ALL · IN · ONE

Civil Engineering PE

EXAM GUIDE
Breadth and Depth

ABOUT THE AUTHOR

Indranil Goswami, Ph.D., P.E., is a consultant engineer in the State of Maryland. He has served on the civil engineering faculty at Morgan State University in Baltimore and is past president of the Baltimore Chapter of the Maryland Society of Professional Engineers. Dr. Goswami was nominated for the Bliss Medal of the Society of American Military Engineers in 1998, and received academic excellence awards from the Maryland Society of Professional Engineers (2003) and the National Society of Professional Engineers (2016). He is a registered Professional Engineer, and teaches an online course geared toward the PE Civil examination.

ALL·IN·ONE

Civil Engineering PE

EXAM GUIDE

Breadth and Depth

FOURTH EDITION

Indranil Goswami, Ph.D., P.E.

New York • Chicago • San Francisco • Athens
London • Madrid • Mexico City • Milan
New Delhi • Singapore • Sydney • Toronto

Library of Congress Cataloging-in-Publication Data

Names: Goswami, Indranil, author.
Title: Civil engineering PE all-in-one exam guide : breadth and depth / Indranil Goswami, Ph.D., P.E.
Description: Fourth edition. | New York : McGraw Hill, [2020] | Includes index.
Identifiers: LCCN 2020020600 | ISBN 9781260457223 (hardcover) | ISBN 9781260457230 (ebook)
Subjects: LCSH: Civil engineering—Examinations—Study guides. | Civil engineers—Certification—
   United States—Study guides.
Classification: LCC TA159 .G58 2020 | DDC 624.076—dc23
LC record available at https://lccn.loc.gov/2020020600

McGraw Hill books are available at special quantity discounts to use as premiums and sales promotions or for use in corporate training programs. To contact a representative, please visit the Contact Us page at www.mhprofessional.com.

Civil Engineering PE All-in-One Exam Guide: Breadth and Depth, Fourth Edition

1 2 3 4 5 6 7 8 9   LCR   25 24 23 22 21 20

ISBN 978-1-260-45722-3
MHID 1-260-45722-2

This book is printed on acid-free paper.

Sponsoring Editor
Ania Levinson

Editing Supervisor
Stephen M. Smith

Production Supervisor
Pamela A. Pelton

Aquisitions Coordinator
Elizabeth M. Houde

Project Manager
Touseen Qadri, MPS Limited

Copy Editor
Yashoda Rawat, MPS Limited

Proofreader
A. Nayyer Shamsi, MPS Limited

Art Director, Cover
Jeff Weeks

Composition
MPS Limited

Information contained in this work has been obtained by McGraw Hill from sources believed to be reliable. However, neither McGraw Hill nor its authors guarantee the accuracy or completeness of any information published herein, and neither McGraw Hill nor its authors shall be responsible for any errors, omissions, or damages arising out of use of this information. This work is published with the understanding that McGraw Hill and its authors are supplying information but are not attempting to render engineering or other professional services. If such services are required, the assistance of an appropriate professional should be sought.

This edition is dedicated with gratitude to my dad Mr. Chitta Goswami,
whose understated efficiency and steadfast principles continue to inspire me every day.

CONTENTS AT A GLANCE

CONTENTS

Contents

XXX

PREFACE TO THE FOURTH EDITION

The fourth edition of the *Civil Engineering PE All-in-One Exam Guide* comes out approximately 5 years after the July 2015 release of the third edition. This new edition appears in concert with the latest updates of codes and standards for the April 2020 exam. The fourth edition has been updated for the codes referenced in the NCEES syllabus effective April 2020, including the following:

- *Highway Capacity Manual* (HCM), 6th edition, 2016
- *A Policy on Geometric Design of Highways and Streets* (AASHTO Green Book), 7th edition, 2018
- *Manual on Uniform Traffic Control Devices* (MUTCD), 2009
- *Minimum Design Loads for Buildings and Other Structures* (ASCE 7), 2010
- *AASHTO Mechanistic-Empirical Pavement Design Guide*, 2nd edition, 2015
- *Building Code Requirements for Structural Concrete* (ACI 318), 2014
- *Building Code Requirements and Specifications for Masonry Structures* (TMS 402/602), 2013
- *AISC Manual of Steel Construction*, 14th edition, 2011
- *AASHTO LRFD Bridge Design Specifications*, 7th edition, 2014
- *PCI Design Handbook—Precast and Prestressed Concrete*, 7th edition, 2010

In the fourth edition, approximately 40 pages of new material have been added, including solved examples. In addition, any errors in the third edition have been corrected.

If you are reading this preface and want a "cookbook" that will evoke some kind of magic at exam time and give you exactly the answer to a particular exam question in a couple of neat sentences, then you should return the book at once. On the other hand, if you want to review and learn civil engineering fundamentals to the extent that you don't need a "cookbook" to pass the PE exam, then you should continue. I won't claim that the book explains each topic better than any other book, but I have tried my best to present the topics in a comprehensive and holistic way.

In addition to studying a book such as this one, it is imperative for readers to create a suite of references to support their exam preparation. In particular, those attempting the construction, structural, or transportation depth area of the PE Civil exam should become intimately familiar with most, if not all, of the design standards relevant to that depth area. Surely, the list of references (that this edition is current with) will be updated and it is up to individuals to make sure that they are using the most current of the standards referenced by NCEES.

As I wrote in the preface to the second edition, a 1200-page book such as this one can only serve as a guide for a field as diverse as civil engineering—it can never be a complete reference. I have tried to strike a balance between instructional aid and exam reference. However, I have no qualms in admitting that the book may well have given too much attention to one area and not enough to another. Overall, I have tried to keep a certain flow to the narrative so that it will be easy enough to read and at the same time be comprehensive enough to serve as an exam reference.

One would hope that the reader would supplement this book with other resources, as deemed appropriate by the official (NCEES) syllabus of the PE exam. I would recommend that readers create a "map" from the current NCEES syllabus to specific pages in this book, thereby familiarizing themselves with both.

I mentioned in the preface to the previous (third) edition that not all of the practice problems at the end of the book have been formatted like actual PE questions. Rather, the intent of these problems is to provide review of the subject matter covered in the corresponding chapter. For questions formatted like the PE exam, the reader is encouraged to seek out a PE practice problems book, one whose primary intent is to present multiple-choice problems similar in content and formatting to the PE Civil examination.

To all who choose to use this book, I wish the very best.

Indranil Goswami, Ph.D., P.E.

PREFACE TO THE FIRST EDITION

This book is the culmination of about a year of writing, but represents experience gathered from teaching an in-class review course for the PE Civil examination over the past 8 years. During that time, the PE exam has gone through several changes and the course has evolved too. Currently, the PE Civil examination has a breadth-depth multiple choice format, with the depth areas being construction, geotechnical, structural, transportation, and water and environmental engineering. As of April 2008, the introduction of construction as a depth specialty area in its own right has meant a significant de-emphasizing of environmental topics compared to the exam's past. Particularly, the topics of solid waste, air quality, and wetlands seem notable omissions in the new exam structure.

A question I get asked all the time is "What is the difference between AM (breadth) and PM (depth) questions?" While there is no easy answer to this question, the most fundamental definition of a breadth-type (AM) question seems to be "these are the things it would be reasonable to expect ALL civil engineers to know, no matter what their specialty happens to be."

Another question is "What types of questions are typically asked on the PE?"

A breadth-depth exam such as the PE Civil examination has, broadly speaking, four categories of questions. These may be described as

1. Basic qualitative questions in the breadth part of the exam

2. Basic quantitative questions in the breadth part of the exam

3. Quantitative—more obscure theoretical areas—more computations possibly required

4. Qualitative—requiring a global viewpoint of the specialty area

A well-designed review process, whether through a review course or individually, can address the first three types of questions quite adequately. Competence in the fourth type of question can only come from long experience and breadth of reading. Even access to good reference materials may not help with this type of question. In fact, having access (during the exam) to a lot of reference materials may actually give one a false sense of "it may be found in one of these books" and result in wasting a lot of time chasing down the answer to a specific question.

Another question that I encourage people to think about, whether they ask it or not, is "What is the best indicator of what the exam will look like?" While a book like this can cover the relevant subject matter adequately, economy of space and, well, let's face it, the plain old human trait of laziness make the statement of problems quite different from the way they are (at least some of them) presented to the examinee on the actual PE exam. The point is, even if you are quite prepared for the content of the question, unexpected details in the presentation of the question can be something that throws off your rhythm during a fast-paced exam such as this. My recommendation to everyone is to look at the sample questions book published by NCEES. If anything can give you an idea about what the actual exam will *look* like, the book published by the exam creators is probably the way to go.

A question that almost everyone will have to ask themselves at some point is "What materials should I take with me?"

During your review, whether you do that on your own or if you are part of a review course, create a folder of notes into which all your work gets inserted, by subject. If you have a set of class notes from a review course, that can be a good start to this folder. As you progress through your review, this folder is going to expand and contract, according to your particular strengths and weaknesses. Over time, this should become your primary reference. Preparing an index to this file is also important. If your depth specialty is one of the code-intensive areas, namely, construction, structural, or transportation, you should of course have the necessary codes and specifications. While there are many texts that are excellent, taking them all with you may actually be detrimental if you are not thoroughly familiar with their contents. I would suggest taking books that fall into one of two categories: (1) you have used it extensively, particularly during your review, and are totally familiar with its contents, or (2) it serves as a "bible" for a particular area, is very comprehensive, and is very well indexed, so that finding specific information in it is relatively easy.

Question: How much time should I dedicate to the review?

Most people who are successful on the PE exam report having spent about 200 to 300 hours on their review, typically distributed over about 3 months. This is about 15 to 20 hours a week.

Question: What should my review schedule look like?

Many are very disciplined when they take on a specific objective such as "passing the PE exam." The typical PE candidate has to balance career and family and this often makes him or her "fall off" plans made in good intentions. For this reason, if not for any other, a review course is recommended to lend structure to the review.

One pitfall of having access to books of practice problems is that some tend to read the solutions to problems and assume that this prepares them to do that problem on their own. This is a huge mistake. I find that unless I am very vigilant, when I am reading someone else's solution to a problem, I typically don't think "Why is step 1 the way it is?" In other words, why does the solution *start* the way it does? This is often the key to the solution. If you start the solution process correctly, the subsequent steps should follow.

Indranil Goswami, Ph.D., P.E.

UNITS AND UNIT CONVERSIONS

Angle

1 radian = 57.29578 degrees

1 gradian = 0.9 degree = 0.015708 radian

Length

1 ft = 0.3048 m = 0.3333 yd

1 m = 3.281 ft = 39.37 in = 1.0936 yd

1 micron = 1 μm = 1.0×10^{-6} m

1 angstrom = 1.0×10^{-10} m

1 mil = 0.001 in

1 mile = 1760 yd = 5280 ft = 1.6093 km

1 yd = 3 ft = 36 in = 0.9144 m

Area

1 acre = 43,560 ft^2 = 1/640 sq mile = 0.40469 hectare (ha)

1 darcy = 9.86923×10^{-13} m^2 (approx. the area of 1 micron $\times$ 1 micron)

1 ha = 10,000 m^2 = 2.471 acres = 107,639 ft^2 = 0.00386 sq mile

1 sq km = 247.105 acres = 100 ha = 1,195,990.05 yd^2

1 sq mile = 640 acres = 259 ha

Volume

1 acre-ft = 43,560 ft^3 = 325,851.43 gal (U.S.)

1 fl oz (U.S.) = 1/128 gal (U.S.) = 29.574 mL = 1.80469 in^3

1 ft^3 = 7.4805 gal (U.S.) = 28.3168 L = 957.5 fl oz

1 gal (Imperial) = 4.5461 L

1 gal (U.S.) = 0.13368 ft^3 = 3.7854 L = 4 quarts = 8 pints = 16 cups = 128 fl oz

1 L = 0.2642 gal (U.S.) = 0.03531 ft^3

1 gal (water) = 8.34 lb

1 ft^3 (water) = 62.45 lb

1 lb (water) = 0.11983 gal = 0.016 ft^3

Velocity

1 mph = 1.46667 ft/s = 0.447 m/s = 1.609 kmph

1 m/s = 3.281 ft/s = 3.6 kmph = 2.237 mph

Mach 1 (speed of sound at standard conditions) = 1225.1 kmph = 761.2 mph

Volumetric Flow Rate

1 in^3/s = 1.63871 × 10^{-5} m^3/s = 5.787 × 10^{-4} ft^3/s

1 ft^3/s (cusec) = 0.02832 m^3/s = 448.83 gal/min = 0.6463 mgd

1 m^3/s (cumec) = 35.3147 ft^3/s = 15,850 gal/min = 22.8245 mgd

1 gal/min = 0.002228 ft^3/s = 0.00144 mgd

1 mgd = 1.5472 ft^3/s = 0.0438 m^3/s = 694.44 gal/min

1 acre-in/h = 1.00833 ft^3/s

Permeability

1 cm/s = 9.225 × 10^8 gal/acre/day

1 in/h = 2 ft/day = 6.517 × 10^5 gal/acre/day

Mass

1 kg = 2.204 lb-m = 1 Ns^2/m

1 slug = 1 lb-s^2/ft = 14.59 kg

1 tonne (metric ton) = 1000 kg

Density

1 g/cm^3 = 1000 kg/m^3 = 62.45 lb-m/ft^3

Runoff or Precipitation Depth

1 acre-ft/sq. mile = 0.01875 in

1 in = 53.333 acre-ft/sq mile

Pressure/Stress

1 std atm = 1.01325 bar = 1.013 × 10^5 Pa = 101.325 kPa = 14.696 lb/in^2 = 2116.2 lb/ft^2

1 std atm = 10.33 m water = 33.9 ft water

1 std atm = 29.92 in mercury (Hg) = 760 mm Hg

1 bar = 1.0 × 10^5 Pa = 100 kPa

1 kip/in^2 (ksi) = 6.895 MPa = 6.895 N/mm^2

1 MPa = 0.14504 ksi = 145.04 psi

1 ton/ft^2 = 13.889 lb/in^2 = 95.76 kPa = 0.945 std atm

Force

1 lb = 4.448 N

1 kip = 4448 N = 4.448 kN = 0.5 ton (U.S.)

1 N = 1 kg-m/s^2 = 0.22481 lb = 1.0 × 10^5 dynes

1 dyne = 1 g-cm/s^2 = 2.2481 × 10^{-6} lb

1 ton (U.S.) = 2000 lb = 2 kips = 8896 N = 8.896 kips

Energy

1 lb-ft = 1.3558 J = 0.001285 Btu = 0.32383 cal (calorie)

1 Btu = 778.169 lb-ft = 1055.056 J = 2.928 × 10^{-4} kW-h

1 cal = 4.1868 J = 3.088 lb-ft

Power

1 hp = 550 lb-ft/s = 745.7 W = 0.746 kW = 42.41 Btu/min = 0.707 Btu/s

1 kW = 56.87 Btu/min = 0.948 Btu/s = 737.6 lb-ft/s = 1.341 hp

Concentration

1 mg/L = 8.3454 lb-m/Mgal

1 mg/L = 1 ppm (if solvent is water, since 1 L of water = 1 kg)

Earth Parameters

Earth radius = 6371 km = 3960 miles (approx.)

Earth mass = 5.9736 × 10^{24} kg

Universal Constants

Acceleration due to gravity (g) = 9.80665 m/s^2 = 32.174 ft/s^2

Atomic mass unit (m_u = 1 u) = 1.66054 × 10^{-27} kg

Avogadro number (N_A) = 6.022141 × 10^{23}/mol

Boltzmann constant (k) = 1.38065 × 10^{-23} J/K

Electric permittivity of free space (ε_o) = 8.854187... × 10^{-12} F/m

Electron charge $(e) = 1.602176 \times 10^{-19}$ C

Electron rest mass $(m_e) = 9.109382 \times 10^{-31}$ kg

Electron volt (eV) $= 1.602176 \times 10^{-19}$ J

Euler's constant $(\gamma) = 0.577215664901532...$

Faraday's constant $(F) = 96485.34$ C/mol

Gravitational constant $(G) = 6.6738 \times 10^{-11}$ N-m^2/kg^2

Magnetic permeability of free space $(\mu_o) = 4\pi \times 10^{-7} = 1.25664 \times 10^{-6}$ H/m

Molar volume of an ideal gas (at 0°C) $= 22.414$ L

Natural (Naperian) logarithm base $(e) = 2.718281828459045...$

Pi $(\pi) = 3.14159265358979...$

Planck constant $(h) = 6.62607 \times 10^{-34}$ J-s

Speed of light in vacuum $(c) = 2.99792458 \times 10^8$ m/s

Stefan-Boltzmann constant $(\sigma) = 5.6704 \times 10^{-8}$ W/m^2-K^4

Universal gas constant $(R) = 8.314462$ J/mol-K

Strength of Materials

This chapter reviews concepts of strength of materials (also known as mechanics of materials). The methods to calculate stresses and strains for various types of loading are a necessary precursor to the process of structural design, wherein those calculated maximum stresses are checked against allowable stresses (allowable stress design). External loads imposed on structures produce internal loads such as axial force (normal to cross section), transverse shear force (in the plane of the cross section), bending moment (about major and minor in-plane axes), and torsion (moment about the axis normal to the cross section). The externally imposed loads and the internal loads are in equilibrium. These internal loads are then responsible for creating internal stresses such as normal stress and shear stress.

Sign Convention for Stresses

The naming convention for stress components on a 3D element uses two subscripts—the first indicating the plane on which the stress component acts and the second indicating the direction of the stress component. Thus, τ_{xy} indicates the (shear) stress component acting on the x face (face whose outward normal is in the x direction, or a face parallel to the yz plane). In Fig. 101.1, this face is shown shaded. In three dimensions, there are three normal stress components (σ_{xx}, σ_{yy}, σ_{zz}) while there are six shear stress components (τ_{xy}, τ_{yx}, τ_{xz}, τ_{zx}, τ_{yz}, τ_{zy}). It can be shown that $\tau_{xy} = \tau_{yx}$, $\tau_{xz} = \tau_{zx}$, $\tau_{yz} = \tau_{zy}$, which leaves only six independent stress components: three normal (σ_{xx}, σ_{yy}, σ_{zz}) and three shear (τ_{xy}, τ_{xz}, τ_{yz}). The stress components on the x plane are shown in Fig. 101.1.

This chapter reviews computation of internal normal and shear stress produced by axial load, normal and shear stress produced by transverse loads, and shear stress produced by torsion. In addition, the membrane stresses produced by internal pressure in thick- and thin-walled pressure vessels are also reviewed.

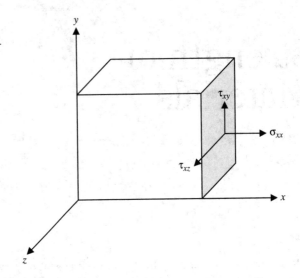

Figure 101.1 Stress components (3D).

Centroid of an Area by Integration

The center of gravity of any object is calculated as the weighted average of the centers of gravity of subobjects that constitute the object. When the area is subdivided into infinitesimal elements, this weighted average is calculated by integration. For an area, the location of the centroid (x_c, y_c) is given by

$$x_c = \frac{\int x\, dA}{\int dA} \qquad\qquad y_c = \frac{\int y\, dA}{\int dA} \qquad\qquad (101.1)$$

where x and y represent the coordinates of the center of the element dA being integrated.

NOTE In practical applications where the centroid location needs to be calculated, e.g., to find section properties of certain built-up shapes, the shape is often composed of geometrically well-known parts. In that case, the weighted average of these parts can be computed, rather than using the much more complex integration approach.

Centroid of a Compound Area—Weighted Average

When one is dealing with a compound shape (which is a combination of a finite number of known shapes), the composite formula may be applied. The coordinates of the centroid are given by the weighted average of the coordinates of individual parts:

$$\bar{x} = \frac{x_1 A_1 + x_2 A_2 + \ldots + x_n A_n}{A_1 + A_2 + \ldots + A_n} = \frac{\sum\limits_{i=1}^{n} x_i A_i}{\sum\limits_{i=1}^{n} A_i} \qquad \text{and} \qquad \bar{y} = \frac{\sum\limits_{i=1}^{n} y_i A_i}{\sum\limits_{i=1}^{n} A_i} \qquad (101.2)$$

Example 101.1 Centroid by Weighted Average

Consider the composite shape shown below. All dimensions are in millimeters. The composite object is broken up into four component parts, each being a well-known shape (rectangle, triangle, and circle) with known expressions for area and centroid location. Note that the circle represents a hole in the object and is therefore treated as a negative area.

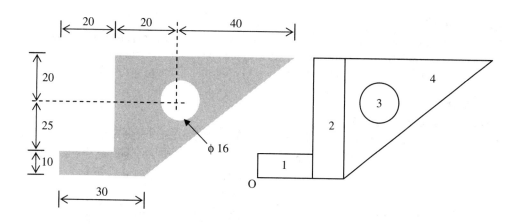

Shape	Area (A)	x_i	y_i	x_iA	y_iA
1	(20)(10) = 200	10	5	2000	1000
2	(10)(55) = 550	25	27.5	13,750	15,125
3	$\pi(8)^2 = -201.1$	40	35	−8042.4	−7037.1
4	1/2(50)(55) = 1375	30 + 1/3(50) = 46.67	2/3(55) = 36.67	64,171.3	50,421.3
	$\Sigma A = 1923.9$			$\Sigma xA = 71,879$	$\Sigma yA = 59,509$

Thus, the centroid location for the composite area is

$$\bar{x} = \frac{\sum x_i A_i}{\sum A_i} = \frac{71,878.9}{1923.9} = 37.36$$

$$\bar{y} = \frac{\sum y_i A_i}{\sum A_i} = \frac{59,509.2}{1923.9} = 30.93$$

NOTE It is vitally important to select an origin and use it consistently throughout as the point of reference for all distances.

The in-plane axes passing through the geometric centroid are also called the elastic neutral axes (ENAs) of the shape. Under the influence of bending moments, if the stresses remain elastic and linear, the bending strain and stress are zero at the neutral axis.

Various Section Properties

Example 101.2

Consider the singly symmetric I-shaped section shown in Fig. 101.2. Calculate (a) the coordinate of the centroid, (b) the moment of inertia, (c) the elastic section modulus, (d) the plastic neutral axis, (e) the plastic section modulus, and (f) the shape factor.

Solution

(a) Coordinate of the Centroid

The y-coordinate of the centroid or the elastic neutral axis is calculated with respect to (an arbitrarily chosen) datum located at the bottom of the bottom flange. This is designated as $y = 0$.

$$\bar{y} = \frac{\sum A_i y_i}{\sum A_i} = \frac{(12 \times 1.5) \times 0.75 + (20 \times 0.75) \times 11.5 + (10 \times 1) \times 22}{12 \times 1.5 + 20 \times 0.75 + 10 \times 1} = \frac{406}{43} = 9.44$$

Since the total depth of the section is 22.5 in., the top fiber is at maximum distance from the neutral axis, at distance $y_{max} = 13.06$ in.

(b) Moment of Inertia

The moment of inertia (I) with respect to any axis is calculated using the *parallel axis theorem* or the *moment of inertia transfer theorem*. This theorem states

$$I_x = I_{xc} + Ad^2$$

where I_{xc} is the moment of inertia with respect to the centroidal axis, A is the area of the cross section, and d is the parallel shift between the centroidal axis and the axis with respect to which

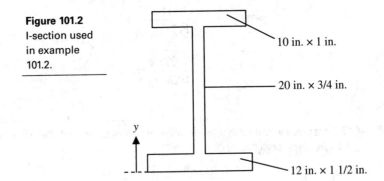

Figure 101.2
I-section used
in example
101.2.

10 in. × 1 in.

20 in. × 3/4 in.

12 in. × 1 1/2 in.

the moment of inertia is desired. Thus, for the section shown in Fig. 101.2, the moment of inertia about the neutral axis is calculated as the sum of the moments of inertia of the three rectangles—top flange, web, and bottom flange. For each of these rectangles, the local centroidal moment of inertia is given by $bh^3/12$.

$$
\begin{aligned}
I_{NA} &= \frac{1}{12} \times 12 \times 1.5^3 + 18 \times (0.75 - 9.44)^2 \\
&+ \frac{1}{12} \times \frac{3}{4} \times 20^3 + 15 \times (11.5 - 9.44)^2 \\
&+ \frac{1}{12} \times 10 \times 1^3 + 10 \times (22 - 9.44)^2 = 3504.7 \ \text{in.}^4
\end{aligned}
$$

NOTE   For any shape, the moment of inertia about its centroidal axis has the minimum value. In other words, any other axis parallel to the centroidal axis would produce a greater magnitude of moment of inertia.

(c) Elastic Section Modulus

The elastic section modulus (S_x) of a section is defined as the ratio I/y_{max}, where y_{max} represents the distance from the elastic neutral axis to the fiber which is furthest from it. Thus, for the I-section shown in Fig. 101.2, the section modulus is given as

$$
S_x = \frac{I_{xc}}{c} = \frac{3504.7}{13.06} = 268.4 \ \text{in.}^3
$$

(d) Plastic Neutral Axis

The plastic neutral axis (PNA) is the line that divides the area of the section into two equal halves. For the section shown, the total area is

$$
A = 18 + 15 + 10 = 43 \ \text{in.}^2
$$

Therefore, half the area is 21.5 in.^{2}.

Working upward from the bottom flange, since the area of the bottom flange is 18 in.^{2}, the needed area from the web is $21.5 - 18.0 = 3.5$ in.^{2}. Therefore, the depth of the web needed is $3.5/0.75 = 4.67$ in.

Measured from the bottom, the coordinate of the PNA is $y_{PNA} = 1.5 + 4.67 = 6.17$ in. See Fig. 101.3.

(e) Plastic Section Modulus

The plastic section modulus (Z) is calculated as the sum of the first moments of the two "halves" of the section. Figure 101.3 shows the parts of the I-section with respect to the PNA ($y = 6.17$). It is convenient to split each T-shaped "half" into its two component rectangles.

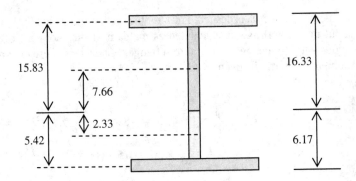

Figure 101.3
Computation of plastic section modulus.

Treating the section as a combination of four rectangles (see Fig. 101.3), the plastic section modulus is given by the first moment of these areas about the PNA:

$$Z_x = (10 \times 1) \times 15.83 + \left(\frac{3}{4} \times 15.33\right) \times \frac{15.33}{2} + \left(\frac{3}{4} \times 4.67\right) \times \frac{4.67}{2} + (12 \times 1.5) \times (4.67 + 0.75)$$

$$= 352.2 \text{ in.}^3$$

(f) Shape Factor

The shape factor for a section is given as the ratio of the plastic section modulus and the elastic section modulus (Z/S). Thus, for the section shown, the shape factor is

$$\frac{Z}{S} = \frac{352.2}{268.4} = 1.31$$

The shape factor can give us some idea of the relative ductility of the member, since it is also the ratio of the plastic (or ultimate) moment capacity of the section to the moment that causes first yield.

Bending Stress

If the material behavior is linear and elastic, and subject to the assumption that *plane sections remain plane after bending*, the bending stress at a plane located at a distance y from the neutral plane, is given by

$$\sigma_b = -\frac{My}{I} \qquad (101.3)$$

NOTE   In this discussion, it has been assumed that the bending moment occurs about the x axis and therefore the relevant properties of the section are I_x, S_x, etc. Similarly, the variable y represents the distance of a plane of interest from the x axis.
The same concept can be extended to bending about the y axis.

This stress distribution is linear, zero at the neutral plane, and has maximum value at the plane furthest from the neutral plane ($y = y_{max}$). Therefore, the maximum bending stress is given by

$$\sigma_{max} = \frac{M_x y_{max}}{I_x} = \frac{M_x}{I_x / y_{max}} = \frac{M_x}{S_x} \qquad (101.4)$$

As the moment is increased, the maximum bending stress reaches the yield stress (F_y). The moment that causes first yield in the section is given by

$$M_{yield} = S_x F_y \qquad (101.5)$$

The basis for the following discussion is the elastoplastic model—commonly used for modeling structural steel behavior. In this model, shown in Fig. 101.4, the stress-strain relationship for steel is assumed to be linear and elastic up to the yield point, and then perfectly plastic ($E = 0$) thereafter. According to this model, the stress remains constant at the yield stress F_y, once the yield strain is exceeded.

Therefore, as the moment is increased beyond M_{yield}, the outer fibers begin to yield and the "plastic behavior zone" gradually progresses inward. The ultimate limit of such a stress distribution is when the plastic zones progress all the way to the neutral axis (all planes at or beyond yield strain). According to the elastoplastic model, there is no more capacity for increased internal stress within the section and therefore the internal moment capacity is at its maximum. Any increase in load beyond this point will cause the formation of an instability which exhibits itself as a "hinge-like" collapse mechanism. This is referred to as *plastic hinge formation*. The moment that causes plastic hinge formation is the ultimate moment capacity of the section. This moment can be calculated as

$$M_P = Z_x F_y \qquad (101.6)$$

Figure 101.4
Elastoplastic
model for steel.

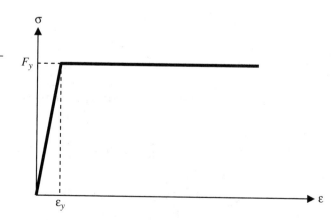

Plastic Collapse Mechanisms

According to the *lower bound theorem*, if a collapse mechanism is visualized for a structure—one that requires the formation of a minimum number of plastic hinges, thereby causing instability—and the load that would cause such a mechanism is computed using equilibrium equations, then that load must be less than the true plastic load that causes collapse. Such a mechanism would then yield a lower bound estimate of the true plastic load capacity. Thus, if several valid collapse mechanisms are visualized, then the highest of the collapse loads is a conservative estimate of the true collapse load for the structure.

Examples 101.3 to 101.7 illustrate the calculation of the collapse load corresponding to assumed collapse mechanisms for various structures.

Example 101.3

Consider a simply supported beam loaded with a concentrated load (P) at midspan as shown below. The beam is statically determinate and therefore needs $0 + 1 = 1$ plastic hinge to form a collapse mechanism (instability). Since the bending moment diagram is, as shown in the third figure, a plastic hinge is going to form at midspan, causing collapse. This is shown in the second figure. Determinate structures require a single plastic hinge for a collapse mechanism to form, so they cannot benefit from ductility as do indeterminate structures, since there is no scope for load redistribution.

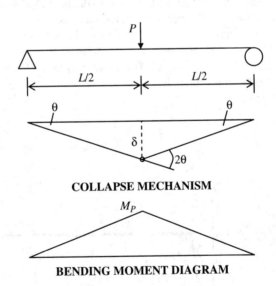

COLLAPSE MECHANISM

BENDING MOMENT DIAGRAM

Equilibrium, expressed in terms of the principle of virtual work, yields

$$W_{ext} = W_{int} \Rightarrow P\delta = M_P 2\theta \Rightarrow P\frac{L}{2}\theta = M_P 2\theta \Rightarrow P = \frac{4M_P}{L}$$

Example 101.4

Consider a "propped cantilever" beam loaded with a concentrated load (P) as shown below. The beam is first-order indeterminate and therefore needs $1 + 1 = 2$ plastic hinges to form a collapse mechanism. Since the bending moment diagram is of the shape as in the third figure, the maximum (and minimum) bending moments form at the fixed support and at the load location. Under elastic conditions, one of these moments is going to reach the plastic moment capacity first. A plastic hinge is going to form at this location and the moment is going to stay at the plastic moment, while with increasing load, redistribution will occur allowing the other point to develop the plastic moment. The beauty of the method outlined here is that we don't have to consider the sequence of these events, but only the final outcome of these events, which is the formation of two plastic hinges, forming the collapse mechanism. This is shown in the second figure.

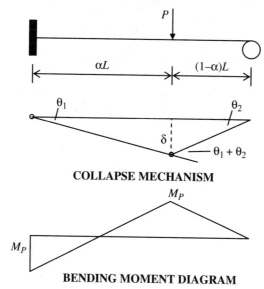

COLLAPSE MECHANISM

BENDING MOMENT DIAGRAM

From compatibility of deformations, for $0 < \alpha < 1$,

$$\delta = \alpha L \theta_1 = (1 - \alpha) L \theta_2 \Rightarrow \theta_2 = \frac{\alpha}{1 - \alpha} \theta_1$$

Equilibrium, expressed in terms of the principle of virtual work, yields

$$W_{ext} = W_{int} \Rightarrow P\delta = M_p \theta_1 + M_p (\theta_1 + \theta_2) \Rightarrow P\alpha L \theta_1 = M_p \left(2\theta_1 + \frac{\alpha}{1 - \alpha} \theta_1 \right)$$

This can be simplified to

$$P = \frac{M_p}{L} \left[\frac{2 - \alpha}{\alpha (1 - \alpha)} \right]$$

The expression above can be used to calculate the collapse load for a specific, given value of $0 < \alpha < 1$.

If the load is a moving load, the minimum load P which causes collapse can be found by minimizing P with respect to the location parameter α, yielding $\alpha = 0.586$.

The corresponding value of P is given by

$$P_{min} = \frac{5.828 M_p}{L}$$

Example 101.5

Consider a propped cantilever beam loaded with a uniformly distributed load (w) as shown below. The beam needs two plastic hinges to form in order to form a collapse mechanism. Since the bending moment diagram is of the shape as in the third figure, the maximum (and minimum) bending moments form at the fixed support and somewhere along the beam span. This location is designated as βL. The final collapse mechanism due to the formation of two plastic hinges is shown in the second figure.

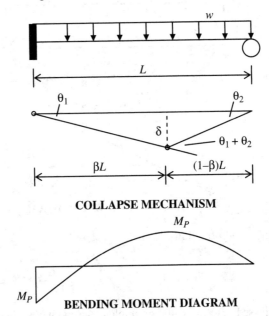

COLLAPSE MECHANISM

BENDING MOMENT DIAGRAM

From compatibility of deformations, we can say

$$\delta = \beta L \theta_1 = (1 - \beta) L \theta_2 \Rightarrow \theta_2 = \frac{\beta}{1 - \beta} \theta_1$$

Equilibrium, expressed in terms of the principle of virtual work, yields

$$W_{ext} = W_{int} \Rightarrow w\left(\frac{1}{2} L \delta\right) = M_p \theta_1 + M_p (\theta_1 + \theta_2) \Rightarrow w = \frac{2 M_p}{L^2}\left[\frac{2 - \beta}{\beta(1 - \beta)}\right]$$

The minimum load w which causes collapse can be found by minimizing w with respect to the location parameter β, yielding $\beta = 0.586$.

The corresponding value of w is given by

$$w_{min} = \frac{11.657 M_P}{L^2}$$

Example 101.6

Now consider a fixed-fixed beam loaded with a uniformly distributed load (w) as shown below. The beam is second-order indeterminate and therefore needs $2 + 1 = 3$ plastic hinges to form a collapse mechanism. Since the bending moment diagram is of the shape as in the third figure, the maximum (and minimum) bending moments form at the two fixed supports and at midspan. The final collapse mechanism due to the formation of three plastic hinges is shown in the second figure.

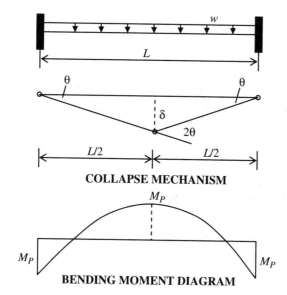

COLLAPSE MECHANISM

BENDING MOMENT DIAGRAM

Equilibrium, expressed in terms of the principle of virtual work, yields

$$W_{ext} = W_{int} \Rightarrow w\left(\frac{1}{2}L\delta\right) = M_P\left(\theta + 2\theta + \theta\right) = 4M_P\theta \Rightarrow w = \frac{16M_P}{L^2}$$

Example 101.7

Now consider a fixed-fixed beam loaded with a point load (P) as shown below. The beam is second-order indeterminate and therefore needs $2 + 1 = 3$ plastic hinges to form a collapse mechanism. Since the bending moment diagram is of the shape as in the third figure, the maximum (and minimum) bending moments form at the fixed support and at the load location. The final collapse mechanism due to the formation of three plastic hinges is shown in the second figure.

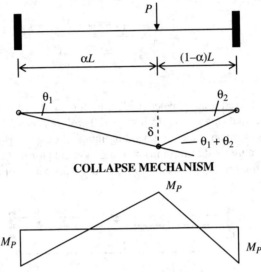

COLLAPSE MECHANISM

BENDING MOMENT DIAGRAM

From compatibility of deformations, we can say

$$\delta = \alpha L \theta_1 = \left(1-\alpha\right) L \theta_2 \Rightarrow \theta_2 = \frac{\alpha}{1-\alpha}\theta_1$$

Equilibrium, expressed in terms of the principle of virtual work, yields

$$W_{ext} = W_{int} \Rightarrow P\delta = M_P\theta_1 + M_P\left(\theta_1 + \theta_1\right) + M_P\theta_2$$

which can be simplified to

$$P = \frac{M_P}{L}\left[\frac{2}{\alpha\left(1-\alpha\right)}\right]$$

The expression above can be used to calculate the collapse load for a specific, given value of $0 < \alpha < 1$.

If the load is a moving load, the minimum load P which causes collapse can be found by minimizing P with respect to the location parameter α, yielding $\alpha = 0.5$.

The corresponding value of P is given by

$$P_{min} = \frac{8M_P}{L}$$

Combined Axial and Bending Stress

For a section subject to an eccentric axial load, the total stress is given by

$$\sigma_{max,min} = \frac{P}{A} \pm \frac{My}{I} = \frac{P}{A} \pm \frac{Pey}{Ar^2} = \frac{P}{A}\left[1 \pm \frac{ey}{r^2}\right] \qquad (101.7)$$

where r is the radius of gyration of the section.

Kern of a Section

For elements such as prestressed concrete beams, the underlying design philosophy is to avoid (or minimize) tensile stress in the cross section. This may be accomplished by keeping the eccentricity of the prestressing force within the "kern" of the cross section. The geometric limits of the kern may be calculated by setting the minimum stress equal to zero.

$$1 \pm \frac{ey}{r^2} = 0 \Rightarrow e_{max} = \frac{r^2}{y_{max}} \qquad (101.8)$$

For example, consider the T-section shown in Fig. 101.5. Let us say that the moment of inertia and area have been calculated and yield a radius of gyration equal to r. The neutral axis is shown, and this defines the distance of the extreme top fiber as y_t and the distance of the extreme bottom fiber as y_b. The outer limits of the kern of the section, shown shaded in the figure on the right, are calculated using

$$k_t = \frac{r^2}{y_b} \qquad \text{and} \qquad k_b = \frac{r^2}{y_t}$$

If the load is applied within the shaded area, the stresses are of the same type everywhere in the section (all compression or all tension).

For a rectangular section, $k_t = k_b = h/6$, leading to the common rule of thumb that stress can be of one type (all tension or all compression) if the load is *within the middle third* of the section width.

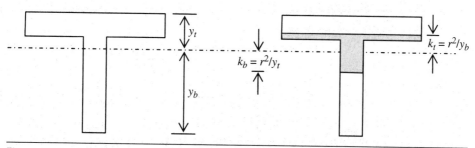

Figure 101.5 Kern limits.

Shear Stress Due to Transverse Load

For a section subject to a transverse shear force V, the shear stress at a longitudinal plane is given by

$$\tau = \frac{VQ}{Ib} \qquad (101.9)$$

where V = transverse shear force
Q = first moment of the area on one side of the longitudinal plane with respect to the elastic neutral plane
I = moment of inertia with respect to the neutral axis
b = width of the section at the plane in question

These are illustrated in Fig. 101.6. The first moment of the area A is calculated as

$$Q = Ay \qquad (101.10)$$

It can be proved that the maximum shear stress occurs at the elastic neutral axis.

For the I-shaped section from Example 101.2, the maximum shear stress (at the elastic neutral axis) is calculated as follows:

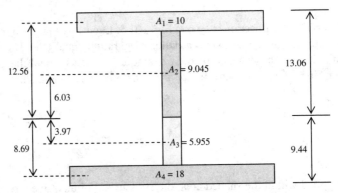

The neutral axis subdivides the section into two "halves"—either the area $A_1 + A_2$ or the area $A_3 + A_4$. Either one can be used to calculate the first moment Q.

Figure 101.6
Parameters for shear
stress calculation.

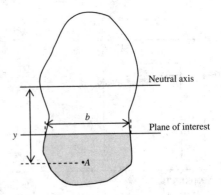

Using the top half, we have $Q = 10 \times 12.56 + 9.045 \times 6.03 = 180.1$ in.^{3}. Using the bottom half, we have $Q = 5.955 \times 3.97 + 18 \times 8.69 = 180.1$ in.^{3}. If this section is subject to a transverse shear force $V = 90$ kips, the shear stress can then be calculated as follows. Note that the moment of inertia was previously calculated as $I = 3504.7$ in.^{4}.

$$\tau = \frac{VQ}{Ib} = \frac{90 \times 180.1}{3504.7 \times 0.75} = 6.17 \, \text{ksi}$$

For I-sections, the flanges play a very small role in resisting shear. Thus, an approximate estimate of the maximum shear stress may be made by distributing the shear force "uniformly" over the web only. In this case, this estimate is given by

$$\tau \approx \frac{V}{A_w} = \frac{90}{20 \times 0.75} = 6.0 \, \text{ksi}$$

This estimate shows an error of 3%.

Typical distributions of shear stress in flanged sections are shown in Fig. 101.7. It can be seen that for I-shaped sections, the approximate estimate obtained by distributing the shear force equally over the area of the web is fairly close to the exact value of the maximum stress at the neutral axis. On the other hand, for a T-section, such an approximation does not yield a good estimate.

For a rectangular section subject to transverse shear (such as is common for timber beams), the shear stress distribution is parabolic and the maximum shear stress (at mid-height) is given by

$$\tau = \frac{3V}{2A} = \frac{3V}{2bd} \tag{101.11}$$

For a circular section subject to transverse shear, the maximum shear stress is given by

$$\tau = \frac{4V}{3A} = \frac{4V}{3\pi r^2} \tag{101.12}$$

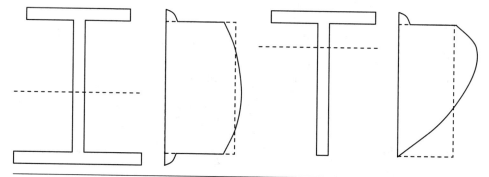

Figure 101.7 Approximations for longitudinal shear stress.

For a hollow circular section subject to transverse shear force V, the maximum shear stress is given by

$$\tau_{max} = \frac{4V}{3A}\left(\frac{r_1^2 + r_1 r_2 + r_2^2}{r_1^2 + r_2^2}\right) \qquad (101.13)$$

where r_1 and r_2 are the inner and outer radii, respectively. If the wall thickness $r_2 - r_1$ is small, then this is approximately equal to

$$\tau_{max} \approx \frac{4V}{3A}\left(\frac{3}{2}\right) = \frac{2V}{A} \qquad (101.14)$$

Example 101.8

A joist is constructed by nailing together two planks as shown in the figure below. If nails are spaced at s = 4 in., what is the shear force (lb) in each nail if the section is subjected to a transverse shear force = 1.2 kips?

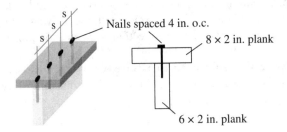

Solution

Assuming a datum at the bottom edge of the section, the y-coordinate of the centroid is

$$\bar{y} = \frac{\sum y_i A_i}{\sum A_i} = \frac{3 \times 12 + 7 \times 16}{12 + 16} = 5.286 \text{ in.}$$

And the centroidal moment of inertia is

$$I_{xc} = \frac{1}{12} \times 2 \times 6^3 + 12 \times (3 - 5.286)^2 + \frac{1}{12} \times 8 \times 2^3 + 16 \times (7 - 5.286)^2 = 151.05 \text{ in.}^4$$

The first moment of the top flange about the centroidal axis is

$$Q = 16 \times (7 - 5.286) = 27.424 \text{ in.}^3$$

Shear stress at interface between two plates: $\tau = \dfrac{VQ}{Ib} = \dfrac{1200 \times 27.424}{151.05 \times 2} = 108.93 \text{ psi}$

Shear force incident to each nail = $108.93 \times 2 \times 4 = 871.5$ lb

Shear Center

When a doubly symmetric section is loaded transversely through the geometric center (centroid) of the section, the load causes only bending of the section. However, when a section (such as the channel shown below) that does not have a vertical (transverse) axis of symmetry is loaded transversely through the centroid, the section is subjected to both bending and torsion. The shear center of such a shape is the point through which the transverse load must pass so that the section undergoes bending only (no torsion). The location of the shear center for a channel is shown below.

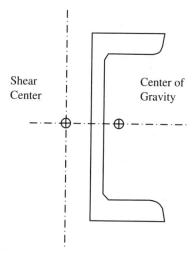

Even if an oblique load is applied through the shear center, the section does not undergo torsion, because the oblique load can be resolved into vertical and horizontal components, both of which pass through the shear center.

Example 101.9

Find the location of the shear center for the C-shaped section shown below.

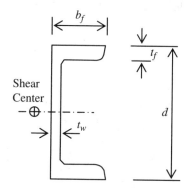

Solution

The shear flow through the channel in response to the vertical shear force is as shown below. The resultant horizontal forces in the two flanges form a clockwise couple, which must be in equilibrium with a counterclockwise couple formed by the vertical forces, one of which is the resultant of the shear flow in the web. For this equilibrating couple to be counterclockwise, the transverse shear force must act to the left of the web, through a point called the shear center.

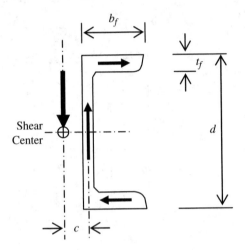

Approximating the flange as a rectangle of width b_f and thickness t_f, the resultant horizontal force in each flange can be derived to be

$$H = \frac{Vdt_f b_f^2}{4I} \qquad (101.15)$$

where I is the moment of inertia about the centroidal x axis.

The clockwise couple created by these two horizontal forces is

$$M = Hd = \frac{Vd^2 t_f b_f^2}{4I} \qquad (101.16)$$

Therefore, in order for the counterclockwise couple to balance the couple M, the distance from the centerline of the web to the shear center is

$$c = \frac{M}{V} = \frac{d^2 t_f b_f^2}{4I} \qquad (101.17)$$

Shear Stress Due to Torsion—Circular Sections

When a solid circular section (radius R) is subjected to torsion, the shear stress between adjacent planes is given by

$$\tau = \frac{Tr}{J} \tag{101.18}$$

where J is the polar moment of inertia of the section, given by

$$J = \frac{\pi R^4}{2} \tag{101.19}$$

Thus, the maximum shear stress in a solid circular shaft due to a torque T is given by

$$\tau_{max} = \frac{TR}{\pi R^4/2} = \frac{2T}{\pi R^3} \tag{101.20}$$

For a hollow cylindrical shaft (inner radius R_1 and outer radius R_2), the polar moment of inertia is given by

$$J = \frac{\pi(R_2^4 - R_1^4)}{2} \tag{101.21}$$

and the maximum shear stress is given by

$$\tau_{max} = \frac{2TR_2}{\pi(R_2^4 - R_1^4)} \tag{101.22}$$

Shear Stress Due to Torsion—Rectangular Sections

The maximum shear stress in a shaft of rectangular section (dimensions $a \times b$, where $b < a$) is given by

$$\tau_{max} = \frac{T}{\alpha a b^2} \tag{101.23}$$

where the parameter α is a function of the aspect ratio a/b as given in Table 101.1.

Table 101.1 Shape Parameter for Torsion-Induced Shear Stress in Rectangular Shafts

a/b	1.0	1.2	1.5	2.0	2.5	3.0	4.0	5.0	10.0	∞
α	0.208	0.219	0.231	0.246	0.258	0.267	0.282	0.291	0.312	0.333

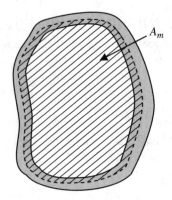

Figure 101.8
Thin-walled tube
in torsion.

A_m

Shear Stress Due to Torsion—Thin-Walled Sections

Economy of weight leads to hollow thin-walled sections being commonly used for structural members subject to significant torsional moments. The thin-walled assumption is considered approximately valid when the wall thickness is less than about 1/20 of the lateral dimension. The maximum shear stress in a thin-walled hollow shaft is given by

$$\tau_{max} = \frac{T}{2A_m t} \tag{101.24}$$

where T = applied torque
A_m = area enclosed by the closed wall centerline of the section (see Fig. 101.8)
t = thickness of the thinnest part of the wall

Stresses in Pressure Vessels

A cylindrical pressure vessel is shown in Fig. 101.9. The inner radius is r_i and the outer radius is r_o.

For a pressure vessel subject to internal pressure (p_i) only, the tangential (hoop) stress is given by

$$\sigma_h = p_i \frac{r_o^2 + r_i^2}{r_o^2 - r_i^2} \tag{101.25}$$

and the radial stress is given by

$$\sigma_r = -p_i \tag{101.26}$$

Figure 101.9
Stresses in pres-
sure vessels.

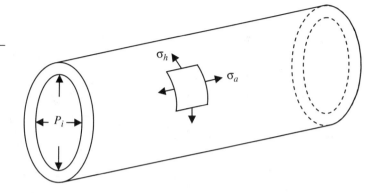

For a pressure vessel subject to external pressure (p_o) only, the tangential (hoop) stress is given by

$$\sigma_h = -p_o \frac{r_o^2 + r_i^2}{r_o^2 - r_i^2} \tag{101.27}$$

and the radial stress is given by

$$\sigma_r = -p_o \tag{101.28}$$

For vessels with end caps, the axial stress is given by

$$\sigma_a = p_i \frac{r_i^2}{r_o^2 - r_i^2} \tag{101.29}$$

Thin-Walled Pressure Vessel

For a cylinder whose wall thickness is less than about one-tenth of the radius, the thin-walled assumption is approximately valid. According to the thin-walled theory, the internal pressure is resisted by hoop stress and axial stress only. They are given by

$$\sigma_h = \frac{p_i r}{t} \tag{101.30}$$

$$\sigma_a = \frac{p_i r}{2t} \tag{101.31}$$

where t is the wall thickness.

Mohr's Circle: Normal (σ) and Shear Stress (τ) Combination

Figure 101.10(a) shows an element in a state of plane stress. The normal stresses are σ_x and σ_y and the shear stresses are τ_{xy} and τ_{yx}. The stresses are given in a framework where the axes are the "traditional" x-y axes—a horizontal x axis and a vertical y axis. Sometimes, the

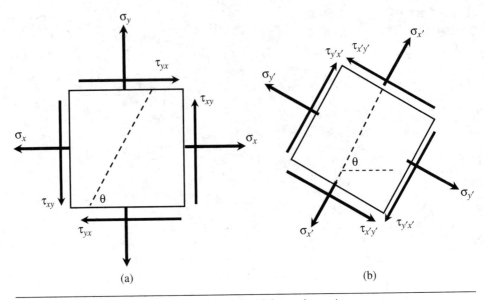

Figure 101.10 Stresses on 2D element (a) original (b) transformed.

engineer might be more interested in stresses oriented along a different set of axes. A "rotated" element (rotated counterclockwise through angle θ) is shown in Fig. 101.10(b). This element has "new" stress components—$\sigma_{x'}$ and $\sigma_{y'}$, $\tau_{x'y'}$ and $\tau_{y'x'}$.

This "new" stress-state is not independent of the original stress-state, but rather a function of it (transformed). The transformation relations are given by Eq. (101.32).

$$\sigma_{x'} = \frac{\sigma_x + \sigma_y}{2} + \frac{\sigma_x - \sigma_y}{2}\cos 2\theta + \tau_{xy}\sin 2\theta$$

$$\sigma_{y'} = \frac{\sigma_x + \sigma_y}{2} - \frac{\sigma_x - \sigma_y}{2}\cos 2\theta - \tau_{xy}\sin 2\theta \qquad (101.32)$$

$$\tau_{x'y'} = -\frac{\sigma_x - \sigma_y}{2}\sin 2\theta + \tau_{xy}\cos 2\theta$$

A very convenient way of (graphically) expressing these results is through Mohr's circle. By eliminating θ from Eq. (101.32), we can write

$$\left(\sigma_{x'} - \frac{\sigma_x + \sigma_y}{2}\right)^2 + \left(\tau_{x'y'} - 0\right)^2 = \left(\frac{\sigma_x - \sigma_y}{2}\right)^2 + \tau_{xy}^2 \qquad (101.33)$$

This represents the equation of a circle with center at

$$\left(\frac{\sigma_x + \sigma_y}{2}, 0\right)$$

and radius

$$R = \sqrt{\left(\frac{\sigma_x - \sigma_y}{2}\right)^2 + \tau_{xy}^2}$$

See Fig. 101.11. Each point on Mohr's circle represents the stress-state along a particular plane.

Maximum and minimum normal stresses are designated σ_{max} and σ_{min} respectively. These are also called the *principal stresses*. Note that these principal directions have only normal stress, i.e., the principal directions are free of shear stress.

$$\sigma_{max}, \sigma_{min} = \sigma_{ave} \pm R = \frac{\sigma_x + \sigma_y}{2} \pm \sqrt{\left(\frac{\sigma_x - \sigma_y}{2}\right)^2 + \tau_{xy}^2} \qquad (101.34)$$

The maximum and minimum values of $\sigma_{x'}$ are calculated by setting $d\sigma_{x'}'/d\theta = 0$. These normal stresses (principal stresses) occur at two specific values of θ, given by

$$\tan 2\theta_p = \frac{2\tau_{xy}}{\sigma_x - \sigma_y} \qquad (101.35)$$

Figure 101.11
Mohr's circle
parameters.

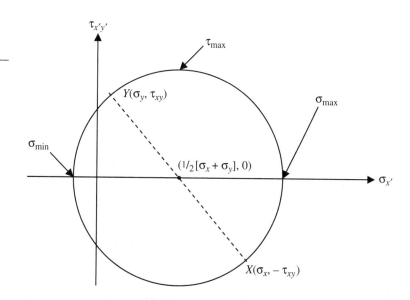

Example 101.10

Consider a 2D element with the stress components as shown in the following figure. Calculate the (a) values of the principal stresses and (b) orientation of the principal planes.

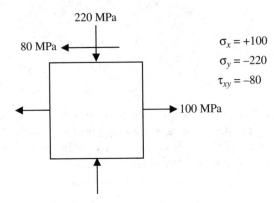

$$\sigma_x = +100$$
$$\sigma_y = -220$$
$$\tau_{xy} = -80$$

Solution From the given stresses, we can calculate the average normal stress σ_{ave} and the radius R of Mohr's circle:

$$\sigma_{ave} = \frac{\sigma_x + \sigma_y}{2} = \frac{100 - 220}{2} = -60$$

$$R = \sqrt{\left(\frac{\sigma_x - \sigma_y}{2}\right)^2 + \tau_{xy}^2} = \sqrt{\left(\frac{100 + 220}{2}\right)^2 + (-80)^2} = 179$$

Therefore, σ_{max}, $\sigma_{min} = -60 \pm 179 = 119, -239$.

The resulting Mohr's circle is shown in the following figure. Each tick mark on the axes is 60 MPa. Point X is located at coordinates (100, 80) and Y is located at (-220, -80).

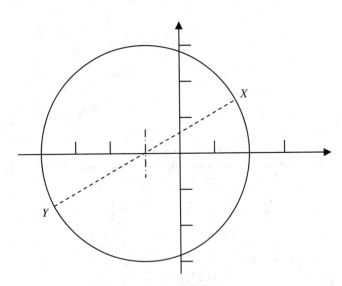

The values of maximum and minimum normal stress can also be calculated by taking the derivative of $\sigma_{x'}$ with respect to θ.

$$\sigma_{x'} = \frac{\sigma_x + \sigma_y}{2} + \frac{\sigma_x - \sigma_y}{2}\cos 2\theta + \tau_{xy}\sin 2\theta$$

$$\frac{d\sigma_{x'}}{d\theta} = -(\sigma_x - \sigma_y)\sin 2\theta + 2\tau_{xy}\cos 2\theta = 0$$

$$\tan 2\theta = \frac{2\tau_{xy}}{(\sigma_x - \sigma_y)} = \frac{2 \times -80}{100 + 220} = \frac{-160}{320} = -0.5$$

This yields $2\theta = \tan^{-1}(-0.5) = -26.6°$ or $153.4°$.

Substituting $2\theta = -26.6°$, we get $\sigma_{x'} = -60 + 160 \times 0.894 + 80 \times 0.447 = 119$. Substituting $2\theta = 153.4°$, we get $\sigma_{x'} = -60 - 160 \times 0.894 - 80 \times 0.447 = -239$. Note that the point with normal stress $\sigma = 119$ is located at a position which is $26.6°$ clockwise (thus negative) with respect to the point X and $\sigma = -239$ is located at a position which is $153.4°$ counterclockwise (thus positive) with respect to X.

Indeterminate Problems in Strength of Materials

A (statically) indeterminate problem is one for which equations of static equilibrium are *necessary but not sufficient* for a complete solution. In such cases, *compatibility equations* must be used in addition to *equilibrium equations*.

Example 101.11

A composite short column has a steel core (diameter 10 in.) surrounded by a snug brass sleeve (inner diameter 10 in, outer diameter 12 in.). The column is loaded uniformly (using loading plates) in compression. If the compressive load is 200 kips, what is the stress in the steel? Assume $E_{steel} = 29,000$ ksi and $E_{brass} = 17,000$ ksi.

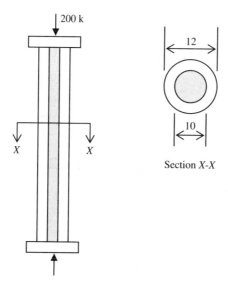

Section X-X

Solution   The only equilibrium equation available is

$$P_{steel} + P_{brass} = 200 \text{ kips}$$

This is not enough to solve for the two unknowns, P_{steel} and P_{brass}. The additional equation is the compatibility equation that states

$$\Delta_{steel} = \Delta_{brass}$$

$$A_{steel} = \pi/4 \, (10)^2 = 78.54 \text{ in.}^2$$

$$A_{brass} = \pi/4 \, (12^2 - 10^2) = 34.56 \text{ in.}^2$$

$$\frac{P_S L}{A_S E_S} = \frac{P_B L}{A_B E_B} \Rightarrow P_S = \frac{A_S E_S}{A_B E_B} P_B = \frac{78.54}{34.56} \times \frac{29,000}{17,000} \times P_B = 3.877 P_B$$

Thus,

$$P_{steel} + P_{brass} = 3.877 \, P_{brass} + P_{brass} =$$

$$4.877 \, P_{brass} = 200 \text{ k} \Rightarrow P_{brass} = 41 \text{ k} \quad \text{and} \quad P_{steel} = 159 \text{ k}$$

Stress in the steel (assuming uniform distribution of load):

$$\sigma_{steel} = 159/78.54 = 2.02 \text{ ksi}$$

Indeterminate Cable Systems

In Fig. 101.12(a), we see a rigid bar suspended by a system of three parallel cables and carrying a load W. The free-body diagram of the rigid bar is shown in Fig. 101.12(b).

At first sight, it might seem as if this system is statically determinate, as it has no more than three unknowns, T_1, T_2, and T_3, and we have three equations of equilibrium. However, note that there are no horizontal forces anywhere in the structure and therefore applying the $\Sigma F_x = 0$

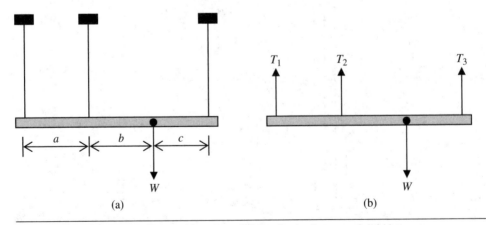

Figure 101.12   (a) Indeterminate cable system (b) free-body diagram of rigid bar.

equation would simply result in a trivial equation of the form $0 = 0$. Thus, we have only two useful equations at our disposal: $\Sigma F_y = 0$ and $\Sigma M_z = 0$. Therefore, the system is first-order indeterminate (two equations, three unknowns).

The two equations of static equilibrium are

$$\sum F_y = 0 \Rightarrow T_1 + T_2 + T_3 - W = 0$$
$$\sum M_1 = T_2 \cdot a - W \cdot (a + b) + T_3 \cdot (a + b + c) = 0$$

The third equation is implicit in a single word in the statement of the problem—"rigid" bar. If the bar is (for all practical purposes) infinitely more rigid than the rest of the components of the system, then we can assume that, as the force **W** pulls down on the assembly, the cables stretch and the bar moves downward, but in such a way that the *bar remains straight* (no bending deformation). As a result, the deflected shape of the assembly will look somewhat as shown in Fig. 101.13.

Applying similar triangles, as shown in Fig. 101.14

$$\frac{\Delta_2 - \Delta_1}{a} = \frac{\Delta_3 - \Delta_2}{b + c}$$
$$a(\Delta_3 - \Delta_2) = (b + c)(\Delta_2 - \Delta_1)$$
$$(b + c)\Delta_1 - (a + b + c)\Delta_2 + a\Delta_3 = 0$$

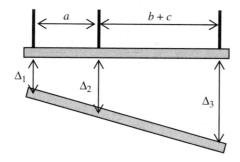

Figure 101.13
Deflection of rigid bar.

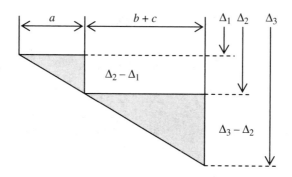

Figure 101.14
Compatibility of deflections using similar triangles.

and recalling from axial deformation

$$\Delta = \frac{TL}{AE}$$

$$(b+c)\frac{T_1 L_1}{A_1 E_1} - (a+b+c)\frac{T_2 L_2}{A_2 E_2} + a\frac{T_3 L_3}{A_3 E_3} = 0$$

This is our missing third equation. The three equations can now be solved simultaneously to calculate T_1, T_2, and T_3.

Example 101.12

A rigid bar ABC is suspended by three cables as shown. Data for the three cables are given in the table following. All cables are steel ($E = 29,000$ ksi).

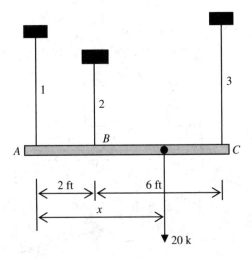

Cable	Length (ft)	Diameter (in.)
1	4.0	0.25
2	3.0	0.40
3	5.0	0.30

Find the distance (x) at which the 20-kip load must be applied such that the bar ABC remains horizontal.

Solution

$$\sum F_y = 0 \Rightarrow T_1 + T_2 + T_3 = 20 \qquad \text{Equilibrium}$$

$$\sum M_A = 0 \Rightarrow 2T_2 + 8T_3 = 20x \qquad \text{Equilibrium}$$

$$\Delta_1 = \Delta_2 = \Delta_3 \Rightarrow \frac{T_1 \times 4}{\dfrac{\pi 0.25^2}{4} E} = \frac{T_2 \times 3}{\dfrac{\pi 0.4^2}{4} E} = \frac{T_3 \times 5}{\dfrac{\pi 0.3^2}{4} E} \qquad \text{Compatibility}$$

The compatibility equation may be simplified to

$$64T_1 = 18.75T_2 = 55.56T_3 \Rightarrow T_1 = 0.293T_2 \quad \text{and} \quad T_3 = 0.338T_2$$

Solving the first equilibrium equation

$$T_1 = 3.59\text{ k}, \quad T_2 = 12.26\text{ k}, \quad T_3 = 4.14\text{ k}$$

Substituting into the second equilibrium equation, we get

$$x = 2.88\text{ ft}$$

Tensile Test

The most common test for steel and other metals is the tensile test, where a sample with an instrumented test section (often called the "gage length") is subjected to axial tension. Because of the narrowing of the test section, most of the deformation (strain) is concentrated in the test section because the stress is much higher than in the end (grip) section. See Fig. 101.15. The gage length L_0 is recorded at the beginning of the test (at zero load), and the extensometer records the elongation (Δ) of the gage length as the load progresses. These measurements (P, Δ) can then be converted to axial stress ($\sigma = P/A$) and strain ($\varepsilon = \Delta/L$).

The typical stress-strain diagrams for ductile and brittle steels are shown in Fig. 101.16. Most grades of steel exhibit a common initial tangent section. Thus, the slope of this tangent section—the modulus of elasticity—is more or less constant across various grades of steel ($E_s = 29,000$ ksi or 200 GPa). If a distinct point of transition (yield point) is not apparent, then the 0.2% offset (line parallel to the initial tangent and passing through a strain intercept of 0.2% or $\varepsilon = 0.002$) is used to identify the yield point. The ratio of the final (fracture) strain to the yield strain is called the ductility.

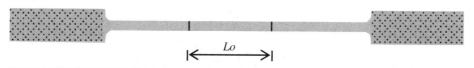

Figure 101.15 Tensile test coupon.

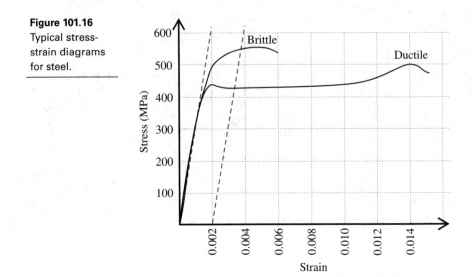

Figure 101.16
Typical stress-
strain diagrams
for steel.

Compression Test

During a compression test of a concrete cylinder (test cylinders may be 6 in diameter × 12 in. height or 4 in. diameter × 8 in. height or 3 in. diameter × 6 in. height), the cylinder is loaded in axial compression as shown in Fig. 101.17. Deformation of the cylinder (Δ) is measured by the travel of the loading crosshead and the resisting load (P) is measured by a load cell. These measurements (P, Δ) can be converted to axial stress ($\sigma = P/A$) and strain ($\varepsilon = \Delta/L$).

The typical stress-strain diagrams for various grades of concrete are shown in Fig.101.18. Unlike steel, for which a tangent modulus is common, the modulus of elasticity of concrete is typically a secant modulus, and as seen in the figure, it is a function of the 28-day

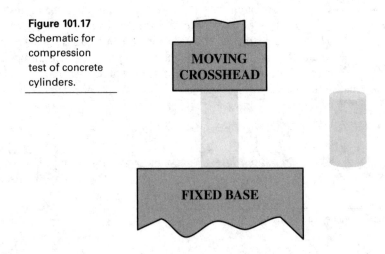

Figure 101.17
Schematic for
compression
test of concrete
cylinders.

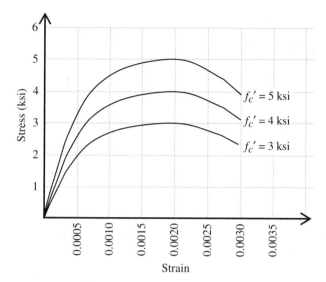

Figure 101.18
Typical stress-
strain diagrams
for concrete.

compressive strength (f_c'). The provisions of ACI 318 say that the modulus of elasticity of concrete E_c (psi) is given by

$$E_c = 33w_c^{1.5}\sqrt{f_c'} \qquad (101.36)$$

where w_c = unit weight of the concrete (lb/ft³). Typically, the maximum compressive strain at which concrete is considered to have failed by "crushing" is $\varepsilon_c = 0.003$.

Split Cylinder Test

In the split cylinder test, a concrete cylinder is placed on its side and loaded transversely in compression as shown in Fig. 101.19. The load is applied diametrically and uniformly along the length of the cylinder until failure occurs along the vertical diameter. This produces a state of tension on the vertical plane and causes the cylinder to "split." The tensile stress on the vertical plane is given by

$$f_{ct} = \frac{2P}{\pi DL} \qquad (101.37)$$

Third-Point Loading Test of a Beam (Flexure)

In this test, a simple span plain concrete beam is loaded at one-third span points. This produces a state of pure bending in the middle third of the beam. See Fig. 101.20. The maximum bending moment ($M_{max} = PL/6$) causes tensile cracking in this zone. The tensile stress corresponding to a load P is given by

$$f_t = \frac{PL}{bd^2} \qquad (101.38)$$

where the rectangular cross section of the beam has width b and depth d.

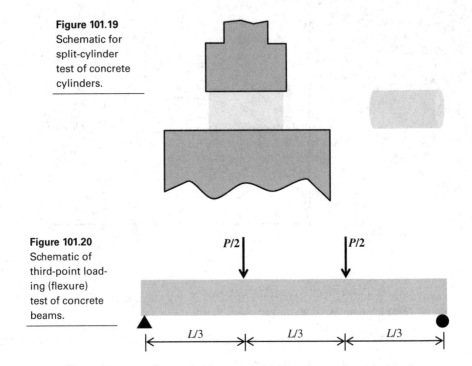

Figure 101.19 Schematic for split-cylinder test of concrete cylinders.

Figure 101.20 Schematic of third-point load-ing (flexure) test of concrete beams.

At the load that causes failure, the value of this tensile stress is called the "modulus of rupture." In ACI 318, the modulus of rupture is represented by f_r. For normal weight concrete, the ACI 318 recommends that the modulus of rupture be taken as

$$f_r = 7.5\sqrt{f_c'} \qquad (101.39)$$

where f_r = modulus of rupture (tensile strength) of concrete (psi)
f_c' = 28-day compression strength (psi)

Hardness

The three most common tests for hardness of a material are (1) the Brinell hardness test, (2) the Rockwell hardness test, and (3) the Vickers hardness test.

Brinell Hardness Test

The Brinell hardness test applies a known load to a surface through a hardened steel ball of known diameter. The diameter of the indentation is measured and the Brinell hardness number (BHN) is reported as

$$BHN = \frac{2P}{\pi D\left(D - \sqrt{D^2 - d^2}\right)} \qquad (101.40)$$

where P = known applied load (kg)
   D = diameter of loading ball (mm)
   d = diameter of indentation (mm)

Rockwell Hardness Test

In the Rockwell hardness test, the depth of indentation under the test load is compared to that made by a preload. The test apparatus is calibrated to directly provide the hardness number (HRA, HRB, etc.) on a prescribed scale, thus eliminating errors involved in measurement. The preload is applied to the sample and allowed to stabilize. This condition is treated as the "zero" point for measurement of the indentation under the actual test load. The Rockwell hardness number (HR) is reported as

$$HR = N - \frac{d}{s} \qquad (101.41)$$

where d = depth of indentation due to test load
   N and s are scale parameters

Vickers Hardness Test

The **Vickers hardness test**, developed as an alternative to the Brinell test, is often easier to use than other hardness tests since the required calculations are independent of the size of the indenter, and the indenter can be used for all materials irrespective of hardness. The Vickers test can be used for all metals and has one of the widest scales among hardness tests. The load is typically applied for between 10 and 15 seconds. Vickers hardness is typically reported as 440HV30/20, where 440 is the Vickers hardness number, HV refers to the hardness scale (Vickers), 30 indicates the load (kgf) and 20 (optional) refers to the loading time (20 s) if it is longer than 10 to 15 seconds.

Toughness

Hardness of a material provides a measure of its strength under load, but not its ability to absorb energy through deformation, which is characteristic of a malleable material. A more appropriate measure of this energy absorbing ability is toughness. The most common test for measuring toughness is the Charpy impact test for notch toughness.

Charpy Impact Test

In the Charpy impact test, a notched specimen is subjected to an impact force from a weighted pendulum. The amount of energy transferred to the specimen can be calculated based on the difference in the height of the hammer before and after the fracture.

Ductility of a material is temperature dependent. There is often a distinct temperature below which the material exhibits brittle fracture as opposed to ductile failure. This temperature is called the ductile-brittle-transition-temperature (DBTT).

Generally high-strength materials have low impact energies due to the fact that fractures easily initiate and propagate in high-strength materials. The impact energies of high-strength materials are usually insensitive to temperature. High-strength BCC steels display a wider variation of impact energy than high-strength metal that do not have a BCC structure because steels undergo microscopic ductile-brittle transition. Regardless, the maximum impact energy of high-strength steels is still low due to their brittleness.

Statically Determinate Structures

Vector

A *vector* is a quantity that has both magnitude and direction, e.g., forces and moments. As an illustration, a force is shown in Fig. 102.1. The orientation of the force is along the line directed from the point A $(-3, 1, 5)$ to the point B $(5, 3, -2)$. The Cartesian components of the line are given by

$$\mathbf{AB} = [5 - (-3)]\mathbf{i} + [3 - 1]\mathbf{j} + [-2 - (5)]\mathbf{k} = 8\mathbf{i} + 2\mathbf{j} - 7\mathbf{k}$$

The length of the line AB is given by

$$|AB| = \sqrt{8^2 + 2^2 + 7^2} = 10.817$$

The unit vector (dimensionless) along the line AB is given by

$$\mathbf{u}_{AB} = \frac{8\mathbf{i} + 2\mathbf{j} - 7\mathbf{k}}{10.817} = 0.740\mathbf{i} + 0.185\mathbf{j} - 0.647\mathbf{k}$$

If the magnitude of the force is 250 kips, then the force $\mathbf{F}$ can be expressed as

$$\mathbf{F} = 250^k (0.740\mathbf{i} + 0.185\mathbf{j} - 0.647\mathbf{k}) = 185\mathbf{i} + 46.3\mathbf{j} - 161.8\mathbf{k}$$

Dot Product

The dot product, also known as the scalar product, between two vectors $\mathbf{A}$ and $\mathbf{B}$ is defined as

$$\mathbf{A} \cdot \mathbf{B} = |\mathbf{A}||\mathbf{B}| \cos\theta \qquad (102.1)$$

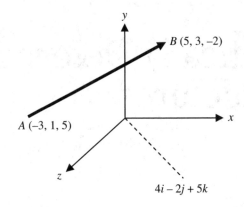

Figure 102.1
Vector in Cartesian coordinates.

B (5, 3, –2)

y

x

A (–3, 1, 5)

z

$4i - 2j + 5k$

where θ is the angle between the vectors A and B. In Cartesian coordinates, the dot product is calculated as

$$\mathbf{A} \cdot \mathbf{B} = A_x B_x + A_y B_y + A_z B_z \tag{102.2}$$

The result of the dot product operation is a scalar, which is why it is also called the scalar product. In the field of mechanics, the most common application for the dot product is the calculation of the component of a force in a particular direction. Note that the dot product only gives the magnitude of this component. For Fig. 102.1, in order to calculate the component of the force F along the dashed line $4i - 2j + 5k$, one must calculate the unit vector (direction only) of the line, which is given by

$$\mathbf{u}_L = \frac{4i - 2j + 5k}{6.708} = 0.596i - 0.298j + 0.716k$$

The component of the force is given by

$$F_L = \mathbf{F} \cdot \mathbf{u}_L = (185i + 46.3j - 161.8k) \cdot (0.596i - 0.298j + 0.716k) = -19.39$$

The (vector) component of the force $\mathbf{F}$ along the line L is then given by

$$\mathbf{F}_L = F_L \mathbf{u}_L = -19.39(0.596i - 0.298j + 0.716k) = -11.56i + 5.78j - 13.88k$$

Cross Product

The cross product, also known as the vector product, between two vectors A and B is calculated as

$$\mathbf{A} \times \mathbf{B} = [A_y B_z - A_z B_y]\mathbf{i} + [A_z B_x - A_x B_z]\mathbf{j} + [A_x B_y - A_y B_x]\mathbf{k} \tag{102.3}$$

This can also be expressed in the form of the determinant

$$\mathbf{A} \times \mathbf{B} = \begin{vmatrix} \mathbf{i} & \mathbf{j} & \mathbf{k} \\ A_x & A_y & A_z \\ B_x & B_y & B_z \end{vmatrix} \tag{102.4}$$

In the field of mechanics, the most common application for the cross product is the calculation of the moment of a force about a point, which is given by

$$\mathbf{M}_o = \mathbf{r} \times \mathbf{F} \tag{102.5}$$

where $\mathbf{r}$ is the (vector) distance from the pivot point O to any point on the line of action of the force.

In the example given in the section Vector in this chapter, the moment of the force about the origin is therefore calculated as the cross product of the distance from O (0, 0, 0) to either (−3, 1, 5) or (5, 3, −2) and the force $\mathbf{F}$.

$$\mathbf{r} = -3\mathbf{i} + \mathbf{j} + 5\mathbf{k}$$

$$\mathbf{F} = 185\mathbf{i} + 46.3\mathbf{j} - 161.8\mathbf{k}$$

$$\mathbf{r} \times \mathbf{F} = \begin{vmatrix} \mathbf{i} & \mathbf{j} & \mathbf{k} \\ -3 & 1 & 5 \\ 185 & 46.3 & -161.8 \end{vmatrix} = -393.3\mathbf{i} + 439.6\mathbf{j} - 323.9\mathbf{k}$$

Equivalent Force System

The analysis of force systems is greatly simplified by reducing a system of forces to an equivalent system of a single force and single couple. An illustrative example is shown below.

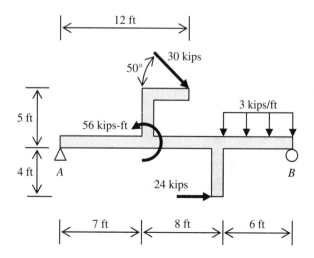

If the force system is to be replaced by a force-couple pair at the left support, the two systems must be equivalent in terms of the resultant force $\Sigma\mathbf{F}$ and the resultant moment about the left support $\Sigma\mathbf{M}_A$.

For the original system, the summation of (vector) forces is

$$\sum\mathbf{F} = 30\cos 50\mathbf{i} - 30\sin 50\mathbf{j} + 24\mathbf{i} - 18\mathbf{j} = 43.28\mathbf{i} - 40.98\mathbf{j}$$

The summation of (vector) moments about support A is

$$\sum\mathbf{M}_A = (12\mathbf{i} + 5\mathbf{j})\times(19.28\mathbf{i} - 22.98\mathbf{j}) + 56\mathbf{k} + (15\mathbf{i} - 4\mathbf{j})\times 24\mathbf{i} + 18\mathbf{i}\times -18\mathbf{j} = -544.16\mathbf{k}$$

 NOTE According to the right-hand rule, which is being used here, the negative **k** component implies that the resultant moment is clockwise.

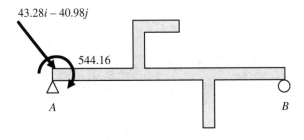

$43.28i - 40.98j$

544.16

A B

Analysis of Trusses

An ideal truss is based on the following assumptions:

1. All members are connected at nodes via frictionless pinned connections.

2. At each node, centroidal axes of connected members are concurrent.

3. Loads are located (lumped) at nodes only—no member forces.

If these assumptions are closely approximated, truss members experience only axial loads (no shear force or bending moment). For an ideal 3D truss, the total number of equilibrium equations available is equal to $3N$, where N is the number of nodes. The total number of unknowns is equal to the number of truss members, M (one axial force per member) plus the number of support reactions, R. For a truss to be classified as statically determinate, the number of equations is equal to the number of unknowns.

Statically determinate (3D) truss:	$3N = M + R$
Statically indeterminate (3D) truss:	$3N < M + R$
Unstable (3D) truss:	$3N > M + R$

Truss Member Forces—Method of Joints

The method of joints involves the listing of all equations of equilibrium, proceeding joint by joint. Each joint of a 2D (plane) truss yields two equilibrium equations—$\Sigma F_x = 0$ and $\Sigma F_y = 0$, where x and y are two mutually perpendicular axes in the plane. On the other hand, the number of unknown quantities for the plane truss is the number of external reactions plus the number of member forces.

Example 102.1   Plane Truss—Method of Joints

For the plane truss shown below, calculate the force in member *EF*.

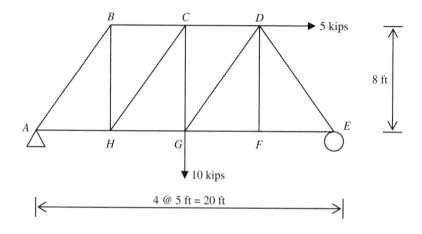

Solution   For the truss shown above, $R = 3$, $M = 13$, and $J = 8$. Thus, we have $R + M = 16 = 2J$. The truss is statically determinate.

Prior to solving for member forces, we need to solve for the external reactions. As long as there are no more than three external reactions, the truss is externally statically determinate.

$$\sum F_x = A_x + 5 = 0 \Rightarrow A_x = -5$$

$$\sum F_y = A_y - 10 + E_y = 0 \Rightarrow A_y + E_y = 10$$

$$\sum M_A = -(10 \times 10) - (8 \times 5) + (20 \times E_y) = 0 \Rightarrow E_y = 7$$

$$\therefore \quad A_y = 10 - E_y = 3$$

The statement of equilibrium of a joint involves two equations of equilibrium. Thus, in order for the equations to be solvable, progressing from node to node, we would have to start at a node where there are no more than two unknowns. *A* and *E*—the two support nodes—are such supports.

Let us start solving from node E.

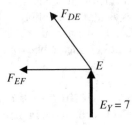

The free-body diagram of node E is shown above. The support reaction E_y has been solved, leaving the two unknowns F_{DE} and F_{EF}. One would therefore solve joints in turn, with no more than two unknown member forces for each free-body diagram.

$$\sum F_x = -F_{EF} - F_{DE}\cos 58 = 0 \Rightarrow F_{EF} = -F_{DE}\cos 58$$

$$\sum F_y = F_{DE}\sin 58 + 7 = 0 \Rightarrow F_{DE} = -\frac{7}{\sin 58} = -8.255$$

$$F_{EF} = -F_{DE}\cos 58 = -(-8.255)\cos 58 = 4.374$$

If faced with a situation such as shown above, where there are only three forces in equilibrium, the solution of the equilibrium equations may also be done geometrically, rather than algebraically. The force polygon formed by three forces in equilibrium is a triangle, and the force triangle must be similar to the geometric triangle. The two triangles are shown in Fig. 102.2.

Note that in the force polygon on the left, the only force which is known *a priori* is the 7-kips vertical reaction at node E. This force is upward. If the forces are drawn "tip-to-tail" so as to complete the force polygon, we have the required directions of the unknown forces F_{DE} and F_{EF}. Comparing these forces with the original free-body diagram, we see that F_{DE} must be compression (acting *into* the node, shown outlined) and F_{EF} must be tension (acting *outward* from the node).

Figure 102.2
Similarity of
force and geo-
metric polygons.

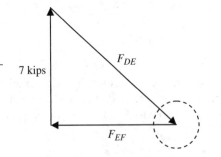

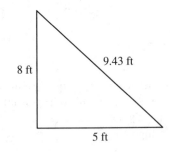

The values of these forces may be obtained by corresponding ratios from the similar triangles. These yield simple equations rather than coupled simultaneous equations.

$$\frac{F_{DE}}{7} = \frac{9.43}{8} \Rightarrow F_{DE} = \frac{7 \times 9.43}{8} = 8.255$$

$$\frac{F_{EF}}{7} = \frac{5}{8} \Rightarrow F_{EF} = \frac{7 \times 5}{8} = 4.375$$

Truss Member Forces—Method of Sections

To apply the method of sections, we want to cut the structure with a section, thus dividing it into two parts. At the location of the cut, we replace each cut member with the member force (which is still unknown). As in the method of joints, we will insert these unknown forces as tension forces to begin with. The algebraic sign of the force, once solved, will tell us whether the actual internal force is tensile or compressive.

There are some rules for choosing a valid section or cut. For a "cut" to be valid—such that the substructures are solvable—one must follow certain rules in choosing the orientation of the cut, i.e., which members it passes through.

1. The "cut" or section must separate the structure into two parts.

2. The "cut" must not pass through more than three unknown members (i.e., whose internal forces are unknown). The substructure is governed by the three equations of static equilibrium. Thus, any more than three unknowns will lead to an unsolvable system.

   Exception: In some situations, there is no single section that meets criteria 1 and 2. In those cases, it may be necessary to make two successive sections, producing a system of six simultaneous equations in six unknown member forces.

3. The unknown forces must not be concurrent. If these forces all pass through a common point, the problem will reduce to equilibrium of that point—producing two equations of equilibrium—$\Sigma F_x = 0$ and $\Sigma F_y = 0$. The equation $\Sigma M = 0$ will be satisfied identically ($0 = 0$) and will not yield any useful information.

Example 102.2   Plane Truss—Method of Sections

For the determinate plane truss shown below, calculate the force in member *CG*.

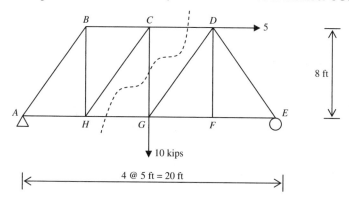

Solution To solve for F_{CG}, consistent with the rule given earlier, we choose a section that passes through the three members CD, CG, and GH. This section is shown as a dashed line in the diagram above. Verifying that this is consistent with the "rules," we note that:

1. The section does separate the structure into two parts.

2. The section passes through (no more than) three members—CD, CG, and GH.

3. The three members mentioned above are not concurrent.

Analyzing the entire truss leads to the following values for the support reactions:

$$A_x = -5 \text{ (left)}$$
$$A_y = +3 \text{ (up)}$$
$$E_y = +7 \text{ (up)}$$

Let us look at the left substructure.

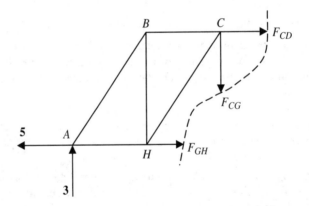

The equations of equilibrium are

$$\sum F_x = -5 + F_{GH} + F_{CD} = 0$$

$$\sum F_y = 3 - F_{CG} = 0 \qquad \Rightarrow F_{CG} = 3$$

$$\sum M_A = -8 \times F_{CD} - 10 \times F_{CG} = 0 \Rightarrow F_{CD} = -\frac{10 \times F_{CG}}{8} = -3.75$$

$$\Rightarrow F_{GH} = 5 - F_{CD} = 8.75$$

In the above case, the section passed through three members—all of which were either vertical or horizontal. As a result, the section equilibrium equations were relatively simple. None of the forces needed to be resolved into components. Let us consider another section of the same truss. Let us consider the vertical section passing through members *CD*, *DG*, and *GF*, as shown below.

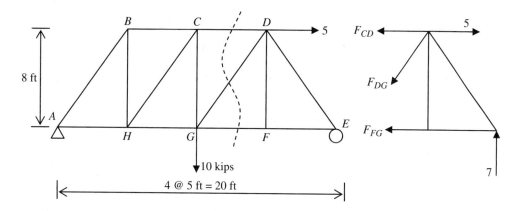

The free-body diagram of the right substructure is shown. The equations of equilibrium for the substructure are

$$\sum F_x = -F_{CD} - F_{DG} \cos 58 - F_{FG} + 5 = 0$$

$$\sum F_y = -F_{DG} \sin 58 + 7 = 0 \Rightarrow F_{DG} = \frac{7}{\sin 58} = 8.255$$

$$\sum M_D = -8 \times F_{FG} + 5 \times 7 = 0 \Rightarrow F_{FG} = \frac{35}{8} = 4.375$$

$$\Rightarrow F_{CD} = -F_{DG} \cos 58 - F_{FG} + 5 = -8.255 \times \cos 58 - 4.375 + 5 = -3.75$$

In many situations, one is asked for only one of the member forces rather than all three. A complete solution of the three (often simultaneous) equations is not necessary. Noting that moment equations give us the flexibility of choosing the moment pivot, we see that:

- If we are interested in solving for F_{CD}, the equation $\Sigma M_G = 0$ is algebraically simple, containing only the variable of interest F_{CD}. We achieve this by noting that we can eliminate the other two (F_{DG} and F_{FG}) unknown forces from our equation by taking the moments about point *G*, which is the intersection of these two forces.

- Similarly, in order to solve for F_{FG}, one would take moments about point *D*.

- If one wanted a simple equation in F_{DG}, one would attempt to eliminate F_{CD} and F_{FG}. This, however, cannot be done with a moment equation, as these two forces are parallel (horizontal). However, note that the $\Sigma F_y = 0$ equation would eliminate these two forces and yield an equation in only one variable F_{DG}.

Identification of Zero-Force Members

For an ideal truss (nodes are pinned, loads act only at nodes), each node is a system of concurrent forces. Some of these forces are collinear, i.e., they share the same line of action. If there are nodes where all forces can be grouped into *only two lines of action*, we may say the following:

1. If one line of action consists of a single force, that force must be zero.

2. If one line of action consists of a pair of forces, they must be equal and opposite.

 NOTE These rules apply if and only if the forces can be grouped into only two lines of action.

Example 102.3

Identify the zero-force members for the plane truss shown. The three panels of the truss are of equal length.

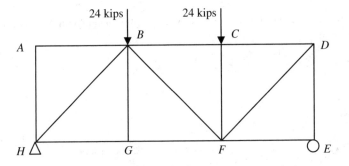

Solution

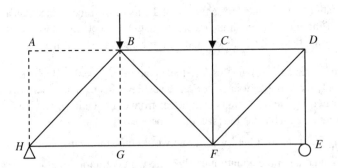

At *A*, there are *two members only*. Since there are no other forces at *A*, both **AB** and **AH** must be zero-force members. Also, at *G*, F_{BG} is the only force along its line of action. So, **BG** must be a zero-force member.

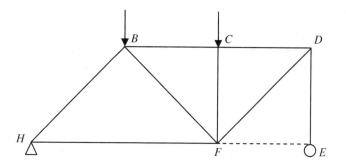

At E, F_{DE} and the vertical reaction at E along one line of action and F_{EF} is the only force along its line of action. Therefore, **EF** must be a zero-force member.

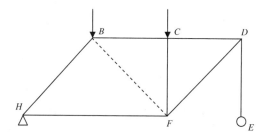

By external symmetry of structure *and* loading, both reactions (at H and E) must be 24 k vertical. Therefore, the panel shear in the middle panel is zero. Only member in the middle panel that can contribute to panel shear is **BF**. Therefore, **BF** is a zero-force member.

Truss Deflection—Method of Virtual Work

The principle of virtual work states that the external virtual work of virtual forces moving through real displacements is equal to the internal virtual work done by the stresses arising from the virtual forces moving through the strains corresponding to the real displacements. The words "real" and "virtual" in this statement may be interchanged.

The *virtual load* is an artificial load which is chosen to be consistent with the desired displacement. For example, if the vertical displacement is desired at point P on the structure, then the virtual load is a unit load applied vertically at point P.

For a truss, this yields the following expression for the deflection at any node (joint) of an ideal truss (axial loads only):

$$1 \cdot \Delta = \sum \mathbf{F} \left(\frac{\mathbf{f}L}{AE} \right) \Rightarrow \Delta = \sum \left(\frac{F \, \mathbf{f}L}{AE} \right) \tag{102.6}$$

where the summation occurs over all members in the truss, $\mathbf{F}$ is the member force due to the *actual load*, f is the member force due to the *virtual load* (a unit load applied in the same sense as the desired deflection), L is member length, A is the member cross-sectional area, and E is the member modulus of elasticity.

Example 102.4

A two-member truss is loaded with a vertical load of 20 kips at node B as shown. The cross sections of the members AB and BC are 2 in² for AB and 3 in² for BC. Modulus of elasticity $E = 29,000$ ksi for both members. Calculate the vertical deflection at B.

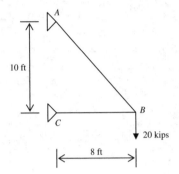

Solution Solve for truss member forces using equilibrium of joint B

$$F_{AB} = +25.6 \text{ k } (T)$$

$$F_{BC} = -16.0 \text{ k } (C)$$

Create the virtual load (consistent with desired deflection)

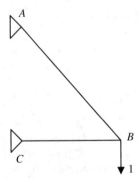

Calculate the member forces due to virtual load.

NOTE In this case, the real load and the virtual load look similar, so we can use scaling.

$$f_{AB} = +1.28 \ (T)$$

$$f_{BC} = -0.80 \ (C)$$

$$\Delta = \sum \left(\frac{FfL}{AE} \right) = \frac{25.6 \times 1.28 \times (12.81 \times 12)}{2 \times 29,000} + \frac{-16 \times -0.8 \times (8 \times 12)}{3 \times 29,000} = 0.10 \text{ in.}$$

NOTE The positive sign of the answer indicates that this deflection is in the same sense as the assumed unit load (downward).

Cables under Point Loads

Cable systems subjected to concentrated loads will result in a configuration of successive straight segments between the points of loading (node). At each node, since we have a system of concurrent forces, there are two or three equations of equilibrium (for 2D and 3D geometry, respectively). These equations can be solved for the forces (tensions in the different cable segments).

Example 102.5

For the cable system shown, find the tension in cables *AB* and *BC*.

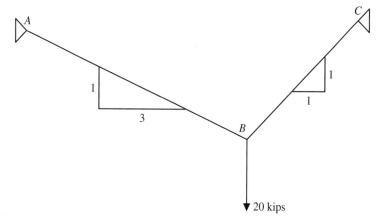

Solution   The free-body diagram (FBD) of joint *B* is shown below. Since this FBD has two unknowns, the two equations of equilibrium may be used to solve for them.

$$\sum F_x = -T_{AB}\frac{3}{\sqrt{10}} + T_{BC}\frac{1}{\sqrt{2}} = 0 \quad \Rightarrow \quad T_{AB} = 0.745T_{BC}$$

$$\sum F_y = T_{AB}\frac{1}{\sqrt{10}} + T_{BC}\frac{1}{\sqrt{2}} = 20 \quad \Rightarrow \quad \left(0.745\frac{1}{\sqrt{10}} + \frac{1}{\sqrt{2}}\right)T_{BC} = 20$$

$$T_{BC} = 21.21 \text{ kips} \quad \text{and} \quad T_{AB} = 15.81 \text{ kips}$$

These cable tensions may then be used to calculate the support reactions at *A* and *C*.

Similarly, a 3D system is statically determinate if no more than three cable tensions are involved. For example, consider a load being lifted by three cables as shown in Example 102.6.

Example 102.6

A crate weighing 900 lb is being lifted by three cables attached as shown. The attachment point for the cables (O) is 8 ft directly above the center of gravity of the load. Point B is at the midpoint of the short side and point C is at the midpoint of the long side. Calculate the tension in each cable.

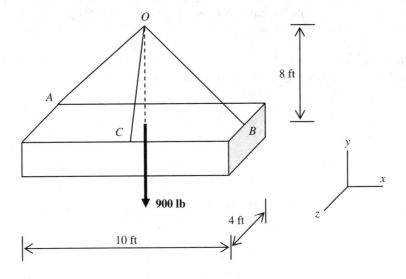

Solution   Since the suspension point is directly above the center of gravity, all four forces (three cable tensions and the weight) pass through it and, therefore, form a system of concurrent forces. The vectors **AO**, **BO**, and **CO** are, respectively

$$\mathbf{AO} = 5\mathbf{i} + 8\mathbf{j} + 2\mathbf{k}$$
$$\mathbf{BO} = -5\mathbf{i} + 8\mathbf{j} + 0\mathbf{k}$$
$$\mathbf{CO} = 0\mathbf{i} + 8\mathbf{j} - 2\mathbf{k}$$

and the unit vectors are

$$\mathbf{u}_{AO} = \frac{5i + 8j + 2k}{\sqrt{5^2 + 8^2 + 2^2}} = 0.52i + 0.83j + 0.21k$$

$$\mathbf{u}_{BO} = \frac{-5i + 8j}{\sqrt{5^2 + 8^2}} = -0.53i + 0.85j$$

$$\mathbf{u}_{CO} = \frac{8j - 2k}{\sqrt{8^2 + 2^2}} = 0.97j - 0.24k$$

The vector equation of equilibrium can be written as

$$T_A \mathbf{u}_{AO} + T_B \mathbf{u}_{BO} + T_C \mathbf{u}_{CO} + \mathbf{W} = \mathbf{0}$$

$$T_A(0.52i + 0.83j + 0.21k) + T_B(-0.53i + 0.85j) + T_C(0.97j - 0.24k) - 900j = 0$$

Separating this vector equation into three scalar equations, we get

$$0.52T_A - 0.53T_B = 0$$

$$0.83T_A + 0.85T_B + 0.97T_C = 900$$

$$0.21T_A - 0.24T_C = 0$$

Solving these equations, $T_A = 358.32$ lb, $T_B = 351.56$ lb, $T_C = 313.53$ lb.

Cables under Uniformly Distributed Load

When cables are subjected to uniformly distributed loads, the cable segments are no longer straight. Cables under distributed loads can be grouped into two categories:

1. Load distributed uniformly along the horizontal axis
2. Load distributed uniformly along the length of the cable

Load Distributed Uniformly along Horizontal Axis (e.g., Bridge Deck)

For a system such as a suspension cable supporting a bridge deck, it is common for the deck load to be specified per unit length along the deck span. For such a loading, the cable assumes a parabolic shape (see Fig. 102.3). The load distribution can be expressed as $dw/dx = $ constant.

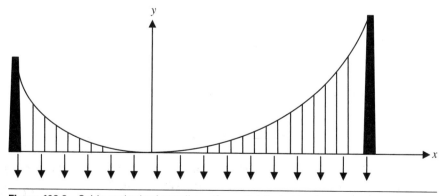

Figure 102.3 Cable under horizontally uniform distributed load.

With the origin at the lowest point on the cable (as shown in Fig. 102.3), the cable has the following parabolic equation:

$$y = \frac{wx^2}{2H} \tag{102.7}$$

where H is (minimum) cable tension at the lowest point $H = T_{min}$.

NOTE If the origin is located elsewhere, the equation of the cable sag shape must be assumed as the general parabola $y = ax^2 + bx + c$ and all coefficients determined before proceeding.

The force H can be found by substituting the coordinates of any point on the cable (x_o, y_o) other than the origin

$$H = \frac{wx_o^2}{2y_o} \tag{102.8}$$

The tension at any point (x, y) on the cable is given by

$$T = \sqrt{H^2 + w^2 x^2} \tag{102.9}$$

Thus, tension in the cable is maximum at the point furthest from the low point. At any point (x, y) on the cable, the angle θ that it makes with the horizontal is given by

$$\tan\theta = \frac{wx}{H} \tag{102.10}$$

Example 102.7

A suspension bridge cable is sprung from towers separated horizontally by 1200 ft. The sag of the cable below the top of the left tower is 190 ft and below the top of the right tower is 110 ft. If the deck load supported by the cable is 6 kips/ft, what is the maximum tension in the cable?

Solution

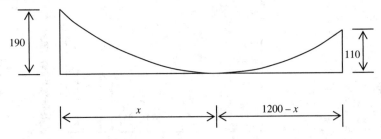

In this case, the cable is not symmetric. The first problem, therefore, is to locate the lowest point. If we assume that the lowest point is at a distance x from the left end, then

$$190 = cx^2 \quad \text{and} \quad 110 = c(1200 - x)^2$$

which can be simplified to

$$\frac{190}{110} = \frac{x^2}{(1200-x)^2} \Rightarrow x = 681.5$$

Thus, the low point of the cable is located 681.5 ft from the left support.

Using the coordinates of the left springing point as x_o, y_o (since maximum tension occurs furthest from low point)

$$H = \frac{wx_o^2}{2y_o} = \frac{6 \times 681.5^2}{2 \times 190} = 7333 \text{ kips}$$

$$T_{max} = \sqrt{7333^2 + 6^2 \times 681.5^2} = 8396 \text{ kips}$$

Load Distributed Uniformly along Cable Length (e.g., Self Weight)

For a cable that carries a load which is uniformly distributed along its length (e.g., self weight), the shape of the deflected cable is called a *catenary*. With the origin below the lowest point as shown in Fig. 102.4, the equation of the deflected cable is

$$y = c\cosh\left(\frac{x}{c}\right) = c\left[1 + \frac{x^2}{2c^2} + \frac{x^4}{2c^4} + ...\right] \qquad (102.11)$$

where $c = H/q_o$ = distance from the origin to the lowest point

H = minimum cable tension (at the lowest point of the cable)

q_o = load per unit length of the cable = dw/ds

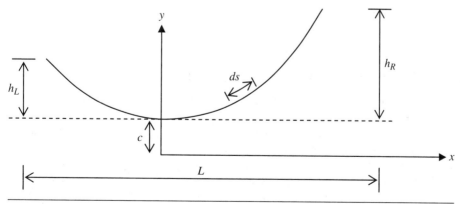

Figure 102.4 Cable under uniformly distributed load (along length).

However, close to the origin, neglecting terms of order higher than x^2, we get

$$y \approx c + \frac{x^2}{2c} \tag{102.12}$$

Thus, close to the origin, where x is small, the catenary can be approximated by a parabola. Given the sag of a catenary (h_L and h_R being the vertical height of the left and right supports above the low point respectively and L being the horizontal distance between supports), the horizontal component of cable tension H can be computed from

$$\frac{q_o L}{H} = \cosh^{-1}\left(\frac{q_o h_L}{H} + 1\right) + \cosh^{-1}\left(\frac{q_o h_R}{H} + 1\right) \tag{102.13}$$

where q_o is the load per unit length of the cable $= dw/ds$. The coordinate s is the distance from the low point measured along the cable. Equations (102.11) and (102.13) can only be solved by trial and error.

Length of arc measured from the low point is given by

$$s = \frac{H}{q_o} \sinh \frac{q_o x}{H} \tag{102.14}$$

At any point on the cable, the tension is given by

$$T = \sqrt{H^2 + q_o^2 s^2} \tag{102.15}$$

Vertical components of support reactions are

$$R_L = H \sinh \frac{q_o a}{H} \quad \text{and} \quad R_R = H \sinh \frac{q_o b}{H} \tag{102.16}$$

where a and b are horizontal distance from the low point (origin) to the left and right supports respectively.

Shear Force and Bending Moment

Figure 102.5 shows a determinate beam loaded with loads that are primarily transverse (to the axis of the beam).

To obtain the internal loads at any intermediate location of the beam, we make a section at some arbitrary point. This section divides our structure into two substructures, as shown in Fig. 102.6. At the "cut" one must insert the set of internal loads as appropriate. There are three internal loads present at an arbitrary location of a 2D beam. These are axial force (longitudinal), shear force (transverse), and bending moment. Note how the internal loads occur in equal and opposite pairs on the two halves of the structure. In other words, the axial force exerted by

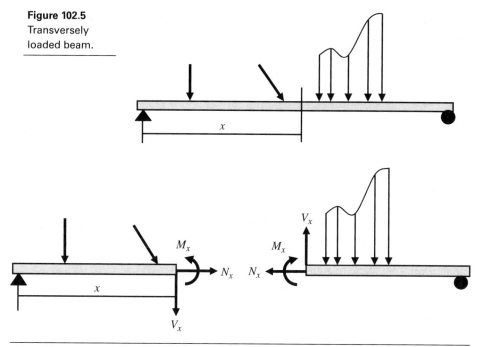

Figure 102.5
Transversely
loaded beam.

Figure 102.6   Internal loads for transversely loaded beam.

the left half on the right is *equal and opposite* to that exerted by the right half on the left. The same holds for the shear force and bending moment. The internal loads are drawn in keeping with a sign convention that is explained in more detail in Fig. 102.6. Notice how the positive sense of an internal load on the left of a cut is exactly opposite to that on the right of the cut.

The sign conventions for internal loads are shown in Fig. 102.7. The dashed line shows the original configuration of an element of the beam, whereas the solid lines show the type of deformation induced by these internal loads.

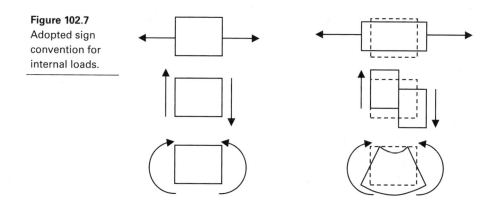

Figure 102.7
Adopted sign
convention for
internal loads.

The following relationships exist between the load function $w(x)$, the shear-force function $V(x)$, and the bending-moment function $M(x)$:

$$\frac{dV}{dx} = -w(x) \tag{102.17}$$

$$\frac{dM}{dx} = V(x) \tag{102.18}$$

Equations (102.17) and (102.18) are in differential form. Geometrically, they have the following significance:

The slope of the shear-force function is equal to the (negative) value of the load function.

The slope of the bending-moment function is equal to the value of the shear force.

In integral form, these relationships may be expressed as

$$V = -\int w(x)dx \Rightarrow V_2 - V_1 = -\int w(x)dx \tag{102.19}$$

$$M = \int V(x)dx \Rightarrow M_2 - M_1 = \int V(x)dx \tag{102.20}$$

The geometric significance of the integral equations may be written as

The change in shear force (between locations 1 and 2) is equal to the (negative) area under the load function between the same points.

The change in bending moment (between locations 1 and 2) is equal to the area under the shear-force function between the same points.

Also, the (maximum or minimum) bending moment occurs at the location of zero shear force.

The following geometric principles may be used to draw shear and bending-moment diagrams, by first following the given load $w(x)$ to create the shear force $V(x)$ and then following the shear diagram to create the bending moment $M(x)$:

- Concentrated (point) loads push shear diagram up/down.
- Clockwise/counterclockwise couples (externally applied or reaction moment) push bending-moment diagram up/down.
- Change in shear force between two locations = negative of the area under the load diagram.
- Change in bending moment between two locations = area under the shear diagram.
- Maximum/minimum bending moment occurs at location of zero shear.

Shear and bending-moment diagrams can be constructed in two ways:

1. To find the shear-force function $V(x)$ or bending-moment function $M(x)$ for a section of the beam, one makes a section or "cut" at that location, inserts the internal loads $V(x)$, $M(x)$ at the cut and solves for these functions.

2. Using the geometric relationships between the load function $w(x)$, the shear-force function $V(x)$ and the bending-moment function $M(x)$, one can progress from the load diagram $w(x)$ to the shear-force diagram $V(x)$ to the bending-moment diagram $M(x)$.

Example 102.8

Draw the shear and bending-moment diagrams for the beam shown below. The beam has an internal hinge at 6 ft from the left support.

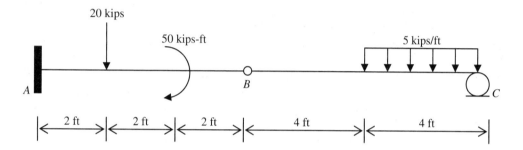

Solution   The first step would be to solve for the support reactions. The beam is statically determinate due to the presence of the internal hinge. Even though a formal approach could be to write the four equations of equilibrium (the standard three equations for a 2D structure plus one hinge moment equation) and solve for the four external reactions (A_x, A_y, M_A, and C_y), it is more convenient and time effective to separate the structure at the hinge B and start solving from the simple span BC (both ends moment free).

Reactions: By lever rule: $C_y = 3/4 \times 20 = 15$ kips   and $B_y = 1/4 \times 20 = 5$ kips
$\qquad A_y = 20 + 5 = 25$ kips
$\qquad M_A = 20 \times 2 + 50 + 5 \times 6 = 120$ kips-ft counterclockwise

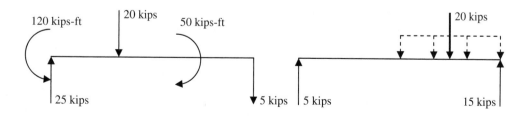

Next, the shear-force diagram can be created by simply following the transverse (shear) loads such as point loads and distributed loads.

Shear-Force diagram

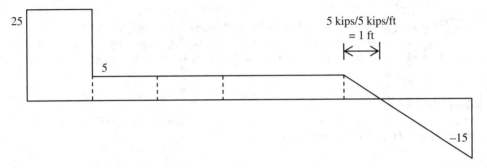

Note that at the internal hinge, there is an internal shear that exists as a pair of equal and opposite forces and, therefore, does not affect the diagram. However, the shear at this location (the hinge shear) is reflected in the diagram. Also, the uniformly distributed load has a resultant (area under the load) of 20 kips. The change in shear force between $x = 10$ and $x = 14$ is the negative of the area under the load. Thus, the shear force changes from $V = +5$ at $x = 10$ to $V = -15$ at $x = 14$. This change occurs at a uniform rate (linear). The location at which the shear force goes through a zero is given by

$$\Delta x = \frac{\Delta V}{w} = \frac{5 \text{ kips}}{5 \text{ kips/ft}} = 1 \text{ ft}$$

Bending-Moment diagram

The bending-moment diagram is constructed by "accumulating areas" from the shear-force diagram. Since the bending-moment function is obtained by integration of the shear function, it is one order more complex than the order of the corresponding shear diagram.

In other words, if the shear function is linear, the corresponding moment function is quadratic. In the Example 102.8, there are two locations where a couple moment exists—at $x = 0$, there is a reaction moment $M = 120$ kips-ft (counterclockwise) and at $x = 4$, there is an applied moment $M = 50$ kips-ft (clockwise). These introduce corresponding discontinuities in the moment diagram at these locations.

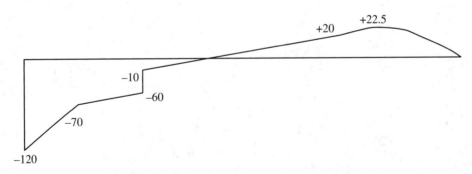

Example 102.9

Draw the shear-force diagram and the bending-moment diagram for the simply supported beam with overhang loaded with a distributed load, as shown below.

Method 1—Geometric Method

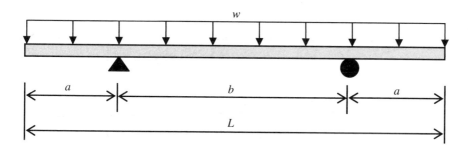

As before, we can write the equations of static equilibrium and solve for the support reactions. However, in this case, we can use symmetry arguments to say that due to symmetry of the structure and the loading, the load wL will be shared equally by the two supports. Next, we will follow the load to draw the shear-force diagram.

The overhang to the left of the first support carries a total load of wa. The shear-force diagram slopes down from a value of zero to $-wa$. At the left support, the reaction ($wL/2$) pushes the diagram up by $wL/2$ to a value $+wb/2$.

Between supports, the total load is wb. Thus, the shear-force diagram goes from a value of $+wb/2$ to $-wb/2$, sloping down continuously at the same rate as the load, i.e., w per unit length.

At the right support, the upward reaction $wL/2$ pushes the diagram up from a value of $-wb/2$ to $+wa$. Over the right overhang, the load wa pushes the diagram down to zero.

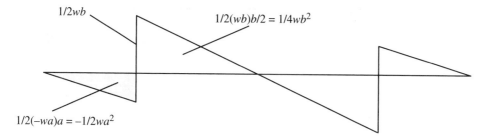

We can then follow the shear-force diagram to construct the bending-moment diagram. The following geometric facts can be used in our construction.

- The value of the bending moment is zero at the free ends.

- The change in the bending moment between any two points is equal to the area under the shear-force diagram between those points.

- The order of the bending-moment curve is one order higher than the shear-force curve. (If the SFD is a straight line, i.e., x^1, then the BMD is a parabola, i.e., x^2.)

- The location of zero shear force, or more correctly, the location of change of sign of the shear force is the location of the maximum (or minimum) bending moment.

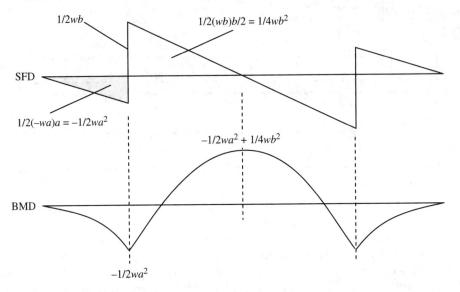

Method 2—Explicit Forms of V(x) and M(x) Functions

As we see from the diagram, the loading on the beam changes at two points—the supports, where the reactions are introduced. These two points divide the beam into three segments— *AB*, *BC*, and *CD*.

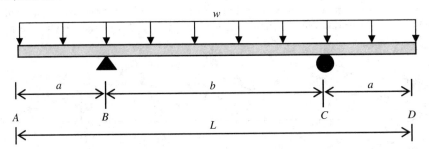

First Section in Segment AB (0 < x < a)

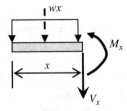

$$\sum F_y = -wx - V_x = 0 \qquad \Rightarrow V_x = -wx$$

$$\sum F_y = M_x + wx\frac{x}{2} = 0 \qquad \Rightarrow M_x = \frac{-wx^2}{2}$$

Second Section in Segment BC (a < x < a + b)

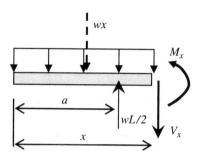

$$\sum F_y = -wx + w\frac{L}{2} - V_x = 0 \qquad \Rightarrow V_x = w\frac{L}{2} - wx$$

$$\sum F_y = M_x + wx\frac{x}{2} - w\frac{L}{2}(x-a) = 0 \qquad \Rightarrow M_x = \frac{-wx^2 + wLx - wLa}{2}$$

Third Section in Segment CD (a + b < x < L)

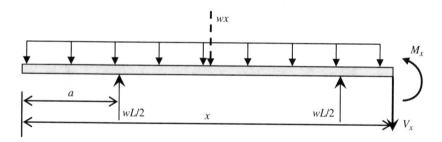

$$\sum F_y = -wx + w\frac{L}{2} + w\frac{L}{2} - V_x = 0 \qquad \Rightarrow V_x = wL - wx$$

$$\sum F_y = M_x + wx\frac{x}{2} - w\frac{L}{2}(x-a) - w\frac{L}{2}(x-a-b) = 0 \qquad \Rightarrow M_x = \frac{-w(L-x)^2}{2}$$

A few more cases that illustrate the geometric principles behind drawing of shear-force and bending-moment diagrams are shown in this section. Note that specific values (of reactions, etc.) are not computed in these examples, but attention is paid to consistency of the diagrams.

Example 102.10

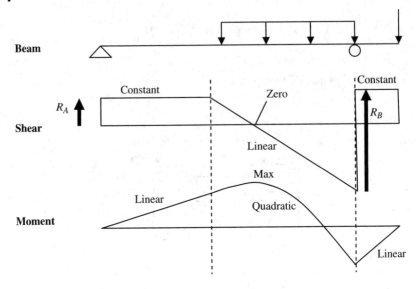

Example 102.11

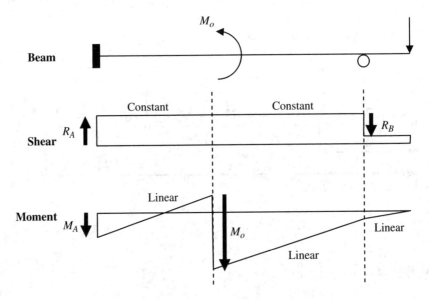

Knowledge of the approximate qualitative shape of the bending-moment diagram allows us to know the nature of the normal stress—tensile or compressive—at various locations of the beam. Consistent with the sign convention adopted here, a location with positive bending moment creates compression above the neutral axis and tension below. For negative bending moment, this pattern is reversed. For elements such as concrete beams in flexure, the inherent weakness of the concrete in tension makes it necessary to place steel reinforcement in the zones with maximum tensile stress.

Example 102.12

For the beam shown below, draw the shear-force diagram and the bending-moment diagram. Also, identify functions $V(x)$ and $M(x)$ for all segments of the beam.

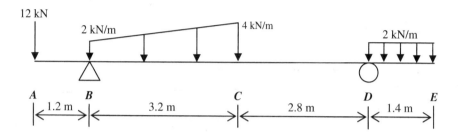

Solution

Step 1   Replace the distributed loads with the resultant forces at the respective centroids. The trapezoid has been separated into a rectangle (area 6.4 kN at a distance of 3.2 m from B) and a triangle (area 3.2 kN at a distance of 1.07 m from C).

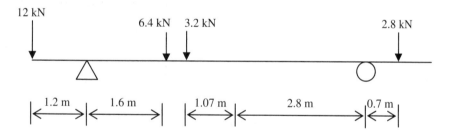

Step 2   Take moments about B, the hinge on the left (clockwise positive)

$$\sum M_B = 0 \Rightarrow -12 \times 1.2 + 6.4 \times 1.6 + 3.2 \times \frac{2}{3} \times 3.2 + 2.8 \times 6.7 - 6D_y = 0 \Rightarrow D_y = 3.571 \text{ kN}$$

Therefore, $B_y = 24.4 - 3.571 = 20.829$ kN

Step 3   The shear-force diagram can be drawn visually by following the transverse loads. Note that the total load in the trapezoidal load is 9.6 kN and, therefore, the change in shear force between B and C is -9.6 kN. At this point, it is known that since the load function between B and C is linear (x^1), the shear-force function must be parabolic (x^2).

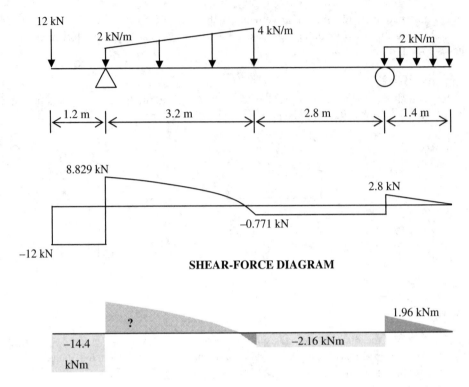

SHEAR-FORCE DIAGRAM

AREAS UNDER SHEAR-FORCE DIAGRAM

Evidently, the areas bounded by the parabola between $x = 1.2$ and $x = 4.4$ are not known yet. The other areas are triangles and rectangles and can be readily calculated.

Step 4   The section $1.2 < x < 4.4$ is shown below:

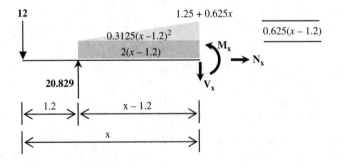

Writing the equilibrium equations, for this section, we get:

$$\sum M_{cut} = 0 \Rightarrow -12x + 20.829(x-1.2) - 2(x-1.2)\frac{x-1.2}{2} - 0.3125(x-1.2)^2\frac{x-1.2}{3} - M_x = 0$$

$$\Rightarrow M_x = -0.1042x^3 - 0.625x^2 + 10.779x - 26.2548$$

and

$$\sum F_y = 0 \Rightarrow -12 + 20.829 - 2(x-1.2) - 0.3125(x-1.2)^2 - V_x = 0$$

$$\Rightarrow V_x = -0.3125x^2 - 1.25x + 10.779$$

Alternatively, in this zone $(1.2 < x < 4.4)$, since the load function is given by a linear function, we can deduce that

$$w(x) = 1.25 + 0.625x \text{ (this function yields the value}$$
$$w = 2 \text{ at } x = 1.2 \text{ and } w = 4 \text{ at } x = 4.4)$$

Since

$$\frac{dV}{dx} = -w(x) \Rightarrow V(x) = -\int w(x)dx = \int(-1.25 - 0.625x)dx = -1.25x - 0.3125x^2 + c_1$$

Since the shear force at $x = 1.2+$ is known to be 8.829 kips, the constant c_1 can be calculated as 10.779.
This completes the shear-force function as

$$V(x) = 10.779 - 1.25x - 0.3125x^2$$

$$\frac{dM}{dx} = V(x) \Rightarrow M(x) = \int V(x)dx = \int(10.779 - 1.25x - 0.3125x^2)dx$$

$$= 10.779x - 0.625x^2 - 0.1042x^3 + c_2$$

Since the bending moment at $x = 1.2+$ is known to be -14.4 kip-ft, the constant c_2 can be calculated as -26.2548.
This completes the bending-moment function as

$$M(x) = -26.2548 + 10.779x - 0.625x^2 - 0.1042x^3$$

Step 5   The point at which the shear is equal to zero is also the point at which the bending moment is maximum (because $dM/dx = V$). Setting the shear function equal to zero, this point is calculated as

$$-0.3125x^2 - 1.25x + 10.779 = 0 \Rightarrow x = 4.204$$

At this location, the bending moment $M = 0.275$ kNm

The area under the shear-force diagram from $x = 1.2$ to 4.204 is calculated as:

$$\int_{1.2}^{4.204}(-0.3125x^2 - 1.25x + 10.779)dx = (-0.1042x^3 - 0.625x^2 + 10.779x)_{1.2}^{4.204} = 14.675$$

This confirms that the bending moment $M_{4.204} = -14.4 + 14.675 = 0.275$ kNm

At $x = 4.4$ m (end of the distributed load), bending moment $M = 0.2$ kNm

The final versions of the shear and moment diagrams are shown on the next page.

The area under the shear-force diagram from $x = 4.204$ to 4.4 is calculated as

$$\int_{4.204}^{4.4}(-0.3125x^2 - 1.25x + 10.779)dx = (-0.1042x^3 - 0.625x^2 + 10.779x)_{4.204}^{4.4} = -0.075$$

This confirms that the bending moment $M_{4.4} = 0.275 - 0.075 = 0.20$ kNm

Step 6   The functions are as follows:

Domain	$V(x)$	$M(x)$
$0 < x < 1.2$	-12	$-12x$
$1.2 < x < 4.4$	$-0.3125x^2 - 1.25x + 10.779$	$-0.1042x^3 - 0.625x^2 + 10.779x - 26.2548$
$4.4 < x < 7.2$	-0.771	$-0.771x + 3.5912$
$7.2 < x < 8.6$	$-2x + 17.2$	$-x^2 + 17.2x - 73.96$

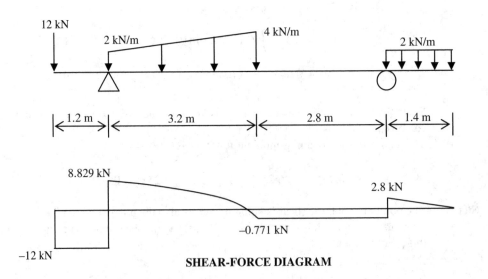

SHEAR-FORCE DIAGRAM

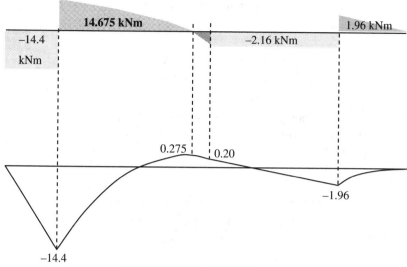

BENDING-MOMENT DIAGRAM

Beam Deflection—the Elastic Curve

If stresses are relatively low, the bending behavior of a beam is likely to be linear and elastic. Moreover, if shear deformations are ignored, the bending deformation is given by the moment-curvature relationship. This relates the curvature at every spanwise location of the beam to the bending moment at that location. The exact form of the moment-curvature relation is given by

$$\text{Curvature } \kappa = \frac{M}{EI} \tag{102.21}$$

If the deflected shape of the beam is described by the function $y(x)$, the curvature κ is given by

$$\text{Curvature } \kappa = \frac{1}{R} = \frac{1}{\text{radius of curvature}} = \frac{y''}{[1 + (y')^2]^{3/2}} \tag{102.22}$$

where y' = the first derivative of the deflection function = dy/dx

$\quad y''$ = the second derivative = d^2y/dx^2

In the moment-curvature relationship [Eq. (102.21)], it has been assumed that shear deformations do not contribute significantly to the total deflection of a beam and therefore can be neglected. According to Rankine, the additional slope of the elastic curve produced by shear is given by

$$\frac{dy_1}{dx} = \frac{VQ}{GIb} = \frac{dM}{dx}\frac{Q}{GIb}$$

(102.23)

As the shear stress varies throughout the cross section, it is very difficult to incorporate this into the deflection equation. However, if we assume a specific case (simply supported beam span L, rectangular section of width b and depth h, carrying a uniformly distributed load w), the additional deflection due to shear compared to the deflection due to the bending moment only is given by

$$\frac{\text{added deflection due to shear}}{\text{deflection due to moment only}} = \frac{6E}{5G}\left(\frac{h}{L}\right)^2$$

If the beam is made of steel, the ratio $E/G = 2.6$. Assuming a typical h/L ratio of 1:20, this added deflection fraction becomes 0.0078 (0.78%), which is small enough to be ignored. On the other hand, for a depth to length (h/L) ratio of 1:4, this fraction becomes nearly 20%. Thus, for deep beams (depth ratio h/L approaching 1:6), the contribution of the shear deformation to the overall deformation of the beam becomes too large to ignore.

For small deformations, assuming slope (y') is small, $(y')^2 \approx 0$, curvature $\kappa \approx d^2y/dx^2$ and the linearized version of the moment curvature equation becomes

$$\frac{d^2y}{dx^2} = \frac{M}{EI}$$

(102.24)

Solution of the equation of the elastic curve yields the deflected shape $y(x)$. This solution may be obtained in one of several ways.

Direct Integration Method

The second-order (linear) differential equation can be integrated twice [see Eqs. (102.26) and (102.27)] to generate the function $y(x)$. This method can work well if the bending-moment diagram is composed of a single function over the entire beam, thereby necessitating a single series of integrations. The disadvantage of this method is that the two integration constants must be evaluated from the boundary conditions to complete the function $y(x)$. This makes the process tedious if the bending-moment diagram is composed of more than one function defined over several subdomains of the beam.

$$y'' = \frac{M(x)}{EI}$$

(102.25)

$$y' = \int \frac{M(x)}{EI} dx + c_1 \qquad\qquad (102.26)$$

$$y = \iint \frac{M(x)}{EI} dx\, dx + c_1 \cdot x + c_2 \qquad\qquad (102.27)$$

Example 102.13

For the simply supported beam shown below, answer the following questions:

(i) What is the equation of the elastic curve?

(ii) Where does the maximum deflection occur?

(iii) What is the maximum deflection in inches?

(iv) What is the rotation at left and right ends (in degrees)?

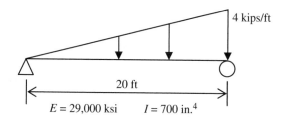

Solution We adopt the following units: kip (force) and inch (length). Therefore, $L = 240$ in., $w = 0.333$ kip/in.

Step 1   Calculate the support reactions.
Resultant load $= 1/2 \times 0.333 \times 240 = 40$ kips
Vertical reactions at left and right ends $= 13.333$ kips and 26.667 kips

Step 2   Make a section at distance x from the left end and solve for the bending-moment function $M(x)$.

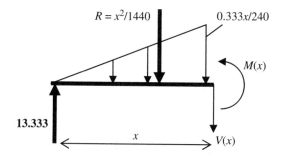

$$\sum M_{cut} = -13.333x + \frac{x^2}{1440}\frac{x}{3} + M(x) = 0 \Rightarrow M(x) = 13.333x - \frac{x^3}{4320}$$

Step 3   Moment curvature relationship

$$y'' = \frac{M}{EI} \Rightarrow EIy'' = M(x) = 13.333x - \frac{x^3}{4320}$$

$$EIy' = 6.6667x^2 - \frac{x^4}{17,280} + c_1$$

$$EIy = 2.2222x^3 - \frac{x^5}{86,400} + c_1x + c_2$$

Step 4   Boundary conditions
(i) At $x = 0$, $y = 0$ $\rightarrow$ $c_2 = 0$
(ii) At $x = 240$, $y = 0$ $\rightarrow$ $c_1 = -89,600$

Step 5   Equation of the elastic curve

$$y = \frac{1}{EI}\left(2.2222x^3 - \frac{x^5}{86,400} - 89,600x\right)$$ **ANSWER (i)**

Step 6   Slope function (used to determine point of zero slope, thus maximum deflection)

$$y' = \frac{1}{EI}\left(6.6667x^2 - \frac{x^4}{17,280} - 89,600\right)$$

Step 7   Solving the equation $6.6667x^2 - \dfrac{x^4}{17,280} - 89,600 = 0$

By making the substitution $u = x^2$, we get the quadratic equation

$$5.787 \times 10^{-5}u^2 - 6.6667u + 89,600 = 0$$

which has the solutions

$$u = 99,669,\ 15,532 \qquad \text{leading to } x = 315.7,\ 124.6$$

The first solution is inadmissible (outside the range 0 to 240). Therefore, deflection is maximum at $x = 124.6$ in.   **ANSWER (ii)**

$$y_{max} = \frac{1}{29,000 \times 700}\left[2.222 \times 124.6^3 - \frac{124.6^5}{86,400} - 89,600 \times 124.6\right] = -0.355 \text{ in.}$$

ANSWER (iii)

Rotation (slope) at left end is $y'\big|_{x=0} = -0.0044 \text{ rad} = -0.25°$ (clockwise).

Rotation (slope) at right end is $y'\big|_{x=240} = +0.005 \text{ rad} = +0.29°$ (counterclockwise).

ANSWER (iv)

Maximum bending moment is obtained by setting $dM/dx = 13.3333 - x^2/1440 = 0 \Rightarrow x = 138.6$.

Therefore, $M_{max} = M\big|_{x=138.6} = 1231.68$ kips-in.

Alternatives to the Direct Integration Method

The tedium of the successive integration method can be alleviated by using one of several graphical techniques that rely on converting one second-order differential equation into two first-order differential equations. Since we have $d^2y/dx^2 = M/EI$, and slope θ may be written as $\theta = dy/dx$ (small slope approximation: $\theta = \tan\theta$), one may write the following first-order equations:

$$\theta = \frac{dy}{dx} \tag{102.28}$$

$$\frac{d\theta}{dx} = \frac{M}{EI} \tag{102.29}$$

Moment-Area Method

The first-order differential equation can also be written in integral form

$$\frac{d\theta}{dx} = \frac{M}{EI} \Rightarrow d\theta = \frac{M}{EI} \cdot dx$$

The integral form of the equation is

$$\theta_2 - \theta_1 = \int_{x_1}^{x_2} \frac{M}{EI} dx \tag{102.30}$$

which has the following geometric significance: change in slope θ = area under M/EI diagram. This may be used to compute slope at any point on the beam if there is at least one point where the slope is known and the M/EI diagram is composed of relatively simple shapes whose areas can be computed easily.

Also, it can be shown that if tangents are drawn at two points A and B on the deflected shape of the beam, then the (vertical) deviation between tangents measured at point B is equal to the product of the area (under M/EI diagram) between A and B and the distance from B to the centroid of this area. This may be written as

$$t_{A/B} = \bar{x} \cdot \left[\int_{x_A}^{x_B} \frac{M}{EI} \cdot dx \right] = \bar{x} \cdot A \tag{102.31}$$

where $t_{A/B}$ is deviation of tangent at A with respect to tangent at B and $\bar{x}$ is distance of centroid of area $\int_{x_A}^{x_B} (M/EI)dx$ from point B.

For a demonstration of the moment area method, see Example 102.14.

Example 102.14   Moment Area Method

What is the slope at left support A for a simply supported beam carrying a point load as shown?

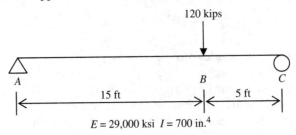

120 kips

A B C

15 ft   5 ft

$E = 29{,}000$ ksi $I = 700$ in.[4]

Solution   The bending-moment diagram is shown below ($M_{max} = 450$ kips-ft). The approximate deflected shape is also shown on the diagram. Note that the tangents at A and C are drawn. The deflection between tangents, measured at C ($t_{A/C}$) is related to the slope θ_A by

$$\tan \theta_A = \frac{t_{A/C}}{L}$$

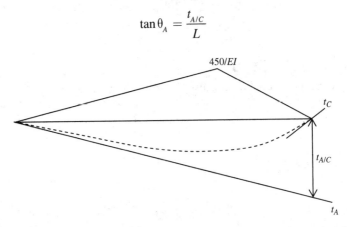

450/EI

t_C

$t_{A/C}$

t_A

According to the moment area method, the deflection between tangents ($t_{A/C}$) is given by the product of the area between A and C (i.e., the entire M/EI diagram) and the distance to the centroid from C. Using the definition of the centroid, we can also calculate this product as the sum of the products of the constituent parts of the diagram.

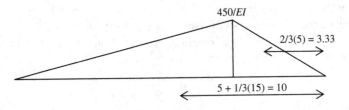

450/EI

2/3(5) = 3.33

5 + 1/3(15) = 10

Thus, $t_{A/C} = \dfrac{1}{2} \times 15 \times \dfrac{450}{EI} \times 10 + \dfrac{1}{2} \times 5 \times \dfrac{450}{EI} \times 3.33 = \dfrac{37{,}500}{EI}$

$\tan \theta_A \approx \theta_A = \dfrac{3750}{EI}$

Similarly, $t_{CIA} = \dfrac{1}{2} \times 15 \times \dfrac{450}{EI} \times 10 + \dfrac{1}{2} \times 5 \times \dfrac{450}{EI} \times 16.67 = \dfrac{52,500}{EI}$

$\tan\theta_C \approx \theta_C = \dfrac{5250}{EI}$

Conjugate Beam Method

The conjugate beam method is a mathematical device which may be used to calculate deflections and slopes in beams from first principles. The method is based on the moment-curvature relationship.

$$K = \frac{1}{\rho} = \frac{d^2 y}{dx^2} = \frac{d\theta}{dx} = \frac{M}{EI}$$

The foundation of the method is in the similarity between the following pairs of equations.

$$\frac{dV}{dx} = -w(x) \qquad \frac{d\theta}{dx} = \frac{M(x)}{EI}$$

$$\frac{dM}{dx} = V(x) \qquad \frac{dy}{dx} = \theta \qquad (102.32)$$

The pair on the left shows the relationships between the load function $w(x)$, the shear-force function $V(x)$ and the bending-moment function $M(x)$.

The pair on the right shows the relationships between the curvature (expressed as the derivative of the slope θ) and the bending-moment function $M(x)$. In this pair, the assumption of small deflections has been used. As a result, curvature is expressed as the linear function d^2y/dx^2 rather than the exact nonlinear function and dy/dx has been approximated by θ rather than $\tan\theta$.

Due to this similarity, one may draw the following analogy:

If an imaginary (conjugate) beam is constructed and loaded with the $-M/EI$ diagram from the real beam, then the slope θ in the real beam is analogous to the shear V in the conjugate beam and the deflection y in the real beam is analogous to the bending moment M in the conjugate beam.

This reduces the complex task of calculating slopes and deflections to the relatively simpler task of calculating shear and bending moments on the conjugate beam.

The steps involved in the conjugate beam method are outlined as follows:

Step 1   Draw the M/EI diagram for the real beam.

Step 2   Create the conjugate beam using these rules for modifying supports:
- Free end becomes fixed support.
- Fixed support becomes free end.

- End hinge/roller support remains hinge/roller support.
- Internal roller becomes internal hinge.
- Internal hinge becomes internal roller.

Step 3   Apply the $-M/EI$ diagram as the load on the conjugate beam (draw load arrows away from the beam).

Step 4   *Slope* at any point on the real beam = *Shear Force* at same point on conjugate beam
Deflection at any point on the real beam = *Bending Moment* at same point on conjugate beam

These steps are illustrated in Fig. 102.8.

$$\sum F_y = 0 \Rightarrow V_B^* = -\frac{PL^2}{2EI} = \theta_B \qquad \text{(clockwise)}$$

$$\sum M_{\text{cut}} = 0 \Rightarrow M_B^* = -\frac{PL^3}{3EI} = \delta_B \qquad \text{(down)}$$

Figure 102.8
Steps in the
conjugate beam
method.

Real beam

Step 1

Bending-moment diagram

Steps 2, 3

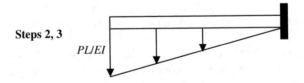

Loaded conjugate beam

Step 4

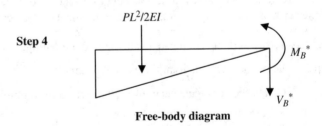

Free-body diagram

Example 102.15   Deflection of a Beam (Conjugate Beam Method)

Calculate the tip deflection of the cantilever beam shown below.

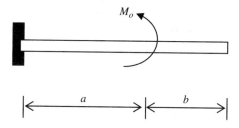

Solution

Step 1   Construct the bending-moment diagram of the beam.

Step 2   Apply the M/EI diagram on the conjugate beam (alter supports).

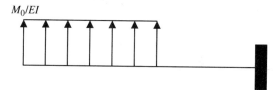

Step 3   Calculate the *internal bending moment* in conjugate beam to get the *deflection* of real beam (internal *shear force* in conjugate beam = *slope* of real beam).

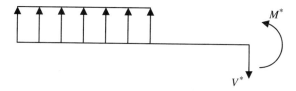

Summing moments about the internal cut, we get

$$\sum M_{cut} = -\frac{M_0 \cdot a}{EI}\left(b + \frac{a}{2}\right) + M^* = 0$$

$$M^* = \frac{M_0}{EI}a\left(b + \frac{a}{2}\right)$$

Therefore, the deflection of the real beam is given by the expression for M^*.

Unit Load Method

The unit load method is derived from the principle of virtual work. Stated in terms of the deformation strain energy, it can be stated as the deflection corresponding to a particular degree of freedom is equal to the partial derivative of the strain energy with respect to the (work) conjugate of the deflection. Thus, if a particular deflection Δ is desired and its work conjugate is the virtual load P.

The bending strain energy is given by $U = \int \dfrac{M^2(x)dx}{2EI}$ and the desired deflection Δ is given by the partial derivative $\partial U/\partial P$

$$\Delta_P = \frac{\partial U}{\partial P} = \int \frac{M(x)\dfrac{\partial M}{\partial P}dx}{EI} = \int \frac{M(x)m(x)dx}{EI}$$

where the function $M(x)$ describes the bending-moment diagram due to the external loads, and the function $m(x)$ describes the bending-moment diagram due to the unit load. The unit load is chosen according to the desired displacement.

Example 102.16   Beam Deflection by Unit Load Method

Calculate the deflection at point B of the beam shown below:

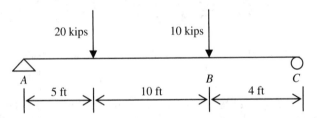

The reaction at A is $20 \times 14/19 + 10 \times 4/19 = 16.842$ kips. The reaction at C is 13.158 kips. The bending-moment diagram for the beam is shown below (labeled M):

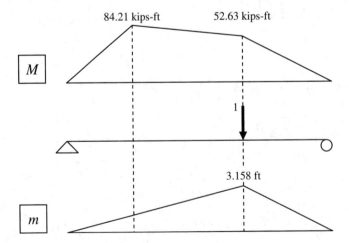

Table 102.1   Values of the Integral $\int mm'dx$

$\int_0^L mm'dx$	rectangle m', L	triangle m', L	trapezoid m_1', m_2', L	parabola m', L
rectangle m, L	$mm'L$	$\frac{1}{2}mm'L$	$\frac{1}{2}m(m_1' + m_2')L$	$\frac{2}{3}mm'L$
triangle m, L	$\frac{1}{2}mm'L$	$\frac{1}{3}mm'L$	$\frac{1}{6}m(m_1' + 2m_2')L$	$\frac{5}{12}mm'L$
trapezoid m_1, m_2, L	$\frac{1}{2}m'(m_1 + m_2)L$	$\frac{1}{6}m'(m_1 + 2m_2)L$	$\frac{1}{6}\left[\begin{array}{l}m_1'(2m_1 + m_2)+\\ m_2'(m_1 + 2m_2)\end{array}\right]L$	$\frac{1}{12}m'(3m_1 + 5m_2)L$
triangle m_1, a L b	$\frac{1}{2}mm'L$	$\frac{1}{6}mm'(L + a)$	$\frac{1}{6}m_1\left[\begin{array}{l}m_1'(L + b)+\\ m_2'(L + a)\end{array}\right]$	$\frac{1}{12}mm'\left(3 + \frac{3a}{L} - \frac{a^2}{L^2}\right)L$
triangle m, L	$\frac{1}{2}mm'L$	$\frac{1}{6}mm'L$	$\frac{1}{6}m(2m_1' + m_2')L$	$\frac{1}{4}mm'L$

The second diagram shows the unit load case (since the answer sought is vertical deflection at B) and the resulting bending-moment diagram (max value $= 1 \times 15 \times 4/19 = 3.158$) is shown in the third diagram. Using the tabular multiplication tables (see Table 102.1), we get

$$\int_0^L Mmdx = \frac{1}{3} \times 84.21 \times 1.053 \times 5$$

$$+ \frac{1}{6} \times [1.053 \times (2 \times 84.21 + 52.63) + 3.158 \times (84.21 + 2 \times 52.63)] \times 10$$

$$+ \frac{1}{3} \times 52.63 \times 3.158 \times 4 = 147.79 + 1385.19 + 221.61 = 1754.59$$

Therefore, the deflection at B is $1754.59/EI$.

Beam Deflection Equations

Origin and coordinate axes: For all diagrams shown in this section, assume that the origin is located at the left end (A) of beam and the right-handed coordinate axes are as follows:

x axis positive to the right

y axis positive upward

positive z axis emerging out of plane. M_z positive counterclockwise

Case 1—Cantilever Beam with Point Load at End

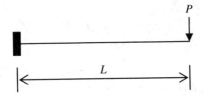

Reactions:	$A_y = +P \qquad M_A = +PL$
Shear Function:	$V(x) = +P$
Bending Moment Function:	$M(x) = P(x - L)$
Maximum Bending Moment:	$M_{max} = -PL$
Deflection Function:	$y(x) = \dfrac{Px^2}{6EI}(x - 3L)$
Maximum Deflection:	$y_{max} = -\dfrac{PL^3}{3EI} \qquad$ at $x = L$

Case 2—Cantilever Beam with Uniformly Distributed Load

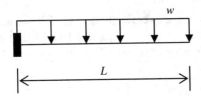

Reactions:	$A_y = +wL \qquad M_A = +\dfrac{wL^2}{2}$
Shear Function:	$V(x) = w(L - x)$
Bending Moment Function:	$M(x) = -\dfrac{w}{2}(L - x)^2$
Maximum Bending Moment:	$M_{max} = -\dfrac{wL^2}{2}$

Deflection Function:
$$y(x) = \frac{wx^2}{24EI}(4Lx - x^2 - 6L^2)$$

Maximum Deflection:
$$y_{max} = -\frac{wL^4}{8EI} \quad \text{at } x = L$$

Case 3—Cantilever Beam with Linearly Varying Distributed Load

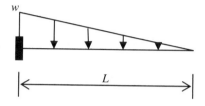

Reactions:
$$A_y = +\frac{wL}{2} \qquad M_A = +\frac{wL^2}{6}$$

Shear Function:
$$V(x) = \frac{w}{2L}(L - x)^2$$

Bending Moment Function:
$$M(x) = -\frac{w}{6L}(L - x)^3$$

Maximum Bending Moment:
$$M_{max} = -\frac{wL^2}{6}$$

Deflection Function:
$$y(x) = -\frac{wx^2}{120EIL}(10L^3 - 10L^2x + 5Lx^2 - x^3)$$

Maximum Deflection:
$$y_{max} = -\frac{wL^4}{30EI} \quad \text{at } x = L$$

Case 4—Cantilever Beam with Linearly Varying Distributed Load

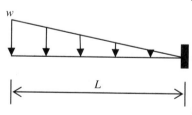

Reactions:
$$B_y = +\frac{wL}{2} \qquad M_B = -\frac{wL^2}{3}$$

Shear Function:
$$V(x) = \frac{wx}{2L}(x - 2L)$$

Bending Moment Function: $M(x) = \dfrac{wx^2}{6L}(x - 3L)$

Maximum Bending Moment: $M_{max} = -\dfrac{wL^2}{3}$

Deflection Function: $y(x) = \dfrac{w}{120EIL}(x^5 - 5Lx^4 + 15L^4x - 11L^5)$

Maximum Deflection: $y_{max} = -\dfrac{11wL^4}{120EI}$ at $x = 0$

Case 5—Cantilever Beam with Moment at One End

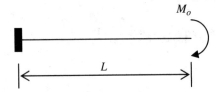

Reactions: $A_y = 0 \qquad M_A = +M_0$

Shear Function: $V(x) = 0$

Bending Moment Function: $M(x) = -M_0$

Maximum Bending Moment: $M_{max} = -M_0$

Deflection Function: $y(x) = -\dfrac{M_0 x^2}{2EI}$

Maximum Deflection: $y_{max} = -\dfrac{M_0 L^2}{2EI}$ at $x = L$

Case 6—Simply Supported Beam with Single Point Load

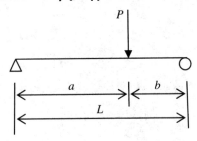

Reactions: $A_y = +\dfrac{Pb}{L} \qquad B_y = +\dfrac{Pa}{L}$

Shear Function:
$$V(x) = +\frac{Pb}{L} \qquad 0 < x < a$$

$$V(x) = -\frac{Pa}{L} \qquad a < x < L$$

Bending Moment Function:
$$M(x) = \frac{Pbx}{L} \qquad 0 < x < a$$

Maximum Bending Moment:
$$M(x) = \frac{Pa(L - x)}{L} \qquad a < x < L$$

$$M_{max} = +\frac{Pab}{L} \qquad \text{at } x = a$$

Deflection Function:
$$y(x) = \frac{Pb}{6EIL}(L^2 x - b^2 x - x^3) \quad 0 < x < a$$

$$y(x) = \frac{Pb}{6EIL}\left[\left(\frac{L}{b}\right)(x - a)^3 + (L^2 - b^2)x - x^3\right] \quad x > a$$

Maximum Deflection:
$$y_{max} = -\frac{0.06415Pb}{EIL}(L^2 - b^2)^{3/2} \quad \text{at } x = \sqrt{\frac{a(L + b)}{3}}$$

Case 7—Simply Supported Beam with Uniformly Distributed Load

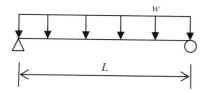

Reactions:
$$A_y = +\frac{wL}{2} \qquad B_y = +\frac{wL}{2}$$

Shear Function:
$$V(x) = \frac{wL}{2} - wx$$

Bending Moment Function:
$$M(x) = \frac{wx(L - x)}{2}$$

Maximum Bending Moment:
$$M_{max} = \frac{wL^2}{8} \qquad \text{at } x = \frac{L}{2}$$

Deflection Function:
$$y(x) = -\frac{wx}{24EI}(x^3 - 2Lx^2 + L^3)$$

Maximum Deflection:
$$y_{max} = -\frac{5wL^4}{384EI} \qquad \text{at } x = \frac{L}{2}$$

Case 8—Simply Supported Beam with Linearly Varying Distributed Load

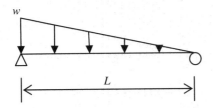

Reactions:
$$A_y = +\frac{wL}{3} \qquad B_y = +\frac{wL}{6}$$

Shear Function:
$$V(x) = +w(L - x)$$

Bending Moment Function:
$$M(x) = \frac{wx}{6L}(L - x)(2L - x)$$

Maximum Bending Moment:
$$M_{max} = 0.0642wL^2 \qquad \text{at } x = 0.423L$$

Deflection Function:
$$y(x) = \frac{wx}{360EIL}(3x^4 - 15Lx^3 + 20L^2x^2 - 8L^4)$$

Maximum Deflection:
$$y_{max} = -0.00652\frac{wL^4}{EI} \qquad \text{at } x = 0.481L$$

Case 9—Simply Supported Beam with Moment at One End

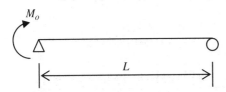

Reactions:
$$A_y = -\frac{M_0}{L} \qquad B_y = +\frac{M_0}{L}$$

Shear Function:
$$V(x) = -\frac{M_0}{L}$$

Bending Moment Function:
$$M(x) = \frac{M_0(L - x)}{L}$$

Deflection Function:
$$y(x) = \frac{M_0}{6EIL}(3Lx^2 - x^3 - 2L^2x)$$

Maximum Deflection:
$$y_{max} = -0.06415\frac{M_0L^2}{EI} \qquad \text{at } x = 0.423L$$

Case 10—Simply Supported Beam Distributed Load Increasing Uniformly to Center

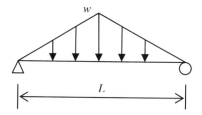

Reactions: $$A_y = +\frac{wL}{4} \qquad B_y = +\frac{wL}{4}$$

Shear Function: $$V(x) = \frac{w}{4L}(L^2 - 4x^2) \quad 0 < x < \frac{L}{2}$$

Bending Moment Function: $$M(x) = \frac{wx}{12L}(3L^2 - 4x^2)$$

Maximum Bending Moment: $$M_{max} = 0.0833wL^2 \qquad \text{at } x = 0.5L$$

Deflection Function: $$y(x) = -\frac{wx}{960EIL}(5L^2 - 4x^2)^2 \quad 0 < x < \frac{L}{2}$$

Maximum Deflection: $$y_{max} = -\frac{wL^4}{120EI} \quad \text{at } x = 0.5L$$

Case 11—Propped Cantilever Beam with Uniformly Distributed Load

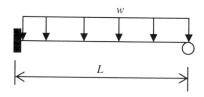

Reactions: $$A_y = +\frac{5wL}{8} \qquad B_y = +\frac{3wL}{8} \qquad M_A = +\frac{wL^2}{8}$$

Shear Function: $$V(x) = \frac{5wL}{8} - wx$$

Bending Moment Function: $$M(x) = \frac{w}{8}(5Lx - L^2 - 4x^2)$$

Maximum Bending Moment: $\qquad M_{max} = -0.125wL^2 \qquad$ at $x = 0$

Maximum Positive Moment: $\qquad M_+ = +\dfrac{9wL^2}{128} \qquad$ at $x = 0.625L$

Deflection Function: $\qquad y(x) = -\dfrac{wx^2}{48EI}(5Lx - 2x^2 - 3L^2)$

Maximum Deflection: $\qquad y_{max} = -\dfrac{wL^4}{185EI} \qquad$ at $x = 0.578L$

Case 12—Propped Cantilever Beam with Concentrated Load

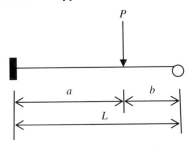

Reactions: $\qquad A_y = \dfrac{Pb}{2L^3}(3L^2 - b^2) \quad B_y = \dfrac{Pa^2}{2L^3}(b + 2L) \quad M_A = \dfrac{Pab}{2L^2}(L + b)$

Shear Function: $\qquad V(x) = \dfrac{Pb^2}{2L^3}(3L^2 - b^2) \quad$ for $0 < x < a$

$\qquad\qquad\qquad\qquad V(x) = -\dfrac{Pa^2}{2L^3}(b + 2L) \quad$ for $0 < x < L$

Bending Moment Function: $\qquad M_x = \dfrac{Pa^2}{2L^3}(b + 2L)(L - x) - P(b - x) \quad$ for $0 < x < a$

$\qquad\qquad\qquad\qquad M_x = \dfrac{Pa^2}{2L^3}(L - x)(b + 2L) \quad$ for $0 < x < L$

Maximum Positive Moment: $\qquad M_+ = \dfrac{Pa^2b}{2L^3}(b + 2L) \quad$ at $x = a$

Maximum Deflection: $\qquad y_{max} = \dfrac{Pa^2b}{6EI}\sqrt{\dfrac{b}{2L + b}} \quad$ at $x = L\sqrt{\dfrac{b}{2L + b}} \quad$ when $a < 0.414L$

$\qquad\qquad\qquad\qquad y_{max} = \dfrac{Pb}{3EI}\dfrac{(L^2 - b^2)^3}{(3L^2 - b^2)^2} \quad$ at $x = L\left(\dfrac{L^2 + b^2}{3L^2 - b^2}\right) \quad$ when $a > 0.414L$

Case 13—Beam Fixed at Both Ends with Uniformly Distributed Load

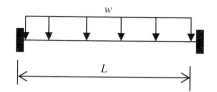

Reactions: $\qquad A_y = B_y = +\dfrac{wL}{2} \qquad M_A = -M_B = +\dfrac{wL^2}{12}$

Shear Function: $\qquad V(x) = \dfrac{wL}{2} - wx$

Bending Moment Function: $\qquad M(x) = \dfrac{w}{12}(6Lx - L^2 - 6x^2)$

Maximum Bending Moment: $\qquad M_{max} = -0.0833wL^2 \qquad$ at $x = 0$ and $x = L$

Maximum Positive Moment: $\qquad M_+ = +\dfrac{wL^2}{24} \qquad$ at $x = 0.5L$

Deflection Function: $\qquad y(x) = -\dfrac{wx^2}{24EI}(L - x)^2$

Maximum Deflection: $\qquad y_{max} = -\dfrac{wL^4}{384EI} \qquad$ at $x = 0.5L$

Case 14—Beam Fixed at Both Ends with Concentrated Load at Midspan

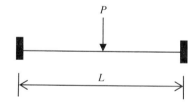

Reactions: $\qquad A_y = B_y = +\dfrac{P}{2} \qquad M_A = -M_B = +\dfrac{PL}{8}$

Shear Function: $\qquad V(x) = +\dfrac{P}{2} \qquad 0 < x < \dfrac{L}{2}$

$\qquad\qquad\qquad\qquad V(x) = -\dfrac{P}{2} \qquad \dfrac{L}{2} < x < L$

Bending Moment Function:
$$M(x) = \frac{P(4x - L)}{8} \qquad 0 < x < \frac{L}{2}$$

$$M(x) = \frac{P(3L - 4x)}{8} \qquad \frac{L}{2} < x < L$$

Maximum Bending Moment:
$$M_{max} = \frac{PL}{8} \quad \text{at } x = 0, \ \frac{L}{2}, \ L$$

Deflection Function:
$$y(x) = \frac{Px^2}{48EI}(4x - 3L) \qquad 0 < x < \frac{L}{2}$$

Maximum Deflection:
$$y_{max} = -\frac{PL^3}{192EI} \quad \text{at } x = 0.5L$$

Case 15—Beam Fixed at Both Ends with Concentrated Load at Any Point

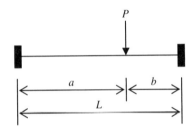

Reactions:
$$A_y = \frac{Pb^2}{L^3}(L + 2a); \qquad B_y = \frac{Pa^2}{L^3}(L + 2b)$$

End Moments:
$$M_A = \frac{Pab^2}{L^2} \ (= M_{max} \text{ when } a < b)$$

$$M_B = \frac{Pa^2b}{L^2} \ (= M_{max} \text{ when } a > b)$$

Shear Function:
$$V(x) = \frac{Pb^2}{L^3}(L + 2a) \qquad 0 < x < a$$

$$V(x) = -\frac{Pa^2}{L^3}(L + 2b) \qquad a < x < L$$

Bending Moment Function:
$$M_x = \frac{Pb^2}{L^3}(-aL + xL + 2ax) \quad \text{for } 0 < x < a$$

$$M_x = \frac{Pa^2}{L^3}(L^2 + bL - xL - 2bx) \quad \text{for } a < x < L$$

Deflection Function:
$$y(x) = \frac{Pb^2x^2}{6EIL^3}(3aL - 3ax - bx) \quad \text{for } x < a$$

Maximum Deflection:
$$y_{max} = -\frac{2Pa^3b^2}{3EI(L+2a)^2} \quad \text{at } x = \frac{2aL}{L+2a} \quad \text{when } a > b$$

Influence Lines

An *influence line* is a plot of the internal force response of a beam (bending moment or shear force) as a function of the location of a single transverse point load on the beam.

Principle of Müller-Breslau

Derived from the principle of virtual work, the Müller-Breslau principle states: *The influence line of a given function (reaction, shear, or bending moment) is proportional to the deflected shape of the beam when the internal force in question is released and a unit displacement or rotation is applied in its place.*

In short, the three steps to apply this principle are as follows:

1. Release the constraint corresponding to the quantity whose influence diagram is sought.

2. Apply a unit displacement corresponding to the released constraint.

3. Visualize the displaced shape of the beam. This is the shape of the influence diagram. In some cases, such as the influence diagram for bending moment, the diagram may require scaling.

Note that for determinate structures, the influence diagram is always composed of straight lines. In other words, applying the unit displacement (step 2 above) is accommodated by the structure such that different parts of the structure deflect as rigid bodies (no bending). The same principle and the outlined procedure is also valid for indeterminate structures, the only difference being that the resulting influence diagram may be composed of nonlinear curves.

Example 102.17 Müller-Breslau Principle

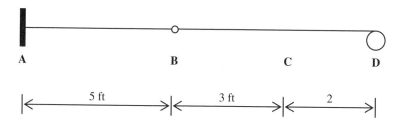

Influence Line for Vertical Reaction at *A*

The support at *A* provides three independent constraints to the beam—it restricts it from translating horizontally, it restricts it from translating vertically, and it restricts it from rotating. By the Müller-Breslau principle, we will first remove the vertical constraint while maintaining the other two constraints, and then apply a unit upward deflection to the beam at *A*. This deflection must not be accompanied by any horizontal translation or rotation of the beam at *A* (since we are supposed to release ONLY the vertical restraint).

The segment *AB* has to remain horizontal so that there is no rotation at *A*. The beam accommodates such a deflected shape without any bending because of the presence of the internal hinge at *B*, which allows the segment *AB* to remain horizontal while rotating segment *BC* in a rigid body mode (without bending). The resulting deflected shape is given by the broken line in the diagram below.

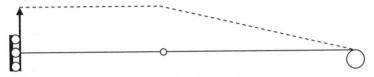

Thus, the *shape* of the influence-line diagram for vertical reaction at *A* is as shown above. Noting that when the unit load is directly above the support *A*, the vertical reaction at *A* is equal to 1.0, the scaling factor for the above diagram is $1.0/1.0 = 1.0$.

Influence Line for Reaction Moment at *A*

For this function (reaction moment at *A*), the beam must be *released* from the rotational constraint while maintaining the other constraints—horizontal and vertical translation. The way to achieve this is to create a hinge at *A* and apply a unit rotation to it. This action is shown in the diagram below. The deflected shape of the beam due to this (rotational) displacement is shown as a broken line.

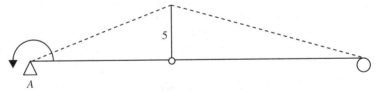

When the (unit downward) load is applied at hinge *B* on the original structure, the support at *C* does not participate in carrying this load and the portion AB of the beam behaves just like a cantilever *AB*. Therefore, the reaction moment at *A* is $+5$ (counterclockwise) when the unit load is at the hinge *B*. Thus, the scaling factor for the influence diagram is given by $+5/+5 = 1.0$. As a result, the scaled-influence-line diagram for the reaction moment at *A* is given by

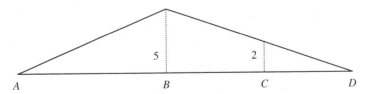

Influence Line for Shear at *C*

For this function (shear at *C*), the beam must be *released* from the internal restraint against shear. This effect is shown in the following diagram. A unit shear is then applied to point *C* (where the beam has been "split open" and therefore released from internal shear resistance).

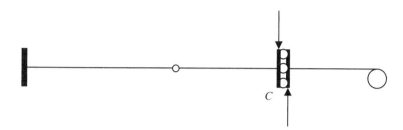

Under the combined action of the force pair, the beam will deflect down on the left of point *C*, and deflect up on the right of point *C*. The total relative deflection at *C* must be equal to 1.0. The theoretical derivation of the equation for the influence diagram for shear indicates that the slopes of the two sides of the influence-line diagram are to be equal. Thus, the total deformation of 1.0 is shared in proportion to the spans *BC* and *CD*. In the above case, *BC*:*CD* = 3:2. Thus, the deflections at *C* must be 0.6 and 0.4. Applying these deformations to point *C* and allowing the rest of the beam to deflect compatibly, we obtain the scaled influence-line diagram for shear at *C*.

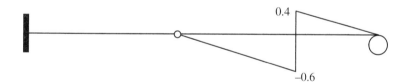

Note that because of the rotational resistance at support *A* and the ability of the hinge *B* to allow rotation, the segment *AB* does not deflect as a result of the deflection of segment *BC*. This implies that at point *C*, there is no internal shear produced if the load is on span *AB* (since the influence diagram has a zero ordinate over *AB*).

Influence Line for Bending Moment at *C*

To apply the Müller-Breslau principle for bending moment at *C*, the internal point *C* must be released from the internal restraint against bending without affecting the internal restraint against axial or shear deformation. This can be accomplished by inserting an internal hinge at *C*. The internal hinge has no resistance to bending. However, it can fully resist axial deformation as well as transverse (shear) deformation. Once the hinge has been inserted, a pair of rotations representing a unit positive bending moment should be applied on either side of *C*. This is shown below. The deflected shape of the beam due to this imposed displacement is shown using dashed lines.

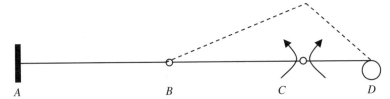

The principle above yields only the *shape* of the diagram. We still need to calculate the *value* of any one point on the diagram in order to *scale* it. The obvious choice for such a point above is point *C*. Thus, placing a unit load at *C*, we calculate the bending moment at *C*. This specific case is shown below:

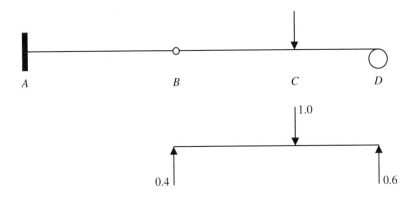

The bending moment under a load *P* on a simple span is given by *Pab/L*. This leads to

$$M = \frac{Pab}{L} = \frac{1 \times 3 \times 2}{5} = 1.2 \text{ ft}$$

Thus, the bending moment at *C* due to a unit load at *C* is 1.2 ft. The influence-line diagram, scaled accordingly, is shown below:

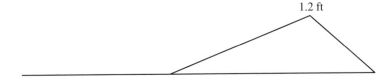

Shear at Midspan of Uniformly Loaded Beams

For a structure on which various load placements are to be investigated, a load envelop must be drawn. This envelop is the plot of the maximum possible load effect (shear force, bending moment, etc.) at all points of the structure. For example, for a simply supported beam subject to uniformly distributed dead and live loads, the shear envelop may be obtained by varying the extent of the live load (the dead load remains fixed) over the span. For example, whereas the maximum shear at the support may be obtained by placing the live load over the entire span, maximum shear at midspan may be obtained by placing the live load over only one-half of the span. The shear envelop is drawn as the plot of the maximum possible shear force at all points along the span.

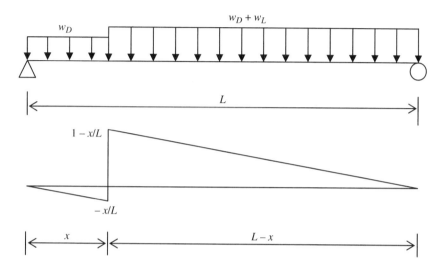

The maximum possible shear at any location along the beam is given by placing the live load partially so as to maximize its contribution. The maximum shear at a distance x from the left support is given by

$$V_x = (w_D + w_L)\left(1 - \frac{x}{L}\right)\frac{L - x}{2} - w_D\left(\frac{x}{L}\right)\frac{x}{2}$$

Setting $x = 0$, the maximum shear at the support is given by

$$V_S = \frac{(w_D + w_L)L}{2}$$

Setting $x = L/2$, the maximum shear at midspan is given by

$$V_x = w_L\frac{L}{8}$$

The approximate shape of the maximum shear envelop over the beam is shown in Fig. 102.9. It can be seen that near the support, the shear envelop is fairly closely approximated by the shear-force diagram obtained by placing the total load (dead + live) over the entire span. This may become relevant for the provisions in ACI 318: *Building Code Requirements for Reinforced*

Figure 102.9
Envelop for
shear at midspan
of simply sup-
ported beam.

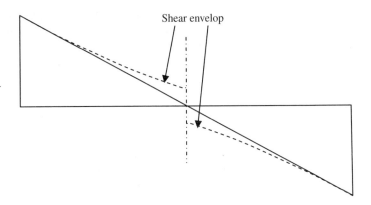

Shear envelop

Concrete, American Concrete Institute, Detroit, Michigan. According to the specifications in ACI 318, the critical section for shear is located at a distance d (effective depth of the beam) from the face of the support. It would then become necessary to calculate the maximum possible shear at this critical section. As may be seen from the diagram above, for all practical purposes, this shear is very close to that obtained by placing the live load over the entire span. For an illustration, see Example 105.3.

Influence of a Series of Concentrated Loads

Highway and railway bridges are designed to carry train and truck loads. The design vehicle, which forms the basis for design, is a sequence of defined wheel loads with specific spacing between wheels. For example, the design of highway bridges is often carried out for the standard AASHTO (American Association of State Highway and Transportation Officials) HS trucks. The axle loads for the HS-20 truck are shown in Fig. 102.10. The designation implies that it is a highway load (H) with a semitrailer (S) and the total design load on the two front axles is 20 tons (40 kips or 40,000 lb). The weight of the third axle is identical to that of the second axle. The distance between the first two axles is fixed at 14 ft but the distance between the second and third axles (in other words the length of the semitrailer) is variable (14 to 30 ft).

For design of railway bridges, AREMA (American Railway Engineering and Maintenance-of-Way Association) has specified similar design railway vehicles. These design vehicles are defined as a sequence of axles, carrying specific magnitudes of wheel load and spacing.

Figure 102.10
AASHTO
standard vehicle
(HS-20).

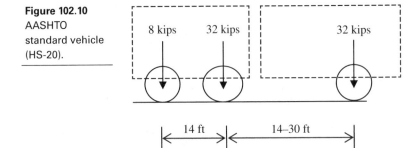

8 kips 32 kips 32 kips

14 ft 14–30 ft

The following discussion extends the discussion on the effect of a single moving load to the effect of a group of wheel loads. Let us consider the simply supported beam shown in the following diagram, subjected to a moving three-axle vehicle as shown.

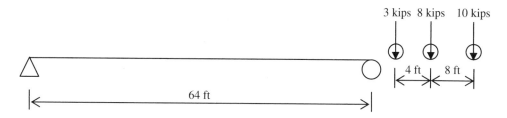

Absolute Maximum Shear

The absolute maximum shear in the beam will occur by maximizing the reaction at one of the supports. This can be accomplished by recognizing that the influence diagram for reaction (at A, for example) is as shown in Fig. 102.11.

The reaction at A can therefore be maximized by placing the heaviest part of the vehicle (rear axle load 10 kips) over A and the rest of the vehicle ahead of it—in other words, a vehicle moving from left to right with the rear axle having just entered the bridge. This load position is shown below.

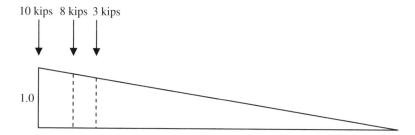

Once the ordinates have been found by similar triangles as 56/64 (under the 8-kip load) and 52/64 (under the 3-kip load), the absolute maximum shear can be found as

$$V_{max} = 1.0 \times 10 + \frac{56}{64} \times 8 + \frac{52}{64} \times 3 = 19.44$$

Figure 102.11
Influence line for
vertical reaction
at A.

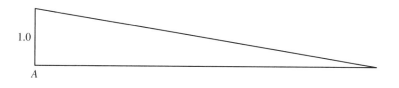

Absolute Maximum Bending Moment

RULE: For a simply supported beam, the absolute maximum bending moment occurs under one of the wheels when the distance between that wheel and the resultant of the wheel group is bisected by the midpoint of the beam.

The wheel group used in the above example is shown below. The resultant is the 21-kip force shown by a broken line. The distance of the resultant from the left (3 kips) wheel is 7.238 ft. In other words, if we were to place this wheel group on the simply supported beam (span = 64 ft) such that the midpoint of the beam ($x = 32$) were halfway between a wheel and the resultant, the maximum bending moment would occur under that wheel.

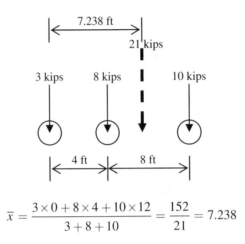

$$\bar{x} = \frac{3 \times 0 + 8 \times 4 + 10 \times 12}{3 + 8 + 10} = \frac{152}{21} = 7.238$$

Since the resultant is between the 8-kip and the 10-kip wheels, let us try two different scenarios.

Scenario 1—Wheel 2 (8 kips) and Resultant (21 kips) Equidistant from Beam Midspan

The distance between wheel 2 and the resultant is 3.238 ft. Since we want to locate the beam midpoint ($x = 32$) midway between these two forces, wheel 2 must be located at $x = 32 - 0.5 \times 3.238 = 30.381$ ft and the resultant at $x = 32 + 0.5 \times 3.238 = 33.619$ ft. This loading is shown in Fig. 102.12. Under this loading, the maximum bending moment should occur directly under wheel 2 (at $x = 30.381$ ft).

Thus, the influence line for the bending moment at $x = 30.381$ is drawn below (Fig. 102.12) and the load placed such that wheel 2 is at $x = 30.381$. (This places wheel 1 at $x = 26.381$ ft and wheel 3 at $x = 38.381$ ft.)

Maximum bending moment (due to this wheel placement)
$$= 3 \times 13.86 + 8 \times 15.96 + 10 \times 12.16 = 290.86 \text{ kips-ft.}$$

Thus, the maximum bending moment of $M = 290.86$ kips-ft occurs at $x = 30.381$ when wheel 2 is located at that location.

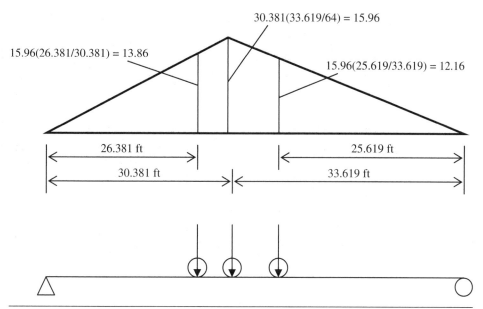

Figure 102.12 Influence line for bending moment under wheel 2.

Scenario 2—Wheel 3 (10 kips) and Resultant (21 kips) Equidistant from Beam Midspan

The distance between wheel 3 and the resultant is 4.762 ft. Since we want to locate the beam midpoint ($x = 32$) midway between these two forces, the resultant must be located at $x = 32 - 0.5 \times 4.762 = 29.619$ ft. and wheel 3 at $x = 32 + 0.5 \times 4.762 = 34.381$ ft. This loading is shown in the Fig. 102.13. Under this loading, the maximum bending moment should occur directly under wheel 3 (at $x = 34.381$ ft).

Thus, the influence line for the bending moment at $x = 34.381$ is drawn in Fig. 102.13 and the load placed such that wheel 3 is at $x = 34.381$. (This places wheel 1 at $x = 22.381$ ft and wheel 2 at $x = 26.381$ ft.)

> Maximum bending moment (due to this wheel placement)
> $= 3 \times 10.36 + 8 \times 12.21 + 10 \times 15.91 = 287.86$ kips-ft

Thus, the maximum bending moment of $M = 287.86$ kips-ft occurs at $x = 34.381$ when wheel 3 is located at that location.

The absolute maximum bending moment in the beam is therefore $M_{max} = 290.86$ kips-ft at $x = 30.381$ ft.

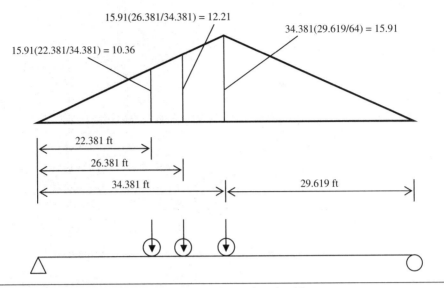

Figure 102.13 Influence line for bending moment under wheel 3.

Calculating Effect of Concentrated and Distributed Loads

Once the influence diagram for a particular effect (reaction, shear, or bending moment) has been constructed and scaled, it can be used to calculate the influence of moving point loads and uniformly distributed loads on that effect. As an example, consider the beam shown in the following diagram:

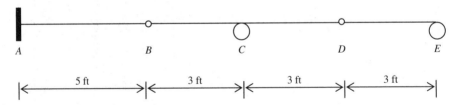

The influence diagram for the hinge shear at *B* is given by

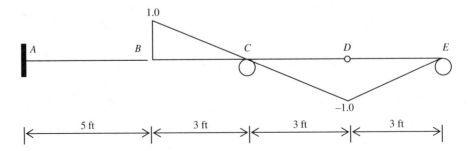

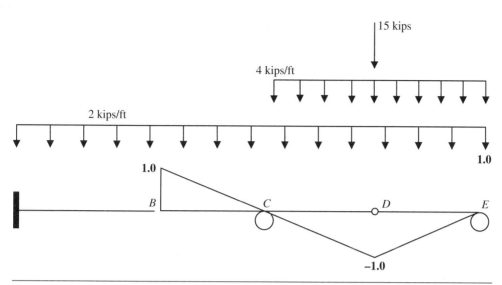

Figure 102.14 Critical placement of loads for maximum hinge shear.

Note that whereas for a simple span, the unit shear at B is distributed according to the span proportions; in this case the rigidity of support A does not allow a rigid body rotation of span AB, and therefore, the entire 1.0 shear displacement is applied to BC. The rigid link BCD then pivots around support C and similar triangles yields the ordinate at D to be -1.0.
Now consider the beam is to be analyzed for the following loads:

Uniformly distributed dead load: 2 kips/ft (FIXED POSITION)

Uniformly distributed live load: 4 kips/ft (VARIABLE POSITION)

One concentrated live load: 15 kips (VARIABLE POSITION)

The dead load must be placed over the entire beam. The live loads must be placed for maximum effect. Since the negative area under the diagram is greater than the positive area, we will try to maximize the negative shear influence. The load placement is then shown in Fig. 102.14.
The effect of a concentrated load is calculated as the product of the load and the ordinate of the influence diagram. The effect of a uniformly distributed load is calculated as the product of the load intensity and the area under the influence diagram. Thus, the maximum hinge shear is calculated as

$$V_{B,\max} = 2 \times \left(\frac{1}{2} \times 3 \times +1.0 + \frac{1}{2} \times 6 \times -1.0 \right) + 4 \times \left(\frac{1}{2} \times 6 \times -1.0 \right) + 15 \times -1.0 = -30 \text{ kips}$$

Example 102.18

Influence Line for Truss Member Force

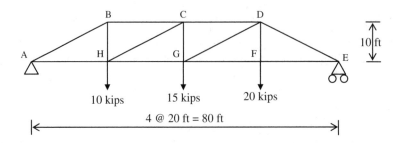

What is the influence-line diagram for the web member *CH* in the truss shown above?

Solution Since the truss is statically determinate, the influence diagram will be composed of straight line segments. When a unit load traverses the truss deck from *A* to *E*, one must therefore consider the two load positions, at either end of the member *H* and *G*, which is directly below *C*. When the load is at the supports *A* and *E*, the member force F_{CH} is zero.

Unit load at H

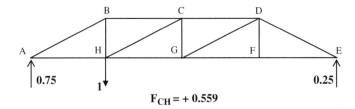

$$F_{CH} = +0.559$$

Using the section shown below:

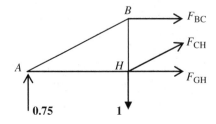

Summing vertical forces, $\quad 0.75 - 1.0 + F_{CH}\dfrac{1}{\sqrt{5}} = 0 \Rightarrow F_{CH} = 0.25\sqrt{5} = +0.559$

Unit load at G

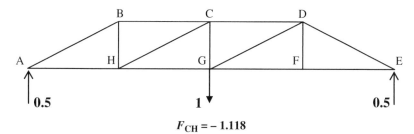

$$F_{CH} = -1.118$$

Using the section shown below:

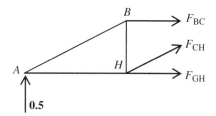

Summing vertical forces, $\qquad 0.5 + F_{CH}\dfrac{1}{\sqrt{5}} = 0 \Rightarrow F_{CH} = -0.5\sqrt{5} = -1.118$

The influence line for F_{CH} is drawn by joining the following ordinates: at A, 0.0; at H, +0.559; at G, −1.118; at E, 0.0.

Influence line for F_{CH}

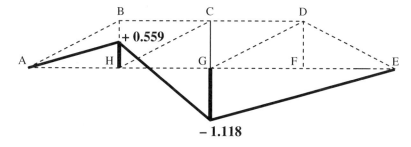

Introduction to Indeterminate Structures

The development of the general method for indeterminate structures may be credited to independent works by James Clerk Maxwell (1864), Otto Mohr (1874), and Heinrich Müller-Breslau (1886).

Stability and Determinacy

Stable structures may be classified into two main groups—statically determinate and statically indeterminate. Statically determinate structures are those for which reactions components and internal stresses can be completely solved using the equations of static equilibrium. For a 2D structure in the x-y plane, these are the familiar equations:

$$\sum F_x = 0 \qquad \sum F_y = 0 \qquad \sum M_z = 0$$

Statically indeterminate structures possess more unknowns than can be solved using the above equations. The degree of indeterminacy is the number of unknowns over and above the number of equilibrium equations available for solution.

Statically indeterminate structures may be externally or internally indeterminate. Two examples of indeterminate truss are shown in Fig. 103.1. In Fig. 103.1(a), we see a truss which is internally determinate but externally indeterminate due to too many external reactions. In Fig. 103.1(b), the truss is externally determinate, but its indeterminacy stems from the number of members in the structure.

Determinate versus Indeterminate Structures

Comparisons may be drawn between determinate and indeterminate structures on the basis of economy, aesthetics, constructability, and safety.

In one sense, determinate structures may be called *marginally stable*. That implies that a determinate structure can become unstable with a single local failure such as a loss of a single

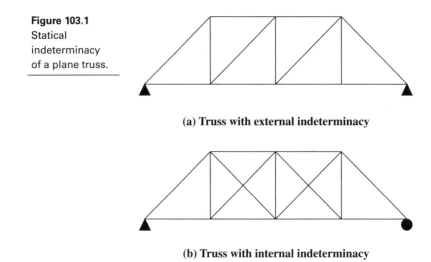

Figure 103.1
Statical
indeterminacy
of a plane truss.

(a) Truss with external indeterminacy

(b) Truss with internal indeterminacy

support restraint. Indeterminate structures are comparatively more redundant, i.e., they possess more constraints than minimally necessary to make the structure stable and determinate. Local "failures" in indeterminate structures, such as formation of a plastic hinge in a steel member, are tolerated via the redistribution of stresses to other parts of the structure.

For determinate structures, imperfections such as initial lack of fit or fabrication errors, settling of supports and temperature change (from design value) do not induce additional stresses in the structure. On the other hand, such situations induce additional stress in indeterminate structures.

The General Force Method

The general force method for indeterminate structures has the following steps:

1. Identify a set of appropriate redundant forces, thereby reducing the structure to a stable, determinate state. This is called the *primary structure*. The external loads are applied to the primary structure. For the primary structure, displacements (deflections and/or rotations) are computed at the locations of the redundants.

2. The primary structure (without loads) is then loaded by unit loads corresponding to the redundants one at a time. These structures are called the *secondary structures*. Thus, we have as many secondary structures as there are redundants, i.e., the same as the degree of indeterminacy of the structure. Deflections or rotations are computed for each of these secondary structures at the locations of the redundants.

3. Compatibility relations are then written for each redundant, where the sum of the deflections (or rotations) at a specific location is set equal to a predetermined value, usually zero. This results in a set of n simultaneous equations in the n unknowns (redundants).

Figure 103.2
Statically
indeterminate
beam.

Force Method Illustration

The beam shown in Fig. 103.2 is indeterminate to the first order, i.e., it has one extra reaction (beyond the three that can be solved using equations of static equilibrium). Consequently, one of the extra reactions is chosen as a redundant force and removed from the structure. For more complex structures, that are higher order indeterminate, the appropriate number of redundants must be chosen. For the demonstration below, let us choose the vertical reaction at the roller support as the redundant. The removal of the redundant from the original structure results in a stable, determinate beam. This is called the *primary structure*. This is shown in Fig. 103.3.

In removing the roller support from the original beam, we have violated a boundary condition there. Whereas the original beam was constrained from deflecting vertically at the roller, the primary beam, due to the removal of the support, is able to deflect. This deflection is shown as Δ^P in Fig. 103.3. Note that, due to the primary beam being determinate, we can easily solve for the deflection Δ^P. Obviously, the one difference between the original beam and the primary beam is that the vertical reaction B_y is missing. We put this back in a *secondary beam*, as shown in Fig. 103.4.

Thus, the original beam has been split into a primary beam and a secondary beam. For the principle of superposition to be valid, the structure must be linear in behavior (e.g., a doubling of the load causes a doubling of the displacements). Also, for the superposition to be valid geometrically, the total deflection at joint B must be zero (constrained by a roller support). Thus, we have

$$\Delta^P + \Delta^S = 0 \tag{103.1}$$

Figure 103.3
Primary beam
(redundant
removed).

Figure 103.4
Secondary beam
(redundant
reinserted).

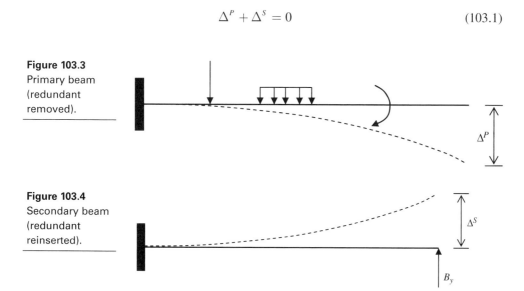

The quantity Δ^P is the end-deflection of a (determinate) cantilever beam, and can be calculated if the loading on the primary beam is specified. The quantity Δ^S, on the other hand, depends on the unknown redundant B_y. However, if the redundant is replaced by a known value (such as 1), the resulting deflection δ^S can be calculated. Due to linearity of the structure, we can say that the deflection Δ^S (due to the full redundant B_y) is a linear multiple of the deflection δ^S (due to the unit load):

$$\Delta^S = B_y \delta^S \qquad (103.2)$$

The total deflection is then given by

$$\Delta^P + \Delta^S = \Delta^P + B_y \delta^S = 0 \qquad (103.3)$$

The value of the redundant force is then given by

$$B_y = -\frac{\Delta^S}{\delta^S} \qquad (103.4)$$

The preceding approach is also called the *unit-load method*.

Castigliano's Method

Castigliano's method or the method of least work is based on Castigliano's second theorem. The theorem is based on the following assumptions—the structure is made from a linear, elastic material, there are no thermal stresses (temperature is constant) or support settlements. Any application of Castigliano's method is restricted to situations where these conditions hold true.

Castigliano's Second Theorem

The deflection component of the point of application of an action on a structure, in the direction of that action, will be obtained by evaluating the first partial derivative of the total internal strain energy of the structure with respect to the applied action.

Castigliano's Method Applied to Trusses

Consider the indeterminate truss shown in Fig. 103.5. The truss is indeterminate to the first order. The indeterminacy in this case is internal, since there are three external reactions, which are solvable using the three equations of static equilibrium. The extra unknown arises due to the presence of one redundant internal member. Remember that any of the truss members could be designated as a redundant. The selection of the redundant member is governed by the requirement to make the primary structure stable. We are going to remove the redundant truss member (CF), thereby creating a stable, determinate truss and then reinsert the "effect" of the removed member into a secondary structure (see Figs. 103.6 and 103.7).

Figure 103.5
First-order inde-
terminate truss.

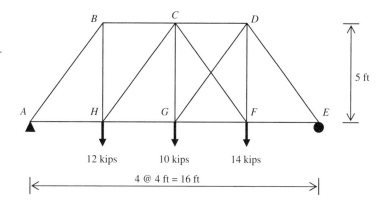

Figure 103.6
Primary
structure: forces.

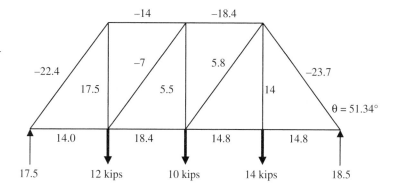

Figure 103.7
Secondary
structure: forces.

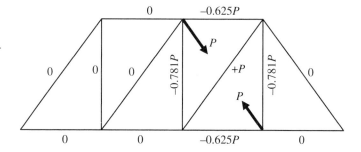

The table below shows a summary of the analysis results from the primary and secondary structures.

Member	Length	F_p	F_s	$F = F_p + F_s$	$\partial F/\partial P$	$F(\partial F/\partial P)$	$F(\partial F/\partial P)L$
AB	6.4	−22.41	0	−22.41	0	0	0
BH	5.0	17.5	0	17.5	0	0	0
HC	6.4	−7.04	0	−7.04	0	0	0
CG	5.0	5.5	−0.781P	5.5 − 0.781P	−0.781	−4.29 + 0.61P	−21.45 + 3.05P
GD	6.4	5.76	+P	5.76 + P	+1.0	5.76 + P	36.86 + 6.4P
DF	5.0	14.0	−0.781P	14.0 − 0.781P	−0.781	−10.93 + 0.61P	−54.67 + 3.05P
DE	6.4	−23.69	0	−23.69	0	0	0
BC	4.0	−14.0	0	−14.0	0	0	0
CD	4.0	−18.4	−0.625P	−18.4 − 0.625P	−0.625	11.5 + 0.39P	46 + 1.56P
AH	4.0	14.0	0	14.0	0	0	0
HG	4.0	18.4	0	18.4	0	0	0
GF	4.0	14.8	−0.625P	14.8 − 0.625P	−0.625	−9.3 + 0.39P	−37.2 + 1.56P
FE	4.0	14.8	0	14.8	0	0	0
FC	6.4	0.0	+P	+P	+1.0	+P	6.4P
						Σ	−30.46 + 22.02P

Castigliano's second theorem may be stated as "in an indeterminate structure, the values of the redundants must be such that the total internal elastic strain energy, resulting from a given system of loads, is minimum." For an ideal truss, where the member forces are purely axial, the axial strain energy is given by a summation of the axial strain energy for all truss members:

$$U = \sum \frac{F^2 L}{2AE} \tag{103.5}$$

For this strain energy to be minimum, its partial derivative with respect to the redundant P is zero:

$$\frac{\partial U}{\partial P} = \sum \frac{2F\left(\frac{\partial F}{\partial P}\right)L}{2AE} = \sum \frac{F\left(\frac{\partial F}{\partial P}\right)L}{AE} = 0$$

If, all members have equal cross-sectional area (A) and the same modulus of elasticity (E), the preceding equation can be simplified to

$$\sum \frac{F\left(\frac{\partial F}{\partial P}\right)L}{AE} = \frac{1}{AE}\sum F\left(\frac{\partial F}{\partial P}\right)L = 0$$

$$\therefore \quad \sum F\left(\frac{\partial F}{\partial P}\right)L = 0$$

Applying the above equation to the specific truss shown, we have

$$-30.46 + 22.02P = 0$$

$$P = \frac{30.46}{22.02} = +1.38$$

This value of P can then be used to calculate the forces in all truss members.

Castigliano's Method Applied to Beams

Consider the following beam. It is indeterminate to the second order.

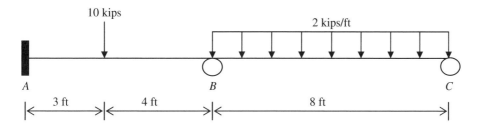

There are various choices of the two redundants. The only condition is that the primary structure remaining after removal of the redundants be stable. For example, of the five external reactions (A_x, A_y, M_A, B_y, and C_y), we can designate M_A and B_y as the redundants. If those two constraints are removed, the primary structure reduces to the simply supported beam shown below.

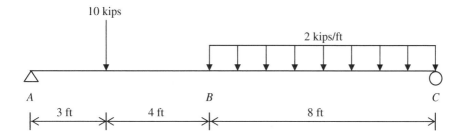

However, if M_A and A_x are designated as the redundants and are removed, the primary structure reduces to an unstable beam (supported by three rollers—no horizontal restraint). This is unacceptable.

For the sake of this example, if the vertical reactions at the two roller supports are taken as the redundants, then the primary structure becomes a cantilever beam. The primary structure is then loaded with the external loads, as well as forces P and Q representing the redundant forces corresponding to support reactions at the rollers.

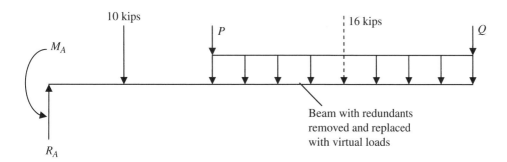

Beam with redundants removed and replaced with virtual loads

From the free-body diagram of the primary beam, one can write the following expressions for the reactions at A. These are, at this stage of the analysis, functions of the loads P and Q.

$$R_A = 10 + 16 + P + Q = 26 + P + Q$$

$$M_A = 30 + 176 + 7P + 15Q$$

Using these expressions for R_A and M_A, one may then further write the following expressions for the internal bending-moment functions in segments 1, 2, and 3.

$$
\begin{aligned}
M_1(x) &= (26 + P + Q)x - (206 + 7P + 15Q) & 0 < x < 3 \\
M_2(x) &= (16 + P + Q)x - (176 + 7P + 15Q) & 3 < x < 7 \\
M_3(x) &= Q(x - 15) - (x - 15)^2 & 7 < x < 15
\end{aligned}
$$

For compatibility with the supports, the vertical deflections at the roller locations must be zero. Turning to Castigliano's second theorem to calculate these deflections, we have

$$\Delta_P = \int \frac{M(x)\dfrac{\partial M}{\partial P}\,dx}{EI} = 0$$

$$\Delta_Q = \int \frac{M(x)\dfrac{\partial M}{\partial Q}\,dx}{EI} = 0$$

Carrying out these integrations, we obtain the following pair of equations:

$$\frac{11{,}002}{3} + \frac{343}{3}P + \frac{931}{3}Q = 0$$

$$\frac{35{,}314}{3} + \frac{931}{3}P + 1125Q = 0$$

$$
\begin{bmatrix} 114.33 & 310.33 \\ 310.33 & 1125 \end{bmatrix}
\begin{Bmatrix} P \\ Q \end{Bmatrix}
=
\begin{Bmatrix} -3667.33 \\ -11{,}771.33 \end{Bmatrix}
\quad \Rightarrow \quad
\begin{Bmatrix} P \\ Q \end{Bmatrix}
=
\begin{Bmatrix} -14.6268 \\ -6.4286 \end{Bmatrix}
$$

Hence, the reactions at B and C are given by 14.63 kips (upward) and 6.43 kips (upward).

Displacement Methods

A broad category of methods known as displacement methods forms a major part of the structural analysis theory useful for indeterminate structures. Two of the most prominent of these methods are the slope-deflection method and the moment-distribution method. These are briefly discussed later. Whereas it is highly unlikely that the current format of the P.E. examination would require these methods to be applied in full, knowledge of key elements of these methods may be useful, particularly for the structural depth examination.

Moment-Distribution Method

The moment-distribution method has been relegated to historical importance only, due to the advent of high-speed computing and the fairly easy application of the stiffness method on the digital computer. The basic idea of the moment-distribution method is to lock all supports so that they behave like fixed supports, and then to unlock them one by one and distribute the resulting "unlock moments" throughout the structures.

Step 1: Fixing All Nodes

Fixing all nodes uncouples the various parts of the structure into independent fixed-fixed spans. The externally applied load then creates fixed-end forces on each span. This also means that at every node, there may exist a moment imbalance. But, since the node is clamped, equilibrium does not need to be satisfied.

Step 2: Releasing Nodes—One at a Time

When a particular node is released, the moment imbalance at that node needs to be corrected by applying a moment of the opposite direction. That correction moment is then distributed to all members connected to the node in proportion to their flexural stiffnesses. The resulting member end moment can also create a carry-over-moment at the far end of the member, if that far end is rigid. The moment-distribution method is thus an iterative, convergent process consisting of the following steps:

1. Calculate fixed-end moments at each node due to the external loads.

2. At each node, apply a correction moment to each member such that the member correction moments are proportional to member bending stiffness, and node satisfies moment equilibrium equation.

3. For any member that has a rigid far node (not pinned), apply a carry-over-moment at the far node equal to half the correction moment at the near node.

4. Repeat steps 2 and 3 until correction moments become acceptably small.

Note that fixed-end moments for various standard loading conditions are summarized at the end of this chapter. These steps are illustrated in Example 103.1.

Example 103.1

Consider the structure shown below. It has a fixed support at A and rollers at B and C. Span AB carries a uniformly distributed load and span BC carries a concentrated load. The free-body diagram with the five unknown support reactions is also shown.

Number of unknowns	$= 5$
Number of equations of static equilibrium	$= 3$
Degree of indeterminacy	$= 2$

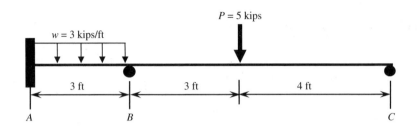

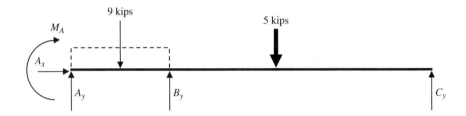

Let EI be constant.

Treating both spans AB and BC as fixed-fixed

Member Stiffnesses ($K = 4EI/L$)

$$K_{AB} = \frac{4EI}{L} = \frac{4EI}{3} = 1.333EI$$

$$K_{BC} = \frac{4EI}{L} = \frac{4EI}{7} = 0.571EI$$

Distribution Factors at Joint B

$$DF_{BA} = \frac{K_{AB}}{K_{AB} + K_{BC}} = \frac{1.333EI}{1.333EI + 0.571EI} = 0.7$$

$$DF_{BC} = \frac{K_{BC}}{K_{AB} + K_{BC}} = \frac{0.571EI}{1.333EI + 0.571EI} = 0.3$$

Fixed-End Moments (Clockwise Positive)

$$MF_{AB} = -\frac{wL^2}{12} = -\frac{3 \times 3^2}{12} = -2.25$$

$$MF_{BA} = +\frac{wL^2}{12} = +\frac{3 \times 3^2}{12} = +2.25$$

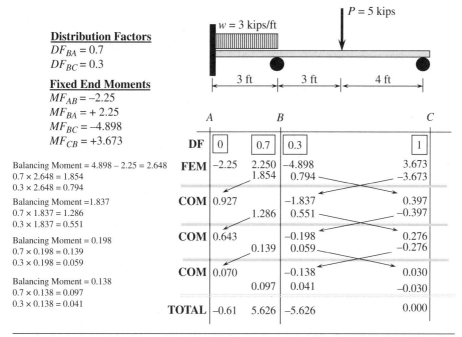

Figure 103.8 Moment-distribution illustration—method one.

$$MF_{BC} = -\frac{Pab^2}{L^2} = -\frac{5 \times 3 \times 4^2}{7^2} = -4.898 \quad \text{(if joint } C \text{ were Fixed)}$$

$$MF_{CB} = +\frac{Pa^2b}{L^2} = -\frac{5 \times 3^2 \times 4}{7^2} = +3.673 \quad \text{(if joint } C \text{ were Fixed)}$$

Figure 103.8 shows the moment-distribution method applied to this structure.

Note that the iterative process is not yet complete—as the next carry-over-moments will continue to diminish, but not equal to zero. Contrast this with the next approach, where the process converges in a single iteration.

Treating span *AB* as fixed-fixed and *BC* as fixed-pinned
 Member Stiffnesses (K = 4EI/L for Fixed-Fixed, K = 3EI/L for Fixed-Pinned)

$$K_{AB} = \frac{4EI}{L} = \frac{4EI}{3} = 1.333EI$$

$$K_{BC} = \frac{3EI}{L} = \frac{3EI}{7} = 0.429EI$$

Note that the member stiffness of *BC* is different from before.

Distribution Factors at Joint B

$$DF_{BA} = \frac{K_{AB}}{K_{AB} + K_{BC}} = \frac{1.333EI}{1.333EI + 0.429EI} = 0.757$$

$$DF_{BC} = \frac{K_{BC}}{K_{AB} + K_{BC}} = \frac{0.429EI}{1.333EI + 0.429EI} = 0.243$$

Fixed-End Moments

$$MF_{AB} = -\frac{wL^2}{12} = -\frac{3 \times 3^2}{12} = -2.25$$

$$MF_{BA} = +\frac{wL^2}{12} = +\frac{3 \times 3^2}{12} = +2.25$$

$$MF_{BC} = -\frac{P}{L^2}\left(ab^2 + \frac{ba^2}{2}\right) = -\frac{5}{7^2}\left(3 \times 4^2 + \frac{4 \times 3^2}{2}\right) = -6.735 \qquad \text{(joint } C \text{ is pinned)}$$

$$MF_{CB} - 0 \qquad \text{(joint } C \text{ is pinned)}$$

The moment distribution process is shown in Fig. 103.9.

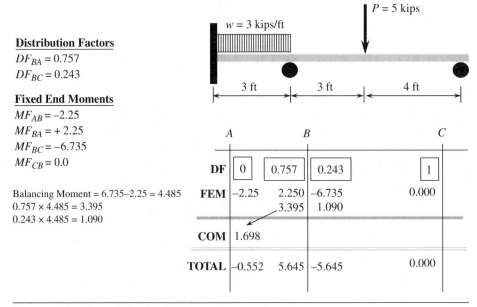

Distribution Factors
$DF_{BA} = 0.757$
$DF_{BC} = 0.243$

Fixed End Moments
$MF_{AB} = -2.25$
$MF_{BA} = +2.25$
$MF_{BC} = -6.735$
$MF_{CB} = 0.0$

Balancing Moment = 6.735−2.25 = 4.485
0.757 × 4.485 = 3.395
0.243 × 4.485 = 1.090

	A		B			C
DF	0	0.757	0.243			1
FEM	−2.25	2.250	−6.735			0.000
		3.395	1.090			
COM	1.698					
TOTAL	−0.552	5.645	−5.645			0.000

Figure 103.9 Moment-distribution illustration—method 2.

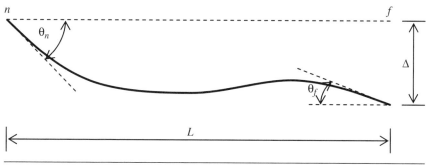

Figure 103.10 Slope-deflection variables.

The Slope-Deflection Method

The slope-deflection equations for a structural member between a "near node" (*n*) and a "far node" (*f*) may be written in terms of the nodal displacements as follows. Here, θ_n and θ_f are nodal rotations and Δ is the transverse displacement of the far node relative to the near node as in Fig. 103.10.

$$M_n = \frac{2EI}{L}\left(2\theta_n + \theta_f - 3\frac{\Delta}{L}\right) + FEM_n$$

$$M_f = \frac{2EI}{L}\left(\theta_n + 2\theta_f - 3\frac{\Delta}{L}\right) + FEM_f$$

These equations show that the moment at a node is composed of contributions of these nodal displacements and the fixed-end moments which are the node-lumped effect of the external loading.

Example 103.2

Let us repeat the same example as considered earlier, using the slope-deflection method.
Fixed-End Moments (Clockwise Positive)

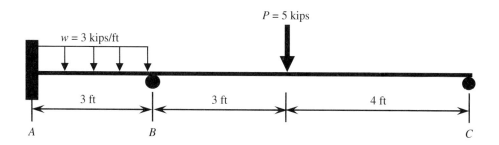

$$MF_{AB} = -\frac{wL^2}{12} = -\frac{3 \times 3^2}{12} = -2.25$$

$$MF_{BA} = +\frac{wL^2}{12} = +\frac{3 \times 3^2}{12} = +2.25$$

$$MF_{BC} = -\frac{Pab^2}{L^2} = -\frac{5 \times 3 \times 4^2}{7^2} = -4.898 \qquad \text{(if joint } C \text{ were fixed)}$$

$$MF_{CB} = +\frac{Pa^2b}{L^2} = -\frac{5 \times 3^2 \times 4}{7^2} = +3.673 \qquad \text{(if joint } C \text{ were fixed)}$$

Slope-Deflection Equations (Clockwise Positive)
Span AB:

$$M_{AB} = \frac{2EI}{3}\left[2\theta_A + \theta_B - 3\frac{0}{3}\right] - 2.25 = \frac{2EI\theta_B}{3} - 2.25$$

$$M_{BA} = \frac{2EI}{3}\left[\theta_A + 2\theta_B - 3\frac{0}{3}\right] + 2.25 = \frac{4EI\theta_B}{3} + 2.25$$

Span BC:

$$M_{BC} = \frac{2EI}{7}\left[2\theta_B + \theta_C - 3\frac{0}{3}\right] - 4.898 = \frac{4EI\theta_B}{7} + \frac{2EI\theta_C}{7} - 4.898$$

$$M_{CB} = \frac{2EI}{7}\left[\theta_B + 2\theta_C - 3\frac{0}{3}\right] + 3.673 = \frac{2EI\theta_B}{7} + \frac{4EI\theta_C}{7} + 3.673$$

Equilibrium equations are

$$M_{BA} + M_{BC} = 0 \Rightarrow 1.905EI\theta_B + 0.286EI\theta_C - 2.648 = 0$$

$$M_{CB} = 0 \Rightarrow 0.286EI\theta_B + 0.571EI\theta_C + 3.673 = 0$$

Solving these equations, we get

$$EI\theta_B = +2.545 \quad EI\theta_C = -7.700$$

Substituting these back into the moment expressions, we get

$$M_{AB} = \frac{2EI\theta_B}{3} - 2.25 = -0.554$$

$$M_{BA} = \frac{4EI\theta_B}{3} + 2.25 = +5.643$$

$$M_{BC} = \frac{4EI\theta_B}{7} + \frac{2EI\theta_C}{7} - 4.898 = -5.644$$

$$M_{CB} = \frac{2EI\theta_B}{7} + \frac{4EI\theta_C}{7} + 3.673 = 0.000$$

These results are fairly close to those obtained from moment distribution, earlier.

Using Results from Moment-Distribution or Slope-Deflection Methods

Step 1: Drawing the Free-Body Diagrams
The solution of the deflection methods consists of the nodal moments M_{AB}, M_{BA}, M_{BC}, and so on. Knowledge of these moments allows the "decoupling" of the individual spans, as shown below.

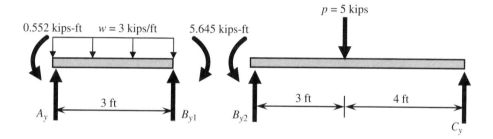

Step 2: Writing the Equilibrium Equations
For the left substructure, the following are the two equilibrium equations:

$$\sum F_y = A_y - 9 + B_{y1} = 0$$

$$\sum M_A = 0.552 - 9 \times 1.5 - 5.645 + 3B_{y1} = 0 \Rightarrow B_{y1} = 6.198 \text{ kips}$$

For the right substructure, the following are the two equilibrium equations:

$$\sum F_y = B_{y2} - 5 + C_y = 0$$

$$\sum M_C = 5.645 + 5 \times 4 - 7B_{y2} = 0 \Rightarrow B_{y2} = 3.664 \text{ kips}$$

Therefore, the total vertical reaction at B is

$$B_y = B_{y1} + B_{y2} = 6.198 + 3.664 = 9.862 \text{ kips.}$$

Reaction at A is $A_y = 2.802$ kips

Reaction at C is $C_y = 1.336$ kips

Fixed-End Moments

Figure 103.11 shows fixed-end moments for a variety of loads on a fixed-fixed beam.

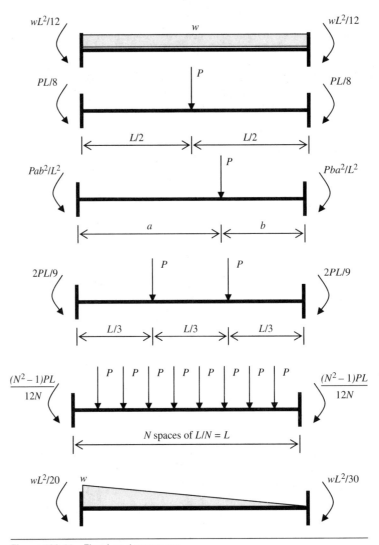

Figure 103.11 Fixed-end moments.

Approximate Methods for Building Frames

Building frames are usually highly indeterminate. Before computer-based analysis methods and tools became ubiquitous, these frames were analyzed using certain approximate methods. Two of the most prevalent of these methods used for laterally loaded frames are: (1) the portal method and (2) the cantilever method. The basic idea behind both of these methods is to make certain

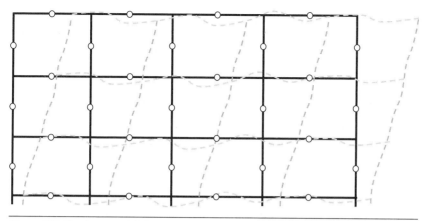

Figure 103.12 Assumed deformed shape of laterally loaded frames.

assumptions about the deformation caused by the loading. Based on the deformed shape, one can make assumptions about location of inflection points. By the moment curvature relationship, the bending moment at these inflection points (zero curvature) must be zero. Therefore, the approximate model works by assuming internal hinges at these locations, as shown in Fig. 103.12. This assumption converts a highly indeterminate structure into a determinate one.

Portal Method

For the portal (doorway) method, the frame is visualized as a collection of portals, each bay representing one portal. The portal method is best suited for frames with a low height:width ratio. The base shear is distributed to the columns using a tributary area concept.

Thus, if bay lengths are equal, the inner columns, which belong to two such portals, receive twice the shear than the exterior columns. This permits a very easy computation of the distribution of the base shear into individual columns.

Example 103.3

Use the portal method to analyze the two-story frame shown below.

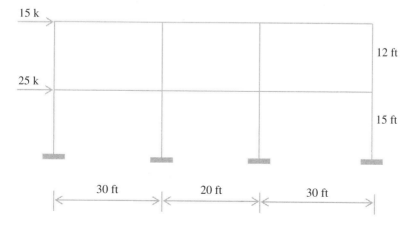

The total base shear = 40 kips. The tributary width for the two exterior columns is 15 ft, while that for the two interior columns is 25 ft. Dividing the total base shear ($V = 40$ k) in proportion, we get $V_1 = V_4 = 7.5$ kips and $V_2 = V_3 = 12.5$ kips.

It is assumed that internal hinges (representing inflection points) exist at the midpoints of the beams and the columns (moment expected to be zero, due to lateral loads). The column shears in the second-story columns, using tributary area, are 2.81 k, 4.69 k, 4.69 k, and 2.81 k, respectively.

Once the column shears at the second level are known, the hinge forces can be obtained as shown below. The process can then be repeated at the next lower level.

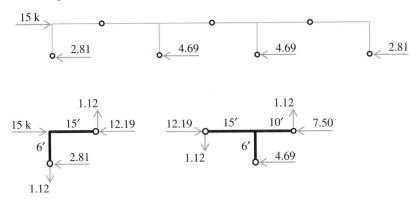

Cantilever Method

Used for lateral loads, the cantilever method models the building as if it were a cantilever beam with the lateral loads causing bending. At a particular location (height) of the frame, the moment caused by all loads above that level is analogous to the internal bending moment in the equivalent beam. Since, under linear, elastic assumptions, the bending stress varies linearly across the cross section, in this approximate method, it is assumed that the *stress* in each column is linearly proportional to the distance from the centerline (vertical) of the structure, which is determined by calculating the centroid of the cross-sectional areas of the columns. See Fig. 103.13.

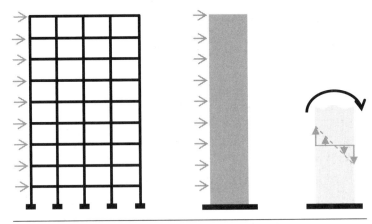

Figure 103.13 Conceptual basis for cantilever method.

As with the portal method, inflection points are assumed to occur at the midpoints of each beam and each column, represented by internal hinges in the structural model. The cantilever method is better-suited model for taller buildings while the portal method works better for low-rise buildings.

Example 103.4

Use the cantilever method to analyze the two-story frame (from Example 103.3) shown below. Assume cross-sectional area of the interior columns is double that of the exterior columns.

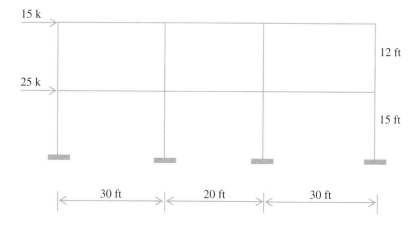

The total base shear $= 40$ kips.

Using principles of statics, taking moments about the outer column on the left, the centroid of the column group (each column represented by its area) is given by:

$$\bar{x} = \frac{A \times 0 + 2A \times 30 + 2A \times 50 + A \times 80}{A + 2A + 2A + A} = 40$$

The stress in each column is proportional to distance from this centroid. Thus, the stress in the exterior columns (located 40 ft from the centroid) is four times the stress in the interior columns (located 10 ft from the centroid). However, since the area of the interior columns is twice that of the exterior columns, the axial force in the exterior columns (F_1 and F_4) is only twice that in the interior columns ($F_1 = F_4 = 2F_2 = 2F_3$).

The moment of these column axial forces, about the centroid of the column group, is given by

$$M = 2F \times 40 + F \times 10 + F \times 10 + 2F \times 40 = 180F$$

Making a section through the hinges assumed at the midpoint of the second-level columns, the overturning moment is:

$$M_{OT} = 15 \times 6 = 90$$

Equating the resisting moment due to the column forces to the overturning moment, $F = 90 \div 180 = 0.5$ kips. Thus, at this level, the axial force in the outer columns $= 1.0$ k and interior columns have $F = 0.5$ k.

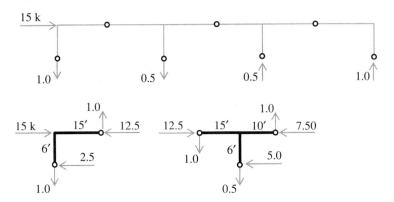

Concrete Fundamentals

Absolute Volume Method

In the past, the design of concrete mixes has been done by empirically based weight ratios of the primary constituents (a cementitious material such as Portland cement, water, and coarse and fine aggregates). The proportions were chosen based on experience and job-specific objectives and limitations. However, it is also possible to use a more methodical process such as the *absolute volume method* as outlined in the book *Design and Control of Concrete Mixtures*[1]. In the methodology outlined in Chap. 9 (Designing and Proportioning Normal Concrete Mixtures) of that book, the design of a concrete mix is done for a unit volume of concrete (1 m^3 or 1 yd^3). The design steps in the method are outlined below. For the sake of brevity, a unit volume of 1 yd^3 has been used in the following discussion:

1. Based on the target compressive strength of the concrete, Table 9.3 gives the desirable water-cement (w/c) ratio.

2. The coarse aggregate size should be part of the specifications for the concrete. The air content is determined based on the anticipated exposure of the concrete. Table 9.5 then gives the approximate mixing water (pounds of water per cubic yard of concrete).

3. Using the unit weight of water, the weight of water may be converted into the volume of water (cubic feet of water per cubic yard of concrete).

4. Knowing the w/c ratio and the weight of water, the weight of cement (pounds of cement per cubic yard of concrete) is calculated.

5. Using the unit weight of cement, the weight of cement may be converted into the volume of cement (cubic feet of cement per cubic yard of concrete).

6. The recommended bulk volume of coarse aggregate is given by Table 9.4.

---

[1] S. Kosmatka, B. Kerkhoff, W. Panarese. Design and Control of Concrete Mixtures, 14th Edition, Portland Cement Association, Skokie, IL, 2008.

7. The bulk volume of coarse aggregate is multiplied by the bulk density to obtain the weight of coarse aggregate (pounds of aggregate per cubic yard of concrete).

8. Using the unit weight of coarse aggregate, the weight of aggregate may be converted into the volume of aggregate (cubic feet of coarse aggregate per cubic yard of concrete).

9. The volume of air in the mix is obtained from the air content determined in step 2.

10. The volume of fine aggregate is obtained as the volume remaining by subtracting air, water, cement, and coarse aggregate volumes from 1 yd^3 (27 ft^3).

11. The weight of fine aggregate is calculated (pounds of aggregate per cubic yard of concrete) based on the known specific gravity.

12. Corrections are made to the mixing water for moisture contained within fine and coarse aggregates according to the following formula:

$$W_{water} = \text{weight of water available from wet aggregate} = W_{wet} \times \left(\frac{MC - MC_{SSD}}{100 + MC} \right)$$

where W_{wet} = weight of wet aggregate
MC = moisture content of wet aggregate (%)
MC_{SSD} = moisture content of SSD aggregate (%) = % absorption

The actual weight of mixing water in the concrete is then given by

Weight of water available = Weight of added water + W_{water} from coarse aggregate + W_{water} from fine aggregate

The weight of SSD (surface saturated dry) aggregate is obtained according to

$$W_{SSD} = W_{wet} - W_{water} = W_{wet} \times \left(\frac{100 + MC_{SSD}}{100 + MC} \right) \tag{104.1}$$

The volume of each component (SSD coarse aggregate, SSD fine aggregate, and cement) can be calculated as

$$V_{coarse} = \frac{W_{SSD} \text{ for coarse aggregate}}{SG_{SSD} \times \gamma_{water}} \tag{104.2a}$$

$$V_{fine} = \frac{W_{SSD} \text{ for fine aggregate}}{SG_{SSD} \times \gamma_{water}} \tag{104.2b}$$

$$V_{cement} = \frac{W_{cement}}{SG_{cement} \times \gamma_{water}} \tag{104.2c}$$

Water-cement ratio = W_{water}/W_{cement} (using total or adjusted weight of mixing water)

Sometimes, in the field, w/c ratio is expressed as gallons per sack instead of a dimensionless weight ratio. The decimal w/c ratio is multiplied by 11.27 to convert to gallons per sack.

Example 104.1   Concrete Mix Design

The following aggregates are used in a concrete mix:

Coarse Aggregate:

Moisture content of wet aggregate	2.8%
Moisture content of SSD aggregate	0.6%
Specific gravity of SSD aggregate	2.65

Fine Aggregate:

Moisture content of wet aggregate	6.0%
Moisture content of SSD aggregate	1.4%
Specific gravity of SSD aggregate	2.60

Cement:

Specific gravity   3.15

Given the following characteristics of the bulk mix:

Weight of wet coarse aggregate	135 lb
Weight of wet fine aggregate	95 lb
Weight of added water	21 lb
Weight of cement	48 lb
Volume of concrete	2.14 ft³

Calculate (1) unit weight of the concrete, (2) volume of coarse aggregate, (3) volume of fine aggregate, and (4) water-cement ratio.

Solution

1. Unit weight $\gamma = \dfrac{135 + 95 + 21 + 48}{2.14} = 139.7 \text{ lb/ft}^3$

2. Weight of SSD coarse aggregate $W_{SSD} = 135 \times \dfrac{100 + 0.6}{100 + 2.8} = 132.11 \text{ lb}$

   Volume of SSD coarse aggregate $V_{coarse} = \dfrac{132.11}{2.65 \times 62.4} = 0.799 \text{ ft}^3$

3. Weight of SSD fine aggregate $W_{SSD} = 95 \times \dfrac{100 + 1.4}{100 + 6.0} = 90.88 \text{ lb}$

   Volume of SSD fine aggregate $V_{fine} = \dfrac{90.88}{2.60 \times 62.4} = 0.560 \text{ ft}^3$

4. Available water from coarse aggregate $= 135 \times \left(\dfrac{2.8 - 0.6}{100 + 2.8} \right) = 2.89$

Available water from fine aggregate $= 95 \times \left(\dfrac{6 - 1.4}{100 + 6} \right) = 4.12$

Actual water weight $W_{water} = 21.0 + 2.89 + 4.12 = 28.01 \, lb$

Actual water-cement ratio $= 28.01/48 = 0.584 = 6.58$ gallons per sack

Cold Weather Concrete

ACI Committee 306 defines "cold weather" as more than 3 days when the average daily temperature is less than 5°C (40°F) and stays below 10°C (50°F) for more than half of a 24-hour period. During periods of cold weather, special considerations must be taken in the mix design, transport, and handling as well as the curing of concrete. Normal concreting operations can resume once the ambient temperature has stayed above 10°C (50°F) for more than half a day.

Cold weather concreting operations involve the use of protective measures such as enclosures, windbreaks, portable heaters, insulated forms, and blankets. The minimum temperature of fresh concrete as mixed should be about 7°C (45°F) for air temperatures above −1°C (30°F), 10°C (50°F) for air temperatures between 0 and 30°F, and 13°C (55°F) for air temperatures below −17°C (0°F).

During the chemical process of hydration, the water demand of the chemical reactions gradually reduces the saturation of the fresh concrete. This occurs at approximately a compressive strength of 500 psi, usually within the first 24 hours. It is during this "early" period that fresh concrete is most susceptible to long-term damage due to cold temperatures.

Concrete which will be exposed to freeze-thaw cycles should be allowed to reach a compressive strength of 4,000 psi before it is exposed to freezing and thawing. Air-entrained concrete is less susceptible to damage by early freezing than non-air-entrained concrete.

Chemical reactions involved in the process of hydration typically completely cease at −10°C (14°F). This is why, in concrete maturity methods (see Nurse-Saul, Arrhenius), this is considered this is considered a baseline, or reference, temperature.

High early strength can be achieved by (1) using Type III (high early strength) cement, (2) using additional cement (100–200 lb/yd³), and (3) using chemical accelerators such as calcium chloride ($CaCl_2$). However, chlorides must not be used where there exists a high potential for corrosion, nor where exposure to air/water containing sulphates or for concrete susceptible to alkali-aggregate reaction.

The objective of using accelerators is to reduce the time of initial setting, but they must not be used as a substitute for proper curing and protection from frost. Low water-cement ratio and low-slump concrete is desirable, especially for flatwork.

Optimum concrete temperature during the curing process is generally considered to be 50 to 60°F. There seems to be very little advantage to having concrete mix temperature above 70°F. Higher concrete temperature increases the potential for thermal shrinkage after hardening and increases the possibility of plastic shrinkage cracking (due to rapid moisture loss through evaporation).

One method to raise the temperature of concrete during mixing is to actually heat up the components. Frozen aggregates must be thawed to avoid frozen pockets and ice lenses as well as to avoid releasing extra water into the concrete during mixing. Even though the quantity of aggregate in a typical concrete mix is much greater than the quantity of water, the latter is easier to heat (specific heat of water is about one-fifth that of aggregates – 0.22 vs. 1.0 Btu/lb/°F). For air temperatures between 30 and 40°F, it is usually necessary only to heat the water to a maximum of about 140°F. For air temperatures below 30°F, the water can be heated to 140 to 212°F and the aggregate to about 45 to 55°F. If both aggregates and the mixing water are preheated, it is recommended that the water be mixed with the aggregates before adding the cement to avoid a flash set.

The temperature (T) of a concrete mix can be calculated as a weighted average of the temperatures (T) and masses (M) of the components.

$$T = \frac{0.22(T_a M_a + T_c M_c) + T_w M_w + T_{wa} M_{wa}}{0.22(M_a + M_c) + M_w + M_{wa}}$$

(104.3)

The subscripts a, c, w, and wa refer to aggregate, cement, added water, and free moisture on aggregates, respectively.

Hot Weather Concrete

Hot weather conditions adversely influence concrete quality, by accelerating the rate of moisture loss and rate of cement hydration. Detrimental hot weather conditions include—high ambient temperature, high concrete temperature, low relative humidity, high wind speed, and solar radiation. These conditions create increased water demand, accelerated slump loss leading to added water in the field, increased rate of setting, increased tendency for plastic cracking, difficulty in controlling entrained air, long- term strength loss, and increased potential for thermal cracking.

Concrete when placed should have a temperature less than 85 to 90°F and precautions and preparations must be taken when concrete temperature begins to approach 80°F. Such precautions include cooling of components, reduction of transport time, rapid placement and consolidation time, adjusting schedules to perform concrete operations during cooler periods of the day, use of moisture retaining films, minimizing moisture loss, etc.

One of the easiest means of lowering concrete temperature is to actually cool the components, of which water is the easiest and most economical to cool down because of its lower specific heat (compared to cement and aggregates). Cooling mixing water 3.5 to 4.0°F usually cools concrete by about 1°F, whereas cooling aggregates 1.5 to 2.0°F lowers concrete temperature by 1°F.

Adding supplementary cementitious materials such as fly ash and other pozzolans and ground granulated blast furnace slag generally slow both the rate of setting as well as the rate of slump loss.

ACI Provisions

In the United States, design of reinforced concrete structures and structural elements is conducted according to the strength design approach, which superseded the allowable stress or working stress method. The current design specifications are entitled *ACI 318: Building Code Requirements for Structural Concrete*[2] and are updated every 3 years.

---

[2]ACI 318, 2014, *Building Code Requirements for Reinforced Concrete*, American Concrete Institute, Detroit, Michigan.

In the strength design method, the strength of the concrete member under an ultimate limit state is compared to the demand created by the governing factored load combination. Unless otherwise specified, the unit weight of normal weight concrete is taken to be $\gamma_c = 150$ lb/ft³. Lightweight structural concrete has unit weight 90–110 lb/ft³. According to the ACI code, the modulus of elasticity is given by

$$E_c = \gamma_c^{1.5} \times 33 \times \sqrt{f_c'} \qquad (104.4)$$

where $f_c' = $ 28-day compressive strength (psi)
$\gamma_c = $ unit weight of the concrete (pcf)
$E_c = $ modulus of elasticity (psi)

The modulus of rupture (tensile strength) is denoted by f_r and is given by ACI as

$$f_r = 7.5\sqrt{f_c'} \qquad (104.5)$$

In the United States, a standard bag of cement contains 94 lb. A concrete mix referred to as a 6-*bag mix* contains 6 standard (94 lb) bags of cement per cubic yard (564 lb cement per cubic yard concrete). Cements are designated as the following types:

Type I   Normal

Type II   Modified for hot weather or large structures

Type III   High early strength

Type IV   Low heat (massive structures)

Type V   Sulphate resistant

Reinforcement

Reinforcement bars are available in various sizes—in the U.S. system of nomenclature, these sizes are designated No. 3, No. 4, and so on, up to No. 18 bars. Nominal diameters and areas of these sizes are given in Table 104.1.

Strength Design Approach

The nominal or theoretical strength (bending, axial, shear, etc.) is denoted R_n (nominal resistance). The design strength ϕR_n is less than the nominal strength by a strength reduction factor ϕ, which is less than unity.

Table 104.1 Properties of Reinforcement Bars

Size	No. 3	No. 4	No. 5	No. 6	No. 7	No. 8	No. 9	No. 10	No. 11	No. 14	No. 18
Diameter (in)	0.375	0.500	0.625	0.750	0.875	1.000	1.128	1.270	1.410	1.693	2.257
Area (in²)	0.11	0.20	0.31	0.44	0.60	0.79	1.00	1.27	1.56	2.25	4.00
Weight (lb/ft)	0.376	0.668	1.043	1.502	2.044	2.670	3.400	4.303	5.313	7.650	13.60

$$\phi R_n \geq P_u \tag{104.6}$$

The strength reduction factor is given by ACI according to Table 104.2.

The factored (ultimate) load U is calculated as

$$P_u = \sum \gamma_i Q_i \tag{104.7}$$

where Q_i = the ith service load level (e.g., dead load and live load).
γ_i = corresponding load factor.

The load combination that produces the maximum load effect for the member is said to "govern."

Load Combinations (ASCE-7)

As updated in the ACI 318-02 specification, the factored load approach should be applied in accordance with Chap. 9 criteria. The load combinations adopted in Chap. 9 are those from ASCE-7.[3] The strength reduction factors ϕ given in Table 104.2 are also from Chap. 9 of ACI 318. The load combinations adopted from ASCE-7 into ACI 318 are shown below:

$$U = 1.4(D + F) \tag{104.8a}$$

$$U = 1.2(D + F + T) + 1.6(L + H) + 0.5(L_r \text{ or } S \text{ or } R) \tag{104.8b}$$

$$U = 1.2D + 1.6(L_r \text{ or } S \text{ or } R) + (1.0L \text{ or } 0.5W) \tag{104.8c}$$

$$U = 1.2D + 1.0W + 1.0L + 0.5(L_r \text{ or } S \text{ or } R) \tag{104.8d}$$

$$U = 1.2D + 1.0E + 1.0L + 0.2S \tag{104.8e}$$

$$U = 0.9D + (1.0W \text{ or } 1.0E) + 1.6H \tag{104.8f}$$

Table 104.2 Strength Reduction Factor for Reinforced Concrete

Stress-State	ϕ
Tension-controlled sections ($\varepsilon_t > 0.005$)	0.90
For all beams ($0.004 < \varepsilon_t < 0.005$)	$0.48 + 83\,\varepsilon_t$
Compression-controlled sections ($\varepsilon_t < 0.002$)	
Members with spiral reinforcement	0.70
Members with tied reinforcement	0.65
Transition sections ($0.002 < \varepsilon_t < 0.005$)	
Members with spiral reinforcement	$0.57 + 67\,\varepsilon_t$
Members with tied reinforcement	$0.48 + 83\,\varepsilon_t$
Shear and torsion	0.75
Bearing on concrete	0.65

Source   ACI 318, 2014, *Building Code Requirements for Reinforced Concrete*, American Concrete Institute, Detroit, Michigan.

Table 104.3   Strength Reduction Factor (ACI Appendix C)

Stress-State	ϕ
Tension-controlled sections ($\varepsilon_t > 0.005$)	0.90
Compression-controlled sections ($\varepsilon_t < 0.002$)	
Members with spiral reinforcement	0.75
Members with tied reinforcement	0.70
Shear and torsion	0.85
Bearing on concrete	0.70

where D = dead load, F = fluid pressure load, T = self-straining force, L = live load, W = wind load, E = earthquake load, L_r = roof live load, S = snow load, R = rain load, H = earth pressure loads.

NOTES
1. The load factor on live load (L) in load combinations in Eqs. (104.8c), (104.8d), and (104.8e) may be reduced to 0.5 except for garages, places of public assembly, and all areas where L is greater than 100 psf.
2. Where W is based on service-level wind loads, 1.6W shall be used in place of 1.0 W in Eqs. (104.8d) and (104.8f), and 0.8 W shall be used in place of 0.5 W in Eq. (104.8c).
3. Where the earthquake load (E) is based on service-level seismic forces, replace 1.0 E with 1.4 E in load combinations in Eqs. (104.8e) and (104.8f).

However, when only dead and live loads are present, the factored load may be taken as

$$U = 1.4D + 1.7L \qquad (104.9)$$

This exception is noted in App. C of the specifications. However, when using this factored load, the strength reduction factors ϕ must be taken as shown in Table 104.3.

Thus, the alternative approaches in Chap. 9 and App. C of the ACI 318 specifications result in significantly different factored loads, but the final result can be quite similar due to different strength-reduction factors. According to the current syllabus for the Structural Depth of the PE Civil exam, "Appendix C does not apply to the Civil Structural examination."

Significant Changes from ACI 318-11 to ACI 318-14

ACI 318-14 has been adopted by reference into the 2015 International Building Code (IBC). There are significant differences – both organizational and technical – between ACI 318-11 and ACI 318-14. For the first time in its development cycle, ACI 318-14 has been reorganized as a member-based document, such that within each chapter devoted to a particular member

[3]*Minimum Design Loads for Buildings and Other Structures*, ASCE/SEI 7-10, American Society of Civil Engineers, 2010.

type, the user would find all the requirements necessary to design that particular member type. For information that belonged to several member types (such as development length), "toolbox" chapters are created, with references made in the member-based chapters.

In Chapter 2 (Notation & Terminology), the following sentence has been added to the definition for the term "hoop": "A closed tie shall not be made up of interlocking headed deformed bars."

In Chapter 5 (Loads), all references to service level earthquake forces have been eliminated.

Also in Chapter 5, recognizing that secondary moments should be considered in member design even when moments are not redistributed, the following has been added to Section 5.3.11 – "Required strength shall include internal load effects due to reaction induced by prestressing with a load factor of 1.0."

In Chapter 6, a new Section 6.9 has provisions to explicitly allow Finite Element Analysis for calculating member loads.

In Chapter 8 (Two-Way Slabs), ACI 318-14 requires that the same minimum bonded reinforcement be provided in slabs with unbonded or bonded tendons, except that only the area of the bonded tendons is considered effective in control of cracking.

A new Section 9.5.4.7, based on research at North Carolina State University, provides the following – For solid precast sections with an aspect ratio $h/b_t \geq 4.5$ (b_t = width of that part of the cross section containing the closed stirrups resisting torsion), it shall be permitted to use open web reinforcement, provided the design procedure produces results that are analytically and empirically justifiable.

Chapter 12 (Diaphragms) of ACI 318-14 has, for the first time, added design provisions for diaphragms in buildings assigned to SDC C and lower.

Chapter 18 (Earthquake Resistant Structures) has some of the most significant technical changes in ACI 318-14. Confinement requirements for columns of special moment frames with high axial load or high concrete compressive strength are significantly different for the plastic hinge regions at the ends of the column. The effectiveness of well-distributed (and laterally supported) reinforcement around the column perimeter is also recognized in ACI 318-14.

For beam-column joints of special moment frames, the new requirements are –

a. restrictions of joint-aspect ratio (Section 18.8.2.4), with the most desirable value being 1.0 and the maximum permitted value being 2.0,

b. requirements for knee joints with headed beam reinforcement (Section 18.8.3.4),

c. hooking of beam reinforcement within a joint (Section 18.8.5.1),

d. requirements for headed longitudinal reinforcement within joints (Section 18.8.5.2).

Chapter 19 (Concrete: Design and Durability Requirements) Table 19.3.1.1 now contains Exposure Categories and Classes. Exposure classes P0 and P1 have been renamed W0 and W1 respectively. In Table 19.3.2.1, the maximum w/c ratio and the minimum compressive strength for exposure classes F1 and F3 are different. Cementitious material types that are allowed in concrete for exposure classes S1, S2, and S3 have changed.

In Chapter 20 (Steel Reinforcement Properties, Durability, and Embedment), the definition of Yield Point for steels without a sharply defined yield point has changed to 0.3% proof stress, to bring it in line with the ASTM definition. Previous versions of ACI 318 used the stress corresponding to a strain of 0.35%.

In Chapter 22 (Sectional Strength), the following has been added to Section 22.6.4.2 – the critical section for two-way shear shall be a polygon selected to minimize the perimeter b_o.

In Chapter 25 (Reinforcement Details), the 3-in. minimum extension of a 90° or 135° hook (previously required of seismic hooks only) has been added to standard hooks as well.

Reinforced Concrete Beams

General

Concrete beams may be reinforced with tension reinforcement only (singly reinforced beams) or both tension and compression steel (doubly reinforced beams). A reinforced concrete beam may be singly reinforced as long as the amount of required reinforcement does not exceed the ACI limits on reinforcement. The reinforcement bars are placed in the tension zone with adequate cover. Clear cover guidelines from ACI 318 (ACI 318, 2014, *Building Code Requirements for Reinforced Concrete,* American Concrete Institute, Detroit, Michigan) are summarized in Table 105.2. For the locations of the beam where the maximum permitted tension reinforcement does not give adequate moment capacity, the deficit may be made up by using additional reinforcement in the compression zone.

Typical progressive behavior of reinforced concrete beams in flexure is summarized in the strain and stress distributions in Fig. 105.1. A rectangular reinforced concrete beam is shown loaded in third-point loading, resulting in a state of pure bending in the middle third of the span. At low loads, when the maximum tensile stress in the concrete is below the rupture strength f_r, the neutral axis is slightly below midheight, and the strain and stress distribution are as shown in the first pair.

As the load increases, the tensile stress in the concrete reaches the rupture strength, causing it to form hairline cracks. The concrete in the tension zone becomes ineffective and the compression in the concrete (possibly still linear or beginning to show some nonlinearity) is balanced solely by the tension in the steel reinforcement. This is shown by the second pair. The location of the neutral axis will depend on the amount of reinforcement and the relative strengths of concrete and steel.

As the load increases further, the compressive stress in the concrete becomes decidedly nonlinear and the steel yields (strain exceeds the yield strain). This is shown by the third pair. The limit of this stress diagram is when the compressive strain in the concrete reaches the limiting value of 0.003, which is defined in ACI 318 as the crushing limit state.

Use of "balanced reinforcement" will cause these limit states—yielding of steel and crushing of concrete—to be reached simultaneously. However, for the failure to be ductile, it is desirable for the beam to be proportioned such that the steel yields well before the crushing

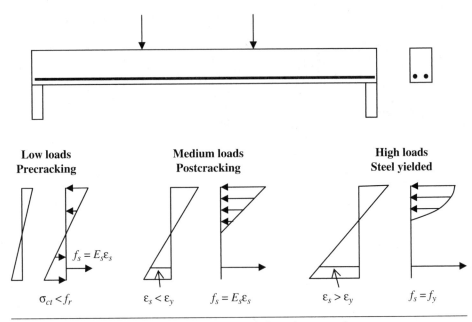

Figure 105.1   Strain and stress progression in a reinforced concrete beam.

strain is reached in the concrete. In the current ACI specifications, this is ensured by imposing a lower limit on the tensile strain in the outermost steel layer ($\varepsilon_t > 0.004$). Prior to the 2002 edition of ACI 318, this criterion was stated as "maximum steel ratio = 75% of the balanced steel ratio ρ_b."

Design Moments at Critical Locations

Table 105.1 can be used to calculate moments and shears at critical locations for continuous beams and one-way slabs.

Cracked Section Characteristics

A doubly reinforced rectangular concrete section is shown in Fig. 105.2. Assume that the flexural stress is tensile on the bottom of the section. When this tensile stress exceeds the modulus of rupture, the concrete cracks and thereby loses any tensile resistance. As a result, the neutral axis shifts upward. In the figure, the compression zone is shown shaded. The steel can be converted to an equivalent area of concrete using the modular ratio (n):

$$n = E_s/E_c \tag{105.1}$$

by multiplying the tension steel area by n and the compression steel area by ($n - 1$). The reason for applying different conversion factors for the steel in the tension zone versus that in the compression zone is that the latter displaces concrete which is still effective

Table 105.1   Critical Shears and Moments for One-Way Beams and Slabs

Design quantity	
Positive moment	
End Spans	
If discontinuous end is unrestrained	$wl^2/11$
If discontinuous end is integral with support	$wl^2/14$
Interior Spans	$wl^2/16$
Negative moment at exterior face of first interior support	
Two spans	$wl^2/9$
More than two spans	$wl^2/10$
Negative moment at other faces of interior supports	$wl^2/11$
Negative moment at face of all supports for slabs with spans less than 10 ft and beams and girders where column stiffness is more than eight times beam stiffness	$wl^2/12$
Negative moment at interior faces of exterior supports for members built integrally with supports	
Where the support is a spandrel beam or girder	$wl^2/24$
Where the support is a column	$wl^2/16$
Shear in end members at first interior support	$1.15wl/2$
Shear at all other supports	$wl/2$

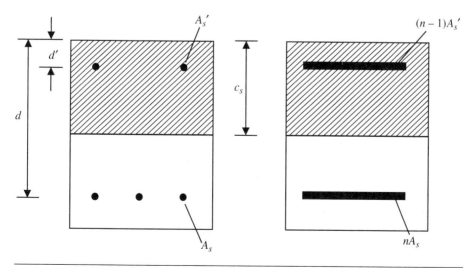

Figure 105.2   Cracked section parameters.

(in compression) whereas the tension steel displaces concrete which is in the cracked zone and therefore ineffective. Taking moments about the presumed neutral axis, one obtains the following quadratic:

$$\frac{bc_s^2}{2} + (n-1)A_s'(c_s - d') = nA_s(d - c_s) \tag{105.2}$$

The quadratic equation can be solved to determine the depth of the neutral axis (c_s) of the cracked section. The cracked moment of inertia about the neutral axis is given by

$$I_{cr} = \frac{bc_s^3}{3} + (n-1)A_s'(c_s - d')^2 + nA_s(d - c_s)^2 \tag{105.3}$$

Effective Moment of Inertia

In computing deflections of concrete beams, the ACI specifications permit the use of an effective moment of inertia (assumed constant along beam length) to capture some cracked and some uncracked sections. This effective moment of inertia is calculated as

$$I_e = \left(\frac{M_{cr}}{M_a}\right)^3 I_g + \left[1 - \left(\frac{M_{cr}}{M_a}\right)^3\right] I_{cr} < I_g \tag{105.4}$$

The cracking moment (M_{cr}) is calculated as

$$M_{cr} = \frac{f_r I_g}{y_t} \tag{105.5}$$

where M_a = maximum bending moment in the beam
f_r = modulus of rupture of concrete
I_g = gross moment of inertia of the section
y_t = distance from the neutral axis to the extreme tension fiber

Clear Cover Guidelines

Based on the guidelines summarized in Table 105.2, the typical minimum clear cover for beams and columns is 1.5 in., for slabs it is 0.75 in, and for footings it is 3.0 in.

ACI Limits on Flexural Reinforcement

As a beam section is subject to increasing bending moment, the concrete in compression progresses toward its crushing limit state (compressive strain $\varepsilon_c = 0.003$) while the steel reinforcement progresses toward its yield limit state (strain $\varepsilon_y = f_y/E_s$). The balanced reinforcement

Table 105.2 Clear Cover Guidelines in ACI 318

Exposure category	Clear cover (in.)
Concrete cast against and permanently exposed to earth	
All members	3.0
Concrete exposed to earth or weather	
No. 6 through no. 18 bars	2.0
No. 5, W31 or D31 wire and smaller	1.5
Concrete not exposed to weather or earth	
Beams and columns	1.5
Slabs, walls, and joists	
No. 11 bar and smaller	0.75
Nos. 14 and 18 bars	1.5
Shells, folded plate members	
No. 6 through no. 18 bars	0.75
No. 5, W31 or D31 wire and smaller	0.5

Source ACI 318, 2014, *Building Code Requirements for Reinforced Concrete*, American Concrete Institute, Detroit, Michigan.

is that for which both materials reach their limit states simultaneously. The ACI 318 specifications prior to 2002 limited the tension steel (A_s) to 75% of the balanced steel. The current ACI 318 specification (ACI 318-14) limits tension reinforcement by imposing a lower limit on the tensile strain ($\varepsilon_t \geq 0.004$). This leads to the following expression for the maximum area of tensile reinforcement for singly reinforced concrete beams:

$$A_{s,\max} = \frac{0.85 f_c' \beta_1 b}{f_y}\left(\frac{3d_t}{7}\right) \tag{105.6}$$

where β_1 = ratio of depth of equivalent (Whitney) stress block to actual depth of compression stress block
b = width of rectangular beam
d_t = distance from compression fiber to center of outermost tension steel layer
f_c' = 28-day compressive strength of concrete
f_y = yield stress of reinforcing steel

For a section whose tensile reinforcement is in a single layer, d_t = effective depth d (see Fig. 105.3). The specifications also require that flexural members be reinforced with minimum reinforcement $A_{s,\min}$ = greater of $3\sqrt{f_c'} b_w d/f_y$ and $200 b_w d/f_y$, where b_w is the width of the beam web (stem). For a statically determinate beam with a flange in tension, the minimum steel area shall be calculated as above, by substituting the smaller of $2b_w$ and the flange width in place of b_w.

Figure 105.3
Rectangular
reinforced
concrete beam—
effective depth.

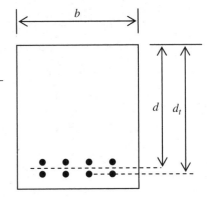

NOTE This minimum reinforcement $A_{s,min}$ is not required if the longitudinal reinforcement provided everywhere is at least 33% more than the reinforcement required by analysis.

Table 105.3 summarizes the minimum and maximum reinforcement ratios for various grades of concrete and steel. The ρ_{max} values listed in Table 105.3 are $A_{s,max}/bd_t$.

Spacing Guidelines

According to ACI 318, the minimum clear spacing between parallel bars in a layer shall be the greater of bar diameter or 1 in. If bars are placed in two or more layers, bars in upper layers shall be placed directly above bars in the bottom layer with clear spacing between layers not less than 1 in. Based on the spacing criteria in ACI specifications, Table 105.4 can be used to determine the minimum beam width required to accommodate a certain number of bars in a single layer.

Table 105.3 Limits on Reinforcement Ratio for Singly Reinforced Beams

f_y	Reinforcement ratios	f_c'			
		3000 psi $\beta_1 = 0.85$	4000 psi $\beta_1 = 0.85$	5000 psi $\beta_1 = 0.80$	6000 psi $\beta_1 = 0.75$
Grade 40 40,000 psi	ρ_{min}	0.0050	0.0050	0.0053	0.0058
	ρ_{max}	0.0232	0.0310	0.0364	0.0410
Grade 50 50,000 psi	ρ_{min}	0.0040	0.0040	0.0042	0.0046
	ρ_{max}	0.0186	0.0248	0.0291	0.0328
Grade 60 60,000 psi	ρ_{min}	0.0033	0.0033	0.0035	0.0039
	ρ_{max}	0.0155	0.0206	0.0243	0.0273

Table 105.4 Minimum Beam Width (inches) Required to Accommodate a Bar Pattern

Bar size	Number of bars in a single layer									Each add'l bar
	2	3	4	5	6	7	8	9	10	
No. 4	6.1	7.6	9.1	10.6	12.1	13.6	15.1	16.6	18.1	1.50
No. 5	6.3	7.9	9.6	11.2	12.8	14.4	16.1	17.7	19.3	1.63
No. 6	6.5	8.3	10.0	11.8	13.5	15.3	17.0	18.8	20.5	1.75
No. 7	6.7	8.6	10.5	12.4	14.2	16.1	18.0	19.9	21.8	1.88
No. 8	6.9	8.9	10.9	12.9	14.9	16.9	18.9	20.9	22.9	2.00
No. 9	7.3	9.5	11.8	14.0	16.3	18.6	20.8	23.1	25.3	2.26
No. 10	7.7	10.2	12.8	15.3	17.8	20.4	22.9	25.4	28.0	2.54
No. 11	8.0	10.8	13.7	16.5	19.3	22.1	24.9	27.8	30.6	2.82
No. 14	8.9	12.3	15.6	19.0	22.4	25.8	29.2	32.6	36.0	3.39
No. 18	10.5	15.0	19.5	24.0	28.6	33.1	37.6	42.1	46.6	4.51

Flexural Capacity of Singly Reinforced Concrete Beams

Figure 105.4 shows a rectangular reinforced concrete beam with tension steel only. Let us assume the beam is simply supported and carrying downward transverse loads, the deflected beam is concave on the top fiber, thus producing compression above the neutral axis and tension below. In such a case, the steel reinforcement is placed near the bottom edge with adequate cover. If the beam is a cantilever beam subject to downward loads, the curvature is concave on the bottom fiber and, therefore, the tension steel is located near the top fiber.

As the load on the beam increases, so does the bending moment. This moment initially (at small loads) creates linear bending stress in the concrete at levels which does not cause

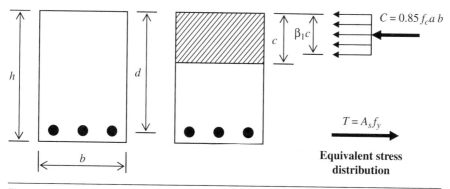

Figure 105.4 Rectangular reinforced concrete beam—stress distribution.

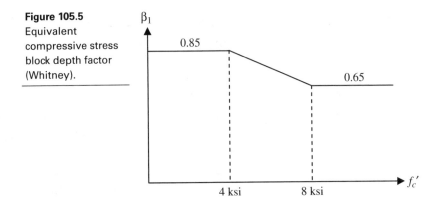

Figure 105.5 Equivalent compressive stress block depth factor (Whitney).

the concrete to crack (the maximum tensile stress in the concrete is lower than the rupture strength of the concrete f_r). When the load increases, the concrete in the tension zone cracks and the steel reinforcement resists the tension force unaided. The concrete progresses toward the compressive limit stress (f_c') while the steel progresses toward yield (f_y). For an under-reinforced section, the steel reaches yield first and at ultimate load, the concrete reaches strain $\varepsilon_c = 0.003$. The actual compression stress diagram (see Fig. 105.1) on the concrete is nonlinear and is replaced by an equivalent stress, as proposed by Whitney (see Fig. 105.4) and adopted by ACI 318. The ultimate stress condition assumes that the tension steel has previously yielded and the limit state corresponds to concrete reaching the strain limit ($\varepsilon_c = 0.003$). The factor β_1 is proposed to be a function of f_c' as shown in Eq. (105.6) and Fig. 105.5.

$$0.65 \leq \beta_1 = 1.05 - \frac{f_c'}{20,000} \leq 0.85 \tag{105.7}$$

The depth of neutral axis may be found by equating the resultant compression to the resultant tension, which yields the following expression:

$$c = \frac{A_s f_y}{0.85\beta_1 f_c' b} \tag{105.8}$$

The depth of the equivalent (Whitney) compression stress block is given by

$$a = \beta_1 c = \frac{A_s f_y}{0.85 f_c' b} \tag{105.9}$$

The nominal (theoretical) moment capacity M_n is given by the strength of the couple formed by the equal and opposite forces C and T.

$$M_n = A_s f_y \left(d - \frac{a}{2} \right) = 0.85 f_c' a b \left(d - \frac{a}{2} \right) \tag{105.10}$$

The design moment capacity is then given by $M_u = \phi M_n$. ACI 318 guidelines on the strength reduction factor (ϕ) is reviewed in section Strength Design Approach in Chap. 104.

Design Problems

Using the dimensionless parameter $w = \rho f_y / f_c'$, we can define the following strength parameter:

$$X = \rho \frac{f_y}{f_c'} \left[1 - \frac{\rho f_y}{1.7 f_c'} \right] = w(1 - 0.588w) \qquad (105.11)$$

The flexural design equation can then be rewritten.

$$\phi M_n = \phi b d^2 f_c' w(1 - 0.588w) = \phi b d^2 f_c' X \qquad (105.12)$$

Thus, since $$\phi M_n \geq M_u \Rightarrow X \geq \frac{M_u}{\phi f_c' b d^2} \qquad (105.13)$$

Table 105.5 shows values of $w = \rho f_y / f_c'$ versus the quantity $X = M_u / (f_c' b d^2 \phi)$.

Table 105.5 can be useful in solving design problems, where all three parameters of the beam (width, depth, area of steel) are not known. In general, in trying to meet a certain design strength M_u, the designer may take the following steps:

1. Presume a reasonable reinforcement ratio (reinforcement area A_s within the permitted range $A_{s,\min}$ to $A_{s,\max}$).
2. Compute the corresponding value of w and use the table above to determine X.
3. Make an assumption about the aspect ratio of the rectangular shape. This may be as simple as stating that, based on aesthetics, the d/b ratio desired is approximately 1.5.
4. Based on these assumptions, calculate the parameters b, d, and A_s.

Example 105.1

A reinforced concrete ($f_c' = 4$ ksi, $f_y = 60$ ksi) beam is simply supported over a span $L = 25$ ft. The beam carries uniformly distributed loads $w_{DL} = 2$ kips/ft and $w_{LL} = 4$ kips/ft over its length. If the width of the beam is 15 in., what is the minimum depth that is satisfactory?

Solution   The factored load is $w_u = 1.2 \, w_{DL} + 1.6 \, w_{LL} = 8.8$ kips/ft. The design moment is $M_u = w_u L^2 / 8 = 687.5$ kips-ft.

We are looking for the *smallest* size of beam, therefore for *maximum allowed reinforcement*.

Table 105.5 Beam Flexural Strength Parameter X versus Reinforcement Parameter w

w	X	w	X	w	X	w	X	w	X	w	X	w	X	w	X
0.001	0.0010	0.051	0.0495	0.101	0.0950	0.151	0.1375	0.201	0.1772	0.251	0.2138	0.301	0.2475	0.351	0.2783
0.002	0.0020	0.052	0.0504	0.102	0.0959	0.152	0.1384	0.202	0.1779	0.252	0.2145	0.302	0.2482	0.352	0.2789
0.003	0.0030	0.053	0.0513	0.103	0.0967	0.153	0.1392	0.203	0.1787	0.253	0.2152	0.303	0.2488	0.353	0.2795
0.004	0.0040	0.054	0.0523	0.104	0.0976	0.154	0.1400	0.204	0.1794	0.254	0.2159	0.304	0.2495	0.354	0.2801
0.005	0.0050	0.055	0.0532	0.105	0.0985	0.155	0.1408	0.205	0.1802	0.255	0.2166	0.305	0.2501	0.355	0.2806
0.006	0.0060	0.056	0.0541	0.106	0.0994	0.156	0.1416	0.206	0.1810	0.256	0.2173	0.306	0.2508	0.356	0.2812
0.007	0.0070	0.057	0.0551	0.107	0.1002	0.157	0.1425	0.207	0.1817	0.257	0.2180	0.307	0.2514	0.357	0.2818
0.008	0.0080	0.058	0.0560	0.108	0.1011	0.158	0.1433	0.208	0.1825	0.258	0.2187	0.308	0.2520	0.358	0.2824
0.009	0.0090	0.059	0.0569	0.109	0.1020	0.159	0.1441	0.209	0.1832	0.259	0.2194	0.309	0.2527	0.359	0.2830
0.010	0.0099	0.060	0.0579	0.110	0.1029	0.160	0.1449	0.210	0.1840	0.260	0.2201	0.310	0.2533	0.360	0.2835
0.011	0.0109	0.061	0.0588	0.111	0.1037	0.161	0.1457	0.211	0.1847	0.261	0.2208	0.311	0.2539	0.361	0.2841
0.012	0.0119	0.062	0.0597	0.112	0.1046	0.162	0.1465	0.212	0.1855	0.262	0.2215	0.312	0.2546	0.362	0.2847
0.013	0.0129	0.063	0.0607	0.113	0.1055	0.163	0.1473	0.213	0.1862	0.263	0.2222	0.313	0.2552	0.363	0.2853
0.014	0.0139	0.064	0.0616	0.114	0.1063	0.164	0.1481	0.214	0.1870	0.264	0.2229	0.314	0.2558	0.364	0.2858
0.015	0.0149	0.065	0.0625	0.115	0.1072	0.165	0.1489	0.215	0.1877	0.265	0.2236	0.315	0.2565	0.365	0.2864
0.016	0.0158	0.066	0.0634	0.116	0.1081	0.166	0.1497	0.216	0.1885	0.266	0.2243	0.316	0.2571	0.366	0.2870
0.017	0.0168	0.067	0.0644	0.117	0.1089	0.167	0.1505	0.217	0.1892	0.267	0.2249	0.317	0.2577	0.367	0.2875
0.018	0.0178	0.068	0.0653	0.118	0.1098	0.168	0.1513	0.218	0.1900	0.268	0.2256	0.318	0.2583	0.368	0.2881
0.019	0.0188	0.069	0.0662	0.119	0.1106	0.169	0.1521	0.219	0.1907	0.269	0.2263	0.319	0.2590	0.369	0.2887
0.020	0.0198	0.070	0.0671	0.120	0.1115	0.170	0.1529	0.220	0.1914	0.270	0.2270	0.320	0.2596	0.370	0.2892
0.021	0.0207	0.071	0.0680	0.121	0.1124	0.171	0.1537	0.221	0.1922	0.271	0.2277	0.321	0.2602	0.371	0.2898
0.022	0.0217	0.072	0.0689	0.122	0.1132	0.172	0.1545	0.222	0.1929	0.272	0.2283	0.322	0.2608	0.372	0.2904
0.023	0.0227	0.073	0.0699	0.123	0.1141	0.173	0.1553	0.223	0.1937	0.273	0.2290	0.323	0.2614	0.373	0.2909
0.024	0.0237	0.074	0.0708	0.124	0.1149	0.174	0.1561	0.224	0.1944	0.274	0.2297	0.324	0.2621	0.374	0.2915
0.025	0.0246	0.075	0.0717	0.125	0.1158	0.175	0.1569	0.225	0.1951	0.275	0.2304	0.325	0.2627	0.375	0.2920

0.026	0.0256	0.076	0.0726	0.126	0.1166	0.176	0.1577	0.226	0.1959	0.276	0.2311	0.326	0.2633	0.376	0.2926
0.027	0.0266	0.077	0.0735	0.127	0.1175	0.177	0.1585	0.227	0.1966	0.277	0.2317	0.327	0.2639	0.377	0.2931
0.028	0.0275	0.078	0.0744	0.128	0.1183	0.178	0.1593	0.228	0.1973	0.278	0.2324	0.328	0.2645	0.378	0.2937
0.029	0.0285	0.079	0.0753	0.129	0.1192	0.179	0.1601	0.229	0.1981	0.279	0.2331	0.329	0.2651	0.379	0.2943
0.030	0.0295	0.080	0.0762	0.130	0.1200	0.180	0.1609	0.230	0.1988	0.280	0.2337	0.330	0.2657	0.380	0.2948
0.031	0.0304	0.081	0.0771	0.131	0.1209	0.181	0.1617	0.231	0.1995	0.281	0.2344	0.331	0.2664	0.381	0.2954
0.032	0.0314	0.082	0.0780	0.132	0.1217	0.182	0.1625	0.232	0.2002	0.282	0.2351	0.332	0.2670	0.382	0.2959
0.033	0.0324	0.083	0.0789	0.133	0.1226	0.183	0.1632	0.233	0.2010	0.283	0.2357	0.333	0.2676	0.383	0.2965
0.034	0.0333	0.084	0.0798	0.134	0.1234	0.184	0.1640	0.234	0.2017	0.284	0.2364	0.334	0.2682	0.384	0.2970
0.035	0.0343	0.085	0.0807	0.135	0.1242	0.185	0.1648	0.235	0.2024	0.285	0.2371	0.335	0.2688	0.385	0.2975
0.036	0.0352	0.086	0.0816	0.136	0.1251	0.186	0.1656	0.236	0.2031	0.286	0.2377	0.336	0.2694	0.386	0.2981
0.037	0.0362	0.087	0.0825	0.137	0.1259	0.187	0.1664	0.237	0.2039	0.287	0.2384	0.337	0.2700	0.387	0.2986
0.038	0.0371	0.088	0.0834	0.138	0.1268	0.188	0.1671	0.238	0.2046	0.288	0.2391	0.338	0.2706	0.388	0.2992
0.039	0.0381	0.089	0.0843	0.139	0.1276	0.189	0.1679	0.239	0.2053	0.289	0.2397	0.339	0.2712	0.389	0.2997
0.040	0.0391	0.090	0.0852	0.140	0.1284	0.190	0.1687	0.240	0.2060	0.290	0.2404	0.340	0.2718	0.390	0.3003
0.041	0.0400	0.091	0.0861	0.141	0.1293	0.191	0.1695	0.241	0.2067	0.291	0.2410	0.341	0.2724	0.391	0.3008
0.042	0.0410	0.092	0.0870	0.142	0.1301	0.192	0.1703	0.242	0.2074	0.292	0.2417	0.342	0.2730	0.392	0.3013
0.043	0.0419	0.093	0.0879	0.143	0.1309	0.193	0.1710	0.243	0.2082	0.293	0.2423	0.343	0.2736	0.393	0.3019
0.044	0.0429	0.094	0.0888	0.144	0.1318	0.194	0.1718	0.244	0.2089	0.294	0.2430	0.344	0.2742	0.394	0.3024
0.045	0.0438	0.095	0.0897	0.145	0.1326	0.195	0.1726	0.245	0.2096	0.295	0.2437	0.345	0.2748	0.395	0.3029
0.046	0.0448	0.096	0.0906	0.146	0.1334	0.196	0.1733	0.246	0.2103	0.296	0.2443	0.346	0.2754	0.396	0.3035
0.047	0.0457	0.097	0.0914	0.147	0.1343	0.197	0.1741	0.247	0.2110	0.297	0.2450	0.347	0.2760	0.397	0.3040
0.048	0.0466	0.098	0.0923	0.148	0.1351	0.198	0.1749	0.248	0.2117	0.298	0.2456	0.348	0.2765	0.398	0.3045
0.049	0.0476	0.099	0.0932	0.149	0.1359	0.199	0.1756	0.249	0.2124	0.299	0.2463	0.349	0.2771	0.399	0.3051
0.050	0.0485	0.100	0.0941	0.150	0.1367	0.200	0.1764	0.250	0.2131	0.300	0.2469	0.350	0.2777	0.400	0.3056

NOTE The facts that the width b has been set and we are looking for maximum reinforcement reduce the number of unknowns from three to one and therefore make this a problem with a unique answer, well suited to a multiple-choice examination.

For $f_c' = 4$ ksi, $f_y = 60$ ksi, $\rho_{max} = 0.0206$

$$w_{max} = \frac{0.0206 \times 60}{4} = 0.309, \; X = 0.2527$$

When $\rho = \rho_{max}$, $\varepsilon_t = 0.004$, and $\phi = 0.48 + 83\varepsilon_t = 0.48 + 83 \times 0.004 = 0.81$

Also, since the section is not fully defined, the tensile strain cannot be calculated and the strength reduction factor is being assumed to be 0.81 (maximum steel → minimum ϕ). Strength reduction factors are summarized in Table 104.2. Thus, the effective depth

$$d = \sqrt{\frac{M_u}{0.81Xbf_c'}} = \sqrt{\frac{687.5 \times 12}{0.81 \times 0.2527 \times 15 \times 4}} = 25.9 \text{ in.}$$

If No. 4 stirrups are used, as well as No. 9 bars as longitudinal reinforcement, and clear cover is 1.5 in., then overall depth $h \approx 25.9 + 1.5 + 0.5 + 1.128/2 \approx 28.46$ in.

$$\text{Use } h = 29 \text{ in.: Effective depth} = 29 - 2.56 = 26.44 \text{ in.}$$

$$\text{Steel area } A_s = 0.0206 \times 15 \times 26.44 = 8.17 \text{ in.}^2$$

$$\text{Depth of neutral axis } c = \frac{8.17 \times 60}{0.85 \times 0.85 \times 4 \times 15} = 11.3 \text{ in.}$$

$$\text{Tensile strain } \varepsilon_t = \frac{0.003(d_t - c)}{c} = \frac{0.003 \times (26.44 - 11.3)}{11.3} = 0.00402$$

This confirms the fact that when close to maximum reinforcement is used, the tensile strain is close to 0.004 and the use of $\phi = 0.81$ is validated.

Example 105.2

Find the design moment capacity of a reinforced concrete beam section with width $b = 12$ in., depth $h = 20$ in., reinforcement 3 No. 9 bars, $f_c' = 3$ ksi, $f_y = 60$ ksi.

Solution All parameters of the beam section are fully specified in this case. The depth of the neutral axis is given by

$$c = \frac{A_s f_y}{0.85\beta_1 f_c' b} = \frac{3.0 \times 60}{0.85 \times 0.85 \times 3 \times 12} = 6.92$$

Assuming bars in a single layer, the effective depth is

$$d = d_t \approx 20 - 2.5 = 17.5$$

The tensile strain is calculated as

$$\varepsilon_t = \frac{0.003(d_t - c)}{c} = \frac{0.003(17.5 - 6.92)}{6.92} = 0.0046$$

The strength-reduction factor is given by $\phi = 0.48 + 83\varepsilon_t = 0.86$.

NOTE   As the tensile stress varies between the minimum permitted value of 0.004 and 0.005, the strength-reduction factor varies from 0.81 to 0.90. So, if the answer choices are well spread, one could avoid the computation of the neutral-axis depth, tensile strain, and strength-reduction factor and can still get a fairly good estimate of the design strength by using $\phi = 0.90$, which is appropriate for $\varepsilon_t > 0.005$.

$$w = \frac{\rho f_y}{f'_c} = 0.286 \Rightarrow X = 0.2377$$
$$\phi M_n = 0.2377 \times bd^2 \phi \times f'_c = 2253.8 \text{ kips-in.} = 187.8 \text{ kips-ft}$$

Doubly Reinforced Rectangular Section

When the tensile reinforcement for a rectangular beam is at the upper limit permitted by ACI 318 and the design moment capacity is still less than the required ultimate moment, it is necessary to either (1) increase the section size or (2) use reinforcement in the compression zone. Figure 105.6 shows a rectangular beam section with compression reinforcement. The tension steel is shown as A_s while the compression steel is shown as A'_s. The distance from the

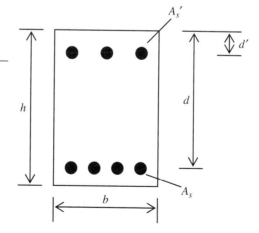

Figure 105.6
Doubly reinforced
rectangular
section.

compression fiber to the centroid of the tension steel is the effective depth d while the distance to the centroid of the compression steel is d'.

For strength design, at the limit state corresponding to ultimate loads, the tension reinforcement is assumed to have yielded. However, whether the compression steel yields or not depends on the applied moment and section geometry. The compression steel yields if

$$A_s - A_s' \geq \frac{0.85\beta_1 f_c' b d'}{f_y}\left(\frac{87,000}{87,000 + f_y}\right) \qquad (105.14)$$

If the compression steel has not yielded, the computation of the moment capacity is further complicated by the additional requirement to calculate the stress in the compression steel.

Compression Steel Yielded

The design moment capacity is then given by the sum of the tension steel-concrete couple and the steel-steel couples. These are the two terms in Eq. (105.15).

$$\phi M_n = \phi\left[(A_s - A_s')f_y\left(d - \frac{a}{2}\right) + A_s' f_y (d - d')\right] \qquad (105.15)$$

where $a = (A_s - A_s')f_y/0.85f_c' b$. The maximum permitted steel area is given by

$$A_{s,\max} = \frac{0.85f_c'\beta_1 b}{f_y}\left(\frac{3d_t}{7}\right) + A_s' \qquad (105.16)$$

Example 105.3

A reinforced concrete ($f_c' = 4\,\text{ksi}, f_y = 60\,\text{ksi}$) beam has a rectangular cross section $b = 12$ in., $h = 25$ in., and is reinforced as shown below. Find the design flexural capacity (kip-ft) of this section.

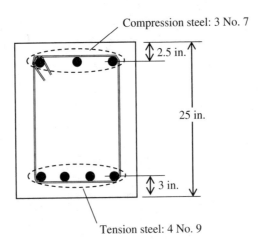

Compression steel: 3 No. 7

2.5 in.

25 in.

3 in.

Tension steel: 4 No. 9

Solution

$$A_s = 4.0 \text{ in.}^2$$

$$A_s' = 2.37 \text{ in.}^2$$

Assume compression steel has yielded.

$$C_c = 0.85 f_c' \beta_1 cb = 0.85 \times 4 \times 0.85 \times c \times 12 = 34.68c$$

$$C_s = A_s' f_y = 2.37 \times 60 = 142.2\,k$$

$$T_s = A_s f_y = 4.0 \times 60 = 240\,k$$

For equilibrium:

$$C_c + C_s = T_s \Rightarrow c = 2.82 \text{ in.}$$

Check yield assumption:

$$\varepsilon_s' = 0.003 \left(\frac{c - d'}{c} \right) = 0.003 \left(\frac{2.82 - 2.5}{2.82} \right) = 0.00034$$

Steel yield stress:

$\varepsilon_y = \dfrac{f_y}{E_s} = \dfrac{50}{29,000} = 0.0017$. Therefore, (since $\varepsilon_s' < \varepsilon_y$) assumption of compression

steel yielding is incorrect.

$$C_s = A_s' f_s' = A_s' \varepsilon_s' E_s = 2.37 \times 0.003 \times \frac{c - 2.5}{c} \times 29,000 = 206.2 \left(\frac{c - 2.5}{c} \right)$$

For equilibrium:

$$C_c + C_s = T_s \Rightarrow 34.68c + 206.2 \left(\frac{c - 2.5}{c} \right) = 240 \Rightarrow c = 4.373$$

For this centroid depth, strain in compressive steel is $0.0013 < 0.0017$ (Unyielded)
ACI code requirement for tension failure

$$\frac{c}{d} = \frac{4.373}{22.0} = 0.199 < 0.375$$

Strain in tensile steel:

$$\varepsilon_s = \frac{d-c}{c} \times 0.003 = \frac{22.0 - 4.373}{4.373} \times 0.003 = 0.012 > 0.005 \quad (\textit{Tension Controlled})$$

Therefore, strength reduction factor $\phi = 0.9$

Forces (shown below) satisfy equilibrium:

$$C_c = 34.68c = 151.66$$

$$C_c = 206.2\left(\frac{c-2.5}{c}\right) = 88.32$$

$$T_s = 4.0 \times 60 = 240$$

Nominal moment capacity:

$$M_n = C_c\left(d - \frac{\beta_1 c}{2}\right) + C_s(d - d') = 151.66 \times \left(22 - \frac{0.85 \times 4.373}{2}\right) + 88.32 \times (22 - 2.5)$$

$$= 4776.9 \text{ kip-in.} = 398.1 \text{ kip-ft}$$

Design moment capacity: $\phi M_n = 0.9 \times 398.1 = 358.3$ kip-ft

Singly Reinforced T-Beams

Floor systems where the floor slab and beam are cast integrally are composed of T-beams where a portion of the slab serves as a compression "flange" for the beam.

Effective Flange Width

For a T-beam with slab flanges on either side (Fig. 105.7), the effective width shall not exceed one-fourth of the span length of the beam. The effective overhanging slab width on either side of the beam stem shall not exceed eight times the slab thickness, nor be greater than one-half the clear distance to the adjacent beam. For a floor system where beams are spaced uniformly, this leads to the following criterion for the effective width of slab:

$$b_{\text{eff}} = \min\left(\frac{L}{4}, b_w + 16t, \text{beam spacing}\right) \tag{105.17}$$

Figure 105.7
Effective flange
width for
T-beams.

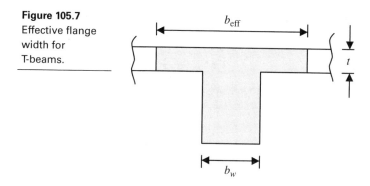

For beams having a slab on one side only (Fig. 105.8), the effective overhang slab width shall not exceed one-twelfth the span length of the beam, six times the slab thickness, or one-half the clear distance to the adjacent beam. This can be expressed as

$$b_{\text{eff}} = \min\left(b_w + \frac{L}{12}, b_w + 6t, b_w + \text{half clear spacing}\right) \tag{105.18}$$

For isolated beams in which the flange is used only to provide additional compression area, the flange thickness shall not be less than half the stem width and the total flange width is limited to four times the stem width (see Fig. 105.9).

Figure 105.8
Effective flange
width for
L-beams.

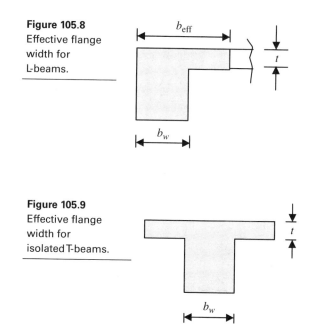

Figure 105.9
Effective flange
width for
isolated T-beams.

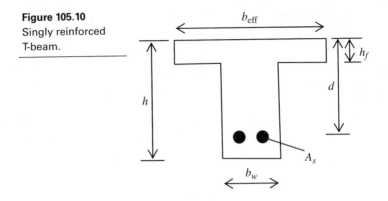

Figure 105.10
Singly reinforced
T-beam.

Flexural Capacity of Singly Reinforced T-Beams

A T-beam with tension steel reinforcement only is shown in Fig. 105.10. The effective width of the slab is shown as b_{eff}. As for rectangular beams, the equivalent (Whitney) depth of the concrete in compression is given by

$$a = \frac{A_s f_y}{0.85 f_c' b_{\text{eff}}}$$

(105.19)

If $a < h_f$, the compression zone is confined to the flange (slab) and the section behaves as a rectangular section (see Fig. 105.11).

Rectangular Beam Behavior

The upper limit for reinforcement is given by

$$A_{s,\max} = \frac{0.85 f_c' \beta_1 b_{\text{eff}}}{f_y} \left(\frac{3 d_t}{7} \right)$$

(105.20)

The nominal moment capacity is given by

$$M_n = 0.85 f_c' a b_{\text{eff}} \left(d - \frac{a}{2} \right)$$

(105.21)

If $a > h_f$, the compression zone extends into the stem (web) and the section behaves as a true T-section (see Fig. 105.12).

Figure 105.11
T-beam exhibiting
rectangular behav-
ior.

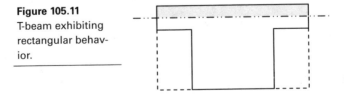

Figure 105.12
T-beam exhibiting
true T-beam
behavior.

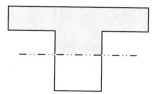

True T-Beam Behavior

The upper limit for reinforcement is given by

$$A_{s,\max} = \frac{0.85 f_c' \beta_1 b_{\text{eff}}}{f_y}\left(\frac{3d_t}{7}\right) + \frac{0.85 f_c' \beta_1 (b_{\text{eff}} - b_w)h_f}{f_y}$$ (105.22)

and the nominal moment capacity is given by

$$M_n = 0.85 f_c'\left[h_f(b_{\text{eff}} - b_w)\left(d - \frac{h_f}{2}\right) + ab_w\left(d - \frac{a}{2}\right)\right]$$ (105.23)

This can also be written as

$$M_n = A_{sf}f_y\left(d - \frac{h_f}{2}\right) + (A_s - A_{sf})f_y\left(d - \frac{a}{2}\right)$$ (105.24)

where A_{sf} is the part of the tensile steel area that is required to balance the compressive force in the overhanging portions of the web, and is given by

$$A_{sf} = \frac{0.85 f_c'(b_{\text{eff}} - b_w)h_f}{f_y}$$ (105.25)

Design of Reinforced Concrete Beams for Shear

Unless the beam qualifies as a deep beam ($L/d < 6$), the effect of shear deformations will be small compared to the flexural deformations. The primary design of the beam is therefore for flexure. Once the beam has been sized so that the flexural capacity is greater than the design moment, the section is checked for shear capacity. The nominal shear capacity of a rectangular reinforced concrete beam section without shear reinforcement is given by

$$V_c = 2\sqrt{f_c'}\,bd$$ (105.26)

where V_c = nominal shear capacity of the concrete section alone (lb)
   f_c' = 28-day compressive strength of the concrete (psi)
   b = beam width (in.)
   d = effective depth (in.).

The shear design statement according to ACI's strength design guidelines is

$$\phi V_n \geq V_u \Rightarrow \phi(V_c + V_s) \geq V_u \tag{105.27}$$

The strength reduction factor ϕ for shear is 0.75. The design shear V_u is then compared to V_c to determine whether the section needs no shear reinforcement, minimum shear reinforcement or calculated shear reinforcement.

Case 1:　　　$V_u < \dfrac{1}{2}\phi V_c$

No shear reinforcement is required.

Case 2:　　　$\dfrac{1}{2}\phi V_c < V_u < \phi V_c$

Use minimum shear reinforcement. Provide stirrups at spacing not to exceed.

$$s = \frac{A_v f_y}{50 b_w} \qquad \text{or} \qquad s = \frac{A_v f_y}{0.75 b_w \sqrt{f_c'}}$$

where A_v is the area of the shear reinforcement normal to the plane of horizontal shear (see Fig. 105.13). If single-loop stirrups are used, A_v is twice the area of the stirrup bar. Maximum permitted spacing of stirrups is not to exceed $s_{max} = d/2$ or 24 in.

Case 3:　　　$V_u > \phi V_c$

Calculate the required shear reinforcement capacity (V_s) following as

$$V_s = \frac{V_u}{\phi} - V_c \leq 8\sqrt{f_c'} bd \tag{105.28}$$

Figure 105.13
Area of stirrups
resisting longitudinal
shear.

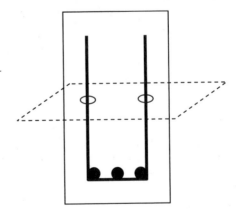

If the magnitude of V_s exceeds $8\sqrt{f_c'}bd$, the beam must be resized. The required spacing of the stirrups to provide this capacity is given by

$$s = \frac{A_v f_y d}{V_s} \tag{105.29}$$

If $V_s \leq 4\sqrt{f_c'}b_w d$, the maximum permitted spacing is not to exceed $d/2$ nor 24 in.

If $V_s > 4\sqrt{f_c'}b_w d$, the maximum permitted spacing is not to exceed $d/4$ nor 12 in.

Shear at Midspan of Uniformly Loaded Beams

The design shear for reinforced concrete beams (according to ACI) is the shear at the critical section V_u^*. The critical section for beam shear is located at a distance d (effective depth) from the face of the support. The shear at this critical section may be found by placing the live load in such a way as to maximize the shear. The shear envelope due to uniformly distributed loads—some permanent (dead loads) and some transient (live loads)—is described by the function

$$V_x = w_D\left(\frac{L}{2} - x\right) + \frac{w_L}{2L}(L - x)^2 \tag{105.30}$$

Example 105.4

A simply supported beam (depth = 20 in.) spans a clear distance of 25 ft between 12-in.-wide supports. The distributed dead load (including self weight) is 2.5 kips/ft and the distributed live load is 4.0 kips/ft. What is the design shear at (a) the critical section and (b) at midspan?

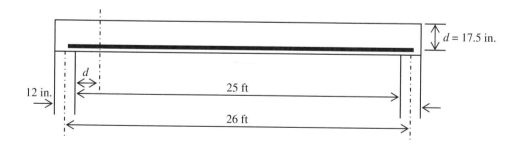

Solution   The effective span = 26 ft. The effective depth is approximately $20 - 2.5 = 17.5$ in. = 1.46 ft. The factored loads are $1.2 \times 2.5 = 3.0$ kips/ft (DL) and $1.6 \times 4.0 = 6.4$ kips/ft (LL). The maximum shear at the critical section is obtained by placing the live load partially as shown.

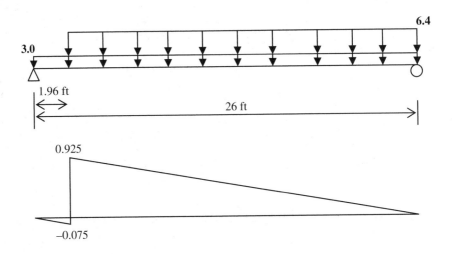

The maximum shear at the critical section is calculated as

$$V_x = 3.0 \times \left(\frac{1}{2} \times 1.96 \times -0.075 \right) + 9.4 \times \left(\frac{1}{2} \times 24.04 \times 0.925 \right) = 104.3 \text{ kips}$$

The maximum shear at midspan is similarly obtained by placing the live load over only half of the span. As a result, the maximum shear at midspan is calculated as $V_x = 20.8$ kips.

$$V_x = 3.0 \times \left(\frac{1}{2} \times 13 \times -0.5 \right) + 9.4 \times \left(\frac{1}{2} \times 13 \times 0.5 \right) = 20.8 \text{ kips}$$

Development Length of Reinforcement

From the location of maximum stress in reinforcement, some length of reinforcement (anchorage) is needed in order to develop the stress through the bond between the reinforcement and the surrounding concrete. This development length is required on both sides of the location of peak stress.

Reinforcement Bars in Tension

For deformed bars and wires in tension, the basic development length is defined by Eq. (105.31). The basic development length in tension is given by

$$l_d = \left\{ \frac{3}{40\lambda} \frac{f_y}{\sqrt{f_c{}'}} \frac{\psi_t \psi_e \psi_s}{\dfrac{c_b + K_{tr}}{d_b}} \right\} d_b \qquad (105.31)$$

where f'_c = 28-day compressive strength of concrete (lb/in.²)

f_y = yield stress of reinforcement steel (lb/in.²)

ψ_t = modifier for reinforcement location

= 1.3 for top bars with more than 12 in. of concrete below, 1.0 for other bars

ψ_e = modifier for coated bars

= 1.5 when re-bars are epoxy coated with cover $< 3d_b$ or clear spacing between bars $< 6d_b$

= 1.2 for all other epoxy coated reinforcing

= 1.0 for non-epoxy coated reinforcing

ψ_s = modifier for bar size

= 0.8 for #6 and smaller bars

= 1.0 for #7 and larger bars

λ = modifier for lightweight aggregate concrete

= 0.85 for sand-lightweight concrete

= 0.75 for all-lightweight

= $f_{ct}/6.7\sqrt{f'_c} \leq 1.0$ when f_{ct} is specified

= 1.0 for normal-weight concrete

The confinement term $\dfrac{c_b + K_{tr}}{d_b}$ is limited to a maximum value of 2.5.

The transverse reinforcement index K_{tr} is given by Eq. (105.32)

$$K_{tr} = \frac{40A_{tr}}{sn} \tag{105.32}$$

where A_{tr} = total cross-sectional area of all transverse reinforcement which is within the spacing s, and which crosses the potential plane of splitting through the reinforcement being developed (in.²)

s = maximum center to center spacing of transverse reinforcement within the development length (in.)

n = number of longitudinal bars or wires being developed along the plane of splitting.

The bar-spacing factor c_b is taken as the smaller of two quantities:

1. Smallest distance from the surface of the concrete to the center of the bar(s) being developed.

2. One-half the center-to-center spacing between bars.

Reinforcement Bars in Compression

For bars in compression, the basic development length is given by Eq. (105.33)

$$l_{dc} = \frac{0.02d_b f_y}{\lambda\sqrt{f'_c}} \geq 0.0003d_b f_y \tag{105.33}$$

Basic development length in compression must not be less than 8 in.

The basic development length in compression may be multiplied by the following factors:

1. If the area of steel is greater than the required area of steel, then the factor $A_{s,reqd}$/ $A_{s,provided}$.

2. For bars confined by a spiral of diameter not less than ¼ in. and pitch not more than 4 in., OR for bars confined by ties with spacing not more than 4 in., the factor = 0.75.

Bundled Reinforcement Bars

For bundles of bars in tension or compression, the diameter d_b in the equations above is replaced by the equivalent diameter of a circular bar having the same area as the total area of the bundle. For a 3-bar bundle, the development length is increased by 20% and for a 4-bar bundle, the development length is increased by 33%.

Example 105.5

A simply supported rectangular reinforced concrete beam has the following reinforcement in a single layer (tension steel at the bottom of the beam). $f_c' = 3000$ psi, $f_y = 60,000$ psi.

Flexural reinforcement: 2 no. 10 bars, 2 no. 8 bars, and 1 no. 6 bars

Shear reinforcement: No. 4 U-shaped stirrups spaced at 6 in. o.c.

At a section where the two no. 8 bars are no longer required, it is determined that they will be terminated. What is the development length of these bars beyond the point of requirement?

Solution   No. 8 bars: Bar diameter $d_b = 1.0$ in.; bottom bars: $\Psi_t = 1.0$; Non-epoxy coated: $\Psi_e = 1.0$; larger than no. 6: $\Psi_s = 1.0$; normal weight concrete: $\lambda = 1.0$.

Area of transverse reinforcement within $s = 6$ in. of the splitting plane, $A_{tr} = 2 \times 0.2 = 0.4$ in.^{2}. Number of bars being developed within splitting plane, $n = 2$

$$K_{tr} = \frac{40A_{tr}}{sn} = \frac{40 \times 0.4}{6 \times 2} = 1.33$$

Without adequate information about bar arrangement, c_b cannot be calculated. The confinement term is taken to be 2.5

$$L_d = \left\{ \frac{3}{40 \times 1} \frac{60,000}{\sqrt{3000}} \frac{1 \times 1 \times 1}{2.5} \right\} \times 1.0 = 32.9 \text{ in.}$$

Reinforced Concrete Slabs

General

Reinforced concrete slabs may be categorized as one- or two-way slabs depending on the prevalent mode of bending of the slab. If the slab is supported on two parallel edges, the slab essentially bends in one direction (between the parallel supports). Bending (curvature) in the other direction is negligible. Such "one-way" slabs can be analyzed and designed in a manner very similar to beams. A unit width of the slab ($b = 1$ ft) is considered as a beam and the thickness of the slab and the required reinforcement are calculated. Since the reinforcement is the area of steel required per foot width of slab, Table 106.1 can be used to select an acceptable pattern. For example, if the required reinforcement is 0.64 in²/ft, then an acceptable pattern is No. 5 bars at 5.5-in spacing (0.669) or No. 6 bars at 8-in spacing (0.663). Since slab thicknesses are small, typically slab reinforcement is accomplished using bars of smaller diameters.

One-Way Reinforced Concrete Slabs

Consider a floor system supported by floor beams as shown in Fig. 106.1. Each slab panel is supported by the beams spaced at a center to center spacing $= S$. The slab primarily bends transversely between these supports, with negligible bending in the longitudinal direction. If a unit strip of the slab (shown shaded) is analyzed, it is equivalent to analyzing a beam with width $= 1$. A simple span between beams would develop a maximum positive moment.

$$M = \frac{wS^2}{8} \tag{106.1}$$

Table 106.1   Reinforcement in Concrete Slabs (in²/ft)

Size													
	Spacing of bars (in)												
	3	**3½**	**4**	**4½**	**5**	**5½**	**6**	**7**	**8**	**9**	**10**	**11**	**12**
3	0.44	0.38	0.33	0.30	0.27	0.24	0.22	0.19	0.17	0.15	0.13	0.120	0.110
4	0.79	0.67	0.59	0.52	0.47	0.43	0.39	0.34	0.30	0.26	0.24	0.214	0.196
5	1.23	1.05	0.92	0.82	0.74	0.67	0.61	0.53	0.46	0.41	0.37	0.335	0.307
6	1.77	1.52	1.33	1.18	1.06	0.96	0.88	0.76	0.66	0.59	0.53	0.482	0.442
7	2.41	2.06	1.80	1.60	1.44	1.31	1.20	1.03	0.90	0.80	0.72	0.656	0.601
8	3.14	2.69	2.36	2.09	1.89	1.71	1.57	1.35	1.18	1.05	0.94	0.857	0.785
9	4.00	3.43	3.00	2.67	2.40	2.18	2.00	1.71	1.50	1.33	1.20	1.090	0.999
10	5.07	4.34	3.80	3.38	3.04	2.76	2.53	2.17	1.90	1.69	1.52	1.382	1.267
11	6.25	5.35	4.68	4.16	3.75	3.41	3.12	2.68	2.34	2.08	1.87	1.703	1.561
14	–	7.72	6.75	6.00	5.40	4.91	4.50	3.86	3.38	3.00	2.70	2.456	2.250
18	–	–	–	10.67	9.60	8.73	8.00	6.86	6.00	5.33	4.80	4.365	4.000

Figure 106.1
Unit width strip
approach for
one-way slabs.

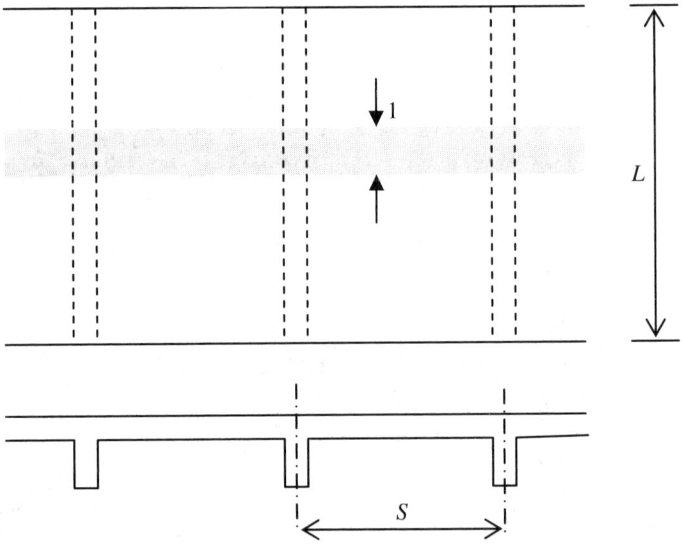

Where w is the uniformly distributed slab load (force/area). With monolithic support from the beam system, if the slab is continuous over three or more supports, the maximum positive bending moment can be reduced to

$$M = \frac{wS^2}{10} \tag{106.2}$$

This is then the design moment for the equivalent beam (strip of unit width).

Table 106.2 Minimum Slab Thickness

Member type	Simply supported	One end continuous	Both ends continuous	Cantilever
	End conditions			
Solid one-way slabs	$L/20$	$L/24$	$L/28$	$L/10$
Beams or ribbed one-way slabs	$L/16$	$L/18.5$	$L/21$	$L/8$

Minimum Slab Thickness

If explicit deflection computations are not performed, the minimum thickness of non-prestressed beams or one-way reinforced concrete slabs is given by ACI (see Table 106.2).

Temperature and Shrinkage Reinforcement

In one-way slabs, the reinforcement for shrinkage and temperature is placed perpendicular to the primary reinforcement according to the criteria in Table 106.3. Spacing of such steel is not to exceed five times the slab thickness nor 18 in.

The reinforcement ratios given in Table 106.3 are computed based on the full depth of the slab, rather than the effective depth. For example, for a 6-in-thick slab using grade-50 reinforcement, the temperature and shrinkage reinforcement is computed as

$$A_s = 0.0020bh = 0.002 \times 12 \times 6 = 0.144\,\text{in}^2/\text{ft}$$

Example 106.1

A reinforced concrete slab is built integrally with its supports and consists of three equal spans, each with a clear span of 16 ft. The ends of the slab are cast integrally with spandrel beams. The service loads are superimposed dead load = 25 psf and live load = 120 psf. Assume $f_c' = 4000$ psi, $f_y = 60,000$ psi. Calculate the moments at (i) exterior support, (ii) interior support, (iii) exterior span, and (iv) interior span.

Solution

For interior spans, minimum slab thickness: $h_{\text{min}} = \dfrac{S}{28} = \dfrac{16 \times 12}{28} = 6.9$ in

For exterior spans, minimum slab thickness: $h_{\text{min}} = \dfrac{S}{24} = \dfrac{16 \times 12}{24} = 8.0$ in

Table 106.3 Temperature and Shrinkage Steel

Minimum ratio of temperature and shrinkage reinforcement in slabs	
Slabs where grade 40 or 50 deformed bars are used	0.0020
Slabs where grade 60 deformed bars or welded wire fabric are used	0.0018
Slabs where reinforcement with $f_y > 60$ ksi is used	$0.0018 \times 60{,}000/f_y \geq 0.0014$

Assume $h = 8$ in

Slab self-weight: $w_{sw} = \dfrac{150 \times 8}{12} = 100$ psf

Total factored load on slab: $w_u = 1.2 \times (25 + 100) + 1.6 \times 120 = 342$ psf

For a unit width (1 ft) of the slab, the factored moments at critical sections (see Table 105.1) are

$$\text{Negative moment exterior support: } M = -\frac{wS^2}{24} = -\frac{0.342 \times 16^2}{24} = -3.65 \text{ kips-ft}$$

$$\text{Positive moment exterior span: } M = +\frac{wS^2}{14} = +\frac{0.342 \times 16^2}{14} = +6.25 \text{ kips-ft}$$

$$\text{Negative moment interior support: } M = -\frac{wS^2}{10} = -\frac{0.342 \times 16^2}{10} = -8.76 \text{ kips-ft}$$

$$\text{Positive moment interior span: } M = +\frac{wS^2}{16} = +\frac{0.342 \times 16^2}{16} = +5.47 \text{ kips-ft}$$

Two-Way Reinforced Concrete Slabs

Slabs that have such proportions or are supported in such a way that their bent shape resembles a dish rather than a cylinder (as with one-way flexure), they are classified as two-way slabs. Two-way slabs may be edge-supported by beams or be primarily column-supported (as in flat plate construction). Slabs exhibit two-way action if the aspect ratio $l_1/l_2 < 2$. Unlike one-way slabs, where the curvature in the minor (long) direction is negligible, there is significant curvature, and therefore, significant moments develop in both directions. The direct design method is a simplified procedure to determine bending moments in a two-way slab. The direct design method may be used when

1. Floor has rectangular panels with aspect ratio $l_1/l_2 < 2$.
2. There are a minimum of three continuous spans in each direction.
3. There is not a more than 33% variation between adjacent spans (based on the longer span).
4. Loading is uniformly distributed gravity loads.
5. Column offsets are less than 10% of the span in the corresponding direction.
6. Loads are gravity loads only and live loads do not exceed twice the dead loads.
7. In the presence of supporting beams, the relative stiffness between beams in the two perpendicular directions is no more than 5:1.

Total Static Moment in Slab Panel

The clear span l_n is defined to extend from face to face of the columns, capitals, brackets, or walls but should not be less than $0.65\,l_1$. For a strip bounded laterally by the centerline

Table 106.4 Distribution Factor α for Bending Moments in Exterior Spans

	Exterior edge unrestrained	Slab with beams between all supports	Slabs without beams between interior supports		Exterior edge fully restrained
			Without edge beam	With edge beam	
Interior negative moment	0.75	0.70	0.70	0.70	0.65
Positive moment	0.63	0.57	0.52	0.50	0.35
Exterior negative moment	0	0.16	0.26	0.30	0.65

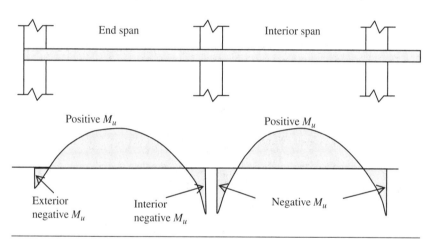

Figure 106.2 Locations of critical moment in continuous slab.

of the panel on each side of the centerline of supports is given by

$$M_o = \frac{w_u l_2 l_n^2}{8} \tag{106.3}$$

where l_n = effective span of slab panel
l_2 = slab dimension perpendicular to span l_n

For interior spans, the total static moment is distributed as follows:

$$\text{Negative factored moment: } M_u = 0.65 M_o$$

$$\text{Positive factored moment: } M_u = 0.35 M_o$$

For exterior spans, moments Fig. 106.2 at various critical sections are given by Eq. (106.4)

$$M_u = \alpha M_o = \alpha \frac{w_u l_2 l_n^2}{8} \tag{106.4}$$

where the distribution factor α is given in Table 106.4.

The terms exterior span, interior span, etc., are as shown in Fig. 106.2.

Slabs on Grade

This section is written in accordance with the provisions of the Army Technical Manual TM 5-809-12: Concrete Floor Slabs on Grade Subjected to Heavy Loads.

Concrete floor slabs poured directly on the subgrade soil are called *slab on grade*. Depending on the intended use of the building, these slabs may be subjected to light or heavy loads such as vehicular loads, stationary live loads, and wall loads.

Loads which consist of forklift axle load of 5 kips or less and stationary live loads (such as due to stored materials) less than 400 lb/ft^2 are typically considered "light." Loads which consist of any one of the following: moving live loads exceeding a forklift axle load of 5 kips, stationary live loads exceeding 400 lb/ft^2, and concentrated wall loads exceeding 600 lb/ft are typically considered "heavy." Soils which exhibit undesirable properties for construction uses such as high compressibility or swell potential are designated as "special soils." A concrete slab resting on grade containing minimal distributed steel, usually of welded wire fabric (WWF), for the purpose of limiting crack width due to shrinkage and temperature change, is designated as an unreinforced slab.

Tests have shown that maximum tensile stresses in floor slabs will occur when vehicle wheels are tangent to a free edge. Stresses for the condition of the vehicle wheels tangent to an interior joint where the two slabs are tied together are less severe than a free edge because of the load transfer across the two adjacent slabs.

The maximum allowable distributed stationary live load w (lb/ft^2) is given by

$$w = 257.876s\sqrt{\frac{kh}{E}} \qquad (106.5)$$

where w = maximum allowable distributed stationary live load (lb/ft^2)
     s = allowable extreme fiber stress in tension (lb/in^2)
     k = modulus of subgrade reaction (lb/in^3). See Table 106.5 for typical values of k.
     h = slab thickness (in)
     E = modulus of elasticity of slab (lb/in^2)

Unless otherwise limited, the limiting stress s is typically taken as half of the 28-day flexural strength, which is typically taken to be $9\sqrt{f_c'}$. Therefore,

$$s = 4.5\sqrt{f_c'}$$

Modulus of Subgrade Reaction

Table 106.5 gives typical values of modulus of subgrade reaction (k) for various types of soil and moisture conditions.

Example 106.2

What is most nearly the required thickness (inches) of a concrete slab ($f_c' = 3500$ psi) supported directly on a compacted silty sand with modulus of subgrade reaction = 200 lb/in^3. Assume the allowable tensile stress is half of the 28-day flexural strength. The design load on the slab (induced by a wheel load) is 900 lb/ft^2.

Table 106.5 Modulus of Subgrade Reaction vs Moisture Content for Various Soils

	Modulus of Subgrade Reaction k (lb/in³) for moisture content w (%)							
	1–4	5–8	9–12	13–16	17–20	21–24	25–28	≥29
Silts and clays with LL > 50 (OH, CH, MH)	–	175	150	125	100	75	50	25
Silts and clays with LL < 50 (OL, CL, ML)	–	200	175	150	125	100	75	50
Silty and clayey sand (SM, SC)	300	250	225	200	150	–	–	–
Gravelly sand (SW, SP)	> 300	300	250	–	–	–	–	–
Silty and Clayey gravels (GM, GC)	> 300	> 300	300	–	–	–	–	–
Gravel and sandy gravels (GW, GP)	> 300	> 300	–	–	–	–	–	–

Solution For an unreinforced slab, the 28-day flexural strength is given by

$$f_r = 9\sqrt{f_c'} = 9 \times \sqrt{3500} = 532 \text{ psi}$$

Allowable stress, $s = 0.5 \times 532 = 266$ psi

Modulus of elasticity of concrete slab,

$$E = 57,000\sqrt{f_c'} = 57,000 \times \sqrt{3500} = 3.37 \times 10^6 \text{ psi}$$

The equation for allowable load can be rewritten as

$$h = 1.5 \times 10^{-5} \frac{w^2 E}{s^2 k} = 1.5 \times 10^{-5} \frac{900^2 \times 3.37 \times 10^6}{266^2 \times 200} = 2.9 \text{ in}$$

Use a 3-in slab.

Subgrade Performance

Load bearing capacity of the subgrade is a function of degree of compaction, moisture content, and soil type. Moisture content tends to affect load bearing capacity and the potential for shrinkage and swelling. Moisture content can be influenced by a number of things such as drainage, groundwater table elevation, infiltration, and pavement porosity (which can be assisted by cracks in the pavement). Generally, excessively wet subgrades will deform excessively under load. Some soils shrink or swell depending upon their moisture content. Additionally, soils with excessive fines may be susceptible to frost heave in colder climates. Shrinkage, swelling, and frost heave will tend to deform and crack any pavement type constructed over them.

Table 106.6 Over-Excavation Recommendations

Subgrade Plasticity Index	Depth of Over-Excavation Below Normal Subgrade Elevation
10–20	2 ft
20–30	3 ft
30–40	4 ft
40–50	5 ft
More than 50	6 ft

Subgrade Rehabilitation Techniques

In construction of any kind of pavement or slab, poor subgrade should be avoided if possible, but when it is necessary to build over weak soils there are several methods available to improve subgrade performance:

Method 1—Removal and Replacement (Over-Excavation)
Poor subgrade soil can simply be removed and replaced with high-quality fill. Although this is simple in concept, it can be expensive. Table 106.6 shows typical over-excavation depths.

Method 2—Stabilization with a Binder
The addition of an appropriate cementitious or asphaltic binder (such as lime, Portland cement, or emulsified asphalt) can increase subgrade stiffness and/or reduce swelling tendencies. The table below provides some typical strategies for subgrade stabilization.

For sandy subgrade soils without excessive soil mass finer than the no. 200 sieve, asphalt emulsification is often recommended as a stabilization technique. On the other hand, for predominantly fine grained soils with plasticity index (PI) less than 10, addition of Portland cement is recommended. Expansive soils are best stabilized with lime.

Method 3—Additional Base Layers
Marginally poor subgrade soils may be compensated for by using additional base layers. These layers (usually of crushed stone—either stabilized or unstabilized) serve to spread pavement loads over a larger subgrade area. A thick pavement structure over a poor subgrade will not necessarily make a good pavement.

Expansive Soils

Expansive soils are those containing sufficient quantities of clay, which tend to swell when they absorb moisture and shrink when they lose moisture. Excessive watering, leaky irrigation systems, and/or poor drainage often accentuates this problem. Poor drainage adjacent to slabs and flatwork is a common problem is expansive soils-related damage. Telltale signs of expansive soils behavior include edge lift at corners and shear cracking near corners. Perimeter heave is common tendency during the first decade after construction, before soil moisture equilibrates beneath the structure. This phenomenon, caused by differential swell beneath the corners, sides, and under the center of a uniformly loaded slab, is sometimes termed "edge curl."

A common problem with lightly loaded structures on expansive soils is differential heave, caused by the ponding of water on the shady side of such structures, leaky water lines, or natural variances in soil moisture content.

Capacity of Anchors in Concrete

Appendix D of ACI 318 provides guidance on calculating the design capacity of anchors embedded in concrete. Figure 106.3 shows three types of anchors—straight, J-, and L-bolts embedded to a depth h_{ef} in concrete. The effective cross-sectional area of an anchor should be provided by the manufacturer.

Effective area of threaded rods is given by Eq. (106.6)

$$A_{se} = \frac{\pi}{4}\left(d_o - \frac{0.9743}{n_t}\right)^2 \qquad (106.6)$$

A_{se} = effective area of threaded rod (in^2)
d_o = outside diameter of threaded rod (inches)
n_t = number of threads per inch

Failure Modes

For steel anchors embedded in concrete, the following types of failure can occur—(1) shear failure of steel section, (2) tensile failure of steel section, (3) concrete breakout, (4) side-face blowout, (5) anchor pullout, and (6) anchor pryout.

There are two kinds of forces acting on anchor rods—transverse shear (V_u) and axial tension (N_u).

Once the design strength in shear (ϕV_n) and tension (ϕN_n) are calculated, the design check is as follows:

1. If $V_u < 0.2\phi V_n$, the full design strength in tension is assumed, and the design requirement is $\phi N_n > N_u$

2. If $N_u < 0.2\phi N_n$, the full design strength in shear is assumed, and the design requirement is $\phi V_n > V_u$

3. If both $V_u > 0.2\phi V_n$ and $N_u > 0.2\phi N_n$, use the interaction Eq. (106.7)

$$\frac{N_u}{\phi N_n} + \frac{V_u}{\phi V_n} \leq 1.2 \qquad (106.7)$$

Figure 106.3
Types of cast-in-place anchors.

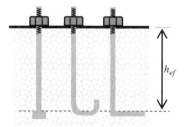

For regions of moderate to high seismicity, the design strengths (for both shear and tension) are reduced by 25% (to $0.75\phi V_n$ and $0.75\phi N_n$, respectively).

Strength Reduction Factor (ϕ)

For anchors governed by strength of steel section:

For tension loads, $\phi = 0.75$ (ductile steel) or 0.65 (brittle steel)

For shear loads, $\phi = 0.65$ (ductile steel) or 0.60 (brittle steel)

For anchors governed by concrete breakout, side-face blowout, anchor pullout, or anchor pryout:

For shear loads, $\phi = 0.75$ (condition A) or 0.70 (condition B)

For tension loads on cast-in headed studs, headed bolts or hooked bolts, $\phi = 0.75$ (condition A) or 0.70 (condition B)

For tension loads on postinstalled anchors:

Category 1—$\phi = 0.75$ (condition A) or 0.65 (condition B)

Category 2—$\phi = 0.65$ (condition A) or 0.55 (condition B)

Category 3—$\phi = 0.55$ (condition A) or 0.45 (condition B)

The three categories of acceptable postinstalled anchors are: category 1 (low sensitivity to installation and high reliability), category 2 (medium sensitivity to installation and medium reliability), and category 3 (high sensitivity to installation and lower reliability).

Condition A: When there is reinforcement in the plane perpendicular to the axis of the anchor.

Condition B: When there is no reinforcement in the plane perpendicular to the axis of the anchor.

Tension

Figure 106.4 shows the breakout cone in tension. The nominal design capacity in tension (N_n) is the smallest of: N_s, N_{cb} or N_{cbg}, N_{pn}, N_{sb}, or N_{sbg}

1. Steel Strength of Anchor(s) in Tension:

$$N_s = nA_{se}f_{ut} \tag{106.8}$$

where f_{ut} = ultimate strength in tension, not to exceed $1.9f_y$, nor 125 ksi

2. N_{cb}—Concrete Breakout Strength (Single Anchor)

$$N_{cb} = \frac{A_N}{A_{No}}\Psi_2\Psi_3 N_b \tag{106.9}$$

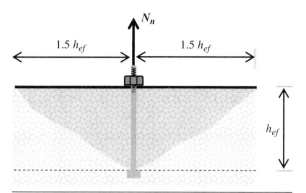

Figure 106.4 Breakout cone for tension.

N_{cbg}—Concrete Breakout Strength (Group of Anchors)

$$N_{cbg} = \frac{A_N}{A_{No}} \Psi_1 \Psi_2 \Psi_3 N_b \qquad (106.10)$$

A_N = anchor or group failure surface in the presence of limiting edges $\leq nA_{No}$
  A_{No}, the projected concrete failure area of one anchor (for calculation of strength in tension), when not limited by edge distance or spacing, is given by (see Fig. 106.4)

$$A_{No} = 9h_{ef}^2 \qquad (106.11)$$

N_b = basic concrete breakout strength: $N_b = k\sqrt{f_c'}h_{ef}^{1.5}$ ($k = 24$ for cast-in anchors; $= 17$ for postinstalled anchors)

Ψ_1 = eccentricity factor (applicable if $e_v' < s/2$): $\Psi_1 = \dfrac{1}{1 + \dfrac{2}{3}\dfrac{e_N'}{h_{ef}}}$

Ψ_2 = edge effect factor (applicable if $c_2 < 1.5c_1$): $\Psi_2 = 0.7 + 0.3\dfrac{c_{min}}{1.5h_{ef}}$

Ψ_3 = 1.25 (cast-in anchor in region where there is no cracking) or 1.4 (postinstalled anchor in region where there is no cracking); 1.0 otherwise.

3. **Pullout Strength of Anchor(s) in Tension:**

$$nN_{pn} = n\Psi_4 N_p \qquad (106.12)$$

For single headed stud: $N_p = A_{brg} 8f_c'$

   For single hooked bolt: $N_p = 0.9f_c'e_h d_o$

Ψ_4 = 1.4 if anchor in region where there is no cracking; = 1.0 otherwise.

4. N_{sb}—Concrete Side Face Blowout Strength of Headed Anchor (Single) in Tension

$$N_{sb} = 160c\sqrt{A_{brg}}\sqrt{f'_c} \tag{106.13}$$

If spacing in the perpendicular direction $c_2 < 3c$, multiply N_{sb} by factor $\frac{1}{4}(1 + c_2/c)$

N_{sbg}—Concrete Side Face Blowout Strength (Group of Anchors with Edge Distance $c < 0.4h_{ef}$)

$$N_{sbg} = \left[1 + \frac{s_o}{6c}\right]N_{sb} \tag{106.14}$$

s_o = spacing of the outer anchors along the edge in the group

Shear

Figure 106.5 shows the breakout cone in shear. The nominal design capacity in shear (V_n) is the smallest of V_s, V_{cb} or V_{cbg}, V_{cp}

1. V_s—Shear Strength of Steel Section (Group of n Anchors)

- For cast-in headed stud anchor, $V_s = nA_{se}f_{ut}$

  f_{ut} = ultimate strength in tension, not to exceed $1.9f_y$, nor 125 ksi
- For cast-in headed bolt and hooked bolt anchors, $V_s = 0.6nA_{se}f_{ut}$
- For postinstalled anchors, $V_s = n(0.6A_{se}f_{ut} + 0.4A_{si}f_{utsi})$

  f_{utsi} = ultimate strength in tension of the anchor sleeve

  A_{si} = cross section of the anchor sleeve

2. V_{cb}—Concrete Breakout Strength (Single Anchor)

$$V_{cb} = \frac{A_v}{A_{Vo}}\Psi_6\Psi_7 V_b \tag{106.15}$$

V_{cbg}—Concrete Breakout Strength (Group of Anchors)

$$V_{cbg} = \frac{A_v}{A_{vo}}\Psi_5\Psi_6\Psi_7 V_b \tag{106.16}$$

V_b = basic concrete breakout strength: $V_b = 7\left(\frac{l}{d_o}\right)^{0.2}\sqrt{d_o}\sqrt{f'_c}c_1^{1.5}$

Ψ_5 = eccentricity factor (applicable if $e'_v < s/2$): $\Psi_5 = \dfrac{1}{1 + \dfrac{2}{3}\dfrac{e'_v}{c_1}}$

Ψ_6 = edge effect factor (applicable if $c_2 < 1.5c_1$): $\Psi_6 = 0.7 + 0.3\dfrac{c_2}{1.5c_1}$

Ψ_7 = 1.4 if anchor in region where there is no cracking; = 1.0 otherwise.

Figure 106.5
Breakout cone for shear.

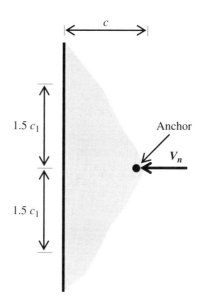

A_V = projected concrete failure area of an anchor or group of anchors $\leq nA_{Vo}$
A_{Vo} = the projected concrete failure area of one anchor (for calculation of strength in shear), when not limited by corner influences, spacing or member thickness, is given by (see Fig. 106.5)

$$A_{Vo} = 4.5c_1^2 \tag{106.17}$$

3. V_{cp}—Concrete Pryout Strength of Anchor

$$V_{cp} = k_{cp}N_{cb} \tag{106.18}$$

k_{cp} = 1.0 when embedment length $h_{ef} < 2.5$ in
k_{cp} = 2.0 when embedment length $h_{ef} \geq 2.5$ in
N_{cb} = concrete breakout strength in tension

Reinforced Concrete Columns

Reinforced concrete columns may be divided into two categories: (1) columns with a purely concentric load (no moment) and (2) columns carrying eccentric loads. A broader name for columns is a compression member—as long as the dominant load effect is axial compression. Columns may be considered "short" or "long," based on the slenderness ratio of the column. For medium and long columns, buckling or the tendency to buckle may diminish the load capacity of the column.

Concrete columns are reinforced with (1) longitudinal reinforcement and lateral ties, (2) longitudinal reinforcement and a continuous spiral, or (3) structural steel shapes, with or without additional longitudinal bars.

Guidelines on Longitudinal Reinforcement

The main reinforcement in concrete columns consists of deformed bars arranged in a rectangular or circular pattern. When longitudinal bars are enclosed by rectangular or circular ties, at least four longitudinal bars should be used, while at least 6 bars are needed if confined by a continuous spiral. The reinforcement ratio, based on the gross area, $\rho_g = (A_s/A_g)$ must be between 1% and 8%.

For tied columns, center-to-center spacing of ties should not be greater than 16 times the longitudinal bar diameter, 48 times the tie bar diameter, nor the least column dimension. Ties must be at least No. 3 bars when longitudinal bars are No. 10 or smaller. Ties must be at least No. 4 size when longitudinal bars are No. 11 or larger. Clear distance between longitudinal bars must be at least 1.5 times the bar diameter or 1.5 in. Clear cover to longitudinal reinforcement must be at least 1.5 in.

For columns whose longitudinal reinforcement is confined by a continuous spiral, the clear distance between spirals must be between 1 in and 3 in. The minimum diameter of a spiral is 3/8 in. The maximum pitch (s) of the spiral is to be calculated according to

$$s = \frac{4A_{sp}}{r_s D_c} \tag{107.1}$$

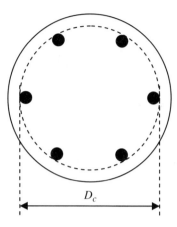

Figure 107.1 Core of a spirally confined reinforced concrete column.

D_c

where A_{sp} = cross-sectional area of the spiral

ρ_s = ratio of volume of spiral steel to volume of core concrete =

$$0.45\left[\frac{A_g}{A_c} - 1\right]\frac{f_c'}{f_y}$$

A_c = area of the column core = $\pi D_c^2/4$

The diameter of the core is shown in Fig. 107.1.

Short versus Long Columns

In braced (nonsway) frames, a column is considered "short" (i.e., effects of slenderness may be ignored) if

$$\frac{Kl_u}{r} \le 34 - 12\left(\frac{M_1}{M_2}\right) \le 40 \tag{107.2}$$

where M_1/M_2 is the ratio of the smaller end moment to the larger end moment (to be taken no smaller than –0.5). This ratio is taken to be positive if column is in single curvature and negative if the column is in reverse curvature. In unbraced frames (sway uninhibited), a column is considered "short" if

$$\frac{Kl_u}{r} \le 22 \tag{107.3}$$

K is the effective length factor and l_u is the unsupported length, taken as the clear distance between floor slabs, beams, or other members providing lateral support.

According to provisions in ACI 318 Chap. 10 (ACI 318, 2014, *Building Code Requirements for Reinforced Concrete*, American Concrete Institute, Detroit, Michigan), the effective length factor must be taken as 1.0 for nonsway frames, unless a lower value can be justified by analysis. For sway frames, K must always be determined by analysis. The radius of gyration for a rectangular column may be taken as $0.30h$, where h is the column dimension in the direction in which stability is being considered. Similarly, the radius of gyration for a circular column may be taken as $0.25h$. Short columns may then be further subdivided into two groups: (1) columns with a purely concentric load (no moment) and (2) columns carrying eccentric loads.

Axial Load Capacity of Short RC Columns

Even for columns carrying loads with "small" eccentricity, the ACI 318 specifications allow the calculation of the axial load capacity as a concentrically loaded column. In other words, it allows the designer to ignore the resulting moment on the column. The eccentricity of the load may be neglected (i.e., column designed for the "axial" load only) if the eccentricity is less than $0.10h$ for tied columns or $0.05h$ for spiral columns. The axial load capacity of short, reinforced concrete column with small eccentricity is given by

$$\phi P_n = \phi\beta P_o = \phi\beta[0.85f_c'(A_g - A_s) + f_y A_s] \qquad (107.4)$$

which can be also written as

$$\phi P_n = \phi\beta A_g[0.85f_c'(1 - \rho_g) + f_y\rho_g] \qquad (107.5)$$

The factors ϕ and β are given in Table 107.1.

Example 107.1

What is the smallest square RC column that can carry the following axial loads?

$P_{DL} = 200$ kips
$P_{LL} = 400$ kips
Use $f_c' = 4000$ psi and $f_y = 60{,}000$ psi.

Solution The factored load is

$$P_u = 1.2P_{DL} + 1.6P_{LL} = 880 \text{ kips.}$$

Table 107.1 Design Factors ϕ and β for Short Columns

	ϕ	β
Tied Column	0.65	0.80
Spiral Column	0.75	0.85

For smallest concrete section, we must use maximum permitted amount of steel. According to ACI, maximum reinforcement ratio $\rho_{max} = 8\%$.

$$P_u = \phi\beta A_g[0.85f_c'(1-\rho_g) + f_y\rho_g]$$

$$A_g = \frac{P_u}{\phi\beta[0.85f_c'(1-\rho_g) + f_y\rho_g]} = \frac{880}{0.65 \times 0.80 \times [0.85 \times 4 \times (1-0.08) + 60 \times 0.08]}$$

$$= 213.5 \ in^2$$

This corresponds to a 14.6 × 14.6 column. If available column sizes are in integer inches, use 15 in × 15 in column ($A_g = 225$ in²).

Example 107.2

What is the largest factored moment that can be ignored in the design of a square reinforced concrete column that carries a factored compression load $P_u = 600$ kips? The column has a 15 in × 15 in cross section. Use $f_c' = 4000$ psi and $f_y = 60,000$ psi.

Solution   A rectangular column will necessarily have lateral confinement in the form of ties. Therefore, according to the ACI, the maximum eccentricity that can be ignored is equal to $0.1h = 0.1 \times 15$ in $= 1.5$ in. Therefore, the maximum factored moment that can be ignored is

$$M_u = P_u e = 600 \times 1.5 = 900 \ \text{kips-in} = 75 \ \text{kips-ft}.$$

Column Interaction Diagrams

The load capacity of short, reinforced concrete columns with significant eccentricity can be calculated using column interaction diagrams. Figures 107.2 through 107.7 show a sampling of column interaction diagrams for three different types of reinforced concrete columns—rectangular with reinforcement on all four faces, rectangular with reinforcement on two parallel faces, and circular. The diagrams for the circular cross section may be used for tied and spiral confinement, as long as the appropriate ϕ is used (0.65 for tied and 0.75 for spiral). The charts are classified by three parameters:

f_c': 28-day compression strength of concrete

f_y: yield stress (tension) of reinforcement steel

γ: factor related to cover to reinforcement

There are three types of problems that may be solved using these diagrams:

1. Given a column cross section (size and reinforcement), what is the maximum eccentricity of a given load?

2. Given an eccentric load and column size, what is the required reinforcement?

3. Given a column cross section (size and reinforcement), what is the maximum load if the eccentricity is predefined?

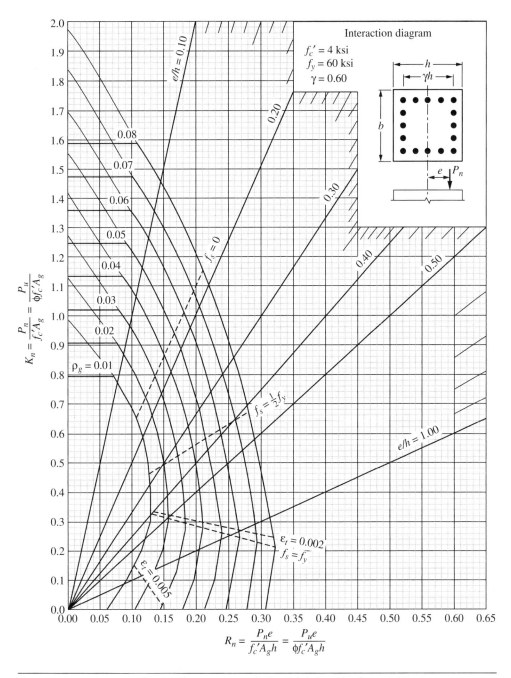

Figure 107.2 Interaction diagram (rectangular 4S, $f_c' = 4$ ksi, $f_y = 60$ ksi, $\gamma = 0.60$).

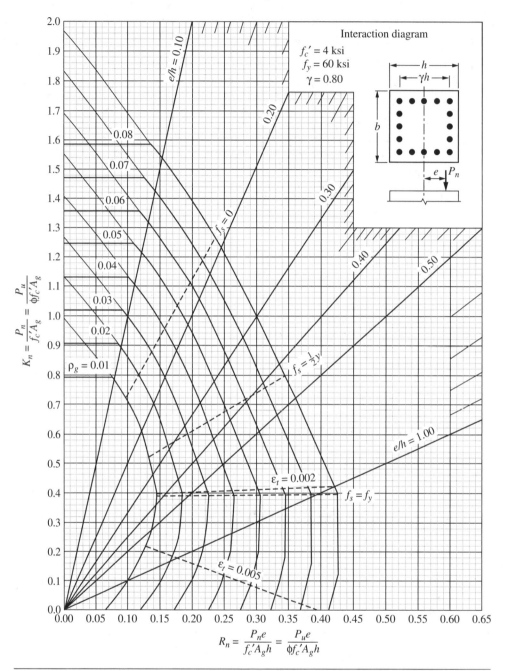

Figure 107.3 Interaction diagram (rectangular 4S, $f_c' = 4$ ksi, $f_y = 60$ ksi, $\gamma = 0.80$).

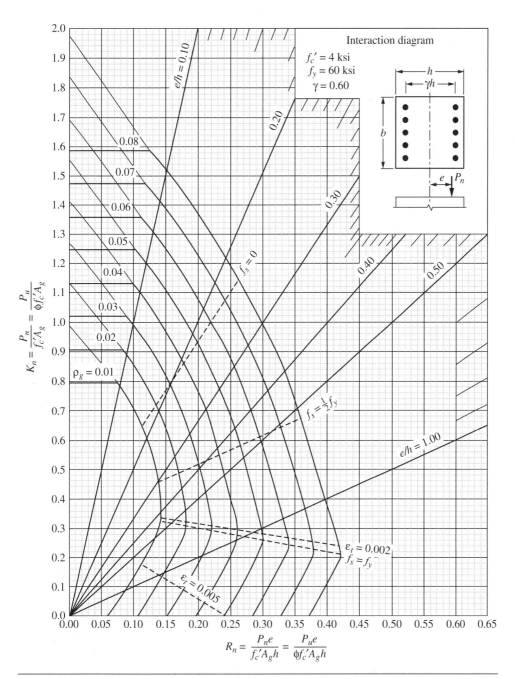

Figure 107.4 Interaction diagram (rectangular 2S, $f_c' = 4$ ksi, $f_y = 60$ ksi, $\gamma = 0.60$).

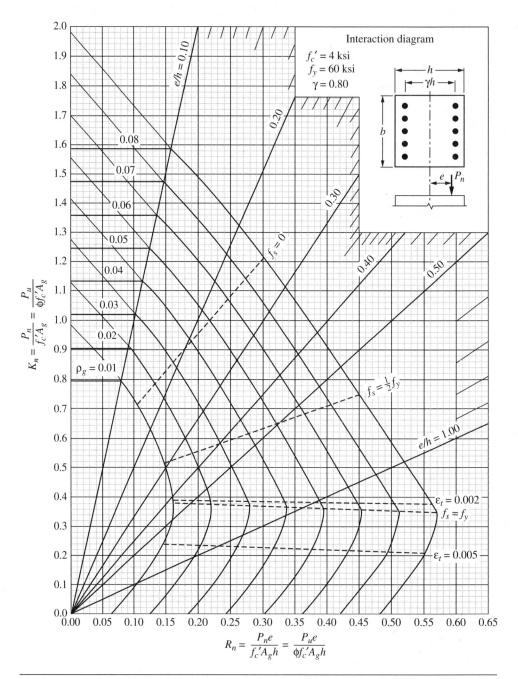

Figure 107.5   Interaction diagram (rectangular 2S, $f_c' = 4$ ksi, $f_y = 60$ ksi, $\gamma = 0.80$).

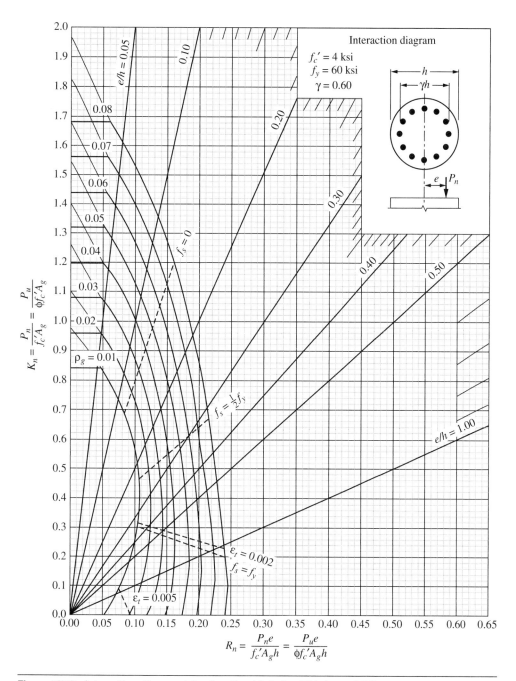

Figure 107.6 Interaction diagram (circular, $f_c' = 4$ ksi, $f_y = 60$ ksi, $\gamma = 0.60$).

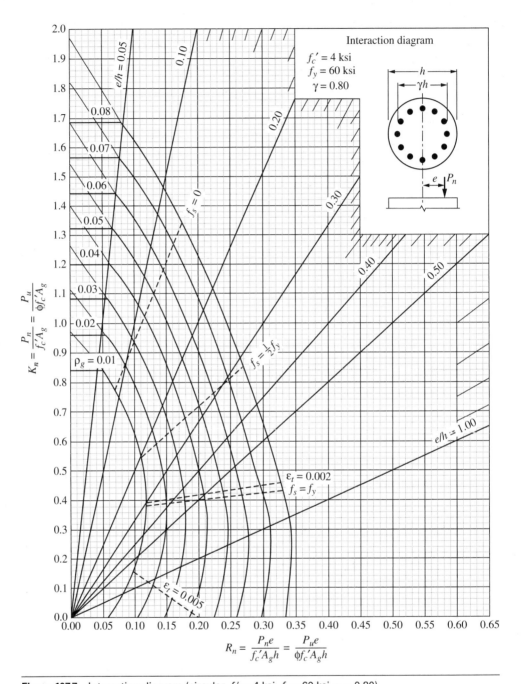

Figure 107.7 Interaction diagram (circular, $f_c' = 4$ ksi, $f_y = 60$ ksi, $\gamma = 0.80$).

Example 107.3

A short reinforced concrete column has an 18 in × 18 in cross section. It is reinforced equally on all four faces with a total of 12 No. 10 bars. The service level axial loads are $P_{DL} = 220$ kips, $P_{LL} = 400$ kips. (Assume $f_c' = 4000$ psi, $f_y = 60,000$ psi.)

What is the maximum eccentricity of the load?

Solution The factored load is $P_u = 1.2 \times 220 + 1.6 \times 400 = 904$ kips. Assuming approximately 2 in of effective cover, the center-to-center distance between parallel lines of reinforcement $= 18 - 2 \times 2 = 14$ in. Parameter $\gamma = 14/18 = 0.78$. Let us use the diagram for $f_c' = 4$ ksi, $f_y = 60$ ksi, $\gamma = 0.80$.

$$\text{Area of steel } A_s = 12 \times 1.27 = 15.24 \text{ in}^2 \quad \text{and} \quad \text{gross area } A_g = 18 \times 18 = 324 \text{ in}^2$$

$$\text{Reinforcement ratio } \rho_g = 15.24/324 = 0.047$$

$$\text{Parameter } K_n = \frac{P_u}{\phi f_c' A_g} = \frac{904}{0.65 \times 4 \times 324} = 1.07$$

Using $K_n = 1.07$ and $\rho_g \approx 0.047$ (visual interpolation between 0.04 and 0.05), we get $e/h \approx 0.15$.

Therefore, the tolerable eccentricity $e = 0.15h = 2.7$ in.

Example 107.4

What is the required reinforcement for a 20 in × 20 in reinforced concrete column ($f_c' = 4000$ psi, $f_y = 60,000$ psi) subjected to the following loads:

$$P_{DL} = 250 \text{ kips}, P_{LL} = 450 \text{ kips. Assume the eccentricity of the load is 4 in.}$$

Solution

$$P_u = 1.2 \times 250 + 1.6 \times 450 = 1020 \text{ kips}$$

Eccentricity is greater than $0.1h$ (4 in > 0.1 × 20 in). Therefore, the column must be designed for a combination of P_u and M_u ($e/h = 4/20 = 0.20$).

$$K_n = \frac{P_u}{\phi_c f_c' A_g} = \frac{1020}{0.65 \times 4 \times 400} = 0.98$$

From the column interaction diagram, using $K_n = 0.98$ and $e/h = 0.2$, $\rho_g = 5.3\%$

$$A_s = 0.053 \times 400 = 21.2 \text{ in}^2$$

Example 107.5

A 20 in × 20 in reinforced concrete column ($f_c' = 4000$ psi, $f_y = 60,000$ psi) is subjected to an eccentric load. The eccentricity of the load is 6 in. The column is reinforced with 12 No. 11 bars distributed equally on all 4 faces. What is the maximum factored load that the column can carry?

Solution   Eccentricity is greater than $0.1h$ (6 in > 0.1×20 in). Therefore, the column must be designed for a combination of P_u and M_u ($e/h = 6/20 = 0.30$).

$$\text{Area of steel } A_s = 12 \times 1.56 = 18.72 \text{ in}^2 \quad \text{and} \quad \text{gross area } A_g = 20 \times 20 = 400 \text{ in}^2$$

$$\text{Reinforcement ratio } \rho_g = 18.72/400 = 0.047$$

Assuming approximately 2 in of effective cover, the center-to-center distance between parallel lines of reinforcement $= 20 - 2 \times 2 = 16$ in.

$$\text{Parameter } \gamma = 16/20 = 0.80$$

Let us use the diagram for $f_c' = 4$ ksi, $f_y = 60$ ksi, $\gamma = 0.80$. From the column interaction diagram, for $e/h = 0.30$ and $\rho_g = 4.7\%$, $K_n = 0.76$:

$$P_u = K_n \phi_c f_c' A_g = 0.76 \times 0.65 \times 4 \times 400 = 790 \text{ kips}$$

Long Columns

When a reinforced concrete column does not meet the criteria for short columns [Eqs. (107.2) and (107.3)], its slenderness ratio must be used to determine the load capacity of the column. The ACI specifications prescribe the use of a moment magnifier to determine the design moment for long columns. The Euler buckling load of the column is given by

$$P_c = \frac{\pi^2 EI}{(Kl_u)^2} = \frac{\pi^2 EI}{(L_{col})^2} \tag{107.6}$$

where $EI = 0.25 E_c I_g$

The analysis and design of long columns can be broadly treated as three separate cases:

(a) Concentrically loaded long columns ($e = 0$)

(b) Eccentrically loaded long columns in non-sway frames

(c) Eccentrically loaded long columns in sway frames

Concentrically Loaded Long Columns ($e = 0$)

For concentrically loaded columns, the ACI specifications require the column be designed for a minimum design eccentricity e_{min}, (in) given by

$$e_{min} = 0.6 + 0.03h \tag{107.7}$$

where h = lateral dimension (inches)

Both end moments are assumed to be

$$M_1 = M_2 = P_u \times e_{min} \qquad (107.8)$$

The magnified design moment for the column is then given by

$$M_u = \frac{P_u \times e_{min}}{1 - \dfrac{P_u}{0.75P_c}} \qquad (107.9)$$

where P_c = Euler critical load given by Eq. (107.6).

P_u and M_u are then used as inputs into an appropriate column strength-interaction diagram.

Eccentrically Loaded Columns in Nonsway Frames

For nonsway frames with end moment M_2 greater than the minimum value $P_u e_{min}$, the magnified moment is calculated as

$$M_c = \delta_{ns} M_2 \qquad (107.10)$$

where the nonsway moment magnifier is given as

$$\delta_{ns} = \frac{C_m}{1 - \dfrac{P_u}{0.75P_c}} \geq 1 \qquad (107.11)$$

$$C_m = 0.6 + 0.4\frac{M_1}{M_2} \geq 0.4 \qquad (107.12)$$

where M_1/M_2 is the ratio of the smaller end moment to the larger end moment (to be taken no smaller than -0.5). This ratio is taken to be positive if column is in single curvature and negative if the column is in reverse curvature.

Eccentrically Loaded Columns in Sway Frames

For sway frames, the two factored end moments for a column are calculated as

$$M_1 = M_{1ns} + \delta_s M_{1s} \qquad (107.13a)$$

$$M_2 = M_{2ns} + \delta_s M_{2s} \qquad (107.13b)$$

The ACI code permits three different methods for calculating these magnified sway moments, all of which are fairly labor intensive, due to the fact that all columns on a particular story must sidesway together in order for the story to sidesway. This group action must therefore be considered in order to calculate the moment magnification factors.

The moment magnification factor δ_s can be calculated as

$$\delta_s = \frac{1}{1-Q} \geq 1 \qquad (107.14)$$

$$Q = \frac{\sum P_u \Delta_o}{V_{us} l_c} \qquad (107.15)$$

Alternatively, δ_s can be calculated as

$$\delta_s = \frac{1}{1 - \dfrac{\sum P_u}{0.75\sum P_c}} \geq 1 \qquad (107.16)$$

where $\sum P_u$ = total factored vertical load on columns in the story being evaluated
$\sum P_c$ = sum of Euler loads for all sway resisting columns in the story (Eq. 107.6)
V_{us} = total horizontal story shear in the story being evaluated
Δ_o = first-order relative lateral deflection between the top and the bottom of that story due to V_{us}

Prestressed Concrete

General

This chapter is dedicated to the design procedures as outlined by the *PCI Design Handbook: Precast and Prestressed Concrete,* 7th ed, 2010, Precast/Prestressed Concrete Institute, Chicago, IL. Its provisions are consistent with those in ACI 318-05 (Building Code Requirements for Structural Concrete), IBC-03 (International Building Code), and ASCE 7-05 and IBC 2006 (Minimum Design Loads for Buildings and Other Structure). The criteria used to determine the safe superimposed load on a prestressed concrete member are based on the provisions of the ACI 318-05 Building Code Requirements for Reinforced Concrete. Thus, the design load is established using the load combinations as prescribed in the ACI 318.

The purpose of prestressing is to introduce a compressive prestress force which produces a stress pattern that is approximately reverse of the stress pattern produced by the external loads. Particularly, for a simply supported beam subject to downward vertical loads, the tensile stresses produced on the bottom surface should be offset (entirely or partially) by the compressive stress produced by the prestress force.

Prestressed concrete sections are classified as one of three classes:

Class C – a prestressed component that is cracked

Class T – a prestressed component that is in transition between uncracked and cracked

Class U – a prestressed component that is uncracked

Strands may be straight or draped. Straight tendons have the same eccentricity at all span locations and therefore induce an identical stress pattern at all span locations. Draped tendons are typically depressed at a single or multiple points. Draped tendons have maximum eccentricity at midspan and therefore produce greater stresses there, in order to compensate for the greater bending stresses at midspan due to transverse loads. For fully developed strands, the critical moment is assumed to occur at midspan for members with straight strands and at $0.4L$ in members with prestressing strands depressed at midspan. If the eccentricity of the strands

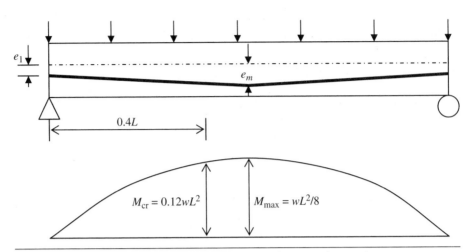

Figure 108.1 Moment at critical section for depressed prestressing tendon.

at the anchorage points is written e_1 and the maximum eccentricity at midspan is written e_m, then the eccentricity at the critical section is given by

$$e_{cr} = \frac{e_1 + 4e_m}{5} \tag{108.1}$$

Figure 108.1 shows a draped prestressing tendon, depressed at a single point (midspan). The eccentricity varies linearly from e_1 at the end anchorage to a maximum value e_m at midspan. The (parabolic) bending-moment diagram due to uniformly distributed transverse load is also shown. The bending moment at the critical section ($x = 0.4L$) is equal to $0.12wL^2$.

Combined Stresses

Figure 108.2 shows a rectangular concrete beam with prestressing strands at an eccentricity $= e$. The prestressing force is P_s. The eccentric load can be treated as the statically equivalent combination of a concentric force P_s and a couple $P_s e$. The concentric force produces a uniform compressive stress equal to

$$\sigma_{ave} = \frac{P_s}{A} = \frac{P_s}{bh} \tag{108.2}$$

The couple produces a linear bending stress which is symmetrical about the centroid and has a maximum value equal to

$$\sigma_b = \pm\frac{My}{I} = \pm\frac{(P_s e) \times h/2}{bh^3/12} = \pm\frac{6P_s e}{bh^2} = \pm\frac{P_s}{bh}\frac{6e}{h} \tag{108.3}$$

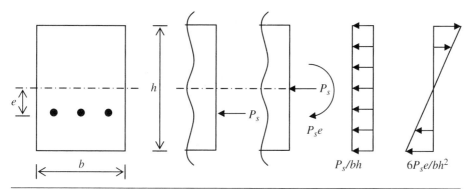

Figure 108.2   Stress pattern due to prestress force.

The combined stress on the top fiber (compression positive) is given by

$$\sigma_{top} = \frac{P_s}{bh} - \frac{6P_s e}{bh^2} = \frac{P_s}{bh}\left[1 - \frac{6e}{h}\right] \tag{108.4}$$

If the eccentricity is significant ($e > h/6$), the resultant stress on the top fiber can be tensile. In that case, the maximum tensile stress must be within the tensile stress limits, as imposed by ACI or AASHTO, whichever is the basis for design. The stress on the bottom fiber (compression positive) is given by

$$\sigma_{bottom} = \frac{P_s}{bh} + \frac{6P_s e}{bh^2} = \frac{P_s}{bh}\left[1 + \frac{6e}{h}\right] \tag{108.5}$$

In addition to these stresses, if the simple span beam is subject to uniformly distributed gravity loads (load per unit length $= w$), the flexural moment (either at $x = 0.5L$ for straight tendons, or at $x = 0.4L$ for depressed tendons) produces compression on the top fiber and tension on the bottom fiber. Incorporating these stresses, we have
Stress on the top fiber (compression positive)

$$\sigma_{top} = \frac{P_s}{bh} - \frac{6P_s e}{bh^2} + \frac{6M_{max}}{bh^2} \tag{108.6}$$

Stress on the bottom fiber (compression positive)

$$\sigma_{bottom} = \frac{P_s}{bh} + \frac{6P_s e}{bh^2} - \frac{6M_{max}}{bh^2} \tag{108.7}$$

Nonrectangular Section

For any section, the section moduli for the top and bottom fibers are given by

$$S_t = \frac{I}{y_t} \quad \text{and} \quad S_b = \frac{I}{y_b} \tag{108.8}$$

The generalized expressions for stress are given by Eqs. (108.9) and (108.10). Stress on the top fiber (compression positive) is given by

$$\sigma_{top} = \frac{P_s}{A} - \frac{P_s e}{S_t} + \frac{M_{max}}{S_t} \tag{108.9}$$

Stress on the bottom fiber (compression positive) is given by

$$\sigma_{bottom} = \frac{P_s}{A} + \frac{P_s e}{S_b} - \frac{M_{max}}{S_b} \tag{108.10}$$

Allowable Stresses (PCI)

Table 108.1 outlines allowable stresses in concrete and prestressing steel, according to the *PCI Design Handbook.*

Table 108.1   Allowable Stresses for Prestressed Concrete (PCI)

Flexural stresses in concrete	
Immediately after prestress transfer	
Extreme fiber stress in compression	$0.6f'_{ci}$
Extreme fiber stress in tension except at the ends of simply supported members	$3\sqrt{f'_{ci}}$
Extreme fiber stress in tension at the ends of simply supported members	$6\sqrt{f'_{ci}}$
At service loads	
Extreme fiber stress in compression due to prestress plus sustained load	$0.45f'_c$
Extreme fiber stress in compression due to prestress plus total load	$0.6f'_c$
Stresses in prestressing steel	
Due to jacking force	$0.94f_{py} \leq 0.8f_{pu}$
Immediately after transfer of prestress	$0.82f_{py} \leq 0.74f_{pu}$
Posttensioning tendons at anchorage devices, immediately after force transfer	$0.7f_{pu}$

Source   PCI Design Handbook

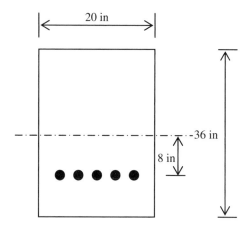

Example 108.1

A 20 in × 36 in prestressed concrete beam is posttensioned with a 1200-kip force using tendons located as shown. If the beam is simply supported over a span of 50 ft, calculate the top fiber stress due to a distributed load of 5 kips/ft and the prestress load.

where f_c' = 28-day compression strength of concrete
f_{ci}' = concrete compression strength at initial transfer of prestress
f_{py} = yield stress of prestressing steel
f_{pu} = ultimate stress of prestressing steel

Solution   Bending moment due to distributed load

$$M = \frac{wL^2}{8} = \frac{5 \times 50^2}{8} = 1562.5 \; kips\text{-}ft = 18,750 \; kips\text{-}in$$

Moment of inertia (neglecting prestressing tendons)

$$I = \frac{1}{12} \times 20 \times 36^3 = 77,760 \; in^4$$

Stress on bottom fiber (compression positive)

$$\sigma = \frac{1200}{20 \times 36} - \frac{18,750 \times 18}{77,760} + \frac{1200 \times 8 \times 18}{77,760} = -0.450 \; ksi$$

Stress on top fiber (compression positive)

$$\sigma = \frac{1200}{20 \times 36} + \frac{18,750 \times 18}{77,760} - \frac{1200 \times 8 \times 18}{77,760} = 3.784 \; ksi$$

Standard Tables from *PCI Design Handbook*

Figure 108.3 shows a typical design table from the *PCI Design Handbook*. The section shown is designated 10DT24, which implies that it is a double-tee (DT) section, which is 10 ft wide and 24 in deep. The strand pattern is designated with the letter S (straight) or D_1 (depressed at 1 point—midspan). The table yields maximum superimposed service loads that can be carried by the beam. The y values listed in column 2 of the table indicate location of center of gravity of the prestressing strands from the bottom edge of the girder.

The values for safe, superimposed, uniform service load are based on the capacity of the component as governed by ACI 318-05 limitations on flexural strength and, in the case of flat deck components, shear strength without shear reinforcement. While ACI 318-05 does not limit tensile stresses, the load tables are based on limits of $12\sqrt{f_c'}$ for double-tees and beams and $7.5\sqrt{f_c'}$ for hollow-core and solid flat slabs.

For untopped deck components, 10 lb/ft² of the capacity shown is assumed as superimposed dead load, which is typical for roof components and for deck components in a parking structure. For deck components with composite topping, 15 lb/ft² of the capacity shown is assumed as superimposed dead load, which is typical for floor components. The capacity shown is in addition to the weight of the topping.

Example: An untopped normal weight 10DT24/88-S (as shown in Fig. 108.3) is prestressed with 88 straight strands with the effective prestressing force at 5 in from the bottom edge of the girder stem. For this section, with a 44-ft span, the capacity shown is 87 lb/ft². The double-tee can safely carry superimposed service loads of 10 lb/ft² dead and 77 lb/ft² live.

The estimated camber at erection is 1.3 in. The estimated long-term camber is 1.5 in.

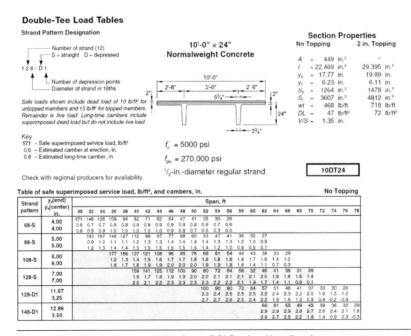

Table of safe superimposed service load, lb/ft², and cambers, in. — No Topping

Strand pattern	y_s(end) / y_s(center) in.	30	32	34	36	38	40	42	44	46	48	50	52	54	56	58	60	62	64	66	68	70	72	74	76	78
68-S	4.00	171	146	126	109	94	82	71	62	54	47	41	35	30	26											
	4.00	0.6	0.7	0.7	0.8	0.8	0.9	0.9	0.9	0.9	0.9	0.8	0.8	0.7	0.6											
		0.8	0.9	0.9	1.0	1.0	1.0	1.0	1.0	0.9	0.8	0.7	0.5	0.3	0.0											
88-S	5.00	193	167	146	127	112	98	87	77	68	60	53	47	41	36	32	27									
	5.00	0.9	1.0	1.1	1.1	1.2	1.3	1.3	1.4	1.4	1.4	1.3	1.3	1.2	1.0	0.9										
		1.2	1.3	1.4	1.4	1.5	1.5	1.5	1.5	1.5	1.4	1.2	1.0	0.8	0.5	0.1										
108-S	6.00								177	156	137	121	108	96	85	76	68	61	54	48	43	38	33	29		
	6.00								1.2	1.3	1.4	1.5	1.6	1.7	1.7	1.8	1.8	1.8	1.8	1.7	1.6	1.4	1.2			
									1.6	1.7	1.8	1.9	2.0	2.0	2.0	1.9	1.9	1.8	1.6	1.4	1.1	0.7	0.3			
128-S	7.00									159	141	125	112	100	90	80	72	64	58	52	46	41	36	31	26	
	7.00									1.6	1.7	1.8	1.9	1.9	2.0	2.0	2.1	2.1	2.1	2.0	1.9	1.8	1.6	1.4		
										2.0	2.1	2.2	2.3	2.3	2.3	2.2	2.2	2.1	1.9	1.7	1.4	1.1	0.6	0.1		
128-D1	11.67											100	90	80	72	64	57	51	46	41	37	33	30	26		
	3.25											2.3	2.4	2.5	2.5	2.5	2.5	2.4	2.3	2.2	2.0	1.6	1.5	1.2		
												2.7	2.7	2.6	2.5	2.4	2.2	1.9	1.6	1.3	0.9	0.4	-0.2	-0.9		
148-D1	12.86																	68	61	55	49	43	39	36	32	29
	3.50																	2.9	2.9	2.9	2.8	2.7	2.6	2.4	2.1	1.8
																		2.9	2.7	2.5	2.2	1.8	1.4	0.9	0.3	-0.3

Figure 108.3 Typical design table from *PCI Design Handbook*.

Prestress Losses

For prestressed concrete members, some of the initial prestress is lost due a combination of several factors. These factors may be broadly classified into (i) instantaneous losses (due to anchorage set, friction, elastic shortening) and (ii) time-dependent losses (due to creep, shrinkage, and relaxation). A report by a joint committee of ASCE and ACI suggested that unless precise data is available, steel stress losses may be assumed to be 35 ksi for pretensioning and 25 ksi for posttensioning.

For pretensioned members, prestress loss is due to elastic shortening, shrinkage, creep of concrete, and relaxation of steel. For members constructed and prestressed in a single stage, relative to the stress immediately before transfer, the loss may be taken as

$$\Delta f_{pT} = \Delta f_{pES} + \Delta f_{pSR} + \Delta f_{pCR} + \Delta f_{pR2} \qquad (108.11)$$

where Δf_{pES} = loss due to elastic shortening (ksi)
Δf_{pSR} = loss due to shrinkage (ksi)
Δf_{pCR} = loss due to creep of concrete (ksi)
Δf_{pR2} = loss due to relaxation of steel after transfer (ksi)

An additional loss occurs during the time between jacking of the strands and transfer. This component is the loss due to the relaxation of steel at transfer, Δf_{pR1}.

Elastic Shortening, Δf_{pES}

$$\Delta f_{pES} = \frac{E_p}{E_{ci}} f_{cgp} \qquad (108.12)$$

where f_{cgp} = sum of concrete stresses at the center of gravity of prestressing tendons due to the prestressing force at transfer and the self-weight of the member at the sections of maximum moment (ksi)
E_p = modulus of elasticity of the prestressing steel (ksi)
E_{ci} = modulus of elasticity of the concrete at transfer (ksi)

Applying this equation requires estimating the stress in the strands after transfer.

Shrinkage Losses

The expression for prestress loss due to shrinkage is a function of the average annual ambient relative humidity, H, and is given by the following equation for pretensioned members:

$$\Delta f_{pSR} = 17 - 0.15H \; (ksi) \qquad (108.13)$$

where H = the average annual ambient relative humidity (%), which can be obtained from local weather statistics.

Creep Losses

The expression for prestress losses due to creep is a function of the concrete stress at the centroid of the prestressing steel at transfer, f_{cgp}, and the change in concrete stress at the centroid of the prestressing steel due to all permanent loads except those at transfer, Δf_{cdp}, and is given by

$$\Delta f_{pCR} = 12 f_{cgp} - 7 \Delta f_{cdp} \geq 0 \qquad (108.14)$$

where f_{cgp} = concrete stresses at the center of gravity of prestressing tendons due to the prestressing force at transfer (ksi)

Δf_{cdp} = change in concrete stress at center of gravity of prestressing steel due to permanent loads (ksi)

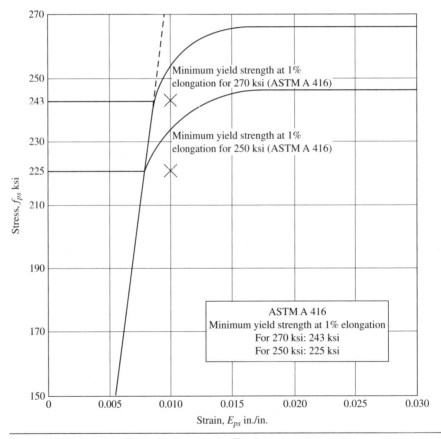

Figure 108.4 Stress vs Strain for Prestressing Tendons.

Camber

Camber is the initial (upward) deflection built into a beam to counter (and limit) the downward deflection caused by live loads. In prestressed concrete beams, the prestressing force is applied below the neutral axis (to create compressive stresses at the bottom fiber, thus countering the tensile stresses that will occur due to live loads).

As a result of this eccentric prestress force, a moment is also applied to the beam cross section. This moment causes an upward deflection, whose magnitude depends on the draping pattern of the prestressing tendon. The simplest case is the one shown in Figure 108.5, where the prestressing tendon is straight, thereby causing a constant moment throughout the length of the beam.

The initial camber (upon transfer of prestress to the beam cross section) is given by

$$\Delta_c = +\frac{PeL^2}{8EI} \tag{108.15}$$

When the beam is simply supported with a span L and allowed to deflect under the self-weight of the beam (w_D), the added downward deflection under the uniformly distributed self-weight is

$$\Delta_D = -\frac{5w_D L^4}{384EI} \tag{108.16}$$

Therefore, before live loads appear on the beam, the net (upward) camber at midspan is given by

$$\Delta_{net} = +\frac{PeL^2}{8EI} - \frac{5w_D L^4}{384EI} \tag{108.17}$$

Design Aid 15.1.4 in the PCI Design Handbook, 7th edition has results for several other tendon draping patterns (such as single point depressed, two point depressed, parabolic, etc.).

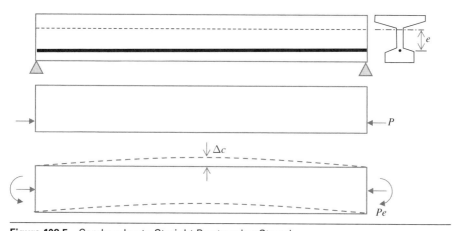

Figure 108.5 Camber due to Straight Prestressing Strand.

Steel Tension Members

This chapter reviews the analysis and design of steel tension members. Specifically, the specification referenced here is the *Steel Construction Manual,* 14th ed. (American Institute of Steel Construction, Inc., 2011), which integrates allowable strength design (ASD) and load and resistance factor design (LRFD) into a combined volume. The load combinations used for design of steel structural elements are reviewed in sections ASD and LRFD.

Allowable Strength Design (ASD)

The design requirement in ASD may be stated as

$$R_a \leq \frac{R_n}{\Omega} \tag{109.1}$$

where R_a is the required strength, R_n is the nominal or theoretical strength, and Ω is the factor of safety. ASD load combinations used to determine the governing required strength are

$$R_a = D + F \tag{109.2a}$$
$$R_a = D + H + F + L + T \tag{109.2b}$$
$$R_a = D + H + F + (L_r \text{ or } S \text{ or } R) \tag{109.2c}$$
$$R_a = D + H + 0.75(L + T) + 0.75(L_r \text{ or } S \text{ or } R) \tag{109.2d}$$
$$R_a = D + H + F + (W \text{ or } 0.7E) \tag{109.2e}$$
$$R_a = D + H + F + 0.75(W \text{ or } 0.7E) + 0.75L + 0.75(L_r \text{ or } S \text{ or } R) \tag{109.2f}$$
$$R_a = 0.6D + (W \text{ or } 0.7E) + H \tag{109.2g}$$

where D = dead load, F = fluid load, H = horizontal earth pressure load, L = live load, T = self-straining load, L_r = roof live load, S = snow load, R = rain load, W = wind load, E = earthquake load

If fluid pressure loads (F), earth pressure loads (H), and self-straining loads (T) are absent, these load combinations simplify to

$$R_a = D \tag{109.3a}$$

$$R_a = D + L \tag{109.3b}$$

$$R_a = D + (L_r \text{ or } S \text{ or } R) \tag{109.3c}$$

$$R_a = D + 0.75L + 0.75(L_r \text{ or } S \text{ or } R) \tag{109.3d}$$

$$R_a = D + (W \text{ or } 0.7E) \tag{109.3e}$$

$$R_a = D + 0.75(W \text{ or } 0.7E) + 0.75L + 0.75(L_r \text{ or } S \text{ or } R) \tag{109.3f}$$

$$R_a = 0.6D + (W \text{ or } 0.7E) \tag{109.3g}$$

Load and Resistance Factor Design (LRFD)

The design requirement in LRFD may be stated as

$$P_u = \sum \gamma_i Q_i \le \phi R_n \tag{109.4}$$

where P_u is the factored load, ϕ is a strength (resistance) reduction factor, γ_i are individual load factors corresponding to service loads Q_i, and R_n is the nominal or theoretical strength (resistance). LRFD (factored) load combinations are

$$P_u = 1.4(D + F) \tag{109.5a}$$

$$P_u = 1.2(D + F + T) + 1.6(L + H) + 0.5(L_r \text{ or } S \text{ or } R) \tag{109.5b}$$

$$P_u = 1.2D + 1.6(L_r \text{ or } S \text{ or } R) + (\alpha L \text{ or } 0.5W) \tag{109.5c}$$

$$P_u = 1.2D + 1.0W + \alpha L + 0.5(L_r \text{ or } S \text{ or } R) \tag{109.5d}$$

$$P_u = 1.2DL + 1.0E + \alpha L + 0.2S \tag{109.5e}$$

$$P_u = 0.9D + (1.0W \text{ or } 1.0E) + 1.6H \tag{109.5f}$$

If fluid pressure loads (F), earth pressure loads (H), and self-straining loads (T) are absent, these load combinations simplify to

$$P_u = 1.4D \tag{109.6a}$$

$$P_u = 1.2D + 1.6L + 0.5(L_r \text{ or } S \text{ or } R) \tag{109.6b}$$

$$P_u = 1.2D + 1.6(L_r \text{ or } S \text{ or } R) + (0.5L \text{ or } 0.5W) \tag{109.6c}$$

$$P_u = 1.2D + 1.0W + 0.5L + 0.5(L_r \text{ or } S \text{ or } R) \tag{109.6d}$$

$$P_u = 1.2D + 1.0E + 0.5L + 0.2S \tag{109.6e}$$

$$P_u = 0.9D + (1.6W \text{ or } 1.0E) \tag{109.6f}$$

NOTES
1. The rain load (R) does not include the effect of *ponding*.
2. For garages, places of public assembly and where the live load exceeds 100 psf, the load factor for live load (L) in load combinations 3, 4, and 5 should be 1.0 instead of 0.5.

Analysis and Design of Tension Members

Steel tension members are designed for two limit states—yielding of the gross section and fracture in the critical net section. In addition, the connection must be checked for the mechanism of block shear.

For ASD, the member design requirement may be stated as

$$P_a \leq \frac{P_n}{\Omega_t} \tag{109.7}$$

where P_a is the required member strength in tension = maximum axial tension calculated from the ASD load combinations. For LRFD, the design requirement may be stated as

$$P_u \leq \phi_t P_n \tag{109.8}$$

where P_u is the required member strength in tension = maximum axial tension calculated from the LRFD load combinations. Both estimations of design strength are based on the same nominal member strength P_n, which is calculated based on the limit states of yield and fracture, as outlined in the next section. In the current specification, it is recommended that the slenderness ratio of a steel tension member (L/r_{min}) should be less than 300. The designer must exercise caution specifying members with high slenderness ratio, as these may exhibit instability problems during erection.

Nominal Member Strength

The nominal strength in tension (P_n) is the lesser of two limit states—yielding of the gross section and fracture of the net section. These are written as
Nominal strength in yielding of the gross area, given by

$$P_n = F_y A_g \tag{109.9}$$

Nominal strength in fracture of the effective net area, given by

$$P_n = F_u A_e = F_u U A_n \tag{109.10}$$

where A_g = gross area of the member cross section
A_n = net area of the member cross section
U = shear lag factor

For ASD, the factor of safety for yield is $\Omega_t = 1.67$, resulting in

$$P_a \leq \frac{P_n}{1.67} = 0.6 F_y A_g \tag{109.11}$$

and the factor of safety for fracture is $\Omega_t = 2.0$, resulting in

$$P_a \leq \frac{P_n}{2.00} = 0.5 F_u A_e \tag{109.12}$$

For LRFD, the strength reduction factor for yield is $\phi_t = 0.90$, resulting in

$$P_u \leq 0.9 F_y A_g \tag{109.13}$$

and the strength reduction factor for fracture is $\phi_t = 0.75$, resulting in

$$P_u \leq 0.75 F_u A_e \tag{109.14}$$

Net Area in Tension

When connections in a tension member consist of bolts, the net area is calculated by subtracting areas removed by drilling for bolts. The usual practice is to drill or punch standard-sized holes with diameter 1/16 in larger than the nominal bolt diameter. To account for roughness around the edges of the hole, AISC specifications require an additional 1/16 in deduction for strength computations. Thus, the diameter of the hole is calculated as the nominal diameter of the bolt plus 1/8 in.

$$d_{\text{hole}} = d_{\text{bolt}} + \frac{1}{8} \text{ in} \tag{109.15}$$

The governing net area is calculated as the net area at the weakest section. For welded members, the net area is equal to the gross area A_g. For bolted connections, this is calculated from the fracture pattern which yields the smallest reduced area after subtracting areas of the holes. This reduction is calculated as the product of the hole diameter (d) and the thickness of the part with the hole (t). For example, for the bolt pattern (line 1) in the plate shown in Fig. 109.1, the net width of the plate is equal to the total width minus two hole diameters. This net width can then be multiplied by the plate thickness to get the net area. Calculating the net area along a line which is not normal to the direction of the load is not as straightforward. The AISC specifications allow the use of a mathematically simple method that closely approximates the strength of such "staggered" bolt lines. This approach is presented in the next section.

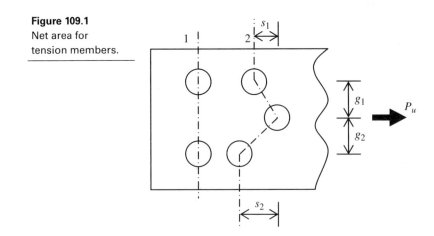

Figure 109.1
Net area for
tension members.

Net Area for Staggered Bolt Lines

Figure 109.1 shows a plate used as a member in tension. The connection is composed of five bolts as shown. The two bolts along line 1 are arranged along a straight line which is normal to the direction of the member force. On the other hand, the three bolts arranged along line 2 are not in such a straight line.

The connection shown in Fig. 109.1 must be checked for two fracture patterns—lines 1 and 2. Line 1 is a net section with two holes without any stagger. However, line 2 is a net section with three holes on a staggered pattern. For such lines, the AISC specifications suggest the following simplified method.

For each staggered line segment, a correction term, given by $s^2/4g$, is added to the net width. Thus, for a pattern that consists of several staggered segments (line 2 above has two), the correction is calculated as the summation of all such terms, calculated individually for each segment. In this calculation, s is the longitudinal spacing between the bolts and g is the transverse or gage distance between the bolts.

$$A_{\text{net}} = A_g - A_{\text{holes}} + \sum \frac{s^2}{4g} t \qquad (109.16)$$

Thus, for the plate shown in Fig. 109.1, the net area along patterns 1 and 2 are

$$A_{\text{net},1} = A_g - \sum A_{\text{holes}} = \left(b - \sum d_i\right) t$$

$$A_{\text{net},2} = \left(b - \sum d_i + \sum \frac{s_i^2}{4g_i}\right) t$$

For the configuration shown, it should be also recognized that line 2 is subject to the full load P_u, whereas line 1 experiences $2/5 \, P_u$ (assuming equal distribution of P_u to the 5 bolts).

Effective Net Area

The effective net area A_e is given by the product of the governing net area A_{net} and the shear lag coefficient U. The shear lag coefficient represents the inefficiency of a connection in transmitting tension when only certain parts of the section are connected. This inefficiency may exist for both bolted and welded connections.

Shear Lag Coefficient

The area used in the calculation of the fracture limit state capacity is called the effective net area (A_e). This area is calculated as a fraction of the minimum net area (A_{net}).

$$A_e = UA_{\text{net}} \qquad (109.17)$$

where U is the shear lag coefficient. For plates connected by bolts, where the connection is in one plane, $U = 1.0$. For other sections, if all parts are connected (such as for a single angle which is connected in both legs), $U = 1.0$. For all other sections, when all parts of the tension member are not directly connected, the shear lag factor U is less than 1.0. The AISC specifications outline several categories of connections for calculating or assigning the shear lag coefficient.

1. *For any type of tension member except plates and round HSS with $l \geq 1.3D$*

   The coefficient is calculated using Eq. (109.18).

   $$U = 1 - \frac{\bar{x}}{L} \qquad (109.18)$$

   where L = length of the connection (distance between outermost bolts)
   $\bar{x}$ = distance from the plane of the connection to the centroid of the section

   If a section has two symmetrically placed planes of connection, then $\bar{x}$ is measured from the centroid of the nearest 1/2 of the area.

2. *Plates*

   For plates where the tension load is transmitted by longitudinal welds (length l and width of plate w) only, as shown in Fig. 109.2.

   for $l \geq 2w$ $U = 1.0$
   for $2w > l \geq 1.5w$ $U = 0.87$
   for $1.5w > l \geq w$ $U = 0.75$

3. *Round HSS with $l \geq 1.3D$*

   $$U = 1.0$$

4. *Single angles*

   In lieu of the equation-based approach, the following values are permitted for single angles. For four or more fasteners in the direction of the load, $U = 0.80$. For two or three fasteners in the direction of the load, $U = 0.60$.

5. *W-, M-, S-, HP-, and T-sections cut from them*

In lieu of the equation-based approach, the following values are permitted for the sections named above:

- Connected through the flange with three or more fasteners in the direction of the load, with a width at least equal to 2/3 of the depth: $U = 0.90$.

- Connected through the flange with three or more fasteners in the direction of the load, with a width less than 2/3 of the depth: $U = 0.85$.

Figure 109.2
Plate tension
member with
longitudinal
welds only.

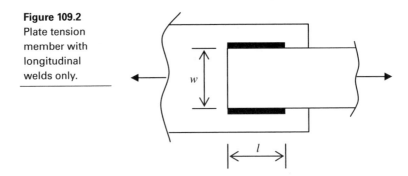

- Connected through the web with four or more fasteners in the direction of the load, with a width less than 2/3 of the depth: $U = 0.70$.

AISC Load Tables

Whereas it is straightforward to list the member strength for yielding ($0.6F_y A_g$ for ASD and $0.9F_y A_g$ for LRFD), since it is based on the gross area A_g, it would be difficult to list the exact member strength for fracture without exact knowledge of the bolt size, thickness of the connected part, and the overall configuration of the connection. For this reason, the fracture member strength listed in the AISC manual is based on the assumption that $A_e = 0.75A_g$. The 0.75 factor represents a combined reduction due to the subtraction of the bolt holes and the shear lag factor. This broad approach may be acceptable for an initial member selection, but the exact value of A_e needs to be calculated once the member selection has been made.

Block Shear

For a poorly proportioned connection, it is possible for a section (or zone) of the member to tear out before one of the member limit states is reached (yield or fracture). For such a failure to occur, shear rupture on certain planes must occur simultaneously with tensile rupture on other planes. Such a failure is termed block shear in the AISC specifications. The AISC specification gives the following nominal strength for block shear:

$$R_n = 0.6F_u A_{nv} + U_{bs} F_u A_{nt} \le 0.6F_y A_{gv} + U_{bs} F_u A_{nt} \qquad (109.19)$$

For example, for the plate-to-plate connection shown in Fig. 109.3, it is possible for block shear to occur in the member or in the gusset plate to which it is connected.

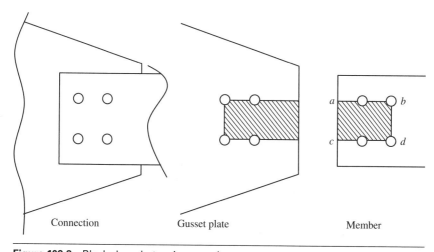

Figure 109.3 Block shear in tension members.

For the member to fail in block shear, shear rupture must occur along planes *ab* and *cd* while tensile rupture occurs along plane *bd*. In the example above, if the thickness of the plate tension member is *t*, then the relevant properties of the block are given by

$$A_{gv} = 2 \times l_{AB} \times t$$
$$A_{nv} = 2 \times \left(l_{AB} - 1.5 \times d\right) \times t$$
$$A_{nt} = \left(l_{BC} - 1.0 \times d\right) \times t$$

The factor $U_{bs} = 1.0$ when the tension stress is uniform and $U_{bs} = 0.5$ when the tension stress is nonuniform.

In ASD, the factor of safety for block shear $\Omega = 2.0$.

In LRFD, the strength reduction factor for block shear $\phi = 0.75$.

Pin-Connected Tension Members

Pin-connected tension members (Fig. 109.4) must be designed for (1) tension on the effective net area, (2) shear on the effective area, (3) bearing on the projected area of the pin, and (4) yielding of the gross area. These are given by

1. Tension on the effective net area

$$P_n = 2tb_{eff}F_u$$
$$\phi_t = 0.75 \quad \Omega_t = 2.0 \tag{109.20}$$

where b_{eff} is the effective width of the plate taken as $2t + 0.63$ in, but not to exceed the actual distance from the edge of the hole to the edge of the connected part, measured perpendicular to the load direction.

Figure 109.4
Pin-connected tension members.

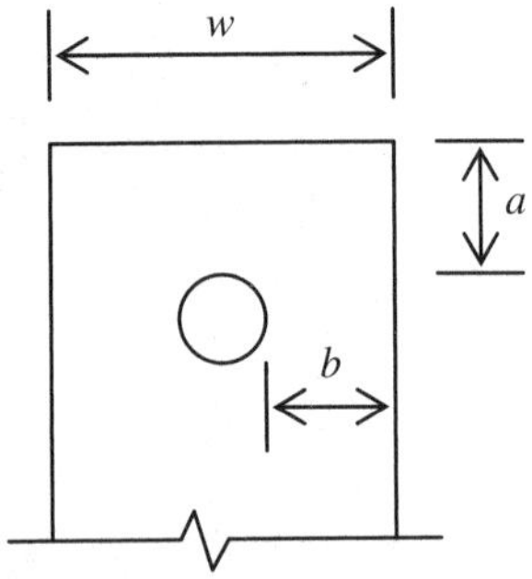

2. Shear on the effective area

$$P_n = 0.6A_{sf}F_u$$
$$\phi_t = 0.75 \quad \Omega_t = 2.0$$

(109.21)

where $A_{sf} = 2t(a + d/2)$

a = shortest distance from the edge of the hole to the edge of the connected part, measured parallel to the load direction

3. Bearing on the projected area of the pin

$$P_n = 1.8tdF_y$$
$$\phi_t = 0.75 \quad \Omega_t = 2.0$$

(109.22)

4. Yielding in the gross section

$$P_n = F_yA_g$$
$$\phi_t = 0.90 \quad \Omega_t = 1.67$$

(109.23)

Example 109.1

A $W8 \times 10$ is used as a tension member as shown below. The connection is made using 3/4 in diameter A325 high strength bolts, with each line of bolts consisting of three bolts. If the steel grade is A992 ($F_y = 50$ ksi, $F_u = 65$ ksi), calculate

(*a*) the design member strength (kips) according to LRFD.

(*b*) the allowable tension force (kips) according to ASD.

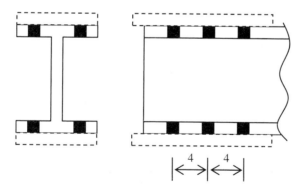

Solution   The following section properties of the $W8 \times 10$ are relevant:

Gross area $A_g = 2.96$ in²

Flange thickness $t_f = 0.205$ in

Flange width $b_f = 3.94$ in

Depth $d = 7.89$ in (therefore, 2/3 $d = 5.26$ in)

Net area $A_n = 2.96 - 4 \times 7/8 \times 0.205 = 2.24$ in²

According to the AISC manual, for W sections connected through the flange with three or more fasteners in the direction of the load, with width less than 2/3 of the depth: $U = 0.85$.

LRFD Solution
Design strength in yielding

$$\phi_t P_n = 0.9 F_y A_g = 0.9 \times 50 \times 2.96 = 133.2 \text{ kips}$$

Design strength in fracture

$$\phi_t P_n = 0.75 F_u U A_n = 0.75 \times 65 \times 0.85 \times 2.24 = 92.8 \text{ kips}$$

Therefore, the design strength (LRFD) is 92.8 kips.

ASD Solution
Allowable load in yielding

$$P_a = 0.6 F_y A_g = 0.6 \times 50 \times 2.96 = 88.8 \text{ kips}$$

Allowable load in fracture

$$P_a = 0.5 F_u U A_n = 0.5 \times 65 \times 0.85 \times 2.24 = 61.9 \text{ kips}$$

Therefore, the allowable load (ASD) is 61.9 kips.

Alternate Definition for the Shear Lag Factor
The shear lag factor may also be calculated using

$$U = 1 - \frac{\overline{x}}{L}$$

$WT\ 4 \times 5$

Since the connection is symmetric about the centroid of the member, the half section is used for determination of the shear lag factor U. For a $W8 \times 10$, the half section is a $WT4 \times 5$, shown above.

$$U = 1 - \frac{0.953}{8} = 0.88$$

For the shear lag factor, the AISC specifications permit the designer to use either the equation based approach or the value tabulated according to member category.

Steel Compression Members

This chapter is dedicated to the design procedures as outlined by American Institute of Steel Construction (AISC). The two design methods described are allowable strength design (ASD) and load and resistance factor design (LRFD). Specifically, the specification referenced here is the *Steel Construction Manual,* 14th ed. (American Institute of Steel Construction, Inc., 2011), which integrates ASD and LRFD into a combined volume.

 NOTE There has been significant realignment of the ASD equations from the ASD, 9th edition, to the *Steel Construction Manual,* 14th edition.

Steel compression members are designed for the instability limit state of buckling. The critical buckling load is based on, and is an extension of, the Euler buckling load which is derived subject to the assumption of linear, elastic behavior.

Stability of Axially Loaded Columns—Euler Buckling

Euler's theory is based on the assumption of elastic behavior, which is acceptable only for very slender columns which buckle at very low loads (i.e., very low stress levels). For a *concentrically loaded column with both ends pinned*, Euler's stability analysis yields the following expression for the critical buckling load P_e:

$$P_e = \frac{\pi^2 EI}{L^2} \tag{110.1}$$

Approximate values of effective-length factor, K

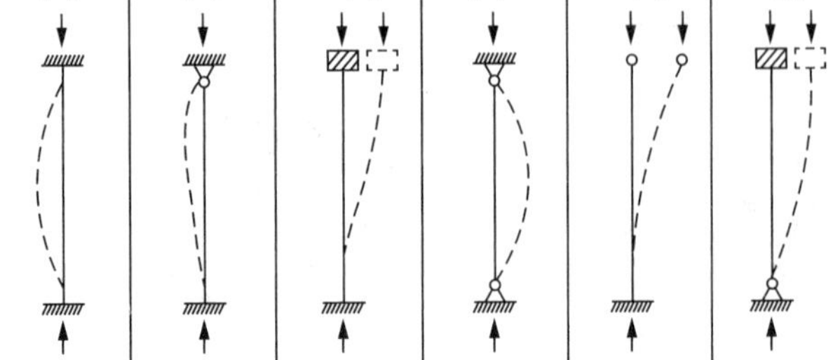

Buckled shape of column is shown by dashed line.	(a)	(b)	(c)	(d)	(e)	(f)
Theoretical K value	0.5	0.7	1.0	1.0	2.0	2.0
Recommended design value when ideal conditions are approximated.	0.65	0.80	1.2	1.0	2.10	2.0
End condition code						

End condition code

Rotation fixed and translation fixed

Rotation free and translation fixed

Rotation fixed and translation free

Rotation free and translation free

Figure 110.1   Effective-length factors for steel columns.

where I is the moment of inertia of the column cross section with respect to the axis about which buckling takes place. For columns with different end conditions, the theory may be extended by replacing the physical length of the column (L) with the effective length (KL).

$$P_e = \frac{\pi^2 EI}{(KL)^2} \tag{110.2}$$

The K value depends on end conditions. Theoretical and AISC-recommended K values are listed in Fig. 110.1. For ASD, the design requirement may be stated as

$$P_a \le \frac{P_n}{\Omega_c} = \frac{P_n}{1.67} \tag{110.3}$$

where P_a is the maximum axial load calculated from the ASD load combinations.

For LRFD, the design requirement may be stated as

$$P_u \le \phi_c P_n = 0.9 P_n \tag{110.4}$$

where P_u is the maximum axial load calculated from the LRFD load combinations.

The nominal axial load capacity (P_n) of steel columns is a function of the *governing (largest) slenderness ratio* of the column. For very slender columns, the behavior is essentially elastic, as the column reaches its critical buckling load at low stress levels. Therefore, for very slender columns, the Euler buckling load is a good indicator of the actual load at which bucking occurs.

An additional factor of safety for initial crookedness of the column is incorporated into the AISC equation for elastic buckling. The nominal load capacity of the column is given by

$$P_n = F_{cr}A_g \qquad (110.5)$$

where F_{cr}, the critical (buckling) stress, is a function of the slenderness ratio.

Critical Buckling Stress for Steel Columns

Euler's critical buckling stress is given by

$$F_e = \frac{\pi^2 E}{(KL/r)^2} \qquad (110.6)$$

The critical buckling stress of a steel column (F_{cr}) depends on whether buckling is classified as inelastic (low slenderness ratio, allowing the steel to reach higher stress levels until buckling occurs) or elastic (more slender columns buckle at low load level, thereby ensuring elastic behavior). The equations outlined in AISC are given below:

Elastic Buckling

$$\text{For } \frac{KL}{r} > 4.71\sqrt{\frac{E}{F_y}} \quad \text{or} \quad F_e < 0.44F_y \quad F_{cr} = 0.877F_e \qquad (110.7)$$

Alternately, the lower limit for elastic behavior can also be expressed in terms of a slenderness parameter λ_c, as shown below:

$$\text{For } \lambda_c = \frac{KL}{r\pi}\sqrt{\frac{F_y}{E}} > 1.5 \quad F_{cr} = 0.877F_e \qquad (110.8)$$

Inelastic Buckling

As the column becomes less slender, the stress level needed to cause buckling increases and the column behaves inelastically, causing significant departure from Euler's (elastic) theory. The following equation is given in the AISC specifications:

$$\text{For } \frac{KL}{r} \leq 4.71\sqrt{\frac{E}{F_y}} \quad \text{or} \quad F_e \geq 0.44F_y \quad F_{cr} = \left(0.658^{F_y/F_e}\right)F_y \qquad (110.9)$$

NOTE As the slenderness ratio KL/r approaches zero, as for a stub column, the critical stress F_{cr} approaches the yield stress F_y. The nominal load corresponding to this limit state is termed the *squash load*, given by $F_y A_g$.

Table 110.1 shows the critical buckling stress F_{cr} as a function of the slenderness ratio KL/r for various grades of steel ($F_y = 35, 36, 42, 46,$ and 50 ksi).

Table 110.1 Critical Buckling Stress for Various Grades of Steel

KL/r	Critical stress F_{cr} (ksi) for various values of yield stress (ksi)				
	35	36	42	46	50
0	35.00	36.00	42.00	46.00	50.00
5	34.96	35.95	41.94	45.92	49.91
10	34.82	35.81	41.74	45.69	49.64
15	34.60	35.58	41.42	45.31	49.18
20	34.29	35.25	40.98	44.78	48.56
25	33.90	34.83	40.42	44.11	47.77
30	33.42	34.33	39.74	43.30	46.82
35	32.87	33.75	38.96	42.36	45.72
40	32.25	33.09	38.07	41.31	44.48
45	31.55	32.36	37.09	40.14	43.12
50	30.80	31.56	36.02	38.88	41.65
55	29.98	30.70	34.88	37.53	40.08
60	29.11	29.78	33.67	36.11	38.43
65	28.19	28.82	32.40	34.62	36.71
70	27.24	27.81	31.08	33.08	34.94
75	26.24	26.77	29.73	31.51	33.14
80	25.22	25.70	28.35	29.91	31.31
85	24.18	24.61	26.95	28.29	29.48
90	23.12	23.50	25.54	26.68	27.65
95	22.05	22.39	24.13	25.07	25.85
100	20.98	21.27	22.73	23.48	24.07
105	19.91	20.15	21.34	21.91	22.33
110	18.84	19.04	19.98	20.38	20.64
115	17.79	17.94	18.64	18.90	18.98
120	16.75	16.87	17.34	17.43	17.43
125	15.73	15.81	16.06	16.06	16.06
130	14.74	14.79	14.85	14.85	14.85
135	13.77	13.77	13.77	13.77	13.77
140	12.81	12.81	12.81	12.81	12.81
145	11.94	11.94	11.94	11.94	11.94
150	11.16	11.16	11.16	11.16	11.16
155	10.45	10.45	10.45	10.45	10.45

(*Continued*)

Table 110.1   Critical Buckling Stress for Various Grades of Steel (*Continued*)

KL/r	Critical stress F_{cr} (ksi) for various values of yield stress (ksi)				
	35	36	42	46	50
160	9.81	9.81	9.81	9.81	9.81
165	9.22	9.22	9.22	9.22	9.22
170	8.69	8.69	8.69	8.69	8.69
175	8.20	8.20	8.20	8.20	8.20
180	7.75	7.75	7.75	7.75	7.75
185	7.33	7.33	7.33	7.33	7.33
190	6.95	6.95	6.95	6.95	6.95
195	6.60	6.60	6.60	6.60	6.60
200	6.28	6.28	6.28	6.28	6.28

Although the AISC specifications do not explicitly place an upper limit on the slenderness ratio KL/r, an upper limit of 200 is recommended.

Braced Columns

The governing slenderness ratio is the largest value of KL/r that can be computed for the column. For a column which is unbraced and for which end conditions (constraints) about both axes are identical, the buckling will of course occur about the so-called *weak axis*, i.e., the axis about which the moment of inertia (and therefore the radius of gyration) is minimum. The lower radius of gyration will result in a higher slenderness ratio and therefore a lower critical load. This may be termed *weak-axis buckling*.

However, one of two situations may cause a stiffening of the weak axis relative to the strong axis—(1) the weak axis may be braced intermittently, so as to reduce the buckling length, or (2) the end conditions may be different, so that different constraints (on translation and rotation) exist with respect to the two axes. In such a case, the KL/r of all buckling scenarios must be investigated, with the largest value chosen as the governing value of the slenderness ratio. Examples 110.1 and 110.2 illustrate the use of AISC column design tables for design of steel columns.

Example 110.1

The steel column shown below is loaded concentrically with the following loads: $P_{DL} = 180$ kips, $P_{LL} = 300$ kips. What is the smallest $W14$ section that can be used? Use $F_y = 50$ ksi.

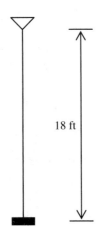

18 ft

Solution

ASD Solution

$$\text{Design load } P = 180 + 300 = 480 \text{ kips}$$
$$KL = 0.80 \times 18 = 14.4 \text{ ft}$$

Using tables, for $P_a = 480$ kips and $KL = 14.4$ ft, the lightest $W14$ section is $W14 \times 82$. Without using tables, one may follow these steps:

(1) Assume $KL/r = 60$.

   a. The critical stress $F_{cr} = 38.43$ ksi.

   b. The allowable stress is $F_a = 38.43/1.67 = 23.01$ ksi.

   c. Therefore, the required area $A_g = 480/23.01 = 20.9$ in². Smallest section is $W14 \times 74$.

(2) Check adequacy of a $W14 \times 74$. $KL/r_y = 14.4 \times 12/2.48 = 69.7$.

   a. $F_a = 20.92$ ksi.

   b. $F_a A_g = 20.92 \times 21.8 = 456.5$ kips < 480 kips. $W14 \times 74$ has insufficient capacity.

(3) Check the next heavier section ($W14 \times 82$). $KL/r_y = 14.4 \times 12/2.48 = 69.7$.

   a. $F_a = 20.92$ ksi, $F_a A_g = 20.92 \times 24.0 = 502.6$ kips. Thus, the $W14 \times 82$ has sufficient capacity.

LRFD Solution

$$\text{Design load } P_u = 1.2 \times 180 + 1.6 \times 300 = 696 \text{ kips}$$
$$KL = 0.80 \times 18 = 14.4 \text{ ft}$$

Using tables, for $P_u = 696$ kips and $KL = 14.4$ ft, the lightest $W14$ section is $W14 \times 82$ (capacity approx. 715 kips). Without using tables, one may use the following steps:

(1) Assume $KL/r = 60$.

    a. The critical stress level is $\phi_c F_{cr} = 0.90 \times 35 = 34.6$ ksi.

    b. Therefore, the required area $A_g = 696/34.6 = 20.1$ in². Smallest section is $W14 \times 74$.

(2) Check adequacy of a $W14 \times 74$. Verify $KL/r_y = 14.4 \times 12/2.48 = 69.7$.

    a. $\phi_c F_{cr} = 0.90 \times 35 = 31.5$ ksi.

    b. $\phi_c F_{cr} A_g = 31.5 \times 21.8 = 686.7$ kips. Not sufficient capacity.

(3) Check the next heavier section ($W14 \times 82$). Verify $W14 \times 82$.
$KL/r_y = 14.4 \times 12/2.48 = 69.7$.

    a. $\phi_c F_{cr} = 0.90 \times 35 = 31.5$ ksi.

    b. $\phi_c F_{cr} A_g = 31.5 \times 24.0 = 756$ kips. Thus, the $W14 \times 82$ has sufficient capacity.

In Example 110.1, the minimum radius of gyration of the section (r_y) has been used to determine the slenderness ratio, since without any restraint (bracing) of the weak axis, buckling first occurs about that axis. For this reason, *the column design strength tables in the AISC manuals are set up in terms of the effective length* (KL) *with respect to the weak axis radius of gyration* (r_y). The exception to this default situation may occur if the weak axis is laterally braced in any way, as shown in Example 110.2.

Example 110.2

The steel column shown in Figs. 110.2 and 110.3 is loaded concentrically with the following loads: $P_{DL} = 180$ kips, $P_{LL} = 300$ kips. The weak axis is laterally braced at point B, as shown in Fig. 110.2. The strong axis buckling mode is shown in Fig. 110.3. What is the smallest $W14$ section that can be used? Use $F_y = 50$ ksi.

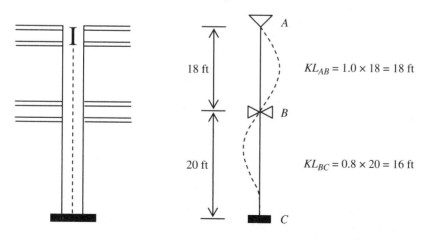

Buckling about minor axis

18 ft $KL_{AB} = 1.0 \times 18 = 18$ ft

20 ft $KL_{BC} = 0.8 \times 20 = 16$ ft

Figure 110.2 Weak axis buckling for a braced column.

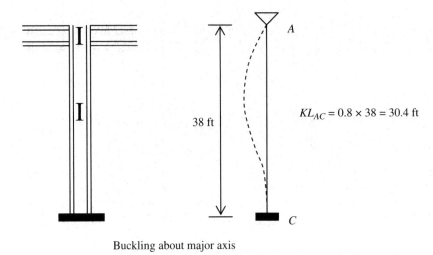

$KL_{AC} = 0.8 \times 38 = 30.4$ ft

38 ft

Buckling about major axis

Figure 110.3 Strong axis buckling for a braced column.

Solution

ASD Solution

$$\text{Design load } P = 180 + 300 = 480 \text{ kips}$$

There are three effective lengths to consider: $KL_{AB} = 1.0 \times 18 = 18$ ft, $KL_{BC} = 0.8 \times 20 = 16$ ft, and $KL_{AC} = 0.8 \times 38 = 30.4$ ft. Of these, KL_{BC} can be ignored since the governing "weak axis" KL is KL_{AB}. However, in terms of slenderness ratio, it is possible that the strong axis KL_{AC} may prove to be more slender. Since the column capacity tables are set up for weak axis buckling, one must use the following approach:

1. Use the weak axis effective length (KL_y) to make a preliminary selection.

2. Use the r_x/r_y ratio for the section selected to convert the KL_x to an equivalent KL_y according to

$$\overline{KL_y} = \frac{KL_x}{r_x/r_y}$$

3. If the converted KL in step 2 is less than the weak axis KL in step 1, the member selection in step 1 stands. Otherwise, the converted (strong axis) KL must be used to select a larger section.

$$KL_{AB} = 18 \text{ ft}$$

Using tables, for $P = 480$ kips and $KL = 18$ ft, the lightest W14 section is W14 × 90. This section has an $r_x/r_y = 1.66$.

$$\overline{KL_y} = \frac{30.4}{1.66} = 18.31 \text{ ft}$$

Since this is greater than 18 ft, check the capacity of W14 × 90 by interpolation between 618 and 600. The capacity is 612.4 kips. This is acceptable, since it is greater than 480.

LRFD Solution

$$\text{Design load } P_u = 1.2 \times 180 + 1.6 \times 300 = 696 \text{ kips}$$
$$KL_{AB} = 18 \text{ ft}, \qquad KL_{BC} = 16 \text{ ft}, \qquad KL_{AC} = 30.4 \text{ ft}$$

Using tables, for $P_u = 696$ kips and $KL = 18$ ft, the lightest W14 section is W14 × 90. This section has an $r_x/r_y = 1.66$.

$$\overline{KL_y} = \frac{30.4}{1.66} = 18.31 \text{ ft}$$

Since this is greater than 18, check the capacity of the W14 × 90 by interpolation between 928 and 902. The capacity is 919.9 kips. This is acceptable, since it is greater than 696.

Effective Length for Columns in a Frame

The predetermination of the effective-length factor (K) is not possible for columns which are not "standalone" but rather are parts of a structural framework. For example, if the column being considered is designated AB, the rotational rigidity of ends A and B depends on the relative bending stiffness of all members (columns and beams) framing into these ends. The AISC specifications outline the following procedure:

1. At either end of the column, calculate a coefficient G given by

$$G = \frac{\left(\sum {I}/{L}\right)_{\text{columns}}}{\left(\sum {I}/{L}\right)_{\text{beams}}} \qquad (110.10)$$

For sidesway inhibited frames, the I/L factors for girders are multiplied by
2.0 if the far end of the girder is fixed and
1.5 if the far end of the girder is pinned.
For column ends supported by, but not rigidly connected to, a footing or foundation, G is theoretically infinite, but the AISC specifications recommend using $G = 10.0$.
For a column end rigidly attached to a properly designed footing, G is theoretically zero, but the AISC specifications recommend using $G = 1.0$.

2. Based on the average compressive stress in the column (service load for ASD, factored load for LRFD), calculate or look up a stiffness reduction factor τ. This factor incorporates the effect of possible inelastic column behavior into the method. The stiffness reduction factor (τ) is given by

$$\text{For } P_n > 0.39 F_y A_g \qquad \tau = -2.724 \frac{P_n}{F_y A_g} \ln\left(\frac{P_n}{F_y A_g}\right) \qquad (110.11)$$

Table 110.2 shows values of the stiffness reduction factor (τ) for various grades of steel for ASD and LRFD.

3. The coefficients G in step 1 are multiplied by the stiffness reduction factor (SRF).

4. The modified G values are then used in one of two alignment charts to obtain the effective-length factor K depending on whether the frame is sidesway inhibited or sidesway uninhibited. These charts are reproduced in Fig. 110.4.

5. The K obtained from the alignment chart is then used to determine the governing slenderness ratio for the column.

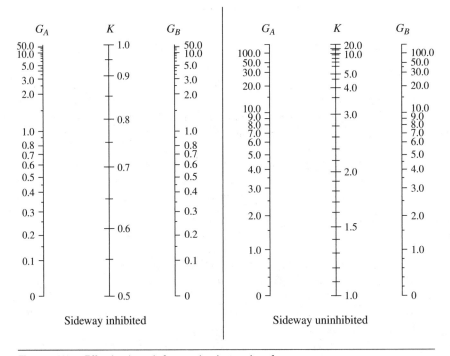

Figure 110.4 Effective length for steel columns in a frame.

Table 110.2 Stiffness Reduction Factor (τ) for LRFD (Shaded) and ASD (White)

P_a/A_g (ksi)	P_u/A_g (ksi)	Yield stress F_y (ksi)								
		35		36		42		46		50	
	45	–	–	–	–	–	–	–	–	–	–
	44	–	–	–	–	–	–	–	–	–	0.060
	43	–	–	–	–	–	–	–	–	–	0.118
	42	–	–	–	–	–	–	–	–	–	0.175
	41	–	–	–	–	–	–	–	0.026	–	0.231
	40	–	–	–	–	–	–	–	0.091	–	0.285
	39	–	–	–	–	–	–	–	0.153	–	0.338
	38	–	–	–	–	–	–	–	0.214	–	0.389
	37	–	–	–	–	–	0.057	–	0.274	–	0.438
	36	–	–	–	–	–	0.127	–	0.331	–	0.486
	35	–	–	–	–	–	0.194	–	0.387	–	0.532
	34	–	–	–	–	–	0.260	–	0.441	–	0.577
	33	–	–	–	–	–	0.323	–	0.492	–	0.620
	32	–	–	–	0.033	–	0.384	–	0.542	–	0.660
	31	–	0.043	–	0.115	–	0.443	–	0.590	–	0.699
30	30	–	0.127	–	0.194	–	0.500	–	0.636	–	0.736
29	29	–	0.207	–	0.270	–	0.554	–	0.679	0.084	0.771
28	28	–	0.285	–	0.344	–	0.606	–	0.720	0.171	0.804
27	27	–	0.360	–	0.414	–	0.655	0.053	0.759	0.254	0.835
26	26	–	0.431	–	0.481	–	0.701	0.148	0.796	0.334	0.863
25	25	–	0.500	–	0.545	0.016	0.745	0.240	0.830	0.410	0.890
24	24	–	0.564	–	0.606	0.122	0.786	0.327	0.861	0.483	0.913
23	23	–	0.626	–	0.663	0.223	0.823	0.410	0.890	0.552	0.934
22	22	–	0.683	–	0.716	0.319	0.858	0.489	0.915	0.617	0.953
21	21	–	0.736	0.069	0.766	0.410	0.890	0.563	0.938	0.678	0.969
20	20	0.122	0.786	0.189	0.811	0.496	0.917	0.633	0.957	0.734	0.982
19	19	0.242	0.831	0.303	0.853	0.577	0.942	0.698	0.974	0.786	0.992
18	18	0.356	0.871	0.410	0.890	0.652	0.962	0.757	0.986	0.833	0.998
17	17	0.462	0.907	0.510	0.922	0.721	0.979	0.811	0.996	0.875	1.000
16	16	0.561	0.937	0.603	0.949	0.784	0.991	0.860	1.000	0.912	1.000

(*Continued*)

Table 110.2 Stiffness Reduction Factor (τ) for LRFD (Shaded) and ASD (White) (*Continued*)

P_a/A_g (ksi)	P_u/A_g (ksi)	Yield stress F_y (ksi)								
		35		36		42		46		50
15	15	0.652	0.962	0.687	0.971	0.840	0.999	0.902	1.000	0.943	1.000
14	14	0.734	0.982	0.764	0.988	0.888	1.000	0.937	1.000	0.968	1.000
13	13	0.807	0.995	0.831	0.998	0.929	1.000	0.965	1.000	0.987	1.000
12	12	0.870	1.000	0.888	1.000	0.962	1.000	0.986	1.000	0.998	1.000
11	11	0.922	1.000	0.935	1.000	0.985	1.000	0.999	1.000	1.000	1.000
10	10	0.962	1.000	0.971	1.000	0.999	1.000	1.000	1.000	1.000	1.000
9	9	0.989	1.000	0.993	1.000	1.000	1.000	1.000	1.000	1.000	1.000
8	8	1.000	1.000	1.000	1.000	1.000	1.000	1.000	1.000	1.000	1.000

Column with Slender Elements

When individual components of a column are so slender as to buckle locally below the stress which would cause the overall column to buckle, the column is said to have slender elements. For these columns, the following procedure is followed:

A reduction factor Q is determined for the slender component(s). This factor represents the fraction of the "effective area" of the components to the total cross-section area. The effective area of each component is based on an effective width b_e, which is calculated from the intrinsic equation:

$$b_e = 1.92t\sqrt{\frac{E}{f}}\left[1 - \frac{0.34}{b/t}\sqrt{\frac{E}{f}}\right] \le b \tag{110.12}$$

The critical stress is then computed using one of the following:

Elastic Buckling

$$\text{For } \frac{KL}{r} > 4.71\sqrt{\frac{E}{QF_y}} \quad \text{or} \quad F_e < 0.44QF_y \qquad F_{cr} = 0.877F_e \tag{110.13}$$

Inelastic Buckling

$$\text{For } \frac{KL}{r} \leq 4.71\sqrt{\frac{E}{QF_y}} \qquad \text{or} \qquad F_e \geq 0.44QF_y \qquad F_{cr} = \left(0.658^{QF_y/F_e}\right)QF_y \qquad (110.14)$$

Single-Angle Compression Elements

For single angles that

1. Are individual members or web members of plane trusses having adjacent web members attached to the same side of a gusset plate or a truss chord
2. Are equal-leg angles or unequal-leg angles connected through the longer leg

The slenderness ratio is calculated as follows:

Buckling is assumed to occur about the y axis, which is the axis parallel to the attached leg.

$$\text{If } \frac{L}{r_x} \leq 80 \qquad \frac{KL}{r} = 72 + 0.75\frac{L}{r_x} \qquad\qquad (110.15a)$$

$$\text{If } \frac{L}{r_x} > 80 \qquad \frac{KL}{r} = 32 + 1.25\frac{L}{r_x} \leq 200 \qquad (110.15b)$$

For single angles that

1. Are web members of box or space trusses having adjacent web members attached to the same side of a gusset plate or a truss chord
2. Are equal-leg angles or unequal-leg angles connected through the longer leg

The slenderness ratio is calculated as follows:

Buckling is assumed to occur about the y axis, which is the axis parallel to the attached leg.

$$\text{If } \frac{L}{r_x} \leq 75 \qquad \frac{KL}{r} = 60 + 0.8\frac{L}{r_x} \qquad\qquad (110.16a)$$

$$\text{If } \frac{L}{r_x} > 75 \qquad \frac{KL}{r} = 45 + \frac{L}{r_x} \leq 200 \qquad (110.16b)$$

Once the slenderness ratio has been computed, the design strength can be computed as before.

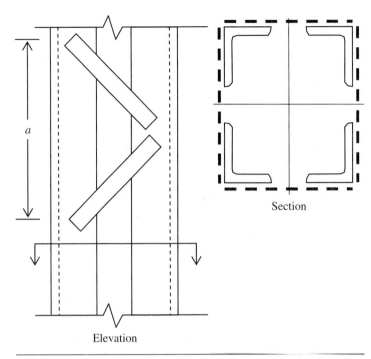

Section

Elevation

Figure 110.5 Built-up compression member.

Built-Up Compression Members

The AISC specifications require that the slenderness ratio of individual parts of a built-up section not be more than 75% of the governing slenderness ratio of the built-up shape. For example, if a tower leg is constructed by "stitching together" four equal-leg angles as shown in Fig. 110.5, the perimeter lattice members must be spaced such that the slenderness ratio of individual angles (KL/a) be less than $0.75(KL/r)$ of the built-up section.

Column Base Plates

Figure 110.6 shows a steel column supported by a steel base plate having dimensions $B \times N$. The area of the base plate is designated $A_1 = BN$ in the following equations. If the area of the concrete or masonry pedestal supporting the base plate is greater than A_1, it is designated A_2 and this greater area leads to a greater bearing strength of the concrete, due to the participation of the adjacent concrete. However, the ratio A_2/A_1 is limited to a maximum value of 4.

Table 110.3 summarizes the steps involved in the design of the base plate. Both ASD and LRFD are summarized. For LRFD, the design load is the factored load P_u, whereas for ASD, the design load is the required strength P_a.

Column Base Plates

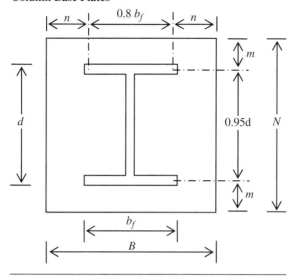

Figure 110.6   Geometry of a column base plate.

Table 110.3   Step-by-Step Procedure for Design of Column Base Plates

LRFD	ASD
Factored axial column load P_u	Largest axial column load P_a
Strength reduction factor $\phi_c = 0.65$	Factor of safety $\Omega_c = 2.31$

$$A_1 = \frac{P_u}{\phi_c(0.85f_c')\sqrt{\dfrac{A_2}{A_1}}} \qquad\qquad A_1 = \frac{P_a\Omega_c}{0.85f_c'\sqrt{\dfrac{A_2}{A_1}}}$$

$$\Delta = 0.5(0.95d - 0.8b_f)$$

$$N = \sqrt{A_1} + \Delta$$

$$B = \frac{A_1}{N}$$

$$m = \frac{N - 0.95d}{2}$$

$$n = \frac{B - 0.8b_f}{2}$$

$$n' = \frac{\sqrt{db_f}}{4}$$

$$l = \max(m, n, n')$$

Required thickness	Required thickness

$$t_{\text{reqd}} = l\sqrt{\frac{2P_u}{0.9F_y BN}} \qquad\qquad t_{\text{reqd}} = l\sqrt{\frac{3.33P_a}{F_y BN}}$$

Example 110.3

A 10×49 column carries the following concentric service loads: $D = 100$ kips, $L = 150$ kips. The column is supported by a 20 in $\times$ 20 in reinforced concrete column ($f'_c = 4{,}000$ psi). The column is grade 50 steel. The base plate is to be A36 steel. What are the required dimensions (length, width, thickness) of the base plate?

Solution

Factored load: $P_u = 1.2P_D + 1.6P_L = 1.2 \times 100 + 1.6 \times 150 = 360$ kips

Bearing on concrete (with supporting area $A_2 = 20 \times 20 = 400$ in^2) with strength reduction factor $\phi = 0.6$

Factored strength in bearing:

$$\phi \left(0.85 f'_c A_1 \sqrt{\frac{A_2}{A_1}} \right) = 0.6 \times 0.85 \times 4 \times A_1 \times \sqrt{\frac{400}{A_1}} = 40.8 \sqrt{A_1} \geq 360$$

This yields: $A_1 \geq 77.85$ in^2

The factored strength in bearing is limited to: $1.7 f'_c A_1$, which yields $A_1 \geq 52.94$ in^2

The footprint of the column envelop $= b_f d = 10 \times 10 = 100$ in^2

It is desirable to have a base plate slightly bigger than the column footprint. Let us choose a base plate 11 in $\times$ 11 in (A_1=121 in^2). $M = N = 11$ in

$$m = \frac{N - 0.95d}{2} = \frac{11 - 0.95 \times 10}{2} = 0.75$$

$$n = \frac{B - 0.8b_f}{2} = \frac{11 - 0.8 \times 10}{2} = 1.5$$

$$n' = \frac{1}{4}\sqrt{db_f} = \frac{1}{4}\sqrt{10 \times 10} = 2.5$$

$$l = max \, (m, n, n') = 2.5$$

Minimum plate thickness: $t_{min} = l \sqrt{\dfrac{2P_u}{0.9 B N F_y}} = 2.5 \sqrt{\dfrac{2 \times 360}{0.9 \times 11 \times 11 \times 36}} = 1.07$ in

Use $11 \times 11 \times 1 \, 1/8$ in base plate.

Steel Beams

General Flexure Theory

For beams having typical L/d ratios (length:depth), the deflection of the beam is predominantly due to the curvature produced by bending moment. This is expressed by the moment-curvature relationship which ignores the contribution of shear deformations to the overall deflection of the beam

$$y'' = \frac{d^2 y}{dx^2} \approx \frac{M}{EI} \tag{111.1}$$

where $y(x)$ = deflection function
$\qquad M(x)$ = bending-moment function
$\qquad\quad E$ = modulus of elasticity of the beam material
$\qquad\quad I$ = moment of inertia of the beam section

For beams having smaller L/d ratios (deep beams), the shear deformation is not negligible in comparison to the flexural deformation and, therefore, produces significant errors if neglected. The internal bending moment in a beam creates bending stresses. Within the proportional limit (stress proportional to strain), the bending stress due to bending moment about the x axis is given by

$$\sigma_b = -\frac{M_x y}{I_x} \tag{111.2}$$

where y = distance to the plane of interest from the centroidal x axis
$\qquad M_x$ = bending moment about the x axis
$\qquad I_x$ = moment of inertia with respect to the x axis.

Progressive Increase of Flexural Stresses

The following discussion is predicated on the elastoplastic stress-strain model for steel shown in Fig. 111.1. It should be noted that in reality, steel begins to show inelastic behavior well before the stress reaches the yield stress F_y, perhaps as soon as $0.5F_y$. At low loads, the bending stresses everywhere in the cross section is within the elastic limit ($\sigma < F_y$) and the elastic stress distribution is as shown in Fig. 111.2(a). The limit of such elastic stress is when the maximum stress reaches exactly F_y. If loaded further, the stresses at the outermost fibers plateau at the value F_y while stresses at internal fibers keep increasing, as shown in Fig. 111.2(b). The limit of such inelastic behavior is when the entire section reaches the stress F_y [Fig. 111.2(c)]. The bending moment that causes first yield of the section (end of elastic behavior) is given by

$$M_{\text{yield}} = F_y S \tag{111.3}$$

where S = elastic section modulus = I/y_{max}.

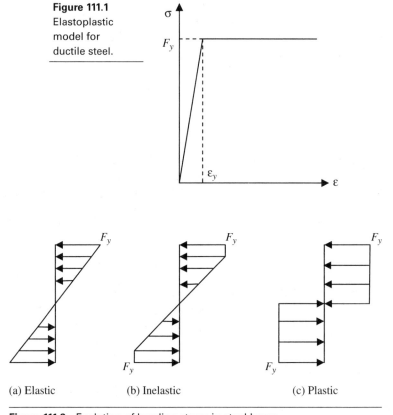

Figure 111.1
Elastoplastic model for ductile steel.

(a) Elastic   (b) Inelastic   (c) Plastic

Figure 111.2 Evolution of bending stress in steel beam.

The bending moment that causes formation of a "plastic hinge" (end of inelastic behavior) and therefore represents the ultimate moment capacity of the cross section is given by

$$M_p = F_y Z \qquad (111.4)$$

where Z is plastic section modulus.

Elastic Section Modulus

The following steps must be followed to calculate the elastic section modulus. For numerical examples (see Chap. 101).

1. Locate elastic neutral axis (ENA) which is the same as geometric centroid. The location of the ENA is the weighted arithmetic average of the centroid locations of the sub-areas that make up the cross section of the beam.

$$\bar{y} = \frac{\sum y_i A_i}{\sum A_i} \qquad (111.5)$$

2. Find moment of inertia I_{NA} with respect to ENA. The contribution of individual parts of the cross section to the overall value of I_{NA} are calculated according to the parallel axis theorem.

$$I_{NA} = I_c + Ad^2 \qquad (111.6)$$

where I_c = moment of inertia with respect to the local centroid
   A = area of the section
   d = distance between the centroidal axis of the shape and the overall centroid of the compound section.

3. The elastic section modulus S is calculated by

$$S = \frac{I_{NA}}{y_{max}} \qquad (111.7)$$

where y_{max} is distance of fiber farthest from ENA.

Plastic Section Modulus

The following steps must be followed to calculate the plastic section modulus:

1. Locate plastic neutral axis (PNA) which is the plane that divides the cross section into two equal halves ($A_{top} = A_{bottom}$).
2. The plastic section modulus Z is calculated as $Z = \Sigma A_i y_i$ which is the first moment of all areas about PNA. These first moments are all taken as positive.

Design of Steel Beams—AISC Specifications

The rest of this chapter is dedicated to the design procedures as suggested by American Institute of Steel Construction (AISC). The two design methods outlined are the allowable strength design (ASD) and load and resistance factor design (LRFD). Specifically, the specification referenced here is the *Steel Construction Manual,* 14th ed. (American Institute of Steel Construction, Inc., 2011), which integrates ASD and LRFD into a combined volume.

For ASD, the design requirement may be stated as

$$M_a \leq \frac{M_n}{\Omega_b} = \frac{M_n}{1.67} \qquad (111.8)$$

where M_a is the required moment strength = maximum moment calculated from the ASD load combinations.

For LRFD, the design requirement may be stated as

$$M_u \leq \phi_b M_n = 0.9 M_n \qquad (111.9)$$

where M_u is the required moment strength = maximum moment calculated from the LRFD load combinations.

Nominal Moment Capacity

The nominal moment capacity (M_n) of steel beams is based on the slenderness of the compression flange and its ability to resist buckling. The full plastic moment capacity M_p can be realized if the unbraced length of the compression flange (L_b) is within a certain limit. For rolled sections, this limit is termed L_p. In other words, for $L_b < L_p$, the nominal moment capacity of the section is equal to the plastic moment capacity. When $L_b > L_p$, the full plastic moment capacity cannot be realized and the section fails due to lateral-torsional buckling (LTB). When $L_p < L_b < L_r$, the section fails due to inelastic LTB. When $L_b > L_r$, the section fails due to elastic LTB.

Figure 111.3
Nominal moment
capacity of steel
beams ($C_b = 1$).

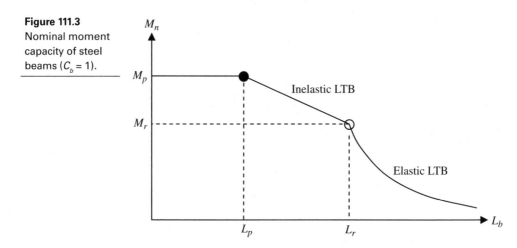

Figure 111.3 shows a plot of the nominal moment capacity M_n versus the unbraced length of the compression flange L_b. The following is a summary of the nominal flexural strength of compact I- and C-shaped sections.

Failure by Reaching Fully Plastic State

$$\text{For } L_b \leq L_p \qquad M_n = M_p = Z_x F_y \qquad\qquad (111.10)$$

Inelastic Lateral Torsional Buckling (LTB)

$$\text{For } L_p \leq L_b \leq L_r \qquad M_n = C_b \left[M_p - (M_p - 0.7 F_y S_x) \left(\frac{L_b - L_p}{L_r - L_p} \right) \right] \leq M_p \quad (111.11)$$

where C_b is the factor to account for the nonuniform bending moment within the unbraced length L_b.

Conservatively, C_b may be taken as 1.0. In fact, the value of C_b depends on the shape of the bending-moment diagram and the AISC design charts are based on $C_b = 1.0$, leaving it up to the user to take advantage of a higher value of C_b if warranted. This is illustrated in section Bending Coefficient C_b.

Elastic Lateral Torsional Buckling (LTB)

$$\text{For } L_b > L_r \qquad M_n = F_{cr} S_x \leq M_p \qquad\qquad (111.12)$$

where F_{cr} = critical stress, given by

$$F_{cr} = \frac{C_b \pi^2 E}{(L_b / r_{ts})^2} \sqrt{1 + 0.078 \frac{Jc}{S_x h_o} \left(\frac{L_b}{r_{ts}} \right)^2} \qquad\qquad (111.13)$$

$$r_{ts}^2 = \frac{\sqrt{I_y C_w}}{S_x} \qquad\qquad (111.14)$$

The torsional constant (J), the warping constant (C_w), and the weak axis moment of inertia (I_y) can be found in Part 1 of the AISC specifications. The other parameters are

$$h_o = d - t_f$$

$c = 1.0$ for doubly symmetric shapes, and

$$c = \frac{h_o}{2}\sqrt{\frac{I_y}{C_w}} \text{ for channels}$$

The critical values of unbraced length which separate these three types of behavior are L_p and L_r. For rolled sections, these values are listed in the AISC specifications, Part 3. These parameters are defined as follows:

$$L_p = 1.76 r_y \sqrt{\frac{E}{F_y}} \tag{111.15}$$

$$L_r = 1.95 r_{ts} \frac{E}{0.7F_y}\sqrt{\frac{Jc}{S_x h_o}}\sqrt{1 + \sqrt{1 + 6.76\left(\frac{0.7F_y S_x h_o}{EJc}\right)^2}} \tag{111.16}$$

The design charts in the AISC manual reflect the $C_b = 1$ case, which is conservative. The effect of incorporating the bending coefficient C_b into the design charts is to increase the plotted values of the moment capacity M_n as shown by the broken line in Fig. 111.4. The only limitation to this increase is that the value of $C_b M_n$ cannot be greater than M_p.

Bending Coefficient C_b

If the bending moment within the unbraced length is nonuniform, the bending coefficient C_b is incorporated into the design to account for the fact that spanwise variations in the bending moment mean that the critical bending moment is not simultaneously reached at all locations,

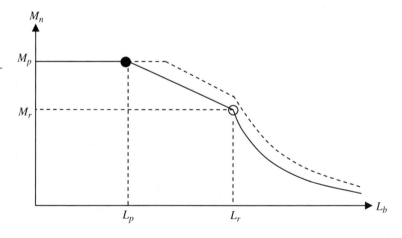

Figure 111.4 Nominal moment capacity of steel beams ($C_b > 1$).

thus making the beam more resistant to compression flange buckling than if C_b were to equal 1.0 (for constant bending moment). The coefficient is defined as

$$C_b = \frac{12.5M_{max}}{2.5M_{max} + 3M_A + 4M_B + 3M_C} \tag{111.17}$$

where M_A, M_B, and M_C are the absolute values of the bending moments at $1/4L_b$, $1/2L_b$, and $3/4L_b$, respectively and M_{max} is the absolute value of the maximum bending moment within the unbraced length L_b (see Fig. 111.5).

For some standard load types, Table 111.1 may be used to look up the value of C_b.

Beam Design Using Z_x Tables

When the unbraced length is very short, such as when the compression flange has "continuous" lateral support from the slab or when relatively closely spaced lateral members frame into the compression zone, L_b may be considered to be zero and the full plastic moment capacity of the beam may be realized.

$$\text{Thus } M_n = M_p = Z_x F_y$$

This translates into the following requirement for minimum plastic section modulus Z_x.

$$\text{ASD}: \quad Z_x \geq \frac{\Omega M_a}{F_y} = \frac{1.67M_a}{F_y} \tag{111.18}$$

$$\text{LRFD}: \quad Z_x \geq \frac{M_u}{\phi F_y} = \frac{M_u}{0.9F_y} \tag{111.19}$$

Figure 111.5
Parameters used
to calculate C_b.

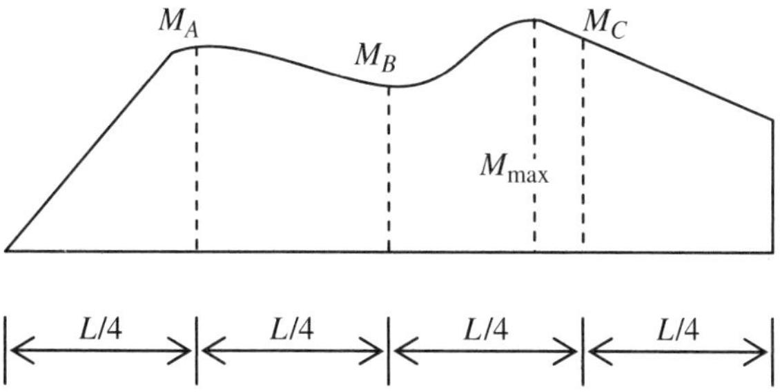

Table 111.1 Values of C_b for Certain Standard Load Cases

Load	Lateral bracing along span	C_b
Midpoint	None	1.32
	At load points	1.67 1.67
Third points	None	1.14
	At load points	1.67 1.00 1.67
Uniform distributed	None	1.14
	At midpoint	1.30 1.30
	At third points	1.45 1.01 1.45

The Z_x tables may then be used to select the section. Figure 111.6 shows a fragment of the Z_x table ($F_y = 50$ ksi) from the AISC manual.

Example 111.1

The top flange of a simply supported beam (span = 20 ft) is continuously braced against buckling by the supported slab. If the service loads are dead load = 2 kips/ft, live load = 3 kips/ft, wind load = 2 kips/ft, what is the lightest W-section which is satisfactory (use $F_y = 50$ ksi).

Shape	Z_x	ASD M_{px}/Ω_b	LRFD $\phi_b M_{px}$	ASD M_{rx}/Ω_b	LRFD $\phi_b M_{rx}$	ASD BF/Ω_b	LRFD $\phi_b BF$	L_p	L_r	I_x
	in³	kip-ft	kip-ft	kip-ft	kip-ft	kips	kips	ft	ft	in⁴
W21 × 55	**126**	**314**	**473**	**192**	**289**	**10.8**	**16.3**	**6.11**	**17.4**	**1140**
W14 × 74	126	314	473	196	294	5.34	8.03	8.76	31.0	795
W18 × 60	123	307	461	189	284	9.64	14.5	5.93	18.2	984
W12 × 79	119	297	446	187	281	3.77	5.67	10.8	39.9	662
W14 × 68	115	287	431	180	270	5.20	7.81	8.69	29.3	722
W10 × 88	113	282	424	172	259	2.63	3.95	9.29	51.1	534
W18 × 55	**112**	**279**	**420**	**172**	**258**	**9.26**	**13.9**	**5.90**	**17.5**	**890**
W21 × 50	**110**	**274**	**413**	**165**	**248**	**12.2**	**18.3**	**4.59**	**13.6**	**984**
W12 × 72	108	269	405	170	256	3.72	5.59	10.7	37.4	597
W21 × 48	**107**	**265**	**398**	**162**	**244**	**9.78**	**14.7**	**6.09**	**16.6**	**959**
W16 × 57	105	262	394	161	242	7.98	12.0	5.65	18.3	758
W14 × 61	102	254	383	161	242	4.96	7.46	8.65	27.5	640
W18 × 50	101	252	379	155	233	8.69	13.1	5.83	17.0	800
W10 × 77	97.6	244	366	150	225	2.59	3.90	9.18	45.2	455
W12 × 65	96.8	237	356	154	231	3.60	5.41	11.9	35.1	533
W21 × 44	**95.4**	**238**	**358**	**143**	**214**	**11.2**	**16.8**	**4.45**	**13.0**	**843**
W16 × 50	92.0	230	345	141	213	7.59	11.4	5.62	17.2	659
W18 × 46	90.7	226	340	138	207	9.71	14.6	4.56	13.7	712
W14 × 53	87.1	217	327	136	204	5.27	7.93	6.78	22.2	541
W12 × 58	86.4	216	324	136	205	3.76	5.66	8.87	29.9	475
W10 × 68	85.3	213	320	132	199	2.57	3.86	9.15	40.6	394
W16 × 45	82.3	205	309	127	191	7.16	10.8	5.55	16.5	586

Figure 111.6 Steel *W*-shapes organized by plastic section modulus (Z_x).

Solution

LRFD Solution

Load combinations are

(i)   LRFD1 $w_u = 1.4DL = 2.8$ kips/ft

(ii) LRFD2 $w_u = 1.2DL + 1.6LL = 1.2 \times 2 + 1.6 \times 3 = 7.2$ kips/ft

(iii) LRFD3 $w_u = 1.2DL + 0.8WL = 1.2 \times 2 + 0.8 \times 2 = 4.0$ kips/ft

(iv) LRFD4 $w_u = 1.2DL + 1.6WL + 0.5LL = 1.2 \times 2 + 1.6 \times 2 + 0.5 \times 3 = 7.1$ kips/ft

$$M_u = \frac{w_u L^2}{8} = \frac{7.2 \times 20^2}{8} = 360 \text{ kips-ft} = 4320 \text{ kips-in}$$

$$Z_{x,\text{reqd}} = \frac{M_u}{0.9 F_y} = \frac{4320}{0.9 \times 50} = 96 \text{ in}^3$$

From Fig. 111.6 (from AISC *Steel Construction Manual*, 13th edition), $W12 \times 65$ is the section with the smallest Z_x (96.8 in^3) that meets the criterion but $W21 \times 48$ is the *lightest* section (top of the group—in bold).

ASD Solution

Load combinations are

(i)   ASD1 $w = DL = 2.0$ kips/ft

(ii) ASD2 $w = DL + LL = 2 + 3 = 5.0$ kips/ft

(iii) ASD5 $w = DL + WL = 2 + 2 = 4.0$ kips/ft

(iv) ASD6 $w = DL + 0.75WL + 0.75LL = 2 + 0.75 \times 2 + 0.75 \times 3 = 5.75$ kips/ft

$$M_a = \frac{w L^2}{8} = \frac{5.75 \times 20^2}{8} = 287.5 \text{ kips-ft} = 3450 \text{ kips-in}$$

$$Z_{x,\text{reqd}} = \frac{\Omega M_a}{F_y} = \frac{1.67 \times 3450}{50} = 115.23 \text{ in}^3$$

From the Z_x table, $W21 \times 55$ is the lightest section according to ASD.

Beam Design Using Charts

When the unbraced length of compression flange is not zero, the beam's flexural strength could be based upon attainment of the plastic moment, or on inelastic or elastic LTB. Without having chosen a section, and therefore not knowing the values of L_p and L_r, it is difficult to make a member selection. However, the M_n versus L_b design charts in the *Manual* may be used as follows:

1. Calculate the required moment capacity from the load combinations.

$$\text{ASD: } M_a = \frac{M_n}{\Omega}$$

$$\text{LRFD: } M_u = \phi M_n$$

2. The unbraced length versus moment capacity chart plots the dual axes ϕM_n and M_n/Ω. Identify the point corresponding to this required capacity.

3. Move vertically upward (keeping L_b constant) from this point until the first solid line is encountered. This is the lightest section. Alternatively, the first dashed line may be used as long as the line plots above the point.

Example 111.2

The top flange of a simply supported beam (span = 20 ft) is laterally braced at the supports and at midspan only. If the service loads are dead load = 2 kips/ft, live load = 3 kips/ft, wind load = 2 kips/ft, what is the lightest *W*-section which is satisfactory (use F_y = 50 ksi)?

Solution

LRFD Solution

As before, the factored load w_u = 7.2 kips/ft. The factored moment is M_u = 360 kips-ft. From the chart (Fig. 111.7) if the point L_b = 10 ft, ϕM_n = 360 kips-ft is plotted, W18 × 55 is the first solid line above that point. This is the lightest section that is satisfactory for the given situation. This happens to be in the second zone.

Verification

$$\phi_b M_n = \phi_b M_p - BF \times (L_b - L_p) = 420 - 13.9 \times (10 - 5.9) = 363$$

Therefore, $\phi_b M_n$ > 360 (OK)

ASD Solution

As before, the ASD load combination w = 5.75 kips/ft. The design moment M_a = 287.5 kips-ft. From the chart, if the point L_b = 10 ft, M_n/Ω = 287.5 kips-ft is plotted, W21 × 62 is the first solid line above that point. This is the lightest section that is satisfactory for the given situation. This happens to be in the second zone.

Compactness Criteria

The provisions of section 5 are for compact shapes. Sections are classified as compact, non-compact, or slender. Most *W*-, *M*-, *S*-, and *C*- shapes are compact. A few have noncompact flanges (webs are compact, without exception) but none are slender. Table 111.2 shows limits for compactness and slenderness based on length to thickness ratios for flanges and webs.

If the slenderness parameter λ is less than λ_p, the section is classified as compact. If $\lambda_p < \lambda < \lambda_r$, the section is classified as noncompact. If λ is greater than λ_r, the section is classified as slender.

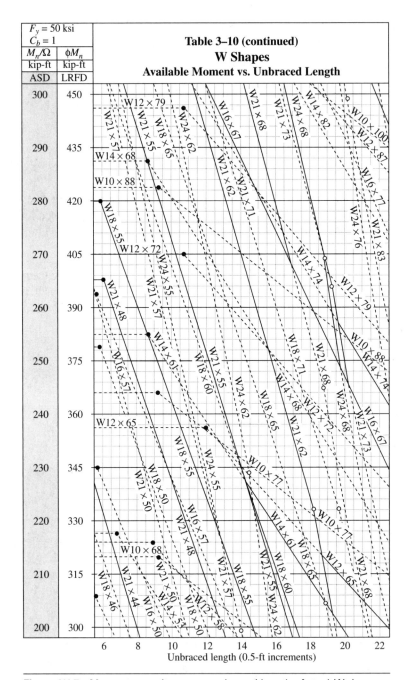

Figure 111.7 Moment capacity versus unbraced length of steel *W*-shapes.

Table 111.2 Compactness Limits on Flanges and Webs in Flexure

Component	λ	λ_p	λ_r
Flange	$\dfrac{b_f}{2t_f}$	$0.38\sqrt{\dfrac{E}{F_y}}$	$1.0\sqrt{\dfrac{E}{F_y}}$
Web	$\dfrac{h}{t_w}$	$3.76\sqrt{\dfrac{E}{F_y}}$	$5.70\sqrt{\dfrac{E}{F_y}}$

Flexural Strength of Noncompact Sections

If the section is noncompact because of the flange, the nominal strength is the smaller of the strengths corresponding to flange local buckling (FLB) and lateral-torsional buckling (LTB).

Flange Local Buckling (FLB)

If $\lambda < \lambda_p$, there is no FLB. If $\lambda_p < \lambda \le \lambda_r$, the nominal moment capacity for FLB is given by

$$M_n = M_p - (M_p - 0.7F_y S_x)\left(\frac{\lambda - \lambda_p}{\lambda_r - \lambda_p}\right) \qquad (111.20)$$

Lateral Torsional Buckling (LTB)

If $L_b \le L_p$, there is no LTB. If $L_p < L_b \le L_r$, the nominal moment capacity for LTB is given by

$$M_n = C_b\left[M_p - (M_p - 0.7F_y S_x)\left(\frac{L_b - L_p}{L_r - L_p}\right)\right] \le M_p \qquad (111.21)$$

$$\text{For } L_b > L_r \qquad M_n = F_{cr} S_x \le M_p \qquad (111.22)$$

where the critical stress is given by

$$F_{cr} = \frac{C_b \pi^2 E}{(L_b/r_{ts})^2}\sqrt{1 + 0.078\frac{Jc}{S_x h_o}\left(\frac{L_b}{r_{ts}}\right)^2} \qquad (111.23)$$

Design for Shear

The exact calculation for shear stress in a transversely loaded member is given by

$$f_v = \frac{VQ}{Ib} \tag{111.24}$$

where f_v is the transverse and longitudinal shear stress at a point in the beam, V is the transverse shear force, Q is the first moment (about the neutral axis) of either "half" of the cross section, I is the moment of inertia of the section about the neutral axis, and b is the width of the beam at the point of interest. For more details, see Fig. 101.7 in Chap. 101.

For wide-flange sections, the average shear stress calculated by distributing the shear force over the area of the web alone yields a good estimate of the actual maximum shear stress (at the neutral axis). This simplified equation [Eq. (111.25)] is adopted by the AISC design specifications. The effective area of the web resisting shear is calculated as $A_w = dt_w$, where $d =$ overall depth and $t_w =$ web thickness (see Fig. 111.8).

$$\text{Average shear stress } f_v = \frac{V}{A_w} \tag{111.25}$$

Taking the shear yield stress to be approximately 60% of the tensile yield stress, the nominal shear capacity of the section is given by

$$V_n = 0.6 F_y A_w = 0.6 F_y dt_w \tag{111.26}$$

To include beams with stiffened and unstiffened webs in a unified specification, the nominal shear capacity is written as

$$V_n = 0.6 F_y A_w C_v \tag{111.27}$$

where C_v is the ratio of the critical web stress to the shear yield stress. For ASD, the shear design requirement may be stated as

$$V_a \le \frac{V_n}{\Omega_v} \tag{111.28}$$

Figure 111.8
W-shape steel beam.

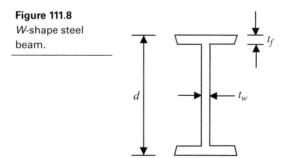

where V_a is the maximum shear calculated from the ASD load combinations, and Ω_v is the factor of safety for shear. For LRFD, the design requirement may be stated as

$$V_u \leq \phi_v V_n \tag{111.29}$$

where V_u is the required shear strength = maximum shear calculated from the LRFD load combinations and ϕ_v is the strength reduction factor for shear. The parameters Ω_v, ϕ_v, and C_v are functions of the slenderness of the web as summarized below.

Case 1—For Rolled Sections with

$$\frac{h}{t_w} \leq 2.24 \sqrt{\frac{E}{F_y}}$$

$$\Omega_v = 1.50, \; \phi_v = 1.0, \text{ and } C_v = 1.0$$

Case 2—For All Other Doubly and Singly Symmetric Sections, Except Round HSS

$$\Omega_v = 1.67, \; \phi_v = 0.90$$

Case 2a

$$\text{For } \frac{h}{t_w} \leq 2.46 \sqrt{\frac{E}{F_y}} \qquad C_v = 1.0$$

Case 2b

$$\text{For } 2.46 \sqrt{\frac{E}{F_y}} \leq \frac{h}{t_w} \leq 3.06 \sqrt{\frac{E}{F_y}} \qquad C_v = \frac{2.46\sqrt{E/F_y}}{h/t_w}$$

Case 2c

$$\text{For } \frac{h}{t_w} > 3.06 \sqrt{\frac{E}{F_y}} \qquad C_v = \frac{7.55E}{(h/t_w)^2 F_y}$$

Floor Framing Systems

A typical floor framing system (shown in Fig. 111.9) has a load path where the floor slab is supported by parallel floor beams which are arranged with constant spacing. The aspect ratio of the slab panels is relatively large, resulting in "one-way" behavior. The tributary width of the slab (to each beam) is therefore the spacing between parallel beams. The beams are supported by, and therefore transfer their loads to, girders. The columns and girders in one vertical plane form a frame. In addition to resisting the vertical loads transmitted

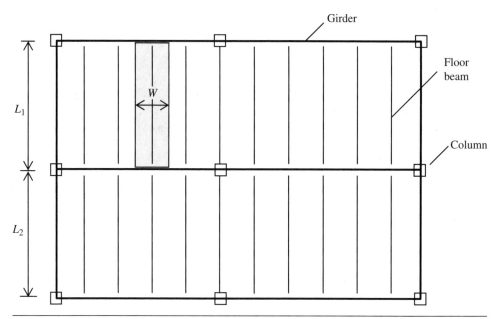

Figure 111.9 Floor system supported by floor beams, girders, and columns.

to them, the frame also acts to resist lateral loads by either acting as a braced frame or a moment-resisting frame.

The first structural component in the load path is the floor slab that is supported by the floor beams. Typically, the aspect ratio of these slab panels is large, since the span L_1 is much greater than the lateral spacing W. The flexural behavior of the slab is therefore predominantly one way, where the slab primarily bends in the short direction (i.e., between supporting beams). If the slab is continuous over three or more supports, it is reasonable to assume that the bending moment per unit width of the slab is given by

$$M_{slab} = \frac{qW^2}{10} \tag{111.30}$$

Assuming the floor beams are simply supported between girders and the uniformly distributed load on them is equal to the floor load (q) multiplied by tributary width (W), the floor beam's free-body diagram can be shown as in Fig. 111.10.

Figure 111.10
Floor beam
with uniformly
distributed load.

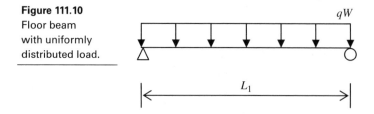

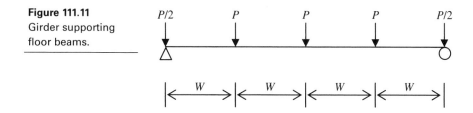

Figure 111.11
Girder supporting
floor beams.

The maximum bending moment in the beams is given by

$$M_{\text{beam}} = \frac{qWL_1^2}{8}$$

(111.31)

The end reactions of the beams are transmitted to the girders as equally spaced point loads as shown below. These loads are the end reactions from the beams. For exterior girders, reaction loads are transmitted from beams on one side only and the load P is given by

$$P = \frac{qWL_1}{2}$$

(111.32)

For interior girders, the load P is equal to

$$P - \frac{qW(L_1 + L_2)}{2}$$

(111.33)

A typical girder free-body diagram is shown in Fig. 111.11.

Composite Construction

Buildings and bridges frequently employ composite construction, wherein the floor or deck is "made composite" with the supporting beams. The term *composite construction* can be applied to any situation where two different materials are used so that they are forced to deform together such as in a reinforced concrete column where the concrete and the longitudinal steel share the compression load while maintaining consistent deformation due to bond. In this chapter, we discuss flexural members—particularly steel I-beams supporting a reinforced concrete deck. When such a system is used in buildings, the criteria from *AISC Steel Construction Manual* apply. When used in bridges, the criteria from *AASHTO LRFD Bridge Design Specifications* apply.

For concrete deck on steel beams, the composite action is achieved by using shear connectors (shear studs, channels, and spirals). The ultimate moment capacity of the member is based on plastic stress distribution with the ductile shear connector transferring shear between the steel beam and the concrete slab.

Advantages and Disadvantages of Composite Construction

Making a flexural member composite results in greater stiffness and leads to smaller deflections. For given values of desired flexural strength and maximum deflection, composite design will always lead to shallower beams. The only significant disadvantage of composite construction is the added cost of installing the shear connectors. Due to the "composite action" a certain width of the concrete slab acts together with the steel beam in forming the composite section. This "effective width" (on one side of the beam centerline) of the slab is given by

$$b' = \min\left(L/8, 1/2 \text{ distance to centerline of adjacent beam, distance to edge of slab}\right)$$

Thus, for the exterior beam in Fig. 111.12, b' is the smaller of one-eighth of the span and the distance to the free edge of the slab. The distance b'_1 is the smaller of one-eighth of the span and half the distance S_1. For the interior beam, the distance b'_2 is the smaller of one-eighth of the span and half the distance S_2 and the distance b'_3 is the smaller of one-eighth of the span and half the distance S_3.

For building systems with steel beams, AISC specifies the total effective width of the slab to be the least of (a) the center-to-center spacing between girders, (b) one-fourth of the girder span, or (c) the girder flange width plus 16 times the slab thickness.

For bridge systems with steel beams, AASHTO-LRFD specifies the total effective width of the slab to be the least of (a) the center-to-center spacing between girders, (b) one-fourth of the girder span, or (c) the girder flange width plus 12 times the slab thickness.

Equivalent Section

Once the effective width of the slab has been calculated, it is reduced to an equivalent width by using the modular ratio (n), which is the ratio of the modulus of elasticity of steel to the modulus of elasticity of concrete.

$$n = \frac{E_s}{E_c} \tag{111.34}$$

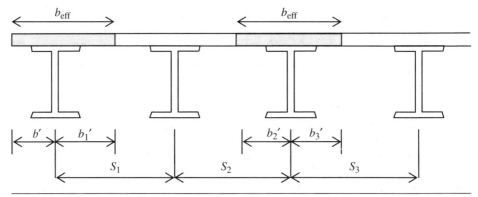

Figure 111.12 Effective width of slab for composite steel beam.

If the *base material* is chosen to be steel, then the concrete slab is converted to an *equivalent steel slab* by dividing the width by the modular ratio *n*.

Elastic Neutral Axis of Composite Section

The converted slab width is shown as b' in Fig. 111.13. The elastic neutral axis of this "steel" section can then be found as

$$\bar{y} = \frac{A_s \dfrac{d}{2} + b't\left(d + \dfrac{t}{2}\right)}{A_s + b't} = y_b \qquad (111.35)$$

The calculation shown in Eq. (111.35) ignores any haunch between the top flange of the beam and the bottom of the slab. Having established the location of the elastic neural axis, the moment of inertia of this equivalent section can be found as

$$I_{NA} = I_s + A_s\left(\bar{y} - \frac{d}{2}\right)^2 + \frac{1}{12}b't^3 + b't\left(d + \frac{t}{2} - \bar{y}\right)^2 \qquad (111.36)$$

where I_s = moment of inertia of the steel beam
A_s = cross-sectional area of the steel beam

The elastic bending stress at the top of the steel due to bending moment *M* is given by

$$\sigma_{ts} = \frac{My_t}{I_{NA}} \qquad (111.37)$$

The stress at the top of the concrete (modified by modular ratio *n*) is given by

$$\sigma_c = \frac{My_c}{n \times I_{NA}} \qquad (111.38)$$

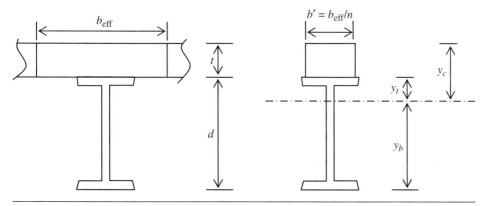

Figure 111.13 Equivalent (steel) section of a composite beam.

The bending stress at the bottom of the steel is given by

$$\sigma_{bs} = \frac{My_b}{I_{NA}}$$ (111.39)

Plastic Neutral Axis of Composite Section

When a composite beam has reached the plastic limit state, stresses will be distributed in one of two ways: either the PNA will be in the concrete slab (in which case, the entire steel is in tension), or the PNA will be in the steel (in which case, the entire slab is in compression). If the entire concrete slab is in compression, the resultant compression force is given by

$$C_c = 0.85 f_c' b_{eff} t$$ (111.40)

If the entire steel (area A_s) is in tension, the resultant tension force is given by

$$T_s = A_s F_y$$ (111.41)

If the shear studs were carrying their full capacity, that shear capacity would be the total shear strength of the shear connectors

$$V_q = \sum Q_n$$ (111.42)

For reasons of economy, shear studs can be chosen or spaced so that they are not providing their full capacity. Such a section is called a partially composite section. For a partially composite section, V_q will control. However, when an adequate number of shear connectors are used, the section can develop its "fully composite" flexural capacity. In a fully composite section, the smaller of T_s and C_c will control. To determine which case governs, calculate the compressive resultant as the *smallest* of

$A_s F_y$	CASE A	PNA in slab
$0.85 f_c' tb$	CASE B	PNA at top of steel
$\sum Q_n$ = total shear strength of shear connectors	CASE B	PNA below top of steel

Plastic Neutral Axis inside Concrete Deck

If $T_s < C_c$, as shown in Fig. 111.14, the implication is that the entire concrete slab does not need to be in compression to balance the tension in the steel shape. The plastic neutral axis is located in the slab. The depth of the slab in compression can be found by equating the tensile resultant force to the compressive resultant force.

$$a = \frac{A_s F_y}{0.85 f_c' b_{eff}} \le t$$ (111.43)

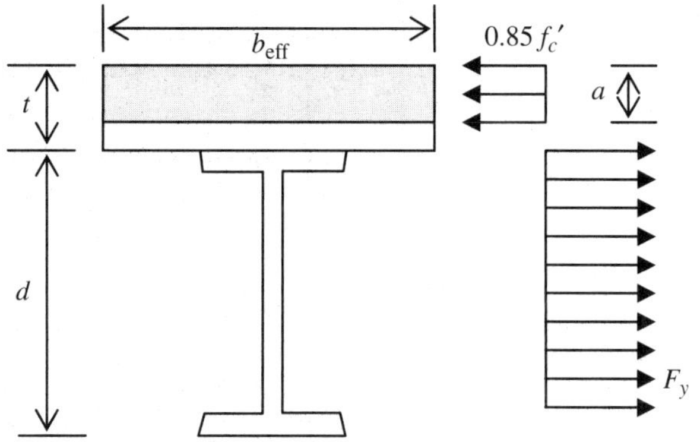

Figure 111.14
Plastic neutral axis
in concrete slab.

The nominal moment capacity of the composite section is given by

$$M_n = A_s F_y \left(\frac{d}{2} + t - \frac{a}{2} \right) \tag{111.44}$$

Plastic Neutral Axis at Top of Steel Shape or Below

If $C_c < T_s$, the implication is that the entire concrete slab does not provide sufficient compression to balance the tension in the steel shape. Some of the steel shape must be in compression to provide equilibrium. The plastic neutral axis is located in the steel shape. This case may be considered as two separate sub-cases: (a) PNA occurs in the top flange and (b) PNA occurs in the web. The force in the flange of the steel shape can be written as

$$P_f = F_y b_f t_f \tag{111.45}$$

and the force in the web can be written as

$$P_w = F_y (A_s - 2b_f t_f) \tag{111.46}$$

If we designate the area of the steel shape in compression as A_{sc}, then we can write

$$0.85 f_c' b_{eff} t + A_{sc} F_y = (A_s - A_{sc}) F_y, \text{ which leads to}$$

$$A_{sc} = \frac{A_s F_y - 0.85 f_c' b_{eff} t}{2F_y} \tag{111.47}$$

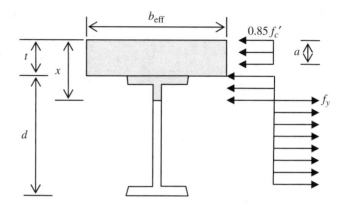

Figure 111.15
Plastic neutral axis in
steel section.

For the case where the PNA is in the top flange $(A_{sc} < b_f t_f)$, the depth of the PNA from the top of the flange is given by (see Fig. 111.15)

$$x = \frac{A_{sc}}{b_f}$$

For the case where the PNA is in the web, $A_{sc} > b_f t_f$, the depth of the PNA from the top of the flange is given by

$$x = t_f + \frac{A_{sc} - b_f t_f}{t_w}$$

Table 3.19 in the AISC *Steel Construction Manual* can be used as a design aid for composite beams.

Use of Composite Beam Design Tables in AISC *Steel Construction Manual*

A composite section subjected to positive moment is shown in Fig. 111.16.

Y_{con} = thickness of the concrete slab

b = effective width of the concrete slab

d = depth of the steel beam

a = depth of the equivalent (Whitney) compression block

The PNA can be located either (a) in the slab (as shown on the left) or (b) in the steel section.

Calculating the plastic moment capacity for the case where the PNA is located in the concrete slab is fairly simple [Eq. (111.44)]. However, the case of PNA located within the steel

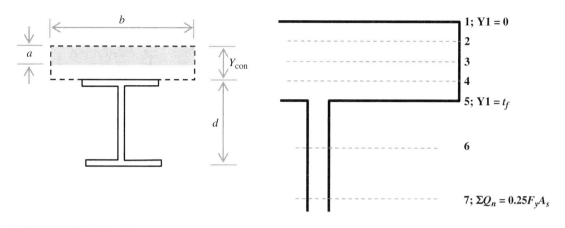

Figure 111.16 Locations of plastic neutral axis for composite beam.

section is more complicated. The AISC SCM (Table 3-19) lists composite section moment capacity for seven different locations of the PNA. These are shown in Fig. 111.16. Five of these locations are in the top flange of the steel section (location 1 at the top of flange, location 5 at the flange-web interface, and locations 2, 3, and 4 located at ¼ flange thickness intervals). Location 7 is located where $\Sigma Q_n = 0.25\, F_y A_s$ and location 6 is midway between 5 and 7.

A typical listing of AISC Table 3-19 is shown below. Only LRFD values are shown. The integrated editions of the SCM (13th and 14th editions) list ASD and LRFD values side by side. The values shown are for $F_y = 50$ ksi.

Shape	$\phi_b M_p$ (k-ft)	PNA	Y1 (in)	ΣQ_n (kip)	Y2 (in) 2	2.5	3	3.5	4	4.5	5	5.5	6	6.5	7
W16x26	166	TFL	0	384	284	298	312	327	341	356	370	385	399	413	428
		2	0.0863	337	276	289	302	314	327	340	352	365	377	390	403
		3	0.173	289	269	280	291	301	312	323	334	345	356	366	377
		4	0.259	242	261	270	279	288	297	306	315	324	334	343	352
		BFL	0.345	194	253	260	267	275	282	289	296	304	311	318	326
		6	2.04	145	241	247	252	258	263	269	274	279	285	290	296
		7	4.00	96	223	227	230	234	237	241	245	248	252	255	259

Y1 is the distance from the top of the compression flange to the PNA (measured down). Thus, when the PNA is at position 1 (marked in table as TFL), Y1 = 0, when PNA is at position 5 (marked in table as BFL), PNA = flange thickness and so on.

Y2 is the distance from the top of the compression flange to the concrete resultant force (measured up).

Example 111.3 illustrates the use of these tables to calculate the design strength of a composite section.

Example 111.3

A floor system is supported by a series of 40-ft simple span W-shape beams, acting compositely with a 4-in-thick concrete slab ($f_c' = 5000$ psi). Beams are spaced at 8 ft on center. Assume a superimposed dead load = 20 psf and live load = 100 psf. $F_y = 50$ ksi. (a) Choose a satisfactory beam size, and (b) calculate the deflection due to unfactored live load.

Solution

Self-weight of the slab tributary to each beam = $0.15 \times 4/12 \times 8 = 0.4$ k/ft
Superimposed dead load on each beam = 20 psf $\times$ 8 ft = 160 plf = 0.16 k/ft
Live load on each beam = 100 psf $\times$ 8 ft = 800 plf = 0.8 k/ft
Assuming a beam self-weight of 80 lb/ft, the total factored load:

$$w_u = 1.2 \times (0.4 + 0.16 + 0.08) + 1.6 \times 0.8 = 2.05 \text{ k/ft}$$

Factored moment (required strength), $M_u = \dfrac{w_u L^2}{8} = \dfrac{2.05 \times 40^2}{8} = 410$ k-ft

Assume $a = 1$ in

Calculate a preliminary value of Y2: $\text{Y2} = Y_{con} - \dfrac{a}{2} = 4 - \dfrac{1}{2} = 3.5$ in

Assuming PNA location at line 4, and with Y2 = 3.5 in, choose (Table 3-19) W16 $\times$ 40 with a capacity of 435 k-ft.
For W16 $\times$ 40 with Y2 = 3.5 in, the composite member has sufficient strength when PNA is at location 4.
Available strength in flexure: $\phi_b M_n = 435$ k-ft and horizontal shear force $\Sigma Q_n = 324$ k
It should also be noted that the plastic moment capacity of the section (W16 $\times$ 40) alone is 274 kip-ft and this must be sufficient to withstand any moments generated on the noncomposite section, during unshored construction.

Calculate depth of equivalent compression block: $a = \dfrac{\Sigma Q_n}{0.85 f_c' b_{eff}} = \dfrac{324}{0.85 \times 5 \times 96} = 0.79$ in
< 1.0 in assumed. OK

Recalculate Y2: $\text{Y2} = Y_{con} - \dfrac{a}{2} = 4 - \dfrac{0.79}{2} = 3.6$ in

Using recalculated Y2, find revised value of available flexural strength ϕM_n (kip-ft) = 437 k-ft > 410 k-ft (actually, the self-weight of the steel beam was assumed to be 80 lb/ft, but is now 40 lb/ft. (This will reduce the required strength slightly less than 410 k-ft).

Table 3-20 has the lower bound moment of inertia (I_{LB}) for composite beams. For the W16 $\times$ 40, with Y2 = 3.6 in and PNA at location 4, $I_{LB} = 1080$ in^4
Live load ($w_{LL} = 0.8$ k/ft = 0.067 k/in) deflection:

$$\Delta_{LL} = \frac{5wL^4}{384EI} = \frac{5 \times 0.067 \times 480^4}{384 \times 29000 \times 1080} = 1.5 \text{ in} = \frac{L}{321}$$

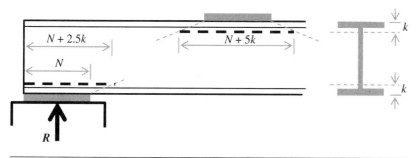

Figure 111.17 Critical sections for web yielding.

Beam Bearing Plate Design

Beam bearing plates are used at locations of large concentrated forces to (1) prevent yielding and crippling of the web of the beam, (2) provide a large enough bearing area to prevent crushing of the supporting material, and (3) have adequate plate thickness for bending strength.

Figure 111.17 shows a schematic of a steel beam transferring the support reaction R to the supporting concrete element via a plate with bearing length N.

The design of beam bearing plates can proceed as follows:

1. Determine bearing dimension N (see Fig. 111.17) such that web yielding and web crippling do not occur.

2. Once N has been chosen, determine dimension B (perpendicular to N) such that the bearing area NB is sufficient to prevent crushing of supporting material (usually concrete).

3. Determine plate thickness so that it has sufficient bending strength.

Web Yielding

Web yielding is calculated at a critical section which is at the base of the web fillet, distance k from the outer edge of the flange (see Fig. 111.17). The dashed lines (sloped at 1V:2.5H) propagate from the edge of the bearing plate, thereby creating a critical section length of $N + 2.5k$ adjacent to the support and $N + 5k$ for an interior load.

The nominal strength for web yielding at the support is

$$R_n = \left(2.5k + N\right)F_y t_w \tag{111.48}$$

For an interior load, the nominal strength for web yielding is

$$R_n = \left(5k + N\right)F_y t_w \tag{111.49}$$

The design strength is ϕR_n ($\phi = 1.0$).

Web Crippling

Web crippling is the buckling of the web caused by the compressive force delivered via the flange.

For an interior load, the nominal strength for web crippling is

$$R_n = 0.8t_w^2 \left[1 + 3 \left(\frac{N}{d} \right) \left(\frac{t_w}{t_f} \right)^{1.5} \right] \sqrt{\frac{EF_y t_f}{t_w}} \qquad (111.50)$$

For a load at or near the support (no greater than half the beam depth from the end), the nominal strength is

$$R_n = 0.4t_w^2 \left[1 + 3 \left(\frac{N}{d} \right) \left(\frac{t_w}{t_f} \right)^{1.5} \right] \sqrt{\frac{EF_y t_f}{t_w}} \qquad \text{for } N/d \leq 0.2 \qquad (111.51a)$$

$$R_n = 0.4t_w^2 \left[1 + \left(\frac{4N}{d} - 0.2 \right) \left(\frac{t_w}{t_f} \right)^{1.5} \right] \sqrt{\frac{EF_y t_f}{t_w}} \qquad \text{for } N/d > 0.2 \qquad (111.51b)$$

The design strength is ϕR_n ($\phi = 0.75$).

Concrete Bearing Strength

The purpose of the bearing plate is to distribute the large reaction force over a large bearing area such that the underlying material (usually concrete) does not experience crushing. According to ACI 318, "the design bearing strength of concrete shall not exceed $\phi(0.85 f_c' A_1)$, except where the supporting surface is wider on all sides than the loaded area, then the design bearing strength of the loaded area shall be permitted to be multiplied by $\sqrt{(A_2/A_1)}$ but not more than 2." The supporting area A_2 must be concentric with A_1 (i.e., the extension must be equal on all sides). In Fig. 111.18, the edge distance from the bearing plate to the edge of the pier is the smallest. Therefore, the supporting area A_2 (dashed line) is constructed by extending the perimeter by this distance on all four sides.

Plate Thickness

The required plate thickness (for adequate flexural capacity) is

$$t \geq \sqrt{\frac{2.222 R_u n^2}{BNF_y}} \qquad (111.52)$$

where $n = B/2 - k$

Figure 111.18
Effective bearing area
on concrete.

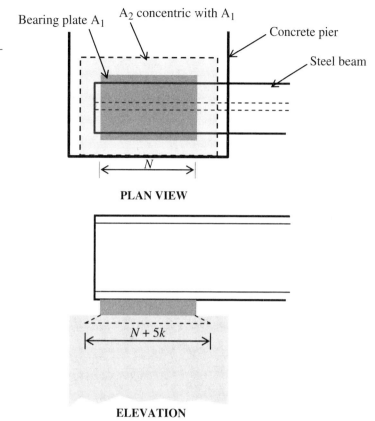

Bearing plate A_1 A_2 concentric with A_1

Concrete pier

Steel beam

N

PLAN VIEW

$N + 5k$

ELEVATION

Example 111.4

A 21 × 68 beam has a simple span $L = 16$ ft. It carries the following uniformly distributed
loads: $w_D = 5$ kip/ft, $w_L = 5$ kip/ft. The beam is supported by reinforced concrete walls (f_c'
= 4,000 psi). The beam is grade 50 steel. The bearing plate is to be A36 steel. What are the
required dimensions (length, width, thickness) of the bearing plate?

Solution

Factored uniform load: $w_u = 1.2w_D + 1.6w_L = 1.2 \times 5 + 1.6 \times 5 = 14$ k/ft
Factored reaction: $R_u = 1/2\ w_u L = 112$ kips
Check web yielding ($\phi = 1.0$) (AISC J10-3) for W21 × 68 ($d = 21.1$ in; $k = 1.19$ in;
$t_w = 0.43$ in)

Factored strength: $\phi(2.5k + N)\ F_y t_w \geq 112$
$$1.0(2.5 \times 1.19 + N)50 \times 0.43 \geq 112 \Rightarrow N \geq 2.23 \text{ in}$$

For web crippling ($\phi = 0.75$), assuming $N/d > 0.2$ (AISC J10-5B)

$$\phi \times 0.4 \times t_w^2 \left[1 + \left(4\frac{N}{d} - 0.2\right)\left(\frac{t_w}{t_f}\right)^{1.5}\right] \sqrt{\frac{EF_y t_f}{t_w}} \geq 112 \Rightarrow N \geq 4.54 \text{ in}$$

Therefore, web crippling governs. Assume bearing plate length, $N = 6$ in

Required bearing area: $\phi_c 0.85 f_c' A_1 \geq R_u \Rightarrow A_1 \geq \dfrac{112}{0.6 \times 0.85 \times 4} = 54.9$ in^2

Use 6 in × 10 in bearing plate. $n = \dfrac{B - 2k}{2} = \dfrac{10 - 2 \times 1.19}{2} = 3.81$ in

Minimum plate thickness: $t_{\min} = \sqrt{\dfrac{2.22 R_u n^2}{BN F_y}} = \sqrt{\dfrac{2.22 \times 112 \times 3.81^2}{10 \times 6 \times 36}} = 1.29$ in

Use 10 × 6 × 1½ in base plate.

Design of Built-Up Beams (Plate Girders)

According to the AISC *Steel Construction Manual*, 14th ed. (2011), a flexural member is classified as a simple or a built-up member depending on the slenderness of the web, measured by the slenderness ratio h/t_w

h = depth of the web (measured between corner fillets)

t_w = thickness of the web

F_y = yield stress (of the web)

If $\dfrac{h}{t_w} \geq 5.7 \sqrt{\dfrac{E}{F_y}}$, design the beam as a built-up section in accordance with AISC chapter G.

Otherwise, design the beam as a simple beam in accordance with AISC chapter F.

For unstiffened girders (i.e., without transverse web stiffeners), the web slenderness ratio h/t_w is limited to 260. Also, a_w which is the ratio of the areas of the web ($A_w = h t_w$) to that of the compression flange ($A_{fc} = b_{fc} t_{fc}$) must be less than or equal to 10.

Figure 111.19
Web "panels" formed by intermediate transverse stiffeners.

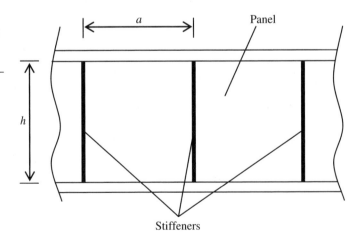

For stiffened girders, the web slenderness has the following limits:

$$\frac{h}{t_w} \leq 11.7\sqrt{\frac{E}{F_y}} \qquad \text{for} \qquad \frac{a}{h} \leq 1.5 \qquad\qquad (111.52a)$$

$$\frac{h}{t_w} \leq 0.42\frac{E}{F_y} \qquad \text{for} \qquad \frac{a}{h} > 1.5 \qquad\qquad (111.52b)$$

where a = clear distance between transverse stiffeners (see Fig. 111.19).

In unstiffened girders, the ratio h/t_w is limited to 260. Also, the ratio of the web area to the compression flange area should not exceed 10.

For a transversely loaded member, the flexural strength (bending moment) and the shear strength (shear force) must be checked for adequacy against the ultimate bending moment and shear force created by the loading.

Flexural Strength of Built-Up Beams

The nominal moment strength M_n shall be the lowest of the following four limit states:

1. Compression Flange Yielding

2. Lateral-Torsional Buckling

3. Compression Flange Local Buckling

4. Tension Flange Yielding

Compression Flange Yielding ($\phi = 0.9$)

For the limit state of compression flange yielding, the nominal moment capacity is given by Eq. (111.53)

$$M_n = R_{pg}F_y S_{xc} \qquad\qquad (111.53)$$

The bending strength reduction factor R_{pg} is given by:

$$R_{pg} = 1 - \frac{a_w}{1200+300a_w}\left(\frac{h_c}{t_w} - 5.7\sqrt{\frac{E}{F_y}}\right) \leq 1.0$$

where R_{pg} = bending strength reduction factor
S_{xc} = elastic section modulus based on compression fiber = I_x/y_c
$a_w = h_c t_w/(b_{fc}t_{fc})$ = ratio of twice the web area in compression to the area of the compression flange elements (a_w is not to exceed 10).

Lateral Torsional Buckling ($\phi = 0.9$)

For the limit state of lateral-torsional buckling, the nominal moment capacity is given by:

$$M_n = R_{pg}F_{cr}S_{xc}$$

The critical stress F_{cr} is determined by the relation of the unbraced length of the compression flange (L_b) to the critical lengths L_p and L_r given by Eqs. (111.54) and (111.55), respectively.

$$L_p = 1.1r_t\sqrt{\frac{E}{F_y}} \tag{111.54}$$

$$L_r = \pi r_t\sqrt{\frac{E}{0.7F_y}} = 3.414L_p \tag{111.55}$$

where $\quad r_t = \dfrac{b_{fc}}{\sqrt{12\left(1+\dfrac{a_w}{6}\right)}}$

When $L_b \leq L_p$, the limit state of lateral-torsional buckling does not apply.

When $L_p \leq L_b \leq L_r \quad$ $F_{cr} = C_b\left[F_y - 0.3F_y\left(\dfrac{L_b - L_p}{L_r - L_p}\right)\right] \leq F_y \tag{111.56}$

When $L_b > L_r \quad$ $F_{cr} = \dfrac{C_b\pi^2 E}{\left(\dfrac{L_b}{r_t}\right)^2} \leq F_y \tag{111.57}$

Compression Flange Local Buckling ($\phi = 0.9$)
For the limit state of compression flange local buckling, the nominal moment capacity is given by Eq. (111.58)

$$M_n = R_{pg}F_{cr}S_{xc} \tag{111.58}$$

For sections with compact flanges ($\lambda < 1_{pf}$), the limit state of compression flange buckling does not apply and $F_{cr} = F_y$

$\lambda = b_{fc}/2t_{fc}$ = slenderness ratio of the "half compression flange"

1_{pf} = limiting slenderness for a compact flange

$$\lambda_{pf} = 0.38\sqrt{\frac{E}{F_y}}$$

1_{rf} = limiting slenderness for a noncompact flange

$$\lambda_{rf} = 0.95\sqrt{\frac{k_c E}{F_L}}$$

where $k_c = \dfrac{4}{\sqrt{h/t_w}}$ $(0.35 \le k_c \le 0.76)$

The stress F_L is determined according to Eqs. (111.59a) and (111.59b).

For $S_{xt}/S_{xc} \ge 0.7$ $\qquad\qquad\qquad F_L = 0.7F_y$ $\hfill$ (111.59a)

For $S_{xt}/S_{xc} < 0.7$ $\qquad\qquad\qquad F_L = F_y \dfrac{S_{xt}}{S_{xc}} \ge 0.5F_y$ $\hfill$ (111.59b)

For section with noncompact flanges $(1_{pf} < \lambda < 1_{rf})$, the critical stress F_{cr} is given by

$$F_{cr} = F_y - 0.3F_y \left(\frac{\lambda - \lambda_{pf}}{\lambda_{rf} - \lambda_{pf}} \right) \tag{111.60}$$

For section with slender flanges $(\lambda > 1_{rf})$, the critical stress F_{cr} is given by

$$F_{cr} = \frac{0.9Ek_c}{\left(b_f / 2t_f \right)^2} \tag{111.61}$$

Tension Flange Yielding ($\phi = 0.9$)
For the limit state of tension flange yielding, the nominal moment capacity is given by

$$M_n = F_y S_{xt} \tag{111.62}$$

When $S_{xt} > S_{xc}$, the limit state of tension flange yielding does not apply.

Shear Strength of Built-Up Beams

The shear strength of a built-up member is affected by the presence (or absence) of stiffeners. One of the primary purposes of having transverse stiffeners is to divide the girder web into smaller panels so that the tension field action can increase the shear strength of the beam. If the stiffeners are absent or are too far apart, there is no tension field action. Otherwise, the shear capacity of the beam is equal to the *shear strength before buckling* plus the *postbuckling strength due to tension field action*.

The design shear strength of stiffened and unstiffened webs (based on the limit states of shear yielding and shear buckling) of built-up sections is:

$$\phi_v V_n = 0.9 \times 0.6 F_y A_w C_v \tag{111.63}$$

Web Plate Buckling Coefficient (k_v)

For unstiffened webs with $h/t_w < 260$, the web plate buckling coefficient $k_v = 1.0$ except for stems of tee shapes for which $k_v = 1.2$.

For stiffened webs, k_v is given by

$$k_v = 5 + \frac{5}{(a/h)^2}$$

The coefficient k_v can be taken as 5 when $a/h > 3.0$ or $\dfrac{a}{h} > \left(\dfrac{260}{h/t_w}\right)^2$.

Web Shear Coefficient (C_v)

The web shear coefficient C_v represents the ratio between the critical shear stress in the web to the yield stress in shear (τ_y).

For $\qquad \dfrac{h}{t_w} \le 1.10 \sqrt{\dfrac{k_v E}{F_{yw}}}$, the web shear coefficient $C_v = 1.0$ $\qquad$ (111.64a)

For $\qquad 1.1 \sqrt{\dfrac{k_v E}{F_{yw}}} \le \dfrac{h}{t_w} \le 1.37 \sqrt{\dfrac{k_v E}{F_{yw}}}$, the coefficient $C_v = \dfrac{1.1\sqrt{\dfrac{k_v E}{F_{yw}}}}{\dfrac{h}{t_w}}$ $\qquad$ (111.64b)

For $\qquad \dfrac{h}{t_w} > 1.37 \sqrt{\dfrac{k_v E}{F_{yw}}}$, the coefficient $C_v = \dfrac{1.51 E k_v}{\left(\dfrac{h}{t_w}\right)^2 F_{yw}}$ $\qquad$ (111.64c)

Tension Field Action

If the web plate is supported on all four sides by flanges or stiffeners, inclusion of tension field action is permitted. Consideration of tension field action is NOT permitted for

a. End panels in all members with transverse stiffeners

b. Members with $a/h > 3.0$ or $\dfrac{a}{h} > \left(\dfrac{260}{h/t_w}\right)^2$

c. $\dfrac{2A_w}{A_{fc} + A_{ft}} > 2.5$

d. $\dfrac{h}{b_{fc}} > 6$ or $\dfrac{h}{b_{ft}} > 2.5$

If tension field action is permitted:

If $\dfrac{h}{t_w} \leq 1.1\sqrt{\dfrac{k_v E}{F_y}}$, nominal shear capacity $V_n = 0.6A_w F_{yw}$ (111.65a)

If $\dfrac{h}{t_w} > 1.1\sqrt{\dfrac{k_v E}{F_y}}$, nominal shear capacity $V_n = 0.6A_w F_{yw}\left[C_v + \dfrac{1 - C_v}{1.15\sqrt{1 + (a/h)^2}}\right]$ (111.65b)

In the previous equation, the first term $0.6A_w F_{yw} C_v$ represents the web-shear buckling strength whereas the second term represents the shear strength due to tension field action.

Example 111.5

For a plate girder, A572 Grade 50 steel is used. The web is ½ in × 70 in and flanges are 3 in × 22 in.

1. Determine the design shear strength of the end panel if the first intermediate stiffener is located 60 in from the support.

2. Compute the design shear strength of an interior panel with a stiffener spacing of 180 in.

3. What is the design shear strength if no intermediate stiffeners are used?

Slenderness ratio of the web: $\dfrac{h}{t_w} = \dfrac{70}{0.5} = 140.$

1. End Panel

Panel length, $a = 60$ in $a/h = 60/70 = 0.85 < 3.$

Since $\dfrac{a}{h} < \left(\dfrac{260}{h/t_w}\right)^2 \Rightarrow k_v = 5 + \dfrac{5}{(a/h)^2} = 5 + \dfrac{5}{0.86^2} = 11.8$

$1.10\sqrt{\dfrac{k_v E}{F_y}} = 1.10\sqrt{\dfrac{11.8 \times 29,000}{50}} = 91$

$1.37\sqrt{\dfrac{k_v E}{F_y}} = 1.37\sqrt{\dfrac{11.8 \times 29,000}{50}} = 113.3$

Since $\dfrac{h}{t_w} > 1.37\sqrt{\dfrac{k_v E}{F_y}}$, the coefficient $C_v = \dfrac{1.51 E k_v}{\left(\dfrac{h}{t_w}\right)^2 F_y} = \dfrac{1.51 \times 29,000 \times 11.8}{140^2 \times 50} = 0.527$

For end panels, design shear strength:

$$\phi_v V_n = 0.9 \times 0.6 \times 50 \times 70 \times 0.5 \times 0.527 = 498 \text{ k}$$

2. Interior Panel

Panel length, $a = 180$ in $\qquad a/h = 180/70 = 2.57 < 3$.

Since $\dfrac{a}{h} < \left(\dfrac{260}{h/t_w}\right)^2 \Rightarrow k_v = 5 + \dfrac{5}{(a/h)^2} = 5 + \dfrac{5}{2.57^2} = 5.76$

$$1.10\sqrt{\dfrac{k_v E}{F_y}} = 1.10\sqrt{\dfrac{5.76 \times 29{,}000}{50}} = 63.6$$

$$1.37\sqrt{\dfrac{k_v E}{F_y}} = 1.37\sqrt{\dfrac{5.76 \times 29{,}000}{50}} = 79.2$$

Since $\dfrac{h}{t_w} > 1.37\sqrt{\dfrac{k_v E}{F_y}}$, the coefficient $C_v = \dfrac{1.51 E k_v}{\left(\dfrac{h}{t_w}\right)^2 F_y} = \dfrac{1.51 \times 29{,}000 \times 5.76}{140^2 \times 50} = 0.257$

For interior panels, including the contribution of tension field action, nominal shear capacity:

$$V_n = 0.6 A_w F_{yw}\left[C_v + \dfrac{1 - C_v}{1.15\sqrt{1 + (a/h)^2}}\right]$$

$$= 0.6 \times 70 \times 0.5 \times 50\left[0.257 + \dfrac{1 - 0.257}{1.15\sqrt{1 + 2.57^2}}\right] = 515.8 \text{ k}$$

Design shear capacity: $\phi V_n = 0.9 \times 515.8 = 464.3 \text{ k}$

3. No Intermediate Stiffeners Used

Assume $a/h > 3$.

$k_v = 5$.

$$234\sqrt{\dfrac{k_v}{F_{yw}}} = 234\sqrt{\dfrac{5}{50}} = 74.0 < \dfrac{h}{t_w} \Rightarrow C_w = \dfrac{44{,}000 k_v}{(h/t_w)^2 F_{yw}} = \dfrac{44{,}000 \times 5}{140^2 \times 50} = 0.224$$

$$V_n = 0.6 A_w F_{yw} C_v = 0.6 \times \left(70 \times \dfrac{1}{2}\right) \times 50 \times 0.224$$

$$= 235.7 \text{ k} \Rightarrow \phi V_n = 0.9 \times 235.7 = 212.1 \text{ k}$$

Bolted and Welded Connections

General

Most failures in steel structures occur due to poorly designed and detailed connections. Member failures are relatively rare. Modern steel structures are connected using welds or bolts. Bolting with high-strength bolts has largely replaced riveting. Welded connections have several advantages over bolted connections. Welded connections require few, if any, holes. Aesthetically, welded connections appear simpler and "cleaner." However, welding requires skilled operators and inspection can be difficult and costly.

Steel bolts are available in several grades:

- A307 bolts are unfinished or ordinary or common bolts. They are available in sizes from 5/8 to 1 1/2 in diameter in 1/8 in increments.

- A325 bolts are high-strength bolts with material properties very similar to A36 steel. In the 14th edition of the *Steel Construction Manual*, these bolts have been grouped with some other bolt designations as "Group A" bolts.

- A490 bolts are high-strength bolts made from an alloy steel. High-strength bolts are available up to a maximum diameter of 1 1/2 in. In the 14th edition of the *Steel Construction Manual*, these bolts have been grouped with some other bolt designations as "Group B" bolts.

Snug-Tight versus Slip-Critical Connections

Bolted connections are either snug-tight or slip-critical. In a snug-tight connection, the bolt is not tensioned. The fitted parts fit snugly but not with enough normal stress at the interface to activate added frictional resistance. In such a connection, the tension or compression in the connected parts leads to development of bearing stress on the shank of the bolt. This is why such connections are called bearing type connections. On the other hand, slip-critical connections use tensioned bolts. A307 bolts are not used in slip-critical connections. The bolt tension

Table 112.1 Minimum Fastener Tension (kips)

	Diameter (in)								
	1/2	5/8	3/4	7/8	1	1 1/8	1 1/4	1 3/8	1 1/2
A325	12	19	28	39	51	56	71	85	103
A490	15	24	35	49	64	80	102	121	148

Table 112.2 Allowable Stress in Fasteners

Fastener grade	Nominal tensile stress F_{nt} (ksi)	Nominal shear stress in bearing type connections F_{nv} (ksi)
A307	45	24
A325 or A325M (where threads are included in the shear plane)	90	48
A325 or A325M (where threads are excluded from the shear plane)	90	60
A490 or A490M (where threads are included in the shear plane)	113	60
A490 or A490M (where threads are excluded from the shear plane)	113	75

causes the connected parts to develop a frictional resistance at the interface. Table 112.1 shows minimum specified tensions for high-strength bolts in slip-critical connections. These tensions correspond to developing at least 70% of the tensile strength of the bolt.

According to *Specification for Structural Joints Using ASTM A325 or A490 Bolts,* Research Council on Structural Connections, Chicago, 2004, the following four methods may be used for bolt tensioning:

1. Turn-of-the-nut method

2. Calibrated wrench tightening

3. Twist-off type bolts

4. Direct tension indicators

Table 112.2 shows nominal stresses in fasteners and threaded parts.

Bearing Type Connections

Besides exceeding allowable shear stress in bolts, bolted connections can also fail due to inadequate distance between bolt centers or inadequate edge distance. The minimum center-to-center spacing between bolts in the direction of the load is 2.67 D and minimum edge distance should be between 1.5D and 2D.

Bearing Strength of Bolts

For bearing type connections using at least 2 bolts per line (in the direction of the load), if the edge distance is greater than or equal to 1.5 times the bolt diameter and distance between bolt centers is greater than or equal to three times the bolt diameter, the nominal bearing strength of a single bolt is given by

$$R_n = 1.2L_c tF_u \leq 2.4dtF_u \qquad (112.1)$$

where L_c = clear distance from the edge of the bolt hole to the edge of the adjacent bolt or the edge of the material (see Fig. 112.1)

t = thickness of the connected part

d = diameter of the bolt

F_u = ultimate tensile stress of the connected part (*not bolt*)

For ASD, the allowable strength is

$$R_a = \frac{R_n}{\Omega} = \frac{R_n}{2.0} \qquad (112.2)$$

For LRFD, the design strength is

$$\phi R_n = 0.75R_n \qquad (112.3)$$

In the discussion above, the limit stress in bearing has been taken as $F_p = 1.2F_u$ (for short-slotted or standard holes). If long-slotted holes are used, the limit stress is to be taken as $F_p = 1.0F_u$. If edge distance L_e is less than 1.5D or for a single bolt, the bearing strength is given by

$$F_p = \frac{L_e F_u}{2D} \qquad (112.4)$$

Figure 112.1
Clear distance
from edge of bolt
hole.

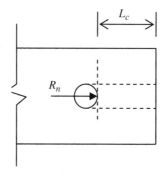

Shear Strength of Bolts

The nominal shear strength of a bolt is given by

$$R_n = F_{nv} A_b \tag{112.5}$$

where F_{nv} = nominal shear stress, given in Table 112.2
A_b = nominal bolt area (unthreaded part)

For ASD, the allowable strength is

$$R_a = \frac{F_{nv} A_b}{\Omega} = \frac{F_{nv} A_b}{2.0} \tag{112.6}$$

For LRFD, the design strength is

$$\phi R_n = 0.75 R_n = 0.75 F_{nv} A_b \tag{112.7}$$

Example 112.1

A 3/8-in-thick × 3-in plate is a tension member and carries concentric axial loads. The plate is connected to a 3/8-in-thick gusset plate using two 3/4-in-diameter bolts, as shown. Check the adequacy of the connection if the following types of bolts are used: (a) A307 normal bolts, (b) A325 high-strength bolts with threads in the plane of shear, and (c) A325 high-strength bolts with threads not in the plane of shear. Assume plate is A36 steel ($F_y = 36$ ksi, $F_u = 58$ ksi).

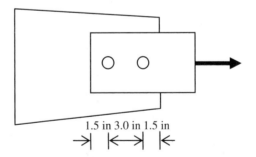

1.5 in   3.0 in   1.5 in

Solution   For bearing strength, AISC recommends using a hole diameter = bolt diameter + 1/16 in (rather than 1/8 in as used for member-strength calculation)

$$h = d + 1/16 = 0.75 + 0.0625 = 0.8125 \text{ in}$$

Clear edge distance for the hole nearest the edge

$$L_c = 1.5 - \frac{0.8125}{2} = 1.094$$

Nominal resistance, $R_n = 1.2L_c tF_u = 1.2 \times 1.094 \times 0.375 \times 58 = 28.6$ kips subject to the upper limit $2.4dtF_u = 2.4 \times 0.75 \times 0.375 \times 58 = 39.2$ kips. *Use* 28.6 *kips.* For the inner hole, the distance L_c is to be taken as $L_c = s - h = 3 - 0.8125 = 2.1875$. Nominal resistance $R_n = 1.2L_c tF_u = 1.2 \times 2.1875 \times 0.375 \times 58 = 57.1$ kips , subject to the upper limit of 39.2 kips. Use 39.2 kips. Therefore, for the two-bolt connection, the nominal strength is $R_n = 28.6 + 39.2 = 67.8$ kips.

- For A307 bolts, the nominal shear stress = 24 ksi. Therefore, the single shear capacity of each bolt ($A_b = 0.44$ in²) is $24 \times 0.44 = 10.6$ kips. The shear strength of two bolts is therefore 21.2 kips. This governs, since it is less than the bearing strength of 67.8 kips. Therefore, for A307 bolts, the nominal strength is $R_n = 21.2$ kips.

  ASD: Allowable strength $R_n/\Omega = 21.2/2.0 = 10.6$ kips

  LRFD: Design strength $\phi R_n = 0.75 \times 21.2 = 15.9$ kips

- For A325 bolts (with threads in the plane of shear), the nominal shear stress = 48 ksi. Therefore, the single shear capacity of each bolt ($A_b = 0.44$ in²) is $48 \times 0.44 = 21.2$ kips. The shear strength of two bolts is therefore 42.4 kips. This governs, since it is less than the bearing strength of 67.8 kips. Therefore, for A325-N bolts, the nominal strength is $R_n = 42.4$ kips.

  ASD: Allowable strength $R_n/\Omega = 42.4/2.0 = 21.2$ kips

  LRFD: Design strength $\phi R_n = 0.75 \times 42.4 = 31.8$ kips

- For A325 bolts (with threads excluded from the plane of shear), the nominal shear stress = 60 ksi. Therefore, the single shear capacity of each bolt ($A_b = 0.44$ in²) is $60 \times 0.44 = 26.4$ kips. The shear strength of two bolts is therefore 52.8 kips. This governs, since it is less than the bearing strength of 67.8 kips. Therefore, for A307-X bolts, the nominal strength is $R_n = 52.8$ kips.

  ASD: Allowable strength $R_n/\Omega = 52.8/2.0 = 26.4$ kips

  LRFD: Design strength $\phi R_n = 0.75 \times 52.8 = 39.6$ kips

Example 112.2

Cover plates are used to reinforce both flanges of a $W18 \times 50$ section as shown. The plates are bolted to both flanges using A325 bolts of 3/4 in diameter as shown. If the transverse shear force $V = 76$ kips, determine the required longitudinal pitch of the bolts.

Solution For $W18 \times 50$, $I_x = 800$ in⁴, $d = 18$ in, $t_f = 0.570$ in, $b_f = 7.5$ in. By symmetry, the neutral axis is at mid-depth, and overall moment of inertia of the reinforced section is given by

$$I = 800 + 2\left[\frac{8 \times 0.5^3}{12} + 8 \times 0.5 \times 9.25^2\right] = 1484.67 \text{ in}^4$$

The first moment of area at the interface between W section and the top plate is

$$Q = 8 \times 0.5 \times 9 = 36 \text{ in}^3$$

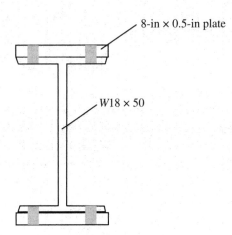

The shear stress (per unit width) developed at the interface between the W section and the top plate is

$$\tau = \frac{VQ}{I} = \frac{76 \times 36}{1484.67} = 1.84 \text{ kips/in}$$

If the allowable shear stress in bolts is 17 ksi, then the allowable shear force in each bolt is given by

$$V_{\text{bolt}} = \frac{\pi \times 0.75^2}{4} \times 17 = 7.5 \text{ kips/bolt}$$

and the maximum required spacing is given by

$$s = \frac{7.5 \times 2}{1.84} = 8.15 \text{ in}$$

Slip-Critical Connections

In slip-critical connections, no slippage between the connected parts is permitted. The resistance to slip is achieved by the friction developed between the surfaces due to the normal force between the connected parts. The nominal slip resistance of a bolt is given by

$$R_n = \mu D_u h_{sc} T_b N_s \tag{112.8}$$

where μ = mean slip coefficient ($= 0.35$ for class A surfaces)
D_u = ratio of mean actual bolt pretension to the specified minimum pretension (default value 1.13)
h_{sc} = hole factor $= 1.0$ for standard bolt holes
T_b = minimum fastener tension
N_s = number of slip planes

Class *A* surfaces are defined as surfaces having clean mill scale. In the AISC specifications, they have the smallest assigned coefficient of friction and, therefore, it is conservative to assume class *A* surfaces if not otherwise specified.

Bolt Group Subject to Shear and Torsion

Figure 112.2 shows a bolt group is subjected to an eccentric load, which causes transverse shear stress in the bolts (which is equally shared among all bolts) as well as torsional shear stress due to the torsional moment (about the center of the bolt group) created by the eccentric load.

The free-body diagram in Fig. 112.3 shows details of the bolt group. The eccentric load of 40 kips produces a torsional moment of $40 \times 6 = 240$ kips-in. In general, for a bolt

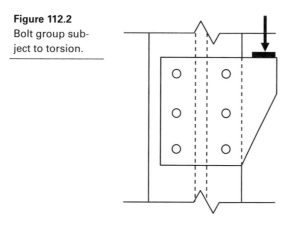

Figure 112.2
Bolt group sub-
ject to torsion.

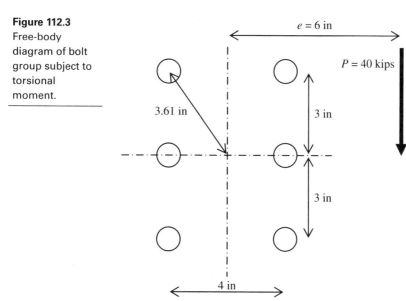

Figure 112.3
Free-body
diagram of bolt
group subject to
torsional
moment.

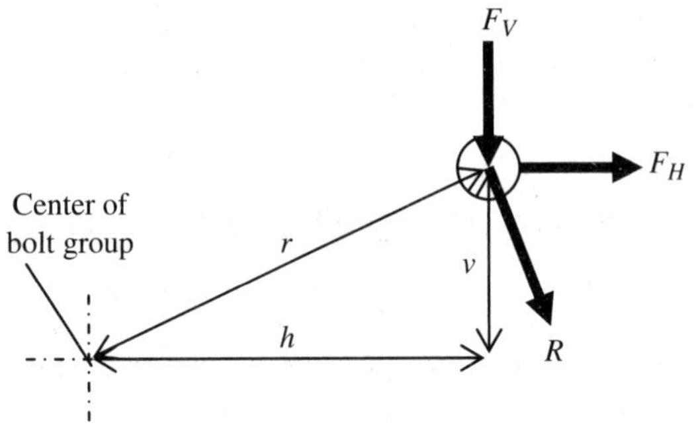

Figure 112.4
Shear forces
acting on bolt at
edge of eccen-
trically loaded
group.

located at horizontal distance h and vertical distance v from the center of the bolt group (see Fig. 112.4), the horizontal and vertical components of the shear force on the bolt are given by

$$F_H = \frac{Mv}{\sum r^2} \tag{112.9}$$

$$F_V = \frac{Mh}{\sum r^2} + \frac{P}{n} \tag{112.10}$$

The quantity Σr^2 is given by

$$\sum r_i^2 = (2 \times 2^2 + 4 \times 3.61^2) = 60 \text{ in}^2$$

The worst vector combination of these forces occurs on the outer bolts on the near line (since the downward component of the torsional shear force combines with the vertical transverse shear force).

$$F_H = \frac{Mv}{\sum r^2} = \frac{240 \times 3}{60} = 12 \text{ kips}$$

$$F_V = \frac{Mh}{\sum r^2} + \frac{V}{n} = \frac{240 \times 2}{60} + \frac{40}{6} = 14.67 \text{ kips}$$

The resultant force on the bolt (combining vectorially)

$$R = \sqrt{12^2 + 14.67^2} = 18.95 \text{ kips}$$

Capacity of Bolted Connections—Design Tables in SCM, 14th Edition

Table 7-1: Available Shear Strength of Bolts (kips)

For group A, group B and A307 bolts in single and double shear, Table 7-1 lists available strength (kips) in shear (ϕr_n for LRFD, $\phi = 0.75$; r_n/Ω for ASD, $\Omega = 2.0$) for bolt diameters 5/8 in to 1.5 in (in 1/8 in increments). For the high-strength bolts (group A and group B), one must also choose whether threads are included (N) or excluded (X) from the shear plane.

Table 7-2: Available Tensile Strength of Bolts (kips)

For group A, group B and A307 bolts in tension, Table 7-2 lists available strength (kips) in tension (ϕr_n for LRFD, $\phi = 0.75$; r_n/Ω for ASD, $\Omega = 2.0$) for bolt diameters 5/8 in to 1.5 in (in 1/8 in increments).

Table 7-3: Available Shear Strength of Slip-Critical Connections (kips)

Table 7-3 lists available shear strength for high-strength bolts (group A and group B) using various types of bolt-holes (standard, short slotted, long slotted, and oversized). Standard (STD) and short-slotted transverse to direction of load (SSLT) holes permit maximum capacity, followed by oversized (OVS) and short-slotted parallel to direction of load (SSLP) holes. Long-slotted holes (LSL) result in lowest bolt capacity. Ordinary (A307) bolts are not used for slip-critical connections. This table is based on the assumption that the faying surface is class A ($\mu = 0.30$).

Table 7-4: Available Bearing Strength (kips/in) Based on Bolt Spacing

For two different values of ultimate stress ($F_u = 58, 65$ ksi), Table 7-4 yields the available bearing strength per unit thickness of the connected part. The results in this table are based on bolt spacing $s = 2.67d_b$ and $s = 3$ in. The required bolt spacing to develop full bearing strength (s_{full}) is also given in this table.

Table 7-5: Available Bearing Strength (kips/in) Based on Bolt Edge Distance

For two different values of ultimate stress ($F_u = 58, 65$ ksi), Table 7-5 yields the available bearing strength per unit thickness of the connected part. The results in this table are based on edge distance $L_e = 1.25$ and 2 in. The required edge distance to develop full bearing strength ($L_{e\,full}$) is also given in this table.

Tables 7-6 through 7-13: Nominal Shear Capacity Coefficient *C* for Eccentrically Loaded Bolt Groups

These tables yield a multiplicative coefficient C that can be used to calculate the capacity of the bolt group (R_n) in terms of the capacity of an individual bolt (r_n).

$$R_n = Cr_n$$

Tables 7-6 through 7-13 can be used for the following scenarios:

Number of vertical rows = 1, 2, 3, or 4

Up to 12 bolts in each vertical row

Range of angles (with vertical) = 0°, 15°, 30°, 45°, 60°, 75°

Vertical spacing between horizontal rows = 3 in, 6 in

Horizontal spacing between vertical lines = 3 in, 5.5 in

Example 112.3

In this example, the applicable table (from 7-6 to 7-13) is used to calculate the ultimate capacity P_u for an eccentrically loaded bolt group as shown below. The group consists of 8 bolts arranged in two vertical lines spaced 5½ in, with 4 bolts per line. The vertical spacing between bolts is 3 in. The resultant load P_u acts at an eccentricity $e_x = 4$ in and is inclined to the vertical by 30°.

Solution   For angle $\theta = 30°$, two vertical lines of bolts spaced 5½ in, we must use Table 7-8 on page 7-44.

Matching all parameters ($n = 4$, $e_x = 4$ in, $s = 3$ in), coefficient $C = 5.30$.
Therefore, $R_n = Cr_n = 5.3r_n$.
If 1-in-diameter group A bolts (with thread included in the shear plane) are used in single shear, Table 7-1 yields $\phi r_n = 31.8$ kips (LRFD) and $r_n/\Omega = 21.2$ kips (ASD).
Therefore, the capacity of the bolt group is obtained as 5.3 times the capacity of a single bolt.

$$P_u = \phi R_n = C\phi r_n = 5.30 \times 31.8 = 168.5 \text{ kips (LRFD)}$$
$$P_a = R_n/\Omega = Cr_n/\Omega = 5.30 \times 21.2 = 112.4 \text{ kips (ASD)}$$

Bolts Subject to Shear and Tension

For bolts subject to shear, Table 112.3 outlines the allowable tension stress (ksi).

Table 112.3   Allowable Tension in Bolts Subject to Shear

Bolt	Threads included in shear plane	Threads included from shear plane
A 307	$26 - 1.8f_v \leq 20$	
A 325	$\sqrt{44^2 - 4.39f_v^2}$	$\sqrt{44^2 - 2.15f_v^2}$
A 490	$\sqrt{54^2 - 3.75f_v^2}$	$\sqrt{54^2 - 1.82f_v^2}$

Basic Weld Symbols

A typical weld symbol is shown in Fig. 112.5. The mandatory reference line is always horizontal. The arrow points to the location of the weld. The tail of the weld symbol (if used) is used to indicate the welding or cutting processes, as well as the welding specs, procedures, or any other information to be used for making the weld. Notations below the reference line refer to the weld on the arrow side, while notations above the reference line refer to the weld on the opposite side. A circle at the junction of the arrow and the reference line indicates that welding must be performed all around. A flag at the junction of the arrow and the reference line indicates a field weld. The flag is always shown flying backward (away from weld arrow). A fillet weld is indicated by a triangle on one or both sides of the reference line, as appropriate. The size of the fillet weld is indicated next to the triangular symbol. For example, in Fig. 112.5, there is a ¼-in-size fillet weld of length 6 in on the arrow side.

The maximum size of fillet welds of connected parts shall be:

a. along edges of material less than ¼ in thick, not greater than the thickness of the material

b. along edges of material ¼ in or more in thickness, not greater than the thickness of the material minus 1/16 in, unless the weld is specifically designated on the drawings to be built out to obtain full-throat thickness

The minimum size of fillet welds shall be:

a. 1/8 in for thickness of thinner part to ¼ in (inclusive)

b. 3/16 in for thinner part thickness over ¼ in to ½ in

c. ¼ in for thinner part thickness over ½ in to ¾ in

d. 5/16 in thinner part thickness over ¾ in

Intermittent welds are indicated by specifying the weld length and the longitudinal pitch (center to center) of the weld lines. An intermittent weld pattern can have welds that are not staggered [Fig. 112.6(a)] or staggered [Fig. 112.6(b)].

Figure 112.7 shows symbols used for different types of welds.

Table 112.4 shows various parts of the nomenclature used for groove welds.

Some examples of common groove welds are shown below. In Figs. 112.8 and 112.9, E is the effective throat thickness. In Fig. 112.8, the symbol "f" is called the land. The land of the weld should be a minimum of 1/8 in.

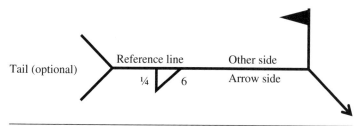

Figure 112.5 Fillet weld.

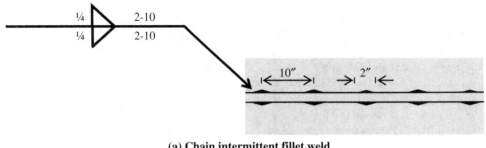

(a) Chain intermittent fillet weld

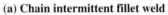

(b) Staggered intermittent fillet weld

Figure 112.6 Intermittent fillet weld.

Table 112.4 Basic Groove Weld Nomenclature

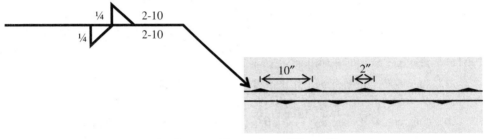

Symbols for weld type	Symbols for joint type	Symbols for weld process	Symbols for base metal thickness and penetration
1 Square Groove 2 Single V-Groove	B = Butt Joint	F = FCAW	U = Unlimited thickness, Complete Joint Penetration
3 Double V-Groove 4 Single Bevel Groove 5 Double Bevel Groove 6 Single U-Groove 7 Double U-Groove 8 Single J-Groove 9 Double J-Groove 10 Flare Groove 11 Flare Groove (open root) 12 Flare Bevel Fillet	C = Corner Joint T = T Joint BC = Butt or Corner Joint TC = T or Corner Joint BTC = Butt, T, or Corner Joint	G = GMAW sc = Short Circuit S = SAW None of the above = SMAW or GTAW	L = Limited thickness, Complete Joint Penetration P = Partial Joint Penetration

Type	Fillet weld	Groove welds					
		Square	V	Bevel	J	U	Plug/Slot
Symbol						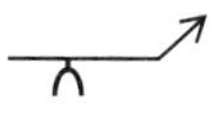	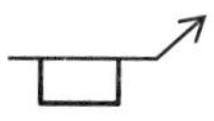

Figure 112.7 Weld symbols.

BTC-P4
Butt, T or Corner Joint. Partial Joint Penetration. Single Bevel Groove Weld

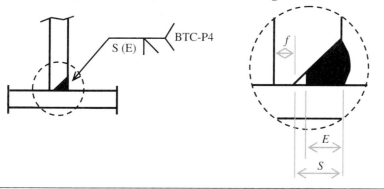

Figure 112.8 Single bevel groove weld.

BTC-P10
Butt, T or Corner Joint. Partial Joint Penetration. Flare Groove Weld

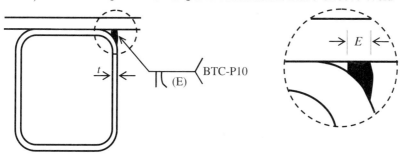

Figure 112.9 Flare groove weld.

Weld Specifications

Some of the best known welding methods are:

- Shielded metal arc welding (SMAW)—also known as arc or stick welding. The electrode slowly melts to form the weld puddle. Slag protects the weld puddle from atmospheric contamination. This very basic welding method is easy to master and is well suited for thicknesses 4 mm and greater. Thinner sheet metals are more suited to the MIG process.

- Gas metal arc welding (GMAW)—commonly termed MIG (metal, inert gas), uses a gun that feeds wire at an adjustable speed and flows an argon-based shielding gas or a mix of argon and carbon dioxide (CO_2) over the weld puddle to protect it from atmospheric contamination.

- Gas tungsten arc welding (GTAW)—also known as TIG (tungsten, inert gas), uses a nonconsumable tungsten electrode to produce the weld. The weld area is protected from atmospheric contamination by an inert shielding gas such as argon or helium. This is considered one of the most difficult and time consuming of welding processes.

- Flux-cored arc welding (FCAW)—also known as "wire welding"—almost identical to MIG welding except it uses a special tubular wire filled with flux; it can be used with or without shielding gas, depending on the filler.

- Submerged arc welding (SAW)—uses an automatically fed consumable electrode and a blanket of granular fusible flux. The molten weld and the arc zone are protected from atmospheric contamination by being "submerged" under the flux blanket.

- Electroslag welding (ESW)—a highly productive, single pass welding process for thicker materials between 1 in and 12 in in a vertical or close to vertical position.

Electrodes

All electrodes must be of the low-hydrogen classification. Shielded metal arc welding (SMAW) is the only preapproved process for welding on bridge members. The most common SMAW electrodes used are E7018 and E8018. E7018 electrodes are used for bridge members that are coated (painted, galvanized, or metalized). E8018 electrodes are used for bridge members that are uncoated.

A typical designation of a SMAW electrode is E7018. The first two digits stand for the minimum tensile strength (ksi) of the welding electrode. Thus, the designation "70" correlates to the electrode having a tensile strength of 70 ksi.

The third digit on a SMAW electrode stands for the positions in which the electrode can be used.

- The number 1 means the electrode can be used in all positions
- The number 2 means the electrode can be used in the flat and horizontal position

The fourth digit on a SMAW electrode indicate the type of coating

- The numbers 6 or 8 classify the electrode as low-hydrogen

Welder Position

Welders must be qualified for the position in which they are welding.

F - Flat position

H - Horizontal position

V - Vertical position

OH - Overhead position

Electrode Storage and Redrying

Hydrogen is one of the major causes for weld defects, so care must be taken to ensure no moisture is picked up in the coating on the electrodes.

Electrodes shall be purchased in hermetically sealed containers or shall be dried for at least 2 hours between 450 and 500°F for E70XX electrodes or between 700 and 800°F for E80XX electrodes.

Immediately after opening of the hermetically sealed container electrodes not being used must be stored in a storage oven (also known as a hot box) and held at a temperature of at least 250°F.

After the electrodes have been removed from the hermetically sealed containers or from the storage oven, the electrodes may be exposed to the atmosphere for a period not to exceed the following:

- E70XX 4 hours maximum
- E80XX 2 hours maximum

If the electrode has been exposed to the atmosphere for a period less than that shown above, then the electrode may be placed back into the storage oven and dried for a period of no less than 4 hours. Electrodes that have been wet shall not be used.

The first step in making a sound weld is to make sure the joint is correctly cleaned and then preheated prior to welding. Cleaning the joint can be accomplished by using a stiff wire brush.

Weld Preheating

Preheat is an important step prior to welding. Preheating the joint helps remove any moisture from the joint and by heating the joint initially before welding commences will allow the joint to cool at a slower rate which will allow for more time for hydrogen to diffuse out of the molten weld metal.

Preheating is the required practice of providing localized heat to the weld zone. The preferred method of preheating is by the use of a manual torch.

Required preheat shall be applied for a distance of 3 in in all directions from the weld joint. Minimum Preheat required is found in Table 4.4 of the AWS/AASHTO D1.5 Bridge Welding Code and is listed below in Table 112.5.

When the base metal temperature falls below 32°F the base metal shall be heated to at least 70°F. No welding shall be done when the ambient temperature around the weld joint is below 0°F.

Table 112.5 Minimum Preheat Temperature (°F)

Base Metal	Thickness of thickest part at point of welding			
	≤ ¾ in	> ¾ in to 1½ in	> 1½ in to 2½ in	> 2½ in
A36, A572, A588 (A 709-Grade 36, 50 50W)	50	70	150	225

Fillet-Welded Joints

Fillet-welded joints such as tee, lap, and corner joints are the most common connections in welded fabrication. Fillets are not only the most frequently used weld joints but also one of the most difficult to weld with any real degree of consistency. Fillet welds require a higher heat input than a butt joint of the same thickness. This can lead to lack of penetration and/or fusion defects that cannot be detected by visual examination and other nondestructive techniques. Inspection methods such as visual inspection, magnetic particle inspection, and penetrant inspection are surface examination techniques only.

Often the fillet welds that are produced are of a poor shape which can adversely influence their performance under load. Some of the deficiencies of fillet welds are shown in Figs. 112.10 to 112.113. Due to the melting of the corner of the upper plate, the vertical leg length is reduced meaning that the design throat is also reduced creating an undersized weld. To prevent this, the weld should be some 0.5 to 1 mm clear of the top corner. See Fig. 112.10.

In addition to the reduction in throat thickness, there is the potential for additional problems such as overlap at the weld toe due to the larger weld pool size or an excessively convex weldface and consequential sharp notches at the weld toe. These situations (illustrated in the

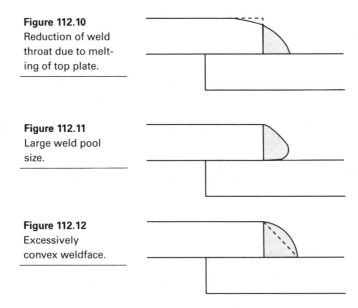

Figure 112.10
Reduction of weld throat due to melting of top plate.

Figure 112.11
Large weld pool size.

Figure 112.12
Excessively convex weldface.

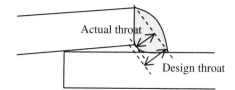

Figure 112.13
Reduction of weld throat due to improper alignment.

Actual throat

Design throat

Figs. 112.11 and 112.12) could adversely influence the fatigue life of the welded joint due to the increased toe angle, which acts as a greater stress concentration. Improper alignment can also reduce the throat thickness as in Fig. 112.13.

Fillet Weld Features

The fillet weld is assumed to have a cross section of a 90-45-45 right triangle. The size of the weld is denoted by w (see Fig. 112.14). Standard weld sizes are specified in increments of 1/16 in. A fillet weld is weakest in shear and failure is assumed to occur on a plane through the throat of the weld. *Throat* is the perpendicular distance from the corner (root) of the weld to the hypotenuse and is equal to 0.707 times the weld size. A deep penetration weld is shown in Fig. 112.15.

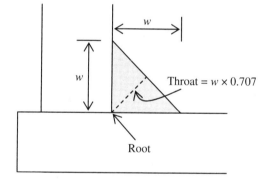

Figure 112.14
Mitre fillet weld parameters.

w

w

Throat = $w \times 0.707$

Root

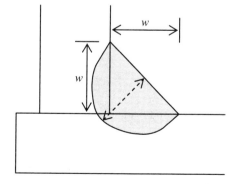

Figure 112.15
Deep penetration weld parameters.

w

w

Strength of a Fillet Weld

For a weld of length L subjected to a force P, the critical shear stress (on the plane inclined along the throat) is given by

$$f_v = \frac{P}{0.707wL} \tag{112.11}$$

If the weld's ultimate shear stress is F_w, the nominal load capacity of the weld is given by

$$R_n = 0.707F_w wL \tag{112.12}$$

The ultimate shearing stress of the weld is 60% of the ultimate tensile strength F_{EXX}. The ultimate tensile strength of the weld metal is defined using the standard notation such as E70XX or E80XX. For example, the E70XX electrode implies that the weld metal has an ultimate tensile strength of 70 ksi. AISC specifications require the designer to use

E70XX electrodes for steel having a yield stress less than 60 ksi

E80XX electrodes for steel having a yield stress of 60 ksi or 65 ksi

When the direction of the load is at an angle to the weld line, as shown in Fig. 112.16, the nominal weld strength (shear) is calculated as

$$F_w = 0.6F_{EXX}\left(1 + 0.5\sin^{1.5}\theta\right) \tag{112.13}$$

The design shear strength of a fillet weld is given by

$$\text{ASD}: \quad R_a \leq \frac{R_n}{\Omega} = \frac{R_n}{2.0} \tag{112.14}$$

$$\text{LRFD}: \quad P_u \leq \phi R_n \tag{112.15}$$

Second Moments of Weld Runs

Table 112.6 shows second moments I_{xx}, I_{yy}, and I_{xy} of several commonly used weld patterns. Example 112.3 illustrates their use.

Figure 112.16
Weld line subject to combined tension and shear.

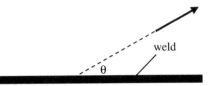

Table 112.6 Second Moment of Weld Runs

Weld pattern	Length L	I_{xx}	I_{yy}	I_{xy}
	d	$\dfrac{d^3}{12}$	0	0
	$2d$	$\dfrac{d^3}{6}$	$\dfrac{b^3 d}{2}$	0
	$2b + d$	$\dfrac{d^2(6b+d)}{12}$	$\dfrac{b^3(b+2d)}{3L}$	0
	$2(b+d)$	$\dfrac{d^2(3b+d)}{6}$	$\dfrac{b^2(b+3d)}{6}$	0
	$b + d$	$\dfrac{d^3(4b+d)}{12L}$	$\dfrac{b^3(b+4d)}{12L}$	$\dfrac{b^2 d^2}{4L}$
	$2\pi R$	πR^3	πR^3	0

Example 112.4

An eccentric load is applied to the flange of a $W14 \times 82$ column as shown. The bracket plate transfers the dead load = 15 kips and live load = 25 kips as shown. What is the minimum weld size needed if E70XX electrode is used?

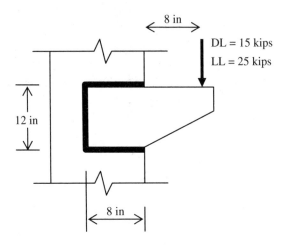

Solution   For the weld profile shown, the centroid is at a distance $b^2/L = 8^2/28 = 2.29$ in from the back edge, causing the eccentricity of the load to be $16 - 2.29 = 13.71$ in.

ASD

The design load $P = 15 + 25 = 40$ kips and moment $M = 40 \times 13.71 = 548.4$ kips-in

$$I_{xx} = \frac{d^2(6b+d)}{12} = \frac{12^2(6 \times 8 + 12)}{12} = 720$$

$$I_{yy} = \frac{b^3(b+2d)}{3L} = \frac{8^3(8 + 2 \times 12)}{3 \times 28} = 195$$

Polar moment of inertia of the weld line = $720 + 195 = 915$ in^4. The torsional shear stress has the following components:

$$\tau_x = \frac{My}{J} = \frac{548.4 \times 5.71}{915} = 3.42 \text{ ksi}$$

$$\tau_y = \frac{Mx}{J} = \frac{548.4 \times 6}{915} = 3.60 \text{ ksi}$$

The direct shear stress is

$$\tau_y = \frac{P}{L} = \frac{40}{28} = 1.43 \text{ ksi}$$

The shear stress on the weld line is $\sqrt{3.42^2 + (3.60 + 1.43)^2} = 6.08$ kips/in. The weld strength per inch is given by (assume E70XX electrode):

$$R_n = 0.707wF_w = 0.707 \times w \times 0.6 \times 70 = 29.69w$$

The allowable load on the weld (per inch) is therefore

$$\frac{R_n}{\Omega} = \frac{29.69w}{2.0} = 14.85w$$

Equating this to the shear stress on the weld (6.08 kips/in), we get $w = 0.41$. Use 7/16 in weld size.

Inspection Criteria for Welds and Bolts

Quality assurance (QA) is *process* oriented and focuses on defect *prevention*, while quality control (QC) is *product* oriented and focuses on defect *identification*. The goal of QA is to improve development and test processes so that defects do not arise when the product is being developed. The goal of QC is to identify defects after a product is developed and before it is released.

Minimum Requirements for Inspection of Structural Steel Buildings

QC inspection tasks shall be performed by fabricator's or erector's QCI (quality control inspector). Applicable construction documents are the shop drawings and the erection drawings and applicable specifications, codes, and standards. QA inspection of fabricated items shall occur at the fabricator's plant. QA inspection of the erected system shall occur at the project site.

Prior to welding, QA and QC inspection must ensure that welding procedure specifications (WPSs) and manufacturer's certifications for welding consumables (rods, etc.) are available for each welded joint or member. After welding, the following must be inspected:

1. Size, length, and location of welds

2. Welds meet the following visual acceptance criteria:

   a. Weld has proper profile and has adequate size as given in job specifications.

   b. Proper fusion of weld and base metal.

   c. Crack prohibition. Crater cracks (occur when a crater is not filled before the arc is broken) can form in longitudinal, transverse, and/or radial directions. An undercut crack, also known as a heat-affected zone (HAZ) crack, is a crack that forms a short distance away from the fusion line. Arc strike cracking occurs when the arc is struck but the spot is not welded due to the spot being heated above the material's upper critical temperature and then essentially quenched.

    d. Porosity of a weld occurs due to gas inclusion, which is the entrapment of gas within the solidified weld due to high sulfur content of the electrode, excessive moisture from the electrode, or due to wrong welding current or polarity.

    e. Cracking in the "k" area, which is the region extending from approximately the midpoint of the radius of the fillet into the web approximately 1 to 1.5 in beyond the point of tangency between the fillet and the web.

   3. Backing removed and weld tabs removed.

Visual Inspection of Welds

A fillet weld gauge is the standard tool to check weld sizes. In addition to checking that the weld size meets specifications, all welds should also be visually inspected for defects. Defects to look for include the following:

Cracks

No cracks in the surface of the welds shall be allowed. If a crack is found, the crack must be removed and magnetic-particle inspection performed to ensure the crack has been removed before rewelding.

Porosity

Porosity is a cavity in the weld that is formed by gas escaping from the molten weld metal during solidification. The AWS D1.5 Code specification for porosity is

- Maximum diameter shall not exceed 3/32 in
- Frequency of any sized porosity shall not exceed one in 4 in or six in 4 ft of weld length

Craters

Craters are the ends of welds where the weld is not filled to its full cross section. The stresses that are caused by the unfilled crater may cause cracks to form because of tension on the weld in the affected area. All welds must have full cross section the entire length of the weld.

Undercut

Undercut occurs at the edge of the weld along the leg. Undercut actually refers more to the base metal adjacent to the weld. Undercut is normally caused by excessive current in the welding operation. Undercut will cause stress risers and should be avoided. The AWS D1.5 Code requirement for undercut is:

- Undercut shall be no more than 0.01 in deep when the weld is transverse to tensile stress.
- Undercut shall be no more than 1/32 in deep for all other cases.

Arc Strikes

Arc strikes are areas where the welding electrode comes into contact with the base metal outside of the final weld. Arc strikes result in heating and very rapid cooling. Arc strikes may result in hardening or fatigue cracking, and serve as potential sites for fracture initiation.

All arc strikes are to be removed by grinding. Grinding to a depth of 1/8 in below the original surface should remove all traces of arc strikes and their hardened heat-affected zones. However, in tension areas of the bridge, the locations where arc strikes were removed shall have magnetic-particle inspection and hardness testing performed per The AWS D1.5 Bridge Welding Code.

Nondestructive Testing (NDT) of Welded Joints

Ultrasonic testing (UT), magnetic particle testing (MT), penetrant testing (PT), and radiographic testing (RT) shall be performed as part of QA, in accordance with AWS D1.1.

For structures in risk categories III or IV (ASCE7), UT shall be performed on all CJP groove welds subject to transversely applied tension loads in butt, T- and corner joints, in material thickness 5/16 in or greater.

Exception   Where the initial rate for UT is 100%, it shall be permitted to be reduced (for an individual welder) to 25% if the reject rate (welds containing unacceptable defects divided by number of welds completed) is less than 5%. This determination is based on a sample of at least 40 completed welds. For continuous welds longer than 36 in where the effective throat is 1 in or less, each 12 in segment or fraction thereof shall be considered as one weld. For welds with effective throat greater than 1 inch, each 6 in segment or fraction thereof shall be considered as one weld.

For structures in risk category II, UT shall be performed on 10% of CJP groove welds subject to transversely applied tension loads in butt, T- and corner joints, in material thickness 5/16 in or greater.

Exception   For structures in risk category II, where the initial rate of UT is 10%, it shall be increased to 100% should the reject rate (for an individual welder) exceeds 5%, based on a minimum sample size of 20 welds. The rate of UT shall be returned to 10% when the reject rate (based on 40 completed welds) falls below 5%.

For structures in risk category I, NDT is not required. For structures in all risk categories, NDT of CJP groove in materials less than 5/16 in thick is not required.

Inspection of High-Strength Bolting

Tables N5.6-1 to N5.6-3 in AISC 360-10 Specifications for Structural Steel Buildings lay out QA and QC inspection requirements prior to, during, and after bolting respectively. Some of these tasks are only random observations made without interrupting operations. Other more critical tasks are mandatory and must be performed for each bolted connection.

1. For snug-tight joints, preinstallation verification testing as specified in Table N5.6-1 and monitoring of the installation procedures as specified in Table N5.6-2 are not applicable. The QCI and QAI need not be present during the installation of fasteners in snug-tight joints.

2. For *pretensioned joints* and slip-critical joints, when the installer is using the *turn-of-nut method* (with match marking), the direct-tension-indicator method, or the twist-off-type tension-control-bolt method, monitoring of bolt pretensioning procedures

shall be as specified in Table N5.6-2. The QCI and QAI need not be present during the installation of fasteners when these methods are used by the installer.

3. For pretensioned joints and slip-critical joints, when the installer is using the calibrated-wrench method or the turn-of-nut method without match marking, monitoring of bolt pretensioning procedures shall be as specified in Table N5.6-2. The QCI and QAI shall be engaged in their assigned inspection duties during installation of fasteners when these methods are used by the installer.

Bridge Design (AASHTO LRFD)

As of the April 2015 exam, the official syllabus for the PE Civil exam references the document *AASHTO LRFD Bridge Design Specifications,* 6th ed., 2012, American Association of State Highway and Transportation Officials, Washington, D.C. These specifications govern the design of highway bridges according to the LRFD design philosophy. The adoption of the LRFD bridge design standards have superseded the *AASHTO Standard Specifications for Highway Bridges,* 16th ed., American Association of State Highway and Transportation Officials, Washington, D.C., which was based on the load factor design method.

In this chapter, we will discuss the specifications regarding the following:

1. Design methods
2. Standard live loadings—truck and lane loads
3. Distribution of deck live loads to superstructure elements
4. Design of concrete deck slabs
5. Design of longitudinal beams (stringers)

Design Philosophy—LRFD

In the working stress design method (AASHTO Standard Specifications), structural members are designed such that specific stresses in the structural member do not exceed a predefined allowable stress, which is defined as a limiting stress divided by a factor of safety. Thus, the allowable stress is a fraction of some kind of "failure stress" for the material.

The main drawbacks of the working stress method are that (a) it designs members for low stress levels (within the elastic limit) and usually results in overdesigned structures and (b) it employs a single factor of safety and therefore makes no distinction between the different degrees of variability of different types of loads.

The load factor design (AASHTO Standard Specifications), conceived to improve upon the stated drawbacks of the working stress method, makes use of the plastic range of material behavior and employs different load factors for different load types. Limit states can be broadly classified into strength limit states (such as yielding, fracture, buckling, etc.) and serviceability limit states (such as fatigue, deflection, vibration, etc.).

The LRFD design philosophy calibrates load factors such that the structure has the same level of reliability for all load types. The LRFD principle may be stated as

$$\sum \eta_i \gamma_i Q_i \leq \phi R_n \tag{113.1}$$

The γ_i is the load factor associated with ith category service load Q_i. The η factor is a load-modification factor, calculated as a product of ductility (η_D), redundancy (η_R), and importance factors (η_I).

The ductility load modifier accounts for the ability of a structure to redistribute stresses from overstressed (inelastic) elements to other parts of the load-resisting system. The redundancy load modifier accounts for the presence of multiple load paths in the structure. Redundancy in a structure increases its factor of safety and this is reflected by the load modifier for redundancy. Bridges are categorized as important if they provide a short route to critical facilities.

LRFD Load Combinations

There are various categories of limit states employed in the LRFD specifications—*service limit states* (restrictions on stress, deformation, and crack width under regular service conditions), *fatigue and fracture limit state, strength limit state,* and *extreme event limit states* (major earthquake or flood, vessel or vehicle collision).

Components and connections of a bridge shall satisfy the design equation for the applicable combinations of factored extreme force effects as specified at each of the following limit states:

Strength I	This is the basic load combination relating to the normal vehicular use of the bridge without wind.
Strength II	Load combination relating to the use of the bridge by owner-specified special design vehicles, evaluation-permit vehicles, or both, without wind.
Strength III	Load combination relating to the bridge exposed to wind velocity exceeding 55 mph.
Strength IV	Load combination relating to very high dead load to live load ratios.
Strength V	Load combination relating to normal vehicular use of the bridge with wind of 55-mph velocity.
Extreme Event I	Load combination including earthquake.
Extreme Event II	Load combination relating to ice load, collision by vessels and vehicles, and certain hydraulic events with a reduced live load other than that which is part of the vehicular collision load CT.

Service I	Load combination relating to the normal operational use of the bridge with a 55-mph wind and all loads taken at their nominal values.
Service II	Load combination intended to control yielding of steel structures and slip of slip-critical connections due to vehicular live load.
Service III	Load combination for longitudinal analysis relating to tension in prestressed concrete superstructures with the objective of crack control and to check principal tension in the webs of segmental concrete girders.
Service IV	Load combination relating only to tension in prestressed concrete columns with the objective of crack control.
Fatigue	Fatigue- and fracture-load combination relating to repetitive gravitational vehicular live load and dynamic responses under a single design truck having the axle spacing specified in Article 3.6.1.4.1.

The LRFD load factors are summarized in Table 113.1.

Table 113.1 AASHTO LRFD Load Factors

Load combination limit state	DC DD DW EH EV ES EL	LL IM CE BR PL LS	WA	WS	WL	FR	TU CR SH	TG	SE	EQ	IC	CT	CV
										Use one at a time			
Strength I	γ_P	1.75	1.0	—	—	1.0	0.5/1.2	γ_{TG}	γ_{SE}	—	—	—	—
Strength II	γ_P	1.35	1.0	—	—	1.0	0.5/1.2	γ_{TG}	γ_{SE}	—	—	—	—
Strength III	γ_P	—	1.0	1.4	—	1.0	0.5/1.2	γ_{TG}	γ_{SE}	—	—	—	—
Strength IV	γ_P	—	1.0	—	—	1.0	0.5/1.2	—	—	—	—	—	—
Strength V	γ_P	1.35	1.0	0.4	1.0	1.0	0.5/1.2	γ_{TG}	γ_{SE}	—	—	—	—
Extreme Event I	γ_P	γEQ	1.0	—	—	1.0	—	—	—	1.0	—	—	—
Extreme Event II	γ_P	0.50	1.0	—	—	1.0	—	—	—	—	1.0	1.0	1.0
Service I	1.0	1.0	1.0	0.3	1.0	1.0	1.0/1.2	γ_{TG}	γ_{SE}	—	—	—	—
Service II	1.0	1.3	1.0	—	—	1.0	1.0/1.2	—	—	—	—	—	—
Service III	1.0	0.8	1.0	—	—	1.0	1.0/1.2	γ_{TG}	γ_{SE}	—	—	—	—
Service IV	1.0	—	1.0	0.7	—	1.0	1.0/1.2	—	1.0	—	—	—	—
Fatigue: LL, IM, & CE only	—	0.75	—	—	—	—	—	—	—	—	—	—	—

Table 113.2 Load Factors for Permanent Loads (γ_p)

Symbol	Name	Load factor	
		Minimum	Maximum
DC	Dead load of structural & nonstructural components	0.90	1.25
DD	Downdrag	0.45	1.80
DW	Dead load of wearing surfaces and utilities	0.65	1.50
EH	Horizontal earth pressure		
	Active	0.90	1.50
	At rest	0.90	1.35
EV	Vertical pressure from dead load of earth fill		
	Overall stability	—	1.00
	Retaining structure	1.00	1.35
	Rigid buried structure	0.90	1.30
	Rigid frames	0.90	1.35
	Flexible buried structures except metal box culverts	0.90	1.95
	Flexible metal box culverts	0.90	1.50
ES	Earth surcharge load	0.75	1.50

Table 113.2 gives the load factor γ_p for various categories of permanent loads. Either the minimum or maximum value of the load factor may produce the most critical condition, and so, both must be considered.

Deflection Limits

The following are the deflection limits prescribed by AASHTO LRFD:

For steel, aluminum, and concrete construction:

Vehicular load, general	$L/800$
Vehicular load and/or pedestrian loads	$L/1000$
Vehicular load on cantilever arms	$L/300$
Vehicular load and/or pedestrian loads on cantilever arms	$L/375$

For wood construction:

Vehicular and pedestrian loads	$L/425$
Vehicular load on wood planks and panels	0.1 in

For orthotropic plates:

Vehicular load on deck plate	$L/300$
Vehicular load on ribs of orthotropic metal decks	$L/1000$
Vehicular load on ribs of orthotropic metal decks (relative deflection between adjacent ribs)	0.1 in

Table 113.3   Minimum Depth of Superstructure Recommended in AASHTO LRFD

Superstructure		Minimum depth	
Material	Type	Simple spans	Continuous spans
REINFORCED CONCRETE	Slabs with main reinforcement parallel to traffic	$\dfrac{1.2(S+10)}{30}$	$\dfrac{S+10}{30} \geq 0.54$ ft
	T-beams	$0.070L$	$0.065L$
	Box beams	$0.060L$	$0.055L$
	Pedestrian structure beams	$0.035L$	$0.033L$
PRESTRESSED CONCRETE	Slabs	$0.030L \geq 6.5$ in	$0.027L \geq 6.5$ in
	CIP box beams	$0.045L$	$0.040L$
	Precast I-beams	$0.045L$	$0.040L$
	Pedestrian structure beams	$0.033L$	$0.030L$
	Adjacent box beams	$0.030L$	$0.025L$
STEEL	Overall depth of composite I-beam	$0.040L$	$0.032L$
	Depth of steel I-beam	$0.033L$	$0.027L$
	Trusses	$0.100L$	$0.100L$

Minimum Depth of Superstructure

Table 113.3 summarizes minimum superstructure depths for various bridge types, as *recommended* in the AASHTO LRFD specifications.

Multiple Presence of Live Load

Unless specified otherwise, the extreme live load force effect shall be determined by considering each possible combination of number of loaded lanes multiplied by a corresponding multiple presence factor to account for the probability of simultaneous lane occupation by the full HL-93 design live load. In other words, the live load force effect on the member being designed shall be computed for multiple scenarios of live load presence (1 lane loaded, 2 lanes loaded, 3 lanes loaded, etc.) multiplied by the corresponding multiple presence factor from Table 113.4.

For the fatigue limit state, only one design truck is used, regardless of the number of design lanes, and therefore the multiple presence factor is not used.

Table 113.4   Multiple Presence Factor for Live Load

Number of loaded lanes	Multiple presence factor
1	1.20
2	1.00
3	0.85
>3	0.65

Vehicular Live Load

Vehicular live loading on the roadways of bridges or incidental structures, designated HL-93, shall consist of three different live loads—a design truck, design tandem, and design lane load. Each design lane under consideration shall be occupied by either the design truck or tandem, coincident with the lane load, where applicable. The loads shall be assumed to occupy 10.0 ft transversely within a design lane.

Design Truck

The design truck is a model load that consists of three axle loads that resemble a typical semitrailer truck. The front axle is 8 kips, the drive axle is 32 kips located 14 ft behind the front axle, and the rear trailer axle is also 32 kips positioned at a variable distance between 14 ft and 30 ft. For most situations, such as to maximize shear and bending moment on simple spans, the compact spacing (14 ft between second and third axle) of the truck axles governs. The design truck is the same configuration that has been used in the Standard Specifications and designated as HS20 (Fig. 113.1).

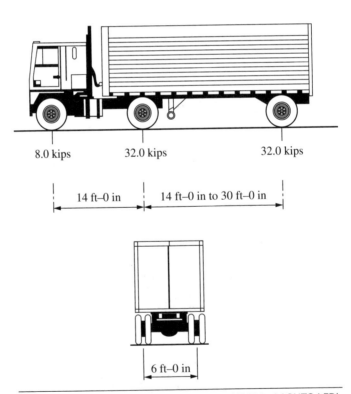

Figure 113.1 AASHTO design truck (former HS20 in AASHTO LFD).

Design Tandem

The design tandem shall consist of a pair of 25.0-kip axles spaced 4.0-ft apart. The transverse spacing of wheels shall be taken as 6.0 ft. A dynamic load allowance shall be considered.

Design Lane Load

The design lane load shall consist of a load of 0.64 klf uniformly distributed in the longitudinal direction. Transversely, the design lane load shall be assumed to be uniformly distributed over a 10.0-ft width. The force effects from the design lane load shall not be subject to a dynamic load allowance.

The load effects of the design truck or design tandem must be superimposed *with the load effect of the design lane, as opposed to the Standard Specifications where the greater of the effects of the truck or lane loading was used for design.*

In addition, a third live load combination is used in the LRFD specifications to model the scenario where a truck is closely followed by another heavily loaded truck. AASHTO LRFD A3.6.1.3.1 specifies

For both negative moment between points of contraflexure under a uniform load on all spans and reaction at interior supports, 90 percent of the effect of two design trucks spaced a minimum of 50 ft between the lead axle of one truck and the rear axle of the other truck, combined with the effect of 90 percent of the design lane load. The distance between the 32 kips axles of each truck shall be taken as 14 ft.

These loads are shown in Figs. 113.2 to 113.4.

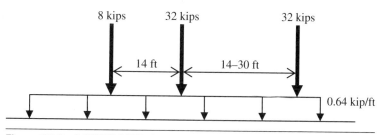

Figure 113.2 AASHTO LRFD design truck + design lane load.

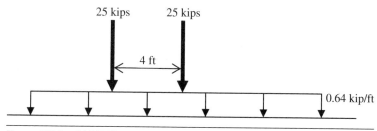

Figure 113.3 AASHTO LRFD design tandem + design lane load.

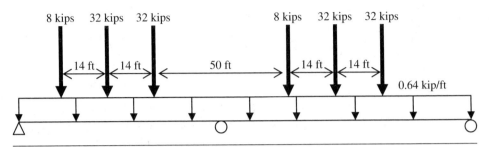

Figure 113.4 AASHTO LRFD dual truck + design lane load.

Pedestrian Loads

A pedestrian load of 75 lb/ft² shall be applied to all sidewalks wider than 2.0 ft and considered simultaneously with the vehicular design live load. Bridges for only pedestrian and/or bicycle traffic shall be designed for a live load of 85 lb/ft².

Dynamic Load Allowance

The static effects of the design truck or tandem, other than centrifugal and braking forces, shall be increased by the percentage specified in Table 113.5 for dynamic load allowance. The factor to be applied to the static load shall be taken as $(1 + IM/100)$. The dynamic load allowance shall not be applied to pedestrian loads or to the design lane load.

Dynamic load allowance need not be applied to retaining walls not subject to vertical reactions from the superstructure, and foundation components that are entirely below ground level.

The dynamic load allowance for culverts and other buried structures shall be taken as

$$IM(\%) = 33(1.0 - 0.125D_E) \geq 0\% \qquad (113.2)$$

where D_E = the minimum depth of earth cover above the structure (ft). The dynamic load allowance is applied to the static live load according to

$$U_{L+I} = U_L(1 + IM) \qquad (113.3)$$

Table 113.5 Dynamic Load Allowance (IM)

Component	IM
Deck joints	
All limit states	75%
All other components	
Fatigue and fracture limit state	15%
All other limit states	33%

Table 113.6 Base Wind Pressure on Superstructure Elements

Skew angle of wind (degrees)	Trusses, columns, and arches		Girders	
	Lateral load (ksf)	Longitudinal load (ksf)	Lateral load (Ksf)	Longitudinal load (Ksf)
0	0.075	0.000	0.050	0.000
15	0.070	0.012	0.044	0.006
30	0.065	0.028	0.041	0.012
45	0.047	0.041	0.033	0.016
60	0.024	0.050	0.017	0.019

Wind on Superstructure (W)

The base wind pressure on superstructure elements shall be taken according to Table 113.6.

Wind on Live Load (WI)

When vehicles are present, the design wind pressure shall be applied to both structure and vehicles. Wind pressure on vehicles shall be represented by an interruptible, moving force of 0.10 klf acting normal to, and 6.0 ft above, the roadway and shall be transmitted to the structure.

Design of Deck Cantilever and Railings

The deck cantilever, located beyond the exterior girder, is designed for a uniform load of 1 kip/ft located 1 ft from the face of the curb or railing. The railing and the deck overhang must sustain the effect of a truck collision. Six test levels are designated based on the momentum characteristics of various vehicles. The design forces are summarized in Table 113.7.

Table 113.7 Design Loads on Railings

Design forces & parameters	Railing test level					
	TL-1	TL-2	TL-3	TL-4	TL-5	TL-6
Transverse force F_t (kips)	13.5	27.0	54.0	54.0	124.0	175.0
Longitudinal force F_L (kips)	4.5	9.0	18.0	18.0	41.0	58.0
Vertical force F_v (kips)	4.5	4.5	4.5	18.0	80.0	80.0
L_t & L_L (ft)	4.0	4.0	4.0	3.5	8.0	8.0
L_v (ft)	18.0	18.0	18.0	18.0	40.0	40.0
Minimum H_e (ft)	18.0	20.0	24.0	32.0	42.0	56.0
Minimum rail height (in)	27.0	27.0	27.0	32.0	42.0	90.0

TL-1: Work zones with low posted speeds and very low-volume, low-speed local streets

TL-2: Work zones and most local and collector roads with favorable site conditions, a small number of heavy vehicles and reduced posted speeds

TL-3: High-speed arterial highways with low heavy vehicle fraction and favorable site conditions

TL-4: Majority of high-speed highways, freeways, expressways, and interstate highways with a mix of heavy vehicles

TL-5: Same as TL-4 category except with a large fraction of heavy vehicles or where unfavorable site conditions justify higher railing design loads

TL-6: Highways where tanker type trucks are expected

where F_t = transverse force (kips) acing on railing
L_t = longitudinal distance for distribution of F_t
F_L = longitudinal force (kips) acing on railing
L_L = longitudinal distance for distribution of F_L
F_v = vertical force (kips) acing on railing
L_v = longitudinal distance for distribution of F_v
H_e = elevation of F_t and F_L above deck
H = height of railing

Deck Design

There are two methods of deck design in the *AASHTO LRFD Bridge Design Specifications:* (1) the approximate method, otherwise known as the equivalent strip method and (2) the empirical method.

Equivalent Strip Method for Design of Reinforced Concrete Decks

Decks can be analyzed by dividing the deck into strips perpendicular to supporting elements. This approach can be used for decks except fully filled and partially filled grids and top slabs of segmental concrete box girders.

Truck axle loads are supported by a transverse strip whose width is given by

1. For overhangs—strip width SW = 45.0 + 10.0X

2. For positive moment—strip width SW = 26.0 + 6.6S

3. For negative moment—strip width SW = 48.0 + 3.0S

where X and S are in feet and SW is in inches.

Minimum deck thickness is 7.0 in. For slabs less than 1/20 of the design span, consideration should be given to prestressing in that direction to control cracking. Effective span of concrete slabs supported by steel stringers or prestressed concrete girders is specified as

- For slabs monolithic with walls or beams: the face-to-face distance
- For slabs supported on steel or concrete girders: the distance between flange tips plus the flange overhang, taken as the distance from the extreme flange tip to the face of the web, disregarding any fillets

Bending Moment in Slab

Table 113.8 may be used to determine design moment in slabs. The values computed in the table are based on the following assumptions:

1. Moments are calculated using the equivalent strip method, applied to concrete deck on parallel girders.

2. Multiple presence factor and dynamic load allowance are included.

3. Deck is supported by at least three girders and has a width of not less than 14.0 ft between centerlines of exterior girders.

4. Moments represent upper bound for moments in interior of slab, calculated assuming different number of girders. For each case, the minimum overhang was assumed to be 21 in and the maximum overhang was assumed to be equal to the smaller of 0.625 times girder spacing and 6 ft.

5. Values in the table do not apply to deck overhangs.

Table 113.8   Design Moment in Concrete Slabs Supported by Steel Beams

Span S (ft)	Positive moment (kip-ft)	Negative moment (kip-ft)						
		Distance from CL of girder to design section for negative moment						
		0 in	3 in	6 in	9 in	12 in	18 in	24 in
4' 0"	4.68	2.68	2.07	1.74	1.60	1.50	1.34	1.25
4' 3"	4.66	2.73	2.25	1.95	1.74	1.57	1.33	1.20
4' 6"	4.63	3.00	2.58	2.19	1.90	1.65	1.32	1.18
4' 9"	4.64	3.38	2.90	2.43	2.07	1.74	1.29	1.20
5' 0"	4.65	3.74	3.20	2.66	2.24	1.83	1.26	1.12
5' 3"	4.67	4.06	3.47	2.89	2.41	1.95	1.28	0.98
5' 6"	4.71	4.36	3.73	3.11	2.58	2.07	1.30	0.99
5' 9"	4.77	4.63	3.97	3.31	2.73	2.19	1.32	1.02
6' 0"	4.83	4.88	4.19	3.5	2.88	2.31	1.39	1.07
6' 3"	4.91	5.10	4.39	3.68	3.02	2.42	1.45	1.13
6' 6"	5.00	5.31	4.57	3.84	3.15	2.53	1.50	1.20
6' 9"	5.10	5.50	4.74	3.99	3.27	2.64	1.58	1.28
7' 0"	5.21	5.98	5.17	4.36	3.56	2.84	1.63	1.37

(Continued)

Table 113.8 Design Moment in Concrete Slabs Supported by Steel Beams (*Continued*)

Span S (ft)	Positive moment (kip-ft)	Negative moment (kip-ft)						
		Distance from CL of girder to design section for negative moment						
		0 in	3 in	6 in	9 in	12 in	18 in	24 in
7′ 3″	5.32	6.13	5.31	4.49	3.68	2.96	1.65	1.51
7′ 6″	5.44	6.26	5.43	4.61	3.78	3.15	1.88	1.72
7′ 9″	5.56	6.38	5.54	4.71	3.88	3.30	2.21	1.94
8′ 0″	5.69	6.48	5.65	4.81	3.98	3.43	2.49	2.16
8′ 3″	5.83	6.58	5.74	4.90	4.06	3.53	2.74	2.37
8′ 6″	5.99	6.66	5.82	4.98	4.14	3.61	2.96	2.58
8′ 9″	6.14	6.74	5.90	5.06	4.22	3.67	3.15	2.79
9′ 0″	6.29	6.81	5.97	5.13	4.28	3.71	3.31	3.00
9′ 3″	6.44	6.87	6.03	5.19	4.40	3.82	3.47	3.20
9′ 6″	6.59	7.15	6.31	5.46	4.66	4.04	3.68	3.39
9′ 9″	6.74	7.51	6.65	5.80	4.94	4.21	3.89	3.58
10′ 0″	6.89	7.85	6.99	6.13	5.26	4.41	4.09	3.77
10′ 3″	7.03	8.19	7.32	6.45	5.58	4.71	4.29	3.96
10′ 6″	7.17	8.52	7.64	6.77	5.89	5.02	4.48	4.15
10′ 9″	7.32	8.83	7.95	7.08	6.20	5.32	4.68	4.34
11′ 0″	7.46	9.14	8.26	7.38	6.50	5.62	4.86	4.52
11′ 3″	7.60	9.44	8.55	7.67	6.79	5.91	5.04	4.70
11′ 6″	7.74	9.72	8.84	7.96	7.07	6.19	5.22	4.87
11′ 9″	7.88	10.01	9.12	8.24	7.36	6.47	5.40	5.05
12′ 0″	8.01	10.28	9.40	8.51	7.63	6.74	5.56	5.21
12′ 3″	8.15	10.55	9.67	8.78	7.90	7.02	5.75	5.38
12′ 6″	8.28	10.81	9.93	9.04	8.16	7.28	5.97	5.54
12′ 9″	8.41	11.06	10.18	9.30	8.42	7.54	6.18	5.70
13′ 0″	8.54	11.31	10.43	9.55	8.67	7.79	6.38	5.85
13′ 3″	8.66	11.55	10.67	9.80	8.92	8.04	6.59	6.01
13′ 6″	8.78	11.79	10.91	10.03	9.16	8.28	6.79	6.16
13′ 9″	8.90	12.02	11.14	10.27	9.40	8.52	6.99	6.30
14′ 0″	9.02	12.24	11.37	10.50	9.63	8.76	7.18	6.45
14′ 3″	9.14	12.46	11.59	10.72	9.85	8.99	7.38	6.58
14′ 6″	9.25	12.67	11.81	10.94	10.08	9.21	7.57	6.72
14′ 9″	9.36	12.88	12.02	11.16	10.30	9.44	7.76	6.86
15′ 0″	9.47	13.09	12.23	11.37	10.51	9.65	7.94	7.02

Concrete Design

In the absence of precise data, the modulus of elasticity of concrete is given by

$$E_c = 33,000 K_l w_c^{1.5} \sqrt{f_c'} \qquad (113.4)$$

where E_c = modulus of elasticity (ksi)
  K_l = parameter related to source of aggregate (default value 1.0)
  w_c = unit weight of the concrete (kcf)
  f_c' = specified compressive strength of the concrete (ksi)

For normal-weight concrete ($w_c = 0.145$ kcf), this reduces to

$$E_c = 1820 \sqrt{f_c'} \qquad (113.5)$$

Poisson's ratio of concrete may be taken as 0.20 if more precise data is not available. For normal-weight concrete, the modulus of rupture f_r (ksi) of concrete is given by

$$f_r = 0.24 \sqrt{f_c'} \qquad (113.6)$$

Minimum reinforcement is provided to ensure that the nominal moment capacity is at least 20% greater than the cracking moment, which is calculated from the modulus of rupture based on Eq. (113.7).

$$f_r = 0.37 \sqrt{f_c'} \qquad (113.7)$$

The following strength reduction factors are specified in AASHTO LRFD:

For tension-controlled reinforced concrete sections	$\phi = 0.90$
For tension-controlled prestressed concrete sections	$\phi = 1.00$
Shear and torsion	
Normal-weight concrete	$\phi = 0.90$
Lightweight concrete	$\phi = 0.70$
Compression-controlled sections with spirals or ties	$\phi = 0.75$
Bearing on concrete	$\phi = 0.70$
Compression in strut and tie models	$\phi = 0.75$

For reinforced concrete decks with only tension reinforcement, the depth of the compression block is given by

$$a = \frac{A_s f_y}{0.85 f_c' b}$$

If the ultimate strength of the deck is written as

$$M_u = \phi k b d^2 \Rightarrow k = \frac{M_u}{\phi b d^2}$$ (113.8)

The strength parameter k may be related to the reinforcement ratio ρ according to

$$\rho = 0.85 \frac{f_c'}{f_y} \left[1 - \sqrt{1 - \frac{k}{1.7 f_c'}} \right]$$ (113.9)

The required reinforcement in the deck is calculated as

$$A_s = \rho b d_e$$ (113.10)

where the effective depth of the slab is given by

$$d_e = \text{total thickness} - \text{bottom cover} - 1/2 \text{ bar diameter} - \text{integral wearing surface}$$

The maximum permitted tensile reinforcement in flexural members, based on a desired level of ductility, corresponds to a lower limit on the tensile strain:

$$\varepsilon_s = 0.003 \left(\frac{d_e - c}{c} \right) > 0.005$$ (113.11)

Distribution Steel in Concrete Slabs

For concrete slabs, distribution reinforcement shall be placed in the secondary direction in the bottom of the slab as the following percentages of the primary reinforcement for positive moment:

For primary reinforcement parallel to traffic

$$\frac{100}{\sqrt{S}} < 50\%$$ (113.12)

For primary reinforcement perpendicular to traffic

$$\frac{220}{\sqrt{S}} < 67\%$$ (113.13)

where S is the effective span length (ft).

Example 113.1

Design a transversely reinforced concrete deck slab for a concrete deck on steel stringer type bridge. The bridge span is 100 ft and the deck is supported by six girders spaced at 8 ft center-to-center spacing. The overall width of the bridge is 46 ft and carries two traffic lanes. The flange width of the steel stringer is 12 in. The design live load is HS20-44. Concrete 28-day compressive strength is 5 ksi and steel reinforcement is grade 60. Provide for a 25 psf future wearing surface.

Solution   For a slab supported continuously over more than two steel stringers, the effective span is given by the sum of the clear distance between flanges and half the flange width:

$$S = 8.0 - 1.0 + 0.5 = 7.5 \text{ ft}$$

Assuming 8-in slab thickness

Dead load per feet width of slab = weight of slab + weight of future wearing surface

$$DC = \frac{8}{12} \times 1 \times 150 = 100 \text{ lb/ft}$$

$$DW = 25 \times 1 = 25 \text{ lb/ft}$$

Factored dead load $D = 1.25 \times 100 + 1.50 \times 25 = 162.5$ lb/ft

The dead load moment for a slab continuous over more than two supports is

$$M_{DL} = \frac{wS^2}{10} = \frac{0.1625 \times 7.5^2}{10} = 0.914 \text{ kip-ft}$$

From the Table 113.8, for $S = 7.5$ ft, live load moment is given by $M_{LL} = 5.44$ kips-ft (includes dynamic allowance). Total factored moment (Strength I):

$$M_u = 0.914 + 1.75 \times 5.44 = 10.43 \text{ kips-ft}$$

Assuming No. 5 main reinforcement bars and 1.0-in clear cover, we get

Effective depth of slab $= 8.0 - 1.0 - 0.625/2 = 6.69$ in

Using Table 105.5

$$X = \frac{M_u}{\phi f_c' bd^2} = \frac{10.43 \times 12}{0.9 \times 5 \times 12 \times 6.69^2} = 0.052 \Rightarrow w = \frac{\rho f_y}{f_c'} = 0.054$$
$$\Rightarrow \rho = 0.0044 \Rightarrow A_s = 0.0044 \times 12 \times 6.69 = 0.35$$

Reinforcement equal to 0.37 in²/ft can be provided by No. 5 bars at 10-in spacing. This reinforcement should be provided on the top and bottom of the slab. Distribution reinforcement is equal to the following percentage of the primary reinforcement: $220/\sqrt{7.5} = 80\%$. Use 67%. Use $0.67 \times 0.37 = 0.25$ in²/ft (use No. 4 at 9-in spacing).

Distribution of Wheel Loads to Girders

Depending on the arrangement of its components, the bridge system may be analyzed as a three-dimensional (3D) system or reduced to equivalent 2D or 1D subsystems for analysis and design of specific components. The validity of such analyses is dependent on the quality of assumptions made in reducing the order of complexity of a subsystem. There are certain analyses that are a hybrid of 1D and 2D systems and are broadly classified as 1.5D level analyses. Similarly, a method which is a hybrid of 2D and 3D analyses could be labeled a 2.5D system.

AASHTO specifications permit the use of so-called distribution factors for analysis.

Bending Moments in Stringers and Longitudinal Beams

The live load bending moment for each interior stringer shall be calculated by applying to the stringer a fraction of the wheel load (front and rear) according to Tables 113.9 and 113.10. The former is to be used to compute distribution factors for concrete deck on steel beams while the latter is for concrete deck on prestressed girders.

This approach of distributing deck loads to girders is limited to bridges that meet the following criteria:

Center-to-center spacing between beams	$3.5 \leq S \leq 16$ ft
Slab thickness	$4.5 \leq t_s \leq 12$ in
Bridge span	$20 \leq L \leq 240$ ft
Longitudinal stiffness parameter	$1 \times 10^4 \leq K_g \leq 7 \times 10^6$ in^4
No. of beams	4 or more

The stiffness parameter K_g is given by

$$K_g = n(I_g + e_g^2 A) \tag{113.14}$$

where n = modular ratio = E_{girder}/E_{deck}
I_g = moment of inertia of the girder
e_g = distance from the centroid of the deck to centroid of the girder

The factors summarized in Tables 113.9 and 113.10 include the effect of the multiple presence factor. The moment or shear is first calculated for a single lane loaded with the most critical load. This load/lane is then multiplied by the lane/girder factor in these tables to obtain the design load/girder. In Tables 113.9 and 113.10, the variable d_e is the distance from the centerline of the exterior girder to the inside face of the curb or barrier.

Effective Width of Flange

When the bridge deck is made composite with the top flange of the girder, a certain width of the deck bends integrally with the girder. This "effective" portion of the deck serves as the compression flange for the composite flexural member.

The slab effective flange width in composite girder and/or stringer systems or in the chords of composite deck trusses may be taken as one-half the distance to the adjacent stringer or girder on each side of the component, or one-half the distance to the adjacent stringer or girder plus the full overhang width.

Table 113.9   Girder Load (Lanes/Girder) for Concrete Deck on Steel Stringers

Internal load	Distribution factor
Moment in interior girder	One design lane loaded
	$$mg = 0.06 + \left(\frac{S}{14}\right)^{0.4}\left(\frac{S}{L}\right)^{0.3}\left(\frac{K_g}{12Lt_s^3}\right)^{0.1}$$
	Multiple design lanes loaded
	$$mg = 0.075 + \left(\frac{S}{9.5}\right)^{0.6}\left(\frac{S}{L}\right)^{0.2}\left(\frac{K_g}{12Lt_s^3}\right)^{0.1}$$
Moment in exterior girder	One design lane loaded Use lever rule
	Multiple design lanes loaded
	$$mg = \left[0.075 + \left(\frac{S}{9.5}\right)^{0.6}\left(\frac{S}{L}\right)^{0.2}\left(\frac{K_g}{12Lt_s^3}\right)^{0.1}\right] \times e$$
	$$e = 0.77 + \frac{d_e}{9.1} \geq 1.0$$
Shear in interior girder	One design lane loaded
	$$mg = 0.36 + \frac{S}{25}$$
	Multiple design lanes loaded
	$$mg = 0.2 + \frac{S}{12} - \left(\frac{S}{35}\right)^2$$
Shear in exterior girder	One design lane loaded Use lever rule
	Multiple design lanes loaded
	$$mg = \left[0.2 + \frac{S}{12} - \left(\frac{S}{35}\right)^2\right] \times e$$
	$$e = 0.6 + \frac{d_e}{10} \geq 1.0$$

Table 113.10 Girder Load (Lanes/Girder) for Concrete Deck on Prestressed Concrete Girders

Internal load	Distribution factor
Moment in interior girder	One design lane loaded $$mg = 0.06 + \left(\frac{S}{14}\right)^{0.4}\left(\frac{S}{L}\right)^{0.3}\left(\frac{K_g}{12Lt_s^3}\right)^{0.1}$$
	Multiple design lanes loaded $$mg = 0.075 + \left(\frac{S}{9.5}\right)^{0.6}\left(\frac{S}{L}\right)^{0.2}\left(\frac{K_g}{12Lt_s^3}\right)^{0.1}$$
Moment in exterior girder	One design lane loaded Use lever rule
	Multiple design lanes loaded $$mg = \left[0.075 + \left(\frac{S}{9.5}\right)^{0.6}\left(\frac{S}{L}\right)^{0.2}\left(\frac{K_g}{12Lt_s^3}\right)^{0.1}\right] \times e$$ $$e = 0.77 + \frac{d_e}{9.1} \geq 1.0$$
Shear in interior girder	One design lane loaded $$mg = 0.36 + \frac{S}{25}$$
	Multiple design lanes loaded $$mg = 0.2 + \frac{S}{12} - \left(\frac{S}{35}\right)^2$$
Shear in exterior girder	One design lane loaded Use lever rule
	Multiple design lanes loaded $$mg = \left[0.2 + \frac{S}{12} - \left(\frac{S}{35}\right)^2\right] \times e$$ $$e = 0.6 + \frac{d_e}{10} \geq 1.0$$

Shear Connector Design

The design guidelines for shear connectors for composite highway bridges are given in AASHTO LRFD 6th ed. Article 6.10.10. For the strength limit state, the resistance factor $\phi_{sc} = 0.85$. For shear connectors in tension, $\phi_{st} = 0.75$.

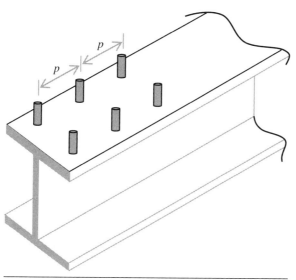

Figure 113.5 Stud-type shear connection ($n = 2$).

AASHTO Article 6.10.10.1 specifies design requirements for stud or channel shear connectors. Simple span bridges shall have shear connectors throughout the length. Straight continuous bridges will normally have shear connectors throughout the length. In negative flexure regions, they shall be provided where longitudinal reinforcement is considered part of the composite section, otherwise not. They help control cracking in these regions. Curved continuous composite bridges shall have shear connectors throughout the length, because of the presence of torsional shear.

Figure 113.5 shows a typical installation of stud-type shear connectors with two shear studs per row. Ratio of height to diameter of shear studs shall not be less than 4. Clear edge distance (between edge of top flange and the first shear stud) shall not be less than 1 in. Clear depth of concrete cover over the top of the shear studs shall not be less than 2 in. Penetration of shear studs into concrete deck shall not be less than 2 in.

Pitch of Shear Connectors

Pitch (longitudinal spacing) shall be determined to satisfy the fatigue limit state. However, the resulting number of shear connectors shall not be less than that required to satisfy strength limit state. Pitch (p) shall satisfy

$$p \leq \frac{nZ_r}{V_{sr}} \tag{113.15}$$

$$V_{fat} = \frac{V_f Q}{l} \tag{113.16}$$

V_{sr} = horizontal fatigue shear range per unit length (resultant of longitudinal V_{fat} and radial F_{fat} components. The latter may be taken as zero for straight bridges with skew less than or equal to 20°).

Z_r = shear fatigue resistance of an individual shear connector.

n = number of shear connectors in a cross section.

The calculated pitch shall not exceed 24 in and shall not be less than six stud diameters.

Transverse c/c spacing shall not be less than four stud diameters.

Fatigue shear resistance of an individual shear connector (Z_r) is calculated as below:

Stud type shear connector

Where 75-year single lane average daily truck traffic (ADTT) $\geq$ 960 trucks per day, Fatigue I load combination shall be used and the fatigue shear resistance for infinite life is

$$Z_r = 5.5d^2 \tag{113.17}$$

For ADTT < 960, Fatigue II load combination shall be used and the fatigue shear resistance for finite life is given by

$$Z_r = \alpha d^2 \tag{113.18}$$

$$\alpha = 34.5 - 4.28 \log N \tag{113.19}$$

where N = number of cycles = $365 \times 75 \times n \times \text{ADTT}_{SL}$
n = number of stress range cycles per truck (Table 6.6.1.2.5.2)

Channel type shear connector

Where 75-year single lane ADTT $\geq$ 1850 trucks per day, Fatigue I load combination shall be used and the fatigue shear resistance for infinite life is

$$Z_r = 2.1w \tag{113.20}$$

ELSE: Fatigue II load combination shall be used and the fatigue shear resistance for finite life is

$$Z_r = Bw \tag{113.21}$$

$$B = 9.37 - 108 \log N \tag{113.22}$$

w = length of channel measure transverse to the direction of the flange.

Factored Shear Resistance of a Single Shear Connector (Q_r)

The available shear strength of a single shear connector is given by

$$Q_r = \phi_{sc} Q_n = 0.85 Q_n \tag{113.23}$$

where Q_n = nominal shear strength of a single shear connector.

Number of shear connectors needed over a region is $n = \dfrac{P}{Q_r}$.

Total nominal shear force (resultant of longitudinal and radial components): $P = \sqrt{P_p^2 = F_p^2}$.

$\quad$ P_p = total longitudinal force in concrete deck is the lesser of: (a) $0.85 f_c' b_s t_s$, which is the maximum compressive resultant when the full depth of the concrete slab is in compression, and (b) $F_{yw} D t_w + F_{yt} b_{ft} t_{ft} + F_{yc} b_{fc} t_{fc}$, which is the maximum tensile resultant when the entire steel beam is in tension yield.

$\quad$ t_w = web thickness

b_{ft} and t_{ft} = full width and thickness of tension flange, respectively.

b_{fc} and t_{fc} = full width and thickness of compression flange, respectively.

Nominal shear resistance of one stud shear connector (Q_n)

The nominal shear resistance of a single stud–type shear connector is given by

$$Q_n = 0.5 A_{sc} \sqrt{f_c' E_c} \le A_{sc} F_u \qquad (113.24)$$

Modulus of elasticity (ksi) of concrete is given by

$$E_c = 33{,}000 K_1 w_c^{1.5} \sqrt{f_c'} \qquad (113.25)$$

w_c = unit weight of concrete (kcf).

f_c' = 28-day compression strength of concrete (ksi).

K_1 = correction factor for source of aggregate to be taken as 1.0 unless otherwise authorized.

A_{sc} = area of a single shear connector

F_u = ultimate strength of shear connector steel

Steps in Shear Connector Design

Step 1: Choose the size (diameter) and transverse layout (n) of shear connectors.

Step 2: Calculate the nominal shear resistance of one shear connector (Q_n).

Step 3: Calculate the design shear resistance of one shear connector ($Q_n = \phi_{sc} Q_n$).

Step 4: Calculate the longitudinal V_{fat} and radial F_{fat} (if present) components of horizontal fatigue shear range V_{sr}. Calculate V_{sr}.

Step 5: Calculate shear fatigue resistance of an individual shear connector (Z_r).

Step 6: Calculate the pitch of shear connectors. Make sure pitch selected satisfies all criteria.

For exterior beams, the effective width of flange may be taken as one-half the effective width of the adjacent interior beam plus the least of (1) one-eighth of the effective span length, (2) six times the slab thickness plus the greater of one-half the web thickness or one-quarter the width of the top flange of the girder, and (3) width of the overhang.

Deflections

Immediate deflections are computed according to standard methods for elastic deflections. The moment of inertia may be taken as either the gross moment of inertia I_g or the effective moment of inertia calculated by Eq. (113.26).

$$I_e = \left(\frac{M_{cr}}{M_a}\right)^3 I_g + \left[1 - \left(\frac{M_{cr}}{M_a}\right)^3\right] I_{cr} \leq I_g \qquad (113.26)$$

where the cracking moment is calculated using

$$M_{cr} = \frac{f_r I_g}{y_t} \qquad (113.27)$$

where f_r = modulus of rupture (ksi) = $0.24\sqrt{f_c'}$
f_c' = 28-day compression strength (ksi)
y_t = distance from centroid to extreme tension fiber

For calculating long-term deflections, the immediate deflection caused by sustained loads is multiplied by one of the following factors:

1. Where the immediate deflection is based on I_g, the factor is 4.

2. Where the immediate deflection is based on I_e, the factor is

$$3 - 1.2\frac{A_s'}{A_s} \geq 1.6$$

where A_s' is the area of the compression reinforcement and A_s is the area of the tension reinforcement.

Timber Design

This chapter reviews some of the provisions of the *National Design Specification for Wood Construction ASD,* 2005 edition and *National Design Specification Supplement, Design Values for Wood Construction, 2005* (American Forest & Paper Association). Specifically, the chapter focuses on the design of sawn timber beams and joists and timber columns.

Note: Only the ASD criteria are summarized in this chapter.

Bending Stress

Using the allowable stress method, the flexural design criterion for timber beams may be stated as

$$f_b = \frac{M}{S} \le F_b' \tag{114.1}$$

where f_b = calculated bending stress
S = section modulus
F_b' = allowable bending stress

The section modulus for a rectangular section (width b, depth d) is given by

$$S = \frac{bd^2}{6}$$

The allowable bending stress parallel to the grain (F_b') is calculated as

$$F_b' = F_b \times C_D \times C_M \times C_t \times C_L \times C_F \times C_V \times C_{fu} \times C_r \times C_c \times C_f \tag{114.2}$$

where F_b = reference design value in the NDS[1] supplement
C_D = load duration factor (NDS Table 2.3.2)
C_M = wet service factor (used when moisture content is > 19%)
C_t = temperature factor (used when timber is used in temperature > 150°F)

---

[1]*National Design Specification for Wood Construction ASD/LRFD,* 2012 edition and *National Design Specification Supplement, Design Values for Wood Construction,* 2012 edition, American Forest and Paper Association, Washington, D.C.

C_L = beam stability factor

C_F = size factor (only for visually graded sawn lumber members and round timber bending members; do not apply simultaneously with C_v for glued laminated timber)

C_V = volume factor (apply only to glued laminated timber bending member)

C_{fu} = flat use factor (when 2- to 4-in timber is loaded on wide face)

C_r = repetitive member factor (apply to dimension bending member 2 to 4 in thick)

C_c = curvature factor (apply to curved glued laminated bending member)

C_f = form factor (for round or diamond section)

Shear Stress

The shear design criterion for timber beams may be stated as

$$f_v = \frac{VQ}{Ib} \leq F_v' \tag{114.3}$$

where f_v = calculated maximum shear stress

F_v' = allowable shear stress

For a rectangular section (width b, depth d), the maximum shear stress due to a transverse shear force V is given by

$$f_v = \frac{3V}{2bd} \tag{114.4}$$

The allowable shear stress is calculated as

$$F_v' = F_v \times C_D \times C_M \times C_t \times C_H \tag{114.5}$$

where F_v = reference design value in the NDS supplement

C_H = shear stress factor depends on length of split and shake. Value of C_H varies from 2 for no split to 1 with 1 1/2 split

Modulus of Elasticity

The elastic modulus shall be calculated as $E' = E \times C_M \times C_t$, where E is the modulus of elasticity in the NDS supplement.

Stress Modification Factors

Values of the load duration factor are summarized in Table 114.1. For a member with nominal depth D and width B, $C_L = 1$ for the conditions summarized in Table 114.2. When the conditions in Table 114.2 are not met, C_L is calculated according to NDS Section 3.3.3.7.

Size Factor C_F

For southern pine 2 in to 4 in thick, size factor need not be applied. For southern pine 4 in thick, 8 in and wider, $C_F = 1.1$. For dimension lumber wider than 12 in, $C_F = 0.9$ except dense

Table 114.1 Load Duration Factor C_D

Load duration	C_D	Design load
Permanent	0.9	Dead load
Ten years	1.0	Occupancy live load
Two months	1.15	Snow load
Seven days	1.25	Construction load
Ten minutes	1.6	Wind/earthquake load
Impact	2.0	Impact load

Table 114.2 Condition for which Beam Stability Factor $C_L = 1$

D/B	Requirements
$D/B \leq 2$	None
$2 < D/B \leq 4$	Solid blocking is provided at both ends of member.
$D/B = 5$	One edge (tension or compression) is fully supported.
$D/B = 6$	Bridge, full depth blocking, cross bracing at 8 ft maximum, and both edges are fully supported or compressive edge is fully supported to prevent lateral displacement, and the ends at the point of bearing are laterally supported to prevent rotation.
$D/B = 7$	Both edges are fully supported.

structural 86, 72, and 65 for which, $C_F = 0.9$. When the depth of dense structural 86, 72, and 65, dimension lumber exceeds 12 in, $C_F = (12/d)^{1/9}$.

Repetitive Member Factor C_r

C_r applies to dimension lumber 2 in to 4 in thick that is subjected to bending. $C_r = 1.15$, when members are used as joists, truss chords, rafters, and the like, and spacing does not exceed 24 in and not less than 3 in.

Wet Service Factor C_M

When the moisture of dimension lumber exceeds 19%, the design value F_b shall be multiplied by $C_M = 0.85$ except that when $F_b \cdot C_M \leq 1500$ psi, $C_M = 1$.

Example 114.1

Floor joists spaced every 16 in carry a floor over a simple span $L = 16$ ft. Assume that the top of the joist is laterally supported by the plywood sheathing. The floor loads are live load $W_L = 40$ psf, dead load $W_D = 10$ psf, superimposed dead load $W_{SD} = 8$ psf.

Timber: Southern pine, moisture less than 19%, used at normal room temperature.

Nominal size of floor joists: 2 in $\times$ 10 in

Calculate the (1) maximum bending stress, (2) allowable bending stress, (3) maximum shear stress, (4) allowable shear stress, and (5) maximum deflection.

Solution

$$\text{Load duration factor for dead load } C_D = 0.9$$

$$\text{Load duration factors for live load } C_D = 1.0$$

$$\text{Calculate design load } W = \left(\frac{W_D + W_{SD}}{0.9} + \frac{W_L}{1.0}\right) \times \frac{16}{12} = 80 \text{ lb/ft}$$

$$\text{Design moment } M = \frac{wL^2}{8} = \frac{80 \times 16^2}{8} = 2560 \text{ lb-ft}$$

$$\text{Nominal dimensions } B = 2 \text{ in}, D = 10 \text{ in}$$

$$\text{Actual dimensions } b = 1.5 \text{ in}, d = 9.25 \text{ in}$$

$$\text{Section modulus } S = 21.39 \text{ in}^3$$

$$\text{Modulus of inertia } I = 98.93 \text{ in}^4 = 4.771 \times 10^{-3} \text{ ft}^4$$

Maximum Bending Stress

$$f_b = \frac{M}{S} = \frac{2560 \times 12}{21.39} = 1436 \text{ psi}$$

For Southern pine No. 2, $F_b = 1500$ psi. The depth-to-width ratio based on nominal dimension, $D/B = 5$. Since compressive edge is fully supported by plywood floor, $C_L = 1$.

$$\text{Repetition factor for joist } C_r = 1.15$$

$$\text{Wet service factor } C_M = 1$$

$$\text{Temperature factor } C_t = 1$$

Other factors are not applicable.

Allowable Bending Stress

$$F_b' = F_b \times C_L \times C_M \times C_t \times C_r = 1500 \times 1 \times 1.15 \times 1 \times 1 = 1725 \text{ psi}$$

Since the maximum bending stress is less than the allowable bending stress, the flexure check is satisfied.

Maximum Shear Force

$$V = \frac{wL}{2} = 640 \text{ lb}$$

Maximum Shear Stress

$$f_v = \frac{3V}{2bd} = \frac{3 \times 640}{2 \times 1.5 \times 9.25} = 69.2 \text{ psi}$$

Conservatively assume shear stress factor $C_H = 1$

Allowable shear stress $F_v = 90 \times C_M \times C_t \times C_H = 90$ psi

Since the maximum shear stress is less than the allowable shear stress, the shear check is satisfied.

Deflection

Elastic modulus $E = 1.6 \times 10^6 \times C_M \times C_t = 1.6 \times 10^6$ psi $= 2.3 \times 10^8$ psf

$$\text{Deflection } \Delta = \frac{5wL^4}{384EI} = \frac{5 \times 80 \times 16^4}{384 \times 2.304 \times 10^8 \times 4.771 \times 10^{-3}} = 0.0621 \text{ ft} = 0.75 \text{ in}$$

Design of Timber Columns

Buckling Limit State

The Euler stress is expressed in the NDS as

$$F_{cr} = \frac{\pi^2 E'_{min}}{(l_e/r)^2} \tag{114.6}$$

For a rectangular column, for which the radius of gyration (r) can be expressed in terms of the least column dimension (d) this relationship for the allowable bucking stress (F_{cE}) becomes

$$F_{cE} = \frac{0.822 E'_{min}}{(l_e/d)^2} \tag{114.7}$$

where l_e is effective unbraced length of the column.

Effective length factors (K) for timber columns are summarized in Fig. 114.1.

Crushing Limit State

At zero slenderness, there is no buckling and the column's limit state is crushing. The allowable stress for this mode of failure is given by

$$F_c^* = F_c \times C_D \times C_M \times C_t \times C_F \times C_i \tag{114.8}$$

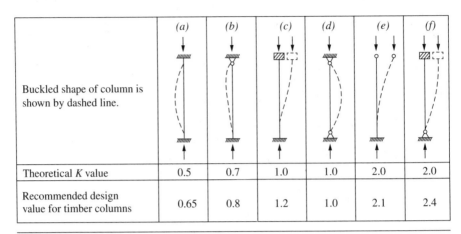

	(a)	(b)	(c)	(d)	(e)	(f)
Buckled shape of column is shown by dashed line.						
Theoretical K value	0.5	0.7	1.0	1.0	2.0	2.0
Recommended design value for timber columns	0.65	0.8	1.2	1.0	2.1	2.4

Figure 114.1 Recommended K values for timber columns.

The modification factors C_D, C_M, and C_t are as defined earlier. C_F is a size factor ($= 1.0$ for timbers). The factor C_i is an incising factor for sawn lumber ($= 0.80$ for incised dimension lumber $= 1.0$ for sawn lumber not incised). F_c is the reference compressive design value parallel to the grain.

Ylinen Column Equation

The Ylinen column equation is an interaction equation between these two limit states. According to this equation, the adjusted compressive design stress in a timber column is given by

$$F_c' = F_c \times C_P \times \text{product of all other relevant factors} \tag{114.9}$$

$$C_P = \frac{1 + F_{cE}/F_c^*}{2c} - \sqrt{\left(\frac{1 + F_{cE}/F_c^*}{2c}\right)^2 - \frac{F_{cE}/F_c^*}{c}} \tag{114.10}$$

where $c =$ buckling and crushing interaction parameter
$= 0.80$ for sawn lumber columns
$= 0.85$ for round timber poles and piles
$= 0.90$ for glulam or structural composite lumber columns

Section Properties of Beams and Joists

Section properties of timber beams and joists are shown in Table 114.3.

Table 114.3 Section Properties of Beams and Joists

Nominal size B × h (in)	Design size b × h (in)	Area A (in²)	Section modulus S (in³)	Moment of inertia I (in⁴)
2 × 2	1.5 × 1.5	2.250	0.563	0.422
2 × 3	1.5 × 2.5	3.750	1.563	1.953
2 × 4	1.5 × 3.5	5.250	3.063	5.359
2 × 6	1.5 × 5.5	8.250	7.563	20.797
2 × 8	1.5 × 7.25	10.875	13.141	47.635
2 × 10	1.5 × 9.25	13.875	21.391	98.932
2 × 12	1.5 × 11.25	16.875	31.641	177.979
2 × 14	1.5 × 13.25	19.875	43.891	290.775
3 × 3	2.5 × 2.5	6.250	2.604	3.255
3 × 4	2.5 × 3.5	8.750	5.104	8.932
3 × 6	2.5 × 5.5	13.750	12.604	34.661
3 × 8	2.5 × 7.25	18.125	21.901	79.391
3 × 10	2.5 × 9.25	23.125	35.651	164.886
3 × 12	2.5 × 11.25	28.125	52.734	296.631
3 × 14	2.5 × 13.25	33.125	73.151	484.626
3 × 16	2.5 × 15.25	38.125	96.901	738.870
4 × 4	3.5 × 3.5	12.250	7.146	12.505
4 × 6	3.5 × 5.5	19.250	17.646	48.526
4 × 8	3.5 × 7.25	25.375	30.661	111.148
4 × 10	3.5 × 9.25	32.375	49.911	230.840
4 × 12	3.5 × 11.25	39.375	73.828	415.283
4 × 14	3.5 × 13.25	46.375	102.411	678.476
4 × 16	3.5 × 15.25	53.375	135.661	1034.419
6 × 6	5.5 × 5.5	30.250	27.729	76.255
6 × 8	5.5 × 7.5	41.250	51.563	193.359
6 × 10	5.5 × 9.5	52.250	82.729	392.964
6 × 12	5.5 × 11.5	63.250	121.229	697.068
6 × 14	5.5 × 13.5	74.250	167.063	1127.672
6 × 16	5.5 × 15.5	85.250	220.229	1706.776
6 × 18	5.5 × 17.5	96.250	280.729	2456.380

(*Continued*)

Table 114.3 Section Properties of Beams and Joists (*Continued*)

Nominal size B × h (in)	Design size b × h (in)	Area A (in²)	Section modulus S (in³)	Moment of inertia I (in⁴)
6 × 20	5.5 × 19.5	107.250	348.563	3398.484
8 × 8	7.5 × 7.5	56.250	70.313	263.672
8 × 10	7.5 × 9.5	71.250	112.813	535.859
8 × 12	7.5 × 11.5	86.250	165.313	950.547
8 × 14	7.5 × 13.5	101.250	227.813	1537.734
8 × 16	7.5 × 15.5	116.250	300.313	2327.422
8 × 18	7.5 × 17.5	131.250	382.813	3349.609
8 × 20	7.5 × 19.5	146.250	475.313	4634.297
8 × 22	7.5 × 21.5	161.250	577.813	6211.484
8 × 24	7.5 × 23.5	176.250	690.313	8111.172
10 × 10	9.5 × 9.5	90.250	142.896	678.755
10 × 12	9.5 × 11.5	109.250	209.396	1204.026
10 × 14	9.5 × 13.5	128.250	288.563	1947.797
10 × 16	9.5 × 15.5	147.250	380.396	2948.068
10 × 18	9.5 × 17.5	166.250	484.896	4242.839
10 × 20	9.5 × 19.5	185.250	602.063	5870.109
10 × 22	9.5 × 21.5	204.250	731.896	7867.880
12 × 12	11.5 × 11.5	132.250	253.479	1457.505
12 × 14	11.5 × 13.5	155.250	349.313	2357.859
12 × 16	11.5 × 15.5	178.250	460.479	3568.714
12 × 18	11.5 × 17.5	201.250	586.979	5136.068
12 × 20	11.5 × 19.5	224.250	728.813	7105.922
12 × 22	11.5 × 21.5	247.250	885.979	9524.276
12 × 24	11.5 × 23.5	270.250	1058.479	12437.130

Section Properties of Planks

Section properties of timber planks are shown in Table 114.4.

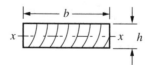

Table 114.4   Section Properties of Planks

Nominal size $b \times h$ (in)	Design size $b \times h$ (in)	Area A (in²)	Section modulus S (in³)	Moment of inertia I (in⁴)
3 × 2	2.5 × 1.5	3.750	0.938	0.703
4 × 2	3.5 × 1.5	5.250	1.313	0.984
6 × 2	5.5 × 1.5	8.250	2.063	1.547
8 × 2	7.25 × 1.5	10.875	2.719	2.039
10 × 2	9.25 × 1.5	13.875	3.469	2.602
12 × 2	11.25 × 1.5	16.875	4.219	3.164
4 × 3	3.5 × 2.5	8.750	3.646	4.557
6 × 3	5.5 × 2.5	13.750	5.729	7.161
8 × 3	7.25 × 2.5	18.125	7.552	9.440
10 × 3	9.25 × 2.5	23.125	9.635	12.044
12 × 3	11.25 × 2.5	28.125	11.719	14.648
14 × 3	13.25 × 2.5	33.125	13.802	17.253
16 × 3	15.25 × 2.5	38.125	15.885	19.857
6 × 4	5.5 × 3.5	19.250	11.229	19.651
8 × 4	7.25 × 3.5	25.375	14.802	25.904
10 × 4	9.25 × 3.5	32.375	18.885	33.049
12 × 4	11.25 × 3.5	39.375	22.969	40.195
14 × 4	13.25 × 3.5	46.375	27.052	47.341
16 × 4	15.25 × 3.5	53.375	31.135	54.487

Section Properties of Decking

Section properties of timber decking are shown in Table 114.5.

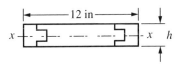

Table 114.5   Section Properties of Decking

Nominal size h (in)	Design size $b \times h$ (in)	Area A (in²)	Section modulus S (in³)	Moment of inertia I (in⁴)
2	12 × 1.5	18.00	4.50	3.375
3	12 × 2.5	30.00	12.50	15.625
4	12 × 3.5	42.00	24.50	42.875

Table 114.6 Section Properties for Class I, Class II, and Structural I Plyforms

Thickness (in)	Weight (psf)	Properties for stress applied parallel to face grain			Properties for stress applied perpendicular to face grain		
		Moment of inertia, I (in⁴/ft)	Section modulus, S (in³/ft)	Rolling shear constant, lb/Q (in²/ft)	Moment of inertia, I (in⁴/ft)	Section modulus, S (in³/ft)	Rolling shear constant, lb/Q (in²/ft)
Class I							
15/32	1.4	0.066	0.244	4.743	0.018	0.107	2.419
½	1.5	0.077	0.268	5.153	0.024	0.130	2.739
19/32	1.7	0.115	0.335	5.438	0.029	0.146	2.834
5/8	1.8	0.130	0.358	5.717	0.038	0.175	3.094
23/32	2.1	0.180	0.430	7.009	0.072	0.247	3.798
¾	2.2	0.199	0.455	7.187	0.092	0.306	4.063
7/8	2.6	0.296	0.584	8.555	0.151	0.422	6.028
1	3.0	0.427	0.737	9.374	0.270	0.634	7.014
1 1/8	3.3	0.554	0.849	10.430	0.398	0.799	8.419
Class II							
15/32	1.4	0.063	0.243	4.499	0.015	0.138	2.434
½	1.5	0.075	0.267	4.891	0.020	0.167	2.727
19/32	1.7	0.115	0.334	5.326	0.025	0.188	2.812
5/8	1.8	0.130	0.357	5.593	0.032	0.225	3.074
23/32	2.1	0.180	0.430	6.504	0.060	0.317	3.781
¾	2.2	0.198	0.454	6.631	0.075	0.392	4.049
7/8	2.6	0.300	0.591	7.990	0.123	0.542	5.997
1	3.0	0.421	0.754	8.614	0.220	0.812	6.987
1 1/8	3.3	0.566	0.869	9.571	0.323	1.023	8.388
Structural I							
15/32	1.4	0.067	0.246	4.503	0.021	0.147	2.405
½	1.5	0.078	0.271	4.908	0.029	0.178	2.725
19/32	1.7	0.116	0.338	5.018	0.034	0.199	2.811
5/8	1.8	0.131	0.361	5.258	0.045	0.238	3.073
23/32	2.1	0.183	0.439	6.109	0.085	0.338	3.780
¾	2.2	0.202	0.464	6.189	0.108	0.418	4.047
7/8	2.6	0.317	0.626	7.539	0.179	0.579	5.991
1	3.0	0.479	0.827	7.978	0.321	0.870	6.981
1 1/8	3.3	0.623	0.955	8.841	0.474	1.098	8.377

Masonry Design

This chapter is dedicated to the ASD design methods for masonry structures as outlined by TMS 402/602 *Building Code Requirements and Specifications for Masonry Structures, 2013* The Masonry Society, Longmont, CO. The official syllabus for the PE Civil (Structural depth) exam states that "Examinees will use only the ASD method, except strength design Section 9.3.5 may be used for walls with out-of-plane loads."

Basic Components of Masonry Structures

Masonry can be used for a wide variety of architectural elements such as bearing walls, shear walls, arches, domes, beams, and columns. The three basic components of masonry structures are masonry units (fired clay or concrete), masonry mortar, and grout. In the United States, three basic types of cementitious mortar are used — cement-lime mortar, masonry-cement mortar and mortar-cement mortar. Grout is fluid concrete (8- to 11-in slump) that is used to fill some or all cells in hollow units, or between wythes. TMS 402-13 permits the use of self-consolidating grout.

Masonry Unit Dimensions

Typically, masonry unit dimensions are specified as thickness × height × length. For example, a typical clay masonry unit has (nominal) dimensions 4 × 2.67 × 8. These nominal dimensions are the actual dimension of the unit plus one half joint thickness on either side. Joint thickness is typically 3/8 in. Thus, the unit named above is 4-in thick, 2.67-in high, and 8-in long. The bed is formed by the thickness × length.

Masonry units can be placed in various bond patterns, such as stack bond, running bond, Flemish bond, etc.

Types of Masonry Mortar

According to ASTM C270, there are four different types of cementitious mortar – M, S, N, and O. The characteristics of these different mortar types are:

- Type M: High compressive (>2500 psi) and tensile bond strength
- Type S: Moderate compressive (1800–2500 psi) and tensile bond strength

307

- Type N: Low compressive (750–1800 psi) and tensile bond strength
- Type O: Very low compressive (350–750 psi) and tensile bond strength

Typical Properties of Concrete Masonry

Compressive strength of concrete masonry is typically between 1,500 and 3,000 psi. ASTM C90 requires a minimum compressive strength of 2,000 psi (measured on the net area). Tensile strength is typically about 10% of compressive strength. Tensile bond strength (strength between mortar and concrete masonry units) is typically 40 to 75 psi (with Portland cement-lime mortar) and about 35 psi or less with masonry-cement mortar. Modulus of elasticity is typically from 1 to 3 × 10^6 psi.

Grout Compressive Strength

For concrete masonry, the specified compressive strength of the grout (f_g') shall equal or exceed the specified compressive strength of the masonry (f_m') but shall not exceed 5,000 psi. For clay masonry, f_g' shall not exceed 6,000 psi.

Strain Compatibility

For any material pair (steel-concrete, steel-masonry, etc.) exhibiting composite behavior, the basic assumption is that the strain in both materials at the same location must be equal.

$$\varepsilon_1 = \varepsilon_2 \Rightarrow \frac{\sigma_1}{E_1} = \frac{\sigma_2}{E_2} \Rightarrow \frac{\sigma_1}{\sigma_2} = \frac{E_1}{E_2} = n \tag{115.1}$$

where n is called the modular ratio.

The modular ratio is used to "transform" one of the materials to an equivalent area of the other. Conventionally, for reinforced masonry, the steel reinforcement is converted to an equivalent area of masonry.

Modulus of Elasticity of Masonry

For clay and concrete masonry, the modulus of elasticity is given by

$$\begin{aligned} E_m &= 700 f_m' \qquad \text{(Clay Masonry)} \\ E_m &= 900 f_m' \qquad \text{(Concrete Masonry)} \end{aligned} \tag{115.2}$$

where E_m = modulus of elasticity of masonry (psi)
f_m' = specified compressive strength of the masonry (psi)

The modulus of rigidity G is given by

$$G = 0.4 E_m \tag{115.3}$$

Poisson's ratio $\nu = 0.25$ is implicit in Eq. (115.3).

Transformed Section

When a masonry member is subjected to bending (such as causes convexity of the bottom surface), the masonry above the neutral axis of the cross section is in compression. The masonry below the neutral axis is assumed to be cracked. The transformed section consists of the area of masonry above the neutral axis and n times the reinforcing steel area below the neutral axis. The transformed area of steel in tension is $A = nA_s$. When the reinforcement and surrounding masonry is in compression, such as a column with a concentric axial load, the transformed area is one of the following:

For long-term loading conditions: $A = (2n - 1)A_s$

For other than long-term loading conditions: $A = (n - 1)A_s$

Using $(n - 1)$ or $(2n - 1)$ rather than n, accounts for the area of masonry in compression being occupied by the actual steel area. Assumed dimensions of several nominal sizes of masonry blocks are shown in Table 115.3. For the purpose of calculating dead loads, normal-weight masonry units are assumed to have an oven-dry weight of 145 pcf.

Allowable Stresses in Masonry

The allowable stress provisions of the 2008 and prior editions of the MSJC code used to permit allowable stresses to be increased by 33% for loading combinations involving wind and earthquake loads. This allowance was removed from the 2011 edition of the code, while at the same time adjusting the allowable stresses.

Unreinforced Masonry

Axial Compression
Under axial compression of unreinforced masonry elements such as walls and columns, the axial stress (f_a) due to applied loads must be less than the allowable compressive stress (F_a).

The axial stress in a masonry wall (f_a) is found as follows:

$$f_a = \frac{P}{A_e} \tag{115.4}$$

where P = the axial load
A_e = the area of the element effective in compression

When determining the capacity of a masonry wall element in compression, the compression reinforcement in the element will be neglected since its contribution is not significant. Only tied compression reinforcement, such as in a column or pilaster, will be considered effective.

Allowable Compressive Stress (F_a)
The allowable stress in compression for an element such as a wall or a column is dependent on the slenderness ratio of the element expressed as (h/r)

where h = effective height of the column = KL

$\quad L$ = actual height

$\quad r$ = least column dimension

For unreinforced masonry compression elements, the allowable axial compressive stress is given by

$$F_a = 0.25 f'_m \left[1 - \left(\frac{h}{140r} \right)^2 \right] \qquad \text{for} \qquad h/r \leq 99 \qquad (115.5a)$$

$$F_a = 0.25 f'_m \left(\frac{70r}{h} \right)^2 \qquad \text{for} \qquad h/r > 99 \qquad (115.5b)$$

As an additional check for stability against an eccentrically applied axial load is given by Eq. (115.6), which limits the axial load to one-fourth of the Euler buckling load (P_e).

$$P \leq 0.25 P_e = 0.25 \left(\frac{\pi^2 E_m I_n}{h^2} \right) \left(1 - 0.577 \frac{e}{r} \right)^3 \qquad (115.6)$$

Allowable Compressive Stress in Flexural Members

For masonry elements subject to an axial compression force P and a bending moment M, the resulting flexural bending stress is calculated as

$$f_b = \frac{Mt}{2I_n} - \frac{P}{A_n} \qquad (115.7)$$

where I_n and A_n are moment of inertia and cross sectional area of the net section.

If the value of the combined stress f_b as given by Eq. (115.7) is positive, the section experiences a net tensile stress, which is limited to the values tabulated in Table 115.1, which shows limits on flexural tension for masonry units (psi).

On the other hand, if the value of the combined stress f_b as given by Eq. (115.7) is negative, the section is in compression, and is subject to the limiting compressive stress, as given by Eq. (115.8).

$$F_b = 0.33 f'_m \qquad (115.8)$$

Combined Axial Compression and Flexure

When unreinforced masonry elements are subjected to a combination of axial load and flexure, the following interaction equation is used to ensure that critical sections remain uncracked under design loads.

$$\frac{f_a}{F_a} + \frac{f_b}{F_b} \leq 1 \qquad (115.9)$$

Table 115.1   Maximum Flexural Tension in Masonry Units

Direction of flexural tensile stress and masonry type	Portland cement/lime or mortar cement		Masonry cement or air entrained Portland cement/lime	
	M or S	N	M or S	N
Normal to bed joints in running or stack bond				
Solid units	53	40	32	20
Hollow units				
Ungrouted	33	25	20	12
Fully grouted	86	84	81	77
Parallel to bed joints in running bond				
Solid units	106	80	64	40
Hollow units				
Ungrouted/partially grouted	66	50	40	25
Fully grouted	106	80	64	40
Parallel to bed joints in masonry not laid in running bond				
Continuous grout section parallel to bed joints	133	133	133	133
Other	0	0	0	0

The allowable masonry stress (compression) due to combined flexure and axial loads has been increased to $0.45f'_m$.

Unreinforced Shear

Shear stress on unreinforced masonry elements are calculated using the net cross-sectional properties in the direction of the applied shear force according to Eq. (115.10).

$$f_v = \frac{VQ}{I_n b} \tag{115.10}$$

where V = the shear load (lbs)
I_n = moment of inertia of net section

The basic check for shear is given by

$$f_v < F_v \tag{115.11}$$

where F_v = allowable shear stress

Allowable In-Plane Shear Stress (F$_v$) in Flexural Members for Unreinforced Masonry

With no shear reinforcement, the allowable shear stress F_v (psi) is the smallest of

(a) $1.5\sqrt{f'_m}$

(b) 120

(c) $37\,\text{psi} + 0.45(N_v/A_v)$ for running bond masonry not fully grouted

(d) $37\,\text{psi} + 0.45(N_v/A_v)$ for masonry not in running bond, constructed of open-end units and fully grouted

(e) $60 + 0.45(N_v/A_v)$ for running bond masonry fully grouted

(f) 15 psi for masonry not laid in running bond, constructed of other than open-end units and fully grouted

A_v = Effective area of shear reinforcement (in^2)
N_v = Compressive force acting normal to the shear surface (lb)

In the MSJC, these limits on allowable shear stress are specified for in-plane shear only. Allowable stresses for out-of-plane shear are not provided. However, the commentary suggests using these same limits for out-of-plane shear.

Reinforced Masonry

For reinforced masonry, the tensile resistance provided by the masonry units, mortar, and grout are neglected while calculating the strength of the section. In other words, the portion of the section subject to net tension are assumed to be cracked.

Allowable Tensile Stress in the Reinforcement

The allowable stresses in the reinforcement steel are:

- For grade 60 bars in tension: $F_s = 32{,}000$ psi
- For grade 40 and grade 50 bars in tension: $F_s = 20{,}000$ psi
- For wire joint reinforcement in tension: $F_s = 30{,}000$ psi

Reinforced Out-of-Plane Flexure

For reinforced masonry, the allowable compressive stress in masonry due to flexure or flexural + axial load is given by

$$f_b \leq F_b = 0.45 f'_m \qquad (115.12)$$

Allowable Bearing Stress

For a member bearing on the full area of a masonry element, the allowable bearing stress is given by

$$F_{br} = 0.33 f_m'$$ (115.13)

For a member bearing on one-third or less of a masonry element, the bearing stress is given by

$$F_{br} = 0.45 f_m'$$ (115.14)

Allowable Flexural Tension for Clay and Concrete Masonry Units

Modulus of Rupture

For strength design, the nominal flexural strength is computed using the values of modulus of rupture values given in Table 115.2. These values are intended to be 2.5 times the corresponding values in Table 115.1. For masonry elements subjected to out-of-plane bending, the modulus of rupture shall be in accordance with Table 115.2. Stress values are given in psi.

Rectangular Beam Analysis

A rectangular masonry section subject to flexure is shown in Fig. 115.1. The following analysis is applicable to fully grouted masonry elements and partially grouted masonry elements with the neutral axis in the compression face shell. The masonry below the neutral axis is assumed cracked (in tension). For allowable stress design, the stresses are expected to remain well within the elastic limit.

In the design of reinforced rectangular sections, the first step is to locate the neutral axis.

This can be accomplished by determining the coefficient, k, which is the ratio of the depth of the compressive stress block to the total depth from the compression face to the reinforcing steel, d. The coefficient k is given by

$$k = \sqrt{(n\rho)^2 + 2n\rho} - n\rho$$ (115.15)

where n = the modular ratio = E_s/E_m
ρ = steel ratio = A_s/bd

The coefficient j, which is the ratio of the distance between the resultant compressive force and the centroid of the tensile force to the effective depth d, is calculated in Eq. (115.10).

$$j = 1 - \frac{k}{3}$$ (115.16)

Table 115.2 Modulus of Rupture for Out-of-Plane Bending

Masonry Type	Portland cement/lime or mortar cement		Masonry cement or air entrained Portland cement/lime	
	M or S	N	M or S	N
Normal to bed joints				
Solid units	133	100	80	51
Hollow units				
Ungrouted	84	64	51	31
Fully grouted	163	158	153	145
Parallel to bed joints in running bond				
Solid units	267	200	160	100
Hollow units				
Ungrouted/partially grouted	167	127	100	64
Fully grouted	267	200	160	100
Parallel to bed joints in stack bond				
Continuous groute section parallel to bed joints	335	335	335	335
Other	0	0	0	0

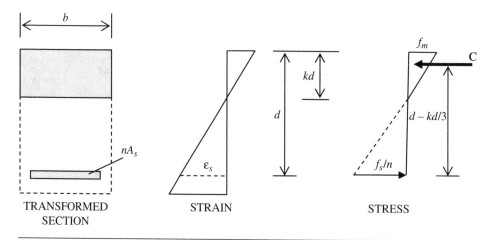

Figure 115.1 Transformed masonry section—strain and stress.

Balanced Steel Ratio

The balanced steel ratio in the working stress design method, (ρ_e) is defined as the reinforcing ratio where the steel and the masonry reach their maximum allowable stresses for the same applied moment. This ratio is calculated using Eq. (115.17).

$$\rho_e = \frac{n}{2\left(\dfrac{F_s}{F_m}\right)\left(n + \dfrac{F_s}{F_m}\right)} \qquad (115.17)$$

where F_s = the allowable tensile stress in the reinforcing steel, psi
$\quad F_m$ = the allowable flexural compressive stress in the masonry, psi.

 NOTE   ACI uses the strength design method for concrete and the definition of the balanced reinforcement for a concrete section is the amount of steel for which both concrete and steel reach their *ultimate* stress or strain states simultaneously. For this reason, the code restricts the maximum permitted reinforcement for concrete sections to be less than this balanced reinforcement (approximately 75%). However, because the balanced reinforcement for the masonry section is based on both masonry and steel reaching their *allowable* stresses simultaneously, there is no such restriction. In fact, to perform an optimized design, one can use the balanced steel ratio, given the allowable stresses on the masonry (F_m) and the steel (F_s).

Flexure of Reinforced Masonry Elements

For a rectangular masonry beam with tension reinforcement, subjected to a moment M, the computed working stresses for the steel (f_s) and the masonry (f_m) are calculated using Eqs. (115.18) and (115.19).

$$\text{Tensile stress in the steel } f_s = \frac{M}{A_s jd} \qquad (115.18)$$

$$\text{Maximum (compression) stress in the masonry } f_m = \frac{2M}{kjbd^2} \qquad (115.19)$$

If $\rho < \rho_e$, the steel stress (f_s) will reach its allowable stress before the masonry ($f_s = F_s$). The resisting moment for the reinforcement, (M_{rs}) can be determined by substituting the allowable steel stress, F_s for the computed steel stress:

$$M_{rs} = \frac{F_s A_s jd}{12} \qquad (115.20)$$

If $\rho > \rho_e$, the masonry stress (f_m) will reach its allowable stress before the steel $(f_m = F_m)$. The resisting moment for masonry (M_{rm}) can be determined by substituting the allowable masonry stress (F_m) for the computed masonry stress:

$$M_{rm} = \frac{1}{2}F_m kjbd^2 \tag{115.21}$$

Allowable Compressive Load for Reinforced Masonry Elements

For reinforced masonry columns, the allowable axial compressive load is given by

$$P_a = \left(0.25f_m'A_n + 0.65A_sF_s\right)\left[1 - \left(\frac{h}{140r}\right)^2\right] \quad \text{for} \quad h/r \leq 99 \tag{115.22a}$$

$$P_a = \left(0.25f_m'A_n + 0.65A_sF_{sc}\right)\left(\frac{70r}{h}\right)^2 \quad \text{for} \quad h/r > 99 \tag{115.22b}$$

where F_s = allowable axial compression in steel = $0.4f_y$, not to exceed 24 ksi
A_s = area of longitudinal steel
A_n = effective or net cross-sectional area of masonry

Equation (115.22) applies only if compression reinforcement with lateral confinement in the form of ties or stirrups is provided.

Allowable In-Plane Shear Stress (F_v) in Flexural Members for Reinforced Masonry

As of the 2011 MSJC code, the shear resistance provided by the masonry (F_{vm}) is added to the shear resistance provided by the shear reinforcement (F_{vs}).

$$F_v = F_{vm} + F_{vs} \tag{115.23}$$

The allowable shear stress (F_v) is limited to the following values:

$$F_v \leq 3\sqrt{f_m'} \quad \text{for } M/Vd \leq 0.25 \tag{115.24}$$

$$F_v \leq 2\sqrt{f_m'} \quad \text{for } M/Vd \geq 1.0 \tag{115.25}$$

When the value of M/Vd falls between 0.25 and 1.0, the maximum value of F_v may be linearly interpolated between the limits given by Eqs. (115.24) and (115.25).

The values of F_{vm} and F_{vs} are calculated using Eqs. (115.26) and (115.27), respectively.

$$F_{vm} = \frac{1}{2}\left[\left(4.0 - 1.75\frac{M}{Vd}\right)f'_m\right] + 0.25\frac{P}{A_n} \tag{115.26}$$

$$F_{vs} = \frac{1}{2}\left(\frac{A_v F_s d}{A_n s}\right) \tag{115.27}$$

When $f_v > F_{vm}$, shear reinforcement must be provided according to the following guidelines:

- The shear reinforcement must be parallel to the direction of the shear force.
- Shear reinforcement spacing (s) must not exceed the lesser of $d/2$ or 48 in.
- Reinforcement must also be provided perpendicular to the shear reinforcement, having an area of at least one-third A_v, must be uniformly distributed and may not be spaced apart more than 8 ft.

Design of Masonry Lintels

A lintel is a beam placed over an opening in a wall to support the loads above. When lintels are reinforced masonry, special U-shaped lintel units can be used that are commonly reinforced with two or more grouted reinforcing bars. Lintels must have a minimum bearing length of 4 in. The effective span of the lintel is the distance between centers of bearings. When 16-in long units are used, it is customary to use 8-in long bearings on either side.

Arching Action

For a lintel supporting a height of wall at least equal to its span, it is assumed that the wall develops "arch action" where a load-carrying arch forms within the wall and the lintel needs to carry only the dead load of the wall defined by the dashed lines at 45 degrees to the horizontal (see Fig. 115.2). The arching action forms only if the masonry on either side of the opening provides sufficient lateral restraint to resist the horizontal thrust from the arching action.

Thus, for the wall shown in Fig. 115.2(a), arching action may be presumed, leading to the type of loading on the lintel shown below the wall. On the other hand, the height of wall above the lintel in Fig. 115.2(b) is not adequate for arch action and, therefore, the lintel must carry the entire wall weight above it as well as any loads transferred to the top of the wall from the roof or ceiling.

Example 115.1

A 7.5 in × 16 in masonry lintel spans a door opening 6-ft wide. The wall above the lintel is 8-in thick with a unit weight of 130 lb/ft³. Assume 8-in bearing on either side of the opening. Assume unit weight of the reinforced lintel to be 140 lb/ft³. What is the maximum bending moment in the lintel, if the height of wall above the lintel is (a) 8 ft, (b) 3 ft? The top of the wall carries joists every 16 in. The joist reaction is 600 lbs.

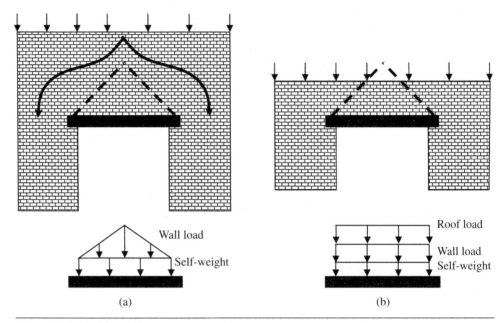

Figure 115.2 Arching action in masonry walls.

Solution

(a) The effective span of the lintel is calculated as

$$S = 6 \text{ ft clear opening width} + 4 \text{ in on either bearing} = 6.67 \text{ ft}$$

Height of wall above lintel = 8 ft > 6.67 ft. Therefore, we can assume that arching action occurs.

Thus, the lintel needs to carry two kinds of loading:

1. Self-weight (rectangular): $w_1 = 140 \times \dfrac{7.5}{12} \times \dfrac{16}{12} = 116.7$ lb/ft

2. Wall weight (triangular): $w_2 = 130 \times \dfrac{7.5}{12} \times 3.33 = 270.6$ lb/ft

Maximum bending moment:

$$M_{max} = M_1 + M_2 = \frac{w_1 L^2}{8} + \frac{w_2 L^2}{12} = 649.0 + 1003.2 = 1652.2 \text{ lb/ft}$$

(b) The effective span of the lintel is $S = 6.67$ ft.

Height of wall above lintel = 3 ft < 6.67 ft. Therefore, no arching action occurs.

Thus, the lintel needs to carry three kinds of loading:

1. Self-weight (rectangular): $w_1 = 140 \times \dfrac{7.5}{12} \times \dfrac{16}{12} = 116.7$ lb/ft

2. Wall weight (rectangular): $w_2 = 130 \times \dfrac{7.5}{12} \times 3 = 243.8$ lb/ft

3. Joist loads (rectangular): $w_3 = 600 \div \dfrac{16}{12} = 450$ lb/ft

Maximum bending moment:

$$M_{max} = M_1 + M_2 + M_3 = \frac{(w_1 + w_2 + w_3)L^2}{8} = 4507.3 \text{ lb/ft}$$

Cracking Moment

The *cracking moment* of a section is defined as the resisting moment of an unreinforced section, calculated as

$$M_{cr} = f_r S \qquad (115.28)$$

where f_r = the modulus of rupture (Table 115.2)
$\quad S$ = section modulus = $bd^2/6$ for a rectangular section

The gross moment of inertia of a rectangular section is calculated as

$$I_g = \frac{bh^3}{12} \qquad (115.29)$$

The cracked moment of inertia is calculated as

$$I_{cr} = \frac{b(kd)^3}{3} + nA_s(d - kd)^2 \qquad (115.30)$$

where $k = \sqrt{(n\rho)^2 + 2n\rho} - n\rho$.

For the purpose of calculating deflections of masonry beams, the "effective" moment of inertia is used, as shown below:

$$I_e = \left(\frac{M_{cr}}{M_a}\right)^3 I_g + \left[1 - \left(\frac{M_{cr}}{M_a}\right)^3\right] I_{cr} < I_g \qquad (115.31)$$

where M_a is the maximum bending moment in the member.

> **NOTE** Section properties of Concrete Masonry Walls are listed in TEK 14-1B (2007) of the National Concrete Masonry Association.

Reinforced Masonry Walls

Lateral Loads

Most masonry walls are designed to span vertically and transfer the lateral loads to the roof, floor, or foundation. Normally, the walls are designed as simple beams spanning between structural supports. Simple beam action is assumed even though reinforcement may be present and will provide at least partial continuity. Under certain circumstances, such as when a system of pilasters is present, the masonry walls may be designed to span horizontally between pilasters, which in turn span vertically to transfer the lateral loads to the horizontal structural support elements above and below.

Axial Loads

Loads enter the wall from roofs, floors, or beams and are transferred axially to the foundation. When the resultant axial force is tension from wind uplift loadings, mortar tension will not be used to resist these uplift forces. Instead, adequate reinforcement will be provided to anchor the top of wall bond beam to the remainder of the wall and on down to the foundation. If the resultant of the vertical loads is not at the center of the wall (not concentric), allowance will be made for the effects of eccentric loading. This includes any moments that are due to eccentric loading as well as any additional moments caused by the rotation of floor or roof elements that frame into the wall.

Uniform loads enter the wall as line loads, stressing the wall uniformly along its length. When concentrated loads (e.g., joists) are not supported by structural elements, such as pilasters, they may be distributed over a length of wall equal to the width of bearing plus four times the wall thickness, but not to exceed the center to center distance between concentrated loads. Concentrated loads will not be distributed across control joints.

Combined Loads

A vertical wall is shown in Figs. 115.3 and 115.4. The wall is subject to a combination of a uniformly distributed lateral load (such as wind) and an eccentric axial load. In Fig. 115.3, the moments due to the eccentric axial load and due to the lateral load are additive, whereas in Fig. 115.4 they are not.

The horizontal reactions at ends of the vertical wall are given by

$$R_a = \frac{wh}{2} \pm \frac{Pe}{12h} \qquad (115.32)$$

where w = lateral distributed load (lb/ft)
h = height of wall (ft)

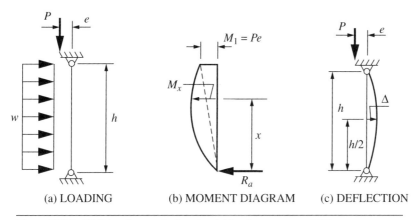

(a) LOADING   (b) MOMENT DIAGRAM   (c) DEFLECTION

Figure 115.3   Wall loading—wind and axial load moments additive.

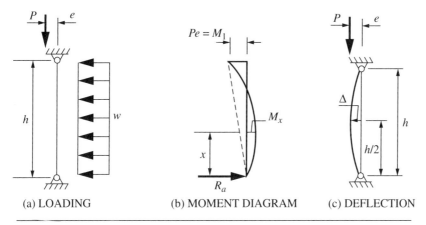

(a) LOADING   (b) MOMENT DIAGRAM   (c) DEFLECTION

Figure 115.4   Wall loading—wind and axial load moments not additive.

P = eccentric axial load (lb/ft of wall length)

e = eccentricity of axial load from centerline of wall (inches)

The location of the maximum moment is given by

$$x = \frac{h}{2} \pm \frac{Pe}{12wh} \tag{115.33}$$

and the maximum moment is given by

$$M_{\max} = \frac{wh^2}{8} \pm \frac{Pe}{2} + \frac{3}{2w}\left(\frac{Pe}{h}\right)^2 \tag{115.34}$$

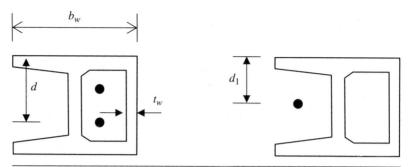

Figure 115.5 Hollow masonry units—dimensions.

Table 115.3 Assumed Dimensions of Hollow Masonry Units (inches)

CMU Nominal Thickness	CMU Design Thickness	Face Shell Thickness	Web Thickness, t_w	d_1	d_2	b_w
6	5 5/8	1	1	2.81	—	7 1/2
8	7 5/8	1 1/4	1	3.81	5.31	7 1/2
10	9 5/8	1 3/8	1 1/8	4.81	7.06	7 1/2
12	11 5/8	1 1/2	1 1/8	5.81	8.81	7 1/2

Figure 115.5 shows hollow masonry units. Table 115.3 gives nominal values of these parameters for 6-, 8-, 10-, and 12-in masonry units.

Design Loads on Buildings and Other Structures

Introduction

The provisions summarized in this chapter are based on ASCE 7-10 (Minimum Design Loads for Buildings and Other Structures, 2010, American Society of Civil Engineers, Reston, VA) and the International Building Code, 2009 edition, International Codes Council, Falls Church, VA.

Loads on building structures can be broadly classified into two classes—gravity (vertical) loads and lateral loads. The most significant gravity loads are dead loads, live loads, and snow loads while the most significant sources of lateral loads are wind and earthquake. Estimating magnitude and distribution of dead loads is usually fairly straightforward once material specifications are complete. On the other hand, calculating the same for snow, wind, and seismic forces are considerably more complicated. In this chapter, the specifications pertaining to live loads, snow loads, wind loads, and seismic loads in the ASCE 7 are summarized.

Risk Category

In ASCE 7-10, buildings and other structures are assigned a risk category based on risk to human life, health, and welfare. This is used for the determination of flood, wind, snow, ice, and earthquake loads. See Table 116.1.

For each type of load, an importance factor (I) is assigned according to Table 116.2.

Live Loads

Chapter 4 of ASCE 7-10 has provisions for determining live loads on buildings based on the occupancy of the structure. Table 4-1 in that document lists minimum design values of uniformly distributed and concentrated live loads.

In buildings where partitions are to be used, regardless of whether or not partitions are shown on the plans, a minimum provision of 15 psf shall be made for partition weight.

Table 116.1 Risk Category Definitions

Risk category	Use or occupancy classification
I	Low risk to human life in the event of failure
II	All structures except those listed in risk categories I, III, and IV
III	Structures whose failure could pose a substantial risk to human life. Buildings and other structures, not included in risk category IV, with potential to cause a substantial economic impact and/or mass disruption of day-to-day civilian life in the event of failure (including, but not limited to, facilities that manufacture, process, handle, store, use, or dispose of hazardous substances).
IV	Buildings and other structures designated as essential facilities. Buildings and other structures, the failure of which could pose a substantial hazard to the community. Buildings and other structures that manufacture, process, handle, store, use, or dispose of hazardous substances. Buildings and other structures required to maintain the functionality of other risk category IV structures.

Table 116.2 Importance Factor for Various Load Types

Risk category	Importance factor			
	Snow (I_s)	Ice (I_i)	Wind (I_w)	Seismic (I_e)
I	0.80	0.80	1.00	1.00
II	1.00	1.00	1.00	1.00
III	1.10	1.25	1.00	1.25
IV	1.20	1.25	1.00	1.50

However, if the minimum specified live load exceeds 80 psf, additional partition load is not required.

Whenever a concentrated live load is investigated for maximum load effect on the building, it shall be distributed over a 2.5 ft by 2.5-ft area.

To allow for impact, the weight of machinery and moving loads shall be increased by the following percentages: (1) for elevator machinery, 100%; (2) for shaft- or motor-driven light machinery, 20%; (3) for reciprocating machinery or power-driven units, 50%; and (4) for hangers for floors or balconies, 33%.

Live Load Reduction

Members for which a factored tributary area ($K_{LL}A_T$) is greater than 400 ft^2 are permitted to be designed for a reduced live load in accordance with Eq. (116.1). (L_o is the minimum uniformly distributed live load in Table 4-1 of ASCE 7.)

$$L = L_o \left(0.25 + \frac{15}{\sqrt{K_{LL}A_T}} \right) \tag{116.1}$$

Table 116.3   Live Load Element Factor K_{LL}

Element type	K_{LL}
For interior columns and exterior columns without cantilever slabs	4
For edge columns with cantilever slabs	3
For corner columns with cantilever slabs, edge beams without cantilever slabs and interior beams	2
All other members: edge beams without cantilever slabs, cantilever beams, one-way slabs, two-way slabs, and members without provision for continuous shear transfer normal to their span	1

where L = reduced design live load per ft² of area supported by the member

L_o = unreduced design live load per ft² of area supported by the member (see Table 4-1 of ASCE 7)

K_{LL} = live load element factor (see Table 116.3)

A_T = tributary area in ft²

The reduced live load L shall not be less than $0.50L_o$ for members supporting one floor and not less than $0.40L_o$ for members supporting two or more floors. Live loads exceeding 100 psf shall not be reduced, except for members supporting two or more floors, for which live loads shall be reduced 20%. Live loads for passenger car garages and places of public assembly shall not be reduced.

The live load element factor (K_{LL}) is given in Table 116.3.

Roof Live Load Reduction

The nominal roof live load L_o shall be reduced by factors R_1 and R_2. The reduced roof live load L_r will be in the range of 12 psf to 20 psf.

The factor R_1 accounts for large tributary area (A_t greater than 200 ft²) according to Eq. (116.2)

$$R_1 = \begin{cases} 1 & A_t \leq 200\,ft^2 \\ 1.2 - 0.01A_t & 200\,ft^2 \leq A_t \leq 600\,ft^2 \\ 0.6 & A_t \geq 600\,ft^2 \end{cases} \qquad (116.2)$$

The factor R_2 accounts for the slope of a pitched roof (F = no. of inches rise per foot) according to Eq. (116.3)

$$R_2 = \begin{cases} 1 & F \leq 4 \\ 1.2 - 0.0F & 4 \leq F \leq 12 \\ 0.6 & F \geq 12 \end{cases} \qquad (116.3)$$

Snow Loads

Chapter 7 of ASCE 7-10 has provisions for determining snow loads on buildings and other structures. The general idea is to start with a mapped value of the ground snow load (p_g) in ASCE 7 Fig. 7-1 and then make adjustments to this value based on characteristics of the site and the roof of the structure.

Ground Snow Loads

Ground snow loads for the contiguous United States are shown in Fig. 7-1 and for Alaska in Table 7-1 of ASCE 7-10. Some special regions are marked CS. Ground snow loads for sites at elevations above the limits indicated in Fig. 7-1 and for all sites within the CS areas shall be approved by the authority having jurisdiction. Ground snow load determination for such sites shall be based on an extreme value statistical analysis of data available in the vicinity of the site using a value with a 2% annual probability of being exceeded (50-year mean recurrence interval).

Flat Roof Snow Load (p_f)

Section 7.3 of ASCE 7 outlines the flat roof snow load for roofs sloped at or less than 5° (1 in/ft = 4.76°). The flat roof snow load p_f is given by Eq. (116.4)

$$p_f = 0.7C_eC_tI_sp_g \tag{116.4}$$

Where $p_g \leq 20$ lb/ft², the design value of p_f is subject to a minimum value $p_f = I_sp_g$
Where $p_g > 20$ lb/ft², the design value of p_f is subject to a minimum value $p_f = 20I_s$

Exposure Factor (C_e)

The exposure factor (C_e) is a function of terrain and roof exposure as shown in Table 116.4.

Table 116.4 Exposure Factor (C_e) for Roof Snow Load

Terrain	Exposure of roof		
	Fully exposed	Partially exposed	Sheltered
B—urban or suburban areas, wooded areas or terrain with closely spaced obstructions the size of single family homes	0.9	1.0	1.2
C—open terrain (such as flat open country and grasslands) with scattered obstructions with heights generally less than 30 ft	0.9	1.0	1.1
D—flat, unobstructed areas and water surfaces.	0.8	0.9	1.0
Above the tree line in windswept mountainous areas	0.7	0.8	N/A
In Alaska in areas without trees within 2-mile radius	0.7	0.8	N/A

Table 116.5 Thermal Factor (C_t) for Roof Snow Load

Thermal condition	C_t
All structures except below:	1.0
Structures kept just above freezing and others with cold, ventilated roofs where the thermal resistance (R) between the ventilated pace and heated space exceeds 25°F hr ft²/Btu (4.4 K m²/W)	1.1
Unheated and open-air structures	1.2
Structures intentionally kept below freezing	1.3
Continuously heated greenhouses with roof having R value less than 2.0°F hr ft²/Btu (0.4 K m²/W)	0.85

Thermal Factor (C_t)

The thermal factor (C_t) is a function of the roof's thermal condition, as shown in Table 116.5.

Sloped Roof Snow Load (p_s)

Snow loads on a sloping surface shall be assumed to act on the horizontal projection of that surface. The balanced snow load on a sloped roof is calculated by multiplying the flat roof snow load (p_f) by a roof slope factor C_s.

$$p_s = C_s p_f \qquad (116.5)$$

Figure 116.1 shows values of C_s as a function of roof temperature and roof slope for warm roofs ($C_t < 1.0$: curves 1 and 2), cold roofs ($C_t = 1.1$: curves 3 and 4 and $C_t = 1.2$: curves 5 and 6).

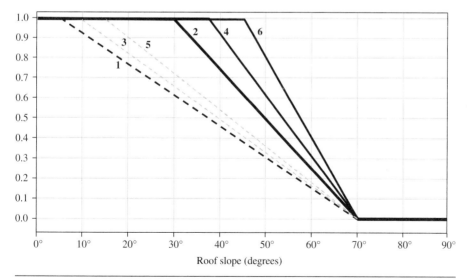

Figure 116.1 Roof slope factor (C_s) for various slope and thermal conditions.

Curve 1: Warm roof ($C_t \leq 1.0$)—unobstructed slippery surface with $R \geq 30°$F h ft²/Btu (5.3 Km²/W) for unventilated roofs or $R \geq 20°$F h ft²/Btu (3.5 Km²/W) for ventilated roofs.

Curve 2: All other warm roofs ($C_t \leq 1.0$) not meeting criteria for curve 1.

Curve 3: Cold roofs with $C_t = 1.1$—unobstructed slippery surfaces.

Curve 4: All other cold roofs with $C_t = 1.1$ not meeting criteria for curve 3.

Curve 5: Cold roofs with $C_t \geq 1.2$—unobstructed slippery surfaces.

Curve 6: All other cold roofs with $C_t \geq 1.2$ not meeting criteria for curve 5.

Unbalanced Roof Snow Loads

Unbalanced snow loads are summarized in the following subsections of Section 7.6 of ASCE 7-10: Hip and gable roofs (7.6.1), curved roofs (7.6.2), multiple folded plate, sawtooth and barrel vault roofs (7.6.3), and dome roofs (7.6.4).

In ASCE 7-10, there is a new exception for required unbalanced snow loads for hip and gable roofs. For hip and gable roofs with a slope exceeding 7/12 (30.2°) or a slope less than 0.5/12 (2.38°), unbalanced snow loads are not required to be applied.

Example 116.1   Snow Load Calculation

A flat roof 450-ft long building located in Milwaukee, WI has 50 ft long × 8 ft high penthouse centered along the length. What is the roof snow load and the drift load?

From snow load map, for Milwaukee, WI, ground snow load $p_g = 30$ psf.

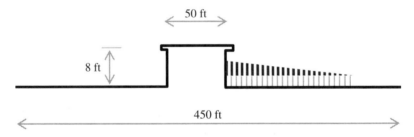

Solution

Balanced Snow Load

Assume exposure B (sheltered) $C_e = 1.2$ (from ASCE 7 Table 7-2)

Heated building, $C_t = 1.0$ (from ASCE 7 Table 7-3)

Importance factor, $I_s = 1.0$ (from ASCE 7 Table 1.5-2 based on risk category in Table 1.5-1)

Flat roof snow load, $p_f = 0.7 C_e C_t I_s p_g = 0.7 \times 1.2 \times 1.0 \times 1.0 \times 30 = 25.2$ psf

Snow unit weight: $\gamma = 0.13 p_g + 14 = 0.13 \times 30 + 14 = 17.9$ pcf

Equivalent height, $h_b = \dfrac{p_g}{\gamma} = \dfrac{25.2}{17.9} = 1.41$ ft

Drift load

$L_u = (450 - 50)/2 = 200$ ft

From ASCE 7 Fig. 7-9, for $L_u = 200$ ft and $p_g = 30$ psf, the drift surcharge height: $h_d = 4.8$ ft. Drift width: $w = 4h_d = 19.2$ ft

Resultant surcharge load: $p_d = \dfrac{1}{2}\gamma h_d w = 0.5 \times 17.9 \times 4.8 \times 19.2 = 825$ lb/ft to be applied as a line load at a distance of $w/3 = 6.4$ ft from the vertical edge.

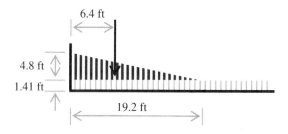

Lateral Force Resisting System (LFRS)

There are essentially three basic forms of LFRS:

1. A moment resisting frame develops resistance to lateral forces via the bending (flexure) of the structural members (beams and columns). Moment frames are further subdivided into ordinary, intermediate, and special moment resisting frames. Special moment frames have specific detailing and proportioning requirements that give them greater ductility and therefore qualify them for higher R-values.

2. A braced frame develops resistance to lateral forces using a system of braces acting as a vertical truss. The braces typically have large slenderness ratio and are therefore effective in tension only.

3. A shear wall is a vertical wall that resists lateral loads through the mechanism of shear. Thus, walls which are parallel to the direction of the lateral force are most effective in resisting these loads. A building may consist of an arrangement of shear walls to provide resistance to lateral forces from several directions.

The basic mechanism of these three types of LFRS is shown in Fig. 116.2.

The two primary sources for lateral load in building frame systems are wind and seismic. The provisions of ASCE 7-10 regarding wind and seismic forces on building structures are briefly summarized in the following sections.

Wind Loads

ASCE 7-10 wind maps are based on "ultimate strength" wind speeds, as opposed to "allowable stress" wind speeds used in ASCE 7-05. Wind speeds in the ASCE 7-10 wind speed maps are

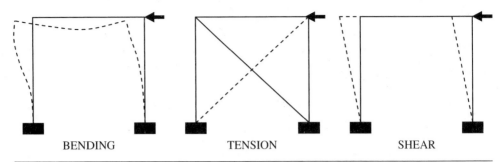

Figure 116.2 Lateral force resisting systems.

significantly higher than in ASCE 7-05, but because the new wind load factor is 1.0 instead of 1.6, the resulting design pressures and uplifts are comparable.

The occupancy category (Importance) factor in ASCE 7-05 has been replaced by the risk category in ASCE 7-10. See Tables 116.1 and 116.2. In ASCE 7-10, wind speed maps are provided for each risk category as opposed to a single map with the importance factor being incorporated in the velocity pressure design equation. An "Exposure D" category was added to the exposure category definitions in ASCE 7-10. Exposure D applies to sites within 600 ft of the coastline or 20 times the building height, whichever is greater.

In ASCE 7, the wind load provisions have been significantly reorganized going from the 2005 edition to the 2010 edition. What previously (ASCE 7-05) was a single wind loads chapter (chap. 6) containing all the provisions that covered lateral load resisting systems, cladding, and testing procedures, is now (ASCE 7-05) a series of six topic-oriented chapters (26–31).

By contrast, in ASCE-05, the design wind loads for the main wind force resisting system (MWFRS) and component and cladding elements were to be determined using one of the following procedures:

Method 1—Simplified Procedure (specified in Section 6.4)

Method 2—Analytical Procedure (specified in Section 6.5)

Method 3—Wind Tunnel Procedure (specified in Section 6.6)

Minimum Wind Force

The design wind load for the MWFRS for an enclosed or partially enclosed building shall not be less than 10 lb/ft^2 multiplied by the area of the building or structure projected onto a vertical plane normal to the assumed wind direction. The design wind pressure for components and cladding of buildings shall not be less than a net pressure of 10 lb/ft^2 acting in either direction normal to the surface.

Definitions

The basic wind speed V is measured as the three-second gust speed at 33 ft above the ground in exposure C.

For the purposes of wind analysis and design, a building is considered flexible if the fundamental frequency is less than 1 Hz. A building is considered low rise if mean roof height is less than or equal to 60 ft AND less than the least plan dimension. A building is considered open if each wall is at least 80% open.

Permitted Procedures

For the MWFRS, the following procedures are permitted:

1. Directional procedure for buildings of all heights (chap. 27)

2. Envelope procedure for low rise buildings (chap. 28)

3. Directional procedure for building appurtenances and other structures (chap. 29)

4. Wind Tunnel procedure for all buildings and other structures (chap. 31)

For components and cladding (C&C), the following procedures are permitted:

1. Analytical procedure (chap. 30)

2. Wind Tunnel procedure (chap. 31)

The basic wind speed V (which is the 3-s gust speed at 33 ft above ground in exposure C) shall be obtained from Fig. 26.5-1A (risk category II), 26.5-1B (risk categories III and IV), or 26.5-1C (risk category I) in ASCE 7-10.

Surface Roughness Categories

For each wind direction considered, the upwind exposure shall be based on ground surface roughness of the two sectors 45 degrees on either side of the design direction. The sector whose upwind exposure would result in the highest wind loads shall be used to represent wind from that direction. The following surface roughness categories are defined:

Surface Roughness B: Urban and suburban areas, wooded areas or other terrain with numerous closely spaced obstructions having the size of single-family dwellings or larger.

Surface Roughness C: Open terrain with scattered obstructions having heights less than 30 ft (including flat open country and grasslands).

Surface Roughness D: Flat unobstructed areas and water surfaces (including smooth mud flats, salt flats, and unbroken ice).

Exposure Categories

Exposure B applies for buildings with mean roof height less than or equal to 30 ft, if surface roughness B prevails for a distance greater than 1500 ft upwind. Exposure B also applies for buildings with mean roof height greater than 30 ft, if surface roughness B prevails for an upwind distance greater than both 2600 ft and 20 times the building height.

Exposure D applies for buildings if surface roughness D prevails for a distance greater than both 5000 ft and 20 times the building height. Exposure D will also be used if the ground surface roughness immediately upwind is B or C and the site is within 600 ft or 20 times building height (whichever is greater) of exposure D.

Exposure C applies wherever exposure B and D do not apply.

Approximate Natural Frequency

The approximate lower-bound natural frequency n_a (Hz) for building height less than 300-ft AND four times effective length (parallel to wind direction) of concrete and steel buildings with mean roof height h (ft) can be calculated using:

For steel moment frame buildings:

$$n_a = \frac{22.2}{h^{0.8}} \tag{116.6}$$

For concrete moment frame buildings:

$$n_a = \frac{43.5}{h^{0.9}} \tag{116.7}$$

For steel and concrete buildings with other lateral force resisting systems:

$$n_a = \frac{75}{h} \tag{116.8}$$

Wind Loads on Buildings—Directional Procedure

The directional procedure is applicable to enclosed, partially enclosed, and open buildings of all heights. The steps involved in the determination of design wind loads are listed below.

Step 1: Determine risk category (Table 1.4-1)

Step 2: Determine basic wind speed V for the risk category (Fig. 26.5-1A, B, or C)

Step 3: Determine wind load parameters—directionality factor (K_d), exposure category, topographic factor (K_{zt}), gust effect factor (G), enclosure classification, internal pressure coefficient (GC_{pi})

Step 4: Determine velocity pressure exposure coefficient (K_z or K_h)

Step 5: Determine velocity pressure (q_z or q_h)

Step 6: Determine external pressure coefficient (C_p or C_N)

Step 7: Calculate wind pressure (p) on each building surface

Example 116.2   Wind Load Calculation

Calculate the wind pressure on the roof of a two-story building shown below. The building is located in the city of Chicago with exposure category is B and risk category is II (nonessential building used as commercial office space). The roof is supported by trusses spaced every 2 ft throughout length of building = 60 ft

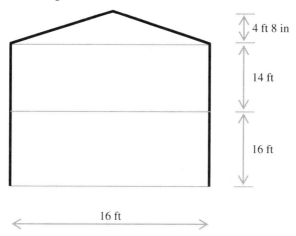

4 ft 8 in

14 ft

16 ft

16 ft

Solution   Since risk category is II, use Fig. 26.5-1A. Basic wind speed (Chicago) is 115 mph.

Importance factor for wind $I_w = 1.0$.

Roof has 30-degree slope. This corresponds to a rise of 6.93 in per foot. $F = 6.93$.

For main LFRS of buildings, directional factor $K_d = 0.85$ (Table 26.6-1)

Mean roof height $= 30 + 4.67/2 = 32.33$ ft

For exposure B

Topographic factor: $K_{zt} = (1 + K_1 K_2 K_3)^2 = 1.0$ (no special topographic features)

For rigid buildings (natural frequency > 1.0 Hz), the gust effect factor, G shall be taken as 0.85 or calculated according to Section 26.9.

Section 26.11 defines internal pressure coefficient $GC_{pi} = \pm 0.18$ for enclosed buildings

Velocity pressure exposure coefficient, $K_h = 2.01\left(\dfrac{h}{z_g}\right)^{2/\alpha} = 2.01\left(\dfrac{32.33}{1200}\right)^{2/7} = 0.716$

From Table 26.9.1, $z_g = 1200$ ft; $\alpha = 7$ (for exposure B)
Using Eq. (27.3-1), velocity pressure:

$$q_h = 0.00256 K_h K_{zt} K_d V^2 = 0.00256 \times 0.716 \times 1 \times 0.85 \times 115^2 = 20.6 \text{ psf}$$

The wind pressure on building surfaces are calculated according to building type—Eqs. (27.4-1) (enclosed and partially enclosed rigid buildings), (27.4-2) (enclosed and partially enclosed flexible buildings), or (27.4-3) (open buildings). Figures 27.4-1 through 27.4-7 show details of how these calculated wind loads are applied to various parts of structures. These figures also provide guidance on the choice of appropriate values of the external pressure coefficient C_p and the net pressure coefficient C_N.

Seismic Loads

Provisions from the following chapters in ASCE 7-10 are summarized below.

Chapter 11: Seismic Design Criteria

Chapter 12: Seismic Design Requirements for Building Structures

Chapter 13: Seismic Design Requirements for Nonstructural Components

Chapter 20: Site Classification Procedure for Seismic Design

The following structures are exempt from the seismic requirements of ASCE 7:

1. Detached one- and two-family dwellings that are located where $S_s < 0.4$ or where the Seismic Design Category is A, B, or C.

2. Detached one- and two-family wood-frame dwellings not included in exception 1 with not more than two stories above grade, designed and constructed in accordance with the International Residential Code.

3. Agricultural storage structures intended only for incidental human occupancy.

4. Structures for which other regulations provide seismic criteria, such as vehicular bridges, electrical transmission towers, hydraulic structures, buried utility lines and their appurtenances, and nuclear reactors.

5. Piers and wharves not accessible to the general public.

Mapped Acceleration Parameters

The parameters S_s (short period 0.2 sec) and S_1 (long period 1.0 sec) shall be determined from maps in chap. 22 (ASCE 7). If $S_1 \leq 0.04$ and $S_s \leq 0.15$, the structure can be assigned to Seismic Design Category A. For Seismic Design Category A, the lateral force at each level of the structure can be taken as 1% of the dead load located at that level.

Site Class

Based on soil properties (averaged over the upper 100-ft depth), the site shall be classified as site class A, B, C, D, E, or F. Classification criteria are in Table 20.3-1 (ASCE 7). The averaging is done as in Example 116.3.

Site Class F

A site is classified as Site Class F if any of the following conditions are satisfied: vulnerable soils such as liquefiable soils, quick and highly sensitive clays and collapsible weakly cemented soils; peats and/or highly organic clay layers thicker than 30 ft; very high plasticity clays (PI > 75) to a depth of 25 ft or more; very thick (> 120 ft) layer of soft/medium stiff clays with undrained shear strength $s_u < 1000$ psf.

Site Class E

If a site does not qualify as Site Class F but has a total thickness of soft clay ($s_u < 500$ psf; moisture content $w > 40\%$ and plasticity index PI > 20%) greater than 10 ft, it shall be classified as Site Class E.

Table 116.6   Seismic Site Characterization

Site class	Description	Average shear wave velocity $\bar{v}_s$	Average std. penetration resistance $\bar{N}$	Average und-rained shear strength $\bar{S}_u$
A	Hard rock	> 5000 ft/s	N/A	N/A
B	Rock	2500–5000 ft/s	N/A	N/A
C	Very dense soil/soft rock	1200–2500 ft/s	> 50	> 2000 psf
D	Stiff soil	600–1200 ft/s	15–50	1000–2000 psf
E	Soft clay soil	< 600 ft/s	< 15	< 1000 psf

Site Classes C, D, E

Site classes C, D, and E are identified by using one of the following parameters, averaged over the top 100 ft of soil at the site: (1) shear wave velocity, (2) field standard penetration resistance, or (3) undrained shear strength. Site classes A and B (hard rock/rock) are classified according to average shear wave velocity only. Table 116.6 highlights the thresholds for the classification of these sites.

Example 116.3

The measured shear wave velocity (v_s) at a site is tabulated below. What is the seismic site classification?

Depth below surface (ft)	Shear wave velocity (ft/s)
0–20	750
20–70	1200
70–125	1800

Solution   The average shear wave velocity is based on the harmonic average of the top 100 ft below the surface.

$$v_s = \frac{100}{\sum \dfrac{d_i}{v_{si}}} = \frac{100}{\dfrac{20}{750} + \dfrac{50}{1200} + \dfrac{30}{1800}} = 1176 \text{ fps}$$

According to Table 116.6, this site has site class D.

Site Coefficient

Using the site class determined above and the previously determined mapped acceleration parameters S_s and S_1, Tables 116.7 and 116.8 are used to determine site coefficients F_a (short period) and F_v (1-s period), respectively. Straight-line interpolation is permitted for intermediate values.

Table 116.7 Site Coefficient F_a

Site class	Mapped risk-targeted maximum considered earthquake spectral response acceleration parameter at short period				
	$S_s \leq 0.25$	$S_s = 0.5$	$S_s = 0.75$	$S_s = 1.0$	$S_s \geq 1.25$
A	0.8	0.8	0.8	0.8	0.8
B	1.0	1.0	1.0	1.0	1.0
C	1.2	1.2	1.1	1.0	1.0
D	1.6	1.4	1.2	1.1	1.0
E	2.5	1.7	1.2	0.9	0.9
F	Site-specific ground motion procedure in chap. 21				

Table 116.8 Site Coefficient F_v

Site class	Mapped risk-targeted maximum considered earthquake spectral response acceleration parameter at 1-s period				
	$S_1 \leq 0.1$	$S_1 = 0.2$	$S_1 = 0.3$	$S_1 = 0.4$	$S_1 \geq 0.5$
A	0.8	0.8	0.8	0.8	0.8
B	1.0	1.0	1.0	1.0	1.0
C	1.7	1.6	1.5	1.4	1.3
D	2.4	2.0	1.8	1.6	1.5
E	3.5	3.2	2.8	2.4	2.4
F	Site-specific ground motion procedure in chap. 21				

It may be noted that the velocity based seismic base shear coefficient (F_v) governs for buildings with long to medium fundamental period (most high rise buildings) whereas the acceleration based seismic base shear coefficient (F_a) governs for buildings with short fundamental periods.

Design Spectral Response Acceleration Parameters

The site coefficients F_a and F_v are then used to adjust the mapped spectral acceleration coefficient into adjusted maximum considered spectral response for each of the two periods (S_{MS} and S_{M1}).

$$S_{MS} = F_a S_s \tag{116.9}$$

$$S_{M1} = F_v S_1 \tag{116.10}$$

Table 116.9 Seismic Design Category Based on Short Period Response Acceleration Parameter S_{DS}

Value of S_{DS}	Risk category	
	I or II or III	IV
$S_{DS} < 0.167$	A	A
$0.167 < S_{DS} < 0.33$	B	C
$0.33 < S_{DS} < 0.50$	C	D
$0.50 < S_{DS}$	D	D

Table 116.10 Seismic Design Category Based on 1-s Response Acceleration Parameter S_{D1}

Value of S_{D1}	Risk category	
	I or II or III	IV
$S_{D1} < 0.067$	A	A
$0.067 < S_{D1} < 0.133$	B	C
$0.133 < S_{D1} < 0.20$	C	D
$0.20 < S_{D1}$	D	D

The maximum spectral response values are then reduced to design values (S_{DS} and S_{D1}) by a simple multiplication by 2/3

$$S_{DS} = \frac{2}{3} S_{MS}$$ (116.11)

$$S_{D1} = \frac{2}{3} S_{M1}$$ (116.12)

Seismic Design Category (SDC)

The design spectral response acceleration parameters (S_{DS} and S_{D1}) and the risk category of the building are then used to determine the seismic design category, as described in Tables 116.9 and 116.10. The worst SDC will be used for design.

Structural Systems

The basic structural system types recognized in ASCE 7-10 are listed below. Table 12.2-1 (ASCE 7) lists response modification factor (R), overstrength factor (Ω_o), deflection amplification factor (C_d), and height limitations for various seismic design categories for these structural systems.

1. Bearing Wall Systems. The walls serve a dual purpose of carrying gravity loads as well as lateral loads. Thus, failure of an element in the LFRS (lateral force resisting system) could compromise the structure's ability to withstand gravity loads.

2. Building Frame Systems. Lateral loads and gravity loads are carried by independent mechanisms in the frame. For example, the gravity loads could be carried by a wood or steel frame, whereas the lateral loads could be carried by a system of non-load-bearing shear walls.

3. Moment Resisting Frames can be subdivided into ordinary, intermediate, and special moment resisting frames, based on specific detailing requirements.

4. Dual Systems with special moment frames capable of resisting at least 25% of seismic forces.

5. Dual Systems with intermediate moment frames capable of resisting at least 25% of seismic forces.

6. Shear Wall-Frame Interactive System with ordinary R.C. moment frame and ordinary R.C. Shear Walls.

7. Cantilevered Column Systems.

8. Steel Systems not specifically detailed for seismic resistance.

Response Modification Factor (*R*)

The response modification factor, *R*, is incorporated to account for the ductility of the structure. If a value of $R = 1$ is used, it signifies that the design force in the structure is at elastic force level. In other words, the design force is such that the element stresses will remain within elastic limits. This results in a very large design force and consequently leads to an overly conservative design. Current seismic design codes recognize the fact that frames exhibit a certain degree of ductility (i.e., the ability to remain stable while undergoing large deformations beyond the elastic limit) and have significant reserve strength that is a combination of element reserve strength as well as system reserve strength. *R*-values for different types of structural configurations are given in *ASCE 7* Table 12.2-1.

The *R*-value reflects the ability of the system to withstand, without collapse, several reversals of cyclic inelastic deformation (ductility). Higher values of *R* represent more ductile systems. In ASCE 7-10, *R*-value varies between 1.0 for the least ductile systems to a value of 8.0 for the most ductile ones (special steel and reinforced concrete moment-resisting frames and steel eccentrically braced frames.) The lowest *R*-values correspond to ordinary plain concrete shear walls, ordinary plain masonry shear walls, and ordinary steel moment frames. Due to their low ductility, these systems are prohibited in Seismic Design Category D.

The overstrength factor Ω_o is applied to design seismic loads to obtain a realistic estimate of the maximum lateral earthquake effects that will develop when the actual lateral capacity of the system is reached. The deflection amplification factor (C_d) amplifies the deflection of the structure based on an elastic analysis. These are shown in Fig. 116.3, where the following notation is used:

V_E = elastic force
V_S = design strength
V_Y = actual strength

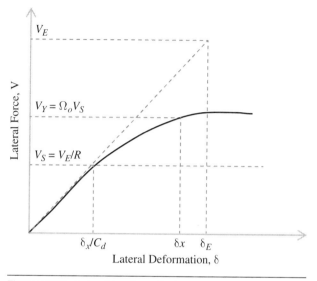

Figure 116.3   Lateral load vs. displacement curve.

Seismic Response Coefficient (C_s)

An essential parameter for the seismic design of building structures—and a useful benchmark for comparison to other projects—is the seismic response coefficient (C_s), which, when multiplied by the weight of the structure, gives the base shear force for seismic design. To derive the seismic response coefficient for the short-period and 1-s period accelerations, *ASCE 7 Section 12.8* provides the following formulas:

$$C_s = \frac{S_{DS}}{(R/I_e)} \qquad (116.13)$$

The value of C_s calculated above is limited to a maximum given by

$$C_s = \frac{S_{D1}}{T(R/I_e)} \qquad \text{for} \qquad T \le T_L \qquad (116.14a)$$

$$C_s = \frac{S_{D1}T_L}{T^2(R/I_e)} \qquad \text{for} \qquad T > T_L \qquad (116.14b)$$

and a minimum value given by

$$C_s = 0.044 S_{DS} I_e \ge 0.01 \qquad (116.14c)$$

where S_{DS} = design spectral response acceleration parameter (short period)
  S_{D1} = design spectral response acceleration parameter (1-s period)
  R = response modification factor
  I_e = importance factor

Table 116.11 Empirical Parameters for Estimating Fundamental Period

Structure type	C_t	x
Steel moment frames	0.028	0.80
Concrete moment frames	0.016	0.90
Steel eccentrically braced frames	0.030	0.75
Steel buckling–restrained braced frames	0.030	0.75
All other systems	0.020	0.75

Fundamental Period (*T*)

T is the fundamental period of the structure. This value can be estimated very accurately using a detailed analytical model of the structure. The eigenvalue analysis of such a model leads to the calculation of the natural frequencies (eigenvalues) and the associated mode shapes (eigenvectors) of the structure. The fundamental period T is the inverse of the fundamental natural frequency f. However, at the preliminary design stage, when structural parameters such as member sizes are not known exactly, such a detailed analysis is infeasible.

In the absence of an exact determination of the fundamental period (T), the following formula may be used to estimate it (T_a):

$$T_a = C_t h_n^x \tag{116.15}$$

where h_n = structural height (ft)
C_t and x are empirical constants defined in Table 116.11.

Seismic Spectra

As shown in Fig. 116.4, the design Response Spectrum developed requires the following parameters—S_{DS}, S_{D1}, and T_L

S_{DS} = design earthquake spectral response acceleration parameter at short period
S_{D1} = design earthquake spectral response acceleration parameter at 1-s period
T_L = long-period transition period (from ASCE Figs. 22-12 through 22-16)
$T_S = S_{D1}/S_{DS}$
$T_o = 0.2T_s$

A simplified pseudostatic design procedure is used for most buildings except for regular structures over 240-ft high or irregular structures greater than five stories or 65-ft height. For buildings falling into the special (tall or irregular) category, estimation of design forces must be based on dynamic analysis.

The general static lateral force procedure consists of the following steps:

- Calculate the Base Shear V
- Calculate the Story Forces F_x
- Calculate the Element Forces

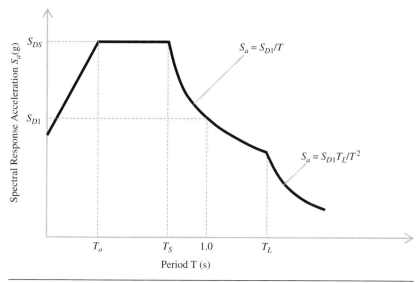

Figure 116.4 Design response spectrum.

Vertical Distribution of Base Shear

The purpose of the vertical force distribution is to translate structure loads to component loads (such as beam and column moments, wall shears, diaphragm shears, etc.). These component loads can then be used as inputs into the design process.

The two primary sources of lateral load in building frames are wind and earthquake. Despite the fact that both wind and seismic loads act laterally with respect to the building frame and are modeled the same way in frame analysis, they are fundamentally different in the way they develop. Wind loads acting on a tributary area of the building are "accumulated" at nodes or joints "collecting" loads from that area. Following the calculation of these nodal loads, the base shear necessary to restore frame equilibrium is calculated. On the other hand, seismic design codes provide means to calculate the design base shear at the frame foundation. The source for this base shear is the inherent inertia of the building frame (its mass) resisting the motion of the ground supporting the structure. This shear force is then distributed to frame nodes as "lumped" lateral forces. In this section, we discuss the procedures for this vertical distribution of the base shear.

According to ASCE 7-10 and IBC 2009, the shear at the base of the structure is given by

$$V = C_s W$$

where W is the effective seismic weight of the structure and includes all of the structural dead load and other loads as follows.

1. In areas used for storage, a minimum of 25% of the live load, except for public garages and open parking structures, and where the inclusion of storage loads adds no more than 5% to the effective seismic weight at a particular level.

2. The larger of the actual partition weight or 10 psf of floor area.

3. Total operating weight of permanent equipment.

4. 20% of the uniform design snow load (regardless of roof slope) in regions where the flat roof snow load exceeds 30 psf.

5. Weight of landscaping and other materials on roof gardens and similar areas.

Distribution of Seismic Base Shear into Story Forces

Once the base shear V has been calculated, we need to distribute it vertically through the structure. The story forces at each level of the structure are essentially inertial forces. When the base or foundation of the structure experiences ground acceleration due to an earthquake, the tendency of the floors (where most of the system weight is concentrated) is to "stay put," as a result of their inertia, while the ground moves underneath. This relative acceleration between the moving ground and the "static" floor creates a "F = ma" type of inertial force on the floor mass. The relative acceleration of each floor mass depends on the "shape" of the vibration of the structure. These are the shapes previously called mode shapes, which are the eigenvectors obtained from an eigenvalue analysis. Typical mode shapes of a prismatic cantilever beam are shown in Fig. 116.5. (The cantilever beam is often used as a simple analogue of a high-rise structure.)

The lowest mode (Mode 1) represents the structure vibrating at the lowest characteristic (natural) frequency. Other higher frequencies are present in the response of the structure and exhibit themselves through the higher modes. Typically, for most structures, the fundamental (first) mode response is a significant fraction of the total response and is used to predict the overall response of the structure. The contributions of the higher modes, though at higher frequencies, are relatively small in magnitude and therefore contribute to the "overall shape of vibration" to a much lesser extent.

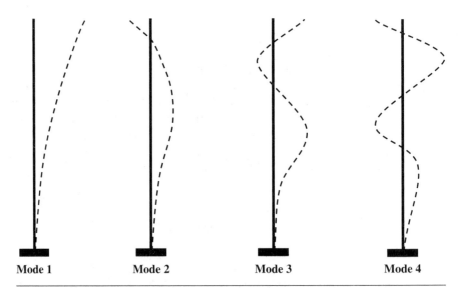

Mode 1   Mode 2   Mode 3   Mode 4

Figure 116.5 Mode shapes of a cantilever structure.

In the following section, the shape of the first mode of vibration serves as the basis for the distribution of the story forces acting on the LFRS. At this point, it may be noted that TWO SETS OF STORY FORCES are calculated for the seismic load. The first set of story forces (F_x) is to be used in the design of the vertical elements of the LFRS. These forces act concurrently and are in equilibrium with the base shear on the structure. A second set of story forces (F_{px}) is used as the design force of the horizontal diaphragms. Except for the roof level, these forces are higher than the corresponding F_x forces. This is due to the fact that during an earthquake, the forces are changing very rapidly. Thus, it is unlikely that the instantaneous maximum value of the force in one diaphragm will occur at the same time as the maximum value is experienced in another. Thus, the F_{px} forces reflect the *largest possible instantaneous force* occurring on each diaphragm, whereas the F_x forces reflect the *concurrent distribution of forces* in the entire structure created due to the base shear V.

Thus F_{px} is used to design the individual horizontal diaphragms, diaphragm collectors (struts) and related connections, whereas F_x is used to design the vertical elements (shear walls) in the primary LFRS.

The procedure is to determine the F_x story forces first and then use them to calculate the F_{px} story forces. The story forces are given by

$$F_x = \frac{w_x h_x^k}{\sum_{i=1}^{n} w_i h_i^k} V$$

where
F_x = story force at level x
h_i and h_x = height from base of structure to level i or level x
w_i and w_x = tributary weights assigned to level i or x
k = an exponent related to structural period ($k = 1$ for $T < 0.5$ s; $k = 2$ for $T > 2.5$ s)

The seismic design story shear at level x is given by

$$V_x = \sum_{i=x}^{n} F_i$$

These loads, when computed, are then applied simultaneously to all levels of the building.

The diaphragm force at any level (x) is calculated as the sum of the story forces acting above level x

$$F_{px} = \frac{\sum_{i=x}^{n} F_i}{\sum_{i=x}^{n} w_i} w_{px}$$

The diaphragm force F_{px} is not to be taken less than $0.2S_{DS}I_e w_{px}$ nor more than $0.4S_{DS}I_e w_{px}$.

Example 116.4   Seismic Load

A hospital building in Memphis, TN (35N, 90W), has a Special Reinforced Concrete Moment Frame as the LFRS. The height of the building is 55 ft. Site class is C. What is the base shear coefficient C_s?

Solution From the USGS website for Memphis TN (35N, 90W), the short period spectral acceleration $S_s = 0.823$ g and the long period spectral acceleration, $S_1 = 0.291$g. These can be also read from Figs. 22.1 and 22.2 of ASCE 7, respectively

From Table 116.7, for site class C and $S_s = 0.823$ g, site coefficient $F_a = 1.07$
From Table 116.8, for site class C and $S_1 = 0.291$ g, site coefficient $F_v = 1.51$

$$S_{MS} = F_a S_s = 1.07 \times 0.823 = 0.882 \text{ g}$$
$$S_{M1} = F_v S_1 = 1.51 \times 0.291 = 0.439 \text{ g}$$
$$S_{DS} = 2/3 S_{MS} = 0.67 \times 0.882 = 0.588 \text{ g}$$
$$S_{D1} = 2/3\ S_{M1} = 0.67 \times 0.439 = 0.293 \text{ g}$$

For concrete moment resisting frame building, fundamental natural frequency,

$$n_a = 43.5/h^{0.9} = 1.18\ Hz$$

Fundamental period: $T_n = 1/n_a = 0.85$ s

Hospital risk category IV

Table 1.5-2, for risk category IV, Seismic importance factor = 1.50
For special reinforced concrete moment frame, from Table 12.2-1, response modification factor $R = 8$

Equation 12.8-2 yields: $C_s = \dfrac{S_{DS}}{R/I} = \dfrac{0.588}{8/1.5} - 0.11$

The upper limit for C_s is given by Eq. (12.8-3): $\dfrac{S_{D1}}{T(R/I)} = \dfrac{0.293}{0.85(8/1.5)} = 0.065$

The lower limit for C_s is given by Eq. (12.8-5): $0.044S_{DS}I = 0.044 \times 0.588 \times 1.5 = 0.039$
Use $C_s = 0.065$.

Example 116.5   Vertical Distribution of Base Shear

Assume a ten-story steel special moment frame building has the following data:

Plan dimensions: 150 ft × 200 ft

Dead load per occupied floor and roof = 40 psf

Live load on floors 1–10 = 60 psf

Roof live load = 25 psf

Flat roof snow load = 22 psf

Exterior walls = 90 psf

Story height = 11 ft

Assume, base shear $= 6\%$ of building's seismic weight

Calculate:

1. Calculate the story forces

2. Calculate the story shear on each level

Solution Effective seismic weight includes dead load and in areas of storage, minimum of 25% of reduced floor live load, partition load of 10 psf, permanent operating equipment, 20% of flat roof snow load (if flat roof snow load exceeds 30 psf)

Building height $= 10 \times 11 = 110$ ft

Roof DL $= 150 \times 200 \times 40 = 1.2 \times 10^6$ lb

Floors 2–10 DL: $150 \times 200 \times 40 \times 9 = 1.08 \times 10^7$ lb

Exterior walls: $2 \times (150 + 200) \times 110 \times 90 = 6.93 \times 10^6$ lb

Partition load on levels 2–10: $150 \times 200 \times 10 \times 9 = 2.7 \times 10^6$ lb

TOTAL $= 2.16 \times 10^7$ lb

Seismic base shear $= 0.06 \times 2.16 \times 10^7 = 1.30 \times 10^6$ lb $= 1300$ kips

Alternatively, it is permitted to determine the approximate fundamental period (T_a), in s, from the following equation for structures *not exceeding 12 stories* above the base as defined in Section 11.2 where the seismic force-resisting system consists entirely of concrete or steel moment resisting frames and the *average story height is at least 10 ft.*

Approximate period $= 0.1 \text{ N} = 1.0$ s

For period T between 0.5 s and 2.5 s, the exponent k for vertical force distribution (by linear interpolation between $k = 1$ and $k = 2$) is $k = 1.25$

$$F_x = \frac{w_x h_x^k}{\sum w_i h_i^k} V$$

Seismic weight at roof level $=$ Roof DL $(1.2 \times 10^6$ lb$)$ + tributary wall − half story $(3.47 \times 10^5$ lb$) = 1.55 \times 10^6$ lb.

Remaining weight can be distributed among nine floors. Each floor has weight $= 2.23 \times 10^6$ lb.

Fractions (of total base shear) at levels 1–10 are: 0.01, 0.03, 0.03, 0.05, 0.07, 0.09, 0.11, 0.14, 0.16, 0.19, and 0.15, respectively.

As a result, the (cumulative from top) story level column shears in levels 10, 9, 8, 7, 6, 5, 4, 3, 2, and 1 are:

1%, 4%, 9%, 16%, 25%, 36%, 50%, 66%, 85%, and 100% of the total base shear, respectively.

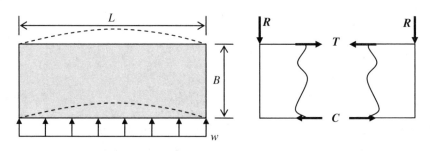

Figure 116.6 Forces in a flexible diaphragm.

Diaphragms

Diaphragms can be considered either flexible or rigid. A flexible diaphragm acts like a deep beam resisting lateral forces. For example, consider a single-story building with rectangular plan as shown in Fig. 116.6.

If the wind pressure on one face of the building is "lumped" at the roof diaphragm level as a uniform load as shown, then the end wall shear forces are each given by

$$R - \frac{wL}{2} \tag{116.16}$$

and the chord forces are

$$C = T = \frac{wL^2}{8B} \tag{116.17}$$

A diaphragm is considered "flexible" is the maximum diaphragm deflection (MDD) is more than twice the average drift of the vertical element (ADVE). For a flexible diaphragm, the shear transfer to the supporting elements is strictly according to tributary area concepts. Torsion is not considered for flexible diaphragms.

Shear Wall System—Center of Rigidity

A system of shear walls resists lateral loads due to the development of shear stresses. The walls which are parallel to the direction of the load are primarily responsible for developing this shear resistance. If the system of walls is asymmetric, the center of rigidity is not in line with the centroid of the load. This causes a torsional moment which induces additional shear in the walls.

In Example 116.1, the center of rigidity of a system of shear walls is calculated.

Example 116.6

The building whose floor plan is shown below has shear walls A, B, C, and D. Calculate the coordinates of the center of rigidity.

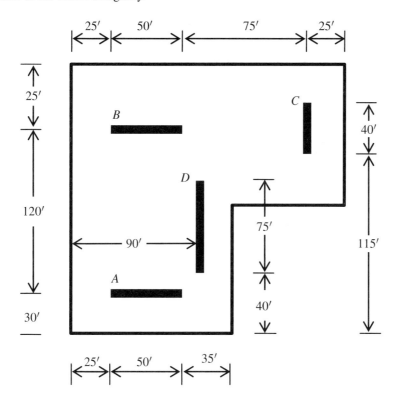

Lengths L and thicknesses of wall t are shown in the table below.

	L (ft)	t (in)
A	50	12
B	50	14
C	40	12
D	75	16

Solution   Shear rigidity of each wall is given by $\dfrac{EA}{L} = \dfrac{ELt}{h}$.

Assume the modulus of elasticity E and the height h is constant for all shear walls. Therefore, wall rigidity is proportional to the product Lt for each segment. Center of rigidity with respect to the southern edge (walls A and B)

$$y_c = \frac{\sum y_i R_i}{\sum R_i} = \frac{\sum y_i L_i t_i}{\sum L_i t_i} = \frac{30 \times 50 \times 12 + 150 \times 50 \times 14}{50 \times 12 + 50 \times 14} = 94.62 \text{ ft}$$

Center of rigidity with respect to the western edge (walls C and D)

$$x_c = \frac{\sum x_i R_i}{\sum R_i} = \frac{\sum x_i L_i t_i}{\sum L_i t_i} = \frac{150 \times 40 \times 12 + 90 \times 75 \times 16}{40 \times 12 + 75 \times 16} = 107.14 \text{ ft}$$

In Example 116.2, the total shear force in the shear walls is computed using a superposition of the direct shear among the participating walls plus the shear force due to the torsion induced by the eccentricity of the load.

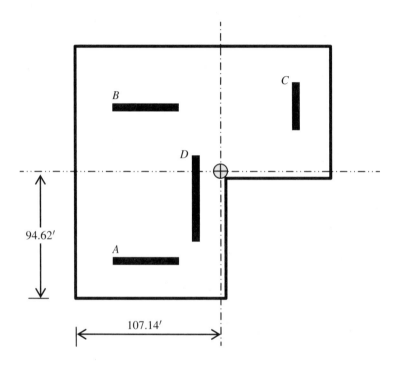

Accidental Torsion

Where diaphragms are not flexible, the design shall include the inherent torsional moment resulting from the location of the structure masses plus the accidental torsional moments caused by an additional eccentricity equal to 5% of the dimension of the structure perpendicular to the direction of the applied forces.

Where earthquake forces are applied concurrently in two orthogonal directions, the required 5% displacement of the center of mass need not be applied in both of the orthogonal directions at the same time, but shall be applied in the direction that produces the greater effect.

Example 116.7

The diagram shows the plan view of a building experiencing a uniform wind load of 200 lb/ft on the south face. The lateral load is to be resisted by a system of shear walls

A, B, C, and D as shown below. All walls are same thickness ($t = 12$ in). Assume that the diaphragm is rigid.

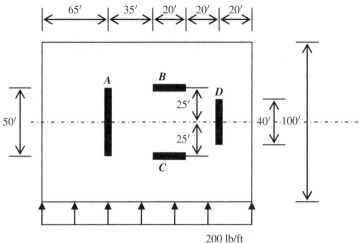

200 lb/ft

Solution The shear in the walls is caused by the wind shear on the building shared by walls A and D and the torsional shear due to the wind resultant not passing through the center of rigidity (since the diaphragm is rigid).

The center of rigidity of the shear wall system is shown as CR. The coordinates of CR are marked (x_c, y_c) on the diagram below.

By symmetry $y_c = 50$ ft

$$x_c = \frac{50 \times t \times 65 + 40 \times t \times 140}{50 \times t + 40 \times t} = 98.33 \text{ ft}$$

Shear in walls A and D is split in proportion to the rigidity of the walls.

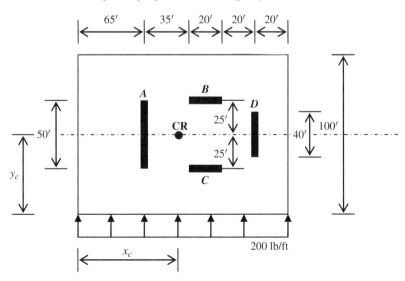

200 lb/ft

Total wind shear = 200 lb/ft × 160 ft = 32,000 lb = 32 kips.

$$V_A = 32 \times \frac{50}{50 + 40} = 17.78 \text{ k}$$

The resultant wind force acts at $x = 80$, whereas the center of rigidity has $x = 98.33$. This causes an eccentricity = 18.33 ft, producing a torsional moment of 32 × 18.33 = 586.7 k-ft.

In addition to the actual eccentricity, ASCE 7 indicates that an accidental eccentricity equal to 5% of the lateral dimension (perpendicular to the direction of the load) must be considered. In this case, with the lateral dimension being 160 ft, the accidental eccentricity = 8 ft.

Therefore, the total eccentricity = 18.33 + 8 = 26.33 ft
Total torsional moment = 32 × 26.33 = 842.56 k-ft
The polar moment of inertia of the shear wall system is given by

$$I_P = I_x + I_y = (2 \times 20 \times 1 \times 25^2) + (50 \times 1 \times 33.33^2 + 40 \times 1 \times 41.67^2) = 25,000 + 125,000 = 150,000 \text{ ft}^4$$

The torsional shear in wall A is given by

$$V_A = \frac{Tr}{I_P} A = \frac{842.56 \times (98.33 - 65)}{150,000} \times (50 \times 1) = 9.36 \text{ k}$$

Total shear in wall A = 17.78 + 9.36 = 27.14 kips.

Phase Relationships for Soils

Processes Leading to the Formation of Soils

Soil is the uncemented aggregate of mineral grains and organic matter. Mineral grains are the product of rock weathering. There are three primary types of rocks, based on the formation process—igneous, sedimentary, and metamorphic.

Igneous rocks are formed by the solidification of molten magma ejected from the earth's mantle. Rate of magma cooling, which depends on a variety of ambient conditions, determines the sequence in which different minerals are formed as magma cools.

Weathering is the breaking down of rocks into smaller particles due to *mechanical and chemical processes.*

Mechanical processes may be due to expansion and contraction from continuous gain and loss of heat. Other physical agents are glacier, wind, ice, rivers and ocean waves. The only change that occurs during such processes is in physical size and not in chemical composition.

In chemical weathering, original rock minerals are transformed by chemical reaction. Water and carbon dioxide from the atmosphere form carbonic acid (H_2CO_3), which in turn reacts with minerals. For example, orthoclase undergoes chemical weathering to turn into kaolinite, which is a clay mineral.

$$2K(AlSi_3O_8) + 2H^+ + H_2O \rightarrow 2K^+ + 4SiO_2 + Al_2Si_2O_5(OH)_4$$
$$\text{(Orthoclase)} \rightarrow \rightarrow \rightarrow \rightarrow \text{(Kaolinite)}$$

Transportation of weathering products occurs due to glaciers, running water in streams (alluvial), deposition in quiet lakes (lacustrine), deposition in seas (marine), wind (aeolian), and soil movement due to gravity (colluvial).

Sedimentary rocks are formed by deposition of soil particles becoming compacted by overburden pressure and cemented by agents such as iron oxide, calcite, dolomite, and quartz. The cementing agents are usually in solution. Soils formed in this way are called *detrital sedimentary rocks.*

Sedimentary rocks can also be formed by chemical processes. Rocks such as limestone, chalk, dolomite, gypsum, anhydrite are called *chemical sedimentary rocks.*

Metamorphic rocks are formed when changes of chemical composition and texture of soils occur due to pressure and heat. New minerals are formed and mineral grains are sheared to give a foliated texture. Granite, diorite, and gabbro transform to gneiss by high-grade metamorphism. Shales and mudstones are transformed into slates and phyllites by low-grade metamorphism.

Diagenesis is also a change in form that occurs in sedimentary rocks. However, diagenetic processes are restricted to those which occur at temperatures below 200°C and pressures below about 300 MPa (approximately 3000 atm). On the other hand, metamorphism occurs at temperatures and pressures higher than 200°C and 300 MPa. Rocks buried deeper in the earth can be subjected to these higher temperatures and pressures. Such burial usually takes place as a result of tectonic processes such as continental collisions or subduction. The upper limit of metamorphism occurs at the pressure and temperature where melting of the rock in question begins. Once melting begins, the process changes to an igneous process rather than a metamorphic process.

Low-grade metamorphism takes place at temperatures between about 200 to 320°C, and relatively low pressure. Low-grade metamorphic rocks are generally characterized by an abundance of hydrous minerals. High-grade metamorphism takes place at temperatures greater than 320°C and relatively high pressure. As grade of metamorphism increases, hydrous minerals become less hydrous, by losing H_2O, and non-hydrous minerals become more common.

The different types of metamorphism are (1) contact with igneous intrusion creating fine-grained rock with little foliation, (2) regional metamorphism over large area showing extensive foliation, (3) cataclastic metamorphism due to large mechanical deformation, such as in the fault zone, (4) hydrothermal, and (5) shock or impact metamorphism.

Most metamorphic textures involve foliation. Foliation is generally caused by a preferred orientation of sheet silicates. Various types of foliation can lead to classifications such as ***slate, phyllite,*** and ***schis***t. A rock that shows a banded texture without a distinct foliation is termed a ***gneiss.***

Schists are a type of metamorphic rock with well-foliated texture. Dolomite and calcite transform into marble by recrystallization. Quartzite is a metamorphic rock formed from quartz-rich sandstones. Silica enters into the void spaces between quartz and sand grains and acts as a cementing agent.

Clay Minerals

Clay minerals are complex aluminum silicates composed of two basic units (1) silica tetrahedron (4 oxygen atoms surrounding a silicon atom) and (2) alumina octahedron (6 hydroxyl ions surrounding an aluminum atom). Clay particles carry a net negative charge on their surfaces.

There are three important clay minerals—Kaolinite, Illite, and Montmorillonite.

Kaolinite (repeating sheets of elemental silica-gibbsite sheets in a 1:1 lattice, each layer about 7.2 Å thick). Specific surface (surface area per unit mass is about 15 m^2/g). The layers are held together by hydrogen bonding.

Illite (sometimes called clay mica) consists of a gibbsite sheet bonded to two silica sheets bonded by potassium ions. Illite particles are typically between 50 and 500 Å thick and have specific area of about 80 m²/g.

Montmorillonite is similar to Illite in structure. Montmorillonite particles are typically between 10 and 50 Å thick and have specific area of about 800 m²/g.

Soil as a Three-Phase System

A soil may be described as a three-phase system consisting of soil solids, voids containing pore water, and empty voids filled with air. A saturated soil may be assumed to have only two of these phases, as voids are completely saturated with water.

The default value of specific weight (weight per unit volume) of water is taken as $\gamma_w = 62.4$ lb/ft³ (9810 N/m³). The weight of water is converted to an equivalent volume by dividing by γ_w. The specific gravity of soil solids (G_s) is typically around 2.60. The weight of solids is converted to an equivalent volume by dividing by $\gamma_s = G_s \gamma_w$.

Figure 201.1 shows a schematic view of the phase distribution in a soil mass. The three-phase system therefore has five unknown quantities (three volumes and two weights). There are two implicit equations—one relating the volume of water to the weight of water and the other relating the volume of solids to the weight of solids.

$$V_w = \frac{W_w}{\gamma_w} \qquad (201.1)$$

$$V_s = \frac{W_s}{G_s \gamma_w} \qquad (201.2)$$

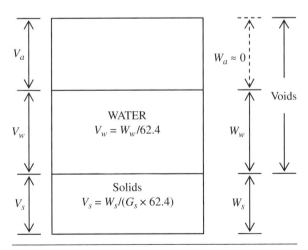

Figure 201.1 Component phases in a soil.

Thus, if three parameters are specified, we have adequate information to solve for the five system parameters. Once all five phase parameters have been calculated, any of the derived parameters—such as void ratio, porosity, degree of saturation—can be calculated.

Fundamental Definitions

The fundamental definition of the phase parameters are given below. In addition, cross relationships between these parameters are also given.

The water content (often expressed as percent) is the dimensionless ratio of the weight of water to the weight of the solids:

$$w = \frac{W_w}{W_s} \tag{201.3}$$

The *void ratio* is the dimensionless ratio of the volume of the voids to the volume occupied by the soil solids:

$$e = \frac{V_{voids}}{V_s} = \frac{V_a + V_w}{V_s} = \frac{n}{1-n} = \frac{wG_s}{S} = \frac{G_s \gamma_w}{\gamma_d} - 1 \tag{201.4}$$

The *porosity* is the dimensionless ratio of the volume of the voids to the total volume:

$$n = \frac{V_{voids}}{V_{total}} = \frac{V_a + V_w}{V_{total}} = \frac{e}{e+1} \tag{201.5}$$

The *degree of saturation* (often expressed as percent) is the fraction of the voids that are saturated:

$$S = \frac{V_w}{V_{voids}} \tag{201.6}$$

Volume occupied by soil solids is

$$V_s = \frac{W_s}{G_s \gamma_w} = V_{total}(1-n) = \frac{V_{total}}{1+e} = \frac{V_{voids}}{e} \tag{201.7}$$

Volume occupied by water is

$$V_w = \frac{W_w}{\gamma_w} = SV_{voids} = nSV_{total} = SV_{total}\frac{e}{1+e} = SV_s e \tag{201.8}$$

Volume occupied by air is

$$V_a = n(1-S)V_{total} = (1-S)V_{total}\frac{e}{1+e} = (1-S)V_{voids} = (1-S)V_s e \tag{201.9}$$

$$\text{Total soil volume} = V_{\text{total}} = V_s(1+e) = V_{\text{voids}}\frac{1+e}{e} = \frac{V_{\text{voids}}}{n} = \frac{V_s}{1-n} \quad (201.10)$$

The total (moist) unit weight is calculated as the total weight of the soil divided by the total volume:

$$\gamma = \frac{W_w + W_s}{V_a + V_w + V_s} = \gamma_d(1+w) = \left(\frac{G_s + Se}{1+e}\right)\gamma_w = \frac{(1+w)SG_s}{wG_s + S}\gamma_w = [(1-n)G_s + nS]\gamma_w$$

$$(201.11)$$

The dry unit weight is calculated as the weight of the soil solids divided by the total volume:

$$\gamma_d = \frac{W_s}{V_a + V_w + V_s} = \frac{\gamma}{1+w} = \frac{G_s\gamma_w}{1+e} = (1-n)G_s\gamma_w = \frac{SG_s\gamma_w}{wG_s + S}$$

$$= \frac{eS\gamma_w}{(1+e)w} = \frac{(\gamma_{sat} - \gamma_w)G_s}{G_s - 1} \quad (201.12)$$

$$\text{Specific Gravity of solids } G_s = \frac{\gamma_s}{\gamma_w} = \frac{Se}{w} = \frac{\gamma_d S}{\gamma_w S - \gamma_d w} = \frac{\gamma}{\gamma_w}(1+e) - Se \quad (201.13)$$

The degree of saturation can also be expressed as

$$S = \frac{\gamma}{\gamma_w}\left(\frac{1+e}{e}\right) - \frac{G_s}{e} = \frac{\gamma wG_s}{(1+w)G_s\gamma_w - \gamma} = \frac{wG_s\gamma_d}{G_s\gamma_w - \gamma_d} \quad (201.14)$$

Buoyant (submerged) unit weight (S = 1)

$$\gamma' = \gamma - \gamma_w = \left(\frac{G_s - 1}{1+e}\right)\gamma_w = \left(\frac{G_s - 1}{1+wG_s}\right)\gamma_w \quad (201.15)$$

To solve all five nonzero components of the three-phase system, it is sometimes convenient to take either a unit-volume or a unit-weight approach. For example, if the water content of the soil (w) is given and we see that water content is defined as the ratio W_w/W_s, it may be convenient to assign $W_s = 1$. This makes the weight of water $W_w = w$. On the other hand, if the porosity (n) is given and we see that porosity is defined as the ratio $V_{\text{voids}}/V_{\text{total}}$, it may be convenient to assign $V_{\text{total}} = 1$. This makes the volume of voids $V_{\text{voids}} = n$. If both options are available, it is often more convenient to use the unit-weight approach rather than unit volume, as there are only two unknown weight components and three volume components.

Example 201.1

What is the dry unit weight of a soil with a porosity $n = 0.23$, specific gravity of solids $= 2.70$, and water content $= 7\%$?

Solution   If the step-by-step approach is followed, since given data have porosity (denominator total volume) we could start with

> Total volume = 1.0 ft³
>
> Volume of voids = 0.23 ft³
>
> Volume of solids = 0.77 ft³
>
> Weight of solids = 0.77 × 2.70 × 62.4 = 129.7 lb
>
> Dry density = 129.7 lb/ft³

Alternative: Using the formula $\gamma_d = (1 - n)G_s\gamma_w = 0.77 \times 2.70 \times 62.4 = 129.7$ pcf.

Example 201.2

A soil has a unit weight of 128 lb/ft³, void ratio = 0.45, and specific gravity of solids = 2.65. What is the degree of saturation of the soil?

Solution   If the step-by-step approach is followed, since given data have void ratio (denominator volume of solids) we could start with

> Volume of solids = 1.0 ft³
>
> Volume of voids = 0.45 ft³
>
> Weight of solids = 1.0 × 2.65 × 62.4 = 165.4 lb
>
> Total weight = 128 × 1.45 = 185.6 lb
>
> Weight of water = 185.6 − 165.4 = 20.2 lb
>
> Volume of water = 20.2 ÷ 62.4 = 0.324 ft³
>
> Saturation = 0.324 ÷ 0.45 = 0.72

Alternative: Using the formula

$$\gamma = \left(\frac{G_s + Se}{1 + e}\right)\gamma_w \Rightarrow 128 = \left(\frac{2.65 + 0.45S}{1.45}\right)62.4 \Rightarrow S = 0.72$$

Pycnometer

A Pycnometer is a device used to calculate the density of a soil. It is a glass flask with a close-fitting ground glass stopper with a capillary hole through it. The following procedure is used:

1. The mass of the empty, dry pycnometer (+ stopper) is measured. (M_1)

2. Fill approximately 1/3 of the pycnometer volume with the soil. Record the mass (M_2)

3. Add water such that the Pycnometer as well as the capillary hole in the stopper is filled. Record the mass (M_3)

4. Empty the Pycnometer. Dry. Fill the Pycnometer with water until capillary hole in stopper is filled. Record the mass (M_4)

From these four masses, specific gravity of the soil is calculated as

$$G_s = \frac{(M_2 - M_1)}{(M_2 - M_1) - (M_3 - M_4)}$$
(201.16)

Shrinkage and Swell

When earth is excavated, its volume is initially greater than its in situ volume, but, after being placed in an embankment and compacted, the earth shrinks so that its volume is less than the volume in the borrow pit. The ratio of the shrunk volume to the in situ volume is termed the shrinkage factor. The approximate shrinkage factors for the various types of soils are sand and gravel 0.92, clay 0.90, loam 0.88, and organic soil 0.85.

When an embankment is made of broken-tip hard rock, it has a volume from 40% to 80% larger than its original volume in the cut, and there is practically no subsequent settling of the embankment. This increase in volume is called the swell of rock. It varies a great deal with conditions, but 50% may be taken as the average swell.

Thus, a unit volume (bank measure) of soil, upon being excavated, swells to a volume = 1 + bulking fraction. This soil, upon being placed in its destined location and compacted, shrinks to a final volume = 1 − shrinkage fraction.

The percent swell is given as

$$\% \text{ swell} = \left[\frac{1}{\text{LF}} - 1\right] \times 100\%$$
(201.17)

where LF is the load factor defined as

$$\text{Load factor} = \frac{\text{loose density}}{\text{bank density}}$$
(201.18)

Example 201.3

A highway project requires 38,000 yd^3 of compacted fill material for construction of an embankment. The contractor has access to a borrow pit of capacity 26,000 yd^3. The soil in this location has 12% shrinkage. The rest of the soil has to come from a borrow pit which has a clay loam with an in-place density of 124 lb/ft^3, shrinkage of 15%, and a bulking of 16%. Trucks used to transport the borrow soil have a net capacity of 24 tons and a volumetric capacity of 15 yd^3. How many truckloads are needed?

Solution

Fill needed	38,000 yd^3
Soil in local borrow pit (*in situ* volume)	26,000 yd^3
Yield volume of local soil (after shrinkage) = 0.88 × 26,000 =	22,880 yd^3
Clay loam yield needed = 38,000 − 22,880 =	15,120 yd^3
In place volume of clay loam = 15,120/0.85 =	17,788 yd^3
In place weight of clay loam = 17,788 × 27 × 124/2000 =	29,777 tons

Therefore, by weight capacity of trucks, we need 1240.7 truckloads. Transported volume of borrow soil $= 17{,}788 \times 1.16 = 20{,}634$ yd^3. Therefore, by volume capacity of trucks, we need 1375.6 truckloads. Therefore, 1376 truckloads (based on volume criterion) will be required.

Example 201.4

An embankment needs 400,000 yd^3 of fill. The fill should be compacted to at least 90% of the standard Proctor maximum dry density obtained in the lab, which is 128 pcf ($\gamma_{d,\max}$). The fill may be obtained from two companies (A and B) with the following profiles. Determine the total cost of haulage (a) by company A and (b) by company B.

	A	B
Distance from borrow pit to embankment	25 miles	30 miles
Cost of haul $/yd³-mile	0.50	0.40
Cost of fill material $/yd³	1.50	1.40
Borrow soil moisture content	16%	18%
Specific gravity of solids	2.70	2.65
Degree of saturation	70%	58%

Solution

At 90% compaction, the dry density $\gamma_d = 0.9 \times 128 = 115.2$ pcf

Weight of solids $= 400{,}000 \times 27 \times 115.2$ lb $= 1.24416 \times 10^9$ lb

$$\text{Void ratio } e = \frac{w \cdot G_s}{S}$$

Company A

$$e_A = \frac{0.16 \times 2.70}{0.70} = 0.617$$

$$V_{\text{solids}} = \frac{1.24416 \times 10^9}{2.70 \times 62.4} = 7.385 \times 10^6 \text{ ft}^3$$

$$V_{\text{total}} = (1 + e)V_s = 11.94 \times 10^6 \text{ ft}^3 = 442{,}222 \text{ yd}^3$$

Material cost for company A $= 400{,}000 \times 1.50 = \$0.60$ million

Haulage cost for company A $= 442{,}222 \times 25 \times 0.5 = \$5.53$ million

Total cost for company A $= \$6.13$ million

Company B

$$e_B = \frac{0.18 \times 2.65}{0.58} = 0.822$$

$$V_{\text{solids}} = \frac{1.24416 \times 10^9}{2.65 \times 62.4} = 7.524 \times 10^6 \text{ ft}^3$$

$$V_{\text{total}} = (1 + e)V_S = 13.71 \times 10^6 \text{ ft}^3 = 507,778 \text{ yd}^3$$

Material cost for company B = 400,000 × 1.40 = $0.56 million

Haulage cost for company B = 507,778 × 30 × 0.4 = $6.09 million

Total cost for company B = $6.65 million

Soil Sampling and Testing

Guidelines for Subsurface Sampling

Soil exploration costs are typically about 0.1% to 0.5% of the ultimate cost of the structure. The number and type of samples to be taken depend on the stratification and material encountered. It is usually recommended to take representative disturbed samples at vertical intervals of not less than 5 ft and at every change in strata. The number and spacing of undisturbed samples depend on the anticipated design problems and the necessary testing program.

Undisturbed samples should comply with the following criteria: they should contain no visible distortion of strata, or opening or softening of materials; specific recovery ratio (length of undisturbed sample recovered divided by length of sampling push) should exceed 95%; and they should be taken with a sampler with an area ratio less than 15%. In cohesive soils, one should obtain undisturbed samples, so that there is at least one representative sample in each boring for each 10 ft depth. If there is danger of caving, use casing or viscous drilling fluid to advance borehole. When sampling above groundwater table, the borehole must remain dry whenever possible. The area ratio of a tubular sampler is defined below:

$$A_r = \frac{D_o^2 - D_i^2}{D_i^2} \times 100\% \tag{202.1}$$

where D_o = outer diameter (O.D.) of tubular sampler
D_i = inner diameter (I.D.) of tubular sampler

The following are considered reasonable guidelines to produce the **minimum** subsurface data needed for a cost-effective geotechnical design and construction.

Structure Foundation

Unless warranted by erratic subsurface conditions, it is recommended to use at least 1 boring per substructure unit under 100 ft (30 m) in width, and 2 per substructure unit over 100 ft (30 m) in width. For spread footings, the minimum depth of borings is equal to *2B* where *L* < *2B* and equal to *4B* where *L* > *2B*. For deep foundations, the minimum depth of boring is the greater of 20 ft below tip elevation or twice the maximum pile group dimension.

Retaining Structures

Borings must be spaced every 100 to 200 ft. Some borings should be at the front of and some behind the wall face. Borings should be extended to a depth of 0.75 to 1.5 times wall height. When stratum indicates potential deep stability or settlement problem, borings should be extended to a hard stratum.

For sand or gravel soils, SPT (split-spoon) samples should be taken at 5 ft (1.5 m) intervals or at significant changes in soil strata. Continuous SPT samples are recommended in the top 15 ft (4.5 m) of borings made at locations where spread footings may be placed in natural soils.

For silt or clay soils, SPT and "undisturbed" thin wall tube samples should be taken at 5 ft (1.5 m) intervals or at significant changes in strata. Take alternate SPT and tube samples in the same boring or tube samples in separate undisturbed boring. Tube samples should be sent to lab to allow consolidation testing (for settlement analysis) and strength testing (for slope stability and foundation bearing capacity analysis). Field vane shear testing is also recommended to obtain in-place shear strength of soft clays, silts, and well-rotted peat.

For rock, continuous cores should be obtained in rock or shales using double- or triple-tube core barrels. In structural foundation investigations, core is a minimum of 10 ft (3 m) into rock to ensure it is bedrock and not a boulder. Core samples should be sent to the lab for possible strength testing (unconfined compression) for foundation investigation. Percent core recovery and rock quality designation (RQD) value should be determined in field or lab for each core run and recorded on boring log.

Groundwater

Water level encountered during drilling, at completion of boring, and at 24 hours after completion of boring should be recorded on the boring log. In low-permeability soils such as silts and clays, a false indication of the water level may be obtained when water is used for drilling fluid and adequate time is not permitted after boring completion for the water level to stabilize. In such soils, an observation well should be installed to allow monitoring of the water level over a period of time. Artesian pressure and seepage zones, if encountered, should also be noted on the boring log.

Interpretation of Boring Logs

ASCE[1] gives the following broad criteria for determining minimum depth of boring:

1. Determine the net increase in vertical stress $\Delta\sigma$ under the foundation.
2. Estimate the effective vertical stress σ'_v.

[1]American Society of Civil Engineers (1972), "Subsurface Investigation for Design and Construction of Foundations of Buildings, Part I," *Journal of the Soil Mechanics and Foundations Division*, ASCE Vol. 98, No. SM5, 481–490.

3. Determine the depth D_1 at which $\Delta\sigma$ is less than 10% of the net stress on the foundation.

4. Determine the depth D_2 at which $\Delta\sigma$ is equal to 5% of the effective vertical stress σ'_v.

Unless bedrock is encountered, the recommended minimum depth of boring is to be the smaller of D_1 and D_2.

Some typical estimates for the minimum depth of borings are given as follows:
For multistory light steel or narrow concrete buildings

$$z_b = 10S^{0.7} \qquad (202.2)$$

For multistory heavy steel or wide concrete buildings

$$z_b = 20S^{0.7} \qquad (202.3)$$

where z_b = minimum depth of boring (ft)
$\quad S$ = number of stories

A typical example of a boring log is shown in Fig. 202.1.

Soil Sampling Techniques

Disturbed Soil Samples

Common sampling techniques for taking disturbed soil samples and rock cores are summarized below:

Split barrel sampler: 2 in O.D. and 1.375 in I.D. are standard. However, penetrometer sizes up to 4 in O.D. to 3.5 in I.D. are available. The split barrel tests provide best results in fine-grained soils. If liners are not used, some soil disturbance is likely with this type of sampling.

Retractable plug sampler: Sampling tubes have 1 in O.D. and 6 in length. Sampling tube is driven by hammer. A maximum of six tubes may be filled in a single penetration. This sampling technique gives good results for silts, clays, and fine and loose sands. These samplers are lightweight and highly portable, but sample disturbance is likely.

Continuous flight helical augers: These augers are available in 3 in to 16 in diameter and are appropriate for most soils above the water table. However, these augers will not penetrate hard soils or those containing cobbles or boulders.

Diamond core barrels: These samplers are available in 1.5 in to 3 in O.D. with a 0.875 in to 2.125 in core. Barrel lengths of 5–10 ft provide best results for hard rock.

A single tube sampler is used primarily for strong, sound rock. Double tube is used primarily for nonuniform fractured and soft erodible rock. The double-tube sampler has an

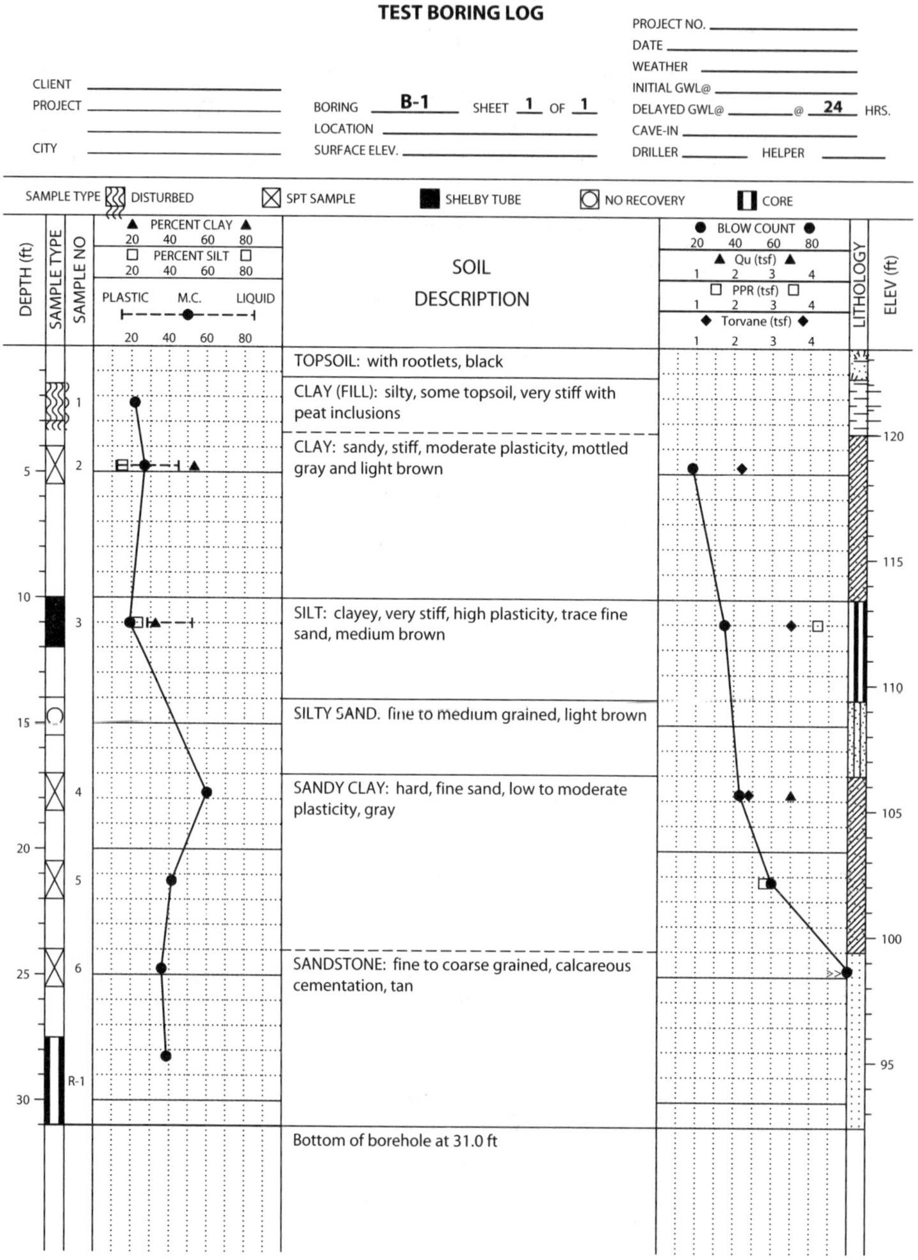

Figure 202.1 Sample boring log.

inner barrel which does not rotate with the outer tube. A triple-tube sampler is similar to the double-tube sampler except that it has an additional inner split tube liner. Intensely fractured rock core is best preserved in this barrel.

Undisturbed Soil Samples

Common sampling techniques for taking undisturbed soil samples and rock cores are summarized below:

Shelby tube: The most common size for the Shelby tube is 3 in O.D., 2.875 in I.D. Available from 2 in to 5 in O.D. Standard sampler length is 30 in. The Shelby tube produces best results for cohesive fine-grained soils. It is not suitable for hard, dense, or gravelly soils.

Stationary piston: The most common size is 3 in O.D. Available from 2 in to 5 in O.D. Standard sampler length is 30 in. Best results for soft to medium clays and fine silts. Usually produces less disturbed samples than Shelby tube.

Hydraulic piston: The most common size is 3 in O.D. Available from 2 in to 4 in O.D. Standard sampler length is 36 in. Best results for silts, clays, and some sandy soils. Piston is usually advanced by hydraulic or compressed air pressure.

Geophysical Methods for Site Characterization

Noninvasive methods such as ground penetrating radar (GPR), resistivity mapping, seismic reflection and refraction can be used instead of taking direct samples. Besides being noninvasive, these methods provide a greater sampling density and beyond the large initial capital expenditure, can result in a lower cost per sample than other more direct sampling methods.

Some applications of geophysical techniques are—characterize hydrogeology, detect and map contaminant plumes, buried wastes, changes in stratification.

Ground Penetrating Radar (GPR)

High-frequency electromagnetic waves are sent out from an emitter and reflected back to a receiver. Reflections occur wherever an interface between two distinct materials occurs. Typical penetration depth of GPR is 15–30 ft. However, it does not yield good data for fine-grained soils (silts and clays) due to low penetration.

Electromagnetic and Resistivity Methods

These methods work on the basis of the current-carrying capacity of the soil. Based on this, spatial variations in the soil can be mapped. Typical sampling depth is 2 ft.

Nuclear Methods

Seismic reflection and refraction methods are based on the fact that wave speed is a function of soil stiffness (quantified by the shear modulus). A source sends calibrated waveforms into the soil and the reflected/refracted waves are detected by a receiver (geophone).

Table 202.1 Rock Mass Classification Based on RQD

RQD	Rock quality classification
< 25%	Very poor
25–50%	Poor
50–75%	Fair
75–90%	Good
90–100%	Excellent

Table 202.2 Recommended Allowable Bearing Pressure for Footings on Rock

Material	Allowable bearing pressure MPa (kip/ft^2)
Igneous and sedimentary rocks (granite, diorite, gneiss, hard limestone, dolomite)	
RQD = 75–100%	11.5 (240)
RQD = 50–75%	6.2 (130)
RQD = 25–50%	2.9 (60)
RQD = 0–25%	1.0 (20)
Metamorphic rocks (schist, slate, bedded limestone)	
RQD > 50%	3.8 (80)
RQD < 50%	1.0 (20)
Sedimentary rocks (hard shales and sandstones)	
RQD > 50%	2.4 (50)
RQD < 50%	1.0 (20)
Soft or broken rock (excluding shale) and soft limestone	
RQD > 50%	1.1 (24)
RQD < 50%	0.8 (16)
Soft shale	0.4 (8)

Rock Quality Designation

Rock cores are obtained by drilling into rock. As soon as possible after drilling, the rock cores are measured to obtain lengths of core pieces longer than 4 in. The RQD is defined as the cumulative length of core pieces longer than 4 in in a run divided by the total length of the core run. The total length of core must include all lost core sections. Any breaks caused by the drilling process or in extracting the core from the core barrel should be ignored. Table 202.1 correlates the qualitative rock mass classification to the numerical value of the RQD. Table 202.2 shows recommended allowable bearing pressure on rock deposits.

Unconfined Compression Strength of Rock

The primary intact rock property of interest for foundation design is the unconfined compressive strength. The strength of jointed rocks is generally less than individual units of the rock mass. Therefore, the unconfined compressive strength provides an upper limit of the rock mass-bearing capacity and an index value for rock classification. In general, samples with unconfined strengths below 250 psi (1.72 MPa) are not considered to behave as rock. As unconfined compressive strength increases, bearing capacity generally increases and scourability decreases.

Rock Mass Rating System

The rock mass rating (RMR) combines the most significant geologic parameters for a rock mass deposit. Each of these parameters is given a rating (score) according to the guidelines of Table 202.3. These are then combined into a single numerical score. The overall RMR (scale of 1–100) is then interpreted according to Table 202.4.

The following five parameters are used for the detailed classification:

1. Strength of intact rock material (based on a point load strength index or uniaxial compressive strength)

2. Drill core quality (in terms of rock quality designation)

3. Spacing of discontinuities

4. Condition of discontinuities

5. Presence of groundwater

Effective Stress

The vertical effective stress at a particular depth affects many aspects of soil behavior, such as consolidation settlement, lateral earth pressure, and shear strength. The effective vertical stress at depth z below the ground surface is calculated as the sum of $\gamma'z$ terms from the ground surface to the depth z.

$$\sigma'_v = \sum \gamma'z = \sum (\gamma - \gamma_w)z \qquad (202.4)$$

In Eq. (202.4), the unit weight γ of a soil layer is replaced by the submerged unit weight γ' only if the soil is below the water table. Equation (202.4) is for the static case when seepage forces are either nonexistent or negligible. When seepage flow takes place through the interparticle voids in the soil, the seepage pressure can either increase or decrease the effective stress. The extent to which this modification occurs depends on the hydraulic gradient which causes the seepage flow.

Effective Stress with Seepage

When an upward hydraulic gradient causes upward seepage in a soil, the effective stress is given by

$$\sigma'_v = \sum \gamma'z - iz\gamma_w = \sum (\gamma - \gamma_w - i\gamma_w)z \qquad (202.5)$$

Table 202.3 Criteria for Rock Mass Rating System

	Parameter	Range of values						
1	Strength of intact rock material (MPa) — Point load strength index	>10	4–10	2–4	1–2	Not used		
	Uniaxial compressive strength	>250	100–250	50–100	25–50	5–25	1–5	<1
	RATING	15	12	7	4	2	1	0
2	Rock quality designation (%)	90–100	75–90	50–75	25–50	<25		
	RATING	20	17	13	8	3		
3	Spacing of discontinuities (m)	>2	0.6–2.0	0.2–0.6	0.06–0.2	<0.06		
	RATING	20	15	10	8	5		
4	Condition of discontinuities	Unweathered wall rock	Slightly weathered walls	Highly weathered walls	Slickensided	Soft gouge		
	RATING	30	25	20	10	0		
5	Groundwater	Completely Dry	Damp	Wet	Dripping	Flowing		
	RATING	15	10	7	4	0		

Table 202.4   Rock Classes and Their Meanings

Class	Rating	Description	Average stand up time	Cohesion (kPa)	Friction angle (deg)
I	81–100	very good rock	20 years for 15 m span	>400	>45
II	61–80	good rock	1 year for 10 m span	300–400	35–45
III	41–60	fair rock	1 week for 5 m span	200–300	25–35
IV	21–40	poor rock	10 hours for 2.5 m span	100–200	15–25
V	<21	very poor rock	30 min for 1 m span	<100	<15

where i = hydraulic gradient in the soil = $\Delta h/L$. When the hydraulic gradient reaches a critical value equal to $i_{cr} = \gamma'/\gamma_w$, the effective vertical stress can approach zero, causing a condition called *boiling* or a *quick* condition.

When a downward hydraulic gradient causes downward seepage in a soil, the effective stress is given by

$$\sigma'_v = \sum \gamma'z + iz\gamma_w = \sum (\gamma - \gamma_w + i\gamma_w)z \tag{202.6}$$

Thus, with downward seepage, the effective stress in the soil is increased by the seepage pressure.

Example 202.1

For the soil layers shown below, calculate the effective vertical stress at the center of the clay layer. Assume no seepage occurs.

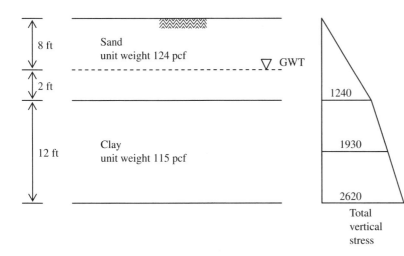

Solution   The center of the clay layer is at a depth of $z = 8 + 2 + 6 = 16$ ft. The effective vertical stress is

$$\sigma'_v = 8 \times 124 + 2 \times (124 - 62.4) + 6 \times (115 - 62.4) = 1430.8\,\text{psf}$$

On the other hand, the total vertical stress is

$$\sigma_v = 10 \times 124 + 6 \times 115 = 1930\,\text{psf}$$

The pore pressure is

$$\sigma_{pw} = 8 \times 62.4 = 499.2\,\text{psf}$$

The total vertical stress σ_v and the pore water pressure diagrams are shown on the figure.

Effective Stress in Capillary Zone in Soil

Depending on the type of soil, a limited zone above the water table may have soil moisture that rises through the soil pores due to capillary action. The pore pressure in this zone is given by

$$u = -S\gamma_w h \tag{202.7}$$

where S = degree of saturation of the soil in the capillary zone
γ_w = unit weight of water = 62.4 lb/ft³ = 9810 N/m³
h = height above water table

Example 202.2

Consider a soil profile as shown below. The sand layer is 13 ft deep and overlies the 12-ft-thick clay layer as shown. The water table elevation is at the top of the clay layer. Capillary forces draw water to an elevation of 3 ft above the water table. The degree of saturation in this capillary zone is 50%. Determine the profiles of (a) total vertical stress, (b) pore pressure, and (c) effective vertical stress in the soil.

Assume the unit weight of the sand above the capillary zone to be 108 pcf, the sand in the capillary zone to be 120 pcf, and that of the clay to be 124 pcf.

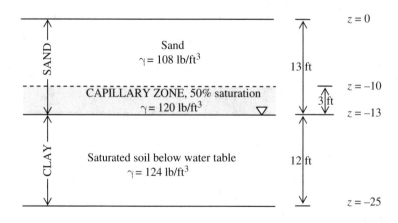

Solution

At an elevation of $z = -10^-$ (just above the top of the capillary zone)
Total (vertical) stress, $p = 108 \times 10 = 1080$ psf
Pore pressure $= 0$
Effective (vertical) stress $= 1080$ psf

At an elevation of $z = -10^+$ (just below the top of the capillary zone)
Total (vertical) stress, $p = 108 \times 10 = 1080$ psf
Pore pressure $= -0.5 \times 62.5 \times 3 = -93.8$ psf
Effective (vertical) stress $= 1080 - (-93.8) = 1173.8$ psf

At an elevation of $z = -13$ (just above the top of the clay layer)
Total (vertical) stress, $p = 1080 + 3 \times 120 = 1440$ psf
Pore pressure $= -0.5 \times 62.5 \times 0 = 0$ psf
Effective (vertical) stress $= 1440 - 0 = 1440$ psf

At an elevation of $z = -25$ (at bottom of the clay layer)
Total (vertical) stress, $p = 1440 + 12 \times 124 = 2928$ psf
Pore pressure $= 62.4 \times 12 = 748.8$ psf
Effective (vertical) stress $= 2928 - 748.8 = 2179.2$ psf

Using these values, the following profiles can be drawn for (a) total vertical stress, (b) pore pressure, and (c) effective vertical stress in the soil.

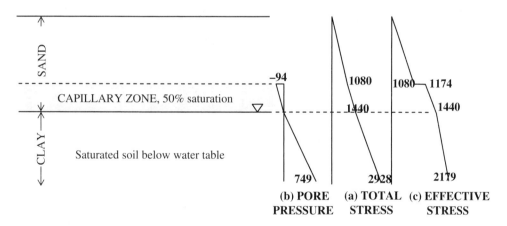

(b) PORE PRESSURE   **(a) TOTAL STRESS**   **(c) EFFECTIVE STRESS**

Soil Consistency

Fine-grained soils containing clay minerals exhibit cohesive properties such as plasticity (the ability to be molded). As the moisture content of such a soil is gradually increased, the soil goes from a brittle solid to a semisolid to a plastic and finally to a liquid state. These transitions occur at specific values of water content known as the shrinkage limit (solid to semisolid), plastic limit (semisolid to plastic state), and liquid limit (plastic to liquid state). These limits have been standardized in terms of Atterberg limits.

Atterberg Limit Tests

Liquid Limit Test (ASTM D-4318)

The apparatus for the liquid limit test consists of a brass cup that is dropped onto a hard rubber base by a cam operated by a crank. An 11-mm-wide groove is cut at the center of the soil paste in the brass cup. The *liquid limit* is defined as the moisture content at which it takes exactly 25 blows to close the groove over a distance of 1/2 in (12.7 mm). Typically, the test is performed with various values of moisture content. The plot of w versus log N is typically linear (see Fig. 202.2). The slope of this plot is called the flow index I_F defined as

$$I_F = \frac{w_1 - w_2}{\log\left(\dfrac{N_2}{N_1}\right)} \qquad (202.8)$$

The water content corresponding to $N = 25$ is interpolated from this plot. This is defined as the water content at which the soil transitions from a plastic state to liquid state (symbol: LL or w_L).

Plastic Limit Test (ASTM D-4318)

The *plastic limit* is defined as the water content at which the soil crumbles when rolled into 1/8-in (3.2-mm) diameter. This is defined as the water content at which the soil transitions from a semisolid to a plastic state (symbol: PL or w_P). Nonplastic soils (sand and silt) have very small values of PL. Clays and plastic silt can have PL from 0 to 100 or more.

Figure 202.3 shows the transition of a soil from a brittle solid to a liquid state. The shrinkage limit is that water content below which removal of water does not result in a reduction of volume. When the moisture content increases beyond the shrinkage limit (symbol: SL or w_S), the volume of the soil increases. The plasticity index (PI) of a cohesive soil is the range of moisture content over which the soil behaves as a plastic material.

$$PI = w_L - w_P \qquad (202.9)$$

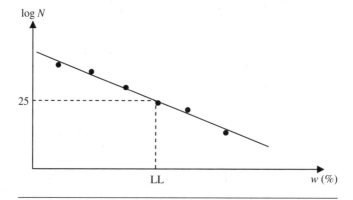

Figure 202.2 Graphical determination of liquid limit.

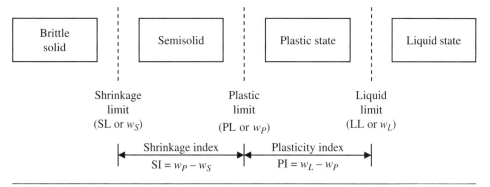

Figure 202.3 Transitions in consistency of a fine-grained soil.

The relative consistency of a cohesive soil is defined by the liquidity index (LI)

$$LI = \frac{w - w_P}{w_L - w_P} \qquad (202.10)$$

Another index that is commonly used is the consistency index (CI)

$$CI = \frac{w_L - w}{w_L - w_P} \qquad (202.11)$$

To find the shrinkage limit of a soil, follow these steps:

Plot the Atterberg limits (LL, PI) of the soil on the Casagrande plasticity chart. The point will be either above (A) or below (B) the A-line. Measure the vertical distance between the point and the A-line. This distance is called Δp_i.

If the point is above the A-line, then the shrinkage limit is given by

$$w_S = 20 - \Delta p_i$$

If the point is below the A-line, then the shrinkage limit is given by

$$w_S = 20 + \Delta p_i$$

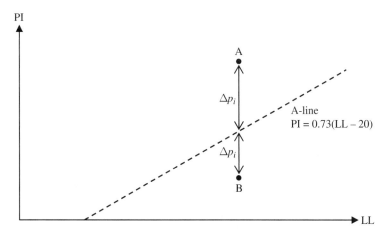

The shrinkage limit w_S is calculated using Eq. (202.12)

$$w_S = 5.4 + 0.73\text{LL} - \text{PI} \qquad (202.12)$$

A plasticity index greater than 25 to 30 may mean troublesome soils with low strength, high compressibility and high shrink-swell potential. Soils with liquid limit greater than 50 are termed highly plastic. Liquidity index is an excellent indicator of geologic history and relative soil properties. Liquidity index = 0 may indicate a heavily over-consolidated soil with high strength while a soil with liquidity index = 1 indicates a normally consolidated soil with much lower strength.

Standard Penetration Test

The standard penetration test (SPT) is one of various standardized tests to measure *in situ* compactness of soils. It can be used for all types of soil, but is preferred for sand deposits. In the standard penetration test, a split-spoon sampler is driven by a 140-lb hammer falling 30 in through three successive 6-in penetrations. The outer diameter of the split spoon is 2 in and the inner diameter is 1.5 in. The field-recorded result for the SPT consists of three numbers—the number of blows required to drive the sampler through 0–6 in, 6–12 in, and 12–18 in penetration or until normal maximum resistance (refusal) is reached, using the standard hammer and drop. *Refusal* is defined as a penetration of less than 6 in for 100 hammer blows. The standard penetration resistance (also known as the SPT N value or blow count) from the field is the sum of the counts (blows per foot) from the second and third intervals.

Corrections to Field *N* Value

The field measured N value needs to be corrected for testing factors such as hammer efficiency, borehole diameter, rod length as well as existing overburden pressure. Skempton[2] suggested the following set of corrections to obtain the corrected N value (N_{60})

$$N_{60} = 1.67 E_m C_b C_r C_s N \qquad (202.13)$$

where E_m = correction for hammer efficiency (0.6 for a safety hammer, 0.45 for a donut hammer)

C_b = correction for borehole diameter (1.0 for 65 mm $< D <$ 115 mm, 1.05 for $D =$ 150 mm, 1.15 for $D =$ 200 mm)

C_r = correction for drill rod length (1.0 for $L >$ 10 m, 0.95 for 6 m $< L <$ 10 m, 0.85 for 4 m $< L <$ 6 m, 0.75 for $L <$ 4 m)

C_s = correction for sampling method (1.0 for standard sampler; 1.2 for sampler without liner)

---

[2]Skempton A. W. (1986), "Standard Penetration Test Procedures and the Effects in Sand of Overburden Pressure, Relative Density, Particle Size, Aging and Overconsolidation," *Geotechnique*, Vol. 36, No. 3, 425–447.

The adjustment due to overburden is applied so that the corrected blow count is normalized to an overburden of 1 ton/ft^2 (which is approximately 1 atmosphere). This is done by multiplying the field N value by a correction factor C_n which corrects the field N value upward (if the overburden is less than 1 tsf) or downward (if the overburden is more than 1 tsf).

$$N' = C_n N \tag{202.14}$$

Various empirical relationships have been suggested for this correction. In these expressions, σ'_v is the effective overburden pressure in tons/ft^2 (1 ton/ft^2 = 2000 lb/ft^2 = 2 kips/ft^2). Some of these empirical equations are presented below:

Liao and Whitman[3]
$$C_n = \sqrt{\frac{1}{\sigma'_v}} \tag{202.15}$$

Peck[4] *et al.*
$$C_n = 0.77 \log_{10}\left(\frac{20}{\sigma'_v}\right) \tag{202.16}$$

Skempton
$$C_n = \frac{2}{1 + \sigma'_v} \tag{202.17}$$

Seed[5] *et al.*
$$C_n = 1 - 1.25 \log_{10} \sigma'_v \tag{202.18}$$

Correlation between *N* Value and Bearing Capacity

For an estimation of the bearing capacity based on settlement in sand, see section Allowable Bearing Pressure in Sand Based on Settlement (Chap. 206).

Table 202.5 shows some correlations between the SPT N value and other soil properties for cohesionless soils. Table 202.6 shows some correlations between the SPT N value and unconfined compression strength (S_{uc}) for cohesive soils.

---

[3]Liao S. and Whitman R. V. (1986), "Overburden Correction Factor for SPT in Sand," *Journal of Geotechnical Engineering*, ASCE, Vol. 112, No. 3, 373–377.

[4]Peck, R. B., Hanson, W. E., and Thornburn, T. H. (1974), *Foundation Engineering*, 2nd ed., Wiley, New York.

[5]Seed, H. B., Arango, I., and Chan, C. K. (1975), "Evaluation of Soil Liquefaction Potential during Earthquakes," Report No. EERC 75–28, Earthquake Engineering Research Center, University of California, Berkeley.

Table 202.5 Correlations between SPT N Value and Strength Parameters (Sand)

Soil type	SPT N	Relative density R_D	Friction angle (Peck)
Very loose sand	<4	<0.20	<29°
Loose sand	4–10	0.20–0.40	29°–30°
Medium sand	10–30	0.40–0.65	30°–36°
Dense sand	30–50	0.65–0.85	36°–41°
Very dense sand	>50	>0.85	>41°

Table 202.6 Correlations between SPT N Value and Strength Parameters (Clay)

SPT (blows/ft)	Estimated consistency	S_{uc} (tons/ft²)
<2	Very soft (extruded between fingers when squeezed)	<0.25
2–4	Soft (molded by light finger pressure)	0.25–0.50
4–8	Medium (molded by strong finger pressure)	0.50–1.00
8–15	Stiff (readily indented by thumb but penetrated with great effort)	1.00–2.00
15–30	Very stiff (readily indented by fingernail)	2.00–4.00
>30	Hard (indented with difficulty by fingernail)	>4.00

Relative Density

The *in situ* density of granular soils is expressed by the relative density R_D. It can be expressed in terms of void ratio (e), porosity (n), or unit weight (γ) as given below:

$$R_D = \frac{e_{max} - e}{e_{max} - e_{min}} = \left(\frac{1 - n_{min}}{1 - n}\right)\left(\frac{n_{max} - n}{n_{max} - n_{min}}\right) = \left(\frac{\gamma_d - \gamma_{d,min}}{\gamma_{d,max} - \gamma_{d,min}}\right)\left(\frac{\gamma_{d,max}}{\gamma_d}\right) \qquad (202.19)$$

where
e, n, γ_d = *in situ* void ratio, porosity, and dry unit weight of the soil, respectively
e_{max} and n_{max} = void ratio and porosity in the loosest state
e_{min} and n_{min} = void ratio and porosity in the densest state
$\gamma_{d,min}$ and $\gamma_{d,max}$ = dry unit weight in the loosest and densest states

It is customary to express the relative density as a percentage. The dry unit weight of granular soils is measured according to ASTM test D-2049. For sand, this test involves a 0.1 ft³ mold. Sand is poured loosely into the mold until the mold is filled without using any compacting effort. The minimum dry unit weight ($\gamma_{d,min}$) is then determined by dividing the net weight of the sand by the mold volume.

The maximum dry unit weight ($\gamma_{d,max}$) is determined by vibrating sand in the mold for 8 minutes (at 60 Hz, amplitude 0.025 in). A surcharge of 2 psi is added to the top of the sand. Soil is added to the mold until full. The value of $\gamma_{d,max}$ is determined at the end of the vibrating period.

The relative density of a soil can have a significant impact on its shear strength, and therefore, on its bearing capacity. For a detailed discussion of how to include the effect of relative density on the bearing capacity of a soil, see the section on bearing capacity.

The relative density of normally consolidated sands can be estimated from the correlation obtained by Marcuson and Bieganousky[6] (1977).

$$R_D = 11.7 + 0.76 \sqrt{222N + 1600 - 53\sigma'_v - 50(C_u)^2} \qquad (202.20)$$

where σ'_v = effective overburden pressure in pounds per square inch

C_u = coefficient of uniformity (D_{60}/D_{10})

N = standard penetration resistance

Cone Penetrometer Test (CPT)

The standard Cone Penetrometer test (ASTM D-3441) involves pushing a 1.41-in-diameter cone with a nose angle of 55°–60° through the ground at a rate of 1 to 2 cm/s. It can produce a continuous record of the tip and sleeve resistance, the ratio between the two, induced pore pressure just behind the tip of the cone, pore pressure ratio (change in pore pressure divided by measured pressure) and changes in stratification. The tapered cone forces failure of the soil about 15 in ahead of the tip.

The tip resistance q_c is measured (tons/ft²) by load cells located just behind the cone. For a saturated cohesive soil, the tip resistance is related to the undrained shear strength. An advantage of the CPT is that it produces a continuous record as the cone is pushed into the soil. A disadvantage is that there is no soil recovery.

Skin Friction

The local friction is measured by tension load cells embedded in the sleeve for a distance of 4 in behind the tip.

Friction Ratio

The friction ratio (FR) (percent) is the ratio of the skin friction to the tip resistance. High values of FR (up to 10%) typically indicate clayey materials while low values are typical of sandy materials (FR < 1%) or dry desiccated clays. Values of FR are rarely higher than 15%.

Pore Pressure

Piezocones measure in situ pore pressure (psi) either in dynamic mode (cone moving through soil) or static mode (cone stationary).

---

[6]Marcuson, W. F. and Bieganousky, W. A. (1977), "SPT and Relative Density in Coarse Sands," *Journal of the Geotechnical Engineering Division*, ASCE, Vol. 103, No. GT11, 1295–1309.

Table 202.7 Consistency Index of Clay vs. SPT *N* value and CPT Tip Resistance

N	Cone tip resistance q_c/P_a	Consistency	Consistency index
<2	<5	Very soft	<0.5
2–8	5–15	Soft to medium	0.5–0.75
8–15	15–30	Stiff	0.75–1.0
15–30	30–60	Very stiff	1.0–1.5
>30	> 60	Hard	>1.5

Temperature Sensor

A temperature shift of about 6°F occurs at the groundwater interface. So, a cone equipped with a temperature sensor can help identify the precise location of saturation zones, which can be very useful for slope stability and consolidation studies.

Table 202.7 shows some correlations between the cone tip resistance obtained from the CPT and the *N* value obtained from the SPT.

$P_a = 1$ atmosphere $= 14.7$ psi $= 101.3$ kPa

Vane Shear Test

The vane shear test consists of inserting a bladed vane into the borehole and then pushing the vane into a clay deposit underlying the borehole. Once the vane is inserted into the clay, the maximum torque required (T_{max}) to shear the clay is recorded. This maximum torque can then be related to the undrained shear strength of the clay (s_u). The precise relation between T_{max} and s_u depends on the configuration of the blades. For example, if the vane carries four rectangular blades, the undrained shear strength is given by

$$s_u = \frac{T_{max}}{\pi(0.5D^2H + 0.167D^3)} \tag{202.21}$$

where *H* and *D* are height and diameter of the vane respectively. There are many variations of the vane shear test equipment, some of them being hand-held devices such as the Torvane device, which is manually inserted into a clay deposit and rotated to induce shear failure of the soil. A calibrated scale on top of the Torvane directly indicates the undrained shear strength s_u.

Direct Shear Test

In the direct shear test, a soil specimen (typical diameter = 2.5 in, height = 1 in) is placed between porous plates in the center of the direct shear apparatus. Vertical load is applied to the specimen and allowed to consolidate fully. A dial gauge is used to measure horizontal deflection. A horizontal force is applied to the upper half of the apparatus to cause shear failure on the middle plane of the sample.

Drained conditions maintained by the porous plates (escape of pore water allowed under slow loading) ensure that total stresses and effective stresses are the same. At various values of the normal force (N), the soil is loaded with increasing horizontal shear forces (P) until the soil fails in shear. The normal stress (σ) is calculated as the vertical force divided by the cross-sectional area, while the shear force at failure (τ) is calculated from the horizontal force at failure. The failure envelope may be obtained from plotting several σ-τ pairs.

This is illustrated in Fig. 202.4. A 4 in × 4 in soil sample is subjected to shear in a direct shear box. Upper and lower surfaces are capped by porous plates, thereby maintaining drained conditions. A normal force (N) is maintained on the sample, while shear forces (P) are applied until failure occurs. Results from tests on three samples are tabulated. Find the cohesion and the angle of internal friction.

N (lb)	P_f (lb)	Normal stress σ (psf)	Shear stress τ (psf)
34.0	63.0	$34/(4/12)^2 = 306$	$63/(4/12)^2 = 567$
56.0	75.0	$56/(4/12)^2 = 504$	$75/(4/12)^2 = 675$
90.0	93.0	$90/(4/12)^2 = 810$	$93/(4/12)^2 = 837$

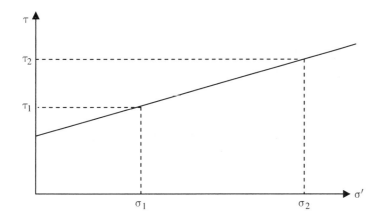

From points 1 and 2 that are on the failure envelope, the angle of internal friction (ϕ) and the cohesion c can be calculated from

$$\tan\phi = \frac{\tau_2 - \tau_1}{\sigma_2 - \sigma_1} = \frac{675 - 567}{504 - 306} = 0.545 \Rightarrow \phi = 28.6^\circ$$

$$c = \frac{\sigma_2\tau_1 - \sigma_1\tau_2}{\sigma_2 - \sigma_1} = \frac{504 \times 567 - 306 \times 675}{504 - 306} = 400 \text{ psf}$$

NOTE Obviously, in a real situation, one would test several samples and draw a line of best fit through these data points. The best-fit line may or may not pass through the data points, and the coordinates of the line should be used rather than coordinates of the test points themselves.

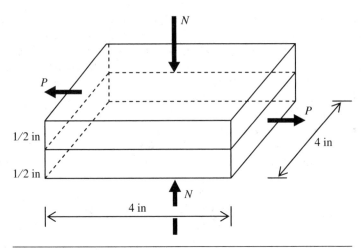

Figure 202.4 Direct shear test.

Unconfined Compression Test

In the unconfined compression test, a cylindrical sample is loaded in axial compression without any radial confining pressure. Generally failure occurs along diagonal planes where the greatest ratio of shear stress to shear strength occurs. Very soft material may not show well-defined diagonal planes of failure. In such cases, failure is generally assumed to have occurred when the axial strain has reached a value of 20%. The unconfined compression test is performed only on cohesive soil samples. For soils with negligible internal friction, cohesion is taken as one-half the unconfined compressive strength.

Compaction

Compaction of soils serves to increase soil shear strength and improve stability, decrease erosion potential as well as decrease permeability and compressibility.

In the field, soil is placed in lifts (layers) and then mechanically compacted using one or more of the following: static loading, impact, vibration, kneading. The *in situ* dry density is often expressed as a percentage of the maximum dry density as obtained in the laboratory. Thus, if the dry density in the field is 95 lb/ft^3 and the maximum dry density is 100 lb/ft^3, then the soil is said to exist at 95% relative compaction.

$$\text{Relative compaction} = \frac{\gamma_d}{\gamma_{d,\max}} \times 100\% \qquad (202.22)$$

where γ_d = dry unit weight achieved on site
$\gamma_{d,\max}$ = maximum dry unit weight from laboratory

The various compaction methods employed can be broadly subdivided into the following categories:

1. Hand methods
   a. Manual weighted tamp
   b. Pneumatic tamp
   c. Vibratory walk-behind sheepsfoot roller
2. Mechanical methods
   a. Drop tamp
   b. Vibratory sheepsfoot roller

Table 202.8 shows a matrix of soil types vs. characteristics relevant to compaction.

Table 202.8 Compaction Characteristics of Various Soils

Soil type	Relevant soil characteristics				
	Permeability	Expansive	Ease of compaction	Foundation support	Pavement subgrade
Gravel	Very high	No	Very easy	Excellent	Excellent
Sand	Medium	No	Easy	Good	Good
Silt	Medium low	Slightly	Moderate	Poor	Poor
Clay	Very low	Yes	Very difficult	Moderate	Poor
Organic	Low	Slightly	Very difficult	Very poor	NA

Table 202.9 shows a matrix of soil types vs. compaction techniques.

Table 202.9 Compaction Techniques for Various Soil Types

Soil type	Lift thickness	Type of energy			
		Impact	Pressure (with kneading)	Vibration	Kneading (with pressure)
		Vibrating Sheepsfoot Rammer	Static sheepsfoot grid roller Scraper	Vibrating plate compactor Vibrating roller Vibrating sheepsfoot	Scraper Rubber-tired roller Loader Grid roller
Gravel	>12 in	Poor	—	Good	Very good
Sand	10 in	Poor	—	Excellent	Good
Silt	6 in	Good	Good	Poor	Excellent
Clay	6 in	Excellent	Very good	—	Good

Standard Proctor Test

The standard Proctor test uses a mold with a diameter of 4 in and a volume of $1/30$ ft^3 (943.3 cm^3)—filled in three lifts, each layer compacted by 25 blows of a hammer (a 5.5-lb weight dropping 12 in). In the modified Proctor test, adapted to better represent current field conditions, the mold volume is still $1/30$ ft^3—however, it is filled in 5 lifts—each layer compacted by 25 blows of a hammer (a 10-lb weight dropping 18 in).

In either test, several samples with varying moisture content are tested. The total unit weight (or density) and moisture content are measured for each sample. The dry unit weight (or density) is calculated as

$$\text{Dry unit weight } \gamma_d = \frac{\gamma}{1 + w} \qquad (202.23)$$

A typical plot (silty-clayey soil) of dry unit weight versus moisture content is shown in Fig. 202.5. Figure 202.5 shows a peak for the dry density at optimum moisture content (w_{opt}). The dry density can also be expressed as the following function of the degree of saturation S:

$$\gamma_d = \frac{G_s \gamma_w}{1 + \dfrac{G_s w}{S}} \qquad (202.24)$$

At any moisture content, the theoretical maximum dry unit weight is obtained when there is no air in the void spaces, i.e., when the degree of saturation $S = 1$. This yields the following formula for the zero air-voids unit weight γ_z (fully compacted):

$$\gamma_z = \frac{G_s \gamma_w}{1 + wG_s} = \frac{\gamma_w}{w + \dfrac{1}{G_s}} \qquad (202.25)$$

Figure 202.5
Variation of dry density with water content.

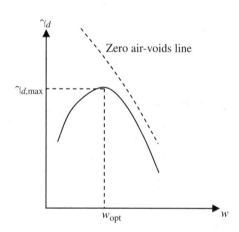

Sandy soil typically exhibits greatest dry density at a moisture content of about 8% while clay compacts best at moisture content of approximately 20%.

Example 202.3

Results from a standard Proctor test on six soil samples are tabulated below. Determine the optimum moisture content, the maximum dry unit weight, the void ratio, and the degree of saturation at the optimum moisture content. Also calculate the range of water content for which 95% compaction can be achieved. Assume specific gravity of solids $= 2.65$.

Sample	Net weight of wet soil (lb)	Moisture content (%)
1	3.45	12
2	3.76	14
3	3.96	16
4	3.98	18
5	3.94	20
6	3.85	22

Solution

Sample	Net Weight of wet soil (lb)	Moisture content (%)	Moist unit weight (pcf)	Dry unit weight (pcf)
1	3.45	12	103.5	92.4
2	3.76	14	112.8	98.9
3	3.96	16	118.8	102.4
4	3.98	18	119.4	101.2
5	3.94	20	118.2	98.5
6	3.85	22	115.5	94.7

In the table above, the shaded cells represent given data. The white cells are calculated. Standard volume of Proctor mold is $1/30$ ft^3. Plotting the above data—moisture content versus dry unit weight—we obtain the following graph. Note that once the maximum dry unit weight has been calculated as 102.4 (approximately) and the 95% level as 97.3, it may be seen from the last column in the table that this would intersect the plot just before the second data point, i.e., at $w = 13.5$ (approximately) and just after the fifth data point, i.e., at $w = 20.5$ (approximately). Thus, unless a very precise read of these values is required, the conclusions can be drawn from the tabulated data itself, without resorting to the much greater effort of plotting.

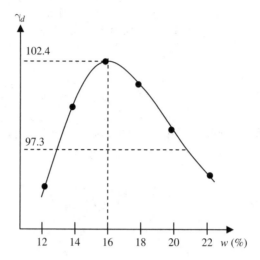

The following results can be obtained:

Optimum moisture content = 16%

Maximum dry unit weight = 102.5 lb/ft³ (approx.)

95% maximum dry unit weight = 97.3 lb/ft³

The range of water content values over which greater than 95% compaction can be achieved is $w_1 = 13.5\%$ to $w_2 = 20.5\%$. At optimum moisture content, the void ratio is given by

$$e = \frac{G_s \gamma_w}{\gamma_d} - 1 = \frac{2.65 \times 62.4}{102.5} - 1 = 0.61$$

At optimum moisture content, the degree of saturation is given by

$$S = \frac{w G_s}{e} = \frac{0.16 \times 2.65}{0.61} = 0.69 = 69\%$$

Compaction Effort

The compaction effort exerted by a falling hammer is proportional to the change in potential energy, which is given by the product of the falling weight and the height of the drop. When soil is placed in multiple lifts, with each lift being compacted by a number of hammer blows, the compaction energy per unit volume of soil is expressed as

$$E = \frac{\text{blows per layer} \times \text{no. of layers} \times \text{weight of hammer} \times \text{drop height}}{\text{volume of soil}} \qquad (202.26)$$

Thus, for the standard Proctor test, this may be calculated as

$$E = \frac{25 \times 3 \times 5.5 \times 1}{1/30} = 12,375 \text{ ft} \cdot \text{lb/ft}^3$$

and for the modified Proctor test, the compaction effort per unit volume is

$$E = \frac{25 \times 5 \times 10 \times 1.5}{1/30} = 56,250 \text{ ft} \cdot \text{lb/ft}^3$$

Dynamic Compaction Techniques

Dynamic and Vibratory Compaction

Vibrodensification as a soil stabilization technique is used primarily for granular soils. It is particularly effective when the relative density is less than 70%. This method is ineffective for partly saturated soils with more than 20% passing the No. 200 sieve.

Dynamic Compaction uses high-energy impact to densify loose granular soils. A heavy weight (30–40 tons or more) is dropped from 100 ft or more to apply compaction energy. Points of compaction are spaced 15–30 ft apart. This method becomes economical when the treatment area is at least 4–8 acres.

Vibrofloatation

Vibrofloatation is another compaction technique suitable for granular soils. A crane suspended cylindrical penetrator (called a vibroflot) about 16 in in diameter and 6 ft long is sunk to the desired treatment depth by using a water jet at the tip. Upon reaching the desired depth, the water jet is shut off and the vibroflot is then withdrawn slowly while a sand or gravel backfill is filled in from the ground level and densified.

Compaction Grouting

This method involves staged injection of low slump mortar-type grout into soils at high pressure into a casing drilled to the desired depth in the targeted area. The casing is gradually withdrawn as the hole is filled.

Field Monitoring of Compaction

In the field, regular monitoring of field compaction is important to know whether a specified unit weight has been achieved. There are several field techniques that are commonly used such as the sand cone method, the rubber balloon method, and nuclear method.

In the sand cone method (ASTM D1556)[7], uniform dry Ottawa sand (of known dry density) is poured into a small hole from which moist soil has been excavated. The volume of the hole

---

[7]American Society for Testing and Materials, "ASTM D1556: Standard Test Method for Density and Unit Weight of Soil in Place by Sand-Cone Method," 2007.

is determined from the net weight of the sand used and its dry density. The moist unit weight and moisture content of the excavated soil are measured in the laboratory. The rubber balloon method (ASTM D-2167) is very similar, except that the volume of the hole is determined by inserting a rubber balloon filled with a measurable amount of water. A nuclear density meter measures the weight of wet soil per unit volume and the weight of water present in a unit volume of soil.

Triaxial Test Fundamentals

In the triaxial test, a soil sample is forced to undergo shear failure by independently controlling a radial confining pressure (σ_R) and an axial stress (σ_A). See Fig. 202.6. The sample fails in shear due to the deviator stress ($\sigma_A - \sigma_R$). The soil specimen is cylindrical, with a diameter of about 1.4 in and length of 3 in, and is encased by a thin rubber membrane. The specimen is placed in a plastic chamber that is equipped with connections that allow either drainage from the soil specimen or measurement of pore pressure when undrained conditions are maintained. Three types of triaxial test are commonly conducted:

1. Consolidated drained (CD) or S test: This test usually takes a very long time because the deviator stress must be applied very slowly to ensure proper drainage. The rate of strain is controlled to prevent the buildup of pore pressure in the specimen. CD tests are generally performed on well-draining soils. For slow-draining soils, several weeks may be required to perform a CD test.

2. Consolidated undrained (CU) or R test: This is the most common type of triaxial test. The soil sample is first consolidated by radial pressure. After the pore pressure generated by this pressure is dissipated, the drainage line is closed and the soil is tested to failure. Pore water pressure and deviator stress are measured simultaneously. Specimens must be completely saturated before application of the deviator stress.

Figure 202.6
Soil sample subject to triaxial test.

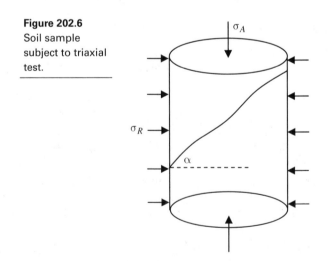

In the absence of pore pressure measurements CU tests can provide only total stress values cohesion (c) and friction angle (ϕ).

3. Unconsolidated undrained (UU) or quick (*Q*) test: In this test, drainage is not permitted during the application of the radial pressure. Because drainage is not allowed during any stage of the test, it can be completed quickly. The shear strength of soil as determined in UU tests corresponds to total stress, and is applicable only to situations where little consolidation or drainage can occur during shearing. It is applicable primarily to soils having permeability less than 10^{-3} cm/s.

Triaxial test results can be affected by added confinement due to membrane stiffness, friction in the loading piston, and other factors. The shear strength of soft sensitive soils is greatly affected by sample disturbance. For UU tests, the laboratory-measured shear strength of disturbed samples will be lower than the in-place strength. For CU or CD tests, the strength may be higher because of the consolidation permitted.

Relevance of Various Types of Triaxial Tests

For clean sands and gravels, undisturbed samples are very difficult to obtain and therefore sophisticated shear tests are usually impractical. For simple foundations, the angle of internal friction can be satisfactorily approximated by correlation with penetration resistance, relative density, and on the basis of soil classification. For earth dam and high embankment work where the soil will be placed under controlled conditions, triaxial compression tests should be conducted.

For clays, the unconfined compression test or UU triaxial test is often adequate. For very soft or sensitive soils, such as in marine deposits, the in-place vane shear test is especially helpful in evaluating the shear strength and its increase with depth. For long-term stability problems requiring effective stress analysis, such as evaluating the safety factor against landslides, CU triaxial tests with pore pressure measurements should be used. Long-term stability problems in some highly overconsolidated clays may require the CD test. For some thinly layered soils, such as varved clay, direct shear tests, or simple shear tests are well suited for determining the strength of the individual layers. Where partial drainage is anticipated, use CU tests with pore water pressure measurements to obtain effective strength parameters.

Overconsolidated soils have defects such as jointing and fissures. The laboratory values of strength which are obtained from a small test specimen are generally higher than the field strength values which are representative of the entire soil mass. In highly overconsolidated soil which may not be fully saturated, unusually high back pressure may be necessary to achieve full saturation, thus making it difficult to perform CU tests. In such cases, CD tests are more appropriate.

For granular soils, full drainage is likely to occur and the drained shear-strength parameters will govern. In contrast, for normally consolidated clays, the time required for dissipation of pore water pressures is too long and undrained conditions are likely to govern. For this reason, $\phi = 0$ is appropriate for such soils. Short- and long-term field conditions are well represented by undrained and drained test results, respectively.

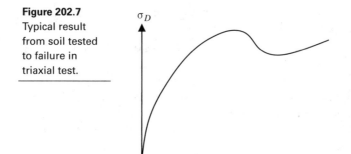

Figure 202.7 Typical result from soil tested to failure in triaxial test.

For a cohesionless soil (clean sand) sample, the failure plane is inclined at an angle $\alpha = 45° + \phi/2$ with the radial plane and the axial stress and the radial stress at failure are related by

$$\frac{\sigma_A}{\sigma_R} = \frac{1 + \sin\phi}{1 - \sin\phi} \qquad (202.27)$$

A typical stress-strain curve obtained from the triaxial test is shown in Fig. 202.7. The deviator stress $\sigma_D = \sigma_A - \sigma_R$ is plotted on the y axis. The deviator stress σ_D is responsible for developing shear stress in the soil.

The radial and axial stresses at failure for a soil sample represent principal stresses. It is conventional to represent them as σ_1 (axial) and σ_3 (radial). Using these normal stresses as the ends of the diameter, a Mohr's circle may be drawn for each stress pair. If several Mohr's circles are drawn based on several samples, a shear failure envelop may be drawn tangent to these circles. Figure 202.8 shows typical (tangent) envelops for various type of soil.

The geometric relationship between the parameters of Mohr's circle (σ_1 and σ_3) and the parameters of the shear envelop (c and ϕ) is shown below for (1) a cohesionless soil and (2) a c-ϕ soil.

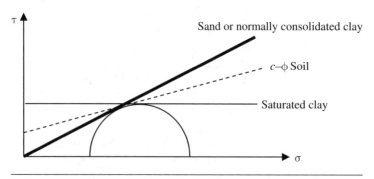

Figure 202.8 Shear-strength envelop and Mohr's circles.

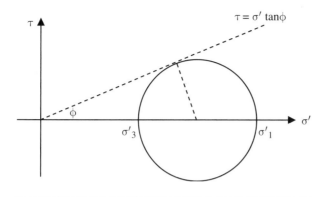

Figure 202.9 Shear-strength envelop for cohesionless soil.

Case 1—Drained Sand (Cohesion c = 0)

For a drained sand sample, the Mohr-Coulomb shear envelop passes through the origin (cohesion = 0), as shown in Fig. 202.9. The radius of Mohr's circle corresponding to a pair of principal stresses (radial stress = σ'_3, axial stress = σ'_1) is given by

$$R = \frac{\sigma'_1 - \sigma'_3}{2} \tag{202.28}$$

The shear-strength envelop has a single parameter (ϕ) which can then be calculated from

$$\sin\phi = \frac{\sigma'_1 - \sigma'_3}{\sigma'_1 + \sigma'_3} \tag{202.29}$$

Case 2—General Case: c-ɸ Soil

The two parameters of the c-ϕ soil can be calculated using results from two triaxial sample tests. Figure 202.10 shows the Mohr-Coulomb envelop for such a soil. The angle of internal friction ϕ is given by

$$\sin\phi = \frac{(\sigma'_{11} - \sigma'_{31}) - (\sigma'_{12} - \sigma'_{32})}{(\sigma'_{11} + \sigma'_{31}) - (\sigma'_{12} + \sigma'_{32})} \tag{202.30}$$

and the cohesion can be calculated using the two following steps:

$$x = \frac{\sigma'_{11} - \sigma'_{31}}{2\sin\phi} - \frac{\sigma'_{11} + \sigma'_{31}}{2} \tag{202.31}$$

$$c = x\tan\phi \tag{202.32}$$

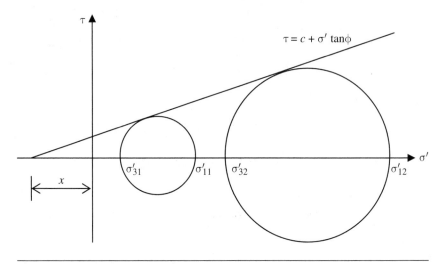

Figure 202.10 Shear-strength envelop for c-ϕ soil.

Example 202.4

A saturated cohesionless soil was tested in a triaxial apparatus. The data at failure are given below:

Pore water pressure	10.0 psi
Vertical total stress	36.5 psi
Horizontal total stress	15.5 psi

Under drained conditions, what is the effective friction angle?

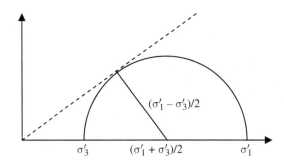

Solution   Under drained conditions, pore pressure is zero. Therefore, we must use effective stresses.

$$\text{Vertical effective stress} = \sigma_1' = 36.5 - 10 = 26.5 \text{ psi}$$

$$\text{Horizontal effective stress} = \sigma_3' = 15.5 - 10 = 5.5 \text{ psi}$$

$$\sin\phi = \frac{\sigma_1' - \sigma_3'}{\sigma_1' + \sigma_3'} = \frac{26.5 - 5.5}{26.5 + 5.5} \Rightarrow \phi = 41°$$

Example 202.5

A saturated cohesionless soil was tested in a triaxial apparatus. The stresses on the failure plane from the test were

$$\text{Effective normal stress} = 1200 \text{ psf}$$

$$\text{Shear stress} = 750 \text{ psf}$$

If the soil is a cohesionless sand, what is the angle of friction?

Solution

$$\tan\phi = \frac{750}{1200} \Rightarrow \phi = 32°$$

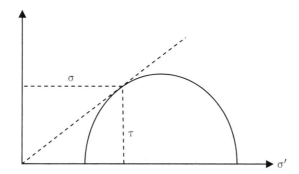

Example 202.6

The following stresses were obtained during triaxial test on two soil samples:

Sample	Effective vertical stress (tons/ft²)	Effective lateral stress (tons/ft²)
1	3	1
2	5	2

Calculate the cohesion and the angle of internal friction for the soil.

Solution If pairs of stresses σ_1 and σ_2 are given, the center coordinate is given by $\sigma_c = (\sigma_1 + \sigma_2)/2$ and the radius is given by $R = (\sigma_2 - \sigma_1)/2$.

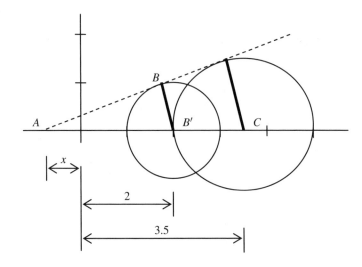

Then for each set, we have $(\sigma_c + x)\sin\phi = R$. The bold lines are radial and therefore perpendicular to the tangent. From the triangles ABB′ and ACC′

$$(2 + x)\sin\phi = 1$$
$$(3.5 + x)\sin\phi = 1.5$$

Solving, we get $x = 1$ and $\phi = \sin^{-1} 0.333 = 19.5°$

Finally, $c = 1.\tan 19.5° = 0.354$ tsf $= 708$ psf

Consolidation

The compression of a soil due to stress increases is caused by deformation of soil particles, expulsion of air and water from the voids and the restructuring of the soil "skeleton." Whereas immediate settlements occur primarily due to the elastic deformation without any accompanying moisture changes, consolidation settlements occur more slowly due to the volume change resulting from expulsion of water from the voids of saturated cohesive soils.

Consolidation Test

The consolidation characteristics of a cohesive soil are determined using a consolidometer (or oedometer), where a soil sample is placed inside a metal ring and loaded with two porous loading plates, one at each end. Specimens are usually 2.5 in in diameter and 1 in thick. The specimen is kept submerged during the test. Each load is usually maintained for 24 hours. Successive loads are usually double the previous load.

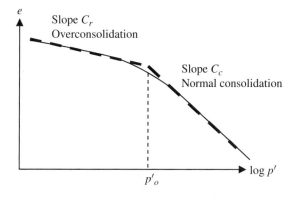

Figure 202.11 Primary consolidation curve.

The volume of the soil specimen consists of a volume occupied by solids and a volume occupied by voidspace. The equivalent height occupied by solids can be calculated as

$$H_s = \frac{W_s}{AG_s\gamma_w} \tag{202.33}$$

where W_s is the dry weight of the specimen, A is the cross-sectional area, G_s is the specific gravity of solids, and γ_w is the unit weight of water. For each load increment, the change in height of the specimen can be converted to an equivalent change in the void ratio, using

$$\Delta e = \frac{\Delta H}{H_s} \tag{202.34}$$

Figure 202.11 shows a typical plot of void ratio versus effective vertical stress (log scale). The plot shows two tangents that have been fitted to the actual curve. The two tangents appear to intersect a point which corresponds to the preconsolidation pressure p_o'. This pressure is the maximum pressure that the soil has experienced in its loading history.

NOTE There are several detailed graphical techniques such as the one by Casagrande[8] that may be used to get an estimation of the preconsolidation pressure. The preconsolidation pressure cannot be determined precisely, but can be estimated from consolidation tests on high-quality undisturbed samples.

---

[8]Casagrande A. (1936), "Determination of the Preconsolidation Load and Its Practical Significance," Proceedings, First International Conference on Soil Mechanics and Foundation Engineering, Cambridge, Massachusetts., Vol. 3, 60–64.

The preconsolidation pressure may be also estimated from

$$p'_o = (q_u/2)/(0.11 + 0.0037\text{PI}) \tag{202.35}$$

where q_u = unconfined compressive strength
PI = soil plasticity index (%)

For pressures less than the preconsolidation pressure, the soil experiences overconsolidation (slope C_r):

$$\Delta e = e_1 - e_2 = C_r \log_{10}\left(\frac{p'_2}{p'_1}\right) \tag{202.36}$$

For pressures exceeding the preconsolidation pressure, the soil experiences normal consolidation (slope C_c):

$$\Delta e = e_1 - e_2 = C_c \log_{10}\left(\frac{p'_2}{p'_1}\right) \tag{202.37}$$

If the pressure on the soil is reduced due to unloading, the soil rebounds and the rebound curve follows a slope C_s. This slope is called the swell index. The swell index is typically 10% to 20% of the compression index C_c except for soils with very high swell potential.

Skempton suggested the following correlations between the liquid limit LL (%) and the compression index C_c:

$$\text{For undisturbed clays: } C_c = 0.009(\text{LL} - 10) \tag{202.38}$$

$$\text{For remolded clays: } C_c = 0.007(\text{LL} - 7) \tag{202.39}$$

The approximate values of C_c for uniform sands in the load range of 1 to 4 tons/ft² may vary from 0.05 to 0.06 (loose condition) and from 0.02 to 0.03 (dense condition).

Consolidation Settlement (Primary)

The settlement of a soil layer of thickness H due to primary consolidation is given by

$$s = H\varepsilon_v = H\left(\frac{\Delta e}{1 + e_0}\right) = \frac{H\Delta e}{1 + e_0} \tag{202.40}$$

where H = layer thickness
e_0 = initial void ratio
Δe = change of void ratio due to the increase of effective vertical stress at the midpoint of the soil layer calculated using Eqs. (202.36 and/or 202.37)

The quantity ε_v is called the vertical strain in the soil.

Example 202.7

A 12-ft-thick clay layer is overlain by 13-ft-thick layer of sand as shown in the diagram. The groundwater table is originally at a depth of 8 ft below the surface. The ground-water is first lowered to the top of the clay layer and then a mat foundation (plan dimensions 200 ft × 150 ft, depth = 3 ft; uniform pressure = 1000 psf) is constructed. Calculate the consolidation settlement (in).

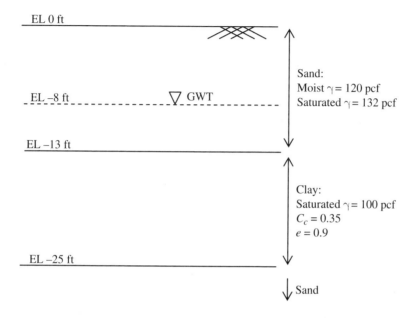

Solution Assuming the behavior of the clay to be normally consolidated (preexisting overburden pressure is the preconsolidation pressure), the effective vertical stress at the center of the clay layer under original conditions is given by

$$p'_1 = \sigma'_v = 120 \times 8 + (132 - 62.4) \times 5 + (100 - 62.4) \times 6 = 1534 \text{ psf}$$

Once the water table has been lowered and the surcharge load is applied, the effective vertical stress at the same location is calculated as

$$p'_2 = \sigma'_v = 1000 + 120 \times 13 + (100 - 62.4) \times 6 = 2786 \text{ psf}$$

Note that because the horizontal extent of the mat foundation load is large compared to the depth (center of clay layer is 16 ft below application of this uniform pressure), uniform pressure of 1000 psf is carried down at full strength (without dissipation).

The change in void ratio is given by

$$\Delta e = C_c \log \frac{p'_2}{p'_1} = 0.35 \log \frac{2786}{1534} = 0.091$$

The consolidation settlement is therefore given by

$$s = \frac{H\Delta e}{1 + e_0} = \frac{12 \times 0.091}{1.9} = 0.575 \text{ ft} = 6.9 \text{ in}$$

Consolidation Rate

The time required for primary consolidation to occur can be expressed as

$$t = \frac{T_v H_d^2}{c_v} \qquad (202.41)$$

where T_v = consolidation time factor (function of the degree of consolidation U_z)
  U_z = degree of consolidation
  H_d = drainage path length, which is equal to half the layer thickness when the layer can drain at both interfaces, or the entire layer thickness if drainage can occur on only one side. For a soil layer at the ground surface, the free surface is assumed to be free draining.
  c_v = coefficient of consolidation, given by

$$c_v = \frac{K(1 + e_0)}{a_v \gamma_w} \qquad (202.42)$$

where a_v = coefficient of compressibility $= -\dfrac{e_2 - e_1}{p_2' - p_1'}$ (202.43)

The following empirical relations have been suggested for the time factor T_v.

$$T_v = \frac{\pi}{4}\left(\frac{U_z}{100}\right)^2 \qquad\qquad 0 \le U_z \le 60$$
$$T_v = 1.781 - 0.933 \log_{10}(100 - U_z) \qquad U_z > 60 \qquad (202.44)$$

Table 202.10 outlines values of the time factor T_v for various degrees of consolidation (U_z).

If the preconsolidation pressure for a soil layer is not specified, it is customary to take it as the existing effective stress due to overburden.

NOTE Units of parameters C_c, C_r, c_v, a_v, U_z, T_v; C_c and C_r are dimensionless; c_v has units of L^2/T (ft²/day typical); a_v has L^2/F dimensions (ft²/lb typical); U_z (degree of consolidation) and T_v (time factor for consolidation) are dimensionless.

Table 202.10 Time Factor versus Degree of Consolidation

U_z (%)	T_v
10	0.008
20	0.031
30	0.071
40	0.126
50	0.197
75	0.477
90	0.848
95	1.129
98	1.500
99	1.781
100	∞

Settlement from Secondary Consolidation

Primary consolidation is caused by dissipation of excess pore water pressure. If the change in void ratio is plotted versus time (log scale), a typical plot may be as shown in Fig. 202.12. Note that the previous plot for determining primary consolidation settlement was for void ratio versus effective vertical pressure (log-scale).

The secondary consolidation in a fine-grained soil occurs due to a plastic restructuring of the soil fabric after the complete dissipation of the excess pore water pressure. The void ratio of the soil at the "end" of primary settlement is denoted e_p and is graphically determined by locating the point of intersection of the tangents from the midsection and the tail of the e-log t curve, as shown in Fig. 202.12. The plot of void ratio against time during secondary

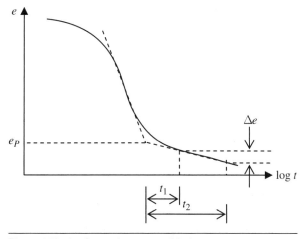

Figure 202.12 Secondary consolidation curve.

consolidation is observed to be practically linear. From the figure, the secondary consolidation index can be defined as

$$C_\alpha = \frac{\Delta e}{\log_{10}\left(\dfrac{t_2}{t_1}\right)}$$

(202.45)

and the secondary consolidation settlement is given by

$$S_s = \frac{C_\alpha}{1+e_p} H \log_{10}\left(\frac{t_2}{t_1}\right)$$

(202.46)

where e_p = void ratio at the end of primary consolidation
H = thickness of soil layer
t_1, t_2 = time markers of interval, measured from the end of primary consolidation

For organic and highly compressible inorganic soils, the settlement due to secondary consolidation can be more than that due to primary consolidation.

Effect of Sample Disturbance

Sample disturbance lowers the void ratio reached under any applied pressure and makes the location of the preconsolidation stress less distinct. Sample disturbance tends to lower the compression index (C_c) and the preconsolidation pressure and coefficient of consolidation. Sample disturbance increases the recompression and swelling indices.

California Bearing Ratio Test

The California Bearing Ratio (CBR) test is conducted to determine the suitability of a soil as a subgrade material for pavements. The results are based on the comparison of the penetration resistance (stress) of the soil to a standard penetration into a crushed stone sample. The resistance loads for a standard 3 in² plunger into a crushed stone sample are given in Table 202.11.

Table 202.11 Standard Penetration Resistance Values for CBR Test

Penetration depth (in)	Standard load (psi)
0.1	1000
0.2	1500
0.3	1900
0.4	2300
0.5	2600

The CBR is given by

$$CBR_{0.1} = \frac{\text{Resistance to 0.1-in penetration into sample}}{1000 \text{ psi}} \times 100 \qquad (202.47)$$

For example, if the load corresponding to a 0.1-in penetration is 1800 lb, then the CBR is

$$CBR_{0.1} = \frac{1800/3 \text{ psi}}{1000 \text{ psi}} \times 100 = 60$$

Similarly, the CBR for 0.2-in penetration is given by

$$CBR_{0.2} = \frac{\text{Resistance to 0.2-in penetration into sample}}{1500 \text{ psi}} \times 100 \qquad (202.48)$$

The test must be repeated if $CBR_{0.2} > CBR_{0.1}$

Hveem Stabilometer Test

This test is used to determine the R-value (resistance) of the soil to the horizontal pressure generated by imposing a vertical load of 160 lb/in on the soil. The test has three steps:

1. Exudation and determination of the exudation pressure. During this phase, each of four cylindrical samples is placed in a steel mold and subjected to a vertical load until water exudes from the soil.

2. At the completion of the exudation test, a perforated brass plate is placed on each sample and covered with water (while in the steel mold). Samples are left to stand for 16 to 20 hours, during which time expansion of the soil is prevented by applying a vertical pressure. This pressure is called the expansion pressure.

3. At the completion of the expansion phase, the specimen is placed in a flexible sleeve in the stabilometer. Vertical displacement is applied at a rate of 0.05 in/min until a pressure of 160 psi reached. The horizontal pressure at this stage is recorded. The vertical pressure is then reduced to 80 psi and the horizontal pressure to 5 psi. The number of turns of a screw-type pump needed to increase the horizontal pressure to 100 psi is then recorded. The soil's resistance value is given by

$$R = 100 - \frac{100}{\dfrac{2.5}{D}\left(\dfrac{P_v}{P_h} - 1\right) + 1} \qquad (202.49)$$

where P_v = the vertical stress (160 psi)
P_h = the horizontal pressure (psi) recorded in step 3
D = number of screw turns needed to increase the horizontal pressure to 100 psi

Shear Strength by Direct Methods

Several devices are available to obtain shear-strength data in the field as a supplement to laboratory tests or where it is not possible to obtain representative samples for testing.

A *pocket penetrometer* is used for obtaining the shear strength of cohesive, nongravelly soils on field exploration or construction sites. Various commercially available penetrometers are able to read unconfined compressive strength directly. The tool is used as a supplement rather than an alternative to other field tests or laboratory tests.

The *Torvane shear device* is used for obtaining rapid approximations of shear strength of cohesive, nongravelly soils. The device can be used on ends of Shelby tubes, penetration samples, block samples from test pits, or sides of test pits. The device is used in uniform soils and does not replace laboratory tests.

In situ vane shear measurements are especially useful in very soft soil deposits where much of the strength may be lost by disturbance during sampling. It should not be used in stiff clays or in soft soils containing gravel, shells, wood, etc.

Deformation Moduli

A number of different methods are available for obtaining values of deformation moduli in soil and rock. Each method has its own advantages or disadvantages and *in situ* testing should only be attempted with a full knowledge of the limitations of the several techniques.

Pressuremeter

The pressuremeter test is an *in situ* lateral loading test performed in a borehole by means of a cylindrical probe. Under increments of pressure, radial expansion is measured, and the modulus of deformation is calculated. If the test is carried to failure, shear strengths can be calculated and are generally higher than those obtained from vane shear tests. Materials difficult to sample (e.g., sands, residual soil, tills, soft rock) can be effectively investigated by the pressuremeter. The tests measure soil compressibility in the radial direction and some assumptions are required on the ratio between the vertical moduli to radial moduli. This may be difficult to interpret and thus of only limited value for stratified soils, for very soft soils, and for soils where drainage conditions during loading are not known. Roughness of the borehole wall affects test results, although the self-boring pressuremeter eliminates some of this disadvantage.

Plate Bearing Test (or Plate Load Test)

In the plate load test, a square or round steel plate is pushed into the soil, while continuously recording the load versus the depth of penetration. The yield point is estimated as the point at which the penetration increases rapidly. The stress (q) and depth of penetration (δ) corresponding to half the yield load are determined from the load versus penetration (logarithmic) plot. The modulus of subgrade reaction (K_v) is calculated as q/δ.

The plate load test can also be used to estimate the settlement under a footing. According to Terzaghi and Peck, the observed settlement of a steel plate (S_1) and the expected settlement below a footing (S)—when both are subjected to the same bearing stress—is given by

$$S = \frac{4S_1}{\left(1 + \dfrac{D_1}{D}\right)^2} \qquad (202.50)$$

where D_1 = smallest lateral dimension of the steel plate
D = smallest lateral dimension of the footing
S_1 = observed settlement below the steel plate
S = expected settlement below the footing

In general, tests should be conducted with groundwater saturation conditions simulating those anticipated under the actual structure. Data from the plate load test is applicable to material only in the immediate zone (say to a depth of two plate diameters) of the plate and should not be extrapolated unless material at greater depth is essentially the same.

Permeability Tests

Some of the most commonly performed permeability tests are described below:

Constant Head Test

This is the most generally applicable permeability test. It may be difficult to perform in materials of either very high or very low permeability since the flow of water may be difficult to maintain or measure.

Rising Head Test

In a saturated zone with sufficiently permeable materials, this test is more accurate than a constant or a falling head test, since plugging of the pores by fines or by air bubbles is less likely to occur. This test should be performed only in the saturated zone.

Falling Head Test

In zones where the flow rates are very high or very low, this test may be more accurate than a constant head test. In an area of unknown permeability the constant head test should be attempted before a falling head test.

Pumping Test

In large-scale seepage investigations or groundwater resource studies, the expense of aquifer or pumping tests may be justified as they provide more useful data than any other type of test. Pump tests require a test well, pumping equipment, observation wells, and lengthy test times achieve steady-state conditions.

In-Place Density

In-place soil density can be measured on the surface by displacement methods to obtain volume and weight, and by nuclear density meters. Density at depth can be measured only in certain soils by the drive cylinder (sampling tube) method.

Displacement Methods

Direct methods of measuring include sand displacement and water balloon methods. The sand displacement and water balloon methods are the most widely used methods because of their applicability to a wide range of material types and good performance. The sand displacement method is the most frequently used surface test and is the reference test for all other methods.

Drive Cylinder Method

The drive cylinder is useful for obtaining subsurface samples from which the density can be ascertained, but it is limited to moist, cohesive soils containing little or no gravel and moist, fine sands that exhibit apparent cohesion.

Nuclear Moisture Density Method

Use ASTM Standard[9] D6938, Density of Soil and Soil-Aggregate in Place by Nuclear Methods (shallow depth). Before nuclear density methods are used on the job, results must be compared with density and water contents determined by displacement methods. Based on this comparison, corrections may be required to the factory calibration curves or a new calibration curve may have to be developed.

---

[9]American Society for Testing and Materials, "ASTM D6938-08a Standard Test Method for In-Place Density and Water Content of Soil and Soil-Aggregate by Nuclear Methods (Shallow Depth)," 2008.

Soil Classification

Sieve Sizes

In the United States, soil classification methods employ the standard sieve sizes listed in Table 203.1. Also in use, but somewhat less common, are metric sieve sizes. For fine-grained soils, plasticity characteristics (Atterberg limits) play a more significant role than the particle size distribution, which is more significant for coarse-grained soils.

Visual description of soils may use the guidelines in Table 203.2.

USDA Textural Classification of Soils

According to the USDA[1] soil classification system, a soil is divided into its gravimetric (weight) fractions into % gravel, % sand, % silt, and % clay using the following thresholds:

Gravel: Particle diameter > 2.0 mm

Sand: 2.0 mm $>$ particle diameter > 0.05 mm

Silt: 0.05 mm $>$ particle diameter > 0.002 mm

Clay: particle diameter < 0.002 mm

If the gravel fraction is separated, the sand, silt, and clay fractions may be redefined in terms of the modified soil fractions shown below:

$$\text{SAND} = \frac{\%\text{ sand}}{1 - \%\text{ gravel}/100}$$

$$\text{SILT} = \frac{\%\text{ silt}}{1 - \%\text{ gravel}/100}$$

$$\text{CLAY} = \frac{\%\text{ clay}}{1 - \%\text{ gravel}/100}$$

---

[1]U.S. Department of Agriculture, Natural Resources Conservation Service. *National Soil Survey Handbook*, 430-VI, 2007.

These modified fractions (which add up to 100) are then entered into the USDA classification chart (Fig. 203.1) to obtain a qualitative textural classification of the soil.

Table 203.1 Standard U.S. Sieve Sizes

Sieve size	Sieve opening (mm)
2 in	50
1 in	25
1/2 in	12.5
No. 4	4.75
No. 10	2.00
No. 20	0.85
No. 40	0.425
No. 60	0.25
No. 100	0.15
No. 140	0.106
No. 200	0.075

Table 203.2 Visual Description of Soil Components by Particle Size

Particle size	Description
>12 in	Boulders
3 in to 12 in	Cobbles
3/4 in to 3 in	Coarse gravel
No. 4 to 3/4 in	Fine gravel
No. 10 to No. 4	Coarse sand
No. 40 to No. 10	Medium sand
No. 200 to No. 40	Fine sand
Passing No. 200	Silt and clay

Example 203.1

According to the USDA Textural Soil Classification System, what is the classification of a soil sample that contains the following particle sizes?

Particle size	Weight (g)
>2 mm	24
0.05–2.00 mm	50
0.002–0.05 mm	64
<0.002 mm	38

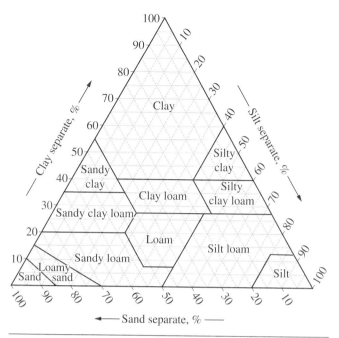

Figure 203.1   USDA textural classification of soils.

(U.S. Department of Agriculture, Natural Resources Conservation Service. *National Soil Survey Handbook*, 2007.)

Solution   The total soil mass is 176 g, of which 24 g is gravel (13.6%). The soil fractions are gravel $= 13.6\%$, sand $= 28.4\%$, silt $= 36.4\%$, and clay $= 21.6\%$. The modified soil fractions are

$$\text{Sand:}\frac{0.284}{1-0.136}=0.33 \qquad \text{Silt:}\frac{0.364}{1-0.136}=0.42 \qquad \text{Clay:}\frac{0.216}{1-0.136}=0.25$$

Plotting these values, we get a classification of *Loam*.

Particle Size Distribution Curves

Figure 203.2 shows a typical particle size distribution curve for a soil aggregate. The shape of the particle size distribution curve is expressed by two parameters—the uniformity coefficient (C_u) and the coefficient of curvature or coefficient of gradation (C_c). These coefficients are defined as

$$\text{Coefficient of uniformity } C_u = \frac{D_{60}}{D_{10}} \tag{203.1}$$

$$\text{Coefficient of curvature } C_c = \frac{D_{30}^2}{D_{10}D_{60}} \tag{203.2}$$

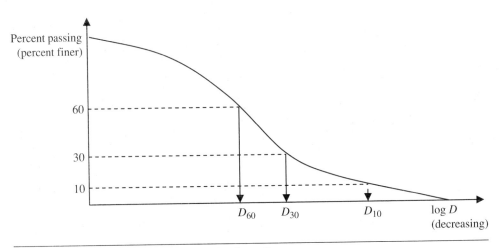

Figure 203.2 Typical particle size distribution for soils.

where D_{10}, D_{30}, and D_{60} are the particle sizes for which 10%, 30%, and 60% (by weight) is finer. These sizes are indicated in Fig. 203.2. D_{10} is called the effective size of the soil (Hazen). The effective size of clean sands and gravels can be correlated to their permeability according to Eq. (203.3).

$$K = Cd_{10}^2 \qquad (203.3)$$

where K(cm/s) = hydraulic conductivity of sandy soil
d_{10}(cm) = effective size of sandy soil
C = 40–80 (very fine sand and very fine to fine sand), 80–120 (medium and coarse sand), 120–150 (clean coarse sand)

Hydrometer Analysis

Particle size distribution for finer soils (silts and clays with particles finer than a No. 200 sieve) can be established by a hydrometer analysis which uses sedimentation of dispersed soil particles (about 50 g of soil in a distilled water suspension filling a 1-L glass cylinder). The hydrometer test is based on type I settling of particles governed by Stokes' Law. The continuous record of settling rates or velocities can be correlated to the particle size distribution.

Unified Soil Classification System (USCS)

The USCS uses letter symbols to describe soils. The parameters used for classification are

1. F_{200} = percent passing the No. 200 sieve (opening size = 0.075 mm)
2. Uniformity coefficient C_u
3. Coefficient of curvature C_c
4. R_4 = percent retained on the No. 4 sieve (opening size = 4.75 mm)

In general, particles retained on a No. 4 sieve ($d > 4.75$ mm) are called gravel and particles retained on a No. 200 sieve ($d > 0.075$ mm) but passing the No. 4 sieve are classified as sand. The USCS classification procedure may be outlined by the following steps:

Step 1   If the soil can be visually identified as peat, classify as Pt.

Step 2   If $F_{200} > 50\%$, the soil is predominantly fine grained. The Atterberg limits liquid limit (LL) and plastic limit (PL) are then used in the Casagrande plasticity chart (Fig. 203.3) to obtain the appropriate classification symbol. The following symbols are used on the plasticity chart:

M   Inorganic silt

C   Inorganic clays

O   Organic silts and clays

L   Low compressibility (LL < 50)

H   High compressibility (LL ≥ 50)

The A-line separates inorganic silts from inorganic clays. The equation of the A-line is

$$PI = 0.73(LL - 20) \tag{203.4}$$

The U-line is approximately the upper limit of the relationship between the plasticity index (PI) and the liquid limit (LL) for any currently known soil. The equation of the U-line is

$$PI = 0.90(LL - 8) \tag{203.5}$$

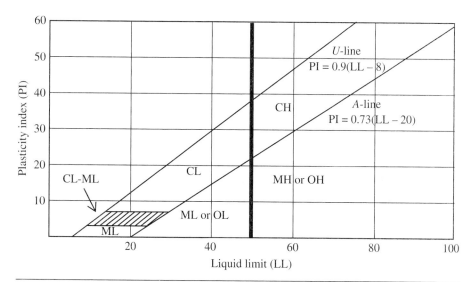

Figure 203.3   Casagrande plasticity chart.

Step 3   If $F_{200} < 50$, the soil is predominantly coarse grained. The first letter of the USCS classification is G or S. Identify cumulative % retained on No. 4 (R_4).

Step 4   If $R_4 > 1/2 (100 - F_{200})$, first letter is G, otherwise S. The following symbols are used for classification of coarse-grained soils:

G   gravel (greater than 50% of coarse fraction retained on No. 4 sieve)

S   sand (less than 50% of coarse fraction retained on No. 4 sieve)

W   well graded, clean

C   significant amounts of clay

P   poorly graded, clean

M   significant amount of silt

Step 5   If first letter is G, use the following guidelines to identify second letter:

If $F_{200} < 5\%$, the soil is classified as GW if $C_u \geq 4$ and $1 < C_c < 3$, GP otherwise.

If $5\% < F_{200} < 12\%$, the soil has a dual classification reflecting both gradation and plasticity. Examples are GP-GM, GP-GC, etc.

If $12\% < F_{200} < 50\%$, clayey gravel (GC) or silty gravel (GM).

Step 6   If first letter is S, use the following guidelines to identify second letter:

If $F_{200} < 5\%$, the soil is classified as SW if $C_u \geq 6$ and $1 < C_c < 3$, SP otherwise.

If $5\% < F_{200} < 12\%$, the soil has a dual classification reflecting both gradation and plasticity. Examples are SP-SM, SP-SC, etc.

If $12\% < F_{200} < 50\%$, clayey sand (SC) or silty sand (SM).

Tables 203.3 and 203.4 summarize properties of various types of gravelly and sandy soils.

Table 203.3 Summary of USCS Classification Parameters for Gravelly Soils

Group symbol	Criteria
GW	$F_{200} < 5$; $C_u \geq 4$ and $1 < C_c < 3$
GP	$F_{200} < 5$; not meeting both criteria for C_u and C_c above
GM	$F_{200} > 12$; PI < 0.73(LL–20) or PI < 4
GC	$F_{200} > 12$; PI > 0.73(LL–20) and PI > 7
GC-GM	$F_{200} > 12$; Atterberg limits in area marked CL-ML on plasticity chart
GW-GM	$5 < F_{200} < 12$; meets criteria for GW and GM
GW-GC	$5 < F_{200} < 12$; meets criteria for GW and GC
GP-GM	$5 < F_{200} < 12$; meets criteria for GP and GM
GP-GC	$5 < F_{200} < 12$; meets criteria for GP and GC

Table 203.4 Summary of USCS Classification Parameters for Sandy Soils

Group symbol	Criteria
SW	$F_{200} < 5$; $C_u \geq 6$ and $1 < C_c < 3$
SP	$F_{200} < 5$; not meeting both criteria for C_u and C_c above
SM	$F_{200} > 12$; PI < 0.73(LL–20) or PI < 4
SC	$F_{200} > 12$; PI > 0.73(LL–20) and PI > 7
SC-SM	$F_{200} > 12$; Atterberg limits in area marked CL-ML on plasticity chart
SW-SM	$5 < F_{200} < 12$; meets criteria for SW and SM
SW-SC	$5 < F_{200} < 12$; meets criteria for SW and SC
SP-SM	$5 < F_{200} < 12$; meets criteria for SP and SM
SP-SC	$5 < F_{200} < 12$; meets criteria for SP and SC

AASHTO Soil Classification

The soil classification protocol defined by AASHTO[2] utilizes the following criteria:

Percent passing No. 10 sieve (F_{10})

Percent passing No. 40 sieve (F_{40})

Percent passing No. 200 sieve (F_{200})

Liquid limit (LL)

Plasticity index (PI) = Liquid limit (LL) – Plastic limit (PL)

The broad distinction between granular materials (groups A-1, A-2, and A-3) and fine-grained materials (groups A-4, A-5, A-6, and A-7) is made on the basis of the percent (by weight) passing the No. 200 sieve (F_{200}).

Less Than or Equal to 35% Passing No. 200 Sieve – Predominantly Granular

Of the three groups meeting this criterion, the A-2 group represents a transition between coarse- and fine-grained soils. Plasticity characteristics such as liquid limit (LL) and plasticity index (PI) are necessary parameters for group A-2.

Is $F_{40} > 50$? If the answer is YES, the soil group is A-3. This group is assigned to fine sands which are nonplastic (NP). If the answer is NO, the soil group is A-1 or A-2.

If F_{200} is less than 35 and Atterberg limits (LL and PL) are given, use Table 203.5 or Fig. 203.4 to determine the subgroup under group A-2 (choices are A2-4, A2-5, A2-6, and A2-7).

Group A-1 is subdivided into A-1-a and A-1-b. Values of F_{10}, F_{40}, and F_{200} are used (see Table 203.5) to differentiate between A1-a and A1-b.

---

[2]*Classification of Soils and Soil-Aggregate Mixtures for Highway Construction Purposes, M-145,* American Association of State Highway and Transportation Officials, Washington DC, 2000.

Table 203.5 AASHTO Soil Classification Criteria

Sieve analysis	Granular materials (35% or less passing No. 200 sieve)							Silt-clay materials (more than 35% passing No. 200 sieve)				
	A-1		A-3	A-2				A-4	A-5	A-6	A-7	A-8
	A-1-a	A-1-b		A-2-4	A-2-5	A-2-6	A-2-7					
% passing No. 10	≤ 50											
No. 40	≤ 30	≤ 50	> 50									
No. 200	≤ 15	≤ 25	≤ 10	≤ 35	≤ 35	≤ 35	≤ 35	> 35	> 35	> 35	> 35	
LL				≤ 40	> 40	≤ 40	> 40	≤ 40	> 40	≤ 40	> 40	
PI		≤ 6	NP	≤ 10	≤ 10	> 10	> 10	≤ 10	≤ 10	> 10	> 10	
General description	Stone, gravel, sand		Fine sand	Silty or clayey gravel and sand				Silty soils		Clayey soils		Peat, highly organic soils
Quality as subgrade material	Good to excellent							Fair to poor				Very poor

More Than 35% Passing No. 200 Sieve—Predominantly Fine Grained

If the fraction passing no. 200 sieve (F_{200}) is greater than 35, the soil is A-4, A-5, A-6, or A-7, and only LL and PI are used for the classification.

Group A-7 is subdivided into A-7-5 and A-7-6, depending on the plastic limit. For PL < 30, classification is A-7-6. For PL ≥ 30, the classification is A-7-5. Figure 203.4 is a graphical representation of classification criteria of the fine-grained soils based on the plasticity characteristics—liquid limit (LL) and plasticity index (PI).

Group Index (GI)

In the AASHTO soil classification system, the group index (GI) is used as an indicator of the suitability of the soil as a highway subgrade material. A low value of GI indicates good subgrade properties. The value of GI is calculated using Eq. (203.6), where F_{200} is the percent weight passing the No. 200 sieve. The value is then reported as the nearest integer. Negative values are reported as zero.

$$GI = (F_{200} - 35)[0.2 + 0.005(LL - 40)] + 0.01(F_{200} - 15)(PI - 10) \qquad (203.6)$$

For groups A-2-6 and A-2-7, only the second term is retained for the calculation. In other words, for these groups, the group index (GI) is defined as

$$GI = 0.01(F_{200} - 15)(PI - 10) \qquad (203.7)$$

For groups A-1-a, A-1-b, A-2-4, A-2-5, and A-3, the group index is zero.

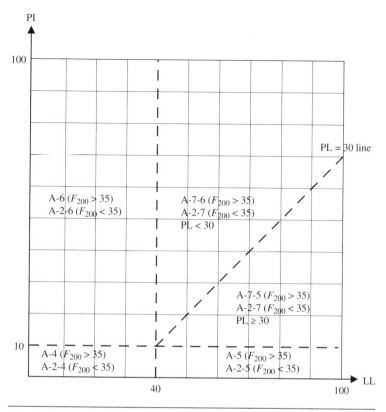

Figure 203.4   Graphical aid to AASHTO soil classification.

Example 203.2

The table on the left shows sieve analysis data for a soil sample (total mass 200 g). The table on the right shows data derived from the raw data.

Sieve size	Opening (mm)	Weight retained (g)	Cumulative weight retained (g)	Cumulative % retained (R)	% Passing (F)
2 in	50	10	10	05.0	95.0
1 in	25	20	30	15.0	85.0
1/2 in	12.5	49	79	39.5	60.5
No. 4	4.75	38	117	58.5	41.5
No. 10	2.00	22	139	69.5	30.5
No. 40	0.425	18	157	78.5	21.5
No. 200	0.075	24	181	90.5	09.5
Pan		19	200	100	00.0

What is the soil classification using (a) USCS and (b) AASHTO guidelines?

Solution   Looking for 10%, 30%, and 60% passing in last column (F) and corresponding sizes, we draw the following conclusions:

$$0.075 < D_{10} < 0.425 \qquad \text{say } D_{10} = 0.08 \text{ mm}$$

$$0.425 < D_{30} < 2.000 \qquad \text{say } D_{30} = 1.9 \text{ mm}$$

$$4.750 < D_{60} < 12.50 \qquad \text{say } D_{60} = 12.0 \text{ mm}$$

$$\text{Uniformity coefficient } C_u = \frac{D_{60}}{D_{10}} = \frac{12.0}{0.08} = 150$$

$$\text{Coefficient of curvature } C_c = \frac{D_{30}^2}{D_{10}D_{60}} = \frac{1.9^2}{0.08 \times 12.0} = 3.76$$

USCS Classification

F_{200} = fraction passing No. 200 sieve = 9.5% (since this percentage is between 5% and 12%, the soil should have dual classification).

Coarse fraction = 100 − 9.5 = 90.5%, which is greater than 50%. Therefore, soil is either G or S (gravel or sand).

Percent retained on No. 4 sieve = 58.5%, which is more than half of coarse fraction (0.5 × 90.5 = 45.3%). Therefore, first letter is G. Since both conditions $C_u > 4$ and $1 < C_c < 3$ are not satisfied, the soil is GP.

Narrowing down the second designation of the soil, i.e., whether it is GP-GM or GP-GC needs further information about the plasticity characteristics of the soil.

AASHTO Classification

Percent passing No. 10 = 30.5

Percent passing No. 40 = 21.5

Percent passing No. 200 = 9.5

Less than 35% passing No. 200 sieve—Predominantly granular

From the table, this soil meets the criteria for A1-a. Group index for group A1-a is zero (0).

Burmister Soil Classification System

The Burmister soil classification system uses a specific nomenclature to describe the soil's texture, color, plasticity, mineralogy, and even geologic origin.

Some of the qualitative descriptors such as *some*, *little*, and *trace* have the following meanings:

Abbrev.	Meaning	Content (% by weight)
A	and	35–50
S	some	20–35
L	little	10–20
T	trace	1–10

Table 203.6 summarizes nomenclature in the Burmister classification system and range of parameters associated with various types of fine-grained and coarse-grained soil fractions.

Further, to describe the gradation of components of coarse-grained soils, the Burmister soil classification system uses the following qualifying labels for coarse-grained soils (see Table 203.7).

The principal component (> 50%) is always listed first. If there is no principal component (i.e., no single component > 50%), sand is listed first.

Table 203.6   Burmister Soil Classification System: Identification of Components

Fine-grained soils

Component	Symbol	Plasticity	Plasticity Index
Silt	$	Nonplastic	0–1
Clayey silt	Cy$	Slight (Sl)	1–5
Silt and clay	$ and C	Low (L)	5–10
Clay and silt	C and $	Medium (M)	10–20
Silty clay	$yC	High (H)	20–40
Clay	C	Very high (VH)	> 40

Coarse-grained soils

Component	Symbol	Sub-class		Particle size
Gravel	G			
		Coarse		1 in–3 in
		Medium	3/8 in–1 in	
		Fine	No. 10–3/8 in	
Sand	S			
		Coarse		No. 30–No. 10
		Medium	No. 60–No. 30	
		Fine	No. 200–No. 60	
Silt	$			
				< No. 200

Table 203.7 Burmister Soil Classification System: Gradation of Components

Qualification	Description	
cmf	Coarse medium to fine	All fractions > 10%
cm	Coarse to medium	Less than 10% fine
mf	Medium to fine	Less than 10% coarse
c	Coarse	Less than 10% medium and fine
m	Medium	Less than 10% coarse and fine
f	Fine	Less than 10% coarse and medium

Two examples of the Burmister soil classification are shown below.

Full description: Coarse$^+$ medium to fine$^-$ sand, some$^-$ medium fine gravel, trace$^+$ Silt

Abbreviated: c$^+$mf$^-$ Sand, s$^-$ mf GRAVEL, t$^+$ Silt

Shorthand: c$^+$mf$^-$ S, s$^-$ mfG, t$^+$\$

Full description: Clay and silt, little$^+$ coarse$^-$ medium to fine$^+$ sand, medium plasticity

Abbreviated: Clay & silt, l$^+$ c$^-$ mf$^+$ S, m Pl

Shorthand: C & \$, l$^+$ c$^-$ mf$^+$ S, m Pl

Vertical Stress Increase at Depth

When surface or near-surface loads are applied to a soil—such as from live loads on a pavement, or loads transmitted through shallow footings or mats—the increment of vertical stress in the soil diminishes with depth. The actual vertical stress experienced by a soil stratum can be calculated using approximate "rule of thumb" approaches, or more sophisticated theories such as those due to Boussinesq and Westergaard.

Approximate Methods

Drawing upon the correlation between observation and the theories such as those by Boussinesq and Westergaard, two common approaches are used for a "quick" estimate of the vertical stress increase at depth due to finitely localized loads. These are (1) using a cone of influence that makes a $2V{:}1H$ slope or (2) using a cone of influence that makes an angle of $60°$ to the horizontal. These are shown in Fig. 204.1.

Thus, using the first approach, if the load P (force per unit length) is supported by a strip footing (width = B, infinite length), the soil pressure directly under the footing is given by P/B while the soil pressure at a depth z below the bottom of footing is given by $P/(B + z)$. Using the second approach, the soil pressure at a depth z below the bottom of footing is given by $P/(B + 1.155z)$.

If the load P (force) is transmitted to a rectangular footing (length L, width B), the zone of influence appears as a truncated pyramid. According to the first approach, the base area at a depth z is given by $(L + z)(B + z)$ and therefore the approximate soil pressure at depth z is given by $P/(L + z)(B + z)$.

Discussion of more detailed methods is outlined below. It should be noted that these methods and the accompanying derivations are based on many idealized assumptions and therefore, while very elegant, should not be perceived as necessarily providing a solution of much greater accuracy than that given by the approximate methods.

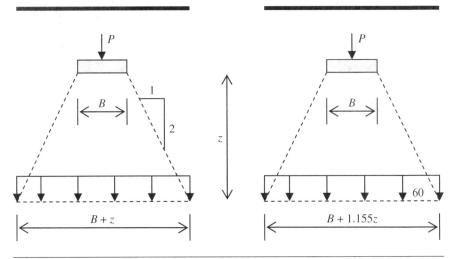

Figure 204.1 Simple rule for soil stress attenuation with depth.

Boussinesq Model for Stress under Uniformly Loaded Area

Figure 204.2 shows a rectangular area loaded with a uniformly distributed vertical load. Boussinesq's theory gives an expression for the stress increase at a point *directly below one corner* of this loaded area (vertical distance $= z$).

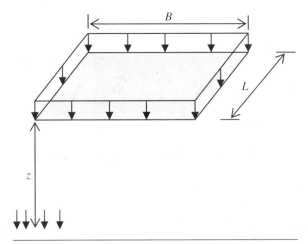

Figure 204.2 Boussinesq's vertical stress increase under a uniformly loaded area.

Vertical stress increase under the *corner* of a rectangular uniformly loaded area is given by the product of the uniform stress directly beneath the loaded area (q_o) and an influence factor I which is a function of the depth z and the geometry of the loaded area

$$\Delta q_v = q_o I \tag{204.1}$$

The influence factor (I) is a function of the ratios

$$m = \frac{B}{z} \quad \text{and} \quad n = \frac{L}{z}$$

and is given by

$$I = \frac{1}{4\pi}\left[\frac{2mn}{\alpha + \alpha_1}\frac{\alpha + 1}{\sqrt{\alpha}} + \tan^{-1}\left(\frac{2mn\sqrt{\alpha}}{\alpha - \alpha_1}\right)\right] \tag{204.2}$$

where the dimensionless parameters α and α_1 are given by

$$\alpha = m^2 + n^2 + 1 \quad \text{and} \quad \alpha_1 = m^2 n^2$$

Figure 204.3 can be used to obtain values of the influence factor I, once m and n values for the loaded area have been calculated.

Newmark's Chart for Graphical Solution of Boussinesq's Equation

In 1942, Newmark developed a graphical technique based on Boussinesq's equation for the vertical stress increase below the corner of a uniformly loaded area. The tool is in the form of a chart which is shown in Fig. 204.4. The procedure for using this chart is

1. Determine the depth z below the uniformly loaded area.
2. Plot the plan of the area scaled such that
   a. the point below which the stress is to be found is at the center of the chart, and
   b. the plan area is scaled such that the "1-unit" mark equals the depth z.
3. Count the number of elements covered by the plan area (N).
4. Calculate the increase in vertical stress as $\Delta p_v = Iq_o N$ where I is the influence value on the chart (0.005 in the sample chart shown) and q_o is the uniform pressure directly under the loaded area.

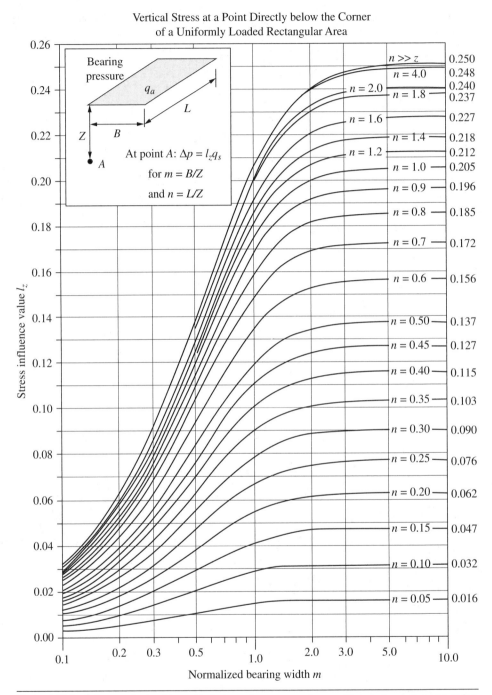

Figure 204.3 Influence factor for vertical stress increase under a uniformly loaded rectangular area (Boussinesq).

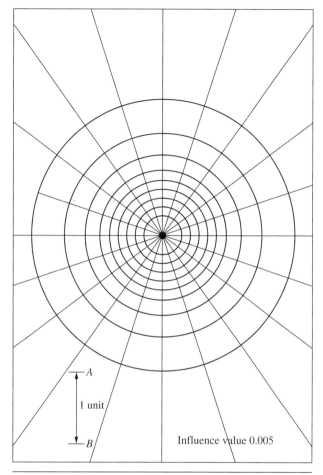

Figure 204.4 Newmark's graphical chart for solution of vertical stress increase under a uniformly loaded area (Boussinesq).

Figure 204.5 shows vertical stress contours under strip and square footings. These represent solutions to Boussinesq's equation. For example, the $0.1p$ stress contour passes very close to the point *at a depth of 2B below the corner of a square footing.* Looking closely, the vertical stress at this point may be estimated as $0.095p$ where p is the uniform pressure under the footing.

Example 204.1

A square footing of width 5 ft carries a concentric column load of 100 kips as shown. The depth of the footing is 3 ft. Find the vertical stress increase at a point 6 ft below the bottom of the footing and at a lateral distance of 10 ft from the center of the footing. Compare the results using (a) load dissipation along 1:2 slope, (b) Boussinesq's influence factor I in Fig. 204.3, and (c) Boussinesq's influence contours in Fig. 204.5.

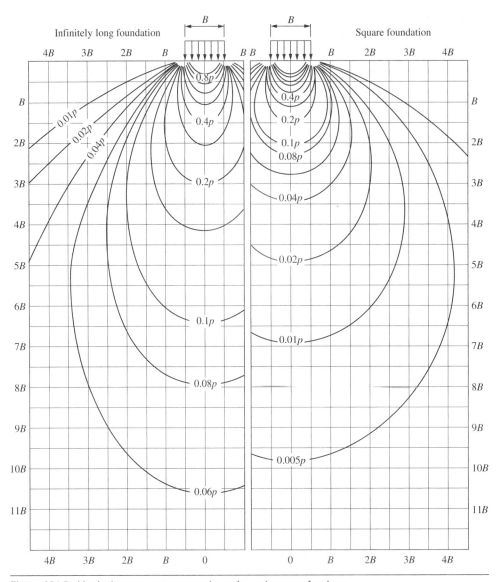

Figure 204.5 Vertical stress contours under strip and square footings.

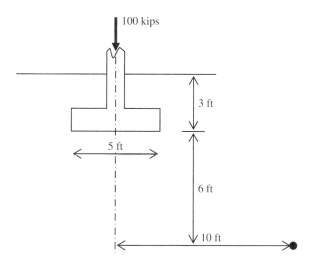

Solution

(a) According to the 2:1 dissipation method, the load must be distributed over a square with dimensions $(B + z)(L + z)$ or 11-ft square. The vertical stress increase is given by

$$\Delta\sigma_v = \frac{100}{121} = 0.826 \text{ ksf}$$

However, this method does not differentiate between a point directly below the footing center and a point offset laterally.

(b) To calculate the Boussinesq influence factor I (which is derived for the stress increase under the *corner* of a loaded area), we must treat the area *abed* as the difference between the areas *acfd* and *bcfe* (both of which have a corner at the point of interest (f).

Area *acfd* has dimensions 12.5 ft × 2.5 ft. $m = 12.5/6 = 2.1$; $n = 2.5/6 = 0.42$. $I = 0.115$

Area *bcfe* has dimensions 7.5 ft × 2.5 ft. $m = 7.5/6 = 1.25$; $n = 2.5/6 = 0.42$. $I = 0.110$

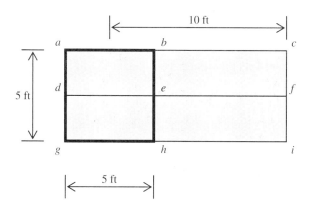

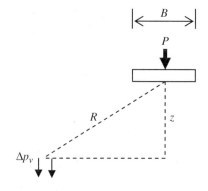

Figure 204.6
Vertical stress under a point load.

Therefore, the composite influence factor $= 2(0.115 - 0.110) = 0.01$ and the stress increase is $\Delta\sigma_v = p_o I = 100/(5 \times 5) \times 0.01 = 0.04$ ksf.

(c) From influence contour chart, for $z = 6$ ft $= 1.2B$ and $x = 10$ ft $= 2B$, the influence factor contour $= 0.01p$ and the stress increase is $\Delta\sigma_v = p_o I = 100/(5 \times 5) \times 0.01 = 0.04$ ksf.

Stress Increase Due to a Point Load

Boussinesq's Theory

A load can be considered a "point load" if the width B over which it is applied is small compared to the depth z at which the pressure increase is to be calculated. The stress increase due to an isolated footing whose lateral dimensions are small compared to the depth z (see Fig. 204.6) can be predicted by the theory by Boussinesq, which gives the stress increase as

$$\Delta p_v = \frac{P}{z^2} \frac{3}{2\pi} \left[1 + (r/z)^2 \right]^{-5/2} = \frac{P}{z^2} I \tag{204.3}$$

The stress increase predicted by this theory yields good results for $z > 2B$. According to this theory, directly under the center of the loaded area, the maximum vertical stress is

$$(\Delta p_v)_{max} = \frac{3P}{2\pi z^2} = 0.478 \frac{P}{z^2} \tag{204.4}$$

Westergaard's Theory

Another theory predicting the stress increase due to a point load is that due to Westergaard[1], who modeled the soil as an elastic medium in which soil layers alternate with thin rigid

---

[1] Westergaard H. M. (1938), "A Problem of Elasticity Suggested by a Problem in Soil Mechanics: Soft Material Reinforced by Numerous Strong Horizontal Sheets" in *Contribution to the Mechanics of Solids, Stephen Timoshenko 60th Anniversary Vol.*, Macmillan, New York.

reinforcements, which can be idealized simulation of alternating clay and silt layers. The vertical stress increase according to Westergaard is given by

$$\Delta p_v = \frac{P}{z^2}\frac{\eta}{2\pi}\left[\eta^2 + (r/z)^2\right]^{-3/2} = \frac{P}{z^2}I \tag{204.5}$$

where $\eta = \sqrt{(1 - 2\mu)/(2 - 2\mu)}$
μ = Poisson's ratio of the soil

According to this theory, directly under the center of the loaded area, the maximum vertical stress is

$$(\Delta p_v)_{max} = \frac{P}{2\pi\eta^2 z^2} = 0.159\frac{P}{\eta^2 z^2} \tag{204.6}$$

Stress Increase Due to a Line Load

When a line load such as that produced by a wall or strip footing acts at a certain depth in the soil, the stress increase at depth z below the footing (see Fig. 204.7) is given by

$$\Delta p_v = \frac{q_o}{z}\frac{2}{\pi}\left[1 + (r/z)^2\right]^{-2} = \frac{q_o}{z}I \tag{204.7}$$

where q_o = force per unit length immediately below the footing
Δq_v = increase in vertical stress at depth z

According to this theory, directly under the centerline of the strip footing, the maximum vertical stress is

$$(\Delta p_v)_{max} = \frac{2}{\pi\,z}q_o = 0.637\frac{q_o}{z} \tag{204.8}$$

Figure 204.7
Vertical stress under an infinitely long line load.

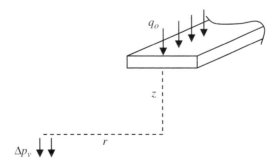

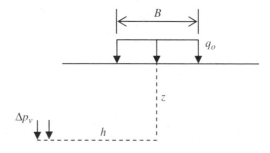

Figure 204.8
Vertical stress under a strip load of finite width.

Stress Increase Due to a Strip Load

The increase in vertical stress due to a strip load, when the load-width B cannot be considered "small" (see Fig. 204.8), is given by

$$\Delta p_v = \frac{q_o}{\pi}\left[\tan^{-1}\left(\frac{z}{h-\dfrac{B}{2}}\right) - \tan^{-1}\left(\frac{z}{h+\dfrac{B}{2}}\right) - \frac{Bz\left(h^2 - z^2 - \dfrac{B^2}{4}\right)}{\left(h^2 + z^2 - \dfrac{B^2}{4} + B^2 z^2\right)}\right] \qquad (204.9)$$

where q_o and Δp_v have units of force/area (stress).

Stress Increase Due to Uniformly Loaded Circular Footing

The vertical stress increase below a uniformly loaded circular area is given by the influence diagram in Fig. 204.9.

Load on Buried Pipes

Conduits laid underground may be either rigid or flexible pipes. Rigid pipes such as those made of steel and concrete have a certain amount of flexural strength, which allows them to withstand the stresses induced by the soil overburden and the loads at the surface, such as due to traffic. Pipes made of concrete or clay do not flex perceptibly when loaded and experience crushing-type failure when their load limit is reached. This mode of failure for rigid pipes has given rise to the terms "crush strength" and "D-Load," but these terms do not apply to flexible pipes.

Pipes made of flexible material (such as PVC) have negligible strength compared to rigid pipes. They flex without breaking when loaded externally from soil overburden and vehicular traffic. When a PVC pipe encounters external loading, its diameter will begin to deflect, meaning its sides will move outward and slightly downward. If the pipe is buried in

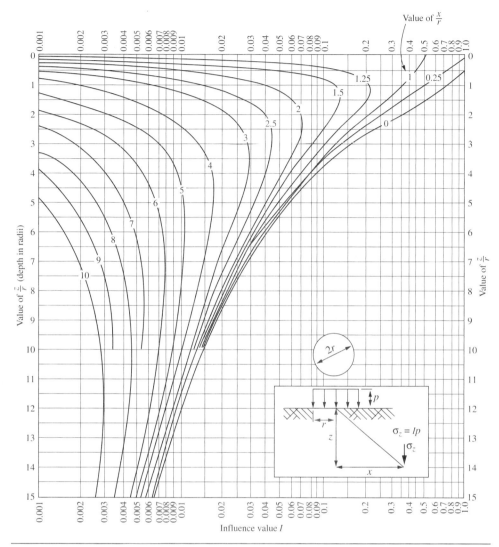

Figure 204.9 Influence factor for vertical stress increase under a uniformly loaded circular area.

supportive soil, the stiffness of the soil will resist the deflection. Thus, the combination of the soil stiffness and the pipe stiffness forms a *system* that acts to support external loads. By itself, the pipe may not support much weight, but the soil/pipe system can have significant load capacity. A PVC pipe's resistance to deflection in an unburied state is measured by its "pipe stiffness."

Deflection of Flexible Pipes

Because PVC pipe flexes rather than breaks when loaded, the failure criterion is not fracture strength. Instead, a limit is placed on pipe diametric deflection. This limit is expressed in terms of a percentage reduction in diameter due to external loading. Industry recommendations for maximum diametric deflection are shown below:

PVC pressure pipes	5.0%
PVC sewer/drain pipes	7.5%
PVC electrical conduits	5.0%

"Failure" of a flexible pipe system from external loading is defined by the point at which the top of the pipe begins to experience inverse curvature. Research has shown this point occurs at a minimum of 30% deflection. This means that the recommendations shown above incorporate safety factors of 4.0 to 6.0.

The deflection of the PVC pipe is estimated by the use of an empirical equation called the *Modified Iowa Equation*. A simplified, conservative version of the equation is presented below.

Modified Iowa Equation

$$\% \text{ deflection} = \frac{0.1(W + P)}{0.149PS + 0.061E} \times 100 \qquad (204.10)$$

where % deflection = predicted percentage of diametric deflection
W = live load (lb/in²) = pressure transmitted to the pipe from surface traffic loads
P = soil prism load (lb/in²) = overburden pressure on the pipe from soil weight
PS = pipe stiffness (lb/in²) = flexible pipe's bending stiffness (unburied state)
E = modulus of soil reaction (lb/in²) = stiffness of the embedment soil

The prism load (soil overburden) can be calculated as $P = \gamma H$ where γ is the unit weight of the soil and H is the height of the soil cover. Thus, for a 5-ft cover of 120-pcf soil, the prism load would be $120 \times 5 = 600$ lb/ft² $= 4.167$ lb/in². Table 204.1 outlines vertical stress increase at various depths for three classes of surface live loads—highway, railway, and airport.

Table 204.1 Live Load as Function of Burial Depth for Various Live Loads

Height of soil cover (ft)	Live load applied to pipe (lb/in²)		
	Highway (*H*20)	Railway (*E*80)	Airport
1	12.50	—	—
2	5.56	26.39	13.14
3	4.17	23.61	12.28
4	2.78	18.40	11.27
5	1.74	16.67	10.09
6	1.39	15.63	8.79
7	1.22	12.15	7.85
8	0.69	11.11	6.93
10	—	7.64	6.09
12	—	5.56	4.76
14	—	4.17	3.06
16	—	3.47	2.29
18	—	2.78	1.91
20	—	2.08	1.53
22	—	1.91	1.14
24	—	1.74	1.05
26	—	1.39	—
28	—	1.04	—
30	—	0.69	—

Table 204.2 gives approximate values of modulus of elasticity for various classes of soil for various degrees of compaction. In this table, the following criteria are used to define the level of compaction:

Slight compaction is defined as < 85% Proctor dry density, < 40% relative density.

Moderate compaction is defined as 85% to 95% Proctor dry density, 40% to 70% relative density.

High compaction is defined as > 95% Proctor dry density, > 70% relative density.

Minimum Burial Depth

The minimum recommended burial depth for PVC pipes beneath a highway is 1 ft. This recommendation assumes proper specification of embedment materials and compaction, and proper installation.

Loads on Rigid Pipes

There are two options for designers of concrete-pipe drainage systems to ensure that the pipe functions as a conduit and a structure. The *indirect design method* determines the required

Table 204.2 Average Values of Modulus of Elasticity for Soil

Pipe bedding material		E (psi) for various degrees of compaction			
Soil class	Soil type (USCS)	Loose	Slight	Moderate	High
Class I	*Crushed rock*	1000	3000	3000	3000
Class II	*Coarse-grained soils* Soils with little or no fines. GW, GP, SW, SP, containing less than 12% fines.	200	1000	2000	3000
Class III	*Fine-grained soils* (LL < 50) Soils with medium to no plasticity. CL, ML, ML-CL, with more than 25% coarse-grained particles. *Coarse-grained soils* Soils with fines. GM, GC, SM, SC containing more than 12% fines.	100	400	1000	2000
Class IV	*Fine-grained soils* (LL < 50) Soils with medium to no plasticity. CL, ML, ML-CL, with less than 25% coarse-grained particles.	50	200	400	1000
Class V	*Fine-grained soils* (LL > 50) Soils with medium to high plasticity. CH, MH, CH-MH.	No data available; consult a competent soils engineer. Otherwise use $E = 0$.			

Source "Soil Reaction for Buried Flexible Pipe" by Amster K. Howard, U.S. Bureau of Reclamation, Denver, Colorado. Reprinted with permission from American Society of Civil Engineers Journal.

strength of the pipe and then a class of pipe is selected that meets that load requirement based on a given bedding design. Using this method, the pipe manufacturer selects the reinforcing steel required in the pipe. *Direct design*, otherwise referred to as standard installations direct design (SIDD), is the design of pipe in the installed condition.

Indirect Design Method

In the indirect design method, the earth pressures and their distribution around the pipe and the resulting moments, thrusts, and shears in the pipe are not calculated. Instead, procedures developed by Marston and Spangler in the 1910s to 1930s are used to calculate bedding factors for the pipe, which relate the *in situ* load to the pipe to the load applied in a three-edge bearing test. Many of the design practices currently in use are based on the research by Marston[2] and Spangler.

The indirect design method was the only standard industry practice through the 1900s, until research by Heger[3] and McGrath[4] in the 1970s and early 1980s led to improvements in

[2]Marston, A. (1930), "The Theory of External Loads on Closed Conduits in the Light of the Latest Experiments," Bulletin 96, Iowa Engineering Experiment Station, Ames Iowa.

[3]Heger, F. J. (1982), "Structural Design Method for Precast Reinforced Concrete Pipe," Transportation Research Record 878, Soil Structure Interaction of Subsurface Conduits, Washington DC.

[4]McGrath, T. J. (1993), "Design of Reinforced Concrete Pipes—A Review of Traditional and Current Methods," The Second Conference on Structural Performance of Pipes, Columbus, Ohio.

understanding the structural behavior of buried pipe in its installed condition. The finite element analysis known as soil pipe interaction design and analysis (SPIDA), identified conservatism in the Marston Spangler work. In the indirect design method, vertical soil pressure due to dead loads is obtained as follows:

$$w = C_w \gamma B^2 \tag{204.11}$$

where w = total dead load on the conduit per unit length
C_w = correction coefficient
B = width of trench at level of top of pipe or pipe outer diameter if buried under an embankment
γ = unit weight of backfill

The coefficient C_w depends on trench depth to width ratio, angle of trench side slopes, friction angle of backfill and trench sides, bedding conditions.

$$\text{Dead load pressure} = w/B$$

Load Factor

The following types of bedding classes and corresponding values of the load factor L_f are tabulated in the NAVFAC 7.1[5] design manual.

Class A: Concrete cradle; $L_f = 4.6$ (for 1% reinforcing steel); $= 3.4$ (for 0.4% steel); $= 2.8$ for plain concrete pipe

Class B: Compacted granular material; $L_f = 1.9$

Class C: Compacted granular material or densely compacted backfill; $L_f = 1.5$

Class D: Flat subgrade; $L_f = 1.1$

To design a rigid conduit, the computed loads (dead and live) are modified to account for bedding conditions and to relate maximum allowable load to the three-edge bearing test load D.

$$D_{0.01} = (P_{DL} + P_{LL}) \frac{N}{L_f} \tag{204.12}$$

where $D_{0.01}$ = allowable load in pound per feet of length of conduit per foot of inside diameter for a crack width of 0.01 in
L_f = load factor
N = safety factor (usually 1.25)

With the specified D load, the supplier is able to provide adequate pipe.

---

[5]Naval Facilities Engineering Command, *Soil Mechanics Design Manual* 7.01, Alexandria, VA, 1986.

Example 204.2

A reinforced concrete pipe culvert is buried 2 ft below a highway pavement. The design load on the highway is a *H*20 truck. Assume the soil unit weight to be approximately 125 pcf. Assume the bedding factor for the pipe to be 1.9 and the overall factor of safety to be 1.5. What is the required allowable load on the pipe (psf)?

Solution

$$P_{DL} = 125 \times 2 = 250 \text{ psf}$$

$$P_{LL} = 5.56 \text{ psi} = 800 \text{ psf}$$

Therefore, the allowable load (pound per feet of length of conduit per foot of inside diameter) is

$$D_{0.01} = (250 + 800)\frac{1.5}{1.9} = 830 \text{ psf}$$

Direct Design Method

The direct design method is a more rational semiempirical approach to reinforced concrete pipe design. Direct design is a limit states design procedure that allows for the design of reinforcing for concrete pipe based on five limit states: (1) reinforcement tension, (2) concrete compression, (3) radial tension, (4) diagonal tension, and (5) crack control. Thus, direct design is much more flexible than indirect design provided that it is used efficiently. According to the direct design method, the required strength of the concrete pipe is determined from the effects of the bending moment, thrust, and shear.

The moment (*M*), thrust (*N*), and shear (*V*) can be computed by using either a computer program or hand calculations with the appropriate coefficients.

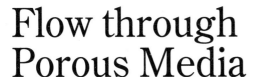
Flow through Porous Media

Groundwater Distribution

Groundwater storage may be divided into the following zones:

1. The soil-water zone extends from the ground surface through the major root zone. The soil in this zone is unsaturated except during heavy infiltration.

2. The intermediate zone extends from the bottom of the soil-water zone down to the top of the capillary fringe.

3. The capillary zone extends from water table up to a height determined by the capillary rise that can be generated by the soil.

4. In the groundwater zone, soil voids are completely saturated with water. Not all of this water can be extracted from the soil because of molecular and surface tension forces. The water that can be drained from the soil by gravity is called the specific yield. The soil water that is held (retained) against gravity drainage is called the specific retention.

A temperature shift of about 6°F occurs at the groundwater interface. So, a stratification study performed with a cone penetrometer equipped with a temperature sensor can help identify the precise location of saturation zones, which can be very useful for slope stability and consolidation studies.

Darcy's Law for Seepage

Seepage, the migration of water through voids in a soil, is due to a soil property called permeability or hydraulic conductivity. Darcy's law specifies that the (average) discharge velocity through soil voids is proportional to the hydraulic gradient (dH/dL).

$$V \alpha \frac{dH}{dL} \Rightarrow V = K \frac{dH}{dL} \qquad (205.1)$$

The constant of proportionality (K) is called the hydraulic conductivity. Thus, the seepage flow rate Q through (and normal to) an area A of the soil is given by

$$Q = K\frac{dH}{dL}A \qquad (205.2)$$

The hydraulic gradient ($i = dH/dL$) is the slope of the energy grade line across the soil boundary. For seepage through soil voids, the velocity is small and therefore the kinetic energy head ($V^2/2g$) can be ignored. The total head H is therefore approximately equal to $H = p/y + z$, which is the piezometric head. The hydraulic gradient is therefore approximated by the slope of the hydraulic grade line. The (actual) seepage velocity through the soil voids is much higher than the average velocity. The seepage velocity may be expressed as

$$V_s = \frac{V}{n} = \frac{K}{n}\frac{dH}{dL} \qquad (205.3)$$

where n is the porosity of the soil.

NOTE   If the seepage loss (flow rate) is to be estimated, it would be appropriate to use the average discharge velocity. But, if the objective is to determine the potential for erosion or scour or the time for seepage flow to travel a certain distance, the seepage velocity should be used, since this is the actual velocity at which water is migrating through the soil.

Hydraulic Conductivity

Since the hydraulic gradient is dimensionless, the hydraulic conductivity K has units of velocity. The value of K is a function of both the soil type as well as the liquid being conveyed through the voids. For example, it is appropriate to state that the conductivity of *water* (hence the name *hydraulic* conductivity) through a fine-grained sand is approximately 0.01 cm/s. However, the soil also has an intrinsic property called absolute permeability (k). The hydraulic conductivity is related to the absolute permeability of the soil as follows:

$$K = \frac{k\gamma_w}{\mu} \qquad \text{or} \qquad k = \frac{K\mu}{\gamma_w} \qquad (205.4)$$

The absolute permeability k has units of L^2. A common unit for the absolute permeability is the Darcy. Converted to SI units, 1 Darcy is equivalent to 9.86923×10^{-13} m² or 0.986923 µm². This conversion is usually approximated as 1 µm². Thus, 1 Darcy is approximately equal to an area of 1 micron $\times$ 1 micron.

Laboratory Measurement of Hydraulic Conductivity

The hydraulic conductivity (K) of a soil can be measured using several laboratory methods. Two of the more common of these methods are

- Constant head test
- Falling head test

Constant Head Test

In the constant head test, the inlet head is controlled to be constant at all times, thereby making the head difference H between inlet and outlet constant. Water is allowed to seep through the soil column and collected at the outlet. The volume of water (V) flowing through the soil in a measured time interval (t) is monitored. A schematic of the constant head test is shown in Fig. 205.1.

The length of the soil column (measured parallel to the direction of flow) is L and the cross section of the soil column (normal to the longitudinal direction) is A. Seepage occurs under constant head H. Therefore, the constant hydraulic gradient is H/L. According to Darcy's law, the average flow rate Q may be written as

$$Q = \frac{V}{t} = K\frac{H}{L}A \tag{205.5}$$

$$K = \frac{VL}{AHt} \tag{205.6}$$

where A = cross section of the soil column
L = length of flow path through soil
V = volume of water collected at outlet
t = duration of water collection
H = constant head

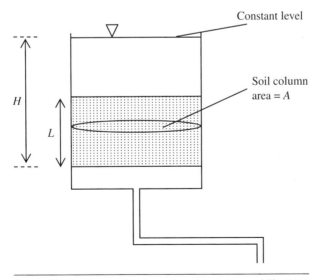

Figure 205.1 Schematic of constant head permeability test.

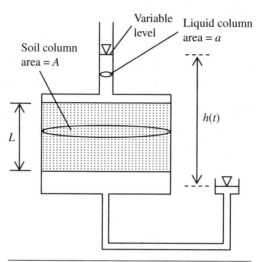

Figure 205.2 Schematic of falling head permeability test.

Falling Head Test

In the falling head test, the liquid is allowed to flow through and out of the system without maintaining a constant inlet reservoir elevation. A schematic view of this type of test is shown in Fig. 205.2. The inlet head is decreasing continuously, as liquid flows through the soil and out at the outlet, making the head difference h (between inlet and outlet) variable with time t. This test requires two-point data, i.e., the head is to be measured at two different instants t_1 and t_2. The head (h) represents the falling head difference between the ends of the soil column. The hydraulic conductivity K is then given by

$$K = \frac{aL \ln\left(\dfrac{h_1}{h_2}\right)}{A(t_2 - t_1)} \tag{205.7}$$

where
a = cross section of the falling liquid column
A = cross section of the soil column
L = length of flow path through soil
$t_2 - t_1$ = time interval
h_1 and h_2 = head at times t_1 and t_2 ($h_1 > h_2$)

Example 205.1

For the apparatus shown in the figure, the volume of water collected at the outlet during a time interval of 40 minutes is 5 cm³. The locations A, B, C, and D are marked with their elevations in centimeters (with respect to a common datum). Assume that reservoirs A and D have constant surface elevation. What is the permeability of the soil (centimeters per hour)?

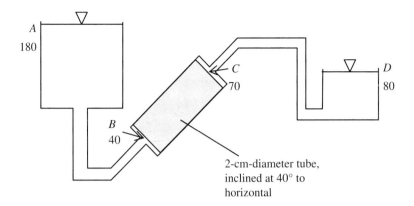

Solution   Since there is no energy loss mechanism between A and B, the total head at B is equal to the total head at $A = 180$ cm. Similarly, the total head at C is equal to the total head at $D = 80$ cm.

$$\text{The length of the soil column } L = \frac{\Delta h}{\sin\theta} = \frac{70-40}{\sin 40} = 46.67 \text{ cm}$$

$$\text{Hydraulic gradient } i = \frac{\Delta H}{\Delta L} = \frac{180-80}{46.67} = 2.14 \text{ cm/cm}$$

$$\text{Cross-sectional area of soil column } A = \frac{\pi D^2}{4} = 3.14 \text{ cm}^2$$

Therefore, from Darcy's law, the hydraulic conductivity is given by

$$K = \frac{Q}{iA} = \frac{5/(40\times60) \text{ cm}^3/\text{s}}{2.14\times3.14 \text{ cm}^2} = 3.1\times10^{-4} \text{ cm/s} = 1.12 \text{ cm/h}$$

Equivalent Hydraulic Conductivity (Layered Soils)

When the soil is not isotropic, flow will occur at different rates in different directions of the soil. Assuming the soil to be made up of a finite number of parallel layers with different values of hydraulic conductivity, this section gives the equivalent hydraulic conductivity for two cases:

1. Seepage flow occurring parallel to soil layers
2. Seepage flow occurring perpendicular to soil layers

Flow Parallel to Soil Layers

Figure 205.3 shows a layered soil of thickness Y, composed of parallel layers with thicknesses Y_1, Y_2, and so on. The seepage flow is parallel to the layer orientation and the corresponding

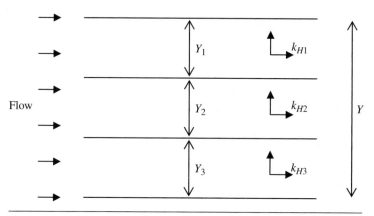

Figure 205.3 Seepage parallel to soil layers.

(horizontal, as shown) values of hydraulic conductivity for each layer are given by k_{H_1}, k_{H_2}, and so on. The equivalent hydraulic conductivity is given by the weighted *arithmetic* average of the hydraulic conductivities of the layers.

$$k_{H,\text{equiv}} = \frac{k_{H_1} \cdot Y_1 + k_{H_2} \cdot Y_2 + \cdots + k_{H_n} \cdot Y_n}{Y} \tag{205.8}$$

Thus, the head loss across each layer is the same, while the total seepage flow rate is equal to the sum of the (different) flow rates in each layer. The seepage flow rate per unit thickness (normal to the plane Fig. 205.3) is given by

$$q = \sum k_{H,i} \frac{\Delta h}{L} Y_i = \frac{\Delta h}{L} \left(k_{H_1} \cdot Y_1 + k_{H_2} \cdot Y_2 + \cdots + k_{H_n} \cdot Y_n \right) \tag{205.9}$$

Flow Transverse to Soil Layers

Figure 205.4 shows a layered soil of thickness Y, composed of parallel layers with thicknesses Y_1, Y_2, and so on. The seepage flow is perpendicular to the layer orientation and the corresponding (vertical, as shown) values of hydraulic conductivity for each layer are given by k_{V_1}, k_{V_2}, etc. The equivalent hydraulic conductivity is given by the weighted *harmonic* average of the conductivities of the layers.

$$k_{V,\text{equiv}} = \frac{Y}{\dfrac{Y_1}{k_{V_1}} + \dfrac{Y_2}{k_{V_2}} + \cdots + \dfrac{Y_3}{k_{V_3}}} \tag{205.10}$$

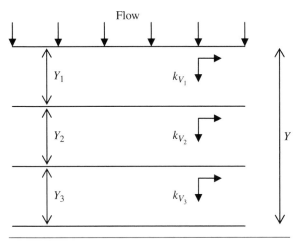

Figure 205.4 Seepage transverse to soil layers.

If the total head dissipated across the layered soil is H, then the head loss across layer i is given by

$$h_i = \frac{\dfrac{Y_i}{k_{V_i}}}{\dfrac{Y_1}{k_{V_1}} + \dfrac{Y_2}{k_{V_2}} + \cdots + \dfrac{Y_n}{k_{V_n}}} H \qquad (205.11)$$

The seepage flow rate through each layer is equal and can be calculated as

$$Q = k_{V,\text{equiv}} \times \frac{H}{Y} \times A = \frac{H \times A}{\dfrac{Y_1}{k_{V_1}} + \dfrac{Y_2}{k_{V_2}} + \cdots + \dfrac{Y_3}{k_{V_3}}} \qquad (205.12)$$

where A is the area of flow normal to the direction of seepage flow.

Example 205.2

In the experimental apparatus shown below, seepage flow occurs due to an elevation difference in the two columns of water. The connecting tube has a diameter of 100 mm and has three soil columns A, B, and C arranged in series as shown. The hydraulic conductivities of the three soils are given in the table. Calculate the seepage flow rate (m^3/s). All dimensions are in mm.

Soil	Hydraulic conductivity K (m/s)
A	1×10^{-4}
B	2×10^{-5}
C	5×10^{-6}

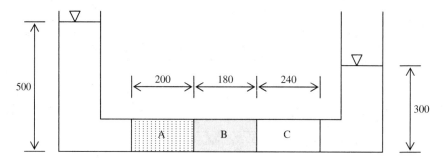

Solution   The difference in static head of the two columns is 200 mm = 0.2 m. The equivalent hydraulic conductivity (flow transverse to layers) is

$$k_{equiv} = \frac{Y}{\dfrac{Y_1}{k_1} + \dfrac{Y_2}{k_2} + \dfrac{Y_3}{k_3}} = \frac{620}{\dfrac{200}{1 \times 10^{-4}} + \dfrac{180}{2 \times 10^{-5}} + \dfrac{240}{5 \times 10^{-6}}} = 1.051 \times 10^{-5}$$

Therefore, $Q = k_{equiv} \times H/L \times A = 1.051 \times 10^{-5} \times 0.2/0.62 \times \pi/4 \times 0.1^2 = 2.66 \times 10^{-8}\,\text{m}^3/\text{s}$.

The head difference across each soil layer must be such that the seepage flow follows the continuity equation $Q_1 = Q_2 = Q_3$. The head loss across layer i is given by

$$h_i = \frac{\dfrac{Y_i}{k_{V_i}}}{\dfrac{Y_1}{k_{V_1}} + \dfrac{Y_2}{k_{V_2}} + \dfrac{Y_3}{k_{V_3}}} H$$

Layer A: $Y/k = 200/1 \times 10^{-4} = 2 \times 10^6$

Layer B: $Y/k = 180/2 \times 10^{-5} = 9 \times 10^6$

Layer C: $Y/k = 240/5 \times 10^{-6} = 48 \times 10^6$

Head loss in layer A: $h_A = \dfrac{2}{2 + 9 + 48} \times 0.2 = 0.0068$ m

Head loss in layer B: $h_B = \dfrac{9}{2 + 9 + 48} \times 0.2 = 0.0305$ m

Head loss in layer C: $h_C = \dfrac{48}{2 + 9 + 48} \times 0.2 = 0.1627$ m

Field Measurement of Hydraulic Conductivity

Hydraulic conductivity of a soil can be estimated using single auger holes. These tests are called slug tests. A hole is drilled to a depth L below the water table. Any water that seeps into the hole is first removed and then the rate of water rise in the hole is recorded. The hydraulic conductivity is given by

$$K = \frac{40}{\left(20 + \dfrac{L}{r}\right)\left(2 - \dfrac{y}{L}\right)} \frac{r}{y} \frac{\Delta y}{\Delta t} \tag{205.13}$$

where r = radius of the auger hole
   y = average depth of water in hole during the time interval Δt
   Δy = rise of water in hole during the time interval Δt
   L = distance measured down from GWT to bottom of auger hole

Flow Nets

The governing equation for seepage transport through a porous medium (soil) is the homogeneous Laplace equation in terms of the potential function $h(x,y)$

$$K_x \frac{\partial^2 h}{\partial x^2} + K_y \frac{\partial^2 h}{\partial y^2} = 0 \tag{205.14}$$

As the groundwater seeps through the soil, driven by a potential difference between the upstream and the downstream boundaries, the potential (head) decreases, according to the solution of Eq. (205.14). For isotropic soils, the hydraulic conductivity is identical in all directions. This leads to

$$\frac{\partial^2 h}{\partial x^2} + \frac{\partial^2 h}{\partial y^2} = 0 \tag{205.15}$$

The graphical solution to the above is a technique known as *flow net*. The flow net is constructed by discretizing the soil domain using two kinds of lines intersecting at right angles. These lines are

1. Equipotential lines—lines of constant total head
2. Flow lines—representing the direction of flow for water

The upstream boundary in the soil domain serves as the first ($i = 0$) equipotential line. The total head at all points along this line is equal to the total head at the free surface of the upstream reservoir. Similarly, the downstream boundary of the soil domain serves as the last ($i = n$) equipotential line. See Fig. 205.5. Rules for constructing a "good" flow net are

1. Equipotential lines and flow lines should intersect at nearly 90 degrees.
2. Cells should have uniform length to width ratio.

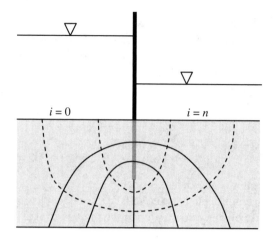

Figure 205.5
Seepage
occurring under
a sheet pile wall.

$i = 0$ $i = n$

Calculation of Seepage Flow from Flow Nets

A typical cell of a flow net is shown in Fig. 205.6. It is bounded by a pair of equipotential lines separated by distance L and a pair of flow lines separated by distance B. The flow (per unit thickness) through one flow channel is given by

$$\Delta q = k\frac{\Delta h}{L}B - k\frac{h/n_e}{L}B = \frac{khB}{n_e L} \tag{205.16}$$

The total seepage flow through all flow channels is given by

$$q = n_f \Delta q = \frac{khn_f B}{n_e L} = k\frac{n_f}{n_e}\frac{B}{L}h$$

If flow net elements are square (i.e., $B = L$), then $q = k(n_f/n_e)h$.

Figure 205.6
Typical cell of a
flow net.

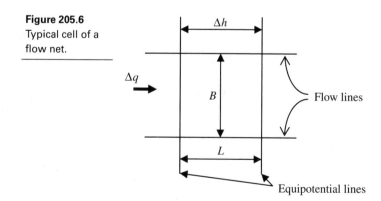

Δh

Δq

B

L

Flow lines

Equipotential lines

NOTE q represents flow rate per unit width (perpendicular to plane of the flow net). It therefore has dimensions L^2/T. To calculate the total seepage flow rate (L^3/T), one must multiply q by the width w ($Q = qw$). This width is normal to the plane in which the flow net is drawn.

Example 205.3

Seepage flow occurs around a sheet pile wall driven to point B (elevation 12 ft) in a soil deposit as shown below. A flow net has been drawn. Points A and C are at elevation 16 ft and 5 ft above the datum shown. Calculate the piezometric pressure at points A, B, and C.

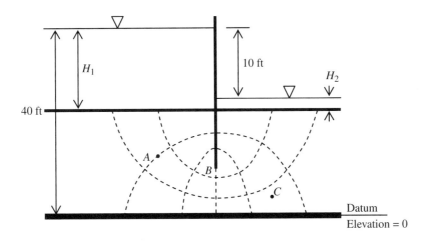

Solution Expressing all pressures as gage pressure (rather than absolute), the pressure head is designated as PH, the velocity head as VH, and the elevation head as EH. The total head is the sum of these three. These quantities are summarized below for the key locations within the soil domain shown above.

At upstream surface:	PH = 0	VH = 0	EH = 40	TH = 40
At ground upstream:	PH = H_1	VH = 0	EH = $40 - H_1$	TH = 40
At downstream surface:	PH = 0	VH = 0	EH = 30	TH = 30
At ground downstream:	PH = H_2	VH = 0	EH = $30 - H_2$	TH = 30

The difference between the total upstream head (TH$_{\text{US}}$) and total downstream head (TH$_{\text{DS}}$) is

$$\text{TH}_{\text{US}} - \text{TH}_{\text{DS}} = 10$$

Since successive equipotential lines represent equal increments of head loss, the total head at points A, B, and C are given by

$$\text{TH}_A = \text{TH}_{\text{US}} - 1/6 \times (\text{TH}_{\text{US}} - \text{TH}_{\text{DS}}) = 40 - 1/6 \times (40 - 30) = 38.33$$
$$\text{TH}_B = \text{TH}_{\text{US}} - 3/6 \times (\text{TH}_{\text{US}} - \text{TH}_{\text{DS}}) = 40 - 3/6 \times (40 - 30) = 35.00$$
$$\text{TH}_C = \text{TH}_{\text{US}} - 4.5/6 \times (\text{TH}_{\text{US}} - \text{TH}_{\text{DS}}) = 40 - 4.5/6 \times (40 - 30) = 32.50$$

Since the total head at A is 38.33 ft and the elevation at A is 16 ft, assuming velocity head to be negligible, the pressure head at A is $38.33 - 16.00 = 22.33$ ft. This pressure head can be then converted to a pressure using

$$p_A = \gamma H = 62.4 \times 22.33 = 1393.4 \text{ psf} = 9.68 \text{ psi}$$

Similarly, the pressure head at B is 23.00 ft and the pressure is 1435.2 psf. The pressure head at C is 27.50 ft and the pressure is 1716.0 psf. The significance of the pressure head at A, B, and C being 22.33 ft, 23.00 ft, and 27.50 ft is that these would be the height of a water column in a fine bore piezometer driven into the soil to these points. In other words, at A, the water column would rise to an elevation $22.33 + 16.00 = 38.33$ ft.

Anisotropic Soils

For anisotropic soils, the equation governing seepage flow is

$$K_x \frac{\partial^2 h}{\partial x^2} + K_y \frac{\partial^2 h}{\partial y^2} = 0 \tag{205.17}$$

This equation can be transformed to Eq. (205.15) using the transformation

$$\bar{x} = x \sqrt{\frac{K_y}{K_x}} \tag{205.18}$$

Procedure for anisotropic soils:

1. The flow net is drawn on a scaled geometry [using the scaling transformation given by Eq. (205.18)].

2. Seepage flow is calculated from the scaled geometry. The equivalent hydraulic conductivity is given by

$$\bar{K} = \sqrt{K_x K_y}$$

3. All calculated results are then scaled back to original geometry.

Uplift Pressure under Hydraulic Structures

Uplift pressures under a hydraulic structure such as a dam can be calculated by constructing a flow net. Hydraulic head difference $H = 41 - 20 = 21$ ft. Number of equipotential drops $= 7$. Thus, drop in total head per cell $= 3$ ft. Total head at A, B, C, D, E, $F = 38, 35, 32, 29, 26, 23$ ft, respectively.

$$\text{Pressure head} = \text{Total head} - \text{elevation head}$$

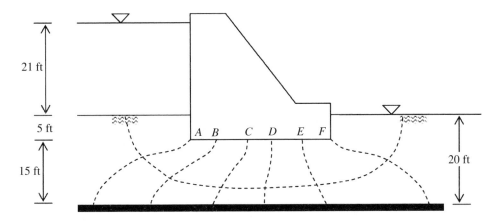

Pressure head at *A, B, C, D, E, F* = 23, 20, 17, 14, 11, 8 ft, respectively.

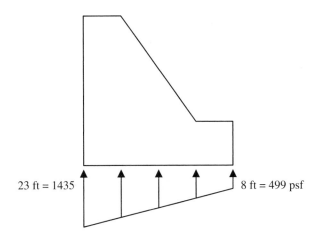

23 ft = 1435   8 ft = 499 psf

Resultant uplift force *R* = (1435 + 499)/2 × length of dam foundation

Factor of safety (uplift) = Weight of dam/*R*

Seepage through Earth Dams

Earth dams are susceptible to the phenomenon of "piping" which is the formation of seepage paths through the body of the dam. Usually, the resulting erosion starts at the downstream toe and works back toward the reservoir, forming a "pipe" or a channel through the dam. This phenomenon can take several years, depending on soil type, particle size distribution, level of compaction, and other factors.

The surface of the water flowing through the dam is called the phreatic surface. To prevent piping, the phreatic surface should be kept below the downstream toe of the dam. The phreatic surface can be controlled by properly designed cores or cutoff walls.

Clays with plasticity index > 15 are resistant to piping. For dams constructed of cohesion-less soils, poor compaction is one of the primary contributors to piping. Uniform (poorly graded) fine sands are also susceptible to piping, even if they are well compacted.

Some of the most common strategies to minimize the risk of piping are listed below. See Fig. 205.7.

1. Lengthening the flow path (and therefore reducing the hydraulic gradient) by using cutoff walls, impermeable cores and impermeable blankets on the upstream face of the dam. To effectively lower the phreatic surface, the location and depth of the cutoff wall has to be near perfect, and therefore this should be combined with other strategies.

2. Internal drain systems to reduce the pore water pressure in the downstream portion of the dam, thereby increasing the stability of the downstream slope.

3. Toe drains, which can be quite successful for low dams, can be used to prevent softening and erosion of the downstream toe.

4. Horizontal drainage blankets are often used for dams of moderate height. Drainage blankets are frequently used over the downstream one-half or one-third of the foundation area. Where pervious material is scarce, the internal strip drains can be placed instead since these give the same general effect.

5. Chimney drains are an attempt to prevent horizontal flow along relatively impervious stratified layers, and to intercept seepage water before it reaches the downstream slope. Chimney drains are often incorporated in high homogeneous dams which have been constructed with inclined or vertical chimney drains.

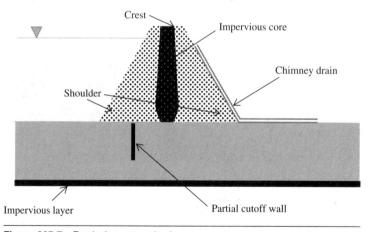

Figure 205.7 Earth dams terminology.

Expansive Soils

Expansive soils containing sufficient quantities of clay tend to swell when they absorb moisture and shrink when they lose moisture. These alternating periods of swelling and shrinkage create a pattern of desiccation or shrinkage cracks. Poor drainage adjacent to slabs and flatwork leads to lifting of slab edges and shear cracking near slab corners. Long-term differential uplift under slabs over time leads to a phenomenon called edge curl. Lightly loaded structures on expansive soils permit more water to penetrate the soil on one side of a structure leading to differential heave. This can be caused by ponding of water on the shady side of such structures, leaky water lines, or natural variances in soil moisture content.

Many natural slopes are covered by residual soils which are expansive. These materials swell and shrink with seasonal regularity, during alternating wet and dry seasons. This cyclic movement allows for plastic creep, or strain under sustained loading. Foundation elements situated on slopes can be expected to move with the slope if they are not deeply embedded.

Gravity Dams

Concrete gravity dams are designed so that the weight of the dam provides adequate safety against overturning due to the forces acting on the dam. These forces are hydrostatic pressure forces on either side of the dam, uplift forces, earthquake forces, and force due to ice buildup. A simple schematic of a gravity dam is shown in Fig. 205.8.

In the Fig. 205.8, $F_{u,v}$ and $F_{u,h}$ are the vertical and horizontal components of the fluid force on the upstream face of the dam. $F_{d,v}$ and $F_{d,h}$ are the corresponding resultant forces on the downstream face of the dam. The ice force, if present, acts at the water surface on the upstream face. Seismic forces, inertial in nature, act at the center of gravity.

Aquifers

Groundwater is extracted principally through wells. Wells may be drilled and installed with or without a casing—the latter situation is appropriate for stable soils where a casing or screen may not be required to stop unwanted inflow. A well typically consists of a drilled hole, a casing that prevents caving in of the sides or unwanted inflow, a pump and a section of perforated casing (the screen) to ensure that water is extracted from a specific elevation within the aquifer, a grout seal at the surface (to prevent contaminated surface runoff from percolating along the sides of the casing and reaching the screen), and a gravel pack around the screen to improve the flow to the well.

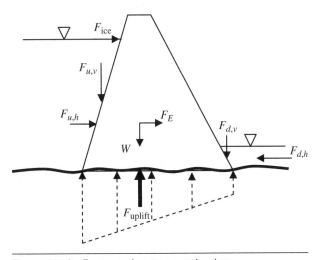

Figure 205.8 Forces acting on a gravity dam.

In a confined aquifer, the hydraulic head is given by the elevation of the piezometric surface, which is above the upper surface of the aquifer. Adding or removing water to the aquifer does not change the saturation of the aquifer, as long as the piezometric surface remains above the upper surface of the aquifer. The change of the hydraulic head is accomplished by a change in pressure. This pressure tries to force individual soil particles apart. Thus, in a confined aquifer, the total storage increases as pressure increases because the granular skeleton expands to create more voidspace.

If water is added to a groundwater reservoir, hydraulic head increases (water level in wells rises) and fluid pressure increases in the aquifer. Porosity increases as the granular skeleton expands.

Storativity (S) is the volume of water that an aquifer will absorb or expel from storage per unit area per unit change in head. Thus, it is a dimensionless property of the aquifer. Thus, the volume of water that will be drained from or added to an aquifer as the head is lowered or raised by Δh is given by

$$V = SA\Delta h \qquad (205.19)$$

where A is the area overlying the aquifer. For a confined aquifer, storativity S is given by

$$S = YS_s \qquad (205.20)$$

where Y = aquifer thickness
S_s = specific storage

In an unconfined aquifer, the hydraulic head is the height of the water table. In an unconfined aquifer, a change in hydraulic head results in both a change in pressure in the saturated portion, as well as a change in thickness of the saturated zone.

The transmissivity (T) is a measure of how much water can be transmitted horizontally through a unit width of a fully saturated aquifer under a hydraulic gradient of 1.0. The transmissivity is calculated as the product of the hydraulic conductivity (K) and the aquifer thickness (Y):

$$T = KY \qquad (205.21)$$

A steady rate of pumping from an aquifer creates a drawdown of the water table. In Fig. 205.9, Y is the thickness of the unconfined aquifer; y is the water table elevation at a radial distance r, s is the drawdown, steady pumping rate is Q; permeability of the soil in the aquifer is K.

$$Y = s + y \qquad (205.22)$$

The steady-state solution, which represents the equilibrium between the rate of pumped flow out of the well and the rate of radial inflow from the cone of depression, is given by Eq. (205.23).

$$Q = \frac{\pi K \left(y_1^2 - y_2^2 \right)}{\ln(r_1 / r_2)} \qquad (205.23)$$

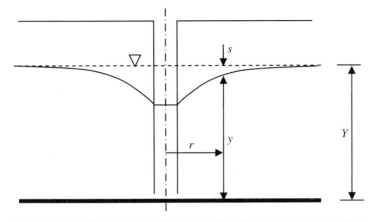

Figure 205.9 Radial drawdown in an unconfined aquifer.

where y_1 = depth of aquifer at radial distance r_1
y_2 = depth of aquifer at radial distance r_2
Q = pumping rate
K = hydraulic conductivity

Equation (205.23) gives reasonably accurate results if the maximum drawdown is less than about half the aquifer depth. For a confined aquifer, the same expression may be used, except that the variable y represents the elevation of the piezometric surface instead of the free surface as in the unconfined aquifer. For a very thick unconfined aquifer, the following approximation may be used:

$$Q \approx \frac{2\pi KY(s_2 - s_1)}{\ln(r_1/r_2)} = \frac{2\pi T(s_2 - s_1)}{\ln(r_1/r_2)} \tag{205.24}$$

The radius of influence of the well is the distance from the centerline of the well to the point where the drawdown becomes negligible. This can be calculated by substituting $s_2 = 0$ in the equation above. This leads to

$$R = r\exp\left(\frac{2\pi Ts}{Q}\right) \tag{205.25}$$

where s is drawdown recorded at a radial distance r.

Example 205.4   Unconfined aquifer
A pumping well is drilled at A and a steady pumping rate of 800 gpm is established. Observation wells are dug at B and C as shown on the map of the region. The original elevation of the

GWT in the underlying soil layer is 295 ft above sea level. The table shows data for the wells. Find the elevation of GWT in the observation well at C.

Well	Type	Diameter (in)	GWT (ft)
A	Pumped	10	288
B	Obs	5	291
C	Obs	6	—

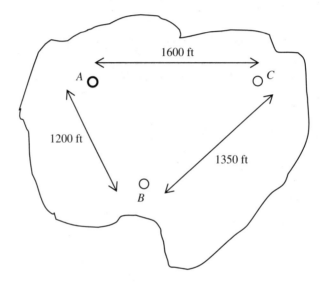

Solution

$$Q = 800 \text{ gpm} = 800 \times 0.002228 = 1.782 \text{ cfs}$$

Assuming aquifer depth Y is large, we can approximate the drawdown equation in terms of the drawdown s instead of the water table elevations y:

$$Q = \frac{\pi K(y_1^2 - y_2^2)}{\ln(r_1/r_2)} \approx \frac{2\pi KY(s_2 - s_1)}{\ln(r_1/r_2)}$$

For well A, $r_1 = 5$ in $= 0.4167$ ft and the drawdown $s_1 = 295 - 288 = 7$ ft

For well B, $r_2 = 1200$ ft and the drawdown $s_2 = 295 - 291 = 4$ ft

Using these values, we get

$$KY = 0.753 \text{ ft}^2/\text{s}$$

Now using this calculated value of KY for the pair A and C, we get

For well A, $r_1 = 5$ in $= 0.4167$ ft and the drawdown $s_1 = 295 - 288 = 7$ ft

For well C, $r_2 = 1600$ ft and the drawdown s_2

$$\text{Therefore, } s_2 - s_1 = -3.11 \text{ ft}$$
$$s_2 = s_1 - 3.11 = 3.89$$

Therefore, elevation of GWT in observation well C is $295 - 3.89 = 291.11$ ft

Example 205.5   Confined Aquifer

A 9-in-diameter well is drilled into a confined aquifer as shown below. The thickness of the aquifer is 32 ft. The impermeable layer above the aquifer is 46 ft thick. The original piezometric surface is at a depth of 30 ft below the ground surface. A steady rate of pumping at 125 gpm is established from the well until water levels in the pumping well and the observation well stabilize. Given the recorded drawdown in the wells

Pumped well   15 ft

Observation well   8 ft

Find the permeability of the soil in the aquifer.

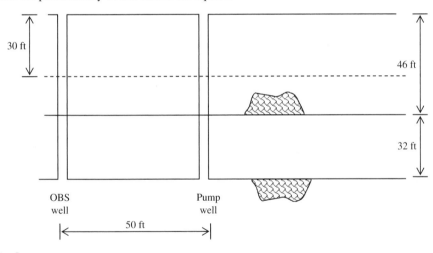

Solution

Initial elevation of piezometric surface $= h_0 = 78$ ft $- 30$ ft $= 48$ ft

Drawn down elevation in pumped well $= 48 - 15$ ft $= 33$ ft

Drawn down elevation in observation well $= 48 - 8$ ft $= 40$ ft

$$Q = 125 \text{ gpm} = 125 \times 0.002228 = 0.2785 \text{ cfs}$$

$$K = \frac{Q \ln\left(\dfrac{r_1}{r_2}\right)}{\pi(h_1^2 - h_2^2)} = \frac{0.2785 \times \ln\left(\dfrac{4.5/12}{50}\right)}{\pi(33^2 - 40^2)} = 8.5 \times 10^{-4} \text{ fps}$$

Example 205.6

A 9-in-diameter well is drilled into a confined aquifer. The thickness of the aquifer is 32 ft. The porosity of the soil within the aquifer is 0.40. A steady rate of pumping at 2500 gpm is established from the well until water levels in the pumping well and the observation well stabilize. What is the radial velocity of groundwater flow at a distance of 3 ft from the centerline of the well?

Solution   The flow rate is $Q = 2500$ gpm $= 5.57$ cfs. The average radial velocity is given by assuming this flow rate to be occurring through the surface area of a cylinder of radius 3 ft and length 32 ft:

$$v = \frac{Q}{2\pi rh} = \frac{5.57}{2\pi \times 3 \times 32} = 0.009235 \text{ fps}$$

However, the actual vseepage velocity is given by

$$v_s = \frac{v}{n} = \frac{0.009235}{0.40} = 0.0231 \text{ fps}$$

Karst

Rainwater that comes in contact with atmospheric carbon dioxide becomes acidic, which makes it a more active dissolving agent. Rocks such as limestone, dolomite, and gypsum are highly soluble in water. A typical landscape caused by the dissolution of such rocks contains features such as sinkholes and underground caves. Features at or near the surface are formed by subsidence and collapse induced by the formation of underground caves.

Water continues to flow through these underground fissures and caves, widening them over time until it reaches the groundwater. Eventually, these features act as aquifers, where the stored water can be extracted for human use. However, one of the concerns is that surface runoff entering the karstified landscape bypasses the normal filtering effect of the soil and bedrock and therefore can be contaminated by pollutants.

Unsteady Well Hydraulics—Theis Method

The conditions described above exist when the cone of depression has reached a steady-state shape and the outflow due to the pumping rate is in equilibrium with the inflow into the well from the groundwater within the radius of influence. Before this equilibrium has been reached, the drawdown is therefore an unsteady (time-dependent) phenomenon.

Assumptions for Transient Drawdown Effects

The nonequilibrium solution for drawdown in an unconfined aquifer (Theis, 1935)[1] is subject to the following assumptions:

- Aquifer is confined top and bottom by impermeable layers.
- There is no source of recharge to aquifer.
- The aquifer is compressible and water is released instantaneously from the aquifer as the head is lowered.
- The well is pumping at a constant rate.

The drawdown (lowering of the piezometric surface) is given by

$$s = h_0 - h = \frac{Q}{4\pi T} \int_U^\infty \frac{e^{-u}}{u} du \qquad (205.26)$$

The dimensionless number u is given by

$$u = \frac{r^2 S}{4Tt} \qquad (205.27)$$

where r = radial distance from the pumping well (L)
   S = storativity of the aquifer (dimensionless)
   T = transmissivity of the aquifer (L^2/T)
   t = time since beginning of pumping (T)

The integral may be replaced by its Taylor series [also known as the well function $W(u)$]

$$s = h_0 - h = \frac{Q}{4\pi T} W(u) \qquad (205.28)$$

Values of the function $W(u)$ for various values of u are given in Table 205.1. The numbers in the table have been truncated to a precision of 0.0001.

When $u < 0.01$ Jacob's approximation may be used:

$$s = \frac{Q}{4\pi T} \ln\left(\frac{2.25Tt}{r^2 S}\right) = \frac{Q}{4\pi T} \ln\left(\frac{0.561}{u}\right) \qquad (205.29)$$

Example 205.7

A confined aquifer is 60 ft thick and consists of a soil with a hydraulic conductivity $K = 2.0$ ft/h. The storativity of the aquifer is 0.004. A fully penetrating well of diameter 10 in is used to pump at a rate of 50 gpm. What is the drawdown in the well after 2 days?

---

[1]Theis, C. V. (1935), "The Relation between the Lowering of the Piezometric Surface and the Rate and Duration of Discharge of a Well Using Groundwater Storage," *American Geophysics Union Transactions*, Vol. 16, 519–524.

Table 205.1 Values of the Well Function

u	$W(u)$	u	$W(u)$	u	$W(u)$	u	$W(u)$
1.00E–10	22.4487	7.00E–08	15.8976	4.00E–05	9.5495	1.00E–02	4.0379
2.00E–10	21.7555	8.00E–08	15.7640	5.00E–05	9.3263	2.00E–02	3.3547
3.00E–10	21.3500	9.00E–08	15.6463	6.00E–05	9.1440	3.00E–02	2.9591
4.00E–10	21.0624	1.00E–07	15.5409	7.00E–05	8.9899	4.00E–02	2.6813
5.00E–10	20.8392	2.00E–07	14.8477	8.00E–05	8.8564	5.00E–02	2.4679
6.00E–10	20.6569	3.00E–07	14.4423	9.00E–05	8.7386	6.00E–02	2.2953
7.00E–10	20.5027	4.00E–07	14.1546	1.00E–04	8.6332	7.00E–02	2.1509
8.00E–10	20.3692	5.00E–07	13.9315	2.00E–04	7.9402	8.00E–02	2.0270
9.00E–10	20.2514	6.00E–07	13.7491	3.00E–04	7.5348	9.00E–02	1.9188
1.00E–09	20.1461	7.00E–07	13.5950	4.00E–04	7.2472	1.00E–01	1.8229
2.00E–09	19.4529	8.00E–07	13.4615	5.00E–04	7.0242	2.00E–01	1.2227
3.00E–09	19.0475	9.00E–07	13.3437	6.00E–04	6.8420	3.00E–01	0.9057
4.00E–09	18.7598	1.00E–06	13.2383	7.00E–04	6.6879	6.00E–01	0.4544
5.00E–09	18.5366	2.00E–06	12.5452	8.00E–04	6.5545	7.00E–01	0.3738
6.00E–09	18.3543	3.00E–06	12.1397	9.00E–04	6.4368	8.00E–01	0.3106
7.00E–09	18.2002	4.00E–06	11.8520	1.00E–03	6.3316	9.00E–01	0.2602
8.00E–09	18.0666	5.00E–06	11.6289	2.00E–03	5.6394	1.00E+00	0.2194
9.00E–09	17.9488	6.00E–06	11.4466	3.00E–03	5.2349	2.00E+00	0.0489
1.00E–08	17.8435	7.00E–06	11.2924	4.00E–03	4.9483	3.00E+00	0.0127
2.00E–08	17.1503	8.00E–06	11.1589	5.00E–03	4.7261	4.00E+00	0.0038
3.00E–08	16.7449	9.00E–06	11.0411	6.00E–03	4.5448	5.00E+00	0.0011
4.00E–08	16.4572	1.00E–05	10.9357	7.00E–03	4.3916	6.00E+00	0.0004
5.00E–08	16.2340	2.00E–05	10.2426	8.00E–03	4.2591	7.00E+00	0.0001
6.00E–08	16.0517	3.00E–05	9.8371	9.00E–03	4.1423	8.00E+00	0.0000

Solution

$$K = 2.0 \text{ ft/h} = 0.033 \text{ ft/min}$$

$$Q = 50 \text{ gal/min} = 6.684 \text{ ft}^3/\text{min}$$

$$S = 0.004$$

$$\text{Transmissivity } T = KY = 0.033 \times 60 = 2.0 \text{ ft}^2/\text{min}$$

$$r = \text{radial distance} = 5 \text{ in} = 0.4167 \text{ ft}$$

$$t = 2 \text{ days} = 2880 \text{ min}$$

$$u = \frac{r^2 S}{4Tt} = \frac{0.4167^2 \times 0.004}{4 \times 2.0 \times 2880} = 3.0 \times 10^{-8}$$

$$W(u) = 16.74 \text{ (See well function } W(u) \text{ in Table 205.1.)}$$

$$\text{Exact: } s = \frac{Q}{4\pi T} W(u) = \frac{6.684 \times 16.74}{4\pi \times 2.0} = 4.452 \text{ ft}$$

Jacob's approximation (may be used for $u < 0.01$) gives

$$s = \frac{Q}{4\pi T} \ln\left(\frac{2.25Tt}{r^2 S} \right) = \frac{6.684}{4\pi \times 2.0} \ln\left(\frac{2.25 \times 2.0 \times 2880}{0.4167^2 \times 0.004} \right) = 4.452 \text{ ft}$$

Groundwater Dewatering

Dewatering makes excavated material lighter and easier to handle, prevents the formation of a quick condition at the bottom of the excavation, and reduces lateral loads on sheeting and bracing. In most soils, the groundwater table must be at least 2 ft and preferably 5 ft below the bottom of the excavation to keep an excavation reasonably dry.

The most common method for dewatering excavations for large projects is the use of well points. Well points are metal well screens about 2 to 3 in in diameter and about 4 ft long. To ensure good drainage in fine and dirty sands or layers of silt or clay, the well point and riser should be surrounded by a sand filter to just below the groundwater table. The space above the filter should be sealed with silt or clay to keep air out of the well point. A pipe connects each well point to a header, from which water is pumped at a certain rate. Spacing of well points usually ranges from 3 ft to 12 ft. Well points are not suitable in very fine-grained soils and soils of low permeability. For dewatering very deep excavations or where the ground-water table must be lowered significantly (more than about 15–20 ft), other methods may become more economical.

Lowering the groundwater table below a construction site may be achieved using well points with plan locations chosen so that their combined field of action results in the desired lowering. For example, consider the dewatering scheme shown in Fig. 205.10. The horizontal dashed line represents the original groundwater table, which must be lowered to a certain minimum vertical clearance below the bottom of the proposed excavation. Dewatering of the site can be achieved by using well points on either side of the excavation. When these wells are pumped at a specified rate, at steady state, each well has a cone of influence as shown in the figure. The dashed line below the bottom of the excavation represents the superposition of these well fields.

Dewatering can also be achieved by installing relief drains (also called extraction drains) as shown in Fig. 205.11. Installing an array of relief drains above a barrier layer (vertical distance $= a$) allows the lowering of the groundwater table to a height b above the barrier layer. The system is described by Donnan's formula, which yields accuracy of $\pm20\%$.

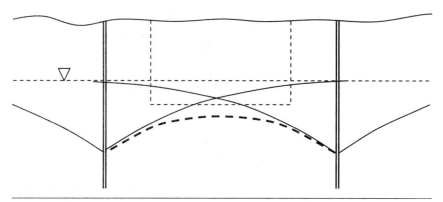

Figure 205.10 Dewatering an excavation site using well points.

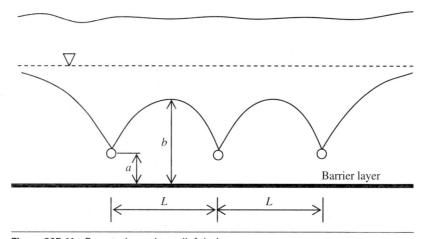

Figure 205.11 Dewatering using relief drains.

The drain spacing calculated from Donnan's formula is typically reduced by 10 to 20% in actual installations. The required spacing of relief drains is given by

$$L^2 = \frac{4K(b^2 - a^2)}{Q_d}$$

(205.30)

where L = drain spacing
K = hydraulic conductivity
a = distance between drain depth and barrier
b = vertical distance between maximum allowable water table height between drains and the barrier
Q_d = recharge rate, also known as the drainage coefficient (L/T)

Wellpoint Systems

Wellpoints consist of 1½ or 2 in diameter pipes with a perforated bottom section protected by screens. Water flows by gravity to the well screen and is then pumped by suction pumps connected to a header pipe. Wellpoints usually produce a drawdown between 15 ft and 18 ft below the center of the header. Discharge capacity is generally 15–30 gallons per minute per point. Typical wellpoint spacing is between 3 ft and 10 ft.

Pumping methods for gravity drainage generally are not effective when the average effective grain size of a soil (D_{10}) is less than 0.05 mm. For fine-grained soils with an effective grain size less than 0.01 mm, a vacuum seal at the ground surface around the wellpoint will improve drainage. Electro-osmosis is a specialized procedure utilized in silts and clays that are too fine grained to be effectively drained by gravity or vacuum methods.

Pumping Wells

These wells are formed by drilling a hole of sufficient diameter to accommodate a pipe column and filter, installing a well casing, and placing filter material in the annular space surrounding the casing.

Pumps may be submersible (placed within the well casing) or have the motor at the surface.

Deep pumping wells are used if (a) dewatering installations must be kept outside the excavation area, (b) large quantities are to be pumped for the full construction period, and (c) pumping must commence before excavation so that there is enough time for the drawdown to reach equilibrium. Deep wells may be used for gravels to silty fine sands, and water bearing rocks.

Bored shallow wells with suction pumps can be used to replace wellpoints where pumping is required for several months or in silty soils where correct filtering is critical.

Relief Wells

These wells are essentially sand columns used to bleed water from underlying strata containing artesian pressures, and to reduce uplift forces at critical locations. Relief wells may be tapped below ground by a collector system to reduce back pressures acting in the well.

Relief wells are frequently used as construction expedients, and in situations where a horizontal drainage course may be inadequate for pressure relief of deep foundations underlain by varved or stratified soils or soils whose permeability increases with depth.

Dewatering Using a Circular Array of Wellpoints

Figure 205.12(a) shows a circular array of n wellpoints. The resulting drawdown of the piezometric surface, calculated using a superposition of the cones of influence of all n wellpoints, is shown in Fig. 205.12(b).

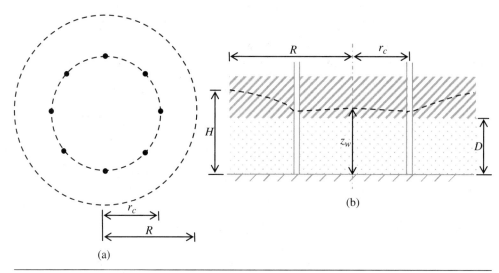

Figure 205.12   Steady-state drawdown with a circular array of wellpoints.

The flow rate (q) in a single well is given by Eq. (205.31)

$$q = \frac{2\pi KD(H - z_w)}{\ln\left(\dfrac{R^n}{nr_w r_c^{\,n-1}}\right)}$$

(205.31)

K = hydraulic conductivity of soil in aquifer
D = depth of aquifer
H = original height of piezometric surface above bottom of aquifer
z_w = lowered (drawn down) height of piezometric surface above bottom of aquifer
R = radius of influence of wellpoints (drawdown = 0)
n = number of wells uniformly spaced in circle of radius r_c
r_w = radius of each well
r_c = radius of circle

Shallow Foundations

Shallow Foundations

Shallow foundations are founded close to the ground surface, generally where the founding depth (D_f) is less than the width of the footing and less than 10 ft. As a rule of thumb, if surface loading or other surface conditions significantly affect the bearing capacity of a foundation it is considered "shallow." Shallow foundations ("spread footings") include pads ("isolated footings"), strip footings, and rafts.

Shallow foundations are used when the soil near the surface is sufficiently strong to support the imposed loads. Usually, they are unsuitable in weak or highly compressible soils, such as poorly compacted fill, peat, lacustrine, and alluvial deposits of insufficient age.

Pad Foundations

Pad foundations are used to support an individual point load such as that due to a structural column. They usually consist of a block or slab of uniform thickness, but they may be stepped if they support the load from a heavily loaded column. Pad foundations are usually shallow, but deep pad foundations can also be used.

Strip Foundations

Strip foundations are used to support a line of loads, either due to a load-bearing wall or if a line of columns need supporting where columns are so close that individual pad foundations would be impractical. The basic form of the bearing capacity equation, such as by Terzaghi, Meyerhof, etc., is defined for strip footings (i.e., for two-dimensional shear failure). All other footing shapes require the inclusion of shape factors in the individual terms of the bearing capacity equation.

Raft Foundations

Raft foundations are used to spread the load from a structure over a large area, typically the entire area of the structure. They are used when column loads or other structural loads are close together and individual pad foundations would interact. A raft foundation normally consists of a concrete slab, which extends over the entire loaded area. It may be stiffened by ribs or beams incorporated into the foundation.

Raft foundations have the advantage of reducing differential settlements as the concrete slab has adequate stiffness to minimize differential settlement between various slab positions. They are often needed on soft or loose soils with low-bearing capacity.

A fully compensated mat foundation is one for which the overburden stress (γD) of the soil removed is equal to the average soil pressure under the mat due to superstructure loads. Thus, for a fully compensated foundation, there is no net increase of soil pressure due to imposition of the superstructure loads.

General

The failure of a substructure can occur in three different ways: (1) shear failure of the soil due to excessive soil pressure, (2) excessive settlement of the foundation due to excessive soil pressure, or (3) structural failure of elements of the footing.

A spread footing simply spreads the load from a wall or column over an area such that the soil pressure is within some prescribed limit. For soils with low bearing capacity, it is more cost-effective to provide a mat foundation to support all walls and columns. If the soil is still too weak to support the structural loads, it may be necessary to transfer the loads to a stronger stratum at greater depth. This may be done using piles and drilled shaft foundations.

Ultimate Bearing Capacity

The ultimate bearing capacity (q_{ult}) is the value of bearing stress which causes a sudden catastrophic settlement of the foundation (general shear failure). By progressive refinement of upper and lower bound solutions (by considering various failure mechanisms and considering more stress regions), the exact solution can be approached.

Terzaghi's Bearing Capacity Theory

Terzaghi's bearing capacity equation predicts the ultimate bearing capacity of a soil medium experiencing a general bearing capacity failure under a strip footing ($L/B \to \infty$). The ultimate bearing capacity is given by

$$q_{ult} = cN_c + \gamma D_f N_q + 0.5\gamma B N_\gamma \qquad (206.1)$$

where c = cohesion
$\quad D_f$ = depth of (bottom of) footing
$\quad B$ = footing width

For a square foundation, this is modified to

$$q_{ult} = 1.3cN_c + \gamma D_f N_q + 0.4\gamma B N_\gamma \qquad (206.2)$$

and for a circular foundation, this is modified to

$$q_{ult} = 1.3cN_c + \gamma D_f N_q + 0.3\gamma BN_\gamma \qquad (206.3)$$

Terzaghi's bearing capacity factors N_c, N_q, and N_γ depend only on the soil friction angle ϕ, as given in Table 206.1.

Table 206.1 Terzaghi's Bearing Capacity Factors

ϕ	N_c	N_q	N_ϕ
0	5.71	1.00	0.00
2	6.30	1.22	0.04
4	6.97	1.49	0.10
6	7.73	1.81	0.20
8	8.60	2.21	0.35
10	9.60	2.69	0.56
12	10.76	3.29	0.85
14	12.11	4.02	1.26
16	13.68	4.92	1.82
18	15.52	6.04	2.59
20	17.69	7.44	3.64
22	20.27	9.19	5.09
24	23.36	11.40	7.08
26	27.09	14.21	9.84
28	31.61	17.81	13.70
30	37.16	22.46	19.13
32	44.04	28.52	26.87
34	52.64	36.50	38.04
36	63.53	47.16	54.36
38	77.50	61.55	78.61
40	95.66	81.27	115.31
42	119.67	108.75	171.99
44	151.95	147.74	261.60
46	196.22	204.19	407.11
48	258.29	287.85	650.67
50	347.51	415.15	1072.80

Example 206.1

What is the smallest isolated square footing to carry a column load of 90 kips? The depth of the footing is 3 ft and the unit weight of the sandy soil ($\phi = 29°$) is 120 pcf. Minimum factor of safety based on ultimate bearing capacity $= 3.0$.

Solution Terzaghi's bearing capacity factors for $\phi = 29°$ are $N_c = 34.4, N_q = 20.1, N_\gamma = 16.4$. According to Terzaghi's theory, the ultimate bearing capacity under a square foundation is given by

$$q_{ult} = 1.3cN_c + \gamma D_f N_q + 0.4\gamma BN_\gamma = 0 + 120 \times 3 \times 20.1 + 0.4 \times 120 \times B \times 16.4$$
$$= 7236 + 787B$$

The allowable soil pressure is $q_{all} = q_{ult}/3 = 2412 + 262B$. The design equation is then

$$\frac{90,000}{B^2} + 120 \times 3 \leq 2412 + 262B \Rightarrow 2052 + 262B \geq \frac{90,000}{B^2}$$

By trial and error, $B = 5.145$ ft $= 5$ ft 2 in.

 NOTE Given the multiple choice format of the PE exam, substituting values of footing size B from the four answer choices is probably more time efficient than trying to solve this cubic equation.

Local Shear Failure

For soils that exhibit local shear failure, Terzaghi's bearing capacity equation can be modified to

Strip foundation: $\quad q_{ult} = 0.667cN_c' + \gamma D_f N_q' + 0.5\gamma BN_\gamma'$ (206.4)

Square foundation: $\quad q_{ult} = 0.867cN_c' + \gamma D_f N_q' + 0.4\gamma BN_\gamma'$ (206.5)

Circular foundation: $\quad q_{ult} = 0.867cN_c' + \gamma D_f N_q' + 0.3\gamma BN_\gamma'$ (206.6)

where the modified bearing capacity factors N_c', N_q', and N_γ' are obtained by replacing the soil friction angle ϕ with

$$\phi' = \tan^{-1}\left(\frac{2}{3}\tan\phi\right) \tag{206.7}$$

General Bearing Capacity Equation

Whereas Terzaghi's bearing capacity equation can be applied to strip, square, and circular footings, it cannot be directly applied to rectangular footings. The generalized expression for ultimate bearing capacity may be written as

$$q_{ult} = cN_c\lambda_{cd}\lambda_{cs}\lambda_{ci} + \gamma D_f N_q\lambda_{qd}\lambda_{qs}\lambda_{qi} + \frac{1}{2}\gamma BN_\gamma\lambda_{\gamma d}\lambda_{\gamma s}\lambda_{\gamma i} \tag{206.8}$$

Terzaghi's analysis was based on an active wedge with angle ϕ' whereas more recent studies have suggested that this angle is closer to $(45 + \phi'/2)$. Equations (206.9) through (206.11) are based on the latter assumption. Values of the general bearing capacity factors listed in Table 206.2 are based on these equations.

$$N_q = e^{\pi \tan \phi'} \tan^2 \left(45 + \frac{\phi'}{2} \right) \tag{206.9}$$

$$N_c = \frac{N_q - 1}{\tan \phi'} \tag{206.10}$$

$$N_\gamma = 2(N_q + 1) \tan \phi' \tag{206.11}$$

Table 206.2 General Bearing Capacity Factors

ϕ	N_c	N_q	N_ϕ
0	5.14	1.00	0.00
2	5.63	1.20	0.15
4	6.19	1.43	0.34
6	6.81	1.72	0.57
8	7.53	2.06	0.86
10	8.34	2.47	1.22
12	9.28	2.97	1.69
14	10.37	3.59	2.29
16	11.63	4.34	3.06
18	13.10	5.26	4.07
20	14.83	6.40	5.39
22	16.88	7.82	7.13
24	19.32	9.60	9.44
26	22.25	11.85	12.54
28	25.80	14.72	16.72
30	30.14	18.40	22.40
32	35.49	23.18	30.21
34	42.16	29.44	41.06
36	50.59	37.75	56.31
38	61.35	48.93	78.02
40	75.31	64.20	109.41
42	93.71	85.37	155.54
44	118.37	115.31	224.63
46	152.10	158.50	330.34
48	199.26	222.30	496.00
50	266.88	319.06	762.86

In many texts and references, the relationship for N_γ may be different from that in Eq. (206.11). The reason is that, there is still some controversy about the variation of N_γ with the soil friction angle ϕ.

The bearing capacity factors N_c, N_q, and N_γ are functions of the angle of internal friction ϕ. The values listed in Table 206.2 are for strip (wall) footings. Each parameter must be multiplied by a shape factor, depth factor, and load inclination factor. In Eq. (206.8), λ_{cs}, λ_{qs}, and $\lambda_{\gamma s}$ are shape correction factors (which approach 1.0 as the aspect ratio of the footing B/L becomes small); λ_{cd}, λ_{qd}, and $\lambda_{\gamma d}$ are depth correction factors (which approach 1.0 as the depth of the footing D_f becomes small); and λ_{ci}, λ_{qi}, and $\lambda_{\gamma i}$ are load inclination correction factors (which approach 1.0 as the angle of inclination of the load becomes small).

The following sections Shape Correction Factors, Depth Correction Factors, and Load Inclination Correction Factor give shape, depth, and load inclination factors for the general bearing capacity equation.

Shape Correction Factors

For shallow footings which are not strip footings, i.e., those with a finite L/B ratio such as rectangular and circular footings, the bearing capacity factors need to be multiplied by shape correction factors outlined below. Note that there are alternate forms suggested by other researchers. According to De Beer (1970)[1] and Hansen (1970)[2], the depth correction factors are given by Eq. (206.12)

$$\lambda_{cs} = 1.0 + \frac{B}{L}\frac{N_q}{N_c} \qquad (206.12a)$$

$$\lambda_{qs} = 1.0 + \frac{B}{L}\tan\phi \qquad (206.12b)$$

$$\lambda_{\gamma s} = 1.0 - 0.4\frac{B}{L} \qquad (206.12c)$$

Depth Correction Factors

As shallow footings get deeper, the bearing capacity mobilized from the supporting soils increases. This is reflected by multiplying the bearing capacity factors by depth correction factors as given by Hansen (1970).

[1]De Beer, E. E. (1970), "Experimental Determination of the Shape Factors and Bearing Capacity Factors of Sand," *Geotechnique*, Vol. 20, No. 4, 387–411.

[2]Hansen, J. B. (1970), "A Revised and Extended Formula for Bearing Capacity," Danish Geotechnical Institute, Bulletin 28, Copenhagen.

For $D_f/B \le 1$

$$\lambda_{cd} = 1.0 + 0.4 \frac{D_f}{B} \qquad (206.13a)$$

$$\lambda_{qd} = 1.0 + 2 \tan \phi (1 - \sin \phi)^2 \frac{D_f}{B} \qquad (206.13b)$$

$$\lambda_{\gamma d} = 1.0 \qquad (206.13c)$$

For $D_f/B > 1$

$$\lambda_{cd} = 1.0 + 0.4 \tan^{-1} \left(\frac{D_f}{B} \right) \qquad (206.14a)$$

$$\lambda_{qd} = 1.0 + 2 \tan \phi (1 - \sin \phi)^2 \tan^{-1} \left(\frac{D_f}{B} \right) \qquad (206.14b)$$

$$\lambda_{\gamma d} = 1.0 \qquad (206.14c)$$

Load Inclination Correction Factors

When the load on the footing is inclined at angle α (degrees) with respect to the vertical, the bearing capacity factors are multiplied by the correction factors outlined below:

$$\lambda_{ci} = \lambda_{qi} = \left(1 - \frac{\alpha}{90} \right)^2 \qquad (206.15a)$$

$$\lambda_{\gamma i} = \left(1 - \frac{\alpha}{\phi} \right)^2 \qquad (206.15b)$$

Factor of Safety for Bearing Capacity

There is some divergence on the application of the factor of safety to the ultimate bearing capacity. The gross allowable load bearing capacity of a footing is determined by dividing the gross ultimate bearing capacity by a factor of safety.

$$q_{all} = \frac{q_{ult}}{FS} \qquad (206.16)$$

On the other hand, sometimes the net allowable stress increase on the soil is seen as the allowable stress that may be imposed upon the soil in addition to the previously existing overburden load due to the soil prism above the bottom of the foundation. This may be written as

$$(q_{all})_{net} = \frac{q_{ult} - \gamma D_f}{FS} \tag{206.17}$$

The factor of safety appropriate for a particular design depends on many factors such as load history, relative density, live-load to dead-load ratio, compaction and rapidity of load application. Typically, a factor of safety of about 1.5 may be desired with respect to shear failure in addition to a minimum factor of safety of about 3 against gross or net ultimate bearing capacity.

Local Shear Failure

For shallow foundations ($D_f < B$), if the relative density R_D is less than about 70%, local or punching shear failure may result. In such cases, the bearing capacity is to be calculated by replacing the friction angle ϕ with an effective friction angle ϕ' given by

$$\phi' = \tan^{-1}\left[(0.67 + R_D - 0.75R_D^2)\tan\phi\right] \quad \text{for} \quad 0 \le R_D \le 0.67 \tag{206.18}$$

Thus, for soils with very low relative density ($R_D \to 0$), the effective angle of internal friction is given by

$$\tan\phi' = 0.67\tan\phi \quad \text{for} \quad R_D = 0 \tag{206.19}$$

Dynamic Loads

When load is applied rapidly to a footing, the ultimate bearing capacity of shallow foundations in sand can be estimated by replacing the friction angle ϕ with ($\phi - 2°$). When foundations should be checked for local bearing capacity failure, the soil properties c (cohesion) and angle of internal friction ϕ are replaced as below. Note that this coincides with the case $R_D = 0$.

$$\bar{c} = \frac{2}{3}c$$

$$\tan\bar{\phi} = \frac{2}{3}\tan\phi \tag{206.20}$$

For foundations subjected to seismic loads, bearing capacity factors N_q and N_γ are modified to values lower than the corresponding values for static loading. According to Richards et al. (1993)[3], the modified values of the bearing capacity coefficients are a function of the seismic inertia angle θ

$$\tan\theta = \frac{k_h}{1 - k_v} \tag{206.21}$$

where k_h and k_v are the horizontal and vertical coefficients of ground acceleration due to an earthquake respectively.

Allowable Bearing Pressure in Sand Based on Settlement

The ultimate bearing capacity of a soil, as given by the Terzaghi-Meyerhof theories, is only mobilized when the settlement of the footing reaches about 10% of the width of the footing. Very often, this amount of settlement is well above the permitted settlement. The settlement criterion, as proposed by Bowles (1977)[4] as a modification to the work done by Meyerhof (1956)[5] is given by Eq. (206.22).

$$q_{net}(\text{ksf}) = \frac{N}{2.5}F_d S \qquad B \le 4' \tag{206.22a}$$

$$q_{net}(\text{ksf}) = \frac{N}{4}\left(\frac{B+1}{B}\right)^2 F_d S \qquad B > 4' \tag{206.22b}$$

where S = allowable settlement (in)
F_d = depth factor = $1 + 0.33D_f/B \le 1.33$
N = corrected standard penetration resistance = $C_N N_{field}$
C_N = overburden correction factor
B = width of the footing (ft)

---

[3]Richards, R., Elms, D. J., and Budhu, M. (1993), "Seismic Bearing Capacity and Settlement of Foundations," *Journal of Geotechnical Engineering*, American Society of Civil Engineers, Vol. 119, No. 4, 662–674.

[4]Bowles, J. E. (1977), *Foundation Analysis and Design*, 2nd ed. McGraw Hill, New York.

[5]Meyerhof, G. G. (1956), "Penetration Tests and Bearing Capacity of Cohesionless Soils," *Journal of the Soil Mechanics and Foundations Division*, American Society of Civil Engineers, Vol. 82, No. SM1, 1–19.

The net allowable soil pressure can also be correlated to the cone penetration resistance q_c, according to Meyerhof (1956). This calculated value of q_{net} (shown in Eq. [206.23]) corresponds to a settlement of 1 in.

$$q_{net} = \frac{q_c}{15} \qquad\qquad B \le 4' \qquad\qquad (206.23a)$$

$$q_{net} = \frac{q_c}{25}\left(\frac{B+1}{B}\right)^2 \qquad\qquad B > 4' \qquad\qquad (206.23b)$$

where q_c = cone penetration resistance (psf)
q_{net} = net allowable bearing pressure (psf)
B = width of the footing (ft)

Effect of Water Table on Bearing Capacity

The preceding formulations for bearing capacity were based on the assumption that the water table is at a significant depth below the bottom of the footing. However, in locations where the water table is shallow, one or more terms in the equation need to be modified. Two cases are discussed below—when the water table is above or below the bottom of the footing.

Case I—Water Table above Bottom of Footing $(0 \le D < D_f)$

When the water table is above the bottom of the footing (Fig. 206.1), the second and third term of the bearing capacity equation need to be modified to Eq. (206.24).

$$q_{ult} = cN_c + \left[\gamma(D_f - D) + (\gamma_{sat} - \gamma_w)D\right]N_q + \frac{1}{2}(\gamma_{sat} - \gamma_w)BN_\gamma \qquad (206.24)$$

Case II—Water Table at or below Bottom of Footing

When the water table is below the bottom of the footing (Fig. 206.2), the last term of the bearing capacity equation is modified by replacing the unit weight of the soil with an average value as shown in Eq. (206.25).

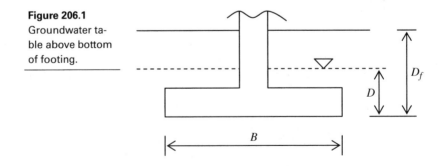

Figure 206.1 Groundwater table above bottom of footing.

Figure 206.2
Groundwater
table at or below
bottom of footing.

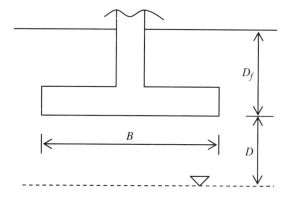

$$q_{ult} = cN_c + \gamma D_f N_q + \frac{1}{2}\gamma_{ave} BN_\gamma$$

$$\gamma_{ave} = \frac{\gamma D + (\gamma - \gamma_w)(B - D)}{B} \qquad D \le B \qquad (206.25)$$

$$= \gamma \qquad\qquad D > B$$

NOTE If the depth D (depth of water table below bottom of footing) is greater than the footing width B, no modification to the original bearing capacity equation is needed.

Coefficient of Subgrade Reaction

If a foundation of width B is supported by a soil and subjected to a uniform pressure q, it will undergo settlement Δ. The coefficient of subgrade reaction (k) is defined as

$$k = \frac{q}{\Delta} \qquad\qquad (206.26)$$

The value of k for a given soil is a function of length, width, and embedment depth of the foundation. In the field, load tests are carried out by loading square plates. The results are then extrapolated to large foundations.

Plate Load Test

The plate load test is conducted in the field to determine the bearing capacity of a soil. The ASTM test uses circular steel bearing plates 6 to 30 in diameter or 12 in $\times$ 12 in square plates. A pit is excavated to the depth of the foundation (D_f) and having a width at least four times the width of the bearing plate to be used. The bearing plate is loaded incrementally until catastrophic settlement indicates failure. The results from the plate bearing test can be extrapolated to the actual footing size using the following guidelines.

For Sandy Soils

The bearing capacity below the footing can be extrapolated to

$$(q_u)_f = (q_u)_p \frac{B_f}{B_p} \tag{206.27}$$

The settlement below the footing is given by

$$S_f = S_p \left(\frac{2B_f}{B_f + B_p}\right)^2 \tag{206.28}$$

The coefficient of subgrade reaction below the footing is given by

$$k_f = k_p \left(\frac{B_f + B_p}{2B_f}\right)^2 \tag{206.29}$$

For Clays

The bearing capacity below the footing is given by

$$(q_u)_f = (q_u)_p \tag{206.30}$$

The settlement below the footing is given by

$$S_f = S_p \frac{B_f}{B_p} \tag{206.31}$$

The coefficient of subgrade reaction below the footing is given by

$$k_f = k_p \left(\frac{B_p}{B_f}\right) \tag{206.32}$$

where $(q_u)_f$ = ultimate bearing capacity below the footing (size B_f)
$(q_u)_p$ = ultimate bearing capacity below the plate (size B_p)
S_f = settlement below the footing
S_p = settlement below the plate
k_f = coefficient of subgrade reaction for the soil underneath the footing
k_p = coefficient of subgrade reaction for the soil underneath the plate

For rectangular foundations having length L and width B ($B < L$), the coefficient of subgrade reaction is given by

$$k_{B \times L} = \frac{k_{B \times B}}{1.5}\left(1 + 0.5\frac{B}{L}\right) \tag{206.33}$$

Thus, for a strip footing ($B/L \to 0$), the coefficient of subgrade reaction is approximately 67% of that below a square footing in the same soil bed. Some typical values of the subgrade reaction (k), as determined from a 1 ft × 1 ft plate load test, are summarized below:

Dry or moist sand:

Loose: 30–90 lb/in³ (8–24 MN/m³)

Medium: 90–460 lb/in³ (24–125 MN/m³)

Dense: 460–1380 lb/in³ (125–375 MN/m³)

Saturated sand:

Loose: 40–55 lb/in³ (11–15 MN/m³)

Medium: 130–150 lb/in³ (35–41 MN/m³)

Dense: 475–550 lb/in³ (130–150 MN/m³)

Clay:

Stiff: 45–90 lb/in³ (12–24 MN/m³)

Very stiff: 90–180 lb/in³ (24–48 MN/m³)

Hard: > 180 lb/in³ (> 48 MN/m³)

The modulus of elasticity of a soil increases with depth. As a result, since increasing values of modulus of elasticity results in decreasing settlement, the value of the coefficient of subgrade reaction also increases with depth.

Combined Footing

When loads on individual columns are large enough, providing individual footings for them may cause them to encroach upon each other or get so close that providing a combined footing may be more economical. Such a combined footing for a pair of columns is shown in the diagram in Example 206.2. The combined footing can be sized to satisfy bearing capacity and settlement criteria in the same fashion as an isolated footing.

Example 206.2

A combined footing is to be designed for columns 1 and 2 as shown. Column loads are shown in the table below:

1. What must be the minimum length of the footing to avoid uplift under the footing?

2. What must be the length of the footing to produce uniform pressure under it?

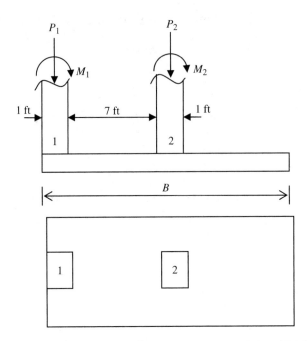

Column	P_{DL} (kips)	P_{LL} (kips)	M_{DL} (kip-ft)	M_{LL} (kip-ft)
1	50	105	80	90
2	70	85	50	100

Solution   Taking moments about the centerline of column 1, the effective location of the resultant is calculated as

$$\bar{x} = \frac{170 + 155 \times 8 + 150}{155 + 155} = 5.03\,\text{ft}$$

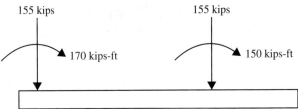

Thus, the resultant of the column loads acts at 5.03 ft from the centerline of column 1, i.e., at 5.53 ft from the left edge of the column.

1. To avoid uplift, resultant must be within the middle third of the footing base. Thus $2B/3 = 5.53$ ft. Therefore $B = 8.3$ ft.

2. To have uniform pressure under footing, resultant should be at footing center, i.e., $B/2 = 5.53$ ft. Therefore $B = 11.06$ ft.

Combined Footing—Design

There are four basic types of combined footings—rectangular combined footing, trapezoidal combined footing, cantilever or strap foundation, and mat foundation. Example 206.3 shows analysis and design of a rectangular combined footing for two columns. A trapezoidal footing is sometimes used in situations where space is limited. A cantilever footing uses a strap beam to connect an eccentrically loaded column to the foundation of an interior column. The strap beam does not provide vertical support for the column loads but serves to transfer some of the moment from the eccentrically loaded column to the adjacent interior column. A mat or raft foundation is a combined footing that covers a large area under a structure (sometimes supporting the entire structure). Mat foundations are preferred for soils having low bearing capacity.

Example 206.3

Design a rectangular combined footing for columns A and B with a center to center spacing of 15 ft. Loads are

Column A	DL = 50 kips	LL = 40 kips
Column B	DL = 100 kips	LL = 80 kips

Assume column A is 1 ft from the edge of the footing. Assume depth of footing = 3 ft and thickness of footing = 2 ft.

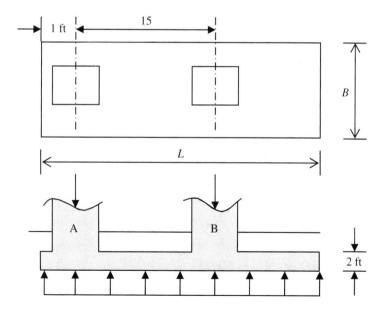

Solution

$$F_A = 50 + 40 = 90 \text{ kips} \qquad F_B = 100 + 80 = 180 \text{ kips}$$

Resultant of soil pressure = 270 kips. Taking moments about centerline of A

$$F_B \times 15 = 270 \left(\frac{L}{2} - 1 \right) \Rightarrow L = 22 \text{ ft}$$

Thus, soil pressure $q = 270 \text{ kips}/22 \text{ ft} = 12.27 \text{ kips/ft}$. If ultimate bearing capacity = 8000 psf and desired factor of safety = 3,

$$q_{\text{net}} = 8000 - 1 \times 120 - 2 \times 150 = 7580 \text{ psf}$$

$$q_{\text{all}} = 7580/3 = 2527 \text{ psf}$$

$$\frac{270{,}000}{LB} \leq 2527 \Rightarrow B \geq \frac{270{,}000}{2527 \times 22} = 4.86 \text{ ft}$$

Mat Foundations

The ultimate bearing capacity of a mat foundation is calculated using the same equation as used for isolated footings. Rafts constructed over sand typically have large factors of safety. For saturated clays with $\phi = 0$, $N_c = 5.14$, $N_q = 1$, and $N_\gamma = 0$, the ultimate bearing capacity under vertical loading is given by

$$q_{\text{ult}} = 5.14 \lambda_{cs} \lambda_{cd} c_u + \gamma D_f \qquad (206.34)$$

where c_u is the undrained cohesion. The shape correction factor is calculated as

$$\lambda_{cs} = 1 + 0.195 \frac{B}{L} \qquad (206.35)$$

The depth correction factor is calculated as

$$\lambda_{cd} = 1 + 0.4 \frac{D_f}{B} \qquad (206.36)$$

Therefore, the net allowable bearing capacity for the mat is given by

$$q_{\text{all}} = \frac{q_{\text{ult}} - \gamma D_f}{FS} = \frac{5.14 c_u}{FS} \left(1 + 0.195 \frac{B}{L} \right) \left(1 + 0.4 \frac{D_f}{B} \right) \qquad (206.37)$$

The net allowable bearing capacity for rafts over granular soils can be correlated to the standard penetration resistance.

$$q_{all} = 0.25N'\left(1 + 0.33\frac{D_f}{B}\right)S \le 0.33N'S \qquad (206.38)$$

where q_{all} = allowable bearing capacity (kips/ft²)
$\quad N'$ = corrected standard penetration resistance
$\quad S$ = settlement (in)

Differential Settlement of Mats

According to the ACI committee 336[6], the following criteria may be used to calculate the differential settlement under a mat foundation. If $K_r > 0.5$, the mat can be considered rigid, and the differential settlement is zero. If $K_r = 0.5$, the differential settlement is approximately 10% of the total settlement. If $K_r = 0$, the differential settlement is 35% of total settlement for square mats ($B/L = 1$) and 50% of total settlement for long foundations ($B/L \approx 0$).

Rigidity factor K_r (dimensionless), which expresses the rigidity of the structure relative to the soil, is defined as

$$K_r = \frac{E'I_b}{E_s B^3} \qquad (206.39)$$

where E' = modulus of elasticity of the material used for the structure
$\quad E_s$ = modulus of elasticity of the soil
$\quad B$ = width of the foundation
$\quad I_b$ = moment of inertia of structure per unit length transverse to the width B.

Compensated Foundations

The settlement of a mat foundation can be reduced by reducing the net pressure increase on the soil, achieved by increasing the depth of embedment (D_f). For a fully compensated foundation, the depth is chosen so that the net pressure increase is zero.

$$q_{net} = \frac{Q}{A} - \gamma D_f = 0 \Rightarrow D_f = \frac{Q}{A\gamma} \qquad (206.40)$$

---

[6]*ACI Committee 336: Committee on Footings, Mats, and Drilled Piers*, American Concrete Institute, Farmington Hills, Michigan.

Strap Footing—Design

The objective of a strap footing is to use a rigid strap to connect two footings so as to achieve equal soil pressure at both footings. The purpose of the strap is to act as a beam that transfers moments from the footing carrying the eccentric load to the other. The strap beam does not provide any bearing resistance.

Example 206.4

Design a strap footing for columns A and B separated by 22 ft. Ultimate bearing capacity = 8000 psf. Depth of each footing = 3 ft. Thickness of each footing = 2 ft. Loads are

Column A	DL = 50 kips	LL = 40 kips
Column B	DL = 100 kips	LL = 80 kips

Assume column A is 1 ft from the edge of the footing.

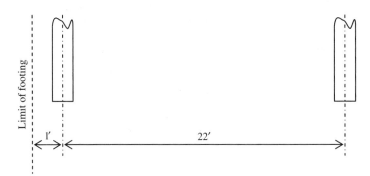

Solution   Preliminary sizes based on individual loads: $F_A = 90$ kips and $F_B = 180$ kips.

$$q_{all} = \frac{q_{ult} - \gamma D}{3} = \frac{8000 - 1 \times 120 - 2 \times 150}{3} = 2527 \text{ psf}$$

Required footing areas: $A_A = 90/2.527 = 35.6$ ft² and $A_B = 180/2.527 = 71.2$ ft². Let width of footings A and B be 6 ft and 8 ft, respectively. Eccentricity of column A = 3 − 1 = 2 ft.

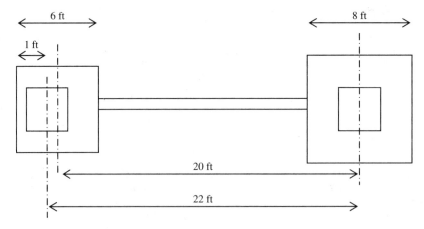

Moment on footing A $\qquad M_A = F_A \times e = 90 \times 2 = 180$ kips-ft

Additional column forces $\qquad V = M_A/d = 180/20 = 9$ kips

Modified footing loads $\qquad F_A = 90 + 9 = 99$ kips $\qquad F_B = 180 - 9 = 171$ kips

Footing A $\qquad$ Required area $= 99 \div 2.527 = 39.2$ ft^2 $\qquad$ Required width $= 6.53$ ft

Footing B $\qquad$ Required area $= 171 \div 2.527 = 67.7$ ft^2 $\qquad$ Required width $= 8.46$ ft

Use a 6.0×6.5 ft footing for column A and a 8.0×8.5 ft footing for column B.

Eccentric Load on a Shallow Footing

Consider a footing of width B as shown in Fig. 206.3. Superstructure loads consist of a vertical load P and a moment (about the footing centerline) M. This represents an eccentricity $= e$. Note that the loads P and M are for unit depth (perpendicular to plane of Fig. 206.3) of footing.

$$e = \frac{M}{P}$$

When the load is located within the kern (commonly referred to as the middle third of the footing), the entire width of the footing is effective (soil in compression). This occurs when the eccentricity of the load is less than one-sixth of the footing width (Fig. 206.4). The resulting soil pressure profile is trapezoidal (ranging from triangular, when the eccentricity is $B/6$ to rectangular, when the eccentricity is zero). The stresses under the footing are given by

$$\sigma_{max,min} = \frac{P}{B} \pm \frac{6M}{B^2} = \frac{P}{B} \pm \frac{6Pe}{B^2} \qquad (206.41)$$

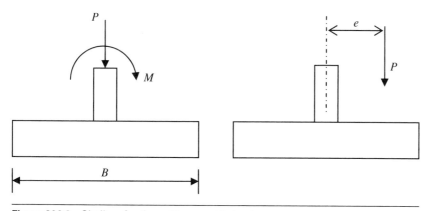

Figure 206.3 Shallow footing with eccentric load.

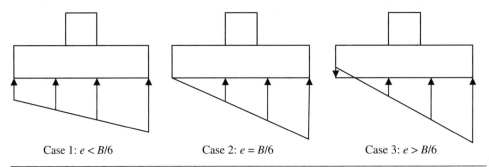

Case 1: $e < B/6$ Case 2: $e = B/6$ Case 3: $e > B/6$

Figure 206.4 Soil pressure predicted by general formula for eccentric load.

Figure 206.5
Soil pressure
under footing
experiencing
partial uplift.

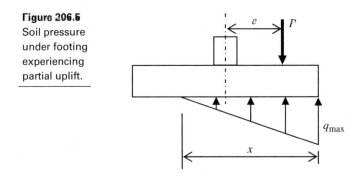

Cases 1 and 2 lead to acceptable diagrams, since the entire footing is in contact with the soil and the soil experiences compressive stresses. However, case 3 leads to an uplift condition on the left side of the footing, which cannot be sustained by the soil. Thus, the total vertical force P at eccentricity e must be resisted by a triangular stress distribution on a partial width of the footing (Fig. 206.5). Solving the above equations gives the following results for maximum soil pressure:

$$q_{max} = \frac{4P}{3(B - 2e)}$$ (206.42)

and effective width of footing (contact width between soil and footing) is given by

$$x = 1.5B - 3e$$ (206.43)

Shear in Footings—One-Way and Two-Way Shear

Footings carrying column or wall loads and supported by soil are subject to shear stresses developed on vertical planes. Wall footings are subject to wide-beam shear or one-way shear, as shown in Fig. 206.6. This type of shear can also occur for isolated rectangular footings.

Figure 206.6
One-way (wide beam) shear under rectangular footing.

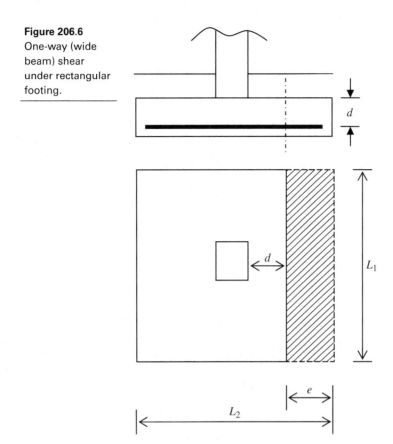

In Fig. 206.6, the plan dimensions of the footing are shown as L_1 and L_2. The shaded area is the area over which the upward soil pressure contributes to one-way shear. The critical plane that defines this area is located a distance d from the face of the column, where d is the effective depth of the footing. For a wall footing, the dimension L_2 is very large compared to L_1.

NOTE   The various design checks for foundations—bearing capacity, settlement, and overall stability—are carried out at service-load level. However, when performing structural design of various elements of a concrete footing, the provisions of the strength design method, as laid out by the ACI specifications[7], should be followed.

---

[7]ACI 318: *Building Code Requirements for Reinforced Concrete*, American Concrete Institute, Detroit, Michigan, 2014.

For a concentric column load, the uniform soil pressure under the footing is given by

$$q_u = \frac{P_u}{A_f}$$ (206.44)

where P_u = strength-level (factored) column load
A_f = area of the footing

The total shear force under the shaded area contributing to shear at the critical section

$$V_u = q_u e L_1$$ (206.45)

This shear is to be resisted by the vertical section at the critical section. The average transverse shear stress on this vertical section is

$$v_u = \frac{V_u}{L_1 d} = \frac{q_u e L_1}{L_1 d} = \frac{q_u e}{d}$$ (206.46)

This ultimate shear must be less than the design shear capacity of the concrete

$$v_u \le \phi v_c = 2\phi \sqrt{f_c'}$$ (206.47)

On the other hand, isolated square and rectangular footings may also be subject to punching or two-way shear as illustrated in Fig. 206.7. The critical section for punching shear is located at a distance $d/2$ from the face of the column. This forms a perimeter with sides b_1 and b_2 around the column as shown. The area of the vertical planes around this perimeter is given by

$$A_p = 2(b_1 + b_2)d$$ (206.48)

For a footing subject to an eccentric load (combination of P_u and M_u), the maximum shear stress on the critical perimeter is

$$v_u = \frac{P_u - R}{A_p} + \frac{\gamma_v M_u (0.5\, b_1)}{J}$$ (206.49)

where

$$R = P_u b_1 b_2 / A_f$$ (206.50)

$$\gamma_v = 1 - \frac{1}{1 + \dfrac{2}{3}\sqrt{\dfrac{b_1}{b_2}}} = \frac{2\sqrt{b_1}}{2\sqrt{b_1} + 3\sqrt{b_2}}$$ (206.51)

The constant J is given by

$$J = \frac{db_1^3}{6}\left[1 + \left(\frac{d}{b_1}\right)^2 + 3\left(\frac{b_2}{b_1}\right)\right]$$ (206.52)

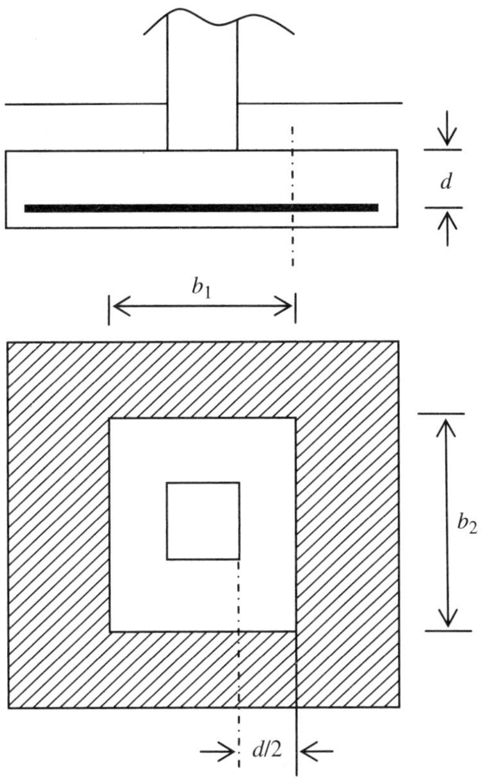

Figure 206.7
Two-way (punch-
ing) shear under
rectangular foot-
ing.

This ultimate shear must be less than the design shear capacity of the concrete in punching shear

$$v_u \leq \phi v_c = \phi(2 + y)\sqrt{f_c'} \qquad (206.53)$$

where the parameter y is given by

$$y = \min \left\{ 2, \frac{4}{\beta}, \frac{40d}{b_0} \right\} \qquad (206.54)$$

Elastic Settlement under Shallow Foundations

Figure 206.8 shows typical profile of settlement experienced by rigid and flexible shallow foundations. The solid line shows the settlement below a rigid footing. The dashed line shows the settlement below a flexible footing. Elastic settlement under a uniformly loaded

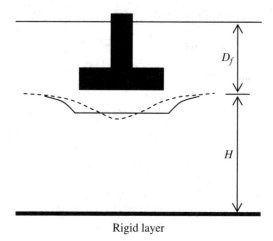

Figure 206.8
Elastic settlement under shallow footings.

Rigid layer

flexible rectangular area on an infinitely thick ($H \rightarrow \infty$) homogeneous elastic soil deposit is given by

$$S_e = \frac{q_o B}{E_s}(1 - \mu^2)\alpha \tag{206.55}$$

where B = least dimension of the loaded area
q_o = pressure on the loaded area
E_s = soil modulus of elasticity
μ = Poisson's ratio of soil
m = aspect ratio = length/width ratio of foundation ($m > 1$)
α = influence factor which is a function of the aspect ratio of the loaded area and the point under which settlement is to be calculated (center or corner). These values are given in Table 206.3.

Table 206.3   Influence Factor α for Elastic Settlement under Shallow Footings

| Shape | Aspect ratio m | Flexible footing | | | Rigid footing |
		Center	Corner	Average	
Circular	–	1.00	–	0.85	0.88
Rectangular	1.0	1.12	0.56	0.95	0.78
	1.5	1.36	0.68	1.20	1.08
	2.0	1.53	0.76	1.30	1.24
	3.0	1.78	0.89	1.52	1.42
	5.0	2.13	1.06	1.82	1.70
	10.0	2.56	1.28	2.25	2.10

Settlement of Granular Soils—Schmertmann and Hartman

The total settlement under a footing situated in granular (sandy) soil is calculated as the sum of contributions from all layers which experience significant influence from the load. The settlement S_e, calculated by a simple extension of Hooke's law of elasticity, is given by

$$S_e = C_1 C_2 (q_0 - \gamma D_f) \sum \frac{I_z}{E_s} \Delta z \qquad (206.56)$$

where C_1 = depth correction factor = $1 - 0.5 \gamma D_f / (q_0 - \gamma D_f)$
  C_2 = correction factor to account for creep in soil = $1.2 + 0.2 \log_{10}$ (time in years)
  q_0 = soil pressure at the base of the footing
  D_f = depth of footing
  I_z = strain influence factor at center of layer i (see Fig. 206.9)
  E_s = modulus of elasticity of layer i
  Δz = thickness of layer i

Strain influence factor I_z is shown below as a function of depth for square, circular, and long rectangular footings.

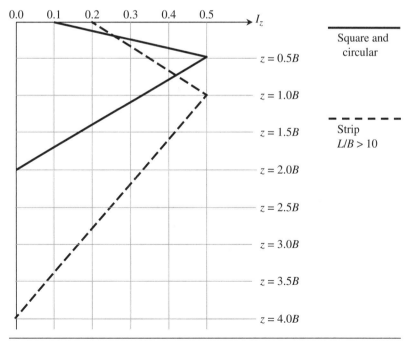

Figure 206.9 Strain influence factor (Schmertmann and Hartmann).

Deep Foundations

Deep Foundations

Deep foundations are those founded so deep below the ground surface that their bearing capacity is not affected by surface conditions, typically at depths greater than 10 ft below finished ground level. They include piles, piers, and caissons or compensated foundations using deep basements and also deep pad or strip foundations. Deep foundations can be used to transfer the loading to a deeper, more competent stratum at depth if unsuitable soils are present near the surface.

Piles are relatively long, slender members that transmit foundation loads through soil strata of low bearing capacity to deeper soil or rock strata having a high bearing capacity. In addition to supporting structures, piles are also used to anchor structures against uplift forces and to assist structures in resisting lateral and overturning forces. Piles get support from both end bearing and skin friction.

Piers are foundations for carrying a heavy structural load which is constructed *in situ* in a deep excavation.

Caissons are a form of deep foundation that are constructed above ground level, then sunk to the required level by excavating or dredging material from within the caisson.

End-bearing piles are those which terminate in hard, relatively impenetrable material such as rock or very dense sand and gravel. They derive most of their carrying capacity from the resistance of the stratum at the toe of the pile.

Friction piles obtain a greater part of their carrying capacity by skin friction or adhesion. This tends to occur when piles do not reach an impenetrable stratum but are driven for some distance into a penetrable soil. Their carrying capacity is derived partly from end bearing and partly from skin friction between the embedded surface of the soil and the surrounding soil.

Settlement-reducing piles are usually incorporated beneath the central part of a raft foundation to reduce differential settlement to an acceptable level. Such piles act to reinforce the soil beneath the raft and help prevent 'dishing' of the raft in the center.

Tension piles: Structures such as tall chimneys and transmission towers can be subject to large overturning moments and so piles are often used to resist the resulting uplift forces at the foundations. In such cases, the resulting forces are transmitted to the soil along the embedded length of the pile. In the design of tension piles, the effect of radial contraction of the pile must be taken into account as this can cause about a 10 to 20% reduction in shaft resistance.

Laterally loaded piles: Almost all piled foundations are subjected to at least some degree of horizontal loading. These lateral loads are generally small compared with the applied vertical axial loading. However, if the horizontal force is relatively large (such as for wharves and jetties subject to impact from ships), then they may prove critical in design.

Batter piles: A vertical pile has some capacity to resist loads applied normally to the axis, although this lateral capacity is much smaller than the axial capacity. If this lateral capacity is sufficient, it is unnecessary to use "raked" or "battered" piles, which are more expensive to install. If needed, battered piles are installed at an angle to the vertical, providing sufficient horizontal resistance by virtue of the horizontal component of axial capacity of the pile.

Negative skip friction: Piles that pass through layers of moderately to poorly compacted fill will be affected by *negative skip friction*, which produces a downward drag along the pile shaft and therefore an additional load on the pile. This occurs as the fill consolidates under its own weight.

Displacement piles cause the soil to be displaced radially as well as vertically as the pile shaft is driven or jacked into the ground. With nondisplacement piles (or replacement piles), soil is removed and the resulting hole filled with concrete or a precast concrete pile is dropped into the hole and grouted in.

Sands and granular soils tend to be compacted by the displacement process, whereas clays tend to heave. Displacement piles themselves can be classified into different types, depending on how they are constructed and how they are inserted. Displacement piles can be totally preformed or driven and cast-in-place. The latter type of pile can be uncased (a temporary steel tube with a closed end is driven into the ground to form a void in the soil, which is then filled with concrete as the tube is withdrawn) or cased (same as other except the steel tube is left in place to form a permanent casing).

Displacement piles are either driven or jacked into the ground. A number of different methods can be used. The *dropping weight or drop hammer* is the most commonly used method of insertion of displacement piles. Variants of the simple drop hammer are the *single-acting and double-acting hammers*. These are mechanically driven by steam, by compressed air, or hydraulically. In the single-acting hammer, the weight is raised by compressed air (or other means), which is then released and the weight allowed to drop. The double-acting hammer is the same except compressed air is also used on the down stroke of the hammer.

Vibratory methods can prove to be very effective in driving piles through cohesionless granular soils. The vibration of the pile excites the soil grains adjacent to the pile making the soil almost free flowing thus significantly reducing friction along the pile shaft. However, the large energy resulting from the vibrations can damage equipment; noise and vibration propagation can also result in the settlement of nearby buildings.

Jacked piles are most commonly used in underpinning existing structures. By excavating underneath a structure, short lengths of pile can be inserted and jacked into the ground using the underside of the existing structure as a reaction.

With *nondisplacement piles* soil is removed and the resulting hole is filled with concrete or sometimes a precast concrete pile is dropped into the hole and grouted in. Clays are especially suitable for this type of pile formation, as in clays the bore hole walls only require support close to the ground surface. When boring through more unstable ground, such as gravels, some form of casing or support such as a bentonite slurry may be required. Alternatively, grout or concrete can be intruded from an auger rotated into a granular soil. This method of construction produces an irregular interface between the pile shaft and surrounding soil, which affords good skin frictional resistance under subsequent loading.

Driven piles can be broadly divided into two types—low displacement piles (such as HP steel piles and open-ended steel pipe piles) and high displacement piles (such as timber and concrete piles and closed-ended pipe piles).

By using a heavy mandrel which is inserted into the pile before installation and driving the mandrel, the impact force is transmitted along the pile shaft, thus reducing the driving stresses experienced by the pile. Another solution is to use corrugated steel piles, which are probably the most common type of mandrel-driven pile in use. The shell is lifted into place in the leaders, the mandrel is inserted into the pile and locked into place, the mandrel and pile are driven into the ground, the mandrel is removed and the shell pile filled with concrete. To prevent soil from plugging the shell, a boot is frequently used at the pile toe.

Driven piles cannot be used economically in ground containing boulders or in clays when ground heave would be detrimental. Similarly, bored piles would not be suitable in loose water-bearing sand, and under-reamed bases cannot be used in cohesionless soils because they are susceptible to collapse before the concrete can be placed.

In coming to the final decision over the choice of pile, cost has considerable importance. The overall cost of installing piles includes the actual cost of the material, the time required for piling in the construction plan, test loading, the cost of the engineer to oversee installation and loading, and the cost of organization and overheads incurred between the time of initial site clearance and the time when construction of the superstructure can proceed.

Site Conditions

Pile foundations, typically composed of steel, concrete, or timber piles, are more expensive than shallow foundations but may still be more cost-effective if one or more of the following conditions exist at the site:

1. Upper soil layers are too weak and too compressible to support the loads from the superstructure.

2. Significant horizontal forces are transmitted to the substructure, as in the foundation for tall structures subject to lateral forces.

3. Soils such as loess which are expansive and collapsible exist to large depths. Piles extend past the "active zone" so as to be unaffected by the swell and shrinkage of these soils.

4. Significant uplift forces exist due to large overturning moments.

5. Possibility of soil erosion exists near the surface.

Materials

Steel piles may be either pipe piles or *H* piles. Pipe piles may be driven with ends open or closed. In most cases, the pipe is filled with concrete after driving. *H* piles are preferred to other rolled steel shapes because their flange and web thicknesses are equal.

Concrete piles are either precast piles or cast *in situ*. Precast piles may be reinforced with either ordinary reinforcement or prestressed using high-strength prestressing cables. Cast-in-place (CIP) concrete piles may be either uncased or cased. A steel casing is driven into the ground to the desired depth and then filled with concrete. For an uncased pile, the casing is withdrawn whereas it is left in place for a cased pile. Timber piles are typically used to maximum lengths of about 60 ft and capacity of about 30 tons.

The following summary lists design characteristics of various types of piles:

Timber piles are best suited as friction piles in granular soils. They can be used in the depth range of 30–60 ft. Design load of timber piles is typically in the 10–50 ton range. Initial cost of timber piles is low. They can suffer significant driving damage unless the pile tip is protected. If timber piles are intermittently submerged, they can be vulnerable to decay. However, permanently submerged timber piles are less susceptible to decay. A major factor that limits length of timber piles is the difficulty of splicing.

Steel *H* piles are best suited where they are end bearing on a rock stratum. They are commonly used in the 40–100 ft depth range. Design load of steel piles is in the 40–120 ton range. Steel piles are susceptible to corrosion. HP section steel piles can suffer damage or deflection during driving operations through soil with obstructions. Advantage of steel HP piles is that splicing is easy and effective.

Precast concrete piles can be of the nonprestressed or the prestressed variety. For the former, 40–50 ft pile lengths are typical, whereas prestressed concrete piles can be used for depths up to about 100 ft. Precast concrete piles can be specifically designed for a wide range of loads. These piles, if not prestressed, are susceptible to damage during handling. However, their capacity to carry large loads, corrosion resistance and tolerance to hard driving makes them a good choice in many situations.

Cast-in-place (CIP) concrete piles are typically used in the depth range of 50–90 ft. They are well suited as friction piles in granular soils for medium loads. CIP concrete piles can be specifically designed for a wide range of loads. These piles cause large soil displacement while driving and the thin shell can be susceptible to damage during driving.

Concrete-filled steel pipe piles can be used in the 40–120 ft depth range. They can have a design capacity of 80–120 tons without a core, but can achieve a capacity of 500–1500 tons with a core. They have high initial cost. If driven open ended, they cause very low soil displacement. They have large bending resistance and are therefore a good choice where there are significant lateral loads.

Auger-placed pressure injected concrete piles can be used for depth ranges of 30–60 ft. They typically have design capacity in the range 35–70 tons. Initial cost of these piles is very low. They function primarily as friction piles and create minimal disturbance to adjacent structures. They are not a good choice in highly compressible soils with a high organic content.

Pile Classification

Piles may be broadly classified into three categories according to the mechanism of load transfer to the soil—point bearing, friction, and compaction piles. If piles extend to bedrock, the load capacity of the pile consists entirely of the point bearing capacity (Q_p). On the other hand, if the pile extends into a hard stratum, the pile capacity is the sum of the point bearing capacity from the hard stratum plus skin friction resistance (Q_s) from all soil layers in contact with the pile.

$$Q_{ult} = Q_p + Q_s \tag{207.1}$$

If Q_p is small, then the pile derives most of its strength from the mechanism of side friction and may be categorized as a friction pile. When piles are driven in granular soils to achieve proper compaction of the soil, they are categorized as compaction piles. Piles are driven using drop hammers, single-acting hammers, double-acting hammers or diesel hammers. Single-acting hammers use air or steam to raise the hammer and gravity to drop it, while double-acting hammers use air or steam pressure for both operations. When a pile group is subjected to high lateral forces, some piles in the group are driven at an angle to the vertical. These are called *batter piles*. Based on the amount of soil displacement caused by the driving of the pile, they are also classified as displacement piles (concrete piles and closed-ended pipe piles) and nondisplacement piles (steel *H* piles, bored piles).

Point Bearing Capacity

In calculating the point bearing capacity (Q_p) of a pile, the Terzaghi-Meyerhof bearing capacity formulation [Eq. (206.1)] may be used. Of the three terms in this formula, the third term may be neglected, since the lateral dimension B is small compared to the depth D.

$$Q_p = A_p \left[cN_c + q'N_q + \frac{1}{2}\gamma BN_\gamma \right] \approx A_p \left[cN_c + q'N_q \right] \tag{207.2}$$

where A_p = area of pile tip
   c = cohesion of the soil at the pile tip
   q' = effective vertical stress at pile tip elevation
N_c, N_q = bearing capacity factors

Point Bearing Capacity for Piles in Cohesionless Soils

For soils such as clean sand, the cohesion is small enough to be neglected (cohesionless assumption). Without the cohesion term, the only surviving term is the "overburden" term. This term is given below. The term is assumed to grow linearly with depth up to a critical depth D_c and then remains constant.

$$Q_p \approx A_p q' N_q \leq A_p q_l \tag{207.3}$$

where the limiting value q_l (psf) is given by $q_l = 1000 N_q \tan \phi$.

The critical depth D_c is approximately equal to $10B$ for loose sands and $20B$ for dense sands. For piles driven to below the critical depth in sand, the bearing capacity is approximately constant and is given by

$$Q_p = 1000 A_p N_q \tan \phi \tag{207.4}$$

where Q_p = ultimate point bearing capacity (pounds)
   A_p = (projected) area at pile tip (square feet)

Table 207.1 Bearing Capacity Factor N_q

Angle of internal friction, ϕ	Bearing capacity factor N_q	
	Driven pile	Drilled pile
26	10	5
28	15	8
30	21	10
31	24	12
32	29	14
33	35	17
34	42	21
35	50	25
36	62	30
37	77	38
38	86	43
39	120	60
40	145	72

Source: NAVFAC DM 7.2, Foundations and Earth Structures, U.S. Department of the Navy, 1984.

According to Meyerhof (1976)[1], the ultimate tip bearing resistance can be correlated to the corrected SPT N value according to

$$q_p = 800 \frac{N'L}{B} \leq 8000 N' \tag{207.5}$$

where q_p = unit tip resistance (per square feet)
N' = corrected SPT N value
L = length of pile
B = diameter of pile

Table 207.1 shows recommended values of the bearing capacity factor N_q as a function of the friction angle ϕ. It is apparent that the N_q value for a drilled pile is about half than that for a driven pile. Since the N_q term dominates the point bearing capacity of a pile in a cohesion-less soil, it follows that a driven pile in sand will have approximately double the capacity of an otherwise identical drilled pile.

---

[1]Meyerhof, G. G. (1976), "Bearing Capacity and Settlement of Pile Foundations," *Journal of the Geotechnical Engineering Division*, American Society of Civil Engineers, Vol. 102, No. GT3, 197–228.

Point Bearing Capacity for Piles in Clay

For cohesive soils such as saturated clays in undrained conditions, it is common to assume that the angle of internal friction $\phi \approx 0$. This leads to the following values for the bearing capacity factors:

$$N_q \approx 1, N_c \approx 9$$

Use of these coefficients in the equation for tip capacity leads to

$$Q_p \approx A_p \left[c_u N_c + \gamma D \right] \approx A_p \left[9c_u + \gamma D \right] \tag{207.6}$$

where c_u is undrained cohesion of the soil below the pile tip.

The net load that can be transmitted to the pile is less than the ultimate load Q_p by the overburden load $A_p \gamma D$, which is the weight of the soil displaced by the pile and approximately compensated by the weight of the pile. This "net" ultimate load is given by

$$(Q_p)_{net} \approx 9A_p c_u \tag{207.7}$$

Point Bearing Capacity for Piles Resting on a Rock Layer

For a group of point bearing piles resting on rock, the group capacity may be taken as the sum of the individual pile capacities as long as the center to center spacing between piles is at least $D + 1$ ft. The ultimate point bearing resistance of piles resting on rock is given by

$$q_p = q_u \sec^2(45 + \phi/2) = \frac{2q_u}{1 - \sin\phi} \tag{207.8}$$

where q_u = unconfined compression strength of the rock
ϕ = drained angle of friction of the rock

If the unconfined compression strength of the rock is determined from small lab specimens, it must be reduced by a factor of safety (about 4 to 5) to represent the scale effect—a phenomenon where the presence of randomly distributed small fractures over a larger area causes the strength to decrease. Also, see Rock Quality Disignation (RQD) in Chap. 202 for correlations between RQD and allowable bearing capacity on rock.

Side Friction Capacity

For piles in sand, the side friction capacity of a pile (Q_s) derives mostly from the friction mobilized between the soil and the surface of the pile, which in turn is proportional to the vertical stress in the soil. On the other hand, for cohesive soils, the side friction derives mostly from the "stickiness" of the soil, which is practically absent for cohesionless soils such as sands. For the side friction to be fully mobilized, the pile tip must go through a displacement of 10% to 25% of the pile width.

The skin friction capacity may be calculated as the product of the unit shaft friction stress f_s and the contact surface area between pile and soil. If the pile is prismatic (constant cross section), then the surface area is equal to the product of the pile perimeter and the contact length.

Table 207.2 Friction Parameters between Various Wall Surfaces and Adjacent Soil

Wall material	Wall friction angle (δ)	Wall adhesion (c_a)
Concrete/brick	20°	If $c_a < 1$ kip/ft² (50 kPa), use $c_a = c_u$
Steel (tar coated)	30°	If sheeting does not penetrate to significant depth, use $c_a = 0$
Steel (uncoated)	15°	If $c_a > 1$ kip/ft² (50 kPa), use $c_a = 1$ kip/ft²

For a pile penetrating through several soil layers, the total side friction resistance is the sum of the resistances from each layer. Thus, the side friction capacity may be expressed as

$$Q_s = \sum p_i f_{s,i} L_i \tag{207.9}$$

where p_i = average perimeter in layer i
$f_{s,i}$ = unit shaft friction in layer i
L_i = contact length in layer i

Table 207.2 shows some typical values of wall friction angle (δ) and wall adhesion (c_a), where enough wall movement occurs to mobilize these values. Where the toe of the wall is founded on hard rock, the wall friction angle and adhesion values quoted in the table should be reduced by 50% for dense granular soils and stiff overconsolidated clays and adhesion should be ignored for loose granular soil.

For granular soils, for which the proximity of machinery or vehicular traffic causes vibrations that may be transmitted to the soil to the wall surface, neither wall friction nor adhesion should be included in calculation of design forces.

Skin Friction Coefficient

The determination of the skin friction coefficient (f_s), also known as unit shaft friction or side friction factor, can be made using one of several methods. Some of the more common methods are outlined below:

1. Value of unit shaft friction given directly. Each soil layer has a specified value of f_s, and the side friction resistance can be calculated based on this given value. The shaft friction factor f_s will have units of stress.

2. α **Method:** Each layer of cohesive soil has an empirical adhesion factor (α) specified. This value is dimensionless and is multiplied by the cohesion of the soil to obtain the unit shaft friction.

$$f_s = \alpha c_u \tag{207.10}$$

3. **β Method:** For piles driven into saturated clays, the unit shaft friction may be correlated to the vertical effective stress.

$$f_s = \beta\sigma'_v = K\tan\phi_R\sigma'_v \qquad (207.11)$$

For normally consolidated clays $K = 1 - \sin\phi_R$ \qquad\qquad (207.12)

For overconsolidated clays $K = (1 - \sin\phi_R)\sqrt{\mathrm{OCR}}$ \qquad\qquad (207.13)

The angle of friction ϕ_R is the drained friction angle of the clay in a remolded state.

4. **λ Method:** Applied mostly to clays, the λ method gives the unit shaft resistance developed by passive pressure of the soil displaced by pile driving. The unit shaft friction is given by

$$f_s = \lambda(\sigma'_v + 2c_u) \qquad (207.14)$$

where the parameter λ varies with embedment depth, c_u is the mean undrained shear strength and σ'_v is the average (over pile length) effective vertical overburden stress.

5. For sands, the unit side friction is due to the effective earth pressure between the pile and the adjacent soil. At some depth z, the unit side friction is given by

$$f_s = K\sigma'_v\tan\delta \qquad (207.15)$$

where K = lateral earth pressure coefficient. For steel piles, K ranges from 0.5 to 1.0; for concrete piles from 1.0 to 2.0 and for timber piles from 1.5 to 2.0
σ'_v = effective vertical stress a depth z
δ = friction angle between the pile and the soil. For steel piles, $\delta = 20$ degrees; for concrete piles, $\delta = 3/4\ \phi$; and for timber piles, $\delta = 2/3\ \phi$

Since the effective vertical stress increases with depth, so does the unit side friction. However, it is observed to remain unchanged for depths approximately greater than $15B$.

Lateral Earth Pressure Coefficient

For bored or jetted piles, the lateral earth pressure coefficient K may be taken as the at-rest coefficient $K_o = 1 - \sin\phi$. For low-displacement driven piles, the lateral force coefficient K may be taken within the range K_o to $1.4K_o$. For high-displacement driven piles, the lateral force coefficient K may be taken within the range K_o to $1.8K_o$.

Capacity of Pile Groups

A pile group is composed of several piles in (most often) a rectangular arrangement and connected at the top by a pile cap (Fig. 207.1). The superstructure transfers its loads to the pile cap. When the piles are close together, their zones of influence overlap, resulting in a group capacity less than the sum of the individual capacities. Typically, the center to center spacing

Figure 207.1
Pile group with
pile cap.

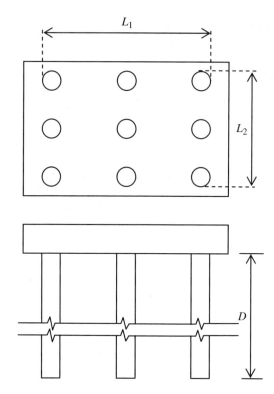

between piles in a group is approximately $3D$, where D is the pile diameter. The tip bearing capacity of the pile group may be calculated as

$$Q_p = 9c_u L_1 L_2 \qquad (207.16)$$

where L_1 and $L_2 =$ plan dimensions of the pile group (see Fig. 207.1)
$\qquad D =$ depth of embedment
$\qquad c_u =$ undrained cohesion of the soil at the pile tip

The side friction capacity is given by

$$Q_s = \sum 2c_u (L_1 + L_2)\Delta L \qquad (207.17)$$

where L_1 and $L_2 =$ plan dimensions of the pile cap
$\qquad \Delta L =$ depth of embedment in each soil layer
$\qquad c_u =$ average undrained cohesion of each layer

Pile group efficiency (η) is given by the ratio of pile group capacity to the sum of individual pile capacities.

$$\eta = \frac{Q_{\text{group}}}{\sum Q_i} \qquad (207.18)$$

NOTE Unlike other definitions of efficiency, the definition of efficiency η may result in a computed value greater than 100%. This simply means that the sum of individual pile capacities is less than the group capacity and therefore is the governing capacity of the pile group.

Special Consideration for Steel *H*-Section Piles

H-section steel piles are common. Hence, in the AISC steel specifications[2], these sections are designated as HP (*H*-pile) sections.

H Piles in Sand

When *H* piles in sand are loaded statically or pseudostatically, the soil does not form a plug at the tip of the pile. The shaft capacity is calculated using the total steel-soil interface, i.e., the total perimeter of the *H*-pile cross section. The base capacity is calculated using the actual pile cross-sectional area.

H Piles in Soft Clay

In soft clays, a plug forms between the flanges of an *H*-pile section. The shaft capacity is calculated using the outside (circumscribed) perimeter of the pile. The base capacity is calculated using the gross area of the pile-soil plug at the tip.

H Piles in Stiff Clay

In stiff clays, some research indicates that there might be significant separation of the adjacent soil from the pile. For this reason, the shaft friction is calculated from only the width of the flanges. The base capacity is calculated using the gross area of the pile-soil plug at the tip.

Example 207.1

What are the shaft perimeter and tip area for an HP12 × 84 pile in (a) soft clay, (b) stiff clay, and (c) sand?

Solution The HP12 × 84 has the following section properties: $d = 12.3$ in; $b_f = 12.3$ in; $t_f = t_w = 0.685$; $A = 24.6$ in^2.

a. Soft clay: Since a plug forms at the tip, the shaft perimeter is considered to be
$P = 2(d + b_f) = 2 \times (12.3 + 12.3) = 49.2$ in and the tip area is considered to be
$A = d \times b_f = 12.3 \times 12.3 = 151.3$ in^2.

---

[2]American Institute of Steel Construction. *Steel Construction Manual*, 13th edition, 2005.

b. Stiff clay: The shaft perimeter is considered to be $P = 2b_f = 2 \times 12.3 = 24.6$ in. The tip area is considered to be $A = d \times b_f = 12.3 \times 12.3 = 151.3$ in².

c. Sand: The shaft perimeter is considered to be $P = 4b_f + 2d - 2t_w = 72.4$ in. The tip area is considered to be cross-sectional area $A = 24.6$ in².

Pile Groups Subject to Overturning Moment

A pile group that has reasonably coherent behavior (piles demonstrating group action) can be analyzed as a vertical member subject to a combination of loads at the pile cap (superstructure loads). These loads may cause uniform compression (if concentric) or a combination of uniform compression (due to equal sharing of the concentric load) plus a bending-stress-type load pattern due to the moment created by the eccentricity of the load. As a result, some piles in the group will be subject to added compression and others will experience some relief (offset) of the uniform compression. If the lateral forces are strong enough, creating large overturning moments, the latter group may even be subject to a net uplift force. For such piles, this uplift must be resisted by the weight of the pile plus the side friction capacity. Typically, approximately 70% of the side friction capacity is assumed to be mobilized during uplift. For piles subject to compression, both point bearing and side friction should be included in determining the ultimate capacity. An illustration is provided in Example 207.2.

Example 207.2

Consider the building supported by two rows of piles as shown in the figure below. Each row of piles consists of five piles separated equally. The individual capacities of the piles are

Tip bearing capacity = 80 kips

Shaft side friction capacity = 40 kips

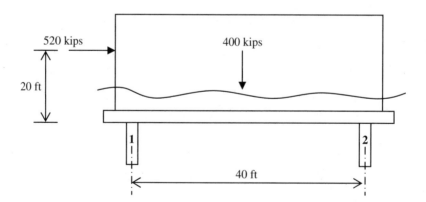

Determine the factor of safety of piles 1 and 2.

Solution The lateral force of 520 kips creates an overturning moment $= 520 \times 20 = 10{,}400$ kips-ft.

$$\text{The vertical reaction on pile row } 1 = \frac{400}{2} - \frac{520 \times 20}{40} = 200 - 260 = -60$$

$$\text{The vertical reaction on pile row } 2 = \frac{400}{2} + \frac{520 \times 20}{40} = 200 + 260 = 460$$

Thus, each pile in row 1 experiences an uplift of $60/5 = 12$ kips and each pile in row 2 experiences a compression force of $460/5 = 92$ kips.

Factor of safety for pile $1 = 40/12 = 3.33$ (note that the tip bearing capacity is neglected since the pile experiences pullout forces)

Factor of safety for pile $2 = 120/92 = 1.30$

Example 207.3 Load Distribution in Pile Groups

Consider a group of 15 piles connected by a pile cap as shown. The superstructure transmits the following loads to the pile cap: $P = 1000$ kips and $M = 1800$ kips-ft. Calculate the load in the worst loaded pile.

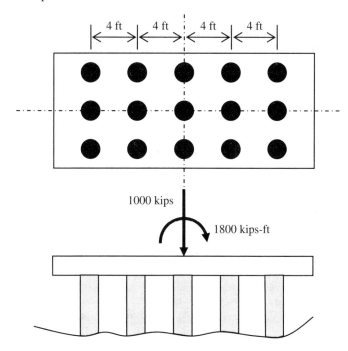

Solution The pile group resists the load by uniform division of load (compression) among the 15 piles plus a bending/overturning effect induced in each pile that causes added compression in the piles on one side (right side in this case) and some tension on the other side of the group centroid. Note that this case is somewhat distinct from the eccentrically loaded shallow footing. In that case, if the resultant stress (pressure) on the left is a tensile stress, the entire footing is "not effective" and there is loss of contact between soil and footing, resulting in uplift. In such a case, an alternate theory must be used that is consistent with such a "partial contact" geometry. In the case of the pile group, even if the resultant load on a pile is negative, the situation may be acceptable as long as the pile can resist such pullout forces. The stress in a pile is

$$\sigma = \frac{P}{15A_P} + \frac{M_y c}{I_y}$$

Moment of inertia of the pile group is calculated by considering pile areas to be lumped at their centers and ignoring their individual centroidal moments of inertia.

$$I = \sum A_P r^2 = 6 \times A_P \times 4^2 + 6 \times A_P \times 8^2 = 480 A_P$$

The maximum compressive load will occur on the easternmost pile ($c = 8$ ft) with axial stress given by

$$\sigma = \frac{1000}{15A_P} + \frac{1800 \times 8}{480A_P} = \frac{96.67}{A_P}$$

Therefore, axial load in worst loaded pile = 96.67 kips. The compressive stress on the piles on the left side of the pile group ($c = -8$ ft) is given by

$$\sigma = \frac{1000}{15A_P} - \frac{1800 \times 8}{480A_P} = \frac{36.67}{A_P}$$

Therefore, axial load in the piles in this line = 36.67 kips.

Caissons

Like piles, the function of caissons is to transmit superstructure loads through weak compressible soil or fill materials onto stiff or dense soil strata or rock at lower levels, in such a manner as to prevent excessive settlement, horizontal displacement or rotation of the supported structure at the caisson cap level. In addition, caissons are required to be founded at sufficient depth to prevent instability due to scour caused by major floods, when located in river environments.

Caissons for bridge foundations are usually cellular reinforced concrete structures comprising one or more excavation compartments, which are wholly or partly constructed at higher level and sunk in stages to the desired founding level, by internal excavation assisted by the application of kentledge.

Kentledge refers to concrete blocks or similar heavy loads repeatedly erected on top of the caisson walls in order to help overcome the frictional resistance of the soil surrounding the caisson during the process of sinking the caisson into the ground.

Small diameter concrete shafts comprising single open cells and constructed in the same manner as caissons are usually called cylinders. Because of their smaller size (usually up to about 8-ft diameter), cylinders lend themselves readily to precast concrete ring elements in their construction.

There are two principal types of caissons:

1. Open caissons may be "well type," with top and bottom open to the air, or "floating type," with open top and closed bottom.

2. Pneumatic caissons, in which during sinking operations the internal air pressure of the closed chamber is maintained by compressed air at greater than atmospheric pressure in order to prevent water from entering the caisson working area.

Caissons are unsuitable when the site has a high incidence of boulders because of the tendency of one side of the caisson to "catch" on boulders during sinking, with the danger of serious local damage to the cutting edge and the adjacent caisson wall.

Batter Piles

When the superstructure transfers lateral loads and moments to the pile group in addition to vertical loads, using vertical piles only would require very large capacity piles or make a very large footprint necessary. Using one or more rows of batter piles can solve this problem.

Figure 207.2 shows a pile cap with a horizontal load H, a vertical load V, and a moment M transferred from the superstructure. The pile cap thickness is D and the horizontal spacing at the top of the piles is B. The equilibrium equations are

$$\sum F_x = 0 \Rightarrow H - F_2 \frac{1}{\sqrt{1+m^2}} = 0 \Rightarrow F_2 = H\sqrt{1+m^2} \qquad (207.19)$$

$$\sum F_y = 0 \Rightarrow -V + F_1 + F_2 \frac{m}{\sqrt{1+m^2}} = 0 \Rightarrow F_1 = V - Hm \qquad (207.20)$$

$$\sum M_O = 0 \Rightarrow Vx_v - H(mB - D) + M = 0 \Rightarrow x_v = \frac{H(mB - D) - M}{V} \qquad (207.21)$$

The problem has nine variables: H, V, M, x_v, B, D, m, F_1, and F_2. Of these, the loads (H, V, and M) will be specified. Of the remaining six variables, if three are assumed, then the remaining three can be calculated using the three equilibrium equations. For example, if the allowable pile loads are substituted for F_1 and F_2 and the pile cap thickness is assumed, then one can solve for the geometric variables B, x_v, and m. These are required pile spacing (B), distance between the vertical load (V) and the vertical pile line, and the slope parameter (m).

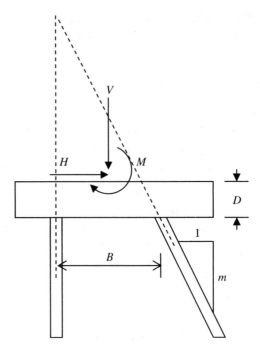

Figure 207.2
Batter piles sub-
ject to generalized
loading.

Laterally Loaded Long Piles

A pile subject to lateral forces undergoes bending, thus mobilizing passive forces in the adjacent soil. The resulting soil pressures depend on the relative stiffness of the pile and the soil. Figure 207.3 shows a long pile subject to a shear Q_g and moment

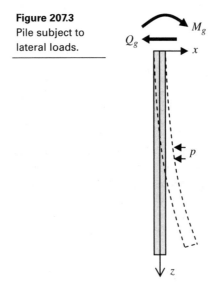

Figure 207.3
Pile subject to
lateral loads.

Mg transferred to it from the superstructure. The characteristic length of the soil-pile system is given by

$$T = \sqrt[5]{\frac{E_p I_p}{n_h}} \qquad (207.22)$$

where E_p = modulus of elasticity of the pile
I_p = moment of inertia of the pile
n_h = constant of the modulus of horizontal subgrade reaction, which is related to
k_z = modulus of horizontal subgrade reaction (at depth z) = $n_h z$

A pile is considered as a long or elastic pile if the length of the pile is more than five times the characteristic length T. When the ratio L/T is less than 2, the pile is considered a rigid pile. The dimensionless depth parameter Z is given by $Z = z/T$. According to Winkler's model, the differential equation governing bending of the pile is

$$E_p I_p \frac{d^4 x}{dz^4} + k_z x = 0 \qquad (207.23)$$

The lateral deflection of the pile at depth z is given by

$$x(z) = A_x \frac{Q_g T^3}{E_p I_p} + B_x \frac{M_g T^2}{E_p I_p} \qquad (207.24)$$

The rotation (slope with respect to the vertical pile axis) at depth z is given by

$$\theta(z) = A_\theta \frac{Q_g T^2}{E_p I_p} + B_\theta \frac{M_g T}{E_p I_p} \qquad (207.25)$$

The bending moment in pile at depth z is given by

$$M(z) - A_m Q_g T + B_m M_g \qquad (207.26)$$

The shear force in pile at depth z is given by

$$V(z) = A_v Q_g + B_v \frac{M_g}{T} \qquad (207.27)$$

Soil reaction against pile at depth z is given by

$$p'(z) = A_p \frac{Q_g}{T} + B_p \frac{M_g}{T^2} \qquad (207.28)$$

where the *A* and *B* coefficients are given in Table 207.3.

Table 207.3   Coefficients for Long Piles

Z	A_x	B_x	A_θ	B_θ	A_m	B_m	A_v	B_v	A_p	B_p
0.0	2.435	1.623	−1.623	−1.750	0.000	1.000	1.000	0.000	0.000	0.000
0.1	2.273	1.453	−1.618	−1.650	0.100	1.000	0.989	−0.007	−0.227	−0.145
0.2	2.112	1.293	−1.603	−1.550	0.198	0.999	0.956	−0.028	−0.422	−0.259
0.3	1.952	1.143	−1.578	−1.450	0.291	0.994	0.906	−0.058	−0.586	−0.343
0.4	1.796	1.003	−1.545	−1.351	0.379	0.987	0.840	−0.095	−0.718	−0.401
0.5	1.644	0.873	−1.503	−1.253	0.459	0.976	0.764	−0.137	−0.822	−0.436
0.6	1.496	0.752	−1.454	−1.156	0.532	0.960	0.677	−0.181	−0.897	−0.451
0.7	1.353	0.642	−1.397	−1.061	0.595	0.939	0.585	−0.226	−0.947	−0.449
0.8	1.216	0.540	−1.335	−0.968	0.649	0.914	0.489	−0.270	−0.973	−0.432
0.9	1.086	0.448	−1.268	−0.87	0.693	0.885	0.392	−0.312	−0.977	−0.403
1.0	0.962	0.364	−1.197	−0.792	0.727	0.852	0.295	−0.350	−0.962	−0.364
1.2	0.738	0.223	−1.047	−0.629	0.767	0.775	0.109	−0.414	−0.885	−0.268
1.4	0.544	0.112	−0.893	−0.482	0.772	0.688	−0.056	−0.456	−0.761	−0.157
1.6	0.381	0.029	−0.741	−0.354	0.746	0.594	−0.193	−0.477	−0.609	−0.047
1.8	0.247	−0.030	−0.596	−0.245	0.696	0.498	−0.298	−0.476	−0.445	0.054
2.0	0.142	−0.070	−0.464	−0.155	0.628	0.404	−0.371	−0.456	−0.283	0.140
3.0	−0.075	−0.089	−0.040	0.057	0.225	0.059	−0.349	−0.213	0.226	0.268
4.0	−0.050	−0.028	0.052	0.049	0.000	−0.042	−0.106	0.017	0.201	0.112
5.0	−0.009	0.000	0.025	−0.011	−0.033	−0.026	0.015	0.029	0.046	−0.002

From *Drilled Pier Foundations*, by R. J. Woodwood, W. S. Gardner, and D. M. Greer, Copyright 1972 by McGraw-Hill Book Company. Used with permission from McGraw-Hill Book Company.

Example 207.4

An 18-in-diameter reinforced concrete pile ($f_c' = 4500$ psi) is subject to a horizontal shear $Q_h = 20$ kips and a moment $M = 15$ kips-ft at the pile cap. Assume the constant of the modulus of subgrade reaction is 3680 lb/ft³. What is the lateral deflection of the pile at a depth of 10 ft?

Solution   28-day compressive strength $f_c' = 4500$ psi.

Modulus of elasticity $E_c = 57,000\sqrt{f_c'} = 3.824 \times 10^6$ psi.

Moment of inertia $I_c = \pi D^4/64 = 5.153 \times 10^3$ in⁴ $E_c I_c = 1.97 \times 10^{10}$ lb-in² $= 1.37 \times 10^5$ kips-ft².

$$n_h = 3680 \text{ lb/ft}^3 = 3.68 \text{ kips/ft}^3$$

Characteristic length of soil-pile system $T = \sqrt[5]{\dfrac{1.37 \times 10^5}{3.68}} = 8.2$ ft

$$Z = 10/8.2 = 1.22$$
$$A_x = 0.721 \qquad B_x = 0.213$$

The lateral deflection is calculated as

$$x = \frac{0.721 \times 20 \times 8.2^3}{1.37 \times 10^5} + \frac{0.213 \times 15 \times 8.2^2}{1.37 \times 10^5} = 0.06 \text{ ft} = 0.72 \text{ in}$$

Pullout Resistance

The net uplift capacity of a pile is given by the gross uplift capacity reduced by the effective weight of the pile.

$$T_{u,\text{net}} = T_{u,g} - W \qquad (207.29)$$

According to a study by Das and Seeley[3], the net uplift capacity of a pile embedded in saturated clay is given by

$$T_{u,\text{net}} = pL\alpha c_u \qquad (207.30)$$

where p = pile perimeter
L = pile length
α = adhesion coefficient at the soil-pile interface
c_u = undrained cohesion of the clay

Negative Skin Friction

Negative skin friction is a term used for a downdrag force created by one of several situations:

1. A clay fill overlies a granular soil layer. When a pile is driven through the clay fill, it will consolidate, thus exerting a downdrag force on the pile.

2. A granular fill of depth H overlies a layer of soft clay. The overburden causes additional consolidation of the clay layer, thus exerting a downdrag force on the pile.

3. If the water table is lowered, the effective stress on soil increases, thus inducing further consolidation settlement. If the pile is located in a clay layer, it will be subjected to a downdrag force.

If a clay fill (effective unit weight γ, fiction angle ϕ, depth H, and friction angle between pile and soil δ) is placed over a granular soil, the resultant downward drag force on a pile with perimeter p can be estimated as

$$Q_{dd} = \frac{1}{2} pK\gamma H^2 \tan\delta = \frac{1}{2} p(1 - \sin\phi)\gamma H^2 \tan\delta \qquad (207.31)$$

---

[3]Das, B. M. and Seeley, G .R. (1982), "Uplift Capacity of Pile Piles in Saturated Clay, Soils and Foundations," *The Japanese Society of Soil Mechanics and Foundation Engineering*, Vol. 22, No. 1, 91–94.

Settlement of Piles

The total settlement of a pile under load is due to the sum of (a) the elastic settlement of the pile, (b) the settlement due to the force at the pile tip, and (c) settlement of the pile due to load transmitted along the pile shaft.

Elastic Settlement

The elastic settlement is given by

$$s_E = \frac{(Q_p + \alpha Q_s)L}{A_p E_p}$$

(207.32)

where Q_p = pile tip load
Q_s = pile shaft load
L = pile length
A_p = pile cross-sectional area
E_p = modulus of elasticity of pile material

The constant α varies from 0.5 (if the unit side friction is constant or parabolic) to 0.67 (if the unit side friction is triangular).

Settlement Due to Tip Load

The settlement caused by the load carried at the tip of the pile can be given by

$$s_p = \frac{Q_p}{A_p}\left(\frac{D}{E_s}\right)(1-\mu^2)I$$

(207.33)

where μ = Poisson's ratio of the soil
E_s = modulus of elasticity of the soil
I = influence coefficient which is approximately equal to 0.85

Settlement Due to Shaft Load

The settlement caused by the load carried by the pile shaft is given by

$$s_F = \frac{Q_s}{pL}\left(\frac{D}{E_s}\right)(1-\mu^2)I$$

(207.34)

The influence coefficient I is given by (Vesic[4])

$$I = 2 + 0.35\sqrt{\frac{L}{D}}$$

(207.35)

---

[4]Vesic, A. S. (1977), "Design of Pile Foundations," *National Cooperative Highway Research Program Synthesis of Practice No. 42,* Transportation Research Board, Washington D.C.

Elastic Settlement of Pile Groups

The simplest model for the elastic settlement of a pile group is that due to Vesic, which states that

$$(s_e)_g = \sqrt{\frac{B}{D}} \times s \qquad (207.36)$$

where $(s_e)_g$ = settlement of the pile group
B = width of the pile group
D = diameter of each pile
s = elastic settlement of an individual pile at a comparable load

Consolidation Settlement of Pile Groups

The approximate procedure for estimating consolidation settlement of a pile group in clay is obtained by distributing the load on 2:1 slope, as shown in Fig. 207.4. The load is assumed to be transferred down from a plane, which is two-thirds of the length of the piles. Any compressible soil layers above this plane are assumed to not contribute significantly to the overall consolidation settlement. In Fig. 207.4, the consolidation settlement of the thickness L_1 of clay layer 2 and thickness L_2 of clay layer 3 will be calculated using the procedures outlined in Chap. 202.

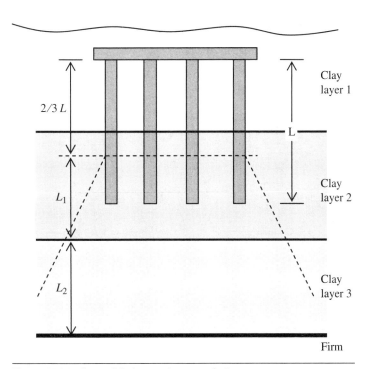

Figure 207.4 Consolidation settlement of pile groups.

Pile-Driving Formula

Driving a pile to a predetermined depth may not always develop the required strength due to variations in the soil profile. For this reason, several equations have been developed to calculate the capacity of a pile during driving. One of these equations is the modified ENR formula, given below:

$$Q_u = \frac{EW_R h}{S + C} \left(\frac{W_R + n^2 W_p}{W_R + W_p} \right) \tag{207.37}$$

where Q_u = ultimate load capacity of the pile
W_R = weight of the ram
W_p = weight of the pile
E = hammer efficiency (typical values: Diesel hammers 0.8 – 0.9; Single and double-acting hammers 0.70 – 0.85; drop hammers 0.7 – 0.9)
$C = 0.1$ in
S = penetration (inches) of pile per hammer blow $= 1/N$
N = number of hammer blows per inch of penetration
n = coefficient of restitution between the ram and the pile cap
h = height of fall of the ram

Retaining Walls

General

Soil accumulates vertical load due to gravity and because of the elasticity of the soil, the vertical pressure creates lateral (horizontal) earth pressure. In the absence of a structural element resisting these horizontal pressures, a mass of soil can become unstable and suffer catastrophic displacements. A naturally stable slope relies on inherent friction and cohesion to resist these lateral destabilizing forces, but lack of inadequate horizontal space often makes such mild slopes infeasible.

Retaining walls "hold back" earth where there is not enough space to provide a stable, gradual slope. By providing stability to the retained earth, retaining walls maximize space utilization by creating a near flat surface at grade level, impacting cost of primary construction as well as cost of secondary features such as drainage elements. Noise and pollution effects of traffic are also minimized by vertical separation between streets and urban residential and commercial development centers.

Retaining Wall Types

When a retaining structure (wall) is used to resist these horizontal forces, it is generally one of the following five categories (see Fig. 208.1):

1. Cantilever walls, which rely on bending strength of the wall stem to provide lateral resistance.

2. Anchored walls use cables or other stays anchored in the rock or soil behind it. Anchors are expanded at the end of the cable, either by mechanical means or often by injecting pressurized concrete (grout), which expands to form a bulb in the soil. Anchors must be placed behind the potential failure plane (Rankine) in the soil.

3. Mechanically stabilized earth (MSE) systems, where the lateral forces are transmitted to the soil using geogrids, which derive strength from the friction mobilized between the geogrid and the soil. The designer must take into account that the geomembranes

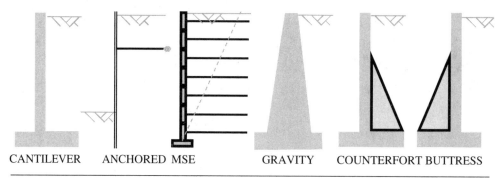

CANTILEVER   ANCHORED MSE   GRAVITY   COUNTERFORT BUTTRESS

Figure 208.1   Retaining wall types.

can only develop the required resistance through the length located behind the potential failure plane (Rankine) in the soil.

4. Gravity walls rely on simply the weight of the wall to have adequate stability against sliding and overturning against horizontal forces. Semigravity walls have a moderate amount of reinforcement.

5. Counterfort walls have ribs on the retained earth side and therefore develop lateral resistance via tension in these ribs. Buttressed walls have ribs (buttresses) on the side opposite the retained earth and therefore develop lateral resistance via compression.

Location geometry most often dictates the selection of a retaining wall system. In recent times, segmental retaining walls have gained favor overpoured-in-place concrete walls or treated-timber walls because of lower cost, easier installment, and better environmental implications. In many situations—especially cuts—MSE may not be the most appropriate wall type. Often the additional excavation and shoring required for installation of MSE walls in cut situations make them uneconomical and difficult to construct.

Typical Damage to Retaining Walls

Damage patterns commonly observed to retaining walls fall into the following categories:

1. Tilting—typically a tilt of about 10 degrees causes a 20% increase in lateral forces.

2. Shifting—usually caused by repeated cycles of freeze and thaw.

3. Bowing—excessive lateral forces behind the wall, usually due to water buildup.

4. Water damage such as green slime, efflorescence, mineral deposits, rust from wall tie.

Drainage behind the wall is critical to the performance of retaining walls. Drainage is typically provided by weep-holes or by French drains behind the wall. Properly designed drainage systems will reduce or eliminate the hydrostatic pressure and increase the stability of the fill material behind the wall. Clayey soil behind the wall can lead to poor draining conditions and cause additional lateral pressure. If not properly designed, such a wall can exhibit structural defects such as bulging and shifting. Installing a drainage membrane with drain tile on the backfilled side during retaining wall construction or during a repair of a retaining wall will help reduce or eliminate most of problems associated with a retaining wall.

Planning

Planning for a retaining wall project usually entails development of a site plan, collection of soils data, and formulation of a construction plan. The site plan is a detailed drawing of the site including wall location, length, relevant elevations, information on grading, layout of underground utilities, erosion control measures, and storm water management. Typically, the wall construction plan has the following elements—wall, toe and crest elevations, reinforcement location and length, soil conditions and parameters, drainage and other wall details, and wall construction specifications.

Recommendations for Design and Construction

Installing a drainage membrane during retaining wall construction will help prevent the effects of water seepage. Freeze and thaw cycles also create extra stress on a retaining wall. The culprit is again water. Unlike footings, which can be located below the frost line, walls usually cannot. So the main strategy is to keep water from accumulating in the backfill zone.

Poor compaction will cause a retaining wall to shift, creating large gaps within the backfill. The soil should be compacted once excavation takes place, again once the gravel base is placed and each time backfill is added. Backfill should be added after each course of the wall has been installed.

When walls are placed on freshly cut slopes, designers should be aware that soil tests conducted on samples may indicate strengths that are higher than long-term properties because freshly excavated material will soften with time. An assessment of long-term strengths must be made in order to get a correct picture of the stability of the wall.

Of particular concern are areas producing groundwater and areas experiencing slope failures during excavation. Each of these indicates potential stability problems and should be brought to the attention of the wall designer. It may be necessary to remove and replace poor soils, install drains, or modify the wall to address such field conditions.

Pressuremeter

A pressuremeter is a device constructed to measure at-rest horizontal earth pressure. It has a probe that is inserted into the borehole and supported at the test depth. The probe has an inflatable flexible membrane which applies even pressure to the walls of the borehole as it expands. As the pressure increases and the membrane expands, the walls of the borehole begin to deform. The pressure inside the probe is held constant for a specific period of time and the increase in volume required to maintain the pressure is recorded at an above-ground unit. There are two types of tests that can be performed with the pressuremeter—the stress-controlled test increases pressure in equal increments while the strain-controlled test increases the volume in equal increments.

The pressuremeter is used to test hard clays, dense sands, and weathered rock which cannot be tested with push equipment. There are three different types of pressuremeter. The borehole pressuremeter is the most common and has a probe that is inserted into a preformed hole (borehole). The self-boring pressuremeter has a probe that is self-bored into the ground to prevent disturbance. The third type, called a cone pressuremeter, has a cone shaped probe that is inserted into the base of the borehole, displacing the soil into the cone of the probe which causes less disturbance to the soil to get a more accurate reading.

Inclinometer

An inclinometer or clinometer is an instrument used for measuring angles of slope (or tilt), elevation, or depression of an object with respect to the direction of gravity. It is also known as a tilt indicator or gradient meter. Clinometers measure both inclines and declines. A tilt sensor can measure the tilting in two axes of a reference plane in two axes. Permanently installed tiltmeters are placed at sites of major earthwork prone structures such as dams to monitor the long-term stability of the structure.

Recommendations for Inspection and Maintenance

Inspection requirements for retaining walls vary by jurisdiction, but a typical code may include the following as situations requiring special inspection—(1) All retaining wall systems with unbalanced fill height greater than 8 ft. (2) All retaining wall system with unbalanced fill height between 4 ft and 8 ft with surcharge either from surcharge or a slope greater than 3H:1V.

Wall height is the measurement from the grade level in front of the wall to the grade level behind the wall. For a tiered wall system, this is the measurement from the grade level in front of the wall located at the lowest elevation to the grade level behind the wall located at the highest elevation.

Walls should be inspected periodically for backfill loss, loss of joint seals, and wall movement. Voids created by loss of backfill can become saturated and create excess pressure on the wall and therefore should be filed with suitable material as soon as it is observed.

Lateral Earth Pressure

For locations where there isn't adequate horizontal space for the provision of a stable slope, earth embankments may be retained using either gravity structures such as gabion walls, flexural structures such as cantilever retaining walls or reinforced earth structures such as mechanically stabilized earth walls. Such elements must be designed to resist the horizontal forces produced as a byproduct of the vertical stress existing within the soil. The effective horizontal stress is related to the effective vertical stress according to

$$\sigma'_h = K\sigma'_v \qquad (208.1)$$

The earth pressure (EP) coefficient (K) will be either the at-rest (K_o), active (K_a), or passive (K_p) coefficient. For surfaces that are not expected to move laterally (such as massive bridge abutments and walls which are adequately restrained), the at-rest earth pressure must be used. For loose sands, an empirical relationship for K_o is a function of the effective friction angle ϕ'

$$K_o = 1 - \sin\phi' \tag{208.2}$$

For dense sands, a better estimation of the at-rest earth pressure coefficient is obtained by

$$K_o = 1 - \sin\phi' + \left(\frac{\gamma_d}{\gamma_{d,\min}} - 1\right) \times 5.5 \tag{208.3}$$

where γ_d and $\gamma_{d,\min}$ are the soil's dry unit weight in situ and in the loosest state, respectively. For fine-grained normally consolidated soils, K_o may be estimated using

$$K_o = 0.44 + 0.42\text{PI} \tag{208.4}$$

where PI = LL − PL = plasticity index of the soil (%). For overconsolidated clays, the coefficient $K_{o,oc}$ may be related to the coefficient $K_{o,nc}$ (normally consolidated) using

$$K_{o,oc} = K_{o,nc} \times \sqrt{\text{OCR}} \tag{208.5}$$

The overconsolidation ratio (OCR) is given by

$$\text{OCR} = \frac{\sigma'_o}{\sigma'} \tag{208.6}$$

where σ'_o = effective preconsolidation pressure
       σ' = effective (vertical) pressure on the soil

Wall Movement Necessary to Develop Lateral Pressures

The amount of wall movement required to develop minimum (active) or maximum (passive) earth pressures depends on the stiffness of the soil and the height of the wall. For stiff soils like dense sands or heavily overconsolidated clays, the required movement is relatively small. For all sands of medium or high density, it can be assumed that the movement required to reach the minimum (active) earth pressure is no more than about 0.4% of the wall height and the movement required to increase the earth pressure to its maximum (passive) value is about 4.0% of the wall height. Thus, it takes about ten times the wall movement to develop full passive resistance as it takes to develop the full active forces. For loose sands, the movement required to reach the minimum active or the maximum passive pressures is somewhat larger.

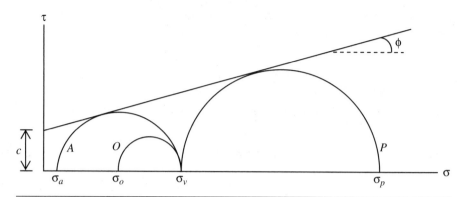

Figure 208.2   Relationship between active, at-rest, and passive pressure.

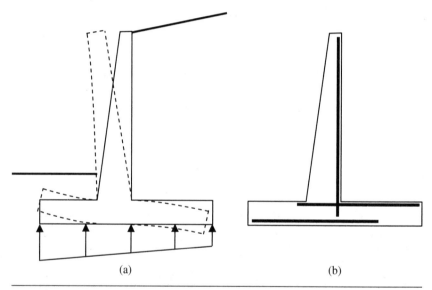

Figure 208.3   Flexure of a cantilever retaining wall.

If the failure envelope for the soil is known, the relation between vertical stress (σ_v) and at-rest horizontal stress (σ_o), and active horizontal stress (σ_a) and passive horizontal stress (σ_p) is shown in Fig. 208.2.

Stability and Strength Checks

The retaining wall must be checked for stability. This includes checks for sliding, overturning, and bearing capacity failure. Once the chosen wall configuration is shown to have adequate factors of safety for these criteria, all components of the wall must also be checked for adequate strength. The typical bending pattern for the various parts of a cantilever retaining wall is shown in Fig. 208.3. As a result, the typical reinforcement pattern for the various parts of the wall is shown in Fig. 208.3(*b*).

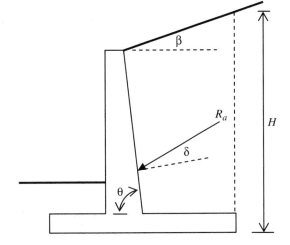

Figure 208.4
Active earth
pressure behind
cantilever
retaining wall.

It is to be noted that the stability and bearing capacity checks are performed at service load levels, i.e., without applying any load factors. However, the structural design of wall components must be performed using the appropriate design method. For example, for a reinforced concrete retaining wall, structural design of wall components must be according to provisions in ACI 318[1], which specifies the strength design approach, using factored loads.

Active Earth Pressure

Figure 208.4 shows a cantilever retaining wall with a sloping backfill (angle with horizontal $= \beta$) behind it. The back of the wall is considered rough (angle of friction between the wall and the backfill $= \delta$). The back of the wall makes an angle θ with the horizontal.

Under the effect of the lateral earth pressure, the stem of the wall rotates and moves away from the backfill. With adequate movement of the wall, the lateral earth pressure developed behind the wall reaches a minimum value called active earth pressure. The active earth pressure (σ_a) at a depth H below the surface is given by

$$\sigma_a = K_a \gamma H - 2c\sqrt{K_a} \qquad (208.7)$$

where K_a = active earth pressure coefficient
c = cohesion of the soil
γ = unit weight of soil

For a cohesive soil, at or near the surface ($H = 0$), the active earth pressure is negative. At these locations, the cohesion of the soil may cause the formation of cracks in the active soil zone.

---

[1]ACI 318: *Building Code Requirements for Structural Concrete*, American Concrete Institute, Detroit, Michigan, 2011.

NOTE For good draining characteristics, backfill soils are often granular soils whose cohesion is negligible.

The maximum depth to which these tension cracks may form can be solved by equating the earth pressure $\sigma_a = 0$ in Eq. (208.7). Coulomb's theory for active earth pressure includes the effect of friction between the backfill soil and the back surface of the wall. According to Coulomb, the active earth pressure coefficient is given by

$$K_a = \frac{\sin^2(\theta + \phi)}{\sin^2\theta \sin(\theta - \delta)\left[1 + \sqrt{\dfrac{\sin(\phi + \delta)\sin(\phi - \beta)}{\sin(\theta - \delta)\sin(\theta + \beta)}}\right]^2} \tag{208.8}$$

Passive Earth Pressure

When the earth retaining structure develops significant movement toward a soil mass, the lateral earth pressure reaches a maximum value called passive earth pressure. Coulomb's theory for passive earth pressure includes the effect of friction between the backfill soil and the back surface of the wall. The passive earth pressure (σ_p) at a depth H below the surface is given by

$$\sigma_p = K_p \gamma H + 2c\sqrt{K_p} \tag{208.9}$$

Coulomb's passive earth pressure coefficient is given by

$$K_p = \frac{\sin^2(\theta - \phi)}{\sin^2\theta \sin(\theta + \delta)\left[1 - \sqrt{\dfrac{\sin(\phi + \delta)\sin(\phi + \beta)}{\sin(\theta + \delta)\sin(\theta + \beta)}}\right]^2} \tag{208.10}$$

Rankine's Theory for Earth Pressure

Rankine's theory for earth pressure ignores the friction between the wall and the backfill ($\delta = 0$). This simplifies the expressions for active and passive earth pressure coefficient:

$$K_a = \frac{\sin^2(\theta + \phi)}{\sin^3\theta\left[1 + \sqrt{\dfrac{\sin\phi\sin(\phi - \beta)}{\sin\theta\sin(\theta + \beta)}}\right]^2} \tag{208.11}$$

$$K_p = \frac{\sin^2(\theta - \phi)}{\sin^3\theta\left[1 - \sqrt{\dfrac{\sin\phi\sin(\phi + \beta)}{\sin\theta\sin(\theta + \beta)}}\right]^2} \quad (208.12)$$

For a simplified situation—horizontal backfill ($\beta = 0$), negligible friction between wall and backfill ($\delta = 0$), and vertical wall stem ($\theta = 90°$)—these expressions become

$$K_a = \frac{1 - \sin\phi}{1 + \sin\phi} \quad (208.13)$$

$$K_p = \frac{1 + \sin\phi}{1 - \sin\phi} \quad (208.14)$$

Table 208.1 shows values of Rankine's earth pressure coefficient K_a for various combinations of angles β, ϕ, and θ. The Rankine theory neglects the friction behind the wall, which is

Table 208.1   Rankine's Active Earth Pressure Coefficient K_a

	Active lateral earth pressure coefficient K_a (Rankine)						
$\varepsilon =$	10°	15°	20°	25°	30°	35°	40°
$\beta = 0°$	0.704	0.589	0.490	0.406	0.333	0.271	0.217
$\beta = 10°$	0.970	0.704	0.569	0.462	0.374	0.300	0.238
$\theta = 90°$ $\quad \beta = 20°$	—	—	0.883	0.572	0.441	0.344	0.267
$\beta = 30°$	—	—	—	—	0.750	0.436	0.318
$\beta = \phi$	0.970	0.933	0.883	0.821	0.750	0.671	0.587
$\beta = 0°$	0.757	0.652	0.559	0.478	0.407	0.343	0.287
$\beta = 10°$	1.047	0.784	0.654	0.550	0.461	0.384	0.318
$\theta = 80°$ $\quad \beta = 20°$	—	—	1.015	0.684	0.548	0.444	0.360
$\beta = 30°$	—	—	—	—	0.925	0.566	0.433
$\beta = \phi$	1.047	1.039	1.015	0.977	0.925	0.860	0.785
$\beta = 0°$	0.833	0.735	0.648	0.569	0.498	0.434	0.375
$\beta = 10°$	1.169	0.895	0.767	0.662	0.572	0.492	0.421
$\theta = 70°$ $\quad \beta = 20°$	—	—	1.205	0.833	0.687	0.576	0.483
$\beta = 30°$	—	—	—	—	1.169	0.740	0.586
$\beta = \phi$	1.169	1.196	1.205	1.196	1.169	1.124	1.064
$\beta = 0°$	0.943	0.851	0.767	0.690	0.619	0.553	0.492
$\beta = 10°$	1.360	1.056	0.926	0.818	0.724	0.639	0.563
$\theta = 60°$ $\quad \beta = 20°$	—	—	1.493	1.047	0.885	0.761	0.657
$\beta = 30°$	—	—	—	—	1.540	0.993	0.808
$\beta = \phi$	1.360	1.436	1.493	1.528	1.540	1.528	1.493

Table 208.2 Coulomb's Active Earth Pressure Coefficient K_a for Vertical Stem ($\theta = 90$)

		Active lateral earth pressure coefficient K_a (Coulomb)						
	ε	10°	15°	20°	25°	30°	35°	40°
δ = 0°	β = 0°	0.704	0.589	0.490	0.406	0.333	0.271	0.217
	β = 10°	0.970	0.704	0.569	0.462	0.374	0.300	0.238
	β = 20°	—	—	0.883	0.572	0.441	0.344	0.267
	β = 30°	—	—	—	—	0.750	0.436	0.318
	β = φ	0.970	0.933	0.883	0.821	0.750	0.671	0.587
δ = 10°	β = 0°	0.635	0.533	0.447	0.373	0.308	0.253	0.204
	β = 10°	0.985	0.664	0.531	0.431	0.350	0.282	0.225
	β = 20°	—	—	0.897	0.549	0.420	0.326	0.254
	β = 30°	—	—	—	—	0.762	0.423	0.306
	β = φ	0.985	0.947	0.897	0.834	0.762	0.681	0.596
δ = 20°	β = 0°	0.607	0.508	0.427	0.357	0.297	0.245	0.199
	β = 10°	1.032	0.654	0.518	0.419	0.340	0.275	0.220
	β = 20°	—	—	0.940	0.547	0.414	0.322	0.250
	β = 30°	—	—	—	—	0.798	0.425	0.305
	β = φ	1.032	0.993	0.940	0.874	0.798	0.714	0.624
δ = 30°	β = 0°	0.606	0.506	0.424	0.356	0.297	0.246	0.201
	β = 10°	1.120	0.669	0.524	0.422	0.343	0.278	0.223
	β = 20°	—	—	1.020	0.565	0.424	0.328	0.256
	β = 30°	—	—	—	—	0.866	0.442	0.315
	β = φ	1.120	1.077	1.020	0.948	0.866	0.775	0.678
δ = 40°	β = 0°	0.631	0.523	0.438	0.368	0.308	0.256	0.210
	β = 10°	1.266	0.712	0.551	0.442	0.358	0.291	0.234
	β = 20°	—	—	1.153	0.605	0.449	0.347	0.270
	β = 30°	—	—	—	—	0.979	0.476	0.337
	β = φ	1.266	1.218	1.153	1.072	0.979	0.876	0.766

The first cluster (δ = 0) represents Rankine's theory.

usually negligible ($\delta \approx 0$). Missing values in the table indicate unstable slope of the backfill ($\beta > \phi$) and therefore need not be considered.

Table 208.2 shows values of Coulomb's earth pressure coefficient K_a for various combinations of angles β, ϕ, and δ. It has been presumed that the wall has a vertical stem ($\theta = 90°$). Coulomb theory accounts for the friction behind the wall (represented by the angle of

Table 208.3 Backfill Rating as a Function of Suitability Number

Range of S_N	Backfill rating
0–10	Excellent
10–20	Good
20–30	Fair
30–50	Poor
> 50	Unsuitable

friction δ). Missing values in the table indicate unstable slope of the backfill ($\beta > \phi$) and therefore need not be considered.

Suitability Number

The grain-size distribution of the backfill material is an important factor that controls the rate of densification. Backfill material is rated by a parameter called the suitability number S_N (Brown 1977)[2] as defined in Eq. (208.15):

$$S_N = 1.7\sqrt{\frac{3}{(D_{50})^2} + \frac{1}{(D_{20})^2} + \frac{1}{(D_{10})^2}}$$ (208.15)

where D_{50}, D_{20}, and D_{10} are diameters (mm) with 50%, 20%, and 10% passing, respectively. Table 208.3 outlines qualitative ratings of a backfill material correlated to the suitability number.

Steps for Evaluating Stability of a Retaining Wall

The following steps must be followed in order to evaluate the overall stability of a retaining wall:

1. Calculate K_a according to either the Rankine or Coulomb theory.

2. The lateral earth pressure may have two components—the triangular (hydrostatic) earth pressure due to overburden and the rectangular (constant) earth pressure due to surcharge load on the backfill, if any. The combined effective horizontal stress at a depth z below the surface is then given by

$$\sigma_h' = K_a(\gamma z + q_s)$$ (208.16)

Note that the right-hand side has been calculated as a total stress, presuming that the water table is at significant depth.

3. If there is groundwater in the backfill zone, add the hydrostatic pore water pressure profile (triangular) to the effective horizontal stress to obtain the total horizontal stress on the wall. (See Example 208.2.)

---

[2]Brown, R. E. (1977), "Vibroflotation Compaction of Cohesionless Soils," *Journal of the Geotechnical Engineering Division*, American Society of Civil Engineers, Vol. 103, No. GT12, 1437–1451.

4. The resultant force (per unit length of the wall) on a wall of height H is then given by (these are shown in the diagram below)

$$R_a = \frac{1}{2} K_a \gamma H^2 + K_a q_s H \qquad (208.17)$$

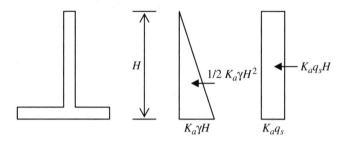

5. The resulting overturning moment (about the toe) is given by

$$M_{\text{OT}} = \frac{1}{2} K_a \gamma H^2 \frac{H}{3} + K_a q H \frac{H}{2} = \frac{1}{6} K_a \gamma H^3 + \frac{1}{2} K_a q H^2 \qquad (208.18)$$

6. The stabilizing moment is due to the gravity forces (vertical)

$$M_s = \sum W_i x_i \qquad (208.19)$$

7. Factor of safety against sliding is obtained by comparing the friction mobilized under the bottom of the footing to the resultant active force.

$$\text{FS} = \frac{\sum W \tan \delta}{R_a} \qquad (208.20)$$

where δ is the friction angle between the base of the wall footing and the supporting soil. Customary factors of safety (wall sliding) used are

   1.5 for retaining walls sliding with active earth pressures

   2.0 for retaining walls sliding with passive earth pressures

 NOTE Designers typically tend to make the conservative decision to ignore the lateral resistance from the passive soil in computing factors of safety.

8. Factor of safety against overturning is obtained by comparing the stabilizing moment of the weight components to the overturning moments produced by the lateral forces.

$$\text{FS} = \frac{M_s}{M_{\text{OT}}} \qquad (208.21)$$

Customary factors of safety (wall overturning) used are

1.5 for retaining walls overturning with granular backfill

2.0 for retaining walls overturning with cohesive backfill

9. The net moment is given by

$$M_{net} = M_s - M_{OT} \tag{208.22}$$

10. The effective location of the resultant force (measured from the toe) is given by

$$\bar{x} = \frac{M_{net}}{\sum W} \tag{208.23}$$

and the eccentricity of this load (with respect to the center of the wall footing) is given by

$$e = \left| \bar{x} - \frac{B}{2} \right| \tag{208.24}$$

11. The maximum soil pressure (as long as the eccentricity is less than $B/6$) is

$$q_{max} = \frac{\sum W}{B} + \frac{6 \times \sum W \times e}{B^2} \tag{208.25}$$

Example 208.1

In the example shown here, assume the unit weight of the backfill soil is $\gamma = 120$ lb/ft^3, unit weight of concrete is 150 lb/ft^3, friction angle between base of wall footing and soil is 26°, and the angle of internal friction for the backfill is 32°. Assume the surcharge to be 400 psf. Calculate the following—(a) factor of safety against sliding, (b) factor of safety against overturning, and (c) maximum soil pressure under wall footing.

Solution   The active earth pressure coefficient is

$$K_a = \frac{1 - \sin 32}{1 + \sin 32} = 0.307$$

The resultant earth pressure has two components—due to overburden and due to surcharge. These are shown as components 4 and 5 in the diagram below:

$$R_4 = \frac{1}{2} K_a \gamma H^2 = \frac{1}{2} \times 0.307 \times 120 \times 17^2 = 5323.4 \text{ lb/ft}$$

$$R_5 = K_a q_s H = 0.307 \times 400 \times 17 = 2087.6 \text{ lb/ft}$$

The weight components are

$$W_1 = 14 \times 1 \times 150 = 2100 \text{ lb/ft (concrete)}$$

$$W_2 = 13 \times 3 \times 150 = 5850 \text{ lb/ft (concrete)}$$

$$W_3 = 7 \times 14 \times 120 = 11{,}760 \text{ lb/ft (soil)}$$

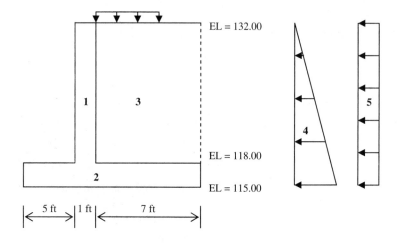

These values are summarized in the table below. Note that the stabilizing moments are designated as positive and the destabilizing (overturning) moments as negative. This is simplified for the purpose of grouping them properly when calculating the factor of safety against overturning and when calculating the net moment.

	Description	F_x (lb)	F_y (lb)	Distance from toe (ft)	Moment about toe (lb-ft)	ΣM
1	Concrete stem	—	2100	5.50	+11,550	
2	Concrete footing	—	5850	6.50	+38,025	+161,295
3	Backfill soil	—	11,760	9.50	+111,720	
4	Lateral EP (overburden)	5323.4	—	5.67	−30,184	−47,929
5	Lateral EP (surcharge)	2087.6	—	8.50	−17,745	

Total vertical load (excluding surcharge) = 19,710

Coefficient of friction (below footing) = $\tan 26° = 0.49$

$$\text{Factor of safety against sliding} = \frac{19,710 \times 0.49}{7411} = 1.30$$

$$\text{Factor of safety against overturning} = \frac{161,295}{47,929} = 3.36$$

Net moment = 161,295 − 47,929 = 113,366

$$\text{Effective distance of resultant from toe} = \frac{M_{net}}{\sum F_y} = \frac{113,366}{19,710} = 5.75$$

Eccentricity of resultant $= 6.50 - 5.75 = 0.75$ ft (left of center)

$$\text{Stress at toe} = \frac{19,710}{13} + \frac{6 \times 19,710 \times 0.75}{13^2} = 1516.2 + 524.8 = 2041 \text{ psf}$$

$$\text{Stress at heel} = \frac{19,710}{13} - \frac{6 \times 19,710 \times 0.75}{13^2} = 1516.2 - 524.8 = 991.4 \text{ psf}$$

Example 208.2 Groundwater behind Wall

For the same wall configuration as in Example 208.1, assume that the drainage behind the wall stops working and the water table rises to a level as shown in the diagram. All other parameters are kept as before.

Solution

$$K_a = \frac{1 - \sin 32}{1 + \sin 32} = 0.307$$

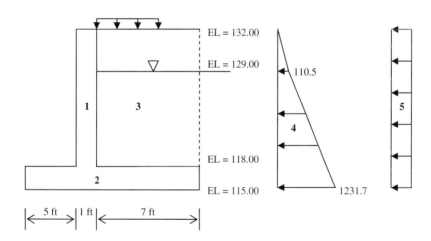

The effective vertical stress is computed as follows: at elevation 129.00, effective vertical stress $= 120 \times 3 = 360$ psf and at elevation 115.00, effective vertical stress $= 360 + (120 - 62.4) \times 14 = 1166.4$ psf.

Therefore, the effective horizontal stress profile is as follows: at elevation 129.00, effective horizontal stress $= 0.307 \times 360 = 110.5$ psf and at elevation 115.00, effective horizontal stress $= 0.307 \times 1166.4 = 358.1$ psf.

In order to obtain the total horizontal stress, we need to add the hydrostatic pore water pressure to the effective horizontal stress. This pore water pressure profile increases linearly from zero at elevation 129.00 to $62.4 \times 14 = 873.6$ at elevation 115.00.

Therefore, the total horizontal pressure on the wall is as shown in the modified component 4—at elevation 132.00, total horizontal pressure $= 0$, at elevation 129.00, total horizontal pressure = 110.5 psf and at elevation 115.00, total horizontal pressure = 358.1 + 873.6 = 1231.7 psf.

The component of the horizontal earth pressure due to surcharge is unchanged and shown as component 5 in the diagram. If the pressure component 4 is subdivided into convenient triangular and rectangular parts, these are shown in the following diagram as 4a, 4b, and 4c:

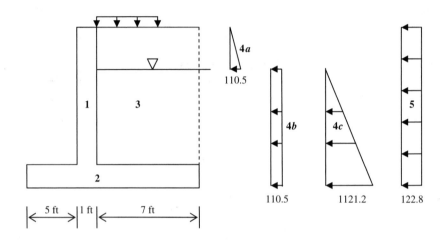

	Description	F_x (lb)	F_y (lb)	Distance from toe (ft)	Moment about toe (lb-ft)	ΣM
1	Concrete stem	—	2100	5.50	+11,550	
2	Concrete footing	—	5850	6.50	+38,025	+161,295
3	Backfill soil	—	11,760	9.50	+111,720	
4a	Lateral EP (overburden)	166	—	15.0	−2490	
4b		1547	—	7.0	−10,829	
4c	Groundwater	7848	—	4.67	−36,650	−67,714
5	Lateral EP (surcharge)	2087.6	—	8.50	−17,745	

Total vertical load (excluding surcharge) = 19,710

Coefficient of friction (below footing) = tan 26° = 0.49

$$\text{Factor of safety against sliding} = \frac{19,710 \times 0.49}{11648.6} = 0.83$$

$$\text{Factor of safety against overturning} = \frac{161,295}{67,714} = 2.38$$

Net moment = 161,295 − 67,714 = 93,581

$$\text{Effective distance of resultant from toe} = \frac{M_{net}}{\sum F_y} = \frac{93,581}{19,710} = 4.75$$

Eccentricity of resultant $= 6.50 - 4.75 = 1.75$ ft (left of center)

$$\text{Stress at toe} = \frac{19{,}710}{13} + \frac{6 \times 19{,}710 \times 1.75}{13^2} = 1516.2 + 1224.6 = 2741 \text{ psf}$$

$$\text{Stress at heel} = \frac{19{,}710}{13} + \frac{6 \times 19{,}710 \times 1.75}{13^2} = 1516.2 - 1224.6 = 291.6 \text{ psf}$$

Retaining Wall with Key

In some locations, the factor of safety against sliding is not adequate. If space limitations exist, this factor of safety cannot be increased by increasing the base width of the footing. In such locations, a base key may be used. Usually, the key is located under the stem and some reinforcement from the stem is continued into the key. The key tends to shear the soil block to its left, thereby mobilizing passive resistance on the vertical surface AB, cohesive forces along line BC, and adhesive forces (between the footing and the soil) along CD and EF (Fig. 208.5). This results in the following expressions for factor of safety against sliding resistance. In cohesive soils

$$\text{FS} = \frac{V \tan \delta + c_a(B - x) + cx + P_p}{H} \tag{208.26}$$

In granular soils

$$\text{FS} = \frac{V \tan \delta + P_p}{H} \tag{208.27}$$

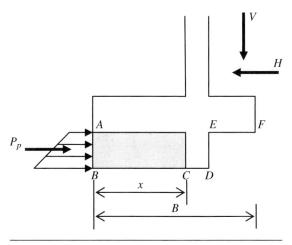

Figure 208.5 Retaining wall with key.

where c = shear strength of the foundation soil

c_a = adhesion between bottom of footing and foundation soil

δ = angle of friction between bottom of footing and foundation soil

V = vertical force (including soil weight and vertical component of lateral force)

H = horizontal component of lateral force

x = width of intact soil wedge in front of key

B = width of footing

P_P = passive resistance of soil in front of key

Horizontal Pressure on Retaining Walls Due to Surface Loads

When the backfill behind a retaining wall is subject to loads on the ground surface, the wall experiences lateral pressure which is a function of the distance of the load from the vertical plane of the wall. For a surface load designated as a surcharge, the extent of the load (in horizontal dimension) is large compared to the depth and therefore is not dissipated with depth. On the other hand, for loads that are localized such as point loads or line loads, the lateral stress decays with depth and distance away from the wall. Two cases presented here are (1) a point load Q (*force*) and (2) a strip or line load q (*force per unit length*).

Point Load on Surface

Figure 208.6 shows a point load Q applied to the backfill at a distance $x = mH$ from the wall. The horizontal stress experienced by the wall at a depth $z = nH$ is given by Eq. (208.28).

$$\sigma_x = \frac{1.77Q}{H^2} \frac{m^2 n^2}{(m^2 + n^2)^3} \qquad m > 0.4 \qquad \text{(208.28a)}$$

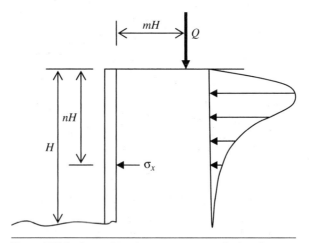

Figure 208.6 Lateral earth pressure on vertical surface due to point load.

Table 208.4 Horizontal Pressure Influence Coefficient due to Point Load on Surface

$n = z/H$	Horizontal distance parameter $m = x/H$					
	≤ 0.4	0.5	0.6	0.7	0.8	1.0
0.10	0.576	0.252	0.126	0.069	0.041	0.017
0.15	1.048	0.492	0.256	0.145	0.088	0.037
0.20	1.416	0.726	0.398	0.233	0.144	0.063
0.25	1.607	0.906	0.528	0.321	0.204	0.092
0.30	1.631	1.013	0.629	0.400	0.262	0.123
0.40	1.383	1.027	0.725	0.505	0.354	0.181
0.50	1.027	0.885	0.702	0.535	0.402	0.227
0.60	0.725	0.702	0.615	0.508	0.408	0.253
0.75	0.422	0.464	0.457	0.418	0.366	0.261
0.90	0.251	0.301	0.322	0.320	0.301	0.242
1.20	0.100	0.132	0.157	0.174	0.181	0.175
1.50	0.046	0.064	0.081	0.095	0.106	0.116
1.80	0.023	0.034	0.044	0.054	0.063	0.075
2.00	0.016	0.023	0.031	0.038	0.045	0.057

$$\sigma_x = \frac{0.28Q}{H^2} \frac{n^2}{(0.16 + n^2)^3} \qquad m \leq 0.4 \qquad (208.28b)$$

These can be expressed as

$$\sigma_x = \frac{Q}{H^2} I \qquad (208.29)$$

where the influence factor (I) is given in Table 208.4.

Line Load on Surface

Figure 208.7 shows a line load q applied to the backfill at a distance $x = mH$ from the wall. The horizontal stress experienced by the wall at a depth $z = nH$ is given by Eq. (208.30).

$$\sigma_x = \frac{4q}{\pi H} \frac{m^2 n}{(m^2 + n^2)^2} \qquad m > 0.4 \qquad (208.30a)$$

$$\sigma_x = \frac{0.203q}{H} \frac{n}{(0.16 + n^2)^2} \qquad m \leq 0.4 \qquad (208.30b)$$

These can be expressed as

$$\sigma_x = \frac{q}{H} I \qquad (208.31)$$

where the influence factor (I) is given in Table 208.5.

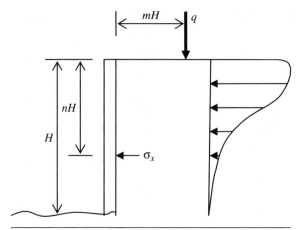

Figure 208.7 Lateral earth pressure on vertical surface due to strip load.

Table 208.5 Horizontal Pressure Influence Coefficient due to Line Load on Surface

	Horizontal distance parameter $m = x/H$					
$n = z/H$	≤ 0.4	0.5	0.6	0.7	0.8	1.0
0.10	11.455	6.592	4.278	3.005	2.230	1.373
0.15	11.851	6.859	4.438	3.102	2.292	1.400
0.20	11.459	6.813	4.456	3.128	2.313	1.413
0.25	10.545	6.519	4.351	3.087	2.296	1.410
0.30	9.371	6.058	4.150	2.990	2.246	1.393
0.40	6.963	4.923	3.579	2.682	2.069	1.325
0.50	4.999	3.820	2.943	2.302	1.832	1.222
0.60	3.579	2.909	2.358	1.921	1.579	1.101
0.75	2.220	1.929	1.661	1.425	1.224	0.913
0.90	1.434	1.303	1.172	1.047	0.933	0.738
1.20	0.676	0.646	0.613	0.578	0.542	0.470
1.50	0.364	0.357	0.348	0.337	0.326	0.301
1.80	0.216	0.214	0.212	0.210	0.206	0.198
2.00	0.159	0.159	0.158	0.157	0.156	0.153

Mechanically Stabilized Earth (MSE) Walls

In Fig. 208.8, a retaining wall supports fill material to a height $D + H$. The wall is laterally restrained using geomembranes that extend into the backfill soil. These layers of fabric develop lateral restraint though the mechanism of friction on both sides (top and bottom) of the geomembrane. The active wedge within the backfill soil forms at an angle of $45° + \phi/2$ to the horizontal, as shown in Fig. 208.8. This (Rankine) failure plane is shown as the dashed

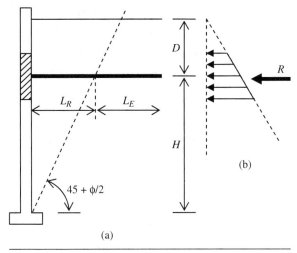

Figure 208.8 Active earth pressure on MSE wall.

line. The soil immediately adjacent to the wall is in an active condition and therefore only the soil beyond this plane is effective in providing the lateral restraint to the wall.

A typical reinforcing strip is shown at depth D below the surface. Depending on the spacing of the strips, each strip serves a limited portion of the wall and provides restraint against the lateral forces acting on it. This "tributary area" of the wall is shown shaded and the (trapezoidal) active earth pressure distribution acting on it is in Fig. 208.8(b).

Thus, each reinforcing strip must have a total length of $L_R + L_E$ = length within the Rankine active zone + effective length required to develop adequate frictional resistance (tension T) against the lateral resultant force R. Using simple trigonometry, the length within the Rankine zone is given as

$$L_R = H\cot\left(45° + \frac{\phi}{2}\right) = H\tan\left(45° - \frac{\phi}{2}\right) \tag{208.32}$$

The total force that can be developed over a length L_E of the membrane beyond the active zone is given by

$$T = 2wL_E \tan\delta\sum\gamma D \tag{208.33}$$

where w = width of the reinforcing strips
    δ = friction angle between the backfill soil and the reinforcing strips
    $\sum\gamma D$ = total vertical overburden pressure at depth D

By equating this tension to the resultant active force R, we get

$$L_E = \frac{R}{2w\tan\delta\sum\gamma D} \tag{208.34}$$

Example 208.3

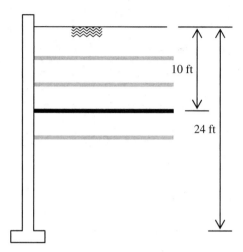

Consider a mechanically stabilized wall as shown in the diagram above. Geotextile strips are spaced at 3-ft vertical spacing as shown. Each strip is 5 in wide and is spaced horizontally every 6 ft. The backfill has the following parameters: $\gamma = 120$ pcf, $\phi = 30°$. The angle of friction between the backfill soil and the reinforcing strips is $\delta = 20°$. Calculate the required length of the reinforcing strip at a depth $D = 10$ ft below the surface.

Solution At a height $H = 14$ ft above the bottom of the footing, the width of the Rankine zone is

$$L_R = 14\cot\left(45° + \frac{30}{2}\right) = 8.08\,\text{ft}$$

At a depth of 10 ft, the total overburden pressure $= 120 \times 10 = 1200$ psf.

$$\text{Active earth pressure coefficient } K_a = \frac{1 - \sin 30}{1 + \sin 30} = 0.333$$

Pressure at centroid of a block at a depth of 10 ft:

$$K_a\gamma H = 0.333 \times 120 \times 10 = 400\,\text{psf}$$

The tributary area for one geotextile strip $= 3\text{ ft} \times 6\text{ ft} = 18\text{ ft}^2$. Resultant active force to be resisted by one reinforcing strip $= 400\text{ psf} \times 18\text{ ft}^2 = 7200$ lb. Therefore, the effective length of reinforcing strip required is given by

$$L_E = \frac{7200}{2 \times \dfrac{5}{12} \times \tan 20 \times 1200} = 19.8 \text{ ft}$$

Therefore, total length of reinforcing strip at depth $D = 10$ ft is $L = 8.1 + 19.8 = 27.9$ ft.

NOTE Factor 2 in the denominator of the calculation for L_E is due to the fact that the reinforcing strip develops friction with the adjacent soil on both upper and lower surfaces.

Support of Excavation

Types of Excavation

There are two basic types of excavations: (1) open excavations where stability is achieved by providing stable side slopes and (2) braced excavations where vertical or sloped sides are stabilized by structural systems that can be restrained laterally by internal or external structural elements. Some examples are skeleton shoring (excavation in most soils up to 20-ft depth), box shoring (depths up to 40 ft), telescopic shoring (very deep trenches).

In selecting and designing the excavation system, the primary controlling factors are (1) soil-type and soil-strength parameters, (2) groundwater conditions, (3) slope protection, (4) side and bottom stability, and (5) vertical and lateral movements of adjacent areas and effects on existing structures.

A *trench shield* is a rigid prefabricated steel unit which extends from the bottom of the excavation to within a few feet of the top of the cut. Pipes are laid within the shield, which is pulled ahead, as trenching proceeds. Typically, this system is useful in loose granular or soft cohesive soils where the excavation depth does not exceed 12 ft.

In *trench timber shoring*, braces and shoring of trench are carried along with the excavation. Braces and diagonal shores of timber should not be subjected to compressive stresses in excess of

$$\sigma_{axial} = 1300 - 20\frac{L}{D} \geq 300 \qquad (209.1)$$

where L = unsupported length (in)
D = least side of the timber (in)
σ_{axial} = allowable compressive stress (psi)

Modes of Failure

The loads exerted on wall/soil system tend to produce a variety of potential failure modes. These failure modes, the evaluation of the loads on the system, and selection of certain system parameters to prevent failure are discussed next.

A *deep-seated failure* causes rotational failure of an entire soil mass containing an anchored or cantilever wall. This type of failure is independent of the structural characteristics of the wall and/or anchor and cannot be remedied by increasing the depth of penetration or by repositioning the anchor. The best option to reduce the likelihood of this failure is to change the geometry of retained material or improve the soil strengths.

Rotational failure due to inadequate pile penetration exhibits itself as large rigid body rotation of a cantilever or anchored wall due to lateral soil and/or water pressures. This type of failure is prevented by adequate penetration of the piling in a cantilever wall or by a proper combination of penetration and anchor position for an anchored wall.

Strength failure of sheet pile or anchor components can occur due to choice of structurally inadequate components. This type of failure can be avoided by designing these components to appropriate strength levels.

Stabilization

During the planning and design stage, if analyses indicate potential slope instability, standard means for slope stabilization or retention should be considered. On occasion, the complexity of a situation may dictate using very specialized stabilization methods. These may include grouting and injection, ground freezing, deep drainage, and stabilization, such as vacuum wells or electro-osmosis.

Bottom Heave in a Cut in Clay

Figure 209.1(a) shows a cut in clay braced with vertical sheet piles. Figure 209.1(b) shows the same cut with the sheet piles driven to an added depth d below the bottom of the cut.

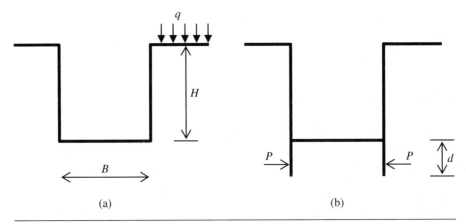

Figure 209.1 Heave at the bottom of a cut in clay.

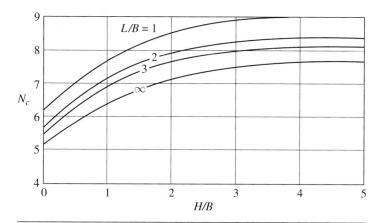

Figure 209.2 N_c factor (Bjerrum and Eide) for evaluating bottom heave in clay. (Based on Bjerrum and Eide's equation.)

According to Bjerrum and Eide[1], the factor of safety against heave at the bottom of a cut in clay is given by the Eq. (209.2). Usually a safety factor of at least 1.5 is desired.

$$FS = \frac{N_c c}{\gamma H + q} \qquad (209.2)$$

where c = cohesion of the clay
H = depth of cut
γ = unit weight of soil
q = uniform surcharge at surface
N_c = factor dependent on the H/B and B/L ratios (see Fig. 209.2)
B = width of trench
L = length of trench

The factor N_c is shown in Fig. 209.2. If the factor of safety is inadequate, then the sheet pile is driven deeper. The force P acting on the buried sheet pile is given by

$$P = 0.7(\gamma HB - 1.4cH - \pi cB) \qquad \text{for } d > 0.47B \qquad (209.3a)$$

$$P = 1.5d\left(\gamma H - \frac{1.4cH}{B} - \pi c\right) \qquad \text{for } d < 0.47B \qquad (209.3b)$$

[1]Bjerrum L. and Eide O. (1956), "Stability of Strutted Excavation in Clay," *Geotechnique*, Vol. 6, No. 1, 32–47.

Typical Plan and Elevation of a Braced Excavation

Figure 209.3 shows a typical schematic of a vertical cut supported by a system of vertical and horizontal structural members. The earth is supported by interlocking sheet piles, which in turn are supported by horizontal members (wales) which are shown spaced a (vertical) distance h_v apart. If the sheet pile is assumed to behave as if it is continuous over several wales, the bending moment per unit width is given by

$$M = \frac{ph_v^2}{10} \tag{209.4}$$

where p is the horizontal soil pressure. Alternatively, a conservative decision may be to assume that the steel pile is pin connected at the wales and use.

$$M = \frac{ph_v^2}{8} \tag{209.5}$$

If the allowable bending stress in the sheet pile is F_a, then the required section modulus (per unit width) of the sheet pile is

$$S \geq \frac{M}{F_a} \tag{209.6}$$

The force per unit length of the wales can be approximately given by

$$w = ph_v \tag{209.7}$$

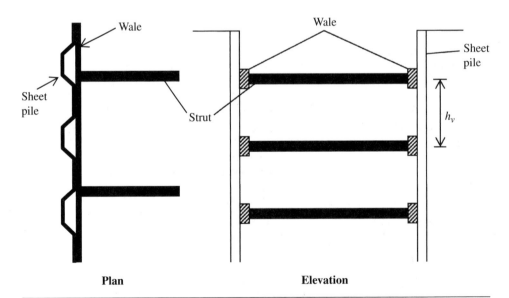

Plan	**Elevation**

Figure 209.3 Plan and elevation of a braced excavation.

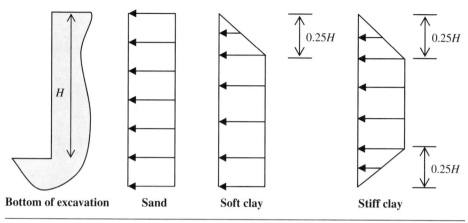

Figure 209.4 Earth pressures in a braced cut (Peck).

Equivalent Pressure Diagrams for Braced Cuts

The soil pressure existing behind structural elements such as sheet piles supporting soil in an excavation can be very complex. One of the commonly used empirical models is that due to Peck[2] which is shown in Fig. 209.4. For a vertical cut of depth H, Peck proposed a lateral soil pressure which is a function of the unit weight, angle of internal friction, and/or the cohesion of the soil.

Sand

For an excavation in cohesionless soil (sand) the active earth pressure is given by

$$p_a = 0.65\gamma H K_a = 0.65\gamma H \tan^2\left(45 - \frac{\phi}{2}\right) \tag{209.8}$$

The resultant active force (per unit length of the cut) is equal to $p_a H$ and may be assumed to act at midheight ($y = 0.5H$ from the bottom of the excavation).

Soft to Medium Clay

For a soft to medium clay, for which the undrained cohesion (c_u) is less than $\gamma H/4$, the active earth pressure is assumed to grow linearly to the maximum value over a depth of $0.25H$ and then remain constant (see Fig. 209.4). The maximum value of the active pressure is given by

$$p_a = \gamma H - 4c \geq 0.3\gamma H \tag{209.9}$$

The resultant active force (per unit length of the cut) is equal to $0.88p_a H$ and may be assumed to act at a height $y = 0.44H$ from the bottom of the excavation.

---

[2]Peck, R.B. (1969), "Deep Excavation and Tunneling in Soft Ground," Proceedings, Seventh International Conference on Soil Mechanics and Foundation Engineering, Mexico City.

Stiff Clay

For stiff clays (unconfined cohesion greater than $\gamma H/4$), the active earth pressure is assumed to grow linearly to the maximum value over a depth of $0.25H$, remain constant over the middle of the depth H and then decay to zero at the bottom of the cut (see Fig. 209.4).

$$p_a = 0.2\gamma H \text{ to } 0.4\gamma H \qquad \text{(with an average of } 0.3\gamma H) \qquad (209.10)$$

The resultant active force (per unit length) is equal to $0.75 p_a H$ and may be assumed to act at midheight ($y = 0.5H$ from the bottom of the excavation).

Example 209.1

A 20-ft-deep, 10-ft-wide, and 60-ft-long trench in a silty clay is braced as shown. Struts are placed every 15 ft (longitudinally). The soil has unconfined compression strength $S_{uc} = 1000$ psf and angle of internal friction $\phi = 0$. The unit weight of the soil is $\gamma = 115$ pcf. Assume that the sheeting is driven 5 ft below the bottom of the cut. Calculate

1. The factor of safety against bottom heave
2. The maximum pressure on the sheet pile from the soil
3. The axial force in the bottom strut

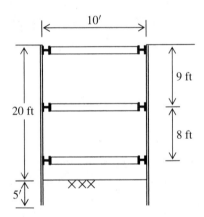

Solution

1. Factor of safety against bottom heave is given by

$$FS = \frac{N_c c}{\gamma H + q}$$

where c = cohesion = $\frac{1}{2} S_{uc}$ = 500 psf
   H = depth of cut = 20 ft
   γ = unit weight of soil = 115 pcf
   q = uniform surcharge at surface = 0 (this case)
   N_c = factor dependent on the H/B and L/B ratios

Using the parameters $H/B = 20/10 = 2.0$ and $L/B = 60/10 = 6.0$, Fig. 209.3 yields $N_c = 7.3$

$$FS = \frac{N_c c}{\gamma H + q} = \frac{7.3 \times 500}{115 \times 20 + 0} = 1.59$$

2. Maximum lateral pressure: Since the cohesion is less than $\gamma H/4 = 575$ psf, the soil may be classified as a soft-to-medium clay. The soil pressure (Peck) is therefore as shown on the figure below. The maximum pressure is calculated as

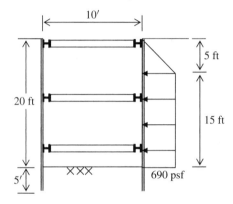

$$p_a = \gamma H - 4c = 115 \times 20 - 4 \times 500 = 300\,\text{psf}$$

However, since this is subject to a minimum value of $0.3\gamma H$, we use

$$p_a = 0.3\gamma H = 0.3 \times 115 \times 20 = 690\,\text{psf}$$

3. Using a tributary area concept, the bottom strut carries load from 4 ft above and 3 ft below the strut. The pressure diagram is uniform ($p = 690$ psf). The resultant load on strut no. 3 is therefore

$$F_2 = 690 \times 7 \times 15 = 72,450 \text{ lb} = 36.2 \text{ tons}$$

Design of Sheet Pile Walls

Sheet pile wall is a row of interlocking, vertical pile segments driven to form an essentially straight wall whose longitudinal dimension is sufficiently large such that its behavior may be based on a typical vertical slice of unit width (usually 1 ft).

Cantilever wall is a sheet pile wall which derives its support solely through interaction with the surrounding soil into which it is embedded. Cantilever walls are usually used as floodwall or as earth retaining walls with low wall heights (10 ft to 15 ft or less). Because cantilever walls derive their support solely from the foundation soils, they may be installed in relatively close proximity (but not less than 1.5 times the overall length of the piling) to existing structures.

Anchored wall is a sheet pile wall which derives its support from a combination of interaction with the surrounding soil and one (or more) mechanical devices which inhibit motion at isolated point(s). An anchored wall is required when the height of the wall exceeds the height suitable for a cantilever or when lateral deflections are a consideration. The proximity of an anchored wall to an existing structure is governed by the horizontal distance required for installation of the anchor.

Retaining wall is a sheet pile wall (cantilever or anchored) which sustains a difference in soil surface elevation from one side to the other. The change in soil surface elevations may be produced by excavation, dredging, backfilling, or a combination.

Dredge side refers to the side of a retaining wall with the lower soil surface elevation. For a floodwall, it refers to the side with the lower water elevation. The *dredge line* is the soil surface on the dredge side of a retaining or floodwall. The wall height is measured from the dredge line. The *retained side* refers to the side of a retaining wall with the higher soil surface elevation or the higher water elevation. The *backfill* is the material on the retained side of the wall.

Anchorage refers to a mechanical assemblage consisting of wales, tie rods, and anchors which supplement soil support for an anchored wall (Fig. 209.5). For a singly anchored wall, anchors are attached to the wall at only one elevation, whereas for a multiply anchored wall, anchors are attached to the wall at more than one elevation. The anchor force is the reaction force (usually expressed per foot of wall) which the anchor must provide to the wall.

Wale is a horizontal beam attached to the wall to transfer the anchor force from the tie rods to the sheet piling (Fig. 209.6).

Tie rods refer to parallel bars or tendons which transfer the anchor force from the anchor to the wales.

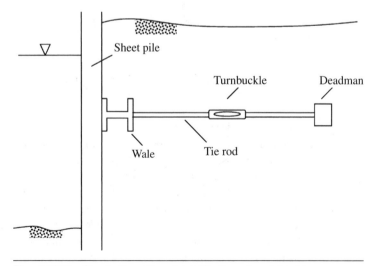

Figure 209.5 Sheet pile wall anchored by tie rod and deadman.

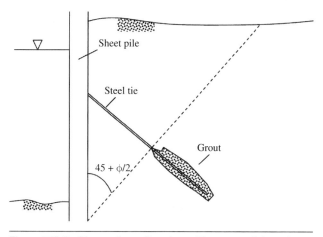

Figure 209.6   Sheet pile wall anchored by grouted tie rod.

Stability Design for Cantilever Walls

It is assumed that a cantilever wall rotates as a rigid body about some point in its embedded length. This assumption implies that the wall is subjected to the net active pressure distribution from the top of the wall down to a point (subsequently called the "transition point") near the point of zero displacement. The design pressure distribution is then assumed to vary linearly from the net active pressure at the transition point to the full net passive pressure at the bottom of the wall. Equilibrium of the wall requires that both the sum of horizontal forces and the sum of moments about any point must be equal to zero. The two equilibrium equations may be solved for the location of the transition point (i.e., the distance z in Fig. 209.7) and the required depth of penetration (distance d). Because the simultaneous equations are nonlinear in z and d, a trial and error solution is required.

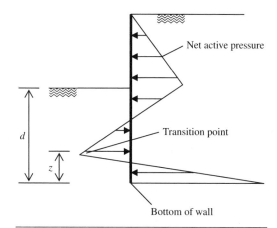

Figure 209.7   Earth pressures on cantilever sheet pile wall.

Ultimate Resistance of Tiebacks

The ultimate resistance of tiebacks in sand is given by

$$T_u = \pi dL \overline{\sigma_v'} K \tan \phi \qquad (209.11)$$

where d = diameter of the grout plug
L = length of the grout plug
$\overline{\sigma_v'}$ = average effective vertical stress for the grout plug
K = earth pressure coefficient
ϕ = angle of friction of soil

In clays, the ultimate resistance of tiebacks can be taken as

$$T_u = \pi dL c_a \qquad (209.12)$$

where c_a = adhesion of the clay

Secant Piles

Secant piles are drilled shafts constructed in such a way that the shafts overlap each other to form a wall. Pile overlap is typically on the order of 3 in. The construction sequence (see Fig. 209.8) involves drilling every other shaft and then returning after the concrete has set in the initial (primary) shafts to drill and pour the shafts for the secondary piles. In Fig. 209.8, the shaded regions show the locations where material from the primary piles is removed to accommodate the secondary piles. Typically, the concrete in the shafts is a combination of low-strength primary and high-strength secondary piles.

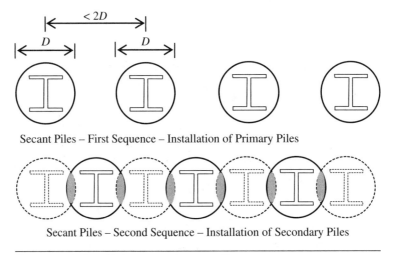

Secant Piles – First Sequence – Installation of Primary Piles

Secant Piles – Second Sequence – Installation of Secondary Piles

Figure 209.8   Installation steps of secant piles.

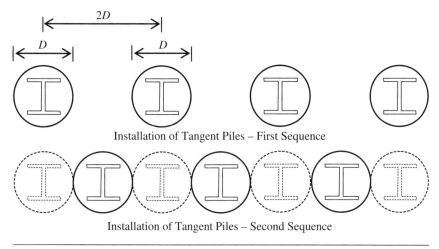

Figure 209.9 Installation steps of tangent piles.

Secant piles are used to build cutoff walls for the control of groundwater inflow and to minimize ground movement in weak and wet soils. They are very stiff retaining walls often used when there are sensitive structures behind the wall. When difficult drilling conditions are anticipated, secant piles can be a very economical method for cutoff wall construction. Another benefit is the relatively low cost of mobilization when compared to other cutoff wall types.

Tangent Pile Wall

Secant pile walls are stiffer than tangent piles walls and are more effective in keeping groundwater out of the excavation. This is because, in a tangent pile wall, there is no pile overlap as the piles are constructed flush to each other. Some of the advantages of tangent and secant pile walls are that they offer increased construction alignment flexibility and faster construction. However, tangent pile walls cannot be used in the presence of high groundwater table without dewatering. The typical installation sequence for tangent pile walls is shown in Fig. 209.9.

OSHA Regulations for Excavations

The OSHA regulations from the 1926 CFR Subpart P (Excavations) are summarized below.

Means of Egress from Trench Excavations

A stairway, ladder, ramp, or other safe means of egress shall be located in trench excavations that are 4 ft (1.22 m) or more in depth so as to require no more than 25 ft (7.62 m) of lateral travel for employees.

Hazardous Atmosphere

Where oxygen deficiency (atmospheres containing less than 19.5% oxygen) or a hazardous atmosphere exists or could reasonably be expected to exist, such as in excavations in landfill areas or excavations in areas where hazardous substances are stored nearby, the atmosphere in the excavation shall be tested before employees enter excavations greater than 4 ft (1.22 m) in depth.

Stability of Adjacent Structures

Where the stability of adjoining buildings, walls, or other structures is endangered by excavation operations, support systems such as shoring, bracing, or underpinning shall be provided to ensure the stability of such structures for the protection of employees. Excavation below the level of the base or footing of any foundation or retaining wall that could be reasonably expected to pose a hazard to employees shall not be permitted except when:

- A support system, such as underpinning, is provided to ensure the safety of employees and the stability of the structure; or
- The excavation is in stable rock; or
- A registered professional engineer has approved the determination that the structure is sufficiently removed from the excavation so as to be unaffected by the excavation activity; or
- A registered professional engineer has approved the determination that such excavation work will not pose a hazard to employees.

Sidewalks, pavements, and appurtenant structure shall not be undermined unless a support system or another method of protection is provided to protect employees from the possible collapse of such structures.

Protection of Employees in Excavations

Employees shall be protected from excavated or other materials or equipment that could pose a hazard by falling or rolling into excavations. Protection shall be provided by placing and keeping such materials or equipment at least 2 ft (.61 m) from the edge of excavations, or by the use of retaining devices that are sufficient to prevent materials or equipment from falling or rolling into excavations, or by a combination of both if necessary.

Each employee in an excavation shall be protected from cave-ins by an adequately designed protective system except when:

- Excavations are made entirely in stable rock; or
- Excavations are less than 5 ft (1.52 m) in depth and examination of the ground by a competent person provides no indication of a potential cave-in.

Walkways

Walkways shall be provided where employees or equipment are required or permitted to cross over excavations. Guardrails which comply with 1926.502(b) shall be provided where walkways are 6 ft (1.8 m) or more above lower levels.

1926 CFR Subpart P Appendix B—Sloping and Benching

Excavations Made in Type A Soil

1. All simple slope excavation 20 ft or less in depth shall have a maximum allowable slope of ¾:1.

   Exception: Simple slope excavations which are open 24 hr or less (short term) and which are 12 ft or less in depth shall have a maximum allowable slope of ½:1.

2. All benched excavations 20 ft or less in depth shall have a maximum allowable slope of ¾ to 1 and maximum bench dimensions such that maximum vertically unsupported height is 4 ft.

3. All excavations 8 ft or less in depth which have unsupported vertically sided lower portions shall have a maximum vertical side of 3½ ft. The portion above the vertical sides can be at a 3/4:1 slope.

4. All excavations more than 8 ft but not more than 12 ft in depth with unsupported vertically sided lower portions shall have a maximum vertical side of 3½ ft. The portion above the vertical sides can be at a 1:1 slope.

5. All excavations 20 ft or less in depth which have vertically sided lower portions that are supported or shielded shall have a maximum allowable slope of ¾:1. The support or shield system must extend at least 18 inches above the top of the vertical side.

Excavations Made in Type B Soil

1. All simple slope excavations 20 ft or less in depth shall have a maximum allowable slope of 1:1.

2. All benched excavations 20 ft or less in depth shall have a maximum allowable slope of 1:1 and maximum bench dimensions such that maximum vertically unsupported height is 4 ft.

3. All excavations 20 ft or less in depth which have vertically sided lower portions shall be shielded or supported to a height at least 18 in above the top of the vertical side. All such excavations shall have a maximum allowable slope of 1:1.

Excavations Made in Type C Soil

1. All simple slope excavations 20 ft or less in depth shall have a maximum allowable slope of 1½:1.

2. All excavations 20 ft or less in depth which have vertically sided lower portions shall be shielded or supported to a height at least 18 in above the top of the vertical side. All such excavations shall have a maximum allowable slope of 1½:1.

3. All other sloped excavations shall be in accordance with the other options permitted in § 1926.652(b).

Slope Stability

Modes of Slope Failure

Embankments and slopes in soil and rock are susceptible to instabilities. Some of the principal modes of failure are (1) rotation on a curved slip surface (often approximated by a circular arc for analytical purposes), (2) translation on a planar surface whose length is large compared to depth below ground, and (3) displacement of a wedge-shaped mass along one or more planes of weakness.

In homogeneous cohesive soils, the critical failure surface is usually deep whereas in homogeneous cohesionless soils, shallow surface sloughing and sliding is more typical. In nonhomogeneous soil, the location and shape of the failure surface depends on the strength parameters of the various soil strata. Failure planes in rock occur along zones of weakness or discontinuities. The orientation and strength of the discontinuities are the most important factors influencing the stability of rock slopes.

Causes of Slope Failure

Natural slopes may fail due to changes in the slope profile (steepening of slope or undercutting of the toe), increase of groundwater pressure causing decrease of shear strength (cohesionless soils) or swell (cohesive soils), weathering, vibrations induced by earthquakes or processes such as blasting or pile driving.

Embankment (fill) slopes may fail due to overstressing of the foundation soil (due to embankment construction), drawdown and subsurface erosion (piping), or vibrations induced by earthquakes, blasting, pile driving, and the like.

Total versus Effective Stress Analysis

The stability analysis makes use of strength parameters which can be found from standard tests. The choice between total stress and effective stress parameters is governed by the expected drainage conditions. Drainage is dependent upon soil permeability, boundary conditions, and time.

Where drainage cannot occur, one should use undrained shear strength parameters determined from tests such as vane shear, unconfined compression, and unconsolidated undrained (UU) triaxial compression tests. Examples where total stress analysis are applicable include cut slopes of normally consolidated clays, embankments on a soft clay stratum, rapid drawdown of water level, and end-of-construction condition for fills built of cohesive soils.

On the other hand, effective shear strength parameters (c', ϕ') should be used for the following cases—long-term stability of clay fills, short-term or end-of-construction condition for fills built of free draining sand and gravel, rapid drawdown condition of slopes in pervious, relatively incompressible, coarse-grained soils, and long-term stability of cuts in saturated clays.

Stability of Infinite Slopes (No Seepage)

Figure 210.1 shows an "infinitely long" slope without the presence of groundwater. The ground surface slopes at an angle of β with the horizontal. The factor of safety for the stability of this slope (i.e., its ability to resist shear failure along a plane parallel to the ground surface) is given by

$$F_s = \frac{c}{\gamma H \cos^2\beta \tan\beta} + \frac{\tan\phi}{\tan\beta} \tag{210.1}$$

where c = cohesion
H = height of embankment
β = slope of plane with horizontal
ϕ = angle of internal friction of soil
γ = unit weight of the soil

The critical depth is obtained by setting $F_s = 1$. This leads to

$$H_{cr} = \frac{c}{\gamma \cos^2\beta (\tan\beta - \tan\phi)} \tag{210.2}$$

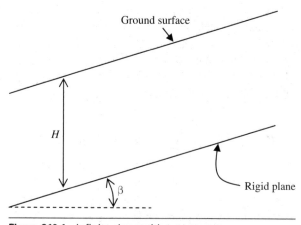

Figure 210.1 Infinite slope without seepage.

For a cohesionless soil ($c = 0$), the factor of safety reduces to

$$F_s = \frac{\tan\phi}{\tan\beta} \tag{210.3}$$

This implies that the maximum stable slope angle (β) in a cohesionless soil is the angle of internal friction (ϕ), leading to the name *angle of repose*. For a cohesive soil for which it is reasonable to assume $\phi = 0$, the factor of safety is

$$F_s = \frac{c}{\gamma H \cos^2\beta \tan\beta} \tag{210.4}$$

Stability of Infinite Slopes (with Seepage)

Figure 210.2 shows an "infinitely long" slope with the groundwater level at the ground surface, resulting in seepage forces acting on the soil. The ground surface slopes at an angle of β with the horizontal. The factor of safety for the stability of this slope (i.e., its ability to resist shear failure along a plane parallel to the ground surface) is given by

$$F_s = \frac{c}{\gamma_{sat} H \cos^2\beta \tan\beta} + \frac{\gamma'}{\gamma_{sat}} \frac{\tan\phi}{\tan\beta} \tag{210.5}$$

where the submerged unit weight is given by $\gamma' = \gamma_{sat} - \gamma_w$. The critical depth is obtained by setting $F_s = 1$. This leads to

$$H_{cr} = \frac{c}{\cos^2\beta(\gamma_{sat} \tan\beta - \gamma' \tan\phi)} \tag{210.6}$$

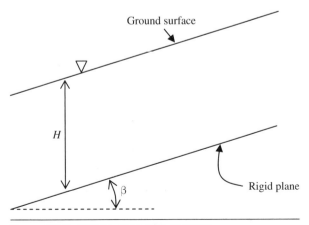

Figure 210.2   Infinite slope with seepage.

Stability of Finite Slopes

There are various methods for the analysis of finite slopes. Some of these are

1. Culmann's method—which employs a plane failure surface

2. Bishop's method—which employs a circular failure surface

3. Bishop and Morgenstern's method—which employs stability coefficients to calculate the factor of safety of a finite slope with seepage

In general, for deep clay subsoils, the critical (minimum factor of safety) failure surface will pass deep into the weakest clay layer. The center of the critical circle will usually lie above the fill slope.

Generalized Method of Slices

The generalized method of slices is illustrated with the aid of Fig. 210.3 in which a finite slope is investigated for slope stability. The method of slices uses the following steps:

1. Create 12 to 15 vertical slices to span the soil domain bounded by an assumed failure circle. In case this domain contains several layers of soil, make sure that the entire bottom boundary of a slice is within a single soil layer.

2. The center of the base of each slice is connected to the center of the circle and the angle between this radius and the vertical radius is measured (α). This angle should be considered counterclockwise positive. (The angle α shown in Fig. 210.3 is clockwise, should be considered negative, and therefore contributes a negative driving force or a stabilizing force.)

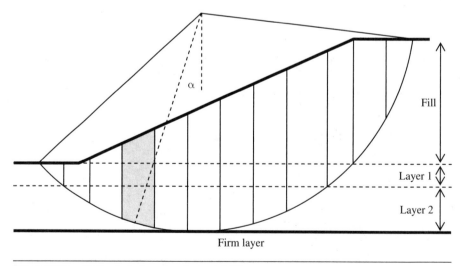

Figure 210.3 Method of slices for slope stability.

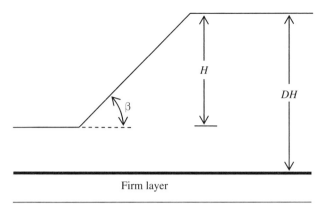

Figure 210.4   Stability of a finite slope.

---

3. The driving force (for shear failure along circle) for each slice is computed as $W_T \sin \alpha$, where W_T is the total weight (soil + water) of the soil in the slice.

4. The resisting force for each slice is computed as $(W_T \cos \alpha - uL) \tan \phi + cL$.

5. The overall factor of safety for the presumed circle is given by

$$F_s = \frac{\sum (W_T \cos \alpha - uL) \tan \phi + \sum cL}{\sum W_T \sin \alpha} \tag{210.7}$$

6. The minimum factor of safety corresponds to the critical circle.

Stability of Finite Slope in Clay (Taylor)[1]

An embankment of height H and slope angle β is shown in Fig. 210.4. The firm soil (bedrock or soil with very high shear strength) is shown at a finite depth DH (below top of embankment). If no such firm stratum can be assumed, assume $D \to \infty$. The factor of safety for slope stability is given by

$$FS = \frac{N_o c}{\gamma H} \tag{210.8}$$

where N_o is Taylor's stability number, shown as a function of slope angle (β) and parameter D in Fig. 210.5. As D approaches infinity, the stability number N_o approaches 5.53. As the slope angle becomes large, the stability number for toe circles approaches the limit 3.83. The critical height of embankment can be found by setting $F_s = 1.0$.

$$H_{cr} = \frac{N_o c}{\gamma} \tag{210.9}$$

---

[1]Taylor, D. W. (1937), "Stability of Earth Slopes," *Journal of the Boston Society of Civil Engineers,* Vol. 24, 197–246.

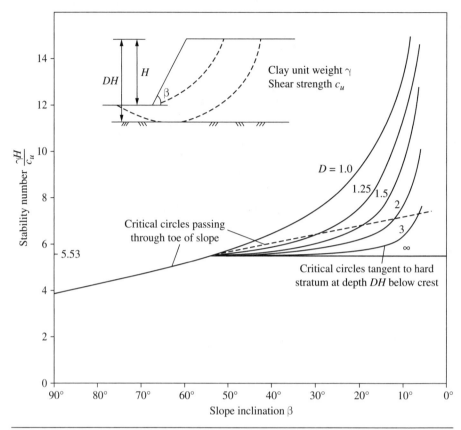

Figure 210.5 Taylor's stability chart.

Example 210.1

An embankment is constructed in clay (unconfined compression strength $S_{uc} = 1400$ psf, unit weight $\gamma = 120$ pcf) to an elevation of 25 ft and at an angle of $40°$ as shown below. There is a stiff layer at a depth of 15 ft below the toe of the slope. Determine the factor of safety for slope stability.

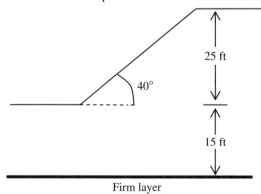

Solution For $\beta = 40°$ and $D = 40/25 = 1.6$, the stability number $N_o = 5.6$. The cohesion of the clay is half the unconfined compression strength $= 700$ psf. The factor of safety for slope stability is given by

$$\text{FS} = \frac{N_o c}{\gamma H} = \frac{5.6 \times 700}{120 \times 25} = 1.30$$

Example 210.2

A 30-ft high embankment exists as shown. ABC is a soil mass with the following properties:
Unit weight $\gamma = 108$ pcf
  Angle of internal friction $\phi = 20°$
  Cohesion $c = 400$ psf.

AC is an interface between the soil and underlying stable rock. For the wedge ABC, determine the factor of safety against sliding along the rock surface.

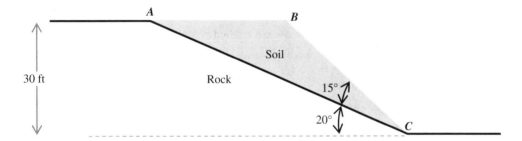

Solution

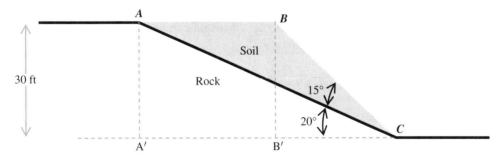

From triangle AA'C, the distance A'C $= 30/\tan 20 = 82.42$ ft
From triangle BB'C, the distance B'C $= 30/\tan 35 = 42.84$ ft
Therefore AB $=$ A'B' $= 39.58$ ft

Length of slope AC $= 30/\sin 20 = 87.71$ ft

Area of soil wedge ABC $= \frac{1}{2} \times 39.58 \times 30 = 593.7$ ft^2

Weight of soil wedge ABC (unit depth) $= 593.7 \times 108 = 64,120$ lb/ft

Component of soil weight parallel to slope: $W_T = 64,120 \sin 20 = 21,930$ lb/ft

Shear stress on slope AB: $\tau = \dfrac{21,930}{87.71 \times 1} = 250$ psf

Component of soil weight perpendicular to slope: $W_N = 64,120 \cos 20 = 60,253$ lb/ft

Normal stress on slope AB: $\sigma = \dfrac{60,253}{87.71 \times 1} = 687$ psf

Shear strength along interface: $\tau_{ult} = c + \sigma \tan \phi = 400 + 687 \tan 20 = 650$ psf

$FS = \tau_{ult}/\tau = 650/250 = 2.6$

Slope Stabilization Methods

Some common methods for slope stabilization are listed below:

1. Flattening and/or benching the slope, or adding material at the toe, as with the construction of an earth berm, increases the stability.

2. Surface control of drainage decreases infiltration to an unstable area. Lowering of groundwater increases effective stresses and eliminates softening of fine-grained soils at fissures.

3. Walls and other retaining structures can be used to stabilize slides of relatively small dimension in the direction of movement or to retain steep toe slopes so that failure will not extend back into a larger mass.

4. Other procedures for stabilizing slopes include grouting, freezing, electro-osmosis, vacuum pumping, and diaphragm walls.

Figure 210.6 presents some common methods for slope protection and stabilization.

Geosynthetics

Geosynthetics, which are synthetic products used to stabilize terrain, may be broadly classified into the following categories:

Geotextiles are permeable synthetic fabrics that can separate, filter, reinforce, protect, or drain soils. They can be either woven or nonwoven.

Geogrids are commonly used to reinforce retaining walls as well as subbases below roads or structures.

A geonet is similar to a geogrid, consisting of integrally connected parallel sets of ribs, usually for in-plane drainage of liquids or gases.

Figure 210.6
Slope stabilization
methods.

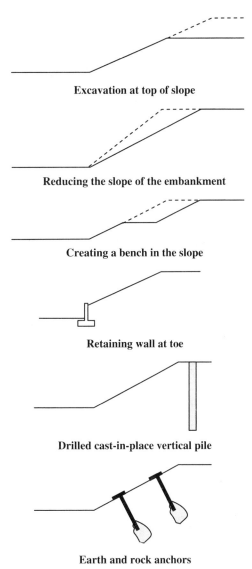

Excavation at top of slope

Reducing the slope of the embankment

Creating a bench in the slope

Retaining wall at toe

Drilled cast-in-place vertical pile

Earth and rock anchors

A geomembrane is a very low-permeability synthetic membrane, made from thin continuous polymeric sheets, used to control fluid or gas migration.

Geosynthetic clay liner (GCL) is a hydraulic barrier consisting of a layer of low permeability material such as bentonite supported by geotextiles and/or geomembranes, held together by needling, stitching, or chemical adhesives.

A geofoam is an expanded polystyrene (EPS) or extruded polystyrene (XPS) material manufactured into large, lightweight blocks, primarily used as a lightweight void fill below a highway, parking lot, bridge approach, etc. Blocks are typically 6 ft × 2.5 ft × 2.5 ft.

Geocells, also known as cellular confinement systems (CCS), are used in construction for erosion control, soil stabilization, channel protection, etc. They are typically made with ultrasonically welded high-density polyethylene (HDPE) strips, expanded on-site to form a honeycomb-like structure and filled with sand, soil, rock, gravel, or concrete.

Geocomposite is a generic name for a need specific design that combines the best features of different materials in such a way that specific applications are addressed in an optimal manner.

Recommended Safety Factors

For slope stability analyses, factor of safety should be no less than 1.5 for permanent or sustained loading conditions. For foundations of structures, a safety factor no less than 2.0 is desirable to limit critical movements at the edge of the foundation. For temporary loading conditions or during construction, safety factors may be reduced to 1.3 or 1.25 if controls are maintained on load application. For transient loads, such as earthquake, safety factors as low as 1.2 or 1.15 may be tolerated.

Slope Protection

Some common techniques for protecting susceptible slopes from the effects of erosion are (1) Covering the slope with a prescribed depth of rocks or cobbles gives adequate protection against wind and rain, (2) Planting and maintaining a regionally suitable variety of grass provides protection against erosion, (3) Rock fragments (riprap) dumped on a properly graded filter provide the best protection against wave action. Rock used should be hard, dense, and durable against weathering and also heavy enough to resist displacement by wave action, (4) Riprap is carefully placed by hand with minimum amount of voids and a relatively smooth top surface. Thickness is typically one-half of the dumped rock riprap but not less than 12 in. A filter blanket must be provided and enough openings should be left in the riprap facing to permit easy flow of water into or out of the riprap, (5) Concrete paving also provides good protection against wave action. Underlying materials should be pervious to prevent development of uplift water pressure. When monolithic construction is not possible, the joints should be kept to a minimum and sealed, and (6) Slopes can be protected by gabions.

Seismic Topics in Geotechnical Engineering

Seismic Stress Waves

There are two primary kinds of seismic waves: body waves and surface waves. The faster body waves move through the earth. Slower surface waves travel along the surface of the earth.

There are two kinds of body waves: (1) compressional waves, also called primary waves (P-waves) and (2) shear waves, also called secondary waves (S-waves). P-waves apply longitudinal (push-pull) stresses. S-waves apply lateral (side-to-side) stresses. P-waves can travel through solids, liquids, or gases, but shear waves can pass only through solids. P-waves are the fastest seismic waves, and they arrive first at a point distant from the epicenter.

Body waves travel faster deep within earth than near the surface. For example, at depths of less than 16 miles, P-waves travel at about 4.2 miles per second, and S-waves travel at 2.4 miles per second. At a depth of 620 miles, the waves travel more than 1.5 times that speed.

There are two kinds of surface waves: (1) Love waves have a horizontal motion that is transverse to the direction of propagation and (2) Rayleigh waves have a retrograde, elliptical motion at the Earth's surface. These are the slowest, but often the largest and most destructive, of the wave types caused by an earthquake. They are usually felt as a rolling or rocking motion and in the case of major earthquakes, can be seen as they approach.

A detector located on the surface first detects the P-waves, then the S-waves, and finally the Rayleigh waves. An approximate depiction of vertical ground disturbance due to these types of waves is shown in Fig. 211.1. Note that the ground displacement caused by Rayleigh waves is significantly greater than that caused by P- and S-waves.

The amplitude of body waves (P- and S-waves) is inversely proportional to the distance ($1/r$) from the rupture point, whereas the amplitude of Rayleigh waves is proportional to $r^{-0.5}$. Transmission velocities of P- and S-waves have been derived from analysis of an

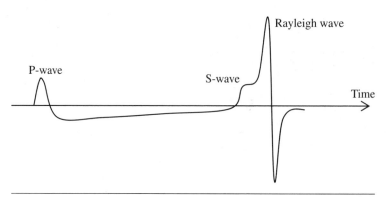

Figure 211.1   Ground motion induced by seismic waves.

elastic half-space and are shown below. Transmission velocity of compression (P) waves is given by

$$v_p = \sqrt{\frac{\lambda + 2G}{\rho}} = \sqrt{\frac{(\lambda + 2G)g}{\gamma}} \qquad (211.1)$$

where the parameter λ is given by

$$\lambda = \frac{\mu E}{(1 + \mu)(1 - 2\mu)} \qquad (211.2)$$

where μ = Poisson's ratio of soil
ρ = density of soil
γ = unit weight of soil

Transmission velocity of shear (S) waves is given by

$$v_s = \sqrt{\frac{G}{\rho}} = \sqrt{\frac{Gg}{\gamma}} \qquad (211.3)$$

The shear modulus G is related to the modulus of elasticity E according to

$$G = \frac{E}{2(1 + \mu)} \qquad (211.4)$$

Shear wave velocity can be measured *in situ* by using an energy source rich in vertical shear component of motion and relatively poor in compressive motion. The three broad categories of such tests are cross-hole, down-hole, and up-hole. In the cross-hole technique, sensors placed in bore-holes are at the same elevation as the triggering device in another bore-hole. In the down-hole method, sensors are placed in various depths and the energy source is located above the sensors, usually at the ground surface. In the up-hole method, the energy source is deep in the boring and the sensors are above it, usually at the surface.

Figure 211.2
Typical shear stress vs. shear strain from a cyclic shear test.

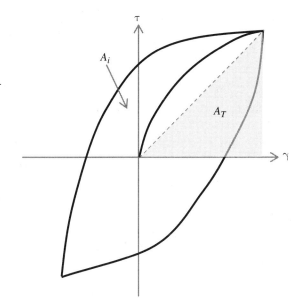

The most widely used laboratory test for determining shear wave velocity is the resonant column method, in which a column of soil is subjected to oscillating longitudinal or torsional load. The frequency is varied until resonance occurs. From the amplitude and frequency at resonance, the modulus and damping of the soil can be calculated.

The most widely used cyclic loading laboratory test is the cyclic triaxial test. Load is cyclic but slowly enough that inertial effects do not occur. A typical shear stress vs. strain diagram is shown in Fig. 211.2. The shear modulus G is calculated as the secant modulus from the shear stress vs. shear strain curve (slope of the broken line on the τ-γ plot) and the damping ratio is calculated as

$$\zeta = \frac{A_i}{4\pi A_T} \qquad (211.5)$$

where A_i = total area included within the hysteretic loop
A_T = shaded area

To avoid resonance of a foundation with periodic loads, such as from oscillatory machinery, one of two strategies can work:

1. For low-speed machinery, increase the natural frequency of the foundation by (1) increasing base area, (2) reducing total static weight, (3) increasing modulus of shear rigidity of the foundation soil by compaction or other stabilization means, and (4) consider using piles to increase stiffness.

2. For high-speed machinery, decrease the natural frequency of the foundation by increasing weight of foundation block.

Vibration Transmission through Soils

The attenuation of vibration as a function of distance from the source is given by Eq. (211.6)

$$A_2 = A_1 \sqrt{\frac{r_1}{r_2}} e^{-\alpha(r_2 - r_1)} \tag{211.6}$$

A_1 = computed or measured amplitude at distance r_1 from source

A_2 = amplitude at distance r_2 from source

α = attenuation coefficient (loose sand 0.06 ft^{-1}; dense sand 0.02 ft^{-1}; silty clay 0.06 ft^{-1}; dense clay 0.003 ft^{-1}; dense weathered rock 0.002 ft^{-1}; competent marble 0.00004 ft^{-1})

For Seismic Site characterization criteria from ASCE 7[1], see Table 116.6.

Example 211.1

The arrivals of P- and S-waves at a seismograph from a remote seismic event are separated by 25 min. Given the following properties for the soil, calculate the distance to the epicenter.

$$E = 3000 \text{ ksf} \qquad \mu = 0.4 \qquad \gamma = 120 \text{ pcf}$$

Solution

$$\lambda = \frac{\mu E}{(1+\mu)(1-2\mu)} = \frac{0.4 \times 3000}{(1+0.4)(1-2 \times 0.4)} = 4286 \text{ ksf}$$

$$G = \frac{E}{2(1+\mu)} = \frac{3000}{2(1+0.4)} = 1071 \text{ ksf}$$

Velocity of P-waves

$$v_p = \sqrt{\frac{\lambda + 2G}{\rho}} = \sqrt{\frac{(4286 + 2 \times 1071) \times 1000 \times 32.2}{120}} = 1313.3 \text{ fps}$$

Velocity of S-waves

$$v_s = \sqrt{\frac{G}{\rho}} = \sqrt{\frac{1071 \times 1000 \times 32.2}{120}} = 536.2 \text{ fps}$$

If distance to epicenter is d, the difference in travel times can be written as

$$t_s - t_p = \frac{d}{536.2} - \frac{d}{1313.3} = 0.0011d = 25 \times 60 \Rightarrow d = 1,359,268 \text{ ft} = 257.4 \text{ miles}$$

Liquefaction

Liquefaction occurs in medium- to fine-grained cohesionless soils due to loss of shear strength. Earthquakes usually create loads characterized by a high strain rate. In fine-grained granular soils, this rapid loading does not permit timely dissipation of pore pressures, effectively

---

[1]Minimum Design Loads for Buildings and Other Structures, ASCE Standard ASCE/SEI 7-10, American Society of Civil Engineers, Reston VA, 2010.

creating an undrained condition, thereby raising pore water pressure. The sudden increase in the pore water pressure causes a sudden decrease in the effective stress. Since the shear strength of a cohesionless soil is solely dependent on the effective stress (little or no cohesion), this causes a sudden decrease in the shear strength, causing the soil to liquefy (behave like a fluid, zero shear strength). Thus, the primary risk factors for liquefaction are

1. Fine-grained soil with little to no cohesion

2. Low relative density

3. Rapid loading

4. Absence of high magnitude loads in loading history

Sand deposits with void ratio greater than a critical void ratio tend to decrease in volume due to seismic load, thus showing susceptibility to liquefaction.

Shear Stress in Soil Due to Ground Acceleration

Consider a rigid block of soil extending from the ground surface to the stratum of interest, as shown in Fig. 211.3. If the peak ground acceleration is a_{max}, then the resulting shear stress at depth h (assuming the soil behaves as a rigid block) is given by

$$\tau = \gamma h \left(\frac{a_{max}}{g} \right) \tag{211.7}$$

Accounting for flexibility of the soil column, the maximum shear stress at depth h is calculated as

$$\tau_{max} = \gamma h \left(\frac{a_{max}}{g} \right) C_D \tag{211.8}$$

where γh = total vertical stress at depth h

C_D = stress reduction factor (related to flexibility of soil column)

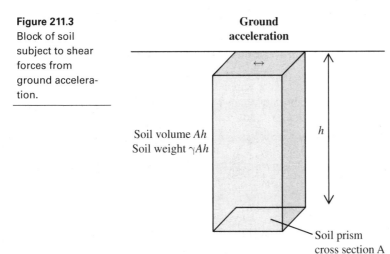

Figure 211.3
Block of soil subject to shear forces from ground acceleration.

Ground acceleration

Soil volume Ah
Soil weight γAh

h

Soil prism cross section A

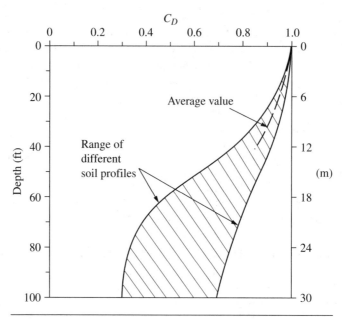

Figure 211.4   Shear stress reduction factor (Seed and Idriss).

The shear stress reduction factor C_D is given by Seed and Idriss[2] and reproduced in Fig. 211.4. The shaded area shows approximate range of values at various depths. The maximum shear stress τ_{max} is then converted to an equivalent average shear stress according to Seed and Idriss.

$$\tau_{ave} = 0.65\gamma h\left(\frac{a_{max}}{g}\right)C_D \tag{211.9}$$

$$\text{Factor of safety against liquefaction} = \frac{\tau_{h,field}}{\tau_{ave}} \tag{211.10}$$

where $\tau_{h,\text{field}}$ = cyclic peak shear stress required to cause initial liquefaction in the field
τ_{ave} = average cyclic shear stress caused by ground acceleration

The shear stress at liquefaction is dependent on the relative density of the soil, and any measured values in the laboratory can be converted to appropriate values in the field according to

$$\frac{\tau_{h,field}}{\tau_{h,lab}} = \frac{R_{D,field}}{R_{D,lab}} \tag{211.11}$$

---

[2]Seed, H. B. and Idriss, I. M. (1971), "Simplified Procedure for Evaluating Soil Liquefaction Potential," *Journal of the Soil Mechanics and Foundations Division,* American Society of Civil Engineers, Vol. 97, SM9, 1249–1273.

where $\quad \tau_{h,\text{field}}$ = cyclic peak shear stress required to cause initial liquefaction in the field
$\quad\quad\quad \tau_{h,\text{lab}}$ = cyclic peak shear stress which causes initial liquefaction in the laboratory
$\quad\quad\quad R_{D,\text{field}}$ = relative density of the soil in the field
$\quad\quad\quad R_{D,\text{lab}}$ = relative density of the soil in the laboratory

Example 211.2

Determine the factor of safety against liquefaction at a depth of 30 ft for a 15-ft-deep fine-sand layer ($\gamma = 122.4$ pcf) overlaid by a 15-ft-deep clay layer ($\gamma = 115$ pcf). The groundwater table is a depth of 15 ft. Use the design earthquake of magnitude 7.0, with a peak ground acceleration of 0.85 g. Let the peak cyclic shear stress required to cause liquefaction be 25 psi.

Solution   At $z = 30$ ft, the total stress $= \gamma h = 15 \times 115 + 15 \times 122.4 = 3561$ psf.
At depth $= 30$ ft, $C_D \approx 0.92$.

$$\tau_{ave} = 0.65 \times 0.92 \times 3561 \times \frac{0.85\,g}{g} = 1810 \text{ psf}$$

$$\tau_{h,\text{max}} = 25 \text{ psi} = 3600 \text{ psf}$$

$$\therefore \text{FS} = \frac{3600}{1810} \approx 2.0$$

Liquefaction Mitigation Techniques

Following are three broad strategies for reducing liquefaction hazards when designing structures.

1. Avoid liquefaction susceptible soils
   Saturated soils formed as sedimentary deposits (fluvial or alluvial deposits in lakes and rivers; colluvial deposits due to eroded materials; aeolian deposits due to wind) can be susceptible to liquefaction. These soils usually have uniform sized, rounded particles that densify easily, causing sudden increases in pore water pressure during earthquakes. Thus, soils with uniform particle size and rounded (rather than angular) particles should be avoided to minimize liquefaction potential.
   Loading history plays an important part in determining liquefaction risk. At a given effective stress, a looser soil is more susceptible than a denser soil. For a given density, a soil under higher effective stress is more susceptible than a soil at low effective stress.

2. Build structures that can tolerate the harmful effects of liquefaction
   Liquefaction has the potential for causing large differential settlements. Two ways to minimize the harmful effects of such differential settlement can be to (a) bridge the "soft spot," as with a stiff mat foundation, or (b) design the structure to be ductile enough so that it can "tolerate" large deformations. If one must build on a soil which is known to be susceptible to liquefaction, incorporating ductility into the design is extremely important.

For deep foundations (piles) liquefaction in a weak layer may cause additional lateral loads to act on the portion of the pile which is embedded in a strong layer. The pile may need to be designed for these addition lateral loads.

3. Improve the soil so that the possibility of liquefaction is reduced/minimized

Soil improvement entails the increase of strength and/or density or the improvement of drainage (to prevent the buildup of pore water pressures). A soil can be densified by several means—vibroflotation, dynamic compaction, compaction piles, etc.

Vibroflotation is an invasive process involving the use of a vibrating probe (vibroflot) that can penetrate a granular soil. The vibration of the probe causes voids in the soil to collapse, causing densification.

Vibroreplacement involves a combination of vibroflotation and gravel backfilling, creating stone columns within the soil. In addition to densifying the soil, these stone columns also improve drainage, thereby preventing the buildup of pore water pressures.

Dynamic compaction involves dropping a heavy weight from a certain height to densify the soil.

Compaction piles—usually constructed from prestressed concrete or timber—serve to densify and reinforce the soil. Compaction grouting can also achieve similar results.

Installing drains of gravel, sand, or synthetic material can improve the drainage ability and reduce the possibility of the buildup of pore water pressures. Sand and gravel drains are usually vertical, but synthetic drains can be installed at other angles, thereby making it easy to improve soil under existing structures.

Bearing Capacity under Dynamic Loading

The soil properties most important for dynamic analysis of foundations and other substructure elements are stiffness, material damping, and unit weight. In addition, location of water table, degree of saturation, and grain size distribution of the soil are also important. For saturated clays, cohesion is calculated from an unconsolidated, or undrained triaxial test. The cohesion obtained under dynamic conditions is higher than that obtained under static conditions. Typically

$$(c_u)_{dynamic} = 1.5(c_u)_{static} \qquad (211.12)$$

On dense sands subject to dynamic loads, the ultimate bearing capacity can be found by replacing the ϕ value as in Eq. (211.16).

$$\phi_{dynamic} = \phi - 2° $$

Cyclic Stress Ratio

The cyclic stress ratio (CSR) is defined as the ratio of the average cyclic shear stress produced in a soil (due to rapid loading) to the effective vertical stress.

$$CSR = \frac{\tau_{ave}}{\sigma'_v} s \qquad (211.13)$$

Thus, if the cyclic stress ratio is specified, it can be used to estimate the average cyclic shear stress due to a particular earthquake event. This shear stress can then be used to calculate the factor of safety given by

$$FS = \frac{\tau_{h,field}}{\tau_{ave}} \tag{211.14}$$

where $\tau_{h,field}$ = cyclic peak shear stress required to cause initial liquefaction in the field.

The cyclic resistance ratio (CRR) is defined as the ratio of the ultimate shear stress (that induces liquefaction) in a soil to the effective vertical stress.

$$CRR = \frac{\tau_{ult}}{\sigma'_v} \tag{211.15}$$

Thus, the factor of safety against liquefaction can also be expressed as

$$FS = \frac{CRR}{CSR} \tag{211.16}$$

Glossary of Earthquake-Related Terms

Deep focus earthquake: One whose focal depth is greater than about 300 km (200 miles).

Epicenter: The point on the ground directly above the focus or hypocenter.

Epicentric distance: The horizontal distance (measured along the ground) from the epicenter to a particular site.

Focus (hypocenter): The point below the ground surface where the rupture of a fault first occurs.

Focal depth: The vertical distance from the ground surface to the focus.

Intermediate focus earthquake: One whose focal depth is between 70 and 300 km (40–200 miles).

Magnitude: According to Richter, the magnitude of an earthquake is based on the amplitude of generated stress waves. It is given by $\log_{10} E = 11.8 + 1.5M$ where E is the energy released (ergs) and M is the Richter magnitude.

Shallow focus earthquake: One whose focal depth is less than 70 km (40 miles).

Earthwork

General

The computation of earthwork volumes requires knowledge of basic geometry—summarized in mensuration charts. It also requires some knowledge of the behavior of soils upon being excavated from their *in situ* condition. For soils that must be compacted to a certain specified state, it is important to specify that state quantitatively in terms of a "percent compaction," which relates the desired in place unit weight and water content to the optimum water content and dry density, as reported by the Proctor test.

Some soils also go through a phenomenon called bulking during handling and transportation. If this is significant, the governing cost of transporting the borrow material may become volume dependent rather than weight dependent. Typically, trucks that are used for transporting the soil will have a maximum weight limit as well as a maximum volumetric capacity defined. The terms struck capacity and heaped capacity are often used in this context.

The struck capacity of a truck is the amount of soil it can carry if the load is flat to the edges. The heaped capacity is the amount of soil that can be carried if the soil load is allowed to form a stable cone above the flat plane formed by the tops of the confining walls. The following is an illustrative example.

"The 25−ton 250C's rated payload is 50,706 lb. Load capacity is 15 yd3 struck and 18 yd3 heaped. Its empty operating weight is 35,534 lb." For this particular earthmover, if a soil with unit weight of 125 pcf is being transported, the struck payload is $15 \times 27 \times 125 = 50,625$ lb, which is very close to the rated payload of 50,706 lb. On the other hand, if the unit weight of the material is about 105 pcf, the truck can transport 18 yd3, while still remaining approximately within the rated limit ($18 \times 27 \times 105 = 51,030$ lb).

Computation of Earthwork Volumes

When calculating area under a curve (with known vertical ordinates) or volume of earthwork (with known end areas), several numerical schemes are available. Of these, the trapezoidal rule and Simpson's rule are very commonly used. A requirement of both formulas is that the stations (marked as $i = 0, 1, 2$, etc., shown in Fig. 212.1) be spaced equal distance apart [shown as Δ in Eqs. (212.1) and (212.2)]. Also, for Simpson's rule, the number of intervals n must be even.

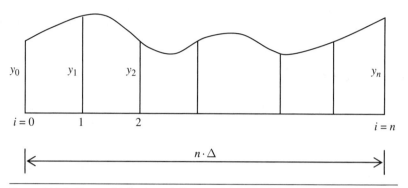

Figure 212.1 Trapezoidal rule and Simpson's rule.

Note that in Fig. 212.1 the first station is marked as $i = 0$ (rather than $i = 1$). This is significant for Simpson's rule because it involves different weighting factors for odd and even i values (4 and 2, respectively).

Using *linear* approximation between (regularly spaced) nodes, we have the *trapezoidal rule:*

$$A = \frac{\Delta}{2}\left[y_0 + y_n + 2\sum_{i=1}^{n-1} y_i \right] \tag{212.1}$$

Using quadratic approximation between (regularly spaced) nodes, we have *Simpson's rule (n must be even):*

$$A = \frac{\Delta}{3}\left[y_0 + y_n + 4\sum_{\substack{odd \\ i}} y_i + 2\sum_{\substack{even \\ i}} y_i \right] \tag{212.2}$$

The same algorithms may also be used to compute earthwork volumes, given end areas at regularly spaced stations (Fig. 212.1). In such a case, the end areas serve as ordinates y, whereas the station coordinates serve as x values. Using the average end area method, the volume between two stations may be expressed as Fig. 212.2

$$V = L\left(\frac{A_1 + A_2}{2}\right) \tag{212.3}$$

Figure 212.2
Earth volume
bounded by two
end areas.

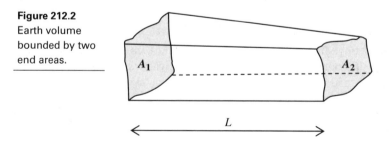

Figure 212.3
Earth volume
bounded by one
end area.

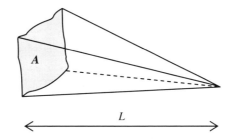

Where the end area at a station can be considered to 'vanish to a point' (Fig. 212.3), the volume should be calculated using the pyramid formula shown in Eq. (212.4). However, for regular earthwork computations, the average end−area approach (trapezoidal rule) is adequate.

$$V = L\left(\frac{A}{3}\right) \tag{212.4}$$

For the data shown in Table 212.1, we apply the trapezoidal rule as follows:

$$V = \frac{50}{2} \times [563 + 410 + 2 \times (457 + 312 + 540 + 256 + 345)]$$
$$= 119{,}825 \text{ ft}^3 = 4438 \text{ yd}^3$$

For the same data, applying Simpson's rule, we have

$$V = \frac{50}{3} \times [563 + 410 + 4 \times (457 + 540 + 345) + 2 \times (312 + 256)]$$
$$= 124{,}617 \text{ ft}^3 = 4615 \text{ yd}^3$$

Table 212.1   Sample Data for Calculating Earthwork Volume

Station i	Distance x_i (ft)	Cut area A_i (ft²)
0	0.0	563.0
1	50.0	457.0
2	100.0	312.0
3	150.0	540.0
4	200.0	256.0
5	250.0	345.0
6	300.0	410.0

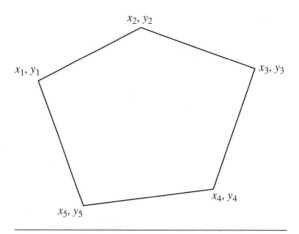

Figure 212.4 Polygon.

Area by Coordinates

The formula to calculate the area of the convex polygon bounded by nodes with coordinates (x_i, y_i) as shown in Fig. 212.4 is given in Eq. (212.5)

$$A = \frac{1}{2}\sum_{i=1}^{n} x_i \left| y_{i+1} - y_{i-1} \right|$$
(212.5)

In this formula, x's and y's are interchangeable. So, the area can also be calculated as

$$A = \frac{1}{2}\sum_{i=1}^{n} y_i \left| x_{i+1} - x_{i-1} \right|$$
(212.6)

A shortcut to calculate this product may be as shown below. The x and y coordinates are written in sequence as below, replicating the first pair (X_1, Y_1) at the end of the sequence. The diagonals in one direction are drawn using the solid lines as shown, and the diagonals in the other direction are drawn using dashed lines. The summation of the series shown in the formulas above is calculated as the absolute value of the sum of one series minus the sum of the other series. Note that the ½ factor is not included in this calculation and must be included to obtain the desired area of the polygon.

$$
\begin{array}{ccccccc}
X_1 & X_2 & X_3 & X_4 & & X_n & X_1 \\
& & & & & & \\
Y_1 & Y_2 & Y_3 & Y_4 & & Y_n & Y_1
\end{array}
$$

$$(X_1 Y_2 + X_2 Y_3 + X_3 Y_4 + \cdots + X_n Y_1) - (X_2 Y_1 + X_3 Y_2 + X_4 Y_3 + \cdots + X_1 Y_n)$$

Shrinkage and Bulking

Materials used in embankments or fills that have been excavated will undergo phenomena that may be broadly labeled as shrinkage and bulking. It is important to determine how these phenomena affect the earthwork volume calculations.

Shrinkage

When earth is excavated and hauled for use as fill, the freshly excavated material generally increases in volume due to the presence of air voids in the uncompacted material. However, when this material is compacted into place in a fill, its volume is normally less than its original condition before it was excavated. This difference is defined as *shrinkage*. The amount of shrinkage will depend on the type of material used and also on the depth of the fill. The shrinkage factor of a soil may be calculated as

$$\text{Shrinkage factor} = \frac{\text{volume after compaction}}{\text{volume before excavation}} \tag{212.7}$$

Bulking

When a particular volume of rock is excavated and placed in an embankment, the material will occupy a larger volume due to the air voids that are introduced into the material. This increase is called *bulking*. The bulking factor of a soil may be calculated as

$$\text{Bulking factor} = \frac{\text{volume after excavation}}{\text{volume before excavation}} \tag{212.8}$$

Typical values of shrinkage and bulking factors are given in Table 212.2.

Table 212.2 Bulking Factor and Shrinkage Factor for Various Soils

Material	Bulking factor	Shrinkage factor
Clay (high PI)	1.40	0.90
Sand	1.05	0.89
Gravel	1.05	0.97
Chalk	1.50	0.97
Shale	1.50	1.33*

*Indicates expansive or collapsible soil. Expansive soil and rock are characterized by clayey material which shrinks and swells as it dries or becomes wet, respectively. The rock type most associated with expansive soils is shale, which can also be expansive.

Using the Mass Diagram

The mass diagram is a useful method to graphically represent the amount of material that will be cut and used for fill on any earthwork job. Some definitions related to mass diagrams are presented below:

Excavation (E)—refers to any excavated material. There are two main categories for excavated material—ordinary material and rock. Both ordinary soil and rock are generally paid for by cubic yards in terms of excavation cost ($/yd³). Rock is generally handled as a separate cost because it is substantially more expensive to excavate.

Free haul (F)—when material is excavated, it will be moved over a certain distance free of charge. This distance is the "free haul" and is normally specified by the contractor.

Overhaul (H)—is defined as any distance over which the excavated material must be hauled less the free haul distance. The rate for overhaul is normally specified by the contractor. This rate is normally given in $/yd³/unit distance.

Borrow (B)—refers to the fill material that must be brought to the proposed highway site from outside the highway cross section. Borrow does not include the material that is excavated on site for use as fill. The borrow cost is normally given in $/yd³, and this rate normally includes cost of excavating borrow. There may be surcharge for borrow if excavated from private property.

Limit of economic overhaul (LEOH)—a distance beyond which it is uneconomic to overhaul.

Limit of economic haul (LEH)—is the distance beyond which it is uneconomic to overhaul plus the free haul distance. It is calculated as the sum of the free haul distance (F) and the LEOH.

$$LEH = F + LEOH$$

(212.9)

Waste—is the excavated material that can not be used for fill on the project site.

Normally, an engineer will try to roughly balance the amount of cut and fill required on a project at the design stage so that the amount of fill that must be hauled in or the amount of waste that must be hauled away are not excessive. Waste can also include excavated materials that are unsuitable for use as a fill because they have unacceptable engineering properties (such as peat, clays, etc.).

Sometimes it is more economical to waste material and use borrow material from a borrow pit within the free haul distance. This occurs where it is necessary to haul excavated material long distances to use as fill.

The mass diagram is a representation of the cumulative volume generated from cuts (positive) and fills (negative), when a proposed profile is overlaid with the existing profile at a site. Figure 212.5 shows the overlay of proposed and existing profiles. The proposed profile is shown as the horizontal line with stations marked using tics along it. Please note that the proposed profile does not have to be horizontal, only so shown in this illustration. For the purpose of the following discussion, let us assume that each of the tics on the horizontal line represents a 100-ft station.

One can see that there is a cut from station 0 + 00.00 to station 4 + 00.00, a fill from station 4 + 00.00 to station 11 + 70.00, and so on.

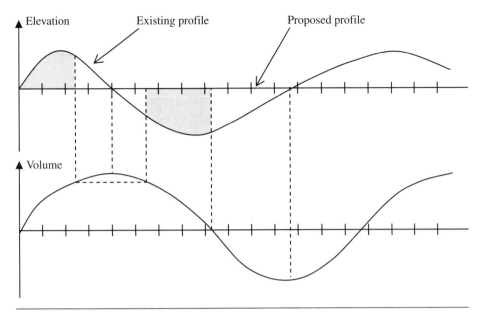

Figure 212.5 Earthwork profile and mass diagram.

Correspondingly, the mass diagram in the second figure, which is the integration of the profile, can be generated by calculating cumulative volumes from the profile, treating cuts as positive mass generation and fills as negative mass generation. When the profile changes from a cut to a fill (as at station $4 + 00.00$), the mass diagram has a positive peak, and when the profile changes from a fill to a cut (as at station $11 + 70.00$), the mass diagram has a negative peak.

Also, in the contracting process for haulage, there is sometimes a free haul limit specified, which is the haulage distance (often specified in stations) for which there is no hauling charge. For all hauls exceeding the limit of free haul, the unit cost is specified in (typical) units of $/yd³-sta. In the illustration, the limit of free haul is taken as 300 ft or 3 stations.

Once the mass diagram has been drawn, a 300-ft-long vector is placed horizontally such that it *fits* the mass diagram. This graphical approach can help identify the stations (approximately $2 + 40.00$ and $5 + 40.00$) which are the limits of free haul. Equal ordinates on the mass diagram imply that the volume of the cut between stations $2 + 40.00$ and $4 + 00.00$ is exactly equal to the volume of the fill between stations $4 + 00.00$ and $5 + 40.00$. The earthwork is balanced between stations $0 + 0.00$ and $8 + 25.00$ (mass diagram ordinate returns to zero). This means that the earthwork volume between stations $0 + 0.00$ and $2 + 40.00$ is exactly equal to the volume between $5 + 40.00$ and $8 + 25.00$. These volumes are shown shaded in Fig. 212.5. Once these stations are identified, volumes outside these stations can be calculated and the associated haulage cost calculated.

Example 212.1   Cut and Fill

The existing and proposed profiles for an embankment project (limits station $10 + 00.00$ to $22 + 00.00$) are shown below. Assume all sections have the same area (perpendicular to the plane of the drawing). The limit of free haul is 4 stations (400 ft).

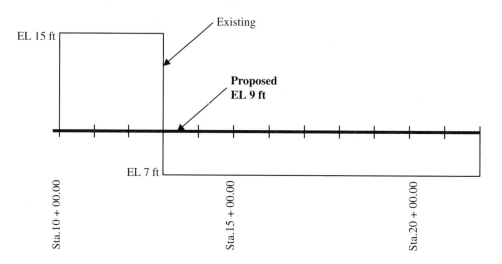

To determine matching volumes of cut and fill, the first task is to calculate the distances x and y such that the volume of the cut is the same as the volume of the fill and $x + y = $ limit of free haul $= 400$ ft.

Since the elevation of the cut is $15 - 9 = 6$ ft and the elevation of the fill is $9 - 7 = 2$ ft, we see that the horizontal distances x and y must be in inverse proportions (2:6). Thus, split 400 ft into two fractions 2/8 and 6/8, which makes $x = 100$ ft and $y = 300$ ft.

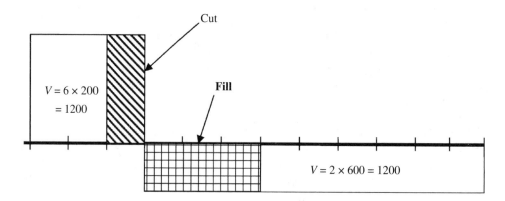

After identifying these matching volumes, one can see that the volume of the cut 1200 ft³ matches the volume of the fill, and their respective centroids are at stations $1 + 00.00$ and $9 + 00.00$. Thus the overhaul may be represented as 1200 ft³ × 8 sta = 9600 ft³-sta = 355.6 yd³-station.

Example 212.2

The table below shows areas of cut and fill at a sequence of 20 stations. Assume shrinkage of 10%. The volume (cut or fill) between two stations is calculated using the average end area method. Fill volumes are adjusted upward using

Adjusted fill volume = volume of hole ÷ 0.9

Station	Area of cut (sf)	Area of fill (sf)	Cut (yd³)	Fill (yd³)	Adjusted fill	Volume of cut	Volume of fill	Mass diagram
0	5	22						0
1	4	45	17	124	138		121	−121
2	8	88	22	246	274		251	−373
3	12	105	37	357	397		360	−733
4	9	44	39	276	307		268	−1000
5	40	32	91	141	156		66	−1066
6	38	22	144	100	111	33		−1033
7	56	3	174	46	51	123		−910
8	105	4	298	13	14	284		−626
9	122	0	420	7	8	412		−214
10	134.2	2	474	4	4	470		256
11	100	11	433	24	27	407		662
12	76	34	326	83	93	233		896
13	66	28	263	115	128	135		1031
14	45	67	206	176	195	10		1041
15	23	56	126	228	253		127	914
16	12	88	65	267	296		231	683
17	6	109	33	365	405		372	310
18	4	88	19	365	405		387	−76
19	23	43	50	243	270		220	−296
20	43	34	122	143	158		36	−332

The figure on the next page shows the mass diagram as computed in the last column of the table. If it is assumed that the free haul distance is 500 ft, then by attempting to fit a horizontal line of length 500 ft between two equal ordinates of the mass diagram, we can identify the limits 282 to 782 ft as limits of free haul (mass diagram ordinate −660 yd³) and 1105 to 1605 ft as limits of free haul (mass diagram ordinate +690 yd³). This essentially means that the earth volume within the cut that exists between 505 and 782 is 660 yd³ and is exactly enough to fill the void between 282 and 505, including compensation for shrinkage. Note that the peaks of the mass diagram are at 505 ft (where it transitions from a fill to a cut) and

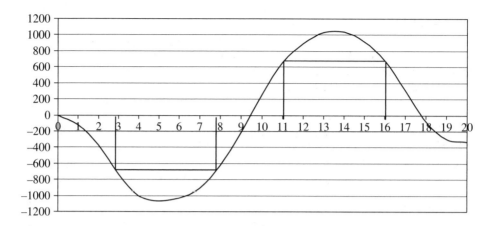

at 1360 ft (where it transitions from a cut to a fill). Similarly, the earth volume within the cut that exists between 1105 and 1360 is 690 yd³ and is exactly enough to fill the void between 1360 and 1605, including compensation for shrinkage.

Earthwork—Borrow Pit Method (Also Known as Grid Method)

The excavation area is referenced to an established grid. The grid density is defined by the size of the project and desired accuracy. An underlying digital terrain model will then yield existing elevations at all points on the grid. For each grid point, final (desired) elevations are established according to the specification of the project. These (proposed) elevations are then subtracted from the existing elevations to generate a depth of cut (if the difference is positive) or fill (if the difference is negative). For each grid square, an average of the depth of cut or fill is multiplied by the grid area to obtain the volume of earthwork associated with that grid square. The volume of earthwork is calculated as follows:

$$V = \left(\frac{D_1 + D_2 + D_3 + D_4}{4} \right) \left(\frac{A_{grid}}{27} \right) \tag{212.10}$$

where V = earthwork volume for the grid square (yd³)
D_i = depths of cut or fill at each grid corner (ft)
A_{grid} = area of grid square (ft²)

Basic Fluid Mechanics

The basic equations governing fluid flow are the conservation of mass, the conservation of energy, and conservation of linear momentum equations. These are derived using a control volume of fluid. By keeping track of the transport of mass, momentum, and energy across the system boundaries, we arrive at the following conservation principles.

Conservation of Mass

The statement of the conservation of mass, better known as the *Continuity equation,* may be stated as—the rate of mass of fluid entering the system must equal that exiting the system. In other words, mass is neither created nor destroyed within the control volume. The mass flow rate through a closed control volume is given by

$$\dot{m} = \rho V A = \text{constant} \tag{301.1}$$

For incompressible fluids (such as water under normal pressures) the density ρ is constant, leading to $Q = VA = \text{constant}$, where Q is the volumetric flow rate (Fig. 301.1).

Conservation of Energy

The statement of the conservation of energy for incompressible fluids, better known as *Bernoulli's equation,* may be stated as—for steady, ideal flow in an incompressible fluid— the energy per unit weight (specific energy) is conserved between two locations, as long as there are no energy loss mechanisms between these locations. The specific energy can be written as

$$E = \alpha \frac{V^2}{2g} + \frac{p}{\gamma} + z \tag{301.2}$$

Figure 301.1
Continuity of
volumetric flow
rate.

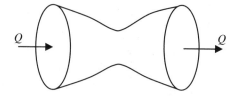

where α is a correction factor for kinetic energy. This factor accounts for the variation of flow velocity across the cross section and is given by

$$\alpha = \frac{\int V^3 \, dA}{V_{avg}^3 \, A}$$

For practical purposes, α may be taken as 1.0. If energy losses between Sections 1 and 2 are ignored, this leads to the ideal form of Bernoulli's equation:

$$\frac{V_1^2}{2g} + \frac{p_1}{\gamma} + z_1 = \frac{V_2^2}{2g} + \frac{p_2}{\gamma} + z_2 \qquad (301.3)$$

If friction and form losses between Sections 1 and 2 are written as h_f and a pump or turbine is located between 1 and 2, this may be extended to

$$\frac{-\eta \dot{W}_S}{\gamma Q} = \frac{V_2^2 - V_1^2}{2g} + \frac{p_2 - p_1}{\gamma} + z_2 - z_1 + h_f \qquad (301.4)$$

where $\dot{W}_S$ = shaft power (positive for turbines, negative for pumps)
η = efficiency of pump or turbine
h_f = head loss between locations 1 and 2

Note that if the equation is applied as shown above, pressures may be expressed as absolute pressure or gage pressure without introducing error.

Example 301.1
Flow occurs in a rectangular open channel (6 ft wide) with velocity of 8 ft/s and depth of 5.8 ft. The elevation of the channel floor is 568.5 ft above sea level. At a downstream location, the channel floor elevation is 564.3 ft. The depth at the downstream end is 6 ft. Determine the head loss due to friction.

Solution Using the continuity equation, we solve for the velocity at the downstream section

$$V_1 A_1 = V_2 A_2 \Rightarrow V_1 d_1 = V_2 d_2 \Rightarrow V_2 = \frac{V_1 d_1}{d_2} = \frac{8 \times 5.8}{6} = 7.73 \text{ ft/s}$$

The energy grade line slopes downward due to the head loss. Thus, the head loss is equal to the decrease of total energy from locations 1 to 2. The surface streamline is at atmospheric pressure (gage pressure = 0). The elevation of the channel surface at the upstream location is $568.5 + 5.8 = 574.3$ ft and at the downstream location is $564.3 + 6.0 = 570.3$ ft.

$$h_f = E_1 - E_2 = \left(\frac{V_1^2}{2g} + \frac{p_1}{\gamma} + z_1\right) - \left(\frac{V_2^2}{2g} + \frac{p_2}{\gamma} + z_2\right)$$

$$= \left(\frac{8^2}{2 \times 32.2} + 0 + 574.3\right) - \left(\frac{7.73^2}{2 \times 32.2} + 0 + 570.3\right) = 4.07 \text{ ft}$$

Conservation of Momentum

The statement of conservation of momentum is derived from Newton's second law which states that the sum of external forces is equal to the rate of change of linear momentum. This principle may be used to calculate the net force acting on a node where a change in momentum occurs. Examples are bends in pipes where the flow changes direction and pipe section enlargements (or contractions) where the flow velocity decreases (or increases).

$$\sum F = \dot{m}(V_2 - V_1) = \rho Q(\beta_2 V_2 - \beta_1 V_1) \tag{301.5}$$

where β is a correction factor for linear momentum. This factor accounts for the variation of flow velocity across the cross section and is given by

$$\beta = \frac{\int V^2 \, dA}{V_{avg}^2 \, A}$$

For practical purposes, β may be taken as 1.0.

The computed net force acts *on* the fluid. Thus, if we are interested in computing the force imparted *by* the fluid, we should reverse the sign of the computed force.

With constant flow rate, force on a rigid surface due to impinging flow $= \rho Q \Delta V$, where ΔV is the change in velocity effected by the static surface.

Example 301.2

A water jet of diameter 1 cm and a flow rate $Q = 1$ L/s is deflected by a fixed plate as shown below. The velocity of the deflected jet is 6 m/s. What is the force on the plate?

(A) 10 N
(B) 15 N
(C) 20 N
(D) 25 N

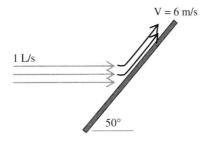

Solution

Flow rate, $Q = 1 \text{ L/s} = 0.001 \text{ m}^3/\text{s}$

Approach velocity: $V = Q/A = \dfrac{0.001}{\pi/4 \times 0.01^2} = 12.73 \text{ m/s}$

After impacting the plate, $V_x = 6 \cos 50 = 3.86 \text{ m/s}$ and $V_y = 6 \sin 50 = 4.60 \text{ m/s}$

Applying conservation of momentum in x and y directions (F_x and F_y represent force acting *on* the fluid jet), respectively,

$$F_x = \Delta(\rho Q V)_x = 1000 \times 0.001 \times (3.86 - 12.73) = -8.88 \text{ N}$$

$$F_y = \Delta(\rho Q V)_y = 1000 \times 0.001 \times (4.6 - 0) = 4.6 \text{ N}$$

Resultant force on plate (equal and opposite to force acting on fluid): $F = \sqrt{8.88^2 + 4.6^2} = 10$ N. Answer is A.

Example 301.3

A pipeline conveying water a flow rate $Q = 32 \text{ ft}^3/\text{s}$ lies in the horizontal plane as shown. At point A, the pipe goes through a 90° bend as well as a reduction in diameter as shown. The pressure at the upstream end is 75 psi. What is the total force experienced by the bend? Ignore the energy losses at the bend.

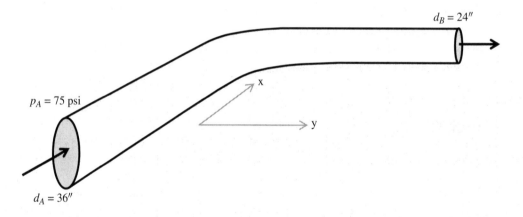

Solution   Pressure at A = 75 psi = 10,800 psf. Therefore, $p_A/\gamma = 10{,}800/62.4 = 173.1$ ft

Velocity at A: $V_A = \dfrac{Q}{A_A} = \dfrac{32}{\dfrac{\pi}{4} \times 3^2} = 4.53 \text{ ft/s}$

Velocity head: $\dfrac{V_A^2}{2g} = \dfrac{4.53^2}{2 \times 32.2} = 0.32 \text{ ft}$

Velocity at B: $V_B = \dfrac{Q}{A_B} = \dfrac{32}{\dfrac{\pi}{4} \times 2^2} = 10.19$ ft/s

Velocity head: $\dfrac{V_B^2}{2g} = \dfrac{10.19^2}{2 \times 32.2} = 1.61$ ft

Neglecting energy losses, Bernoulli's equation yields:

$$\frac{p_A}{\gamma} + \frac{V_A^2}{2g} + z_A = \frac{p_B}{\gamma} + \frac{V_B^2}{2g} + z_B \Rightarrow \frac{p_B}{\gamma} = \left(\frac{p_A}{\gamma} + \frac{V_A^2}{2g} - \frac{V_B^2}{2g} \right)$$
$$= 173.1 + 0.32 - 1.61 = 171.81 \text{ ft}$$

Pressure at B $= 171.81 \times 62.4 = 10{,}721$ psf.

Summation for forces (acting on the fluid) in the original flow direction:

$$F_x + p_A A_A = \rho Q(0 - 4.53) = -\frac{62.4}{32.2} \times 32 \times 4.53 = -281 \Rightarrow F_x = -10{,}800 \times \frac{\pi}{4} \times 3^2 - 281$$
$$= -76{,}622 \text{ lb}$$

Summation for forces (acting on the fluid) in the final flow direction:

$$F_y - p_B A_B = \rho Q(10.19 - 0) = \frac{62.4}{32.2} \times 32 \times 10.19 = 632 \Rightarrow F_y = 10{,}721 \times \frac{\pi}{4} \times 2^2 + 632$$
$$= 34{,}313 \text{ lb}$$

Resultant force: $F = \sqrt{F_x^2 + F_y^2} = \sqrt{76622^2 + 34313^2} = 83{,}954$ lb

Force acting from the fluid on the pipe bend is the equal and opposite of this force.

Energy Grade Line and Hydraulic Grade Line

With respect to a horizontal datum, the energy grade line (EGL) and the hydraulic grade line (HGL) may be plotted as the functions.

$$\begin{aligned} \text{EGL} \qquad & z + \frac{p}{\gamma} + \frac{V^2}{2g} \\[2mm] \text{HGL} \qquad & z + \frac{p}{\gamma} \end{aligned} \qquad (301.6)$$

For example, if a horizontal pipe is tapped with a Pitot tube as shown in Fig. 301.2, the stagnation pressure at the head of the tube causes the liquid column to rise to the EGL elevation, whereas a tap into the same central streamline (but which doesn't stagnate the flow) causes the liquid column to rise to the elevation of the HGL. If these two pressure taps were connected to a differential manometer, the height difference recorded on the manometer is proportional to the quantity $V^2/2g$, where $V =$ flow velocity.

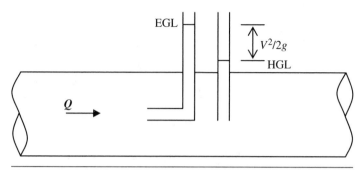

Figure 301.2 Elevation of HGL and EGL in a Pitot tube.

These lines are plotted for the simple system shown in Fig. 301.3. A system of pipes carries flow from the reservoir on the left to the one on the right. The flow is aided by a pump and regulated by a valve. In the figure, the dark solid line plots the EGL while the dashed line plots the HGL. The following observations may be made about Fig. 301.3:

1. At free surfaces of large reservoirs, since the gage pressure is zero and the velocity is effectively zero, both EGL and HGL coincide with the reservoir free surface.

2. The slope of the EGL and HGL along each pipe segment represents friction loss per unit length along that segment.

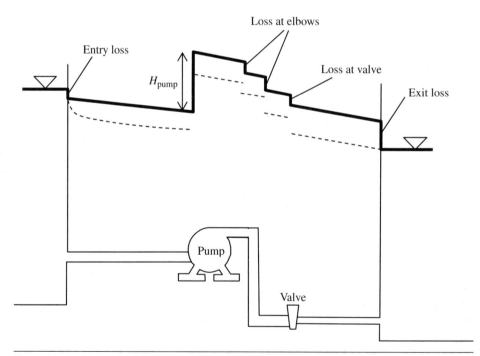

Figure 301.3 Elevation of HGL and EGL in a water delivery system.

3. The vertical separation between EGL and HGL is proportional to the square of the velocity. Thus, with the same flow rate through various pipes, this separation will be greater for smaller pipe diameters.

4. At the location of the pump, there is a discontinuous jump in both HGL and EGL.

5. At all locations where form losses occur (such as entry into the pipe system, exit from the pipe system, pipe bends, etc.), there is a discontinuous jump in both HGL and EGL.

The fluid power (power needed to transmit a flow rate Q of a fluid with unit weight γ and provide it with a lift H) is given by $P = \gamma QH$. In order that a pump operating at an efficiency η provides this fluid power, it must be rated at a higher power given by

$$\dot{W}_p = \frac{\gamma QH}{\eta} \tag{301.7}$$

Example 301.4

Five manholes (A to E) with characteristics shown in the table exist along a storm sewer as shown below. During extreme rainfall events, the pipe flows full with the pressures given in the fourth column. At which manholes does the storm water overflow through the manhole onto the street above?

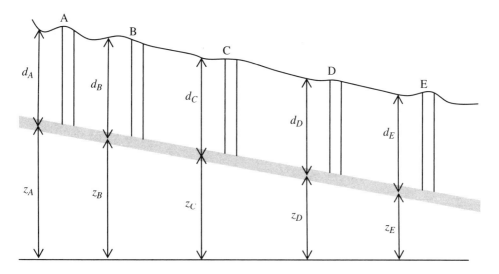

Location	Pipe C.L depth d (ft)	Pipe C.L elevation z (ft)	Pressure (lb/in²)
A	8.2	100	0
B	10.5	86	4.2
C	12.4	71	6.3
D	6.8	54	3.1
E	6.2	43	1.2

Solution For each location, the pressure is converted to pressure head (listed in the third column in the table below). For example,

$$4.2 \text{ psi} = 4.2 \times 144 \text{ psf} = \frac{4.2 \times 144}{62.4} = 9.7 \text{ ft head}$$

The elevation of the hydraulic grade line is given by HGL $= p/\gamma + z$ (listed in the fourth column of the table).

Location	Pressure (lb/in²)	Pressure head p/γ (ft)	HGL elevation $p/\gamma + z$ (ft)	Ground elevation (ft)
A	0	0	100	108.2
B	4.2	9.7	95.7	96.5
C	6.3	14.5	85.5	83.4
D	3.1	7.2	61.2	60.8
E	1.2	2.8	45.8	49.2

Comparing HGL elevation to surface elevation, manholes C and D will overflow. In other words, if the pressure head at a certain location exceeds the burial depth of the sewer, the manhole will overflow.

Viscosity

The viscosity of a fluid is a measure of its ability to resist shear stresses. When a shear stress, no matter how small, is applied to any fluid, a velocity gradient develops transverse to the plane on which the shear stress is applied. For Newtonian fluids, the applied shear stress causes a proportional velocity gradient. The constant of proportionality is termed the viscosity, or dynamic viscosity, μ.

$$\tau = \mu \frac{dV}{dy} \tag{301.8}$$

Viscosity has units of FT/L² (N-s/m² or lb-s/ft²) or M/LT (kg/m-s or slug/ft-s).

For non-Newtonian fluids, the relationship between the applied shear stress and the resulting velocity gradient is nonlinear and is most commonly expressed as a power law (Eq. 301.9).

$$\tau = K \left(\frac{dV}{dy} \right)^n \tag{301.9}$$

Raw sewage flowing through sewers is mostly water, containing a small amount of solids. For all practical purposes, this behaves as a Newtonian fluid. However, once sewage reaches a treatment plant and is concentrated into a sludge, its behavior becomes distinctly no-Newtonian and may require more advanced modeling to determine its viscosity (to calculate pumping costs, etc.).

The kinematic viscosity v is given by

$$v = \frac{\mu}{\rho} \tag{301.10}$$

For a Newtonian fluid, the Reynolds number, which is a dimensionless parameter expressing the ratio between the inertial forces and the viscous forces acting on the fluid, is given by

$$\text{Re} = \frac{\rho VD}{\mu} = \frac{VD}{\nu} \tag{301.11}$$

A non-Newtonian fluid, where the relationship between the applied shear stress and the resulting velocity gradient is given by Eq. (301.9), can be further subclassified into pseudoplastic ($n < 1$) and dilatant ($n > 1$) fluids. For non-Newtonian fluids, the Reynolds number is given by

$$\text{Re}' = \frac{\rho V^{2-n} D^n}{K\left(\dfrac{3n+1}{4n}\right)^n 8^{n-1}} \tag{301.12}$$

Example 301.5

A Newtonian fluid with viscosity $\mu = 0.01$ lb-s/ft^2 fills the annular space between a fixed inner cylinder diameter $= 2.0$ in and an outer rotating drum (inner diameter $= 2.04$ in). The length of the assembly is 35 in. The drum is rotated at a constant speed of 2000 rpm. What is the torque required to maintain the rotation of the drum?

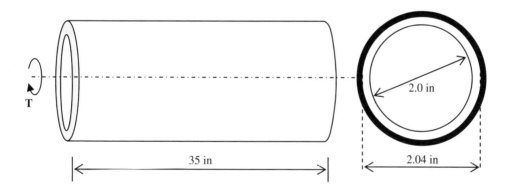

Solution

Rotational speed $\omega = 2000$ rpm $= \dfrac{2000 \times 2\pi}{60} = 209.44$ rad/s.

The tangential velocity of the drum $v = r\omega = 1.02 \times 209.44 = 213.6$ in/s.

Thickness of fluid film $= 0.04 \div 2 = 0.02$ in.

Velocity gradient in the thin fluid film (in the annular space)

$$\frac{dV}{dy} = \frac{213.6}{0.02} = 10681 \text{ s}^{-1}.$$

Shear stress at the surface of the rotating drum

$$\tau = \mu \frac{dV}{dy} = 0.01 \times 10{,}681 = 106.8 \text{ lb/ft}^2 .$$

The (cylindrical) surface on which this shear stress acts is

$$A = 2\pi r L = 2\pi \times 1.0 \times 35 = 220 \text{ in}^2 = 1.527 \text{ ft}^2$$

Therefore, the torque needed is the moment of the resultant surface force about the axis

$$T = \tau \times A \times r = 106.8 \times 1.527 \times 1.0 \text{ lb-in} = 14 \text{ lb-ft}$$

Table 301.1 summarizes some of the fundamental properties of water as a function of temperature.

Table 301.2 summarizes some of the fundamental properties of atmospheric air as a function of elevation above mean sea level.

Table 301.1 Properties of Water (U.S. Units)

Temp (°F)	Specific weight (lb/ft³)	Density (slug/ft³)	Dynamic viscosity (lb-s/ft²) × 10⁻⁵	Kinematic viscosity (ft²/s) × 10⁻⁵	Surface tension (lb/in) × 10⁻⁴	Vapor pressure (lb/in²)
32	62.42	1.940	3.746	1.931	4.32	0.09
40	62.43	1.940	3.229	1.664	5.12	0.12
50	62.41	1.940	2.735	1.410	4.24	0.18
60	62.37	1.938	2.359	1.217	4.20	0.26
70	62.30	1.936	2.050	1.059	4.15	0.36
80	62.22	1.934	1.799	0.930	4.10	0.51
90	62.11	1.931	1.595	0.826	4.05	0.70
100	62.00	1.927	1.424	0.739	4.00	0.95
110	61.86	1.923	1.284	0.667	3.94	1.27
120	61.71	1.918	1.168	0.609	3.89	1.69
130	61.55	1.913	1.069	0.558	3.83	2.22
140	61.38	1.908	0.981	0.514	3.78	2.89
150	61.20	1.902	0.905	0.476	3.73	3.72
160	61.00	1.896	0.838	0.442	3.68	4.74
170	60.80	1.890	0.780	0.413	3.62	5.99
180	60.58	1.883	0.726	0.385	3.56	7.51
190	60.36	1.876	0.678	0.362	3.50	9.34
200	60.12	1.868	0.637	0.341	3.44	11.52
212	59.83	1.860	0.593	0.319	3.37	14.70

Table 301.2 Properties of Air (U.S. Units)

Elevation (ft)	Temperature (°F)	Density (slug/ft³ × 10⁻⁵)	Kinematic viscosity (ft²/s × 10⁻⁵)	Pressure (lb/ft²)
0	59.0	237	15.6	2116
1,000	55.4	231	16.0	2041
2,000	51.9	224	16.4	1968
5,000	41.2	205	17.7	1760
10,000	23.4	176	20.0	1455
15,000	5.54	150	22.8	1194
20,000	−12.3	127	26.1	973
25,000	−30.1	107	30.0	785
30,000	−48.0	89	34.7	628
35,000	−65.8	74	40.4	498
40,000	−67.6	59	50.6	392
50,000	−67.6	36	81.8	242
100,000	−67.6	3.3	89.5	22.4
150,000	−113.5	0.3	1.32	3
200,000	−160.0	0.06	6.84	0.665

Static Pressure on Submerged Surfaces

A submerged static object is subject to hydrostatic pressure acting on the various surfaces defining the boundary of the object. The resultant of these static pressures is the buoyancy force acting on the object. For an object of negligible volume, these forces are in static equilibrium and the buoyancy is zero. Archimedes' principle on buoyancy may be stated as

1. The buoyancy force on a submerged object is equal to the weight of the displaced fluid.

2. An object in neutral equilibrium (floating) displaces a volume of fluid whose weight equals the weight of the object.

Thus, an ideal submerged plate (zero thickness, zero volume) develops equal and opposite pressure profiles on either side, and therefore has no buoyancy. The following section outlines the procedure for calculating the static pressure on one side of a plane surface.

Static Pressure on Plane Area of Arbitrary Shape

Figure 301.4 shows a plane surface which is completely submerged and inclined at an angle θ with the horizontal. Z_C and Z_R are distance from the free surface, measured parallel to the plane of the submerged object. The center of gravity of the object is marked as CG and the "effective

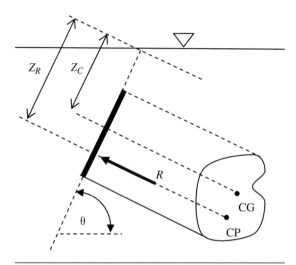

Figure 301.4 Static fluid pressure on a submerged plane surface.

location" of the resultant force is marked as CP (center of pressure). The resultant force on one side of the plane object is given by

$$R = \gamma h_c . A \tag{301.13}$$

The depth of the center of gravity (measured vertically) is

$$h_c = Z_c \sin \theta \tag{301.14}$$

The distance Z_R is given by

$$Z_R = Z_C + \frac{I}{AZ_C} \tag{301.15}$$

where I is the second moment of area (moment of inertia) of the submerged plane about a horizontal axis passing through its center of gravity, and A is the cross-sectional area of the submerged plane.

Example 301.6

A triangular plate serves as a gate for a water storage tank as shown on the next page. The gate is hinged at the top (point A in the tank cross section or line AA in the side view of the gate) and at the bottom (point B). Calculate (1) the resultant force (lbs) acting on the gate, (2) the effective depth of the resultant force (ft), and (3) the force exerted at the hinge B.

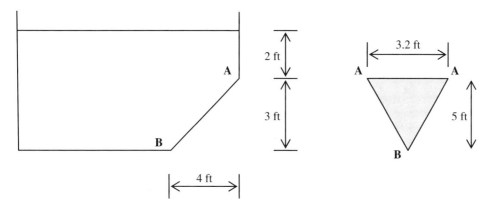

Solution Length of the triangle (perpendicular distance from the vertex B to the base AA) = 5 ft.

Centroidal moment of inertia of a triangle $I = \dfrac{bh^3}{36} = \dfrac{3.2 \times 5^3}{36} = 11.11$ ft^4.

Area of triangle $I = \dfrac{bh}{2} = \dfrac{3.2 \times 5}{2} = 8$ ft^2.

The depth of the centroid is $h_c = 2 + \dfrac{1}{3} \times 3 = 3$ ft.

The inclined distance to the centroid is $Z_c = h_c \times \dfrac{5}{3} = 5$ ft.

The inclined distance to the center of pressure is $Z_R = Z_c + \dfrac{1}{AZ_c} = 5 + \dfrac{11.11}{8 \times 5} =$ 5.278 ft.

1. The resultant force is $R = \gamma h_c . A = 62.4 \times 3 \times 8 = 1497.6$ lb.
2. The (vertical) depth to the center of pressure is $h_R = Z_R \times \dfrac{3}{5} = 3.167$ ft.
3. Since the top of the gate is at depth 2 ft, the bottom is at depth 5 ft, and the depth of the center of pressure is at 3.167 ft, the reaction at B, using the Lever rule, is given by

$$F_B = \frac{3.167 - 2}{5 - 2} \times 1497.6 = 582.6 \text{ lb}$$

Static Pressure on Compound Area (Curved or Multiple Linear Segments)

Calculating the resultant force on a nonplanar surface can be very complex if performed as an integration of the pressure profile acting normal to infinitesimal elements. An easier and equivalent procedure is to perform the calculation in two steps as follows:

1. Draw a vertical plane through the toe of the surface and calculate the horizontal force as the resultant of the pressure acting on the projection of the surface onto this vertical plane. In Fig. 301.5, the toe of the surface is at the point d and the vertical projection of the surface onto the vertical plane is de.

2. Calculate the vertical force as the weight of the fluid above the surface. In Fig. 301.5, this would be the weight of the fluid in the region $abcde$.

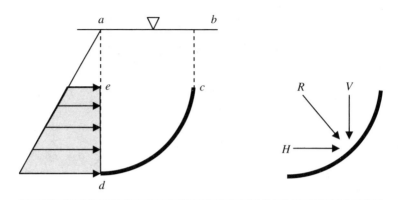

Figure 301.5 Static fluid pressure on a submerged curved surface.

To calculate total force on the curved surface *cd*, draw a vertical plane *de* through the toe. The horizontal force *H* = resultant of the hydrostatic pressure on *de*. The vertical force *V* = weight of fluid in area *abcde*. The resultant force acting on the surface can then be calculated from

$$R = \sqrt{H^2 + V^2}$$

Dynamic Similarity

In conducting model tests on fluid behavior, certain dimensionless groups are used to demonstrate that dynamic similarity (or dissimilarity) exists between different fluid flow fields. For example, if model tests are carried out in a laboratory experiment (such as in a wind tunnel or water tunnel) to study a full-scale phenomenon, one must ensure that the model and prototype flow regimes have similar characteristics. This may be ensured by maintaining the same Reynolds number (Re) or Froude number (Fr) or Mach number (M) in both flow fields.

There are two aspects to maintaining similarity between model and prototype. The first is *geometric similarity*, which is to maintain the same geometric proportions (model is true-to-scale in length, area, and volume). The second is *dynamic similarity*, which is to maintain equal ratios of all types of forces for the model and prototype.

There are various types of forces at work in a fluid field. Some of these are inertial forces, viscous forces, surface tension forces, and gravitational forces. For the field of civil engineering, the two most important of these dimensionless numbers are the Reynolds number and the Froude number.

Case 1—Inertial and Viscous Forces Dominate (Reynolds Number Similarity)

For systems where inertial and viscous forces dominate, the relative effect of these forces is expressed in terms of the Reynolds number (Re). Examples of systems where Reynolds number similarity is desirable are subsonic aircraft, closed pipe-flow (turbulent), pumps, submarines, turbines, and drainage through tank orifices.

Reynolds number

$$\text{Re} = \frac{VD}{v} = \frac{\rho VD}{\mu} \tag{301.16}$$

Thus, if the model and prototype are to have equal Reynolds number,

$$\text{Re}_m = \frac{\rho_m \cdot V_m \cdot L_m}{\mu_m} = \frac{\rho_p \cdot V_p \cdot L_p}{\mu_p} = \text{Re}_p \tag{301.17}$$

This can be rearranged to relate the velocity ratio (or scale) to the length scale as given below. The length scale, velocity scale, and viscosity scale are expressed as λ_L, λ_V, and λ_μ, respectively.

$$\lambda_V = \frac{V_m}{V_p} = \frac{\rho_p}{\rho_m} \frac{L_p}{L_m} \frac{\mu_m}{\mu_p} = \frac{1}{\lambda_\rho} \frac{1}{\lambda_L} \lambda_\mu \tag{301.18}$$

Thus, if the fluid surrounding the prototype structure and the model are the same (density and viscosity are the same), this would reduce to the simpler relation

$$\lambda_V = \frac{V_m}{V_p} = \frac{1}{\lambda_L} \tag{301.19}$$

Example 301.7

A 1:50 scale model of a suspension bridge is tested in a water tunnel. The model is completely immersed in the fluid. If the temperature of the water is 70°F, what is the velocity of the water needed to replicate wind effects (air temperature 60°F) due to a 60 mph wind?

Viscosity of water at 70°F = 1.059×10^{-5} ft²/s.

Viscosity of air at 60°F = 1.58×10^{-4} ft²/s.

Solution   This is a case of submerged flow. Therefore, Reynolds number similarity must be maintained.

$$\text{Re}_m = \frac{V_m \cdot L_m}{v_m} = \frac{V_p \cdot L_p}{v_p} = \text{Re}_p \Rightarrow V_m = V_p \times \frac{D_p}{D_m} \times \frac{v_m}{v_p}$$

$$V_m = 60 \text{ mph} \times \frac{50}{1} \times \frac{v_{\text{water } 70°F}}{v_{\text{air } 60°F}} = 60 \times 50 \times \frac{1.059 \times 10^{-5}}{15.8 \times 10^{-5}} = 201 \text{ mph}$$

Case 2—Inertial and Gravitational Forces Dominate (Froude Number Similarity)

For systems where inertial and gravitational forces dominate, the relative effect of these forces is expressed in terms of the Froude number (Fr). Examples of systems where Froude number similarity is desirable are flows over spillways, weirs, open channel flow with varying surface levels, surface ships, and surge and flood waves.

Froude number

$$\text{Fr} = \frac{V}{\sqrt{gL}} \qquad (301.20)$$

Thus, if the model and prototype are to have equal Froude number,

$$\text{Fr}_m = \frac{V_m}{\sqrt{g_m L_m}} = \frac{V_p}{\sqrt{g_p L_p}} = \text{Fr}_p \qquad (301.21)$$

This can be rearranged to relate the velocity ratio (or scale) to the length scale as given below. The length scale, velocity scale, and gravity scale are expressed as λ_L, λ_V, and λ_g, respectively.

$$\lambda_V = \frac{V_m}{V_p} = \sqrt{\frac{L_m}{L_p} \frac{g_m}{g_p}} = \sqrt{\lambda_L} \sqrt{\lambda_g} \qquad (301.22)$$

Thus, if the gravity field strength for the prototype structure and the model are the same, this would reduce to the simpler relation

$$\lambda_V = \sqrt{\lambda_L} \qquad (301.23)$$

Example 301.8

A 1:50 scale model of a spillway is constructed to replicate flow phenomena for a flow rate of 500 cfs. What must be the flow rate in the model?

Solution This is a case of free surface flow. Therefore, Froude number similarity must be maintained.

$$\text{Fr}_m = \text{Fr}_p \Rightarrow \frac{V_m^2}{L_m g_m} = \frac{V_p^2}{L_p g_p}$$

which can be used to solve for the velocity scale

$$\frac{V_m}{V_p} = \sqrt{\frac{L_m}{L_p} \frac{g_m}{g_p}} = \sqrt{\frac{1}{50} \frac{1}{1}} = \sqrt{0.02} = 0.1414$$

Flow rate is given by $Q = VA$. From a point of view of dimensional analysis, $Q = VL^2$. Therefore, the flow rate scale is given by

$$\frac{Q_m}{Q_p} = \frac{V_m}{V_p} \frac{A_m}{A_p} = \frac{V_m}{V_p} \left(\frac{L_m}{L_p} \right)^2 = 0.1414 \times 0.02^2 = 0.0000566$$

Therefore, $Q_m = 0.0000566 \times 500 = 0.0283$ cfs.

Laminar versus Turbulent Flow

Reynolds number is the flow parameter used to distinguish laminar and turbulent flows. The term *laminar flow* indicates that individual streamlines remain parallel and distinct and mixing does not occur. For turbulent flows, slight perturbations upstream cause adjacent streamlines to cross, thereby inducing mixing. For flow through closed conduits, flow is laminar when the Reynolds number is less than approximately 1000 and turbulent if the Reynolds number is greater than approximately 3000. Between these limits, the flow regime goes through a transition from laminar to turbulent flow.

Figure 301.6 shows laminar and turbulent flow profiles in circular pipes and between parallel plates. The boundary layer for turbulent flow is of insignificant thickness, as seen from the Fig. 301.6(b). As a result, for fully turbulent flow, the average velocity is a good indicator of the true maximum velocity, which occurs at the center streamline. On the other hand, for laminar flow, a parabolic velocity profile develops, given by

$$V(r) = V_{max} \left[1 - \left(\frac{r}{R} \right)^2 \right] \tag{301.24}$$

where r = radial distance from centerline
R = radius (circular pipe) or half the distance between parallel plates

The nominal velocity of flow in a conduit is often calculated as the average velocity which is given by dividing the flow rate by the cross-sectional area. A direct result of the parabolic profile is that for laminar flow in a circular pipe, the maximum velocity is twice the average velocity and for laminar flow between parallel plates, the maximum velocity is 1.5 times the average velocity. For fully turbulent flow in a circular conduit, the maximum velocity is approximately equal to 1.18 times the average velocity.

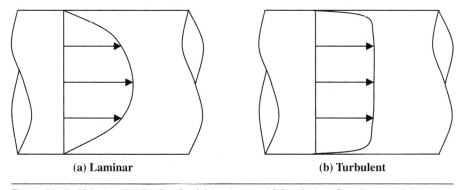

(a) Laminar   **(b) Turbulent**

Figure 301.6 Velocity distribution for (a) laminar and (b) turbulent flow in a conduit.

Example 301.9

The velocity is measured along the centerline of a circular pipe (internal diameter 2 in) experiencing laminar flow is $V = 1.5$ in/s. What is the flow rate (cfs)?

Solution Since the flow is laminar flow in a circular pipe, the theoretical velocity distribution is parabolic and the average velocity is half of the maximum velocity (along centerline).

Thus, average velocity $= 0.75$ in/s.

Flow rate $Q = VA = 0.75 \times \dfrac{\pi}{4} \times 2^2 = 2.36$ in^3/s $= 0.0014$ ft^3/s.

Closed Conduit Hydraulics

General

Pipes conveying flow under the influence of a pressure gradient experience friction losses which are dependent on pipe roughness, flow regime (laminar or turbulent), and flow velocity. The two most prevalent models for quantifying friction loss in pipes are those due to Darcy and Weisbach and to Hazen and Williams. The Darcy-Weisbach model gives a dimensionally consistent formula which requires inputs about Reynolds number and pipe roughness. It is applicable to all fluids. The Hazen-Williams model leads to an empirical equation which yields good results for water at a temperature of 60°F. It is not applicable to other fluids.

Pipelines may need gate valves, check valves, air-release valves, surge control devices, pumping stations, manholes, and so on. Gate valves are provided at regular intervals so that specific sections of the pipeline can be drained for cleaning or repair. Check valves are provided to prevent back flow due to adverse head conditions. Air-release valves are needed at high points to vent the pipeline in order to prevent vacuum formation. Drains are located at low points to allow sediment removal.

For conduits or channels transporting water-containing fine sediment (silt), a minimum velocity of 2 to 3 ft/s is recommended. Factors such as channel erosion and hydraulic surge limit velocities to about 10 to 20 ft/s.

Darcy-Weisbach Equation

According to the Darcy-Weisbach model, the head loss due to friction h_f is given by

$$h_f = f \cdot \frac{L}{D} \cdot \frac{V^2}{2g}$$

(302.1)

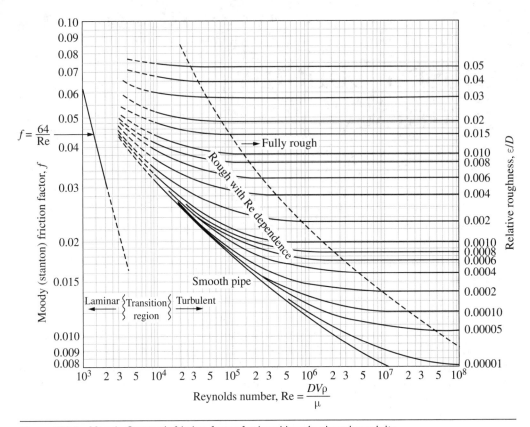

Figure 302.1 Moody-Stanton's friction factor for head loss in closed conduits.

where f = friction factor
L = length of pipe
D = pipe diameter
V = average flow velocity in the pipe

The Darcy friction factor f is a function of relative roughness, ε/D of the pipe, and Reynolds number (Re). The Moody-Stanton diagram for Darcy friction factor is shown in Fig. 302.1.

If a *fully turbulent* flow assumption is made, the flat portion of the Moody diagram may be used. This makes determination of the Reynolds number unnecessary. For example, if the relative roughness is 0.002, friction factor for fully turbulent flow is 0.024.

For Reynolds numbers less than about 2000, laminar flow typically occurs in pipes. The friction factor for laminar flow can be approximated by

$$f_{\text{LAMINAR}} = \frac{64}{\text{Re}} = \frac{64v}{VD} = \frac{64\mu}{\rho VD} \tag{302.2}$$

where v is the kinematic viscosity and μ is the absolute viscosity of the fluid.

Table 302.1 Absolute Roughness for Various Pipe Materials

Pipe material	Absolute roughness ε	
	× 10⁻⁶ ft	× 10⁻⁶ m
Plastic	5	1.5
Brass	5	1.5
Copper	5	1.5
Steel	150	45
Wrought iron	150	45
Asphalted cast iron	400	120
Galvanized iron	500	150
Cast iron	850	260
Concrete	4,000	1,200
Riveted steel	10,000	3,000

Thus, for laminar flow, the friction head loss in a circular pipe is given by the Hagen-Poiseulle equation

$$h_f = \frac{64}{\mathrm{Re}} \cdot \frac{L}{D} \cdot \frac{V^2}{2g} = \frac{32\mu L V}{\rho g D^2} = \frac{128\mu L Q}{\pi \rho g D^4} \tag{302.3}$$

Typical values of absolute roughness of various types of pipe material are given in Table 302.1.

The Moody diagram (Fig. 302.1) is based on the Colebrook-White[1] equation which requires a trial and error solution. An explicit form approximation of the Colebrook equation is given by Swamee and Jain[2].

Example 302.1

Estimate the head loss (ft) due to friction for a cast iron pipe with nominal diameter = 8 in transporting water at a flow rate = 3.0 cfs over a distance of 2450 ft. Use the Darcy-Weisbach equation.

Solution

Flow velocity $V = \dfrac{Q}{A} = \dfrac{3.0}{\pi \times (4/12)^2} = 8.6$ fps

Assuming water temperature = 60°F, viscosity = 1.217×10^{-5} ft²/s

Reynolds number $\mathrm{Re} = \dfrac{VD}{v} = \dfrac{8.6 \times 0.667}{1.217 \times 10^{-5}} = 4.7 \times 10^5$

---

[1]Colebrook C. F. (1939), "Turbulent Flow in Pipes, with Particular Reference to the Transition Region between Smooth and Rough Pipe Laws," *Journal of the Institute of Civil Engineers*, London, Vol. 11.
[2]Swamee, P. K. and Jain, A. K. (1976), "Explicit Equations for Pipe-Flow Problems," *Journal of the Hydraulics Division,* American Society of Civil Engineers, Vol. 102 (5), 657–664.

For cast iron pipe, the absolute roughness is $\varepsilon = 850 \times 10^{-6}$ ft. Therefore, the relative roughness $= \varepsilon/D = 850 \times 10^{-6} = 0.0013$.

Using these values of Re and ε/D in the Moody diagram, the friction factor is approximately $f = 0.021$. Therefore, the Darcy-Weisbach head loss is

$$h_f = f \cdot \frac{L}{D} \cdot \frac{V^2}{2g} = 0.021 \times \frac{2450}{0.667} \times \frac{8.6^2}{2 \times 32.2} = 88.6 \text{ ft}$$

Hazen-Williams Equation

The Hazen-Williams equation should be used only for turbulent flow. It yields good results for liquids having kinematic viscosity around 1.2×10^{-5} ft²/s. Thus, it yields good results for water around 60°F.

The flow velocity according to the Hazen-Williams equation is given by Eq. (302.4)

$$V = 1.318 C R^{0.63} S^{0.54} \quad \text{(US)} \tag{302.4a}$$

$$V = 0.849 C R^{0.63} S^{0.54} \quad \text{(SI)} \tag{302.4b}$$

where $S =$ (decimal) slope of the energy grade line and is equal to h_f/L
$R =$ hydraulic radius (ft or m)
$C =$ Hazen-Williams roughness coefficient. Some typical values of C are given in the Table 302.2.

Table 302.2   Hazen-Williams Roughness Coefficient for Various Pipe Materials

Pipe material	Hazen-Williams constant C	
	Range	Design
Plastic	150–120	130
Brass	150–120	130
Copper	150–120	130
Steel	80–150	100
Wrought iron	80–150	100
Asphalted cast iron	50–140	100
Galvanized iron	80–150	100
Cast iron	80–150	100
Concrete	80–150	100
Riveted steel	80–150	100

Using the expression for average flow velocity in Eq. (302.4), we can write the following expression for head loss in terms of hydraulic radius:

$$h_f = \frac{0.6V^{1.85}L}{C^{1.85}R^{1.165}} \qquad \text{(where } h_f, L, \text{ and } R \text{ are in ft; } V \text{ is in ft/s)} \qquad (302.5a)$$

$$h_f = \frac{1.35V^{1.85}L}{C^{1.85}R^{1.165}} \qquad \text{(where } h_f, L, \text{ and } R \text{ are in m; } V \text{ is in m/s)} \qquad (302.5b)$$

Head Loss in Circular Conduits

For a circular conduit flowing full, substituting $R = D/4$ leads to

$$h_f = \frac{3.022V^{1.85}L}{C^{1.85}D^{1.165}} \qquad \text{(where } h_f, L, \text{ and } D \text{ are in ft; } V \text{ is in ft/s)} \qquad (302.6a)$$

$$h_f = \frac{6.81V^{1.85}L}{C^{1.85}D^{1.165}} \qquad \text{(where } h_f, L, \text{ and } D \text{ are in m; } V \text{ is in m/s)} \qquad (302.6b)$$

To apply the Hazen-Williams equation to noncircular ducts, the hydraulic diameter D_h must be used in place of the variable D. The hydraulic diameter is calculated as four times the hydraulic radius. Alternate versions of Eq. (302.6) are shown in Table 302.3.

The head loss h_f may also expressed as the overall lowering of the energy grade line, given as the product of the slope of the EGL (S) and the length of pipe L

$$h_f = SL \qquad (302.7)$$

Table 302.3 Versions of Hazen-Williams Equation for Head Loss in Circular Conduits

Find	Given parameters	Equation
h_f (ft)	V (ft/s), L (ft), D (ft)	$h_f = \dfrac{3.022V_{fps}^{1.85}L_{ft}}{C^{1.85}D_{ft}^{1.165}}$
h_f (ft)	Q (ft³/s), L (ft), D (ft)	$h_f = \dfrac{4.725Q_{cfs}^{1.85}L_{ft}}{C^{1.85}D_{ft}^{4.865}}$
h_f (ft)	Q (gal/min), L (ft), D (ft)	$h_f = \dfrac{5.862\times10^{-5}\times Q_{gpm}^{1.85}L_{ft}}{C^{1.85}D_{ft}^{4.865}}$
h_f (ft)	Q (gal/min), L (ft), D (in)	$h_f = \dfrac{10.429Q_{gpm}^{1.85}L_{ft}}{C^{1.85}D_{in}^{4.865}}$
h_f (ft)	Q (mgd), L (ft), D (ft)	$h_f = \dfrac{10.63Q_{mgd}^{1.85}L_{ft}}{C^{1.85}D_{ft}^{4.865}}$

Table 302.4 Flow Rate in Closed Conduits (Hazen-Williams)

Find	Given parameters	Equation
Q (mgd)	S (decimal), D (ft)	$Q_{mgd} = 0.279CD^{2.63}S^{0.54}$
Q (ft³/s)	S (decimal), D (ft)	$Q_{cfs} = 0.432CD^{2.63}S^{0.54}$
Q (m³/s)	S (decimal), D (m)	$Q = 0.278CD^{2.63}S^{0.54}$

The equation for the flow rate Q can be also written in terms of the slope of the energy grade line, as shown in Table 302.4.

Example 302.2

Estimate the head loss (ft) due to friction for a cast iron pipe with nominal diameter = 8 in transporting water at a flow rate = 3.0 cfs over a distance of 2450 ft. Use the Hazen-Williams equation.

Solution

Flow velocity $\quad V = \dfrac{Q}{A} = \dfrac{3.0}{\pi \times \left(4/12\right)^2} = 8.6$ fps

Roughness coefficient design value $C = 100$

$$h_f = \frac{3.022V_{fps}^{1.85}L_{ft}}{C^{1.85}D_{ft}^{1.165}} = \frac{3.022 \times 8.6^{1.85} \times 2450}{100^{1.85} \times 0.667^{1.165}} = 126.8 \text{ ft}$$

This was previously (in Example 302.1) calculated using the Darcy-Weisbach equation as 88.6 ft.

Minor Losses

When flow passes through regulating devices such as valves and gates, through path changes such as bends, or through change in flow area such as pipe entrances and exits, there is a loss of flow energy at that location. For typical systems, the magnitude of these losses is small compared to the losses that occur due to friction from the pipe walls. For this reason, these losses are called minor losses. However, like pipe friction loss, these losses are also proportional to the kinetic head or velocity head $V^2/2g$. Thus, minor losses are expressed in a manner very similar to friction losses, except that the source of energy loss is discrete (located at the specific locations of these devices) rather than occurring continuously along the length of the pipe. Each location is assigned a standardized value or coefficient K and the total system minor loss may be expressed in terms of the K values for all devices.

$$h_{minor} = K\frac{V^2}{2g} \tag{302.8}$$

Table 302.5 Typical Minor Loss Coefficients

Type of component or fitting		Form loss factor K
Regular elbows		
	Flanged 90° elbow	0.3
	Threaded 90° elbow	1.5
	Threaded 45° elbow	0.4
Long radius elbows		
	Flanged 90° elbow	0.2
	Threaded 90° elbow	0.7
	Flanged 45° elbow	0.4
Tees		
	Flanged tees, line flow	0.2
	Threaded tees, line flow	0.9
	Flanged tees, branched flow	1.0
	Threaded tees, branched flow	2.0
Globe valve, fully open		10.0
Gate valves		
	Fully open	0.15
	1/4 closed	0.26
	1/2 closed	2.1
	3/4 closed	17.0
Ball valves		
	Fully open	0.05
	1/3 closed	5.5
	2/3 closed	200
Sudden enlargement (use upstream velocity as V)		$K = \left[1 - \left(\dfrac{D_1}{D_2} \right)^2 \right]^2$
Sudden contraction (use downstream velocity as V)		$K = \dfrac{1}{2} \left[1 - \left(\dfrac{D_1}{D_2} \right)^2 \right]$
Pipe exit		1.0
Borda inlet		0.75

Alternatively, by analogy with the Darcy-Weisbach friction loss formula, the loss coefficient K may also be expressed in terms of an equivalent length L_e. This is the length of pipe (of specific material and diameter) which would cause friction loss equivalent to that caused by the device.

$$f\frac{L}{D}\cdot\frac{V^2}{2g} = K\frac{V^2}{2g} \Rightarrow K = \frac{f\cdot L_e}{D} \quad \text{or} \quad L_e = \frac{KD}{f} \qquad (302.9)$$

Thus, the critical parameters for the pipeline are f (friction factor), L (length), and D (diameter), whereas the devices (valves, etc.) and geometry changes (enlargements, bends, etc.) are assigned K values and represented as ΣK. Some typical minor loss coefficients are given in Table 302.5.

Example 302.3

A fully open globe valve ($K = 10$) is installed on a 4-in-diameter cast iron pipe. If the nominal friction factor for the pipe is 0.022, what is the equivalent length (ft) of the valve?

Solution   The equivalent length is given by

$$L_{eq} = \frac{KD}{f} = \frac{10 \times 4}{0.022} = 1818 \text{ in} = 152 \text{ ft}$$

Pipe Deflection

Water mains are typically available in 16 or 20 ft lengths. In locations where the alignment of the pipeline has significant curvature, commensurate bends are utilized. Moderate curvature can be achieved by using sleeves and high deflection couplings. If no such fittings are used, the pipe manufacturer prescribes a maximum allowable deflection at each bell-spigot joint. The figure below shows a deflection angle θ between two successive pipe lengths (L). The broken line approximates a curve that can be created by laying several successive pipe length, each one deflected by the angle θ.

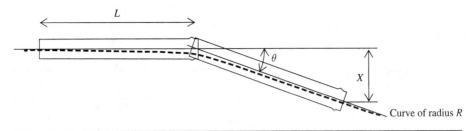

Figure 302.2   Curve radius achievable with pipe deflection.

The lateral deflection (X) between successive pipe axes is given by

$$x = L\sin\theta \qquad (302.10)$$

And the effective radius of curvature (R) is given by

$$R = \frac{L}{2\sin\left(\frac{\theta}{2}\right)}$$

(302.11)

Example 302.4

PVC pipes with 8 in diameter have nominal length of 20 ft. If the manufacturer's prescribed limit on joint deflection is 1°, what are (1) the maximum lateral axis shift permitted and (2) the minimum radius of curvature achievable through joint deflection?

Solution

$$X = L\sin\theta = 20 \times \sin 1° = 0.349\,\text{ft} = 4.19\,\text{in}$$

$$R = \frac{L}{2\sin\left(\frac{\theta}{2}\right)} = \frac{20}{2\sin(0.5°)} = 1145\,\text{ft}$$

Pipe Networks

Pipe networks such as those for municipal water distribution systems are composed of thousands of nodes (connection or branching points) and links (pipe length between two nodes). Additionally, these systems have one or more sources (such as water flowing into the system from a distribution station) and a large number of loads (such as household water connections). The final solution of such a system must satisfy the following conditions:

1. The sum of flow rates at each node must be zero. In other words, total flow rate into each node must be equal to the total flow rate out of that node.

2. The sum of head losses in all links in any closed loop around the network must be zero.

Obtaining such a solution requires an iterative approach. One of the most prominent methods for such a problem is the Hardy-Cross method. The following is an outline of the Hardy-Cross method:

1. A flow pattern is initially assumed and distributed in the network. This initially assumed flow pattern has to satisfy the continuity equation, that is, the first condition stated above.

2. These flow rates are then used to calculate the head loss in each link in the network. The typical relationship between the flow rate Q and the head loss h_f may be written as $h_f = kQ^n$.

Hazen-Williams model: $h_f = \left(\dfrac{4.727L}{C^{1.85}D^{4.865}}\right)Q^{1.85}$ (L in ft, D in ft, and Q in cfs)

Darcy-Weisbach model: $h_f = \left(\dfrac{0.0252fL}{D^5}\right)Q^2$ (L in ft, D in ft, and Q in cfs)

3. For each loop, a correction to the discharge pattern is calculated as

$$\Delta Q = -\frac{\sum kQ^n}{\sum \left| nkQ^{n-1} \right|}$$ (302.12)

In the summation shown above, head losses due to clockwise flows and counterclockwise flows are considered with opposite signs. The flow rates in each loop are adjusted and the method is repeated until it converges.

Alternatively, values of H may be presumed and then the flows balanced by correcting the assumed heads, according to

$$\Delta H = -\frac{n\sum Q}{\sum \left| \frac{Q}{H} \right|}$$ (302.13)

Two-Node Network

While the process described above may be "computationally tedious," a simple two node, two link "network" such as the one in Example 302.5 can be solved without iteration.

Example 302.5

A 6-in-diameter steel pipe is in place to carry a flow rate of 2.4 cfs. The pipe is deemed inadequate and a second 4-in-diameter steel pipe is laid in parallel to the first pipe. The length of both pipes is 1200 ft. What is the velocity of flow in the original 6-in pipe after connecting the new 4-in pipe?

Solution   The flow will split in the pipes such that the head loss h_f in each branch is equal.

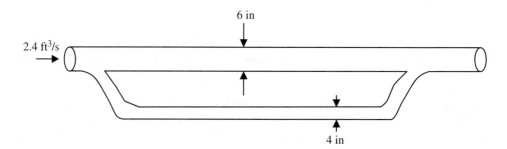

Using the Darcy-Weisbach approach for friction loss in pipes, we can write

$$h_f = f\frac{L}{D}\frac{V^2}{2g} = f\frac{L}{D}\frac{Q^2}{2gA^2} = f\frac{L}{D}\frac{Q^2}{2g\left(\frac{\pi D^2}{4}\right)^2} \Rightarrow h_f \propto \frac{fLQ^2}{D^5}$$

For an approximate solution, let us assume that the friction factor f is the same for both pipes. Since the lengths are also equal,

$$\frac{Q_1^2}{D_1^5} = \frac{Q_2^2}{D_2^5} \Rightarrow \frac{Q_1}{Q_2} = \left(\frac{D_1}{D_2}\right)^{5/2} = \left(\frac{6}{4}\right)^{5/2} = 2.8$$

Split the total flow of $Q = 2.4$ cfs into fractions 2.8/3.8 and 1.0/3.8. Therefore,

$$Q_1 = \frac{2.8}{3.8} \times 2.4 = 1.77 \text{ cfs} \qquad V_1 = \frac{Q_1}{\frac{\pi D^2}{4}} = 9.0 \text{ fps}$$

Note that since the relative roughness of pipe 2 is greater than that of pipe 1 (due to smaller diameter of pipe 2), pipe 1 will actually carry slightly higher flow than 1.77 fps. Thus, the velocity in pipe 1 will be slightly greater than 9 fps.

Also, note that if the Hazen-Williams friction loss equation were used

$$h_f = \frac{4.725 L Q^{1.85}}{C^{1.85} D^{4.865}}$$

and from $h_{f1} = h_{f2}$, we get

$$\frac{Q_1^{1.85}}{D_1^{4.865}} = \frac{Q_2^{1.85}}{D_2^{4.865}} \Rightarrow \frac{Q_1}{Q_2} = \left(\frac{D_1}{D_2}\right)^{4.865/1.85} = \left(\frac{6}{4}\right)^{2.63} = 2.9$$

These two solutions are fairly close.

Fire Hydrant Flow Testing
Hydrant Classifications

Fire hydrants are classified at 20 psi residual pressure and the hydrant tops and nozzle caps are color coded in accordance with NFPA 291 (Recommended Practice for Fire Flow Testing and Marking of Hydrants). The 20 psi pressure is recommended to prevent backflow and the ensuing contamination of the public water supply.

Table 302.6 Fire Hydrant Classifications

Classification	Capacity (gpm)	Color
Class AA	≥1,500	Light Blue
Class A	1,000–1,499	Green
Class B	500–999	Orange
Class C	≤500	Red

Rated Capacity at 20 psi from Fire Hydrant

$$Q_R = Q_F \times \left(\frac{H_R}{H_F}\right)^{0.54}$$

(302.14)

Q_R = rated capacity (gpm) at 20 psi
Q_F = total test flow (gpm)
$H_R = P_s - 20$ (psi)
$H_F = P_s - P_R$ (psi)
P_s = static pressure (psi)
P_R = residual pressure (psi)

Fire Hydrant Discharging to Atmosphere

For a hydrant freely discharging through an outlet, the discharge is related to the stagnation pressure measured according to

$$Q = 29.8 D^2 C_d \sqrt{P}$$

(302.15)

Q = discharge (gpm)
D = outlet diameter (in)
C_d = hydrant coefficient (0.9 for smooth and well-rounded outlet, 0.8 for square and sharp-edged outlet, 0.7 for square outlet projecting into barrel)
P = pressure detected by Pitot tube (psi)

Example 302.6

The static pressure at a hydrant with no flow occurring is 60 psi and when flow is allowed to occur, the pressure drops to 45 psi. The Pitot pressure at the hydrant (smooth and rounded outlet diameter = 2.5 in) is 25 psi. What is the rated flow (gpm) at the desired residual pressure of 20 psi?

Solution   At the given flow condition, discharge Q_F is

$$Q_F = 29.8 \times 2.5^2 \times 0.9 \times \sqrt{25} = 838 \, \text{gpm}$$

At the desired residual pressure of 20 psi
$H_R = 60 - 20 = 40$ psi
$H_F = 60 - 45 = 15$ psi

$$Q_R = Q_F \times \left(\frac{P_s - 20}{P_s - P_R}\right)^{0.54} = 838 \times \left(\frac{40}{15}\right)^{0.54} = 1423 \text{ gpm}$$

Flow Measurement Devices

Flow measurement techniques can be broadly classified as either direct or indirect. Direct methods involve the actual measurement of the volume of fluid passing a point in a specific length of time. Indirect methods usually involve the measurement of pressure changes which are related to the flow rate. For example, an orifice meter restricts the flow by forcing it through an orifice. The resulting pressure drop is measured and since it is related to the velocity of flow, it can also be related to the flow rate.

Pitot Tube

Possibly the simplest flow measurement device is the Pitot tube. It is an open-end tube of very fine bore that is placed parallel to the flow such that the flow impinges on the

Figure 302.3
Pitot tube.

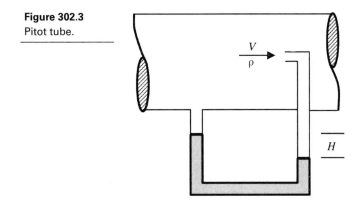

open end (see Fig. 302.3). The flow stagnates at the tip of the tube, thereby converting the kinetic energy to pressure energy. If the pressures within the free stream and at the tip of the tube are compared, they show a pressure differential. This pressure rise can then be recorded on a manometer. The fundamental physics involved in flow measuring devices of this philosophy is the conversion of velocity head ($V^2/2g$) to pressure energy (p/γ). Thus,

$$\frac{V^2}{2g} = \frac{p}{\gamma} \Rightarrow V = \sqrt{\frac{2gp}{\gamma}} = \sqrt{2gH} \qquad (302.16)$$

However, if the monitored fluid and the manometer fluid are different, the pressure head recorded H would change by a factor dependent on the densities of the two fluids. This results in the following expression for the stream velocity V:

$$V = \sqrt{2gH\left(\frac{\rho_m - \rho}{\rho}\right)} \qquad (302.17)$$

where ρ = density of the fluid in the conduit
   ρ_m = density of the manometer fluid
   H = manometer differential column height

Note that this term reappears in the expression for flow rate for the orifice meter and the Venturi meter as well.

Orifice Meter

The orifice meter uses a sharp-edged orifice (diameter D_o) to constrict the flow in a pipe (see Fig. 302.4). Pressure taps are installed at distances D and $1/2\ D$ upstream and downstream of the orifice, respectively. The flow narrows to its smallest diameter (*vena contracta*) downstream of the orifice. The area of the contracted jet A_j is smaller than the area of the orifice A_o by a constant called C_c (coefficient of contraction) given by

$$C_c = \frac{A_j}{A_o} = \left(\frac{d_j}{d_o}\right)^2 \qquad (302.18)$$

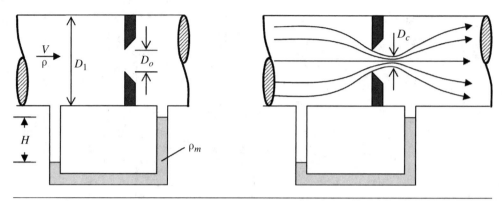

Figure 302.4   Orifice meter.

The flow rate in the conduit is then given by

$$Q = C_f A_o \sqrt{\frac{2gH(\rho_m - \rho)}{\rho}}$$

(302.19)

where the discharge coefficient is given by

$$C_f = \frac{C_v C_c}{\sqrt{1 - C_c^2 A_o^2 / A_1^2}}$$

(302.20)

where A_o and A_1 are the areas of the orifice and the upstream section, respectively. C_v is a velocity coefficient which is less than 1.0 due to significant viscous effects at low to moderate Reynolds numbers. At high Reynolds numbers, $C_v \approx 1.0$.

Venturi Meter

The Venturi meter constricts the flow in a conduit through a small diameter "throat" before allowing the flow to expand to the original diameter. The advantage of the Venturi meter is that while very similar to the orifice meter in principle, it imposes a much smaller head loss because of the more streamlined flow. The two pressure taps (marked 1 and 2 in Fig. 302.5) are located upstream of and in the throat section, respectively. The flow rate for a Venturi meter is given by Eq. (302.21):

$$Q = \frac{A_2 C_d}{\sqrt{1 - \beta^4}} \sqrt{\frac{2gH(\rho_m - \rho)}{\rho}}$$

(302.21)

where $\beta = D_2 / D_1$ and $C_d =$ discharge coefficient (approaches 1.0 for large Reynolds numbers and small β).

In all three equations, if direct pressure measurements are made, then $\Delta p / \gamma$ can take the place of H.

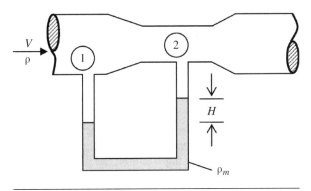

Figure 302.5 Venturi meter.

Electromagnetic Flow Meter

An electromagnetic flow meter works on the basis of the principle that a conductor moving in a magnetic field experiences an electromotive force. Thus, liquids having some conductivity will produce a voltage between electrodes placed transverse to the flow. The EM flow meter produces an output which is linearly proportional to the flow rate and offers no resistance to the flow and therefore causes no energy losses.

Sonic Flow Meter

Sonic flow meters may operate by either (1) recording the difference in travel times of a sound wave traveling upstream and downstream between two monitoring stations, or (2) by recording the difference in frequency (Doppler shift) of a wave reflected back from the flow when an ultrasonic beam is projected into the fluid. Using either of these approaches, it is possible to calculate the flow velocity and therefore the flow rate.

Pumps

When water needs to be conveyed to a community through several miles of pipelines while maintaining adequate water pressure, energy losses due to friction and due to flow obstructions and modifications (valves, bends, etc.) may be significant. To compensate for these losses, pumps are often needed to provide the needed boost in energy.

Pump Selection

There are two broad classifications of pumps—*kinetic* and *positive displacement*. A kinetic pump imparts velocity energy to the fluid, which is converted to pressure energy upon exiting the pump casing. A positive displacement pump moves a fixed volume of fluid within the pump casing by applying a force to moveable boundaries containing the fluid volume.

Kinetic pumps can be further divided into two categories—*centrifugal* and *special effect*. Special effect pumps include jet pumps, reversible centrifugal, gas lift, electromagnetic, and hydraulic ram.

Positive displacement pumps are also divided into two major categories—reciprocating and rotary. *Reciprocating pumps* transfer a volume of fluid by a crankshaft, eccentric cam or an alternating fluid pressure acting on a piston, plunger, or a diaphragm in a reciprocating motion. *Rotary pumps* operate by transferring a volume of fluid in cavities located between rotating and stationary components inside the pump casing.

Very broadly, it can be stated that centrifugal pumps have the lowest cost, followed by rotary and then reciprocating pumps. Pump selection can be affected by head and capacity (flow) requirements, flow variability, and fluid characteristics (viscosity, specific gravity, particulate content, abrasiveness, entrained vapors, and gases). The relative features of reciprocating and rotary pumps, as well as centrifugal pumps, are summarized in Table 302.7.

Table 302.7 Typical Operating Parameters for Various Types of Pumps

Parameter	Centrifugal pumps	Reciprocating pumps	Rotary pumps
Optimum flow rates	Medium/high	Low	Low/medium
Optimum pressures	Low/medium	High	Low/medium
Maximum flow rate (gal/min)	>100,000	>10,000	>10,000
Low flow rate capability	No	Yes	Yes
Maximum pressure (psi)	>6,000	>100,000	>4,000
Requires relief valve	No	Yes	Yes
Smooth or pulsating flow	Smooth	Pulsating	Smooth
Variable or constant flow	Variable	Constant	Constant
Self priming?	No	Yes	Yes
Space requirements	Less space	More space	Less space
COST			
Initial	Low	High	Low
Maintenance	Low	High	Low
Power	High	Low	Low
Fluid characteristics			
Abrasive	Suitable for a wide range including clean, clear, nonabrasive fluids to fluids with abrasive, high-solid content	Suitable for clean, clear, nonabrasive fluids. Specially fitted pumps suitable for abrasive-slurry service	Requires clean, clear, nonabrasive fluid due to close tolerances
High viscosity	Not suitable for high-viscosity fluids	Suitable for high-viscosity fluids	Optimum performance with high-viscosity fluids
Entrained gases	Lower tolerance for entrained gases	Higher tolerance for entrained gases	Higher tolerance for entrained gases

Positive displacement pumps are self-priming and can be used for metering. A self-priming pump, by definition, is a pump which will clear its passages of air if it becomes air bound and resume delivery of the pumpage without external intervention. This is accomplished by the retention of a charge of liquid sufficient to prime the pump in the casing or in an accessory priming chamber. When the pump starts, the rotating impeller creates a partial vacuum, drawing in air from the suction piping and entraining it in the liquid drawn from the priming chamber. The air is separated from the liquid and expelled through the discharge piping, whereas the liquid returns to the priming chamber. This cycle is repeated until all of the air from the suction piping has been expelled and replaced by pumpage. At this point, the pump is "primed."

Kinetic pumps can be higher in efficiency than all but the most efficient reciprocating pumps. Reciprocating pump losses are the least, rotary pumps a little more, and centrifugal pumps show the greatest losses.

Flow Rate and Pressure Head

The centrifugal pump has varying flow depending on the system pressure or head. The positive displacement pump has more or less a constant flow regardless of the system pressure or head. Positive displacement pumps generally give more pressure than centrifugal pumps.

Capacity and Viscosity

In the centrifugal pump, the flow is reduced when the viscosity is increased, whereas in the positive displacement pump, the flow is increased when viscosity is increased due to a higher volumetric efficiency and a positive displacement pump is better suited for high-viscosity applications. A centrifugal pump becomes very inefficient at even modest values of fluid viscosity.

Mechanical Efficiency

Changing the system pressure or head has little or no effect on the flow rate in the positive displacement pump, whereas changing the system pressure or head has a dramatic effect on the flow rate in the centrifugal pump.

Net Positive Suction Head (NPSH)

In a centrifugal pump, net positive suction head (NPSH) varies as a function of flow determined by pressure. In a positive displacement pump, NPSH varies as a function of flow determined by speed. Reducing the speed of the positive displacement pump reduces the NPSH.

Specific Speed

Specific speed is a number characterizing the type of impeller in a unique and coherent manner. Specific speed is determined independent of pump size and can be useful in comparing different pump designs. The specific speed identifies the geometrically similarity of pumps.

Specific speed N_s is dimensionless and is given by

$$N_s = \frac{\omega Q^{1/2}}{H^{3/4}} \tag{302.22}$$

where ω = pump shaft rotational speed (rpm)
    Q = flow rate (gal/min)
    H = head rise (ft)

Typical values for specific speed—N_s—for different designs in U.S. units (see above) are as follows:

- Radial flow—$500 < N_s < 4000$—typical for centrifugal impeller pumps with radial vanes—double and single suction. Francis vane impellers in the upper range.
- Mixed flow—$2000 < N_s < 8000$—more typical for mixed impeller single suction pumps.
- Axial flow—$7000 < N_s < 20{,}000$—typical for propellers and axial fans.

Pump Affinity Laws

The following dimensionless ratios are applicable for analysis of pumps. They are called Affinity Laws.

$$\frac{q_1}{q_2} = \left(\frac{n_1}{n_2}\right)\left(\frac{d_1}{d_2}\right) \tag{302.23}$$

$$\frac{\Delta p_1}{\Delta p_2} = \left(\frac{n_1}{n_2}\right)^2 \left(\frac{d_1}{d_2}\right)^2 \tag{302.24}$$

$$\frac{P_1}{P_2} = \left(\frac{n_1}{n_2}\right)^3 \left(\frac{d_1}{d_2}\right)^3 \tag{302.25}$$

where q = volumetric flow capacity
    n = wheel velocity (rpm)
    d = wheel diameter
    Δp = head or pressure
    P = power

Example 302.7   Pump Affinity Laws—Changing Pump Speed

The pump speed is changed when the impeller size is constant. The initial flow is 100 gpm, the initial head is 100 ft, the initial power is 5 bhp, the initial speed is 1750 rpm, and the final speed is 3500 rpm.

Using Affinity Law 1: The final flow capacity q_2 is given by

$$q_2 = q_1 \left(\frac{n_2}{n_1}\right)\left(\frac{d_2}{d_1}\right) = 100 \times \frac{3500}{1750} \times 1 = 200 \text{ gpm}$$

Using Affinity Law 2: The final head Δp_2 is given by

$$\Delta p_2 = \Delta p_1 \left(\frac{n_2}{n_1}\right)^2 \left(\frac{d_2}{d_1}\right)^2 = 100 \times \left(\frac{3500}{1750}\right)^2 \times 1 = 400 \text{ ft}$$

Using Affinity Law 3: The final power consumption P_2 is given by

$$P_2 = P_1 \left(\frac{n_2}{n_1}\right)^3 \left(\frac{d_2}{d_1}\right)^3 = 5 \times \left(\frac{3500}{1750}\right)^3 \times 1 = 40 \text{ bhp}$$

Example 302.8   Pump Affinity Laws—Changing Impeller Diameter

The diameter of the pump impeller is reduced when the pump speed is constant. The initial flow is 100 gpm, the initial head is 100 ft, the initial power is 5 bhp. The diameter is changed from 8 to 6 in.

Using Affinity Law 1: The final flow capacity q_2 is given by

$$q_2 = q_1 \left(\frac{n_2}{n_1}\right)\left(\frac{d_2}{d_1}\right) = 100 \times 1 \times \frac{6}{8} = 75 \text{ gpm}$$

Using Affinity Law 2: The final head Δp_2 is given by

$$\Delta p_2 = \Delta p_1 \left(\frac{n_2}{n_1}\right)^2 \left(\frac{d_2}{d_1}\right)^2 = 100 \times 1 \times \left(\frac{6}{8}\right)^2 = 56.3 \text{ ft}$$

Using Affinity Law 3: The final power consumption P_2 is given by

$$P_2 = P_1 \left(\frac{n_2}{n_1}\right)^3 \left(\frac{d_2}{d_1}\right)^3 = 5 \times 1 \times \left(\frac{6}{8}\right)^3 = 2.1 \text{ bhp}$$

When a pump is used to deliver a volumetric flow rate Q from location 1 to 2, the modified Bernoulli equation for energy balance is given by

$$H_p = \frac{\eta \dot{W}_s}{\gamma Q} = \frac{p_2 - p_1}{\gamma} + z_2 - z_1 + \frac{V_2^2 - V_1^2}{2g} + h_f \qquad (302.26)$$

where H_p = pump head
$\quad\quad \eta$ = pump efficiency
$\quad\quad \gamma$ = fluid unit weight
$\quad\quad Q$ = volumetric flow rate

The rated power of the pump (also called *the rate of doing shaft work* $\dot{W}_s$) is given by

$$\dot{W}_s = \frac{\gamma Q H_P}{\eta} = \frac{\dot{m} g H_P}{\eta} \qquad (302.27)$$

The mass flow rate (typical units kg/s or slugs/s or lb-s/ft) is given by

$$\dot{m} = \rho Q = \frac{\gamma Q}{g} \qquad (302.28)$$

In U.S. units, typical units of pump power are lb-ft/s, which may be converted to horsepower (hp) or kilowatts (kW) using

$$1 \text{ hp} = \frac{Q_{\text{gpm}} \times H_{\text{ft}} \times \text{SG}}{3960} = 550 \text{ lb-ft/s} = 0.746 \text{ kW} = 746 \text{ N-m/s}$$

where SG is the specific gravity of the fluid.

Improving Pipeline Efficiency

Pressure drops or head losses in liquid pumping systems increase the energy requirements of these systems. Pressure drops are caused by resistance or friction in piping and in bends, elbows, or joints, as well as by throttling across the control valves. The power required to overcome a pressure drop is proportional to both the fluid flow rate and the magnitude of the pressure drop. For example, for water (SG = 1), a pressure drop of 1 psi is equivalent to a head loss of approximately 2.3 ft.

The friction loss and pressure drop caused by fluids flowing through valves and fittings depend on the size and type of pipe and fittings used, the roughness of interior surfaces, and the fluid flow rate and viscosity.

Pumping system controls should be evaluated to determine the most economical control method. High-head-loss valves (such as globe valves) are commonly used for control purposes. Significant losses occur with these types of valves, however, even when they are fully open. If the evaluation shows that a control valve is needed, choose the type that minimizes pressure drop across the valve.

Adjustable speed drives (ASDs) are often recommended for pumping systems that have variable flow rate requirements. When systems are retrofitted with ASDs, the control valve can be removed from the system to eliminate unnecessary pressure drops. The control valve can be replaced with a low-loss replacement valve.

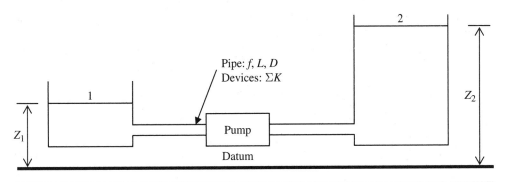

Figure 302.6 Pump delivering flow between two reservoirs.

System Curve

Consider a pump which is used to deliver a rate of flow Q from reservoir 1 (surface elevation Z_1) to reservoir 2 (surface elevation Z_2), as shown in Fig. 302.6. The head loss in the piping system is taken as the sum of friction losses and minor losses. The friction loss may be calculated using either the Darcy-Weisbach equation

$$h_f = f \frac{L}{D} \frac{V^2}{2g}$$

or the Hazen-Williams equation

$$h_f = \frac{LV^{1.85}}{C^{1.85} D^{1.165}}$$

Since both locations 1 and 2 are at reservoir surface, they are under equal pressure ($p_1 = p_2 = p_{atm}$) and since the reservoirs are of large extent compared to the pipe, the continuity equation dictates that velocity is negligible ($V_1 = V_2 \approx 0$). The equation then simplifies to

$$H_p = \left(\frac{p_2 - p_1}{\gamma}\right) + (Z_2 - Z_1) + \left(\frac{V_2^2 - V_1^2}{2g}\right) + h_f + h_M = (Z_2 - Z_1) + f \frac{L}{D} \frac{V_{pipe}^2}{2g} + \Sigma K \frac{V_{pipe}^2}{2g}$$

Since $V_{pipe} = Q/A_{pipe}$, this can also be written in terms of the flow rate Q.

$$H_p = (Z_2 - Z_1) + \left(\frac{fL}{D2gA^2} + \frac{\Sigma K}{2gA^2}\right) Q^2 \qquad (302.29)$$

The elevation difference to be overcome by the pump is called the static head.

$$H_s = Z_2 - Z_1$$

The second term in H_p is due to the flow conveyed by the pipe and is called the dynamic head $H_d = CV^2$ or CQ^2. Thus, the total dynamic head TDH = static head + dynamic head.

Note that if the Hazen-Williams equation for head loss were used, the dynamic head H_d would be of the form

$$H_d = CV^{1.85} \quad \text{or} \quad CQ^{1.85}$$

Example 302.9

The pump shown in the diagram below operates at efficiency of 85%. The suction line is 12 in in diameter and 500-ft long. The discharge line is 8 in in diameter and 2200-ft long. To pump water at a discharge of 1200 gpm, what is the required power rating of the pump (hp)? Assume all pipes to be cast iron. Assume minor losses totaling 15 ft of head loss.

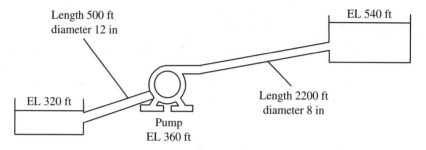

Solution   Flow rate $Q = 1200$ gpm $= 2.674$ ft³/s

Velocity of flow in suction line $= 2.674 \div 0.785 = 3.4$ fps

Reynolds number of flow in suction line $\text{Re} = \dfrac{3.4 \times 1.0}{1.2 \times 10^{-5}} = 2.8 \times 10^5$

Relative roughness of suction line $\dfrac{\varepsilon}{D} = 0.00085$

Friction factor $f = 0.02$

Velocity of flow in discharge line $= 2.674 \div 0.349 = 7.66$ fps

Reynolds number of flow in discharge line $\text{Re} = \dfrac{7.66 \times 0.67}{1.2 \times 10^{-5}} = 4.3 \times 10^5$

Relative roughness of discharge line $\dfrac{\varepsilon}{D} = 0.00128$

Friction factor $f = 0.02$

Note that depending on the relative spread between answer choices, an assumption of $f = 0.02$ for both pipes would have been a very good time-saving assumption.

Head loss (friction) $h_f = \displaystyle\sum f \dfrac{L}{D}\dfrac{V^2}{2g} = \dfrac{0.02 \times 500 \times 3.4^2}{1 \times 2 \times 32.2} + \dfrac{0.02 \times 2200 \times 7.66^2}{0.667 \times 2 \times 32.2} = 62$ ft

Total dynamic head = static head + head loss $= (540 - 320) + 15 + 62 = 297$ ft

Rate power of pump $W_s = \dfrac{\gamma Q H_P}{\eta} = \dfrac{62.4 \times 2.674 \times 297}{0.85} = 58302$ lb-ft/s $= 106$ hp

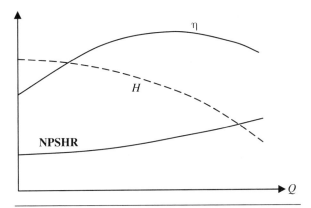

Figure 302.7 Typical pump manufacturer's curves.

Pump Curves

The manufacturer of the hydraulic machine (pump) provides calibrated "pump curves" which plot various pump parameters (efficiency, required suction head, operating head) against pump discharge. These curves may be developed at a factory or in the field. When the system curve is overlaid with the pump curves, the intersection of the H-Q plots yields the operating point for the pump-system combination. A typical set of pump curves is shown in Fig. 302.7. The operating head (H), pump efficiency (η), and net positive suction head required (NPSHR) are shown plotted versus discharge (Q) in this figure.

Every pump has a shutoff head, which is the maximum operating head for the pump. When the head exceeds this value, the discharge becomes zero.

When a particular pump is used in a specific system, the intersection of the pump's H versus Q curve and the system's H versus Q curve gives us the operating point for the pump (see Example 302.10). Once this operating point is identified, parameters such as operating discharge, efficiency, and NPSHR can be estimated from the corresponding manufacturer's curves.

Example 302.10

The manufacturer's curves (H vs. Q and η vs. Q) are given below. The pump is used to deliver water from reservoir A (surface elevation 100 ft above sea level) to reservoir B (surface elevation 140 ft above sea level) through 800 ft of 8-in-diameter cast iron pipe ($C = 100$). Determine the flow rate and the efficiency at the operating point of the pump.

Solution   Using the Hazen-Williams equation to determine the frictional head loss

$$h_f = \frac{3.022\,LV^{1.85}}{C^{1.85}D^{1.165}} = \frac{4.725\,LQ^{1.85}}{C^{1.85}D^{4.865}} = \frac{4.725 \times 800 \times Q^{1.85}}{100^{1.85}0.667^{4.865}} = 5.422Q_{\text{cfs}}^{1.85} = 6.73 \times 10^{-5}Q_{\text{gpm}}^{1.85}$$

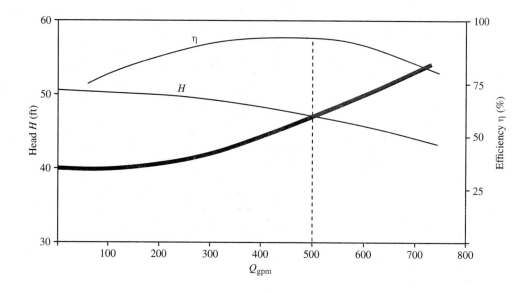

NOTE To convert flow rate in ft³/s to gal/min, multiply by 448.83 ($Q_{gpm} = 448.83 \times Q_{cfs}$).
To convert flow rate in gal/min to ft³/s, multiply by 0.002228 ($Q_{cfs} = 0.002228 \times Q_{gpm}$).

The system curve equation, obtained by applying Bernoulli's equation to reservoir surfaces, is shown below:

$$H_p = h_f + \left(\frac{p_2 - p_1}{\gamma}\right) + (z_2 - z_1) + \left(\frac{V_2^2 - V_1^2}{2g}\right) = h_f + (z_2 - z_1) = 40 + 6.73 \times 10^{-5} Q_{gpm}^{1.85}$$

This is used to plot the system curve (heavy line in the diagram above).

The point of intersection of the system curve and the *H-Q* pump curve is at approximately $Q = 510$ gpm. This is called the operating discharge. At this discharge, the efficiency of the pump is $\eta_{op} = 90\%$.

Pumps in Series

When two pumps operate in series, the discharge line of the first serves as the suction line for the second. Thus, the discharge through each pump is the same and so the total head for their combined operation is the sum of *H* for each pump. This may be accomplished graphically by adding the *H-Q* curves vertically, as shown in Fig. 302.8.

NOTE When an array of pumps A and B (in series) are used instead of using either pump by itself, the intersection with the system curve moves up and to the right, thereby increasing both operating head and discharge.

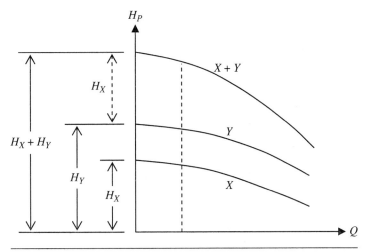

Figure 302.8 Pumps in series—manufacturer's curve.

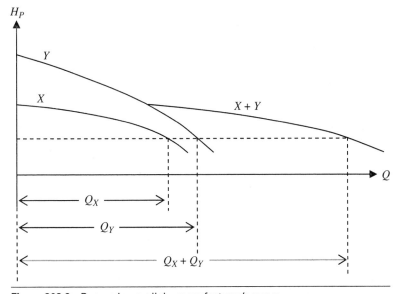

Figure 302.9 Pumps in parallel—manufacturer's curve.

Pumps in Parallel

When pumps are needed to serve a system with highly variable demands, it may be economical to install several pumps in a parallel configuration while using only those needed to satisfy the demand at a particular time. These pumps may be discharging into a common collector pipe that supplies water and the head across all the pumps will be the same; no matter how many pumps are operating. If two dissimilar pumps are operated in parallel, the H-Q curves of the two pumps are combined graphically by adding them horizontally as shown in Fig. 302.9.

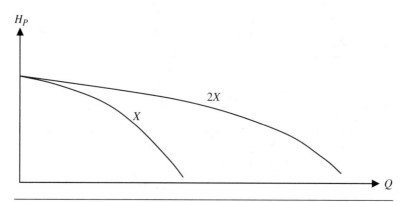

Figure 302.10 Two identical pumps in parallel—manufacturer's curve.

Note that each pump has a shutoff head, which is the maximum head under which it can operate. The pump discharge goes to zero when the head is equal to the shutoff head. Therefore, on the typical *H-Q* curve shown above (Fig. 302.9), the shutoff head is the *y*-intercept for the pump curve. If the operating head is greater than the shutoff head, the pump needs to be equipped with a check valve which prevents flow from passing back through the pump.

If two *identical* pumps are used in parallel, the *H* versus *Q* curve for the array looks as shown in Fig. 302.10. It is apparent that using identical pumps in parallel does not increase the shutoff head.

Cavitation

Liquids can be made to boil either by raising the temperature to the boiling point while pressure remains constant, or by lowering the pressure to the saturation pressure at a particular temperature. For hydraulic machines such as pumps and turbines, the fluid can reach very high velocities. As a result, as predicted by Bernoulli's law of energy conservation, pressures can fall to very low values. When the hydrodynamic pressure is less than the vapor pressure (temperature dependent), small pockets of vapor (cavities) can form spontaneously. These vapor pockets collapse, causing noise, vibration, and pitting corrosion of surfaces. This phenomenon is called cavitation.

Cavitation can occur when the available head is less than the head required for satisfactory operation. The minimum fluid energy required at the pump inlet is the net positive suction head required (NPSHR). This is a specification of the pump manufacturer. The dependence of NPSHR on the pump discharge is quadratic. Thus, one may write

$$\frac{NPSHR_2}{NPSHR_1} = \left(\frac{Q_2}{Q_1}\right)^2 \qquad (302.30)$$

The actual fluid energy (head) at the pump inlet is called the positive suction head available (PSHA). This is a function of the specific parameters of the system. For example, consider flow from a reservoir (free surface elevation Z_1) through a pipe to a pump whose suction line inlet is at elevation Z_2. A schematic of this system is shown in Fig. 302.11. The head losses in the suction line (friction losses plus minor losses) may be written as h_f and the (positive) static head $h_s = Z_1 - Z_2$.

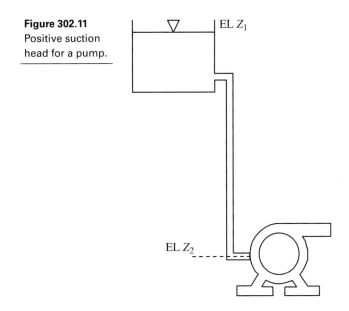

Figure 302.11
Positive suction
head for a pump.

The PSHA at the pump inlet is the sum of all pressure sources contributing to positive pressure minus all the sources of pressure (or head) loss.

$$\text{PSHA} = h_{\text{atm}} + h_s - h_f \qquad (302.31)$$

where h_{atm} = atmospheric pressure head (approximately 33.9 ft)
h_s = elevation head at the pump inlet = $Z_1 - Z_2$
h_f = friction loss in the suction line

Since cavitation becomes possible when the pressure falls below the vapor pressure, the net positive suction head available (NPSHA) is given by

$$\text{NPSHA} = h_{\text{atm}} + h_s - h_f - h_{\text{vp}} \qquad (302.32)$$

where h_{vp} is the vapor pressure (temperature dependent). The vapor pressure increases with water temperature, reaching 1 atmosphere (101.3 kPa) at 100°C (boiling point of water at standard atmospheric pressure). Values of vapor pressure of water at various temperatures are given in Table 301.1. To increase the NPSHA, the following steps may be used:

1. Increase h_s by increasing the height of the fluid source (usually difficult, if not impossible).

2. Increase h_s by lowering the pump (not difficult).

3. Reduce h_f by shortening suction line length.

4. Reduce h_f by increasing suction line diameter (thereby reducing velocity).

5. Reduce the pump speed.

6. Locate the pump in the coldest part of the system, so that the vapor pressure is low.

Where cavitation cannot be eliminated, its effects can be mitigated by

1. The boundaries may be designed so that cavities form and collapse out in the flow field away from the boundaries.

2. Introducing air into the fluid changes cavitation from vaporous to gaseous. This eliminates the explosive growth and complete collapse of cavities, which reduces the potential for damage.

3. The damaging effects of cavitation can be reduced by using materials such as stainless steels.

Cavitation Index

A dimensionless parameter called the cavitation index is often used to quantify the potential for cavitation. A high value of this index signifies low potential for cavitation.

$$\sigma = \frac{p_o - p_{vp}}{\dfrac{\rho V_o^2}{2}} \tag{302.33}$$

where p_o = hydrodynamic pressure
p_{vp} = vapor pressure
ρ = fluid density
V_o = fluid velocity

Example 302.11

A pump is used to pump water from reservoir A (surface elevation 50.0 ft) to reservoir B (surface elevation 220.0 ft). The pump is located at elevation 60 ft as shown. All pipes are 3-in-diameter steel pipe (assume friction factor 0.02). The water temperature is 60°F. What is the maximum flow rate permissible in the system?

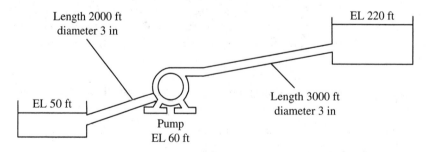

Solution The maximum flow rate is determined by the potential for cavitation, which occurs when the hydrodynamic pressure within the pump falls below the vapor pressure, which is 0.26 psi at 60°F (equivalent to 0.6 ft of water).

Therefore, setting the (absolute) pressure at pump location to be 0.6 ft, we can obtain the maximum flow velocity in the system. Note that the parameters of the system downstream of the pump do not affect this calculation.

$$\frac{p_{atm}}{\gamma} + 50 + \frac{0^2}{2g} - f\frac{L}{D}\frac{V^2}{2g} = h_{vp} + 60 + \frac{V^2}{2g}$$

$$33.9 + 50 + 0 - 0.02\left(\frac{2000}{0.25}\right)\frac{V^2}{2 \times 32.2} = 0.6 + 60 + \frac{V^2}{2 \times 32.2}$$

$$2.5V^2 = 33.9 + 50 + 0 - 0.6 - 60 = 0 \Rightarrow V = 3.05 \text{ fps}$$

$$\text{Flow rate } Q = \left(\frac{\pi D^2}{4}\right)V = 0.15 \text{ cfs} = 67.2 \text{ gpm}$$

Water Hammer (Surge)

Rapidly closing or opening a valve causes pressure transients in pipelines, known as water hammer. Valve closure can result in pressures well over the steady-state values, while valve opening can cause seriously low pressures, possibly so low that the flowing liquid vaporizes (cavitation) inside the pipe.

Mass conservation and momentum conservation are the fundamental equations used to analyze hydraulic transients (water hammer). The equations, subject to the boundary conditions, are not readily solved analytically—a numerical solution is required. Quick-closing valves, positive displacement pumps, and vertical pipe runs can create damaging pressure spikes, leading to blown diaphragms, seals and gaskets, and damage to meters and gauges.

Liquids are virtually incompressible; therefore, any energy that is applied to it is instantly transmitted. This energy becomes dynamic in nature when a force such as quick-closing valve or a pump applies velocity to the fluid.

Surge or water hammer is the result of a sudden change in liquid velocity. Water hammer usually occurs when a transfer system is quickly started, stopped, or is forced to make a rapid change in direction. Any of these events can lead to catastrophic component failures. The primary cause of water hammer in process applications is the quick-closing valve, whether manual or automatic. A valve closing in 1.5 s or less, depending upon valve size and system conditions, can cause an abrupt stoppage of flow. The pressure spike (acoustic wave) created at rapid valve closure can be high as five times the system working pressure.

The equation below can be used to estimate the pressure increase due to the rapid closure of a valve in a pipeline.

$$P = P_1 + \frac{0.07VL}{t} \tag{302.34}$$

where P = increased pressure (psi)
P_1 = inlet pressure (psi)

V = flow velocity (ft/s)
t = valve closing time (s)
L = upstream pipe length (ft)

The critical time for valve closure in a buried ductile pipe of length L is given by

$$t_{min} = \frac{2L}{V} \qquad (302.35)$$

where V is the velocity (fps) of the induced pressure wave given by

$$V = \frac{4720}{\sqrt{1 + (1 - \mu^2)\dfrac{K}{E}\dfrac{D}{t}}} \qquad (302.36)$$

where E and μ = modulus of elasticity and Poisson's ratio of the pipe material
D and t = nominal diameter and the wall thickness of the pipe
K = bulk modulus of the fluid (water)

NOTE   For rigid pipe the ratio K/E approaches zero, and the velocity of the pressure wave approaches 4720 ft/s. Bulk modulus of water is approximately 2.2 GPa. Modulus of elasticity of steel is approximately 200 GPa.

The bulk modulus is related to the compressibility of the fluid and is given by

$$K = -\frac{dp}{\dfrac{dV}{V}} = \frac{dp}{\dfrac{d\rho}{\rho}} \qquad (302.37)$$

where dp = pressure change
dV = resulting volume change
$d\rho$ = change in density

Variation of Atmospheric Pressure with Elevation

As a result of decreasing density of air molecules at higher elevations, atmospheric pressure decreases with increasing altitude. The following empirical relationship is a first-order approximation for atmospheric pressure:

$$\log_{10} P \approx 5 - \frac{h}{15,500} \qquad (302.38)$$

where P = pressure in Pa (N/m^2)
h = elevation (meters)

Table 302.8 Atmospheric Pressure and Air Temperature versus Elevation

Elevation			Atmospheric pressure				Temp
Feet	Meters	Psia	atm	kPa	in Hg	mm Hg	(°C)
0	0	14.7	1.00	101	29.9	760	15.0
328	100	14.5	0.99	100	29.5	752	14.6
500	150	14.4	0.98	99.4	29.4	747	14.0
656	200	14.3	0.97	98.8	29.2	743	13.7
1000	300	14.2	0.96	97.6	28.9	734	13.0
1312	400	14.0	0.95	96.4	28.5	725	12.4
1500	450	13.9	0.94	95.6	28.3	719	12.0
2000	600	13.7	0.93	93.9	27.8	706	11.0
2500	750	13.4	0.91	92.3	27.3	694	10.0
3000	900	13.2	0.89	90.6	26.8	681	9.0
3500	1070	12.9	0.88	88.8	26.3	668	8.0
4000	1220	12.7	0.86	87.1	25.8	655	7.0
4500	1370	12.4	0.85	85.8	25.4	645	6.0
5000	1520	12.2	0.83	84.2	24.9	633	5.0
5500	1680	12.0	0.81	82.5	24.4	620	4.0
6000	1830	11.8	0.80	81.1	24.0	610	3.0
6500	1980	11.5	0.78	79.4	23.5	597	2.0
7000	2130	11.3	0.77	78.1	23.1	587	1.0
7500	2290	11.1	0.76	76.7	22.7	577	0.0
8000	2440	10.9	0.74	75.0	22.2	564	− 0.8
8500	2590	10.7	0.73	73.7	21.8	554	− 1.8
18000	5486	7.35	0.50	50.5	15.0	380	− 20.0
52929	16132	1.47	0.10	10.1	3.0	76	− 56.5

A rough approximation valid for the first few kilometers above the surface is that pressure decreases by 10 kPa/km. Table 302.8 shows atmospheric pressure up to an elevation of 16 km. The last column of the table gives approximate air temperature (which is highly variable with location—an approximate average ground level air temperature of 15°C has been assumed in this table). The "lapse rate" for standard atmosphere is approximately 6.5°C for every 1000 m. The nominal ground level air density is approximately 1.225 kg/m³.

1 standard atmosphere = 14.7 lb/in² = 101.3 kPa = 1.013 bar = 29.9 in of mercury = 760 mm of mercury = 33.9 ft of water = 10.34 m of water. 1 bar = 100 kPa.

Open Channel Hydraulics

Fundamentals

The primary driving force for flow in open channels is gravity, rather than pressure gradients (as in pipes flowing full). The longitudinal slope of the channel bottom is therefore a key parameter in determining flow rate and flow velocity in an open channel. The specific energy of the open channel (total energy per unit weight) is described as the energy of the free surface streamline with respect to the channel bottom.

$$E = d + \frac{V^2}{2g} \tag{303.1}$$

where d = depth of flow
V = average flow velocity
g = acceleration due to gravity

The key dimensionless parameter that defines flow characteristics in an open channel is the Froude number (Fr), given by

$$\text{Fr} = \frac{V}{\sqrt{gd_h}} \tag{303.2}$$

where d_h is the hydraulic depth, given by

$$d_h = \frac{\text{area of flow}}{\text{width at free surface}} \tag{303.3}$$

For a rectangular channel, the hydraulic depth is equal to the actual depth of flow.

The expression $\sqrt{gd_h}$ represents the speed of propagation of a small disturbance wave on the channel surface.

If the Froude number is greater than 1.0, the flow is termed supercritical (shallow depth, high velocity flows), while if the Froude number less than 1.0, the flow is termed subcritical (high depth, low velocity). When the Froude number is equal to 1.0, the channel experiences critical flow. At critical flow, the specific energy of the channel is minimized.

For uniform flow, the depth of flow remains constant and the free surface is parallel to the channel bottom.

Velocity in Open Channels

The flow rate in the channel is given by the product of the average discharge velocity and the flow area [as given by Eq. (303.4)]. Note that the velocity of flow is typically not constant with depth, but reaches a maximum value at some depth below the free surface. The average discharge velocity (given by either the Chezy-Manning equation or the Hazen-Williams equation) does not reflect this variation.

$$Q = VA_f \qquad (303.4)$$

While the driving force for the flow is gravity, induced by the slope of the bottom of the channel S_o, the resisting force is provided by the frictional resistance from the sides of the channel. The side friction from the walls is quantified in terms of the Manning constant n or Hazen-Williams constant C. The equilibrium depth flowing in an open channel is called the normal depth d_n. The average discharge velocity in the channel is given by either the Chezy-Manning empirical formula

$$V = \frac{k}{n} R_h^{2/3} \sqrt{S} \qquad k = 1 \text{ (SI units) or } k = 1.486 \text{ (U.S. units)} \qquad (303.5)$$

or the Hazen-Williams empirical formula

$$V = kCR_h^{0.63} S^{0.54} \qquad k = 0.849 \text{ (SI units) or } k = 1.318 \text{ (U.S. units)} \qquad (303.6)$$

The discharge Q, according to the Chezy-Manning equation, is given by

$$Q = VA = \frac{k}{n} AR_h^{2/3} \sqrt{S} = K\sqrt{S} \qquad (303.7)$$

where K is called the conveyance of the channel. The conveyance is a function of the flow cross section (which determines the area A and the hydraulic radius R_h) and the channel material (which determines the roughness coefficient n).

In both models, the hydraulic radius R_h reflects a combined effect of the flow area and the wetted perimeter and is given by

$$R_h = \frac{\text{area of flow}}{\text{wetted perimeter}} \qquad (303.8)$$

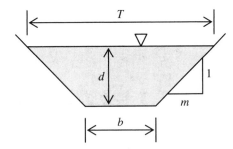

Figure 303.1
Parameters of
a straight-sided
open channel.

The hydraulic diameter D_h is given by four times the hydraulic radius

$$D_h = 4R_h \tag{303.9}$$

Hydraulic Parameters of Straight-Sided Open Channels

An open channel with straight sides is shown in Fig. 303.1. If the side slope parameter $m = 0$, the channel is rectangular. If the bottom width $b = 0$, the channel is triangular. For $b \neq 0$ and $m \neq 0$, the channel is trapezoidal. The hydraulic parameters of a straight-sided channel are given by Eqs. (303.10) to (303.13). Width at the free surface is given by

$$T = b + 2md \tag{303.10}$$

The area of flow is given by

$$A = (b + md)d \tag{303.11}$$

The wetted perimeter is given by

$$P = b + 2d\sqrt{1 + m^2} \tag{303.12}$$

The hydraulic radius is given by

$$R_h = \frac{d(b + md)}{b + 2d\sqrt{1 + m^2}} \tag{303.13}$$

Table 303.1 lists normalized values of top width (T), wetted perimeter (P), flow area (A), and hydraulic radius (R_h) as prescribed by Eqs. (303.10) through (303.13). Only the cases $m = 0$ (rectangular channel) and $m = 1, 2$ are shown.

Table 303.1 Hydraulic Parameters of Symmetric Open Channels with Straight Sides

d/b	m = 0				m = 1				m = 2			
	T/b	P/b	A/b²	R_h/b	T/b	P/b	A/b²	R_h/b	T/b	P/b	A/b²	R_h/b
0.00	1.00	1.0000	0.0000	0.0000	1.00	1.0000	0.0000	0.0000	1.00	1.0000	0.0000	0.0000
0.05	1.00	1.1000	0.0500	0.0455	1.10	1.1414	0.0525	0.0460	1.20	1.2236	0.0550	0.0449
0.10	1.00	1.2000	0.1000	0.0833	1.20	1.2828	0.1100	0.0857	1.40	1.4472	0.1200	0.0829
0.15	1.00	1.3000	0.1500	0.1154	1.30	1.4243	0.1725	0.1211	1.60	1.6708	0.1950	0.1167
0.20	1.00	1.4000	0.2000	0.1429	1.40	1.5657	0.2400	0.1533	1.80	1.8944	0.2800	0.1478
0.25	1.00	1.5000	0.2500	0.1667	1.50	1.7071	0.3125	0.1831	2.00	2.1180	0.3750	0.1771
0.30	1.00	1.6000	0.3000	0.1875	1.60	1.8485	0.3900	0.2110	2.20	2.3416	0.4800	0.2050
0.35	1.00	1.7000	0.3500	0.2059	1.70	1.9899	0.4725	0.2374	2.40	2.5652	0.5950	0.2319
0.40	1.00	1.8000	0.4000	0.2222	1.80	2.1314	0.5600	0.2627	2.60	2.7889	0.7200	0.2582
0.45	1.00	1.9000	0.4500	0.2368	1.90	2.2728	0.6525	0.2871	2.80	3.0125	0.8550	0.2838
0.50	1.00	2.0000	0.5000	0.2500	2.00	2.4142	0.7500	0.3107	3.00	3.2361	1.0000	0.3090
0.55	1.00	2.1000	0.5500	0.2619	2.10	2.5556	0.8525	0.3336	3.20	3.4597	1.1550	0.3338
0.60	1.00	2.2000	0.6000	0.2727	2.20	2.6971	0.9600	0.3559	3.40	3.6833	1.3200	0.3584
0.65	1.00	2.3000	0.6500	0.2826	2.30	2.8385	1.0725	0.3778	3.60	3.9069	1.4950	0.3827
0.70	1.00	2.4000	0.7000	0.2917	2.40	2.9799	1.1900	0.3993	3.80	4.1305	1.6800	0.4067
0.75	1.00	2.5000	0.7500	0.3000	2.50	3.1213	1.3125	0.4205	4.00	4.3541	1.8750	0.4306
0.80	1.00	2.6000	0.8000	0.3077	2.60	3.2627	1.4400	0.4413	4.20	4.5777	2.0800	0.4544
0.85	1.00	2.7000	0.8500	0.3148	2.70	3.4042	1.5725	0.4619	4.40	4.8013	2.2950	0.4780
0.90	1.00	2.8000	0.9000	0.3214	2.80	3.5456	1.7100	0.4823	4.60	5.0249	2.5200	0.5015
0.95	1.00	2.9000	0.9500	0.3276	2.90	3.6870	1.8525	0.5024	4.80	5.2485	2.7550	0.5249
1.00	1.00	3.0000	1.0000	0.3333	3.00	3.8284	2.0000	0.5224	5.00	5.4721	3.0000	0.5482

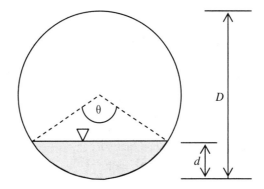

Figure 303.2
Parameters of
a circular open
channel.

Hydraulic Parameters of Circular Open Channels

A circular conduit flowing partially full is shown in Fig. 303.2. The central angle subtended by the free surface is θ (radians) ranging from 0 to 2π. The hydraulic parameters of a circular channel are given by Eqs. (303.14) to (303.17). The width at the free surface is given by

$$T = D\sin\frac{\theta}{2} = 2\sqrt{d(D-d)} \tag{303.14}$$

The area of flow is given by

$$A = \frac{D^2}{8}[\theta - \sin\theta] \qquad \text{where} \quad \theta \ (\text{radians}) = 2\cos^{-1}\left(1 - \frac{2d}{D}\right) \tag{303.15}$$

The wetted perimeter is given by

$$P = \frac{\theta D}{2} \tag{303.16}$$

The hydraulic radius is given by

$$R_h = \frac{D}{4}\left[1 - \frac{\sin\theta}{\theta}\right] \tag{303.17}$$

Some of these hydraulic parameters are listed in Table 303.3.

Friction Loss in Open Channels

Friction head loss in an open channel carrying normal depth flow is given by $h_f = LS$. This is the geometric change in elevation of the channel floor over a distance L. If this is combined with the (Manning) equation for flow velocity, we get

$$h_f = LS = \frac{Ln^2V^2}{k^2R^{4/3}} = \frac{Ln^2Q^2}{k^2R^{4/3}A^2} \qquad k = 1 \ (\text{SI units}) \ \text{or} \ k = 1.486 \ (\text{U.S. units}) \tag{303.18}$$

Table 303.2 Typical Values of Manning's Roughness Coefficient n

Lined canals	
Cement plaster	0.011
Wood, planed	0.012
Wood, unplaned	0.013
Concrete, troweled	0.012
Asphalt, smooth	0.013
Asphalt, rough	0.016
Natural channels	
Gravel beds, straight	0.025
Gravel beds with large boulders	0.040
Earth, straight (some grass)	0.026
Earth, winding (no vegetation)	0.030
Earth, winding	0.050

Using the Hazen-Williams equation, the expression for head loss in an open channel is

$$h_f = \frac{0.6V^{1.85}L}{C^{1.85}R^{1.165}} \qquad \text{where } h_f, L, \text{ and } R \text{ are in ft; } V \text{ is in ft/s} \qquad (303.19)$$

$$h_f = \frac{1.35V^{1.85}L}{C^{1.85}R^{1.165}} \qquad \text{where } h_f, L, \text{ and } R \text{ are in } m; V \text{ is in m/s} \qquad (303.20)$$

Typical values of Manning's roughness coefficient n for various channel conditions are listed in Table 303.2.

Rectangular Open Channels

For a rectangular channel, if the flow rate is Q and the (constant) width is b, then the flow rate per unit width may be written $q = Q/b$. The average velocity of flow is given by

$$V = \frac{Q}{bd} = \frac{q}{d} \qquad (303.21)$$

Therefore, the specific energy of the channel can be written as

$$E = d + \frac{V^2}{2g} = d + \frac{q^2}{2gd^2} \qquad (303.22)$$

The plot of this function is shown in Fig. 303.3. (Only the quadrant $E > 0, d > 0$ is shown.)

Table 303.3 Velocity and Flow Ratios for a Circular Open Channel

d/D	P/D	A/D²	R/D	n/n_full	$n = n_{full}$ V/V_full	$n = n_{full}$ Q/Q_full	$n \neq n_{full}$ V/V_full	$n \neq n_{full}$ Q/Q_full
0.00	0.0006	0.0000	0.0000	1.0000	0.0000	0.0000	0.0000	0.0000
0.02	0.2838	0.0037	0.0132	1.1118	0.1407	0.0007	0.1266	0.0006
0.04	0.4027	0.0105	0.0262	1.1548	0.2219	0.0030	0.1922	0.0026
0.06	0.4949	0.0192	0.0389	1.1847	0.2889	0.0071	0.2439	0.0060
0.08	0.5735	0.0294	0.0513	1.2074	0.3478	0.0130	0.2880	0.0108
0.10	0.6435	0.0409	0.0635	1.2253	0.4009	0.0209	0.3272	0.0170
0.12	0.7075	0.0534	0.0755	1.2397	0.4496	0.0306	0.3627	0.0247
0.14	0.7670	0.0668	0.0871	1.2514	0.4949	0.0421	0.3955	0.0337
0.16	0.8230	0.0811	0.0986	1.2608	0.5372	0.0555	0.4261	0.0440
0.18	0.8763	0.0961	0.1097	1.2684	0.5770	0.0707	0.4549	0.0557
0.20	0.9273	0.1118	0.1206	1.2744	0.6146	0.0876	0.4823	0.0687
0.22	0.9764	0.1281	0.1312	1.2790	0.6502	0.1061	0.5084	0.0830
0.24	1.0239	0.1449	0.1416	1.2823	0.6839	0.1263	0.5334	0.0985
0.26	1.0701	0.1623	0.1516	1.2846	0.7160	0.1480	0.5574	0.1152
0.28	1.1152	0.1800	0.1614	1.2858	0.7465	0.1712	0.5806	0.1332
0.30	1.1593	0.1982	0.1709	1.2862	0.7756	0.1958	0.6030	0.1522
0.32	1.2025	0.2167	0.1802	1.2857	0.8033	0.2217	0.6248	0.1725
0.34	1.2451	0.2355	0.1891	1.2845	0.8296	0.2489	0.6459	0.1938
0.36	1.2870	0.2546	0.1978	1.2825	0.8548	0.2772	0.6665	0.2161
0.38	1.3284	0.2739	0.2062	1.2799	0.8787	0.3066	0.6866	0.2395
0.40	1.3694	0.2934	0.2142	1.2767	0.9015	0.3369	0.7061	0.2639
0.42	1.4101	0.3130	0.2220	1.2729	0.9232	0.3682	0.7253	0.2893
0.44	1.4505	0.3328	0.2295	1.2685	0.9438	0.4002	0.7440	0.3155
0.46	1.4907	0.3527	0.2366	1.2637	0.9633	0.4329	0.7623	0.3426
0.48	1.5308	0.3727	0.2435	1.2583	0.9818	0.4662	0.7803	0.3705
0.50	1.5708	0.3927	0.2500	1.2525	0.9993	0.4999	0.7978	0.3992
0.52	1.6108	0.4127	0.2562	1.2462	1.0157	0.5341	0.8150	0.4285
0.54	1.6509	0.4327	0.2621	1.2396	1.0312	0.5684	0.8319	0.4586
0.56	1.6911	0.4526	0.2676	1.2325	1.0457	0.6029	0.8484	0.4892
0.58	1.7315	0.4724	0.2728	1.2250	1.0591	0.6374	0.8646	0.5203
0.60	1.7722	0.4920	0.2776	1.2172	1.0716	0.6718	0.8804	0.5519
0.62	1.8132	0.5115	0.2821	1.2090	1.0831	0.7059	0.8959	0.5839
0.64	1.8546	0.5308	0.2862	1.2005	1.0936	0.7396	0.9110	0.6161
0.66	1.8965	0.5499	0.2900	1.1916	1.1031	0.7729	0.9257	0.6486
0.68	1.9391	0.5687	0.2933	1.1825	1.1116	0.8054	0.9400	0.6811

(*Continued*)

Table 303.3 Velocity and Flow Ratios for a Circular Open Channel (*Continued*)

d/D	P/D	A/D²	R/D	n/n_full	V/V_full (n = n_full)	Q/Q_full (n = n_full)	V/V_full (n ≠ n_full)	Q/Q_full (n ≠ n_full)
0.70	1.9823	0.5872	0.2962	1.1730	1.1190	0.8371	0.9539	0.7137
0.72	2.0264	0.6054	0.2987	1.1632	1.1253	0.8679	0.9674	0.7461
0.74	2.0715	0.6231	0.3008	1.1532	1.1305	0.8975	0.9803	0.7783
0.76	2.1176	0.6405	0.3024	1.1429	1.1345	0.9257	0.9927	0.8100
0.78	2.1652	0.6573	0.3036	1.1323	1.1374	0.9524	1.0045	0.8412
0.80	2.2143	0.6736	0.3042	1.1214	1.1389	0.9774	1.0156	0.8716
0.82	2.2653	0.6893	0.3043	1.1103	1.1391	1.0003	1.0260	0.9009
0.84	2.3186	0.7043	0.3038	1.0989	1.1378	1.0209	1.0354	0.9290
0.86	2.3746	0.7186	0.3026	1.0873	1.1349	1.0390	1.0438	0.9555
0.88	2.4341	0.7320	0.3007	1.0755	1.1303	1.0541	1.0509	0.9801
0.90	2.4981	0.7445	0.2980	1.0635	1.1235	1.0657	1.0564	1.0021
0.92	2.5681	0.7560	0.2944	1.0512	1.1143	1.0732	1.0600	1.0209
0.94	2.6467	0.7662	0.2895	1.0387	1.1019	1.0756	1.0608	1.0355
0.96	2.7389	0.7749	0.2829	1.0260	1.0851	1.0713	1.0576	1.0441
0.98	2.8578	0.7816	0.2735	1.0131	1.0610	1.0566	1.0473	1.0429
1.00	3.1416	0.7854	0.2500	1.0000	0.9993	0.9999	0.9993	0.9999

Note that the function is cubic in nature, exhibiting a minimum for E (called the critical point). The minimum value of E is called the critical specific energy E_c and the depth corresponding to it is called the critical depth d_c.

The critical depth for a rectangular open channel is given by minimizing E with respect to the depth d, leading to the following expressions for critical depth d_c, critical velocity V_c, and the critical specific energy E_c.

$$d_c = \sqrt[3]{\left(\frac{q^2}{g}\right)} = \sqrt[3]{\left(\frac{Q^2}{gb^2}\right)} \tag{303.23}$$

$$V_c = \sqrt[3]{qg} = \sqrt[3]{\left(g\frac{Q}{b}\right)} \tag{303.24}$$

$$E_c = \frac{3}{2}d_c \tag{303.25}$$

Flows with depth less than the critical depth are called supercritical flow (shallow, rapid flow) and flows with depth greater than the critical depth are called subcritical flow (deep, tranquil flow).

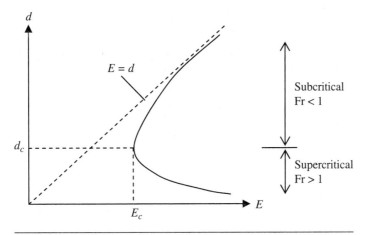

Figure 303.3 Specific energy versus depth for rectangular open channel.

Alternate Depths

Two depths with the same value of specific energy E are called alternate depths. One of these is on the subcritical branch ($d_1 > d_c$) and the other on the supercritical branch ($d_2 < d_c$) as shown in Fig. 303.4.

Close to the critical point, flow in an open channel is very sensitive to channel conditions. When the specific energy is close to its minimum value, a very small perturbation to such flows (such as small changes in the channel bottom profile) can cause large changes to the type of flow. The effect of a localized perturbation, such as an abrupt change in the channel floor, can be illustrated in Example 303.1.

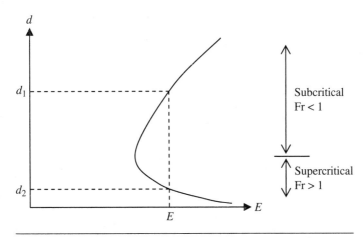

Figure 303.4 Alternate depths for rectangular open channel.

Example 303.1

The diagram shows a rectangular (width = 8 ft) open channel with an approach flow velocity of 6.25 fps and a depth of 4 ft. At location A, there is an abrupt rise of the channel bottom by 4 in. What is the effect on the channel surface due to the "bump" on the channel bottom?

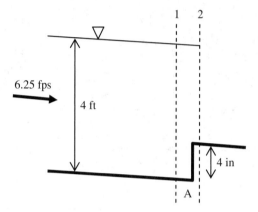

Solution   Considering two planes 1 and 2 with negligible separation, we may assume that the absolute energy (with respect to a horizontal datum) remains constant. However, if the specific energy is used as defined earlier (i.e., measured with respect to the channel bottom), then the specific energy at location 2 (E_2) is less than that at location 1 (E_1) by the magnitude of the channel bottom "bump" (4 in = 0.333 ft).

$$E_2 = E_1 - 0.333$$

Approach flow rate $Q = 6.25 \times 4 \times 8 = 200$ cfs

For a rectangular channel, the actual depth of flow is equal to the hydraulic depth. The approach Froude number is given by

$$\text{Fr}_1 = \frac{6.25}{\sqrt{32.2 \times 4}} = 0.55$$

The specific energy E_1 is given by

$$E_1 = d_1 + \frac{V_1^2}{2g} = 4 + \frac{6.25^2}{2 \times 32.2} = 4.607 \text{ ft}$$

Therefore, $E_2 = E_1 - 0.333 = 4.274\,\text{ft}$.

Using the continuity equation, we have $V_1 d_1 = V_2 d_2 \Rightarrow V_2 = \dfrac{V_1 d_1}{d_2} = \dfrac{6.25 \times 4}{d_2} = \dfrac{25}{d_2}$

$$E_2 = d_2 + \frac{V_2^2}{2g} = d_2 + \frac{25^2}{2 \times 32.2 \times d_2^2} = d_2 + \frac{9.705}{d_2^2} = 4.274$$

Solving this by trial and error, we get $d_2 \approx 2.125$ ft.

Thus, the water surface dips by $4.0 - 0.333 - 2.125 = 1.542$ ft $= 18.5$ in.

The Froude number at location 2 is

$$\mathrm{Fr}_2 = \frac{11.77}{\sqrt{32.2 \times 2.125}} = 1.42$$

Thus, if a subcritical flow approaches a bump in the channel floor, the flow surface dips.

As a point of reference, it may be noted that the critical depth for this rectangular channel is

$$d_c = \sqrt[3]{\left(\frac{q^2}{g}\right)} = \sqrt[3]{\left(\frac{25^2}{32.2}\right)} = 2.687\,\text{ft}$$

In this case, the introduction of a 4-in bump in the channel floor causes the flow to change from subcritical to supercritical.

Supercritical Approach Flow

Results are summarized for an approach velocity of 12.0 fps and approach depth 2.08 ft (same flow rate as before $Q = 12.0 \times 2.08 \times 8 = 200$ cfs).

Approach flow rate $Q = 12.0 \times 2.08 \times 8 = 200$ cfs, and approach Froude number

$$\mathrm{Fr}_1 = \frac{12}{\sqrt{32.2 \times 2.08}} = 1.47$$

$$E_1 = d_1 + \frac{V_1^2}{2g} = 2.08 + \frac{12^2}{2 \times 32.2} = 4.316 \text{ ft} \Rightarrow E_2 = 3.983$$

$$V_1 d_1 = V_2 d_2 \Rightarrow V_2 = \frac{V_1 d_1}{d_2} = \frac{12 \times 2.08}{d_2} = \frac{25}{d_2}$$

$$E_2 = d_2 + \frac{v_2^2}{2g} = d_2 + \frac{25^2}{2 \times 32.2 \times d_2^2} = d_2 + \frac{9.705}{d_2^2} = 3.983$$

Solving this by trial and error, we get $d_2 \approx 2.688$ ft.

Thus, the water surface rises by $2.688 + 0.333 - 2.08 = 0.941$ ft $= 11.3$ in. Thus, if a supercritical flow approaches a bump in the channel floor, the flow surface rises.

Example 303.2

A rectangular channel (width = 10 ft) conveys a flow rate $Q = 60$ cfs at a depth of 0.5 ft. A smooth step of height 0.2 ft obstructs the flow. What is the depth of flow above the step?

Solution For a rectangular channel, the actual depth of flow is equal to the hydraulic depth. Therefore, Froude number is given by

$$Fr = \frac{V}{\sqrt{gd}}$$

Due to the raising of the channel floor, the specific energy above the step is

$$E_2 = E_1 - 0.2$$

Approach flow rate $Q = 60$ cfs and approach Froude number

$$Fr_1 = \frac{12}{\sqrt{32.2 \times 0.5}} = 2.99$$

The specific energy E_1 is given by

$$E_1 = d_1 + \frac{V_1^2}{2g} = 0.5 + \frac{12^2}{2 \times 32.2} = 2.736 \text{ ft}$$

Therefore, $E_2 = E_1 - 0.2 = 2.536$ ft.

Using the continuity equation, we have $V_1 d_1 = V_2 d_2 \Rightarrow V_2 = \frac{V_1 d_1}{d_2} = \frac{12 \times 0.5}{d_2} = \frac{6}{d_2}$

$$E_2 = d_2 + \frac{V_2^2}{2g} = d_2 + \frac{6^2}{2 \times 32.2 \times d_2^2} = d_2 + \frac{0.559}{d_2^2} = 2.536$$

Solving this by trial and error, we get $d_2 \approx 0.53$ ft.

Effect of Variations in the Channel Floor

The results obtained in the previous section can also be explained by the following differential equation relating the depth of flow d and the elevation of the channel floor z.

$$(1 - Fr^2)\frac{dd}{dx} + \frac{dz}{dx} = 0 \tag{303.26}$$

Thus, when the approach flow is subcritical (Fr < 1), a bump in the channel floor ($dz/dx > 0$) causes a lowering ($dd/dx < 0$) of the water surface. When the approach flow is supercritical (Fr > 1), a bump in the channel floor ($dz/dx > 0$) causes a rising ($dd/dx > 0$)

of the water surface. The effect of a lowering of the channel floor (trench) is exactly the opposite in either case.

Momentum in Open Channels

In general, the principle of conservation of momentum can be stated as

Net force acting on a fluid domain between two boundaries is equal to the rate of change of momentum between those boundaries.

Thus, when there are no forces acting on the fluid between sections 1 and 2, the momentum is conserved. The generalized expression for specific force (force per unit weight acting on the water) is given by the sum of the pressure force and the momentum flow through the channel.

$$\frac{F}{\gamma} = A\bar{d} + \frac{Q^2}{gA} \tag{303.27}$$

where A = area of flow
  $\bar{d}$ = distance from centroid to water surface
  Q = flow rate
  g = acceleration due to gravity

Applied to a rectangular open channel, the principle of conservation of momentum can be written as

$$\frac{d_1^2}{2} + \frac{qV_1}{g} = \frac{d_2^2}{2} + \frac{qV_2}{g} \tag{303.28}$$

where the momentum per unit width m may be written in terms of the flow rate per unit width q as

$$m = \frac{d^2}{2} + \frac{qV}{g} \tag{303.29}$$

Thus, if the depth and velocity of flow are known at a location upstream of a channel transition, then the depth and/or velocity at a location downstream of the transition can be solved.

Example 303.3

A rectangular channel (width = 5 ft) conveys a flow rate $Q = 300$ cfs at a depth of 10 ft. The channel smoothly transitions to a 4-ft-wide channel with no change in elevation. What is the depth of flow after the width transition?

Solution   Flow rate per unit width $q = 300/5 = 60$ ft^2/s

The critical depth is calculated as $d_c = \left(\dfrac{60^2}{32.2}\right)^{1/3} = 4.82$

Approach depth = 10 ft (subcritical)

$$M = \frac{d^2}{2} + \frac{q^2}{gd} = \frac{10^2}{2} + \frac{60^2}{32.2 \times 10} = 61.2$$

For narrow channel, critical depth $d_c = \left(\frac{75^2}{32.2}\right)^{1/3} = 5.59$

Using the principle of conservation of momentum, we may write

$$\frac{d^2}{2} + \frac{75^2}{32.2 \times d} = 61.2 \Rightarrow d = 3.1, 9.2$$

Of these, since the approach flow is subcritical, the depth of 9.2 ft is accessible through the transition.

Most Efficient Channel Section

The most efficient open channel (best hydraulic section) minimizes wetted perimeter for a given flow area. For a rectangular section, the best hydraulic section is one for which the depth of flow is exactly half the channel width. For a trapezoidal section, the best hydraulic section is half a hexagon. For a circular section, it is the half circle. For a triangular section, it is half a square. Of all channel sections, the semicircular section is the most efficient. These four types of open channel sections are shown in Fig. 303.5. The variable T is the width at the top (free) surface and b is the width of the base (for the rectangular and trapezoidal channels).

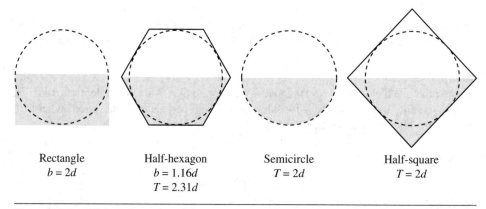

Rectangle
$b = 2d$

Half-hexagon
$b = 1.16d$
$T = 2.31d$

Semicircle
$T = 2d$

Half-square
$T = 2d$

Figure 303.5 Efficient channel sections.

Example 303.4

A rectangular concrete channel is to be built to convey water. It will be placed on a 1% grade and will discharge into a natural earth channel with a slope of 5.28 ft/min. The design discharge is 250 cfs and Manning's roughness for the concrete channel is 0.015. What is the width of the most hydraulically efficient concrete channel, allowing 1 ft freeboard?

Solution   Most efficient rectangular channel section has $d = b/2$.

$$R = \frac{bd}{b+2d} = \frac{b\dfrac{b}{2}}{b+2\dfrac{b}{2}} = \frac{\dfrac{b^2}{2}}{2b} = \frac{b}{4} \qquad \text{and} \qquad A = bd = b\frac{b}{2} = \frac{b^2}{2}$$

$$Q = \frac{1.486}{n} AR^{2/3} S^{1/2} = \frac{1.486}{n} \frac{b^2}{2} \left(\frac{b}{4}\right)^{2/3} S^{1/2} = \frac{0.296}{n} b^{8/3} S^{1/2} \Rightarrow b = \left[\frac{nQ}{0.296\sqrt{S}}\right]^{3/8}$$

If $n = 0.015$

$$b = \left|\frac{nQ}{0.296\sqrt{S}}\right|^{3/8} = \left|\frac{0.015 \times 250}{0.296\sqrt{0.01}}\right|^{3/8} = 6.15 \text{ ft}$$

Normal Depth of Flow in Circular Open Channels

Figure 303.2 shows a circular conduit flowing partially full as an open channel. The key hydraulic parameters for circular open channels, such as wetted perimeter, flow area, and hydraulic radius are outlined in the section "Hydraulic Parameters of Circular Open Channels."
   Using the Chezy-Manning equation for average velocity

$$V = \frac{k}{n} R^{2/3} S^{1/2} \qquad k = 1.486 \text{ (U.S.)} \qquad k = 1.0 \text{ (SI)} \tag{303.30}$$

The flow rate is given by

$$Q = VA = \frac{0.0496k}{n} \left[\frac{(\theta - \sin\theta)^5}{\theta^2}\right]^{1/3} D^{8/3} S^{1/2} \tag{303.31}$$

The Manning coefficient n varies with depth according to the empirical relationship

$$\frac{n}{n_{\text{full}}} = 1 + \left(\frac{d}{D}\right)^{0.54} - \left(\frac{d}{D}\right)^{1.20} \tag{303.32}$$

If the pipe is flowing full, the velocity and flow rate are given by

$$V_f = \frac{0.397k}{n_{full}} D^{2/3} S^{1/2} \tag{303.33}$$

$$Q_f = \frac{0.312k}{n_{full}} D^{8/3} S^{1/2} \tag{303.34}$$

Given a flow rate, the required diameter of a circular conduit (flowing full) is given by

$$D_{reqd} = 1.548 \left(\frac{nQ}{k\sqrt{S}} \right)^{3/8} \qquad k = 1.486 \text{ (U.S.)} \qquad k = 1.0 \text{ (SI)} \tag{303.35}$$

Table 303.3 yields key parameters for a circular conduit for both situations—when depth related changes in Manning's n are ignored (i.e., n is taken as the nominal, constant value; $n = n_{full}$) as well as when Manning's n is assumed to vary with depth and this change is incorporated into the results ($n \neq n_{full}$).

In Table 303.3, the cells which represent the maximum value of the velocity series or the flow rate series are highlighted in dark gray. Thus, we may say that a circular pipe conveys the maximum (free-surface) flow when it is flowing at a depth which is approximately 95% of the diameter. This is due to the fact that at this depth, the slight loss of flow area is more than compensated by the decrease in wall friction and the resulting increase in flow velocity. The flow ratios in Table 303.3 are graphically represented in Fig. 303.6.

Example 303.5

What is the minimum size of a circular sewer pipe to convey a flow rate of 18 cfs if the bottom of the pipe slopes at 0.2% and the pipe lining has a Manning $n = 0.013$?

Solution   Using Eq. (303.35)

$$D_{reqd} = 1.548 \left[\frac{0.013 \times 18}{1.486\sqrt{0.002}} \right]^{3/8} = 2.48 \text{ ft} = 29.8 \text{ in} \qquad \text{Use } D = 30 \text{ in.}$$

Example 303.6

What is the depth of flow if a flow rate of $Q = 6$ cfs is conveyed by a 24-in-diameter pipe? Bottom slope $= 0.3\%$ and nominal value of Manning's $n = 0.017$.

Solution   This is an example of a pipe flowing partially full. First, using Eq. (303.34), the flow rate for a circular pipe flowing full is given by

$$Q_f = \frac{0.463}{0.017} \times 2^{8/3} \times 0.003^{1/2} = 9.47 \text{ cfs}$$

$$Q/Q_f = 6.0/9.47 = 0.63$$

Hydraulic-Elements Graph for Circular Sewers

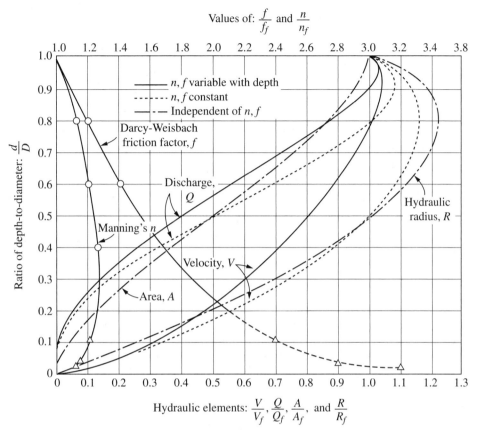

Values of: $\dfrac{f}{f_f}$ and $\dfrac{n}{n_f}$

Hydraulic elements: $\dfrac{V}{V_f}, \dfrac{Q}{Q_f}, \dfrac{A}{A_f},$ and $\dfrac{R}{R_f}$

Design and Construction of Sanitary and Storm Sewers, Water Pollution Control Federation and American Society of Civil Engineers, 1970.

Figure 303.6 Hydraulic elements of circular open channels.

Using this ratio in Fig. 303.6, we obtain the depth ratio d/D.

$d/D = 0.58$, if Manning's n considered constant with depth $d = 13.92$ in

$d/D = 0.65$, if Manning's n considered variable with depth $d = 15.60$ in

Thus, the assumptions of constant n and variable n can lead to significantly different (12% difference in this example) solutions. Typically, on the P.E. exam, the question indicates which option is to be used.

Values in Table 303.3 are more accurate than those read from Fig. 303.6.

Normal Depth of Flow in Channels with Straight Sides

The trapezoidal open channel shown in Fig. 303.1 has the following parameters:

m = side slope parameter

b = bottom width

d = depth of flow

If the geometric parameters of the flow section (depth d or width b) are not known, the Chezy-Manning equation for flow velocity is highly nonlinear. The material presented in this section facilitates the solution of this equation without iteration.

Case 1—Depth of Flow Unknown

The flow rate in an open channel with straight sides may be written as

$$Q = \frac{Kkb^{8/3}S^{1/2}}{n} \qquad k = 1.486 \text{ (U.S.) and } k = 1.0 \text{ (SI)} \qquad (303.36)$$

Table 303.4 summarizes the values of the parameter K for various d/b ratios and side slopes. If the normal depth of flow is unknown, this table greatly facilitates the solution of what would otherwise be the trial and error solution of a nonlinear equation. The following procedure must be used to solve for the unknown depth of flow for a particular flow rate Q:

1. Compute the value of the parameter

$$K = \frac{Qn}{kb^{8/3}S^{1/2}} \qquad k = 1.486 \text{ (U.S.) and } k = 1.0 \text{ (SI)}$$

2. In the appropriate column of the table (given m) search for this computed value of K.
3. The leftmost column in Table 303.4 gives the d/b ratio.
4. The depth of flow is calculated as this d/b ratio multiplied by the known bottom width.

Case 2—Depth of Flow Known

The flow rate in an open channel with straight sides may also be written

$$Q = \frac{K'kd^{8/3}S^{1/2}}{n} \qquad k = 1.486 \text{ (U.S.) and } k = 1.0 \text{ (SI)} \qquad (303.37)$$

Thus, if the normal depth of flow is known but the corresponding bottom width needs to be calculated, Table 303.5 may be used. The following procedure must be used to solve for the unknown width of flow for a particular flow rate Q.

Table 303.4 Values of Parameter *K* for Straight-Sided Open Channels

d/b	Horizontal projection *m*										
	0	0.25	0.5	0.75	1.0	1.5	2.0	2.5	3.0	3.5	4.0
0.01	0.0005	0.0005	0.0005	0.0005	0.0005	0.0005	0.0005	0.0005	0.0005	0.0005	0.0005
0.02	0.0014	0.0014	0.0015	0.0015	0.0015	0.0015	0.0015	0.0015	0.0015	0.0015	0.0015
0.04	0.0044	0.0045	0.0046	0.0046	0.0047	0.0047	0.0048	0.0048	0.0049	0.0049	0.0050
0.06	0.0085	0.0087	0.0089	0.0090	0.0091	0.0093	0.0095	0.0096	0.0098	0.0099	0.0101
0.08	0.0135	0.0139	0.0142	0.0145	0.0147	0.0152	0.0155	0.0159	0.0162	0.0165	0.0168
0.1	0.0191	0.0198	0.0204	0.0209	0.0214	0.0221	0.0228	0.0234	0.0241	0.0247	0.0253
0.12	0.0253	0.0265	0.0275	0.0283	0.0290	0.0303	0.0314	0.0324	0.0334	0.0345	0.0355
0.14	0.0320	0.0338	0.0352	0.0365	0.0376	0.0395	0.0412	0.0428	0.0444	0.0459	0.0475
0.16	0.0392	0.0416	0.0437	0.0455	0.0471	0.0498	0.0523	0.0546	0.0569	0.0591	0.0614
0.18	0.0467	0.0500	0.0529	0.0553	0.0575	0.0612	0.0646	0.0678	0.0710	0.0741	0.0772
0.2	0.0547	0.0589	0.0627	0.0659	0.0687	0.0737	0.0783	0.0826	0.0868	0.0910	0.0952
0.22	0.0629	0.0683	0.0731	0.0772	0.0809	0.0874	0.0932	0.0989	0.1043	0.1098	0.1152
0.24	0.0714	0.0781	0.0841	0.0893	0.0939	0.1021	0.1096	0.1167	0.1237	0.1306	0.1374
0.26	0.0801	0.0884	0.0957	0.1021	0.1078	0.1180	0.1273	0.1361	0.1448	0.1534	0.1619
0.28	0.0891	0.0990	0.1078	0.1156	0.1226	0.1350	0.1464	0.1572	0.1678	0.1783	0.1888
0.3	0.0983	0.1100	0.1205	0.1298	0.1382	0.1532	0.1669	0.1800	0.1928	0.2055	0.2181
0.35	0.1220	0.1392	0.1547	0.1686	0.1812	0.2038	0.2246	0.2446	0.2641	0.2834	0.3025
0.4	0.1468	0.1705	0.1922	0.2118	0.2297	0.2621	0.2919	0.3206	0.3486	0.3763	0.4038
0.45	0.1723	0.2038	0.2330	0.2596	0.2840	0.3283	0.3693	0.4086	0.4472	0.4852	0.5231
0.5	0.1984	0.2390	0.2770	0.3119	0.3440	0.4027	0.4571	0.5094	0.5606	0.6112	0.6614
0.55	0.2251	0.2761	0.3243	0.3688	0.4100	0.4856	0.5558	0.6234	0.6896	0.7550	0.8199
0.6	0.2523	0.3150	0.3748	0.4304	0.4822	0.5773	0.6660	0.7514	0.8350	0.9176	0.9996
0.65	0.2799	0.3557	0.4286	0.4968	0.5605	0.6781	0.7880	0.8938	0.9974	1.0998	1.2015
0.7	0.3079	0.3980	0.4856	0.5681	0.6453	0.7884	0.9222	1.0513	1.1777	1.3027	1.4266
0.75	0.3361	0.4421	0.5460	0.6442	0.7367	0.9083	1.0692	1.2245	1.3766	1.5268	1.6760
0.8	0.3646	0.4879	0.6096	0.7254	0.8347	1.0383	1.2293	1.4138	1.5946	1.7732	1.9505
0.85	0.3934	0.5354	0.6767	0.8118	0.9397	1.1785	1.4030	1.6199	1.8325	2.0426	2.2511
0.9	0.4223	0.5846	0.7471	0.9033	1.0516	1.3292	1.5907	1.8433	2.0911	2.3358	2.5787
0.95	0.4514	0.6354	0.8210	1.0002	1.1708	1.4908	1.7927	2.0846	2.3708	2.6536	2.9343
1	0.4807	0.6879	0.8984	1.1024	1.2973	1.6636	2.0095	2.3442	2.6725	2.9968	3.3187
1.25	0.6292	0.9758	1.3390	1.6981	2.0452	2.7038	3.3294	3.9359	4.5313	5.1199	5.7041
1.5	0.7800	1.3062	1.8733	2.4432	2.9993	4.0625	5.0771	6.0626	7.0306	7.9879	8.9380
1.75	0.9324	1.6804	2.5070	3.3492	4.1778	5.7718	7.2985	8.7836	10.2433	11.6872	13.1206
2	1.0858	2.0998	3.2459	4.4277	5.5985	7.8623	10.0373	12.1556	14.2389	16.3001	18.3465
2.25	1.2399	2.5658	4.0955	5.6898	7.2784	10.3636	13.3356	16.2330	19.0840	21.9052	24.7067
2.5	1.3947	3.0800	5.0614	7.1461	9.2341	13.3042	17.2337	21.0681	24.8425	28.5783	32.2882

Table 303.5 Values of Parameter K' for Straight-Sided Open Channels

d/b	\multicolumn{11}{c}{Horizontal projection m}
	0	0.25	0.5	0.75	1.0	1.5	2.0	2.5	3.0	3.5	4.0
0.01	98.6885	99.0602	99.3590	99.6000	99.7991	100.1201	100.3845	100.6206	100.8416	101.0542	101.2621
0.02	48.7096	49.0775	49.3748	49.6159	49.8163	50.1424	50.4146	50.6605	50.8934	51.1196	51.3426
0.04	23.7497	24.1102	24.4045	24.6456	24.8483	25.1831	25.4682	25.7304	25.9821	26.2292	26.4750
0.06	15.4539	15.8074	16.0988	16.3397	16.5440	16.8859	17.1814	17.4561	17.7219	17.9842	18.2460
0.08	11.3224	11.6693	11.9578	12.1983	12.4039	12.7516	13.0554	13.3400	13.6166	13.8904	14.1638
0.1	8.8555	9.1961	9.4817	9.7217	9.9282	10.2806	10.5912	10.8838	11.1689	11.4514	11.7336
0.12	7.2200	7.5546	7.8374	8.0767	8.2840	8.6402	8.9565	9.2555	9.5473	9.8366	10.1256
0.14	6.0590	6.3879	6.6679	6.9064	7.1142	7.4737	7.7945	8.0988	8.3961	8.6908	8.9850
0.16	5.1939	5.5174	5.7948	6.0324	6.2406	6.6026	6.9274	7.2360	7.5377	7.8368	8.1352
0.18	4.5259	4.8441	5.1188	5.3556	5.5639	5.9282	6.2562	6.5684	6.8738	7.1765	7.4783
0.2	3.9953	4.3086	4.5807	4.8166	5.0250	5.3910	5.7218	6.0371	6.3455	6.6511	6.9557
0.22	3.5645	3.8730	4.1427	4.3776	4.5860	4.9534	5.2866	5.6044	5.9155	6.2235	6.5304
0.24	3.2083	3.5122	3.7795	4.0134	4.2217	4.5904	4.9255	5.2455	5.5588	5.8689	6.1777
0.26	2.9094	3.2088	3.4738	3.7067	3.9148	4.2845	4.6213	4.9432	5.2583	5.5702	5.8806
0.28	2.6552	2.9504	3.2131	3.4450	3.6529	4.0234	4.3616	4.6851	5.0018	5.3152	5.6270
0.3	2.4367	2.7279	2.9883	3.2192	3.4269	3.7980	4.1374	4.4624	4.7804	5.0951	5.4081
0.35	2.0059	2.2875	2.5427	2.7711	2.9780	3.3501	3.6920	4.0197	4.3404	4.6575	4.9728
0.4	1.6895	1.9625	2.2127	2.4387	2.6447	3.0173	3.3608	3.6904	4.0131	4.3320	4.6489
0.45	1.4486	1.7137	1.9593	2.1829	2.3880	2.7606	3.1052	3.4363	3.7603	4.0806	4.3987
0.5	1.2599	1.5177	1.7590	1.9804	2.1844	2.5568	2.9023	3.2343	3.5594	3.8806	4.1996
0.55	1.1087	1.3597	1.5970	1.8163	2.0193	2.3913	2.7372	3.0700	3.3959	3.7179	4.0376
0.6	0.9853	1.2300	1.4635	1.6807	1.8827	2.2543	2.6005	2.9339	3.2604	3.5829	3.9031
0.65	0.8829	1.1218	1.3519	1.5671	1.7681	2.1391	2.4855	2.8193	3.1462	3.4692	3.7898
0.7	0.7969	1.0304	1.2571	1.4705	1.6705	2.0408	2.3874	2.7215	3.0487	3.3721	3.6930
0.75	0.7238	0.9522	1.1758	1.3874	1.5865	1.9562	2.3027	2.6370	2.9646	3.2882	3.6094
0.8	0.6611	0.8847	1.1054	1.3153	1.5135	1.8825	2.2289	2.5634	2.8912	3.2150	3.5364
0.85	0.6067	0.8259	1.0438	1.2521	1.4494	1.8177	2.1641	2.4987	2.8266	3.1507	3.4722
0.9	0.5593	0.7742	0.9895	1.1964	1.3928	1.7604	2.1067	2.4413	2.7694	3.0936	3.4153
0.95	0.5176	0.7286	0.9414	1.1468	1.3424	1.7094	2.0555	2.3902	2.7183	3.0426	3.3644
1	0.4807	0.6879	0.8984	1.1024	1.2973	1.6636	2.0095	2.3442	2.6725	2.9968	3.3187
1.25	0.3470	0.5382	0.7385	0.9366	1.1280	1.4912	1.8363	2.1708	2.4992	2.8238	3.1460
1.5	0.2646	0.4430	0.6354	0.8287	1.0173	1.3779	1.7220	2.0563	2.3846	2.7093	3.0316
1.75	0.2096	0.3778	0.5637	0.7531	0.9394	1.2978	1.6411	1.9750	2.3032	2.6279	2.9502
2	0.1710	0.3307	0.5112	0.6973	0.8817	1.2382	1.5808	1.9144	2.2425	2.5671	2.8894
2.25	0.1426	0.2952	0.4711	0.6545	0.8373	1.1922	1.5341	1.8674	2.1954	2.5200	2.8422
2.5	0.1211	0.2675	0.4396	0.6207	0.8021	1.1556	1.4969	1.8300	2.1579	2.4823	2.8046
∞	0.0000	0.0612	0.1842	0.3361	0.5000	0.8359	1.1696	1.4989	1.8247	2.1479	2.4694

1. Compute the value of the parameter

$$K' = \frac{Qn}{kd^{8/3}S^{1/2}} \qquad k = 1.486 \text{ (U.S.) and } k = 1.0 \text{ (SI)}$$

2. In the appropriate column of the table (given m) search for this computed value of K'.

3. The leftmost column in Table 303.5 gives the d/b ratio.

4. The width b is calculated as this known depth d divided by this d/b ratio.

NOTE Table 303.5 must be used for triangular (V-notch) channels ($b = 0$, $d/b \rightarrow \infty$), drawing the K' parameter from the very last row.

Example 303.7

What is the depth of flow in a trapezoidal channel with a bottom width = 10 ft and side slopes 1V:3H conveying a flow rate $Q = 20$ cfs. Assume $n = 0.018$ and bottom slope = 0.002 ft/ft.

Solution   Using Eq. (303.36), the value of the parameter K is given by

$$K = \frac{Qn}{kb^{8/3}\sqrt{S}} = \frac{20 \times 0.018}{1.486 \times 10^{8/3}\sqrt{0.002}} = 0.0117$$

Side slope parameter $m = 3.0$.

Interpolating from Table 303.4, we get $d/b = 0.0659$.

Therefore, $d = 0.0659 \times 10 \text{ ft} = 0.659 \text{ ft} = 7.91 \text{ in} \approx 8 \text{ in}.$

Example 303.8

What is the flow rate (cfs) in a trapezoidal open channel with 1:1 side slopes in which water flows at a normal depth = 4 ft? The bottom width of the channel is 10 ft. Assume $n = 0.015$ and bottom slope = 0.003 ft/ft.

Solution   Note that the traditional approach may be taken in this case (without resorting to the tables) since depth of flow is known. This would proceed as follows: (1) compute the flow area, (2) calculate the wetted perimeter, (3) calculate the hydraulic radius, (4) using the Chezy-Manning equation, calculate the normal velocity, and (5) calculate the flow rate = product of velocity and flow area.

However, either table may be used (since both depth of flow and bottom width are known). Using Table 303.4, for $m = 1$ and $d/b = 0.40$, $K = 0.2297$.

$$Q = \frac{Kkb^{8/3}S^{1/2}}{n} = \frac{0.2297 \times 1.486 \times 10^{8/3} \times 0.003^{1/2}}{0.015} = 578.6 \text{ cfs}$$

As may be seen, using the table reduces the solution time significantly.
   Alternatively, using Table 303.5, for $m = 1$ and $d/b = 0.40$, $K' = 2.6447$.

$$Q = \frac{K'kd^{8/3}S^{1/2}}{n} = \frac{2.6447 \times 1.486 \times 4^{8/3} \times 0.003^{1/2}}{0.015} = 578.6 \text{ cfs}$$

Critical Depth of Flow in Open Channels

In general, for any open channel (any cross section), the following holds true when flow is occurring at critical depth (Froude number Fr = 1).

$$Q^2T = A^3g \tag{303.38}$$

where Q = volumetric flow rate
   T = width of the channel at the free surface
   A = cross-sectional flow area
   g = acceleration due to gravity

   For natural streams that do not have a regular cross section, bathymetry data may be used to represent the depth versus cross-sectional area and depth versus surface width relationships. These are shown for a typical stream in Fig. 303.7.
   If the depth of flow is known, such curves can then be used to determine the corresponding values of T and A.

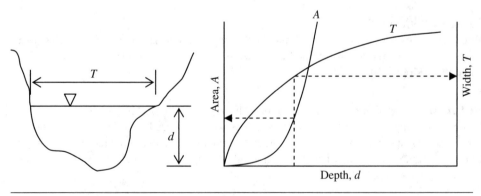

Figure 303.7   Hydraulic parameters of a natural open channel.

Example 303.9

The table below shows flow data for an open channel with an irregular cross section. At which flow rate is the channel most nearly conveying flow at critical depth? What is the corresponding critical velocity?

Depth of flow (ft)	Width at top (ft)	Area of flow (ft²)	Flow rate (ft³/s)
2	3.52	4.12	52
4	4.95	6.75	108
6	7.10	24.90	265
8	10.34	36.22	430

Solution Since at critical depth of flow, the relation $Q^2T = A^3g$ must hold, we calculate the expression Q^2T/A^3 for all four data points.

Point 1: Value $= 136.1$

Point 2: Value $= 187.7$

Point 3: Value $= 32.3$

Point 4: Value $= 40.2$

Of these, 32.3 is numerically closest to $g = 32.2$ ft/s². Thus, when the flow rate is 265, the channel is most nearly conveying critical flow. Therefore, the critical velocity is

$$V_c = \frac{265}{24.9} = 10.6 \text{ fps}$$

Critical Depth in Circular Conduits

If the general principle [Eq. (303.38)] is applied to a circular conduit (diameter D) flowing at critical depth (d_c), the following must hold true:

$$\frac{Q}{\sqrt{gD^5}} = \frac{(\theta - \sin\theta)^{3/2}}{32\left[\frac{d_c}{D} - \left(\frac{d_c}{D}\right)^2\right]^{1/4}} \qquad (303.39)$$

where Q = volumetric flow rate
d_c = critical depth of flow
D = pipe diameter
θ = angle (radians) subtended by the free surface at the center of the conduit

Based on Eq. (303.39), Fig. 303.8 shows a plot of the depth ratio d_c/D versus the dimensionless parameter $Q/\sqrt{gD^5}$.

Note that being dimensionless, this approach can be used in the same way for U.S. and SI units.

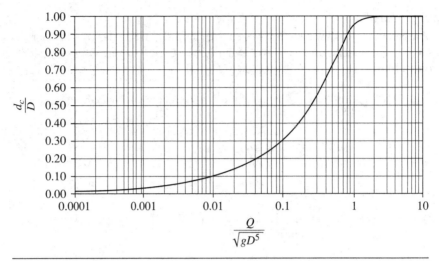

Figure 303.8 Critical depth of flow in circular conduits.

Example 303.10

What is the critical depth of flow in a circular sewer (diameter 7 ft) conveying a flow rate of 500 cfs?

Solution Calculate the parameter $\dfrac{Q}{\sqrt{gD^5}} = \dfrac{500}{\sqrt{32.2 \times 7^5}} = 0.680$

From Fig. 303.8, the critical depth ratio is given by $d_c/D = 0.84$. Critical depth of flow $d_c = 0.84 \times 7 = 5.88$ ft = 5 ft 10.5 in.

Critical Depth in Straight-Sided Channels

Rectangular and Trapezoidal Channels

Using the concept outlined in the section "Critical Depth of Flow in Open Channels," the following relation may be derived for the dimensionless flow parameter for rectangular and trapezoidal open channels:

$$\frac{Q}{\sqrt{gb^5}} = \frac{(\alpha + m\alpha^2)^{3/2}}{(1 + 2m\alpha)^{1/2}} \tag{303.40}$$

where m = side-slope parameter
α = dimensionless parameter d_c/b

Table 303.6 shows values of the parameter $Q/\sqrt{gb^5}$ for various values of α and m.

Table 303.6 Critical Depth Parameter $Q/\sqrt{gb^5}$ for Straight-Sided Open Channels

$\alpha = d_c/b$	Side-slope parameter m				
	$m = 0$	$m = 1$	$m = 2$	$m = 3$	$m = 4$
0.01	0.0010	0.0010	0.0010	0.0010	0.0010
0.02	0.0028	0.0029	0.0029	0.0029	0.0029
0.03	0.0052	0.0053	0.0054	0.0054	0.0055
0.04	0.0080	0.0082	0.0083	0.0085	0.0087
0.05	0.0112	0.0115	0.0118	0.0121	0.0124
0.06	0.0147	0.0152	0.0156	0.0162	0.0167
0.07	0.0185	0.0192	0.0199	0.0207	0.0215
0.08	0.0226	0.0236	0.0246	0.0257	0.0268
0.09	0.0270	0.0283	0.0297	0.0311	0.0327
0.10	0.0316	0.0333	0.0351	0.0371	0.0390
0.20	0.0894	0.0994	0.1104	0.1220	0.1340
0.30	0.1643	0.1925	0.2242	0.2572	0.2908
0.40	0.2530	0.3124	0.3789	0.4477	0.5175
0.50	0.3536	0.4593	0.5774	0.6988	0.8216
0.60	0.4648	0.6342	0.8225	1.0153	1.2098
0.70	0.5857	0.8379	1.1170	1.4018	1.6887
0.80	0.7155	1.0717	1.4638	1.8627	2.2641
0.90	0.8538	1.3363	1.8652	2.4020	2.9417
1.00	1.0000	1.6330	2.3238	3.0237	3.7268
2.00	2.8284	6.5727	10.5409	14.5285	18.5218
3.00	5.1962	15.7117	26.6905	37.6969	48.7110
4.00	8.0000	29.8142	52.3877	74.9955	97.6127
5.00	11.1803	49.5434	89.0091	128.5151	168.0320
6.00	14.6969	75.4922	137.7755	200.1043	262.4453
7.00	18.5203	108.2023	199.7952	291.4385	383.0951
8.00	22.6274	148.1748	276.0905	404.0610	532.0460
9.00	27.0000	195.8786	367.6151	539.4106	711.2216
10.00	31.6228	251.7557	475.2663	698.8398	922.4297

Example 303.11

A trapezoidal open channel has equal side slopes 1H:1V, base width $= 20$ ft, longitudinal slope of 0.001 ft/ft, Manning's $n = 0.017$, and conveys a flow rate $Q = 325$ cfs. What is the critical depth and critical velocity of flow in the channel?

Solution Computing the parameter $\dfrac{Q}{\sqrt{gb^5}} = \dfrac{325}{\sqrt{32.2 \times 20^5}} = 0.0320$

For $m = 1$, linear interpolation in Table 303.6 yields $d_c/b = 0.097$. Therefore, $d_c = 0.097 \times 20 = 1.94$ ft.

  With depth of flow $d = 1.94$ ft

  Top width $T = 23.88$ ft

  Area of flow $A = 21.94 \times 1.94 = 42.56$ ft^2

  Critical velocity $V = \dfrac{Q}{A} = \dfrac{325}{42.56} = 7.64$ fps

  Verification $\dfrac{Q^2 T}{A^3 g} = \dfrac{325^2 \times 23.88}{42.56^3 \times 32.2} = 1.02$

Therefore, the calculated depth $d = 1.94$ ft approximately satisfies the criterion for critical depth.

Triangular Open Channels

For a triangular channel, the bottom width is equal to zero. The critical depth of flow in a V-shaped open channel with equal side slopes (H:V ratio $= m$) is given by

$$d_c = \left(\frac{2Q^2}{gm^2} \right)^{1/5} \tag{303.41}$$

and the critical velocity is given by

$$V_c = \left(\frac{Qg^2}{4m} \right)^{1/5} \tag{303.42}$$

Example 303.12

A triangular open channel has equal side slopes 2H:1V, longitudinal slope of 0.001 ft/ft, Manning's $n = 0.017$ and conveys a flow rate $Q = 45$ cfs. What is the critical velocity of flow in the channel?

Solution The longitudinal slope $S = 0.001$ and $n = 0.017$ are parameters that can be used to calculate the *normal depth* of flow. However, they are unnecessary for calculation of the *critical depth*, which is a function of the channel cross section and the flow rate only.

$$d_c = \left(\frac{2Q^2}{gm^2} \right)^{1/5} = \left(\frac{2 \times 45^2}{32.2 \times 2^2} \right)^{1/5} = 1.99 \text{ ft}$$

and the critical velocity is given by

$$V_c = \left(\frac{Qg^2}{4m}\right)^{1/5} = \left(\frac{45 \times 32.2^2}{4 \times 2}\right)^{1/5} = 5.66 \text{ fps}$$

Verification: With a depth of flow $d = 1.99$ ft
Top width $T = 7.96$ ft
Area of flow $A = 7.92$ ft^2

$$\frac{Q^2 T}{A^3 g} = \frac{45^2 \times 7.96}{7.92^3 \times 32.2} = 1.008$$

Thus, the critical depth criterion has been satisfied to reasonable accuracy.

Occurrence of Critical Depth in Open Channels

Critical depth occurs at a distance approximately three to four times the critical depth (y_c) upstream of a free outfall at the end of a channel with mild slope. At the outfall, the depth is supercritical (approximately $0.7y_c$):

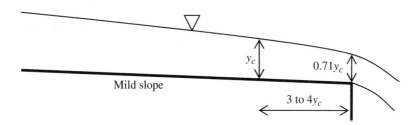

Critical flow occurs when water passes over a broad-crested weir:

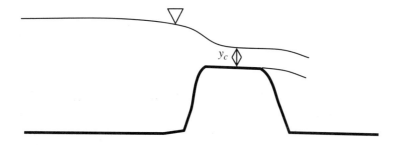

At slope transitions (when channel slope changes from mild to steep), the depth of flow passes through the critical stage. The critical depth occurs a very short distance upstream of the transition. A mild slope is one for which the normal depth of flow is greater than the critical depth. A steep slope is one for which the normal depth of flow is less than the critical depth.

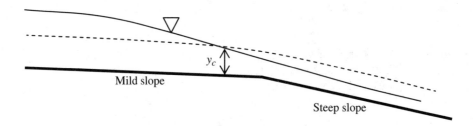

Critical Slope in Open Channels

For any open channel of known geometry and carrying a given flow rate, a specific longitudinal slope exists for which the normal depth of flow is equal to the critical depth for that geometry. This slope is called the critical slope and can be calculated by the following steps. Knowledge of the critical slope is necessary to classify the flow type in the channel (see Table 303.8).

For *circular conduit* (diameter D), follow these steps:

1. Calculate the parameter $\dfrac{Q}{\sqrt{gD^5}}$. Use the Fig. 303.8 to determine the ratio $\dfrac{d_c}{D}$.

2. Using the depth ratio d_c/D from step 1, use the normal depth table (Table 303.3) to look up the ratio Q/Q_f. Use either constant n ($n = n_{\text{full}}$) or variable n ($n \neq n_{\text{full}}$) based on the problem statement.

3. Since Q is known, calculate Q_f.

4. Use the formula for Q_f [Eq. (303.34)] to calculate the slope S.

For *straight-sided channels* (bottom width b), follow these steps:

1. Calculate the parameter $\dfrac{Q}{\sqrt{gb^5}}$. Use Table 303.6 to determine the ratio $\dfrac{d_c}{b}$.

2. Use the normal depth table (Table 303.4) to look up the appropriate K parameter.

3. Since flow rate Q is known, use Eq. (303.36) to calculate corresponding slope S.

Open Channels Having Compound Cross Sections

Many natural watercourses have irregular cross sections which may be best described as a compound cross section which is a composite of various known shapes. The hydraulic parameters of each section are calculated separately. Note that the common interface between adjacent areas is not included in the calculation of the wetted perimeter. As an example, consider the channel shown in Example 303.13. Assume a longitudinal slope of 0.4% and Manning's $n = 0.035$. Given the water surface as shown, calculate the flow rate (cfs).

Example 303.13

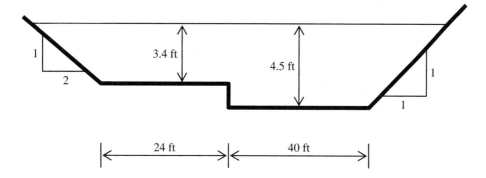

Solution   Dividing the channel flow area into two trapeziums (A and B), we calculate

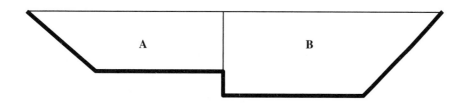

For area A: bottom width = 24 ft, top width = 24 + 6.8 = 30.8 ft, depth = 3.4 ft

Flow area $A = \dfrac{24 + 30.8}{2} \times 3.4 = 93.16 \text{ ft}^2$

Wetted perimeter $P = 3.4 \times \sqrt{5} + 24 = 31.60 \text{ ft}$

Hydraulic radius $R = \dfrac{93.16}{31.60} = 2.95 \text{ ft}$

Velocity $V = \dfrac{1.486}{0.035} \times 2.95^{2/3} \times 0.004^{1/2} = 5.52 \text{ ft/s}$

Flow rate $Q = VA = 5.52 \times 93.16 = 514.55 \text{ ft}^3/\text{s}$

For area B: bottom width = 40 ft, top width = 40 + 4.5 = 44.5 ft, depth = 4.5 ft

Flow area $A = \dfrac{40 + 44.5}{2} \times 4.5 = 190.13 \text{ ft}^2$

Wetted perimeter $P = 4.5 \times \sqrt{2} + 40 + 1.1 = 47.46 \text{ ft}$

Hydraulic radius $R = \dfrac{190.13}{47.46} = 4.01 \text{ ft}$

Velocity $V = \dfrac{1.486}{0.035} \times 4.01^{2/3} \times 0.004^{1/2} = 6.77$ ft/s

Flow rate $Q = VA = 6.77 \times 190.13 = 1287.70$ ft^3/s

For the entire channel, flow rate $Q = 514.55 + 1287.70 = 1802.25$ cfs

Flow in Gutters and Swales

The triangular-shaped area defined by the curb, gutter, and the spread onto the pavement creates an open-channel flow section for conveying runoff. The typical geometry is shown in Fig. 303.9. The cross slope S_x is significantly greater than the normal cross slope of the pavement, in order to reduce spread to the pavement. Since the cross section is typically very shallow compared to the top width, the hydraulic radius is not a very representative hydraulic parameter.

Thus, modification of Manning's equation is necessary for use in computing flow in triangular channels. The equation in terms of cross slope S_x and spread T

$$Q = \frac{0.56}{n} S_x^{5/3} S^{1/2} T^{8/3} \tag{303.43}$$

where Q = flow rate in the gutter (cfs)
n = Manning's roughness coefficient for the gutter
S_x = cross slope of the gutter (decimal)
S = longitudinal slope (decimal)
T = top width (spread) (ft)

Pavement Drainage Inlets

When the capacity of the curb-gutter-pavement section has been exceeded, typically as a result of spread considerations, runoff must be diverted from the roadway surface. A common solution is often interception of all or a portion of the runoff by drainage inlets that are connected to a storm drain pipe. Inlets used for intercepting runoff from highway surfaces can be divided into four major classes: (1) grate inlets, (2) curb-opening inlets, (3) combination inlets, and (4) slotted-drain inlets. See Fig. (303.10)

Curb opening and slotted-drain inlet length for total interception is given by

$$L_T = 0.6 \frac{Q^{0.42} S^{0.3}}{(nS_x)^{0.6}} \tag{303.44}$$

Figure 303.9
Typical gutter
cross section.

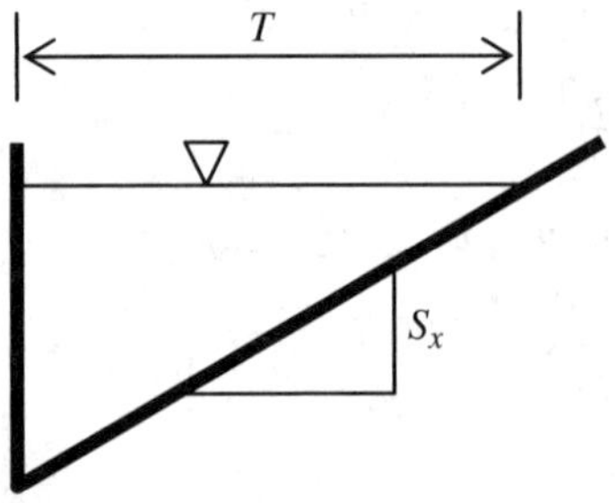

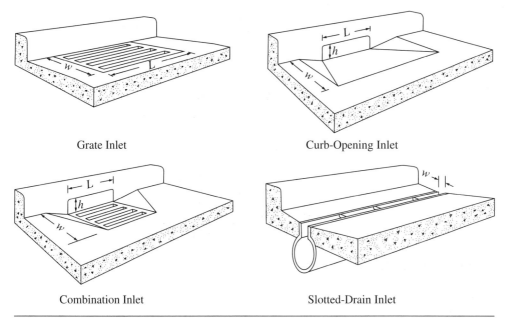

Grate Inlet   Curb-Opening Inlet

Combination Inlet   Slotted-Drain Inlet

Figure 303.10   Various types of pavement drainage inlets.

For an inlet length $L < L_T$, the inlet interception efficiency is given by

$$E = 1 - \left(1 - \frac{L}{L_T}\right)^{1.8}$$
(303.45)

Example 303.14

A street has a cross slope of 2%. The longitudinal slope of the street and the adjoining gutter is 4%. Assume Manning's n value is 0.015.

1. What is the required curb opening width for an intercepted discharge of 1.65 ft³/s?

2. If a 15-ft-long curb opening is provided, how much of the discharge will be bypassed to the next inlet?

Solution

1. The total interception curb opening length is given by

$$L_T = 0.6 \times \frac{1.65^{0.42} \times 0.04^{0.3}}{(0.015 \times 0.02)^{0.6}} = 36.6 \text{ft}$$

2. For $L = 15$ ft and $L_T = 36.6$ ft, $E = 0.61$

Thus, the discharge intercepted by a 15-ft-long inlet = 0.61 × 1.65 = 1.01 ft³/s
Flow bypassed to the next inlet = 1.65 − 1.01 = 0.64 ft³/s

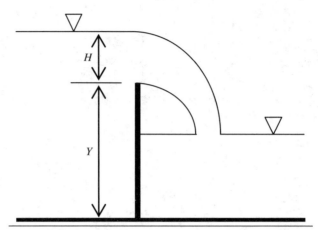

Figure 303.11   Discharge over a sharp-crested weir.

Flow Measurement with Weirs

Sharp-Crested Weirs

Measurement weirs are vertical, sharp-crested plates obstructing flow (Fig. 303.11). If the weir opening does not span the entire width of the channel, the flow width is contracted and the effective weir length is less than the actual weir length. In such a case, the weir is said to be *contracted*. If a weir has no end walls, or if contractions are suppressed by flaring the walls, it is called a *suppressed* weir. For example, the Cipoletti weir which is commonly used in irrigation channels, uses a trapezoidal section with the side walls inclined at an angle of 28° to the vertical to accomplish suppression of end contractions.

The discharge equation for a sharp-crested rectangular weir is given by

$$Q = Kb\sqrt{2g}H^{3/2} \tag{303.46}$$

where K is the weir coefficient and is given by (Kindsvater and Carter, 1959)[1]

$$K \approx 0.40 + 0.05\frac{H}{Y} \tag{303.47}$$

For contracted ends, the physical width of the weir b is replaced by the effective width b_{eff} given by $b_{\text{eff}} \approx b - 0.1NH$, where N = number of contractions.

For submerged weirs (Fig. 303.12), where a nappe does not form downstream of the weir, the discharge is corrected according to Eq. (303.48).

$$Q_{\text{submerged}} = Q_{\text{free flow}}\left[1 - (H_2/H_1)^{3/2}\right]^{0.385} \tag{303.48}$$

[1] Kindsvater, C. E. and Carter, R. W. (1959), "Discharge Characteristics of Rectangular Thin-Plate Weirs," *Transactions*, American Society of Civil Engineers, Vol. 24, Paper No. 3001.

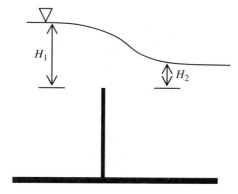

Figure 303.12
Discharge over a fully submerged sharp-crested weir.

Triangular Weirs

For low flow rates, triangular weirs are commonly used (Fig. 303.13). The discharge equation for a triangular weir is given by

$$Q = \frac{8}{15} K \sqrt{2g} \tan (\theta/2) H^{5/2} \qquad (303.49)$$

where K is the weir coefficient which is a function of the head H. For weirs with θ between 60° and 90°, K is between 0.60 and 0.57 as the head H varies between 0.2 and 2.0 ft.

Specifically, for a 90° V-notch weir, the equation becomes

$$Q = 2.54H^{5/2} \qquad (303.50)$$

Equation (303.50) is for USCS units. For SI units, the weir coefficient is 1.40.

Broad-Crested Weirs

For a broad-crested weir, the discharge equation is given by

$$Q = 0.385Kb\sqrt{2g}H^{3/2} \qquad (303.51)$$

where K is the weir coefficient which depends on several factors, such as approach velocity, slope of the upstream face, and longitudinal length of the crest. As H/Y approaches zero, the coefficient K approaches 0.85.

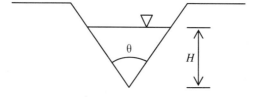

Figure 303.13
Discharge through a triangular weir.

Table 303.7 Weir Coefficient for Broad-Crested Weirs

Measured head (ft)	Weir crest breadth w (ft)										
	0.5	0.75	1.0	1.5	2.0	2.5	3.0	4.0	5.0	10.0	15.0
0.2	2.80	2.75	2.69	2.62	2.54	2.48	2.44	2.38	2.34	2.49	2.68
0.4	2.92	2.80	2.72	2.64	2.61	2.60	2.58	2.54	2.50	2.56	2.70
0.6	3.08	2.89	2.75	2.64	2.61	2.60	2.68	2.69	2.70	2.70	2.70
0.8	3.30	3.04	2.85	2.68	2.60	2.60	2.67	2.68	2.68	2.69	2.64
1.0	3.32	3.14	2.98	2.75	2.66	2.64	2.65	2.67	2.68	2.68	2.63
1.2	3.32	3.20	3.08	2.86	2.70	2.65	2.64	2.67	2.66	2.69	2.64
1.4	3.32	3.26	3.20	2.92	2.77	2.68	2.64	2.65	2.65	2.67	2.64
1.6	3.32	3.29	3.28	3.07	2.89	2.75	2.68	2.66	2.65	2.64	2.63
1.8	3.32	3.32	3.31	3.07	2.88	2.74	2.68	2.66	2.65	2.64	2.63
2.0	3.32	3.31	3.30	3.03	2.85	2.76	2.27	2.68	2.65	2.64	2.63
2.5	3.32	3.32	3.31	3.28	3.07	2.89	2.81	2.72	2.67	2.64	2.63
3.0	3.32	3.32	3.32	3.32	3.20	3.05	2.92	2.73	2.66	2.64	2.63
3.5	3.32	3.32	3.32	3.32	3.32	3.19	2.97	2.76	2.68	2.64	2.63
4.0	3.32	3.32	3.32	3.32	3.32	3.32	3.07	2.79	2.70	2.64	2.63
4.5	3.32	3.32	3.32	3.32	3.32	3.32	3.32	2.88	2.74	2.64	2.63
5.0	3.32	3.32	3.32	3.32	3.32	3.32	3.32	3.07	2.79	2.64	2.63
5.5	3.32	3.32	3.32	3.32	3.32	3.32	3.32	3.32	2.88	2.64	2.63

Source Brater, E. F. and King, H. W. (1976), *Handbook of Hydraulics*, 6th ed., New York, McGraw-Hill Book Company.

The discharge equation for a broad-crested weir is written as

$$Q = CbH^{3/2} \qquad (303.52)$$

where C = weir coefficient varies with weir crest breadth (w) as given in Table 303.7
  b = weir length (ft)
  H = measured head (ft)

Proportional Weirs

The proportional weir is shaped so that the discharge area varies nonlinearly with head. This results in the weir having a linear head-discharge relationship. A typical proportional weir is shown in Fig. 303.14. The discharge (cfs) for a proportional weir is given by

$$Q = 4.97b\sqrt{a}\left(H - \frac{a}{3}\right) \qquad (303.53)$$

where the dimensions a, b, and H are in feet and are shown in Fig. 303.14.

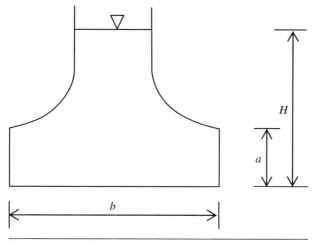

Figure 303.14 Discharge through a proportional weir.

A proportional weir has a cross section such that the flow rate Q is a linear function of the weir head (H).

$$Q = C_d K \frac{\pi}{2} \sqrt{2g} H \qquad (303.54)$$

Parshall Flume

The Venturi flume, developed and calibrated by Parshall (1926)[2], introduces a contracted section where the width is reduced and bottom slope is increased. A head-discharge relationship can be established as long as the downstream depth is low enough to allow free flow through the contracted section. Under these conditions, the Parshall flume is a critical depth meter. For a flume with a throat width (W, ft) of at least 1 ft but less than 8 ft, the flow rate (Q, cfs) can be calculated using the empirical relationship.

$$Q = 4WH^{1.522W^{0.026}} \qquad (303.55)$$

where H (ft) is the head at a specified upstream location.

Alternately, the discharge is given by

$$Q = K\sqrt{2g}WH^{1.5} \qquad (303.56)$$

K is the discharge coefficient which primarily depends on depth of flow. At higher depths, the coefficient K approaches 0.486.

---

[2]Parshall, R. L. (1926), "The Improved Venturi Flume," *Transactions*, American Society of Civil Engineers, Vol. 89, 841–851.

Gradually Varied Flow in Open Channels

Denoting the bottom slope of an open channel as S_o and the friction slope h_f/L as S_f, one may write the following differential equation for gradually varied flow along an open channel. Both S_o and S_f are considered positive for a channel sloping downward. The variable y is the elevation of the free surface with respect to the channel bottom.

$$\frac{dy}{dx} = \frac{S_o - S_f}{1 - Fr^2}$$

(303.57)

The friction slope S can be estimated using the Chezy-Manning equation

$$S_f = \frac{0.453n^2V^2}{R^{4/3}}$$

(303.58)

Classification of Surface Profiles

Water surface profiles in open channels are classified according to two criteria—the slope of the channel and the depth of flow. Slopes are designated as horizontal (H), mild (M), critical (C), steep (S), or adverse (A). Depth of flow is designated as zone 1 (if actual depth of flow is greater than both normal and critical depths), zone 2 (if actual depth of flow is between normal and critical depths), or zone 3 (if actual depth of flow is less than both normal and critical depths).

The types of water profiles can be obtained by analyzing Eq. (303.59)

$$\frac{dy}{dx} = S_o \frac{\left[1 - \left(\dfrac{n}{n_o} \right)^2 \left(\dfrac{y_o}{y} \right)^{10/3} \right]}{\left[1 - \left(\dfrac{y_c}{y} \right)^3 \right]}$$

(303.59)

Table 303.8 summarizes characteristics of various flow profiles.

If the slope is so small that the normal depth of flow is greater than the critical depth for the given discharge, then the slope is characterized as *mild* (M). Similarly, if the slope is so steep that the normal depth is less than the critical depth, the channel is *steep* (S). For horizontal (H) and adverse (A) slopes, normal depth does not exist. An adverse slope indicates an upward slope in the flow direction. Therefore, H1 and A1 categories of flow are not defined. These flow profiles are also shown in Fig. 303.15.

If the approach flow in an open channel is an M3 profile (actual depth of flow less than critical depth), then the Froude number is greater than 1 and if the S_f is greater than S_o, then according to Eq. (303.59), both the numerator and denominator are negative, thus dy/dx is positive (i.e., the depth increases in the direction of flow). For such an approach flow, as the upstream depth approaches the critical depth, the Froude number approaches 1, denominator approaches zero, and dy/dx approaches infinity.

Table 303.8 Classification Parameters for Gradually Varied Flow Profiles

Characteristics of water surface profiles				
Mild	$S_o > 0$	$y_c < y_o < y$	1	M1
Mild	$S_o > 0$	$y_c < y < y_o$	2	M2
Mild	$S_o > 0$	$Y < y_c < y_o$	3	M3
Critical	$S_o > 0$	$Y > y_o = y_c$	1	C1
Critical	$S_o > 0$	$Y < y_o = y_c$	3	C3
Steep	$S_o > 0$	$y_o < y_c < y$	1	S1
Steep	$S_o > 0$	$y_o < y < y_c$	2	S2
Steep	$S_o > 0$	$Y < y_o < y_c$	3	S3
Horizontal	$S_o = 0$	$y > y_c$	2	H2
Horizontal	$S_o = 0$	$y < y_c$	3	H3
Adverse	$S_o < 0$	$y > y_c$	2	A2
Adverse	$S_o < 0$	$y < y_c$	3	A3

When the approach flow is of very large depth, the approach velocity is very low and the Froude number approaches zero and the denominator approaches 1. For such a condition, the water surface approaches the horizontal. This situation occurs for M1, S1, and C1 approach profiles.

Hydraulic Jump

When a supercritical flow (Froude number greater than 1) in an upstream section of the channel is forced to become subcritical, there may occur a sudden increase in depth, accompanied by a significant loss of energy. This phenomenon is known as a hydraulic jump and is often the preferred mechanism of dissipating energy at the bottom of spillways. By controlling the depth of the tailwater or receiving reservoir, the jump may be forced to occur within the apron of the spillway, thereby reducing the erosion of the channel further downstream. All following results are derived for a horizontal channel, but are fairly applicable to channels of moderate slope (less than approximately 2%).

The existence of a jump assumes adequate tailwater conditions exist. Without adequate tailwater, the jump will be swept downstream out of the culvert, causing a potentially large scour hole at the culvert outlet.

The principal parameter that affects hydraulic jump performance is the upstream Froude number. Reynolds number and channel geometry play a minor role. For Froude number less than about 2.5, weak jumps form, causing energy dissipation around 15%. For Froude number between 2.5 and 4.5, an unstable, oscillating jump forms that can travel for long distance downstream, resulting in total energy dissipation between 15% and 45%. When the Froude number is greater than 4.5, a stable jump forms that is insensitive to downstream conditions and causes significant energy dissipation (45% to 70%).

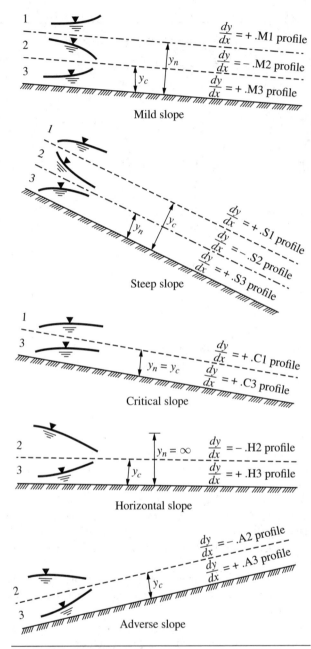

Figure 303.15 Classification of water surface profiles for gradually varied flow.

(From Chow V. T. (1959), Open Channel Hydraulics, McGraw-Hill, New York.)

Depths on either side of a hydraulic jump are called conjugate depths. The momentum is conserved in a hydraulic jump. For a rectangular channel, the conjugate depths d_1 and d_2 are related by

$$d_1 = -\frac{1}{2}d_2 + \sqrt{\frac{2V_2^2 d_2}{g} + \frac{d_2^2}{4}}$$ (303.60)

$$d_2 = -\frac{1}{2}d_1 + \sqrt{\frac{2V_1^2 d_1}{g} + \frac{d_1^2}{4}}$$ (303.61)

where d_1 is the upstream depth and d_2 is the downstream depth. This may also be expressed as a ratio d_2/d_1 which is a function of the upstream Froude number.

The relation between the supercritical depth and the sequent depth for a rectangular channel is

$$\frac{d_2}{d_1} = \frac{1}{2}\left[\sqrt{1 + 8\mathrm{Fr}_1^2} - 1\right]$$ (303.62)

where the upstream Froude number is $\mathrm{Fr}_1 = V_1/\sqrt{gd_1}$.

The length of a hydraulic jump (L_j) increases with upstream Froude number Fr_1. L_j is approximately $6d_2$ for $4 < \mathrm{Fr}_1 < 20$. For Froude numbers outside this range, the length of the jump is slightly less than $6d_2$.

If depths, upstream and downstream, of a hydraulic jump are known as d_1 and d_2, the upstream velocity is given by

$$V_1 = \sqrt{\frac{gd_2}{2d_1}(d_1 + d_2)}$$ (303.63)

The specific energy lost (per unit weight) in the hydraulic jump is given by the difference between the upstream specific energy E_1 and the downstream energy E_2

$$\Delta E = E_1 - E_2 = \left(d_1 + \frac{V_1^2}{2g}\right) - \left(d_2 + \frac{V_2^2}{2g}\right) \approx \frac{(d_2 - d_1)^3}{4d_1 d_2}$$ (303.64)

Rate of power dissipation in a hydraulic jump is given by

$$P = \gamma Q \Delta E$$ (303.65)

Figure 303.16 shows d_2/d_1, length of hydraulic jump L, and head loss h_L as a function of upstream Froude number.

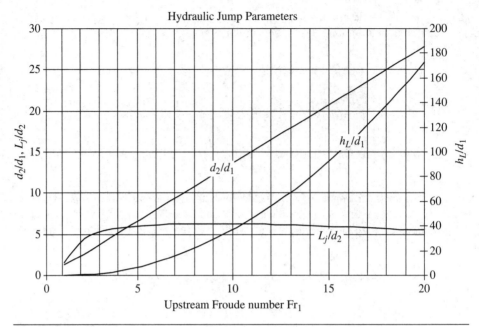

Figure 303.16 Parameters of a hydraulic jump.

Example 303.15 Hydraulic Jump in the Apron of a Spillway

A spillway operates with 3 ft of head as shown. The top of the spillway is 40 ft above the toe. Assume the spillway discharge coefficient to be 3.45.

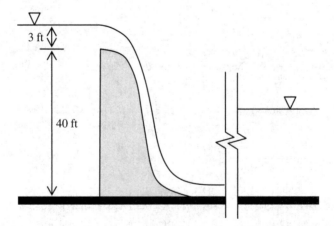

1. What is the discharge per foot width of the spillway?

2. What is the depth of flow at the toe?

3. What must be the tailwater depth in order to force a hydraulic jump to occur at the toe?

4. What is the energy loss (in hp) per foot width?

Solution

1. Discharge per foot width $Q = C_s bH^{3/2} = 3.45 \times 1 \times 3^{3/2} = 17.93$ cfs/ft

2. Equating total energy upstream of the spillway crest and at the toe (i.e., ignoring energy dissipation over the spillway)

$$43 + \frac{V^2}{2g} = d_{toe} + \frac{V^2_{toe}}{2g}$$

Neglecting upstream velocity (V) and depth at the toe (d_{toe}), we get

$$V_{toe} = \sqrt{2 \times 32.3 \times 43} = 52.62 \,\text{ft/s}$$

Using continuity equation $V_1 A_1 = V_2 A_2$ to calculate d_{toe}, we get

$$d_{toe} = \frac{17.93}{52.62} = 0.34 \,\text{ft}$$

3. $d_1 = d_{toe} = 0.34 \qquad V_1 = 52.62$ ft/s

$$d_2 = -\frac{1}{2}d_1 + \sqrt{\frac{2V_1^2 d_1}{g} + \frac{d_1^2}{4}} = 7.48 \,\text{ft}$$

4. $V_2 = \dfrac{V_1 d_1}{d_2} = \dfrac{52.62 \times 0.34}{7.48} = 2.39$ fps

$$E_1 = 0.34 + \frac{52.62^2}{2 \times 32.2} = 43.33$$

$$E_2 = 7.48 + \frac{2.39^2}{2 \times 32.2} = 7.57$$

$$\Delta E = E_1 - E_2 = 35.76 \,\text{ft}$$

The rate of energy dissipation is calculated using the fluid power equation.

$$P = \gamma Q \Delta E = 62.4 \times 17.93 \times 35.76 = 40{,}009 \,\text{lb-ft/s} = 72.74 \,\text{hp}$$

NOTE The depth of flow at the toe, calculated in step 3, is small enough to justify neglecting it in step 2.

Conservation of Momentum—Nonrectangular Open Channel

If the open channel cross section is rectangular, then the simplified hydraulic jump equations [Eqs. (303.60) and (303.61)] for conjugate depths can be used to apply the conservation of momentum. However, those equations are not valid for other cross sections. The following example applies the momentum conservation concept to a triangular channel.

Example 303.16

A roadside gutter is triangular in cross section (1:1 side slopes). The longitudinal slope changes abruptly from 12% to 2%. The normal depth of flow immediately upstream of the transition section is 10 cm. What is the depth of flow (m) immediately downstream of the transition?

    A. 0.20

    B. 0.28

    C. 0.46

    D. 1.20

Solution

Upstream flow occurs at depth $d = 0.1$ m

Since the side slopes are 1:1, the top width $T = 0.2$ m and area $A = 0.01$ m²

Assuming upstream flow to be at normal depth,

Slope = 12% = 0.12, $b = 0$, $m = 1$, $K = 0.5$ (Table 303.5)

Flow rate: $Q = \dfrac{Kkd^{8/3}S^{1/2}}{n} = \dfrac{0.5 \times 1.0 \times 0.1^{8/3} \times 0.12^{1/2}}{0.013} = 0.0287\,\dfrac{m^3}{s}$

For triangular channel, hydraulic depth: $d_h = d/2 = 0.1/2 = 0.05$ m

Upstream velocity: $V = Q/A = 0.0287/0.01 = 2.87$ m/s

Froude number: $Fr = \dfrac{V}{\sqrt{gd_h}} = \dfrac{2.87}{\sqrt{9.81 \times 0.05}} = 4.1$ (supercritical approach flow)

For triangular channel: $\bar{d} = \dfrac{d}{3}$ and $A = md^2 = d^2$

At the upstream section, specific force: $\dfrac{F}{\gamma} = A\bar{d} + \dfrac{Q^2}{gA} = \dfrac{d^3}{3} + \dfrac{Q^2}{gd^2}$

$$= \dfrac{0.1^3}{3} + \dfrac{0.0287^2}{9.81 \times 0.1^2} = 0.00873\,m^3$$

For $d = 0.20$, $F/\gamma = 0.00477$

For $d = 0.28$, $F/\gamma = 0.00839$ Closest match (exact solution $d = 0.2847$)

For $d = 0.46$, $F/\gamma = 0.03284$

For $d = 1.2$, $F/\gamma = 0.57606$

Shear Stress on Channel Bed

Channel linings are used whenever needed to protect the channel bed and sides from scouring due to flow of the water. The average shear stress on the wetted perimeter is given by

$$\tau_o = \gamma R S_o \qquad (303.66)$$

where γ = unit weight of the fluid (water)
R = hydraulic radius of the channel
S_o = longitudinal slope of the channel bed

Another approach to calculate the channel bed shear stress is based on the Karman-Prandtl velocity equation

$$\tau_o = \frac{\rho V^2}{\left[5.75 \log\left(12.27\dfrac{y_0}{k_s}\right)\right]^2} \qquad (303.67)$$

where ρ = fluid density (= 1.937 lb-s^2/ft for water)
V = average channel velocity
y_0 = normal depth of flow
k_s = roughness height (typically 3.5 times D$_{84}$ for natural coarse bed channels)

Example 303.17

A gravel-lined trapezoidal open channel has the following characteristics:

Flow rate = 650 ft^3/s

Side slopes = 3H:1V

Base width = 15 ft

D$_{84}$ size for channel lining = 6 in

Manning's roughness coefficient $n = 0.030$

Longitudinal slope = 0.5%

Calculate the shear stress along the channel bed and along the wetted perimeter.

Solution

$$K = \frac{nQ}{1.486b^{8/3}S^{1/2}} = \frac{0.030 \times 650}{1.486 \times 15^{8/3} \times \sqrt{0.005}} = 0.1356$$

For side slope parameter $m = 3$, $K = 0.1356$, and depth ratio $d/b = 0.25$.

   Depth of flow $= 0.25 \times 15 = 3.75$ ft

   Top width $= 37.5$ ft

   Area of flow $= 98.44$ ft^2

   Wetted perimeter $= 38.72$ ft

   Hydraulic radius $= 2.54$ ft

   Velocity of flow $= 650 \div 98.44 = 6.6$ ft/s

Shear stress along wetted perimeter

$$\tau_o = \gamma R S_o = 62.4 \times 2.54 \times 0.005 = 0.79 \, \text{psf}$$

Shear stress along channel bed

$$\tau_o = \frac{\rho V^2}{\left[5.75 \log\left(12.27 \dfrac{y_0}{k_s}\right)\right]^2} = \frac{1.937 \times 6.6^2}{\left[5.75 \log\left(12.27 \dfrac{3.75}{3.5 \times 0.5}\right)\right]^2} = 1.27 \, \text{psf}$$

Flow in Culverts

A culvert is a short conduit that is typically used to accommodate flow of water underneath a roadway or other embankment. The primary components of a culvert are the entrance (square edged, angled wingwalls, beveled edge, mitered, etc.), the outlet, and the barrel. Common barrel shapes are circular pipe, rectangular box, elliptical, and arched.

   Culverts are usually laid on a downward slope, but in some cases the barrel may be horizontal or on an adverse slope. The most common culvert materials are concrete, corrugated metal, and plastic. While most culverts employ a closed conduit, many culvert installations utilize three-sided culverts. This type of structure uses one material for the top and sides, while the bottom of the culvert is the natural channel bottom. For such installations, a composite Manning's roughness coefficient must be computed. While there may be several ways to obtain the composite roughness as a weighted average, one method (Chow, 1959)[3] is based on the assumption that the each part of the section has the same average velocity. This results in the following expression for the composite n:

$$n_{\text{ave}} = \frac{\left(P_1 n_1^{1.5} + P_2 n_2^{1.5}\right)^{2/3}}{(P_1 + P_2)^{2/3}} \tag{303.68}$$

---

[3]Chow, V. T. (1959), *Open Channel Hydraulics*, McGraw-Hill, New York.

where P_1 = wetted perimeter of the top and side material

$\quad\quad P_2$ = wetted perimeter of the natural channel

$\quad\quad n_1$ = Manning's roughness coefficient of the top and side material

$\quad\quad n_2$ = Manning's roughness coefficient of the natural channel

A culvert barrel may flow full over all of its length or partly full. Full flow in a culvert barrel is rare. Generally, at least part of the barrel flows partly full. A water surface profile calculation is the only way to determine how much of the barrel flows full.

Full Flow

The hydraulic condition in a culvert flowing full is called pressure flow. High downstream water surface (tailwater) elevation or high upstream water surface (headwater) elevation may produce full flow in a culvert barrel. The capacity of a culvert operating under pressure flow is affected by upstream and downstream conditions and by the hydraulic characteristics of the culvert.

Partly Full (Free Surface) Flow

When the flow in a channel is subcritical, the velocity is less than that of disturbances (perturbations) traveling along the channel and therefore depth and velocity can be affected by downstream disturbances. In the supercritical flow regime, flow characteristics are not affected by downstream disturbances.

For a steep culvert flowing partly full, critical depth would occur at the culvert inlet, subcritical flow could exist in the upstream channel, and supercritical flow would exist in the culvert barrel.

A special type of free surface flow is called "just-full flow." This is a special condition where a pipe flows full with no pressure. The water surface just touches the crown of the pipe. The analysis of this type of flow is the same as for free surface flow.

Types of Flow Control

According to research sponsored by the Federal Highway Administration (FHWA), flow in a culvert occurs under two kinds of conditions—inlet control and outlet control. Inlet control occurs when the culvert barrel is capable of conveying more flow than the inlet will accept. The control section of a culvert operating under inlet control is located just inside the entrance. Critical depth occurs at or near this location, and the flow regime immediately downstream is supercritical.

The FHWA method for culvert design involves first assuming that inlet control governs and computing the corresponding headwater elevation. This is followed by computing the headwater elevation for outlet control. The two headwater values are then compared and the higher of the two is used as the basis for the design.

Inlet Control

For many culverts, the inlet control condition governs. It is hydraulically complex and cannot be modeled exactly to yield headwater depth. However, empirical procedures such as in

Table 303.9 Factors Affecting Culvert Performance

Factor	Inlet control	Outlet control
Headwater elevation	X	X
Inlet area	X	X
Inlet edge configuration	X	X
Inlet shape	X	X
Barrel roughness		X
Barrel area		X
Barrel shape		X
Barrel length		X
Barrel slope	Negligible	X
Tailwater elevation		X

the FHWA HDS-5: Hydraulic Design of Highway Culverts[4] can be used with a great degree of success.

Hydraulic characteristics downstream of the inlet control section do not affect the culvert capacity. The upstream water surface elevation and the inlet geometry represent the major flow controls.

Outlet Control

Outlet control flow occurs when the culvert barrel is not capable of conveying as much flow as the inlet opening will accept. The control section for outlet control flow in a culvert is located at the barrel exit or further downstream. Either subcritical or pressure flow exists in the culvert barrel under these conditions. All of the geometric and hydraulic characteristics of the culvert play a role in determining its capacity. Table 303.9 gives a list of hydraulic parameters that affect culvert performance under inlet control and outlet control.

Headwater

The depth of the upstream water surface measured from the invert at the culvert entrance is generally referred to as headwater depth.

Tailwater

Tailwater is defined as the depth of water downstream of the culvert measured from the outlet invert. It is an important factor in determining culvert capacity under outlet control conditions. Tailwater may be caused by an obstruction in the downstream channel or by the hydraulic

---

[4]Norman, J. M., Houghtalen, R. J., and Johnston, W. J. (2001), *Hydraulic Design of Highway Culverts, Second Edition,* Hydraulic Design Series No. 5, National Highway Institute, FHWA.

resistance of the channel. In either case, backwater calculations from the downstream control point are required to precisely define tailwater.

Outlet Velocity

Increased velocities in the culvert, caused by the constriction of the flow, can cause streambed scour and bank erosion in the vicinity of the culvert outlet. Minor problems can occasionally be avoided by increasing the barrel roughness. Energy dissipaters and outlet protection devices are sometimes required to avoid excessive scour at the culvert outlet. When a culvert is operating under inlet control and the culvert barrel is not operating at capacity, it is often beneficial to flatten the barrel slope or add a roughened section to reduce outlet velocities.

Performance Curves

A performance curve is a plot of headwater depth or elevation versus flow rate. In developing a culvert performance curve, both inlet and outlet control curves must be plotted. This is necessary because the dominant control at a given headwater is hard to predict. Also, control may shift from the inlet to the outlet or vice versa over a range of flow rates. Headwater versus discharge curves for pipe culverts are shown in Figs. 303.17 (inlet control) and 303.18 (full flow).

Scour at Inlets

A culvert barrel usually forces the flow in the channel through a reduced opening, creating vortices which may tend to cause scouring. In many cases, a scour hole also forms upstream of the culvert floor as a result of the acceleration of the flow as it leaves the natural channel and enters the culvert. Upstream slope paving, channel paving, headwalls, wingwalls, and cutoff walls help protect the slopes and channel bed at the upstream end of the culvert.

Scour at Outlets

Increased velocity through the culvert barrel causes erosion. At the outlet, turbulence and erosive eddies form as the flow expands to conform to the natural channel. Scour in the vicinity of a culvert outlet can be classified into two types. The first type is called local scour, where a scour hole is produced at the culvert outlet due to high exit velocities. Coarse material scoured from the circular or elongated hole is deposited immediately downstream, often forming a low bar. Finer material is transported further downstream. The second type of scour is classified as general stream degradation. This phenomenon is independent of culvert performance. Natural causes produce a lowering of the stream bed over time.

   Protection against scour at culvert outlets varies from limited riprap placement to complex and expensive energy dissipation devices. At some locations, use of a rougher culvert material or a flatter slope alleviates the need for a special outlet protection device. Riprapped channel expansions and concrete aprons protect the channel and redistribute or spread the flow. More complex energy dissipation devices, when warranted, include hydraulic jump basins, impact basins, drop structures, and stilling wells.

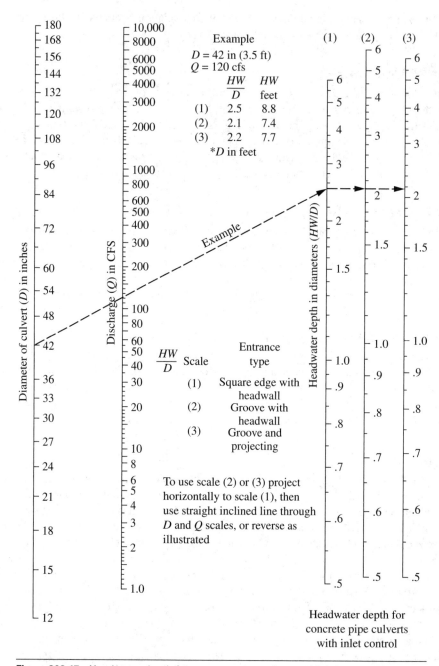

Figure 303.17 Headwater depth for concrete pipe culverts with inlet control.

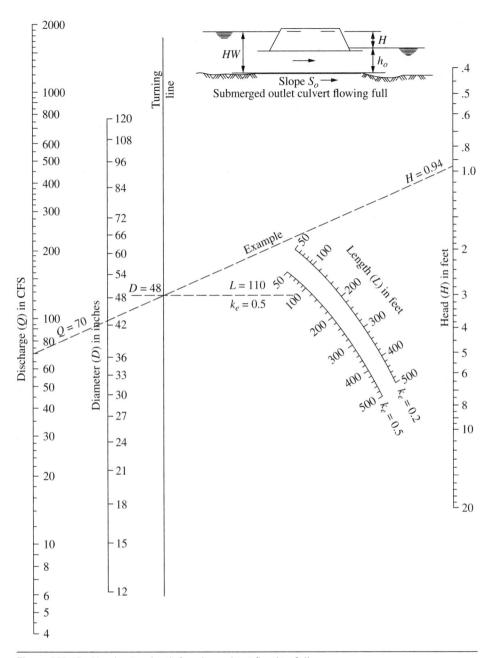

Figure 303.18 Headwater depth for pipe culvert flowing full.

Sedimentation

Most streams carry a sediment load and tend to deposit this load when their velocities decrease. Barrel slope and roughness are key indicators of potential problems at culvert sites. Other important factors in sedimentation processes are the discharge and the characteristics of the channel material. Culverts which are located on and aligned with the natural channel generally do not have a sedimentation problem. Storm events tend to cleanse culverts of sediment when increased velocities are experienced. Helical corrugations tend to promote this cleansing effect if the culvert is flowing full. In a degrading channel, erosion is a greater potential problem than sedimentation.

Hydrology

Hydrologic Balance

Hydrology is the study of the distribution and movement of water throughout the earth (hydrologic cycle). The hydrologic balance for the watershed involves positive flows into the system (such as from precipitation) and negative flows out of the system (such as through evaporation). The overall water balance (hydrologic budget) may be written as

$$R = P - \text{ET} - \Delta S \qquad (304.1)$$

where R = stream flow
P = precipitation (gross rainfall)
ET = evapotranspiration
ΔS = change in storage

This equation uses the principles of conservation of mass in a closed system, whereby any water entering a system (via precipitation) must be transferred into either evaporation, surface runoff (eventually reaching the channel and leaving in the form of river discharge), or stored in the ground. This equation requires the system to be closed, and where it isn't (for example when surface runoff contributes to a different basin), this must be taken into account.

Precipitation

The average precipitation on a watershed may be written as

$$\text{Average precipitation} = \frac{\text{total rainfall volume}}{\text{watershed area}} \qquad (304.2)$$

The total rainfall volume may be estimated from an examination of isohyetal contours. An *isohyet* is a line drawn through geographical points recording equal amounts of precipitation during a specific period. These lines are labeled using precipitation depth (such as 0.5 in, 1.0 in rainfall, etc.). If one considers a specific region for study, the area enclosed by an isohyet within the region may be estimated using standard techniques. Figure 304.1 shows isohyets for a watershed. Typical precipitation data may be as shown in Table 304.1.

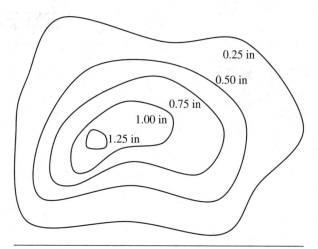

Figure 304.1 Isohyetal contours.

Using the data in Table 304.1, one may calculate the incremental areas as shown in Table 304.2. Each incremental area is then represented by the average precipitation between the two adjacent isohyets. The average precipitation over the entire watershed can be calculated as the weighted average for all these incremental areas that make up the watershed.

Average precipitation (gross rainfall depth) over the entire watershed is then given by

$$P_{ave} = \frac{15.656 \times 12}{321} = 0.59 \text{ in}$$

Rainfall Intensity

Rainfall intensity, on the other hand, is calculated from the temporal (rather than spatial) distribution of precipitation. Typical units of rainfall intensity are centimeters per hour or inches per hour. Rain gauge data may be used to determine rainfall intensity during specific time bands of a precipitation event. A rain gauge basically records cumulative precipitation depth over a period of time. Typical data collected in a rain gauge may appear, as in shown Table 304.3.

Table 304.1 Typical Precipitation Data

Line	Precipitation (in)	Enclosed area (ac)
1	0.25	321
2	0.50	126
3	0.75	85
4	1.00	47
5	1.25	12

Table 304.2   Precipitation Volume Calculation from Precipitation Data

Area	Precipitation range	Precipitation (in)	Enclosed area (ac)	Volume (ac–ft)
1	<0.25			
2	0.25–0.50	0.375	195	6.094
3	0.50–0.75	0.625	41	2.135
4	0.75–1.00	0.875	38	2.771
5	1.00–1.25	1.125	35	3.281
6	>1.25	1.375	12	1.375
Total				15.656

Table 304.3   Precipitation Depth versus Time

Time (min)	Precipitation (in)
0.00	0.0
10.00	0.4
20.00	0.6
30.00	1.2
40.00	1.5

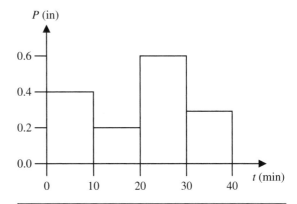

Figure 304.2   Storm hyetograph for data in Table 304.3.

The storm hyetograph is the bar graph of instantaneous rainfall versus time. For the data shown in Table 304.3, the hyetograph is shown in Fig. 304.2. Based on this data, one may say that the rainfall intensity during the time interval 0 to 10 min is 0.4 in/10 min = 2.4 in/h, the intensity during the time interval 10 to 20 min is 0.2 in/10 min = 1.2 in/h, and so on. The average rainfall intensity during the entire 40 min duration is 1.5 in/40 min = 2.25 in/h.

Recorded Precipitation Data

To quantitatively describe rainfall distribution over a region, there are several methods which may be used.

Station Average Method

For a drainage area containing a relatively large number of uniformly distributed gages, one may simply calculate the arithmetic average of the precipitation depth recorded at all gauges.

$$P_{ave} = \frac{\sum_{i=1}^{N} P_i}{N} \tag{304.3}$$

where P_i is the precipitation at station i (total number of stations = N).

Figure 304.3
Thiessen polygon
method—tributary
area.

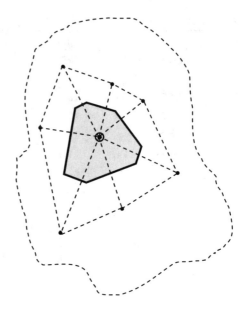

Thiessen Polygon Method

In this method, the precipitation recorded at each gauge is weighted with the area closest to it. Once these areas have been defined and calculated or measured, the average precipitation is calculated as the weighted average of the precipitation depths recorded at each station.

$$P_{ave} = \frac{\sum_{i=1}^{N} P_i A_i}{\sum_{i=1}^{N} A_i}$$

(304.4)

where P_i = precipitation at station i (total number of stations = N)
  A_i = area tributary to station i

The graphical method employed to associate each station with its adjacent area is shown in Fig. 304.3. The following are the major steps in this method:

- On a map showing the location of rain gauges and the boundary of the area of concern, draw straight lines connecting adjacent rain gauge locations and identify the midpoint on each line. These lines are shown as dashed lines in Fig. 304.3.

- Draw lines that are perpendicular bisectors of the dashed lines. Darken portions of these lines to create polygons around each rain gauge, such that each polygon contains land closest to the enclosed rain station.

- Determine the area of each polygon.

Intensity-Duration-Frequency Curves

Estimation of peak flow can be separated into two categories: (1) sites with measured stream gage data and (2) sites without gauged data. When reliable gauged data are available, statistical analysis of the flow record can be used to estimate flood peaks for various return periods.

When gauged data are not available, estimates are made by empirical equations such as the rational method or by regional regression equations. Regional regression equations are typically appropriate for larger drainage areas, and methods such as the rational method are commonly used for smaller areas, less than about 200 ac.

The recurrence interval is the period during which, on an average, a given event (e.g., rainfall or runoff) is equaled or exceeded. The (annual) exceedance probability is the reciprocal of recurrence interval. Thus, the "100-year flood" is defined as the event which has an annual probability of exceedance $= 1/100 = 0.01$.

If the recurrence interval is too short, the design event is of low intensity, resulting in a low initial cost, but the life-cycle cost is high because of frequently occurring damage. On the other hand, if the recurrence interval is too long, the initial cost is high, but annual maintenance costs are low. Somewhere between these limits lies the design frequency that will produce a reasonable balance of construction cost, annual maintenance cost, and risk of flooding.

Based on regional precipitation data recorded over a long period of time, intensity-duration-frequency (I-D-F) curves are plotted. A typical example of a set of I-D-F curves is shown in Fig. 304.4.

To select the governing rainfall intensity for a watershed, the duration is set equal to the time of concentration and the appropriate curve (based on return period) is used to predict the intensity. Alternatively, the intensity can be calculated using empirical models such as Steel's formula, which gives intensity as

$$I = \frac{K}{t_c + b} \tag{304.5}$$

where t_c = time of concentration (minutes).

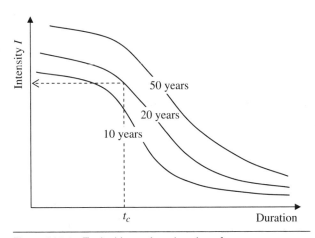

Figure 304.4 Typical intensity–duration–frequency curves.

Table 304.4 Coefficients K and b for Steel's Formula

Return period (y)	Coefficients	Region						
		1	2	3	4	5	6	7
2	K	206	140	106	70	70	68	32
	b	30	21	17	13	16	14	11
5	K	247	190	131	97	81	75	48
	b	29	25	19	16	13	12	12
10	K	300	230	170	111	111	122	60
	b	36	29	23	16	17	23	13
25	K	327	260	230	170	130	155	67
	b	33	32	30	27	17	26	10
50	K	315	350	250	187	187	160	65
	b	28	38	27	24	25	21	8
100	K	367	375	290	220	240	210	77
	b	33	36	31	28	29	26	10

The coefficients K and b are listed in Table 304.4 according to storm return period and geographical region. The map of the continental United States shows the hydrologic regions, relevant to the Steel's formula (Fig. 304.5).

Time of Concentration

The time of concentration for an inlet is the time required for water to flow from the most remote (in terms of time of flow) point of the area to the inlet once the soil has become saturated and minor depressions have been filled. It is assumed that when the duration of the storm equals the time of

Figure 304.5 Geographical regions (continental U.S.) for Steel's formula.

concentration, all parts of the watershed are contributing simultaneously to the discharge at the inlet. For this reason, the time of concentration is chosen as the duration while using the intensity-duration-frequency (IDF) curves. Time of concentration is calculated as the sum of overland flow, swale or ditch flow, and channel flow.

$$t_c = t_{\text{sheet flow}} + t_{\text{ditch flow}} + t_{\text{channel flow}} \tag{304.6}$$

Several empirical models are available for estimating the time of concentration for a watershed. All of these models employ the following parameters-the length of the hydraulically longest path, the average slope along the path, and some parameter that quantifies the land use of the terrain. Some of the better known models are presented below. Unless mentioned otherwise, time of concentration tc is specified in minutes in the models mentioned below.

Kirpich's Equation

Kirpich's equation was developed from data obtained in seven rural watersheds in Tennessee. The watersheds had well-defined channels and steep slopes of 3% to 10% and areas of 1 to 112 ac. It is used widely in urban areas for both overland flow and channel flow and for agricultural watersheds up to 200 ac.

$$t_c = 0.0078L^{0.77}S^{-0.385} \tag{304.7}$$

where tc = time of concentration (min)
L = length if the flow path (ft)
S = average slope (decimal) along the flow path

FAA Formula (Applicable to Urban Areas)

The FAA method was developed from data obtained from airport runoff but has been successfully applied to overland flow in urban areas.

$$t_c = \frac{1.8(1.1 - C)\sqrt{L}}{S^{1/3}} \tag{304.8}$$

where C = overall rational coefficient for the watershed
L = length of the drainage path (ft)
S = average slope along the hydraulic path (percent)

Manning's Kinematic Wave Formula (Applicable to Paved Areas)

According to the kinematic wave formula (Manning), the time of concentration in minutes is given by

$$t_c = \frac{0.938}{i^{0.4}} \left(\frac{nL_o}{\sqrt{S}} \right)^{0.6} \tag{304.9}$$

where L_o = length of the drainage path (ft)

$\quad\;\; n$ = Manning's roughness coefficient

$\quad\;\; i$ = intensity (in/h)

$\quad\;\; S$ = slope (decimal) of the hydraulic path

Values of n used in this formula are

Smooth impervious surfaces	0.035
Smooth soil surfaces free of stones	0.05
Poor grass surfaces	0.10
Pasture or medium grass cover	0.20
Dense grass or forested areas	0.40

NRCS Lag Equation (Applicable to Small Urban Watersheds)

The Natural Resources Conservation Service (NRCS)[1] gives the following formula for the time of concentration (hours):

$$t_c = \frac{1.67 L_0^{0.8} \left(\frac{1000}{CN} - 9 \right)^{0.7}}{1900 S^{0.5}} \tag{304.10}$$

where $\quad L_o$ = length of the drainage path (ft)

$\quad\;\; CN$ = NRCS curve number

$\quad\;\;\; S$ = slope of hydraulic path (percent)

Kerby's Equation

The Kerby equation was developed from data obtained in watersheds having watercourses less than 1200 ft, slopes less than 1%, and areas less than 10 ac.

$$t_c = 0.67 \left(\frac{nL_o}{\sqrt{S}} \right)^{0.467} \tag{304.11}$$

where t_c = time of concentration (min)

$\quad\;\; S$ = overland slope (decimal)

$\quad\;\; L_o$ = length of overland flow path (ft)

$\quad\;\; n$ = overland roughness coefficient (values in Table 304.5)

Sheet Flow

Manning's kinematic equation (valid for sheet flow $l_o < 300$ ft) is given by

$$t_{\text{sheet flow}} = \frac{0.007(nl_o)^{0.8}}{P_2^{0.5} S^{0.4}} \tag{304.12}$$

---

[1] *Urban Hydrology for Small Watersheds*, Natural Resources Conservation Service, USDA, Washington D.C., 1986.

Table 304.5 Kerby's Roughness Coefficient

Surface	Kerby *n* coefficient
Smooth, impervious surface	0.02
Smooth, packed bare soil	0.10
Poor grass, cultivated row crops of moderately rough bare soil	0.20
Pasture or average grass	0.40
Deciduous timberland	0.60
Timberland with deep forest litter or dense grass	0.80

where l_o = length of overland flow path (ft)
$\quad p_2$ = 2-year storm's 24-year rainfall (in)
$\quad S$ = slope of hydraulic slope line (decimal)
$\quad n$ = Manning's roughness, specified earlier in this section

Shallow Concentrated Flow

For hydraulic path length exceeding 300 ft, sheet flow changes to shallow concentrated flow. The time for ditch flow is calculated using an average velocity *V*.

$$t_{\text{ditch flow}} = \frac{L}{V} \qquad (304.13)$$

According to the Natural Resources Conservation Service (NRCS) *Urban Hydrology for Small Watersheds*, Technical Report No. 55, the velocity *V* is given by
For unpaved surfaces:

$$V_{\text{fps}} = 16.1345\sqrt{S_{\text{decimal}}} \qquad (304.14)$$

For paved surfaces:

$$V_{\text{fps}} = 20.3282\sqrt{S_{\text{decimal}}} \qquad (304.15)$$

Channel Flow

The relatively coherent flow that occurs in channels that collect runoff from the watershed is governed by empirical models such as the Chezy-Manning equation or the Hazen-Williams equation for open channel flows. The time of flow in channel flow will therefore be given by

$$t_{\text{storm drain}} = \frac{L}{V} \qquad (304.16)$$

where the average velocity of flow in the channel is given by either the Chezy-Manning equation or the Hazen-Williams equation.

Chezy-Manning

$$V = \frac{k}{n} R^{2/3} S^{1/2} \qquad k = 1.486 \text{ (U.S.)} \quad \text{and} \quad k = 1.0 \text{ (SI)} \qquad (304.17)$$

Hazen-Williams

$$V = 1.318 C R^{0.63} S^{0.54} \qquad C = 1.318 \text{ (U.S.)} \quad \text{and} \quad C = 0.849 \text{ (SI)} \qquad (304.18)$$

where V = average velocity (ft/s or m/s)
R = hydraulic radius (ft or m)
S = longitudinal channel slope (decimal)

The roughness of the channel is expressed in terms of the Manning's roughness parameter (n) or the Hazen-Williams coefficient (C). For more details, see Chap. 303.

Rainfall Distribution by Storm Type

The intensity of rainfall varies considerably during a storm as well as by geographic region. To represent the various regions of the United States, NRCS developed four synthetic 24-h rainfall distributions (I, IA, II, and III) from available National Weather Service (NWS) duration-frequency data or local storm data. These are shown in Fig. 304.6. Type IA is the least

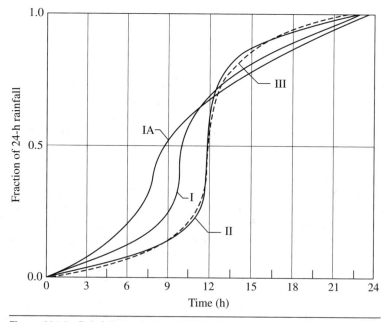

Figure 304.6 Rainfall distribution (NRCS).

intense and type II the most intense short-duration rainfall. The four distributions are shown in Fig. 304.6. Types I and IA represent the Pacific maritime climate with wet winters and dry summers. Type III represents Gulf of Mexico and Atlantic coastal areas where tropical storms bring large 24-h rainfall amounts. Type II represents the rest of the country.

Flow rate in a stream is often measured by measuring river stage. Previously established rating curves that relate stage (surface elevation) to flow rate can then be used to estimate the flow rate. A stream hydrograph is a plot of stream flow rate versus time. When interception and infiltration of incident rainfall (precipitation) is complete, the runoff (excess precipitation) begins to occur and the stream carries more discharge than the previously existing flow in the stream (base flow). Thus, there is a time lag between the start of precipitation and the start of excess precipitation.

Once the base flow is separated from the stream flow (several methods are discussed below), the excess precipitation is plotted against time. This represents the direct runoff due to the rainfall event. Once the total depth of runoff is calculated, the unit hydrograph for the rainfall event can be derived by dividing the direct runoff discharge values by the runoff depth.

Thus, a unit hydrograph of time T represents a hypothetical case of 1 unit (1 mm, 1 cm, 1 in) of runoff occurring from the catchment area during a time interval T.

Hydrograph Separation

Net runoff lags rainfall since the soil has a finite initial abstraction, based upon its infiltration rate. Runoff will only occur when the rain falls on a very wet watershed that is unable to absorb more.

Stream monitoring stations record the discharge in a stream that collects runoff flow from a contributing watershed. The recorded discharge includes the preexisting flow (baseflow) that must be separated to identify the discharge produced by the storm runoff. Many baseflow separation techniques are graphical which tend to focus on defining the points where baseflow intersects the rising and falling limbs of the quickflow response.

Graphical Separation Methods

Graphical methods are commonly used to plot the baseflow component of a flood hydrograph event, including the point where the baseflow intersects the falling limb.

Stream flow subsequent to this point is assumed to be entirely baseflow, until the start of the hydrographic response to the next significant rainfall event. Some of these graphical approaches to partitioning baseflow are discussed below.

The constant discharge method assumes that baseflow is constant during the storm hydrograph. The minimum streamflow immediately prior to the rising limb is used as the constant value. This is illustrated in Fig. 304.7. This method does not require plotting of the curve to identify points of inflection and the like and therefore is the most likely one to appear on the PE exam.

The constant slope method connects the start of the rising limb with the inflection point on the receding limb. This assumes an instant response in baseflow to the rainfall event. This method is illustrated in Fig. 304.8.

The concave method attempts to represent the assumed initial decrease in baseflow during the climbing limb by projecting the declining hydrographic trend evident prior to the rainfall

Figure 304.7
Constant discharge
method of base-
flow separation.

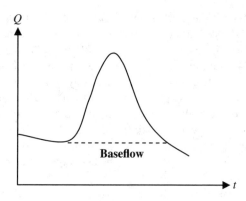

Figure 304.8
Constant slope
method of base-
flow separation.

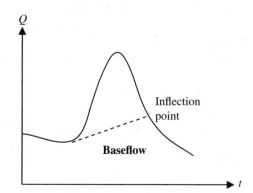

event to directly under the crest of the flood hydrograph. This point is then connected to the inflection point on the receding limb of storm hydrograph to model the delayed increase in baseflow. This method is illustrated in Fig. 304.9.

The excess precipitation due to the storm is calculated by subtracting the baseflow from the stream discharge, resulting in a plot of Q_{net} versus time. The volume of excess rainfall from the

Figure 304.9
Concave method
of baseflow
separation.

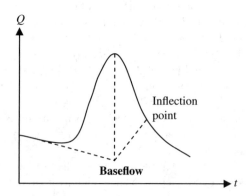

watershed is calculated as the area under the Q_{net} versus time curve. If this volume is spread evenly over the watershed area, the average depth of excess precipitation is given by

$$P_{av} = \frac{V}{A_d} \qquad (304.19)$$

where A_d = drainage area.

Unit Hydrograph

If the overland flow hydrograph is divided everywhere by P_{av}, we get the unit hydrograph of excess precipitation. Typical units of Q_{unit} are cubic feet per second per inch or cubic meter per second per centimeter. For example, a 1-h unit hydrograph for a watershed is the pattern of discharge produced by a 1-h period of uniform excess precipitation equal to 1 unit (1 in or 1 cm, for example).

Example 304.1

The 1-h unit hydrograph for a watershed is given below:

Time (h)	0.0	1.0	2.0	3.0	4.0	5.0
Discharge (cfs/in)	0.0	0.7	3.3	1.9	0.5	0.0

This data is shown below graphically. The time $t = 0$ marks the beginning of excess precipitation from the storm (after interception and infiltration are complete).

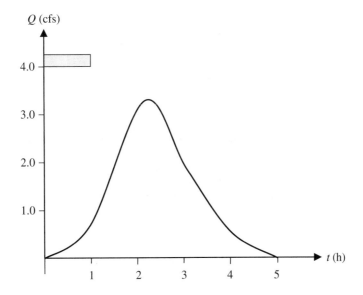

What is the runoff at $t = 2$ h due to a storm that is described as an hour of uniform excess precipitation (average intensity $i = 0.6$ in/h) followed by a second hour of uniform excess precipitation (average intensity $i = 0.3$ in/h)?

Solution Thus, at time $t = 2$ h, the contribution of the two periods of uniform precipitation (0.6 in in the first hour followed by 0.3 in during the second hour) can be calculated as

$$Q = 0.6 \times 3.3 + 0.3 \times 0.7 = 2.19 \text{ cfs}$$

This represents the contribution of the first hour of excess precipitation due to a lag time of 2 h plus the contribution of the second hour of excess precipitation due to a lag time of 1 h. Similarly, at time $t = 3$ h, the contribution of the two periods can be calculated as

$$Q = 0.6 \times 1.9 + 0.3 \times 3.3 = 2.13 \text{ cfs}$$

This represents the contribution of the first hour of excess precipitation due to a lag time of 3 h plus the contribution of the second hour of excess precipitation due to a lag time of 2 h.

A unit hydrograph developed from one storm can be used as design basis for other storms. This is subject to the following assumptions:

- All storms in the watershed have the same duration.
- Shape of the rainfall curve is the same.
- Only the total rainfall amount varies.

NRCS Synthetic Unit Hydrograph

Natural Resources Conservation Service (NRCS) developed a synthetic, unit hydrograph based on the curve number (CN) of the watershed. It is applicable for urban watersheds up to approximately 5000 ac. This hydrograph is shown in Fig. 304.10. The hydrograph has two parameters: t_p (time to peak discharge) and Q_p (peak discharge). These are given by

$$t_p = 0.5t_r + t_1 \tag{304.20}$$

$$t_1 = \frac{L_o^{0.8}(S+1)^{0.7}}{31.67\sqrt{G_{percent}}} \tag{304.21}$$

where t_r = storm duration (min)
t_1 = time from centroid of rainfall distribution to peak discharge (lag time, min)
$S = \dfrac{1000}{CN} - 10$ is the storage capacity of the soil
L_o = longest distance to collection point (ft)
G = average longitudinal grade (percent) of hydraulic path

The peak discharge (cfs/in) of the unit hydrograph is given by

$$Q_p = \frac{0.756A_{d,ac}}{t_p} = \frac{484A_{d,mi^2}}{t_p} \tag{304.22}$$

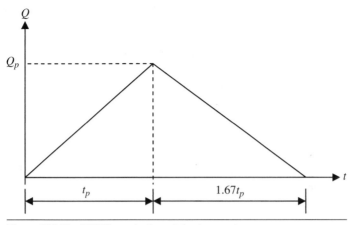

Figure 304.10 NRCS synthetic unit hydrograph.

Hydrograph Synthesis

The unit hydrograph can be used to predict runoff from a storm only if the duration of the storm is the same as that of the unit hydrograph. Hydrograph synthesis techniques can be used to construct the hydrograph of a longer storm from the unit hydrograph of a shorter storm.

Method 1 — Lagging Storm Method

This method is applicable when the duration of the synthesized storm is an exact multiple of the duration of the unit hydrograph ($t_r' = nt_r$). The procedure may be summarized as follows:

- Draw the n hydrographs, each separated by the duration t_r.
- Add the ordinates of the overlapping hydrographs.
- Divide resulting hydrographs ordinates by n.

Method 2 — S-Curve Method

This method is applicable when the synthesized hydrograph duration (t_r') is not an exact multiple of the duration of the base hydrograph (t_r). The procedure may be summarized as follows:

- Draw many ($n \to \infty$) hydrographs, the first at $t = 0$ and successive ones lagging the previous by time t_r.
- Draw the cumulative curve by adding together all of the hydrographs. If n is sufficiently large, the cumulative curve will be S-shaped and level off.
- Draw another S-curve repeating the two steps described above by starting at time t_r'.
- The difference between the two S-curves is multiplied by the factor (t_r'/t_r). This is the synthesized unit hydrograph.

Example 304.2   Lagging Storm Method

Ordinates for a unit hydrograph for a 2-h storm are shown below. Create the unit hydrograph for a 4-h storm synthesized from this storm.

Time t (h)	Discharge Q (cfs)
0.0	0.0
1.0	423.7
2.0	805.1
3.0	1016.9
4.0	296.6
5.0	42.4

Solution   The table later shows the application of the lagging storm method. This method is applicable since the synthesized storm duration (4 h) is an integer multiple of the basis storm (2 h).

Time t (h)	Q_1 (cfs)	Q_2 (cfs)	$Q_1 + Q_2$	$(Q_1 + Q_2)/2$
0.0	0.0		0.0	0.0
1.0	423.7		423.7	211.9
2.0	805.1	0.0	805.1	402.6
3.0	1016.9	423.7	1440.6	720.3
4.0	296.6	805.1	1101.7	550.9
5.0	42.4	1016.9	1059.3	529.7
6.0		296.6	296.6	148.3
7.0		42.4	42.4	21.2

NOTE   Since the original hydrograph was a unit hydrograph and the 4-h storm was formed by combining two such storms, the sum is then divided by two to obtain the new unit hydrograph.

Example 304.3   S-Curve Method

Ordinates for a unit hydrograph for a 2-h storm are shown below:

Time t (h)	Discharge Q (cfs)
0.0	0.0
1.0	423.7
2.0	805.1
3.0	1016.9
4.0	296.6
5.0	42.4

Create the unit hydrograph for a 3-h storm synthesized from this storm. The procedure may be summarized as follows:

1. Starting at $t = 0$, create a set of a large number of lagging storms (lag = duration of basis storm = 2 h). Calculate the cumulative runoff from this set. This cumulative curve should be approximately S-shaped. For future reference, let us call this curve S_1.

2. Starting at $t = 3$ h, create a set of a large number of lagging storms (lag = duration of basis storm = 2 h). Calculate the cumulative runoff from this set. Let us call this curve S_2.

3. Obtain the difference between the curves $S_1 - S_2$.

4. Scale the data obtained in step 3 by the ratio $2/3$. This is approximately the unit hydrograph of the 3-h storm.

As may be obvious, this procedure may be performed fairly easily using a spreadsheet type of program, but is too tedious to perform manually. Alternatively, if one had constructed the 4-h unit hydrograph (with a peak runoff of 720.3), one could scale it to approximately predict the peak of the 3-h unit hydrograph as

$$Q_{\text{peak,3h}} = Q_{\text{peak,4h}} \times \frac{4}{3} = \frac{720.3 \times 4}{3} = 960.4$$

Note that this approximation assumes that since both (unit) hydrographs have the same total precipitation (area under the curve), a scaling of the time axis would result in the inverse scaling of the vertical (runoff) axis.

Runoff Estimation by NRCS Curve Number

The Natural Resources Conservation Service (NRCS), formerly the Soil Conservation Service (SCS), specifies an approach for calculating storm runoff as specified in the document TR55. This method is also referred to as the soil cover complex method. In this approach, a curve number (CN) is used as the basis for calculating the runoff. It is applicable to any size homogeneous watershed with a known percentage of imperviousness. SCS method assumes a type II storm. This is not representative of storms that drop most of their precipitation early as in Hawaii, Alaska, and coastal regions in the Gulf of Mexico.

The curve number for a watershed is a single number representing the aggregate effect of land use and soil type. The following soil types are used:

Type A   Deep sands, deep loess, and aggregated silts are classified as type A soils. These soils exhibit high infiltration rate (>0.3 in/h) and therefore low runoff potential.

Type B   Shallow loess and sandy loam are classified as type B soils. These soils exhibit moderate infiltration rates (0.15–0.3 in/h) if thoroughly wetted.

Type C   Clay loams, shallow sandy loam, and soils low in organic content are classified as type C soils. These soils exhibit low infiltration rates (0.05–0.15 in/h) if thoroughly wetted.

Type D   Soils that swell significantly when wet. These soils exhibit low infiltration rates (less than 0.05 in/h) and therefore have a high runoff potential. Examples are heavy plastic clays and saline soils.

Table 304.6 Antecedent Moisture Condition

		5-day antecedent rainfall total (in)	
AMC	Description	Dormant season	Growing season
I	Dry soils after drought condition	<0.5	<1.4
II	Typical conditions	0.5–1.1	1.4–2.1
III	Saturated soil due to heavy rainfall	>1.1	>2.1

TR55 tables give curve numbers (CNs) for AMC II. The antecedent moisture condition (AMC) describes the moisture condition of the soil. AMC definitions are given in Table 304.6.

For AMC I and AMC III conditions, one must convert CN_{II} values to CN_I and CN_{III} using the following empirical relationships:

$$CN_I = \frac{4.2CN_{II}}{10 - 0.058CN_{II}} \qquad (304.23)$$

$$CN_{III} = \frac{23CN_{II}}{10 + 0.13CN_{II}} \qquad (304.24)$$

Paved watersheds are assigned $CN = 98$. For a composite area that can be partitioned into separable sections, each qualifying as a homogeneous watershed, a composite curve number may be used as in Eq. (304.25).

$$\overline{CN} = \frac{\sum (A_i \times CN_i)}{\sum A_i} \qquad (304.25)$$

Table 304.7 is reproduced from TR55 with permission. It shows curve numbers for AMC−II for urban areas. For a complete listing, the reader is referred to the complete document—Natural Resources Conservation Service, *Technical Release 55: Urban Hydrology for Small Watersheds*, USDA, Washington D.C., 1986.

Procedure for the TR55 Method

The procedure for the methodology in this document is outlined below:

1. Choose soil type and land use for each region.
2. Use the tables to look up CN_{II} for each region.
3. Correct to CN_I or CN_{III} if necessary (AMC not "normal" conditions).
4. Calculate composite CN for entire watershed.
5. Calculate storage capacity of soil $S = \frac{1000}{CN} - 10$.
6. Estimate time to concentration for the watershed.

Table 304.7 Curve Numbers for Urban Watersheds

Cover description		Curve numbers for hydrologic soil group			
Cover type and hydrologic condition	Average percent impervious area $\cong$	A	B	C	D
Fully developed urban areas (vegetation established)					
Open space (lawns, parks, golf courses, cemeteries, etc.):					
Poor condition (grass cover <50%)		68	79	86	89
Fair condition (grass cover 50% to 75%)		49	69	79	84
Good condition (grass cover >75%)		39	61	74	80
Impervious areas:					
Paved parking lots, roofs, driveways, etc. (excluding right-of-way)		98	98	98	98
Streets and roads:					
Paved; curbs and storm sewers (excluding right-of-way)		98	98	98	98
Paved; open ditches (including right-of-way)		83	89	92	93
Gravel (including right-of-way)		76	85	89	91
Dirt (including right-of-way)		72	82	87	89
Western desert urban areas:					
Natural desert landscaping (pervious areas only) $\cong$		63	77	85	88
Artificial desert landscaping (impervious weed barrier, desert shrub with 1- to 2-in sand or gravel mulch and basin borders)		96	96	96	96
Urban districts:					
Commercial and business	85	89	92	94	95
Industrial	72	81	88	91	93
Residential districts by average lot size:					
1/8 ac or less (town houses)	65	77	85	90	92
1/4 ac	38	61	75	83	87
1/3 ac	30	57	72	81	86
1/2 ac	25	54	70	80	85
1 ac	20	51	68	79	84
2 ac	12	46	65	77	82

(*Continued*)

Table 304.7 Curve Numbers for Urban Watersheds[a] (*Continued*)

Cover description		Curve numbers for hydrologic soil group			
Cover type and hydrologic condition	Average percent impervious area $\cong$	A	B	C	D
Developing urban areas					
Newly graded areas: (pervious areas only, no vegetation) $\cong$		77	86	91	94

Idle lands (CNs are determined using cover types similar to those in table 2-2c).

---

1 Average run off condition, and $I_a = 0.2S$.
2 The average percent impervious area shown was used to develop the composite CNs. Other assumptions are as follows; Impervious areas are directly connected to the drainage system. Impervious areas have a CN of 98, and pervious areas are considered equivalent to open space in good hydrologic condition. CNs for other combinations of conditions may be computed using figure 2-3 or 2-4.
3 CNs shown are equivalent to those of pasture. Composite CNs may be computed for other combinations of open space cover type.
4 Composite CNs for natural desert landscaping should be computed using figures 2-3 or 2-4 based on the impervious area percentage (CN-98) and the pervious area CN. The pervious are CNs are assumed equivalent to desert shrub in poor hydrologic condition.
5 Composite CNs to use for the design of temporary measures during grading and construction should be computed using figure 2-3 or 2-4 based on the degree of development (impervious are percentage) and the CNs for the newly graded pervious areas.

7. Using the time of concentration as the duration and a given recurrence interval, determine total rainfall (P_g).

8. Modify the gross rainfall for large areas.

9. Total runoff (net rain or excess rain) is calculated as

$$Q_{in} = \frac{(P_g - I_a)^2}{P_g - I_a + S} = \frac{(P_g - 0.2S)^2}{P_g + 0.8S} \qquad (304.26)$$

The results of Eq. (304.26) are tabulated in Table 304.8 and also plotted in Fig. 304.11, reproduced from TR55. For very low values of curve number CN and gross rainfall (P_g), this model does not yield reliable values. In the upper left section of the table, these are the cells populated by zero estimated runoff.

Peak Discharge

Once the runoff depth Q has been computed, the graphical peak discharge method is used to compute the peak discharge, according to the following equation:

$$q_p = q_u A_d Q F_p \qquad (304.27)$$

Table 304.8 Runoff Depth for Selected CNs and Rainfall Amounts*

Rainfall	Runoff depth for curve number of—												
	40	45	50	55	60	65	70	75	80	85	90	95	98
						inches							
1.0	0.00	0.00	0.00	0.00	0.00	0.00	0.00	0.03	0.08	0.17	0.32	0.56	0.79
1.2	.00	.00	.00	.00	.00	.00	.03	.07	.15	.27	.46	.74	.99
1.4	.00	.00	.00	.00	.00	.02	.06	.13	.24	.39	.61	.92	1.18
1.6	.00	.00	.00	.00	.01	.05	.11	.20	.34	.52	.76	1.11	1.38
1.8	.00	.00	.00	.00	.03	.09	.17	.29	.44	.65	.93	1.29	1.58
2.0	.00	.00	.00	.02	.06	.14	.24	.38	.56	.80	1.09	1.48	1.77
2.5	.00	.00	.02	.08	.17	.30	.46	.66	.89	1.18	1.53	1.96	2.27
3.0	.00	.02	.09	.19	.33	.51	.71	.96	1.25	1.59	1.98	2.45	2.77
3.5	.02	.08	.20	.35	.53	.75	1.01	1.30	1.64	2.02	2.45	2.94	3.27
4.0	.06	.18	.33	.53	.76	1.03	1.33	1.67	2.04	2.46	2.92	3.43	3.77
4.5	.14	.30	.50	.74	1.02	1.33	1.67	2.05	2.46	2.91	3.40	3.92	4.26
5.0	.24	.44	.69	.98	1.30	1.65	2.04	2.45	2.89	3.37	3.88	4.42	4.76
6.0	.50	.80	1.14	1.52	1.92	2.35	2.81	3.28	3.78	4.30	4.85	5.41	5.76
7.0	.84	1.24	1.68	2.12	2.60	3.10	3.62	4.15	4.69	5.25	5.82	6.41	6.76
8.0	1.25	1.74	2.25	2.78	3.33	3.89	4.46	5.04	5.63	6.21	6.81	7.40	7.76
9.0	1.71	2.29	2.88	3.49	4.10	4.72	5.33	5.95	6.57	7.18	7.79	8.40	8.76
10.0	2.23	2.89	3.56	4.23	4.90	5.56	6.22	6.88	7.52	8.16	8.78	9.40	9.76
11.0	2.78	3.52	4.26	5.00	5.72	6.43	7.13	7.81	8.48	9.13	9.77	10.39	10.76
12.0	3.38	4.19	5.00	5.79	6.56	7.32	8.05	8.76	9.45	10.11	10.76	11.39	11.76
13.0	4.00	4.89	5.76	6.61	7.42	8.21	8.98	9.71	10.42	11.10	11.76	12.39	12.76
14.0	4.65	5.62	6.55	7.44	8.30	9.12	9.91	10.67	11.39	12.08	12.75	13.39	13.76
15.0	5.33	6.36	7.35	8.29	9.19	10.04	10.85	11.63	12.37	13.07	13.74	14.39	14.76

*Interpolate the values shown to obtain runoff depths for CNs or rainfall amounts not shown.

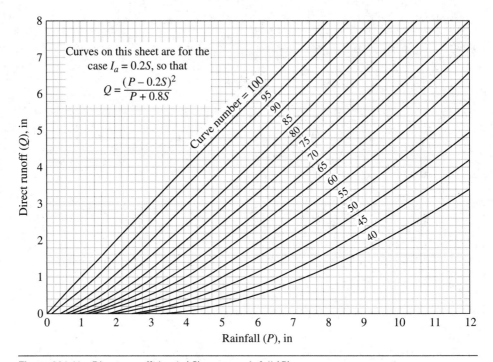

Figure 304.11 Direct runoff depth (Q) versus rainfall (P).

(*Urban Hydrology for Small Watersheds*, Natural Resources Conservation Service, USDA, Washington D.C., 1986.)

where q_p = peak discharge (cfs)
  q_u = unit peak discharge (csm/in)
  A_d = drainage area (mi^2)
  Q = runoff (in)
  F_p = pond and swamp adjustment factor

The unit peak discharge (q_u) for various types of rainfall distribution (types I, IA, II, and III) are given in Chap. 4 of the document *Urban Hydrology for Small Watersheds* (TR55). A representative figure (for type II rainfall distribution) is shown in Fig. 304.12.

The pond and swamp adjustment factor (F_p) is given in Table 304.9.

Example 304.4

A 30-ac wooded lot is to be converted into a community of 1/4-ac residential lots. The 2-year precipitation depth is 3.2 in. The predevelopment curve number is 65 and the time of concentration is 50 min. It is estimated that after construction, the curve number will become 80 and the time of concentration will become 20 min. What is the size of the infiltration basin needed to maintain the runoff at predevelopment level.

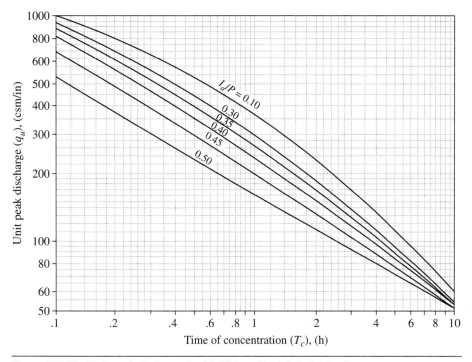

Figure 304.12 Unit peak discharge for NRCS type II rainfall distribution.
(*Urban Hydrology for Small Watersheds*, Natural Resources Conservation Service, USDA, Washington D.C., 1986.)

Table 304.9 Pond and Swamp Adjustment Factor F_p

Percentage of pond and swamp areas	F_p
0.0	1.00
0.2	0.97
1.0	0.87
3.0	0.75
5.0	0.72

Solution Drainage area $= 30$ ac $= 30 \div 640 = 0.047$ mi^2
Predevelopment

$$\text{CN} = 65 \Rightarrow S = \frac{1000}{65} - 10 = 5.385 \Rightarrow I_a = 0.2S = 1.08 \Rightarrow \frac{I_a}{P} = \frac{1.08}{3.2} = 0.34$$

For CN = 65 and gross rainfall = 3.2 in, $Q_1 = 0.6$ in
Using exhibit 4-II from TR-55, for $I_a/P = 0.34$ and $t_c = 0.83$ h, $q_u = 280$ csm/in
   2-year peak discharge, $q_p = q_u A_d Q F_p = 280 \times 0.047 \times 0.6 = 7.9$ cfs
   Postdevelopment

$$\text{CN} = 80 \Rightarrow S = \frac{1000}{80} - 10 = 2.5 \Rightarrow I_a = 0.2S = 0.50 \Rightarrow \frac{I_a}{P} = \frac{0.50}{3.2} = 0.16$$

For CN = 80 and gross rainfall = 3.2 in, $Q_2 = 1.4$ in
Using exhibit 4-II from TR-55, for $I_a/P = 0.16$ and $t_c = 0.33$ h, $q_u = 620$ csm/in
   2-year peak discharge, $q_p = q_u A_d Q F_p = 620 \times 0.047 \times 1.4 = 40.8$ cfs
   The increased upland runoff depth for the 2-year storm is given by

$$\Delta Q = Q_2 - (q_{u1}/q_{u2})Q_1 = 1.4 - (280/620) \times 0.6 = 1.13 \text{ in}$$

The infiltration basin is sized to store 1.13 in (123,057 ft^3 = 0.92 Mgal) of runoff.

Rational Method for Predicting Runoff

The error in the runoff estimate increases as the size of the drainage area increases. For these reasons, the rational method should not be used to determine the rate of runoff from large drainage areas. For the design of highway drainage structures, the use of the rational method should be restricted to drainage areas less than 80 hectares (200 ac). The assumptions involved in using the rational method are

1. The peak flow occurs when the entire watershed is contributing.

2. The rainfall intensity is uniform over a duration equal to the time of concentration.

3. The frequency of the computed peak flow is equal to the frequency of the rainfall intensity. In other words, the 10-year rainfall intensity is assumed to produce the 10-year flood.

Most urban watersheds served by storm drains contribute flow to these large drainage channels well before they reach significant size and therefore the rational method can be applied to these tributary areas. The rational method uses Eq. (304.28) to predict peak runoff flow rate Q_p from a watershed of area A_d due to a storm of intensity I.

$$Q_p = CIA_d \tag{304.28}$$

The intensity (typical units inches/per hour) may be estimated from historical data presented in the form of intensity-duration-frequency curves. The rational coefficient C is a single number representing land use and land cover for the watershed. Table 304.10 shows typical values of the runoff coefficient.

Table 304.10   Values of Runoff Coefficient C, for Use in the Rational Method

Surface type	Runoff coefficient (C)
Rural areas	
Concrete or sheet asphalt pavement	0.8–0.9
Asphalt macadam pavement	0.6–0.8
Steep grassed areas	0.5–0.7
Gravel roadways	0.4–0.6
Bare earth	0.2–0.9
Cultivated fields	0.2–0.4
Turf meadows	0.1–0.4
Forested areas	0.1–0.3
Urban areas	
Residential, flat, 30% impervious	0.40
Residential, flat, 60% impervious	0.55
Residential, steep, 50% impervious	0.65
Built—up area, steep, 70% impervious	0.80
Commercial, flat, 90% impervious	0.80

Example 304.5

Land use classification for a watershed in the western United States is summarized in the table below. The longest flow path in the watershed to the storm sewer is 337 yd and the average slope is 1.3%. What is the flow to the storm sewer resulting from a 10-year storm?

Land use	Area (ac)	Rational coefficient
Shingle roof	0.8	0.85
Concrete surface	1.1	0.88
Asphalt surface	2.6	0.83
Poorly drained lawn	10.0	0.15

Solution   The runoff coefficient for the watershed is calculated as the weighted average.

$$\overline{C} = \frac{\sum C_i A_i}{\sum A_i} = \frac{0.85 \times 0.8 + 0.88 \times 1.1 + 0.83 \times 2.6 + 0.15 \times 10}{0.8 + 1.1 + 2.6 + 10} = 0.366$$

Using the FAA formula, the time of concentration is calculated as

$$t_c = \frac{1.8(1.1 - 0.366)\sqrt{1011}}{1.3^{1/3}} = 38.5 \, \text{min}$$

Using Steel's formula, $K = 60$, $b = 13$, we get

$$I = \frac{K}{t_c + b} = \frac{60}{38.5 + 13} = 1.17 \, \text{in/h}$$

Runoff is given by

$$Q = CIA_d = 0.366 \times 1.17 \times 14.5 = 6.21 \, \text{ac-in/h} = 6.26 \, \text{cfs}$$

Example 304.6

A community has 45 houses, with average occupancy of four persons per house. Each roof has dimensions 65 ft $\times$ 40 ft. Assuming an average sewage flow of 115 gpcd, what size sewer is needed to handle the peak domestic flow rate plus drainage from the roofs from a storm with rainfall intensity = 1.5 in/h? Assume the sewer pipe has $n = 0.015$ and longitudinal slope = 2%.

Solution   Domestic sewage flow = $45 \times 4 \times 115 = 20,700 \, \text{gpd} = 14.38 \, \text{gpm} = 0.032 \, \text{cfs}$ (average)

Assuming the maximum hour to carry three times the average, $Q = 0.1 \, \text{cfs}$

$$\text{Roof area} = 45 \times 65 \times 40 = 117,000 \, \text{ft}^2 = 2.69 \, \text{ac}$$

Assuming a rational $C = 1.0$ (impervious roof), the roof runoff is

$$Q = CIA = 1.0 \times 1.5 \times 2.69 = 4.04 \, \text{ac-in/h} = 4.04 \, \text{cfs}$$

The combined flow rate is 4.14 cfs.
Using the tables for circular conduits,

$$Q_f = \frac{0.312k}{n_{\text{full}}} D^{8/3} S^{1/2} \qquad\qquad (k = 1.0 \text{ for SI units}, k = 1.486 \text{ for U.S. units})$$

$$D = \left(\frac{2.157nQ}{\sqrt{S}}\right)^{3/8} = \left(\frac{2.157 \times 0.015 \times 4.14}{\sqrt{0.02}}\right)^{3/8} = 0.98 \, \text{ft} \qquad\qquad (\text{Use 12-in pipe})$$

Modified Rational Method

The modified rational method allows the development of a storm hydrograph based only on the storm duration t_r, the time of concentration t_c, and the peak discharge Q_p (Fig. 304.13). The peak discharge is calculated using

$$Q_p = CIA_d \qquad\qquad (304.29)$$

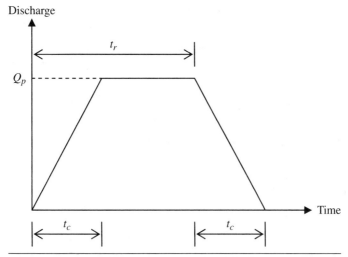

Figure 304.13 Modified rational method unit hydrograph.

Reservoir Sizing Using the Rippl Diagram

A reservoir may be used to stabilize/equalize "variable" inflows into a watershed so as to meet a "constant" demand for public use. If the reservoir size is known, a tabular approach may be used to construct a continuous record of quantities in the reservoir. The storage at the end of one period is simply the storage in the previous period + inflows − outflows and other losses (such as evaporation and infiltration). If this value becomes negative, it implies that the reservoir is empty and the value is reset to zero. If the value becomes greater than the reservoir capacity, it implies a spill, and the value is reset to the reservoir capacity. However, if the reservoir capacity is unknown, construction of such a table is not possible. Several alternative approaches may be used to determine the minimum required storage of a reservoir. One of these is the use of the mass diagram as proposed by Rippl (1883).[2]

For example, Table 304.11 shows a typical annual record of inflow into an impounding reservoir from the contributing watershed. Inflow values are in 1000 gal/mi^2. The reservoir is expected to meet a demand of 7.56×10^5 gal/mi^2/y. This is equivalent to a monthly draft of 63,000 gal/mi^2.

For each time interval, the deficiency is calculated as the difference between the outflow (O) and the inflow (I). A negative value of this deficiency indicates a surplus for that interval. In the next column of the table, the cumulative deficiency is computed, using only positive values of the deficiency and resetting the column every time a surplus month is encountered. The maximum value of the cumulative deficiency is an estimate of the minimum required reservoir capacity.

---

[2]Rippl, W. (1883), "The Capacity of Storage Reservoirs for Water Supply," Proceedings of the Institution of Civil Engineers, 71.

Table 304.11 Inflow–Outflow Data for a Reservoir

Month	Inflow (*I*)	Draft (*O*)	Cumul. inflow	Deficiency (*O* – *I*)	Cumul. deficiency $\Sigma|O-I|$	Surplus (*I* – *O*)	Cumul. surplus $\Sigma|I-O|$
Jan	43	63.0	43	20.0	20.0	−20.0	0
Feb	58	63.0	101	5.0	25.0	−5.0	0
Mar	113	63.0	214	−50.0	0	50.0	50.0
Apr	45	63.0	259	18.0	18.0	−18.0	0
May	43	63.0	302	20.0	38.0	−20.0	0
Jun	12	63.0	314	51.0	89.0	−51.0	0
Jul	11	63.0	325	52.0	141.0	−52.0	0
Aug	56	63.0	381	7.0	148.0	−7.0	0
Sep	143	63.0	524	−80.0	0	80.0	80.0
Oct	121	63.0	645	−58.0	0	58.0	138.0
Nov	86	63.0	731	−23.0	0	23.0	161.0
Dec	50	63.0	781	13.0	13.0	−13.0	0

Similarly, a surplus column and a cumulative surplus column are constructed. The maximum value of the cumulative surplus is an estimate of the required reservoir capacity to avoid a spill.

The cumulative inflow column is used to plot Fig. 304.14. The cumulative deficiency column may be used to determine the required reservoir capacity without having to resort to the graphical solution.

Thus, for the pattern of inflows, the required volume of the water supply reservoir that can meet a monthly demand of 63,000 gal/mi² is 148,000 gal/mi² or approximately 70 days of demand.

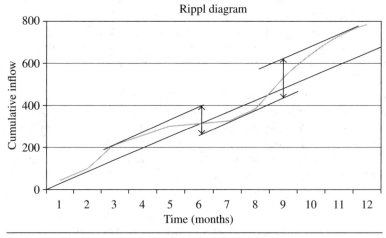

Figure 304.14 Cumulative inflow for data in Table 304.11.

Table 304.12 Procedure for Sizing Equalization Basin

Month	Inflow (I)	Draft (O)	Cumulative inflow	Surplus (I − O)	Cumulative surplus Σ|I − O|
Jan	43	65.0	43	−22.0	0.0
Feb	58	65.0	101	−7.0	0.0
Mar	113	65.0	214	48.0	48.0
Apr	45	65.0	259	−20.0	0.0
May	43	65.0	302	−22.0	0.0
Jun	12	65.0	314	−53.0	0.0
Jul	11	65.0	325	−54.0	0.0
Aug	56	65.0	381	−9.0	0.0
Sep	143	65.0	524	78.0	78.0
Oct	121	65.0	645	56.0	134.0
Nov	86	65.0	731	21.0	155.0
Dec	50	65.0	781	−15.0	0.0

The cumulative demand line is drawn (straight line if demand is constant with respect to time). Tangents with the same slope are drawn at each crest and the following trough. The distance between each pair is measured vertically.

The maximum separation between all tangents drawn from peaks to following troughs is the minimum reservoir capacity required to *serve the demand without a drought.* This is seen as approximately 150 in Fig. 304.14. That estimate coincides with the 148 obtained using the tabular approach, which is the maximum cumulative of the deficiency column (outflow−inflow).

The maximum separation between all tangents drawn from troughs to following peaks is the minimum reservoir capacity required to *serve the demand without a spill.* This is seen as approximately 160 in Fig. 304.14. That estimate coincides with the 161 obtained using the tabular approach, which is the maximum cumulative of the surplus column (inflow−outflow).

Flow Equalization

On the other hand, if the objective is to design an equalization basin, such as used for stabilizing fluctuating wastewater flows, the draft is equal to the entire cumulative inflow divided equally into the number of flow periods. In this problem, the draft is then equal to $781/12 = 65$. Instead of tabulating positive values of the deficiency $(O − I)$, we need to tabulate positive values of the surplus $(I − O)$. The results of the analysis for size of the equalization basin are presented in Table 304.12.

Thus, for the pattern of inflows, the required volume of the equalization basin is 155,000 gal/mi².

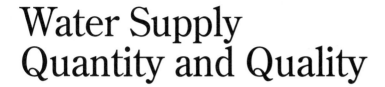
Water Supply Quantity and Quality

Determination of Needed Fire Flow

Population-Based Fire Flow Requirements

The National Board of Fire Underwriters[1] has proposed the following formula for computing needed fire flow for communities of less than 200,000:

$$Q = 1020\sqrt{P}(1 - 0.01\sqrt{P}) \qquad (305.1)$$

where Q is in gal/min and P is the population in thousands.

Fire Flow Requirements Based on Structure Type

The Insurance Services Office (New Jersey, 2008) has published *Guide for the Determination of Needed Fire Flow*, which outlines the procedures for calculating water requirements for municipal fire protection. The amount of water needed for fighting a fire in an individual, nonsprinklered building is given by

$$\text{NFF} = CO[1 + (X + P)] \qquad (305.2)$$

where NFF = needed fire flow (gal/min)
      C = factor related to type of construction
      O = factor related to type of occupancy
      X = factor related to exposure of buildings
      P = factor related to communication between buildings

---

[1]National Board of Fire Underwriters (1974) now merged into Insurance Services Office (ISO), New Jersey.

The step-by-step procedure for determining the needed fire flow is

1. Determine the predominant construction type and the associated factor (F).
2. Determine the effective area (A_{eff}) in square feet.
3. Calculate the construction factor $C = 18F\sqrt{A_{eff}}$.
4. Round the construction factor (C) to the nearest 250 gpm.
5. Determine the predominant occupancy type and the associated factor (O).
6. Determine if there is an exposure charge (factor X) by identifying the construction type and length-height value of the exposure building as well as the distance (in feet) to the exposure building. Also, make note of any openings and protection of those openings in the wall facing the subject building (the building the needed fire flow is being calculated for).
7. Determine if there is a communication charge (factor P) by identifying the combustibility of the passageway, whether the passageway is open or closed, the length, and a description of any protection provided in the passageway openings.
8. Substitute the values for the factors in the formula for NFF to determine the needed fire flow.

Type of Construction (*C*)

$$C = 18F\sqrt{A_{eff}}$$ (305.3)

where $F = 1.5$ for class 1 (wood frame)
$F = 1.0$ for class 2 (joisted masonry)
$F = 0.8$ for class 3 (noncombustible)
$F = 0.8$ for class 4 (masonry, noncombustible)
$F = 0.6$ for class 5 (modified fire resistive)
$F = 0.6$ for class 6 (fire resistive)

For construction classes 1 and 2, the calculated value of C shall not exceed 8000 gpm nor be less than 500 gpm. For construction classes 3, 4, 5, and 6 and for one-story buildings of any class, the calculated value of C shall not exceed 6000 gpm nor be less than 500 gpm.

Effective Area (*A*eff)

The effective area (ft²) shall be the total square feet area of the largest floor plus the following percentage of the total area of the other floors:

1. For construction classes 1 to 4: 50%.
2. For construction classes 5 and 6: If all vertical openings are protected, 25% of the area not exceeding the two other largest floors. If any vertical openings are unprotected, 50% of the area not exceeding eight other floors with unprotected openings.

Some areas such as sprinklered floors, vacant areas, roof structures, and the like may be disregarded in the calculation of the effective area.

Occupancy Factor (*O*)

The occupancy factor *O* ranges from 0.75 to 1.25 based on the combustibility characteristics of building contents. The values for the five classes C-1 through C-5 are given below:

Occupancy combustibility class	Occupancy factor
C-1 (noncombustible)	0.75
C-2 (limited combustible)	0.85
C-3 (combustible)	1.00
C-4 (free burning)	1.15
C-5 (rapid burning)	1.25

Exposure and Communication Factor (*X* + *P*)

These factors reflect the influence of adjoining and connected buildings on the NFF.

- An exposure building has a wall 100 ft or less from a wall of the subject building.
- A communicating building has a passageway to the subject building.

$$(X + P) = 1.0 + (X_i + P_i) \leq 1.60 \tag{305.4}$$

The factor X_i depends on the construction and the length-height value (product of length of wall in feet and the height in stories) of the exposure building and the distance between facing walls of the subject building and the exposure building. When there is no exposure on a side, $X_i = 0$.

Depending on proximity of the exposure building, the exposure factor *X* may be as high as 0.25. When the height of the facing wall of the exposure building is the same or less than the height of the facing wall of the subject building, use $X = 0$.

Communication Factor P_i

Depending on the fire classification of the doors on either side of the passageway communicating between the subject building and another connected building, the communication factor may range from 0 to 0.5. With a single class A fire door at each end of the passageway, or double class A fire doors at one end, the communication factor is zero.

Communication charges from adjacent buildings are exempt when they are rated sprinklered or habitational, buildings of construction class 5 or 6, buildings of construction class 3 or 4 with C-1 or C-2 contents combustibility.

Table 305.1 Needed Fire Flow for Small Residential Buildings

Distance between buildings (ft)	NFF (gpm)
>100	500
31–100	750
11–30	1000
<10	1500

Other guidelines

1. When a subject building or exposure buildings have wood-shingle roof that could contribute to spreading fires, add 500 gpm to needed fire flow.
2. Maximum NFF is 12,000 gpm. Minimum NFF is 500 gpm.
3. For NFF less than 2500 gpm, round to nearest 250 gpm. For NFF greater than 2500 gpm, round to nearest 500 gpm.
4. For one and two family dwellings not exceeding two stories in height, use the criteria in Table 305.1.
5. For other types of habitational buildings, maximum NFF is 3500 gpm.

Example 305.1

Calculate the needed fire flow in gal/min for a one-story wood frame building. The primary use of the building is for contractor equipment storage. The floor area is 2100 ft². There are no other buildings closer than 100 ft from any of the faces of the building.

Solution   Wood frame building—construction class 1—construction class factor $F = 1.5$

$$C = 18F\sqrt{A_{\text{eff}}} = 18 \times 1.5 \times \sqrt{2100} = 1237.3 = 1250 \text{ (rounded to nearest 250)}$$

Occupancy Combustibility class C-3 (combustible) $O = 1.00$

No exposures or communications $X + P = 0.0$

NFF $= 1250 \times 1.0 \times 1.0 = 1250$ gpm (greater than 500, less than 6000 O.K.)

Example 305.2

Calculate the needed fire flow in gal/min for a two-story building with masonry walls, wood-joisted roof and floors. The primary use of the building is for manufacturing paper products. The ground floor has a footprint of 150 ft × 80 ft. There are no other buildings closer than 100 ft from any of the faces of the building.

Solution

Construction class 2—construction class factor $F = 1.0$

Effective floor area = ground floor + half of second floor = 18,000 ft^2

$$C = 18F\sqrt{A_{\text{eff}}} = 18 \times 1.0 \times \sqrt{18,000} = 2415 = 2500 \text{ (rounded to nearest 250)}$$

Occupancy combustibility class C-4 (free-burning) $O = 1.15$

No exposures or communications $X + P = 0.0$

NFF $= 2500 \times 1.15 \times 1.0 = 2875$ gpm (specify NFF $= 3000$ gpm)

For NFF greater than 2500 gpm, round to nearest 500 gpm.

Drinking Water Treatment and Distribution

Public Water Systems

Public water systems (PWSs) may be publicly or privately owned and maintained. All PWSs must have at least 15 service connections or serve at least 25 people per day for 60 days of the year. Drinking water standards apply to water systems differently based on their type and size. In spite of significant differences of scale, the common goal of all PWSs is to provide safe, reliable drinking water to the communities they serve. To do this, most water systems must treat their water. The types of treatment provided by a specific PWS varies depending on the size of the system and whether the source they use is ground water or surface water.

A community water system (CWS) is a PWS that serves the same people year-round. Most residences including homes, apartments, and condominiums in cities, small towns, and mobile home parks are served by CWSs.

A non-community water system (NCWS) is a PWS that serves the public but does not serve the same people year-round. There are two types of non-community systems:

- Non-transient non-community water system (NTNCWS)—A NCWS that serves the same people more than 6 months per year, but not year-round. For example, a school with its own water supply is considered a non-transient system.

- Transient non-community water system (TNCWS)—A NCWS that serves the public but not the same individuals for more than 6 months. For example, a rest area or campground may be considered a transient water system.

Sources of Drinking Water

Large-scale water supply systems tend to rely on surface water sources, while smaller systems tend to rely on ground water. Around 32% of the population served by CWSs drinks water that originates as ground water. Ground water is usually pumped from wells ranging from shallow to deep (50–1000 ft). The remaining 68% of the population served by CWSs receive water taken primarily from surface water sources like rivers, lakes, and reservoirs.

Lakes as an example of an ecosystem

Limnology is the study of the ecology of freshwater biotic communities such as lakes, rivers, creeks, and ponds. Specifically, within the field of limnology, the study of the stratification and productivity of freshwater lakes is useful for civil engineers.

Stratification of Lakes

Nearly all deep lakes in temperate climates become stratified during summer and experience "turnover" in the fall. The turnover occurs due to changes in water temperature resulting from annual cycle of air temperature changes.

During the summer, the surface of the lake is heated both by direct sunlight and due to contact with warm air. This warm water, being less dense than the cold water below, stays near the surface unless mixed downward by turbulence from wind, waves, and other forces. As the effect of the turbulence extends only to shallow depths, the result is an upper layer of well-mixed, warm water (called the epilimnion) which floats on a poorly mixed, cooler layer called the hypolimnion. The transition zone between the epilimnion and the hypolimnion is called the metalimnion. Within the metalimnion, in a layer known as the thermocline, the temperature and density of the water changes rapidly with depth, whereas they are relatively stable within the epilimnion and the hypolimnion. The thermocline may be defined as the zone within which the temperature gradient is more than $1°C/m$ ($0.55°F/ft$). Once formed, thermal stratification of lakes is fairly stable.

In the fall, the epilimnion cools until it is denser than the hypolimnion. It then sinks, causing turnover. The hypolimnion rises to the surface, where it cools, causing it to sink again. This causes a complete mixing of the water in the lake. In colder climates, this process stops when the deeper water temperature reaches $4°C$, at which temperature it has the highest density. Further cooling or freezing causes winter stratification, with a frozen surface layer and progressively warmer (up to $4°C$) waters at depth. As the water warms again in spring, it overturns again and becomes completely mixed. Thus, lakes in temperate climates undergo at least one, if not two, cycle of complete mixing every year.

The epilimnion is aerobic (higher dissolved oxygen) due to oxygen absorption at the surface and good mixing.

Biological Zones in Lakes

Lakes contain several distinct zones of biological activity, based on the availability of light and oxygen. Vertically, the lake may be divided into the euphotic zone, the profundal zone, and the benthic zone. The euphotic zone is the upper layer of water through which sunlight can penetrate. The euphotic zone is considered to extend to the depth where the light intensity is about 1% of the intensity at the surface. Below the euphotic zone lies the profundal zone. The interface between the euphotic zone and the profundal zone is called the light compensation point. At this depth, the CO_2 conversion rate (due to photosynthesis) is approximately equal to the CO_2 production due to respiration. The benthic zone is the lowermost layer occupied by sediments formed by the deposition and decomposition of dead organic matter. Bacteria and fungi are always present in the benthic zone.

Laterally, the lake may be divided into the littoral zone and the limnetic zone. The shallow water near the shore in which rooted (emergent) plants can grow is called the littoral zone. The littoral zone cannot extend deeper than the euphotic zone. The limnetic zone is the area of open water where photosynthesis can occur. The producers in this zone are planktonic algae.

The primary consumers are zooplankton such as crustaceans and rotifers. Secondary consumers are insects and fish.

Lake Productivity

A lake can also be classified according to its productivity, which is the measure of its ability to support aquatic life. A more productive lake will have greater biomass concentration due to the availability of nutrients. A high concentration of biomass is often undesirable, resulting in taste and odor problems, low dissolved oxygen levels, and many other issues. Productivity of a water body is controlled by a limiting factor, which is typically phosphorus, because, of all the nutrients, only phosphorus is not readily available from the atmosphere or the natural water supply. Productivity of a lake is often determined by the amount of algal growth that can be supported by the lake. It has been estimated that phosphorus concentration should be below approximately 0.015 mg/L to limit algal blooms.

Oligotrophic lakes have a low level of productivity due to lack of nutrients. In such lakes, the euphotic zone often extends into the hypolimnion, which is aerobic. These lakes support cold water game fish. On the other hand, eutrophic lakes have high productivity due to abundant supply of nutrients. Because of the overproduction of algae, the water becomes highly turbid, resulting in the euphotic zone to extend only partially into the epilimnion. Eutrophic lakes support only warm water fish as most cold water fish species require a dissolved oxygen concentration of at least 5 to 6 mg/L.

Lakes intermediate between oligotrophic and eutrophic are called mesotrophic. In these lakes, the oxygen levels in the hypolimnion may be depleted, but still at aerobic levels. On the other hand, dystrophic lakes receive a large quantity of organic material and have low productivity due to low nutrient concentrations. These lakes have zero oxygen concentration near the bottom and aerobic conditions exist only in the shallow regions.

Hypereutrophic lakes have high algal productivity and intense algal blooms. These lakes are usually shallow with a lot of organic sediment in the benthic zone. Dissolved oxygen concentration in these lakes can go through extreme variations, sometimes down to anaerobic levels. This can result in frequent occurrences of "fish kills."

Treating Raw Water

The type of treatment used by a public water system varies with the source type and quality. Ground water goes through "natural" treatment due to the filtering action of the media it trickles through. As a result, many ground water systems can satisfy all federal requirements without applying any treatment, while others need to add chlorine or additional treatment. Because surface water systems are exposed to direct wet weather runoff and to the atmosphere and are therefore more easily contaminated, federal and state regulations require that these systems treat their water. Disinfection of drinking water is a very important step in reaching desired quality standards. However, the disinfectants can react with constituents in the water to form unintended byproducts which may pose health risks. The dosage of disinfectant must be carefully chosen to balance the risk from microbial pathogens and disinfection byproducts. The Stage 1 Disinfectants and Disinfection Byproducts Rule and the Interim Enhanced Surface Water Treatment Rule together address these risks.

The most commonly used water treatment processes include filtration, flocculation and sedimentation, and disinfection for surface water. Depending on the quality of the source water, some treatment plants also include ion exchange and adsorption.

Types of Treatment

Flocculation/Sedimentation: Flocculation refers to water treatment processes that coagulate small suspended particles into larger particles ("floc"), which settle out of the water as sediment. Alum and iron salts or synthetic organic polymers (used alone or in combination with metal salts) are generally used to enhance/accelerate coagulation. Settling occurs naturally as flocculated particles settle out of the water.

Filtration: Many water treatment facilities use filtration to remove particles (clays and silts, natural organic matter, precipitates from other treatment processes in the facility, iron and manganese, and microorganisms) from the water. Filtration clarifies water and enhances the effectiveness of disinfection.

Ion Exchange: Ion exchange processes are used to remove inorganic contaminants if they cannot be removed adequately by filtration or sedimentation. It can be used to treat hard water. It can also be used to remove arsenic, chromium, excess fluoride, nitrates, radium, and uranium.

Adsorption: A substance such as granulated activated carbon (GAC) can be used to remove organic contaminants, color, taste, and odor-causing compounds which are adsorbed ("stick to") the surface of the adsorbent and are thus removed from the drinking water.

Disinfection: Water is often disinfected before it enters the distribution system to ensure that potentially dangerous microbes are killed. Chlorine, chloramines, or chlorine dioxide are most often used because they are very effective disinfectants, not only at the treatment plant but also in the pipes that distribute water to our homes and businesses. Ozone is a powerful disinfectant, and ultraviolet radiation is an effective disinfectant and treatment for relatively clean source waters, but neither of these is effective in controlling biological contaminants in the distribution pipes.

Monitoring Water Quality

Water systems monitor for a wide variety of contaminants to verify that the water they provide to the public meets all federal and state standards. Currently, the nation's CWSs and NTNCWSs must monitor for more than 83 contaminants. The major classes of contaminants include volatile organic compounds (VOCs), synthetic organic compounds (SOCs), inorganic compounds (IOCs), radionuclides, and microbial organisms (including bacteria). Testing for these contaminants takes place on varying schedules and at different locations throughout the water system.

TNCWSs may monitor less frequently and for fewer contaminants than CWSs. Because these types of systems serve a constantly changing population, it is not critical that these systems monitor for contaminants that cause chronic conditions due to cumulative exposure, but it is necessary that they screen for contaminants such as microorganisms and nitrate that can cause an immediate, acute public health effect.

Water Distribution Systems

Reservoirs must be located such that gravity flow can be utilized for much of the water distribution to a community. In establishing a proper reservoir elevation, the rule of thumb is that every foot of head produces approximately 0.434 psi. Therefore, to generate about 60 psi in the water distribution system, the storage reservoir needs to be located approximately 150 ft above the service area. For certain low-lying areas, regulators must be installed to protect systems from overpressure.

Pumps should be installed at locations where gravity flow does not result in adequate pressures. These "booster" locations and the required pumping power should be chosen such that they have optimal effect on the entire network.

Reservoirs should have adequate capacity to provide continuous domestic flow and anticipated fire flows for a reasonable duration. A reasonable rule of thumb is that storage should be sufficient to provide at least 2 days of peak domestic consumption plus required fire flows. For example, in a residential area, if fire codes call for 2 hours of fire flow at 1000 gpm, then additional fire flow storage is approximately 1.44 million gallons. For industrial areas, fire flow of 4000 to 5000 gpm for 3 hours is more typical.

A water distribution system must have some redundancy built into it so that adequate flow and pressure can be maintained in spite of closure of one or more branches of the network. Each point of the distribution system should be supplied by primary feeders which are typically 16 in in diameter. Secondary feeders (typically 12 in diameter) are used to supply specific neighborhoods. Distributor mains of 8 in diameter are typically used to supply water to individual streets.

Valves should be provided at every junction for mains branching from those junctions. Operation of these valves should allow individual sections of the main to be isolated for service or repair, so that "failure" of one section does not result in loss of water delivery to other parts of the network.

National Primary Drinking Water Standards (EPA)

"National Primary Drinking Water Regulations (NPDWRs or primary standards)[2] are legally enforceable standards that apply to public water systems. Primary standards protect public health by limiting the levels of contaminants in drinking water."

Surface Water Treatment Rule

The Surface Water Treatment Rule (effective December 31, 1990) seeks to prevent waterborne diseases caused by viruses, *Legionella*, and *Giardia lamblia*. These disease-causing microbes are present at varying concentrations in most surface waters. The rule requires that water systems filter and disinfect water from surface water sources to reduce the occurrence of unsafe levels of these microbes. This rule governs water supplies whose source of drinking water is surface water, which it defines as "all water which is open to the atmosphere and subject to surface runoff." Surface water is particularly susceptible to microbial contamination from sewage treatment plant discharges and runoff from storm water and snow melt. These sources often contain high levels of fecal microbes that originate in livestock wastes or septic systems.

Maximum Contaminant Level Goal

The rule mentioned above sets nonenforceable health goals, or maximum contaminant level goals (MCLGs) for *Legionella*, *Giardia lamblia,* and viruses at zero because any amount of exposure to these contaminants represents some health risk. The MCLG is an ideal limit, which may not be realizable based on economics and current technology and therefore is not legally enforceable. In establishing legal limits for contaminants in drinking water, Environmental Protection Agency (EPA) can set either a legal limit (MCL) and require monitoring for the contaminant in drinking water, or, for those contaminants that are difficult to measure, EPA can establish a

---

[2]U.S. Environmental Protection Agency, Washington DC. http://www.epa.gov/safewater/contaminants/index.html.

treatment technique requirement. Since measuring disease-causing microbes in drinking water is not considered to be feasible, EPA established a treatment technique in this rule.

All systems must filter and disinfect their water to provide a minimum of 99.9% combined removal and inactivation of *Giardia* and 99.99% of viruses. The adequacy of the filtration process is established by measuring turbidity (a measure of the amount of particles) in the treated water and determining if it meets EPA's performance standard. Some public water supplies that have pristine sources may be granted a waiver from the filtration requirement. These supplies must provide the same level of treatment as those that filter; however, their treatment is provided through disinfection alone. The great majority of water supplies in the United States that use a surface water source filter their water.

Contaminants in drinking water are subdivided into six functional groups: disinfectants (D), disinfectant byproducts (DBP), inorganic compounds (IOC), microorganisms (M), organic compounds (OC), and radionuclides (R). Some of the most important of these contaminants are presented in Table 305.2. For the complete listing of all contaminants, see the EPA Web site (www.epa.gov).

Table 305.2 Selected Criteria from National Primary Drinking Water Standards

Contaminant	MCL	MCLG
Arsenic	0.01 mg/L as of 1/23/06	0
Asbestos (fibers >10 μm)	7 million fibers per liter (MFL)	7 MFL
Benzene	0.005 mg/L	0
Chloramines (as Cl$_2$)	MRDL = 4.0 mg/L	MRDLG = 4.0 mg/L
Chlorine (as Cl$_2$)	MRDL = 4.0 mg/L	MRDLG = 4.0 mg/L
Copper	Action level = 1.3 mg/L (more than 10% of samples exceeding level)	1.3 mg/L
Cryptosporidium	99% removal	0
Dioxin	0.00003 μg/L	0
Fluoride	4.0 mg/L	4.0 mg/L
Giardia lamblia	99.9% removal/inactivation	0
Lead	Action level = 0.015 mg/L (more than 10% of samples exceeding level)	0
Mercury (inorganic)	0.002 mg/L	0
Nitrate (measured as nitrogen)	10 mg/L	10 mg/L
Nitrite (measured as nitrogen)	1 mg/L	1 mg/L
Total coliforms (including fecal coliform and *Escherichia coli*)	No more than 5.0% samples total coliform positive in a month.	0
Turbidity	No samples over 5 NTU. For systems that filter: no more than 5% of samples over 1 NTU. For conventional or direct filtration: no more than 5% of samples over 0.5 NTU. As of January 1, 2002, turbidity may never exceed 1 NTU, and must not exceed 0.3 NTU in 95% of daily samples in any month.	
Viruses (enteric)	99.99% removal/inactivation	0

Table 305.3 National Secondary Drinking Water Standards

Contaminant	Secondary standard
Aluminum	0.05–0.2 mg/L
Chloride	250 mg/L
Color	15 color units
Copper	1.0 mg/L
Corrosivity	Noncorrosive
Fluoride	2.0 mg/L
Foaming agents	0.5 mg/L
Iron	0.3 mg/L
Manganese	0.05 mg/L
Odor	3 threshold odor number
pH	6.5–8.5
Silver	0.10 mg/L
Sulfate	250 mg/L
Total dissolved solids	500 mg/L
Zinc	5 mg/L

Source U.S. Environmental Protection Agency, Washington DC.
https://www.epa.gov/ground-water-and-drinking-water/national-primary-drinking-water-regulations

National Secondary Drinking Water Standards (EPA)

"National Secondary Drinking Water Standards[3] are non-enforceable guidelines regulating contaminants that may cause cosmetic effects (such as skin or tooth discoloration) or aesthetic effects (such as taste, odor, or color) in drinking water. EPA recommends secondary standards to water systems but does not require systems to comply. However, states may choose to adopt them as enforceable standards." These secondary standards are given in Table 305.3.

Taste and Odor in Water

Drinking water often has unpleasant taste and odor. Even though the judgment is subjective, some of the most common complaints are sewer smell, chlorine smell and/or taste, smell of rotten eggs, petroleum smell or taste, metallic smell or taste, and earthy or fishy smell and taste.

Very often, the contaminants that contribute to these problems are classified as secondary drinking water contaminants, meaning that water treatment systems aren't required by law to address them.

---

[3]U.S. Environmental Protection Agency, Washington D.C. http://www.epa.gov/safewater/contaminants/index.html.

Sewer Smell

If water tastes or smells stale like a sewer, it is often stale water caused by water standing in dead-end lines or service lines that are too long or little-used lines in a new development. The best way to address this problem is through routine maintenance procedures such as regular flushing of these water lines.

Chlorine Smell

If tap water only smells of chlorine, but there is no chlorine taste, it could be due to the fact that chloro-organics and chloramines are present in the water. These compounds may result in taste and odor problems. Adding more chlorine neutralizes these compounds, taking the treated water beyond the breakpoint and creating free residual chlorine.

Chlorine Smell and Taste

If tap water has a distinct taste *and* smell of chlorine it usually means there is a high chlorine residual in the water. The maximum residual disinfectant level (MRDL) of free available chlorine is 4.0 mg/L but levels below that can also be offensive to taste. A system that chlorinates heavily so that customers at the end of long lines receive properly disinfected water will have customers who are close to the chlorination point receiving water with higher chlorine residuals, leading to complaints of chlorine taste and odor.

One way to help reduce the heavy dosage would be to add a booster chlorination system in the distribution system. If this is economically infeasible, installation of an activated carbon filter by the customer can eliminate, or at least reduce the chlorine residual.

Rotten Egg Smell

In groundwater systems, a smell like rotten eggs is caused by hydrogen sulfide (H_2S), which is not regulated by the national primary drinking water standards but may be regulated at the state level. Hydrogen sulfide does not usually pose a health risk at low levels (1 to 2 ppm).

The best way to treat water containing H_2S is to aerate the water. This can be done mechanically (stack aeration) or chemically (with oxidizing chemicals such as potassium permanganate or chlorine). Other treatment methods include ion exchange, oxidizing filters, and carbon filtration.

If the complaint is with a surface water system, the problem could be contamination from a sewer and should be thoroughly investigated. If the complaint is only with the hot water in a customer's house, the problem could be due to the magnesium rod used in heaters (for corrosion control) chemically reducing sulfates to H_2S. Replacing the water heater's magnesium corrosion control rod with one made of aluminum or other metal may improve the situation.

Petroleum Smell or Taste

Petroleum smell or taste sometimes occurs in groundwater systems that may be near areas of naturally occurring gas and oil. Very often, Benzene (C_6H_6) is the main contaminant. The U.S. Environmental Protection Agency's National Primary Drinking Water Standards sets the maximum contaminant level (MCL) for benzene at 0.005 mg/L or ppm with a public health goal (MCLG) of zero. The best way to treat groundwater with this problem is to use activated carbon filters immediately after pumping followed by using chlorine as an oxidant. Typical contact time for the chlorine is about 20 min.

Metallic Smell or Taste

If the water is acidic (low pH), it can corrode the inside of metal distribution and copper service lines, causing the material to leach into the water. If pH adjustment does not work, corrosion can be prevented by adding a coating on the inside of the line to protect the metals from corroding and leaching into the water.

Earthy or Fishy Smells

These problems can be seasonal as they are often connected to algal growth. Algae are usually blue-green in color and are easy to detect if present in the source water. If algae bloom is present in the source water, copper sulfate can be added to kill the algae. Some states do not allow copper sulfate to be added to rivers or streams because it kills fish and other aquatic life. This problem can also be addressed by using mechanical aerators. In addition to eliminating algae, aerators can also reduce H_2S, iron, manganese, and phosphorus. Potassium permanganate can also be used for control of taste and odor problems caused by decaying vegetation. Activated carbon is another treatment option that can be used in conjunction with potassium permanganate.

Example 305.3

A drinking water sample shows the following results:

Ca^{2+}	45 mg/L
Mg^{2+}	15 mg/L
Na^+	4 mg/L
Alkalinity	160 mg/L as $CaCO_3$
Total coliform	1.7 MPN
Turbidity	1.8 NTU
Odor	2.4 TON
Nitrate [NO_3-N]	6 mg/L
Nitrite [NO_2]	3 mg/L
Total dissolved solids	350 mg/L

Which of the constituents are in violation of primary or secondary drinking water standards?

Solution The metal ions (Ca^2+, Mg^2+, and Na+) are not contaminants. Neither is alkalinity. Table 305.2 lays out criteria for batch of several samples to pass (no more than 5% coliform positive). But for a single sample, it is considered coliform positive if the MPN is greater than 1. Therefore, this sample is coliform positive.

The current standard for turbidity is "no samples exceeding 1.0 NTU." Therefore, this sample violates the standard for turbidity.

Odor = 2.4 TON <3.0 TON. Therefore, this standard is not violated.

The standard for NO_3-N is 10 mg/L. Therefore, this standard is not violated.

The standard for NO_2-N is 1 mg/L. However, before evaluating this constituent, the given concentration must be converted to "as nitrogen." This can be done using

$$[NO_2\text{-}N] = [NO_2] \times \frac{\text{eq.wt. of N}}{\text{eq.wt. of } NO_2} = 3 \times \frac{14}{46} = 0.91 \text{ mg/L as } NO_2\text{-N}$$

Since this is less than the primary standard of 1 mg/L, this constituent is not in violation.

Total dissolved solids 350 mg/L <500 mg/L. Therefore, this constituent is not in violation.

Therefore, the constituents that violate primary or secondary drinking water standards are total coliform and turbidity.

Dissolved Oxygen in Water

Solubility of Gases in Water

The concentration of a gas that can be present in solution is governed by (1) the solubility of the gas as defined by Henry's law, (2) temperature, (3) presence of impurities (such as salinity), and (4) the partial pressure of the gas in the atmosphere.

Henry's Law for Dissolved Gases

The saturation concentration of a gas dissolved in a liquid is a function of the partial pressure of the gas. Henry's law states that

$$p_g = \frac{H}{P_T} x_g \tag{305.5}$$

where p_g = mole fraction of the gas in air (moles gas/moles air)

x_g = mole fraction of the gas in the liquid (moles gas/moles liquid)

H = Henry's law constant (atm)

P_T = total pressure (atm)

Henry's law constant is a function of the type of gas, temperature, and type of liquid. When the liquid is water, the values in Table 305.4 may be used.

Example 305.4

What is the saturation concentration of oxygen (mg/L) in water at a pressure of 1 atm?

Solution

Henry's law constant for oxygen in water = H = 41,100 atm

Atmospheric air contains approximately 21% (mole fraction) of oxygen p_g = 0.21.

The mole fraction of (saturated) oxygen in water is therefore

$$x_g = \frac{P_T}{H} p_g = \frac{1}{41,100} \times 0.21 = 5.1 \times 10^{-6} \qquad \text{mol } O_2/\text{mol } H_2O$$

Table 305.4 Henry's Law Constant *H* When Water Is the Solvent

Gas	Henry's constant, *H* (atm)
Air	66,400
Carbon dioxide	1,420
Chlorine	579
Hydrogen	68,300
Nitrogen	80,400
Oxygen	41,100
Ozone	5,300

One liter of water has nominal mass $= 1$ kg $= 1 \times 10^6$ mg $= 55.556$ g-mol (1 g-mol of water $= 18$ g $= 18,000$ mg). Therefore, it contains 0.000283 g-mol of oxygen gas $= 0.000283 \times 32,000 = 9.07$ mg/L.

Dissolved Oxygen

Dissolved oxygen (DO) is measured in standard solution units such as millimoles (of oxygen) per liter (mmol/L), milligrams per liter (mg/L), milliliter per liter (mL/L), or parts per thousand (ppt). Oxygen saturation is calculated as the percent of DO relative to a theoretical maximum concentration given the temperature, pressure, and salinity of the water. Well-aerated water (in free interchange with the air) will usually be 100% saturated. In general, lower temperatures, lower salinity, and higher atmospheric pressures lead to higher values of dissolved oxygen.

Regimes of low concentrations (between 0% and 30% DO) are often called hypoxic (low oxygen). In estuaries, lakes, and coastal waters, low oxygen usually means a concentration of less than 2 parts per million (ppm). Hypoxic waters do not have enough oxygen to support fish and other aquatic animals. Hypoxia can be caused by a process called eutrophication which is due to the presence of excess nutrients in water. Nutrients can come from many sources, such as agricultural runoff, deposition of nitrogen from the atmosphere, erosion of soil containing nutrients, and sewage treatment plant discharges. Excess nutrients can cause intensive growth of algae causing reduced sunlight penetration, decreased dissolved oxygen, loss of habitat for aquatic animals and plants. The decrease in dissolved oxygen is caused by the degradation of dead plant material (algae), which consumes available oxygen.

The state of 0% saturation (no DO) is called *anoxia*. Most fish species cannot live in water once saturation falls below 30%. Supersaturation can sometimes be harmful for organisms and cause gas bubble disease. Percent saturation values of 80% to 120% are considered to be excellent, and values less than 60% or over 125% are considered to be poor.

Measurement of Dissolved Oxygen

Dissolved oxygen is measured using a DO probe. The DO probe consists of small silver anodes and a gold cathode. These electrodes are separated from the surrounding lake water by a Teflon membrane. Dissolved oxygen diffuses across the membrane and is reduced to hydroxyl ions (OH^-) at the cathode and silver chloride (AgCl) is formed at the anode. The current associated with this process is proportional to the DO in the surrounding water.

Saturated Dissolved Oxygen

Currently used values for saturation dissolved oxygen are based on an equation by Weiss[4] (1970) which fits data by Carpenter (1966)[5]. Carpenter's values are, at the present time, widely accepted as the most accurate determinations of saturation DO available. Table 305.5 shows saturated dissolved oxygen concentrations in mg/L for various degrees of salinity S (measured in g/kg) at a pressure of 1 standard atmosphere.

Table 305.5 Saturated Dissolved Oxygen as a Function of Temperature and Salinity

Temperature (°C)	Saturated dissolved oxygen (g/kg)				Vapor pressure (mmHg)
	$S = 0$	$S = 5$	$S = 10$	$S = 20$	
0	14.60	14.11	13.64	12.74	4.6
1	14.20	13.73	13.27	12.40	4.9
2	13.81	13.36	12.91	12.07	5.3
3	13.45	13.00	12.58	11.76	5.7
4	13.09	12.67	12.25	11.47	6.1
5	12.76	12.34	11.94	11.18	6.5
6	12.44	12.04	11.65	10.91	7.0
7	12.13	11.74	11.37	10.65	7.5
8	11.83	11.46	11.09	10.40	8.0
9	11.55	11.19	10.83	10.16	8.6
10	11.28	10.92	10.58	9.93	9.2
11	11.02	10.67	10.34	9.71	9.8
12	10.77	10.43	10.11	9.50	10.5
13	10.53	10.20	9.89	9.30	11.2
14	10.29	9.98	9.68	9.10	12.0
15	10.07	9.77	9.47	8.91	12.8
16	9.86	9.56	9.28	8.73	13.6
17	9.65	9.36	9.09	8.55	14.5
18	9.45	9.17	8.90	8.39	15.5

[4]Weiss, R. (1970), "The Solubility of Nitrogen, Oxygen, and Argon in Water and Seawater," *Deep-Sea Res.*, Vol. 17, 721–735.

[5]Carpenter, J. H. (1966), "New Measurements of Oxygen Solubility in Pure and Natural Water," *Limnology and Oceanography*, Vol. 11, No. 2, 264–277.

Table 305.5 Saturated Dissolved Oxygen as a Function of Temperature and Salinity (*Continued*)

Temperature (°C)	Saturated dissolved oxygen (g/kg)				Vapor pressure (mmHg)
	S = 0	**S = 5**	**S = 10**	**S = 20**	
19	9.26	8.99	8.73	8.22	16.5
20	9.08	8.81	8.56	8.07	17.5
21	8.90	8.64	8.39	7.91	18.6
22	8.73	8.48	8.23	7.77	19.8
23	8.56	8.32	8.08	7.63	21.1
24	8.40	8.16	7.93	7.49	22.4
25	8.24	8.01	7.79	7.36	23.8
26	8.09	7.87	7.65	7.23	25.2
27	7.95	7.73	7.51	7.10	26.7
28	7.81	7.59	7.38	6.98	28.4
29	7.67	7.46	7.26	6.87	30.0
30	7.54	7.33	7.14	6.75	31.8
31	7.41	7.21	7.02	6.65	33.7
32	7.29	7.09	6.90	6.54	35.7
33	7.17	6.98	6.79	6.44	37.7
34	7.05	6.86	6.68	6.33	39.9
35	6.93	6.75	6.58	6.24	42.2
36	6.82	6.65	6.47	6.14	44.6
37	6.72	6.54	6.37	6.05	47.1
38	6.61	6.44	6.28	5.96	49.7
39	6.51	6.34	6.18	5.87	52.4
40	6.41	6.25	6.09	5.79	55.3

NOTE 1 g/kg = 1000 mg/kg = 1000 ppm = 1000 mg/L (approximately, for water).

Higher barometric pressures lead to higher values of saturated dissolved oxygen. The correction for barometric pressure may be written as

$$DO' = DO_o \frac{P - u}{760 - u} \qquad (305.6)$$

where DO' = saturation DO at barometric pressure P (mmHg)
  DO_o = saturation DO at barometric pressure 760 mmHg
  u = vapor pressure of water (mmHg)

The vapor pressure of water may be calculated from one of several empirical equations in terms of temperature T. A representative profile is included in the last column of Table 305.5.

$$\ln u(\text{mmHg}) = 8.10765 - \frac{1750.286}{235 + T} \tag{305.7}$$

where T = temperature in °C.

Biochemical Oxygen Demand

Microorganisms in water digest organic material (substrate or food). During this process, they use oxygen and thus place an *oxygen demand* on the water. This oxygen demand is biochemical oxygen demand (BOD). The oxygen demand on a water body varies in accordance with the population dynamics of the microbial population. A typical model for such a population model includes a lag phase, a constant growth phase, a stationary phase, and a decay phase. A mathematical model for the growth of BOD is the exponential model where the BOD grows asymptotically to the so-called ultimate BOD of the water, which is its oxygen demand potential at $t = 0$. This model does not account for the lag or decay phases. According to this model, BOD is given by the function shown below:

$$\text{BOD}(t) = \text{BOD}_u(1 - e^{-kt}) \tag{305.8}$$

where BOD_u = ultimate BOD = oxygen equivalent of organics at $t = 0$
k = reaction rate constant (base e, day^{-1}), also called the deoxygenation rate constant or the endogenous decay constant (alternate symbol k_d)

Equation (305.8) employs the Naperian or natural logarithmic base e (2.71828 . . .). The same model could also be expressed as

$$\text{BOD}(t) = \text{BOD}_u(1 - 10^{-kt}) \tag{305.9}$$

where k = reaction rate constant (base 10, day^{-1}).
The growth constants in the two models are related by

$$k(\text{base} - e) = 2.303 \times k \,(\text{base} - 10)$$

A representative plot of BOD is shown in Fig. 305.1. Table 305.6. shows some typical values of the rate constant k at 20°C.

Table 305.6 Typical Values for k (day^{-1} at 20°C)

Sewage	Base e	Base 10
Raw	0.12–0.46	0.28–1.06
Treated	0.12–0.23	0.28–0.53

Figure 305.1
Biochemical
oxygen demand.

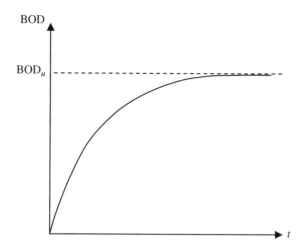

Temperature Dependence of BOD Rate Constant

The reaction rate constant is temperature dependent, due to the fact that metabolism progresses faster at higher temperatures. It should be noted, however, that the ultimate BOD (the oxygen demand potential at $t = 0$) remains the same while the rate of growth increases for higher temperatures. Commonly, the base or reference temperature is taken as 20°C. The rate constant at other temperatures is given by

$$k_T = k_{20}\, \theta^{T-20} \tag{305.10}$$

where θ is the temperature coefficient. According to current research, the following are the recommended values of θ for typical ranges of temperature encountered in water treatment.

$$\theta = 1.135 \qquad 4°C < T < 20°C$$
$$\theta = 1.056 \qquad 20°C < T < 30°C$$

Some references quote a universal $\theta = 1.047$ for all temperature ranges.

Example 305.5

A BOD sample was incubated at 27°C for 7 days and found to be 100 mg/L. Given $K_{20} = 0.23$ day^{-1} (base $- e$). What would be the BOD of this sample recorded at 5 days, if incubated at 20°C?

Solution   The reaction rate constant at 27°C is given by

$$k_T = k_{20}\theta^{T-20} \Rightarrow k_{27} = 0.23 \times 1.047^{27-20} = 0.32 \text{ day}^{-1}$$

Since BOD_7 at 27°C is 100 mg/L, the ultimate BOD is given by

$$BOD_u = \frac{BOD_7}{1 - e^{-0.32 \times 7}} = \frac{100}{1 - e^{-0.32 \times 7}} = 112 \text{ mg/L}$$

Based on this ultimate BOD, the BOD_5 at 20°C is given by

$$BOD_5 = BOD_u \left(1 - e^{-0.23 \times 5}\right) = 112 \times 0.683 = 76.5 \text{ mg/L}$$

The results are summarized in the diagram below:

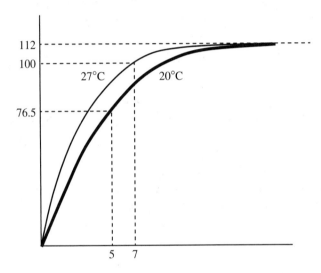

Standard BOD Test

The standard BOD test uses a BOD incubation bottle of volume 300 mL. A wastewater sample is added to the bottle and diluted by filling the bottle with aerated dilution water, which contains a phosphate buffer and inorganic nutrients. Untreated municipal wastewaters and unchlorinated effluents usually have adequate microbial populations and don't require addition of seed material. In a *seeded BOD test,* seed material—usually derived from aged wastewater—is added to industrial wastewaters and chlorinated effluents to accelerate biological activity. The dissolved oxygen of the diluted sample is measured at $t = 0$. The diluted sample is then incubated at a specified temperature for a specified time (commonly 5 days at 20°C). The dissolved oxygen of the diluted sample is measured at the end of the incubation period. For the standard BOD test, the BOD is given by

$$BOD = \frac{DO_i - DO_f}{\dfrac{V_{\text{sample}}}{V_{\text{sample}} + V_{\text{dilution}}}} \qquad (305.11)$$

where DO_i and DO_f are the measured values of dissolved oxygen at the beginning and at the end of the incubation period, respectively. The denominator is the dilution fraction that extrapolates the oxygen depletion from the sample to full volume.

Some of the recorded oxygen demand is due to a contribution from the dilution water itself. If this contribution is to be excluded, thus reflecting the "true" BOD of the wastewater sample, then the BOD must be calculated as

$$\text{BOD} = \frac{(DO_i - DO_f) - x(DO_i^* - DO_f^*)}{\dfrac{V_{sample}}{V_{sample} + V_{dilution}}} \qquad (305.12)$$

where x = volume fraction of dilution water in the diluted sample. Thus, in such a test, the diluted sample is tested at the beginning and end of the dilution period to yield the values DO_i and DO_f, respectively. A separate bottle containing dilution water only is tested to yield the values DO_i^* and DO_f^*, respectively.

Example 305.6
Calculate the BOD of a wastewater sample, given the following data:

Sample volume = 10 mL

BOD incubation bottle volume = 300 mL

Initial DO of diluted sample = 9 mg/L

Final DO of diluted sample = 2 mg/L

Initial DO of dilution water = 9 mg/L

Final DO of dilution water = 8 mg/L

Solution

Volume of sample = 10 mL

Sample volume fraction = 10/300 = 0.0333

Volume of dilution water = 290 mL

Volume fraction of dilution water = 290/300 = 0.9667

$$\delta DO_{sample} = 7 \text{ mg/L}$$

$$\delta DO_{dilution\ water} = 1 \text{ mg/L}$$

$$\text{BOD} = \frac{(9-2) - 0.9667 \times (9-8)}{0.0333} = 181 \text{ mg/L}$$

NOTE If the dilution water had been excluded from the calculations, the BOD would have been overestimated as 210 mg/L.

Exceptions

Whenever multiple tests are performed to determine parameters that show stochastic variation, statistical analysis of the data yields estimates of mean value and variance. Also, such analysis can be used as a basis for rejecting "statistical outliers" based on a specified confidence level.

Results obtained from a single sample can be corrupted by one of two phenomena. First, the dissolved oxygen in the incubation bottle may become so depleted that it may be unclear whether the oxygen demand process stagnated at an early time. In such a case, *if the concentration of DO in incubation bottle at the end of the incubation period be less than 0.5 mg/L, the sample may be rejected.* On the other hand, if the actual depletion of dissolved oxygen, i.e., *the difference between the initial and final values of DO be less than 2 mg/L,* it is likely that, for unknown reasons, the metabolism processes of that particular batch of microorganisms never progressed satisfactorily. Such results should also be excluded from the data set.

Seeded BOD Test

For very weak wastewater samples, the microbial population is not potent enough to exert a measurable and reliable oxygen demand. In such cases, a seeded test is conducted. A seed material is added to the sample to accelerate metabolism of the microorganisms. The volume of seed material added to the sample is recorded. An independent test is performed on a known volume of the seed material and initial and final DO values are recorded. The BOD from the seeded BOD tests is given by

$$BOD = \frac{DO_i - DO_f - x(DO_i^* - DO_f^*)}{\dfrac{V_{sample}}{V_{sample} + V_{dilution}}}$$

(305.13)

where DO_i and DO_f = initial and final DO of seeded sample
DO_i^* and DO_f^* = initial and final DO of seed material
x = ratio of seed volume used in seeded sample to the seed volume used to determine DO_i^* and DO_f^*

Example 305.7

A 12-mL water sample is tested for dissolved oxygen. Three milliliters of seed material is added to the water sample. The test results are summarized as follows:

Total volume of BOD bottle = 300 mL

Initial dissolved oxygen concentration of diluted, seeded sample = 9.5 mg/L

Dissolved oxygen of diluted, seeded sample after 5 days incubation at 20°C = 7.3 mg/L

An independent test is carried out on the seed material. The results from that test are

Volume of seed material $= 10$ mL

Initial dissolved oxygen concentration of seed material $= 4.5$ mg/L

Dissolved oxygen of seed material after 5 days incubation at $20°C = 2.2$ mg/L

What is most nearly the BOD_5 of the water sample?

Solution

$$BOD_5 = \frac{(9.5 - 7.3) - \dfrac{3}{10}(4.5 - 2.2)}{\dfrac{12}{300}} = 38 \text{ mg/L}$$

Carbonaceous versus Nitrogenous Oxygen Demand

The initial depletion of dissolved oxygen (approximately first-order reaction, resulting in the exponential equation) is due to the carbonaceous oxygen demand of bacteria and protozoa that reduce organic matter to carbon dioxide. If present in sufficient numbers, nitrifying bacteria exert an additional oxygen demand, oxidizing ammonia nitrogen to nitrites and nitrates. This additional oxygen demand is exerted after a lag of several days, when nitrifying bacteria become competitive due to reduction in dissolved oxygen and nutrients. See Fig. 305.2. This is the reason the standard BOD test is typically terminated at 5 days, before the nitrogenous contribution to BOD can become significant.

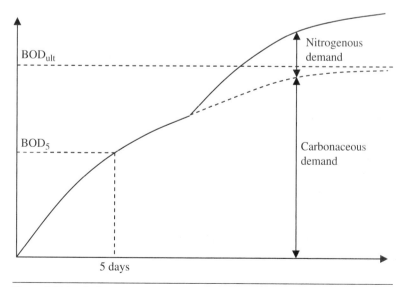

Figure 305.2 Carbonaceous versus nitrogenous oxygen demand.

Limitations of the BOD Test

The BOD test has the following limitations:

1. The test duration is relatively long (typically 5 days).

2. Only biodegradable organics are measured.

3. The effect of nitrifying bacteria is dependent on the nature of the waste and is therefore unpredictable.

4. For weak wastewaters, a high concentration of seed bacteria is required.

Chemical Oxygen Demand

The chemical oxygen demand (COD) of a wastewater sample is a measure of the oxygen equivalent of the organic matter susceptible to oxidation by a strong chemical oxidant (mixture of chromic acid and sulphuric acid). The reagent used for the standard COD test is a mixture of a standard quantity of potassium dichromate and sulphuric acid containing silver sulphate. The mixture is added to a known volume of wastewater sample and refluxed for 2 h. A blank sample is tested identically to compensate for any oxygen demand exerted by any organic matter contained in the reagents. The COD is then calculated as

$$\text{COD} = \frac{(a - b) \times \text{normality of Fe(NH}_4)_2(\text{SO}_4)_2}{V_{\text{sample}}} \times 8000 \text{ mg/L} \qquad (305.14)$$

where a = amount of ferrous ammonium sulphate titrant added to the blank (mL)
b = amount of ferrous ammonium sulphate titrant added to the sample (mL)

Total Organic Carbon

The total organic carbon (TOC) test is conducted to determine the total organic carbon in a water sample. Various methods are utilized to convert the organic carbon to carbon dioxide, which is then measured. The TOC of a wastewater is an indication of its degree of pollution. The advantage of this test is the short time (5 to 10 min) required to get results. TOC may be divided into particulate TOC (coarser than 0.45 μm) and soluble TOC (finer than 0.45 μm).

Relationship between BOD, COD, and TOC

For typical municipal wastewater, the ratio between BOD and COD is between 0.3 and 0.8 and the ratio between BOD and TOC is between 1.2 and 2.0.

Dilution Purification of Wastewater Streams

When a stream of untreated or partially treated sewage is discharged into a large body of water, the wastewater is diluted. Assuming instantaneous mixing at the point of discharge, the parameters of the wastewater-river mixture can be calculated using a weighted average approach.

This can be applied for calculating parameters such as temperature/dissolved oxygen/BOD/ suspended solids, and the like. If the resulting levels are within water quality standards for the stream, the wastewater does not require pretreatment prior to discharge.

Consider a wastewater flow rate Q_w with ultimate BOD L_w, and dissolved oxygen DO_w mixing with a river with flow rate Q_r, ultimate BOD L_r, and dissolved oxygen DO_r. The initial (immediately after mixing) ultimate BOD of the river-wastewater mix is

$$L_o = \frac{Q_w L_w + Q_r L_r}{Q_w + Q_r} \tag{305.15}$$

Initial dissolved oxygen immediately after mixing (DO_o) is given by

$$DO_o = \frac{Q_w DO_w + Q_r DO_r}{Q_w + Q_r} \tag{305.16}$$

Temperature immediately after mixing (T_o) is given by

$$T_o = \frac{Q_w T_w + Q_r T_r}{Q_w + Q_r} \tag{305.17}$$

The saturation dissolved oxygen (DO_{sat}) for this temperature can be found from the Table 305.5. Initial oxygen deficit after mixing (D_o) is given by

$$D_o = DO_{sat} - \frac{Q_w DO_w + Q_r DO_r}{Q_w + Q_r} \tag{305.18}$$

NOTE Exercise care in distinguishing between dissolved oxygen (often abbreviated as DO) and the initial oxygen deficit (often abbreviated as D_o). They are both typically expressed in units of mg/L.

Example 305.8

A wastewater flow of 1200 gpm is discharged into a stream with flow rate of 12 cfs. The ultimate BOD of the wastewater is 1800 lb/day. The stream has negligible BOD upstream of the mixing point. Other parameters are

Upstream temperature of river = 15°C

Upstream dissolved oxygen of river = 5.2 mg/L

Temperature of wastewater = 38°C

Dissolved oxygen of wastewater = 1.2 mg/L

(a) What is the ultimate BOD (mg/L) immediately downstream of the mixing point?
(b) What is the oxygen deficit (mg/L) immediately downstream of the mixing point?

Solution The wastewater flow rate is 1200 gpm = 2.674 cfs. The BOD load of the wastewater needs to be converted to a concentration using one of the following expressions:

$$mg/L \times MGD \times 8.3454 = lb/day$$
$$mg/L \times gpm \times 0.012 = lb/day$$
$$mg/L \times cfs \times 5.393 = lb/day$$

Therefore, the ultimate BOD of the wastewater can be calculated as

$$L_w = \frac{1800}{1200 \times 0.012} = 125 \text{ mg/L}$$

The ultimate BOD of the stream-wastewater mix is given by the weighted average

$$L_o = \frac{2.674 \times 125 + 12 \times 0}{2.674 + 12} = 22.78 \text{ mg/L}$$

The dissolved oxygen of the stream-wastewater mix is given by the weighted average

$$DO_o = \frac{2.674 \times 1.2 + 12 \times 5.2}{2.674 + 12} = 4.47 \text{ mg/L}$$

The temperature of the stream-wastewater mix is given by the weighted average

$$T_o = \frac{2.674 \times 38 + 12 \times 15}{2.674 + 12} = 19.2°C$$

At a temperature of 19.2°C, the saturated dissolved oxygen = 9.22 mg/L (Table 305.5). Therefore, oxygen deficit immediately downstream of mixing point = 9.22 − 4.47 = 4.75 mg/L.

Example 305.9

A factory produces chromium at a concentration of 3000 µg/L in a wastewater stream of 500 gpm. If the upstream chromium content of the receiving stream (flow rate 7.5 cfs) is 30 µg/L and the EPA limit on chromium is 50 µg/L, what must be the removal efficiency of chromium the factory must achieve before discharging into the stream?

Solution

$$Q_w = 500 \text{ gal/min} = 500 \times 0.002228 = 1.114 \text{ cfs}$$

If the concentration of chromium discharged into the stream = C_w

The weighted average of the stream and wastewater = $\dfrac{C_w \times 1.114 + 30 \times 7.5}{1.114 + 7.5} \leq 50$

Therefore, $C_w < 184.65$ µg/L

Therefore, percent removal during pretreatment = $\dfrac{3000 - 184.65}{3000} \times 100\% = 93.85\%$

Streeter-Phelps Equations

The Streeter-Phelps equations describe the dissolved oxygen "sag curve" obtained as a result of simultaneous deoxygenation and reoxygenation that occurs when wastewater is discharged into receiving stream. The organic load of the effluent (represented by the ultimate BOD or BOD potential) brings in microorganisms, which continuously exert an oxygen demand on the receiving water, resulting in a lowering of the dissolved oxygen. The rate of deoxygenation is expressed in terms of the coefficient k_d. At the same time, depending on the extent of stream turbulence, surface aeration entraps atmospheric oxygen and tends to increase the dissolved oxygen. The rate of reoxygenation is expressed in terms of the coefficient k_r.

Under the influence of these two effects, the dissolved oxygen follows a sag curve, showing a minimum value at the critical time t_c. The dissolved oxygen sag curve is shown in Fig. 305.3. The maximum oxygen deficit at t_c is designated D_c. Far from the point of mixing (shown as $t = 0$ in the Fig. 305.3), the dissolved oxygen asymptotes to the saturation value of the dissolved oxygen corresponding to the stream temperature.

Oxygen Deficit

According to the Streeter-Phelps model, the oxygen deficit after time t ($t = 0$ at the instant of mixing) is given by

$$D_t = \frac{k_d L_o}{k_r - k_d}(e^{-k_d t} - e^{-k_r t}) + D_o e^{-k_r t} \qquad (305.19)$$

where L_o = ultimate BOD immediately after mixing (mg/L)
D_o = oxygen deficit immediately after mixing (mg/L)
k_d = deoxygenation rate (day^{-1})
k_r = reoxygenation rate (day^{-1})

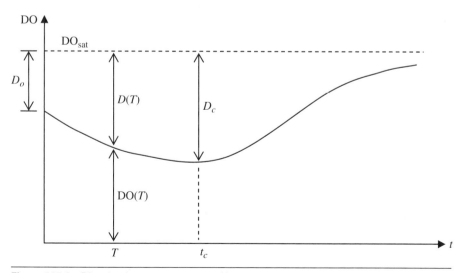

Figure 305.3 Dissolved oxygen sag curve (Streeter-Phelps).

Note that in these equations, dissolved oxygen concentration, oxygen deficit, and ultimate BOD should all be expressed in similar units (mg/L). However, the BOD L_o is often expressed as a daily load (as in lb/day). In that case, it must be converted to a concentration using

$$L_o \text{ (mg/L)} = \frac{L_o \text{ (lb/day)}}{(Q_r + Q_w)\text{(MGD)} \times 8.3454} \quad (305.20)$$

The flow rate in the denominator $(Q_r + Q_w)$ is the combined flow rate of the stream and wastewater.

Critical (Maximum) Oxygen Deficit (Minimum Dissolved Oxygen)

The time at which the dissolved oxygen is critical (minimum) is given by

$$t_c = \frac{1}{k_r - k_d} \ln\left(\frac{k_d L_o - k_r D_o + k_d D_o}{k_d L_o}\frac{k_r}{k_d}\right) \quad (305.21)$$

At this time, the critical (maximum) oxygen deficit is given by

$$D_c = \frac{k_d L_o}{k_r} e^{-k_d t_c} \quad (305.22)$$

The distance (from the point of mixing) to the critical location is given by

$$x_c = v t_c \quad (305.23)$$

where v = average stream velocity.

 NOTE   In the Streeter-Phelps model, the Eqs. (305.19), (305.21), and (305.22) are written with $e = 2.81828\ldots$ as the base for the exponential terms. This is consistent with the rate constants $(k_d$ and $k_r)$ being defined with respect to the same base. It is just as legitimate for these equations to be written with base 10, as long as the rate constants are also defined similarly.

Hardness in Water

Hardness in water is quantified as the sum of the concentration of all polyvalent cations such as Ca^{++}, Mg^{++}, Fe^{++}, and the like. Each of these species has a different potential for causing hardness and therefore cannot be combined unless expressed in equivalent terms such as milliequivalents per liter (meq/L). For practical purposes these concentrations are expressed in consistent units mg/L as $CaCO_3$. This is accomplished by converting concentrations of individual species by a multiplicative factor given by

$$\frac{\text{Eq. wt. CaCO}_3}{\text{Eq. wt. of species}} = \frac{50}{\text{Eq. wt. of species}}$$

Table 305.7 shows relevant data for the most dominant cations causing hardness in water.

Table 305.7 Properties of Anions, Cations, and Compounds

Anions					
Name	Formula	Ionic weight	Valence	Equivalent weight	Factor
Bicarbonate	HCO_3^-	61.0	1	61.0	0.82
Carbonate	CO_3^{2-}	60.0	2	30.0	1.67
Chloride	Cl^-	35.5	1	35.5	1.41
Hydroxide	OH^-	17.0	1	17.0	2.94
Nitrate	NO_3^-	62.0	1	62.0	0.81
Sulphate	SO_4^{2-}	96.1	2	48.1	1.04

Cations					
Name	Formula	Ionic weight	Valence	Equivalent weight	Factor
Aluminum	Al^{3+}	27.0	3	9.0	5.56
Calcium	Ca^{2+}	40.1	2	20.0	2.50
Cupric	Cu^{2+}	63.5	2	31.8	1.57
Cuprous	Cu^+	63.5	1	63.5	0.79
Ferric	Fe^{3+}	55.8	3	18.6	2.69
Ferrous	Fe^{2+}	55.8	2	27.9	1.79
Magnesium	Mg^{2+}	24.3	2	12.2	4.11
Manganese	Mn^{2+}	54.9	2	27.4	1.82
Zinc	Zn^{2+}	65.4	2	32.7	1.53

Compounds					
Name	Formula	Molecular weight	Valence	Equivalent weight	Factor
Aluminum sulphate	$Al_2(SO_4)_3$	342.1	6	57.0	0.88
Aluminum sulphate	$Al_2(SO_4)_3 \cdot 18H_2O$	666.1	6	111.0	0.45
Calcium bicarbonate	$Ca(HCO_3)_2$	162.1	2	81.1	0.62
Calcium carbonate	$CaCO_3$	100.1	2	50.1	1.00
Calcium hydroxide	$Ca(OH)_2$	74.1	2	37.1	1.35
Calcium sulphate	$CaSO_4$	136.2	2	68.1	0.74
Carbon dioxide	CO_2	44.0	2	22.0	2.27
Magnesium bicarbonate	$Mg(HCO_3)_2$	146.3	2	73.2	0.68
Magnesium carbonate	$MgCO_3$	84.3	2	42.2	1.19
Magnesium hydroxide	$Mg(OH)_2$	58.3	2	29.2	1.71
Sodium carbonate	Na_2CO_3	106.0	2	53.0	0.94
Sodium hydroxide	$NaOH$	40.0	1	40.0	1.25

Of the total hardness, the carbonate hardness (or temporary hardness) is the lesser of alkalinity or total hardness. If the hardness is greater than alkalinity, the difference between the total hardness and the alkalinity is called the noncarbonate hardness (or permanent hardness).

Milliequivalents per Liter Bar Graph

Results of a water analysis are usually reported as concentrations of various species in mg/L. These may be converted to milliequivalents per liter (meq/L) by dividing by the equivalent weight of the species. In order to identify chemical compounds in the water sample, these ions are arranged in a bar graph, where the top row consists of the major cations arranged in the order of calcium (Ca^{2+}), magnesium (Mg^{2+}), sodium (Na^+), and potassium (K^+). In the second row, the major anions are arranged in the order carbonate (CO_3^{2-}), bicarbonate (HCO_3^-), sulphate (SO_4^{2-}), and chloride (Cl^-). For a water sample in ionic equilibrium, the total positive milliequivalents per liter must equal the total negative milliequivalents per liter.

Example 305.10

Given the following results of a water sample analysis, construct the milliequivalents per liter bar graph to determine the constituent compounds.

Ca^{++}	50 mg/L
Mg^{++}	15 mg/L
Na^+	12 mg/L
K^+	9 mg/L
HCO_3^-	112 mg/L
CO_3^{2-}	21 mg/L
SO_4^{2-}	79 mg/L
Cl^-	11 mg/L

Solution   The table below shows each concentration converted to meq/L.

Ion	Conc. mg/L	Equivalent weight	meq/L
Ca^{2+}	50	20.040	2.50
Mg^{2+}	15	12.156	1.23
Na^+	12	22.990	0.52
K^+	9	39.102	0.23
		Sum	4.49
CO_3^{2-}	21	30.005	0.70

HCO_3^-	112	61.017	1.84
SO_4^{2-}	79	48.031	1.64
Cl^-	11	35.453	0.31
		Sum	4.49

If this data is plotted on a bar graph, we see the following:

0		2.5		3.73		4.25		4.49
	Ca			Mg	Na		K	

0	0.70		2.54			4.18		4.49
CO_3^{2-}		HCO_3^-			SO_4^{2-}		Cl^-	

The following compounds are therefore identified. Note that the meq/L can be multiplied by the equivalent weight to obtain the concentration in mg/L.

$CaCO_3$	0.70 meq/L	35.0 mg/L
$Ca(HCO_3)_2$	1.80 meq/L	145.9 mg/L
$Mg(HCO_3)_2$	0.04 meq/L	2.9 mg/L
$MgSO_4$	1.19 meq/L	71.6 mg/L
Na_2SO_4	0.45 meq/L	32.0 mg/L
NaCl	0.07 meq/L	4.1 mg/L
KCl	0.24 meq/L	17.9 mg/L

Note that the total concentration of ions (309 mg/L) matches the total concentration of compounds (309.4 mg/L).

Hardness Removal

The following chemical compounds are commonly used for removal of hardness from municipal water:

Lime (calcium oxide)	CaO
Slaked lime (hydrated lime)	$Ca(OH)_2$
Soda ash (sodium carbonate)	Na_2CO_3
Caustic soda (sodium hydroxide)	NaOH

For each molecule of calcium bicarbonate hardness removed, one molecule of lime is used. For each molecule of magnesium bicarbonate hardness removed, two molecules of lime are used. For each molecule of non-carbonate calcium hardness removed, one molecule of soda ash is used. For each molecule of non-carbonate magnesium hardness removed one molecule of lime plus one molecule of soda ash is used.

The different stages in hardness removal are outlined below:

Carbon Dioxide Demand

CO_2 demand is met first before any softening occurs. Removal of carbon dioxide raises the pH and makes hardness removal more efficient

$$CO_2 + Ca(OH)_2 \rightarrow CaCO_3 \downarrow + H_2O$$

Removal of Carbonate Hardness

Next, lime or caustic soda reacts with carbonate hardness. This category of hardness in water is caused primarily by calcium and magnesium bicarbonates. This can be accomplished by the addition of slaked lime or caustic soda. The chemical reactions shown below describe the removal of calcium and magnesium carbonate hardness using lime [$Ca(OH)_2$] or caustic soda ($NaOH$).

Calcium Carbonate Hardness Removal with Ca(OH)₂

$$Ca(HCO_3)_2 + Ca(OH)_2 \rightarrow 2CaCO_3 \downarrow + 2H_2O$$

Calcium Carbonate Hardness Removal with NaOH

$$Ca(HCO_3)_2 + 2NaOH \rightarrow CaCO_3 \downarrow + 2Na^+ + CO_3^{2-} + H_2O$$

Magnesium Carbonate Hardness Removal with Ca(OH)₂

$$Mg(HCO_3)_2 + 2Ca(OH)_2 \rightarrow 2CaCO_3 \downarrow + Mg(OH)_2 \downarrow + 2H_2O$$

Magnesium Carbonate Hardness Removal with NaOH

$$Mg(HCO_3)_2 + 4NaOH \rightarrow Mg(OH)_2 \downarrow + 4Na^+ + 2CO_3^{2-} + 2H_2O$$

Removal of Noncarbonate Hardness

To remove noncarbonate hardness (e.g., $MgSO_4$), soda ash is needed as a secondary treatment after lime is added.

Calcium Noncarbonate Hardness Removal

$$CaSO_4 + Na_2CO_3 \rightarrow CaCO_3 \downarrow + Na_2SO_4$$

Magnesium Noncarbonate Hardness Removal

$$MgSO_4 + Ca(OH)_2 + Na_2CO_3 \rightarrow CaCO_3 \downarrow + Mg(OH)_2 \downarrow + Na_2SO_4$$

Destruction of Excess Alkalinity

$$2HCO_3^- + Ca(OH)_2 \rightarrow CaCO_3 \downarrow + CO_3^{2-} + 2H_2O$$

Recarbonation

Recarbonation (injection with carbon dioxide) is then used to lower the pH.

$$Ca^{2+} + 2OH^- + CO_2 \rightarrow CaCO_3 \downarrow + H_2O$$

Excess Lime Treatment

When magnesium hardness is more than about 40 mg/l as $CaCO_3$, magnesium hydroxide scale deposits in household hot-water heaters operated at normal temperatures of 140 to 150° F. To reduce magnesium hardness, more lime must be added to the water. Extra lime will raise pH above 10.6 to help magnesium hydroxide precipitate out of the water.

To identify individual species contributing to hardness, one may also use the flowchart in Fig. 305.4. One must make sure that all concentrations are in equivalent units such as meq/L or mg/L as $CaCO_3$ (see Fig. 305.4). Note that the flowchart should not be used if species other than those identified below are present in significant concentrations.

- Ca = calcium ion (Ca^{++}) concentration
- Mg = magnesium ion (Mg^{++}) concentration
- M = total alkalinity
- H = total hardness
- S = sulphates (SO_4^{2-}) concentration
- O = hydroxides (OH^-) concentration

Example 305.11

A water sample shows the following results:

Ca^{++}	25 mg/L
Mg^{++}	30 mg/L
Na^+	25 mg/L
Alkalinity	150 mg/L as $CaCO_3$

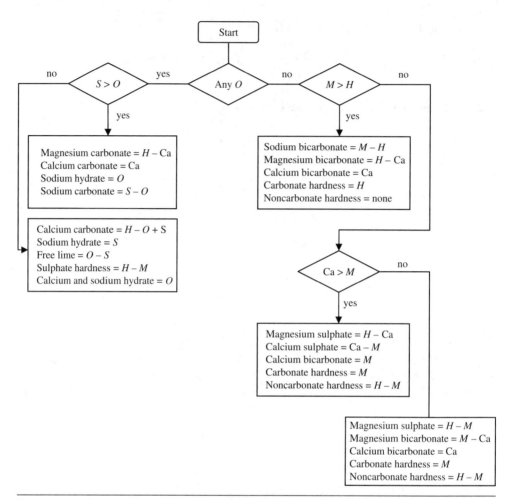

Figure 305.4 Flowchart for determining hardness constituents.

What is the dosage of 88% pure lime $Ca(OH)_2$ and 90% pure Na_2CO_3 needed to treat a flow rate of 2 MGD of this water?

Solution Total hardness $= 25 \times 2.5 + 30 \times 4.1 = 185.5$ mg/L as $CaCO_3$
Alkalinity $= 150$ mg/L as $CaCO_3$
From the flowchart, answering the following questions:

Any O?	Answer: NO
$M > H$?	Answer: NO ($M = 150$, $H = 185.5$)
$Ca > M$?	Answer: NO ($Ca = 62.5$, $M = 150$)

We have the following constituents:

Magnesium sulphate $= H - M = 185.5 - 150 = 35.5$ mg/L as $CaCO_3$
Treatment of 1 mol of $MgSO_4$ requires 1 mol of lime and 1 mol of Na_2CO_3

Magnesium bicarbonate $= M - Ca = 150 - 62.5 = 87.5$ mg/L as $CaCO_3$
Treatment of 1 mol of $Mg(HCO_3)_2$ requires 2 mol of lime and 0 mol of Na_2CO_3

Calcium bicarbonate $= Ca = 62.5$ mg/L as $CaCO_3$
Treatment of 1 mol of $Ca(HCO_3)_2$ requires 1 mol of lime and 0 mol of Na_2CO_3

Carbonate hardness $= M = 150$ mg/L as $CaCO_3$

Noncarbonate hardness $= H - M = 185.5 - 150 = 35.5$ mg/L as $CaCO_3$

Pure lime required $= 1 \times 35.5 + 2 \times 87.5 + 1 \times 62.5 = 273$ mg/L as $CaCO_3$

88% purity lime required $= 273/0.88 = 310.23$ mg/L as $CaCO_3$

$$= 310.23/1.35 = 229.8 \text{ mg/L as } Ca(OH)_2$$

Lime dosage $- 2.0 \times 229.8 \times 8.3454 = 3835$ lb/day

Pure Na_2CO_3 required $= 1 \times 35.5 = 35.5$ mg/L as $CaCO_3$

90% purity Na_2CO_3 required $= 35.5/0.90 = 39.4$ mg/L as $CaCO_3$

$$= 39.4/0.94 = 41.2 \text{ mg/L as } Na_2CO_3$$

Na_2CO_3 dosage $= 2.0 \times 41.2 \times 8.3454 = 700$ lb/day

Example 305.12

The analysis of a water sample yields the results shown below. Hardness removal is to be accomplished by a two stage lime-soda ash process. Excess lime and soda ash will be added to achieve a residual hardness of 30 mg/L as $CaCO_3$ and 10 mg/L $Mg(OH)_2$ as $CaCO_3$. Assume excess lime addition of 35 mg/L

Ca^{++}	80 mg/L
Mg^{++}	30 mg/L
Na^+	19 mg/L
Cl^-	18 mg/L
SO_4^-	64 mg/L
HCO_3^-	336 mg/L
CO_2	15 mg/L

What is the amount of lime (CaO) required per million gallons of water?
What is the amount of soda ash (Na_2CO_3) required per million gallons of water?

Solution Each ionic concentration needs to be converted from mg/L to "mg/L as $CaCO_3$."

Ion	mg/L	Factor	mg/L as CaCO$_3$	
Ca^{++}	80	2.50	200.0	
Mg^{++}	30	4.12	123.6	364.9
Na$^+$	19	2.17	41.3	
Cl$^-$	18	1.41	25.4	
SO$_4^-$	64	1.04	66.7	367.5
HCO$_3^-$	336	0.82	275.4	
CO$_2$	15	2.27	34.1	

Based on these concentrations, the following bar plots of cations and anions can be constructed:

| 200 | 124 | 41 | = 365 |

| 275 | 67 | 25 | = 367 |

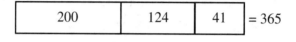

200   75   49 18 25

Ca(HCO$_3$)$_2$	200 mg/L as CaCO$_3$
Mg(HCO$_3$)$_2$	75 mg/L as CaCO$_3$
MgSO$_4$	49 mg/L as CaCO$_3$
Na$_2$SO$_4$	18 mg/L as CaCO$_3$
NaCl	25 mg/L as CaCO$_3$

LIME is used to treat Ca(HCO$_3$)$_2$ (1:1) and Mg(HCO$_3$)$_2$ (1:2) and MgSO$_4$ (1:1)
LIME is also needed to neutralize CO$_2$ (1:1)
Total lime needed = $200 + 2 \times 75 + 49 + 34 = 433$ mg/L as CaCO$_3$ = $433 \div 1.79 = 241.9$ mg/L lime
Excess lime (CaO) = 35 mg/L
Total lime = 276.9 mg/L as CaCO$_3$
For every 1 mg, lime dosage required = $1 \times 276.9 \times 8.34 = 2309$ lb (as CaO)

SODA ASH is used to treat MgSO$_4$ (1:1)
Total soda ash needed = 49 mg/L as CaCO$_3$
For every 1 mg, Na$_2$CO$_3$ dosage required = $1 \times 49 \times 8.34 = 409$ lb (as CaCO$_3$) = $409 \div 0.94 = 435$ lb (as Na$_2$CO$_3$)

pH and pOH

pH (potential hydrogen) is a measure of the activity of hydrogen ions (H$^+$) in a solution. pH is a quantification of the acidity or alkalinity of a solution. In aqueous systems, the hydrogen ion activity is dictated by the dissociation constant of water ($K_w = 1.011 \times 10^{-14}$ at 25°C). The pH is defined as the negative logarithm of the ionic activity (a_H^+) of H^+ ions.

$$pH = -\log\left[a_H^+\right] \tag{305.24}$$

In dilute solutions, the activity is approximately equal to the numeric value of the concentration of the H^+ ion, denoted as $[H^+]$ measured in mol/L. Thus, approximately,

$$pH = -\log_{10}[H^+] \tag{305.25}$$

$$pOH = -\log_{10}[OH^-] \tag{305.26}$$

where $[H^+]$ = concentration of H^+ ions in mol/L

For a neutral solution (H^+ ion activity equals OH^- ion activity), the pH is approximately 7. Aqueous solutions with pH values lower than 7 are considered acidic, while pH values higher than 7 are considered alkaline.

At temperatures other than 25°C, the pH of pure water will not be 7. Note also that pure water, when exposed to the atmosphere, will absorb carbon dioxide, some of which reacts with water to form carbonic acid, thereby lowering the pH to about 5.7. Most substances have a pH in the range 0 to 14, although extremely acidic or basic substances may have pH less than 0 or greater than 14.

pOH measures the concentration of OH^- ions. Since water self-ionizes, and denoting $[OH^-]$ as the concentration of hydroxide ions, we have

$$K_w = a_H a_{OH} = 10^{-14}$$

where K_w is the ionization constant of water. This leads to

$$pH + pOH = 14 \tag{305.27}$$

Calculation of pH for Weak and Strong Acids

Values of pH for weak and strong acids can be approximated using certain assumptions. A *strong acid* may be defined as a species which is a much stronger acid than the hydronium (H_3O^+) ion. In that case the dissociation reaction (strictly $HX + H_2O \leftrightarrow H_3O^+ + X^-$ but simplified as $HX \leftrightarrow H^+ + X^-$) goes to completion, i.e., no unreacted acid remains in solution. Dissolving the strong acid HCl (hydrochloric acid) in water can therefore be expressed as

$$HCl \leftrightarrow H^+ + Cl^-$$

This means that in a 0.01-mol/L solution of HCl, it is approximated that there is a concentration of 0.01 mol/L dissolved hydrogen ions. Thus, since $pH = -\log_{10}[H^+]$:

$$pH = -\log(0.01) = 2.0$$

Equilibrium Constant, Dissociation, and pH

For weak acids, the dissociation reaction does not go to completion, but rather to an equilibrium point dictated by the equilibrium constant. It is necessary to know the value of the equilibrium constant of the reaction for each acid in order to calculate its pH.

$$K_a = \frac{[H^+][A^-]}{[HA]} \tag{305.28}$$

When calculating the pH of a weak acid, it is usually assumed that the water does not provide any hydrogen ions.

Example 305.13

What is the pH of a solution of methanoic acid (HCOOH) having molarity $= 0.01$? The acidity constant for methanoic acid is equal to 1.6×10^{-4}.

Solution The dissociation constant for methanoic acid is defined as

$$K_a = \frac{[H^+][HCOO^-]}{[HCOOH]} = 1.6 \times 10^{-4}$$

Given that an unknown amount of the acid has dissociated, [HCOOH] will be reduced by this amount, while $[H^+]$ and $[HCOO^-]$ will each be increased by this amount. Therefore, [HCOOH] may be expressed as $(0.01 - x)$, and $[H^+]$ and $[HCOO^-]$ may each be expressed as x, giving us the following equation:

$$\frac{x^2}{0.01 - x} = 1.6 \times 10^{-4}$$

Solving this for x yields 1.187×10^{-3}, which is the concentration of hydrogen ions after dissociation. Therefore, the pH $= -\log(1.187 \times 10^{-3})$ or about 2.93.

Example 305.14

A $0.01M$ solution of hydrochloric acid is ionized partially in an aqueous solution. If the pH of the solution is 4.8, what is the fraction ionized?

Solution The ionization reaction for hydrochloric acid is

$$HCl \rightarrow H^+ + Cl^-$$

Concentration of HCl is 0.01 mol/L.
If the fraction ionized is x, then the concentration of ionized HCl $= 0.01x$ mol/L.

$$\text{Therefore } [H^+] = 0.01x$$
$$pH = -\log_{10} [H^+] = 4.8$$
$$\text{Therefore } [H^+] = 10^{-4.8} = 1.6 \times 10^{-5}$$

Thus $\qquad\qquad 0.01x = 1.6 \times 10^{-5} \Rightarrow x = 0.0016 = 0.16\%$

Example 305.15

A river carrying a flow rate of 14,000 gal/min has a natural pH = 6.8. If a wastewater stream of 3000 gal/min having a pH = 4.7 mixes into the stream, assuming perfect and instantaneous mixing, what is the stream pH immediately downstream of the mixing point?

Solution   Since pH $= -\log_{10}[H^+]$, where the hydrogen ion concentration $[H^+]$ is expressed in moles per liter, the number of moles (per minute) the river and the wastewater flow are, respectively

$14,000 \times 3.7854 \times 10^{-6.8} = 0.0084$ moles and

$3000 \times 3.7854 \times 10^{-4.7} = 0.2266$ moles

Thus, the concentration of $[H^+]$ in the combined flow of 17,000 gallons is

0.235 moles/$17,000 \times 3.7854 = 3.652 \times 10^{-6}$

Which results in a pH $= {}^-\log_{10}[3.652 \times 10^{-6}] = 5.44$

Note that the unit conversion (from gallons to liters) is not necessary, as follows:

Step 1: Weighted average of the ion concentrations

$$\left[H^+\right] = \frac{14,000 \times 10^{-6.8} + 3000 \times 10^{-4.7}}{14,000 + 3000} = 3.652 \times 10^{-6}\, moles\, per\, liter$$

Step 2: Calculate pH

$$pH = -\log_{10}(3.652 \times 10^{-6}) = 5.44$$

Alkalinity

Alkalinity is a measure of the ability of water to absorb hydrogen ions without significant change in pH. In municipal water supply, alkalinity is primarily due to hydroxyl (OH^-), carbonate (CO_3^{2-}) and bicarbonate (HCO_3^-) ions, and traces of nitrate (NO_3^-). As with hardness computations, these species must be converted to a common basis (commonly "mg/L as $CaCO_3$"). Table 305.8 summarizes the chemical properties of these anions.

Below a pH of 4.5, no alkalinity exists. Between pH = 4.5 and pH = 8.3, the alkalinity is entirely due to bicarbonate, and above pH = 8.3, the bicarbonate ions are converted to carbonate. Hydroxyl ions appear above a pH = 9.5 and react with carbon dioxide to produce carbonates and bicarbonates. Alkalinity is measured by titrating a water sample with a standard acidic titrant (sulphuric acid). The volume of acid needed to lower the pH to 8.3 (phenolphthalein end point) and the volume needed to further lower the pH to 4.5 (bromocresol red/methyl red end point) are both recorded. Alkalinity is expressed in units **mg/L as $CaCO_3$**.

where V_P = volume of acidic titrant needed to reach pH 8.3

V_M = volume of acidic titrant needed to reach pH 4.5

Let P = phenolphthalein alkalinity and M = total alkalinity. These are calculated as

$$P = \frac{V_P \times N_{titrant}}{V_{sample}} \times 50 \times 1000 \qquad (305.29)$$

$$M = \frac{V_M \times N_{titrant}}{V_{sample}} \times 50 \times 1000 \qquad (305.30)$$

Table 305.8 Properties of Primary Anions Responsible for Alkalinity in Water

Cation	Atomic weight	Valence	Equivalent weight	Factor
OH^-	17.008	1	17.008	2.94
CO_3^{2-}	60.010	2	30.005	1.67
HCO_3^-	61.018	1	61.018	0.82
NO_3^-	62.01	1	62.01	0.81

The initial pH determines the concentration of hydroxyl ions. If the initial pH is x, then the initial pOH is $14 - x$ and the hydroxyl ion concentration (converted from mol/L to mg/L as $CaCO_3$) is given by

$$[OH^-] = 10^{x-14} \times 50,000$$

As the titration progresses, the removal (neutralization) of the hydroxyl ions is followed by the neutralization of the carbonate ions. The phenolphthalein alkalinity (corresponding to a pH = 8.3) corresponds to the sum of the hydroxyl ions plus half the carbonate ions. The methyl red alkalinity (corresponding to a pH = 4.5) corresponds to the sum of the hydroxyl ions plus carbonate ions plus bicarbonate ions. Expressed in terms of milliequivalents per liter (meq/L), one can also express the alkalinity as

$$\text{Alk (meq/L)} = [HCO_3^-] + 2[CO_3^{2-}] + [OH^-] - [H^+] \qquad (305.31)$$

Example 305.16

A 200-mL water sample has an initial pH of 10.3 and is titrated using a 0.05 N sulphuric acid solution. A pH of 8.3 (phenolphthalein end point) is reached after 5.5 mL of acid is added and a pH of 4.5 (methyl red end point) is reached after addition of a further 10.5 mL of acid. What are the alkalinity species present and in what concentrations (mg/L as $CaCO_3$)?

Solution

$$pH = 10.3$$
$$pOH = 14 - pH = 3.7 = -\log_{10} [OH^-]$$

$[OH^-] = 10^{-3.7} = 1.995 \times 10^{-4}$ mol/L $= 1.995 \times 10^{-4} \times 17 \times 1000$ mg/L $= 3.39$ mg/L
$= 3.39 \times 2.94 = 9.97$ mg/L as $CaCO_3$ (say 10 mg/L as $CaCO_3$)

The phenolphthalein alkalinity is given by

$$P = \frac{V_P \times N_{titrant}}{V_{sample}} \times 50 \times 1000 = \frac{5.5 \times 0.05 \times 50,000}{200} = 68.75 \text{ mg/L as } CaCO_3$$

and the total alkalinity as (note: total acid added $= 16$ mL)

$$M = \frac{V_M \times N_{titrant}}{V_{sample}} \times 50 \times 1000 = \frac{16 \times 0.05 \times 50,000}{200} = 200 \text{ mg/L as } CaCO_3$$

In reaching pH of 8.3, all the hydroxide alkalinity and half of the carbonate alkalinity must be neutralized.

$$1/2 \, CO_3 + OH = 68.75$$

Since $OH^- = 10$ mg/L as $CaCO_3$

$$CO_3^{2-} = 2 \times 58.75 = 117.5 \text{ mg/L as } CaCO_3$$
$$HCO_3^- = 200 - 10 - 117.5 = 72.5 \text{ mg/L as } CaCO_3$$

Analysis of Solids Data

The solids in a water sample can be broadly classified into suspended solids and dissolved solids. If the water sample is filtered using a standard pore-size filter, this process separates these solids—suspended solids remain on the filter and the dissolved solids pass into the filtrate. Each of these solid fractions can then be further separated into a volatile fraction and a fixed fraction by igniting them at a standard temperature (500°C). The solids remaining after volatilization represent the fixed fraction and the fraction "burnt off" represents the volatile fraction. Table 305.9 summarizes these processes.

Figure 305.5 shows a schematic of the filtration, evaporation, and ignition processes used to separate these solids.

Example 305.17

The following test results were obtained using a 50-mL wastewater sample. All mass measurements are given in grams. Determine TS, TVS, TSS, VSS, TDS, and VDS.

Data

Mass of evaporating dish	52.6480
Mass of evaporating dish plus residue after evaporation at 105°C	52.6845
Mass of evaporating dish plus residue after ignition at 550°C	52.6575
Mass of Whatman GF/C filter (dry)	1.5386
Mass of filter plus residue after drying	1.5435
Mass of Whatman filter plus residue after ignition at 550°C	1.5418

Table 305.9   Definitions of Various Solid Fractions in Water

Test	Description
Total solids (TS)	The residue remaining after a wastewater sample has been evaporated and dried at a specified temperature (103 to 105°C).
Total volatile solids (TVS)	Solids that can be volatilized and burned off when the TS are ignited at 500 ± 50°C.
Total fixed solids (TFS)	The residue that remains after TS are ignited (TS −TVS).
Total suspended solids (TSS)	Portion of the TS retained on a filter (most commonly the Whatman glass fiber filter, with a nominal pore size of about 1.58 microns), measured after being dried at a specified temperature (105°C).
Volatile suspended solids (VSS)	Solids that can be volatilized and burned off when the TSS are ignited at 500 ± 50°C.
Fixed suspended solids (FSS)	The residue that remains after TSS are ignited (TSS − VSS).
Total dissolved solids (TDS) (TS −TSS)	Solids that pass through the filter, and are then evaporated and dried at specified temperature. These solids include colloidal and dissolved solids.
Total volatile dissolved solids (VDS)	Solids that can be volatilized and burned off when the TDS are ignited (500 ± 50°C).
Fixed dissolved solids (FDS)	The residue that remains after TDS are ignited (500 ± 50°C).
Settleable solids	Suspended solids, expressed as milliliters per liter, that will settle out of suspension within a specified period of time.

Solution

$$TS = \frac{(\text{dish} + \text{residue}) - (\text{dish})}{\text{sample size}} = \frac{(52.6845 - 52.6480)\,10^3}{0.05} = 730 \text{ mg/L}$$

$$TVS = \frac{[(\text{dish} + \text{dry residue}) - (\text{dish} + \text{ignited residue})]10^3}{0.05}$$
$$= \frac{(52.6845 - 52.6575) \times 1000}{0.05} = 540 \text{ mg/L}$$

$$TFS = TS - TVS = 730 - 540 = 190 \text{ mg/L}$$

$$TSS = \frac{(1.5435 - 1.5386)10^3}{0.05} = 98 \text{ mg/L}$$

$$VSS = \frac{(1.5435 - 1.5418)10^3}{0.05} = 34 \text{ mg/L}$$

$$TDS = TS - TSS = 730 - 98 = 632 \text{ mg/L}$$
$$VDS = TVS - VSS = 540 - 34 = 506 \text{ mg/L}$$

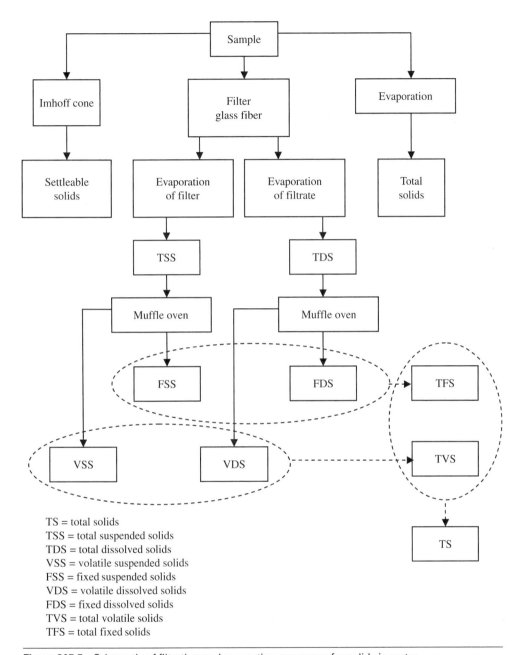

Figure 305.5 Schematic of filtration and separation processes for solids in water.

Toxicity

Trace metals in water originate from residences, groundwater infiltration, and discharge from industrial facilities. Whereas many of these metals are necessary nutrients for biological life, excessive concentrations can cause toxicity and thus affect the usefulness of a water source.

EPA Defaults for Calculating Intake

EPA recommends the following default values for estimating intake of a substance:

Average (adult) body weight	70 kg
Average (child) body weight	
0–1.5 years	10 kg
1.5–5 years	14 kg
5–12 years	26 kg
Volume of water ingested per day	
Adult	2.0 L
Child	1.0 L
Volume of air breathed per day	
Adult	20 m^3
Child	5 m^3
Exposure duration	
Lifetime (for carcinogens)	70 years
At one residence (90th percentile)	30 years
National median	5 years

The *chronic daily intake* (CDI) (mg/kg-day) of a substance is defined as

$$CDI = \frac{\text{total dose (mg)}}{\text{body weight (kg)} \times \text{lifetime (days)}} \tag{305.32}$$

The *total dose* is calculated as

$$\text{Total dose (mg)} = \text{concentration (mg/L)} \times \text{daily intake (L/day)} \times \text{duration (days)} \\ \times \text{absorption factor}$$

The *absorption factor* (dimensionless) is the mass of a contaminant absorbed into the cell tissue per unit mass inhaled or ingested.

The lifetime risk is defined as the product of CDI and the potency factor.

Potency Factor

The potency factor (also known as the slope factor) is the slope of the dose-response curve at very low doses. This is shown in Fig. 305.6. Note that for a carcinogenic constituent, the NOAEL is zero. In other words, there is no chronic dose, no matter how small, that can be tolerated without causing an increase in lifetime risk. On the other hand, for a noncarcinogenic

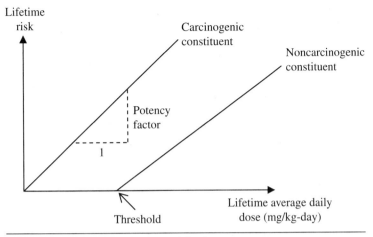

Figure 305.6 Lifetime risk versus daily dose.

constituent, there is usually a threshold value (indicated on the diagram). If the daily intake is less than this threshold, there is no observable increase in the lifetime risk from exposure to this constituent.

The acceptable daily intake (ADI) of a noncarcinogenic chemical is the dose that poses no lifetime risk when taken daily. The typical units used for the ADI are mg/kg body weight/day. In expressing various thresholds of contaminants, the following definitions are used:

NOAEL (no observed adverse effect level) is defined as the concentration of a substance below which, there are no observable adverse impacts to a subject.

LOAEL (lowest observed adverse effect level) is defined as the lowest tested dose of a substance that has been reported to cause harmful (adverse) health effects on people or animals.

The reference dose (R_fD) may be based on the LOAEL or NOAEL and is lower than this level by a uncertainty factor (safety factor). The U.S. EPA has developed reference doses for various constituents based on the assumption that the threshold ADI exists for some toxic effects but not for others (such as carcinogens).

$$R_fD = \frac{\text{NOAEL or LOAEL}}{\text{safety factor}} \qquad (305.33)$$

In order to determine the concentration (mg/L) in drinking water corresponding to a particular reference dose, the *drinking water equivalent level* (DWEL) is given by

$$\frac{\text{DWEL}}{\text{(mg/L)}} = \frac{R_fD \times \text{body weight (kg)}}{\text{drinking water consumption (L/day)}} \qquad (305.34)$$

The *hazard ratio* is defined as the ratio of the exposure dose and the reference dose (the threshold for no observable health effects).

The LC_{50} is defined as the median *lethal concentration* in air which, based on laboratory tests, is expected to kill 50% of a group of test animals, when administered as a single exposure over a duration of 1 or 4 h.

The LD_{50} is defined as the median *lethal dose* of a material, which based on laboratory tests, is expected to kill 50% of a group of test animals, usually by skin or oral exposure.

Example 305.18

GenX ($C_6HF_{11}O_3$) is a clear, colorless liquid with high water solubility. The No-observed-adverse-effect-level (NOAEL) of GenX = 0.1 mg/kg body weight/day. Calculate the MCLG (ng/L) for GenX for bottle-fed infants (body weight 7.8 kg, daily water intake = 1.1 L/day. RSC (relative source contribution), which is the proportion of total GenX exposure from drinking water = 0.2.

The following uncertainty factor are to be applied:

- Intraspecies UF = 10 (to account for the variation insensitivity among members of human population)

- Interspecies UF = 10 (to account for the uncertainty involved in extrapolating from animal data to humans)

- Subchronic-chronic UF = 10 (to account for the uncertainty involved in extrapolating from less than chronic NOAELs to chronic NOAELs)

Solution

Composite uncertainty factor (UF) $= 10 \times 10 \times 10 = 1000$

Reference dose: $R_f D = \dfrac{NOAEL}{UF} = \dfrac{0.1}{1000} = 0.0001 \dfrac{\text{mg}}{\text{kg}}$ /day

Formula: $MCLG\left(\dfrac{\text{ng}}{\text{L}}\right) = 10^6 \dfrac{\text{ng}}{\text{mg}} \times 0.0001 \dfrac{\text{mg}}{\text{kg} - \text{day}} \div 1.1 \dfrac{\text{L}}{\text{day}} \times 0.2 \times 7.8 \text{kg} = 140 \dfrac{\text{ng}}{\text{L}}$

Water and Wastewater Treatment

Total Maximum Daily Load (TMDL)

Total Maximum Daily Load (TMDL) is the maximum amount of a pollutant that a water body can receive and still meet water quality standards. The load must also be allocated to the various sources of the pollutant. TMDLs must account for seasonal variations in water quality and incorporate a margin of safety to account for uncertainty in predicting how well pollutant reductions will result in meeting water quality standards. TMDL is calculated as

$$TMDL = \sum WLA + \sum LA + MOS$$

where $\sum WLA$ = sum of wasteload allocations (point sources)
$\sum LA$ = sum of load allocations (nonpoint sources and background)
MOS = margin of safety

The goal of developing a TMDL is to end up with an implementation plan or a watershed plan designed to meet water quality standards and restore impaired water bodies. These plans can build on information in one or more TMDLs.

Pollutant sources are characterized as either point sources that receive wasteload allocations (WLA), or nonpoint sources that receive a load allocation (LA). Point sources include all sources subject to regulation under the National Pollutant Discharge Elimination System (NPDES) program, e.g., wastewater treatment facilities, some stormwater discharges, and concentrated animal feeding operations (CAFOs). Nonpoint sources include all remaining sources of the pollutant as well as anthropogenic and natural background sources.

A TMDL is developed for each water body/pollutant combination. For example, if one water body is impaired or threatened by three pollutants, three TMDLs will be developed for the water body. However, in many cases, the word TMDL is used to describe a document that addresses several water body/pollutants combinations (i.e., several TMDLs exist in one TMDL document). States have been developing TMDLs on a water body/pollutant basis or have grouped several water body/pollutant combinations in larger watershed-scale analyses. Some states are also doing very large-scale TMDLs that cover multiple watersheds.

According to the Clean Water Act (CWA), each state must develop TMDLs for all the waters on the list in Article 303(d). It is at the discretion of states to set priorities for developing TMDLs for waters on the 303(d) list.

Under the CWA, states are responsible for developing TMDLs and submitting them to the Environmental Protection Agency (EPA) for approval and many states have taken the lead in developing all of their TMDLs. In some states, nonprofit statewide environmental groups or watershed organizations have taken on significant responsibility in the development of the TMDL document and supporting analysis, known as a third-party TMDL. These TMDLs must still be submitted to EPA by the states. The EPA reviews and issues approval or disapproval of all TMDLs. The EPA has developed many TMDLs itself in response to court orders or upon requests from states.

National Pollutant Discharge Elimination System

Polluted stormwater runoff is a leading cause of impairment to U.S. water bodies which do not meet water quality standards. Polluted surface runoff is often discharged directly into local water bodies. Left uncontrolled, this water pollution can result in the destruction of fish, wildlife, and aquatic life habitats, a loss in aesthetic value and adverse impacts on public health.

Under the auspices of the Clean Water Act, the National Pollutant Discharge Elimination System (NPDES) is a comprehensive national program for addressing the nonagricultural sources of stormwater discharges, which adversely affect the quality of our nation's waters. The NPDES program uses the permitting mechanism to require the implementation of controls, designed to prevent harmful pollutants from being carried by stormwater runoff into local water bodies.

The Clean Water Act prohibits anybody from discharging pollutants through a point source into a water of the United States unless they have an NPDES permit. The permit contains limits on what can be discharged, monitoring and reporting requirements, and other provisions to ensure that the discharge does not hurt water quality or people's health. The permit translates general requirements of the Clean Water Act into specific provisions tailored to the operations of each entity discharging pollutants.

NPDES permits make sure that a state's mandatory standards for clean water and the federal minimums are being met. There are various methods used to monitor NPDES permit conditions. The permit will require the facility to sample its discharges and notify the EPA and the state regulatory agency of these results. In addition, the permit will require the facility to notify EPA and the state regulatory agency when the facility determines that it is not in compliance with the requirements of the permit. In addition, EPA and state regulatory agencies will send inspectors to companies in order to determine if they are in compliance with the conditions imposed under their permits. EPA and state regulatory agencies may issue administrative orders, which require facilities to correct violations and that assess monetary penalties.

EPA and state agencies to pursue civil and criminal actions that may include mandatory injunctions or penalties, as well as jail sentences for persons found willfully violating requirements and endangering the health and welfare of the public or environment.

Members of the general public can also enforce permit conditions. If any member of the general public, upon reviewing the monitoring documents, finds that a facility is violating its NPDES permit, he/she can independently start a legal action, unless EPA or the state regulatory agency has taken an enforcement action.

The term *pollutant* includes any type of industrial, municipal, and agricultural waste discharged into water. Some examples of pollutants covered by the NPDES requirements are dredged soil, solid waste, incinerator residue, sewage, garbage, sewage sludge, munitions, chemical wastes, biological materials, radioactive materials, heat, wrecked or discarded equipment, rock, sand, and industrial, municipal, and agricultural waste.

The term *point source* means any discernible, confined, and discrete conveyance, such as a pipe, ditch, channel, tunnel, conduit, discrete fissure, or container. It also includes vessels or other floating craft from which pollutants may be discharged. By law, agricultural stormwater discharges and return flows from irrigated agriculture are not considered point sources.

For discharging pollutants into a municipal *sanitary* sewer system, an NPDES permit is not required. For discharging pollutants into a municipal *storm* sewer system, a permit may be required depending on the constituents.

The Clean Water Act limits the length of NPDES permits to 5 years. NPDES permits can be renewed (reissued) at any time after the permit holder applies. Federal laws provide EPA and authorized state regulatory agencies with various methods of taking enforcement actions against violators of NPDES permit requirements.

Municipal Separate Storm Sewer System (MS4)

The regulatory definition of an MS4 is a conveyance or system of conveyances (including roads with drainage systems, municipal streets, catch basins, curbs, gutters, ditches, manmade channels, or storm drains) which is (1) owned or operated by a state, city, town, borough, county, parish, district, association, or other public body (created to or pursuant to state law) or a designated and approved management agency under section 208 of the Clean Water Act that discharges into waters of the United States; (2) designed or used for collecting or conveying stormwater; (3) not a combined sewer; and (4) not part of a Publicly Owned Treatment Works (POTW).

In practical terms, operators of MS4s can include municipalities and local sewer districts, state and federal departments of transportation, public universities, public hospitals, military bases, and correctional facilities. The Stormwater Phase II Rule added federal systems, such as military bases and correctional facilities by including them in the definition of small MS4s.

EPA's NPDES Stormwater Program regulates medium, large, and regulated small MS4s. A medium MS4 is a system that is located in an incorporated place or county with a population between 100,000 and 249,999. A large MS4 is a system that is located in an incorporated place or county with a population of 250,000 or more. A regulated small MS4 is any small MS4 located in an "urbanized area," as defined by the Bureau of the Census, or located outside an urbanized area and brought into the program by the NPDES permitting authority. An NPDES permit must be obtained by the operator of an MS4 covered by the NPDES Stormwater Program.

Water Quality–Based Effluent Limits (WQBEL)

Water quality goals for bodies of water within state boundaries are set by individual states. For ecologically important water bodies that abut several states, such as the Chesapeake Bay, these water quality standards are set by agreement between the affected states, with EPA having overall jurisdiction. Section 303(c) of the Clean Water Act (CWA) requires every state to develop water quality standards at or above EPA standards.

Water quality standards have three components:

1. *Use classification:* The defining authority (the state) must make clear the intended use classification of the water body—water supply, recreation, or propagation of fish and wildlife.

2. *Water quality criteria:* The defining authority must establish numerical and narrative criteria for all identified pollutants. These will be in the form of specific limits on the concentration of each pollutant.

   Water quality criteria consist of three components:
   a. *Magnitude:* This is a specific level (concentration) of the pollutant allowable in the water body.
   b. *Duration:* This is the period over which the concentration is averaged. For example, the limiting concentration of pollutant A may be on the 1-h average, while that for pollutant B may be on the 8-h average.
   c. *Frequency:* This is how often the criteria may be exceeded.

3. *Antidegradation policy:* The defining authority must adopt policies and identify methods used to implement these policies to prevent/remediate degradation of the water bodies below their defined minimum quality standards. There are three tiers of protection provided, depending on the importance of the water body, with tier 3 protection being the most stringent.

Wastewater Flow Rates for Various Sources

Table 306.1 summarizes wastewater flow rates from various sources in the United States.

Approximately 70% to 80% of a community's domestic and industrial water supply returns as wastewater. Sanitary sewer sizing is commonly based upon a flow rate of 100 to 125 gpcd. Hourly variation in this number is the most significant for designing sewer size.

According to the "Ten States Standards" (Recommended Standards for Sewage Works),[1] new sanitary sewer *systems* should be designed to carry the average flow of 100 gpcd,

---

[1]"Recommended Standards for Sewage Works. Policies for the Review and Approval of Plans and Specifications for Sewage Collection and Treatment." *A Report of Committee of the Great Lakes—Upper Mississippi River Board of State Sanitary Engineers*, Health Education Service, Albany, N.Y., 1978.

"Recommended Standards for Water Works. Policies for the Review and Approval of Plans and Specifications for Public Water Supplies." *A Report of the Water Supply Committee of the Great Lakes—Upper Mississippi River Board of State and Provincial Public Health and Environmental Managers*, Health Education Service, Albany, N.Y., 2007.

Table 306.1 Average Flow Rates from Various Sources in the United States

Category	Unit	Flow rate [gpcd (Lpcd)]
Residential	2 persons/household	65–80 (225–385)
	4 persons/household	40–70 (155–270)
	6 persons/household	40–65 (145–250)
Commercial	Airport (per passenger)	3–5 (11–19)
	Conference center (per person)	6–10 (40–60)
	Industrial building (per employee)	15–35 (57–130)
	Office (per employee)	7–16 (26–60)
	Restaurant (per customer)	7–10 (26–40)
	Shopping center (per employee)	7–13 (26–50)
Institutions	Hospital (per bed)	175–400 (660–1500)
	Prison (per inmate)	80–150 (300–570)
	School (per student)	15–30 (60–120)

but sewer *pipes* should be sized to carry the peak flow. The peak factor for a community (P = population in thousands of people) is

$$\frac{Q_{\text{peak}}}{Q_{\text{ave}}} = \frac{18 + \sqrt{P}}{4 + \sqrt{P}} \tag{306.1}$$

The following guidelines about Sewers are given in the *Recommended Standards for Wastewater Facilities: Policies for the Design, Review, and Approval of Plans and Specifications for Wastewater Collection and Treatment Facilities, 2004* (also known as "Ten States Standards"):

No public gravity sewer conveying raw wastewater shall be less than 8 in (200 mm) in diameter.

All sewers shall be designed and constructed to give mean velocities, when flowing full, of not less than 2.0 ft/s, based on Manning's formula using an "*n*" value of 0.013.

Sewers 48 in or larger should be designed and constructed to give mean velocities, when flowing full, of not less than 3.0 ft/s, based on Manning's formula using an "*n*" value of 0.013.

Slopes which are slightly less than the recommended minimum slopes may be permitted. Such decreased slopes may be considered where the depth of flow will be 0.3 of the diameter or greater for design average flow.

Where velocities greater than 15 ft/s are attained, special provision shall be made to protect against displacement by erosion and impact.

Sewers on 20% slopes or greater shall be anchored securely with concrete, or equal, anchors spaced as follows:

    a. Not over 36 ft center to center on grades 20% and up to 35%;

    b. Not over 24 ft center to center on grades 35% and up to 50%;

    c. Not over 16 ft center to center on grades 50% and over.

No pipe shall exceed a deflection of 5%.

The leakage exfiltration or infiltration shall not exceed 100 gallons per inch of pipe diameter per mile per day for any section of the system. An exfiltration or infiltration test shall be performed with a minimum positive head of 2 ft.

Manholes shall be installed: at the end of each line; at all changes in grade, size, or alignment; at all intersections; and at distances not greater than 400 ft for sewers 15 in or less in diameter, and 500 ft for sewers 18–30 in in diameter, except that distances up to 600 ft may be approved in cases where adequate modern cleaning equipment for such spacing is provided.

The minimum diameter of manholes shall be 48 in; larger diameters are preferable for large diameter sewers. A minimum access diameter of 24 in shall be provided.

The top of all sewers entering or crossing streams shall be at a sufficient depth below the natural bottom of the stream bed to protect the sewer line. In general, the following cover requirements must be met:

    a. 1.0 ft of cover where the sewer is located in rock;

    b. 3.0 ft of cover in other material. In major streams, greater cover may be required; and

    c. In paved stream channels, the top of the sewer line should be placed below the bottom of the channel pavement.

Inflow, Infiltration, and Exfiltration

Inflow is stormwater that enters into sanitary sewer systems at points of direct connection to the systems from various sources such as roof leaders, cellar and yard area drains, foundation drains, commercial and industrial "clean-water" discharges, drains from springs and swampy areas, etc. Inflow does not include infiltration, which is groundwater that enters sanitary sewer systems through cracks and/or leaks in the sanitary sewer pipes.

The EPA states that a maximum allowable test section rate of 200 gallons per inch of diameter per mile of pipe per day can normally be achieved with little effect on construction costs. This limit is appropriate when the average depth of the groundwater is between 2 ft and 6 ft above the crown of the pipe. With heads of more than 6 ft, the permissible limit should be multiplied by the square root of the ratio of the actual head to the base head of 6 ft.

For example, if the average head is 10 ft above the crown of the pipe, then the multiplier is $\sqrt{(10/6)} = 1.29$ and the acceptable limit may be raised to $200 \times 1.29 = 258$ gpd/inch diameter/mile.

Health and Economic Impacts

Inflow and infiltration add clear water to sewer systems increasing the load on the systems. Once the sanitary sewer system reaches capacity or becomes overloaded, wastewater flows at much higher water level than normal and creates the potential for backflow into households and/or overflows through manholes into streets. This poses a public health hazard and also creates the potential for environmental fines and litigation costs. Pumps that transport wastewater operate round the clock and must work harder as the sewer system's water level load increases, putting additional strain on the pumps and shortening their life expectancy. Failure on the part of a municipality to mitigate these overflow events can compromise their rating and reduce property values in affected areas.

Collection and Evaluation of Systemwide data

Quantification of the inflow and infiltration into a sanitary sewer system requires the recording of instances of observed overflows, surcharges and bypasses, customer backup complaints, and chronic maintenance activities. Typically, it is recommended that the system has one flow meter for every 30,000 to 50,000 ft of sanitary sewer pipe. The flow meter should collect intermittent data, typically at about 15-minute intervals. Rain gauges must be used throughout the system to record several rainfall events. The system should be monitored during a period of high seasonal groundwater.

Once the flow monitoring data has been collected it should be carefully evaluated. The corrected data should be tabulated and analyzed to make comparisons between the measured inflow and infiltration and the corresponding rainfall intensity. Data under surcharge conditions should be avoided for analysis purposes. The analysis will provide two essential parameters that are used to quantify the inflow and infiltration problem. The first parameter is a comparison between different basins so that basins can be prioritized for future studies and potential inflow and infiltration reduction. The second parameter is information that will be useful if subsequent relief or replacement sewer systems are necessary to reduce or eliminate overflow or bypass conditions.

Municipal Wastewater Treatment

In municipal wastewater treatment plants, the sewage first passes through a *screen*, where large floating objects such as rags and sticks are removed. The screened sewage then passes through a *grit chamber*, where the heavier particles sink to the bottom and are removed from the through flow. The through flow still contains organic and inorganic matter along with other suspended solids. Minute suspended solids can be removed in a *sedimentation tank* in the form of sludge. Primary treatment removes a significant fraction of the inorganic (fixed) solids, but it reduces volatile solids to a much lesser extent.

In the United States, some of the most common secondary treatment processes are trickling filters, activated sludge process, and rotating biological contactors. Filter media for recently installed trickling filters comprise of interlocking pieces of corrugated plastic or other synthetic materials, to increase surface area for biological populations. These synthetic media filters can be significantly taller than rock media trickling filters and therefore have a much smaller footprint.

The activated sludge process is favored both for economy of space and time over trickling filters. The fundamental idea of the activated sludge process is to bring air and highly concentrated (activated) sludge into contact with sewage for several hours in an aeration tank. From the aeration tank, the partially treated sewage is sent to another sedimentation tank (secondary clarifier) to remove excess bacteria. The process usually involves recirculation of the activated sludge in order to maintain a design concentration of suspended solids in the aeration tank.

The EPA requires that secondary treatment should reduce BOD_5 below 30 mg/L and suspended solids below 30 mg/L. The secondary effluent is usually disinfected with chlorine, typically killing more than 99% of the harmful bacteria. Excess chlorine can lead to the formation of toxic byproducts such as trihalomethanes (THMs) and therefore should be removed by dechlorination. Ultraviolet radiation and ozone are alternatives to chlorine disinfection.

Depending on the pollutants found in the wastewater, advanced treatment processes that target-specific pollutants (such as ammonia removal by air stripping or phosphorus removal by precipitation) may be used following secondary treatment.

Lift Stations

Lift stations are hydraulic structures that are designed to move wastewater from a lower elevation to a higher one, so that it can be transported through municipal sewer lines for eventual processing at a wastewater treatment plant. A lift station has the following functional components—pumps, motors, power source and controls for mechanical equipment, a wet-well, and discharge conduits.

Lift Stations Operation

When influent liquid sewage rises to a preset trigger level in the wet-well, a monitoring mechanism sends a signal to the station's control panel to start pumping. Once the liquid has been evacuated through the discharge conduits to the main sewer, the level of liquid in the wet-well drops to a shutoff trigger level and the pump ceases operation and waits for the chamber to fill again.

Types of Lift Stations

There are various types of lift stations in common use. These are:

1. Wet-Well, Submersible Lift Station
   Lift stations using submersible, nonclogging, centrifugal pumps are the most frequently specified systems for pumping raw sewage. They are equipped with pumps that operate while submerged in the wet-well. They are used frequently because of their high efficiency and capacity and for overall economy because of low maintenance costs.

2. Wet-Well/Dry-Well, Suction Lift Station
   This type of station places mechanical equipment in a dry-well that is separated by a divider wall from the wet-well. This arrangement provides good equipment accessibility and because of the more favorable operating conditions, there is a high degree of reliability.

3. Wet-Well, Suction Lift Station

In a wet-well mounted lift station, the pumps are located at ground level above the wet-well. The water rises up from the wet-well to the pump through a vertical suction pipe. The pumps can be self-priming or primed with a small vacuum pump. The pumps, valves, and control panel are located above grade.

Wet-Well Design

According to "Design of Wastewater and Stormwater Pumping Stations," Water Pollution Control Federation, Manual of Practice No. FD-4, 1981, the wet-well is sized so that the cycle time for each pump shall not be less than 5 min, or that the average cycle time shall not be more than 30 min. The pump cycle time is equal to the sum of the time to fill the well and the pump runtime.

The shortest operating cycle time occurs when the inflow equals half the pump discharge rate. The required volume of the wet-well is given by

$$V = \frac{\theta q}{4} \qquad (306.2)$$

where V = required capacity (gallons)

θ = minimum time of one pumping cycle or time between successive starts (minutes) and

q = pump capacity (gpm)

Once the volume V is calculated, the vertical distance between the conditions "lead pump on" and "all pumps off" can be calculated based on the wet-well geometry.

Reactors Used for Wastewater Treatment

The various unit operations and processes of wastewater treatment—physical, chemical, and biological—take place in the following types of reactors: batch reactor, complete mix reactor, complete-mix reactors in series, plug-flow reactor, packed bed reactor, and fluidized bed reactor.

In a *batch reactor*, the wastewater enters the reactor, is treated and once this batch is discharged, it is replaced by a new batch of untreated wastewater. An example of an operation using a batch reactor is flocculation. In an ideal *complete-mix reactor*, it is assumed that complete mixing occurs instantaneously as the wastewater enters the reactor. An example of a complete-mix reactor is the aeration tank of the activated sludge process. A large number of complete-mix reactors in series approximate a plug-flow reactor. In the *plug-flow reactor*, fluid packets pass through the reactor with no mixing. Longitudinal dispersion is minimal. An example of a plug-flow reactor is a chlorination contact chamber. In a *packed-bed reactor*, the fluid moves either in upflow or downflow past a bed of packed media. An example of a packed-bed reactor is a trickling filter. A *fluidized-bed reactor* is similar to a packed-bed reactor except that the packed media is expanded by the upward movement of the fluid through the media. An example of a fluidized-bed reactor is air stripping.

Reaction Kinetics and Reactor Hydraulics

In complete mixing, the chemical (such as a coagulant) entering the reactor is immediately distributed throughout by impellers. If the mixing basin is small, it may be known as a rapid mixer. In plug-flow reactors, the water flows through a long chamber, the chemical is added at the inlet (as a "plug"), and there is no mechanical agitation to aid mixing.

In the treatment of wastewater, operations in which some transformation is effected through the use of chemical reactions are termed chemical unit operations. The rate at which a chemical reaction progresses depends on the nature of the reaction. Broadly, chemical reactions may be classified as zero order, first order, second order, and so on.

The basic characteristics of different types of reactions are given below.

Zero-Order Reactions

Zero-order reactions proceed at a rate which is independent of the concentration of the reactants. This may be expressed as

$$\frac{dC}{dt} = -k \qquad \text{leading to} \qquad C(t) = C(0) - kt \qquad (306.3)$$

If concentration data is available at various times, a zero-order reaction would exhibit the following property: equal intervals of time would lead to equal changes in concentration. The plot of concentration $C(t)$ versus time is linear (slope k).

First-Order Reactions

A reaction is classified as a first-order reaction if the rate of reaction is directly proportional to the concentration of the reactant. This may be expressed as

$$\frac{dC}{dt} = -kC \qquad \text{leading to} \qquad C(t) = C(0)e^{-kt} \qquad (306.4)$$

If concentration data is available at various times, a first-order reaction would exhibit the following property: equal intervals of time would lead to equal changes in $\ln C$. The plot of $\ln C$ versus time is linear (slope k).

Second-Order Reactions

A reaction is classified as a second-order reaction if the rate of reaction is proportional to the square of the concentration of the reactant. This may be expressed as

$$\frac{dC}{dt} = -kC^2 \qquad \text{leading to} \qquad \frac{1}{C(t)} = \frac{1}{C(0)} + kt \qquad (306.5)$$

If concentration data is available at various times, a second-order reaction would exhibit the following property: equal intervals of time would lead to equal changes in $1/C$. The plot of $1/C$ versus time is linear (slope k).

Table 306.2   Required Hydraulic Detention Time for Various Orders of Reaction

Reaction order	Required hydraulic detention time	
	Complete mix reactor	**Plug flow reactor**
Zero order	$\dfrac{C_0 - C_e}{K}$	$\dfrac{C_0 - C_e}{K}$
First order	$\dfrac{1}{K}\left(\dfrac{C_0}{C_e} - 1\right)$	$\dfrac{1}{K}\ln\left(\dfrac{C_0}{C_e}\right)$
Second order	$\dfrac{1}{K}\left(\dfrac{C_0 - C_e}{C_e^2}\right)$	$\dfrac{1}{K}\left(\dfrac{1}{C_e} - \dfrac{1}{C_0}\right)$
Saturation	$\dfrac{(C_0 - C_e)(K_s + C_e)}{KC_e}$	$\dfrac{[K_s \ln(C_0/C_e) + C_0 - C_e]}{K}$

Saturation Reactions

A reaction is classified as a saturation reaction if the rate of reaction saturates as the reaction progresses. This may be expressed as

$$\frac{dC}{dt} = \frac{kC}{a + C} \qquad \text{leading to} \qquad a\ln\frac{C(0)}{C(t)} + C(0) - C(t) = kt \qquad (306.6)$$

The plot of $1/t \ln(C_o/C_t)$ versus $(C_o - C_t)/t$ is linear (slope k). Table 306.2 shows a summary of the required hydraulic detention time for complete-mix and plug-flow reactors for various reaction orders.

The variables in Table 306.2 have the following meaning:

C_o = concentration in the influent

C_e = required concentration in the effluent

K = reaction coefficient

K_s = half saturation coefficient

PHYSICAL UNIT OPERATIONS IN WASTEWATER TREATMENT

In the treatment of wastewater, operations in which some transformation is effected through the use of physical forces are termed *physical unit operations*. Some of the physical unit operations used for wastewater treatment are (1) screening; (2) reduction of coarse solids—comminution, maceration, and grinding; (3) flow equalization; (4) mixing and flocculation; (5) grit removal; (6) sedimentation; (7) high rate clarification; (8) accelerated gravity separation; (9) floatation; (10) oxygen transfer; (11) aeration; and (12) stripping of VOC's.

Screening

A screen is a device with openings used to retain coarse solids in the influent wastewater, in order to protect equipment such as pumps, improve overall treatment process effectiveness, and prevent contamination of waterways receiving the effluent. Screens may be broadly classified into two types—coarse screens having clear openings ranging from 6 to 150 mm (0.25 to 6 in) and fine screens having openings less than 6 mm. Microscreens, having openings less than 50 μm are primarily used to remove fine solids from treated effluents.

The screen channel floor should either be level or sloping down toward the screen. The channel should preferably be straight and perpendicular to the plane of the screen. The screens themselves should be periodically cleaned, either by hand or mechanically. In hand-cleaned installations, the approach velocity should be limited to about 1.5 ft/s (0.5 m/s) at average flow. For mechanically cleaned installations, two or more units should be installed so that one unit may be shut down for maintenance.

As the trapped solids accumulate on the screen, head losses increase. For the system using mechanical cleaning, the cleaning mechanism may be actuated by the buildup of headloss. The head loss through coarse screens may be estimated by the following formula:

$$h_L = \frac{1}{C} \frac{V_2^2 - V_1^2}{2g} \tag{306.7}$$

where V_1 = velocity in the upstream channel
V_2 = velocity through the screens
C = an empirical loss coefficient, typically 0.7 for a clean screen and 0.6 for a clogged screen

The calculation of headloss through fine screens differs from that for coarse screens. The clear water headloss through fine screens may be calculated from

$$h_L = \frac{1}{2g}\left(\frac{Q}{CA}\right)^2 \tag{306.8}$$

where Q = discharge through the screen
A = effective open area of the submerged screen
C = loss coefficient typically = 0.6 for a clean screen

Coarse Solids Reduction

As an alternative to screens, comminutors and macerators can be used to intercept and grind them to finer sizes. Use of these devices can eliminate the difficult and expensive task (especially in colder climates) of handling and disposal of the screenings. A comminutor (also known as a grinding pump) is used most commonly in small plants (less than about 5 MGD capacity). It houses a rotating cutting screen that shreds material to sizes from ¼ in to ¾ in without removing the shredded material from the flow stream. Because of high operational and maintenance costs associated with comminutors, newer plants often use a screen and macerator.

Macerators are slow-speed grinders, typically consisting of two sets of counter-rotating assemblies with blades. High-speed grinders (hammermills) receive screened materials from bar screens, pulverizing them by a high-speed rotating assembly.

Flow Equalization

Flow equalization is simply the damping of flow-rate variations to achieve a more even flow rate. Flow equalization has the following advantages: (1) Due to the elimination of "shock" loadings, operational parameters for biological treatment can be stabilized, leading to better system performance; (2) filtration area requirements are reduced; and (3) in chemical treatment, chemical feed control and process reliability are improved. Equalization can be performed either in-line or off-line, with in-line equalization achieving significantly better flow damping than off-line.

For sizing of equalization basins, see the section *Reservoir Sizing Using the Rippl Diagram* in Chap. 304.

Mixing and Flocculation

Mixing is employed in many parts of wastewater treatment to improve the efficiency of unit processes. Most mixing operations in wastewater treatment can be classified as continuous-rapid mixing (less than 30 s) or continuous mixing.

Power Requirements for Mixing

The energy imparted by an impeller causes turbulence in the fluid, thereby inducing mixing. According to studies by Camp and Stein (1943),[2] the power requirement of an impeller in a mixing operation is given by

$$P = G^2 \mu V \tag{306.9}$$

where G = the desirable value of the velocity gradient (T^{-1})
μ = dynamic viscosity of the fluid (FTL^{-2})
V = volume of the tank (L^3)

Impeller power and pumping capacity of mixers can be also written as

$$P = N_p \rho n^3 D^5 \tag{306.10}$$

$$Q = N_Q n D^3 \tag{306.11}$$

where P = power requirement for mixer
Q = pumping capacity of mixer
N_p = power number (dimensionless, provided by manufacturer)
N_Q = flow number (dimensionless, provided by manufacturer)
ρ = fluid density

---

[2]Camp, T. R. and Stein, P. C. (1943), "Velocity Gradient and Internal Work in Fluid Motion," *Journal of the Boston Society of Civil Engineers*, Vol. 30, No. 4, 219–237.

n = revolutions per second of impeller blades
D = diameter of impeller

The power requirement for a paddle system can be written as

$$P = \frac{1}{2}C_D A\rho v_P^3 \tag{306.12}$$

where C_D = drag coefficient of the paddle moving perpendicular to the fluid
A = paddle area
ρ = fluid density
v_P = relative velocity of the paddles with respect to the fluid, typically assumed to be about 70% of the tip speed of the paddles

Flocculation

Flocculation typically follows rapid mixing where chemicals (coagulants) have been added to destabilize the repulsive forces between particles. Flocculation may be divided into (1) microflocculation, for particles in the size range 0.001 to 1 μm, which is the aggregation of particles induced by random thermal motion of the fluid, also known as Brownian motion and (2) macroflocculation, for particles larger than about 1 μm, which is induced by velocity gradients or differential settling. A velocity gradient imparted to the fluid causes faster moving particles to collide with slower moving particles, thus creating larger particles that are easier to remove. In differential settling, larger particles tend to settle faster than smaller particles and therefore collide and coalesce with the smaller particles, thereby forming larger particles.

Settling of Particles

Particle settling can be subdivided into four types—discrete particle settling, flocculent settling, hindered or zone settling, and compression settling.

Type I Settling (Discrete Particle Settling)

Particles that settle discretely without any flocculation settle according to the laws of sedimentation by Newton and Stokes. For spherical particles, the terminal settling velocity is calculated by equating the gravitational force to the drag force. This yields Newton's law for type I settling:

$$V_s = \sqrt{\frac{4g(\rho_s - \rho_w)d_p}{3C_D\rho_w}} \tag{306.13}$$

where V_s = settling velocity of the particle
d_p = particle diameter
ρ_s = density of the particle
ρ_w = density of the fluid (water)
C_D = drag coefficient (dependent on Reynolds number)

Type I settling occurs commonly in presedimentation for sand removal prior to coagulation, settling of sand and other particles during cleaning of rapid sand filters, and grit chambers in wastewater treatment plants. While Newton's law is strictly valid for spherical particles, it can be adapted for nonspherical particles as shown below:

$$V_s = \sqrt{\frac{4g(\rho_s - \rho_w)}{3C_D\rho_w} \frac{d_p}{SF}}$$

(306.14)

The shape factor SF is 1.0 for spheres, 2.0 for sand grains, and 20 or more for fractal floc.

Laminar Region
For low Reynolds number (Re < 1.0), we have Stokes flow, for which the drag coefficient is given by $C_D \approx 24/Re$ leading to

$$V_s = \frac{g(\rho_s - \rho_w)d_p^2}{18\mu}$$

(306.15)

Transition Region
For intermediate values of Reynolds number ($1.0 < Re < 10^4$), the drag coefficient is given by

$$C_D = \frac{24}{Re} + \frac{3}{\sqrt{Re}} + 0.34$$

(306.16)

Turbulent Region
In the turbulent region (Re > 10^4), the coefficient of drag is often taken as 0.4, leading to

$$V_s = \sqrt{\frac{10g(\rho_s - \rho_w)d}{3\rho_w}}$$

(306.17)

Percent Removal of a Particular Size by Settling
For discrete particle settling, all particles having settling velocity greater than the overflow rate $V^* = Q/A$ will be removed entirely. Finer particles will be partially removed according to

$$\%removal = \frac{V_s}{V^*} \times 100\% \leq 100\%$$

(306.18)

where V_s = settling velocity of chosen particle size
V^* = overflow rate = Q/A
Q = influent flow rate
A = surface area of the sedimentation tank

The hydraulic detention time t_d is given by

$$t_d = \frac{\text{tank volume}}{\text{flow rate}} \qquad (306.19)$$

The design depth of the sedimentation tank is given by

$$d = V^* t_d \qquad (306.20)$$

Example 306.1

What is the settling velocity for sand particles with nominal diameter $= 0.10$ mm. Assume specific gravity SG $= 2.65$ and drag coefficient $C_D = 10.0$.

Solution   The settling velocity is

$$V_s = \sqrt{\frac{4gd}{3C_D} \frac{\rho_s - \rho_w}{\rho_w}} = \sqrt{\frac{4 \times 9.81 \times 10^{-4}}{3 \times 10} \times 1.65} = 0.0147 \text{ m/s}$$

Example 306.2

What is the percent removal of the 0.10-mm particle (in Example 306.1) if a flow rate 0.10 m³/s were flowing through a grit chamber of diameter 2 m?

Solution   The overflow rate is calculated as

$$V^* = \frac{0.1}{\dfrac{\pi}{4} \times 2^2} = 0.0318 \text{ m/s}$$

From the Example 306.1, the settling velocity for the 0.1-mm size is 0.0147 m/s. Therefore, the percent removal for this particle size is given by

$$\%R = \frac{0.0147}{0.0318} = 0.46 = 46\%$$

Type II Settling (Flocculent Settling)

For particles which are too fine to settle by discrete settling, addition of a coagulant causes particles to flocculate as they settle. Particle size changes continuously due to flocculation. As a result, settling rates also change continuously. In wastewater treatment plants, floc from ferric chloride or alum coagulation in primary sedimentation tanks and secondary settling tanks follows type II settling. Settling characteristics of a suspension of flocculent particles can be established from a settling column test (see Settling Column Data—Analysis). The settling column should be equal in height to the depth of the sedimentation tank.

Type III Settling (Hindered or Zone Settling)

Particles present in high concentration tend to settle as a mass, creating a clear zone above the sludge mass. In water treatment, sludge from lime softening exhibits type III settling. In wastewater treatment plants, type III settling occurs in activated sludge and gravity thickeners. Particles form a blanket or a zone, which settles as water moves upward through the interstices in the blanket. A distinct interface develops between the upper (clarified) region and the hindered settling region below. The rate of settling in the hindered zone is a function of the solids concentration.

Type IV Settling (Compression Settling)

As settling progresses, a compressed layer of particles forms on the bottom of the tank. There is continuous variation of solids concentration in the compression zone. If the interface height is plotted versus time, a typical plot may be, as shown in Fig. 306.1. The tangents from either end of the plot are extended to meet as shown. The points of departure from these tangents can be approximately taken as the points of transition between hindered, transition, and compression settling.

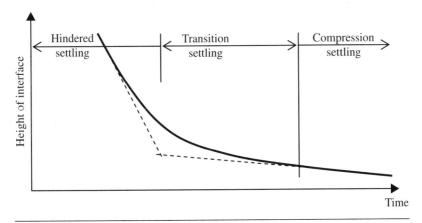

Figure 306.1   Different regimes of settling.

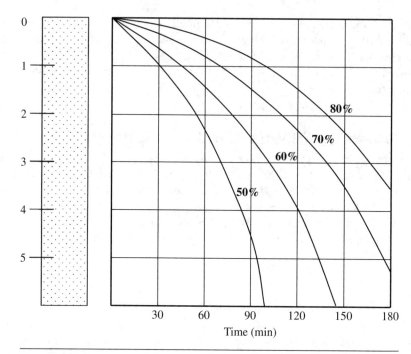

Figure 306.2 Isoconcentration curves from settling column data.

Settling Column Data—Analysis

A water sample containing suspended solids is put into a vertical column instrumented such that at various discretely spaced depths, the concentration of solids can be measured optically. If we refer to the (uniform) concentration at time $t = 0$ as C_o (arbitrary units), and the concentration at time t as C_t, then the percent removal may be expressed as

$$R\% = \left(1 - \frac{C_t}{C_o}\right) \times 100\% \tag{306.21}$$

Figure 306.2 shows a schematic of a settling column with isoconcentration curves (line joining points of equal concentration).

Example 306.3

Given the settling column percent removal curves in Fig. 306.2, calculate the removal efficiency at a depth of 4 ft and a settling time of 90 min.

Solution   The curves are reproduced below:

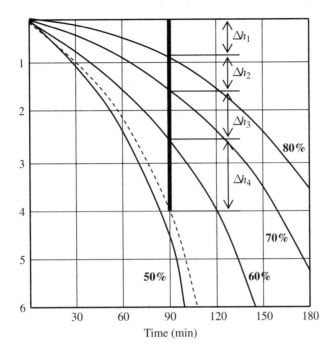

The overall removal efficiency at 90 min at a depth of 4 ft is obtained by drawing a vertical line from that point (90 min, 4 ft) to the 100% removal line. Visual interpolation indicates that the isoconcentration line passing through the point is the 52% removal line. The removal efficiency is given by the weighted average of removal efficiencies over the depth range 0 to D, using the depth increments as weights.

$$R\% = \sum \left(\frac{\Delta h_i}{D} \frac{R_i + R_{i+1}}{2} \right) \tag{306.22}$$

In this example,

$$R = \frac{0.8}{4.0} \times \frac{100 + 80}{2} + \frac{0.8}{4.0} \times \frac{80 + 70}{2} + \frac{0.9}{4.0} \times \frac{70 + 60}{2} + \frac{1.5}{4.0} \times \frac{60 + 52}{2} = 68.63\%$$

Grit Removal

In wastewater treatment plants, grit chambers are usually located after the bar screens and before the primary sedimentation tanks. There are three basic types of grit chambers—horizontal flow, aerated, or vortex type.

Horizontal Flow Grit Chamber

In the horizontal-flow-type grit chamber, flow passes through the rectangular or square chamber horizontally. The basis for design is the complete removal of a specified size of grit by settling. The length of the channel is determined from the depth required for this settling velocity. The cross-sectional area is determined by the rate of flow being treated.

The horizontal velocity in grit chambers and sedimentation tanks must be less than the velocity that would carry the particles along the bottom of the tank. Camp (1946)[3] derived this "scour velocity" as

$$V_H = \sqrt{\frac{8C(\rho_s - \rho_w)gd}{f\rho_w}} = \sqrt{\frac{8C(G_s - 1)gd}{f}} \tag{306.23}$$

where C = Camp constant (0.04 for a unigranular material and 0.06 for a sticky interlocking material)
   f = Darcy-Weisbach friction factor (typically in the range 0.02–0.03)
   G_s = specific gravity of particles
   g = gravitational acceleration
   d = particle size

Thus, the surface loading rate (or the overflow rate) for a grit chamber must be less than the settling velocity of the particle size that needs to be removed. At the same time, the horizontal velocity must be more than the scouring velocity for organic material, but less than that for the grit particles. As a result, grit particles will not be scoured and transported longitudinally, but organic material will be transported, leaving the grit reasonably clean (free from organic material). In these computations, in the absence of more specific data, we may assume the specific gravity of organic material as 1.0 and that of the inorganic grit particles as 2.65.

Aerated Grit Chamber

The aerated grit chamber consists of a spiral-flow aeration tank where the spiral velocity is induced by the geometry of the tank and the quantity of air supplied to the flow. The liquid particles follow a helical pattern while advancing longitudinally through the tank. Heavier particles settle to the bottom of the tank while lighter particles (mostly organic) pass through.

Vortex-Type Grit Chamber

The vortex type consists of a cylindrical tank in which the flow enters tangentially, thereby creating a vortex. At the center of the vortex, effluent is carried out from the eye of the vortex while centrifugal and gravitational forces are responsible for the heavier, primarily inorganic particles to settle and be removed.

---

[3]Camp, T. R. (1946), "Sedimentation and the Design of Settling Tanks," ASCE Transactions 1946, page 895.

Primary Sedimentation Tanks

The objective of primary sedimentation is to remove approximately 50% to 70% of suspended solids and approximately 25% to 40% of the BOD. Primary sedimentation tanks are rectangular or circular concrete tanks that are mechanically cleaned. Typical dimensions of rectangular tanks are 80 to 120 ft long, 15 to 30 ft wide, and 15 ft deep. Typical diameter of circular primary sedimentation tanks is between 40 and 150 ft. The overflow rate is defined as $V^* = Q/A$ where A is the surface area of the tank.

The settling time for a particular size particle is given by $t_s = H/V_s$ where V_s is the settling velocity and H is the depth of the tank. Particles whose settling velocity is greater than the overflow rate (or whose settling time is shorter than the detention time $t_d = V_{tank}/Q$) are presumed to be removed completely. Finer sizes are removed at a removal rate given by Eq. (306.18).

For example, for the rectangular tank shown in Fig. 306.3, the surface area, $A = WL$.

The surface loading rate (or overflow rate), which is relevant for the rate of particle settling, is given by Q/A.

The detention time is given by $t = AH/Q$.

Also, the horizontal flow-through velocity is given by $v_H = Q/WH$. This velocity is important because excessive horizontal velocity will cause scouring and resuspension of settled sludge.

Length to width ratio of rectangular sedimentation tanks ranges from 3 to 6, with a value of 4 being common,

The weir loading rate is important for rectangular tanks. The maximum weir loading rate should be about 1000 gal/ft-h (with 330–650 more typical). Usually, a single weir across the end of the tank is too short to limit weir loading rates to these values. The weir length can be increased by providing multiple suspended weir troughs, which take the shape of square fingers projecting into the tank facing the oncoming flow. In circular radial flow tanks, the weir loading rate on a single perimeter weir is usually within acceptable limits.

The following guidelines about Sedimentation Tanks are given in the *Recommended Standards for Wastewater Facilities: Policies for the Design, Review, and Approval of Plans and Specifications for Wastewater Collection and Treatment Facilities, 2004* (also known as "Ten States Standards"):

For settling tanks, the minimum length of flow from inlet to outlet shall be 10 ft. Sizing shall be calculated for both design average and design peak hourly flow conditions, and the larger surface area determined shall be used.

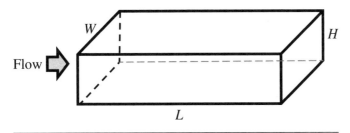

Figure 306.3 Rectangular Sedimentation Tank Geometry.

Surface overflow rates for intermediate settling tanks following series units of fixed film reactor processes shall not exceed 1500 gpd/ft^2 based on design peak hourly flow.

For average plant capacity not exceeding 1 MGD, weir loading at design peak hourly flow shall not exceed 20,000 gpd/ft. For plant capacity greater than 1 MGD, weir loading rate should not exceed 30,000 gpd/ft.

Example 306.4

A primary clarifier receives a wastewater flow rate of 2 MGD containing 350 mg/L of total suspended solids. The suspended solids removal efficiency of the clarifier is 65%. The primary sludge contains 3.5% solids. Calculate (1) the volumetric flow rate of primary sludge (in gpd) and (2) the mass of suspended solids removed (lb/day).

Solution

Approach 1

If 65% of incoming solids are being removed, then the solids removed is $0.65 \times 350 = 227.5$ mg/L. The daily solids removed is given by

$$W_{\text{solids}} = 227.5 \times 2 \times 8.3454 = 3797 \text{ lb/day}$$

Since the solids represent 3.5% of the primary sludge, the weight of the sludge is

$$W_{\text{sludge}} = \frac{3797}{0.035} = 108,490 \text{ lb/day}$$

At 3.5% solids, the sludge is mostly water, and therefore the unit weight is 8.34 lb/gal.

$$V_{\text{sludge}} = \frac{108,490 \text{ lb/day}}{8.34 \text{ lb/gal}} = 13,008 \text{ gal/day}$$

Approach 2

If 65% of incoming solids are being removed, then the concentration of solids going through is $0.35 \times 350 = 122.5$ mg/L. Solids fraction $= 3.5\% = 35$ g/1000 g which is approximately equal to 35 g/L (since the sludge is mostly water) $= 35,000$ mg/L. One may then draw the following quantities flowing into and out of the clarifier:

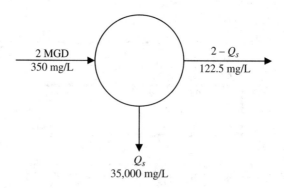

Balancing solids (entering versus leaving), we get

$$2 \times 350 = (2 - Q_s) \times 122.5 + Q_s \times 35,000 \Rightarrow Q_s = 0.01305 \text{ MGD} = 13,050 \text{ gal/day}$$

The two approaches have less than 0.3% disagreement.

Absorption versus Adsorption

Separation operations involve the transfer of material from one homogeneous phase to another, thereby reducing the concentration (of the substance to be removed) in one stream and increasing the concentration in another. Mass transfer of the substance relies on the inequilibrium between these streams and therefore ceases when equilibrium is reached. Two examples of such mass transfer processes are (1) absorption—the transfer of a substance across a gas-liquid interface as in air stripping and (2) adsorption—the transfer of a substance across a liquid-solid interface as in adsorption onto granulated-activated carbon (GAC).

Absorption of Gases

The mass transfer of a gas into a liquid (such as the increase of dissolved oxygen concentration of a water body through the gas-liquid interface at the surface) is dictated by the following relationship:

$$\frac{C_s - C_t}{C_s - C_o} = e^{-K_L at} \tag{306.24}$$

where C_s = saturation concentration of the gas in the liquid
C_t = concentration at time t
C_o = initial concentration
a = interface area per unit volume of the liquid (L^{-1})
K_L = liquid mass transfer coefficient (LT^{-1})

The quantity $K_L a$ is called the volumetric mass transfer coefficient (T^{-1}). At equilibrium, the concentration C_t asymptotes to the saturation value C_s.

Desorption of Gases

Similarly, the process of desorption (the volatilization of a gas from a liquid supersaturated with the gas) may be described by the following relationship:

$$\frac{C_t - C_s}{C_o - C_s} = e^{-K_L at} \tag{306.25}$$

At equilibrium, the concentration C_t asymptotes to the saturation value C_s.

Activated Carbon Adsorption

Adsorption, the binding of molecules or particles to a surface, is different from *absorption*, the filling of pores in a solid. The binding to the surface is usually weak and reversible. The most common industrial adsorbents are activated carbon, silica gel, and alumina because they present very large surface areas per unit weight. Activated carbon is produced by roasting organic material (such as wood, coconut shell) to decompose it to granules of carbon. Silica gel is a matrix of hydrated silicon dioxide. Alumina is mined or precipitated aluminum oxide and hydroxide. The black color of activated carbon adds a gray tinge if even trace amounts are left after treatment, which can be mitigated by using filters with fine pores.

Adsorption progresses in four distinct stages: (1) movement of the organic material to be adsorbed through the bulk liquid to the fixed film of liquid surrounding the adsorbent, (2) diffusion of the organic material through the film, (3) transport of the organic material through the pores of the adsorbent, and (4) the attachment of the material to an available adsorption site.

The quantity of adsorbate that can be adsorbed by an adsorbent is a function of the type and concentration of the adsorbate as well as the temperature and this function is called an adsorption isotherm. Two such functions—the Freundlich isotherm and the Langmuir isotherm—are discussed below.

Freundlich Isotherm

The Freundlich isotherm is used most commonly to describe the adsorption characteristics of activated carbon used for wastewater treatment. It is defined by

$$\frac{x}{m} = K_f C_e^{1/n} \tag{306.26}$$

where x/m = mass of adsorbate adsorbed per unit mass of adsorbent (mg adsorbent/g activated carbon)

K_f = Freundlich capacity factor

C_e = equilibrium concentration of adsorbate in solution after adsorption, mg/L

$1/n$ = Freundlich intensity parameter

The ratio x/m is also related to the adsorbent dose by

$$\frac{x}{m} = \frac{V}{M}(C_o - C_e) \tag{306.27}$$

where V = volume of liquid in the reactor (L)

M = mass of adsorbent (g)

C_o = initial concentration of adsorbate (mg/L)

C_e = equilibrium concentration of adsorbate in solution after adsorption (mg/L)

Example 306.5

A chemical formed as a byproduct of wastewater treatment is to be removed/mitigated by GAC adsorption. The Freundlich isotherm coefficients are $K_f = 3.6$ and $1/n = 0.70$. If the original concentration of the chemical is 0.10 mg/L and the concentration in the effluent is to be 0.03 mg/L, how much GAC is required (lb/day) to treat 2.6 MGD of wastewater?

Solution

$$C_o = 0.10 \text{ mg/L} \qquad C_e = 0.03 \text{ mg/L}$$

$$\frac{x}{m} = K_f C_e^{1/n} = 3.6 \times 0.03^{0.7} = 0.309$$

$$\frac{x}{m} = \frac{V}{M}(C_o - C_e) = 0.309 \Rightarrow \frac{V}{M} = \frac{0.309}{0.10 - 0.03} = 4.414 \text{ L/g} \Rightarrow \frac{M}{V} = 0.227 \text{ g/L}$$
$$= 227 \text{ mg/L}$$

Thus, the quantity of adsorbent (GAC) required $= 2.6 \text{ MGD} \times 227 \text{ mg/L} \times 8.3454 = 4925 \text{ lb/day}$.

Langmuir Isotherm

The Langmuir isotherm, also used to describe the phenomenon of adsorption, is given by

$$\frac{x}{m} = \frac{abC_e}{1 + bC_e} \tag{306.28}$$

where x/m = mass of adsorbate adsorbed per unit mass of adsorbent (mg adsorbent/g activated carbon)

a, b = Langmuir empirical constants

C_e = equilibrium concentration of adsorbate in solution after adsorption (mg/L)

CHEMICAL PROCESSES FOR WASTEWATER TREATMENT

Coagulation

Untreated surface waters contain many inorganic and organic materials such as clay, minerals, bacteria, inert solids, microbiological organisms, oxidized metals, and other suspended solids. All of these particles can cause cloudiness (turbidity) of the water. Besides lowering the aesthetic quality of the water, turbidity is associated with the formation of cells that can contain pathogens and shield them from treatment processes. The purpose of coagulation is to remove these particles. While larger particles may be removed physically by gravity settling within a reasonable length of time, colloidal particles ranging in size from 0.001 to 1 μm in diameter require long periods for complete settling. Since detention times in water treatment processes are typically less than 12 h, the rate of settling of these colloidal particles must be increased. This is accomplished in the coagulation process when tiny particles agglomerate into larger, denser "superparticles" which settle more quickly.

Coagulation involves three basic steps: charge destabilization, flocculation, and sedimentation. Colloidal particles in suspension have negative charges. These charges produce electrical repulsion forces that keep them in suspension. Addition of an appropriate cation (such as Al^{3+})

reduces the charge on these particles—causing formation of floc. This process is called *flocculation*.

The Shultz-Hardy rule states that the destabilization of a colloid by an indifferent electrolyte is brought about by ions of opposite charge to that of the colloid, and that coagulation effectiveness increases exponentially with charge. Recent studies show departures from the Shultz-Hardy rule. Based on these studies, the coagulation power for monovalent, divalent, and trivalent species is taken as 1:60:700.

The conventional choices for appropriate coagulants with appropriate characteristics are alum $[Al_2(SO_4)_3 \cdot 14H_2O]$, ferric chloride ($FeCl_3$), or ferric sulfate $[Fe_2(SO_4)_3 \cdot H_2O]$.

Example 306.6

If 0.15 mol/L of sodium is sufficient to destabilize a colloid, what is the required dosage of alum for equal effectiveness?

Solution Sodium is monovalent, whereas alum contains the Al^{3+} ion (trivalent). Therefore, the coagulation power is in the ratio 1:700. Therefore, the alum dose corresponding to 0.15 mol/L sodium is

$$\text{Concentration of alum} = \text{concentration of sodium} \div 700 = (0.15 \text{ mol/L})/700$$
$$= 0.00022 \text{ mol/L}$$

Since the molecular weight of alum is 594 g/mol, this concentration is equivalent to 0.13 g/L = 130 mg/L.

Properties of a Good Coagulant

A good coagulant must be nontoxic, be insoluble in water within the expected pH range, and have a high charge density. Some compounds that have these properties are aluminum sulphate, $Al_2(SO_4)_3 \cdot xH_2O$ (effective for water containing organic matter); ferrous sulphate, $FeSO_4 \cdot 7H_2O$; ferric chloride, $FeCl_3$; and ferric sulphate, $Fe_2(SO_4)_3 \cdot H_2O$.

Factors Influencing Coagulation

The pH range in which a coagulation process occurs is the single most important factor in satisfactory progress of the coagulation process. The presence of some anions such as sulfate (SO_4^{2-}) or phosphate (PO_4^{3-}) in a water supply typically causes the optimum pH range to shift lower. Thus, abrupt changes in the concentrations of these anions can disrupt the coagulation process. Generally, higher turbidity levels require higher coagulant dosages. On the other hand, water with low turbidity can be very difficult to coagulate due to the difficulty in inducing collision between the colloids. Low water temperatures can cause retarded rates of coagulation, due to lower chemical reaction rates and slower floc settling velocities at lower temperatures. Poor or inadequate mixing, resulting in an uneven dispersion of the coagulant, can affect overall coagulation effectiveness. The choice of the proper coagulant for the given conditions is of critical importance for efficient coagulation. The chemicals most commonly used in the coagulation process are aluminum sulfate, ferric chloride, ferric sulfate, and cationic polymers. Table 306.3 shows chemicals commonly used for chemical coagulation and precipitation.

Table 306.3 Chemicals Commonly Used for Chemical Coagulation

Chemical	Formula	Molecular weight	Equivalent weight
Alum	$Al_2(SO_4)_3 \cdot 14H_2O$	594.4	99.1
Aluminum chloride	$AlCl_3$	133.3	44.4
Calcium hydroxide	$Ca(OH)_2$	74.1	37.0
Ferric chloride	$FeCl_3$	162.2	54.1
Ferric sulphate	$Fe_2(SO_4)_3$	399.9	66.6
Ferrous sulphate	$FeSO_4 \cdot 7H_2O$	278.0	139.0
Sodium aluminate	$Na_2Al_2O_4$	163.9	82.0

Alum (Aluminum Sulphate) as a Coagulant

On a theoretical basis, 1.0 mg/L of dry alum will react with

- 0.50 mg/L of natural alkalinity as calcium carbonate
- 0.33 mg/L of 85% quicklime as calcium oxide
- 0.39 mg/L of 95% hydrated lime as calcium hydroxide
- 0.54 mg/L of soda ash as sodium carbonate

Alum can be effective in the pH range of 5.5 to 8.0, but seems to work best in a pH range of 6.8 to 7.5. Alum is typically shipped and fed in dry granular form, although it is also available as a powder or liquid slurry. Aluminum sulphate reacts with natural alkalinity to form aluminum hydroxide as the precipitate (floc).

$$Al_2(SO_4)_3 \cdot 14.3H_2O + 3Ca(HCO_3)_2 \rightarrow 2Al(OH)_3 + 3CaSO_4 + 6CO_2 + 14.3H_2O$$

Polymers

Synthetic polymers are high-molecular-weight organic compounds that have multiple electrical charges along a long chain of carbon atoms. Depending on whether the ionic groups have positive or negative charges, the polymer may be classified as a cationic or anionic polymer, respectively. In the absence of electrical charges, the polymer is called a nonionic polymer.

In water and wastewater treatment, anionic and nonionic polymers are effective aids to coagulation. Polymers promote the formation of larger and tougher floc by a mechanism known as particle bridging. If used in conjunction with alum, a polymer can significantly reduce the alum dosage. In order for the polymer to be effective, thorough mixing (without hindering particle bridging) is necessary. Polymers are available dry (powdered or granular), emulsion (in oil), solution (in water), or Mannich polymers (water soluble). If dry polymers are used, the solution must be aged before use, so that the long molecular chains of the polymer "uncoil" and become effective.

Chemical Neutralization

Wastewater treatment has many operations and processes that may create either excess acidity or excess alkalinity in the water. All waters that are either too acidic (low pH) or too alkaline (high pH) require neutralization. Some of the most common chemicals used for raising the pH (i.e., for treatment of acidic waters) are calcium carbonate, lime (available as quicklime or slaked hydrated lime), sodium hydroxide (caustic soda), and sodium carbonate (soda ash).

Some chemicals used for lowering the pH (i.e., for treatment of alkaline waters) are carbon dioxide, carbonic acid, hydrochloric acid, and sulphuric acid.

Disinfection

The Clean Water Act was passed as federal law in 1972. In 1973, EPA adopted specific regulations for secondary treatment. In 1976, the fecal coliform requirement established in 1973 was deleted and replaced with a provision that allowed individual states to develop site-specific water quality standards including disinfection standards for waste water treatment plants. As a result, these standards vary. The most common standard for receiving waters is 200 MPN fecal coliform/100 mL.

An ideal disinfectant must have the following characteristics:

1. Should be available in large quantities and economically

2. Should form a homogeneous solution

3. Should have enough toxicity (to microorganisms) to be effective at high dilutions

4. Should be effective in ambient temperature ranges

5. Must not be absorbed by organic matter other than bacterial cells

6. Should possess deodorizing ability

7. Should be noncorrosive (metals), nonstaining (clothes), and nontoxic (humans/animals)

8. Should be safe to handle, transport, store, and use

Physical disinfection methods can use heat, light, and sound waves to achieve disinfection. Heating water to its boiling point destroys major pathogens, but is not economically feasible for the treatment of large quantities of wastewater.

Sunlight is a good disinfectant, because of the UV radiation. Special UV lamps have been developed for wastewater disinfection applications, but contact geometry between the UV radiation source and the water is extremely important because suspended matter, dissolved organic molecules, and the water itself will absorb the radiation, reducing its effectiveness.

Mechanism of Various Disinfectants

A disinfectant destroys microorganism cells through the following basic mechanisms: (1) damage to the cell wall, (2) alteration of cell permeability, (3) changes to the colloidal nature of the protoplasm, (4) alteration of the cell RNA, or DNA, and (5) inhibition of enzyme activity.

Table 306.4 summarizes characteristics of various disinfectants. Specifically, the mechanisms of disinfection of chlorine, ozone, and UV radiation are summarized below:

Chlorine:

(1) Oxidation

(2) Reactions with available chlorine

(3) Protein precipitation

(4) Modification of cell wall permeability

(5) Hydrolysis and mechanical disruption

Ozone:

(1) Direct oxidation/destruction of cell wall with leakage of cellular constituents

(2) Reactions with radical byproducts of ozone decomposition

(3) Damage to the constituents of the nucleic acids

(4) Breakage of carbon-nitrogen bonds leading to depolymerization

UV Radiation:

(1) Photochemical damage to DNA and RNA (thereby also affecting reproduction)

(2) Nucleic acids in microorganisms are most important absorbers of the electromagnetic energy in the 240–280 nm range

The following guidelines about Disinfection are given in the *Recommended Standards for Wastewater Facilities: Policies for the Design, Review, and Approval of Plans and Specifications for Wastewater Collection and Treatment Facilities, 2004* (also known as "Ten States Standards"):

Design shall consider meeting both bacterial standards as well as disinfectant residual limit in effluent. Chlorine is the most commonly used chemical for wastewater disinfection. The most common forms are liquid chlorine, sodium or calcium hypochlorite. Some other disinfectants are chlorine dioxide, ozone, bromine, and UV radiation.

For normal domestic wastewater, the following chlorine doses are suggested:

Trickling filter plant effluent	10 mg/L
Activated sludge plant effluent	8 mg/L
Tertiary filtration effluent	6 mg/L
Nitrified effluent	6 mg/L

150 lb cylinders are typically used for storage of chlorine gas where consumption is less than 150 lb/day. Where consumption is higher, use of one ton (907 kg = 2000 lb) containers should be considered. For liquid hypochlorite solutions, use sturdy nonmetallic lined containers equipped with secure tank tops and pressure relief and overflow piping. Protect from light or extreme temperatures.

Table 306.4 Comparison of Characteristics of Common Disinfectants

Characteristic	Chlorine	Sodium hypochlorite	Calcium hypochlorite	Chlorine dioxide	Ozone	UV radiation
Availability & cost	Low	Moderate/low	Moderate/low	Moderate/low	Moderate/high	Moderate/high
Deodorizing ability	High	Moderate	Moderate	High	High	No
Homogeneity	Homogeneous	Homogeneous	Homogeneous	Homogeneous	Homogeneous	No
Interaction with extraneous material	Oxidizes organic matter	Active oxidizer	Active oxidizer	High	Oxidizes organic matter	Absorbance of UV radiation
Corrosion/staining	Highly corrosive	Corrosive	Corrosive	Highly corrosive	Highly corrosive	No
Toxicity to higher forms of life	Highly toxic	Toxic	Toxic	Toxic	Toxic	Toxic
Penetration	High	High	High	High	High	Moderate
Safety concern	High	Moderate	Moderate	High	Moderate	Low
Solubility	Moderate	High	High	High	High	No
Stability	Stable	Slightly unstable	Relatively stable	Unstable	Unstable	No
Toxicity to microorganisms	High	High	High	High	High	High
Toxicity at ambient temp	High	High	High	High	High	High

Disinfectant shall be positively mixed as rapidly as possible (complete mix in less than 3 s)

Minimum contact period of 15 min at design peak hourly flow is recommended.

Leak detection: Bottle of 56% ammonium hydroxide solution shall be available for detecting chlorine leaks.

Ventilation: One complete fresh air change per minute when room is occupied.

Respiratory air-pac protection equipment, meeting the requirements of the National Institute for Occupational Safety and Health (NIOSH), shall be available where chlorine gas is handled, and shall be stored at a convenient location, but not inside any room where chlorine is used or stored.

Dechlorination of effluent may be necessary to reduce toxicity due to chlorine residuals. Most common chemicals used for dechlorination are sulfur compounds such as sulfur dioxide gas.

Theoretical dosage (mg/L) of chemical needed to neutralize 1 mg/L of chlorine

Sodium thiosulfate (solution)	0.56
Sodium sulfite (tablet)	1.78
Sulfur dioxide (gas)	0.90
Sodium meta bisulfite (solution)	1.34
Sodium bisulfite (solution)	1.46

Minimum contact time of 30 s is recommended for dechlorination.

Chlorine as a Disinfectant

Chlorine is universally favored as a disinfectant because it possesses many of the properties mentioned above. It is available at low cost in large quantities, it is moderately soluble in water (7g/L at 1 atmosphere and 15°C) and highly toxic to microorganisms if used in adequate concentration. However, while it is transported as a liquid under high pressure, it vaporizes if unconfined and can pose a health risk if released. Some of the byproducts of chlorination can be harmful.

The principal chlorine compounds used for disinfection are chlorine (Cl_2), sodium hypochlorite (NaOCl), calcium hypochlorite [$Ca(OCl)_2$], chlorine dioxide (ClO_2). The available chlorine in a compound is an indication of its oxidizing power. For example, for the compound sodium hypochlorite (NaOCl), the actual chlorine content is given by

$$\frac{\text{Weight of chlorine}}{\text{Molecular weight}} = \frac{35.5}{74.5} = 0.4765$$

Available Chlorine

However, the available chlorine is given by the actual chlorine multiplied by the chlorine equivalent, which is the change in valence of chlorine from what it is in the compound to a value of -1. Thus, since the valence of Cl in the compound NaOCl is $+1$, the change of valence from $+1$ to -1 is 2. Therefore, the available chlorine is given by $0.4765 \times 2 = 0.953$. When chlorine gas is used as a disinfectant for water, the reactions that occur are hydrolysis and ionization.

$$Cl_2 + H_2O \longleftrightarrow HOCl + H^+ + Cl^-$$

When sodium or calcium hypochlorite is used, the hydrolysis reactions are given by

$$NaOCl + H_2O \rightarrow HOCl + NaOH$$
$$Ca(OCl)_2 + 2H_2O \rightarrow 2HOCl + Ca(OH)_2$$

Stages in the Chlorination Process

The various stages in chlorination can be generally described as follows:

1. The initial chlorine added is destroyed by reacting with reducing compounds.

2. As the demand exerted by the reducing compounds is met, the chlorine residual increases in the form of chloro-organics and chloramines.

3. Chlorine residual decreases as chloro-organics and chloramines are partly destroyed. This will continue until the breakpoint, when no ammonia remains.

4. Chlorine added after breakpoint will contribute to the free chlorine residuals (such as Cl_2, HOCl, OCl^-).

Chlorine Demand

Added chlorine can be "used up" in many ways without producing significant disinfection. The term breakpoint chlorination is used for the process where enough chlorine is added to the water to react with all the oxidizable compounds so that any additional chlorine appears as free available chlorine. For example, chlorine reacts with ammonia (NH_3) in the water to form chloramines. Above a pH of 8.5, monochloramine (NH_2Cl) is formed. Dichloramine ($NHCl_2$) is formed at a pH of 4.5 and nitrogen trichloride (NCl_3) is formed below a pH of 4.4. The formation of chloramines depends on the pH of the water, the amount of ammonia available, and the temperature.

The appearance of *free chlorine* is a signal that all chemical reactions (using chlorine) and disinfection are complete. Chlorine demand or dose is the amount of chlorine required to leave a desired chlorine residual (0.5 mg/L typical) 15 min after mixing. Using a very high chlorine dose can lead to the formation of trihalomethane (THM).

Chlorination effectiveness depends on, and is often tabulated by the CT value, which is the product of the concentration of free available chlorine (C) and the contact time (T). Typical units are mg/L and minutes. Some typical CT values are

99% removal (2 log inactivation) of *Escherichia coli* by free chlorine at 5°C and pH of 6 requires an average $CT = 0.045$ mg/L-min.

99% removal of *Giardia lamblia* requires CT of 65 to 150 mg/L-min.

Inactivation of Giardia Lamblia by Chlorination

The empirical model shown below Eq. (306.29) yields the required CT value for a certain level of inactivation (L) of *Giardia Lamblia* by using free chlorine (C) at various values of water pH and temperature (T).

$$CT = 0.2828 \times pH^{2.69} \times C^{0.15} \times 0.933^{T-5} \times L \qquad (306.29)$$

CT (mg/L-min) = Product of residual chlorine concentration C (mg/L) and contact time T (min)

pH = pH of water

C = Residual chlorine concentration at or before the first customer (mg/L)

T = Temperature (°C)

L = Log removal

Table 306.6 is based on the model above.

Example 306.7

The flow rate through a water treatment plant is 2 MGD. The hypochlorite ion (OCl^-) dose is 20 mg/L. If the purity of calcium hypochlorite $Ca(OCl)_2$ is 92.8%, what is the feed rate in lb/day of calcium hypochlorite needed?

Solution

$$Ca(OCl)_2 \rightarrow Ca^{2+} + 2(OCl)^-$$

Molecular weight of $Ca(OCl)_2$ is $40 + 2x(16 + 35.5) = 143$ g
Ionic weight of OCl^- is $16 + 35.5 = 51.5$ g

$$\frac{2[OCl^-]}{[Ca(OCl)_2]} = \frac{2 \times 51.5}{143} = 0.72$$

To obtain 20 mg/L of OCl^-, we need $20/0.72 = 27.78$ mg/L of *pure* calcium hypochlorite, which is equivalent to $27.78/0.928 = 29.93$ mg/L of available $Ca(OCl)_2$ stock (92.8% purity). Therefore, feed rate $= 29.93 \times 2 \times 8.3454 = 499.6$ lb/day.

Baffling Factor

The calculated (or effective) CT value can be less than the product of the chlorine concentration (C) and the hydraulic detention time t_d. This is expressed by a baffling factor (see Table 306.5).

$$CT_{calc} = C \times t_{10} \qquad (306.30)$$

$$t_{10} = t_d \times BF \qquad (306.31)$$

Table 306.5 Baffling Factor for Different Types of Reactors

Baffle condition	Description	Baffling factor (BF)
Unbaffled (Mixed Flow)	No baffles. Agitated basin. Very low length to width ratio. High inlet and outlet velocities	0.1
Poor	Single or multiple unbaffled inlets and outlets. No intra-basin baffles	0.3
Average	Baffled inlet or outlet with some intra-basin baffles	0.5
Superior	Perforated inlet baffles. Serpentine or perforated intra-basin baffles. Outlet weir or perforated launders	0.7
Perfect (Plug Flow)	Very high length to width ratio. Perforated inlet and outlet. Intra-basin baffles	1.0

where CT_{calc} = calculated CT value (mg/L-min)

C = residual chlorine concentration during peak flow (mg/L)

t_{10} = time it takes for 10% of the water to flow through the reactor during peak flow (min)

t_d = hydraulic detention time = V/Q

BF = baffling factor

Log Inactivation

Log inactivation levels are related to removal efficiency. For example, for 99% removal efficiency, the remaining fraction is 0.01. This corresponds to 2-log inactivation since

$$-\log_{10} 0.01 = 2$$

Chick's law expresses the dependence of number of organisms as a function of time. The differential equation is given by

$$\frac{dN}{dt} = -kN \qquad (306.32)$$

Solution of Eq. (306.28) leads to the function

$$\frac{N_t}{N_0} = e^{-kt} \Rightarrow -\log \frac{N_t}{N_0} = kt \qquad (306.33)$$

Thus, according to Chick's law, the log inactivation is equal to the product kt where k is the inactivation rate constant and t is the contact time.

Table 306.6 *CT* Values for 3-log Inactivation of *Giardia* Cysts by Free Chlorine

Chlorine Concentration (mg/L)	Temperature ≤ 0.5°C pH							Temperature = 5°C pH							Temperature = 10°C pH						
	≤6.0	6.5	7.0	7.5	8.0	8.5	9.0	≤6.0	6.5	7.0	7.5	8.0	8.5	9.0	≤6.0	6.5	7.0	7.5	8.0	8.5	9.0
≤0.4	137	163	195	237	277	329	390	97	117	139	166	198	236	279	73	88	104	125	149	177	209
0.6	141	168	200	239	286	342	407	100	120	143	171	204	244	291	75	90	107	128	153	183	218
0.8	145	172	205	246	295	354	422	103	122	146	175	210	252	301	78	92	110	131	158	189	226
1.0	148	176	210	253	304	365	437	105	125	149	179	216	260	312	79	94	112	134	162	195	234
1.2	152	180	215	259	313	376	451	107	127	152	183	221	267	320	80	95	114	137	166	200	240
1.4	155	184	221	266	321	387	464	109	130	155	187	227	274	329	82	98	116	140	170	206	247
1.6	157	189	226	273	329	397	477	111	132	158	192	232	281	337	83	99	119	144	174	211	253
1.8	162	193	231	279	338	407	489	114	135	162	196	238	287	345	86	101	122	147	179	215	259
2.0	165	197	236	286	346	417	500	116	138	165	200	243	294	353	87	104	124	150	182	221	265
2.2	169	201	242	297	353	426	511	118	140	169	204	248	300	361	89	105	127	153	186	225	271
2.4	172	205	247	298	361	435	522	120	143	172	209	253	306	368	90	107	129	157	190	230	276
2.6	175	209	252	304	368	444	533	122	146	175	213	258	312	375	92	110	131	160	194	234	281
2.8	178	213	257	310	375	452	543	124	148	178	217	263	318	382	93	111	134	163	197	239	287
3.0	181	217	261	316	382	460	552	126	151	182	221	268	324	389	95	113	137	166	201	243	292

(*Continued*)

Table 306.6 CT Values for 3-log Inactivation of *Giardia* Cysts by Free Chlorine (*Continued*)

Chlorine Concentration (mg/L)	Temperature = 15°C pH							Temperature = 20°C pH							Temperature = 25°C pH						
	≤6.0	6.5	7.0	7.5	8.0	8.5	9.0	≤6.0	6.5	7.0	7.5	8.0	8.5	9.0	≤6.0	6.5	7.0	7.5	8.0	8.5	9.0
≤0.4	49	59	70	83	99	118	140	36	44	52	62	74	89	105	24	29	35	42	50	59	70
0.6	50	60	72	86	102	122	146	38	45	54	64	77	92	109	25	30	36	43	51	61	73
0.8	52	61	73	88	105	126	151	39	46	55	66	79	95	113	26	31	37	44	53	63	75
1.0	53	63	75	90	108	130	156	39	47	56	67	81	98	117	26	31	37	45	54	65	78
1.2	54	64	76	92	111	134	160	40	48	57	69	83	100	120	27	32	38	46	55	67	80
1.4	55	65	78	94	114	137	165	41	49	58	70	85	103	123	27	33	39	47	57	69	82
1.6	56	66	79	96	116	141	169	42	50	59	72	87	105	126	28	33	40	48	58	70	84
1.8	57	68	81	98	119	144	173	43	51	61	74	89	106	129	29	34	41	49	60	72	86
2.0	58	69	83	100	122	147	177	44	52	62	75	91	110	132	29	35	41	50	61	74	88
2.2	59	70	85	102	124	150	181	44	53	63	77	93	113	135	30	35	42	51	62	75	90
2.4	60	72	86	105	127	153	184	45	54	65	78	95	115	138	30	36	43	52	63	77	92
2.6	61	73	88	107	129	156	188	46	55	66	80	97	117	141	31	37	44	53	65	78	94
2.8	62	74	89	109	132	159	191	47	56	67	81	99	119	143	31	37	45	54	66	80	96
3.0	63	76	91	111	134	162	195	47	57	68	83	101	122	146	32	38	46	55	67	81	97

BIOLOGICAL WASTEWATER TREATMENT

Biological processes used for wastewater treatment can be broadly divided into two categories: suspended growth and attached growth processes. Examples of suspended growth processes are activated sludge, aerated lagoons, aerobic digestion, and the like. Examples of attached growth processes are trickling filters, rotating biological contactors, packed bed reactors, and the like.

In suspended growth processes, the microorganisms are maintained in liquid suspension at sufficient concentration to effect treatment. These processes can be either aerobic or anaerobic. The most common suspended growth process of the aerobic variety is the activated sludge process where the wastewater with the microbial suspension—known as the mixed liquor suspended solids (MLSS) or mixed liquor volatile suspended solids (MLVSS)—is aerated in an aeration tank. Mechanical equipment is used to facilitate mixing and air entrainment. The activated sludge process produces floc particles that can be removed by gravity settling in a secondary clarifier. Removal efficiency for the secondary clarifier is typically around 99%.

In attached growth processes, the microorganism population resides on the surface of an inert packing material. Packing materials include rock, gravel, sand to a wide range of plastic, and other synthetic materials. The microbe layer is called the *biofilm*. The most common attached growth process is the trickling filter, in which the water is distributed (typically using a radial rotating arm) over the top of a bed of packing material. Air circulates in the void spaces in the bed and provides oxygen for the microorganisms to facilitate their metabolic processes and "consume" the organic material in the influent wastewater.

Bacterial Biochemistry

Wastewater engineering involves designing processes and facilities that utilize microorganisms to destroy organic and inorganic substances. Metabolism is a term used to describe all chemical activities in the cell. Those processes that allow the bacterium to synthesize new cells from stored energy (ATP: adenosine triphosphate) are called *anabolic*. All processes where the cells convert substrate into useful energy and waste products are called *catabolic*.

To grow and reproduce, microorganisms need an energy source, a carbon source for the synthesis of new cellular material and inorganic nutrients such as nitrogen, phosphorus, potassium, calcium, magnesium, and organic nutrients (growth factors). Carbon is present in large quantities in wastewater. Treatment of wastewater involves converting this carbon into cellular material in microorganisms. This cellular material then forms a sludge that can be removed by settling. Thus, organisms that use organic material are encouraged.

Microorganisms derive their carbon from either organic matter (heterotrophs or organotrophs) or from carbon dioxide (autotrophs). Organisms that use inorganic material as a carbon source are also called *lithotrophic*. The energy needed for metabolic processes is derived from light (phototrophs) or from chemical oxidation-reduction reactions that transfer electrons from an electron donor to an electron acceptor (chemotrophs). In these reactions, the electron donor gets oxidized and the electron acceptor gets reduced. When oxygen is the electron acceptor, the reaction is termed *aerobic*. Chemical end products of aerobic decomposition are carbon dioxide, water, and cellular material. Organisms that can exist only in an environment that is devoid of oxygen generate energy by fermentation and are called *obligate anaerobes*. On the other hand, facultative anaerobes have the ability to grow in either the presence or absence of molecular oxygen. The term *anoxic* is used for organisms that use nitrite and nitrate ions as the electron acceptor. These ions get reduced to nitrogen gas, a process known as *denitrification*. *Microaerophiles* are organisms that may use oxygen, but only at low concentration. Microaerophiles carry out aerobic respiration, and some of them can also do anaerobic respiration. Aerotolerant organisms can survive in the presence of oxygen, but they are anaerobic because they cannot use it. Almost all animals, most fungi and several bacteria are obligate aerobes. Most anaerobic organisms are bacteria. An example of a facultative aerobe is yeast.

Obligate (strict) anaerobes die in presence of oxygen. Obligate anaerobes may use fermentation or anaerobic respiration. In the presence of oxygen, facultative anaerobes use aerobic respiration; without oxygen some of them ferment, some use anaerobic respiration. Aerotolerant organisms are strictly fermentative.

For dilute wastewater (BOD$_5$ <500 mg/L), aerobic oxidation is preferred because decomposition is rapid and efficient. For strong (BOD$_5$ >1000 mg/L) wastewater, aerobic decomposition requires large quantities of oxygen and sludge production is very high. For this reason, aerobic decomposition may not be the preferred mechanism for strong wastewaters. Conditions in a wastewater treatment plant are adjusted so that chemoheterotrophs (such as bacteria) predominate.

Effect of Temperature and pH

Temperature and pH have an important effect on the growth and survival of microorganisms. Psychrophiles or cryophiles grow in the temperature range of 10 to 30°C (optimum range 12–18°C), mesophiles grow in the temperature range of 20 to 50°C (optimum range 25–40°C), thermophiles grow in the temperature range of 35 to 75°C (optimum range 55–65°C), while ultrathermophiles grow in the temperature range of 60 to 80°C. For example, *E. coli*, a mesophile, operates best in the range 20 to 50°C, but can continue to reproduce at temperatures around 0°C. Most bacteria cannot tolerate pH values in excess of 9.5 or less than 4.0. Generally, the optimum pH for bacteria is in the range 6.5 to 7.5.

Fungi are aerobic, multicellular, nonphotosynthetic, heterotrophic eukaryotes. Most fungi are obligate or facultative aerobes. Molds (true fungi) produce a filamentous mass called the mycelium. Yeasts are unicellular fungi that cannot form a mycelium. Fungi are able to grow in low moisture, low nitrogen environments and are tolerant of low pH (acidic) environments. Fungi release CO_2 and N_2 during breakdown of organic material. Fungi use half the amount of nitrogen as bacteria for formation of cell material. (For this reason, they may replace bacteria as the dominant species in nitrogen-deficient wastewater.)

Bacteria are single-celled prokaryotes which reproduce primarily by binary fission. Their cytoplasm contains ribonucleic acid (RNA), which is responsible for protein synthesis, and deoxyribonucleic acid (DNA), which contains all the information necessary for reproduction.

Algae are autotrophic, photosynthetic eukaryotes. They can be unicellular or multicellular. Algae and bacteria have a symbiotic relationship in aquatic systems where algae produce the oxygen used by bacteria. In the presence of sunlight, photosynthetic production of oxygen is more than the amount used in respiration. At night, algae use up oxygen for respiration. Thus, long summer days result in a net production of oxygen. Excessive algal growth (algal blooms) can cause supersaturated oxygen conditions in daytime and anaerobic conditions at night. Algae produce turbidity in water. Some algae cause a smell and taste in natural water.

Table 306.7   Summary of Various Types of Bacteria

Type of bacteria	Reaction	Carbon source	Substrate	Electron acceptor	Products
Aerobic Autotrophic	Nitrification	CO_2	NH_3, NO_2	O_2	NO_2, NO_3
	Iron oxidation	CO_2	Fe (II)	O_2	Fe (III)
	Sulphur oxidation	CO_2	H_2S, S, S_2O_3	O_2	SO_4
Aerobic Heterotrophic	Aerobic oxidation	Organic compounds	Organic compounds	O_2	CO_2, H_2O
Facultative Heterotrophic	Denitrification anoxic reaction	Organic compounds	Organic compounds	NO_2, NO_3	N_2, CO_2, H_2O
Anaerobic Heterotrophic	Acid fermentation	Organic compounds	Organic compounds	Organic compounds	Volatile fatty acids
	Iron reduction	Organic compounds	Organic compounds	Fe (III)	Fe (II), CO_2, H_2O
	Sulphate reduction	Organic compounds	Organic compounds	SO_4	H_2S, CO_2, H_2O
	Methanogenesis	Organic compounds	Volatile fatty acids	CO_2	Methane

Protozoa are usually single-celled heterotrophic eukaryotes. Most of protozoa are aerobic heterotrophs, while some are aerotolerant anaerobes and a few are anaerobic. They are desirable in wastewater effluent, as they act as polishers by consuming bacteria and particulate matter.

Viruses are parasitic organisms—pass through filters that retain bacteria—grow and reproduce inside a host but can survive outside. They are not cells, but a nucleic acid core (DNA or RNA) surrounded by a protein sheath. Their size is 10 to 25 nm (typical size found in water supply). Viruses are host specific. Bacteriophages are viruses that infect bacteria as the host. Table 306.7 summarizes microorganisms by function, carbon source, electron donor, substrate, and end products.

Aerobic versus Anaerobic Biological Treatment

The primary advantages of anaerobic wastewater treatment (compared to aerobic processes) are low-energy requirement, less sludge production, smaller reactor volumes, and often a net production of energy. Also, an anaerobic process requires fewer nutrients than a comparable aerobic process and for this reason, is often a better option for industrial wastewaters.

The primary disadvantages of anaerobic processes are (1) need for addition of alkalinity, (2) nitrogen and phosphorus removal not possible, (3) potential for production of odors, longer start-up times required to build up the necessary biomass, and (4) and the possibility of needing aerobic treatment to polish the effluent.

For anaerobic processes, the fundamental design parameter is the solids retention time (SRT). By definition, the SRT is the average time the solids in the activated sludge spend in the system. The SRT is calculated as the mass of solids in the system divided by the mass of solids removed per day.

In general, at temperatures around 30°C, SRT values are typically 20 days or more. Three types of anaerobic suspended growth treatment processes are (1) the complete-mix suspended growth anaerobic digester, (2) the anaerobic contact process, and (3) the anaerobic sequencing batch reactor. For the complete-mix process, the solids retention time (SRT) and the hydraulic detention time (t_d) are equal, on the order of 15 to 30 days. This leads to very large reactor volumes compared to the other two types of processes.

Population Growth of Bacteria

Bacterial growth patterns in a batch reactor are closely related to the quantity of available substrate and nutrients and the bacterial population (biomass). Four distinct phases of bacterial growth can be identified as

1. The lag phase represents the time needed for the organisms to acclimate to the environment in the reactor. Neither substrate nor biomass changes significantly during the lag phase.

2. During the exponential growth phase, the bacterial population increases exponentially while substrate decreases at the maximum rate. If unlimited substrate and nutrients are available, the only factor that affects the rate of exponential growth is temperature.

3. The stationary phase is characterized by equilibrium between the bacterial growth rate and the death rate of bacterial cells, resulting in a relatively constant biomass concentration.

4. The death phase is marked by completely depleted substrate, so that no growth occurs and the change in bacterial population is due to death of bacterial cells.

These stages are shown in Fig. 306.4, which plots the number of microorganisms in a population as a function of time.

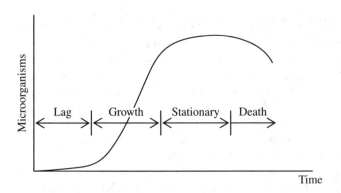

Figure 306.4 Different stages of microorganism life cycle.

Hydraulic Detention Time

The hydraulic detention time is purely a function of the flow rate being treated by the reactor and the volume of the reactor. It does not represent the organic load being imposed on the reactor. The hydraulic detention time is given by

$$t_d = \frac{V}{Q} \tag{306.34}$$

where V = volume of the reactor
$\quad\;\; Q$ = influent flow rate

Example 306.8

What is the hydraulic detention time for a circular tank (diameter = 8 ft, height = 4 ft) that treats a flow Q = 15,000 gpd. The hydraulic efficiency of the tank is 75%.

Solution Since the tank has a hydraulic efficiency of 75%, the detention time is less (by a factor of 0.75) of the ideal hydraulic detention time.

Tank volume $V = \pi R^2 H = \pi \times 4^2 \times 4 = 201 \,\text{ft}^3$

Flow rate $Q = 15,000$ gpd = 0.0232 cfs

At 100% efficiency, detention time $t_d = \dfrac{201}{0.0232} = 8664 \,\text{s} = 2.4 \,\text{h}$

At 75% efficiency, detention time = $0.75 \times 2.4 = 1.8 \,\text{h}$

Indicator Organisms

Indicator organisms are used as a measure of the quality of the water. Because pathogens of interest may be present in small numbers leading to difficulties in their isolation and identification, microorganisms that are more numerous and easily detected and quantified are used as surrogate indicators. An ideal indicator organism must have the following characteristics [Maier et al. (2000)].[4]

Characteristics of a Good Indicator Organism

1. The indicator organism must always be present when fecal contamination is present.

2. The indicator organism must be present in large numbers, equal to or greater than the numbers of the pathogenic organism.

3. The indicator organism must have the same survival characteristics and response to the environment as the pathogen.

---

[4]Maier, R. M., Pepper, I. L., and Gerba, C. P. (2000), *Environmental Microbiology*, Academic Press, A. Harcourt Science and Technology Company, San Diego, CA.

4. In order that it not pose a health risk to laboratory technicians, the indicator must not be able to reproduce outside the host organism.

5. The isolation, identification, and enumeration of the indicator organism must be easier and cheaper than the pathogen itself.

6. Indicator and pathogen must come from the same source. Thus, detection of the indicator is a strong indication that the pathogen is present. The organism should be a member of the intestinal microflora of warm-blooded animals.

To date, no organism has been found that has all of these characteristics. However, total coliform and fecal coliform have been used, as they possess several of the desirable characteristics. However, it has not been demonstrated that presence of total and fecal coliform necessarily indicates the presence of enteric viruses and protozoa. Some pathogens such as pathogenic *Escherichia coli (E. coli) Cryptosporidium*, and *G. lamblia* may arise from nonhuman sources. More recently, bacteriophages have been used as indicators of enteric viruses.

Identification, Isolation, and Enumeration of Bacteria

Bacteria can be enumerated by (1) direct counting under a microscope, (2) pour plate and spread plate counts, (3) membrane filtration, and (4) multiple tube fermentation.

Direct counting can be accomplished using a Petroff-Hauser counting chamber. In the pour plate method, a wastewater sample is diluted serially. Each dilution is mixed with a culture medium and incubated. Following incubation, bacterial colonies are counted and reported as colony-forming units per unit volume (cfu/mL). In the spread plate method, a small amount of diluted wastewater is spread on the surface of a culture dish and incubated, followed by counting bacterial colonies. In the membrane-filter method, a known volume of wastewater is passed through a membrane with small pore size (typically 0.45 μm). The membrane, with retained bacteria, is then placed in contact with a culture medium and incubated. After incubation, the bacterial colonies are counted. The multiple tube fermentation method is based on the principle of dilution to extinction in which successively more and more dilute samples are tested for the presence of microorganisms. The test is based on the Poisson distribution for extreme values and employs multiple samples of equal volume and in concentrations decreasing in a geometric series. It reports results as an MPN (most probable number, which is not an absolute concentration of organisms present, but only a statistical estimate of that concentration). The table below shows typical data from multiple-tube fermentation.

Portion size (mL)	Number of tubes	Number of positive tubes	Number of negative tubes
10.0	8	8	0
1.0	8	6	2
0.1	8	5	3
0.01	8	3	5
0.001	8	0	8

The Thomas equation can be used to analyze results from multiple-tube fermentation.

$$\text{MPN/100 mL} = \frac{\text{number of positive tubes}}{\sqrt{\text{mL of sample in negative tubes} \times \text{mL of sample in all tubes}}} \times 100$$

(306.35)

The data that must be considered for this analysis starts from the dilution level at which some tubes yield negative results (in this example, the 10-mL dilution does not yield any negatives; therefore one must start counting at the 1.0-mL dilution). For the data shown in the table, this results in

$$\text{MPN/100 mL} = \frac{14}{\sqrt{2.35 \times 8.88}} \times 100 = 306/100\,\text{mL}$$

In addition, a simple presence-absence test can be conducted to determine if coliform bacteria are present in a water sample, without indicating the number of microorganisms in the sample. This test is based on the concept that the MCL for coliform bacteria is zero. The water sample is treated with a reagent containing indicators and organic nutrients and incubated at 35°C for 24 h.

Activated Sludge Process

The most fundamental of aerobic secondary treatment processes is the activated sludge process, which consists of the following basic components: (1) an aeration tank where a concentration of microorganisms are exposed to the influent wastewater and air, (2) a clarifier where solids (primarily organic solids) are settled, and (3) a recycle system to separate waste-activated sludge from the return-activated sludge, which is then returned to blend with the primary effluent. A schematic of the activated sludge process, showing its relative position in a wastewater treatment plant, is shown in Fig. 306.5.

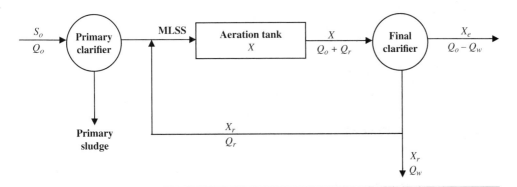

Figure 306.5 Schematic of the activated sludge process.

Most activated sludge plants receive wastewaters that have been treated with primary sedimentation, where much of the settleable solids are removed. The secondary treatment is then responsible for removing soluble, colloidal, and suspended organic solids and nutrients such as nitrogen and phosphorus. The primary design elements for the activated sludge process are (1) the volume of the aeration tank, (2) the sludge production rate, and (3) oxygen requirements.

In the schematic of the activated sludge process shown in Fig. 306.4, the blending of the return-activated sludge with the primary effluent produces the "mixed liquor" and the suspended solids concentration in it is called the "mixed liquor suspended solids" (MLSS). The organic component of the MLSS is called the "mixed liquor volatile suspended solids" (MLVSS). In the process diagram, the MLVSS is designated as X.

X = volatile solids concentration in the aeration tank = MLVSS
V_a = volume of aeration tank
S_o = influent biomass concentration (effluent from primary clarifier)
S_e = growth-limiting substrate concentration
Q_o = incoming flow rate
MLSS = mixed liquor suspended solids
MLVSS = mixed liquor volatile suspended solids = X
μ_m = maximum specific growth rate (T^{-1}) = kY
μ = specific growth rate
k_d = endogenous decay coefficient (T^{-1})
k = maximum rate of substrate utilization (T^{-1})
K_s = half-velocity constant (ML^{-3})
Y = maximum cell yield ratio (mg VSS/mg BOD_5)
Y_{obs} = observed cell yield ratio (mg VSS/mg BOD_5)
f_d = fraction of biomass that remains as biomass (g VSS/g VSS)

The following critical parameters of the activated sludge process may be defined using the nomenclature shown previously. Food arrival rate is given by

$$F = S_o Q_o \qquad (306.36)$$

Mass of microorganisms is given by

$$M = V_a X \qquad (306.37)$$

Food to microorganism ratio is given by

$$F:M = \frac{S_o Q_o}{V_a X} \qquad (306.38)$$

Hydraulic detention time for the aeration tank is given by

$$t_d = \frac{V_a}{Q_o + Q_r} \qquad (306.39)$$

Solids retention time (SRT) is given by

$$\text{SRT} = \frac{V_a X}{(Q_o - Q_w)X_e + Q_w X_r} \qquad (306.40)$$

Volumetric organic loading rate is given by

$$L_{org} = \frac{S_o Q_o}{V_a} \qquad (306.41)$$

Observed cell yield is given by

$$Y_{obs} = \frac{Y}{1 + k_d \text{SRT}} \qquad (306.42)$$

Heteromorphic biomass is given by

$$P = \left[\frac{Q_o Y (S_o - S_e)}{1 + k_d \text{SRT}} \right] = Q_o Y_{obs} (S_o - S_e) \qquad (306.43)$$

Cell debris produced by decay is given by

$$P = f_d k_d \text{SRT} \left[\frac{Q_o Y (S_o - S_e)}{1 + k_d \text{SRT}} \right] \qquad (306.44)$$

Biomass concentration (MLVSS) is given by

$$X = \frac{\text{SRT}}{V_a} \left[\frac{Q_o Y (S_o - S_e)}{1 + k_d \text{SRT}} \right] \qquad (306.45)$$

Rate of soluble substrate utilization is given by

$$r_{su} = -\frac{kXS_e}{K_s + S_e} = -\frac{\mu_m S_e}{Y(K_s + S_e)} X \qquad (306.46)$$

Growth-limiting substrate concentration is given by

$$S_e = \frac{K_s (1 + k_d \text{SRT})}{\text{SRT}(Yk - k_d) - 1} \qquad (306.47)$$

Net biomass production rate is given by

$$r_g = Y \frac{kXS}{K_s + S} - k_d X \qquad (306.48)$$

Note that, as a result of the preceding definitions, one may also write the following relationship between the sludge production rate (P) and other key variables such as volume of aeration tank (V_a), solids retention time (SRT), and mixed liquor volatile suspended solids (X).

$$P \times \text{SRT} = V_a \times X \qquad (306.49)$$

Bacteria oxidize a portion of the bCOD (bacterial carbonaceous oxygen demand) to provide energy and the rest of the bCOD for cell growth. Since a portion of the biomass is wasted, the oxygen requirement may be written as

$$R_o = Q_o(S_o - S_e) - 1.42 P_{X,\text{bio}} \tag{306.50}$$

Typically, for BOD removal, the oxygen requirement varies from 0.9- to 1.3-kg O_2 per kg BOD removed for SRT between 5 and 20 days, respectively.

Mass Balance for a Closed System Boundary

Performing a mass balance around the aeration tank (boundary 1 in Fig. 306.6) yields

$$X_r Q_r + S_o Q_o = (Q_o + Q_r)X \tag{306.51}$$

Performing a mass balance around the secondary clarifier (boundary 2 in Fig. 306.6) yields

$$(Q_o + Q_r)X = (Q_r + Q_w)X_r + (Q_o - Q_w)X_e \tag{306.52}$$

Assuming the concentration S_e to be negligible, we have the following approximate relationships for the recycle ratio and the waste sludge:

$$R = \frac{Q_r}{Q_o} = \frac{1 - t_d/\text{SRT}}{X_r/X - 1} \tag{306.53}$$

$$Q_w = \frac{V_a}{\text{SRT}} \tag{306.54}$$

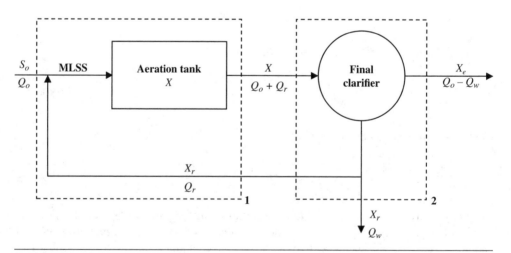

Figure 306.6 Mass balance for closed volumes.

Table 306.8 Permissible Aeration Tank Capacities and Loadings

Process	Aeration tank Organic loading lb BOD_5/day/1000 ft³	F:M ratio lb BOD_5/day/lb MLVSS	MLSS mg/L
Conventional step aeration complete mix	40	0.2–0.5	1,000–3,000
Contact stabilization	50	0.2–0.6	1,000–3,000
Extended aeration single-stage nitrification	15	0.05–0.10	3,000–5,000

According to Eq. (306.53), if sludge recycling is not used in a biological process ($R = 0$), the solids retention time (SRT) and the hydraulic retention time (t_d) are identical. This leads to a very large reactor volume for typical BOD removal goals. The purpose of using recycled sludge is to increase the SRT to be significantly greater than the hydraulic retention time, thus keeping the reactor volume within practical limits.

The following guidelines about aeration tanks are given in the *Recommended Standards for Wastewater Facilities: Policies for the Design, Review, and Approval of Plans and Specifications for Wastewater Collection and Treatment Facilities, 2004* (also known as "Ten States Standards"):

Table 306.8 shows permissible aeration tank capacities and loadings as suggested in the "Ten States Standards."

Aeration equipment shall be capable of maintaining a minimum of 2.0 mg/L of dissolved oxygen in the mixed liquor at all times and provide thorough mixing of the mixed liquor. In the absence of experimentally determined values, the design oxygen requirements for all activated sludge processes shall be 1.1 lb O_2/lb design peak hourly BOD_5 applied to the aeration tanks, with the exception of the extended aeration process, for which the value shall be 1.5 to include endogenous respiration requirements.

Sludge Bulking

When sludge shows poor settling characteristics in the clarifier, it may lead to poor system performance due to high levels of suspended solids in the effluent. With good settling characteristics, the sludge levels are typically 0.5 to 1.0 ft at the bottom of the clarifier. Two primary types of sludge bulking are filamentous bulking and viscous bulking. Filamentous organisms thrive in substrate depleted conditions. This includes low DO, low organic substrate, and low nutrients. Viscous bulking is caused by excessive amounts of extracellular biopolymer, producing sludge with a slimy jellylike consistency. Viscous sludge is usually found in nutrient-limited systems or systems with very high organic load.

Two methods commonly used for quantifying sludge settling characteristics are—the sludge volume index (SVI) and the zone settling velocity (ZSV).

The SVI is defined as the volume of 1 g of sludge after 30 min of settling. It is determined by placing a mixed liquor sample in a 1-L cylinder and measuring the settled volume after 30 min and the corresponding sample MLSS concentration. The SVI is calculated as

$$\text{SVI} = \frac{\text{settled volume (mL/L)}}{\text{MLSS (mg/L)}} \times 1000 \text{ mg/g} = \frac{\text{mL}}{\text{g}} \qquad (306.55)$$

Typically an upper limit of 100 mL/g is considered desirable for good settling characteristics. SVI values above 150 mL/g are typically associated with filamentous bacteria.

The zone settling velocity is typically a negative exponential function of the MLSS. Once the ZSV is calculated, the overflow rate V/A is then established as a fraction of it.

Specific Gravity of Sludge

The specific gravity of solid matter in sludge can be computed using

$$\frac{1}{S} = \frac{P_f}{S_f} + \frac{P_v}{S_v}$$

(306.56)

where S = specific gravity of sludge solids
S_f = specific gravity of fixed (inorganic) solids, usually assumed 2.6 (approx.)
S_v = specific gravity of volatile (organic) solids, usually assumed 1.0
P_f = fraction of fixed solids
P_v = fraction of volatile solids = $1 - P_f$

Example 306.9

A biological sludge contains 15% solids, with a volatile fraction of 60%. What is the specific gravity of the sludge?

Solution The wet sludge may be considered 85% water (SG = 1.0), 6% (40% of 15%) fixed solids (SG = 2.6), and 9% (60% of 15%) volatile solids (SG = 1.0). Therefore, the specific gravity of the sludge is

$$\frac{1}{S} = \frac{0.85}{1.0} + \frac{0.06}{2.6} + \frac{0.09}{1.0} = 0.963 \Rightarrow S = 1.038$$

Completely Mixed Aerated Lagoons

Aerated ponds are commonly used for treatment of industrial wastes or first-stage treatment of municipal wastewater. Complete mixing and adequate aeration are essential requirements. No provisions are made for settling and recycling activated sludge. Aeration is achieved by using either floating or platform-mounted aerators. Aerated lagoons can reliably produce an effluent with both biological oxygen demand (BOD) and TSS < 30 mg/L if provisions for settling are included at the end of the system. BOD removal efficiency depends on temperature, detention time, and nature and strength of the waste. Pond depths range from 6 to 20 ft, with 10 ft being the most typical. Detention times range from 10 to 30 days, with 20 days the most typical (shorter detention times use higher intensity aeration). A typical relationship for the BOD removal achieved by a series of n equal-sized cells is given by

$$\frac{L_e}{L_o} = \frac{1}{\left[1 + \dfrac{K_T t}{n}\right]^n}$$

(306.57)

where L_e = effluent BOD (mg/L)
       L_o = influent BOD (mg/L)
       K_T = BOD removal rate constant (base e, day^{-1})
       t = detention time (days)

The temperature-dependent rate constant K_T is given by

$$K_T = K_{20}\theta^{T-20} \qquad (306.58)$$

where K_T = temperature-dependent rate constant
       K_{20} = rate constant at 20°C (typical value 0.276, base e, day^{-1})
       θ = temperature coefficient (1.036)

An aerated lagoon may be classified as either partial-mix or complete-mix. A partial-mix system provides only enough aeration to satisfy the oxygen requirements of the system and does not provide energy to keep all total suspended solids (TSS) in suspension. A complete-mix system typically uses about ten times the amount of energy needed for an equally sized partial mix system to treat municipal wastes. A complete-mix aerated lagoon is similar to the activated sludge treatment process except that it does not include sludge recycling, which requires a longer hydraulic detention time than activated sludge treatment to achieve the same level of treatment. Partial mix lagoons are also called facultative aerated lagoons.

Advantages

An aerated lagoon requires less land than facultative lagoons. An aerated lagoon can usually discharge throughout the winter while discharge may be prohibited from an ice-covered facultative lagoon in the same climate. Sludge disposal may be necessary but the quantity is relatively small compared to other secondary treatment processes.

Disadvantages

Aerated lagoons are not as effective as facultative ponds in removing ammonia nitrogen or phosphorous, unless designed for nitrification. Aerated lagoons may experience surface ice formation. Reduced rates of biological activity occur during cold weather. Mosquito and similar insect vectors can be a problem if vegetation on the dikes and berms is not properly maintained. Sludge accumulation rates will be higher in cold climates because low temperature inhibits anaerobic reactions. The use of surface or submerged aerators requires energy input.

Typical Performance Parameters

For an aerated lagoon, BOD removal can range up to 95%. Effluent TSS can range from 20 to 60 mg/L, depending on the design of the settling basin and the concentration of algae in the effluent. Removal of ammonia nitrogen in aerated lagoons is usually less effective than in facultative lagoons because of shorter detention times. Phosphorus removal is also less effective than in facultative lagoons because of more stable pH and alkalinity conditions. Phosphorus removal of about 15 to 25% can be expected with aerated lagoons. Removal of coliforms and fecal coliforms can be effective, depending on detention time and temperature. Disinfection may be necessary if effluent limits are less than 200 MPN/100 mL.

Rotating Biological Contactors

Rotating biological contactors (RBC) is a remediation technology used in the secondary treatment of wastewater. This technology involves allowing wastewater to come in contact with a biological medium in order to facilitate the removal of contaminants.

In its simplest form a rotary biological contactor (RBC) consists of a series of closely spaced circular disks mounted on a shaft which is driven so that the disks rotate (1 to 2 rpm) at right angles to the flow of settled sewage. The disks are usually made of plastic (polythene, PVC, expanded polystyrene) and are contained in a trough so that about 40% of their area is immersed. The disks are arranged in groups or packs with baffles between each group to minimize surging or short circuiting. With small units the trough is covered and large units are often housed within buildings. This is to reduce the effect of weather on the active biofilm which becomes attached to the disk surfaces.

The biological growth that becomes attached to the disks assimilates the organic materials in the wastewater. Aeration is provided by the rotating action, which exposes the disks to the air after contacting them with the wastewater. Excess biomass is sheared off in the tank, where the rotating action of the disks maintains the solids in suspension. Eventually, the flow of the wastewater carries these solids out of the system and into a clarifier, where they are separated. By arranging several sets of disks in series, it is possible to achieve a high degree of organic removal and nitrification.

Advantages of Rotating Biological Contactors

Some advantages of the RBC method of wastewater treatment are short contact periods, good settling characteristics of the sloughed biomass, low operating costs, low power requirements and sludge production, and excellent process control.

Disadvantages of Rotating Biological Contactors

Some disadvantages of the RBC method of wastewater treatment are requirement for covering RBC units in northern climates to protect against freezing and frequent maintenance of shaft bearings and mechanically driven units.

Performance of Biological Contactors

The performance of RBC systems depends on the temperature, the concentration of the pollutants, and the rate at which the treatment is expected to proceed. Studies have shown that, in terms of BOD removal, there is a critical hydraulic retention time of 3 h and that any further increase in the retention results in little or no improvement in performances. Also, for most designs, there is usually a critical organic loading rate above which the BOD removal efficiency decreases rapidly. Removal efficiency of RBC units is given by empirical equations. A typical equation is given below:

$$\frac{S_e}{S_o} = \left[1 + \frac{KA}{Q}\right]^{-n}$$

(306.59)

where S_e = effluent (following RBC process) BOD_5 (mg/L)
   S_o = influent (preceding RBC process) BOD_5 (mg/L)
   K = empirical process constant (typical units gpd/ft^2)
   A = RBC disk area (ft^2)
   Q = hydraulic flow rate (gpd)

Trickling Filters

A *trickling filter* is a biological reactor, which uses rock or plastic packing media over which wastewater is distributed continuously, usually with recirculation. A biofilm attaches to the surface of the media and treatment occurs as the water flows over this biofilm. The depth of trickling filters varies between 3 to 8 ft for rock media and 14 to 40 ft for plastic media. For rock media filters, the most common material is rounded river rock, uniformly graded so that 95% is within 3- to 4-in-diameter range. Corrugated plastic packing offers greater surface area per unit volume and therefore plastic media filters require less land area due to the ability to use greater loading rates and taller filters. The wastewater is distributed over the media bed using rotating distributor arms. Facultative bacteria are primarily responsible for water treatment in trickling filters, although aerobic and anaerobic bacteria are also present. Both BOD removal and nitrification are accomplished in a trickling filter with low organic load. With high-strength wastewaters, a two-stage filter system with an intermediate clarifier is most commonly used.

The following guidelines about Trickling Filters are given in the *Recommended Standards for Wastewater Facilities: Policies for the Design, Review, and Approval of Plans and Specifications for Wastewater Collection and Treatment Facilities, 2004* (also known as "Ten States Standards").

Trickling filter media shall have a minimum depth of 6 ft above the underdrains. Rock and/or slag filter media depths shall not exceed 10 ft and manufactured filter media depths should not exceed the recommendations of the manufacturer.

Underdrains for trickling filters shall have a minimum slope of 1%. Effluent channels shall be designed to produce a minimum velocity of 2 ft/s at design average flow rates of application to the filter including recirculated flows.

The piping system shall be designed for recirculation as required to achieve the design efficiency. The recirculation rate shall be variable and subject to plant operator control at the range of 0.5:1 up to 4:1 (ratio of recirculation rate versus design average flow). A minimum of two recirculation pumps shall be provided.

Single-Stage Trickling Filter

Trickling filters can remove up to 90% suspended solids and up to 85% to 90% of BOD. Recirculation of flow through the filter accomplishes long contact times.

$$\text{Hydraulic loading} = \text{total flow rate} \div \text{plan area} = (Q_w + Q_r)/A$$

The BOD loading rate (lb/1000 ft^3-day) is given by

$$L_{BOD} = \frac{Q \times S \times 8345.4}{V} \tag{306.60}$$

where Q = wastewater flow rate (MGD)
$\quad\quad S$ = influent BOD concentration (mg/L)
$\quad\quad V$ = trickling filter volume (ft^3)

NRC (National Research Council) Equation

The NRC equation was developed (Mohlman et al., 1946)[5] for BOD removal by rock media filters based on data from wastewater treatment plants at military installations. The BOD removal efficiency for a single-stage rock filter is given by

$$\eta = \frac{1}{1 + 0.0561\sqrt{\dfrac{L_{BOD}}{F}}} \tag{306.61}$$

$$F = \frac{1+R}{(1+0.1R)^2} \tag{306.62}$$

where L_{BOD} = organic loading (lb/1000 ft^3-day)
$\quad\quad F$ = recirculation factor = effective number of passes of organic material through the filter
$\quad\quad R$ = ratio of the filter discharge returned to the inlet to the raw influent, typically varying between 0 and 2

For a two-stage trickling filter system, the efficiency of BOD removal of the second filter is given by

$$\eta_2 = \frac{1}{1 + \dfrac{0.0561}{1-\eta_1}\sqrt{\dfrac{L_{BOD}}{F}}} \tag{306.63}$$

where $\quad\eta_1$ = BOD removal efficiency of the first stage
$\quad\quad L_{BOD}$ = BOD loading applied to the second-stage filter
$\quad\quad F$ = recirculation factor for the second stage

Trickling filters and rotating biological contactors (RBC) have been used as aerobic attached growth processes for BOD removal only, combined BOD removal and nitrification, and tertiary nitrification following secondary treatment.

Example 306.10

A municipal wastewater flow rate Q = 2 MGD and a BOD$_5$ = 265 mg/L is treated with a two-stage rock media trickling filter. The first-stage filter has diameter = 72 ft and depth = 6 ft. The second-stage filter has diameter = 120 ft and depth = 6 ft. Calculate the BOD$_5$ in the effluent. Assume that the recirculation rate R = 2 for both filters

---

[5]Mohlman, F. W. et al. (1946), "Sewage Treatment at Military Installations," National Research Council, Subcommittee Report, *Sewage Works Journal*, Vol. 18, No. 5, 787–1028.

For the first filter, volume $V = \dfrac{\pi}{4} \times 72^2 \times 6 = 24{,}429 \text{ ft}^3$

BOD load $= 265 \times 2 \times 8.3454 = 4423 \text{ lb/day}$

BOD loading rate $L_{BOD} = 4423 / 24.429 = 181.1 \text{ lb/1000 ft}^3\text{-day}$

Recirculation factor $F = \dfrac{1+2}{(1+0.1\times 2)^2} = 2.08$

Efficiency of the first-stage $\eta_1 = \dfrac{1}{1 + 0.0561\sqrt{\dfrac{181.1}{2.08}}} = 0.656$

For the second filter, volume $V = \dfrac{\pi}{4} \times 120^2 \times 6 = 67{,}858 \text{ ft}^3$

BOD_5 transmitted to the second stage $= (1 - 0.656) \times 265 = 91.05 \text{ mg/L}$

This leads to a BOD load $= 1521.5 \text{ lb/day}$ and a BOD loading rate

$L_{BOD} = 1521.5/67.858 = 22.4 \text{ lb/1000 ft}^3\text{-day}$

Efficiency of the second-stage $\eta_2 = \dfrac{1}{1 + \dfrac{0.0561}{1 - 0.656}\sqrt{\dfrac{22.4}{2.08}}} = 0.651$

BOD_5 transmitted to the second-stage effluent $= (1 - 0.651) \times 91.05 = 31.8 \text{ mg/L}$

Overall efficiency $= \dfrac{265 - 31.8}{265} \times 100\% = 88\%$

Formulations for Plastic Packing

According to Schulze (1960),[6] the BOD removal for BOD removal in a trickling filter with plastic media is given by

$$\frac{S_e}{S_o} = \exp\left(-\frac{kD}{Q^n}\right) \tag{306.64}$$

where S_e = BOD in the filter effluent (mg/L)
$\quad\quad\; S_o$ = BOD in the influent wastewater (mg/L)
$\quad\quad\; k$ = packing coefficient
$\quad\quad\; D$ = packing depth (ft)

---

[6]Schulze, K. L. (1960), "Load and Efficiency in Trickling Filters," *Journal Water Pollution Control Federation*, Vol. 33, No. 3, 245–260.

Q = hydraulic application rate excluding recirculation (gpm/ft^2)

n = constant characteristic of the packing material

The maximum recommended application rate is 0.75 gpm/ft^2. For trickling filters, the commonly used correction for temperature is given as

$$k_T = k_{20} 1.035^{T-20} \qquad (306.65)$$

Nutrient Removal from Wastewater Streams

Nutrients such as nitrogen and phosphorus need to be removed beyond what can be accomplished by conventional secondary treatment processes to limit eutrophication. Eutrophication is the process by which a body of water acquires a high concentration of nutrients, especially phosphates and nitrates, promoting excessive growth of algae. As the algae die and decompose, high levels of organic matter and the decomposing organisms deplete the water of available oxygen, causing the death of other organisms, such as fish.

Phosphorus Removal Technologies

Treatment technologies available for phosphorus removal can fall into three categories:

Physical: (1) filtration for particulate phosphorus (2) membrane technologies

Chemical: (1) precipitation (2) other techniques such as physical-chemical adsorption

Biological: (1) assimilation (2) enhanced biological phosphorus removal (EBPR)

The most recent progress has been made in EBPR, which has the potential to remove phosphorus down to very low levels at relatively low costs. The question of sludge handling and treatment of phosphorus in side streams is also being addressed.

Physical Treatment

Filtration for particulate phosphorus Sand filtration or other methods of TSS removal (e.g., membrane, chemical precipitation) are necessary for plants with low effluent total phosphorus permits.

Membrane technologies

Membrane technologies have been of growing interest for phosphorus removal. In addition to removing the phosphorus in the TSS, membranes can also remove dissolved phosphorus. Membrane bioreactors (MBRs, which incorporate membrane technology in a suspended growth secondary treatment process), tertiary membrane filtration (after secondary treatment), and reverse osmosis (RO) systems have all been used in full-scale plants with good results. Current reliable limits of technology are 0.04 mg/L for MBRs and tertiary membrane filtration, and 0.008 mg/L for reverse osmosis.

Chemical Treatment

Precipitation   Chemical precipitation has long been used for phosphorus removal. The chemicals most often employed are compounds of calcium, aluminum, and iron. Chemical addition can occur at various locations—prior to primary settling, during secondary treatment, or as part of a tertiary treatment process. Full-scale systems may perform better (0.005–0.04 mg/L) than the 0.05 mg/L limit predicted in laboratory tests.

One disadvantage of using chemical precipitation for phosphorus removal is the production of additional sludge, especially if the method selected is lime application during primary treatment. Use of alum after secondary treatment tends to produce much less sludge.

Biological Treatment

Assimilation   Biological assimilation is the incorporation of the phosphorus as an essential element in biomass, through growth of photosynthetic organisms such as plants and algae. Assimilation occurs in treatment ponds containing planktonic or attached algae, rooted and floating plants. During the growing season, land application of effluent has also been used. In more recent years, using constructed wetlands as a site for bioassimilation is becoming common practice. In all of these cases, however, it is necessary to remove the net biomass growth in order to prevent eventual decay of the biomass and rerelease of the phosphorus.

EBPR   Enhanced biological phosphorus removal has the potential to achieve very low (< 0.1 mg/L) effluent phosphorus levels at low cost and without excessive sludge production. Removal of BOD, nitrogen, and phosphorus can all be achieved in a single system, although it can be challenging to achieve very low concentrations of both total nitrogen and phosphorus in such systems. The phosphate in EBPR is removed in the waste activated sludge, which might have 5% or more phosphorus as opposed to only 2% to 3% in non-EBPR sludge. Simultaneous biological nutrient removal (SBNR) has also been observed in oxidation ditches not specifically designed for nutrient removal.

Cold weather can provide a challenge for many biological treatment processes.

The following guidelines about Phosphorus Removal are given in the *Recommended Standards for Wastewater Facilities: Policies for the Design, Review, and Approval of Plans and Specifications for Wastewater Collection and Treatment Facilities, 2004* (also known as "Ten States Standards"):

Addition of aluminum salts, iron salts, or lime may be used for the chemical removal of soluble phosphorus. The phosphorus reacts with the aluminum, iron, or calcium ions to form insoluble compounds. Those insoluble compounds may be flocculated with or without the addition of a coagulant aid such as a polyelectrolyte to facilitate separation by sedimentation, or sedimentation followed by filtration.

Dosage   The design chemical dosage shall include the amount needed to react with the phosphorus in the wastewater, the amount required to drive the chemical reaction to the desired state of completion, and the amount required due to inefficiencies in mixing or dispersion. Excessive chemical dosage should be avoided.

When lime is used, it may be necessary to neutralize the high pH prior to subsequent treatment in secondary biological systems or prior to discharge in those flow schemes where lime treatment is the final step in the treatment process. Problems associated with lime usage, handling, and sludge production and dewatering shall be evaluated.

Each chemical must be mixed rapidly and uniformly with the flow stream. Where separate mixing basins are provided, they should be equipped with mechanical mixing devices. The detention period should be at least 30 s.

The particle size of the precipitate formed by chemical treatment may be very small. Consideration should be given in the process design to the addition of synthetic polyelec- trolytes to aid settling. The flocculation equipment should be adjustable in order to obtain optimum floc growth, control deposition of solids, and prevent floc destruction.

Effluent filtration shall be considered where effluent phosphorus concentrations of less than 1 mg/L must be achieved.

Each dry chemical feeder shall be equipped with a dissolver which is capable of provid- ing a minimum 5-min retention at the maximum feed rate. Polyelectrolyte feed installations should be equipped with two solution vessels and transfer piping for solution make-up and daily operation.

All chemical feed equipment and storage facilities shall be constructed of materials resis- tant to chemical attack by all chemicals normally used for phosphorus removal.

Precautions shall be taken to prevent chemical storage tanks and feed lines from reach- ing temperatures likely to result in freezing or chemical crystallization at the concentrations employed. A heated enclosure or insulation may be required. Consideration shall be given to temperature, humidity, and dust control in all chemical feed room areas.

Nitrogen Removal Technologies

Nitrogen is present in wastewater in various forms—organic nitrogen (both soluble and particulate), ammonia/ammonium, and nitrate. All of the biological nitrogen removal processes (such as anoxic/aerobic process, intermittent aeration, postanoxic denitrification with methanol addition) include an aerobic zone in which biological nitrification occurs. Some anoxic volume must also be included in the process design to allow for biologi- cal denitrification. This can help achieve the objective of total nitrogen removal by both NH_4-N oxidation and NO_2-N to nitrogen gas.

Suspended growth biological nitrogen removal processes can be divided into single-sludge and two-sludge systems. A single-sludge system, which is more common, has only one solids separation device (normally a secondary clarifier). The activated sludge tank may be divided into different zones of aerobic and anoxic conditions and the mixed liquor may be pumped from one zone to another (internal recycle). In the two-sludge system, an aerobic process (for nitrification) is followed by an anoxic process (for denitrification), each with its own clarifier, thus producing two sludges.

Biological Nitrification and Denitrification

Biological Nitrification

Nitrification is the process in which ammonia nitrogen (NH_4-N) is oxidized to nitrite (NO_2-N) and then to nitrate (NO_3-N). The EPA primary drinking water maximum contaminant level (MCL) for nitrate is 10 mg/L as nitrogen (or 45 mg/L as nitrate). Excessive nitrogen levels can cause eutrophication (an increase in chemical nutrients such as nitrogen or phosphorus in an ecosystem, leading to excessive plant growth and decay, lack of oxygen, and severe reductions in water quality and in fish and other animal populations). Excessive ammonia levels are also toxic to fish populations.

Nitrification processes can be suspended growth or attached growth. Bacteria responsible for nitrification grow much more slowly than heterotrophic bacteria. Therefore, systems designed for nitrification typically have much greater hydraulic and solids residence time than those for systems designed for only BOD removal.

In attached growth systems designed for nitrification, most of the BOD must be removed before nitrifying organisms can be established as a competitive population. In the first stage of nitrification, autotrophic bacteria (*Nitrosomonas, Nitrosococcus, Nitrosospira, Nitrosolobus,* and the like) convert (oxidize) ammonia to nitrite. In the second stage, another group of autotrophic bacteria (*Nitrobacter, Nitrococcus, Nitrospira, Nitrospina,* and the like) oxidize nitrite to nitrate. The two-step oxidation can be summarized as

Nitrosobacteria: $$2NH_4^+ + 3O_2 \rightarrow 2NO_2^- + 4H^+ + 2H_2O$$

Nitrobacteria: $$2NO_2^- + O_2 \rightarrow 2NO_3^-$$

Which may be combined as $$NH_4^+ + 2O_2 \rightarrow NO_3^- + 2H^+ + H_2O$$

Factors Affecting Nitrification

Nitrifying bacteria need carbon dioxide and phosphorus for cell growth. Trace element concentrations that have been observed to stimulate nitrifying bacteria are Ca (0.5 mg/L), Cu (0.01 mg/L), Mg (0.03 mg/L), Mo (0.001 mg/L), Ni (0.1 mg/L), and Zn (1.0 mg/L). However, higher concentrations of certain metal ions can cause remarkable decrease in nitrification rates. Low levels of dissolved oxygen (less than 05 mg/L) and pH below 6.8 can cause inhibition of the nitrification process.

Biological Denitrification

Denitrification is the biological process where nitrate (NO_3-N) is reduced to nitric oxide, nitrous oxide, and nitrogen gas. This is a significant part of the process of biological nitrogen removal. There are two modes in which nitrate reduction takes place, termed *assimilating* and *dissimilating* nitrate reduction.

Assimilating nitrate reduction occurs when NH_4-N is not available and is independent of dissolved oxygen concentration.

Dissimilating nitrate reduction or biological denitrification occurs when nitrate or nitrite ions serve as the electron acceptor for the purpose of oxidation of a variety of inorganic and organic electron donors. This process is coupled to the respiratory process of microorganisms.

The nitrate reduction process progresses from nitrate (NO_3^-) to nitrite (NO_2^-) to nitric oxide (NO) to nitrous oxide (N_2O) to nitrogen gas (N_2).

Capacity Analysis

Transportation Planning

Planning plays a major role in large transportation projects. Lack of proper planning can lead to severe traffic congestion, dangerous travel patterns, undesirable land use patterns, adverse environmental impact, and waste of resources. Large transportation projects require a long lead time for their design and construction. The basic steps in the transportation planning process are the following:

- Problem definition
- Define goals, objectives, and criteria
- Data collection
- Forecasts (modeling)
- Develop alternatives
- Evaluation
- Implementation plan

A major part of transportation planning is the use of empirical models developed from economic data. The term "model" is used to refer to a series of mathematical equations that are used to represent how choices are made when people travel. The coefficients and parameters in the model are set (calibrated) to match existing data. Usually, these relationships are assumed to be valid and to remain constant in the future.

Models are created based on population forecasts, economic forecasts, and land use patterns. They are then used to estimate the number of trips that will be made on a network (trip generation), how these trips will originate and terminate (trip distribution), how these trips will distribute across various modes of travel (modal split), and what routes will be available or need to be made available for these forecast trips (route assignment). These estimates are the basis for transportation plans and are used in major investment analysis, environmental impact statements, and in setting priorities for investments.

For purposes of simulations, an urban area is represented as a series of small geographic areas (zones). Zones are characterized by their population, employment, and other factors and are the places where trips begin (trip producers) or end (trip attractors). Trip making is first estimated at

807

the household level and then aggregated to the zone level. Zones can be as small as a single block but typically are one-fourth to 1 mi^2 in area.

Trip Generation

The first step in the traffic forecasting process—trip generation—predicts the number of trips originating from or destined for a particular traffic analysis zone. Typically, trip generation analysis focuses on residential communities, based on an economic model which is a function of the social and economic attributes of households. Residential trip generation analysis is often undertaken using statistical regression using variables such as household size, number of workers in the household, persons in an age group, type of residence (single family, apartment, etc.), auto-ownership per household, and so on. Usually, measures on five to seven independent variables are available.

There are two kinds of trip generation models: production models and attraction models. Trip production models estimate the number of home-based trips created from zones where trip makers reside. Information from land use, population, and economic forecasts are used to estimate how many person trips will be made to and from each zone. Traffic analysis zones are also destinations of trips or trip attractors. The analysis of attractors focuses on nonresidential land uses. Trip attractions are typically based on the level of employment in a zone. Trip attraction models estimate the number of home-based trips to each zone at the nonhome end of the trip. Different production and attraction models are used for each trip purpose.

Gravity Model for Trip Distribution

Once the trip generation model predicts production (P_i) and attraction (A_j) values for each zone within the analysis area, these could be out of balance. In other words, the total number of trips produced within the area may not be equal to the total number of trips attracted. An iterative method must then be used to balance the system.

The most commonly used procedure for trip distribution is the "gravity model." The gravity model takes the trips produced at one zone and distributes then to other zones based on the size of the other zones (as measured by their trip attractions) and on the basis of the distance to other zones. According to the gravity model, the number of trips between two zones is inversely proportional to the square of the time that it takes to travel between those zones.

$$T_{ij} = P_i \left[\frac{A_j F_{ij} K_{ij}}{\sum_k A_k F_{ik} K_{ik}} \right] \tag{401.1}$$

where T_{ij} = number of trips produced in zone i and attracted to zone j
P_i = total number of trips produced in zone i (produced by trip generation model)
A_j = total number of trips attracted to zone j (produced by trip generation model)
F_{ij} = value representing impedance (usually the inverse of travel time) between zones i and j

K_{ij} = socioeconomic adjustment factor for route between i and j (these factors are used to adjust model predictions when there is a significant difference between predicted and observed values due to unique characteristics of certain zones and/or routes).

The sum of all the trip productions (P_i) must equal the sum of all the trip attractions (A_j).

Example 401.1

A study area consists of three zones—A, B, and C. The trip generation model predicts the following trip productions and attractions:

Zone	A	B	C	Total
Trip production	170	400	330	900
Trip attraction	350	325	225	900

Travel time (minutes) between zones is given in the matrix below. For this example, the F factor is taken as the inverse of the corresponding travel time. Also, for this example, all K_{ij} values are taken as 1.0.

Zone	A	B	C
A	7	4	5
B	4	5	8
C	5	8	3

The first iteration of trip distribution yields the following results:

Zone	A	B	C	Production (P)
A	48	78	43	170
B	194	144	62	400
C	124	72	133	330
Computed A	366	295	239	
Given A	350	325	225	

For instance, the distribution of number of trips (170) produced in zone A is calculated as

$$T_{AA} = P_A \frac{A_A/t_{AA}}{\sum_j A_j/t_{Aj}} = 170\left(\frac{350/7}{350/7 + 325/4 + 225/5}\right) = 48$$

$$T_{AB} = P_A \frac{A_B/t_{AB}}{\sum_j A_j/t_{Aj}} = 170\left(\frac{325/4}{350/7 + 325/4 + 225/5}\right) = 78$$

$$T_{AC} = P_A \frac{A_C/t_{AC}}{\sum_j A_j/t_{Aj}} = 170\left(\frac{225/5}{350/7 + 325/4 + 225/5}\right) = 43$$

As may be seen, this produces an imbalance in the attractions for each zone. The computed attractions to zones A, B, and C are 366, 295, and 239, respectively—which are different from the initial values of 350, 325, and 225. The next step will be to apply a correction to the attraction values and reiterate through the distribution process. The process is continued until the values converge.

Modal Split by the Logit Model

Mode choice is one of the most critical parts of the travel demand modeling process. It is the step where trips between a given origin and destination are split into trips using transit, trips by car pool or as automobile passengers, and trips by automobile drivers. The most commonly used process for mode split is to use the "Logit" model. This involves a comparison of the "disutility" of travel between two points for the different modes that are available.

Disutility is a term used to represent a combination of the travel time, cost, and convenience of a mode between an origin and a destination. Once disutilities are known for the various mode choices between an origin and a destination, the trips are split among various modes based on the relative differences between disutilities. The logit equation is used in this step. A large advantage in disutility will mean a high percentage for that mode. Traffic assignment is typically done for peak hour travel, while forecasts of trips are done on a daily basis. A ratio of peak hour travel to daily travel is needed to convert daily trips to peak hour travel.

The probability of choosing a particular mode (p) out of all available modes (n) is then given by

$$P(p) = \frac{e^{U_P}}{\sum_1^n e^{U_i}} \tag{401.2}$$

where $U(i)$ is the value of the utility function for mode i.

Automobile trips must be converted from person trips to vehicle trips with an auto-occupancy model. Mode split and auto-occupancy analysis can be two separate steps or can be combined into a single step, depending on how a forecasting process is set up.

Once trips have been split into highway and transit trips, the specific path that they use to travel from their origin to their destination must be found. These trips are then assigned to that path in the step called traffic assignment.

Example 401.2

Citizens of a small community have three choices for commuting to the employment zone defined as "downtown," which is a distance of 17 miles away. The choices are (1) driving, (2) 5 commuter rail, and (3) a suburban bus line. The utility values for the three modes, based on a factored sum of attributes such as time, cost, etc., are -1.23 for driving, -2.11 for commuter rail, and -1.89 for the bus line. If the total number of commuters originating from the community is 560, how many are expected to take the bus?

Solution   The probability of selecting mode C (bus line) can be written as

$$P(C) = \frac{e^{U_C}}{e^{U_A} + e^{U_B} + e^{U_C}} = \frac{e^{-1.89}}{e^{-1.23} + e^{-2.11} + e^{-1.89}} = \frac{0.151}{0.292 + 0.121 + 0.151} = 0.268$$

Number of commuters expected to use the bus line $= 0.268 \times 560 = 149.9$.

Travel Speed

Average running speed is based on the observation of vehicle travel times traversing a section of highway of known length. It is the length of the segment divided by the average running time of vehicles to traverse the segment. Running time includes only time that vehicles are in motion.

Average travel speed is based on travel time observed on a known length of highway. It is the length of the segment divided by the average travel time of vehicles traversing the segment, including all stopped delay times. It is also called space mean speed.

Time mean speed is the arithmetic average of speeds of vehicles observed passing a point on a highway, also referred to as the average spot speed.

Space mean speed is always less than time mean speed, but the difference decreases as the absolute value of speed increases. According to Drake et al.,[1] the space mean speed S_R and the time mean speed S_T are empirically related according to

$$S_R = 1.026 S_T - 1.890 \qquad (401.3)$$

where S_R = space mean speed (mph)
S_T = time mean speed (mph)

Free-flow speed is the average speed of vehicles on a given facility, measured under low-volume conditions, when drivers tend to drive at their desired speed and are not constrained by control delay.

Operating speed is the speed at which drivers are observed operating their vehicles under free-flow conditions. The 85th percentile of the distribution of observed speeds is most frequently used as the operating speed.

Design speed is defined as "the maximum safe speed that can be maintained over a specified section of highway when conditions are favorable such that the design features of the highway govern" (*AASHTO Green Book*).[2] The design speed assumed for a highway should be based on functional classification, topography, anticipated operating speeds, and adjacent land use. Design speeds range from 20 to 70 mph, usually in 10 mph increments. Usually, freeways should be designed for 60 to 70 mph whereas design speeds for other arterial highways range from 30 to 60 mph. A common rule of thumb for the design speed is to add 10 to 15 mph (15 to 25 kmph) to the posted speed.

Design Traffic Volume

Design hourly volume (DHV) is the 30th highest hourly volume of the year, based on hourly vehicular counts taken throughout a full year.

---

[1]Drake, J. S., Schofer, J. L., and May, A. D. (1967), "A Statistical Analysis of Speed Density Hypotheses." *Highway Research Record 154*, Highway Research Board, NRC, Washington, D.C.:53–87.

[2]*AASHTO Green Book—A Policy on Geometric Design of Highways and Streets*, 7th ed., American Association of State and Highway Transportation Officials, 2018.

Average annual daily traffic (AADT) is the average 24-h traffic volume at a given location over a full 365-day year.

Average annual weekday traffic (AAWT) is the average 24-h traffic volume occurring on weekdays over a full 365-day year.

Average daily traffic (ADT) is an average 24-h volume at a given location for some period of time less than a year, but more than 1 day.

Average weekday traffic (AWT) is an average 24-h traffic volume occurring on weekdays for some period less than 1 year.

K is the proportion of AADT on a roadway segment or link that occurs during the design hour. Thus the K factor is given by Eq. (401.4)

$$DHV = K \times AADT \qquad (401.4)$$

D is the proportion of DHV occurring in the heavier direction, and is called the directional split.

The directional design hourly volume, denoted by DDHV, is given by Eq. (401.5)

$$DDHV = D \times DHV = D \times K \times AADT \qquad (401.5)$$

Some typical range of values of K and D factors are listed below.

Facility type	K factor	D factor
Rural	0.15–0.25	0.65–0.80
Suburban	0.12–0.15	0.55–0.65
Urban		
Radial Route	0.07–0.12	0.55–0.60
Circumferential Route	0.07–0.12	0.50–0.55

Expansion Factors

To make reasonable estimates of annual traffic volume characteristics on an area-wide basis, different types of periodic counts, with count durations ranging from 15 min to continuous, are conducted.

Periodic volume counts are used to calculate expansion factors needed to estimate the annual traffic volume.

Hourly expansion factor (HEF) is defined as

$$HEF = \frac{\text{Total Volume for 24 h period}}{\text{Volume for particular h}} \qquad (401.6)$$

So, if hourly volume counts are available for a continuous 24-h day, the HEF can be computed for each hour. This factor can be used to extrapolate the hourly count to a daily count.

Similarly, a daily expansion factor (DEF) is defined as

$$DEF = \frac{\text{Average Total Volume for a week}}{\text{Volume for particular day}} \qquad (401.7)$$

Similarly, a monthly expansion factor (MEF) is defined as

$$\text{MEF} = \frac{\text{AADT}}{\text{Volume (ADT) for particular month}} \qquad (401.8)$$

For example, if the following hourly data is given for a rural road:

Hour	Volume	HEF	Hour	Volume	HEF
6–7 am	312	41.13	6–7 pm	782	16.41
7–8 am	456	28.14	7–8 pm	702	18.28
8–9 am	554	23.16	8–9 pm	603	21.28
9–10 am	678	18.93	9–10 pm	511	25.11
10–11 am	723	17.75	10–11 pm	432	29.70
11–12 pm	682	18.82	11–12 pm	421	30.48
12–1 pm	654	19.62	12–1 am	249	51.53
1–2 pm	765	16.77	1–2 am	189	67.89
2–3 pm	843	15.22	2–3 am	156	82.26
3–4 pm	892	14.39	3–4 am	102	125.80
4–5 pm	932	13.77	4–5 am	98	130.94
5–6 pm	908	14.13	5–6 am	188	68.26

The 24-h volume = 12,832. The HEF values are generated as follows: for 6–7 am, HEF = 12,832 ÷ 312 = 41.13. This means that the count during the 6–7 am hour should be multiplied by 41.13 to extrapolate to a 24-h period.

Weekly traffic counts are summarized in the following table:

Day of the week	Daily count	DEF
Sunday	7900	12.80
Monday	15,600	6.48
Tuesday	17,120	5.91
Wednesday	17,340	5.83
Thursday	16,890	5.99
Friday	14,250	7.09
Saturday	12,000	8.43

Weekly volume = 101,100. The DEF values are generated as follows: for Sunday, DEF = 101,100 ÷ 7900 = 12.80. This means that the count taken on Sunday should be multiplied by 12.80 to extrapolate to a weekly count.

Further, if monthly ADT values are recorded as shown in the table below,

Month	ADT	MEF
January	14,200	1.73
February	12,500	1.97
March	15,400	1.60
April	16,650	1.48
May	18,100	1.36
June	25,460	0.96
July	43,200	0.57
August	48,900	0.50
September	39,200	0.63
October	23,350	1.05
November	21,300	1.15
December	16,500	1.49

Mean average daily volume (AADT) = 24,563. The MEF values are generated as follows: for January, MEF = 24,563 ÷ 14,200 = 1.73. This means that the ADT based on January data should be multiplied by 1.73 to extrapolate to AADT.

Example 401.3

The following traffic counts were observed on a Thursday in April on a rural primary road. If the data in the previous three tables (showing hourly, daily, and monthly traffic counts) is from a roadway of similar classification, what is the AADT of the road?

Hour	Volume
8–9 am	294
9–10 am	426
10–11 am	560
11–12 pm	657
10–11 am	722

Using the HEF factors calculated previously, the 24-h volume for Thursday is given by

$$\frac{294 \times 23.16 + 426 \times 18.93 + 560 \times 17.75 + 657 \times 18.82 + 722 \times 19.62}{5} = 10,269$$

Since the data was collected on a Thursday (DEF = 5.99), the average (daily) volume for the week is

$$\frac{10,269 \times 5.99}{7} = 8787 \text{ vpd}$$

Since the data was collected in the month of April (MEF = 1.48), the AADT is given by

$$\text{AADT} = 8787 \times 1.48 = 13{,}005 \text{ vpd}$$

Speed-Volume-Density Relationships

Many traffic flow models are based on relationships between the critical mobility variables—speed, volume, and density. Fundamental definitions of some of these parameters are given below.

The flow rate is the equivalent hourly rate at which vehicles pass a point on a highway during a time period less than 1 h. The flow rate q (vph) is equal to the product of speed S (mph) and density D (veh/mile)

$$q = SD \tag{401.9}$$

The average spacing d_a (ft/veh) between corresponding points on successive vehicles is given by

$$d_a = \frac{5280}{D} \tag{401.10}$$

The average headway h_a (s/veh) between successive vehicles is given by

$$h_a = \frac{3600}{q} \tag{401.11}$$

The saturation headway—the average time gap between successive vehicles discharging at capacity of a roadway—is a concept often used for the design of signal timing at an intersection.

Example 401.4

The average vehicular speed on a highway lane is 45 mph. If the peak flow rate in this lane is 1200 vph, what is most nearly the density of vehicles (veh/mile) on this stretch of highway? Also, what is the average spacing between vehicles (ft)?

Solution

Density $D = \dfrac{q}{S} = \dfrac{1200}{45} = 26.7$ veh/mi

Average spacing $d_a = \dfrac{5280}{D} = \dfrac{5280}{26.7} = 198$ ft

Example 401.5

The two-way hourly volume on a bicycle path is 242 bicycles per hour. The peak hour factor is 0.85. The typical directional split is 60:40. If the average speed is 14 mph, what is the maximum density of bicycle in the design direction? What is the corresponding minimum spacing between bicycles?

Solution The bidirectional flow rate is given by

$$v_p = \frac{V}{PHF} = \frac{242}{0.85} = 285$$

The directional flow rate $= 0.6 \times 285 = 171$ bicycles per hour
If the average speed $S = 14$ mph, the corresponding density is

$$D = \frac{v_p}{S} = \frac{171}{14} = 12.2 \text{ bicycles per mile}$$

This translates into an average spacing $= 5280 \div 12.2 = 433$ ft.

Time-Mean Speed versus Space-Mean Speed

There are two types of mean speed—time-mean speed and space-mean speed. The time-mean speed is the arithmetic mean of the speeds of vehicles passing a specific point on a highway during an interval of time. Thus, the time-mean speed of n vehicles is given by

$$\bar{u}_t = \frac{1}{n}\sum_{i=1}^{n} u_i \tag{401.12}$$

where $u_i =$ instantaneous speed of the ith vehicle.

On the other hand, the space mean speed is the harmonic mean of the speeds of vehicles passing through a stretch of the highway during an interval of time.

$$\bar{u}_s = \frac{n}{\displaystyle\sum_{i=1}^{n} \frac{1}{u_i}} \tag{401.13}$$

Example 401.6

The speeds of five cars are recorded using a radar gun as follows: 48, 34, 42, 39, and 45 mph. What is the (a) time mean speed and (b) space mean speed?

Solution The time mean speed is given by Eq. (401.12)

$$\bar{u}_t = \frac{1}{n}\sum_{i=1}^{n} u_i = \frac{48 + 34 + 42 + 39 + 45}{5} = 41.6 \text{ mph}$$

The space mean speed is given by Eq. (401.13)

$$\bar{u}_s = \frac{n}{\displaystyle\sum_{i=1}^{n} \frac{1}{u_i}} = \frac{5}{\dfrac{1}{48} + \dfrac{1}{34} + \dfrac{1}{42} + \dfrac{1}{39} + \dfrac{1}{45}} = 41.0 \text{ mph}$$

For comparison, Eq. (401.3) would have given space mean speed $= 40.8$ mph.

Speed-Volume-Density Models

There have been several analytical models proposed for traffic flow. Two of these models—the first due to Greenshields and the second due to Greenberg are detailed below:

Greenshields Linear Model

Greenshields' model is appropriate for both light and dense traffic situations. In this model, it is hypothesized that a linear relationship exists between space mean speed S and density D. This is expressed as

$$S = S_f\left(1 - \frac{D}{D_j}\right) \tag{401.14}$$

where S = average travel speed (space mean)
   S_f = free flow speed
   D = actual density
   D_j = jam density, the density that (theoretically) causes speed and flow rate to become zero

According to this model, the travel speed varies linearly from the free flow speed S_f at zero density to zero at the jam density D_j.
   Flow rate q is given by the product of the space mean speed and density.

$$q = SD \tag{401.15}$$

Speed S and flow rate q are related by

$$S^2 = S_f S - \frac{S_f q}{D_j} \tag{401.16}$$

Flow rate q and density D are related by

$$q = S_f D - \frac{S_f}{D_j}D^2 \tag{401.17}$$

In the Greenshields model, maximum flow occurs when the speed S is exactly half the free-flow speed S_f and when the density D is exactly half the jam density D_j.
   D_o is the critical or optimum density, which corresponds to the critical speed. This maximum flow rate (capacity) is given by

$$q_{max} = \frac{D_j S_f}{4} \tag{401.18}$$

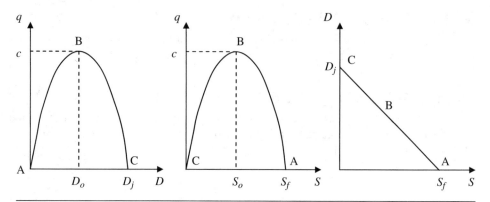

Figure 401.1 Speed-volume-density relationships. (Greenshields.)

Figure 401.1 shows these relationships graphically. The corresponding points on the graphs are marked as A, B, and C.

At point A, density $D \to 0$, speed $S \to$ free-flow speed S_f, and flow rate $q \to 0$

At point C, density $D \to$ jam density D_j, speed $S \to 0$, and flow rate $q \to 0$

At point B, density $D =$ optimum density D_o, speed $S =$ optimum speed S_o, and flow rate $q =$ maximum flow rate q_{max}

Greenberg Model

Using a fluid flow analogy for traffic flow, Greenberg proposed the following model:

$$S = C \ln\left(\frac{D_j}{D}\right) \tag{401.19}$$

where $C =$ speed at maximum flow rate.

Because of inconsistencies with boundary conditions, this model is appropriate only for dense traffic conditions. Some other results from this model are

$$\text{Flow rate } q = CD \ln\left(\frac{D_j}{D}\right) \tag{401.20}$$

The maximum flow rate is given by

$$q_{max} = \frac{CD_j}{e} \tag{401.21}$$

where $e = 2.8182818 \ldots$

Other Models

Two other analytical models for traffic flow are by Underwood and Drake. The development of these models can be carried out in a manner similar to the Greenshields and the Greenberg models. In each case, for maximum flow, the derivative of flow rate q with respect to density is equated to zero.

$$S = C \exp\left(\frac{D_j}{D}\right) \qquad \text{Underwood} \qquad (401.22)$$

$$S = U_f \exp\left(\frac{1}{2}\left(\frac{D_j}{D}\right)^2\right) \qquad \text{Drake} \qquad (401.23)$$

Shockwave in Traffic Stream

In highways with relatively large flow and density, a bottleneck is a location where a sudden reduction in capacity and a sudden increase in density occur. The bottleneck location then moves *upstream* as traffic continues to approach the bottleneck. This migration of the point of discontinuity is referred to as a shock wave.

The shock wave velocity is given by

$$v = \frac{q_2 - q_1}{D_2 - D_1} \qquad (401.24)$$

where q_1 = flow rate immediately upstream of the discontinuity
  q_2 = flow rate immediately downstream of the discontinuity ($q_2 > q_1$)
  D_1 = density immediately upstream of the discontinuity
  D_2 = density immediately downstream of the discontinuity ($D_2 < D_1$)

As the shock wave progresses upstream, there are two possible scenarios—if the location of the discontinuity is static (such as due to an accident), the impacted zone spreads with the velocity of the shock wave. On the other hand, if the discontinuity moves at a constant velocity (such as a slow-moving vehicle in a single lane with no opportunity for passing), the impacted zone spreads with the relative velocity of the shock wave with respect to the speed of the discontinuity. For an application of such a case, see Example 401.7.

Example 401.7

The volume at a section of a two-lane highway is 1800 vph in each direction and the density is approximately 30 veh/mi. A slow-moving truck joins the traffic stream and travels for a length of 3 mi at 25 mph. There is no option for following vehicles to pass the truck due to large opposing traffic volume. The platoon forming behind the truck has a density of 100 veh/mi and a flow rate of 1100 vph. How many vehicles are in the platoon before the truck leaves the highway?

Solution   The velocity of the shock wave is

$$V = \frac{q_2 - q_1}{D_2 - D_1} = \frac{1800 - 1100}{30 - 100} = -10 \text{ mph}$$

Since the truck moves forward (downstream) with speed = 25 mph and the shockwave moves backward (upstream) with speed = 10 mph, the relative velocity between the truck and the wave = 35 mph

Time spent by the truck on the highway = 3 mi ÷ 25 mph = 0.12 h

Length of the platoon forming behind the truck = 0.12 × 35 = 4.2 mi

Using the density behind the truck, the number of vehicles in the platoon = 4.2 mi × 100 veh/mi = 420 vehicles

Discharge of a Queue on a Highway

When an incident such as an accident or lane blockage affects traffic flow, the resulting flow patterns can be extremely complex, with a variety of merging and weaving patterns. An oversimplified analysis may consider, whereas the flow upstream of the incident location (the arriving flow) is at the normal rate of flow on the highway, the flow downstream (the discharging flow) occurs at capacity. The incident may be perceived to affect traffic flow as long as it takes for these two flows to reach equilibrium. This concept is illustrated in Example 401.8.

Example 401.8

Two lanes of a freeway have a capacity of 4400 veh/h. The normal flow in the two lanes is 3200 veh/h. An incident blocks both lanes for 15 min. After 15 min, both lanes are opened to full traffic flow. How long after the incident does it take the queue to dissipate?

Solution   Assume $t = 0$ to be the instant at which the incident occurs and $t = T$ to be the time for the resulting queue to dissipate.

Capacity = 4400 veh/h = 73.33 veh/min

Flow rate = 3200 veh/h = 53.33 veh/min

Total time for arrivals = T

Total time for departures = $T - 15$

Assuming arrivals occur at 53.33 veh/min and departures occur at capacity (73.33 veh/min) until the queue is dissipated.

$$53.33T = 73.33(T - 15)$$

Therefore, $T = 55$ min.

Queueing Theory—M/M/1 Queue

A M/M/1 queue represents the length of a queue in a system with a single server, where arrivals follow a Poisson distribution (arrival rate λ) and service times (service rate μ) follow an exponential distribution. The system is considered stable only if $\lambda < \mu$. If the arrivals occur faster than they can be served ($\lambda > \mu$), the queue grows indefinitely.

The ratio $\rho = \lambda/\mu$ is called the utilization factor.

The probability of finding n customers in the system (those being served + those in queue) is given by

$$P(n) = \rho^n(1 - \rho) \tag{401.25}$$

The expected length of the queue is given by: $\dfrac{\rho^2}{1 - \rho}$ (401.26)

The average time spent waiting is: $\dfrac{\rho}{\mu - \lambda}$ (401.27)

Average number of vehicles in the system (those being served + those in queue) $= \dfrac{\rho}{1 - \rho}$ (401.28)

Example 401.9

A toll facility has a single booth. The hourly volume is 500 vph and the peak hour factor is 0.9. If the average service time per driver is 5 s and the average vehicle length is 16 ft.,

1. What is the average length of the queue (number of vehicles) during the peak flow period?
2. What is the average number of vehicles in the system?
3. What is the average wait time for each vehicle?
4. What is the 95th-percentile number of vehicles in the system?

Solution

Average arrival rate $= 500$ vph

Average arrival rate during peak flow, $\lambda = V/PHF = 500/0.9 = 556$ vph

Service rate, $\mu = 3600/5 = 720$ vph. Since this is more than the peak arrival rate, we do not have instability.

According to the Transportation Engineer's Handbook (ITE),

Utilization factor: $\rho = \lambda/\mu = 556/720 = 0.77$

1. Expected queue length $= \dfrac{\rho^2}{1 - \rho} = \dfrac{0.77^2}{1 - 0.77} = 2.6$. So, on average, we expect an average of 2.6 (waiting) vehicles in the queue.

2. Average number of vehicles in the system (those being served + those in queue) =

$$\frac{\rho}{1-\rho} = \frac{0.77}{1-0.77} = 3.38$$

3. Average wait time = $\dfrac{\rho}{\mu - \lambda} = \dfrac{0.77}{720 - 556} = \dfrac{1}{213}\dfrac{h}{veh} = 17\dfrac{s}{veh}$

4. Probability of finding n vehicles in the system: $P(n) = \rho^n(1 - \rho)$. Using this formula, we build the table below. $P(n)$ is the probability of finding n vehicles in the system. $F(n)$ is the cumulative of $P(n)$. It represents the cumulative probability that there will be n or less vehicles in the system

n	$P(n)$	$F(n)$
0	0.228	0.228
1	0.176	0.404
2	0.136	0.540
3	0.105	0.644
4	0.081	0.725
5	0.063	0.788
6	0.048	0.836
7	0.037	0.874
8	0.029	0.902
9	0.022	0.925
10	0.017	0.942
11	0.013	0.955

Therefore, the 90th percentile number of vehicles in the system = 8 and the 95th percentile number of vehicles in system = 11.

Constant Acceleration and Deceleration

For applications where motion of a vehicle may be assumed to be either at constant speed (zero acceleration) or constant acceleration (uniform acceleration from rest to maximum speed or uniform braking from initial travel speed to final speed), Newton's laws of motion adapted for uniform acceleration may be used. These are summarized in Eq. (401.29).

$$V_f = V_i + at \tag{401.29a}$$

$$s = s_o + V_i t + \frac{1}{2}at^2 \tag{401.29b}$$

$$V_f^2 = V_i^2 + 2a(s - s_o) \tag{401.29c}$$

where V_i = initial speed (at time $t = 0$)
V_f = final speed (at time t)
a = constant acceleration
s_o = initial position (at time $t = 0$)
s = final position (at time t)

On the P.E. exam, the most common applications for these equations are calculations of stopping sight distance, which involve travel at uniform speed during the perception-reaction time and uniform deceleration during the braking time. Example 401.10 illustrates the application of these equations.

Example 401.10

A commuter train has stations spaced 1 mi apart. The top speed of the train is 80 mph. The train accelerates at 5.5 ft/s^2 and decelerates at 4.5 ft/s^2. It dwells at each platform for 20 s to pick up and discharge passengers. What is the average running speed of the train?

Solution   The motion of the train occurs in three phases: (1) acceleration from rest up to peak speed, (2) travel at peak speed, and (3) deceleration from peak speed to rest, as the train enters the destination station.

$$\text{Peak speed} = 80 \text{ mph} = 80 \times 1.47 = 117.6 \text{ fps}$$

Phase 1   From Rest to 117.6 fps @ 5.5 ft/s^2

Time to reach peak speed: $t_1 = \dfrac{V_f - V_i}{a} = \dfrac{117.6 - 0}{5.5} = 21.4$ s

Distance traveled during this phase: $s_1 = \dfrac{at_1^2}{2} = \dfrac{5.5 \times 21.4^2}{2} = 1259$ ft

Phase 3   From 117.6 fps to Rest @ 4.5 ft/s^2

Time to decelerate to rest: $t_3 = \dfrac{V_f - V_i}{a} = \dfrac{0 - 117.6}{-4.5} = 26.1$ s

Distance traveled during this phase: $s_3 = V_i t_3 + \dfrac{at_3^2}{2} = 117.6 \times 26.1 - \dfrac{4.5 \times 26.1^2}{2}$
$$= 1536 \text{ ft}$$

Phase 2   Constant Velocity at 117.6 fps

Distance traveled during this phase: $s_2 = s - s_1 - s_3 = 5280 - 1259 - 1536 = 2485$ ft

Time of travel at constant speed: $t_2 = \dfrac{s_2}{V_2} = \dfrac{2485}{117.6} = 21.1$ s

Therefore, total travel time between stations $= t_1 + t_2 + t_3 = 68.6$ s
Average running speed (not including dwell time) $= 5280 \div 68.6 = 77$ fps $= 52.3$ mph
   On the other hand, space mean speed is calculated by including the dwell time (train traverses 1 mi in $68.6 + 20 = 88.6$ s)
Space mean speed $= 5280 \div 88.6 = 59.6$ fps $= 40.5$ mph

Example 401.11

A commuter train has a travel time of 70 s between uniformly spaced stations. During rush hour, the uniform headway between trains is 5 min. Trains have a dwell time of 15 s at each station. Each train has four coaches, each with a capacity of 240 passengers. What is the peak capacity of the line (passengers per hour)?

Solution Note that the headway (time gap between successive trains) is 5 min. This means that every 5 min, 960 passengers are being moved. No dwell time needs to be added to this. The difference between the travel time for one train (85 s) and the headway between trains (5 min) simply means that approximately 3 1/2 min of every 5 min is "dead time" along the commuter line.

$$\text{The capacity is therefore } \frac{960}{5} \times 60 = 11{,}520 \text{ passengers/h}$$

Example 401.12

For the commuter line described in Example 401.11, what is the minimum headway between trains?

Solution In order to avoid conflicts on the time-space diagram for successive trains, the minimum headway between trains is equal to the dwell time. If trains are spaced every 15 s, as soon as one train departs a station, the next one pulls in behind it.

The minimum headway is therefore 15 s.

Example 401.13

A car engine exerts a thrust of 1800 lb to accelerate the car (weight = 3400 lb) up a 5% grade from rest to a peak speed of 80 mph. Rolling resistance on all wheels is estimated to be 40 lb/ton. Once at peak speed, the engine generates just enough thrust to maintain constant velocity. If the total distance traveled is 0.5 mile, what is the average running speed (mph) of the car?

A. 64

B. 68

C. 71

D. 74

Solution During acceleration phase, total resistance factor = grade resistance + rolling resistance = 0.05 + 40lb/ton = 0.05 + 40/2000 = 0.07.

Total resistance = 0.07 × 3400 = 238 lb

Net thrust = 1800 – 238 = 1562 lb. This produces acceleration = 1562 × 32.2/3400 = 14.8 ft/s²

Time required to go from rest to 80 mph (117.33 ft/s) at this acceleration is 7.93 s

Distance traveled during acceleration = 0.5 × 14.8 × 7.93² = 465 ft

Distance at cruising speed = 0.5 mile – 465.3 ft = 2640 – 465 = 2175 ft

Time for cruising speed travel = 2175/117.33 = 18.54 s

Total travel time (for 0.5 miles) = 26.47 s

Average running speed = 2640/26.47 = 99.7 fps = 68 mph

Table 401.1   Level of Service on Transit Facilities

	Bus		Rail		
LOS	ft²/p	p/seat	ft²/p	p/seat	Comments
A	>12.9	0.00–0.50	>19.9	0.00–0.50	No passenger need sit next to another.
B	8.6–12.9	0.51–0.75	14.0–19.9	0.51–0.75	Passengers can choose where to sit.
C	6.5–8.5	0.76–1.00	10.2–13.9	0.76–1.00	All passengers can sit.
D	5.4–6.4	1.01–1.25	5.4–10.1	1.01–2.00	Comfortable standee load for design.
E	4.3–5.3	1.26–1.50	3.2–5.3	2.01–3.00	Maximum schedule load.
F	<4.3	>1.50	<3.2	>3.00	Crush loads.

Source   Highway Capacity Manual, Transportation Research Board, Washington DC, 2010.

Level of Service for Transit Facilities

Table 401.1 from the *Highway Capacity Manual*[3] outlines guidelines for level of service on transit facilities such as bus and rail. The thresholds between different levels of service are either in terms of available space (sq ft per passenger) or the average number of riding passengers per seat.

Choice to Use Public Transit

The public's choice to use public transit depends on the "convenience factor." If transit facilities are well-planned in layout and capacity, they can offer a motorist a viable alternative to driving. The potential user will often base the decision to switch to public transit based on the total extra time a particular trip takes by using public transit as opposed to driving. Table 401.2 outlines some guidelines that planners can use to determine the effectiveness of an existing transit system based on time of travel.

Table 401.2   Choice to Use Transit

LOS	Travel time difference (min)	Comments
A	≤0	Faster by transit than by automobile.
B	1–15	About as fast by transit as by automobile.
C	16–30	Tolerable to choice riders.
D	31–45	Round-trip at least an hour longer by transit.
E	46–60	Tedious for all riders. May be best possible outcome in small cities.
F	>60	Unacceptable to most riders.

[3]*Highway Capacity Manual*, Transportation Research Board, Washington D.C. 2000.

Table 401.3 Level of Service for Urban Scheduled Transit Service

	Exhibit 27-1 service frequency LOS for urban scheduled transit service		
LOS	Headway (min)	Veh/h	Comments
A	<10	>6	Passengers don't need schedules.
B	≥10–14	5–6	Frequent service; passengers consult schedules.
C	≥14–20	3–4	Maximum desirable time to wait if bus/train missed.
D	≥20–30	2	Service unattractive to choice riders.
E	≥30–60	1	Service available during hour.
F	≥60	<1	Service unattractive to all riders.

Source   Highway Capacity Manual, Transportation Research Board, Washington D.C., 2010.

Table 401.3 shows service frequency LOS criteria for urban scheduled transit service.

Parking Facilities

The following terms are commonly used for analysis of parking requirements:

A space-hour defines the occupation of a single parking space for the duration of 1 h.

A parking accumulation curve shows the cumulative accumulation of parked vehicles in a defined area over a period of time.

Parking turnover is the rate of use of a parking space. It is calculated by dividing the parking volume by the number of parking spaces within the study area.

On street parking facilities may have various configurations, ranging from parking stalls aligned parallel to the curb to perpendicular stalls. The parallel and angled configurations are shown in Fig. 401.2. Table 401.4 lists the dimensions L_1, L_2, and L_3 for various angles of

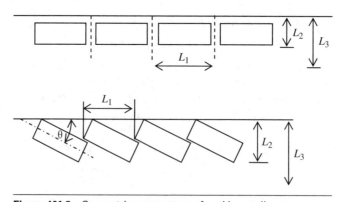

Figure 401.2 Geometric parameters of parking stalls.

Table 401.4 Typical Geometry of Parking Stalls

Curb angle (π)	Stall length L_1 (ft)	Stall width L_2 (ft)	Stall width + aisle L_3 (ft)	Number of spaces in length L	Area per vehicle (ft^2)
0	22	8	18	$N = \dfrac{L}{22}$	396
30	17	16.4	26	$N = \dfrac{L-2.8}{17}$	460
45	12	18.7	30	$N = \dfrac{L-6.7}{12}$	A
60	9.8	19.8	37	$N = \dfrac{L-6.6}{9.8}$	425
90	8.5	18	43	$N = \dfrac{L}{8.5}$	365

inclination θ between the edge of the street and the longitudinal axis of the vehicle. Equations summarized in Table 401.4 give the number of vehicles that may utilize a curb length L.

It is to be noted, however, that parking standards are usually determined by local authorities and therefore values in this table are merely a rough indication.

Example 401.14

Based on entry-exit records in a gated parking lot, 275 cars parked during a typical day between 9 a.m. and 6 p.m. Of these cars, approximately 10% of the cars were parked for 1 h, 35% for 2 h, 25% for 3 h, and the remaining for 4 h. On average, about 12% of the bays are vacant throughout the day. Assume an efficiency factor of 75%. What are (a) the space-hour demand and (b) the number of parking spaces in the lot?

Solution   Space hours of demand = sum of number of cars × hours parked.
Utilized space-hour demand = $275 \times (0.10 \times 1 + 0.35 \times 2 + 0.25 \times 3 + 0.30 \times 4) = 756$
Since this represents 88% occupancy, the available space-hour demand = $756 \div 0.88 = 859$ space hours.
If there are N parking spaces in the lot and the overall efficiency is 75%, then $0.75 \times N \times 9 = 859 \Rightarrow N = 127$.

Highway Safety

Postimpact Behavior of Vehicles

Analysis of the after effects of a collision can give clues about travel speeds of vehicles immediately prior to impact. The following principles of physics must hold:

1. Conservation of linear momentum

2. Coefficient of restitution relating preimpact and postimpact relative velocity

Conservation of Linear Momentum

Total momentum is conserved in all directions (vector):

$$\sum (m_j \mathbf{V}_j)_i = \sum (m_j \mathbf{V}_j)_f \tag{402.1}$$

where the subscripts i and f refer to initial and final conditions, respectively. This means that the total linear momentum of the system (both colliding vehicles) before the impact, with speeds written as vector quantities, is equal to the total momentum immediately after the impact. The direction of travel immediately after impact can be estimated from the orientation of the skid marks and distance traveled by the wreckage.

Coefficient of Restitution

Also, the nature of the impact is specified in terms of a coefficient of restitution (e), which is in the range $0 \le e \le 1$, with $e = 0$ implying a *perfectly plastic collision* (objects stick together after impact) and $e = 1$ implying a *perfectly elastic collision* (relative velocity of objects unchanged by impact). The coefficient of restitution is defined as

$$e = \frac{(V_1)_f - (V_2)_f}{(V_2)_i - (V_1)_i} \tag{402.2}$$

In Eq. (402.2), the velocity components affected by the nature of the impact (i.e., the value of e) are those that are parallel to the line of impact (the line joining the centers of mass at the instant of impact). The velocity components perpendicular to this line are unaffected by the impact. If the vehicles get stuck together as a result of the collision, the collision is plastic ($e = 0$) and the two vehicles have a common velocity.

Example 402.1

At a right-angle intersection, two cars collide and the wreckage is seen to follow the path as shown in the following figure. The distance traveled by the wreckage before coming to a stop is 52 ft. It has been established that the effective coefficient of friction of the level ground over which the wreckage travels is 0.72. The weight of the westbound vehicle (B) was 3200 lb and that of the northbound vehicle (A) was 5100 lb. What were the speeds (mph) of the two vehicles at the time of impact?

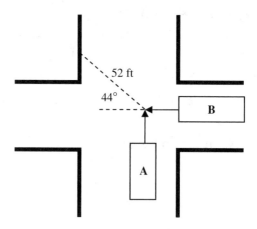

Solution   Coefficient of friction $f = 0.72$

Therefore, deceleration rate $a = fg = 0.72 \times 32.2$ ft/s^2 = 23.2 ft/s^2

The speed at which the wreckage moves immediately after impact is thus

$$v = \sqrt{2as} = \sqrt{2 \times 23.2 \times 52} = 49.1 \text{ fps} = 33.5 \text{ mph}$$

Conservation of momentum in the y direction gives

$$5100 \times v_A = (3200 + 5100) \times 33.5 \times \sin 44 \Rightarrow v_A = 37.9$$

$$\text{Speed of vehicle A} \qquad v_A = 37.9 \text{ mph}$$

Conservation of momentum in the x direction gives

$$3200 \times v_B = (3200 + 5100) \times 33.5 \times \cos 44 \Rightarrow v_B = 62.5$$

$$\text{Speed of vehicle B} \qquad v_B = 62.5 \text{ mph}$$

Design of Crash Cushions

An elementary design for crash cushion design is based on the energy absorption capacity of the cushion. The critical design variables are—mass of design vehicle, impact speed, cushion length, and deceleration rate.

Example 402.2

A crash cushion must be chosen to safely stop a runaway vehicle. The following data are given:

Weight of design vehicle = 4300 lb

Approach speed = 70 mph

Equivalent spring stiffness of barrier system = 1200 lb/ft

What is the minimum length of crash cushion required?

Solution

Mass of the vehicle: $m = \dfrac{w}{g} = \dfrac{4300}{32.2} = 133.54$ lb-s^2/ft

Impact speed: $V = 70$ mph $= 70 \times 1.47 = 102.9$ fps

Kinetic Energy at the time of impact: $E = \dfrac{1}{2} mV^2 = \dfrac{1}{2} \times 133.54 \times 102.9^2 = 706{,}988$ lb-ft

Energy absorbed by cushion (which is idealized as a spring): $E = \dfrac{1}{2} kx^2$

Equating impact energy to absorbed energy, we get: $\dfrac{1}{2} kx^2 = 706{,}988 \Rightarrow x = 34.3$ ft

Thus, the crash cushion must be at least 35-ft long to absorb the energy of the impact.

Crash Cushion Calculations

When a vehicle (mass M_v, speed V_o) strikes a stationary crash cushion (mass M_c), based on the Principle of conservation of momentum, the speed following impact (assuming a plastic impact) is given by

$$V_1 = \frac{M_v V_o}{M_v + M_c} \tag{402.3}$$

During the impact, both vehicle and the crash cushion move together with speed V_1 and travel through the distance D (diameter of the first barrel) before the vehicle hits the second barrel (outline of the vehicle just before the second impact is shown as a dashed line in the figure below).

Assuming uniform deceleration during this phase, the average deceleration is given by Newton's laws of motion.

$$a = \frac{V_o^2 - V_1^2}{2D} \tag{402.4}$$

The time taken to travel through the distance D is calculated as

$$t = \frac{V_o - V_1}{a} \tag{402.5}$$

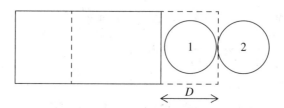

Example 402.3

A 4000 lb vehicle traveling at 60 mph (88 ft/sec) strikes a crash barrier system constructed of 2 ft diameter, 3-ft-high water filled drums. Each drum has a weight of 300 lb. Calculate (1) the vehicle speed after striking the first two barrels, (2) the time taken for collisions 1 & 2 respectively, and (3) the percent reduction in the car's kinetic energy as a result of the first two collisions.

Solution

Before collision 1, vehicle speed = 88 fps

After collision 1, vehicle speed $= \dfrac{4000 \times 88}{4000 + 300} = 81.86$ fps

Deceleration rate $= \dfrac{88^2 - 81.86^2}{2 \times 2} = 260.7\dfrac{\text{ft}}{\text{s}^2} = 8.1$ g

Time to go from 88 fps to 81.86 fps $= \dfrac{88 - 81.86}{260.7} = 0.024$ second

Kinetic energy ratio $= \dfrac{V_1^2}{V_o^2} = \dfrac{81.86^2}{88^2} = 0.865$

Before collision 2, vehicle speed = 81.86 fps

After collision 2, vehicle speed $= \dfrac{4000 \times 81.86}{4000 + 300} = 76.15$ fps

Deceleration rate $= \dfrac{81.86^2 - 76.15^2}{2 \times 2} = 225.6\dfrac{\text{ft}}{\text{s}^2} = 7.0$ g

Time to go from 81.86 fps to 76.15 fps $= \dfrac{81.86 - 76.15}{225.6} = 0.025$ second

Kinetic energy ratio $= \dfrac{V_2^2}{V_1^2} = \dfrac{76.15^2}{81.86^2} = 0.865$

After striking the second barrel, velocity = 76.15 fps = 51.9 mph

Total time needed for the two collisions = 0.049 second

Energy reduction factor $= 0.865^2 = 0.748$ (25% reduction)

Accident Rates

Accident rates for a particular location are a critical parameter for decision making about possible improvements to the location or facility. Once the crash rate for the location (stretch of roadway or intersection) has been measured, it is then compared to the critical crash rate for the surrounding region (district, state, and the like).

The *equivalent number of accidents* for a facility is often calculated by adopting weighing factors for various types of accidents depending on their severity. These are not economic weighting factors, but are meant to represent the accident risk level for that facility.

Critical Crash Rate

The critical crash rate is calculated using various approaches. For example, in Kentucky, Eq. (402.3) is used to calculate the critical crash rate.

$$C_c = C_a + K\sqrt{\frac{C_a}{M}} + \frac{1}{2M} \tag{402.6}$$

where C_c = critical crash rate
$\quad\;\; C_a$ = average statewide crash rate for type of intersection
$\quad\;\; K$ = a constant related to statistical significance level
$\quad\;\; M$ = exposure (e.g., for intersections M is in terms of million entering vehicles, for highway segments it is hundred million vehicle miles)

The actual crash rate for each intersection is calculated using Eq. (402.6). The critical rate factor is then defined as the actual crash rate divided by the critical crash rate.

Example 402.4

A 0.25-mi segment of an urban arterial is being studied for its accident history. The AADT is 12,000 vehicles per day. During the 3-year study period, there have been a total of 43 "equivalent" accidents (fatal + injury + property damage) per year. The statewide average for similar arterials is 725 equivalent accidents per year. Use a 95% confidence level. What is the accident ratio for this arterial?

Solution The exposure M is calculated as

$$M = \frac{N \times \text{AADT} \times L \times 365}{10^8} = \frac{3 \times 12,000 \times 0.25 \times 365}{10^8} = 0.0329\,\text{HMVM}$$

For 95% confidence level, the constant $K = 1.645$.

$$\text{Critical crash rate } C_c = 725 + 1.645\sqrt{\frac{725}{0.0329} + \frac{1}{2 \times 0.0329}} = 984$$

$$\text{Segment accident history} = \frac{43}{0.0329} = 1307$$

$$\text{Accident ratio} = \frac{1307}{984} = 1.33$$

Since this ratio exceeds 1.0, this highway segment may have a safety problem.

Example 402.5

The example below outlines various ways for normalizing and reporting accident rates for a study jurisdiction. The table below shows some gross accident statistics for a relatively small urban jurisdiction in 2003.

Fatalities	75
Fatal accidents	60
Injury accidents	300
PDO accidents	2000
Total involvement	4000
Vehicle miles traveled	1.5×10^9
Registered vehicles	100,000
Licensed drivers	150,000
Population of area	300,000

The following "accident rates" may then be calculated and reported:

1. Accidents per hundred million miles

$$\frac{2360}{1.5 \times 10^9} \times 10^8 = 157/\text{HMVM}$$

2. Deaths per 100,000 population

$$\frac{75 \times 10^5}{3 \times 10^5} = 25$$

3. Deaths per 10,000 registered vehicles

$$\frac{75 \times 10^4}{10^5} = 7.5$$

For Highway Segment (Length L)

For a segment of highway of a finite length, the rate of accidents over a study period of N years is reported as number of accidents per hundred million vehicle miles (number per HMVM) and is defined as

$$\text{Accidents/hundred million vehicle miles} = \frac{\text{No. of accidents } \times 10^8}{L \times \text{ADT} \times 365 \times N} \qquad (402.7)$$

where ADT is the average daily traffic for the study period. If ADT values for these N years are reported individually, then the average ADT is calculated as

$$\text{ADT} = \frac{\sum_1^N \text{ADT}_i}{N} \tag{402.8}$$

For Intersections

For an intersection, the rate of accidents over a study period of N years is reported as number of accidents per million entering vehicles (number per MEV) and is defined as

$$\text{Accidents/million entering vehicles} = \frac{\text{No. of accidents} \times 10^6}{\text{ADT} \times 365 \times N} \tag{402.9}$$

The number of entering vehicles is counted for all approaches into the intersection.

Accident Countermeasures

If accident rates at a particular site are unacceptable, then several countermeasures can be implemented. Some effective countermeasures are lane widening, sightline improvements, parking restrictions, speed control, etc. If empirical data is available about these countermeasures and their impact on accident rates, then a particular site can be evaluated for the overall impact (reduction in number of accidents) of implementing several countermeasures.

The *Highway Safety Manual*, 2010, American Association of State Highway & Transportation Officials, Washington, D.C., includes several countermeasures grouped into categories and subcategories. Some examples of countermeasures and the category and subcategory they are nested under are listed below.

Category: Delineation

Subcategory: Supplemental delineation

Countermeasure: Install postmounted delineators

Countermeasure: Install snowplowable, permanent raised pavement markers

Subcategory: On-pavement markings

Countermeasure: Install edgelines, centerlines, and postmounted delineators

Each countermeasure has a crash modification factor (CMF) which is a multiplicative factor used to compute the expected number of crashes after implementing a given countermeasure at a specific site. The crash reduction factor (CRF) is equal to $1 - \text{CMF}$.

The overall crash reduction achieved by implementing a series of countermeasures $(1 \ldots n)$ is given by Eq. (402.7).

$$CR = CR_1 + (1 - CR_1)CR_2 + (1 - CR_1)(1 - CR_2)CR_3 + \ldots + (1 - CR_1) \ldots (1 - CR_{n-1})CR_n \tag{402.10}$$

Example 402.6

An urban intersection had 23 crashes during the planning year 2001–2002. The ADT (all entering vehicles) for this year was 6400. Two specific countermeasures were implemented. They are described below:

Countermeasure 1

Widening of lanes

Expected reduction of crashes = 12%

Countermeasure 2

Eliminating curbside parking

Expected reduction of crashes = 32%

If the ADT is expected to increase by 3% every year, what is the number of crashes expected during the year 2011–2012?

A. 12

B. 19

C. 26

D. 31

Solution

Growing at the rate of 3% every year, ADT after interval of 10 years = $6400 \times 1.03^{10} = 8601$

Crash reduction factor for countermeasure 1: $CR_1 = 0.88$

Crash reduction factor for countermeasure 2: $CR_2 = 0.68$

Overall crash reduction factor: $CR = CR_1 + 1 - CR_1\,CR_2 = 0.12 + 0.88 \times 0.32 = 0.4016$

Therefore, number of crashes prevented in 2011–2012:

$$N \times CR \times \frac{ADT_2}{ADT_1} = 23 \times 0.4016 \times \frac{8601}{6400} = 12.4$$

It may be assumed that without the countermeasures, the number of crashes would increase in proportion to the ADT. This would mean $23 \times 8601/6400 = 30.9$

With the countermeasures, expected number of crashes = $30.9 - 12.4 = 18.5$

Answer is B.

Clear Zone

The Roadside Design Guide (AASHTO Roadside Design Guide, 4th edition, 2011, American Association of State Highway & Transportation Officials, Washington, D.C.) outlines minimum required width of recovery zones adjacent to highways. The required inputs are—(1) design speed, (2) ADT, and (3) severity of slopes adjacent to the roadway. Slopes are classified as either foreslopes (or fill-slopes) or backslopes (or cut-slopes). Table 402.1 gives the basic value of the clear zone width L_c.

On a curved section, the design width of the clear zone is obtained by multiplying the clear-zone width from the straight section by a correction factor K_{cz}, which is a function of design speed and curve radius. The clear-zone correction factor is applied to the outside of curves only. The clear zone on the outside of curvature (CZ_c) is then given by Eq. (402.8)

$$CZ_c = L_c \times K_{cz}$$ (402.11)

Table 402.1 Clear-Zone Distance L_c (ft) from Edge of Through Traveled Way

Design Speed	Design ADT	Foreslopes			Backslopes		
		1V:6H or flatter	1V:5H to 1V:4H	1V:3H	1V:3H	1V:5H to 1V:4H	1V:6H or flatter
40 mph or less	UNDER 750	7–10	7–10	**	7–10	7–10	7–10
	750–1500	10–12	12–14	**	10–12	10–12	10–12
	1500–6000	12–14	14–16	**	12–14	12–14	12–14
	OVER 6000	14–16	16–18	**	14–16	14–16	14–16
45–50 mph	UNDER 750	10–12	12–14	**	8–10	8–10	10–12
	750–1500	12–14	16–20	**	10–12	12–14	14–16
	1500–6000	16–18	20–26	**	12–14	14–6	18–20
	OVER 6000	18–20	24–28	**	14–16	18–20	20–22
55 mph	UNDER 750	12–14	14–18	**	8–10	10–12	10–12
	750–1500	16–18	20–24	**	10–12	14–16	16–18
	1500–6000	20–22	24–30	**	14–16	16–18	20–22
	OVER 6000	22–24	26–32*	**	16–18	20–22	22–24
60 mph	UNDER 750	16–18	20–24	**	10–12	12–14	14–16
	750–1500	20–24	26–32*	**	12–14	16–18	20–22
	1500–6000	26–30	32–40*	**	14–18	18–22	24–26
	OVER 6000	30–32*	36–44*	**	20–22	24–26	26–28
65–70 mph	UNDER 750	18–20	20–26	**	10–12	14–16	14–16
	750–1500	24–26	28–36*	**	12–16	18–20	20–22
	1500–6000	28–32*	34–42*	**	16–20	22–24	26–28
	OVER 6000	30–34*	38–46*	**	22–24	26–30	28–30

*Where a site specific investigation indicates a high probability of continuing crashes, or such occurrences are indicated by crash history, the designer may provide clear-zone distances greater than the clear-zone shown in the above table. Clear zone may be limited to 30 ft for practicality and to provide a consistent roadway template if previous experience with similar projects or design indicates satisfactory performance.

** Since recovery is less likely on the unshielded, traversable 1V:3H slopes, fixed objects should not be present in the vicinity of the toe of the slopes. Recovery of high-speed vehicles that encroach beyond the edge of the shoulder may be expected to occur beyond the toe of slope. Determination of the width of the recovery area at the toe of slope should take into consideration right-of-way availability, environmental concerns, economics factors, safety needs, and crash histories. Also the distance, between the edge of the through traveled lane and the beginning of the 1V:3H slope should influence the recovery area provided at the top of slope.

Table 402.2 Horizontal Curve Adjustment Factor K_{cz}

Radius (ft)	Design Speed (mph)					
	40	45	50	55	65	70
2950	1.1	1.1	1.1	1.2	1.2	1.2
2300	1.1	1.1	1.2	1.2	1.2	1.3
1970	1.1	1.2	1.2	1.2	1.3	1.4
1640	1.1	1.2	1.2	1.3	1.3	1.4
1475	1.2	1.2	1.3	1.3	1.4	1.5
1315	1.2	1.2	1.3	1.3	1.4	—
1150	1.2	1.2	1.3	1.4	1.5	—
985	1.2	1.3	1.4	1.5	1.5	—
820	1.3	1.3	1.4	1.5	—	—
660	1.3	1.4	1.5	—	—	—
495	1.4	1.5	—	—	—	—
330	1.5	—	—	—	—	—

Example 402.7

The design ADT for a roadway is 5000. Design speed is 55 mph. A paved shoulder width of 8 ft is available adjacent to the edge of the traveled way, as shown. At the location shown, the roadway curves to the left with a radius = 1485 ft. A culvert headwall is to be located on the foreslope of 1:4 at a distance X as shown. What is the minimum required distance X?

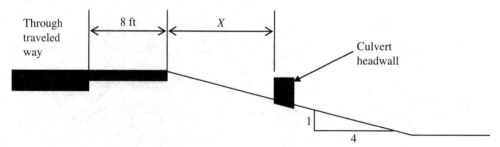

Solution

For ADT = 5000, design speed = 55 mph and foreslope = 1:4, the base value of the clear zone is 30 ft.

Since the roadway curves to the left, the clear zone on the right is on the outside of the curve and therefore needs to be modified for curvature. From Table 402.2, for $V = 55$ mph and $R = 1485$ ft, the correction factor $K_{cz} = 1.3$.

Therefore, the required clear zone width = $30 \times 1.3 = 39$ ft. The shoulder is considered to be part of the recovery width. Therefore, minimum required $X = 31$ ft.

Use of Roadside Barriers

Figure 402.1 shows the spatial variables involved in the layout of a roadside barrier, which may be required if an obstruction is located closer than the clear zone requirement based on design speed, ADT, sideslope, and roadway curvature. According to the Roadside Design Guide (AASHTO), roadside barriers perform most effectively when they are installed on slopes of 1V:10H or flatter.

Typically, a barrier is flared (suggested flare rates b/a in Table 5-9 of the RDG), with the barrier terminal located farther from the roadway (to minimize driver reaction) and then gradually tapered ("flared") to a parallel configuration. Table 5-10 gives suggested runout lengths (L_R) as a function of design speed and ADT.

The lateral extent of the area of concern (L_A) is the distance from the edge of traveled way to outside edge of obstacle. The lateral offset to the barrier (L_2) is chosen to be greater than the shy line offset (L_s), which is the minimum distance beyond which presence of a barrier does not affect driver behavior. Table 5-7 of the Roadside Design Guide provides guidance on Shy-Line Offset values as a function of speed. The distance L_1 (chosen by designer) = longitudinal distance from upstream edge of obstruction to the start of the barrier flare. If no flare is used, $L_1 = 0$.

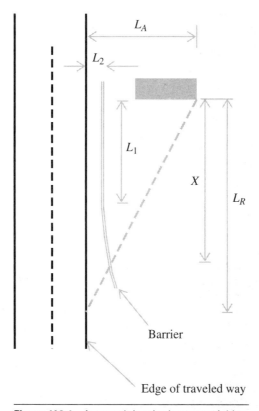

Figure 402.1 Approach barrier layout variables.

Once these variables (L_A, L_R, L_1, L_2, b/a) are defined, the required length of need (X) can be calculated as

$$X = \frac{L_A + \dfrac{b}{a}L_1 - L_2}{\dfrac{b}{a} + \dfrac{L_A}{L_R}}$$

(402.12)

Sight Distance

AASHTO Recommendations

The stopping sight distance (SSD) is the sum of the distance traveled during perception-reaction time (t_R) and the distance traveled during braking to a stop from the design speed. The recommended value for t_R is 2.5 s. The default value of the deceleration rate in the *AASHTO Green Book*[1] is 11.2 ft/s². The total distance traveled during the stopping maneuver, from the first sighting of the obstruction is given by

$$\text{SSD} = Vt_R + \frac{V^2}{2(a \pm gG)} \tag{403.1}$$

where V = design speed
t_R = perception-reaction time
a = deceleration rate
g = acceleration due to gravity
G = longitudinal grade

In Eq. (403.1), all quantities are in consistent units (e.g., if V is fps, t_R is seconds, a is ft/s², and G is decimal). However, another form of this equation in AASHTO is presented with the more commonly used U.S. units

$$\text{SSD} = 1.47 \times S_{\text{mph}} \times t_R + \frac{1.075 S_{\text{mph}}^2}{a \pm 0.32 G_{\%}} \tag{403.2}$$

[1]*AASHTO Green Book—A Policy on Geometric Design of Highways and Streets*, 7th ed., American Association of State and Highway Transportation Officials, 2018.

This can also be expressed as

$$\text{SSD} = 1.47 \times S_{mph} \times t_R + \frac{S_{mph}^2}{30\left(\dfrac{a}{32.2} \pm \dfrac{G_\%}{100}\right)} \qquad (403.3)$$

Table 403.1 shows the stopping distance for various combinations of design speed (mph) and grade G (%). The default values of $t_R = 2.5$ s and $a = 11.2$ ft/s² have been used in generating the table.

In Table 403.1, the braking distance may be obtained by subtracting the reaction distance from the total stopping distance. For example, for a vehicle traveling at 40 mph on a downgrade of 4%, the braking distance is equal to $320.3 - 146.7 = 173.6$ ft.

The braking distance traveled to slow a vehicle from an initial speed V_i (mph) to a final speed V_f (mph) at a uniform rate of deceleration a (ft/s²) is given by

$$S_B = \frac{1.075(V_i^2 - V_f^2)}{a \pm 0.32 G_\%}$$

Table 403.1 Stopping Sight Distance for Various Speeds and Grades

Design speed (mph)	Reaction distance (ft)	Stopping distance $S_R + S_B$ (ft) for various gradients G (%)						
		Downgrades			Level	Upgrades		
		−6	−4	−2	0	+2	+4	+6
15	55.0	81.1	79.4	77.9	76.6	75.4	74.4	73.4
20	73.3	119.8	116.7	114.1	111.7	109.7	107.8	106.1
25	91.7	164.2	159.5	155.3	151.7	148.4	145.5	142.9
30	110.0	214.4	207.7	201.7	196.4	191.7	187.5	183.7
35	128.3	270.5	261.3	253.1	246.0	239.6	233.8	228.7
40	146.7	332.3	320.3	309.7	300.3	292.0	284.5	277.7
45	165.0	400.0	384.7	371.3	359.5	348.9	339.4	330.9
50	183.3	473.5	454.6	438.1	423.4	410.4	398.7	388.1
55	201.7	552.7	529.9	509.9	492.2	476.4	462.2	449.4
60	220.0	637.8	610.6	586.8	565.7	546.9	530.1	514.9
65	238.3	728.6	696.8	668.8	644.1	622.0	602.2	584.4
70	256.7	825.3	788.4	755.9	727.2	701.6	678.7	658.0
75	275.0	927.8	885.4	848.1	815.2	785.8	759.5	735.7
80	293.3	1036.1	987.8	945.4	907.9	874.5	844.5	817.5

Example 403.1

Calculate and verify the stopping distance on a rural highway with a downgrade of 4% and a design speed of 70 mph.

Solution   Assuming a reaction time $t_R = 2.5$ s, $a = 11.2$ ft/s^2, and $G = -4$, we get
Reaction distance $S_R = 1.47 \times 2.5 \times 70 = 257.3$ ft

Braking distance $S_B = \dfrac{1.075 \times 70^2}{11.2 - 0.32 \times 4} = 531$ ft

Therefore, total stopping distance $= 788.3$ ft (Table 403.1 shows 788.4 ft).

Example 403.2

A car traveling at 55 mph is followed by another traveling at 65 mph. The lead car suddenly brakes to a stop within a distance of 270 ft. Assuming the following car to have a perception-reaction time of 2 s, what must be the minimum following distance in order to avoid a collision?

Solution   The lead car travels a distance of 270 ft during the braking maneuver alone. (The braking to a stop is *not* in response to the sighting of an obstruction.) The key assumption in this case is that under identical road conditions, the same deceleration rate and grade should be used for both vehicles.

For the lead car, equating the braking distance to 270 ft, we have

$$270 = \frac{1.075 \times 55^2}{(a \pm gG)} \Rightarrow a \pm gG = 12.044$$

Therefore, for the following car, the stopping sight distance is given by

$$SSD = 1.47 \times 65 \times 2 + \frac{1.075 \times 65^2}{12.044} = 568.2\,\text{ft}$$

During the time that car 1 travels 270 ft, car 2 will travel 568 ft. Therefore, to avoid a collision, the minimum following distance $= 568 - 270 = 298$ ft.

Acceleration and Deceleration

Table 403.2 shows typical acceleration rates for passenger cars (approximately 30 lb/hp) versus trucks (approximately 200 lb/hp) as suggested by the *Traffic Engineering Handbook*, 5th edition, Institute of Transportation Engineers.

In the absence of other data, these values of acceleration can be used to determine the distance of travel associated with maneuvers such as acceleration from a stopped condition and merging into a stream of vehicles.

Table 403.2 Acceleration Rates for Typical Cars and Trucks

Speed range (mph)	Acceleration rate (ft/s²)	
	Typical car	Typical truck
0–20	7.5	1.6
20–30	6.5	1.3
30–40	5.9	0.7
40–50	5.2	0.7
50–60	4.6	0.3

Source Traffic Engineers Handbook, 5th edition, Institute of Transportation Engineers.

Sight Distance on Vertical Curves

Stopping Sight Distance on Crest Curves

Simple (parabolic) vertical curves are either crest curves or sag curves. On crest curves the vertical profile itself obstructs the line of sight and the length of curve should provide adequate stopping sight distance based on height of driver's eye (h_1) and height of the obstruction being sighted (h_2). AASHTO defines Eq. (403.4) for stopping sight distance on a crest vertical curve.

$$L = \frac{AS^2}{100\left(\sqrt{2h_1} + \sqrt{2h_2}\right)^2} \qquad \text{for} \qquad S \leq L \qquad (403.4a)$$

$$L = 2S - \frac{200\left(\sqrt{h_1} + \sqrt{h_2}\right)^2}{A} \qquad \text{for} \qquad S > L \qquad (403.4b)$$

For stopping and passing sight distance calculations for passenger cars, the height of driver's eye above the roadway is assumed to be $h_1 = 3.5$ ft. For large trucks, the *AASHTO Green Book* recommends a value of $h_1 = 7.6$ ft.

For stopping sight distance calculations, the recommended value for height of object $h_2 = 2.0$ ft. For passing sight distance calculations, the recommended value of $h_2 = 3.5$ ft.

Thus, for stopping sight distance for a passenger car, adopting $h_1 = 3.5$ ft and $h_2 = 2.0$ ft yields Eq. (403.5).

$$L = \frac{AS^2}{2158} \qquad \text{for} \qquad S \leq L \qquad (403.5a)$$

$$L = 2S - \frac{2158}{A} \qquad \text{for} \qquad S > L \qquad (403.5b)$$

where S = stopping sight distance (ft)
L = length of vertical curve (ft)
A = absolute grade difference = $|G_1 - G_2|$ (%)

Example 403.3

The alignment of a roadway contains a parabolic curve which transitions from a grade of $+5\%$ to a grade of -3%. Based on the AASHTO guidelines for stopping sight distance, what is the minimum length of curve required for a design speed of 60 mph?

Solution   For a design speed of 60 mph, the stopping sight distance $S = 566$ ft. Using the default values for $h_1 = 3.5$ ft, $h_2 = 2.0$ ft, and $A = 8$, Eq. (403.5a) gives

$$L = \frac{AS^2}{2158} = \frac{8 \times 566^2}{2158} = 1187.6 \text{ ft}$$

This satisfies the criterion $S < L$.
Therefore, the required length of curve $= 1188$ ft.

Passing Sight Distance on Crest Curves

For passing sight distance for a passenger car, adopting $h_1 = 3.5$ ft and $h_2 = 3.5$ ft yields Eq. (403.6).

$$L = \frac{AS^2}{2800} \qquad \text{for} \qquad S \leq L \qquad \qquad (403.6a)$$

$$L = 2S - \frac{2800}{A} \qquad \text{for} \qquad S > L \qquad \qquad (403.6b)$$

NOTE   This passing sight distance is purely a function of the geometry of the vertical curve. It relates the length of vertical curve to the potential of the crest curve to obstruct the line of sight. For passing sight distance on two-lane highways, where passing involves the maneuver of crossing over into the opposite lane, see the section "Passing Sight Distance."

Stopping Sight Distance on Sag Curves

On sag curves, the vertical profile does not obstruct the line of sight. However, the length of curve should provide adequate stopping sight distance based on headlight illumination distance. AASHTO defines Eq. (403.7) for stopping sight distance on a sag vertical curve.

$$L = \frac{AS^2}{400 + 3.5S} \qquad \text{for} \qquad S \leq L \qquad \qquad (403.7a)$$

$$L = 2S - \frac{400 + 3.5S}{A} \qquad \text{for} \qquad S > L \qquad \qquad (403.7b)$$

Example 403.4

A parabolic vertical curve of length 875 ft connects a tangent grade $= +4\%$ to a grade of -6%. What is the safe design speed based on stopping sight distance?

Solution Length of curve $= 875$ ft. Difference of grades $A = 10\%$.
According to Eq. (403.5a),

$$S = \sqrt{\frac{2158L}{A}} = \sqrt{\frac{2158 \times 875}{10}} = 435 \text{ ft} \qquad \text{(This satisfies the criterion } S \leq L.)$$

Corresponding to SSD $= 435$ ft, the design speed (on level grade) is 50.8 mph (Table 403.1). Specify safe design speed $= 50$ mph.

Comfort Criterion

Another criterion for minimum length of curve is based on rider comfort. Physiologically, discomfort is produced by the perception of acceleration, rather than velocity. The vertical acceleration a_v is equal to the rate of change of vertical velocity (the product of the velocity V and the gradient G).

$$a_v = \frac{\Delta v_v}{\Delta t} = \frac{\Delta v_v}{\dfrac{L}{V}} = \frac{V\Delta v_v}{L} = \frac{V(VG_2 - VG_1)}{L} = \frac{V^2(G_2 - G_1)}{L} = \frac{V^2 A}{L} \qquad (403.8)$$

If the upper limit of vertical acceleration is $a_{v,max}$, the minimum length of curve may be given by

$$L = \frac{AV^2}{a_{v,max}} \qquad \text{(consistent units: e.g., } L \text{ ft, } A \text{ decimal, } V \text{ fps, } a_{max} \text{ ft/s}^2) \qquad (403.9)$$

The AASHTO default for $a_{v,max} = 1$ ft/s^2.
  Thus, on sag vertical curves, the length of the vertical curve must be greater than

$$L_{min} = \frac{AV^2}{46.5} \qquad (403.10)$$

where L_{min} = minimum length of sag curve (ft)
    A = absolute difference in grades $= |G_2 - G_1|$ (%)
    V = design speed (mph)

  This formula may be generalized to Eq. (403.11) (for situations where the specified value of maximum vertical acceleration is different from 1 ft/s^2, such as for railroad alignments).

$$L_{min} = \frac{AV^2}{46.5a_{v,max}} \qquad (403.11)$$

Limited Sight Distance under Obstructing Object over Sag Curve

On sag curves, drivers of high vehicles such as trucks may have limited sight distance due to the presence of obstructing structures (such as an overpass). This situation is illustrated in Fig. 403.1.

The minimum length of curve is given by

$$L = \frac{AS^2}{800\left[C - \dfrac{h_1 + h_2}{2}\right]} \qquad \text{for} \qquad S \le L \qquad (403.12a)$$

$$L = 2S - \frac{800}{A}\left[C - \frac{h_1 + h_2}{2}\right] \qquad \text{for} \qquad S > L \qquad (403.12b)$$

Example 403.5

A parabolic curve of length 800 ft is used to transition from a grade of –4% to a grade of +5%. The minimum vertical clearance under an overpass is 13 ft 8 in. What is the maximum safe speed for a truck (assume height of driver's eye = 7.6 ft and height of obstruction = 2.0 ft) in order to satisfy the AASHTO criterion for sight distance?

Solution   The data for the problem are length of curve $L = 800$ ft, vertical clearance $C = 13.67$ ft, height of driver's eye $h_1 = 7.6$ ft, height of obstruction $h_2 = 2.0$ ft, algebraic grade difference $A = 5 - (-4) = 9$. Using the second Eq. (403.12b)

$$2S = L + \frac{800}{A}\left[C - \frac{h_1 + h_2}{2}\right] = 800 + \frac{800}{9}\left(13.67 - \frac{7.6 + 2.0}{2}\right) \Rightarrow S = 794 \text{ ft}$$

This does not satisfy the criterion $S > L$. Therefore, we use Eq. (403.12a)

$$S = \sqrt{\frac{800\left[C - \dfrac{h_1 + h_2}{2}\right]L}{A}} = \sqrt{\frac{800\left(13.67 - \dfrac{7.6 + 2.0}{2}\right)800}{9}} = 794 \text{ ft}$$

This does satisfy the criterion $S \le L$.

The design speed (for level roadway) corresponding to a sight distance $S = 794$ ft is 73.8 mph.

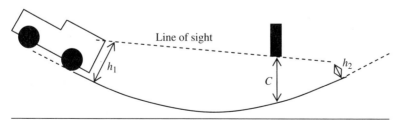

Figure 403.1 Sight distance on a vertical curve under an obstruction.

Table 403.3 Stopping Sight Distance on Vertical Curves (AASHTO)

	$S \leq L$	$S > L$
Crest vertical curve general:	$L = \dfrac{AS^2}{100\left(\sqrt{2h_1} + \sqrt{2h_2}\right)^2}$	$L = 2S - \dfrac{200\left(\sqrt{h_1} + \sqrt{h_2}\right)^2}{A}$
Stopping sight distance Crest vertical curve $h_1 = 3.5$ ft and $h_2 = 2.0$ ft	$L = \dfrac{AS^2}{2158}$	$L = 2S - \dfrac{2158}{A}$
Passing sight distance Crest vertical curve $h_1 = 3.5$ ft and $h_2 = 3.5$ ft	$L = \dfrac{AS^2}{2800}$	$L = 2S - \dfrac{2800}{A}$
Sag vertical curve general:	$L = \dfrac{AS^2}{200(2 + S\tan 1°)}$	$L = 2S - \dfrac{200(2 + S\tan 1°)}{A}$
Stopping sight distance Sag vertical curve (headlight criterion)	$L = \dfrac{AS^2}{400 + 3.5S}$	$L = 2S - \dfrac{400 + 3.5S}{A}$
Sag vertical curve (comfort criterion: centripetal acceleration $a_{v,max} = 1$ ft/s²)	$L = \dfrac{AV^2}{46.5}$	
Sag vertical curve (limited vertical clearance under overhead structure) C = vertical clearance (ft)	$L = \dfrac{AS^2}{800\left(C - \dfrac{h_1 + h_2}{2}\right)}$	$L = 2S - \dfrac{800}{A}\left(C - \dfrac{h_1 + h_2}{2}\right)$
For $h_1 = 8$ ft and $h_2 = 2$ ft	$L = \dfrac{AS^2}{800(C - 5)}$	$L = 2S - \dfrac{800}{A}(C - 5)$

Table 403.3 shows a summary of the stopping sight distance criteria in the section "Sight Distance on Vertical Curves."

Horizontal Curves

On horizontal curves, the obstruction to sight lines is due to proximity of adjacent structures (signposts, buildings, etc.). In Fig. 403.2, a two-lane roadway is shown. The dashed line shows the centerline of the roadway (usually the baseline curve). The centerlines of the inner and outer lanes are also shown. A sight obstructing object is shown located adjacent to the inner edge of the roadway. The obstructing effect of this object is more pronounced on the driver traveling on the inside lane.

The lateral offset of the object from this curve (centerline of the inside lane) is shown as M in the figure. The design speed of the vehicle can be directly related to the required stopping sight distance S. This distance must be minimally available with the sight obstruction at distance M as shown.

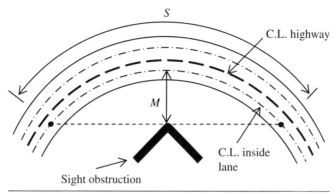

Figure 403.2 Sight distance on a horizontal curve around an obstruction.

For a horizontal circular curve of radius R, the middle ordinate distance M that corresponds to a particular arc length S is given by

$$M = R\left[1 - \cos\left(\frac{28.65S}{R}\right)\right] \tag{403.13}$$

This may also be written as

$$S = \frac{R}{28.65}\left[\cos^{-1}\left(1 - \frac{M}{R}\right)\right] \tag{403.14}$$

The middle ordinate distance M should be measured to the centerline of the traveled lane. In Eqs. (403.13) and (403.14), angles are expressed in degrees, not radians.

Example 403.6

The centerline of the local horizontal alignment of a two-lane highway is a circular curve with degree of curve = 3° (based on 100-ft arc). A sight obstructing object is located at a clear distance of 10 ft from the inside edge of the highway. What is the safe design speed (mph) on the curve? Assume standard lane width = 12 ft.

Solution   The radius of curve (highway centerline) is given by

$$R = \frac{5729.578}{3} = 1909.86 \text{ ft}$$

The radius of the inside lane = $1909.86 - 6.0 = 1903.86$ ft.

The middle ordinate distance is calculated as the offset between the obstruction and the centerline of the inside lane $= 10.0 + 6.0 = 16$ ft. The corresponding stopping sight distance is calculated as

$$S = \frac{R}{28.65}\left[\cos^{-1}\left(1 - \frac{M}{R}\right)\right] = \frac{1903.86}{28.65}\left[\cos^{-1}\left(1 - \frac{16}{1903.86}\right)\right] = 493.96 \, \text{ft}$$

Using the Table 403.1 for level grade, the corresponding design speed is 55 mph.

Decision Sight Distance

Some sections of a highway should provide greater (than safe stopping sight distance) sight distance to allow drivers to react to more complex situations than a simple stop. Providing only the stopping sight distance may preclude drivers from performing evasive maneuvers.

This may be applicable to interchanges, intersections where unusual or unexpected maneuvers are required, changes in cross section (lane drops or lane additions), toll plazas, and intense demand areas with a lot of "visual noise" from multiple sources of information such as roadway elements, traffic control devices, and advertising signs. Because decision sight distance gives drivers sufficient length to maneuver their vehicle at the same or reduced speed, rather than just to stop, its value is significantly more than the stopping sight distance.

The *AASHTO Green Book* provides recommendations for five avoidance maneuvers. These are as follows:

Avoidance Maneuver A: Stop on rural road ($t_R = 3.0$ s)

Avoidance Maneuver B: Stop on urban road ($t_R = 9.1$ s)

Avoidance Maneuver C: Speed/path/direction change rural road ($t = 10.2$ to 11.2 s)

Avoidance Maneuver D: Speed/path/direction change suburban road ($t = 12.1$ to 12.9 s)

Avoidance Maneuver E: Speed/path/direction change urban road ($t = 14.0$ to 14.5 s)

For avoidance maneuvers A and B, the decision sight distance is calculated in the same way as for stopping sight distance, using the reaction times as listed above, instead of 2.5 s.

For example, decision sight distance for maneuver B at design speed of 40 mph is given by

$$S = 1.47 \times 40 \times 9.1 + \frac{1.075 \times 40^2}{11.2} = 535.1 + 153.6 = 688.7 \, \text{ft}$$

In Table 403.4, this has been rounded to 690 ft.

For avoidance maneuvers C, D, and E, the decision sight distance is calculated by replacing the second term (braking distance) with a maneuver time that varies between 4.5 s at low speeds to 3.5 s at higher speeds. This results in a total (premaneuver and maneuver) time for maneuver C to vary between 10.2 and 11.2 s, for maneuver D to vary between 12.1 and 12.9 s, and for maneuver E to vary between 14.0 and 14.5 s. The values of DSD listed in Table 403.4 are based on

Table 403.4 Decision Sight Distance by Maneuver Type

Design speed (mph)	Decision sight distance (ft) by maneuver type				
	A	B	C	D	E
30	220	490	450	535	620
35	275	590	525	625	720
40	330	690	600	715	825
45	395	800	675	800	930
50	465	910	750	890	1030
55	535	1030	865	980	1135
60	610	1150	990	1125	1280
65	695	1275	1050	1220	1365
70	780	1410	1105	1275	1445
75	875	1545	1180	1365	1545
80	970	1685	1260	1455	1650
85	1070	1830	1340	1565	1785

considering both the premaneuver and maneuver times. For example, DSD for maneuver E at design speed of 40 mph is given by

$$S = 1.47 \times 40 \times (9.5 + 4.5) = 558.6 + 264.6 = 823.2 \text{ ft}$$

In the table, this has been rounded to 825 ft.

Passing Sight Distance

In the matter of calculating passing sight distance, there has been a significant change from the 5th edition (2004) of the "Policy on Geometric Design of Highways and Streets" to the 6th edition (2011). In the 5th edition of the Green Book, the passing sight distance on two-lane highways was calculated as the sum of travel distance during four separate maneuvers:

1. The overtaken vehicle travels at constant speed.

2. During the passing maneuver, once the passing driver perceives a sufficient clear distance, the passing vehicle accelerates to a speed which is 10 mph higher than the vehicle being passed.

3. When the passing vehicle returns to its lane, there is suitable clearance between it and the opposing vehicle.

4. While the passing vehicle traverses distance d_1, d_2, and d_3, the opposing vehicle travels a distance d_4.

The minimum passing sight distance for two-lane highways was then calculated as the sum of distances d_1, d_2, d_3, and d_4. The passing sight distances calculated from this method are shown in Table 403.5 and are significantly higher than those in the MUTCD (Manual on

Table 403.5 Passing Sight Distance for Design of Two-Lane Highways (Green Book, 5th ed.)

Speed range (mph)	Assumed speeds (mph)		Passing sight distance (ft)
	Passed Vehicle	**Passing Vehicle**	
30–40	24.9	34.9	1040
40–50	33.8	43.8	1468
50–60	42.6	52.6	1918
60–70	52.0	62.0	2383

Uniform Traffic Control Devices, 2009) guidelines, which are aimed toward providing guidance for demarcating no-passing zones (solid lines separating opposing traffic on two-lane highways). The MUTCD guidelines are shown in Table 403.7.

Table 403.6 shows the passing sight distance according to the guidelines of the 6th edition of the Green Book.

Table 403.6 Passing Sight Distance for Design of Two-Lane Highways (Green Book, 6th ed.)

Design speed (mph)	Assumed speeds (mph)		Passing sight distance (ft)
	Passed Vehicle	**Passing Vehicle**	
20	8	20	400
25	13	25	450
30	18	30	500
35	23	35	550
40	28	40	600
45	33	45	700
50	38	50	800
55	43	55	900
60	48	60	1000
65	53	65	1100
70	58	70	1200
75	63	75	1300
80	68	80	1400

Table 403.7 Minimum Passing Sight Distances for No-Passing Zone Markings (MUTCD 2009)

85th Percentile speed/posted speed/ statutory speed limit (mph)	Minimum passing zone length (ft)
25	450
30	500
35	550
40	600
45	700
50	800
55	900
60	1000
65	1100
70	1200

Methods for Increasing Passing Opportunities on Two-Lane Roads

The Green Book provides the following guidelines for providing passing sections on two-lane highways:

1. Horizontal and vertical alignment should be designed to provide as much of the highway as possible with the passing sight distance in Table 403.6.

2. Where the design volume approaches capacity, the effect of the lack of passing opportunities on the level of service must be recognized.

3. Where the physical length of an upgrade is greater than the critical length of grade (Green Book Figure 3-29), addition of added climbing lanes must be considered.

4. Where strategies 1–3 still do not provide adequate passing opportunities, construction of passing lane sections should be considered.

Passing lane sections are particularly advantageous in rolling terrain, especially where alignment is winding or the profile includes critical lengths of grade. A minimum length of 1000 ft, excluding tapers, is needed so that delayed vehicles have an opportunity to complete at least one pass in the added lane. The optimal length is usually 0.5–2.0 miles, with longer lengths appropriate were traffic volumes are greater.

Highway Curves

Elements of Surveying

Azimuth

The azimuth of a line is the horizontal angle measured clockwise to the line from a specific meridian (usually north). Figure 404.1 shows a line *AB* with the azimuthal angles shown at ends *A* and *B*. The back azimuth of a line is the azimuth of the line running in the reverse direction. For example, in Fig. 404.1, the back azimuth of line *AB* is equal to the azimuth of line *BA*. When the azimuth is less than 180°, the back azimuth equals the azimuth plus 180° and when the azimuth is greater than 180°, the back azimuth equals the azimuth minus 180°.

Bearings

Bearings of lines are directional (horizontal) angles with respect to a meridian (north or south) measured at the originating point on the line. The bearing angle is conventionally chosen to be the acute angle. For example, the bearing of a line headed in the north-west direction can be written as N45°W ("45° west of north").

The angle between two lines whose bearings are given may be calculated as the difference between their azimuths. For example, if the two lines are N50°E and S70°W, then the first step will be to describe them as azimuths. It is helpful to visualize the angles: S70°W is in the third quadrant; 70° west of south, as shown in the following figure. The azimuth of N50°E is 50° and the azimuth of S70°W is 70° + 180° = 250°. Therefore, the angle between the two lines is 250° − 50° = 200°.

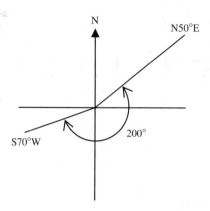

Distance Measurements

The distance between two points on a plane can be calculated if the coordinates of the points are known with respect to some origin. These coordinates, usually described in a coordinate system with axes running north-south and east-west, are called northings and eastings. On the other hand, orthogonal projections of a line segment are called latitude and departure.

Latitude and Departure

In the rectangular coordinate system, where the north-south meridian serves as the y-axis and the east-west line serves as the x-axis, the projections of a line are termed *latitude* (y projection) and *departure* (x projection), as shown in Fig. 404.2. The departure is considered positive to the east (i.e., the line shows an *increase in easting*) and the latitude is considered positive to the north (i.e., the line shows an *increase in northing*).

If the azimuth of the line shown is θ, then the latitude is given by $L\cos\theta$ and the departure by $L\sin\theta$, where L is the length of line AB.

Figure 404.1
Azimuth angle.

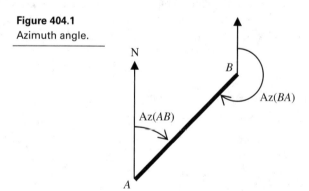

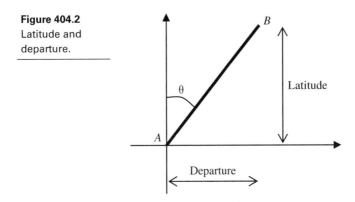

Figure 404.2
Latitude and departure.

Northings and Eastings

While latitudes and departures are north-south and east-west projections, respectively, of a line segment, northings and eastings are coordinates of points in a traverse. Thus, for line segment *AB*, the latitude is the difference of the northings of points *A* and *B* and the departure is the difference of the eastings.

Horizontal Curves

The geometry of a horizontal circular curve is shown in Fig. 404.3. PC is the *point of curvature,* PT the *point of tangent,* and PI the *point of intersection* (of tangents). The *internal angle* subtended by the major chord (PC to PT) is *I*. This is also the *deflection angle between*

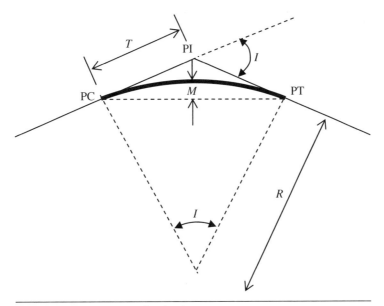

Figure 404.3 Geometry of a circular curve.

tangents at PI. For the sense of the curve as shown in the figure, one would say that the deflection angle is "to the right." E is called the *external distance* and M the *middle ordinate distance*. There are two different ways to specify the curvature of a circular curve:

1. Using a 100-ft arc, specifying a circular curve as "degree of curve = $2°$" means that a 100-ft arc along the curve subtends an angle of $2°$ at the center. In this case, the radius of the curve is given by

$$R = \frac{5729.578}{D} \qquad (404.1)$$

   Note that the factor 5729.578 is the product of 100 and 57.29578...—the radian-to-degree conversion factor—and is shown truncated to a precision of 0.001.
   The length of the curve is then given by

$$L = \frac{100I}{D} \qquad (404.2)$$

2. Using a 100-ft chord, specifying a circular curve as "degree of curve = $2°$" means that a 100-ft *chord* subtends an angle of $2°$ at the center. In this case, the radius and length of the curve are given by

$$R = \frac{50}{\sin(D/2)} \qquad (404.3)$$

$$L = \frac{RI}{57.29578} \qquad (404.4)$$

In Eqs. (404.2) through (404.4), all angles are in degrees.

$$\text{The length of the major chord (from PC to PT)} = 2R\sin\left(\frac{I}{2}\right)$$

The tangent length T is the distance from the PC to the PI (also the distance from the PI to the PT, by symmetry) and is given by

$$T = R\tan\left(\frac{I}{2}\right) \qquad (404.5)$$

The middle ordinate distance M is the radial distance from the midpoint of the major chord (which connects the PC to the PT) to the curve.

$$M = R\left[1 - \cos\left(\frac{I}{2}\right)\right] = E\cos\left(\frac{I}{2}\right) \qquad (404.6)$$

The external distance E is the radial distance from the PI to the curve.

$$E = R\left[\sec\left(\frac{I}{2}\right) - 1\right] = M \sec\left(\frac{I}{2}\right) \tag{404.7}$$

Staking out of a circular curve may be done in one of several ways:

By deflection angle from the back tangent

By offsets from the back tangent

By offsets from the major chord

Horizontal Curve Layout—by Deflection Angle

In this method, the horizontal curve is constructed by laying out chord distances from the PC using corresponding deflection angles from the back tangent. In Fig. 404.4, the point A is a point on the curve, α is the deflection angle from the back tangent, and PC-A is the chord length. In order to lay out the curve, one needs pairs of values of angle α and distance PC-A. Note that the deflection angle (α) is half the angle at the center (β). The deflection angle for any point A on the curve may be calculated using

$$\alpha = (\text{arc PC} - A) \times \left(\frac{I}{2L}\right) \tag{404.8}$$

In Eq. (404.8), both the length of curve and the length of the arc should be measured in the same units (either feet or stations) and the angles (α and I) are both in degrees. Since the arc length PC-A is the difference in stations between station A and station PC,

$$\text{sta. } A_{\text{ft}} = \text{sta. } \text{PC}_{\text{ft}} + \frac{200\alpha}{D} \tag{404.9}$$

where α is the deflection angle (degrees) and D is the degree of the curve.

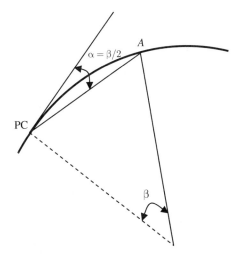

Figure 404.4
Layout of a circular curve by deflection angle.

Example 404.1

A circular curve is laid out by deflection angles. The degree of curve (based on 100-ft arc) is $3°$. The PC is at station $12 + 30.65$ and the PI is at station $15 + 43.76$. What is the deflection angle for station $13 + 00.00$ on the curve?

Solution Tangent length T is the distance from the PC to the PI.

$$T = 1543.76 - 1230.65 = 313.11 \text{ ft}$$

$$\text{Radius of curve is calculated as } R = \frac{5729.578}{D} = \frac{5729.578}{3} = 1909.86$$

Using $T = R\tan(I/2)$, the deflection angle between tangents is calculated as $I = 18.62°$.

$$\text{Length of curve } L = \frac{100I}{D} = \frac{100 \times 18.62}{3} = 620.67 \text{ ft}$$

The arc length from PC to station $13 + 00.00 = 69.35$ ft. The corresponding deflection angle is given by

$$\alpha = 69.35 \times \left(\frac{18.62}{2 \times 620.67}\right) = 1.0403° = 1°02'25''$$

Horizontal Curve Layout—by Chord Offset

In this method, the horizontal curve is constructed by laying out points along the major chord (connecting PC to PT) and then laying out perpendicular offsets from these points. In Fig. 404.5, the point B is a point on the major chord and BA is the offset distance to the point on the curve (A),

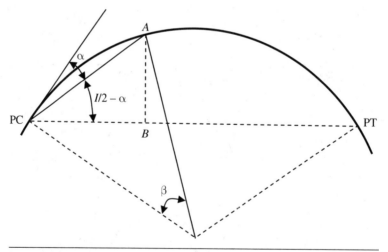

Figure 404.5 Layout of a circular curve by chord offset.

α is the deflection angle from the back tangent, and PC-A is the chord length. In order to lay out the curve, one needs pairs of values of distance PC-B and distance BA. Note that the deflection angle (α) is half the angle at the center (β).

$$\text{The length of the major chord (distance from PC to PT)} = 2R\sin\left(\frac{I}{2}\right) \qquad (404.10)$$

$$\text{The length of the chord PC-A} = 2R\sin\alpha$$

$$\text{Distance along the chord PC-B} = 2R\sin\alpha\cos\left(\frac{I}{2}-\alpha\right) \qquad (404.11)$$

$$\text{Offset from the chord } BA = 2R\sin\alpha\sin\left(\frac{I}{2}-\alpha\right) \qquad (404.12)$$

If pairs of PC-B and BA values are given, squaring and adding them yields the square of the quantity $2R\sin\alpha$. The ratio BA:PC-B yields the quantity $\tan(1/2 - \alpha)$. Thus, if the intersection angle (I) is known, one may solve for angle (α) and then for radius of curve (R).

Example 404.2

A circular horizontal curve is being constructed by the method of chord offsets. The degree of the curve is $4°$. The curve joins tangents with a deflection angle $23°30'$ (right). A point P is located on the major chord at a distance of 352.00 ft from the PC. Determine the length of the chord offset at point P.

Solution   Given $I = 23.5°, D = 4°$

Radius of the curve is $R = \dfrac{5729.578}{4} = 1432.4$ ft

The distance along the chord is given by

$$2R\sin\alpha\cos\left(\frac{I}{2}-\alpha\right) = 2864.8\sin\alpha\cos\left(11.75 - \alpha\right)$$

If this is set equal to 352, a trial and error solution yields the value of the deflection angle $\alpha = 7.1°$. Using this value of the deflection angle, we get the offset distance as

$$2R\sin\alpha\sin\left(\frac{I}{2}-\alpha\right) = 2864.8 \times \sin 7.1 \times \sin\left(11.75 - 7.1\right) = 28.7 \text{ ft}$$

Horizontal Curve Layout—by Tangent Offset

In this method, the horizontal curve is constructed by laying out points along the back tangent and then laying out perpendicular offsets from these points. In Fig. 404.6, the point B is a point on (an extension of) the back tangent and BA is the offset distance to the point on the curve (A). To lay out the curve, one needs pairs of values of distance PC-B and distance BA.

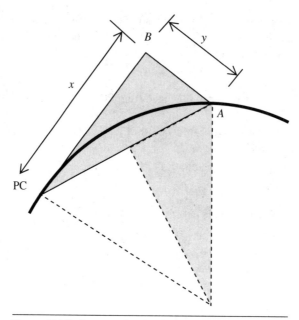

Figure 404.6 Layout of a circular curve by tangent offset.

Since the angle between the tangent and the chord (deflection angle) is half the central angle subtended by the chord, the two triangles shown shaded in Fig. 404.6 are similar triangles. This results in the following relationship:

$$R = \frac{x^2 + y^2}{2y} \qquad (404.13)$$

This form is useful for estimating the radius of the curve if tangent distance (x) and tangent offset (y) are known. This can also be written as $y = R - \sqrt{R^2 - x^2}$ (useful for estimating the tangent offset (y) if radius of the curve (R) and tangent distance (x) are known).

Example 404.3

Two tangents intersecting at an angle of 36°42′ are to be joined by a 6°30′ curve. If the degree of curve is based on a 100-ft arc, compute the tangent distance and the radius of the curve.

Solution The deflection angle for the curve is $I = 36°42′$ and the degree of curve $D = 6°30′$. The radius of the curve is calculated as

$$R = \frac{5729.578}{6.5} = 881.47 \text{ ft}$$

The tangent distance is

$$T = R\tan\frac{I}{2} = 881.47 \times \tan 18°21' = 292.37 \text{ ft}$$

Example 404.4

Two tangents intersecting at an angle of 36°42′ are to be joined by a 6°30′ curve. The PI for the curve is at 9 + 31.00 and the degree of curve is based on a 100-ft chord. Determine the station of PC and PT.

Solution   If the degree of curve is based on a 100-ft *chord*, the relationship is

$$100 = 2R\sin\frac{6.5°}{2} \Rightarrow R = \frac{100}{2\sin 3.25°} = 881.95 \text{ ft}$$
$$T = R\tan\frac{I}{2} = 881.95 \times \tan 18°21' = 292.53 \text{ ft} = 2 + 92.53 \text{ sta}$$

PI station = PC sta + T = 9 + 31.00

Therefore, PC station = (9 + 31.00) − (2 + 92.53) = 6 + 38.47. Length of curve L is the *arc length* corresponding to R = 881.95 ft and intersection angle = 36°42′

$$L = 881.95 \times \frac{36.7}{57.29578} = 564.92 \text{ ft} = 5 + 64.92 \text{ sta}$$

Therefore, PT station = (6 + 38.47) + (5 + 64.92) = 12 + 03.39.

Example 404.5

The PC of a 6° curve is at station 15 + 27.00. Determine the deflection angle from the tangent at the PC to each of the following points on the curve: (a) station 16 + 00.00, (b) station 17 + 00.00, (c) station 20 + 98.00.

Solution   Degree of curve = 6

$$\text{Using the equation} \quad \text{sta.}A_{ft} = \text{sta. PC}_{ft} + \frac{200\alpha}{D}$$

$$\alpha = \frac{D}{200} \times (\text{sta.}A - \text{sta. PC}) \quad \text{we obtain}$$

For station 16 + 00 $\alpha = 2.19° = 2°11'24''$
For station 17 + 00 $\alpha = 5.19° = 5°11'24''$
For station 20 + 98 $\alpha = 17.13° = 17°7'48''$

Example 404.6

The PC of a 3° curve, whose degree is based on a chord of 100 ft, is at station 19 + 39.60 and the PT is at station 25 + 64.60. Calculate the deflection angle from the tangent at the PC to station 22 + 00.00 on the curve and to the PT.

Solution   Degree of curve (based on 100-ft chord) = 3

$$R = \frac{50}{\sin 1.5} = 1910.08 \text{ ft}$$

Length of curve = distance from PC to PT = 625 ft

$$L = \frac{RI}{57.29578} \Rightarrow I = \frac{57.29578 \times 625}{1910.08} = 18.75$$

$$\text{Central angle } \beta = \frac{\text{arc length}}{\text{radius}}$$

For station 22 + 00.00, arc length = 260.4 ft, central angle = 0.1363 radian, deflection angle = 0.0682 radian = 3.908° = 3°54′22″.

For station 25 + 64.60, central angle = 18.75°, deflection angle = 9°22′30″.

Example 404.7

The PC of a 14° curve, whose degree is based on an arc of 100 ft, is at station 49 + 27.36 and the angle of intersection of the tangents is 60°58′. Calculate the deflection angle from the tangent at the PC to station 49 + 50.00 on the curve.

Solution

$$\text{Length of curve } L = \frac{100I}{D} = \frac{100 \times 60.967}{14} = 435.48$$

Therefore, PT is at station 53 + 62.84.

$$\text{Radius } R = \frac{5729.578}{14} = 409.26 \text{ ft}$$

For station 49 + 50.00, arc length = 22.64 ft, central angle = 0.0553 radian, and deflection angle = 0.0277 radian = 1.585° = 1°35′05″.

The following templates can be used to determine bearings or azimuths of specific lines for a horizontal (circular) curve. Some of these lines are (1) the back tangent, (2) the forward tangent, (3) the major chord from PC to PT, (4) the radius to the PC, and (5) the radius to the PT, etc. The following letter subscripts are used below to identify these lines:

F, forward tangent; B, back tangent; C, major chord from PC to PT; R_1, outward radius to PC; R_2, outward radius to PI; R_3, outward radius to PT.

The first set of results is for a curve that deflects right and the next set is for a curve that deflects left.

All these relationships are expressed in terms of azimuth because the azimuth is *always* expressed consistently (as in "clockwise from the north meridian").

Curve Deflects Right

$Az_F = Az_B + \Delta$

$Az_C = Az_B + \Delta/2$

$Az_{R1} = Az_B - 90$

$Az_{R2} = Az_C - 90 = Az_B + \Delta/2 - 90$

$Az_{R3} = Az_F - 90 = Az_B + \Delta - 90$

Deflection angle of any point P on curve $\alpha = arc\ PC - P \times \dfrac{l}{2L}$

$Az_{chord\ to\ P} = Az_B + \alpha/2$

$Az_{radius\ thru\ P} = Az_B + \alpha/2 - 90$

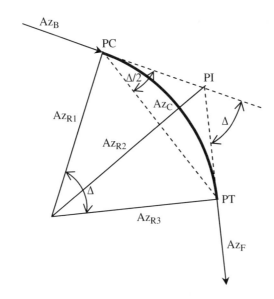

Curve Deflects Left

$Az_F = Az_B - \Delta$

$Az_C = Az_B - \Delta/2$

$Az_{R1} = Az_B + 90$

$Az_{R2} = Az_C + 90 = Az_B - \Delta/2 + 90$

$Az_{R3} = Az_F + 90 = Az_B - \Delta + 90$

Deflection angle of any point P on curve $\alpha = arc\ PC - P \times \dfrac{l}{2L}$

$Az_{chord\ to\ P} = Az_B - \alpha/2$

$Az_{radius\ thru\ P} = Az_B + \alpha/2 - 90$

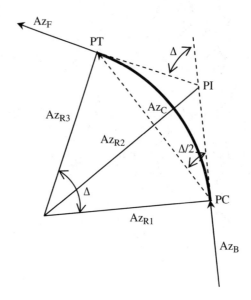

Example 404.8

A horizontal circular curve has the following data:

Radius = 1250 ft

Bearing of the back tangent = N 23°34′45″W

Bearing of the forward tangent = N 34°46′30″E

PC at sta. 23 + 32.67

What is the bearing of the (a) major chord, (b) radial line to the PI, and (c) the chord from the PC to the station 30 + 0.00 on the curve?

Solution

Bearing of the back tangent = N 23.579° W

The azimuth of the back tangent = 360 – 23.579 = 336.421°

Bearing of the forward tangent = N 34.775° E

The azimuth of the forward tangent = 34.775°

Therefore, curve deflects by $\Delta = 34.775 - 336.421 = -301.646$ (which is the same as 58.354)

Degree of curve: $D = \dfrac{5729.578}{R} = \dfrac{5729.578}{1250} = 4.584$

Length of curve: $L = \dfrac{100\Delta}{D} = \dfrac{100 \times 58.354}{4.584} = 1273.09$ ft

(a) Azimuth of major chord, $Az_C = Az_B + \Delta/2 = 336.421 + 58.354/2 = 365.598$, which is the same as 5.598

Bearing of major chord $= $ N 5°35'53"E

(b) Azimuth of the radial line (from center to PI), $Az_{R_2} = Az_B + \Delta/2 - 90 = 275.598$

Bearing of this radial line (from center to PI) $= 360 - 275.598 = 84.402° = $ N 84°24'07"W

(c) Arc length to point on curve $= 30.0 - 23.3267 = 6.6733$ sta $= 667.33$ ft

Deflection angle to this point: $\alpha = arc \times \dfrac{\Delta}{2L} = 667.33 \times \dfrac{58.354}{2 \times 1273.09} = 15.294°$

Azimuth of chord from PC to this point on the curve, $Az_P = Az_B + \alpha/2 = 336.421 + 15.294/2 = 344.068°$

Bearing of this chord $= $ N 15°55'55"W

Offtracking and Pavement Widening

Offtracking is the phenomenon, more prevalent in larger vehicles, in which the rear wheels do not follow the same path as the front wheels when the vehicle is traveling on a curve. If the effect is severe enough, the AASHTO Green Book recommends widening of the pavement. The recommendations for pavement widening (one-way and two-way two-lane highways) are summarized in Table 404.1. The base values are for a WB-62 design vehicle (interstate semi-trailer). The last two columns provide adjustments for WB-100T (triple trailer combination) and WB-109D (turnpike-double combination) vehicles. Because widening is expensive, value lower than 2.0 ft from Table 404.1 may be ignored.

Turning Roadways at Intersections

There are three typical types of right turning roadways at intersections:

1. minimum edge of traveled way design (where it is appropriate to provide for turning vehicles within minimum space
2. a design with a corner triangular island, and
3. a free-flow design using a simple radius or compound radii

For curves at intersections—either a simple curve or a simple curve with taper—Table 9-15 of the AASHTO Green Book provides recommended curve radius for various standard vehicle types and various turn angles (ranging from 30° to 180°).

Table 9-16 provides guidance for three-centered curves.

Table 9-17 gives guidance on cross-street width occupied by turning vehicles for two scenarios:

Case A: Where vehicle turns from proper lane and swings wide on cross street, and

Case B: Where turning vehicle swings equally wide on both streets.

Table 404.1 Traveled Way Widening on Open Highway Curves (Two-Lane Highways) for WB-62 Design Vehicle

Curve radius (ft)	Design Values for Pavement Widening on Highway Curves (ft): Base Values are for WB-62 Vehicle												Δ for WB-100T	Δ for WB-109D
	Roadway width = 24 ft				Roadway width = 22 ft				Roadway width = 20 ft					
	Design speed (mph)				Design speed (mph)				Design speed (mph)					
	30	40	50	60	30	40	50	60	30	40	50	60		
7000	0.0	0.0	0.0	0.0	0.7	0.8	0.9	1.0	1.7	1.8	1.9	2.0	−0.1	0.2
6500	0.0	0.0	0.0	0.1	0.7	0.8	1.0	1.1	1.7	1.8	2.0	2.1	−0.1	0.2
6000	0.0	0.0	0.0	0.1	0.7	0.9	1.0	1.1	1.7	1.9	2.0	2.1	−0.1	0.2
5500	0.0	0.0	0.1	0.2	0.8	0.9	1.1	1.2	1.8	1.9	2.1	2.2	−0.1	0.2
5000	0.0	0.0	0.1	0.3	0.9	1.0	1.1	1.3	1.9	2.0	2.1	2.3	−0.1	0.3
4500	0.0	0.1	0.2	0.4	0.9	1.1	1.2	1.4	1.9	2.1	2.2	2.4	−0.1	0.3
4000	0.0	0.2	0.3	0.5	1.0	1.2	1.3	1.5	2.0	2.2	2.3	2.5	−0.1	0.3
3500	0.1	0.3	0.5	0.6	1.1	1.3	1.5	1.6	2.1	2.3	2.5	2.6	−0.1	0.4
3000	0.3	0.4	0.6	0.8	1.3	1.4	1.6	1.8	2.3	2.4	2.6	2.8	−0.1	0.5
2500	0.5	0.7	0.9	1.1	1.5	1.7	1.9	2.1	2.5	2.7	2.9	3.1	−0.1	0.5
2000	0.7	1.0	1.2	1.4	1.7	2.0	2.2	2.4	2.7	3.0	3.2	3.4	−0.2	0.7
1800	0.9	1.1	1.4	1.6	1.9	2.1	2.4	2.6	2.9	3.1	3.4	3.6	−0.2	0.8
1600	1.1	1.3	1.6	1.8	2.1	2.3	2.6	2.8	3.1	3.3	3.6	3.8	−0.2	0.8
1400	1.3	1.6	1.9	2.1	2.3	2.6	2.9	3.1	3.3	3.6	3.9	4.1	−0.3	1.0
1200	1.7	1.9	2.1	2.5	2.7	2.9	3.2	3.5	3.7	3.9	4.2	4.5	−0.3	1.1
1000	2.1	2.4	2.7	3.0	3.1	3.4	3.7	4.0	4.1	4.4	4.7	5.0	−0.4	1.4
900	2.4	2.7	3.1		3.4	3.7	4.1		4.4	4.7	5.1		−0.4	1.5
800	2.7	3.1	3.5		3.7	4.1	4.5		4.7	5.1	5.5		−0.4	1.7
700	3.2	3.6	4.0		4.2	4.6	5.0		5.2	5.6	6.0		−0.5	1.9
600	3.8	4.2	4.6		4.8	5.2	5.6		5.8	6.2	6.6		−0.6	2.3
500	4.6	5.1			5.6	6.1			6.6	7.1			−0.7	2.7
450	5.2	5.7			6.2	6.7			7.2	7.7			−0.8	3.0
400	5.9	6.4			6.9	7.4			7.9	8.4			−0.9	3.4
350	6.8	7.3			7.8	8.3			8.8	9.3			−1.0	3.9
300	7.9				8.9				9.9				−1.2	4.6
250	9.6				10.6				11.6				−1.4	5.5
200	12.0				13.0				14.0				−1.8	7.0

For three-lane roadways, multiply values in the table by 1.5; for four-lane roadways, multiply by 2.0

Compound Curves

A compound horizontal curve consists of two or more circular curves in succession, turning in the same direction and successive curves sharing a common tangent. These curves are sometimes necessary at at-grade intersections, interchange ramps, and locally difficult topographic conditions. For successive curves, AASHTO recommends that ratio of the curve radii be not more than 1.5. The maximum desirable ratio of the radii is given by AASHTO as 1.75. Figure 404.7 shows a typical layout of a compound curve composed of two circular curves sharing a common tangent at PCC. The total deflection angle is the sum of the deflection angles of curves 1 and 2.

$$\Delta = \Delta_1 + \Delta_2 \qquad (404.14)$$

The tangent lengths of the simple circular curves are given by

$$t_1 = R_1 \tan\frac{\Delta_1}{2} \qquad (404.15)$$

$$t_2 = R_2 \tan\frac{\Delta_2}{2} \qquad (404.16)$$

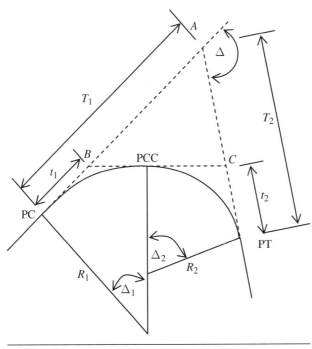

Figure 404.7 Geometry of a compound horizontal curve.

From triangle ABC, the law of sines gives

$$\frac{AB}{\sin \Delta_2} = \frac{AC}{\sin \Delta_1} = \frac{t_1 + t_2}{\sin \Delta} \qquad (404.17)$$

and the tangent lengths of the compound curve are given by

$$T_1 = t_1 + AB = t_1 + (t_1 + t_2)\frac{\sin \Delta_2}{\sin \Delta} \qquad (404.18)$$

$$T_2 = t_2 + AC = t_2 + (t_1 + t_2)\frac{\sin \Delta_1}{\sin \Delta} \qquad (404.19)$$

Alternate forms of the relationships for a two-centered compound (circular) curve are:

$$T_1 = \frac{R_2 - R_1 \cos \Delta + (R_1 - R_2)\cos \Delta_2}{\sin \Delta} \qquad (404.20)$$

$$T_2 = \frac{R_1 - R_2 \cos \Delta + (R_2 - R_1)\cos \Delta_1}{\sin \Delta} \qquad (404.21)$$

$$\sin \Delta_1 = \frac{T_1 + T_2 \cos \Delta - R_2 \sin \Delta}{R_1 - R_2} \qquad (404.22)$$

$$\sin \Delta_2 = \frac{T_2 + T_1 \cos \Delta - R_1 \sin \Delta}{R_2 - R_1} \qquad (404.23)$$

Example 404.9

A compound horizontal curve is shown in the figure below. The radius (ft) of the second curve (R2) is most nearly:

A. 1890
B. 2240
C. 2650
D. 2940

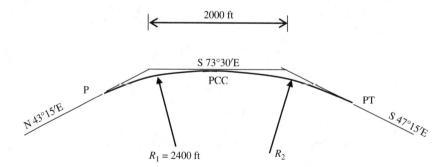

Solution

Forward tangent of curve 1 has bearing S73°30′E, which corresponds to an azimuth of $106°30′ = 106.5°$

Deflection angle for curve 1: $\Delta_1 = 106.5 - 43.25 = 63.25°$

Back tangent of curve 2 has bearing S47°15′E, which corresponds to an azimuth of $132°45′ = 132.75°$

Deflection angle for curve 2: $\Delta_2 = 132.75 - 106.5 = 26.25°$

Tangent length for curve 1: $t_1 = R_1 \tan\left(\dfrac{\Delta_1}{2}\right) = 2400 \times \tan\left(\dfrac{63.25}{2}\right) = 1477.93$

Tangent length for curve 1: $t_2 = 2000 - 1477.93 = 522.07 = R_2 \tan\left(\dfrac{\Delta_2}{2}\right) = R_2 \tan\left(\dfrac{26.25}{2}\right) =$

$R_2 = 2238.02$ ft

Answer is (B).

Vertical Curves

Vertical highway curves are parabolic curves connecting an approach tangent (gradient G_1) and an exit tangent (gradient G_2). If the x coordinate is measured horizontally (rather than along the curve), the equation of a simple parabolic vertical curve is

$$y(x) = y_{\text{PVC}} + G_1 x + \frac{1}{2}Rx^2 \qquad (404.24)$$

where $R = $ the rate of gradient change is given by

$$R = \frac{G_2 - G_1}{L} \qquad (404.25)$$

In Eqs. (404.25) and (404.26), gradient G and distance x may be expressed in percent (%) and stations (sta) respectively only if 100-ft (or 100-m) stations are employed. The effects of the two factors of 100 implicit in such a choice of units will cancel each other and elevation (y) will have units of feet (or meters). Thus, for 100-ft (m) stations, the rate of gradient change (R) will have units of %/sta. If the stationing interval is not 100, gradients G must be expressed in decimal (rather than percent) and distance x must be expressed in feet (or meters) rather than stations. As a result, the units of R will be ft^{-1} or m^{-1}.

 NOTE For crest curves, the rate of gradient change (R) is negative, resulting in a negative (downward) offset distance, while for sag curves the offset is positive.

The PVC is the beginning of the vertical curve, PVT is the end of the vertical curve, and PVI is the point of intersection of the two tangents. If the stationing of points (x-coordinate) is expressed horizontally, the PVI is exactly midway between the PVC and the PVT.

The offset from the tangent at any location grows as a parabolic function. This offset distance is the last term in the second-order equation for $y(x)$. Since this term is $1/2\,Rx^2$, we can say that if two locations x and x_1 are considered, the ratio of the tangent offsets at these two locations will be equal to the square of the ratio of the distances x and x_1.

$$\frac{AB}{A'B'} = \left(\frac{x}{x_1}\right)^2 \tag{404.26}$$

The vertical offset at the midpoint of the curve is given by $H_{L/2} = (G_2 - G_1)L/8$ and at the PVT is given by $H_L = (G_2 - G_1)L/2$. The location of the high or low point (turning point, marked TP in Fig. 404.8) of the curve is given by

$$x_{\mathrm{TP}} = -\frac{G_1}{R} \tag{404.27}$$

For a sag curve, the low point would be the optimum point for locating a drainage outlet.

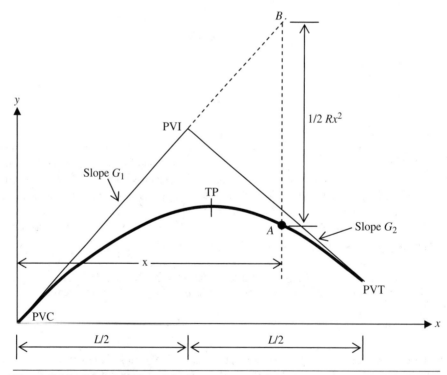

Figure 404.8 Geometry of a vertical curve.

Example 404.10

A –4% grade meets a +5% grade at station 34 + 00. If a vertical curve of length 600 ft is used to connect the two tangents, find the station of the low point.

Solution

$$\text{sta PVC} = \text{sta PVI} - 1/2\,L = 34.00 - 6/2 = 31.00$$

$$R = (G_2 - G_1)/L = [5 - (-4)]/6 = 1.5\ \%/\text{sta}$$

$$x_{TP} = -G_1/R = 4/1.5 = 2.67\ \text{sta}$$

Thus, the low point is at station $(31 + 00) + (2 + 66.677) = 33 + 66.67$.

Example 404.11

In the figure shown below, vertical curve no. 1 is followed by a tangent section, which is followed by vertical curve no. 2. What is the vertical clearance under the bridge structure at station 40 + 55.00?

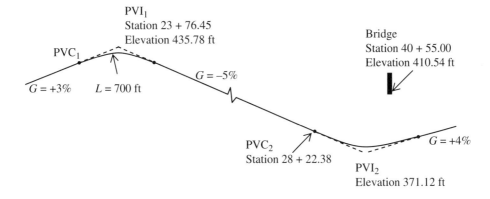

Solution   The distance between PVI_1 and $PVC_2 = 2822.38 - 2376.45 = 445.93$ ft $= 4.4593$ sta.
The elevation at $PVC_2 = 435.78 - 5 \times 4.4593 = 413.48$ ft
The elevation difference between PVC_2 and $PVI_2 = 413.48 - 371.12 = 42.36$ ft
The distance between PVC_2 and $PVI_2 = 42.36 \div 5 = 8.472$ sta
Therefore, the length of curve no. 2 $= 16.944$ sta

The rate of gradient change for curve no. 2 $= R = \dfrac{4 - (-5)}{16.944} = 0.5312\ \%/\text{sta}$

The elevation of the point on curve no. 2 at sta 40 + 55.00 (12.3262 sta. ahead of PVC) is calculated as

$$y = 413.48 + (-5) \times 12.3262 + \frac{1}{2} \times 0.5312 \times 12.3262^2 = 392.20$$

Vertical clearance $= 410.54 - 392.20 = 18.34$ ft

Parabolic Curve to Pass through a Given Point

The application of the vertical curve parabolic equation becomes problematic if the PVC is not predefined (location, elevation). This is because that formulation utilizes the PVC as the origin for the parabolic curve (i.e., in the previous formulation, the distance x is measured horizontally from the PVC).

When the position of a point on the curve is known with respect to the PVI (rather than the PVC) in terms of horizontal and vertical offsets h and v (Fig. 404.9), the required length of curve L is given by

$$\frac{L + 2h}{L - 2h} = \sqrt{\frac{v - G_1 h}{v - G_2 h}} \tag{404.28}$$

Example 404.12

A grade of −1.6% is followed by a grade of +3.8%, the grades intersecting at station 42 + 0.00. The elevation of the PVI is 210.00 ft. A parabolic vertical curve connects these two grades and passes through station 42 + 60.00 at elevation 213.70 ft. Calculate the required length of curve.

Solution   The offsets of the point on the curve from the PVI are given by

$$h = 42.6 - 42.0 = 0.6 \text{ sta}$$
$$v = 213.70 - 210.00 = 3.70 \text{ ft}$$

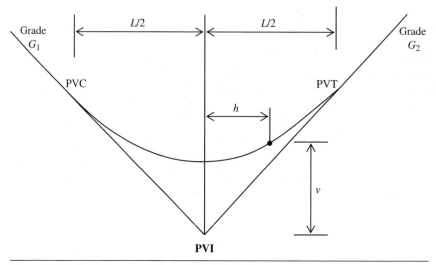

Figure 404.9   Vertical curve defined in terms of offsets from PVI.

The gradients are $G_1 = -1.6$ and $G_2 = +3.8$. Therefore, using Eq. (404.28)

$$\frac{L + 2 \times 0.6}{L - 2 \times 0.6} = \sqrt{\frac{3.7 + 1.6 \times 0.6}{3.7 - 3.8 \times 0.6}} = 1.81154$$

Solving this linear equation, $L = 4.157$ stations $= 415.7$ ft.

Example 404.13

A vertical curve connects tangents of $G_1 = +2\%$ and $G_2 = -5\%$ that intersect at station $58 + 25.60$ having elevation $= 128.54$ ft and the vertical curve is to be 800-ft long, what should be the elevation at station $59 + 00.00$?

Solution The conditions in this case represent a crest vertical curve. Here, the PVI is at station $58 + 25.60$ and, since the curve extends 400 ft on each side of that point, it starts at station $54 + 25.60$ and ends at station $62 + 25.60$.

$$G_1 = +2, G_2 = -5, L = 8 \text{ sta}$$

Therefore, the rate of gradient change $R = \dfrac{-5 - 2}{8} = -0.875$ %/sta

The elevation at PVC (station $54 + 25.60$) is given by

$$y_{PVC} = y_{PVI} - G_1 \frac{L}{2} = 128.54 - 2 \times 4 = 120.54 \text{ ft}$$

The elevation at station $59 + 00.00$ (4.744 sta. ahead of PVC) is given by

$$y = y_{PVC} + G_1 x + \frac{1}{2} R x^2 = 120.54 + 2 \times 4.744 + \frac{1}{2} \times -0.875 \times 4.744^2 = 120.18 \text{ ft}$$

Example 404.14

A slope with a grade of -3.2% and one with a grade of $+2.6\%$ intersect at station $34 + 28.00$ at an elevation of 49.20 ft. If a sag curve is formed by the slopes and they are to be connected by a vertical curve 600-ft long, what is the elevation at station $35 + 0.00$?

Solution The conditions in this case represent a sag curve. Here, the PVI is at station $34 + 28$ and, since the curve extends 300 ft on each side of that point, the PVC is at station $31 + 28$ and the PVT is at station $37 + 28$.

$$G_1 = -3.2, G_2 = +2.6, L = 6 \text{ sta}$$

Therefore, the rate of gradient change $R = \dfrac{2.6 - (-3.2)}{6} = +0.967$ %/sta

The elevation at PVC (station 31 + 28.00) is given by

$$y_{\text{PVC}} = y_{\text{PVI}} - G_1 \frac{L}{2} = 49.20 - (-3.2) \times 3 = 58.8 \text{ ft}$$

The elevation at station 35 + 00.00 (3.72 sta. ahead of PVC) is given by

$$y = y_{\text{PVC}} + G_1 x + \frac{1}{2} R x^2 = 58.8 - 3.2 \times 3.72 + \frac{1}{2} \times 0.967 \times 3.72^2 = 53.58 \text{ ft}$$

Unsymmetrical Vertical Curves

When a simple vertical curve creates unusual constraints on sight distance, two separate parabolic vertical curves may be used with a common point directly under the PVI as shown in Fig. 404.10. The length e may be calculated if the elevations of PVC, PVI, and PVT are known. According to Eq. (404.29),

$$2e = \text{EL}_{\text{PVI}} - \frac{L_2 \times \text{EL}_{\text{PVC}} + L_1 \times \text{EL}_{\text{PVT}}}{L} \tag{404.29}$$

The offsets from each tangent can then be calculated using

$$y_1 = e \left(\frac{x_1}{L_1} \right)^2 \tag{404.30}$$

$$y_2 = e \left(\frac{x_2}{L_2} \right)^2 \tag{404.31}$$

$$e = \frac{L_1 L_2}{2(L_1 + L_2)} A \tag{404.32}$$

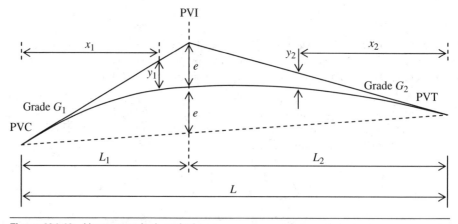

Figure 404.10 Unsymmetrical vertical curve.

Spiral Curves

A horizontal spiral or transition curve is often used to produce a gradual transition from tangents to circular curves. The radius of curvature of a spiral gradually decreases from infinity (on the tangent) to the radius of the circular curve (R_c). There are various forms of transition curves, the most common of them being a clothoid, for which the curvature at any point is directly proportional to the distance (measured along the curve) from the beginning of the transition curve.

The length of the spiral must be adequate to achieve the transition from a normal crown section to a fully superelevated section. In Fig. 404.11, a spiral curve is introduced at each end of the circular curve. The station TS marks the transition from the tangent to the spiral while SC marks the transition from the spiral to the circular curve. Length of spiral L_s is the distance from TS to SC.

There are various methods to calculate the required length of a spiral or transition curve. One of them is the Shortt[1] equation (1909) which was originally developed for gradual attainment of lateral acceleration on railroad curves. It gives the length of the transition spiral as

$$L_s = \frac{3.15S^3}{RC} \tag{404.33}$$

where S = speed (mph)
R = curve radius (ft)
C = rate of increase of lateral acceleration (ft/s^3)

The value of C is between 1 and 3 ft/s^3 for highways. If C is taken as 2, the length of transition spiral becomes

$$L_s = \frac{1.6S^3_{mph}}{R_{ft}} \tag{404.34}$$

The spiral angle is calculated from Eq. (404. 35)

$$\text{Spiral angle } \theta_s = \frac{L_s D_c}{200} \tag{404.35}$$

where D_c = degree of curve for circular curve. With an identical spiral on either side of the circular curve, the deflection angle of circular curve $\Delta_c = \Delta - 2\theta_s$. If, as shown in Fig. 404.11, the x-axis is oriented along the main tangent, with origin at the TS, x_c and y_c are coordinates of SC.

$$x_c = L_s\left(1 - \frac{\theta_s^2}{10}\right) \tag{404.36}$$

$$y_c = L_s\left(\frac{\theta_s}{3} - \frac{\theta_s^3}{42}\right) \approx \frac{L_s^2}{6R_c} \tag{404.37}$$

[1]Shortt, W. H. (1909), "A Practical Method for Improvement of Existing Railroad Curves," *Proceedings, Institution of Civil Engineering*, Vol. 76, pp. 97–208.

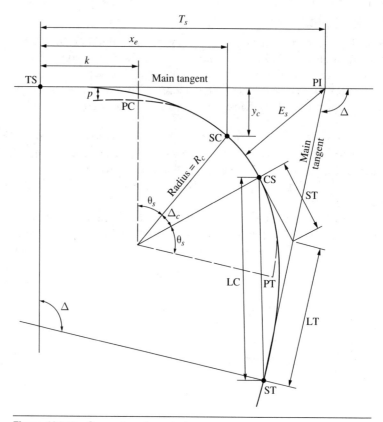

Figure 404.11   Geometry of a spiral curve.

In Eqs. (404.36) and (404.37), θ_s should be entered in radians. If θ_s is in degrees, the equations are

$$x_c = L_s \left(1 - \frac{\theta_s^2}{32828} \right) \tag{404.38}$$

$$y_c = L_s \left(\frac{\theta_s}{172} - \frac{\theta_s^3}{7.9 \times 10^6} \right) \tag{404.39}$$

Likewise, k and p are the coordinates of the PC (which is the station at which the tangent to the circular curve is parallel to the main tangent). In Eqs. (404.40) and (404.41), θ_s should be entered in radians.

$$k = x_c - R_c \sin\theta_s = L_s \left(\frac{1}{2} - \frac{\theta_s^2}{60} \right) \tag{404.40}$$

$$p = y_c - R_c \left(1 - \cos\theta_s \right) = L_s \left(\frac{\theta_s}{12} - \frac{\theta_s^3}{336} \right) \cong \frac{L_s^2}{24R} \tag{404.41}$$

The deflection angle of the SC from the main tangent is given by

$$\delta_s = \frac{L_s}{6R_c} = \frac{\theta_s}{3} \qquad (404.42)$$

For small values of the spiral angle, the tangent offset of any point on the spiral is given by

$$y \approx \frac{xL_s}{6R_c} \qquad (404.43)$$

The tangent length for the spiral is given by

$$T_s = (R_c + p)\tan\left(\frac{\Delta}{2}\right) + k = (y_c + R_c\cos\theta_s)\tan\left(\frac{\Delta}{2}\right) + x_c - R_c\sin\theta_s \qquad (404.44)$$

The original tangent length for the circular curve (T_c) is given by

$$T_c = (R_c + p)\tan\frac{\Delta}{2} \qquad (404.45)$$

The external distance for the spiral is given by

$$E_s = \left(R_c + p\right)\sec\left(\frac{\Delta}{2}\right) - R_c \qquad (404.46)$$

Example 404.15

A 4° curve is to be designed on a highway with two 12-ft lanes and design speed 60 mph. Maximum superelevation is 6%. Normal drainage cross slope is 1%. If a transition spiral is to be used, find the length of spiral and the stations for TS, SC, CS, and ST. Angle for deflection for the original tangents is 38° and PI is at 11 + 62.00.

Solution

$$\text{Circular curve radius } R = \frac{5729.578}{4} = 1432.4 \text{ ft}$$

$$\text{According to Shortt equation, length of spiral } L_s = \frac{1.6 \times 60^3}{1432.4} = 241.3 \text{ ft}$$

For $S = 60$ mph, side friction coefficient $f = 0.12$

$$\text{Design superelevation } e = \frac{S^2}{15R} - f = \frac{60^2}{15 \times 1432.4} - 0.12 = 0.048\,(4.8\%)$$

For $S = 60$ mph, the maximum gradient $\Delta = 0.45$ (see Table 405.1).

Therefore, the minimum length for superelevation attainment is

$$L_r = \frac{12 \times 2 \times 4.8 \times 0.75}{0.45} = 192 \text{ ft}$$

Therefore, the length of the spiral is $L_s = 241.3$ ft. The angle for the spiral is

$$\theta_s = \frac{L_s D}{200} = \frac{241.3 \times 4}{200} = 4.824°$$

$$\Delta - 2\theta_s = 38 - 2 \times 4.824 = 28.35°$$

$$L_c = 100 \frac{\Delta - 2\theta_s}{D} = 100 \times \frac{28.35}{4} = 708.75 \text{ ft}$$

$$T_s = 1432.4 \times \tan\left(\frac{38}{2}\right) + \left[1432.4\cos 4.824 - 1432.4 + \frac{241.3^2}{6 \times 1432.4}\right]\tan\left(\frac{38}{2}\right)$$
$$+ (241.3 - 1432.4\sin 4.824) = 614.64 \text{ ft}$$

$$\text{TS} = \text{PI} - T_s = 1162.00 - 614.64 = 547.36 = 5 + 47.36$$

$$\text{SC} = \text{TS} + L_s = 547.36 + 241.30 = 788.66 = 7 + 88.66$$

$$\text{CS} = \text{SC} + L_c = 788.66 + 708.75 = 1497.41 = 14 + 97.41$$

$$\text{ST} = \text{CS} + L_s = 1497.41 + 241.30 = 1738.71 = 17 + 38.71$$

Superelevation

Forces Acting on a Turning Vehicle

Centrifugal forces on curves are proportional to the local curvature κ which is the inverse of the radius of curvature. The centrifugal force acts along the outward normal and tends to destabilize a vehicle by tipping it about the outside wheel line. One way to counter this destabilizing effect is to "bank" the roadway. Superelevation is the banking (rotation) of a highway to counter some of the lateral force. In Fig. 405.1, the components of the two body forces—the gravity force and the centrifugal force—are shown as solid lines. As shown in the figure, the banking causes a portion of the lateral acceleration to act normal to the pavement ($mV^2/R \sin \phi$). This is felt as a downward force by the vehicle occupants. The remaining portion of the lateral force may act one of three ways depending on the banking and speed of the vehicle.

If the speed is balanced for the banking, the lateral force acting outward on the vehicle will be countered by the forces pushing the vehicle down the slope of the banking. This is a neutral equilibrium condition.

If the vehicle is traveling faster than the equilibrium speed, the resultant lateral force acts outward on the vehicle and occupants. At excessive speeds, the vehicle will skid or roll off the road.

If the speed is lower than the equilibrium speed, the vehicle and occupants are forced inward. Extreme banking can cause top heavy vehicles to roll over toward the inside of the curve. Additionally, icy conditions can cause the vehicle to slide down the pavement.

Maximum Superelevation Rates

High rates of superelevation may cause slow moving vehicles to slide down the banking in snow and ice. High superelevation rates can be difficult to attain in urban settings due to closely spaced intersections, numerous driveways, and limited right of way. Maximum superelevation

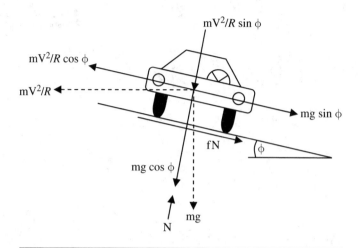

Figure 405.1 Forces on a vehicle on a superelevated roadway.

rates are chosen to limit the adverse effects of superelevation. The following recommendations are from the *AASHTO Green Book:*[1]

1. For urban areas, the maximum superelevation rates of 4% and 6% are recommended.

2. For areas that have frequent ice and snow, maximum superelevation rates of 6% and 8% are recommended.

3. In rural areas without ice or snow, maximum superelevation rates of 10% and 12% are recommended. Higher values should be used with caution, particularly considering the accommodation of occasional slow moving vehicles, construction equipment, and maintenance equipment.

Coefficient of Side Friction

The wheel on wet pavement side friction factor is designated f. Note that this is different from the coefficient of friction (a combination of rolling and skidding friction) which is used to calculate stopping distance. (Note that the current version of the *AASHTO Green Book* no longer uses a friction coefficient approach for computing stopping distance. Instead it uses a deceleration rate a, default value $a = 11.2$ ft/s².)

The superelevation factor is given by $e = \tan \phi$, where ϕ is the banking angle (with the horizontal).

[1] *A Policy on Geometric Design of Highways and Streets*, 7th ed., American Association of State Highway and Transportation Officials, Washington, D.C., 2018.

Solving the equilibrium equations for the vehicle traveling on a banked curve, to prevent the outward slide of the vehicle, the required superelevation factor e for a design speed V, curve radius R, and side friction coefficient f is given by

$$e = \tan\phi = \frac{V^2 - fgR}{fV^2 + gR} \qquad (405.1)$$

On the other hand, if superelevation e and side friction coefficient f are known, the dimensionless centrifugal acceleration (normalized with respect to the gravitational acceleration) is given by

$$\frac{V^2}{gR} = \frac{e+f}{1-ef} \qquad (405.2)$$

For most practical values, the product $(e \times f)$ is negligible and therefore the centrifugal factor is given by the sum of the superelevation factor e and the side friction factor f. For stability, the centrifugal factor $e + f$ must be greater than or equal to the factor v^2/gR.

$$e + f \geq \frac{V^2}{gR} \qquad (405.3)$$

Given maximum superelevation e_{max} and design side friction factor f, the minimum safe radius (R_{min}) and maximum degree of curve (D_{max}) for a horizontal curve are given by

$$R_{min} = \frac{S^2}{15(0.01e_{max} + f_{design})} \qquad (405.4)$$

$$D_{max} = \frac{5729.578}{R_{min}} \qquad (405.5)$$

where S = design speed (mph)
e = superelevation (%)
f = recommended side friction factor (decimal)
R = curve radius (ft)
D = degree of curve (degrees)

Example 405.1

The exit ramp from a highway has 4% superelevation. The design speed for the ramp is 35 mph. If the side friction factor is 0.18, what is the maximum radius (ft) of the horizontal curve for the ramp?

Solution   The minimum radius is calculated from

$$R_{min} = \frac{S^2}{15(0.01e_{max} + f_{design})} = \frac{35^2}{15(0.04 + 0.18)} = 371.2\,\text{ft}$$

Use a radius of 375 ft for the ramp.

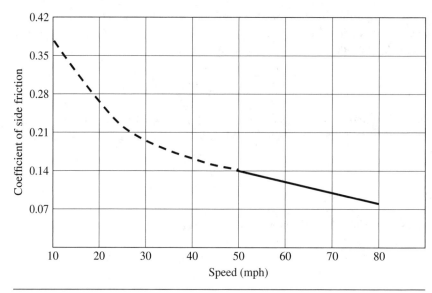

Figure 405.2   Side friction factors assumed for design. (*AASHTO Green Book.*)

Side Friction Coefficient

Figure 405.2 summarizes the recommendations from the *AASHTO Green Book* about side friction factor as a function of design speed.

Distribution of *e* and *f* over a Range of Curves

According to the *AASHTO Green Book*, there are five methods for sustaining centripetal acceleration on curves by use of superelevation *e* or side friction *f*:

Method 1	Superelevation and side friction vary in proportion (linear) to the curvature (inverse of curve radius) from zero to their respective maximum values.
Method 2	For a vehicle traveling at design speed all lateral acceleration is sustained by side friction until f_{max} is reached. Then, side friction is held constant while superelevation is increased. Variation of both parameters is proportional to the curvature.
Method 3	For a vehicle traveling at design speed, all lateral acceleration is sustained by superelevation until e_{max} is reached. Then, superelevation is held constant while side friction is increased.
Method 4	Same as method 3, except that average running speed is used instead of design speed.
Method 5	Superelevation and side friction vary in nonlinear fashion with the curvature (inverse of curve radius). Resulting values are between those produced by methods 1 and 3.

Superelevation—AASHTO Recommendation

The *AASHTO Green Book* (A Policy on Geometric Design of Highways and Streets, 7th ed., 2018) has the following design aids based on Method 5:

1. Figures 3-9 to 3-13: Design Superelevation Rates for Maximum Superelevation Rates (e_{max}) ranging from 4% to 12%
2. Tables 3-8 to 3-12: Minimum Radii for Design Superelevation Rates, Design Speeds, and e_{max} ranging from 4% to 12%

Example 405.2

For maximum superelevation rate $e_{max} = 8\%$, what is the recommended (*AASHTO Green Book*) superelevation for a horizontal curve ($R = 4000$ ft) on a highway with a design speed $= 70$ mph?

Solution

From Figure 3-11 ($e_{max} = 8\%$), choosing the curve for $V = 70$ mph, $R = 4000$ ft yields $e = 4.9\%$

From Table 3-10b ($e_{max} = 8\%$), choosing the column for $V = 70$ mph, $R_{min} = 4100$ ft for $e = 4.8\%$ and $R_{min} = 3{,}910$ ft for $e = 5.0\%$. Interpolation yields, for $R_{min} = 4000$ ft, $e = 4.905\%$

Recommendation: Use $e = 5\%$

NOTE Using this superelevation, the implicit side-friction factor is

$$f = \frac{V^2}{15R} - e = \frac{70^2}{15 \times 4000} - 0.05 = 0.032, \text{ which is well within the maximum}$$

$f = 0.10$ for $=$ design speed $= 70$ mph and $e_{max} = 8\%$ from Table 3-7.

Transition to Superelevation

The normal cross section (for drainage) of a roadway is usually crowned along the centerline and has a nominal cross slope of 1.5% to 2.0%. While this is acceptable over a tangent section, it can cause problems when the roadway is curved. The problem is more severe on the outer lane, where the adverse cross slope may be considered "negative superelevation." To satisfy the stability criterion stated above, it is necessary to bank or superelevate the roadway. The transition from a tangent section with a normal crown to a superelevated horizontal curve occurs in two parts:

Tangent Runoff Length L_t

Using a portion of the tangent leading into the curve, the outside lane(s) are rotated from a crown cross slope to a level condition over a length known as the tangent runoff length. The rate at which the adverse cross slope is removed is called the relative gradient for superelevation runoff, or the superelevation runoff rate. The tangent length required to achieve this change is called the tangent runoff length or tangent runout L_t. The tangent runout is computed from

$$L_t = \frac{w \times n \times e_{NC} \times b_w}{\Delta} \tag{405.6}$$

where L_t = length of tangent runoff
w = lane width
n = no. of lanes being rotated
b_w = a reduction factor = $(1+n)/2n$
b_w = 1.0 for $n = 1$, $b_w = 0.75$ for $n = 2$, $b_w = 0.67$ for $n = 3$
e_{NC} = normal cross slope (%)
Δ = superelevation runoff rate (%)

Superelevation Runoff Length L_r

The length over which outside lane(s) are rotated from level to the fully superelevated condition is known as the superelevation runoff length.

$$L_r = \frac{w \times n \times e_d \times b_w}{\Delta}$$ (405.7)

where L_r = length of superelevation runoff
e_d = design superelevation rate (%)

Ideally, the curve should have no superelevation on the tangent section and be fully superelevated on the curve. One option is obviously to achieve full superelevation exactly at the point of curve (PC) and provide the distance L_r entirely on the tangent. However, this practice is by no means universal. Many agencies distribute the distance L_r to lie partially on the tangent and partially within the curve. Thereby, the curve is not fully superelevated at the PC. The distribution of the superelevation runoff varies from agency to agency. For example, if the distance L_r is partitioned into 2:1 parts, then two-thirds of L_r is located on the tangent (i.e., before the PC of the curve) and one-third of L_r is located within the curve. At the PC the curve has only two-thirds of the design superelevation. This situation is illustrated in Fig. 405.3. The pavement cross sections are shown to the right. On the tangent, upto point A, the pavement has a normal

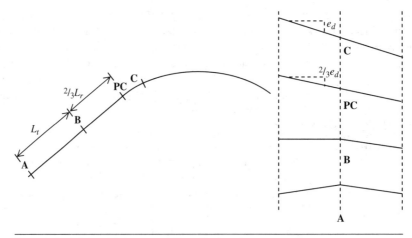

Figure 405.3 Transition to superelevation.

Table 405.1 Maximum Relative Gradient for Superelevation Runoff (AASHTO)

Design speed (mph)	Maximum relative gradient (%)	Equivalent relative slope	Design speed (mph)	Maximum relative gradient (%)	Equivalent relative slope
15	0.78	1:128	50	0.50	1:200
20	0.74	1:135	55	0.47	1:213
25	0.70	1:143	60	0.45	1:222
30	0.66	1:152	65	0.43	1:233
35	0.62	1:161	70	0.40	1:250
40	0.58	1:172	75	0.38	1:263
45	0.54	1:185	80	0.35	1:286

Source   Adapted in part from Exhibit 3-30 (*AASHTO Green Book*).

crowned section. At B the adverse cross slope has been rotated to the horizontal profile. The distance AB to achieve this rotation is the tangent runout L_t. From B to C the cross section is rotated until it is fully superelevated at C. The distance BC is the superelevation runoff (L_r).

Maximum Gradient for Superelevation Runoff

Table 405.1 shows the maximum relative gradients recommended by the *AASHTO Green Book*. For example, for a design speed of 50 mph, the maximum Δ is 0.5% (which may also be expressed as 1:200). Commonly adopted rates are significantly less than this value and for $V = 50$ mph, the SRR adopted may be between 1:400 and 1 : 300.

Example 405.3
The alignment of a roadway is composed of the following sections:

Segment	Length (ft)	Radius (ft)	Speed (mph)	Deflection	Section
AB	655	765	40	12°30′ Rt	Superelevated, rotated 6% about centerline
BC	1560	∞	60	Tangent	Crowned at centerline, cross slopes = 2%
CD	800	750	50	8°15′ Lt	Superelevated, rotated 4% about edge
DE	1200	850	40	15°45′ Rt	Crowned

Which segment has the lowest factor of safety for lateral stability?

Solution   For lateral stability, the superelevation factor $e + f$ must be greater than V^2/gR. Using speed S (mph), radius R (ft), and g (ft/s^2), this factor becomes

$$\frac{S^2_{mph}}{15R_{ft}}$$

Assuming the coefficient of side friction f according to design speed, the segment having the greatest value of $V^2/gR - e - f$ will be most critical. In fact, theoretically, if this factor is positive, the location is unsafe with regard to lateral stability. Calculating the value of this parameter for each segment, we get

Segment	Coefficient of side friction (f)	$S^2/15R - e - f$
AB	0.17	$0.14 - 0.06 - 0.17 = -0.09$
BC	0.12	$0.00 - (-0.02) - 0.12 = -0.14$
CD	0.14	$0.22 - 0.04 - 0.14 = +0.04$
DE	0.17	$0.13 - (-0.02) - 0.17 = -0.02$

According to this table, the segment CD has the lowest factor of safety for lateral stability.

Example 405.4

The alignment of a highway includes a circular curve with radius $R = 3000$ ft. The maximum permitted superelevation is 10%. If the design speed on the highway is 65 mph, what is the superelevation recommended by the *AASHTO Green Book*?

Solution

Using Table 3-11 from the 7th edition of the *AASHTO Green Book* ($e_{max} = 10\%$), for $R = 3000$ ft and $V = 65$ mph, the recommended superelevation is $e = 6\%$ (based on minimum radius of 2980 ft).

Spiral Curves

A spiral curve or transition curve is often used to produce a gradual transition from tangents to circular curves. While the cross section is being rotated from the crowned to the fully superelevated section, the spiral may be used to gradually change the radius of curvature (from infinite on the tangent to the specified curve radius).

For more details on the geometry of spiral curves, see Chap. 404.

Freeways

Level of Service

According to the *Highway Capacity Manual*[1], the density of passenger car equivalents per mile per lane of a freeway segment (pcpmpl or pc/mi/ln) of a freeway segment is used to define the level of service (LOS). These LOS boundaries can also be expressed in terms of the following proxy variables—speed (mph), volume to capacity ratio, and service flow rate (pcphpl or pc/mi/ln). Figure 406.1 shows the boundaries for the various regimes of level of service (A through E) based on these parameters. This figure can be used once the free-flow speed (FFS) is determined.

 NOTE   The upper limit for LOS E corresponds to a *v/c* ratio = 1.0. Therefore, the corresponding value of service flow rate represents freeway lane capacity. See Eq. (406.3).

Default Values of Parameters

For basic freeway segments, the following default values of key input parameters are listed in Exhibit 12-6 of the HCM and may be used when field measured data are not available.

Lane width = 12 ft

Right-side lateral clearance = 10 ft

Base free-flow speed = speed limit + 5 mph

PHF = 0.94

Percentage of heavy vehicles, P_T = 12% (rural), 5% (urban)

Driver population factor, f_p = 1.00.

CAF (capacity adjustment factor) = 1.0

SAF (speed adjustment factor) = 1.0

---

[1]*Highway Capacity Manual*, 6th ed., Transportation Research Board, Washington, DC, 2016.

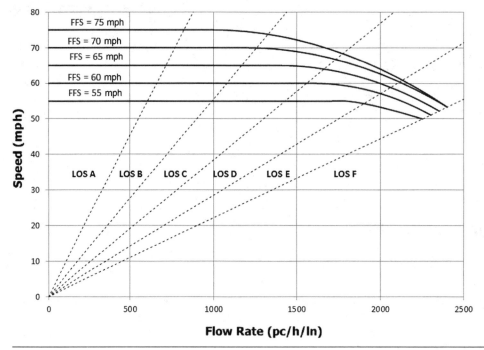

Figure 406.1 LOS criteria for basic freeway segments.

FFS is not equal to posted speed limit or 85th percentile speed. It is the mean field measured speed when volumes are less than 1300 pcphpl.

Free-Flow Speed

The Highway Capacity Manual outlines the following procedure for determining free-flow speed of a basic freeway segment based on the base free-flow speed (BFFS) and certain site-specific adjustments (speed deductions).

$$FFS = 75.4 - f_{LW} - f_{LC} - 3.22 \times TRD^{0.84} \qquad (406.1)$$

Speed Adjustments to BFFS

The parameters f_{LW} and f_{LC} are adjustments for lane width and right shoulder lateral clearance, respectively. These are given by Tables 406.1 and 406.2, respectively.

TRD, the total ramp density (ramps per mile), is defined as the total number of ramps (on- and off-ramps in one direction) located between points 3 miles upstream and 3 miles downstream of the midpoint of the basic freeway segment under study, divided by 6 miles.

Table 406.1 Free-Flow Speed Adjustment for Lane Width (f_{LW})

Lane width (ft)	Reduction in free-flow speed (mph)
$>= 12$	0.0
$>= 11-12$	1.9
$>= 10-11$	6.6

Table 406.2 Free-Flow Speed Adjustment for Lateral Clearance (f_{LC})

Right shoulder lateral clearance (ft)	Reduction in free-flow speed (mph)			
	Lanes in one direction			
	2	3	4	$\geq$5
$\geq$6	0.0	0.0	0.0	0.0
5	0.6	0.4	0.2	0.1
4	1.2	0.8	0.4	0.2
3	1.8	1.2	0.6	0.3
2	2.4	1.6	0.8	0.4
1	3.0	2.0	1.0	0.5
0	3.6	2.4	1.2	0.6

Source *Highway Capacity Manual*, Transportation Research Board, Washington, DC, 2010.

Adjusted Free-Flow Speed

The FFS estimated by Eq. (406.1) does not account for variable conditions such as weather, incidents, work zone, and driver population. The speed adjustment factor (SAF) is a calibration parameter used to adjust for these factors.

The adjusted FFS is then given by

$$FFS_{adj} = FFS \times SAF \leq FFS \qquad (406.2)$$

Capacity

The base lane capacity (pcphpl) of a freeway segment for $55 \leq FFS \leq 75$ mph is given by

$$c = 2200 + 10\,(FFS - 50) \leq 2400 \qquad (406.3)$$

The capacity adjustment factor (CAF) is a calibration parameter used to adjust for local conditions and to account for nonrecurring sources of congestion such as weather, incidents, work zone, and driver population.

The adjusted segment capacity is then given by

$$c_{adj} = c \times \text{CAF} \leq c \qquad (406.4)$$

Flow Rate

The peak flow rate (v_p) is calculated from the peak hourly volume V (vph) using

$$v_p = \frac{V}{\text{PHF} \times N \times f_{HV} \times f_p} \qquad (406.5)$$

where PHF is the peak hour factor based on peak 15-minute vehicular count
N is the number of lanes
f_p is a driver population factor and
f_{HV} is a heavy vehicle factor

Peak Hour Factor

The peak hour factor is an indicator of the variability of flow within the design hour. The greater the variability in the flow, the lower is the PHF. A lower value of PHF results in a higher value of peak flow rate v_p.

Example 406.1   Calculation of Peak-Hour Factor (PHF)

Fifteen minute traffic counts for a highway were conducted over a contiguous 2-h window. These are shown in the table below. What is the peak-hour factor?

Time (hh:mm)	Vehicles
09:00–09:15	234
09:15–09:30	256
09:30–09:45	245
09:45–10:00	231
10:00–10:15	237
10:15–10:30	266
10:30–10:45	245
10:45–11:00	210

Solution The sliding window (1-h width) captures the hourly volumes for the 9:00–10:00, 9:15–10:15, 9:30–10:30, 9:45–10:45, and 10:00–11:00 windows as 966, 969, 979, 979, and 958, respectively.

Time (hh:mm)	Vehicles	Consecutive hourly counts
09:00–09:15	234	
09:15–09:30	256	966
09:30–09:45	245	969
09:45–10:00	231	979
10:00–10:15	237	979
10:15–10:30	266	958
10:30–10:45	245	
10:45–11:00	210	

The peak hourly volume is then $V = 979$ veh/h and the peak 15-min flow is 266 veh/15 min. The PHF is calculated as

$$PHF = \frac{979}{4 \times 266} = 0.92 \quad \text{and the peak flow rate } v_p \text{ is given by}$$

$$v_p = \frac{979}{0.92} = 1064 \text{ vph} \quad \text{which is also the same as } 4v_{15} = 4 \times 266 = 1064 \text{ vph}$$

Heavy Vehicle Factor

The heavy vehicle factor converts the mixed traffic (passenger cars, trucks, buses, and recreational vehicles) into an equivalent number of passenger cars. The factor is calculated in terms of equivalence factors E_T (trucks and buses) and E_R (RVs), which in turn are dependent primarily on terrain. The factor f_{HV} is given by

$$f_{HV} = \frac{1}{1 + P_T(E_T - 1)} \tag{406.6}$$

where P_T is the proportion (decimal) of heavy vehicle in the traffic stream. For general terrain segments, the equivalence factor E_T is given in Table 406.3. Freeway segments longer than 0.5 miles with grades between 2% and 3% or longer than 0.25 miles with grades of 3% or greater should be considered as separate segments.

Table 406.3 Passenger Car Equivalents for General Terrain Segments

Passenger Car Equivalent	Terrain Type	
	Level	Rolling
E_T	2.0	3.0

Source Highway Capacity Manual, 6th ed.
Transportation Research Board, Washington, DC, 2016.

The factors given in Table 406.3 apply *when analyzing extended sections* of freeways that have been broadly classified as level/rolling. In the HCM 2016, heavy vehicles are broadly classified as one of two types—SUT (single unit trucks, RVs, and buses) and TT (tractor trailers). HCM Exhibits 12-26 through 12-28 are PCE tables for the following splits—30% SUTs + 70% TTs, 50% SUTs + 50% TTs, and 70% SUTs + 30% TTs, respectively.

Driver Population Factor

The driver population factor (f_p) accounts for the effect of driver familiarity with the roadway on level of service. This factor varies between 0.85 and 1.0. Unless there is sufficient reason for using a lesser value, the default value is 1.0, which represents weekday commuter traffic.

Mean Speed

The mean speed of the traffic stream under base conditions is given by

$$S = \text{FFS}_{adj} \qquad\qquad for\ v_p \le \text{BP} \qquad\qquad (406.7a)$$

$$S = \text{FFS}_{adj} - \frac{\left(\text{FFS}_{adj} - \dfrac{c_{adj}}{D_c}\right)(v_p - BP)^2}{(c_{adj} - BP)^2} \qquad for\ v_p > \text{BP} \qquad (406.7b)$$

where D_c = density at capacity = 45 pc/mi
and BP is the breakpoint (beyond which flow rate varies parabolically with speed), given by

$$\text{BP} = \left[1000 + 40 \times (75 - \text{FFS}_{adj})\right] \times \text{CAF} \qquad (406.8)$$

Determining Level of Service

A basic freeway segment can be characterized by three performance measures: density D (pcpmpl), space mean speed S (mph), and the ratio of demand flow rate to capacity (v/c). Each of these measures is an indication of how well traffic is being accommodated by the basic freeway segment. Because speed is constant through a broad range of flows and the v/c ratio is not directly discernible to road users (except at capacity), the service measure for basic freeway segments is density. The LOS criteria in terms of density are given in Table 406.4. The density D (pc/mi/ln) is calculated from the flow rate v_p (pc/h/ln) using

$$D = \frac{v_p}{S} \qquad\qquad (406.9)$$

where S is the mean speed (mph).

Table 406.4 Level of Service Criteria for Basic Freeway Segments

LOS	Density (pcphpl)
A	≤11
B	>11–18
C	>18–26
D	>26–35
E	>35–45
F	Demand exceeds capacity OR density >45

Source Highway Capacity Manual, 6th ed. Transportation Research Board, Washington, DC, 2016.

Example 406.2

Determine the level of service for an extended stretch of a four-lane (two lanes in each direction) rural freeway in rolling terrain. Lanes are 11 ft wide. Shoulder width = 2 ft. Assume a peak hour factor *PHF* = 0.92. The traffic consists of approximately 5% trucks, 2% RVs, and 3% tractor-trailers. The one way peak hour volume is 2300 vph. There are four full-cloverleaf interchanges within the 6-mile study area. Assume that the traffic primarily consists of commuters. Assume CAF = SAF = 1.0

Solution   Lane width = 11 ft. Table 406.1 gives $f_{LW} = 1.9$

Right shoulder clearance = 2 ft. Table 406.2 gives $f_{LC} = 2.4$

Each cloverleaf interchange has two on-ramps and two off-ramps in each direction, i.e., 16 such ramps in the 6-mile stretch. Therefore TRD = 16/6 = 2.67 ramps per mile.

$$FFS = 75.4 - 1.9 - 0.0 - 3.22 \times 2.67^{0.84} = 63.8 \text{ mph}$$

$$FFS_{adj} = FFS \times SAF = 63.8 \text{ mph}$$

Heavy vehicles: SUT = 7%, TT = 3%.

For rolling terrain, $E_T = 3.0$

Heavy vehicle factor:

$$f_{HV} = \frac{1}{1 + P_T(E_T - 1)} = \frac{1}{1 + 0.1(3.0 - 1)} = 0.833$$

Commuter traffic implies population factor $f_p = 1.0$

$V = 2300$ vph

$$v_p = \frac{V}{PHF \times N \times f_{HV} \times f_p} = \frac{2300}{0.92 \times 2 \times 0.833 \times 1.0} = 1500$$

For a $FFS_{adj} = 63.8$ mph and flow rate $v_p = 1500$ pcphpl, LOS C

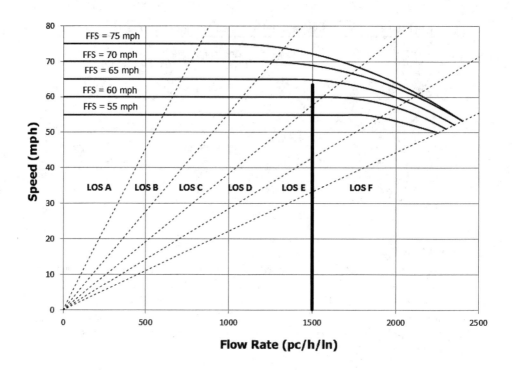

Flow Rate (pc/h/ln)

Example 406.3

On a freeway serving traffic to and from coastal beach resorts, the minimum level of service during the peak hour is to be *D*. Determine the number of lanes required if:

  One directional hourly volume = 4200 vph

  PHF = 0.90

  Lane width = 12 ft

  Rolling terrain

  Average spacing of interchanges = 2 miles

  8% trucks and 2% RVs

  Right shoulder lateral clearance = 6 ft

Solution For a suburban freeway, with a FFS = 65 mph, maximum flow rate for LOS D is approximately 2000 pcphpl. Based on the $V = 4200$ vph, which will be inflated by PHF and f_{HV} to approximately 6000 pcphpl, let us assume three lanes.

Assume BFFS = 70 mph

Lane width = 11 ft. Table 406.1 gives $f_{LW} = 0.0$
Right shoulder clearance = 6 ft. Table 406.2 gives $f_{LC} = 0.0$

Interchange spacing = 2 miles. Therefore, we have three interchanges in 6 miles. Six ramps in 6 miles for each direction. TRD = 1.0

$$FFS = 75.4 - 0.0 - 0.0 - 3.22 \times 2^{0.84} = 72.2 \text{ mph}$$

From Fig. 406.1, for FFS = 72.2 mph, the maximum flow rate for LOS D is approximately 2100.

Recreational destination, implying population factor $f_p = 0.9$.

Heavy vehicle factor (assuming $E_T = 3.0$).

$$f_{HV} = \frac{1}{1 + P_T(E_T - 1)} = \frac{1}{1 + 0.1(3.0 - 1)} = 0.833$$

$V = 4500$ vph

$$v_p = \frac{V}{\text{PHF} \times N \times f_{HV} \times f_p} = \frac{4200}{0.92 \times 2 \times 0.833 \times 1.0} \leq 2100$$

$N \geq 2.96$. Therefore, three lanes are okay.

Weaving

Weaving length is measured from a point at the merge gore where the right edge of the freeway lane and the left edge of the merging lane are 2 ft apart to a point on the diverge gore where the two edges are 12 ft apart. The length of the weaving segment constrains the time and space within which the driver must make all lane changes. Typically, weaving segments are limited to a length of 2500 ft. Longer weaving segments are possible, but they typically have merging and diverging movements separated. Table 406.5 defines the LOS criteria for weaving segments.

Table 406.5 LOS Criteria for Weaving Segments

	Density (pc/mi/ln)	
LOS	Freeway weaving segment	Multilane and collector distributor weaving segments
A	≤10	≤12
B	>10–20	>12–24
C	>20–28	>24–32
D	>28–35	>32–36
E	>35–43	>36–40
F	>43 (or demand > capacity)	>40 (or demand > capacity)

Source *Highway Capacity Manual*, 6th ed. Transportation Research Board, Washington, DC, 2016.

Table 406.6 Weaving Configuration Types

Number of lane changes required by movement v_{w2}	Number of lane changes required by movement v_{w1}		
	0	1	≥ 2
0	Type B	Type B	Type C
1	Type B	Type A	N/A
≥ 2	Type C	N/A	N/A

Source Highway Capacity Manual, 6th ed. Transportation Research Board, Washington, DC, 2016.

Weaving Segment Configuration

The configuration of a weaving segment is based on the number of lane changes required of each weaving movement. Three configurations—type A, type B, and type C are used. The applicability of these three types is summarized in the Table 406.8, where v_{w1} is the larger of the two weaving flow rates (pc/h) and v_{w2} is the smaller of the two weaving flow rates (pc/h).

The three types of geometric configurations are defined below:

Type A   Weaving vehicles in both directions must make one lane change to successfully complete a weaving maneuver.

Type B   Weaving vehicles in one direction may complete a weaving maneuver without making a lane change, whereas other vehicles must make one lane change to successfully complete a weaving maneuver.

Type C   Weaving vehicles in one direction may complete a weaving maneuver without making a lane change, whereas other vehicles must make at least two lane changes to successfully complete a weaving maneuver.

Weaving Parameters

Three geometric characteristics affect a weaving segment's operating characteristics: length, width, and configuration.

Weaving Configuration

Figure 406.2 shows a weaving segment of a highway segment between two closely spaced on- and off-ramps. Movements A to C and B to D (shown as solid lines) are called non-weaving movements, while the movements from A to D and from B to C (shown as broken lines) have to make one or more lane changes and cross each others' paths in the process.

In the analysis of LOS in weaving segments, the following flow rates are defined:

v_{FF} = freeway to freeway demand flow rate (pc/h) (AC: non-weaving)

v_{RF} = ramp to freeway demand flow rate (pc/h) (BC: weaving)

v_{FR} = freeway to ramp demand flow rate (pc/h) (AD: weaving)

v_{RR} = ramp to ramp demand flow rate (pc/h) (BD: non-weaving)

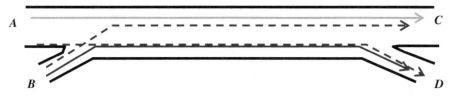

Figure 406.2 Weaving movements.

All quantities used in the computational steps are flow rates, derived from the corresponding hourly volumes and adjusted for heavy vehicle presence, as below:

$$v = \frac{v}{\text{PHF} \times f_{HV}} \qquad (406.10)$$

v_W = weaving demand flow rate in the weaving segment = $v_{RF} + v_{FR}$

v_{NW} = non-weaving demand flow rate in the weaving segment = $v_{FF} + v_{RR}$

Volume ratio, $VR = \dfrac{v_w}{v_w + v_{Nw}}$ $\qquad (406.11)$

N_{WL} = Number of lanes from which a weaving maneuver may be made with one or no lane changes. (For two-sided weaving segments, $N_{WL} = 0$.)

LC_{RF} = Minimum number of lane changes that must be made by a single-weaving vehicle moving from the on-ramp to the freeway.

LC_{FR} = Minimum number of lane changes that must be made by a single-weaving vehicle from the freeway to the off-ramp.

LC_{MIN} = Minimum rate of lane changing that must exist for all weaving vehicles to complete their weaving maneuvers successfully, in lane changes per hour.

For one-sided weaving segments,

$$LC_{MIN} = LC_{RF} \times v_{RF} + LC_{FR} \times v_{FR} \qquad (406.12)$$

For two-sided weaving segments,

$$LC_{MIN} = LC_{RR} \times v_{RR} \qquad (406.13)$$

Weaving Length

There are two weaving lengths—L_B and L_S. These are shown in Fig. 406.3.

L_S = Short length is the distance between the end points of any barrier markings (solid white lines) that discourage lane changing.

L_B = Base length is the distance between points in the two gore areas where the edge of the raveled way intersects with the edge of the ramp traveled way.

The maximum length of a weaving segment is given by

$$L_{MAX} = 5728(1 + VR)^{1.6} - 1566 N_{WL} \qquad (406.14)$$

If the short length $L_S \geq L_{MAX}$, HCM recommendation is to analyze the merge and diverge areas as separate segments.

Freeway Ramps

For freeway ramps, in the absence of local field measured data, the following defaults may be assumed:

Acceleration lane length = 590 ft

Deceleration lane length = 140 ft

Ramp free-flow speed = 35 mph

PHF = 0.88 (rural), 0.92 (urban)

Heavy vehicles = 10% (rural), 5% (urban)

Driver population factor = 1.0

The methodology for analyzing the LOS in the ramp influence area (shown highlighted in Fig. 406.3) involves predicting the number of approaching freeway vehicles that remain in lanes 1 and 2 immediately upstream of the ramp–freeway junction.

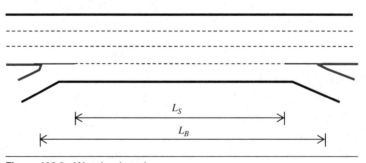

Figure 406.3 Weaving length.

Freeway Ramp Analysis

The critical variables for ramp junctions are shown in Fig. 406.4.

On-Ramps (Merge Areas)

The methodology to determine LOS on on-ramps has three major steps.

First, flow entering lanes 1 and 2 immediately upstream of the merge influence area (v_{12}) or at the beginning of the deceleration lane at diverge is determined as a fraction of the approach flow (v_F) using Eq. (406.15).

$$v_{12} = v_F P_{FM}$$

(406.15)

The procedure to determine the fraction P_{FM} for on-ramps is summarized in Table 406.8.

Second, several capacity values are determined to determine the likelihood of congestion. These are as follows:

- Maximum total flow approaching a major diverge area on the freeway (v_F)
- Maximum total flow departing from a merge or diverge area on the freeway (v_{FO})
- Maximum total flow entering the ramp influence area (v_{R12} for merge areas and v_{12} for diverge areas)
- Maximum flow on a ramp (v_R)

The density of flow within the ramp influence area (D_R) and the level of service based on this variable are determined, as shown in Table 406.7. For some situations, the average speed of vehicles within the influence area (S_R) may also be estimated.

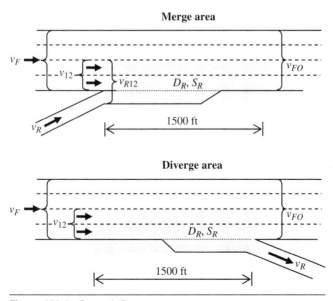

Figure 406.4 Ramp influence area.

All relevant freeway and ramp flows must be converted to equivalent pc/h under base conditions during the peak 15 minutes of the hour. This is accomplished by using the peak hour factor (PHF) and heavy vehicle factor (f_{HV}) as shown in Eq. (406.6).

Ramps on four-lane, eight-lane, and ten-lane freeways are always analyzed as isolated merge or diverge areas. The nature of the procedure for predicting v_{12} makes the four-lane case trivial, and there is insufficient data to determine the effects of adjacent ramps on eight-lane and ten-lane freeways.

Six-Lane Freeways

For six-lane freeways, the effect of adjacent ramps on lane distribution at a subject ramp can be calculated. When nearby ramps inject vehicles into or remove them from lane 1, the lane distribution may be seriously altered. Important variables determining this impact include the total flow on the upstream (v_U) or downstream (v_D) ramp (or both), in pc/h, and the distance from the subject ramp to the adjacent upstream (L_{up}) or downstream (L_{down}) ramp (or both), in feet. For ramps on six-lane freeways, therefore, an additional analysis step is required to determine whether adjacent ramps are close enough to affect traffic distribution in the subject ramp.

With all of these variables, the total approaching freeway flow has the most dominant influence on flow in lanes 1 and 2. Longer acceleration lanes lead to lower densities in the influence area and higher flows in lanes 1 and 2. When the ramp has a higher free-flow speed, vehicles tend to enter the freeway at higher speeds, and approaching freeway vehicles tend to move further left to avoid the possibility of high speed turbulence.

Table 406.7 LOS Criteria for Merge and Diverge Areas

LOS	Density (pc/mi/ln)
A	≤ 10
B	>10–20
C	>20–28
D	>28–35
E	>35
F	Demand exceeds capacity

Table 406.8 HCM Models for Predicting v_{12} at Off-Ramps

Four-lane freeways (two lanes in each direction)	$P_{FM} = 1.000$	
Six-lane freeways (three lanes in each direction)	$P_{FM} = 0.5775 + 0.000028L_A$ $P_{FM} = 0.7289 - 0.0000135(v_F + v_R) - 0.03296$ $\qquad + 0.000063L_{up}$ $P_{FM} = 0.5487 + 0.2628v_D/L_{down}$	Eq. (406.16) Eq. (406.17) Eq. (406.18)
Eight-lane freeways (four lanes in each direction)	For $v_F/S_{FR} \leq 72$, $P_{FM} = 0.2178 - 0.000125v_R + \dfrac{0.01115L_A}{S_{FR}}$ For $v_F/S_{FR} > 72$, $P_{FM} = 0.2178 - 0.000125v_R$	Eq. (406.19a) Eq. (406.19b)

v_{12} = flow rate in lanes 1 and 2 of freeway immediately upstream of merge (pc/h)

v_F = freeway demand flow rate immediately upstream of merge (pc/h)

v_R = on-ramp demand flow rate (pc/h)

v_D = demand flow rate on adjacent downstream ramp (pc/h)

P_{FM} = fraction of approaching freeway flow remaining in lanes 1 and 2 immediately upstream of merge

L_A = length of acceleration lane (ft)

S_{FR} = free-flow speed of ramp (mi/h)

L_{up} = distance to adjacent upstream ramp (ft)

L_{down} = distance to adjacent downstream ramp (ft)

Selecting equations for P_{FM} for six-lane freeways				
Adjacent upstream ramp	Subject ramp	Adjacent downstream ramp	Equation used	Remarks
None	On	None	Eq. (406.16)	
None	On	On	Eq. (406.16)	
None	On	Off	Eq. (406.16) or (406.18)	Use Eq. (406.16) if distance between ramps $\geq L_{EQ}$, Eq. (406.18) otherwise
On	On	None	Eq. (406.16)	
Off	On	None	Eq. (406.16) or (406.17)	Use Eq. (406.16) if distance between ramps $\geq L_{EQ}$, Eq. (406.17) otherwise
On	On	On	Eq. (406.16)	
On	On	Off	Eq. (406.16) or (406.18)	Use Eq. (406.16) if distance between ramps $\geq L_{EQ}$, Eq. (406.18) otherwise
Off	On	On	Eq. (406.16) or (406.17)	Use Eq. (406.16) if distance between ramps $\geq L_{EQ}$, Eq. (406.17) otherwise
Off	On	Off	Eqs. (406.16), (406.17), or (406.18)	Use Eq. (406.16) if distance between ramps $\geq L_{EQ}$. Else, try both Eqs. (406.17) and (406.18) and choose larger value of P_{FM}

L_{EQ} is that distance for which Eqs. (406.16) and (406.17) or (406.18), as appropriate, yield the same value of P_{FM}.

Where an adjacent upstream off-ramp exists, calculate L_{EQ} according to Eq. (406.20).

$$L_{EQ} = 0.214(v_F + v_R) + 0.44L_A + 52.32S_{FR} - 2403 \qquad (406.20)$$

If $L_{up} > L_{EQ}$, use Eq. (406.16), otherwise use Eq. (406.20).

Similarly, when an adjacent downstream off-ramp exists, Eq. (406.21) is used to compute L_{EQ}.

$$L_{EQ} = \frac{v_D}{0.1096 + 0.000107L_A} \qquad (406.21)$$

If $L_{down} > L_{EQ}$, use Eq. (406.16), otherwise use Eq. (406.18).

Two-Lane On-Ramps

For two-lane on-ramps, however, the following values of P_{FM} are used:

- Four-lane freeways, $P_{FM} = 1.000$
- Six-lane freeways, $P_{FM} = 0.555$
- Eight-lane freeways, $P_{FM} = 0.209$

Off-Ramps

Flow rate in lanes 1 and 2 immediately upstream of the diverge (v_{12}) is determined as the sum of the flow rate in ramp plus a fraction of the rest of approach flow ($v_F - v_R$) using Eq. (406.20).

$$v_{12} = v_R + (v_F - v_R)P_{FD} \qquad (406.22)$$

The procedure to determine the fraction P_{FD} for off-ramps is summarized in Table 406.9. The variables used in Table 406.9 are defined as follows:

v_{12} = flow rate in lanes 1 and 2 of freeway immediately upstream of diverge (pc/h)

v_F = freeway demand flow rate immediately upstream of diverge (pc/h)

v_R = off-ramp demand flow rate (pc/h)

v_U = demand flow rate on adjacent upstream ramp (pc/h)

Table 406.9 HCM Models for Predicting v_{12} at Off-Ramps

Four-lane freeways (two lanes in each direction)	$P_{FD} = 1.000$	
Six-lane freeways (three lanes in each direction)	$P_{FD} = 0.760 - 0.000025v_F - 0.000046v_R$	Eq. (406.23)
	$P_{FD} = 0.717 - 0.000039v_F + 0.0604v_U/L_{up}$	Eq. (406.24)
	$P_{FD} = 0.616 - 0.000021v_F + 0.124v_D/L_{down}$	Eq. (406.25)
Eight-lane freeways (four lanes in each direction)	$P_{FD} = 0.436$	

v_D = demand flow rate on adjacent downstream ramp (pc/h)

P_{FD} = proportion of through freeway flow remaining in lanes 1 and 2 immediately upstream of diverge

L_{up} = distance to adjacent upstream ramp (ft)

L_{down} = distance to adjacent downstream ramp (ft)

Selecting equations for P_{FD} for six-lane freeways				
Adjacent upstream ramp	Subject ramp	Adjacent downstream ramp	Equation used	Remarks
None	Off	None	Eq. (406.23)	
None	Off	On	Eq. (406.23)	
None	Off	Off	Eq. (406.23) or (406.25)	Use Eq. (406.23) if distance between ramps $\geq L_{EQ}$, Eq. (406.25) otherwise
On	Off	None	Eq. (406.23) or (406.24)	Use Eq. (406.23) if distance between ramps $\geq L_{EQ}$, Eq. (406.24) otherwise
Off	Off	None	Eq. (406.23)	
On	Off	On	Eq. (406.23) or (406.24)	Use Eq. (406.23) if distance between ramps $\geq L_{EQ}$, Eq. (406.24) otherwise
On	Off	Off	Eqs. (406.23), (406.24), or (406.25)	Use Eq. (406.23) if distance between ramps $\geq L_{EQ}$. Else, try both Eqs. (406.24) and (406.25) and choose larger value of P_{FD}
Off	Off	On	Eq. (406.23)	
Off	Off	Off	Eq. (406.23) or (406.25)	Use Eq. (406.23) if distance between ramps $\geq L_{EQ}$, Eq. (406.25) otherwise

L_{EQ} is that distance for which Eqs. (406.23) and Eq. (406.24) or (406.25), as appropriate, yield the same value of P_{FD}.

Where an adjacent upstream on-ramp exists, L_{EQ} is calculated using Eq. (406.26).

$$L_{EQ} = \frac{v_U}{0.071 + 0.000023v_F - 0.000076v_R} \qquad (406.26)$$

Where an adjacent downstream off-ramp exists, L_{EQ} is calculated using Eq. (406.27).

$$L_{EQ} = \frac{v_D}{1.15 + 0.000023v_F - 0.0000369v_R} \qquad (406.27)$$

Two-Lane On-Ramps

P_{FD} for two-lane off-ramps is found as follows:

- Four-lane freeways, $P_{FD} = 1.000$
- Six-lane freeways, $P_{FD} = 0.450$
- Eight-lane freeways, $P_{FD} = 0.260$

Multilane Highways

In the *Highway Capacity Manual*, 6th edition (2016), it has been recognized that the analysis methods for basic freeway segments and multilane highway segments are very similar and, therefore, their treatment is unified under one chapter (Chap. 12).

Level of Service

According to the HCM, the density of passenger car equivalents per mile per lane of a multilane highway segment (pcpmpl) is used to define the level of service. These level of service (LOS) boundaries can also be expressed in terms of the following proxy variables—speed (mph), volume to capacity ratio, and service flow rate (pcphpl). Figure 407.1 summarizes the boundaries for the various regimes of level of service (A through E) based on these parameters. This figure can be used once the free-flow speed (FFS) has been determined.

Default Values of Parameters

For multilane highway segments, the following default values of certain input parameters are listed in Exhibit 12-6 of the HCM and may be used when field measured data are not available.

Lane width = 12 ft (range 10–12 ft)

Right-side lateral clearance = 6 ft (range 0–6 ft)

Median (left-side) lateral clearance = 6 ft (range 0–6 ft)

Access point density: 8 per mile (rural); 16 per mile (low-density suburban); 25 per mile (high-density suburban)

Base free-flow speed = speed limit + 5 mph (range 45–70 mph)

PHF = 0.95 (urban); 0.88 (rural)

Percentage of heavy vehicles = 12% (rural), 5% (urban)

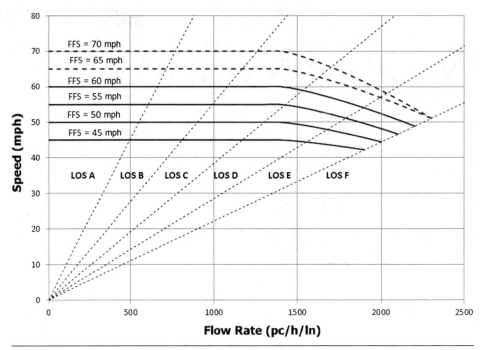

Figure 407.1 Level of service criteria for multilane highways.

Driver population factor $= 1.00$.

Unlike freeway segments, capacity and speed adjustment factor (CAF and SAF, respectively) are not used for multilane highway segments.

Note that the upper limit for LOS E corresponds to a v/c ratio $= 1.0$. Therefore, the corresponding value of service flow rate represents highway lane capacity. Thus, at a free-flow speed of 60 mph, a highway lane has a capacity of 2200 pc/h. See Eq. (407.3).

Free-Flow Speed

The *Highway Capacity Manual* outlines the following procedure for determining free-flow speed based on the base free-flow speed (BFFS) and certain site-specific adjustments (speed deductions).

$$\text{FFS} = \text{BFFS} - f_{LW} - f_{LC} - f_M - f_A \tag{407.1}$$

Where field measurements are unavailable, BFFS $= 60$ mph may be assumed for rural or suburban multilane highways.

Speed Adjustments to BFFS

The parameters $f_{LW}, f_{LC}, f_M,$ and f_A are adjustments for lane width, total lateral clearance, median type, and access-points density, respectively. These are given by Tables 407.1 through 407.4, respectively.

Table 407.1 Free-Flow Speed Adjustment for Lane Width

Lane width (ft)	Reduction in free-flow speed (mph)
12	0.0
11	1.9
10	6.6

Table 407.2 Free-Flow Speed Adjustment for Lateral Clearance

Four-lane highways		Six-lane highways	
Total lateral clearance (ft)	Reduction in FFS (mph)	Total lateral clearance (ft)	Reduction in FFS (mph)
12	0.0	12	0.0
10	0.4	10	0.4
8	0.9	8	0.9
6	1.3	6	1.3
4	1.8	4	1.7
2	3.6	2	2.8
0	5.4	0	3.9

Table 407.3 Free-Flow Speed Adjustment for Median Type

Median type	Reduction in free-flow speed (mph)
Undivided highways	1.6
Divided highways (including TWLTL)	0.0

Table 407.4 Free-Flow Speed Adjustment for Access Point Density

Number of access points per mile	Reduction in free-flow speed (mph)
0	0.0
10	2.5
20	5.0
30	7.5
≥ 40	10.0

The adjustment for lateral clearance is based on total lateral clearance, which is the sum of left and right lateral clearances

$$TLC = LC_L + LC_R$$

(407.2)

LC_R = lateral clearance from the right edge of the travel lanes to the roadside obstructions. The value of LC_R should be limited to 6 ft.

LC_L = lateral clearance from the left edge of the travel lanes to obstructions in the roadway median. The value of LC_L should be limited to 6 ft.

NOTE: For undivided highways, the lack of left clearance is taken into account by the median adjustment. So that the design is not subject to a double penalty, use $LC_L = 6$ ft for the undivided design. Also, for roadways with two-way left turn lanes, $LC_L = 6$ ft.

Access points are defined as intersections and driveways on the right side of the roadway that influence (impede) traffic flow. Access points which are unnoticed by the driver or with minimal activity should not be considered in calculating access point density.

Capacity

The base lane capacity (pcphpl) of a multilane highway segment for $45 \leq FFS \leq 70$ mph is given by

$$c = 1900 + 20(FFS - 45) \leq 2300$$

(407.3)

Flow Rate

The peak hourly volume V (vph) is converted to an equivalent peak flow rate v_p (pcphpl) using

$$v_p = \frac{V}{PHF \times N \times f_{HV} \times f_P}$$

(407.4)

where   PHF is the peak hour factor based on peak 15-minute vehicular count
N is the number of lanes
f_P is a driver population factor
f_{HV} is a heavy vehicle factor

Peak Hour Factor

The peak hour factor is an indicator of the variability of flow. Larger fluctuations in the traffic flow result in lower computed values of PHF. A lower value of PHF results in a higher value of peak flow rate v_p. For an example of PHF calculation from traffic counts, see Chap. 406.

Table 407.5 Passenger Car Equivalents for General Terrain Segments

Passenger Car Equivalent	Type of terrain	
	Level	Rolling
E_T	2.0	3.0

For multilane highways, the default values of PHF as given in Chap. 12 of the *Highway Capacity Manual* are: 0.88 (rural) and 0.95 (urban).

Heavy Vehicle Factor

The heavy vehicle factor converts the mixed traffic (passenger cars, trucks, buses, and recreational vehicles) into an equivalent number of passenger cars. The factor is calculated in terms of equivalence factor E_T, which in turn is dependent primarily on terrain. The factor f_{HV} is given by

$$f_{HV} = \frac{1}{1 + P_T(E_T - 1)} \qquad (407.5)$$

where P_T is the proportion (decimal) of heavy vehicle in the traffic stream. Equivalence factor E_T is given in Table 407.5. Highway segments longer than 0.5 miles with grades between 2% and 3% or longer than 0.25 miles with grades of 3% or greater should be considered as separate segments.

The factors given in Table 407.5 apply *when analyzing extended sections* of multilane highways that have been broadly classified as level or rolling. In the HCM 2016, heavy vehicles are broadly classified as one of two types—SUT (single unit trucks, RVs, and buses) and TT (tractor trailers). HCM Exhibits 12-26 through 12-28 are PCE tables for the following splits—30% SUTs + 70% TTs (typical for rural highways), 50% SUTs + 50% TTs, and 70% SUTs + 30% TTs, respectively.

Driver Population Factor

The driver population factor (f_p) accounts for the effect of driver familiarity with the roadway on level of service. This factor varies between 0.85 and 1.0. Unless there is sufficient reason for using a lesser value, the default value is 1.0, which represents weekday commuter traffic.

Mean Speed

The mean speed of the traffic stream under base conditions is given by

$$S = FFS_{adj} \qquad \qquad for\ v_p \leq 1400 \qquad (407.6a)$$

$$S = FFS_{adj} - \frac{\left(FFS_{adj} - \dfrac{c}{D_c}\right)(v_p - 1,400)^{1.31}}{(c - BP)^{1.31}} \qquad for\ v_p > 1400 \qquad (407.6b)$$

where D_c = density at capacity = 45 pc/mi and the breakpoint BP = 1400.

Table 407.6 Level of Service Criteria for Basic Freeway Segments

LOS	Density (pc/mi/ln)
A	≤11
B	>11–18
C	>18–26
D	>26–35
E	>35–45
F	Demand exceeds capacity OR density >45

Determining Level of Service

Either density or peak flow rate can be used in conjunction with the free-flow speed to determine LOS using Fig. 407.1. The density D (pc/mi/ln) is calculated from the flow rate v_p (pc/h/ln) using

$$D = \frac{v_P}{S} \tag{407.7}$$

where S is average speed (mph).

Table 407.6 gives LOS thresholds for multilane highway segments.

Example 407.1

Determine the level of service for a 3000 ft segment of a four-lane highway (two lanes in each direction) with a 3% grade. The field measured FFS is 45 mph. Lanes are 11 ft wide. Assume a peak hour factor PHF = 0.88. The traffic consists of approximately 10% single unit trucks and 5% tractor-trailers. The one way peak hour volume is 2200 vph. Assume traffic primarily consists of commuters.

Solution Since the field measured FFS is reported, no adjustments are necessary. FFS = 45 mph

Commuter traffic, implying population factor $f_p = 1.0$

The heavy vehicle mix is 67% SUT + 33% TT. From Exhibit 12-28 (for 70% SUT and 30% TT), using 3% grade, 3000 ft (0.57 mile) segment length and 15% trucks, $E_T = 2.35$. Heavy vehicle factor is

$$f_{HV} = \frac{1}{1 + 0.15(2.35 - 1)} = 0.832$$

$V = 2200$ vph

$$v_P = \frac{2200}{0.88 \times 2 \times 0.832 \times 1.0} = 1502$$

For a free-flow speed = 45 mph and flow rate = 1502 pcphpl, LOS D (range 1170–1550)

Example 407.2

Determine the level of service for a 3 mile segment of a four-lane undivided highway (two lanes in each direction) with a 3% grade. The field measured FFS is 48 mph. Lanes are 11 ft wide. Assume a peak hour factor PHF = 0.91. The traffic consists of approximately 12% trucks, buses and RVs, and 4% tractor-trailers. The one way peak hour volume is 2450 vph. Assume weekend traffic.

Solution Since the field measured FFS is reported, no adjustments are necessary.

FFS = 48 mph
Non-commuter traffic, implying population factor $f_p = 0.9$
Because of the long segment length, considering it to be a rolling terrain, $E_T = 3.00$.
Heavy vehicle factor

$$f_{HV} = \frac{1}{1 + 0.16(3.0 - 1)} = 0.758$$

$V = 2450$ vph

$$v_p = \frac{2450}{0.91 \times 2 \times 0.832 \times 0.9} = 1798 \text{ pcphpl}$$

For a free-flow speed, FFS = 48 mph (interpolating between 50 mph and 45 mph), the maximum flow rates for LOS A through E are 528, 864, 1248, 1740, 1960 (approx.).
Therefore, for flow rate = 1798 pcphpl, LOS E.

Urban Streets
Free-Flow Speed

The base free-flow speed (S_{fo}) for urban streets is defined to be the free-flow speed on longer segments. It includes the influence of speed limit, access point density, median type, curb presence, and on-street parking presence. The corresponding adjustment factors are listed in Exhibit 18-11 of the HCM 2016.

Shorter segment length (defined by spacing of signals) affects the driver's choice of free-flow speed (S_f)

$$S_f = S_{fo} f_L \geq S_{pl}$$

where the adjustment factor f_L is given by

$$f_L = 1.02 - 4.7 \frac{S_{fo} - 19.5}{\max(L_s, 400)} \leq 1.0 \qquad (407.9)$$

L_s = segment length = spacing between signals (ft)
S_{fo} = free-flow speed (mph)
S_{pl} = posted speed limit (mph)

Level of Service Criteria for Urban Street Segments

LOS criteria for urban street segments are based on multimodal approach—motorized vehicles (Table 407.6), pedestrians (Table 407.7), and bicycles and transit vehicles (Table 407.8).

LOS is determined separately for both directions of travel along the segment. HCM Exhibit 18-1 lists the LOS thresholds established for this purpose. As indicated in this exhibit, LOS is defined by two performance measures. One measure is the travel speed for through vehicles. The second is the volume-to-capacity ratio for the through movement at the downstream boundary intersection.

Table 407.6 LOS Criteria: Motorized Vehicle Mode

LOS	Travel Speed Threshold by Base Free-Flow Speed (mph)							Volume-to-Capacity Ratio
	55	50	45	40	35	30	25	
A	>44	>40	>36	>32	>28	>24	>20	
B	>37	>34	>30	>27	>23	>20	>17	
C	>28	>25	>23	>20	>18	>15	>13	≤1.0
D	>22	>20	>18	>16	>14	>12	>10	
E	>17	>15	>14	>12	>11	>9	>8	
F	≤17	≤15	≤14	≤12	≤11	≤9	≤8	
F	Any							>1.0

Table 407.7 LOS Criteria: Pedestrian Mode

Segment-Based Pedestrian LOS Score	Segment-Based LOS by						Link-Based Pedestrian LOS	
	Average Pedestrian Space (ft²/p)							
	>60	>40–60	>24–40	>15–24	>8–15	≤8	Link-Based LOS Score	LOS
≤2.00	A	B	C	D	E	F	≤1.50	A
>2.00–2.75	B	B	C	D	E	F	>1.50–2.50	B
>2.75–3.50	C	C	C	D	E	F	>2.50–3.50	C
>3.50–4.25	D	D	D	D	E	F	>3.50–4.50	D
>4.25–5.00	E	E	E	E	E	F	>4.50–5.50	E
>5.00	F	F	F	F	F	F	>5.50	F

Table 407.8   LOS Criteria: Bicycle and Transit Mode

LOS	Segment-Based Bicycle LOS Score	Link-Based Bicycle LOS Score	Transit LOS Score
A	$\leq$2.00	$\leq$1.50	$\leq$2.00
B	>2.00–2.75	>1.50–2.50	>2.00–2.75
C	>2.75–3.50	>2.50–3.50	>2.75–3.50
D	>3.50–4.25	>3.50–4.50	>3.50–4.25
E	>4.25–5.00	>4.50–5.50	>4.25–5.00
F	>5.00	>5.50	>5.00

CHAPTER 408

Two-Lane Highways

General

The methodology for the analysis of two-lane highways underwent significant change from the HCM 2000 to the HCM 2010. First of all, the new analysis method is for one-directional traffic as opposed to analyzing both directions. There were two classes (Class I and Class II) of two-lane highway defined in the HCM 2000. There is an additional class (Class III) defined in the HCM 2010. In the transition from HCM 2010 to HCM 2016, the methodology for two-lane highways has essentially remained unchanged.

Because of the restrictions on passing maneuvers, level of service for two-lane highways is determined based on average travel speed (ATS), percent time spent following (PTSF), and percent free-flow speed (PFFS). This is in contrast to multilane highways and freeways for which level of service is determined by density.

The free-flow speed is determined from either field measured speed (S_{FM}) or base free-flow speed (BFFS). If S_{FM} is used, then adjustments for flow rate and heavy vehicles are made to obtain the free-flow speed (FFS). On the other hand, BFFS needs to be adjusted for lane width, shoulder width, and access point density to obtain the free-flow speed.

Next, the hourly volume is converted to peak flow rate using peak-hour factor, heavy vehicle factor, and grade adjustment factor. These computations must be carried out twice—one sequence to obtain the flow rate for ATS ($v_{i,ATS}$) and another to obtain the flow rate for PTSF ($v_{i,PTSF}$).

Default Parameters

The following default parameters are specified in the *Highway Capacity Manual, 2016*:

Lane width = 12 ft
Shoulder width = 6 ft
Access point density (Class I and II): 8 per mile
Access point density (Class III): 16 per mile
Percent no-passing zone: Level terrain: 20%
Rolling terrain: 40%
More extreme: 80%
BFFS: Speed Limit + 10 mph
PHF = 0.88
Heavy vehicles: 6%

917

Classification of Two-Lane Highways

The three classes of two-lane highways are defined as follows:

Class I two-lane highways are those for which motorists expect to travel at relatively high speeds. These are highways that serve mostly long-distance trips or provide connections between facilities that serve long-distance trips. Some examples of Class I two-lane highways are major intercity routes or major links in state or national highway networks.

Class II two-lane highways are those on which motorists do not necessarily expect to travel at high speeds. Class II highways most often serve relatively short trips. Some typical examples are two-lane highways serving as access routes to Class I facilities, serving as scenic or recreational routes.

Class III two-lane highways are those serving moderately developed areas. They may be portions of Class I and Class II highways that pass through small towns or developed recreational areas. On these highways, local traffic often mixes with through traffic and density of unsignalized access points is significantly higher than in a strictly rural area. Most rural two-lane arterials and trunk roads would be considered Class I highways, while most two-lane collectors and local roads would be considered Class II or Class III.

Level of Service

Level of service criteria for Classes I, II, and III two-lane highways are summarized in Table 408.1 (from Exhibit 15-3 of HCM 2016). For Class I highways, LOS is determined by the worse of ATS-based LOS and PTSF-based LOS.

Average Travel Speed (ATS)

ATS is defined as the highway segment length divided by the average travel time taken by vehicles to traverse it during a designated time interval.

Percent Time Spent Following (PTSF)

PTSF represents the freedom to maneuver and the comfort and convenience of travel. It is the average percentage of time that vehicles must travel in platoons behind lower vehicles due to the inability to pass.

Table 408.1 Automobile LOS Criteria for Two-Lane Highways

| LOS | Class I highways | | Class II highways | Class III highways |
	ATS (mph)	PTSF (%)	PTSF (%)	PFFS (%)
A	>55	≤35	≤40	>91.7
B	>50–55	>35–50	>40–55	>83.3–91.7
C	>45–50	>50–65	>55–70	>75.0–83.3
D	>40–45	>65–80	>70–85	>66.7–75.0
E	≤40	>80	>85	≤66.7

Percent Free-Flow Speed (PFFS)

PFFS represents the ability of vehicles to travel at or near the posted speed limit. It is calculated by dividing the average travel speed (ATS) by the free-flow speed (FFS).

The base conditions for two-lane highways are:

- Lane widths 12 ft or greater
- Clear shoulders 6 ft or wider
- Zero no-passing zones
- No heavy vehicles
- Level terrain
- No impediments to through traffic (signals, turning vehicles)

Capacity

Under base conditions, the capacity of a two-lane highway is 1700 pc/h in one direction, with a limit of 3200 pc/h for both directions combined. For short lengths, such as tunnels and bridges, a capacity of 3200 to 3400 pc/h for both directions may be attained.

Bicycle Mode

Bicycle levels of service are based on a bicycle LOS (BLOS) score, which is based on (1) average effective width of the outside through lane, (2) motorized vehicle volumes, (3) motorized vehicle speeds, (4) heavy vehicle volumes, and (5) pavement condition. Table 408.2 gives bicycle LOS criteria for two-lane highways.

Free-Flow Speed

The free-flow speed is determined from either field measured speed (S_{FM}) or base free-flow speed (BFFS). If S_{FM} is used, then adjustments for flow rate and heavy vehicles are made to obtain the free-flow speed. On the other hand, BFFS needs to be adjusted for lane width, shoulder width, and access point density to obtain the free-flow speed.

Table 408.2 Bicycle LOS Criteria for Two-Lane Highways

LOS	BLOS score
A	≤ 1.5
B	>1.5–2.5
C	>2.5–3.5
D	>3.5–4.5
E	>4.5–5.5
F	>5.5

If the field-measured speed (S_{FM}) is available, the free-flow speed FFS is calculated from Eq. (408.1).

$$FFS = S_{FM} + 0.00776\frac{v}{f_{HV,ATS}} \qquad (408.1)$$

v = total demand flow rate (both directions) during period of speed measurements
$f_{HV,ATS}$ = heavy vehicle adjustment factor for ATS

The equivalence factors E_T and E_R used for calculating $f_{HV,ATS}$ are in Exhibits 15-11 through 15-13 in HCM 2010.

If field data are not available, FFS can be estimated from the base free-flow speed (BFFS) according to Eq. (408.2).

$$FFS = BFFS - f_{LS} - f_A \qquad (408.2)$$

f_{LS} = adjustment for lane and shoulder width (mph) in Table 408.3
f_A = adjustment for access point density (mph) in Table 408.4

Table 408.4 shows the adjustment for access point density, which can be calculated by dividing the number of unsignalized intersections and driveways on both sides of the roadway segment by the length of the segment (miles). Access points which are not noticed by the driver and therefore have no influence traffic flow should not be counted.

Table 408.3 Adjustment to Free-flow Speed for Lane and Shoulder Width (f_{LS})

	Reduction in FFS (mph)			
	Shoulder width SW (ft)			
Lane Width (ft)	0 ≤ SW < 2	2 ≤ SW < 4	4 ≤ SW < 6	6 ≤ SW
9 ≤ LW < 10	6.4	4.8	3.5	2.2
10 ≤ LW < 11	5.3	3.7	2.4	1.1
11 ≤ LW < 12	4.7	3.0	1.7	0.4
12 ≤ LW	4.2	2.6	1.3	0.0

Table 408.4 Adjustment to Free-flow Speed for Access Point Density, f_A

Access points per mile	Reduction in FFS (mph)
0	0.0
10	2.5
20	5.0
30	7.5
40	10.0

Heavy Vehicle Factor

Normally, the heavy vehicle factor is calculated as:

$$f_{HV,ATS} = \frac{1}{1 + P_T(E_T - 1) + P_R(E_R - 1)} \qquad (408.3)$$

P_T = proportion of trucks in the traffic stream
E_T = passenger car equivalent for trucks (HCM Exhibit 15-11)
P_R = proportion of RVs in the traffic stream
E_R = passenger car equivalent for RVs (HCM Exhibit 15-11)

However, on specific downgrades (3% or steeper AND 0.6 miles or longer) where trucks travel at crawl speed, the heavy vehicle factor is calculated as

$$f_{HV,ATS} = \frac{1}{1 + P_{TC}P_T(E_{TC} - 1) + (1 - P_{TC})P_T(E_T - 1) + P_R(E_R - 1)} \qquad (408.4)$$

P_{TC} = proportion of trucks operating at crawl speed
E_{TC} = passenger car equivalent for trucks operating at crawl speed (HCM Exhibit 15-14)

Methodology for Class I Two-Lane Highways

Average Travel Speed (ATS)

For both directions, the demand flow rate is calculated using Eq. (408.5).

$$v_{i,ATS} = \frac{V_i}{PHF \times f_{g,ATS} \times f_{HV,ATS}} \qquad (408.5)$$

V_i = demand hourly volume for direction i (vph)
i = "d" (analysis direction) or "o" (opposing direction)
$f_{g,ATS}$ = grade adjustment factor for ATS (HCM Exhibit 15-9 or 15-10)
$f_{HV,ATS}$ = heavy vehicle adjustment factor for ATS [(Eq. (408.3) or (408.4)]

Table 408.5 gives values for the grade adjustment factor $f_{g,ATS}$ for extended segments of level and rolling terrain, as well as for specific downgrades.

For specific upgrades, HCM Exhibit 15-10 gives the ATS grade adjustment factor $f_{g,ATS}$.

Next, the average travel speed (ATS) is calculated according to

$$ATS_d = FFS - 0.00776(v_{d,ATS} + v_{o,ATS}) - f_{np,ATS} \qquad (408.6)$$

ATS_d = average travel speed in analysis direction (mph)
FFS = free-flow speed (mph)
$v_{d,ATS}$ = demand flow rate for ATS calculation in the analysis direction (pc/h)
$v_{o,ATS}$ = demand flow rate for ATS calculation in the opposing direction (pc/h)
$f_{np,ATS}$ = ATS adjustment factor for no passing zones (mph) is given by Table 408.6

Table 408.5 Grade Adjustment Factor to Determine Speeds $(f_{g,ATS})$

One direction demand flow rate (vph)	Type of terrain	
	Level terrain & specific downgrades	Rolling terrain
≤ 100	1.00	0.67
200	1.00	0.75
300	1.00	0.83
400	1.00	0.90
500	1.00	0.95
600	1.00	0.97
700	1.00	0.98
800	1.00	0.99
≥ 900	1.00	1.00

Percent Time Spent Following (PTSF)

The PTSF is required only for Class I and Class II highways. It is calculated from Eq. (408.7)

$$\text{PTSF}_d = \text{BPTSF}_d + f_{np,PTSF}\left(\frac{v_{d,PTSF}}{v_{d,PTSF} + v_{o,PTSF}}\right) \tag{408.7}$$

The base percent time spent following is estimated by

$$\text{BPTSF}_d = 100\left[1 - \exp\left(av_d^b\right)\right] \tag{408.8}$$

Where the empirical coefficients a and b are given in Table 408.7.
Table 408.8 gives values of $f_{np,PTSF}$ to be used in Eq. (408.7).

Methodology for Class II Two-Lane Highways

The LOS for Class II two-lane highways is based solely on the calculation of PTSF [Eq. (408.7)].

Methodology for Class III Two-Lane Highways

Once the FFS and ATS have been calculated for the design direction, the PFFS is given by

$$\text{PFFS} = \frac{\text{ATS}_d}{\text{FFS}} \tag{408.9}$$

Table 408.6 ATS Adjustment Factor for No-Passing Zones ($f_{np,ATS}$)

Opposing demand flow rate v_o (pc/h)	Percent no-passing zones (%)				
	≤ 20	40	60	80	100
FFS ≥ 65 mph					
≤100	1.1	2.2	2.8	3.0	3.1
200	2.2	3.3	3.9	4.0	4.2
400	1.6	2.3	2.7	2.8	2.9
600	1.4	1.5	1.7	1.9	2.0
800	0.7	1.0	1.2	1.4	1.5
1000	0.6	0.8	1.1	1.1	1.2
1200	0.6	0.8	0.9	1.0	1.1
1400	0.6	0.7	0.9	0.9	0.9
≥1600	0.6	0.7	0.7	0.7	0.8
FFS = 60 mph					
≤100	0.7	1.7	2.5	2.8	2.9
200	1.9	2.9	3.7	4.0	4.2
400	1.4	2.0	2.5	2.7	3.9
FFS = 60 mph					
600	1.1	1.3	1.6	1.9	2.0
800	0.6	0.9	1.1	1.3	1.4
1000	0.6	0.7	0.9	1.1	1.2
1200	0.5	0.7	0.9	0.9	1.1
1400	0.5	0.6	0.8	0.8	0.9
≥1600	0.5	0.6	0.7	0.7	0.7
FFS = 55 mph					
≤100	0.5	1.2	2.2	2.6	2.7
200	1.5	2.4	3.5	3.9	4.1
400	1.3	1.9	2.4	2.7	2.8
600	0.9	1.1	1.6	1.8	1.9
800	0.5	0.7	1.1	1.2	1.4
1000	0.5	0.6	0.8	0.9	1.1
1200	0.5	0.6	0.7	0.9	1.0
1400	0.5	0.6	0.7	0.7	0.9
≥1600	0.5	0.6	0.6	0.6	0.7
FFS = 50 mph					
≤100	0.2	0.7	1.9	2.4	2.5
200	1.2	2.0	3.3	3.9	4.0
400	1.1	1.6	2.2	2.6	2.7
600	0.6	0.9	1.4	1.7	1.9
800	0.4	0.6	0.9	1.2	1.3

(Continue)

Table 408.6 ATS Adjustment Factor for No-Passing Zones ($f_{np,ATS}$) (*Continued*)

Opposing demand flow rate v_o (pc/h)	Percent no-passing zones (%)				
	≤ 20	40	60	80	100
1000	0.4	0.4	0.7	0.9	1.1
1200	0.4	0.4	0.7	0.8	1.0
1400	0.4	0.4	0.6	0.7	0.8
≥1600	0.4	0.4	0.5	0.5	0.5
FFS ≤ 45 mph					
≤100	0.1	0.4	1.7	2.2	2.4
200	0.9	1.6	3.1	3.8	4.0
400	0.9	0.5	2.0	2.5	2.7
600	0.4	0.3	1.3	1.7	1.8
800	0.3	0.3	0.8	1.1	1.2
1000	0.3	0.3	0.6	0.8	1.1
1200	0.3	0.3	0.6	0.7	1.0
1400	0.3	0.3	0.6	0.6	0.7
≥1600	0.3	0.3	0.4	0.4	0.6

Table 408.7 Values of Coefficients *a* and *b* Used for Estimating $BPTSF_d$

Opposing demand flow rate v_o (pc/h)	a	b
≤200	−0.0014	0.973
400	−0.0022	0.923
600	−0.0033	0.870
800	−0.0045	0.833
1000	−0.0049	0.829
1200	−0.0054	0.825
1400	−0.0058	0.821
≥1600	−0.0062	0.817

Example 408.1

Determine the level of service for a 5-mile length of a two-lane Class I highway with the following characteristics:

Two-way hourly volume = 1800 vph

Directional split (during analysis period) = 60/40

12% trucks and buses, 5% RVs

PHF = 0.90

Table 408.8 No-Passing-Zone Adjustment Factor ($f_{np,PTSF}$) for Determination of PTSF

Two-way flow rate $v_d + v_o$ (pc/h)	No passing zones (%)					
	0	20	40	60	80	100
Directional Split 50/50						
≤200	9.0	29.2	43.4	49.4	51.0	52.6
400	16.2	41.0	54.2	61.6	63.8	65.8
600	15.8	38.2	47.8	53.2	55.2	56.8
800	15.8	33.8	40.4	44.0	44.8	46.6
1400	12.8	20.0	23.8	26.2	27.4	28.6
2000	10.0	13.6	15.8	17.4	18.2	18.8
2600	5.5	7.7	8.7	9.5	10.1	10.3
3200	3.3	4.7	5.1	5.5	5.7	6.1
Directional split 60/40						
≤200	11.0	30.6	41.0	51.2	52.3	53.5
400	14.6	36.1	44.8	53.4	55.0	56.3
600	14.8	36.9	44.0	51.1	52.8	54.6
800	13.6	28.2	33.4	38.6	39.9	41.3
1400	11.8	18.9	22.1	25.4	26.4	27.3
2000	9.1	13.5	15.6	16.0	16.8	17.3
2600	5.9	7.7	8.6	9.6	10.0	10.2
Directional split 70/30						
≤200	9.9	28.1	38.0	47.8	48.5	49.0
400	10.6	30.3	38.6	46.7	47.7	48.8
600	10.9	30.9	37.5	43.9	45.4	47.0
800	10.3	23.6	28.4	33.3	34.5	35.5
1400	8.0	14.6	17.7	20.8	21.6	22.3
2000	7.3	9.7	15.7	13.3	14.0	14.5
Directional split 80/20						
≤200	8.9	27.1	37.1	47.0	47.4	47.9
400	6.6	26.1	34.5	42.7	43.5	44.1
600	4.0	24.5	31.3	38.1	39.1	40.0
800	4.8	18.5	23.5	28.4	29.1	29.8
1400	3.5	10.3	13.3	16.3	16.9	32.2
2000	3.5	7.0	8.5	10.1	10.4	10.7
Directional split 90/10						
≤200	4.6	24.1	33.6	43.1	43.4	43.6
400	0.0	20.2	28.3	36.3	36.7	37.0
600	−3.1	16.8	23.5	30.1	30.6	31.1
800	−2.8	10.5	15.2	19.9	20.3	20.7
1400	−1.2	5.5	8.3	11.0	11.5	11.9

Rolling terrain

Lane width = 11 ft

Shoulder width = 5 ft

Access points spacing 300 ft

40% no-passing zones

BFFS = 60 mph

Solution

Estimate the Free-Flow Speed (FFS)

For 11-ft lanes and 5-ft shoulders, the adjustment $f_{LS} = 1.7$ mph (Table 408.3)

For $5280/300 = 17.6$ access points per mile, $f_A = 4.4$ mph (Table 408.4)

$$\text{FFS} = 60 - 1.7 - 4.4 = 53.9 \text{ mph}$$

Demand adjustments for ATS

Since split is 60:40, the dominant flow, $V_d = 0.6 \times 1800 = 1080$ vph. Corresponding flow rate $v = 1080/0.9 = 1200$ vph

For rolling terrain and $v = 1200$ vph, Table 408.5 gives $f_{g,ATS} = 1.00$

HCM Exhibit 15-11, $E_T = 1.3$ and $E_R = 1.1$

$$f_{HV,ATS} = \frac{1}{1 + P_T(E_T - 1) + P_R(E_R - 1)} = \frac{1}{1 + 0.12 \times (1.3 - 1) + 0.05 \times (1.1 - 1)} = 0.96$$

$$v_{d,ATS} = \frac{V_d}{\text{PHF} \times f_{g,ATS} \times f_{HV,ATS}} = \frac{1080}{0.9 \times 1.00 \times 0.96} = 1250 \text{ pc/h}$$

Estimate ATS

$$\text{ATS}_d = \text{FFS} - 0.00776(v_{d,ATS} + v_{o,ATS}) - f_{np, ATS}$$

$$v_{d,ATS} = 1250 \text{ pc/h}$$

$$v_{o,ATS} = 833 \text{ pc/h}$$

From Table 408.6, for $vd = 1250$, FFS = 54 mph, 40% no-passing zones $f_{np,ATS} = 0.6$ mph

$$\text{ATS}_d = 53.9 - 0.00776(1250 + 833) - 0.6 = 37.7 \text{ mph}$$

Table 408.9 Grade Adjustment Factor to Determine Speeds ($f_{g,ATS}$)

Directional demand flow rate (vph)	Type of terrain	
	Level terrain & specific downgrades	Rolling terrain
≤ 100	1.00	0.73
200	1.00	0.80
300	1.00	0.85
400	1.00	0.90
500	1.00	0.96
600	1.00	0.97
700	1.00	0.99
800	1.00	1.00
≥ 900	1.00	1.00

Demand adjustments for PTSF

For rolling terrain and $v = 1200$ vph, Table 408.9 gives $f_{g,OTSF} = 1.00$

HCM Exhibit 15-18, $E_T = 1.0$ and $E_R = 1.0$. Therefore, $f_{HV,PTSF} = 1.0$

$$v_{d,PTSF} = \frac{V_i}{PHF \times f_{g,PTSF} \times f_{HV,PTSF}} = \frac{1080}{0.9 \times 1.00 \times 1.00} = 1200 \text{ pc/h}$$

Estimate PTSF

Opposing demand flow rate $= 0.4 \times 1800/0.9 = 800$ pc/h

From Table 408.7, $a = -0.0045$, $b = 0.833$

$$BPTSF_d = 100\left[1 - \exp\left(av_d^b\right)\right] = 100 \times [1 - \exp(-0.0045 \times 1200^{0.833})] = 80.8$$

Two-way flow rate $= 2000$ pc/h. Adjustment factor for no-passing zone, from Table 408.8, $f_{np,PTSF} = 15.6$

$$PTSF_d = BPTSF_d + f_{np,PTSF}\left(\frac{v_{d,PTSF}}{v_{d,PTSF} + v_{o,PTSF}}\right) = 80.8 + 15.6 \times \frac{1200}{2000} = 90.2\%$$

Using ATS $= 37.7$ mph and PTSF $= 90.2\%$ leads to LOS E.

If the capacity had been exceeded (1700 one-way or 3200 two-way) LOS would be F.

Note: If this had been a Class II highway, the LOS would be based only on PTSF $=$ 90.2 and the LOS would be E (PTSF > 85).

If this had been a Class III highway, the LOS would be based only on PFFS $=$ ATS/FFS $= 37.7/53.9 = 70\%$ and the LOS would be D (66.7% $>$ PFFS $>$ 75%).

Signalization Warrants

The *Manual for Uniform Traffic Control Devices* (MUTCD)[1] outlines the following nine warrants for signalization at an intersection.

Warrant 1—Eight-Hour Vehicular Volume

Warrant 1 is satisfied if *one* of the following conditions exists for *each of any 8 h of an average day*:

1. The vehicles per hour given in *both* of the 100% columns of condition A in Table 409.1 exist on the major-street (both directions combined) and the higher-volume minor-street approaches.

2. The vehicles per hour given in *both* of the 100% columns of condition B in Table 409.1 exist on the major-street (both directions combined) and the higher-volume minor-street.

In applying each condition the major-street and minor-street volumes shall be for the same 8 h. On the minor street, the higher volume shall not be required to be on the same approach during each of these 8 h.

If the posted or statutory speed limit or the 85th-percentile speed on the major street exceeds 70 km/h (40 mph), or if the intersection lies within the built-up area of an isolated community having a population of less than 10,000, the traffic volumes in the 70% columns in Table 409.1 should be used in place of the 100% columns.

---

[1]*Manual of Uniform Traffic Control Devices*, Federal Highway Administration, Washington, DC, 2009.

Table 409.1 Warrant 1 Criteria (MUTCD)

Condition A: minimum vehicular volume								
Number of lanes of traffic by approach		Vehicles per hour on major street (total both approaches)				Vehicles per hour on higher volume minor street (one direction)			
Major	Minor	100%	80%	70%	56%	100%	80%	70%	56%
1	1	500	400	350	280	150	120	105	84
≥ 2	1	600	480	420	336	150	120	105	84
≥ 2	≥ 2	600	480	420	336	200	160	140	112
1	≥ 2	500	400	350	280	200	160	140	112

Condition B: interruption of continuous traffic								
Number of lanes of traffic by approach		Vehicles per hour on major street (total both approaches)				Vehicles per hour on higher volume minor street (one direction)			
Major	Minor	100%	80%	70%	56%	100%	80%	70%	56%
1	1	750	600	525	420	75	60	53	42
≥ 2	1	900	720	630	504	75	60	53	42
≥ 2	≥ 2	900	720	630	504	100	80	70	56
1	≥ 2	750	600	525	420	100	80	70	56

Source   Manual of Uniform Traffic Control Devices, Federal Highway Administration, Washington, DC, 2009.

Warrant 1 is also satisfied if *both* of the following conditions exist for *each of any 8 h of an average day*:

1. The vehicles per hour given in *both* of the 80% columns of condition A in Table 409.1 exist on the major-street and the higher-volume minor-street approaches.

2. The vehicles per hour given in *both* of the 80% columns of condition B in Table 409.1 exist on the major-street and the higher-volume minor-street approach.

These major-street and minor-street volumes shall be for the same 8 h for each condition; however, the 8 h satisfied in condition A shall not be required to be the same 8 h satisfied in condition B. On the minor street, the higher volume shall not be required to be on the same approach during each of the 8 h.

If the posted or statutory speed limit or the 85th-percentile speed on the major street exceeds 70 km/h (40 mph), or if the intersection lies within the built-up area of an isolated community having a population of less than 10,000, the traffic volumes at the 56% level may be used in place of the 80% columns.

Warrant 2—Four-Hour Vehicular Volume

Warrant 2 is satisfied if, *for each of any 4 h of an average day,* the plotted points representing the vehicles per hour on the major street (total of both approaches) and the corresponding vehicles per hour on the higher-volume minor-street approach (one direction only) all fall above the applicable curve in Fig. 409.1 for the existing combination of approach lanes. On

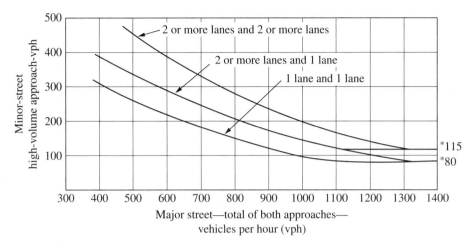

*Note: 115 vph applies as the lower threshold volume for a minor-street approach with two or more lanes and 80 vph applies as the lower threshold volume for a minor-street approach with one lane.

Figure 409.1 MUTCD warrant 2: 4-h vehicular volume.

Source Manual of Uniform Traffic Control Devices, Federal Highway Administration, Washington, DC, 2009.

the minor street, the higher volume shall not be required to be on the same approach during each of these 4 h.

If the posted or statutory speed limit or the 85th-percentile speed on the major street exceeds 70 km/h (40 mph) or if the intersection lies within the built-up area of an isolated community having a population of less than 10,000, warrant 2 is satisfied with 70% of the numbers in Fig. 409.1.

Warrant 3—Peak-Hour Volume

This signal warrant shall be applied only in unusual cases. Such cases include, but are not limited to, office complexes, manufacturing plants, industrial complexcs, or high-occupancy vehicle facilities that attract or discharge large numbers of vehicles over a short time.

Warrant 3 is satisfied if the criteria in *either* of the following two categories are met:

1. If all three of the following conditions exist for the same 1 h (any four consecutive 15-min periods) of an average day:
   a. The total stopped time delay experienced by the traffic on one minor-street approach (one direction only) controlled by a *stop* sign equals or exceeds: 4 vph for a one-lane approach; or 5 vph for a two-lane approach.
   b. The volume on the same minor-street approach (one direction only) equals or exceeds 100 vph for one moving lane of traffic or 150 vph for two moving lanes.
   c. The total entering volume serviced during the hour equals or exceeds 650 vph for intersections with three approaches or 800 vph for intersections with four or more approaches.

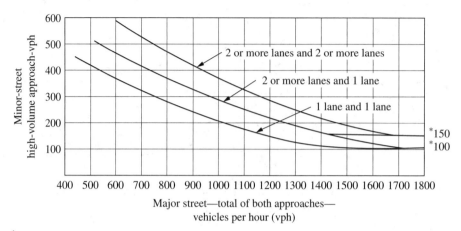

*Note: 150 vph applies as the lower threshold volume for a minor-street approach with two or more lanes and 100 vph applies as the lower threshold volume for a minor-street approach with one lane.

Figure 409.2 MUTCD warrant 3: peak-hour volume.

Source Manual of Uniform Traffic Control Devices, Federal Highway Administration, Washington, DC, 2009.

2. The plotted point representing the vehicles per hour on the major street (total of both approaches) and the corresponding vehicles per hour on the higher-volume minor-street approach (one direction only) for 1 h (any four consecutive 15-min periods) of an average day falls above the applicable curve in Fig. 409.2 for the existing combination of approach lanes.

If the posted or statutory speed limit or the 85th-percentile speed on the major street exceeds 70 km/h (40 mph), or if the intersection lies within the built-up area of an isolated community having a population of less than 10,000, warrant 4 is satisfied with 70% of the numbers in Fig. 409.2.

Warrant 4—Pedestrian Volume

Warrant 4 is met if one of the following criteria is met:

A. For each of any 4 h of an average day, the plotted points representing the vehicles per hour on the major street (total of both approaches) and the corresponding pedestrians per hour crossing the major street (total of all crossings) all fall above the curve in Fig. 409.3; or

B. For 1 h (any four consecutive 15-min periods) of an average day, the plotted point representing the vehicles per hour on the major street (total of both approaches) and the corresponding pedestrians per hour crossing the major street (total of all crossings) falls above the curve in Fig. 409.4.

If the posted or statutory speed limit or the 85th-percentile speed on the major street exceeds 35 mph, or if the intersection lies within the built-up area of an isolated community having a population of less than 10,000, the thresholds may be lowered by a factor of 70%.

The Pedestrian volume signal warrant shall not be applied at locations where the distance to the nearest traffic control signal or STOP sign controlling the street that pedestrians desire to cross is less than 300 ft, unless the proposed traffic control signal will not restrict the progressive movement of traffic.

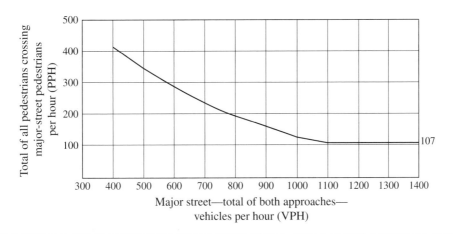

Figure 409.3 MUTCD warrant 4 condition A: pedestrian 4-h volume.

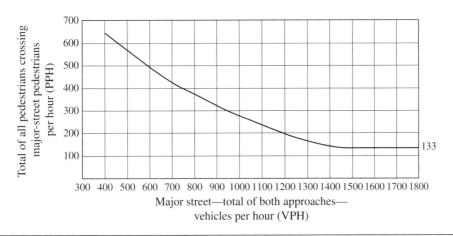

Figure 409.4 MUTCD warrant 4 condition B: pedestrian peak-hour volume.

The criterion for the pedestrian volume crossing the major street may be reduced as much as 50 percent if the 15th-percentile crossing speed of pedestrians is less than 3.5 ft/s.

A traffic control signal may not be needed at the study location if adjacent coordinated traffic control signals consistently provide gaps of adequate length for pedestrians to cross the street.

Warrant 5—School Crossing

Warrant 5 is satisfied when an engineering study of the frequency and adequacy of gaps in the vehicular traffic stream as related to the number and size of groups of school children at an established school crossing across the major street shows that the number of adequate gaps in the traffic stream during the period when the children are using the crossing is less than the number of minutes in the same period and there are a minimum of 20 students during the highest crossing hour. Before a decision is made to install a traffic control signal, consideration shall be given to the implementation of other remedial measures, such as warning signs and flashers, school speed zones, school crossing guards, or a grade-separated crossing.

The school crossing signal warrant shall not be applied at locations where the distance to the nearest traffic control signal along the major street is less than 90 m (300 ft), unless the proposed traffic control signal will not restrict the progressive movement of traffic.

Warrant 6—Coordinated Signal System

Warrant 6 is satisfied if *one* of the following criteria is met:

1. On a one-way street or a street that has traffic predominantly in one direction, the adjacent traffic control signals are so far apart that they do not provide the necessary degree of vehicular platooning.

2. On a two-way street, adjacent traffic control signals do not provide the necessary degree of platooning and the proposed and adjacent traffic control signals will collectively provide a progressive operation.

Warrant 7—Crash Experience

Warrant 7 is satisfied if *all* of the following criteria are met:

1. Adequate trial of alternatives with satisfactory observance and enforcement has failed to reduce the crash frequency.

2. Five or more reported crashes, of types susceptible to correction by a traffic control signal, have occurred within a 12-month period, each crash involving personal injury or property damage apparently exceeding the applicable requirements for a reportable crash.

3. For each of any 8 h of an average day, the vehicles per hour (vph) given in both of the 80% columns of condition A in Table 409.1, or the vph in both of the 80% columns of condition B in Table 409.1 exists on the major-street and the higher-volume minor-street approach, or the volume of pedestrian traffic is not less than 80% of the requirements specified in the pedestrian volume warrant. These major-street and minor-street volumes shall be for the same 8 h. On the minor street, the higher volume shall not be required to be on the same approach during each of the 8 h.

If the posted or statutory speed limit or the 85th-percentile speed on the major street exceeds 70 km/h (40 mph), or if the intersection lies within the built-up area of an isolated community having a population of less than 10,000, the traffic volumes in the 56% columns in Table 409.1 may be used in place of the 80% columns.

Warrant 8—Roadway Network

Warrant 8 is satisfied if the common intersection of two or more major routes meets *one* or *both* of the following criteria:

1. The intersection has a total existing, or immediately projected, entering volume of at least 1000 veh/h during the peak hour of a typical weekday and has 5-year projected traffic volumes, based on an engineering study, that meet one or more of Warrants 1, 2, and 3 during an average weekday.

2. The intersection has a total existing or immediately projected entering volume of at least 1000 veh/h for each of any 5 h of a nonnormal business day (Saturday or Sunday).

A major route as used in this signal warrant shall have *one* or *more* of the following characteristics:

1. It is part of the street or highway system that serves as the principal roadway network for through traffic flow.

2. It includes rural or suburban highways outside, entering, or traversing a city.

3. It appears as a major route on an official plan, such as a major-street plan in an urban area traffic and transportation study.

Warrant 9—Signalization of an Intersection near a Grade Crossing

Warrant 9 (new in the 2009 MUTCD) is met if both of the following conditions are met:

A. A grade crossing exists on an approach controlled by a STOP or YIELD sign and the center of the track nearest to the intersection is within 140 ft of the stop line or yield line on the approach; and

B. During the highest traffic volume hour during which rail traffic uses the crossing, the plotted point representing the vehicles per hour on the major street (total of both approaches) and the corresponding vehicles per hour on the minor-street approach that crosses the track (one direction only, approaching the intersection) falls above the applicable curve in Fig. 409.5 or 409.6 (Figures 4C-9 or 4C-10 from the MUTCD 2009) for the existing combination of approach lanes over the track and the distance D, which is the clear storage distance as defined in Section 1A.13.

The minor-street approach volume may be multiplied by up to three adjustment factors as follows:

Because the curves are based on an average of four occurrences of rail traffic per day, the vehicles per hour on the minor-street approach may be multiplied by the adjustment factor shown in Table 4C-2 for the appropriate number of occurrences of rail traffic per day. The factor ranges from 0.67 (for daily rail traffic = 1) to 1.33 (for daily rail traffic = 12).

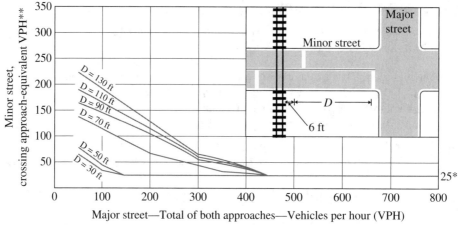

*25 vph applies as the lower threshold volume
**VPH after applying the adjustment factors in Tables 4C-2, 4C-3, and/or 4C-4, if appropriate

Figure 409.5 Warrant 9, intersection near a grade crossing (one approach lane at the track crossing).

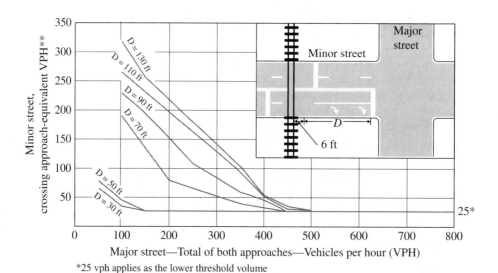

*25 vph applies as the lower threshold volume
**VPH after applying the adjustment factor in Tables 4C-2, 4C-3, and/or 4C-4, if appropriate

Figure 409.6 Warrant 9, intersection near a grade crossing (two or more approach lanes at the track crossing).

Because the curves are based on typical vehicle occupancy, if at least 2% of the vehicles crossing the track are (high occupancy) buses carrying at least 20 people, the vehicles per hour on the minor-street approach may be multiplied by the adjustment factor, shown in Table 4C-3, for the appropriate percentage of high-occupancy buses. For 6% or more of the minor street approach traffic composed of high-occupancy buses, this adjustment factor = 1.32.

Because the curves are based on tractor-trailer trucks comprising 10% of the vehicles crossing the track, the vehicles per hour on the minor-street approach may be multiplied by the adjustment factor shown in Table 4C-4 for the appropriate distance and percentage of tractor-trailer trucks. This factor ranges from 0.50 to 4.18 for distance D less than 70 ft and from 0.50 to 2.09 for distance D greater than 70 ft.

Clear storage distance—when used in Part 8, the distance available for vehicle storage measured between 6 ft from the rail nearest the intersection to the intersection stop line or the normal stopping point on the highway. At skewed grade crossings and intersections, the 6-ft distance shall be measured perpendicular to the nearest rail either along the center line or edge line of the highway, as appropriate, to obtain the shorter distance. Where exit gates are used, the distance available for vehicle storage is measured from the point where the rear of the vehicle would be clear of the exit gate arm. In cases where the exit gate arm is parallel to the track(s) and is not perpendicular to the highway, the distance is measured either along the center line or edge line of the highway, as appropriate, to obtain the shorter distance.

Example 409.1

A two-lane minor street intersects a four-lane major street as shown in the diagram below.

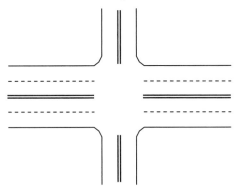

Given the following data, determine which MUTCD warrants for signalization are satisfied at this location.

On an average day, for each of the 8 h, total traffic volume on major street (sum of both directions) is 450 vph. For the higher-volume minor-street approach, the volume is 130 vph.

85th-percentile speed on the major-street traffic is 47 mph.

The peak pedestrian volume crossing the major street is 230 pedestrians per hour. During peak pedestrian periods, the average number of gaps per hour in the traffic stream is 48.

The nearest traffic signal is located 800 ft from this intersection.

In the last 12 months, there have been two PDO accidents and one injury accident (nonfatal) at this intersection.

Solution   With the limited data, only the following relevant warrants are examined:

  Warrant 1—8-h volume

  Warrant 4—pedestrian volume

  Warrant 7—crash experience

Warrant 1

For two lanes on major street and one lane on minor street, the minimum volume requirements are 600 (major street total both approaches) and 150 (larger of minor volumes). The given volumes (450 and 130) are less than these levels. Conditions A or B are not met at the 100% level. However, since the 85th-percentile speed is greater than 40 mph, check these levels at 70% levels (420 and 105). Condition A for warrant 1 is therefore met.

Warrant 4

In order for this warrant to be met, both conditions of warrant 4 need to be met—pedestrian volume greater than 190 pedestrians per hour during any 1 h, or greater than 100 pedestrians per hour during any 4 h and number of acceptable (to pedestrians) gaps in traffic stream less than 60 per hour. The given peak-hour volume of 230 pedestrians per hour is greater than 190 and number of gaps (48) is less than 60. Therefore, this warrant is met.

Warrant 7

Number of crashes during the 12-month period is less than 5. Therefore, this warrant is not met.
  Warrants 1 and 4 for signalization of the intersection are satisfied.

Example 409.2

A one-lane major street intersects with a one-lane minor street at an at-grade intersection. Traffic counts were collected during 12 consecutive hours of a typical day. Investigate whether vehicular warrants 1 and 2 of MUTCD are met.

Time	Major street		Minor street	
	NB	SB	EB	WB
6 (a.m.)–7 (a.m.)	410	135	45	55
7 (a.m.)–8 (a.m.)	725	320	110	160
8 (a.m.)–9 (a.m.)	650	430	100	180
9 (a.m.)–10 (a.m.)	570	500	80	100
10 (a.m.)–11 (a.m.)	450	530	90	110
11 (a.m.)–12 (p.m.)	450	550	80	110
12 (p.m.)–1 (p.m.)	550	600	120	110
1 (p.m.)–2 (p.m.)	410	620	160	180
2 (p.m.)–3 (p.m.)	390	640	130	150
3 (p.m.)–4 (p.m.)	350	710	160	190
4 (p.m.)–5 (p.m.)	400	750	200	190
5 (p.m.)–6 (p.m.)	420	720	170	150

Solution

WARRANT 1: EIGHT-HOUR VEHICULAR VOLUME

Warrant 1: Condition A thresholds (100% level) for one lane major −one lane minor are 500 vph total on both approaches major AND 150 vph on higher-volume minor.

Table 4C-1. Warrant 1, eight-hour vehicular volume
Condition A—minimum vehicular volume

Number of lanes for moving traffic on each approach		Vehicles per hour on major street (total of both approaches)				Vehicles per hour on higher-volume minor-street approach (one direction only)			
Major Street	Minor Street	100%[a]	80%[b]	70%[c]	56%[d]	100%[a]	80%[b]	70%[c]	56%[d]
1	1	500	400	350	280	150	120	105	84
2 or more	1	600	480	420	336	150	120	105	84
2 or more	2 or more	600	480	420	336	200	160	140	112
1	2 or more	500	400	350	280	200	160	140	112

The intervals during which these thresholds are exceeded are highlighted in the table below.

	Major street		Minor street	
Time	NB	SB	EB	WB
6 (a.m.)–7 (a.m.)	410	135	45	55
7 (a.m.)–8 (a.m.)	725	320	110	160
8 (a.m.)–9 (a.m.)	650	430	100	180
9 (a.m.)–10 (a.m.)	570	500	80	100
10 (a.m.)–11 (a.m.)	450	530	90	110
11 (a.m.)–12 (p.m.)	450	550	80	110
12 (p.m.)–1 (p.m.)	550	600	120	110
1 (p.m.)–2 (p.m.)	410	620	160	180
2 (p.m.)–3 (p.m.)	390	640	130	150
3 (p.m.)–4 (p.m.)	350	710	160	190
4 (p.m.)–5 (p.m.)	400	750	200	190
5 (p.m.)–6 (p.m.)	420	720	170	150

As shown, the two criteria for Condition A of Warrant 1 are met during 7 h. Warrant 1 requires the thresholds met during any 8 h. So, this condition of warrant 1 is not met.

If this condition was met at the 100% level, there would be no reason to evaluate condition B.

Warrant 1: Condition B thresholds (100% level) for one lane major—one lane minor are 750 vph total on both approaches major AND 75 vph on higher-volume minor.

Condition B—interruption of continuous traffic

Number of lanes for moving traffic on each approach		Vehicles per hour on major street (total of both approaches)				Vehicles per hour on higher-volume minor-street approach (one direction only)			
Major Street	Minor Street	100%[a]	80%[b]	70%[c]	56%[d]	100%[a]	80%[b]	70%[c]	56%[d]
1	1	750	600	525	420	75	60	53	42
2 or more	1	900	720	630	504	75	60	53	42
2 or more	2 or more	900	720	630	504	100	80	70	56
1	2 or more	750	600	525	420	100	80	70	56

Time	Major street		Minor street	
	NB	SB	EB	WB
6 (a.m.)–7 (a.m.)	410	135	45	55
7 (a.m.)–8 (a.m.)	725	320	110	160
8 (a.m.)–9 (a.m.)	650	430	100	180
9 (a.m.)–10 (a.m.)	570	500	80	100
10 (a.m.)–11 (a.m.)	450	530	90	110
11 (a.m.)–12 (p.m.)	450	550	80	110
12 (p.m.)–1 (p.m.)	550	600	120	110
1 (p.m.)–2 (p.m.)	410	620	160	180
2 (p.m.)–3 (p.m.)	390	640	130	150
3 (p.m.)–4 (p.m.)	350	710	160	190
4 (p.m.)–5 (p.m.)	400	750	200	190
5 (p.m.)–6 (p.m.)	420	720	170	150

As shown, the two criteria for condition B of Warrant 1 are met during 11 h. Warrant 1 requires the thresholds met during any 8 h. So, this condition of warrant 1 is met.

Overall, Warrant 1 is met if EITHER condition A or condition B is met at the 100% level.

If neither condition A nor condition B were met at the 100% level, then meeting BOTH conditions A and B at the 80% level would have also satisfied Warrant 1.

WARRANT 2: FOUR-HOUR VEHICULAR VOLUME

Figure 4C-1. Warrant 2, four-hour vehicular volume

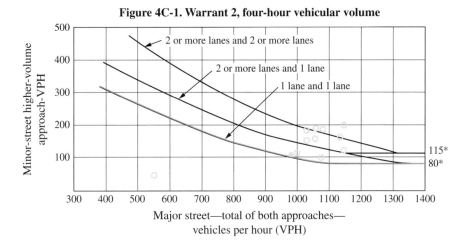

The data in the table is plotted on Fig. 4C-1. As may be seen, all but one data point plots above the red line (one lane major and one lane minor). The warrant requires that this threshold be exceeded for at least 4 h on a single day. Therefore, this warrant is met.

Intersections

General

An *intersection* may be defined as an area shared by two or more roads, thus allowing drivers an option to change route direction. Intersections may be divided into three broad categories: (1) grade separated without ramps (which simply allows the two routes to share the same space without making any provision for route changes), (2) grade separated with ramps (more commonly known as an interchange), and (3) at-grade intersections.

At-grade intersections may be divided into three categories—T or three-leg intersections, cross or four-leg intersections, and multileg intersections. The two primary aspects of an at-grade intersection that impact design are the potential for conflict between different streams of traffic flow and the delay experienced by motorists using the intersection.

Figure 410.1 shows the conflict points between the various traffic streams at a four-approach unsignalized intersection. The 32 conflict points represent the maximum number possible at such an intersection, if all movements are permitted, i.e., through left turn and right turn from all four approaches. If certain movements are prohibited, the number of conflict points will be less than 32.

Signalization of an intersection is a very costly design alternative and should be recommended in those locations where it is fully warranted. Occasionally, a very inexpensive option to improve the safety of an unsignalized intersection is to prohibit certain movements (e.g., no left turns from a particular approach). This eliminates a certain number of conflict points between traffic streams, thereby reducing or eliminating certain types of accidents. This recommendation can only be made if the prohibited movements can be accommodated at a nearby location.

Level of Service

The level of service for two-way stop control (TWSC) and all-way stop control (AWSC) intersections is determined by the measured or computed control delay and is defined for each minor movement. Table 410.1 shows level of service criteria for TWSC and AWSC intersections, as given in the *Highway Capacity Manual*[1].

---

[1]*Highway Capacity Manual*, Transportation Research Board, Washington, DC, 2000.

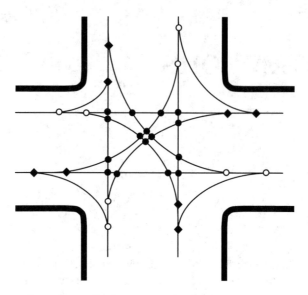

● Crossing conflict points = 16
○ Merging conflict points = 8
◆ Diverging conflict points = 8

Figure 410.1 Points of conflict between traffic streams at a four-way intersection.

Table 410.1 Level of Service Criteria for Unsignalized Intersections

LOS criteria for unsignalized intersections	
LOS	**Average control delay (s/veh)**
A	≤10
B	>10–15
C	>15–25
D	>25–35
E	>35–50
F	>50

The average delay experienced by motorists at an intersection, coupled with data about number and types of accidents experienced over a specified study period, may lead to decisions about signalization of the intersection.

Sight Distance at Intersections

At-grade intersections either have: (1) no control, (2) yield control, (3) stop control, or (4) signalized control. At each intersection, one may draw an approach sight triangle with one side along each approach road, intersecting at the potential point of conflict between the two vehicle paths.

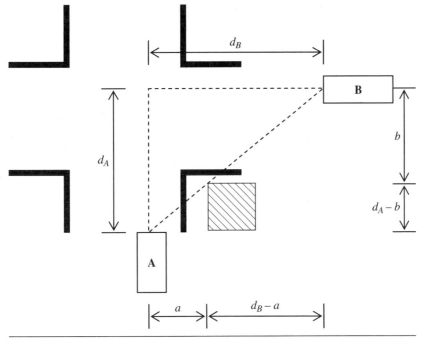

Figure 410.2 Sight triangle at an intersection with no control.

Figure 410.2 shows the sight triangle at an intersection with no control. For the discussion, let us consider the east-west roadway (the one on which vehicle B is traveling) to be the major road and the north-south roadway (path of vehicle A) to be the minor road.

At the instant shown, vehicle A is at a distance d_A from the "point of conflict" while vehicle B is at distance d_B. The travel paths of the vehicles A and B are offset from the edge of the sight obstruction by distances a and b, respectively. Using similar triangles, one may write

$$\frac{b}{d_B - a} = \frac{d_A - b}{a} \qquad (410.1)$$

which can be simplified to

$$d_B = \frac{ad_A}{d_A - b} \quad \text{or} \quad d_A = \frac{bd_B}{d_B - a}$$

Thus, when the position of one vehicle is known, the position of the other when they first become visible to each other can be computed. Similarly, one may construct a departure sight triangle for a vehicle stopped on a minor road, waiting to make a left or right turn onto a major road.

Table 410.2 Length of Sight Triangle Leg for Intersections with No Control

Design speed (mph)	Length of approach leg (ft)
15	70
20	90
25	115
30	140
35	165
40	195
45	220
50	245
55	285
60	325
65	365
70	405
75	445
80	485

Although desirable at high-volume approaches, approach sight triangles are not necessary for intersection approaches controlled by stop signs or traffic signals. The following cases are identified by AASHTO *Geometric Design of Highways and Streets*[2].

Case A—Intersections with No Control

At intersections where right-of-way is not explicitly assigned through signals, stop, or yield signs, the limitations are based on the gap acceptance problem. In other words, what is the minimum acceptable (safe) gap in the traffic flow in one direction as perceived by drivers traveling in the other direction? Table 410.2 shows the distance traveled during these 3.0 s at various speeds. This must be the minimum sight triangle distance on that approach. These distances do not give drivers adequate time to stop, but only time to change the vehicle's speed.

When the approach grade exceeds 3%, the distance given in Table 410.2 should be adjusted by about 10% (increased for downgrades and decreased for upgrades).

NOTE If the sight triangle distance for a safe stopping maneuver is to be computed, then *stopping sight distances* must be used rather than the values from Table 410.2.

[2]*A Policy on Geometric Design of Highways and Streets*, 7th ed., American Association of State Highway and Transportation Officials, Washington, DC, 2018.

Example 410.1

A sight obstructing structure exists on the north-west corner of the unsignalized intersection of third street and main street as shown. Distances to the face of the structure are from the centerlines of right travel lanes. The speed limit on main street is 40 mph. What should be the speed limit on third street in order to allow drivers to avoid collision by adjusting their speeds?

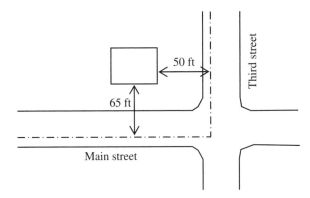

Solution   The length of the sight triangle leg on main street (corresponding to its speed of 40 mph) is 195 ft. Designating this distance as d_A, we get the length of the sight triangle leg along third street as

$$d_B = \frac{65 \times 195}{195 - 50} = 87.4 \text{ ft}$$

From Table 410.2, the allowable speed on third street is nearly 20 mph.

Case B—Intersections with Stop Control on the Minor Road

Departure sight triangle distance may be calculated for three situations:

   B1: Left turns from minor road

   B2: Right turns from minor road

   B3: Crossing the major road from a minor road approach

Case B1—Left Turn from the Minor Road

The vertex of the departure sight triangle on the minor road should be 14.5 ft from the edge of the major road traveled way. Where practical, this distance should be increased to 18.0 ft. The length of the sight triangle along the minor road is the sum of the distance from the major road plus half lane width for vehicles approaching from the left, or 1 1/2 lane widths for vehicles approaching from the right. The intersection sight distance along the major road is given by

$$\text{ISD} = 1.47 V_{\text{major}} t_g \qquad (410.2)$$

where t_g is the time gap acceptable to motorists making this maneuver. Table 410.3 shows the time gap for a stopped vehicle to turn left into a two-lane highway with no median and grades

Table 410.3 Time Gap for Stopped Vehicle to Turn Left into Two-Lane Highway

Design vehicle	Time gap t_g (s)
Passenger car	7.5
Single unit truck	9.5
Combination truck	11.5

3% or less. For left turns onto two-way highways with more than two lanes, add 0.5 s for passenger cars and 0.7 s for trucks for each additional lane to be crossed by the turning vehicle. If the minor road approach grade is an upgrade exceeding 3%, add 0.2 s for each percent grade on the approach.

Example 410.2

Traffic from a new commercial development exits onto a T-junction with a four-lane, undivided highway. The design speed on the highway is 50 mph. The commercial development will generate a significant amount of combination truck traffic. What is the required intersection sight distance (ft) for left turns from the new development onto the highway?

(A) 700

(B) 800

(C) 900

(D) 1000

Solution Case B1: Left turns from minor road
For combination trucks, from Table 9-5 of the *Green Book*, time gap = 11.5 s

For multilane highways, the merging vehicle will have to cross two lanes. Thus, we should add 0.7 s for the additional lane that must be crossed. Therefore, adjusted time gap = 11.5 + 0.7 = 12.2 s.

$$\text{ISD} = 1.47Vt = 1.47 \times 50 \times 12.2 = 896.7 \text{ ft}$$

Answer is C.

Case B2—Right Turn from the Minor Road

The ISD along the major road should be provided in the same way as for case B1, except that the time gap t_g is 1.0 s less than that for case B1. Table 410.4 shows the time gap for a stopped vehicle to turn right onto or cross a two-lane highway with no median and grades 3% or less. For crossing a major road with more than two lanes, add 0.5 s for passenger cars and 0.7 s for trucks for each additional lane to be crossed by the turning vehicle. If the minor road approach grade is an upgrade exceeding 3%, add 0.1 s for each percent grade on the approach.

Table 410.4 Time Gap for Stopped Vehicle to Turn Right onto or Cross Two-Lane Highway

Design vehicle	Time gap t_g (s)
Passenger car	6.5
Single unit truck	8.5
Combination truck	10.5

Case B3—Crossing Maneuver from the Minor Road

In most cases, providing adequate ISD according to Cases B1 and B2 will also provide more than adequate sight distance for minor road traffic to cross the major road. The calculated ISD for Case B3 is the same as that for Case B2.

Case C—Intersections with Yield Control on the Minor Road

Minor road drivers approaching yield signs are permitted to enter and cross the major road without stopping, if there are no potentially conflicting vehicles on the major road. Sight distance needed by drivers on yield-controlled approaches is greater than that required on stop-controlled approaches. Sight triangle distance may be calculated for two situations:

C1: Crossing maneuver from minor road

C2: Left and right turn maneuvers

Critical Gap

As may be seen, the sight distance problem at an unsignalized intersection is one of gap acceptance, i.e., what is the minimum gap length (measured in seconds) that a driver will accept while waiting to merge into a stream of traffic. This "critical gap" has been defined in several ways—one of which (Greenshields) defines the critical gap as the gap accepted by 50% of the drivers. Raff defined it as the gap for which the number of accepted gaps shorter than it is equal to the number of gaps longer than it.

For example, the following table outlines gap acceptance data for an intersection. It provides a summary of accepted and rejected gaps of various lengths for a particular maneuver (e.g., crossing from a minor approach).

Length of gap (s)	No. of gaps accepted	No. of gaps rejected
0.0	0	123
1.0	4	105
2.0	15	90
3.0	45	54
4.0	65	31
5.0	94	12
6.0	137	0

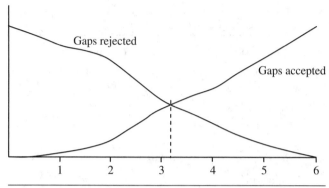

Figure 410.3   Critical gap determination (Raff).

It is apparent that the critical gap (Raff)[3] is somewhere in the interval 3.0 to 4.0 s. At $t = 3.0$ s, the gap differential is $45 - 54 = -9$. At $t = 4.0$ s, the gap differential is $65 - 31 = +34$. Thus, using linear interpolation, the gap differential is zero at approximately 3.2 s. This conclusion is also validated by plotting the data, as shown in Fig. 410.3.

Criteria for Installing Multiway Stop Control (MUTCD)

The solutions presented above apply to cases where (1) a clear major-minor relationship exists, i.e., the volume in one direction is significantly greater than the other and (2) adequate sight distances may be provided. When local conditions do not permit adequate sight distances, it is safer to install stop control on both approaches. This is likely to occur in an urban environment where the two intersecting roads carry equivalent traffic volumes. The following criteria should be considered in an engineering study for a multiway STOP sign installation:

A. The multiway STOP is an interim measure for an intersection where traffic control signals are justified.

B. A crash problem exists (five or more reported crashes in a 12-month period) including right and left turn collisions as well as right-angle collisions.

C. Minimum volumes

   1. The vehicular volume entering the intersection from both major-street approaches averages at least 300 vph for any 8 h of an average day.

   2. Combined vehicular, pedestrian, and bicycle volume entering the intersection from the minor-street approaches averages at least 200 units per hour for the same 8 h and average delay to minor street vehicular traffic is at least 30 s per vehicle during the highest hour.

---

[3]Raff, M. S. and Hart, J. W. (1950), "A Volume Warrant for Urban Stop Sign," *Traffic Engineering and Control*, 5/1983, pp. 255–258.

NOTE If the 85th-percentile approach speed of major-street traffic exceeds 40 mph, then minimum vehicular volume warrants are 70% of the values in criteria C-1 and C-2.

D. Eighty percent level satisfaction of criteria B, C-1, and C-2 (i.e., four or more reported crashes and entering volume at least 240 vph for 8 h and combined volume at least 160 units per hour and average delay to minor street traffic at least 24 s per vehicle).

Geometric Characteristics of Intersections

The following criteria may be considered in the preliminary design of intersections. If left-turn volume is greater than 100 vph, an exclusive left turn lane should be considered if space is available. Double exclusive left turn lanes should be considered when left turn volumes are greater than 300 vph. When right turn volumes are higher than 300 vph and the adjacent mainline volume is also higher than 300 vph, provide an exclusive right turn lane. Maximum volume on any through or through-plus-right turn lane should not exceed 450 vph. Lane widths should be 12 ft if possible.

Capacity and Level of Service at Signalized Intersections

Whereas the level of service for highway segments is computed based on the through flows only, turning movements at intersections can significantly affect their level of service. The capacity of an intersection approach does not have a strong correlation to the level of service. For freeway segments, the volume to capacity (v/c) ratio is 1.00 at the upper limit of level of service E. If this ratio exceeds 1.00, the LOS is automatically F. However, for intersections, it is possible to have the v/c ratio below 1.00 and still have LOS F. This may be due to unacceptable levels of delay due to a combination of one or more of the following factors—long cycle lengths, disproportionately distributed green times, or poor signal progression. Similarly, it is possible to have an approach to have a v/c ratio equal to 1.00 and still have short delays due to the following conditions—short cycle length, and favorable signal progression.

Therefore, LOS F does not necessarily mean that the intersection is oversaturated and therefore level of service and capacity must be calculated separately when signalized intersections are being evaluated.

An isolated intersection, i.e., one whose timing is not coordinated with other nearby intersections, should have a relatively short cycle length—preferably below 60 s, although high approach volumes may necessitate longer cycle lengths. However, cycle lengths should be kept below 120 s, since very long cycles will lead to excessive delay.

Table 410.5   Level of Service Criteria for Signalized Intersections

LOS	Control delay per vehicle (s)	Comments
A	≤ 10	Typically short cycle lengths. Vehicles arrive mainly during green phase.
B	>10–20	Short cycle lengths. Few vehicles stopped. Good vehicle progression.
C	>20–35	Significant number of vehicles stopped at signal. Some vehicles may not clear the intersection during first cycle. Fair progression. Relatively long cycle lengths.
D	>35–55	Noticeable number of cycle failures. Long cycle lengths, high v/c ratios, unfavorable progression.
E	>55–80	Frequent cycle failures. Long cycle lengths, high v/c ratios, poor progression.
F	>80	Long cycle lengths, poor progression. High v/c ratios. Often, but not necessarily oversaturation occurs (v/c >1).

Source   Highway Capacity Manual, Transportation Research Board, Washington, DC, 2010.

Table 410.5 (*Highway Capacity Manual*) gives the level of service criteria for signalized intersections based on average control delay per vehicle. The total control delay for a particular lane group is calculated as

$$d = d_1 \times \text{PF} + d_2 + d_3 \tag{410.3}$$

where d_1 = uniform delay (assuming uniform arrivals)
   PF = adjustment factor for quality of progression
   d_2 = incremental delay, assuming no residual demand at the start of the period
   d_3 = residual demand delay

The intersection delay is calculated as the weighted average of all approaches, where the weights are the lane group volumes for each approach.

$$d_I = \frac{\sum d_A v_A}{\sum v_A} \tag{410.4}$$

The various delay components are affected by the cycle length as follows: the uniform delay shows a linear increase with the cycle length, as expected from the uniform delay equation. The incremental delay is constant with respect to cycle length because the v/c ratio is kept constant. The total delay is the sum of the uniform and incremental components and is therefore linear. Table 410.5 summarizes LOS criteria for signalized intersections.

Critical Lane Volumes at an Intersection

The first step in doing an analysis of a signalized intersection is to calculate critical lane volumes from the design hourly volumes obtained from traffic counts. An illustrative example is shown below.

Example 410.3

A right-angle intersection of two 4-lane highways is shown below. Northbound and southbound left turn lanes have a protected left turn signal. Eastbound and westbound left turning vehicles yield on green. The peak-hour factor is 0.92. Assume moderate pedestrian volumes on both crosswalks. Pedestrian walking speed = 4.0 fps and speed limit = 50 mph both approaches. Level grades. Crosswalk width = 10 ft.

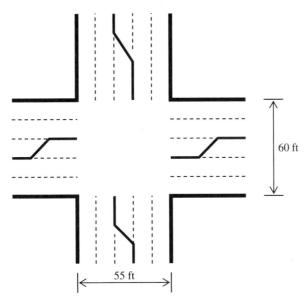

The table below shows all 12 components of approach volumes at the intersection

Approach	Movement	Volume (vph)	Equivalent	Volume	Lane group volume	Volume/ lane*
EB	Left	35	4.00	140	140	140
	Through	610	1.00	610	702	351
	Right	70	1.32	92		
WB	Left	25	5.15	129	129	129
	Through	500	1.00	500	566	283
	Right	50	1.32	66		
NB	Left	220	1.05	231	231	231
	Through	700	1.00	700	944	472
	Right	185	1.32	244		
SB	Left	250	1.05	263	263	263
	Through	800	1.00	800	1031	516
	Right	175	1.32	231		

Table 410.6 Fraction of Lane Group Volume in Design Lane (Typical)

Number of lanes in group	Volume fraction assigned to design lane
2	0.55
3	0.40
4	0.33

In Example 410.3, in calculating the critical lane volume, the volume of the lane group has been equally distributed to the lanes within the lane group. However, it is common practice to distribute a greater fraction to the design lane, as shown in Table 410.6.

Equivalence Factors for Turning Movements

Tables 410.7 and 410.8 give equivalence factors that can be used to convert left and right turning volumes to an equivalent through volume. It should be noted that this procedure is equivalent to, but not the one outlined in the *Highway Capacity Manual*.

Table 410.7 Through Vehicle Equivalents for Left Turns E_{LT}

Opposing flow (vph)	No. of opposing lanes		
	1	2	3
0	1.1	1.1	1.1
200	2.5	2.0	1.8
400	5.0	3.0	2.5
600	10.0	5.0	4.0
800	13.0	8.0	6.0
1000	15.0	13.0	10.0
$\geq$1200	15.0	15.0	15.0

E_{LT} for all protected left turns = 1.05

Table 410.8 Through Vehicle Equivalents for Right Turns E_{RT}

Pedestrian volume in conflicting crosswalk (ped/h)	Equivalent
None	1.18
Low (50)	1.21
Moderate (200)	1.32
High (400)	1.52
Extreme (800)	2.14

Length of Storage Bay in a Dedicated Left Turn Lane

The length of the median cutout storage bay that is needed to accommodate left turns is a function of the left turn volume anticipated. The *AASHTO Green Book*[4] makes the following recommendation:

> "At unsignalized intersections, the storage length, exclusive of taper, may be based on the number of turning vehicles likely to arrive in an average 2-minute period during the peak hour. Space for at least two passenger cars must be provided; with over 10% truck traffic, provision should be made for at least one car and one truck . . . At signalized intersections, the storage length needed depends on the signal cycle length, the signal phasing arrangement and the rate of arrivals and departures of left-turning vehicles. The storage length is a function of the probability of occurrence of events and should usually be based on one and one-half to two times the average number of vehicles that would store per cycle."

The Institute of Transportation Engineers (ITE) provides for calculating the storage length Eq. (410.5).

$$\text{Storage length (ft)} = \frac{25VK(1 + P_T)}{N_c} \qquad (410.5)$$

where K = factor to capture randomness of arrivals (usually 2.0)
V = peak hourly left turning volume (veh/h)
N_c = number of cycles per hour
P_T = proportion of trucks (decimal)

An alternate formulation gives

$$\text{Storage length (ft)} = \frac{Vks}{N_c} \qquad (410.6)$$

where k = factor to capture randomness of arrivals (usually 1.5 to 1.8 for minor and 2.0 for major roadways)
V = peak hourly left turning volume (veh/h)
N_c = number of cycles per hour
s = average vehicle length, given by Table 410.9

Table 410.9 Average Vehicle Length Used to Determine Length of Storage Bay

% trucks	Average vehicle length s ft (m)
<4%	25 (7.6)
5%	27 (8.4)
10%	30 (9.0)
15%	35 (10.0)

---

[4]*A Policy on Geometric Design of Highways and Streets*, 5th ed., American Association of State Highway and Transportation Officials, Washington, DC, 2004.

Storage Length at Unsignalized Intersections

According to the *AASHTO Green Book*, the storage length at unsignalized intersections should be based on the number of turning vehicles likely to arrive in an average 2-min period within the peak hour. Space for at least two passenger cars must be provided, unless the turning truck traffic exceeds 10%, in which case provision should be made for at least one passenger car and one truck. Where turning lanes are designed for two-lane operation, the storage length can be reduced to approximately half of that required for single lane operation.

Auxiliary Lanes

Auxiliary lanes are used preceding median openings and at intersections preceding left and right turn movements. They should be at least 10-ft wide, if not as wide as the through lanes. On rural high-speed roadways, a minimum 6-ft-wide shoulder is preferred adjacent to auxiliary lanes. On auxiliary lanes with heavy truck traffic, a paved shoulder 2–4 ft wide may be needed.

Deceleration lanes are desirable on high-speed roadways, because the driver leaving the highway is forced to slow down to on the through-traffic lane. Acceleration lanes are not always desirable at stop-controlled intersections, but on roads without stop control and on high-volume roadways where there are infrequent and short gaps between vehicles, they are advantageous (with or without stop control).

The deceleration lane length consists of a perception-reaction distance (L_1), the full deceleration length ($L_2 + L_3$) and the storage length (L_4). These are shown in Fig. 410.4. Table 410.10 lists the full deceleration length for various approach speeds. The table is based on the assumptions that (1) the speed differential between the turning vehicle and the following through vehicle is 10 mph and (2) the average deceleration rate while moving from the through lane into the turn lane is 5.8 ft/s².

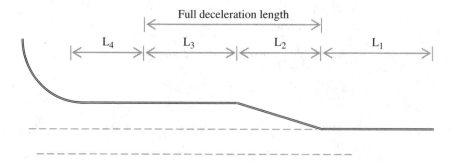

Figure 410.4 Area upstream of an intersection showing components of deceleration length.

Table 410.10 Desirable Full Deceleration Lengths

Speed (mph)	Deceleration distance (ft)
20	70
30	160
40	275
50	425
60	605
70	820

Change and Clearance Intervals at an Intersection

The change and clearance interval of a signal cycle is the sum of the yellow (amber) and all-red intervals that are provided between phases to allow pedestrian and vehicular traffic to clear the intersection before the release of conflicting movements. The all-red interval must provide enough time for a vehicle that has entered the intersection on yellow, to clear the intersection.

Length of Yellow Phase

A poor choice of the yellow interval can create a dilemma zone—an area near the intersection where a vehicle can neither stop safely, nor clear the intersection without speeding before the signal turns red. According to the Institute of Transportation Engineers (ITE), the yellow interval may be calculated as

$$y = t_R + \frac{W + L}{1.47V_{mph}} + \frac{1.47V_{mph}}{2a + 64.4G} \quad\quad (410.7)$$

where y = length of the yellow interval (s)
V = approach speed (mph). This may be an 85th-percentile speed of approach
a = deceleration rate (commonly assumed to be 10 ft/s^2)
t_R = perception-reaction time (commonly assumed as 1.0 s)
G = grade (decimal)
W = width of the intersection (ft)
L = length of the vehicle (20 ft typical)

The *Manual on Uniform Traffic Control Devices (MUTCD)* states that a yellow change interval should be approximately 3 to 6 s, and the *Traffic Engineering Handbook*[5] states that a maximum of 5 s is typical for the yellow change interval. The red clearance interval, if used, should not exceed 6 s (MUTCD), but 2 s or less is typical (*Traffic Engineering Handbook*). The traffic laws in each state may vary from these suggested practices. ITE recommends that the yellow interval not exceed 5 s, so as not to encourage driver disrespect for signals. If the calculated value of the yellow interval is greater than 5.0 s, an all-red phase is inserted following the yellow interval, so that the required clearance interval is still available.

---

[5]*Traffic Engineering Handbook*, 5th ed., Institute of Transportation Engineers.

Length of All-Red Phase

The all-red interval is determined according to the following equations (ITE):
For intersections with no pedestrian traffic

$$ar = \frac{w + L}{1.47 S_{15}} \qquad (410.8)$$

For intersections with significant pedestrian traffic

$$ar = \frac{P + L}{1.47 S_{15}} \qquad (410.9)$$

For intersections with some pedestrian traffic

$$ar = \max\left[\frac{w + L}{1.47 S_{15}}, \frac{P}{1.47 S_{15}}\right] \qquad (410.10)$$

where w = distance from departure stop line to far side of furthest conflicting traffic lane
L = length of standard vehicle (18–20 ft)
S_{15} = 15th-percentile speed of approaching traffic, or speed limit
P = distance from departure stop line to far side of furthest conflicting crosswalk

Effective Green Time

During a signal cycle, start up time is lost when the first few drivers in the queue react to the change from red to green and the traffic flow gradually increases to the saturation flow rate. For the first few vehicles in the queue, the headway, or time between successive vehicles, is higher than the saturation headway h. The analysis for signal timing is done using the saturation headway, but after reducing the actual green time to the effective green time.

Effective green time g_i is the time during which vehicles are moving at the rate of one vehicle every h s. The effective green time g_i for a signal phase is given by

$$g_i = G_i + Y_i - t_{Li} \qquad (410.11)$$

where G_i = actual green time for phase i
Y_i = sum of yellow and all red intervals = $y_i + ar_i$
t_{Li} = total lost time in phase i

Saturation Flow Rate

The saturation flow rate for each lane group is calculated based on a base saturation flow rate s_o (1900 pcphpl) and 13 site-specific multipliers as defined in Eq. (410.12) (*Highway Capacity Manual*). These adjustment factors are shown in Table 410.11.
Adjusted saturation flow rate per lane in the subject lane group is given by

$$s = s_o N f_w f_{HV,\,g} f_p f_{bb} f_a f_{LU} f_{LT} f_{RT} f_{Lpb} f_{Rpb} f_{wz} f_{ms} f_{sp} \qquad (410.12)$$

Table 410.11 Adjustment Factors for Saturation Flow Rate

Factor	Formula	Variables	Notes
Lane width	For $W < 10.0$, $f_w = 0.96$ For $W \geq 10.0$–12.9, $f_w = 1.00$ For $W \geq 12.9$, $f_w = 1.04$	W = lane width (ft)	Factors apply to average lane widths of 8.0 ft or more
Heavy vehicle	For downhill ($P_g < 0$) $$f_{HVg} = \frac{100 - 0.79P_{HV} - 2.07P_g}{100}$$ For level or uphill ($P_g \geq 0$) $$f_{HVg} = \frac{100 - 0.78P_{HV} - 0.31P_g^2}{100}$$	P_{HV} = percent heavy vehicles in lane group P_g = approach grade (%)	
Parking	$$f_p = \frac{N - 0.1 - \dfrac{18N_m}{3600}}{N} \geq 0.050$$	N = no. of lanes in lane group N_m = no. of parking maneuvers/h	$0 \leq N_m \leq 180$ $f_P = 1.0$ for no parking
Bus blockage	$$f_{bb} = \frac{N - \dfrac{14.4N_b}{3600}}{N} \geq 0.050$$	N = no. of lanes in lane group N_b = no. of buses stopping/h	$0 \leq N_b \leq 250$ $f_{bb} = 1$ for bus stop
Type of area	$f_a = 0.9$ in CBD $f_a = 1.0$ all other areas		
Lane utilization	$$f_{LU} = \frac{v_g}{N_e v_{g1}}$$	v_g = unadjusted demand flow rate for lane group (vph) v_{g1} = unadjusted demand flow rate on heaviest lane in group (vph) N_e = no. of lanes in lane group	
Protected right turns	$$f_{RT} = \frac{1}{E_R}$$		E_R = equivalent no. of through cars for a protected right-turning vehicle = 1.18

If the right-turn movement shares a lane with another movement or has permitted operation, the procedure in Section 3 of HCM Chap. 31 should be used to compute the adjusted saturation flow rate for the shared-lane lane group.

Factor	Formula	Variables	Notes
Pedestrian/ Bicycle blockage factors f_{Lpb} and f_{Rpb}	LT adjustment: $$f_{Lpb} =$$ $$1.0 - P_{LT}\left(1 - A_{pbT}\right)\left(1 - P_{LTA}\right)$$ RT adjustment: $$f_{Rpb} =$$ $$1.0 - P_{RT}\left(1 - A_{pbT}\right)\left(1 - P_{RTA}\right)$$	P_{LT} = proportion of left turns in lane group A_{pbT} = permitted phase adjustment P_{LTA} = proportion of LT protected green over total LT green P_{RT} = proportion of right turns in lane group P_{RTA} = proportion of RT protected green over total RT green	If no pedestrians or bicycles present, f_{Lpb} = f_{Rpb} = 1.0

Work zone presence factor f_{wz}	If some or all of a work zone is located between the STOP line and a point 250 ft upstream of the STOP line		If no work zone present, $f_{wz} = 1.0$
Downstream lane blockage factor f_{ms}	A downstream lane closure is a closure located downstream of the subject intersection. The factor is applied only to those lane groups entering the segment on which the closure is present. The lane closure can be associated with a work zone or special event.	A procedure for computing this factor is provided in Section 3 of Chap. 30, Urban Street Segments: Supplemental.	The factor has a value of 1.0 if no downstream lane blockage is present.
Sustained spillback factor f_{sp}	Sustained spillback occurs as a result of oversaturation (i.e., more vehicles discharging from the upstream intersection than can be served at the subject downstream intersection	A procedure for calculating the sustained spillback factor is provided in Chap. 29, Urban Street Facilities: Supplemental.	

Example 410.4

The intersection used in Example 410.2 is located in a commercial district. Through lanes for the east-west roadway are 11-ft wide and the left turn lanes are 9-ft wide. For the north-south roadway, through lanes are 12-ft wide and the left turn lanes are 9-ft wide. For the east-west roadway, there are 5% left turns and 12% right turns. For the north-south roadway, there are 3% left turns and 8% right turns. The signal cycle consists of the following phases:

Green on east-west roadway = 34 s

Green on north-south roadway = 30 s

A 3-s yellow phase separates these phases. There is no parking on either approach. The north-south roadway has a local gradient of 4% and the east-west roadway is level. For both approaches, assume 6% trucks. What is the saturation flow rate for the east-west approach (pc/h)?

Solution   For the east-west approach, the following conditions apply:

Lane width = 11 ft

Number of lanes = 2

Proportion of left turns = 0.05

Proportion of right turns = 0.12

Grade = 0

Proportion of heavy vehicles = 6%

$$f_w = 1 + \frac{11 - 12}{30} = 0.967$$

$$f_{HV} = \frac{100}{100 + 6(2.0 - 1)} = 0.943$$

$$f_a = 0.9$$

Exclusive left turn lane $f_{LT} = 0.95$

Shared right turn lane $f_{RT} = 1.0 - 0.15 \times 0.12 = 0.982$

Saturation flow rate for east-west approach

$s = 1900 \times 2 \times 0.967 \times 0.943 \times 0.9 \times 0.95 \times 0.982 = 2909$ pc/h

Capacity of a Lane Group

The capacity of a lane group is calculated from the saturation flow rate (vehicles per hour green) using the ratio of the effective green time (in the direction) to the total length of the cycle.

$$c_i = s_i \times \frac{g_i}{C} \qquad (410.13)$$

where c_i = capacity of lane group (veh/h)
  s_i = saturation flow rate for lane group (veh/hg) (veh/h green)
  g_i = effective green time (s)
  C = signal cycle length (s)

The ratio of flow to capacity for a particular approach is given by the v/c ratio for that approach. This is also called the degree of saturation X.

$$X_i = \left(\frac{v}{c}\right)_i = \frac{v_i}{c_i} = \frac{v_i}{s_i \times \frac{g_i}{C}} = \frac{v_i C}{s_i g_i} \qquad (410.14)$$

To evaluate the overall intersection, the critical volume-to-capacity ratio X_c is used. This is obtained by considering only the critical lane groups, which have the highest flow ratio (v/s) for each signal phase. The critical v/s ratio X_c for the entire intersection is given by

$$X_c = \sum \left(\frac{v}{s}\right)_{ci} \frac{C}{C - L} \qquad (410.15)$$

where $(v/s)_{ci}$ is the critical (maximum) flow ratio v/s for the ith phase
  C = cycle length
  L = total lost time per cycle

Example 410.5

A three-phase cycle has 3 s of lost time per phase. The actual volume and saturation flow for the critical movements during each phase are given in the table below.

	Phase A	Phase B	Phase C
Volume (vph)	70	400	550
Saturation flow (vph)	500	2400	3800

1. What is the shortest cycle time that will avoid oversaturation?
2. What is the critical v/c ratio if cycle time $C = 40$ s?

Solution

Phase A: $v/s = 0.140$

Phase B: $v/s = 0.167$

Phase C: $v/s = 0.145$

Sum of v/s ratios $= 0.452$
Total lost time $= 3 \times 3 = 9$ s

1. Setting $X_c = 1$, we get

$$X_c = 1 = \sum \left(\frac{v}{s}\right)_{ci} \frac{C}{C-L} = 0.452 \times \frac{C}{C-9} \Rightarrow 0.548C = 9 \Rightarrow C = 16.4 \text{ s (Say 17 s)}$$

2. If $C = 40$ s, $X_c = \sum \left(\frac{v}{s}\right)_{ci} \frac{C}{C-L} = 0.452 \times \frac{40}{40-9} \Rightarrow 0.583$

Optimal Signal Cycle Length—Webster's Theory

According to Webster's theory, the minimum intersection delay is obtained if cycle length C is given by

$$C = \frac{1.5L + 5}{1 - \sum_{i=1}^{N} Y_i} \tag{410.16}$$

where $L =$ total lost time per cycle
$Y_i =$ maximum value of approach flow to saturation flow ratio of all traffic streams using phase i
$N =$ number of phases in signal cycle

The total lost time during a cycle is the sum of startup lost time and the clearance lost time. The startup lost time is the sum of headways in excess of the saturation headway and applies to the first few vehicles in the discharging queue. The clearance lost time is the time lost when the signal changes to yellow and the flow going through the intersection gradually decreases to zero.

It is possible, indeed likely, for Webster's equation to yield unreasonably long cycle times, if the denominator of Eq. (410.16) approaches zero. Typically, for most jurisdictions, the maximum cycle time is set to about 120 s.

Example 410.6

What is the recommended length of cycle for a four-phase signal with 3 s lost time per phase, with the following critical movements?

Phase	Critical lane volume	Saturation volume	$Y_i = (q/s)_i$
A	438	2000	0.219
B	120	800	0.150
C	540	1920	0.281
D	90	800	0.113
		TOTAL	0.763

Solution The critical flow to saturation flow ratios are calculated and shown in the last column of the table. The sum of the Y_i ratios is equal to 0.763. Webster's optimal cycle length is then calculated as

$$C = \frac{1.5 \times 12 + 5}{1 - 0.763} = 97\,\text{s}$$

Time-Space Diagram

The time-space diagram is a visual tool for engineers to analyze a coordinated traffic signal system. To construct the time-space diagram, one needs the following information—intersection locations, cycle length, splits, signal offsets, left turn phasing (on the arterial in the direction of the diagram), and speed limit. In actuated systems, the phase timings change on a cycle-by-cycle basis. The outputs of a time-space diagram include bandwidth (or vehicle progression opportunities), estimates of vehicle delay, stops, queuing, and queue spillback.

A time-space diagram is usually drawn with time on the horizontal axis and distance on the vertical axis. Vehicle trajectories are plotted on the time-space diagram and the difference in distance over time represents the speed or a sloped line on the diagram. The trajectories always move left to right along with time, and the distance traversed can be either northbound (bottom to top of the diagram) or southbound (top to bottom). Vehicles can have a positive or negative slope that indicates the movements on a street network. Stopped vehicles (no change

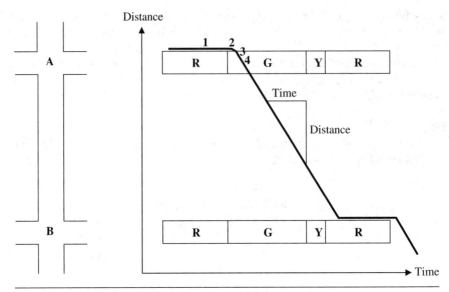

Figure 410.5 Time-space diagram showing vehicle progression.

in distance) are shown as horizontal lines. The assumed speed for coordination on the corridor may be the speed limit, the 85th-percentile speed, or a desired speed. The resulting speeds on the corridor are affected by the presence of other traffic, the signal timing settings, the acceleration and deceleration rates of the vehicles, and other elements within the immediate vicinity of the intersection. The acceleration rates are especially important considering the departures of standing queues at intersections.

Figure 410.5 shows an idealized time-space diagram for a vehicle traveling southbound between intersection 1 and 2. The blocks marked R, G, Y represent the signal phase timings (red, green, yellow) at each intersection. A motorist experiences four different conditions in moving from a stop to the progression speed:

1. Vehicles stopped at intersection,

2. Driver perception—reaction time at the onset of green,

3. Vehicle acceleration, and

4. Running speed of the vehicle (often assumed to be the speed limit or an estimated progression speed).

In this example, there is no offset between the two signal cycles, i.e., they are perfectly in phase with each other. On the other hand, if these signals were offset by a time lag, it may allow for better progression of vehicles through this block. Taking a very narrow view, a vehicle will experience a continuous series of green phases if the signal offset between successive intersections is calculated as

$$t = \frac{d}{1.47V}$$

where d = length of the block (ft)

$\quad V$ = posted speed or 85th percentile speed (mph)

$\quad t$ = ideal signal offset (s)

Bandwidth

Bandwidth is the length of time available for vehicles to travel through a system at a determined progression speed. This is an outcome of the signal timing that is determined by the offsets between intersections and the allotted green time for the coordinated phase at each intersection. The bandwidth is calculated as the time difference between the first and last vehicle trajectory that can travel at the progression speed without impedance. In Fig. 410.6, the bandwidth (even though shown only for a single block bounded by two intersections) is shown as BW. Bandwidth is a parameter that is commonly used to describe capacity or maximized vehicle throughput, but in reality it is only a measure of progression opportunities. Bandwidth is independent of traffic flows and travel paths and for that reason it may not necessarily be used by travelers. In other words, on an arterial with 10 signalized intersections, a bandwidth solution would be established to allow vehicles to travel through the entire system. In reality, one must consider how many vehicles desire to travel through all intersections without stopping.

A few important points related to bandwidth are given below:

- Bandwidth is different for each direction of travel on the arterial and dependent on the assumed speed on the time-space diagram;

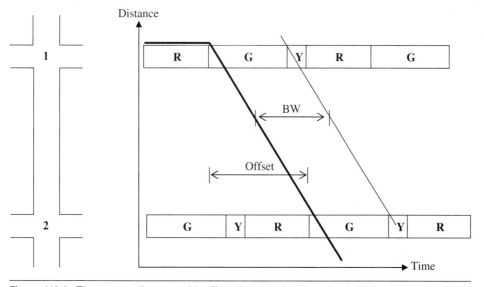

Figure 410.6 Time-space diagram with offset signal timing showing vehicle progression.

- As additional intersections are added to the system, it is increasingly difficult to achieve and measure the impact of an additional signal;

- During periods of oversaturated conditions, bandwidth solutions may result in poor performance, often simultaneous offsets are more effective; and

- Timing plans that seek the greatest bandwidth increase network delay and fuel consumption.

Guidelines for Pedestrian Facilities

This section summarizes provisions of the Guide for the Planning, Design, and Operation of Pedestrian Facilities, American Association of State Highway and Transportation Officials, 2004 (hereafter referred to as the *Pedestrian Guide*).

Pedestrian Characteristics

According to the Pedestrian Guide, pedestrian walking speeds range from 2.5 to 6.0 fps. MUTCD2009, HCM2010, and GDHS 6th ed. indicate an average walking speed (for street crossings) of 3.5 fps. This may be increased to 5.0 fps for an open sidewalk/walkway.

Two people walking side by side need a minimum width of 4.67 ft. Two wheelchairs side by side need a width of 5.0 ft.

"Spatial bubble" refers to the preferred distance of unobstructed forward vision while walking. The recommended values are—for a public event 6 ft, for shopping 9–12 ft, for a normal walk 15–18 ft, for a leisurely walk > 35 ft.

Traffic Calming

Traffic calming devices are typically spaced 300–500 ft apart.

Chicanes are curb extensions or other features that alternate from one side of the street to the other and serve to modify the straight line, wide open look of residential streets, thereby slowing down through vehicles.

Speed humps are typically located on local or neighborhood collector streets with volume between 300 and 3000 vpd. Well-designed speed humps allow vehicles to cross at 20 mph without causing discomfort to occupants. Speed bumps (not speed humps) are smaller raised areas 1–3 ft wide, not suitable for public roads.

Sloped curbs allow a vehicle to climb them and park and interfere with pedestrian space. To prevent this, vertical curb faces are recommended. However, vertical curbs are not recommended adjacent to streets with design speeds over 45 mph (*AASHTO Green Book*).

Parking and Other Sight Impediments

On urban streets with speed limit 20–30 mph, a no parking zone extending for a minimum 20 ft from the crosswalk on both near and far side of the intersection should be provided on all intersection legs (50 ft desirable for speed limit 35–45 mph). A no parking zone extending 30 ft should be provided in advance of each signal, stop sign or yield sign. In areas where speed limit exceeds 45 mph, on street parking should be prohibited.

Sidewalk Widths

Minimum recommended clear width of sidewalks is 4 ft. Where sidewalk width is less than 5 ft, passing spaces 5-ft wide should be provided intermittently, to allow two wheelchairs to pass each other. Along arterials not in the CBD, 6–8 ft width is desirable where a planting strip exists between sidewalk and curb and 8–10 ft where the sidewalk is flush with the curb. In CBD areas, minimum recommended sidewalk width is 10 ft.

Sidewalk Grades

Adjacent to a public right of way, there is no limit to sidewalk grade if it follows the grade of the street. Maximum cross slope of the sidewalk is 2%. Not adjacent to PROW, maximum sidewalk grade without railings is 5%; maximum ramp grade with handrails and landings is 8.33%.

Intersections

At intersections, curb radii should be appropriate for largest design vehicle likely to use the intersection. However, providing larger curb radii causes a lengthening of the pedestrian cross-walk, thereby increasing exposure of pedestrians to hazard and reducing the storage space for pedestrians behind the curb.

At signalized intersections, median islands provide storage space for pedestrians that cannot cross in a single cycle. They should be considered when crossing distance exceeds 60 ft. Width of newly constructed median island should be 6 ft or more (wheelchair).

Sidewalk and Curb Treatments at Pedestrian Crossings

A typical curb ramp has the following components—approach, landing, ramp, and flares. These are shown in Fig. 410.7.

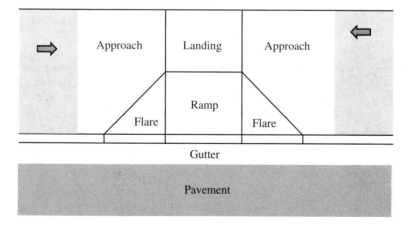

Figure 410.7   Standard curb ramp components.

For new construction, curb ramp slope should not exceed 8.33% (equivalent to a 6-in rise over a typical 6-ft ramp. If such a flat slope is not feasible, a 12.5% slope (3-in rise over 2 ft) is permitted for a 2-ft long ramp, or a 10% slope (6-in rise over a 5 ft) is permitted for a 5-ft long ramp. The cross slope of the ramp must not be greater than 2%. For new construction, curb ramp should have a minimum width of 4 ft.

Curb ramps are of three types—perpendicular (as shown in Fig. 410.7), parallel (installed where the available space between the curb and property line is too tight to permit the use of both a ramp and a landing) or diagonal (single perpendicular curb ramp located at the apex of the corner).

The gutter requires a counter slope at the point where the ramp meets the street. This counter slope should not exceed 5%. The algebraic grade difference between the gutter counter slope and the ramp slope should not exceed 11%.

Level of Service for Pedestrians

The following data may be taken as default values for pedestrian movements:

Average walking speed = 4.0 ft/s (85th percentile)

For older persons, walking speed = 3.5 ft/s

85th percentile gap accepted by crossing pedestrian = 125 ft

Table 410.12 (*Highway Capacity Manual*) outlines LOS criteria for pedestrian facilities.

At unsignalized intersections, pedestrians do not have a designated crossing interval and therefore their crossing behavior is correlated to gap acceptance. Table 410.13 shows LOS criteria for pedestrians at unsignalized intersections.

Pedestrians at Signalized Intersections

The average delay per pedestrian at a crosswalk is given by

$$d_p = \frac{0.5(C-g)^2}{C} \qquad (410.17)$$

Table 410.12 Level of Service for Pedestrian Facilities

LOS	Space (ft²/p)	Flow rate (p/min/ft)	Speed (ft/s)	v/c ratio
A	>60	≤5	>4.25	≤0.21
B	>40–60	>5–7	>4.17–4.25	>0.21–0.31
C	>24–40	>7–10	>4.00–4.17	>0.31–0.44
D	>15–24	>10–15	>3.75–4.00	>0.44–0.65
E	>8–15	>15–23	>2.50–3.75	>0.65–1.00
F	≤8	Variable	≤2.50	Variable

Source   Highway Capacity Manual, Transportation Research Board, Washington, DC, 2010.

Table 410.13 LOS Criteria for Pedestrians at Unsignalized Intersections

LOS	Average delay per pedestrian (s)	Likelihood of risk-taking behavior (accepting short gaps)
A	<5	Low
B	≥5–10	
C	>10–20	Moderate
D	>20–30	
E	>30–45	High
F	>45	Very high

Source *Highway Capacity Manual*, Transportation Research Board, Washington, DC, 2010.

where d_p = average pedestrian delay (s)
g = effective green time for pedestrians (s)
C = cycle length (s)

Table 410.14 shows LOS criteria for pedestrians at signalized intersections.

Effective Walkway Width

The effective width (W_E) of the walkway that is used by pedestrians is given by

$$W_E = W_T - W_O \qquad (410.18)$$

where W_T and W_O are the total width and the sum of all width obstructions, respectively. The effective length of an occasional obstruction is assumed to be five times its actual width. The average effect of such obstructions is calculated as the product of their effective width and the ratio of their effective length to the average distance between them.

For example, if a walkway has benches spaced every 300 ft and the actual length of a typical bench is 6 ft, the effective length of each bench is to be taken as 30 ft. Therefore, the average

Table 410.14 LOS Criteria for Pedestrians at Signalized Intersections

LOS	Pedestrian delay (s/ped)	Likelihood of noncompliance
A	<10	Low
B	≥10–20	
C	>20–30	Moderate
D	>30–40	
E	>40–60	High
F	>60	Very high

Source *Highway Capacity Manual*, Transportation Research Board, Washington, DC, 2010.

effect of these benches on a long segment of the sidewalk is to be taken as the product of the effective length (30 ft) and the ratio of the effective length (30 ft) to the average spacing (300 ft). Thus, the average effect of the benches is

$$30 \times \frac{30}{300} = 3.0 \text{ ft}$$

Table 410.15 shows values of width adjustments recommended in the *Highway Capacity Manual*.

NOTE To account for the avoidance distance between pedestrians and obstacles, 1.0 to 1.5 ft must be added to the preemption distance in Table 410.15.

Example 410.7

A 15-ft-wide curbed sidewalk segment has regularly spaced parking meters on one side and is bordered on the other side by storefront windows. The average pedestrian volume is 1000 ped/h. The peak-hour factor is 0.85. What is the LOS during the peak 15-min period on average and within platoons?

Solution   Effective width of sidewalk $= 15.0 - 1.5$ (curb) $- 2.0$ (parking meters) $- 3.0$ (window shopping) $= 8.5$ ft.
    The 15-min flow rate is

$$v_{15} = \frac{1000}{0.85} = 1176 \text{ ped/h}$$

The flow rate is given by

$$v_p = \frac{v_{15}}{15 \times W_E} = \frac{1176}{15 \times 8.5} = 9.2 \text{ p/min/ft}$$

Table 410.12 gives LOS C for average conditions.

Minimum Green Time for Pedestrians

Total crossing time or effective green time required by a pedestrian to cross an intersection is given by

$$\begin{aligned} G_p &= 3.2 + \frac{L}{S_p} + 2.7\frac{N_{\text{ped}}}{W_E} \qquad & W_E > 10\,\text{ft} \\[2mm] G_p &= 3.2 + \frac{L}{S_p} + 0.27\,N_{\text{ped}} \qquad & W_E \leq 10\,\text{ft} \end{aligned}$$

$$(410.19)$$

Table 410.15 Preemption of Walkway Width due to Various Obstacles

Obstacle	Preempted width (ft)
Street furniture	
Light pole	2.5–3.5
Traffic signal poles and boxes	3.0–4.0
Fire alarm boxes	2.5–3.5
Fire hydrants	2.5–3.0
Traffic signs	2.0–2.5
Parking meters	2.0
Mail boxes (1.7 ft × 1.7 ft)	3.2–3.7
Telephone booths (2.7 ft × 2.7 ft)	4.0
Waste baskets	3.0
Benches	5.0
Public underground access	
Subway stairs	5.5–7.0
Subway ventilation gratings	6.0 +
Transformer vault ventilation gratings	5.0 +
Landscaping	
Trees	2.0–4.0
Planter boxes	5.0
Commercial uses	
News stands	4.0–13.0
Vending stands	Variable
Advertising displays	Variable
Store displays	Variable
Sidewalk café	7.0
Building protrusions	
Columns	2.5–3.0
Stoops	2.0–6.0
Cellar doors	5.0–7.0
Standpipe connections	1.0
Awning poles	2.5
Truck docks	Variable
Garage entrance/exit	Variable
Driveways	Variable

Source   Highway Capacity Manual, Transportation Research Board, Washington, DC, 2010.

where L = crosswalk length
W_E = effective crosswalk width
S_p = average pedestrian walking speed (ft/s)
N_{ped} = number of pedestrians crossing during interval

Example 410.8

For the intersection shown below, each pedestrian walkway is 9-ft wide. The signal phasing is as follows:

N-S green time = 36 s

E-W green time = 28 s

Yellow time between phases = 3 s

Overall cycle length = 72 s

Peak-hour pedestrian volumes:

On north-south walkways   500 ped/h

On east-west walkways   300 ped/h

What is the minimum green time in the N-S direction, based on pedestrian volume?

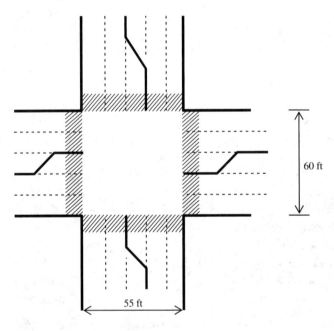

Solution

Cycle length = 72 s

Number of cycles (intervals) per hour $= 3600 \div 72 = 50$ cycles/h

For the N-S crosswalk, number of pedestrians per cycle $N_{ped} = 500 \div 50 = 10$

Assuming a walking speed $S_p = 4$ fps

$$G_p = 3.2 + \frac{60}{4.0} + 0.27 \times 10 = 20.9\,s$$

Highway Interchanges

An interchange is a road junction that utilizes grade separation and one or more ramps to permit traffic on at least one roadway to pass through the junction without crossing any other traffic stream. A complete interchange has enough ramps to provide access to and from any direction. Some of the most common types of interchanges are discussed below.

Diamond

The basic diamond is often the design of choice for lower-traffic interchanges without special constraints. It does work well when there is heavy traffic on the surface street or ramps, or if there is heavy left-turning traffic. Traffic signals can be installed at the two points where the ramps meet the surface street, but high enough traffic volumes can cause backups on the street and the ramps—even resulting in stopped traffic on the freeway. All ramps function to connect the freeway to the surface street, as well as transition traffic from low speeds, or a dead stop, to freeway speeds. If a ramp also has the task of storing queued-up traffic, its length becomes a critical design factor.

For higher traffic volumes, the surface street will need left turn lanes for the entrance ramps. In a tight diamond, there's not much length between ramps available for turn lanes. Having turn lanes for each direction in parallel forces the roadway to be wider. If the surface road is on a bridge, where lanes are expensive to add, each left turn lane takes away a potential through lane.

SPUI—Single Point Urban Interchange

The single-point urban interchange (also known as "urban interchange" and "single-point diamond") is a relatively new variant of the diamond. The SPUI's advantages (for appropriate situations) include compact layout, less right-of-way acquisition, allows concurrent left turns for greater capacity.

Disadvantages include (1) complex intersection and signal phases may be unfamiliar to drivers, (2) multilane ramps or surface streets can lead to very large areas of uncontrolled pavement, (3) distance between stop bars on surface street creates problems for bicycles, which need more time to clear the area between them, (4) more free-flow vehicle movements (which increases the SPUI's capacity) makes it harder for pedestrians to safely cross, (5) standard traffic signal timing does not include a phase for pedestrian crossing, (6) vehicle clearance time (where all lights must be red) is longer, and (7) longer overpasses can require larger bridge girders.

From high enough up, the SPUI resembles a slim classical diamond. However, where a diamond has two ramp intersections at the surface street (one on each side of the freeway), an SPUI's ramps are placed close together to make them effectively part of the same intersection. This allows one traffic signal to control all crossing movements, and enables concurrent opposing left turns, which increases the capacity of the interchange.

SPUIs are new to some areas, and drivers can be unfamiliar with how they work. Often special signals and channeling (using curbs and barriers to guide traffic) are set in place to help.

Cloverleaf

The cloverleaf type interchange is appropriate for the intersection of two busy roads since it provides nonstop flow of traffic as traffic streams don't cross. Tight right of way may force ramps to have small radius, thereby reducing the safe speed on these ramps. In situations where there are heavy left turning volumes, this can affect capacity.

A disadvantage to the plain cloverleaf is the "weaving" process, where drivers exiting one loop have to merge and cross other drivers entering the next one. Weaving, which causes bottlenecks and accidents, is the primary reason cloverleafs are being superseded by other types of interchanges. For more details on weaving, see Chap. 406. Many cloverleafs have been replaced with either signalized interchanges or higher-capacity directional interchanges with flyovers.

One way to improve a cloverleaf is to add collector/distributor (C/D) roads, running parallel to the freeway and isolating it from the weaving action at the loops. Traffic exits the freeway onto the C/D road, and then can split into the various directions. Likewise, onramps from both directions of the other road merge together first, then merge onto the primary road. This moves the weaving maneuvers to a lighter-traffic side road, and the freeway now has two ramps (entrance and exit) to deal with instead of four. Where two freeways intersect, each one may have C/D roads.

The analysis of a particular site for the ideal type of interchange uses complex data about many relevant factors such as land use, right of way, proposed alignments, traffic volume and distribution, and soil type. Such analysis is well beyond the scope of the PE exam. However, some examples are presented here where only traffic distribution data are used as the basis for optimal interchange type selection.

Example 410.9

The intersection of two arterials is being evaluated for the optimal interchange type. The traffic volume distribution is given in the figure below. What is the most suitable interchange type?

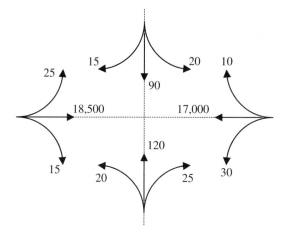

Solution   The traffic distribution shows large through-volume in one direction (E-W) compared to the other. The number of turns from all approaches is insignificant. Therefore, a diamond-type interchange is indicated.

Example 410.10

The intersection of two arterials is being evaluated for the optimal interchange type. The traffic volume distribution is given in the figure below. What is the most suitable interchange type?

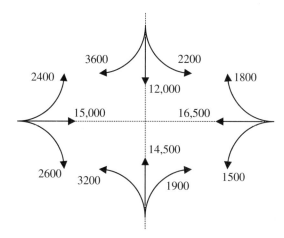

Solution   The traffic distribution shows competing through volume in both directions. Turn volumes significant but not excessive. A full cloverleaf type interchange is indicated.

Example 410.11

The intersection of two arterials is being evaluated for the optimal interchange type. The traffic volume distribution is given in the figure below. What is the most suitable interchange type?

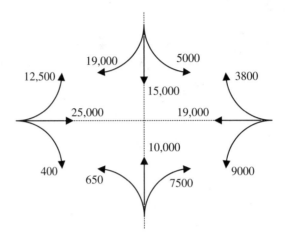

Solution   In this case, there is competing through volume in both directions accompanied by large turn volumes. A full cloverleaf would have capacity problems on the 270° ramps for the heavy left turn volumes from the EB and WB approaches.

In this case, a directional type interchange is indicated.

Roundabout Design

In volume 3 (Interrupted Flow) of the HCM 2016, Chap. 22 outlines the methodology for the analysis of roundabouts, which are usually circular and characterized by circulation of traffic around a central island (counterclockwise in the United States). A typical example is shown in figure 410.8. The number of circulating lanes may be one or more and the number of lanes entering the roundabout can also be one or more. Some roundabouts have right turning bypass

Figure 410.8
Traffic movements on an approach to a roundabout.

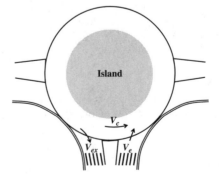

lanes—classified as either Type I (yielding bypass lane) or Type 2 (nonyielding bypass lane). These are not shown in figure 410.8

Roundabouts share the same basic control delay configuration as TWSC (two-way stop control) and AWSC (all-way stop control) intersections, adjusted for yield control. This is because the circulating flow around the island has the right of way and the entering flow must yield to it before entering the circulation. In Fig. 410.7, the circulating (or conflicting) flow rate is shown as v_c, the entering flow rate (northbound) is v_e, and the flow rate exiting (before v_e enters) the roundabout is v_{ex}.

The following data are required for analaysis of a roundabout:

1. 1. Number and configuration of lanes on each approach
2. Demand volume for each entering vehicular movement and each pedestrian crossing movement during the peak 15-min period
3. Percentage of heavy vehicles
4. Lane utilization for multilane approaches

The following default values may be used for a roundabout analaysis:

Peak hour factor, PHF $= 0.92$

Percent heavy vehicles, $P_T = 3\%$

Equivalence factor for heavy vehicles, $E_T = 2.0$

For two-lane roundabout approaches, where the approach lane configuration is either left-through + through-right or left-through-right + right, the default lane utilization values are 47% in the left lane and 53% in the right lane.

For two-lane roundabout approaches where the approach lane configuration is left + left-through-right, the default lane utilization values are 53% in the left lane and 47% in the right lane.

Table 410.16 Level of Service of Roundabouts as a Function of Control Delay per Vehicle

Control delay (s/veh)	LOS by *v/c* ratio	
	v/c ≤ 1.0	*v/c* > 1.0
0–10	A	F
>10–15	B	F
>15–25	C	F
>25–35	D	F
>35–50	E	F
>50	F	F

Capacity

The capacity of an approach decreases as the conflicting flow increases. If the conflicting flow rate approaches zero, the maximum entry flow is given by $3600/h$ where h is the follow-up headway (seconds/vehicle).

For a single-lane roundabout (single entry lane conflicted by one circulating lane, the capacity of the entering lane, adjusted for heavy vehicles (pc/h), is given by Eq. (410.20)

$$c_{c,pce} = 1{,}130e^{-1.0 \times 10^{-3} v_{c,pce}} \tag{410.20}$$

where $c_{e,pce}$ = lane capacity, adjusted for heavy vehicles (pc/h), and
$v_{c,pce}$ = conflcting flow rate, adjusted for heavy vehicles (pc/h)

When there is a single entry lane conflicting with two circulating lanes, the capacity of the entering lane, adjusted for heavy vehicles (pc/h), is given by Eq. (410.21)

$$c_{c,pce} = 1{,}130e^{-0.7 \times 10^{-3} v_{c,pce}} \tag{410.21}$$

$v_{c,pce}$ = conflcting flow rate (total of both lanes), pc/h
When there are two entry lanes conflicting with two circulating lanes, the capacity of the entering lanes L and R, adjusted for heavy vehicles (pc/h), are given by Eq. (410.22)

$$c_{e,R,pce} = 1{,}130e^{-0.7 \times 10^{-3} v_{c,pce}} \tag{410.22}$$

where $c_{e,R,pce}$ = capacity of the right entry lane, adjusted for heavy vehicles (pc/h)
$c_{e,L,pce}$ = capacity of the left entry lane, adjusted for heavy vehicles (pc/h)
$v_{c,pce}$ = conflcting flow rate (total of both lanes), pc/h

When calculating the circulating flow, contributions from all approaches must be considered (see Fig. 410.9) For example, when analyzing the northbound entering flow ($v_{e,NB}$), we see that the conflicting flow ($v_{c,NB,pce}$) has six components—the westbound vehicles making a U-turn at the roundabout (v_{WBU}), the southbound vehicles making U and left turns (v_{SBU} and v_{SBL}) and the eastbound vehicles making U-turns, left turns, and through (v_{EBU}, v_{EBL}, v_{EBT}). This is shown in Eq. (410.24).

$$v_{c,NB,pce} = v_{WBU,pce} + v_{SBL,pce} + v_{SBU,pce} + v_{EBU,pce} + v_{EBL,pce} + v_{EBT,pce} \tag{410.23}$$

Provisions of the AASHTO Guide for the Planning, Design, and Operation of Pedestrian Facilities

Rounabout islands have outside radii ranging from 45 ft to 200 ft and yield signs for entering traffic are used as the primary traffic control mechanism. The design of the roundabout should slow down entering and exiting motorists to an operating speed of 20–35 mph and to safely move the pedestrians across the entry and exiting intersections. Crosswalks to the splitter

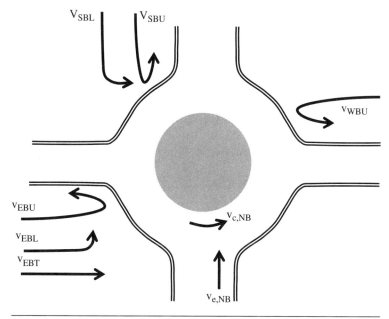

Figure 410.9 Contributions to conflicting flow from all approaches.

islands should be offset a minimum of 20 ft (approximately one car length) from the yield line of each of the approach intersections

Provisions of the AASHTO Policy on Geometric Design of Highways and Streets (Green Book)

The *AASHTO Green Book*, 6th edition, covers roundabout in Section 9.3.4. The following guidance about choice of roundabout type is given in Table 9-2 of that document.

Design element	Mini roundabout	Single lane roundabut	Multilane roundabout
Recommended maximum entry design speed	15–20 mph	20–25 mph	25–30 mph
Maximum number of lanes per approach	1	1	2+
Typical inscribed circle diameter	45–90 ft	90–150 ft	140–250 ft
Central island treatment	Mountable	Raised	Raised
Typical daily volume on four-leg roundabout (vpd)	0–15,000	0–20,000	20,000+

Design of Highway Pavements

Equivalent Single Axle Load

For design of highway pavements—rigid and flexible—one of the most significant inputs into the design process is the traffic volume and composition. Heavier vehicle axles have significantly higher potential for causing damage to pavements than lighter axles. Passenger cars and pickup trucks are excluded from the calculation of wheel load impact on pavements. To standardize the vehicular impact on the condition and life expectancy of a pavement, *AASHTO Guide for Design of Pavement Structures* converts all vehicle axles (single, double, and triple axles) of various axle loads to an equivalent number of 18-kip single axles. The load equivalence factor (LEF) is based on damage potential. For example, if the LEF for a 12-kip single axle on a flexible pavement with a pavement structural number (SN) of 4 is 0.213, this means that on a flexible pavement with SN = 4, a 12,000-lb single axle has the potential to cause about 21% of the damage that would be caused by an 18-kip single axle.

The equivalent single axle load (ESAL) is the cumulative 18-kip equivalent for a pavement over its entire design life. It is calculated as the summation of the LEF values for the total number of axles expected to use the pavement over the plan duration. If the axles are classified by type (single-axle, tandem-axle, triple axle, etc.) and load, then the ESAL is calculated as

$$\text{ESAL} = \sum (N_i \times \text{LEF}_i) \tag{411.1}$$

where N_i is the number of axles in a particular category and LEF_i is the load equivalence factor for that category.

The truck factor (TF) is defined as

$$\text{TF} = \frac{\text{ESAL}}{\text{No. of trucks}} \tag{411.2}$$

Load Equivalence Factors

The load equivalence factor (LEF) values were generated based on AASHTO road tests conducted in Ottawa, Illinois. Test traffic consisted of thousands of single axles (ranging from 2 to 30 kips) and tandem axles (ranging from 24 to 48 kips) being driven over pavements composed of an asphalt surface course (three different thicknesses ranging from 1 to 6 in), a well-graded crushed limestone base course (three different thicknesses ranging from 0 to 9 in) and a uniformly graded sand-gravel subbase (three different thicknesses ranging from 0 to 6 in).

Tables 411.6 to 411.11 show load equivalence factors for single, double, and triple axles on flexible and rigid pavements for terminal serviceability index $p_t = 2.5$.

The serviceable life of a pavement is related to the difference in present serviceability index (PSI) between construction and end-of-life. Typical values used for PSI are:

Postconstruction: 4.0 to 5.0 depending upon construction quality, smoothness, etc.

End-of-life (called "terminal serviceability" and designated "p_t"): 1.5 to 3.0 depending upon road use (e.g., interstate highway, urban arterial, residential). This chapter tabulates load equivalence factors for terminal serviceability $p_t = 2.5$.

Example 411.1

Traffic data for a section of two-lane, bidirectional roadway shows the following truck axle loads. Number of average daily trips (heavy vehicles only) is 8500 with a 60/40 directional split. What is the annual equivalent single axle load for the design lane? Assume all trucks are two-axle vehicles.

Axle type	Gross load (lb)	LEF	% ADT
Single	6000	0.017	67
	10,000	0.118	21
Tandem	14,000	0.042	6
	22,000	0.229	6

Solution Since the ADT is 8500, the number of axles = 17,000. Of these, 67% + 21% are single axles and 6% + 6% are tandem axles. The ESAL is calculated as

$$\text{ESAL} = \sum N_i \text{LEF}_i = 17,000 \times (0.017 \times 0.67 + 0.118 \times 0.21 + 0.042 \times 0.06 + 0.229 \times 0.06)$$
$$= 891/\text{day}$$

$$\text{Annual ESAL} = 891 \times 365 = 325,328 \text{ ESAL/yr} \qquad \text{(bidirectional)}$$

$$\text{For design lane: ESAL} = 0.6 \times 325,328 = 195,197 \text{ ESAL/yr} \qquad \text{(directional)}$$

Example 411.2

The annual ESAL (W_{18}) for a highway is calculated to be 280,000 in 2005. Expecting a 4% growth in traffic per year over the next 10 years, what is the cumulative ESAL for this highway over the 10-year period?

Solution   The situation described is a geometric series with the first term $a = 280,000$ and the rate of increase $r = 1.04$. The sum of the first 10 terms is given by

$$S_n = \frac{a(r^n - 1)}{r - 1} = \frac{280,000 \times (1.04^{10} - 1)}{1.04 - 1} = 280,000 \times 12.0061 = 3,361,710$$

Note that the factor 12.0061 in the example above may be called a growth factor and it is numerically identical to the F/A factor in the engineering economics tables (Chap. 501) for $n = 10$ and $i = 4\%$.

Flexible Pavements

Flexible pavements are composed of a wearing surface, usually composed of bituminous materials, underlain by a layer of granular material (base course) and a layer of blended aggregates (subbase). This arrangement is underlain by a well-compacted subgrade that serves as the foundation of the pavement. Flexible pavements are also classified as high-type, intermediate-type, and low-type. High-type pavements support the traffic loads without any visible distress and are not susceptible to weather conditions. Low-type pavements have wearing surfaces that range from untreated to loose materials to surface treated earth. Intermediate-type pavements, as the name suggests, have qualities between those of high- and low-type pavements.

The subbase course is the portion of the flexible pavement structure between the roadbed soil and the base course. It usually consists of a compacted layer of granular material or of a layer of soil treated with an admixture. The subbase may be omitted from the pavement cross-section design if the underlying soil bed is of high quality. In addition to contributing to the overall structural strength of the pavement, the subbase course may have the following secondary functions—improve drainage characteristics, prevent intrusion of fines into the base course, and minimize frost damage.

The base course is that portion of the pavement structure, which is immediately beneath the surface course. It lies either above the subbase course or if no subbase course is used, it may lie directly on the roadbed soil. It usually consists of aggregates such as crushed stone, crushed gravel, and sand, which may be untreated or treated with stabilizing admixtures such as asphalt, lime, Portland cement, etc.

Stress Distribution within the Pavement Thickness

Modeling the surface layer as a flexible beam subject to the wheel load, Fig. 411.1 shows typical distribution of vertical and horizontal stresses through the pavement thickness.

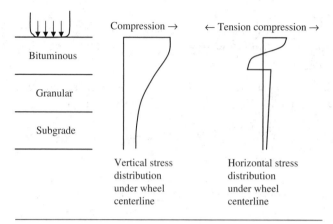

Bituminous

Granular

Subgrade

Vertical stress distribution under wheel centerline

Horizontal stress distribution under wheel centerline

Figure 411.1 Stresses in a flexible pavement due to wheel loads.

Structural Number

Flexible pavements are characterized by the structural number (SN), which is calculated as

$$SN = a_1 D_1 + a_2 m_2 D_2 + a_3 m_3 D_3 \qquad (411.3)$$

where a_1, a_2, and a_3 are strength coefficients for layers 1, 2, and 3, and m_2 and m_3 are drainage coefficients for layers 2 and 3.

Default values for layer strength coefficients are

Asphalt concrete surface course	$a_1 = 0.44$
Crushed stone base course	$a_2 = 0.14$
Sandy gravel subbase	$a_3 = 0.11$

Coefficients m_2 and m_3 represent drainage coefficients of base course and subbase, respectively. Values of these coefficients range from 0.4 to 1.4. Values of m higher than 1.0 are assigned where these courses have very good drainage characteristics.

Minimum pavement layer thicknesses recommended by AASHTO are shown in Table 411.1.

Example 411.3

The design structural number of a flexible pavement is 5. The pavement cross section consists of an asphalt surface course (minimum thickness = 4.0 in) underlain by a granular base course (maximum thickness 18 in). The following layer coefficients are given:

Asphalt concrete surface course	$a_1 = 0.45$ in^{-1}
Crushed stone base course	$a_2 = 0.15$ in^{-1}

What is the minimum required thickness of the surface course (in)?

Table 411.1 Minimum Recommended Thickness of Flexible Pavement Components (AASHTO)

ESAL	Minimum thickness (in)	
	Asphalt concrete	Aggregate base
<50,000	1.0	4.0
50,001–150,000	2.0	4.0
150,001–500,000	2.5	4.0
500,001–2,000,000	3.0	6.0
2,000,001–7,000,000	3.5	6.0
>7,000,000	4.0	6.0

Solution   In order for the surface course to have minimum thickness, we must use the maximum permissible thickness for the base course, $D_2 = 18$ in. Assuming drainage coefficient $m_2 = 1.0$, we can write

$$SN = a_1 D_1 + a_2 m_2 D_2 \Rightarrow D_1 = \frac{SN - a_2 m_2 D_2}{a_1} = \frac{5 - 0.15 \times 18}{0.45} = 5.11 \, \text{in}$$

Since this is greater than the minimum requirement of 4 in, the required thickness of the asphalt surface course is 5.5 in.

Flexible Pavement Design

The following criteria are important guideposts for the overall design of asphalt pavements:

- Sufficient asphalt to ensure a durable pavement
- Sufficient stability under traffic loads
- Sufficient air-voids—lower limit to allow room for initial densification due to traffic and upper limit to prevent excessive environmental damage
- Sufficient workability

Purposes of Compaction

- To prevent further compaction and settlement
- To increase shear strength
- To improve water tightness of mixture
- To prevent excessive oxidation of the asphalt binder

The basic design equation for flexible pavements is given by

$$\log_{10} W_{18} = Z_R S_o + 9.36 \log_{10} (\text{SN} + 1) - 0.20 + \frac{\log_{10}[\Delta \text{PSI}/(4.2 - 1.5)]}{0.40 + [1094/(\text{SN} + 1)^{5.19}]}$$
$$+ 2.32 \log_{10} M_r - 8.07 \qquad (411.4)$$

where W_{18} = number of equivalent single axle load applications over design life (ESAL)
   Z_R = standard normal deviation corresponding to a given reliability
   S_o = overall standard deviation
   SN = structural number of pavement
   ΔPSI = loss of serviceability index = $p_i - p_t$
   M_r = resilient modulus of subgrade soil (lb/in²)

Asphalt

Hot mix asphalt (HMA) is a mixture of asphalt binder and well-graded, high-quality aggregate which is heated and compacted. Asphalt is placed in multiple lifts (layers). Using deep lifts has the following advantages: (1) thicker layers hold heat longer and it is therefore easier to roll the layers to the required density, (2) deeper lifts can be placed in cooler weather, (3) one deep lift is more economical than multiple lifts, and (4) deep lifts suffer less distortion due to rolling than thin lifts.

Asphalt mix design may be performed using the Hveem, Marshall, or Superpave mix design methods.

Asphalt Grading

In the past, asphalt cement (AC) was graded by either penetration resistance or viscosity.

Penetration Grading

Penetration graded asphalts were specified by a measurement by a standardized penetrometer needle (mass = 100 g) under a standard load at a standard temperature. Penetration graded asphalts were typically expressed as "Penetration Grade 85-100," meaning that the needle penetration was between 85 and 100 mm. Higher penetration signifies a softer AC. Five different penetration grades ranging from hard (40–50 mm penetration) to soft (200–300 mm penetration) are specified in this classification system.

Penetration grading describes only the consistency at an intermediate temperature (25°C). Low-temperature properties are not directly measured by this grading system.

Viscosity Grading

Viscosity-graded asphalts were specified by determining the viscosity of asphalt cement. A temperature of 60°C (140°F) was considered to be a typical summer pavement temperature, and at this temperature, the unit of viscosity used was the poise. Standard terminology referred to AC-10 and AC-20, meaning that the viscosity of the AC was 1000 or 2000 poise, respectively. AC-20 was thicker or harder than AC-10. A temperature of 135°C (275°F) was considered the mixing and handling control point. At this temperature, different laboratory equipment was used and the unit of viscosity used was the centistoke (cS).

Although viscosity is a fundamental measure of flow, it only provides information about high temperature viscous behavior, not about the low or intermediate temperature elastic behavior.

Performance Grading

In 1994, a new system of design for asphalt paving materials known as Superpave, which introduced a new concept called performance grading, was introduced based on research done under the Strategic Highway Research Program (SHRP). The performance grading (PG) system of specifying binder is based on a complex series of performance-based tests.

The new system for specifying asphalt binders is based on performance at specified temperatures. Physical property requirements are the same, but the temperature at which the binder must attain the properties changes. For example, the high temperature, unaged binder stiffness (G^*/sin δ) is required to be at least 1.0 kPa, but this must be achieved at higher temperatures if the binder is to be adequate in a hot climate.

Binder physical properties are measured using four devices:

1. *Dynamic shear rheometer*   The dynamic shear rheometer is used to characterize the viscoelastic properties of the binder. It measures the complex shear modulus (G^*) and phase angle (δ). For totally elastic materials, there is no lag ($\delta = 0$) between the applied shear stress and the shear strain response of the sample. For totally viscous materials, $\delta = 90°$. The binder specification uses either G^*/sin δ at higher temperatures ($T > 46°C$) or G^* sin δ at intermediate temperatures ($7°C < T < 34°C$) as a means of controlling asphalt stiffness.

2. *Rotational viscometer*   This test characterizes the stiffness of the asphalt at 135°C, at which temperature it behaves almost entirely as a viscous fluid. The RTV is a rotational coaxial cylinder that measures viscosity by the torque required to rotate a spindle submerged in a sample of hot asphalt at a constant speed. The binder specification requires that binders have a viscosity of less than 3 Pa-s.

3. *Bending beam rheometer*   The Bending beam rheometer (BBR) measures the creep stiffness (S) and the logarithmic creep rate (m) by measuring the response of a small binder beam specimen to a creep load at low temperatures. Binders with low creep stiffness and/or higher m values will not crack in cold weather.

4. *Direct tension tester*   A high creep stiffness (at low temperatures) may be acceptable if a direct tension test shows that the binder is sufficiently ductile at low temperatures.

Superpave

The new specification system no longer refers to asphalt cement, but rather to binder, which includes modified and unmodified asphalts. It specifies asphalt binders as PG followed by two numbers, for example PG 66-20. The first number is always higher and positive, while the second number is smaller and negative. The first number represents the *high pavement temperature* and is based on the 7-day average high air temperature of the surrounding area, while the second number represents the *low pavement temperature* and is based on the 1-day low air temperature of the surrounding area. Both numbers referred to are in degrees Celsius. PG asphalt binders are specified in 6°C increments. If the sum of the two numbers (absolute value) >90, then use of polymer-modified asphalt is indicated.

The Superpave software calculates high pavement temperature 20 mm below the pavement surface and low temperature at the pavement surface. Design pavement temperature calculations are based on HMA pavements subjected to fast-moving traffic. Pavements subject to slow traffic, such as at intersections, toll booths, and bus stops should contain a stiffer asphalt binder than that which would be used for fast-moving traffic. Superpave allows the high-temperature grade to be increased by one grade (6°C) for slow transient loads and by two grades (12°C) for stationary loads. Additionally, the high-temperature grade should be increased by one grade for anticipated 20-year loading in excess of 30 million ESALs. For pavements with multiple conditions that require grade increases, only the largest grade increase should be used. For example, for a pavement intended to experience slow loads (one grade increase) and greater than 30 million ESALs (one grade increase), the asphalt binder high-temperature grade should be increased by only one grade. Table 411.2 shows performance grading criteria.

In Superpave, the high pavement design temperature at a depth of 20 mm is calculated using Eq. (411.5)

$$T_{20mm} = 0.9545(T_{air} - 0.00618 \times lat^2 + 0.2289 \times lat + 42.2) - 17.78 \qquad (411.5)$$

where T_{air} = 7-day average high air temperature (°C) of the surrounding area
lat = latitude (degrees) of location

The pavement low temperature (°C) is calculated using Eq. (411.6)

$$T_{min} = 0.859 \, T_{air} + 1.7 \qquad (411.6)$$

Table 411.2 shows performance grading criteria based on test results from the dynamic shear rheometer, rotational viscometer, bending beam rheometer, and direct tension tester.

Example 411.4

For Topeka, KS (39.02N, 95.687W), the average 7-day maximum air temperature is 36°C with a standard deviation of 2°C. The average coldest air temperature is −23°C, with a standard deviation of 4°C.

Table 411.2 Performance Graded (PG) Binder Grading System

Performance grade	PG 52							PG 58					PG 64					
	−10	−16	−22	−28	−34	−40	−46	−16	−22	−28	−34	−40	−10	−16	−22	−28	−34	−40
Average 7-day maximum pavement design temperature, °C	<52							<58					<64					
Minimum pavement design temperature, °C	>−10	>−16	>−22	>−28	>−34	>−40	>−46	>−16	>−22	>−28	>−34	>−40	>−10	>−16	>−22	>−28	>−34	>−40
Original binder																		
Minimum flash point temperature, °C			230															
Maximum viscosity 3 Pa-s, test temperature, °C			135															
Dynamic shear G*/sinδ, minimum 1.0 kPa, test temperature, °C	52							58					64					
Rolling thin film oven (RTFO) residue																		
Maximum mass loss %								1.00										
Dynamic shear G*/sinδ, minimum 2.2 kPa, test temperature, °C	52							58					64					
Pressure aging vessel (PAV) residue																		
PAV aging temp, °C	90							100					100					
Dynamic shear G*/sinδ, maximum 5.0 MPa, test temperature, °C	25	22	19	16	13	10	7	25	22	19	16	13	31	28	25	22	19	16
Physical hardening report																		
Creep stiffness, S maximum 300 MPa / M-value, minimum 0.3, test temperature, °C	0	−6	−12	−18	−24	−30	−36	−6	−12	−18	−24	−30	0	−6	−12	−18	−24	−30
Direct tension failure strain, minimum 1%, test temperature, °C	0	−6	−12	−18	−24	−30	−36	−6	−12	−18	−24	−30	0	−6	−12	−18	−24	−30

Source Federal Highway Administration Report FHWA-SA-95-03, "Background of Superpose Asphalt Mixture Design and Analysis," Nov. 1994.

Solution According to Eqs. (411.5) and (411.6), for a high air temperature of 36°C, the mean pavement high temperature is expected to be 56°C and for a low air temperature of −23°C, the mean pavement low temperature is expected to be −18°C. For 98% reliability (2 standard deviations away from the mean), these should be adjusted to $56 + 2 \times 2 = 60$°C and to $-18 - 2 \times 4 = -26$°C. Therefore, the performance grading should be PG 64-34. To account for slow transient loads, the designer should select one grade higher binder, a grade of PG 70-34.

Mixture Volumetric Requirements

Voids in mineral aggregate (VMA) is the sum of the volume of air voids and effective (unabsorbed) binder in a compacted sample. It represents the void space between aggregate spaces. Table 411.3 shows Superpave VMA requirements.

Voids filled with asphalt (VFA) is defined as the percentage of the VMA containing asphalt binder. Table 411.4 shows Superpave VFA requirements.

Table 411.3 Voids in Mineral Aggregate (VMA) Requirements in Superpave

Nominal maximum aggregate size (mm)	Minimum VMA (%)
9.5	15
12.5	14
19.0	13
25.0	12
37.5	11

Table 411.4 Voids Filled with Asphalt (VFA) Requirements in Superpave

Traffic (ESALs)	Design VFA (%)
$<3 \times 10^5$	70–80
$<1 \times 10^6$	65–78
$<3 \times 10^6$	65–78
$<1 \times 10^7$	65–75
$<3 \times 10^7$	65–75
$<1 \times 10^8$	65–75
$>1 \times 10^8$	65–75

Dust Proportion

Dust proportion is computed as the ratio of the percentage (by weight) of aggregate finer than no. 200 sieve (0.075 mm) to the effective asphalt content, expressed as a percent of the total mix. Effective asphalt content is the total asphalt content less the percentage of absorbed asphalt. In the Superpave guidelines, acceptable dust proportion should be in the range 0.6 to 1.2.

The Superpave mix design method consists of the following seven basic steps:

1. Aggregate selection
2. Asphalt binder selection
3. Sample preparation (including compaction)
4. Performance tests
5. Density and voids calculations
6. Optimum asphalt binder content selection
7. Moisture susceptibility evaluation

Aggregates in Asphalt Mix

Desirable properties of aggregates in hot mix asphalt are toughness, soundness, and good gradation.

The *nominal maximum aggregate size* is defined as one size larger than the first sieve to retain more than 10%. The *maximum aggregate size* is defined as one size larger than the nominal maximum aggregate size.

Coarse aggregate is that designated as retained on the 4.75-mm sieve and fine aggregate is that passing the 4.75-mm sieve.

Combined Specific Gravity of Aggregates

When various aggregates $(1, ..., n)$ with specific gravities $G_1, ..., G_n$ are combined in proportions (percentages) $P_1, ..., P_n$ (where $P_1 + ... + P_n = 100$), the overall specific gravity of the mix is given by

$$G = \frac{100}{\dfrac{P_1}{G_1} + \dfrac{P_2}{G_2} + \cdots + \dfrac{P_n}{G_n}} \qquad (411.7)$$

Apparent specific gravity of an aggregate is designated G_{sa}. It is calculated as

$$G_{sa} = \frac{\text{Mass of oven dry aggregate}}{\text{Volume of aggregate}} \qquad (411.8)$$

The bulk specific gravity of an aggregate mixture is calculated using bulk specific gravity values for each component aggregate. Similarly, the apparent specific gravity of an aggregate mixture is calculated using apparent specific gravity values for each component aggregate.

Coarse Aggregate Specific Gravity Calculations (ASTM C127)

The standard test procedure ASTM C127 outlines the following steps for determining parameters of a coarse aggregate:

Steps: Dry aggregate
Soak in water for 24 h
Decant water
Use dampened cloth to obtain surface saturated dry (SSD) condition
Determine weight of SSD aggregate (B)
Submerge and determine weight of submerged aggregate (C)
Dry to constant mass
Determine oven dry weight (A)

From the test, the following measurements are made:

A = weight of oven dry aggregate

B = weight of SSD aggregate

C = weight of submerged (water) aggregate

The bulk specific gravity is given by

$$G_{sb} = \frac{A}{B - C} \qquad (411.9)$$

The specific gravity of the SSD aggregate is given by

$$G_{SSD} = \frac{B}{B - C} \qquad (411.10)$$

The apparent specific gravity of the aggregate is given by

$$G_{sa} = \frac{A}{A - C} \qquad (411.11)$$

$$\text{Absorption (\%)} = \frac{B - A}{A} \times 100 \qquad (411.12)$$

Fine Aggregate Specific Gravity Calculations (ASTM C128)

The standard test procedure ASTM C128 outlines the following steps for determining parameters of a fine aggregate:

Steps: Dry aggregate
Soak in water for 24 h
Spread out and dry to SSD condition
Add 500 g of SSD aggregate to pycnometer of known volume
Add water and agitate to remove all air

Fill to line and determine mass of pycnometer, aggregate, and water (C)

Empty aggregate into pan and dry to constant mass

Determine oven dry mass (A)

From the test, the following measurements are made:

A = weight of oven dry aggregate

B = weight of pycnometer filled with water

C = weight of pycnometer, SSD aggregate and water

S = weight of SSD aggregate (standard value 500 g)

The bulk specific gravity is given by

$$G_{sb} = \frac{A}{B + S - C} \qquad (411.13)$$

The specific gravity of the SSD aggregate is given by

$$G_{SSD} = \frac{S}{B + S - C} \qquad (411.14)$$

The apparent specific gravity of the aggregate is given by

$$G_{sa} = \frac{A}{A + B - C} \qquad (411.15)$$

$$\text{Absorption } (\%) = \frac{S - A}{A} \times 100 \qquad (411.16)$$

Hot-Mix Asphalt Volumetric Relationships

The various constituents of total mass and total volume of an asphalt mix are shown in Fig. 411.2.

The maximum specific gravity of the paving mixture is calculated using

$$G_{mm} = \frac{100}{\dfrac{P_s}{G_{se}} + \dfrac{P_b}{G_b}} \qquad (411.17)$$

where P_s = percentage of aggregate in the mixture

G_{se} = effective specific gravity of the aggregate

P_b = percentage of asphalt in the mixture

G_b = specific gravity of the asphalt

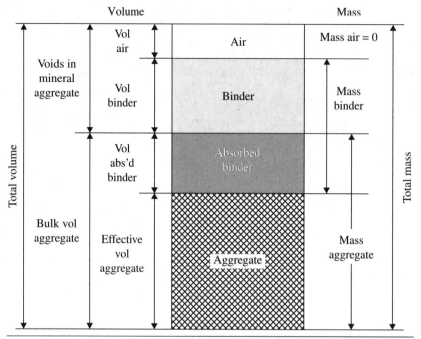

Volume   Mass

Figure 411.2   Composition of an asphalt mix.

This can be also written as

$$G_{se} = \frac{100 - P_b}{\dfrac{100}{G_{mm}} - \dfrac{P_b}{G_b}}$$ (411.18)

The total air voids (%) is given by

$$\text{VTM} = \frac{G_{mm} - G_{mb}}{G_{mm}} \times 100$$ (411.19)

where G_{mb} = maximum bulk specific gravity
  G_{mm} = maximum possible specific gravity
  G_{sb} = bulk specific gravity of aggregate mixture
  P_s = percentage of aggregate in the mixture

Voids in mineral aggregate (VMA) is an indication of the film thickness on the surface of the aggregate. The VMA (%) is given by

$$\text{VMA} = 100 - \frac{G_{mb}P_s}{G_{sb}}$$ (411.20)

Voids filled with asphalt (VFA) is the percentage of VMA that is filled with asphalt. The VFA (%) is given by

$$\text{VFA} = 100 \times \frac{\text{VMA} - \text{VTM}}{\text{VMA}} \qquad (411.21)$$

Unit Volume Approach to Calculating Asphalt Properties

For a total volume = 1.0, the following relationships are useful in calculating all components of the asphalt mix:

Volume of bulk aggregate = mass of aggregate ÷ bulk SG of aggregate

Effective volume of aggregate = mass of aggregate ÷ effective SG of aggregate

Volume of absorbed asphalt = volume of bulk aggregate − effective volume of aggregate

Effective volume of asphalt = total volume of asphalt − volume of absorbed asphalt

Volume of air = total volume − total volume of asphalt − effective volume of aggregate which can also be expressed as:

Volume of air = total volume − total volume of asphalt + volume of absorbed asphalt − volume of bulk aggregate

or

Volume of air = total volume − (effective volume of asphalt + volume of bulk aggregate)

VMA = volume of air + effective volume of asphalt = total volume − volume of bulk aggregate

VFA = effective volume of asphalt ÷ VMA

Example 411.5

A sample of compacted hot mix asphalt is known to have the following properties at 25°C:

Mix bulk specific gravity = 2.329

Bulk specific gravity of aggregate = 2.705

Effective specific gravity of aggregate = 2.731

Asphalt binder specific gravity = 1.015

Asphalt content = 5% by weight

What are the (a) VMA, (b) VFA, and (c) maximum theoretical specific gravity?

Solution

Assume a total volume $= 1.0\ \text{ft}^3$

Weight of asphalt mix $= 2.329 \times 62.4 \times 1.0 = 145.33\ \text{lb}$

Total weight of asphalt $= 0.05 \times 145.33 = 7.27\ \text{lb}$

Total volume of asphalt $= 7.27/(1.015 \times 62.4) = 0.115\ \text{ft}^3$

Total weight of aggregate $= 145.33 - 7.27 = 138.06\ \text{lb}$

Volume of bulk aggregate $= 138.06/(2.705 \times 62.4) = 0.818\ \text{ft}^3$

Effective volume of aggregate $= 138.06/(2.731 \times 62.4) = 0.810\ \text{ft}^3$

Volume of absorbed asphalt $= 0.818 - 0.810 = 0.008\ \text{ft}^3$

Effective volume of asphalt $= 0.115 - 0.008 = 0.107\ \text{ft}^3$

Effective mass of asphalt $= 0.107 \times 1.015 \times 62.4 = 6.78\ \text{lb}$

Effective asphalt content $= 6.78/145.33 = 4.66\%$

Absorbed asphalt content $= 5.0 - 4.66 = 0.34\%$

Volume of air $= 1.0 - 0.115 - 0.810 = 0.075\ \text{ft}^3$

(a) VMA $= 1.0 - 0.818 = 0.182\ \text{ft}^3$

(b) VFA $= 0.107/0.182 = 0.588 = 58.8\%$

Maximum theoretical unit weight $=$ (Weight of asphalt $+$ weight of aggregate)/ (Effective volume of asphalt $+$ Bulk aggregate volume) $= (7.27 + 138.06)/(0.107 + 0.818) = 157.11\ \text{lb/ft}^3$

(c) Maximum theoretical specific gravity $= 157.11/62.4 = 2.518$

Example 411.6

An asphalt mix contains the following constituents (see the table below). The bulk specific gravity of the mixture is 2.34. The specific gravity of a voidless mixture (i.e., the maximum specific gravity) is 2.55.

Calculate the following: (a) bulk specific gravity of the aggregate, (b) effective specific gravity of the aggregate, (c) asphalt absorption, (d) air void content of the asphalt mixture, (e) VMA of the asphalt mixture, and (f) the effective asphalt content of the mixture.

Component	Percentage (by weight)	Specific gravity	Apparent specific gravity
Asphalt	5.4		1.02
Limestone dust	14.2	2.66	2.80
Sand	29.5	2.61	2.68
Gravel	50.9	2.62	2.65

Solution

(a) The aggregate is composed of the three components – limestone dust, sand, and gravel.

The bulk specific gravity of the aggregate is calculated as

$$G_{sb} = \frac{\sum P_i}{\sum \dfrac{P_i}{G_i}} = \frac{14.2 + 29.5 + 50.9}{\dfrac{14.2}{2.66} + \dfrac{29.5}{2.61} + \dfrac{50.9}{2.62}} = 2.62$$

(b) The effective specific gravity of the aggregate is calculated from Eq. (411.18)

$$G_{sc} = \frac{100 - P_b}{\dfrac{100}{G_{mm}} - \dfrac{P_b}{G_b}} = \frac{100 - 5.4}{\dfrac{100}{2.55} - \dfrac{5.4}{1.02}} = 2.79$$

(c) The asphalt absorption is calculated from

$$P_{ba} = \frac{G_b}{G_{sb}} - \frac{G_b}{G_{sc}} = \frac{1.02}{2.62} - \frac{1.02}{2.79} = 0.023 = 2.32\%$$

(d) The air void content (VTM) is calculated from Eq. (411.19)

$$VTM = \frac{G_{mm} - G_{mb}}{G_{mm}} = \frac{2.55 - 2.34}{2.55} = 0.082 = 8.2\%$$

(e) The voids in mineral aggregate (VMA) is calculated from Eq. (411.20)

$$VMA = 100 - \frac{G_{mb} P_s}{G_{sb}} = 100 - \frac{2.34 \times 94.6}{2.62} = 15.6\%$$

(f) The effective asphalt content (%) is calculated using

$$P_{be} = P_b - P_{ba} P_s = 5.4 - \frac{2.32 \times 94.6}{100} = 3.21\%$$

Rigid Pavement Design

The design of a rigid pavement involves design of the thickness of the concrete slab, choice of reinforcement, and load transfer devices for joints. The basic materials in the pavement slab are—Portland cement concrete, reinforcement steel, either in the form of reinforcement bars or welded wire fabric, joint transfer devices, and joint sealing materials. There are four primary types of concrete pavement. They are (1) jointed plain concrete pavement (JPCP), (2) jointed reinforced concrete pavement (JRCP), (3) continuous reinforced concrete pavement (CRCP), and (4) prestressed concrete pavement (PCP).

AASHTO Method: Rigid Pavement Design

According to *AASHTO Guide for Design of Pavement Structures*, the basic design equation for flexible pavements is given by

$$\log_{10} W_{18} = Z_R S_o + 7.35\log_{10}(D+1) - 0.06 + \frac{\log_{10}[\Delta PSI/(4.5-1.5)]}{1+[1.624\times10^7/(D+1)^{8.46}]}$$

$$+ (4.22 - 0.32 P_t)\log_{10}\left\{\frac{S_c'C_d}{215.63J}\left(\frac{D^{0.75}-1.132}{D^{0.75}-[18.42/(E_c/k)^{0.25}]}\right)\right\} \quad (411.22)$$

where W_{18} = predicted number of equivalent single axle load applications over design life
Z_R = standard normal deviation corresponding to a given reliability
S_o = combined standard error of traffic prediction and performance prediction
D = thickness of the concrete pavement (in)
ΔPSI = loss of serviceability index = $p_i - p_t$
E_c = modulus of elasticity of concrete (lb/in²)
S_c' = modulus of rupture of concrete (lb/in²)
J = load transfer coefficient = 3.2
C_d = drainage coefficient
k = effective modulus of subgrade reaction (lb/in³); the k-value, similar to modulus of elasticity, is the primary performance indicator of the soil

The tensile strength of concrete is expressed as the modulus of rupture (S_c'). This is similar to, but is not exactly f_r (as defined by ACI) but rather is specified by AASHTO T97 or ASTMC78.

The modulus of elasticity of concrete E_c (psi) is given by

$$E_c = 57,000\sqrt{f_c'}$$

where f_c' is the 28-day compressive strength (psi) of the concrete.

California bearing ratio (CBR) is correlated with subgrade modulus k.

Reinforcement

The purpose of reinforcement in a rigid pavement slab is to hold cracks together, thus maintaining the overall integrity of the pavement. Cracking in a slab-on-grade is caused by differential between the temperature and moisture related contraction of the slab and the frictional resistance from the material underlying the slab. For such slabs, the maximum tensile stresses occur at mid-depth. If this maximum stress exceeds the tensile strength of the concrete, cracks form, and the stress transfers to the reinforcement.

Short slabs ($L < 15$ ft) are not expected to develop transverse cracks and therefore do not require reinforcement.

For jointed reinforced concrete pavements (JRCP), the required reinforcement (%) is calculated as

$$P_s = \frac{L \times F \times 100}{2 f_s} \tag{411.23}$$

where L = slab length (ft)
F = friction factor between bottom of slab and top of subgrade
f_s = steel working stress (psi)

Joint Sealing Materials

Three different types of joint sealing materials are used currently: (1) liquid sealants, such as asphalt, silicone, hot rubber and polymers; (2) cork expansion joint fillers; and (3) preformed elastomeric (neoprene) seals.

The purpose of using longitudinal joints in a concrete pavement is that cracks form at known locations, so that such cracks may be sealed properly. The maximum recommended spacing of longitudinal joints is 16 ft.

Table 411.5 shows Z_R values for various reliability levels.

Table 411.6 shows overall standard deviation recommended for flexible and rigid pavements.

Figure 411.3 shows a nomograph, reproduced from *AASHTO Guide for Design of Pavement Structures*, which can be used to solve Eq. (411.4).

Table 411.5 Standard Normal Deviation (Z_R)

Reliability (%)	Z_R
90	−1.282
95	−1.645
99	−2.327
99.9	−3.090
99.99	−3.750

Table 411.6 Standard Deviation Recommended for Pavement Design

Pavement type	Standard deviation, S_o
Flexible	0.40–0.50
Rigid	0.30–0.40

Nomograph solves:

$$\log_{10}W_{18} = Z_R{*}S_o + 9.36{*}\log_{10}(SN+1) - 0.20 + \frac{\log_{10}\left[\dfrac{\Delta PSI}{4.2 - 1.5}\right]}{0.40 + \dfrac{1094}{(SN + 1)^{5.19}}} + 2.32{*}\log_{10}M_r - 8.07$$

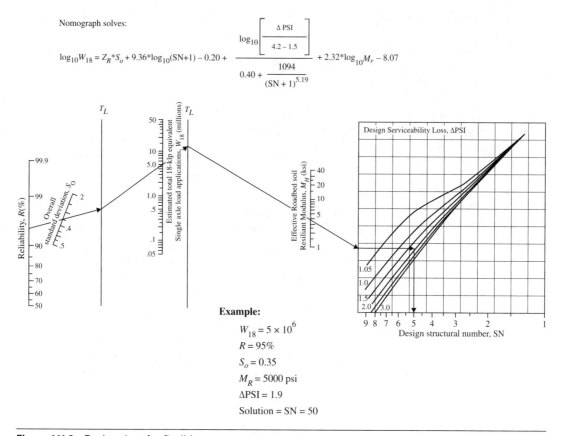

Example:

$W_{18} = 5 \times 10^6$

$R = 95\%$

$S_o = 0.35$

$M_R = 5000$ psi

$\Delta PSI = 1.9$

Solution = SN = 50

Figure 411.3 Design chart for flexible pavements.

Resilient Modulus M_R

As adopted by *AASHTO Design of Pavement Structures*, the definitive material property used to characterize roadbed soil is the resilient modulus (M_R), which is determined using AASHTO Test Method T 274. The resilient modulus can be used directly for the design of flexible pavements but must be converted to the modulus of subgrade reaction (k) for the design of rigid pavements. A correlation between the CBR value and the resilient modulus has been established (Huekelom and Klomp, 1962) as

$$M_R\,(\text{psi}) = 1500\,\text{CBR} \tag{411.24}$$

Similarly, the Asphalt Institute has developed the following correlation between M_R and the R-value:

$$M_R\,(\text{psi}) = A + BR \tag{411.25}$$

where $A = 772$ to 1155
$B = 369$ to 555

Nomograph solves:

$$\log_{10}W_{18} = I_R*S_O + 7.35*\log_{10}(D+1) - 0.06 + \frac{\log_{10}\left[\dfrac{\Delta PSI}{4.5-1.5}\right]}{1+\dfrac{1.624*10^7}{(D+1)^{8.46}}} + (4.22 - 0.32P_t)*\log_{10}\frac{S_c^1 * C_d\left[D^{0.75} - 1.132\right]}{215.63*0\left[D^{0.75} - \dfrac{18.42}{(E_c/k)^{0.25}}\right]}$$

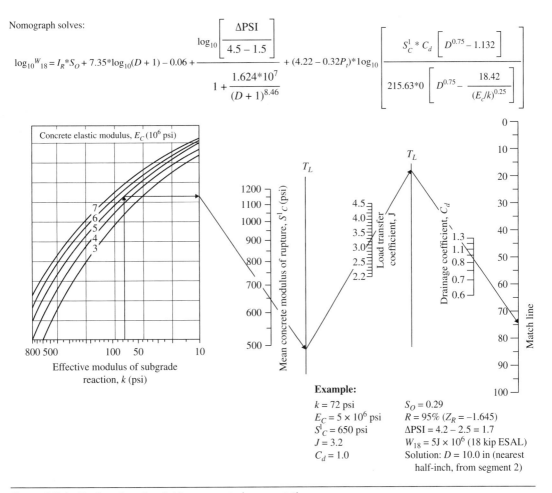

Example:

$k = 72$ psi	$S_O = 0.29$
$E_C = 5 \times 10^6$ psi	$R = 95\%$ $(Z_R = -1.645)$
$S_C^1 = 650$ psi	$\Delta PSI = 4.2 - 2.5 = 1.7$
$J = 3.2$	$W_{18} = 5J \times 10^6$ (18 kip ESAL)
$C_d = 1.0$	Solution: $D = 10.0$ in (nearest half-inch, from segment 2)

Figure 411.4 Design chart for rigid pavements (segment 1).

The pavement design guide recommends using

$$M_R \text{ (psi)} = 1000 + 555R \qquad (411.26)$$

In the approach using the design charts and nomographs, the variables R, S_o, and ESAL are used to determine the x-coordinate, whereas the variables k, E_c, S_c', J, C_d, and ΔPSI are used to determine the y-coordinate. A design chart in *AASHTO Guide for Design of Pavement Structures* is then used to plot a point with these coordinates and estimate the required pavement thickness.

Tables 411.7 to 411.9 show load equivalence factors for single, double, and triple axles on flexible pavements of various structural number, and $p_t = 2.5$.

Tables 411.10 to 411.12 show load equivalence factors for single, double, and triple axles on rigid pavements of various thicknesses, and $p_t = 2.5$.

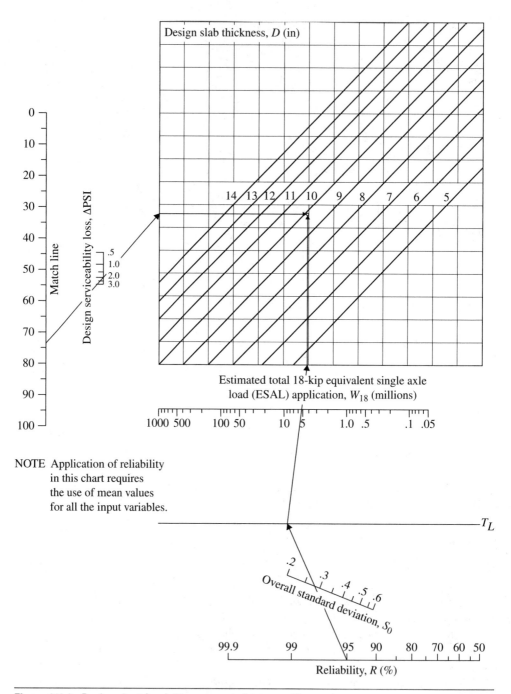

Figure 411.4 Design chart for rigid pavements (segment 2).

Table 411.7 Axle Load Equivalence Factors for Flexible Pavements, Single Axles, $p_t = 2.5$

Axle load (kips)	Pavement structural number (SN)					
	1	2	3	4	5	6
2	0.0004	0.0004	0.0003	0.0002	0.0002	0.0002
4	0.003	0.004	0.004	0.003	0.002	0.002
6	0.011	0.017	0.017	0.013	0.010	0.009
8	0.032	0.047	0.051	0.041	0.034	0.031
10	0.078	0.102	0.118	0.102	0.088	0.080
12	0.168	0.198	0.229	0.213	0.189	0.176
14	0.328	0.358	0.399	0.388	0.360	0.342
16	0.591	0.613	0.646	0.645	0.623	0.606
18	1.00	1.00	1.00	1.00	1.00	1.00
20	1.61	1.57	1.49	1.47	1.51	1.55
22	2.48	2.38	2.17	2.09	2.18	2.30
24	3.69	3.49	3.09	2.89	3.03	3.27
26	5.33	4.99	4.31	3.91	4.09	4.48
28	7.49	6.98	5.90	5.21	5.39	5.98
30	10.3	9.5	7.9	6.8	7.0	7.8
32	13.9	12.8	10.5	8.8	8.9	10.0
34	18.4	16.9	13.7	11.3	11.2	12.5
36	24.0	22.0	17.7	14.4	13.9	15.5
38	30.9	28.3	22.6	18.1	17.2	19.0
40	39.3	35.9	28.5	22.5	21.1	23.0
42	49.3	45.0	35.6	27.8	25.6	27.7
44	61.3	55.9	44.0	34.0	31.0	33.1
46	75.5	68.8	54.0	41.4	37.2	39.3
48	92.2	83.9	65.7	50.1	44.5	46.5
50	112.0	102.0	79.0	60.0	53.0	55.0

Table 411.8 Axle Load Equivalence Factors for Flexible Pavements, Tandem Axles, $p_t = 2.5$

Axle load (kips)	Pavement structural number (SN)					
	1	2	3	4	5	6
2	0.0001	0.0001	0.0001	0.0000	0.0000	0.0000
4	0.0005	0.0005	0.0004	0.0003	0.0003	0.0002
6	0.002	0.002	0.002	0.001	0.001	0.001
8	0.004	0.006	0.005	0.004	0.003	0.003
10	0.008	0.013	0.011	0.009	0.007	0.006
12	0.015	0.024	0.023	0.018	0.014	0.013
14	0.026	0.041	0.042	0.033	0.027	0.024
16	0.044	0.065	0.070	0.057	0.047	0.043
18	0.070	0.097	0.109	0.092	0.077	0.070
20	0.107	0.141	0.162	0.141	0.121	0.110
22	0.160	0.198	0.229	0.207	0.180	0.166
24	0.231	0.273	0.315	0.292	0.260	0.242
26	0.327	0.370	0.420	0.401	0.364	0.342
28	0.451	0.493	0.548	0.534	0.495	0.470
30	0.611	0.648	0.703	0.695	0.658	0.633
32	0.813	0.843	0.889	0.887	0.857	0.834
34	1.06	1.08	1.11	1.11	1.09	1.08
36	1.38	1.38	1.38	1.38	1.38	1.38
38	1.75	1.73	1.69	1.68	1.70	1.73
40	2.21	2.16	2.06	2.03	2.08	2.14
42	2.76	2.67	2.49	2.43	2.51	2.61
44	3.41	3.27	2.99	2.88	3.00	3.16
46	4.18	3.98	3.58	3.40	3.55	3.79
48	5.08	4.80	4.25	3.98	4.17	4.49
50	6.12	5.76	5.03	4.64	4.86	5.28
52	7.33	6.87	5.93	5.38	5.63	6.17
54	8.72	8.14	6.95	6.22	6.47	7.15
56	10.3	9.6	8.1	7.2	7.4	8.2
58	12.1	11.3	9.4	8.2	8.4	9.4
60	14.2	13.1	10.9	9.4	9.6	10.7
62	16.5	15.3	12.6	10.7	10.8	12.1
64	19.1	17.6	14.5	12.2	12.2	13.7

(Continued)

Table 411.8 Axle Load Equivalence Factors for Flexible Pavements, Tandem Axles, $p_t = 2.5$ (*Continued*)

Axle load (kips)	Pavement structural number (SN)					
	1	2	3	4	5	6
66	22.1	20.3	16.6	13.8	13.7	15.4
68	25.3	23.3	18.9	15.6	15.4	17.2
70	29.0	26.6	21.5	17.6	17.2	19.2
72	33.0	30.3	24.4	19.8	19.2	21.3
74	37.5	34.4	27.6	22.2	21.3	23.6
76	42.5	38.9	31.1	24.8	23.7	26.1
78	48.0	43.9	35.0	27.8	26.2	28.8
80	54.0	49.4	39.2	30.9	29.0	31.7
82	60.6	55.4	43.9	34.4	32.0	34.8
84	67.8	61.9	49.0	38.2	35.3	38.1
86	75.7	69.1	54.5	42.3	38.8	41.7
88	84.3	76.9	60.6	46.8	42.6	45.6
90	93.7	85.4	67.1	51.7	46.8	49.7

Table 411.9 Axle Load Equivalence Factors for Flexible Pavements, Triple Axles, $p_t = 2.5$

Axle load (kips)	Pavement structural number (SN)					
	1	2	3	4	5	6
2	0.0000	0.0000	0.0000	0.0000	0.0000	0.0000
4	0.0002	0.0002	0.0002	0.0001	0.0001	0.0001
6	0.0006	0.0007	0.0005	0.0004	0.0003	0.0003
8	0.001	0.002	0.001	0.001	0.001	0.001
10	0.003	0.004	0.003	0.002	0.002	0.002
12	0.005	0.007	0.006	0.004	0.003	0.003
14	0.008	0.012	0.010	0.008	0.006	0.006
16	0.012	0.019	0.018	0.013	0.011	0.010
18	0.018	0.029	0.028	0.021	0.017	0.016
20	0.027	0.042	0.042	0.032	0.027	0.024
22	0.038	0.058	0.060	0.048	0.040	0.036
24	0.053	0.078	0.084	0.068	0.057	0.051
26	0.072	0.103	0.114	0.095	0.080	0.072
28	0.098	0.133	0.151	0.128	0.109	0.099

(*Continued*)

Table 411.9 Axle Load Equivalence Factors for Flexible Pavements, Triple Axles, $p_t - 2.5$ (*Continued*)

Axle load (kips)	Pavement structural number (SN)					
	1	2	3	4	5	6
30	0.129	0.169	0.195	0.170	0.145	0.133
32	0.169	0.213	0.247	0.220	0.191	0.175
34	0.219	0.266	0.308	0.281	0.246	0.228
36	0.279	0.329	0.379	0.352	0.313	0.292
38	0.352	0.403	0.461	0.436	0.393	0.368
40	0.439	0.491	0.554	0.533	0.487	0.459
42	0.543	0.594	0.661	0.644	0.597	0.567
44	0.666	0.714	0.781	0.769	0.723	0.692
46	0.811	0.854	0.918	0.911	0.868	0.838
48	0.979	1.015	1.072	1.069	1.033	1.005
50	1.17	1.20	1.24	1.25	1.22	1.20
52	1.40	1.41	1.44	1.44	1.43	1.41
54	1.66	1.66	1.66	1.66	1.66	1.66
56	1.95	1.93	1.90	1.90	1.91	1.93
58	2.29	2.25	2.17	2.16	2.20	2.24
60	2.67	2.60	2.48	2.44	2.51	2.58
62	3.09	3.00	2.82	2.76	2.85	2.95
64	3.57	3.44	3.19	3.10	3.22	3.36
66	4.11	3.94	3.61	3.47	3.62	3.81
68	4.71	4.49	4.06	3.88	4.05	4.30
70	5.38	5.11	4.57	4.32	4.52	4.84
72	6.12	5.79	5.13	4.80	5.03	5.41
74	6.93	6.54	5.74	5.32	5.57	6.04
76	7.84	7.37	6.41	5.88	6.15	6.71
78	8.83	8.28	7.14	6.49	6.78	7.43
80	9.92	9.28	7.95	7.15	7.45	8.21
82	11.1	10.4	8.8	7.9	8.2	9.0
84	12.4	11.6	9.8	8.6	8.9	9.9
86	13.8	12.9	10.8	9.5	9.8	10.9
88	15.4	14.3	11.9	10.4	10.6	11.9
90	17.1	15.8	13.2	11.3	11.6	12.9

Table 411.10 Axle Load Equivalence Factors for Rigid Pavements, Single Axles, $p_t = 2.5$

Axle load (kips)	Slab thickness, D (in)								
	6	7	8	9	10	11	12	13	14
2	0.0002	0.0002	0.0002	0.0002	0.0002	0.0002	0.0002	0.0002	0.0002
4	0.003	0.002	0.002	0.002	0.002	0.002	0.002	0.002	0.002
6	0.012	0.011	0.010	0.010	0.010	0.010	0.010	0.010	0.010
8	0.039	0.035	0.033	0.032	0.032	0.032	0.032	0.032	0.032
10	0.097	0.089	0.084	0.082	0.081	0.080	0.080	0.080	0.080
12	0.203	0.189	0.181	0.176	0.175	0.174	0.174	0.173	0.173
14	0.376	0.360	0.347	0.341	0.338	0.337	0.336	0.336	0.336
16	0.634	0.623	0.610	0.604	0.601	0.599	0.599	0.599	0.598
18	1.00	1.00	1.00	1.00	1.00	1.00	1.00	1.00	1.00
20	1.51	1.52	1.55	1.57	1.58	1.58	1.59	1.59	1.59
22	2.21	2.20	2.28	2.34	2.38	2.40	2.41	2.41	2.41
24	3.16	3.10	3.22	3.36	3.45	3.50	3.53	3.54	3.55
26	4.41	4.26	4.42	4.67	4.85	4.95	5.01	5.04	5.05
28	6.05	5.76	5.92	6.29	6.61	6.81	6.92	6.98	7.01
30	8.16	7.67	7.79	8.28	8.79	9.14	9.35	9.46	9.52
32	10.8	10.1	10.1	10.7	11.4	12.0	12.3	12.6	12.7
34	14.1	13.0	12.9	13.6	14.6	15.4	16.0	16.4	16.5
36	18.2	16.7	16.4	17.1	18.3	19.5	20.4	21.0	21.3
38	23.1	21.1	20.6	21.3	22.7	24.3	25.6	26.4	27.0
40	29.1	26.5	25.7	26.3	27.9	29.9	31.6	32.9	33.7
42	36.2	32.9	31.7	32.2	34.0	36.3	38.7	40.4	41.6
44	44.6	40.4	38.8	39.2	41.0	43.8	46.7	49.1	50.8
46	54.5	49.3	47.1	47.3	49.2	52.3	55.9	59.0	61.4
48	66.1	59.7	56.9	56.8	58.7	62.1	66.3	70.3	73.4
50	79.4	71.7	68.2	67.8	69.6	73.3	78.1	83.0	87.1

Table 411.11 Axle Load Equivalence Factors for Rigid Pavements, Tandem Axles, $p_t = 2.5$

Axle load (kips)	Slab thickness, D (in)								
	6	7	8	9	10	11	12	13	14
2	0.0001	0.0001	0.0001	0.0001	0.0001	0.0001	0.0001	0.0001	0.0001
4	0.0006	0.0006	0.0005	0.0005	0.0005	0.0005	0.0005	0.0005	0.0005
6	0.002	0.002	0.002	0.002	0.002	0.002	0.002	0.002	0.002
8	0.007	0.006	0.006	0.005	0.005	0.005	0.005	0.005	0.005
10	0.015	0.014	0.013	0.013	0.012	0.012	0.012	0.012	0.012
12	0.031	0.028	0.026	0.026	0.025	0.025	0.025	0.025	0.025
14	0.057	0.052	0.049	0.048	0.047	0.047	0.047	0.047	0.047
16	0.097	0.089	0.084	0.082	0.081	0.081	0.080	0.080	0.080
18	0.155	0.143	0.136	0.133	0.132	0.131	0.131	0.131	0.131
20	0.234	0.220	0.211	0.206	0.204	0.203	0.203	0.203	0.203
22	0.340	0.325	0.313	0.308	0.305	0.304	0.303	0.303	0.303
24	0.475	0.462	0.450	0.444	0.441	0.440	0.439	0.439	0.439
26	0.644	0.637	0.627	0.622	0.620	0.619	0.618	0.618	0.618
28	0.855	0.854	0.852	0.850	0.850	0.850	0.849	0.849	0.849
30	1.11	1.12	1.13	1.14	1.14	1.14	1.14	1.14	1.14
32	1.43	1.44	1.47	1.49	1.50	1.51	1.51	1.51	1.51
34	1.82	1.82	1.87	1.92	1.95	1.96	1.97	1.97	1.97
36	2.29	2.27	2.35	2.43	2.48	2.51	2.51	2.52	2.53
38	2.85	2.80	2.91	3.03	3.12	3.16	3.18	3.20	3.20
40	3.52	3.42	3.55	3.74	3.87	3.94	3.98	4.00	4.01
42	4.32	4.16	4.30	4.55	4.74	4.86	4.91	4.95	4.96
44	5.26	5.01	5.16	5.48	5.75	5.92	6.01	6.06	6.09
46	6.36	6.01	6.14	6.53	6.90	7.14	7.28	7.36	7.40
48	7.64	7.16	7.27	7.73	8.21	8.55	8.75	8.86	8.92
50	9.11	8.50	8.55	9.07	9.68	10.14	10.42	10.58	10.66
52	10.8	10.0	10.0	10.6	11.3	11.9	12.3	12.5	12.7
54	12.8	11.8	11.7	12.3	13.2	13.9	14.5	14.8	14.9
56	15.0	13.8	13.6	14.2	15.2	16.2	16.8	17.3	17.5
58	17.5	16.0	15.7	16.3	17.5	18.6	19.5	20.1	20.4
60	20.3	18.5	18.1	18.7	20.0	21.4	22.5	23.2	23.6

(*Continued*)

Table 411.11 Axle Load Equivalence Factors for Rigid Pavements, Tandem Axles, $p_t = 2.5$ (*Continued*)

Axle load (kips)	Slab thickness, D (in)								
	6	7	8	9	10	11	12	13	14
62	23.5	21.4	20.8	21.4	22.8	24.4	25.7	26.7	27.3
64	27.0	24.6	23.8	24.4	25.8	27.7	29.3	30.5	31.3
66	31.0	28.1	27.1	27.6	29.2	31.3	33.2	34.7	35.7
68	35.4	32.1	30.9	31.3	32.9	35.2	37.5	39.3	40.5
70	40.3	36.5	35.0	35.3	37.0	39.5	42.1	44.3	45.9
72	45.7	41.4	39.6	39.8	41.5	44.2	47.2	49.8	51.7
74	51.7	46.7	44.6	44.7	46.4	49.3	52.7	55.7	58.0
76	58.3	52.6	50.2	50.1	51.8	54.9	58.6	62.1	64.8
78	65.5	59.1	56.3	56.1	57.7	60.9	65.0	69.0	72.3
80	73.4	66.2	62.9	62.5	64.2	67.5	71.9	76.4	80.2
82	82.0	73.9	70.2	69.6	71.2	74.7	79.4	84.4	88.8
84	91.4	82.4	78.1	77.3	78.9	82.4	87.4	93.0	98.1
86	102	92	87	86	87	91	96	102	108
88	113	102	96	95	96	100	105	112	119
90	125	112	106	105	106	110	115	123	130

Table 411.12 Axle Load Equivalence Factors for Rigid Pavements, Triple Axles, $p_t = 2.5$

Axle load (kips)	Slab thickness, D (in)								
	6	7	8	9	10	11	12	13	14
2	0.0001	0.0001	0.0001	0.0001	0.0001	0.0001	0.0001	0.0001	0.0001
4	0.0003	0.0003	0.0003	0.0003	0.0003	0.0003	0.0003	0.0003	0.0003
6	0.001	0.001	0.001	0.001	0.001	0.001	0.001	0.001	0.001
8	0.003	0.002	0.002	0.002	0.002	0.002	0.002	0.002	0.002
10	0.006	0.005	0.005	0.005	0.005	0.005	0.005	0.005	0.005
12	0.011	0.010	0.010	0.009	0.009	0.009	0.009	0.009	0.009
14	0.020	0.018	0.017	0.017	0.016	0.016	0.016	0.016	0.016
16	0.033	0.030	0.029	0.028	0.027	0.027	0.027	0.027	0.027
18	0.053	0.048	0.045	0.044	0.044	0.043	0.043	0.043	0.043
20	0.080	0.073	0.069	0.067	0.066	0.066	0.066	0.066	0.066
22	0.116	0.107	0.101	0.099	0.098	0.097	0.097	0.097	0.097
24	0.163	0.151	0.144	0.141	0.139	0.139	0.138	0.138	0.138

(*Continued*)

Table 411.12 Axle Load Equivalence Factors for Rigid Pavements, Triple Axles, $p_t = 2.5$ (*Continued*)

Axle load (kips)	Slab thickness, D (in)								
	6	7	8	9	10	11	12	13	14
26	0.222	0.209	0.200	0.195	0.194	0.193	0.192	0.192	0.192
28	0.295	0.281	0.271	0.265	0.263	0.262	0.262	0.262	0.262
30	0.384	0.371	0.359	0.354	0.351	0.350	0.349	0.349	0.349
32	0.490	0.480	0.468	0.463	0.460	0.459	0.458	0.458	0.458
34	0.616	0.609	0.601	0.596	0.594	0.593	0.592	0.592	0.592
36	0.765	0.762	0.759	0.757	0.756	0.755	0.755	0.755	0.755
38	0.939	0.941	0.946	0.948	0.950	0.951	0.951	0.951	0.951
40	1.14	1.15	1.16	1.17	1.18	1.18	1.18	1.18	1.18
42	1.38	1.38	1.41	1.44	1.45	1.46	1.46	1.46	1.46
44	1.65	1.65	1.70	1.74	1.77	1.78	1.78	1.78	1.79
46	1.97	1.96	2.03	2.09	2.13	2.15	2.16	2.16	2.16
48	2.34	2.31	2.40	2.49	2.55	2.58	2.59	2.60	2.60
50	2.76	2.71	2.81	2.94	3.02	3.07	3.09	3.10	3.11
52	3.24	3.15	3.27	3.44	3.56	3.62	3.66	3.68	3.68
54	3.79	3.66	3.79	4.00	4.16	4.26	4.30	4.33	4.34
56	4.41	4.23	4.37	4.63	4.84	4.97	5.03	5.07	5.09
58	5.12	4.87	5.00	5.32	5.59	5.76	5.85	5.90	5.93
60	5.91	5.59	5.71	6.08	6.42	6.64	6.77	6.84	6.87
62	6.80	6.39	6.50	6.91	7.33	7.62	7.79	7.88	7.93
64	7.79	7.29	7.37	7.82	8.33	8.70	8.92	9.04	9.11
66	8.90	8.28	8.33	8.83	9.42	9.88	10.17	10.33	10.42
68	10.1	9.4	9.4	9.9	10.6	11.2	11.5	11.7	11.9
70	11.5	10.6	10.6	11.1	11.9	12.6	13.0	13.3	13.5
72	13.0	12.0	11.8	12.4	13.3	14.1	14.7	15.0	15.2
74	14.6	13.5	13.2	13.8	14.8	15.8	16.5	16.9	17.1
76	16.5	15.1	14.8	15.4	16.5	17.6	18.4	18.9	19.2
78	18.5	16.9	16.5	17.1	18.2	19.5	20.5	21.1	21.5
80	20.6	18.8	18.3	18.9	20.2	21.6	22.7	23.5	24.0
82	23.0	21.0	20.3	20.9	22.2	23.8	25.2	26.1	26.7
84	25.6	23.3	22.5	23.1	24.5	26.2	27.8	28.9	29.6
86	28.4	25.8	24.9	25.4	26.9	28.8	30.5	31.9	32.8
88	31.5	28.6	27.5	27.9	29.4	31.5	33.5	35.1	36.1
90	34.8	31.5	30.3	30.7	32.2	34.4	36.7	38.5	39.8

Mechanistic-Empirical Pavement Design Guide (AASHTO) Principles

Nomenclature

CRCP	Continuously reinforced concrete pavement
CTE	Coefficient of thermal expansion
IRI	International roughness index
JPCP	Jointed plain concrete pavement
LTE	Load transfer efficiency
PCC	Portland cement concrete

The MEPDG[1] (AASHTO) represents an evolution in the design of continuously reinforced concrete pavements (CRCP) from the methods presented in the AASHTO Pavement Design Guide (1993). It has been developed to represent the state of the art in rigid pavement stress calculations, fatigue damage analysis and performance prediction.

The procedure in the MEPDG allows the engineer significant control on the various (approximately 150) inputs and features. Most of the inputs have default values, but the following are typically reviewed and input by the design engineer for each project—slab thickness, base type, soil type, type, size and location of reinforcement steel, shoulder type, climate zone, elastic and strength properties of concrete, lane width, and traffic.

The first step in the design process is gathering the required inputs and selecting the desired design features. Based on these inputs, the program first predicts the mean crack spacing that will eventually develop as a result of the steel restraint, concrete properties, base friction, and climatic conditions. An age-dependent prediction of crack width growth is subsequently calculated.

Performance studies have shown that crack spacing between 3 and 6 ft and crack width less than 0.02 in have generally resulted in successful performance of continuously reinforced concrete pavements. Once the crack spacing and crack width are calculated, the modeling of the development of a classic punchout can begin.

Classic punchouts in CRCP are directly linked to fatigue produced by repeated traffic loading. The critical tensile stresses for punchout development are located at the top of the slab between the wheels. The slab tensile stresses are calculated at various times as significant variables (traffic load, climate, etc.) change. Incremental concrete fatigue damage is then calculated at the critical stress location for each month in the design life. The cumulative fatigue damage is then related to the number of expected punchouts through a field-calibrated performance model.

The design engineer limits the allowable number of punchouts at the end of the design life to an acceptable level (typically between 10 and 20 per mile) at a given level of reliability.

---

[1]Mechanistic-Empirical Pavement Design Guide, Interim Edition: A Manual of Practice (AASHTO 2008).

Lastly, CRCP smoothness at any time increment is determined based on the calculated punchouts, initial CRCP roughness (IRI), and site factors such as pavement age, soil type, and climate. For most CRCP designs, the trigger value for IRI roughness failure is 172 in/mile.

AASHTOware Pavement ME Design Software

The provisions of the MEPDG are encoded in the AASHTOware Pavement ME Design software, the step by step input process for which is summarized below.

1. Choose design type (new/overlay/restoration/rehabilitation) and pavement type (JPCP/CRCP/Flexible). When CRCP is chosen, user must also provide PCC thickness and material properties.

2. Define supporting layers under the PCC layer (six different choices)—PCC, asphalt-concrete, sandwiched granular, non-stabilized base, subgrade, or bedrock. Within each of these six general layer categories, several material options exist. When the user has built the trial section to be analyzed, the Pavement ME Design software graphically displays the pavement section to confirm the user's choices.

3. After the proposed CRCP cross section has been input with accompanying material properties, the user must specify several critical design input parameters to the "CRCP Design Property" category. The user specifies the percentage of steel in the cross section, bar diameter, and steel cover depth. The user must also specify the shoulder type, base/slab friction level, and whether the crack spacing will be predicted using the program's algorithm or directly input by the user.

4. In the MEPDG approach, one significant change relative to the 1993 AASHTO Pavement Design Guide is that traffic is no longer characterized in terms of an equivalent single-axle load (ESAL). Instead, load spectra information is utilized in the fatigue analysis by defining the FHWA vehicle class distributions and expected axle load distribution for single, tandem, tridem, and quad axles. These data can also be uploaded from standard AVC (automatic vehicle classification) outputs from weigh-in-motion systems. To characterize the volume, the total amount of truck traffic is input as average annual daily truck traffic (AADTT), including the expected lane and directional distribution factor for the facility.

5. A significant improvement to the CRCP design process is accounting for site-specific climate. The Pavement ME Design program models account for daily and seasonal fluctuations in temperature and moisture profiles in the CRCP and soil layer, respectively, through site-specific factors such as percent sunshine, air temperature, precipitation, wind, and water table depth.

For CRCP, the software predicts only two performance criteria that can be used for assessing the validity of the CRCP design at a given level of reliability. These are IRI and the number of CRCP punchouts per mile. Three other quantities affect the performance prediction of punchouts and IRI are crack spacing, crack width, and crack LTE. To achieve and maintain good performance, crack spacing should generally be within 3 to 6 ft, crack width should remain less than 0.02 in, and crack LTE should be greater than 80% to 90%. The Pavement ME Design program also utilizes a design reliability level to account for uncertainty in the inputs, model predictions, as-constructed pavement materials, and construction process. The IRI and punchout thresholds as well as the reliability level selected are related to the roadway's functional classification.

Once the traffic, pavement cross section, material properties, and climate inputs have been entered, the program can be run to either predict the number of punchouts and smoothness at the end of the design life or until an appropriate thickness is found to the nearest ¼ in that does not exceed the user-defined CRCP performance criteria (maximum number of punchouts per mile and maximum value of IRI).

Sensitivity of CRCP Performance Criteria to Design Parameters

Continuously reinforced concrete pavements are evaluated using the performance criteria—punchouts per mile and terminal IRI. The most sensitive design inputs have been found to be slab thickness, climate, shoulder type, concrete strength, base properties, steel content and depth, and construction month.

In some cases, a reduction of only ¼ in in PCC thickness results in a near-doubling of the CRCP punchouts. Similarly, reduction of steel content from 0.7% to 0.6% in one design resulted in CRCP punchouts to increase from about 8 per mile to about 32 per mile. Increasing steel content also decreases the spacing between cracks, leading to tighter crack widths and more sustained load transfer between slabs. Location of the steel reinforcement is also critical. It is common practice to not place the steel below the slab mid-depth, as there is significant increase in punchouts and terminal IRI with an increased depth of steel from the slab surface.

Changing the type of coarse aggregate type from granite, which has a coefficient of thermal expansion value $= 5.7 \times 10^{-6}/°F$ to a low-expansion coarse aggregate such as basalt (CTE $= 4.9 \times 10^{-6}/°F$) can reduce punchouts and maintain a high ride quality on the CRCP. As the concrete CTE is lowered for a given crack spacing, the crack width is reduced, leading to better load transfer across these cracks. The choice of using aggregates with low CTE values is subject to the overall economics of availability of local resources. Increasing the steel content in the slab can be used as a potential strategy to offset higher concrete CTE without increasing the slab thickness.

A concrete shoulder, whether monolithically paved or paved separately, can be used to significantly reduce the slab bending stresses and deflections and subsequently punchouts relative to an asphalt or gravel shoulder. The terminal IRI is comparatively insensitive to the choice of shoulder type.

The base type selected for support in a CRCP is a critical factor impacting projected performance both in the development of cracks and tight crack widths as well as in resisting foundation layer erosion from repeated loading. The base type can have a pronounced impact on the computed crack spacing, crack width, crack LTE, and, ultimately, performance of the CRCP section. In addition, the use of a stabilized material as the base type can assist in reducing both the bending stresses in the PCC and the creation of erosion-induced voids, thereby increasing the fatigue life of the CRCP section. Stabilized base materials, such as a cement-treated base or asphalt-treated base, significantly reduce the projected number of punchouts in comparison to a granular base material, as the resulting crack spacing and subsequent widths are significantly affected. This reduction in punchouts also leads to a significant improvement in ride quality.

The construction month has been shown to impact the temperature development at early ages and zero-stress temperature in CRCP. Cooler months of construction such as March and October (relative to warmer months such as June) produce smaller crack widths, which promote a high load transfer between adjacent CRCP panels, reducing bending stresses and deflections from axle loads and achieving a lower number of predicted punchouts at the end of the design life.

Frost Action

Frost action, which can be quite detrimental to pavements because of its effect on the underlying subgrade, can be divided into "frost heave" and "thaw weakening." "Frost heave" is an upward movement of the subgrade resulting from the expansion of accumulated soil moisture as it freezes, while "thaw weakening" is a weakened subgrade condition resulting from soil saturation as ice within the soil melts.

Frost Heave

Frost heaving of soil is caused by the formation of ice crystals within the soil voids and the tendency of this ice to form continuous ice lenses, layers, veins, or other ice masses. As the ice lens grows/thickens, the overlying soil and pavement will "heave" up potentially resulting in a rough, cracked pavement.

Frost heave occurs primarily in soils containing fine particles ("frost susceptible" soils), while clean sands and gravels (small amounts of fine particles) are non-frost susceptible (NFS). Thus, the degree of frost susceptibility is mainly a function of the percentage of fine particles within the soil. Many agencies classify materials as being frost susceptible if 10% or more passes a No. 200 sieve (0.075-mm opening size) or 3% or more passes a No. 635 sieve (0.02-mm opening size).

The following rule-of-thumb criterion is widely used for identifying potentially frost susceptible soils (Casagrande 1932):

"Under natural freezing conditions and with sufficient water supply one should expect considerable ice segregation in non-uniform soils containing more than 3 percent of grains smaller than 0.02 mm, and in very uniform soils containing more than 10 percent smaller than 0.02 mm. No ice segregation was observed in soils containing less than 1 percent of grains smaller than 0.02 mm, even if the groundwater level is as high as the frost line."

Thaw Weakening

Thaw weakening occurs when the ice contained within the subgrade melts. As the ice melts, the water cannot drain out of the soil fast enough and thus the subgrade becomes substantially weaker and loses bearing capacity. Therefore, loading that would not normally damage a given pavement may cause significant damage during spring thaw.

Mitigating Frost Action

Frost action mitigation generally involves structural design considerations as well as other techniques applied to the base and subgrade to limit its effects. The basic methods used can be broadly categorized into the following techniques:

Frost Heave

- *Limit the depth of frost into the subgrade soils.* This is typically accomplished by specifying the depth of pavement to be some minimum percentage of the frost depth. By extending the pavement section well into the frost depth, the depth of frost-susceptible subgrade under the pavement (between the bottom of the pavement structure and frost depth) is reduced, causing correspondingly less damage.

- *Removing and replacing frost-susceptible subgrade.* Ideally the subgrade will be removed at least down to the typical frost depth. Removing frost-susceptible soils removes frost action.

- *Providing a capillary break.* By breaking the capillary flow path, frost action will be less severe because frost heaving requires substantially more water than is naturally available in the soil pores.

Thaw Weakening

- *Design the pavement structure based on reduced subgrade support.* This method simply increases the pavement thickness to account for the damage and loss of support caused by frost action.

- *Restrict pavement loading during thaw conditions.* Permanent pavement damage can be limited by limiting pavement loading while the subgrade support is weak. Typically, a load reduction in the range of 40% to 50% should accommodate a wide range of pavement conditions.

Engineering Economics

Types of Cash Flow

The primary objective of engineering economic analyses is to evaluate various economic alternatives based on cost. The four types of idealized "cash flow" discussed in this chapter are:

1. Present value (P)—a one-time cash flow that occurs now (i.e., at $t = 0$)
2. Future value (F)—a one-time cash flow that occurs after a finite duration ($t = n$)
3. Annuity (A)—a constant value cash flow that starts at $t = 1$ and recurs for n time periods
4. Gradient series (G)—a uniformly increasing (arithmetic) finite cash flow series that has value zero at $t = 1$, value G at $t = 2$, $2G$ at $t = 3$, etc., ending at a value $(n - 1)G$ at $t = n$.

These cash flows are shown graphically in Fig. 501.1. The analytical relationships to transform one type of cash flow to another are summarized in Eqs. (501.1) through (501.9).

The Year-End Accounting Convention

Equations and tables in this chapter are based on the year-end convention of accounting, in which all expenses incurred in a particular year are represented as cash flows at the end of that year.

Equations in this section are based on the concept of *time value of money*, which is the anticipated or expected growth of an investment, due to being compounded at the prevalent or expected rate of return. This rate of return is called the MARR—the minimum attractive rate of return. In the formulations in this chapter, this rate of return is represented by the variable i. MARR is one of the most important concepts in economic analysis. It is the lowest rate of return that investors will accept by balancing the investment risk or the opportunity to invest elsewhere for possibly greater returns.

Two sets of formulas are presented—one for discrete compounding (where the compounding interval is a finite length of time) and the other for continuous compounding (the limit as the number of compounding intervals $n \rightarrow \infty$). Note that the gradient series loses meaning for continuous compounding, hence those factors are missing.

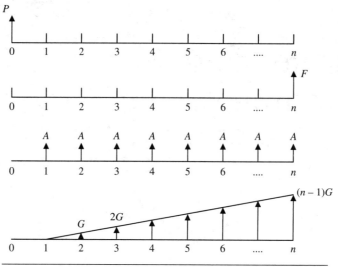

Figure 501.1 Types of cash flow.

The equations representing the various conversions for the time value of money are given below. In these equations, the following variables are used:

P—present worth (single lump sum).

A—an annuity (constant installment at the end of every period).

F—future worth (single lump sum).

G—gradient series starting at zero for the first period and increasing by constant increment *G* every period.

i—nominal interest rate per period (MARR). This is often expressed as a percentage, but must be used as a decimal value in the equations.

n—the number of compounding intervals.

Single Payment Compounded—Symbol (F/P, i%, n)—Converts to F, Given P

Discrete compounding:
$$(F/P, i, n) = (1 + i)^n \qquad (501.1a)$$

Continuous compounding:
$$(F/P, i, n) = e^{in} \qquad (501.1b)$$

Single Payment Present Worth—Symbol (P/F, i%, n)—Converts to P, Given F

Discrete compounding:
$$(P/F, i, n) = (1 + i)^{-n} \qquad (501.2a)$$

Continuous compounding:
$$(P/F, i, n) = e^{-in} \qquad (501.2b)$$

Uniform Series Sinking Fund—Symbol (A/F, i%, n)—Converts to A, Given F

Discrete compounding:
$$(A/F, i, n) = \frac{i}{(1+i)^n - 1}$$
(501.3a)

Continuous compounding:
$$(A/F, i, n) = \frac{e^i - 1}{e^{in} - 1}$$
(501.3b)

Capital Recovery—Symbol (A/P, i%, n)—Converts to A, Given P

Discrete compounding:
$$(A/P, i, n) = \frac{i(1+i)^n}{(1+i)^n - 1}$$
(501.4a)

Continuous compounding:
$$(A/P, i, n) = \frac{e^i - 1}{1 - e^{-in}}$$
(501.4b)

Uniform Series Compounded—Symbol (F/A, i%, n)—Converts to F, Given A

Discrete compounding:
$$(F/A, i, n) = \frac{(1+i)^n - 1}{i}$$
(501.5a)

Continuous compounding:
$$(F/A, i, n) = \frac{e^{in} - 1}{e^i - 1}$$
(501.5b)

Uniform Series Present Worth—Symbol (P/A, i%, n)—Converts to P, Given A

Discrete compounding:
$$(P/A, i, n) = \frac{(1+i)^n - 1}{i(1+i)^n}$$
(501.6a)

Continuous compounding:
$$(P/A, i, n) = \frac{1 - e^{-in}}{e^i - 1}$$
(501.6b)

Uniform Gradient Present Worth—Symbol (P/G, i%, n)—Converts to P, Given G

Discrete compounding:
$$(P/G, i, n) = \frac{(1+i)^n - 1}{i^2(1+i)^n} - \frac{n}{i(1+i)^n}$$
(501.7)

Continuous compounding: Undefined

Uniform Gradient Future Worth—Symbol (F/G, i%, n)—Converts to F, Given G

Discrete compounding:

$$(F/G, i, n) = \frac{(1+i)^n - 1}{i^2} - \frac{n}{i}$$

(501.8)

Continuous compounding: Undefined

Uniform Gradient Uniform Series—Symbol (A/G, i%, n)—Converts to A, Given G

Discrete compounding:

$$(A/G, i, n) = \frac{1}{i} - \frac{n}{(1+i)^n - 1}$$

(501.9)

Continuous compounding: Undefined

Based on these relations, Tables 501.1 to 501.12 are presented in the section "Engineering Economics Factors" for interest rates ranging from $i = 1\%$ to $i = 12\%$.

Nonannual Compounding

Note that these interest rates are per period. For example, if the duration is $n = 3$ years, and the effective annual interest rate is 6%, then the table yields $(P/F, i = 6\%, n = 3) = 0.8396$. However, if the time period is measured in months (e.g., $n = 36$ months) and the interest rate is still given as an effective annual interest rate of 6%, then the tables cannot be used directly. This becomes a case of nonannual compounding. The equivalent monthly interest rate i_M is that rate of increase (monthly) which, compounded over a year (12 periods) will yield 6%. Thus, the value of i_M can be solved from

$$(1 + i_M)^{12} = 1.06 \Rightarrow i_M = 0.00487 \quad (0.487\%)$$

Noting the absence of tables for this interest rate, the formula for $(P/F, i = 0.00487, n = 36)$ yields the value 0.8396. This value is identical to that obtained using three 1-year periods. This is because the 6% is designated as the *effective* annual rate.

However, if the annual interest rate of 6% were quoted as a *nominal* rate (instead of an effective rate), then the equivalent monthly interest rate would be $6\%/12 = 0.5\% = 0.005$. This would have yielded a different P/F factor $(P/F, i = 0.005, n = 36) = 0.8356$.

Engineering Economics Factors

In this section, conversion factors are tabulated for interest rate $i = 1\%$ to $i = 12\%$.

Table 501.1 Economics Factors for $i = 1\%$

n	A/P	A/F	A/G	P/A	P/F	P/G	F/A	F/P	F/G	G/A	G/P	G/F
1	1.0100	1.0000	0.0000	0.9901	0.9901	0.0000	1.0000	1.0100	0.0000			
2	0.5075	0.4975	0.4975	1.9704	0.9803	0.9803	2.0100	1.0201	1.0000	2.0100	1.0201	1.0000
3	0.3400	0.3300	0.9934	2.9410	0.9706	2.9215	3.0301	1.0303	3.0100	1.0067	0.3423	0.3322
4	0.2563	0.2463	1.4876	3.9020	0.9610	5.8044	4.0604	1.0406	6.0401	0.6722	0.1723	0.1656
5	0.2060	0.1960	1.9801	4.8534	0.9515	9.6103	5.1010	1.0510	10.1005	0.5050	0.1041	0.0990
6	0.1725	0.1625	2.4710	5.7955	0.9420	14.3205	6.1520	1.0615	15.2015	0.4047	0.0698	0.0658
7	0.1486	0.1386	2.9602	6.7282	0.9327	19.9168	7.2135	1.0721	21.3535	0.3378	0.0502	0.0468
8	0.1307	0.1207	3.4478	7.6517	0.9235	26.3812	8.2857	1.0829	28.5671	0.2900	0.0379	0.0350
9	0.1167	0.1067	3.9337	8.5660	0.9143	33.6959	9.3685	1.0937	36.8527	0.2542	0.0297	0.0271
10	0.1056	0.0956	4.4179	9.4713	0.9053	41.8435	10.4622	1.1046	46.2213	0.2264	0.0239	0.0216
11	0.0965	0.0865	4.9005	10.3676	0.8963	50.8067	11.5668	1.1157	56.6835	0.2041	0.0197	0.0176
12	0.0888	0.0788	5.3815	11.2551	0.8874	60.5687	12.6825	1.1268	68.2503	0.1858	0.0165	0.0147
13	0.0824	0.0724	5.8607	12.1337	0.8787	71.1126	13.8093	1.1381	80.9328	0.1706	0.0141	0.0124
14	0.0769	0.0669	6.3384	13.0037	0.8700	82.4221	14.9474	1.1495	94.7421	0.1578	0.0121	0.0106
15	0.0721	0.0621	6.8143	13.8651	0.8613	94.4810	16.0969	1.1610	109.6896	0.1467	0.0106	0.0091
16	0.0679	0.0579	7.2886	14.7179	0.8528	107.2734	17.2579	1.1726	125.7864	0.1372	0.0093	0.0079
17	0.0643	0.0543	7.7613	15.5623	0.8444	120.7834	18.4304	1.1843	143.0443	0.1288	0.0083	0.0070
18	0.0610	0.0510	8.2323	16.3983	0.8360	134.9957	19.6147	1.1961	161.4748	0.1215	0.0074	0.0062
19	0.0581	0.0481	8.7017	17.2260	0.8277	149.8950	20.8109	1.2081	181.0895	0.1149	0.0067	0.0055
20	0.0554	0.0454	9.1694	18.0456	0.8195	165.4664	22.0190	1.2202	201.9004	0.1091	0.0060	0.0050
21	0.0530	0.0430	9.6354	18.8570	0.8114	181.6950	23.2392	1.2324	223.9194	0.1038	0.0055	0.0045
22	0.0509	0.0409	10.0998	19.6604	0.8034	198.5663	24.4716	1.2447	247.1586	0.0990	0.0050	0.0040
23	0.0489	0.0389	10.5626	20.4558	0.7954	216.0660	25.7163	1.2572	271.6302	0.0947	0.0046	0.0037
24	0.0471	0.0371	11.0237	21.2434	0.7876	234.1800	26.9735	1.2697	297.3465	0.0907	0.0043	0.0034
25	0.0454	0.0354	11.4831	22.0232	0.7798	252.8945	28.2432	1.2824	324.3200	0.0871	0.0040	0.0031
26	0.0439	0.0339	11.9409	22.7952	0.7720	272.1957	29.5256	1.2953	352.5631	0.0837	0.0037	0.0028
27	0.0424	0.0324	12.3971	23.5596	0.7644	292.0702	30.8209	1.3082	382.0888	0.0807	0.0034	0.0026
28	0.0411	0.0311	12.8516	24.3164	0.7568	312.5047	32.1291	1.3213	412.9097	0.0778	0.0032	0.0024
29	0.0399	0.0299	13.3044	25.0658	0.7493	333.4863	33.4504	1.3345	445.0388	0.0752	0.0030	0.0022
30	0.0387	0.0287	13.7557	25.8077	0.7419	355.0021	34.7849	1.3478	478.4892	0.0727	0.0028	0.0021
31	0.0377	0.0277	14.2052	26.5423	0.7346	377.0394	36.1327	1.3613	513.2740	0.0704	0.0027	0.0019
32	0.0367	0.0267	14.6532	27.2696	0.7273	399.5858	37.4941	1.3749	549.4068	0.0682	0.0025	0.0018
33	0.0357	0.0257	15.0995	27.9897	0.7201	422.6291	38.8690	1.3887	586.9009	0.0662	0.0024	0.0017
34	0.0348	0.0248	15.5441	28.7027	0.7130	446.1572	40.2577	1.4026	625.7699	0.0643	0.0022	0.0016
35	0.0340	0.0240	15.9871	29.4086	0.7059	470.1583	41.6603	1.4166	666.0276	0.0626	0.0021	0.0015
36	0.0332	0.0232	16.4285	30.1075	0.6989	494.6207	43.0769	1.4308	707.6878	0.0609	0.0020	0.0014
37	0.0325	0.0225	16.8682	30.7995	0.6920	519.5329	44.5076	1.4451	750.7647	0.0593	0.0019	0.0013
38	0.0318	0.0218	17.3063	31.4847	0.6852	544.8835	45.9527	1.4595	795.2724	0.0578	0.0018	0.0013
39	0.0311	0.0211	17.7428	32.1630	0.6784	570.6616	47.4123	1.4741	841.2251	0.0564	0.0018	0.0012
40	0.0305	0.0205	18.1776	32.8347	0.6717	596.8561	48.8864	1.4889	888.6373	0.0550	0.0017	0.0011
∞	0.0100	0.0000	99.9523	99.9952	0.0000	9994.7517	∞	∞	∞	0.0100	0.0001	0.0000

Table 501.2 Economics Factors for $i = 2\%$

n	A/P	A/F	A/G	P/A	P/F	P/G	F/A	F/P	F/G	G/A	G/P	G/F
1	1.0200	1.0000	0.0000	0.9804	0.9804	0.0000	1.0000	1.0200	0.0000			
2	0.5150	0.4950	0.4950	1.9416	0.9612	0.9612	2.0200	1.0404	1.0000	2.0200	1.0404	1.0000
3	0.3468	0.3268	0.9868	2.8839	0.9423	2.8458	3.0604	1.0612	3.0200	1.0134	0.3514	0.3311
4	0.2626	0.2426	1.4752	3.8077	0.9238	5.6173	4.1216	1.0824	6.0804	0.6779	0.1780	0.1645
5	0.2122	0.1922	1.9604	4.7135	0.9057	9.2403	5.2040	1.1041	10.2020	0.5101	0.1082	0.0980
6	0.1785	0.1585	2.4423	5.6014	0.8880	13.6801	6.3081	1.1262	15.4060	0.4095	0.0731	0.0649
7	0.1545	0.1345	2.9208	6.4720	0.8706	18.9035	7.4343	1.1487	21.7142	0.3424	0.0529	0.0461
8	0.1365	0.1165	3.3961	7.3255	0.8535	24.8779	8.5830	1.1717	29.1485	0.2945	0.0402	0.0343
9	0.1225	0.1025	3.8681	8.1622	0.8368	31.5720	9.7546	1.1951	37.7314	0.2585	0.0317	0.0265
10	0.1113	0.0913	4.3367	8.9826	0.8203	38.9551	10.9497	1.2190	47.4860	0.2306	0.0257	0.0211
11	0.1022	0.0822	4.8021	9.7868	0.8043	46.9977	12.1687	1.2434	58.4358	0.2082	0.0213	0.0171
12	0.0946	0.0746	5.2642	10.5753	0.7885	55.6712	13.4121	1.2682	70.6045	0.1900	0.0180	0.0142
13	0.0881	0.0681	5.7231	11.3484	0.7730	64.9475	14.6803	1.2936	84.0166	0.1747	0.0154	0.0119
14	0.0826	0.0626	6.1786	12.1062	0.7579	74.7999	15.9739	1.3195	98.6969	0.1618	0.0134	0.0101
15	0.0778	0.0578	6.6309	12.8493	0.7430	85.2021	17.2934	1.3459	114.6708	0.1508	0.0117	0.0087
16	0.0737	0.0537	7.0799	13.5777	0.7284	96.1288	18.6393	1.3728	131.9643	0.1412	0.0104	0.0076
17	0.0700	0.0500	7.5256	14.2919	0.7142	107.5554	20.0121	1.4002	150.6035	0.1329	0.0093	0.0066
18	0.0667	0.0467	7.9681	14.9920	0.7002	119.4581	21.4123	1.4282	170.6156	0.1255	0.0084	0.0059
19	0.0638	0.0438	8.4073	15.6785	0.6864	131.8139	22.8406	1.4568	192.0279	0.1189	0.0076	0.0052
20	0.0612	0.0412	8.8433	16.3514	0.6730	144.6003	24.2974	1.4859	214.8685	0.1131	0.0069	0.0047
21	0.0588	0.0388	9.2760	17.0112	0.6598	157.7959	25.7833	1.5157	239.1659	0.1078	0.0063	0.0042
22	0.0566	0.0366	9.7055	17.6580	0.6468	171.3795	27.2990	1.5460	264.9492	0.1030	0.0058	0.0038
23	0.0547	0.0347	10.1317	18.2922	0.6342	185.3309	28.8450	1.5769	292.2482	0.0987	0.0054	0.0034
24	0.0529	0.0329	10.5547	18.9139	0.6217	199.6305	30.4219	1.6084	321.0931	0.0947	0.0050	0.0031
25	0.0512	0.0312	10.9745	19.5235	0.6095	214.2592	32.0303	1.6406	351.5150	0.0911	0.0047	0.0028
26	0.0497	0.0297	11.3910	20.1210	0.5976	229.1987	33.6709	1.6734	383.5453	0.0878	0.0044	0.0026
27	0.0483	0.0283	11.8043	20.7069	0.5859	244.4311	35.3443	1.7069	417.2162	0.0847	0.0041	0.0024
28	0.0470	0.0270	12.2145	21.2813	0.5744	259.9392	37.0512	1.7410	452.5605	0.0819	0.0038	0.0022
29	0.0458	0.0258	12.6214	21.8444	0.5631	275.7064	38.7922	1.7758	489.6117	0.0792	0.0036	0.0020
30	0.0446	0.0246	13.0251	22.3965	0.5521	291.7164	40.5681	1.8114	528.4040	0.0768	0.0034	0.0019
31	0.0436	0.0236	13.4257	22.9377	0.5412	307.9538	42.3794	1.8476	568.9720	0.0745	0.0032	0.0018
32	0.0426	0.0226	13.8230	23.4683	0.5306	324.4035	44.2270	1.8845	611.3515	0.0723	0.0031	0.0016
33	0.0417	0.0217	14.2172	23.9886	0.5202	341.0508	46.1116	1.9222	655.5785	0.0703	0.0029	0.0015
34	0.0408	0.0208	14.6083	24.4986	0.5100	357.8817	48.0338	1.9607	701.6901	0.0685	0.0028	0.0014
35	0.0400	0.0200	14.9961	24.9986	0.5000	374.8826	49.9945	1.9999	749.7239	0.0667	0.0027	0.0013
36	0.0392	0.0192	15.3809	25.4888	0.4902	392.0405	51.9944	2.0399	799.7184	0.0650	0.0026	0.0013
37	0.0385	0.0185	15.7625	25.9695	0.4806	409.3424	54.0343	2.0807	851.7127	0.0634	0.0024	0.0012
38	0.0378	0.0178	16.1409	26.4406	0.4712	426.7764	56.1149	2.1223	905.7470	0.0620	0.0023	0.0011
39	0.0372	0.0172	16.5163	26.9026	0.4619	444.3304	58.2372	2.1647	961.8619	0.0605	0.0023	0.0010
40	0.0366	0.0166	16.8885	27.3555	0.4529	461.9931	60.4020	2.2080	1020.0992	0.0592	0.0022	0.0010
∞	0.0200	0.0000	50.0000	50.0000	0.0000	2499.9999	∞	∞	∞	0.0200	0.0004	0.0000

Table 501.3 Economics Factors for $i = 3\%$

n	A/P	A/F	A/G	P/A	P/F	P/G	F/A	F/P	F/G	G/A	G/P	G/F
1	1.0300	1.0000	0.0000	0.9709	0.9709	0.0000	1.0000	1.0300	0.0000			
2	0.5226	0.4926	0.4926	1.9135	0.9426	0.9426	2.0300	1.0609	1.0000	2.0300	1.0609	1.0000
3	0.3535	0.3235	0.9803	2.8286	0.9151	2.7729	3.0909	1.0927	3.0300	1.0201	0.3606	0.3300
4	0.2690	0.2390	1.4631	3.7171	0.8885	5.4383	4.1836	1.1255	6.1209	0.6835	0.1839	0.1634
5	0.2184	0.1884	1.9409	4.5797	0.8626	8.8888	5.3091	1.1593	10.3045	0.5152	0.1125	0.0970
6	0.1846	0.1546	2.4138	5.4172	0.8375	13.0762	6.4684	1.1941	15.6137	0.4143	0.0765	0.0640
7	0.1605	0.1305	2.8819	6.2303	0.8131	17.9547	7.6625	1.2299	22.0821	0.3470	0.0557	0.0453
8	0.1425	0.1125	3.3450	7.0197	0.7894	23.4806	8.8923	1.2668	29.7445	0.2990	0.0426	0.0336
9	0.1284	0.0984	3.8032	7.7861	0.7664	29.6119	10.1591	1.3048	38.6369	0.2629	0.0338	0.0259
10	0.1172	0.0872	4.2565	8.5302	0.7441	36.3088	11.4639	1.3439	48.7960	0.2349	0.0275	0.0205
11	0.1081	0.0781	4.7049	9.2526	0.7224	43.5330	12.8078	1.3842	60.2599	0.2125	0.0230	0.0166
12	0.1005	0.0705	5.1485	9.9540	0.7014	51.2482	14.1920	1.4258	73.0677	0.1942	0.0195	0.0137
13	0.0940	0.0640	5.5872	10.6350	0.6810	59.4196	15.6178	1.4685	87.2597	0.1790	0.0168	0.0115
14	0.0885	0.0585	6.0210	11.2961	0.6611	68.0141	17.0863	1.5126	102.8775	0.1661	0.0147	0.0097
15	0.0838	0.0538	6.4500	11.9379	0.6419	77.0002	18.5989	1.5580	119.9638	0.1550	0.0130	0.0083
16	0.0796	0.0496	6.8742	12.5611	0.6232	86.3477	20.1569	1.6047	138.5627	0.1455	0.0116	0.0072
17	0.0760	0.0460	7.2936	13.1661	0.6050	96.0280	21.7616	1.6528	158.7196	0.1371	0.0104	0.0063
18	0.0727	0.0427	7.7081	13.7535	0.5874	106.0137	23.4144	1.7024	180.4812	0.1297	0.0094	0.0055
19	0.0698	0.0398	8.1179	14.3238	0.5703	116.2788	25.1169	1.7535	203.8956	0.1232	0.0086	0.0049
20	0.0672	0.0372	8.5229	14.8775	0.5537	126.7987	26.8704	1.8061	229.0125	0.1173	0.0079	0.0044
21	0.0649	0.0349	8.9231	15.4150	0.5375	137.5496	28.6765	1.8603	255.8829	0.1121	0.0073	0.0039
22	0.0627	0.0327	9.3186	15.9369	0.5219	148.5094	30.5368	1.9161	284.5593	0.1073	0.0067	0.0035
23	0.0608	0.0308	9.7093	16.4436	0.5067	159.6566	32.4529	1.9736	315.0961	0.1030	0.0063	0.0032
24	0.0590	0.0290	10.0954	16.9355	0.4919	170.9711	34.4265	2.0328	347.5490	0.0991	0.0058	0.0029
25	0.0574	0.0274	10.4768	17.4131	0.4776	182.4336	36.4593	2.0938	381.9755	0.0954	0.0055	0.0026
26	0.0559	0.0259	10.8535	17.8768	0.4637	194.0260	38.5530	2.1566	418.4347	0.0921	0.0052	0.0024
27	0.0546	0.0246	11.2255	18.3270	0.4502	205.7309	40.7096	2.2213	456.9878	0.0891	0.0049	0.0022
28	0.0533	0.0233	11.5930	18.7641	0.4371	217.5320	42.9309	2.2879	497.6974	0.0863	0.0046	0.0020
29	0.0521	0.0221	11.9558	19.1885	0.4243	229.4137	45.2189	2.3566	540.6283	0.0836	0.0044	0.0018
30	0.0510	0.0210	12.3141	19.6004	0.4120	241.3613	47.5754	2.4273	585.8472	0.0812	0.0041	0.0017
31	0.0500	0.0200	12.6678	20.0004	0.4000	253.3609	50.0027	2.5001	633.4226	0.0789	0.0039	0.0016
32	0.0490	0.0190	13.0169	20.3888	0.3883	265.3993	52.5028	2.5751	683.4253	0.0768	0.0038	0.0015
33	0.0482	0.0182	13.3616	20.7658	0.3770	277.4642	55.0778	2.6523	735.9280	0.0748	0.0036	0.0014
34	0.0473	0.0173	13.7018	21.1318	0.3660	289.5437	57.7302	2.7319	791.0059	0.0730	0.0035	0.0013
35	0.0465	0.0165	14.0375	21.4872	0.3554	301.6267	60.4621	2.8139	848.7361	0.0712	0.0033	0.0012
36	0.0458	0.0158	14.3688	21.8323	0.3450	313.7028	63.2759	2.8983	909.1981	0.0696	0.0032	0.0011
37	0.0451	0.0151	14.6957	22.1672	0.3350	325.7622	66.1742	2.9852	972.4741	0.0680	0.0031	0.0010
38	0.0445	0.0145	15.0182	22.4925	0.3252	337.7956	69.1594	3.0748	1038.6483	0.0666	0.0030	0.0010
39	0.0438	0.0138	15.3363	22.8082	0.3158	349.7942	72.2342	3.1670	1107.8078	0.0652	0.0029	0.0009
40	0.0433	0.0133	15.6502	23.1148	0.3066	361.7499	75.4013	3.2620	1180.0420	0.0639	0.0028	0.0008
∞	0.0300	0.0000	33.3333	33.3333	0.0000	1111.1111	∞	∞	∞	0.0300	0.0009	0.0000

Table 501.4 Economics Factors for $i = 4\%$

n	A/P	A/F	A/G	P/A	P/F	P/G	F/A	F/P	F/G	G/A	G/P	G/F
1	1.0400	1.0000	0.0000	0.9615	0.9615	0.0000	1.0000	1.0400	0.0000			
2	0.5302	0.4902	0.4902	1.8861	0.9246	0.9246	2.0400	1.0816	1.0000	2.0400	1.0816	1.0000
3	0.3603	0.3203	0.9739	2.7751	0.8890	2.7025	3.1216	1.1249	3.0400	1.0268	0.3700	0.3289
4	0.2755	0.2355	1.4510	3.6299	0.8548	5.2670	4.2465	1.1699	6.1616	0.6892	0.1899	0.1623
5	0.2246	0.1846	1.9216	4.4518	0.8219	8.5547	5.4163	1.2167	10.4081	0.5204	0.1169	0.0961
6	0.1908	0.1508	2.3857	5.2421	0.7903	12.5062	6.6330	1.2653	15.8244	0.4192	0.0800	0.0632
7	0.1666	0.1266	2.8433	6.0021	0.7599	17.0657	7.8983	1.3159	22.4574	0.3517	0.0586	0.0445
8	0.1485	0.1085	3.2944	6.7327	0.7307	22.1806	9.2142	1.3686	30.3557	0.3035	0.0451	0.0329
9	0.1345	0.0945	3.7391	7.4353	0.7026	27.8013	10.5828	1.4233	39.5699	0.2674	0.0360	0.0253
10	0.1233	0.0833	4.1773	8.1109	0.6756	33.8814	12.0061	1.4802	50.1527	0.2394	0.0295	0.0199
11	0.1141	0.0741	4.6090	8.7605	0.6496	40.3772	13.4864	1.5395	62.1588	0.2170	0.0248	0.0161
12	0.1066	0.0666	5.0343	9.3851	0.6246	47.2477	15.0258	1.6010	75.6451	0.1986	0.0212	0.0132
13	0.1001	0.0601	5.4533	9.9856	0.6006	54.4546	16.6268	1.6651	90.6709	0.1834	0.0184	0.0110
14	0.0947	0.0547	5.8659	10.5631	0.5775	61.9618	18.2919	1.7317	107.2978	0.1705	0.0161	0.0093
15	0.0899	0.0499	6.2721	11.1184	0.5553	69.7355	20.0236	1.8009	125.5897	0.1594	0.0143	0.0080
16	0.0858	0.0458	6.6720	11.6523	0.5339	77.7441	21.8245	1.8730	145.6133	0.1499	0.0129	0.0069
17	0.0822	0.0422	7.0656	12.1657	0.5134	85.9581	23.6975	1.9479	167.4378	0.1415	0.0116	0.0060
18	0.0790	0.0390	7.4530	12.6593	0.4936	94.3498	25.6454	2.0258	191.1353	0.1342	0.0106	0.0052
19	0.0761	0.0361	7.8342	13.1339	0.4746	102.8933	27.6712	2.1068	216.7807	0.1276	0.0097	0.0046
20	0.0736	0.0336	8.2091	13.5903	0.4564	111.5647	29.7781	2.1911	244.4520	0.1218	0.0090	0.0041
21	0.0713	0.0313	8.5779	14.0292	0.4388	120.3414	31.9692	2.2788	274.2300	0.1166	0.0083	0.0036
22	0.0692	0.0292	8.9407	14.4511	0.4220	129.2024	34.2480	2.3699	306.1992	0.1118	0.0077	0.0033
23	0.0673	0.0273	9.2973	14.8568	0.4057	138.1284	36.6179	2.4647	340.4472	0.1076	0.0072	0.0029
24	0.0656	0.0256	9.6479	15.2470	0.3901	147.1012	39.0826	2.5633	377.0651	0.1036	0.0068	0.0027
25	0.0640	0.0240	9.9925	15.6221	0.3751	156.1040	41.6459	2.6658	416.1477	0.1001	0.0064	0.0024
26	0.0626	0.0226	10.3312	15.9828	0.3607	165.1212	44.3117	2.7725	457.7936	0.0968	0.0061	0.0022
27	0.0612	0.0212	10.6640	16.3296	0.3468	174.1385	47.0842	2.8834	502.1054	0.0938	0.0057	0.0020
28	0.0600	0.0200	10.9909	16.6631	0.3335	183.1424	49.9676	2.9987	549.1896	0.0910	0.0055	0.0018
29	0.0589	0.0189	11.3120	16.9837	0.3207	192.1206	52.9663	3.1187	599.1572	0.0884	0.0052	0.0017
30	0.0578	0.0178	11.6274	17.2920	0.3083	201.0618	56.0849	3.2434	652.1234	0.0860	0.0050	0.0015
31	0.0569	0.0169	11.9371	17.5885	0.2965	209.9556	59.3283	3.3731	708.2084	0.0838	0.0048	0.0014
32	0.0559	0.0159	12.2411	17.8736	0.2851	218.7924	62.7015	3.5081	767.5367	0.0817	0.0046	0.0013
33	0.0551	0.0151	12.5396	18.1476	0.2741	227.5634	66.2095	3.6484	830.2382	0.0797	0.0044	0.0012
34	0.0543	0.0143	12.8324	18.4112	0.2636	236.2607	69.8579	3.7943	896.4477	0.0779	0.0042	0.0011
35	0.0536	0.0136	13.1198	18.6646	0.2534	244.8768	73.6522	3.9461	966.3056	0.0762	0.0041	0.0010
36	0.0529	0.0129	13.4018	18.9083	0.2437	253.4052	77.5983	4.1039	1039.9578	0.0746	0.0039	0.0010
37	0.0522	0.0122	13.6784	19.1426	0.2343	261.8399	81.7022	4.2681	1117.5562	0.0731	0.0038	0.0009
38	0.0516	0.0116	13.9497	19.3679	0.2253	270.1754	85.9703	4.4388	1199.2584	0.0717	0.0037	0.0008
39	0.0511	0.0111	14.2157	19.5845	0.2166	278.4070	90.4091	4.6164	1285.2287	0.0703	0.0036	0.0008
40	0.0505	0.0105	14.4765	19.7928	0.2083	286.5303	95.0255	4.8010	1375.6379	0.0691	0.0035	0.0007
∞	0.0400	0.0000	25.0000	25.0000	0.0000	625.0000	∞	∞	∞	0.0400	0.0016	0.0000

Table 501.5 Economics Factors for $i = 5\%$

n	A/P	A/F	A/G	P/A	P/F	P/G	F/A	F/P	F/G	G/A	G/P	G/F
1	1.0500	1.0000	0.0000	0.9524	0.9524	0.0000	1.0000	1.0500	0.0000			
2	0.5378	0.4878	0.4878	1.8594	0.9070	0.9070	2.0500	1.1025	1.0000	2.0500	1.1025	1.0000
3	0.3672	0.3172	0.9675	2.7232	0.8638	2.6347	3.1525	1.1576	3.0500	1.0336	0.3795	0.3279
4	0.2820	0.2320	1.4391	3.5460	0.8227	5.1028	4.3101	1.2155	6.2025	0.6949	0.1960	0.1612
5	0.2310	0.1810	1.9025	4.3295	0.7835	8.2369	5.5256	1.2763	10.5126	0.5256	0.1214	0.0951
6	0.1970	0.1470	2.3579	5.0757	0.7462	11.9680	6.8019	1.3401	16.0383	0.4241	0.0836	0.0624
7	0.1728	0.1228	2.8052	5.7864	0.7107	16.2321	8.1420	1.4071	22.8402	0.3565	0.0616	0.0438
8	0.1547	0.1047	3.2445	6.4632	0.6768	20.9700	9.5491	1.4775	30.9822	0.3082	0.0477	0.0323
9	0.1407	0.0907	3.6758	7.1078	0.6446	26.1268	11.0266	1.5513	40.5313	0.2721	0.0383	0.0247
10	0.1295	0.0795	4.0991	7.7217	0.6139	31.6520	12.5779	1.6289	51.5579	0.2440	0.0316	0.0194
11	0.1204	0.0704	4.5144	8.3064	0.5847	37.4988	14.2068	1.7103	64.1357	0.2215	0.0267	0.0156
12	0.1128	0.0628	4.9219	8.8633	0.5568	43.6241	15.9171	1.7959	78.3425	0.2032	0.0229	0.0128
13	0.1065	0.0565	5.3215	9.3936	0.5303	49.9879	17.7130	1.8856	94.2597	0.1879	0.0200	0.0106
14	0.1010	0.0510	5.7133	9.8986	0.5051	56.5538	19.5986	1.9799	111.9726	0.1750	0.0177	0.0089
15	0.0963	0.0463	6.0973	10.3797	0.4810	63.2880	21.5786	2.0789	131.5713	0.1640	0.0158	0.0076
16	0.0923	0.0423	6.4736	10.8378	0.4581	70.1597	23.6575	2.1829	153.1498	0.1545	0.0143	0.0065
17	0.0887	0.0387	6.8423	11.2741	0.4363	77.1405	25.8404	2.2920	176.8073	0.1461	0.0130	0.0057
18	0.0855	0.0355	7.2034	11.6896	0.4155	84.2043	28.1324	2.4066	202.6477	0.1388	0.0119	0.0049
19	0.0827	0.0327	7.5569	12.0853	0.3957	91.3275	30.5390	2.5270	230.7801	0.1323	0.0109	0.0043
20	0.0802	0.0302	7.9030	12.4622	0.3769	98.4884	33.0660	2.6533	261.3191	0.1265	0.0102	0.0038
21	0.0780	0.0280	8.2416	12.8212	0.3589	105.6673	35.7193	2.7860	294.3850	0.1213	0.0095	0.0034
22	0.0760	0.0260	8.5730	13.1630	0.3418	112.8461	38.5052	2.9253	330.1043	0.1166	0.0089	0.0030
23	0.0741	0.0241	8.8971	13.4886	0.3256	120.0087	41.4305	3.0715	368.6095	0.1124	0.0083	0.0027
24	0.0725	0.0225	9.2140	13.7986	0.3101	127.1402	44.5020	3.2251	410.0400	0.1085	0.0079	0.0024
25	0.0710	0.0210	9.5238	14.0939	0.2953	134.2275	47.7271	3.3864	454.5420	0.1050	0.0075	0.0022
26	0.0696	0.0196	9.8266	14.3752	0.2812	141.2585	51.1135	3.5557	502.2691	0.1018	0.0071	0.0020
27	0.0683	0.0183	10.1224	14.6430	0.2678	148.2226	54.6691	3.7335	553.3825	0.0988	0.0067	0.0018
28	0.0671	0.0171	10.4114	14.8981	0.2551	155.1101	58.4026	3.9201	608.0517	0.0960	0.0064	0.0016
29	0.0660	0.0160	10.6936	15.1411	0.2429	161.9126	62.3227	4.1161	666.4542	0.0935	0.0062	0.0015
30	0.0651	0.0151	10.9691	15.3725	0.2314	168.6226	66.4388	4.3219	728.7770	0.0912	0.0059	0.0014
31	0.0641	0.0141	11.2381	15.5928	0.2204	175.2333	70.7608	4.5380	795.2158	0.0890	0.0057	0.0013
32	0.0633	0.0133	11.5005	15.8027	0.2099	181.7392	75.2988	4.7649	865.9766	0.0870	0.0055	0.0012
33	0.0625	0.0125	11.7566	16.0025	0.1999	188.1351	80.0638	5.0032	941.2754	0.0851	0.0053	0.0011
34	0.0618	0.0118	12.0063	16.1929	0.1904	194.4168	85.0670	5.2533	1021.3392	0.0833	0.0051	0.0010
35	0.0611	0.0111	12.2498	16.3742	0.1813	200.5807	90.3203	5.5160	1106.4061	0.0816	0.0050	0.0009
36	0.0604	0.0104	12.4872	16.5469	0.1727	206.6237	95.8363	5.7918	1196.7265	0.0801	0.0048	0.0008
37	0.0598	0.0098	12.7186	16.7113	0.1644	212.5434	101.6281	6.0814	1292.5628	0.0786	0.0047	0.0008
38	0.0593	0.0093	12.9440	16.8679	0.1566	218.3378	107.7095	6.3855	1394.1909	0.0773	0.0046	0.0007
39	0.0588	0.0088	13.1636	17.0170	0.1491	224.0054	114.0950	6.7048	1501.9005	0.0760	0.0045	0.0007
40	0.0583	0.0083	13.3775	17.1591	0.1420	229.5452	120.7998	7.0400	1615.9955	0.0748	0.0044	0.0006
∞	0.0500	0.0000	20.0000	20.0000	0.0000	400.0000	∞	∞	∞	0.0500	0.0025	0.0000

Table 501.6 Economics Factors for $i = 6\%$

n	A/P	A/F	A/G	P/A	P/F	P/G	F/A	F/P	F/G	G/A	G/P	G/F
1	1.0600	1.0000	0.0000	0.9434	0.9434	0.0000	1.0000	1.0600	0.0000			
2	0.5454	0.4854	0.4854	1.8334	0.8900	0.8900	2.0600	1.1236	1.0000	2.0600	1.1236	1.0000
3	0.3741	0.3141	0.9612	2.6730	0.8396	2.5692	3.1836	1.1910	3.0600	1.0404	0.3892	0.3268
4	0.2886	0.2286	1.4272	3.4651	0.7921	4.9455	4.3746	1.2625	6.2436	0.7007	0.2022	0.1602
5	0.2374	0.1774	1.8836	4.2124	0.7473	7.9345	5.6371	1.3382	10.6182	0.5309	0.1260	0.0942
6	0.2034	0.1434	2.3304	4.9173	0.7050	11.4594	6.9753	1.4185	16.2553	0.4291	0.0873	0.0615
7	0.1791	0.1191	2.7676	5.5824	0.6651	15.4497	8.3938	1.5036	23.2306	0.3613	0.0647	0.0430
8	0.1610	0.1010	3.1952	6.2098	0.6274	19.8416	9.8975	1.5938	31.6245	0.3130	0.0504	0.0316
9	0.1470	0.0870	3.6133	6.8017	0.5919	24.5768	11.4913	1.6895	41.5219	0.2768	0.0407	0.0241
10	0.1359	0.0759	4.0220	7.3601	0.5584	29.6023	13.1808	1.7908	53.0132	0.2486	0.0338	0.0189
11	0.1268	0.0668	4.4213	7.8869	0.5268	34.8702	14.9716	1.8983	66.1940	0.2262	0.0287	0.0151
12	0.1193	0.0593	4.8113	8.3838	0.4970	40.3369	16.8699	2.0122	81.1657	0.2078	0.0248	0.0123
13	0.1130	0.0530	5.1920	8.8527	0.4688	45.9629	18.8821	2.1329	98.0356	0.1926	0.0218	0.0102
14	0.1076	0.0476	5.5635	9.2950	0.4423	51.7128	21.0151	2.2609	116.9178	0.1797	0.0193	0.0086
15	0.1030	0.0430	5.9260	9.7122	0.4173	57.5546	23.2760	2.3966	137.9328	0.1687	0.0174	0.0072
16	0.0990	0.0390	6.2794	10.1059	0.3936	63.4592	25.6725	2.5404	161.2088	0.1593	0.0158	0.0062
17	0.0954	0.0354	6.6240	10.4773	0.3714	69.4011	28.2129	2.6928	186.8813	0.1510	0.0144	0.0054
18	0.0924	0.0324	6.9597	10.8276	0.3503	75.3569	30.9057	2.8543	215.0942	0.1437	0.0133	0.0046
19	0.0896	0.0296	7.2867	11.1581	0.3305	81.3062	33.7600	3.0256	245.9999	0.1372	0.0123	0.0041
20	0.0872	0.0272	7.6051	11.4699	0.3118	87.2304	36.7856	3.2071	279.7599	0.1315	0.0115	0.0036
21	0.0850	0.0250	7.9151	11.7641	0.2942	93.1136	39.9927	3.3996	316.5454	0.1263	0.0107	0.0032
22	0.0830	0.0230	8.2166	12.0416	0.2775	98.9412	43.3923	3.6035	356.5382	0.1217	0.0101	0.0028
23	0.0813	0.0213	8.5099	12.3034	0.2618	104.7007	46.9958	3.8197	399.9305	0.1175	0.0096	0.0025
24	0.0797	0.0197	8.7951	12.5504	0.2470	110.3812	50.8156	4.0489	446.9263	0.1137	0.0091	0.0022
25	0.0782	0.0182	9.0722	12.7834	0.2330	115.9732	54.8645	4.2919	497.7419	0.1102	0.0086	0.0020
26	0.0769	0.0169	9.3414	13.0032	0.2198	121.4684	59.1564	4.5494	552.6064	0.1070	0.0082	0.0018
27	0.0757	0.0157	9.6029	13.2105	0.2074	126.8600	63.7058	4.8223	611.7628	0.1041	0.0079	0.0016
28	0.0746	0.0146	9.8568	13.4062	0.1956	132.1420	68.5281	5.1117	675.4685	0.1015	0.0076	0.0015
29	0.0736	0.0136	10.1032	13.5907	0.1846	137.3096	73.6398	5.4184	743.9966	0.0990	0.0073	0.0013
30	0.0726	0.0126	10.3422	13.7648	0.1741	142.3588	79.0582	5.7435	817.6364	0.0967	0.0070	0.0012
31	0.0718	0.0118	10.5740	13.9291	0.1643	147.2864	84.8017	6.0881	896.6946	0.0946	0.0068	0.0011
32	0.0710	0.0110	10.7988	14.0840	0.1550	152.0901	90.8898	6.4534	981.4963	0.0926	0.0066	0.0010
33	0.0703	0.0103	11.0166	14.2302	0.1462	156.7681	97.3432	6.8406	1072.3861	0.0908	0.0064	0.0009
34	0.0696	0.0096	11.2276	14.3681	0.1379	161.3192	104.1838	7.2510	1169.7292	0.0891	0.0062	0.0009
35	0.0690	0.0090	11.4319	14.4982	0.1301	165.7427	111.4348	7.6861	1273.9130	0.0875	0.0060	0.0008
36	0.0684	0.0084	11.6298	14.6210	0.1227	170.0387	119.1209	8.1473	1385.3478	0.0860	0.0059	0.0007
37	0.0679	0.0079	11.8213	14.7368	0.1158	174.2072	127.2681	8.6361	1504.4686	0.0846	0.0057	0.0007
38	0.0674	0.0074	12.0065	14.8460	0.1092	178.2490	135.9042	9.1543	1631.7368	0.0833	0.0056	0.0006
39	0.0669	0.0069	12.1857	14.9491	0.1031	182.1652	145.0585	9.7035	1767.6410	0.0821	0.0055	0.0006
40	0.0665	0.0065	12.3590	15.0463	0.0972	185.9568	154.7620	10.2857	1912.6994	0.0809	0.0054	0.0005
∞	0.0600	0.0000	16.6667	16.6667	0.0000	277.7778	∞	∞	∞	0.0600	0.0036	0.0000

Table 501.7 Economics Factors for $i = 7\%$

n	A/P	A/F	A/G	P/A	P/F	P/G	F/A	F/P	F/G	G/A	G/P	G/F
1	1.0700	1.0000	0.0000	0.9346	0.9346	0.0000	1.0000	1.0700	0.0000			
2	0.5531	0.4831	0.4831	1.8080	0.8734	0.8734	2.0700	1.1449	1.0000	2.0700	1.1449	1.0000
3	0.3811	0.3111	0.9549	2.6243	0.8163	2.5060	3.2149	1.2250	3.0700	1.0472	0.3990	0.3257
4	0.2952	0.2252	1.4155	3.3872	0.7629	4.7947	4.4399	1.3108	6.2849	0.7064	0.2086	0.1591
5	0.2439	0.1739	1.8650	4.1002	0.7130	7.6467	5.7507	1.4026	10.7248	0.5362	0.1308	0.0932
6	0.2098	0.1398	2.3032	4.7665	0.6663	10.9784	7.1533	1.5007	16.4756	0.4342	0.0911	0.0607
7	0.1856	0.1156	2.7304	5.3893	0.6227	14.7149	8.6540	1.6058	23.6289	0.3662	0.0680	0.0423
8	0.1675	0.0975	3.1465	5.9713	0.5820	18.7889	10.2598	1.7182	32.2829	0.3178	0.0532	0.0310
9	0.1535	0.0835	3.5517	6.5152	0.5439	23.1404	11.9780	1.8385	42.5427	0.2816	0.0432	0.0235
10	0.1424	0.0724	3.9461	7.0236	0.5083	27.7156	13.8164	1.9672	54.5207	0.2534	0.0361	0.0183
11	0.1334	0.0634	4.3296	7.4987	0.4751	32.4665	15.7836	2.1049	68.3371	0.2310	0.0308	0.0146
12	0.1259	0.0559	4.7025	7.9427	0.4440	37.3506	17.8885	2.2522	84.1207	0.2127	0.0268	0.0119
13	0.1197	0.0497	5.0648	8.3577	0.4150	42.3302	20.1406	2.4098	102.0092	0.1974	0.0236	0.0098
14	0.1143	0.0443	5.4167	8.7455	0.3878	47.3718	22.5505	2.5785	122.1498	0.1846	0.0211	0.0082
15	0.1098	0.0398	5.7583	9.1079	0.3624	52.4461	25.1290	2.7590	144.7003	0.1737	0.0191	0.0069
16	0.1059	0.0359	6.0897	9.4466	0.3387	57.5271	27.8881	2.9522	169.8293	0.1642	0.0174	0.0059
17	0.1024	0.0324	6.4110	9.7632	0.3166	62.5923	30.8402	3.1588	197.7174	0.1560	0.0160	0.0051
18	0.0994	0.0294	6.7225	10.0591	0.2959	67.6219	33.9990	3.3799	228.5576	0.1488	0.0148	0.0044
19	0.0968	0.0268	7.0242	10.3356	0.2765	72.5991	37.3790	3.6165	262.5566	0.1424	0.0138	0.0038
20	0.0944	0.0244	7.3163	10.5940	0.2584	77.5091	40.9955	3.8697	299.9356	0.1367	0.0129	0.0033
21	0.0923	0.0223	7.5990	10.8355	0.2415	82.3393	44.8652	4.1406	340.9311	0.1316	0.0121	0.0029
22	0.0904	0.0204	7.8725	11.0612	0.2257	87.0793	49.0057	4.4304	385.7963	0.1270	0.0115	0.0026
23	0.0887	0.0187	8.1369	11.2722	0.2109	91.7201	53.4361	4.7405	434.8020	0.1229	0.0109	0.0023
24	0.0872	0.0172	8.3923	11.4693	0.1971	96.2545	58.1767	5.0724	488.2382	0.1192	0.0104	0.0020
25	0.0858	0.0158	8.6391	11.6536	0.1842	100.6765	63.2490	5.4274	546.4148	0.1158	0.0099	0.0018
26	0.0846	0.0146	8.8773	11.8258	0.1722	104.9814	68.6765	5.8074	609.6639	0.1126	0.0095	0.0016
27	0.0834	0.0134	9.1072	11.9867	0.1609	109.1656	74.4838	6.2139	678.3403	0.1098	0.0092	0.0015
28	0.0824	0.0124	9.3289	12.1371	0.1504	113.2264	80.6977	6.6488	752.8242	0.1072	0.0088	0.0013
29	0.0814	0.0114	9.5427	12.2777	0.1406	117.1622	87.3465	7.1143	833.5218	0.1048	0.0085	0.0012
30	0.0806	0.0106	9.7487	12.4090	0.1314	120.9718	94.4608	7.6123	920.8684	0.1026	0.0083	0.0011
31	0.0798	0.0098	9.9471	12.5318	0.1228	124.6550	102.0730	8.1451	1015.3292	0.1005	0.0080	0.0010
32	0.0791	0.0091	10.1381	12.6466	0.1147	128.2120	110.2182	8.7153	1117.4022	0.0986	0.0078	0.0009
33	0.0784	0.0084	10.3219	12.7538	0.1072	131.6435	118.9334	9.3253	1227.6204	0.0969	0.0076	0.0008
34	0.0778	0.0078	10.4987	12.8540	0.1002	134.9507	128.2588	9.9781	1346.5538	0.0952	0.0074	0.0007
35	0.0772	0.0072	10.6687	12.9477	0.0937	138.1353	138.2369	10.6766	1474.8125	0.0937	0.0072	0.0007
36	0.0767	0.0067	10.8321	13.0352	0.0875	141.1990	148.9135	11.4239	1613.0494	0.0923	0.0071	0.0006
37	0.0762	0.0062	10.9891	13.1170	0.0818	144.1441	160.3374	12.2236	1761.9629	0.0910	0.0069	0.0006
38	0.0758	0.0058	11.1398	13.1935	0.0765	146.9730	172.5610	13.0793	1922.3003	0.0898	0.0068	0.0005
39	0.0754	0.0054	11.2845	13.2649	0.0715	149.6883	185.6403	13.9948	2094.8613	0.0886	0.0067	0.0005
40	0.0750	0.0050	11.4233	13.3317	0.0668	152.2928	199.6351	14.9745	2280.5016	0.0875	0.0066	0.0004
∞	0.0700	0.0000	14.2857	14.2857	0.0000	204.0816	∞	∞	∞	0.0700	0.0049	0.0000

Table 501.8 Economics Factors for $i = 8\%$

n	A/P	A/F	A/G	P/A	P/F	P/G	F/A	F/P	F/G	G/A	G/P	G/F
1	1.0800	1.0000	0.0000	0.9259	0.9259	0.0000	1.0000	1.0800	0.0000			
2	0.5608	0.4808	0.4808	1.7833	0.8573	0.8573	2.0800	1.1664	1.0000	2.0800	1.1664	1.0000
3	0.3880	0.3080	0.9487	2.5771	0.7938	2.4450	3.2464	1.2597	3.0800	1.0540	0.4090	0.3247
4	0.3019	0.2219	1.4040	3.3121	0.7350	4.6501	4.5061	1.3605	6.3264	0.7123	0.2150	0.1581
5	0.2505	0.1705	1.8465	3.9927	0.6806	7.3724	5.8666	1.4693	10.8325	0.5416	0.1356	0.0923
6	0.2163	0.1363	2.2763	4.6229	0.6302	10.5233	7.3359	1.5869	16.6991	0.4393	0.0950	0.0599
7	0.1921	0.1121	2.6937	5.2064	0.5835	14.0242	8.9228	1.7138	24.0350	0.3712	0.0713	0.0416
8	0.1740	0.0940	3.0985	5.7466	0.5403	17.8061	10.6366	1.8509	32.9578	0.3227	0.0562	0.0303
9	0.1601	0.0801	3.4910	6.2469	0.5002	21.8081	12.4876	1.9990	43.5945	0.2864	0.0459	0.0229
10	0.1490	0.0690	3.8713	6.7101	0.4632	25.9768	14.4866	2.1589	56.0820	0.2583	0.0385	0.0178
11	0.1401	0.0601	4.2395	7.1390	0.4289	30.2657	16.6455	2.3316	70.5686	0.2359	0.0330	0.0142
12	0.1327	0.0527	4.5957	7.5361	0.3971	34.6339	18.9771	2.5182	87.2141	0.2176	0.0289	0.0115
13	0.1265	0.0465	4.9402	7.9038	0.3677	39.0463	21.4953	2.7196	106.1912	0.2024	0.0256	0.0094
14	0.1213	0.0413	5.2731	8.2442	0.3405	43.4723	24.2149	2.9372	127.6865	0.1896	0.0230	0.0078
15	0.1168	0.0368	5.5945	8.5595	0.3152	47.8857	27.1521	3.1722	151.9014	0.1787	0.0209	0.0066
16	0.1130	0.0330	5.9046	8.8514	0.2919	52.2640	30.3243	3.4259	179.0535	0.1694	0.0191	0.0056
17	0.1096	0.0296	6.2037	9.1216	0.2703	56.5883	33.7502	3.7000	209.3778	0.1612	0.0177	0.0048
18	0.1067	0.0267	6.4920	9.3719	0.2502	60.8426	37.4502	3.9960	243.1280	0.1540	0.0164	0.0041
19	0.1041	0.0241	6.7697	9.6036	0.2317	65.0134	41.4463	4.3157	280.5783	0.1477	0.0154	0.0036
20	0.1019	0.0219	7.0369	9.8181	0.2145	69.0898	45.7620	4.6610	322.0246	0.1421	0.0145	0.0031
21	0.0998	0.0198	7.2940	10.0168	0.1987	73.0629	50.4229	5.0338	367.7865	0.1371	0.0137	0.0027
22	0.0980	0.0180	7.5412	10.2007	0.1839	76.9257	55.4568	5.4365	418.2094	0.1326	0.0130	0.0024
23	0.0964	0.0164	7.7786	10.3711	0.1703	80.6726	60.8933	5.8715	473.6662	0.1286	0.0124	0.0021
24	0.0950	0.0150	8.0066	10.5288	0.1577	84.2997	66.7648	6.3412	534.5595	0.1249	0.0119	0.0019
25	0.0937	0.0137	8.2254	10.6748	0.1460	87.8041	73.1059	6.8485	601.3242	0.1216	0.0114	0.0017
26	0.0925	0.0125	8.4352	10.8100	0.1352	91.1842	79.9544	7.3964	674.4302	0.1186	0.0110	0.0015
27	0.0914	0.0114	8.6363	10.9352	0.1252	94.4390	87.3508	7.9881	754.3846	0.1158	0.0106	0.0013
28	0.0905	0.0105	8.8289	11.0511	0.1159	97.5687	95.3388	8.6271	841.7354	0.1133	0.0102	0.0012
29	0.0896	0.0096	9.0133	11.1584	0.1073	100.5738	103.9659	9.3173	937.0742	0.1109	0.0099	0.0011
30	0.0888	0.0088	9.1897	11.2578	0.0994	103.4558	113.2832	10.0627	1041.0401	0.1088	0.0097	0.0010
31	0.0881	0.0081	9.3584	11.3498	0.0920	106.2163	123.3459	10.8677	1154.3234	0.1069	0.0094	0.0009
32	0.0875	0.0075	9.5197	11.4350	0.0852	108.8575	134.2135	11.7371	1277.6692	0.1050	0.0092	0.0008
33	0.0869	0.0069	9.6737	11.5139	0.0789	111.3819	145.9506	12.6760	1411.8828	0.1034	0.0090	0.0007
34	0.0863	0.0063	9.8208	11.5869	0.0730	113.7924	158.6267	13.6901	1557.8334	0.1018	0.0088	0.0006
35	0.0858	0.0058	9.9611	11.6546	0.0676	116.0920	172.3168	14.7853	1716.4600	0.1004	0.0086	0.0006
36	0.0853	0.0053	10.0949	11.7172	0.0626	118.2839	187.1021	15.9682	1888.7768	0.0991	0.0085	0.0005
37	0.0849	0.0049	10.2225	11.7752	0.0580	120.3713	203.0703	17.2456	2075.8790	0.0978	0.0083	0.0005
38	0.0845	0.0045	10.3440	11.8289	0.0537	122.3579	220.3159	18.6253	2278.9493	0.0967	0.0082	0.0004
39	0.0842	0.0042	10.4597	11.8786	0.0497	124.2470	238.9412	20.1153	2499.2653	0.0956	0.0080	0.0004
40	0.0839	0.0039	10.5699	11.9246	0.0460	126.0422	259.0565	21.7245	2738.2065	0.0946	0.0079	0.0004
∞	0.0800	0.0000	12.5000	12.5000	0.0000	156.2500	∞	∞	∞	0.0800	0.0064	0.0000

Table 501.9 Economics Factors for $i = 9\%$

n	A/P	A/F	A/G	P/A	P/F	P/G	F/A	F/P	F/G	G/A	G/P	G/F
1	1.0900	1.0000	0.0000	0.9174	0.9174	0.0000	1.0000	1.0900	0.0000			
2	0.5685	0.4785	0.4785	1.7591	0.8417	0.8417	2.0900	1.1881	1.0000	2.0900	1.1881	1.0000
3	0.3951	0.3051	0.9426	2.5313	0.7722	2.3860	3.2781	1.2950	3.0900	1.0609	0.4191	0.3236
4	0.3087	0.2187	1.3925	3.2397	0.7084	4.5113	4.5731	1.4116	6.3681	0.7181	0.2217	0.1570
5	0.2571	0.1671	1.8282	3.8897	0.6499	7.1110	5.9847	1.5386	10.9412	0.5470	0.1406	0.0914
6	0.2229	0.1329	2.2498	4.4859	0.5963	10.0924	7.5233	1.6771	16.9259	0.4445	0.0991	0.0591
7	0.1987	0.1087	2.6574	5.0330	0.5470	13.3746	9.2004	1.8280	24.4493	0.3763	0.0748	0.0409
8	0.1807	0.0907	3.0512	5.5348	0.5019	16.8877	11.0285	1.9926	33.6497	0.3277	0.0592	0.0297
9	0.1668	0.0768	3.4312	5.9952	0.4604	20.5711	13.0210	2.1719	44.6782	0.2914	0.0486	0.0224
10	0.1558	0.0658	3.7978	6.4177	0.4224	24.3728	15.1929	2.3674	57.6992	0.2633	0.0410	0.0173
11	0.1469	0.0569	4.1510	6.8052	0.3875	28.2481	17.5603	2.5804	72.8921	0.2409	0.0354	0.0137
12	0.1397	0.0497	4.4910	7.1607	0.3555	32.1590	20.1407	2.8127	90.4524	0.2227	0.0311	0.0111
13	0.1336	0.0436	4.8182	7.4869	0.3262	36.0731	22.9534	3.0658	110.5932	0.2075	0.0277	0.0090
14	0.1284	0.0384	5.1326	7.7862	0.2992	39.9633	26.0192	3.3417	133.5465	0.1948	0.0250	0.0075
15	0.1241	0.0341	5.4346	8.0607	0.2745	43.8069	29.3609	3.6425	159.5657	0.1840	0.0228	0.0063
16	0.1203	0.0303	5.7245	8.3126	0.2519	47.5849	33.0034	3.9703	188.9267	0.1747	0.0210	0.0053
17	0.1170	0.0270	6.0024	8.5436	0.2311	51.2821	36.9737	4.3276	221.9301	0.1666	0.0195	0.0045
18	0.1142	0.0242	6.2687	8.7556	0.2120	54.8860	41.3013	4.7171	258.9038	0.1595	0.0182	0.0039
19	0.1117	0.0217	6.5236	8.9501	0.1945	58.3868	46.0185	5.1417	300.2051	0.1533	0.0171	0.0033
20	0.1095	0.0195	6.7674	9.1285	0.1784	61.7770	51.1601	5.6044	346.2236	0.1478	0.0162	0.0029
21	0.1076	0.0176	7.0006	9.2922	0.1637	65.0509	56.7645	6.1088	397.3837	0.1428	0.0154	0.0025
22	0.1059	0.0159	7.2232	9.4424	0.1502	68.2048	62.8733	6.6586	454.1482	0.1384	0.0147	0.0022
23	0.1044	0.0144	7.4357	9.5802	0.1378	71.2359	69.5319	7.2579	517.0215	0.1345	0.0140	0.0019
24	0.1030	0.0130	7.6384	9.7066	0.1264	74.1433	76.7898	7.9111	586.5535	0.1309	0.0135	0.0017
25	0.1018	0.0118	7.8316	9.8226	0.1160	76.9265	84.7009	8.6231	663.3433	0.1277	0.0130	0.0015
26	0.1007	0.0107	8.0156	9.9290	0.1064	79.5863	93.3240	9.3992	748.0442	0.1248	0.0126	0.0013
27	0.0997	0.0097	8.1906	10.0266	0.0976	82.1241	102.7231	10.2451	841.3682	0.1221	0.0122	0.0012
28	0.0989	0.0089	8.3571	10.1161	0.0895	84.5419	112.9682	11.1671	944.0913	0.1197	0.0118	0.0011
29	0.0981	0.0081	8.5154	10.1983	0.0822	86.8422	124.1354	12.1722	1057.0595	0.1174	0.0115	0.0009
30	0.0973	0.0073	8.6657	10.2737	0.0754	89.0280	136.3075	13.2677	1181.1949	0.1154	0.0112	0.0008
31	0.0967	0.0067	8.8083	10.3428	0.0691	91.1024	149.5752	14.4618	1317.5024	0.1135	0.0110	0.0008
32	0.0961	0.0061	8.9436	10.4062	0.0634	93.0690	164.0370	15.7633	1467.0776	0.1118	0.0107	0.0007
33	0.0956	0.0056	9.0718	10.4644	0.0582	94.9314	179.8003	17.1820	1631.1146	0.1102	0.0105	0.0006
34	0.0951	0.0051	9.1933	10.5178	0.0534	96.6935	196.9823	18.7284	1810.9149	0.1088	0.0103	0.0006
35	0.0946	0.0046	9.3083	10.5668	0.0490	98.3590	215.7108	20.4140	2007.8973	0.1074	0.0102	0.0005
36	0.0942	0.0042	9.4171	10.6118	0.0449	99.9319	236.1247	22.2512	2223.6080	0.1062	0.0100	0.0004
37	0.0939	0.0039	9.5200	10.6530	0.0412	101.4162	258.3759	24.2538	2459.7328	0.1050	0.0099	0.0004
38	0.0935	0.0035	9.6172	10.6908	0.0378	102.8158	282.6298	26.4367	2718.1087	0.1040	0.0097	0.0004
39	0.0932	0.0032	9.7090	10.7255	0.0347	104.1345	309.0665	28.8160	3000.7385	0.1030	0.0096	0.0003
40	0.0930	0.0030	9.7957	10.7574	0.0318	105.3762	337.8824	31.4094	3309.8049	0.1021	0.0095	0.0003
∞	0.0900	0.0000	11.1111	11.1111	0.0000	123.4568	∞	∞	∞	0.0900	0.0081	0.0000

Table 501.10 Economics Factors for $i = 10\%$

n	A/P	A/F	A/G	P/A	P/F	P/G	F/A	F/P	F/G	G/A	G/P	G/F
1	1.1000	1.0000	0.0000	0.9091	0.9091	0.0000	1.0000	1.1000	0.0000			
2	0.5762	0.4762	0.4762	1.7355	0.8264	0.8264	2.1000	1.2100	1.0000	2.1000	1.2100	1.0000
3	0.4021	0.3021	0.9366	2.4869	0.7513	2.3291	3.3100	1.3310	3.1000	1.0677	0.4294	0.3226
4	0.3155	0.2155	1.3812	3.1699	0.6830	4.3781	4.6410	1.4641	6.4100	0.7240	0.2284	0.1560
5	0.2638	0.1638	1.8101	3.7908	0.6209	6.8618	6.1051	1.6105	11.0510	0.5524	0.1457	0.0905
6	0.2296	0.1296	2.2236	4.3553	0.5645	9.6842	7.7156	1.7716	17.1561	0.4497	0.1033	0.0583
7	0.2054	0.1054	2.6216	4.8684	0.5132	12.7631	9.4872	1.9487	24.8717	0.3814	0.0784	0.0402
8	0.1874	0.0874	3.0045	5.3349	0.4665	16.0287	11.4359	2.1436	34.3589	0.3328	0.0624	0.0291
9	0.1736	0.0736	3.3724	5.7590	0.4241	19.4215	13.5795	2.3579	45.7948	0.2965	0.0515	0.0218
10	0.1627	0.0627	3.7255	6.1446	0.3855	22.8913	15.9374	2.5937	59.3742	0.2684	0.0437	0.0168
11	0.1540	0.0540	4.0641	6.4951	0.3505	26.3963	18.5312	2.8531	75.3117	0.2461	0.0379	0.0133
12	0.1468	0.0468	4.3884	6.8137	0.3186	29.9012	21.3843	3.1384	93.8428	0.2279	0.0334	0.0107
13	0.1408	0.0408	4.6988	7.1034	0.2897	33.3772	24.5227	3.4523	115.2271	0.2128	0.0300	0.0087
14	0.1357	0.0357	4.9955	7.3667	0.2633	36.8005	27.9750	3.7975	139.7498	0.2002	0.0272	0.0072
15	0.1315	0.0315	5.2789	7.6061	0.2394	40.1520	31.7725	4.1772	167.7248	0.1894	0.0249	0.0060
16	0.1278	0.0278	5.5493	7.8237	0.2176	43.4164	35.9497	4.5950	199.4973	0.1802	0.0230	0.0050
17	0.1247	0.0247	5.8071	8.0216	0.1978	46.5819	40.5447	5.0545	235.4470	0.1722	0.0215	0.0042
18	0.1219	0.0219	6.0526	8.2014	0.1799	49.6395	45.5992	5.5599	275.9917	0.1652	0.0201	0.0036
19	0.1195	0.0195	6.2861	8.3649	0.1635	52.5827	51.1591	6.1159	321.5909	0.1591	0.0190	0.0031
20	0.1175	0.0175	6.5081	8.5136	0.1486	55.4069	57.2750	6.7275	372.7500	0.1537	0.0180	0.0027
21	0.1156	0.0156	6.7189	8.6487	0.1351	58.1095	64.0025	7.4002	430.0250	0.1488	0.0172	0.0023
22	0.1140	0.0140	6.9189	8.7715	0.1228	60.6893	71.4027	8.1403	494.0275	0.1445	0.0165	0.0020
23	0.1126	0.0126	7.1085	8.8832	0.1117	63.1462	79.5430	8.9543	565.4302	0.1407	0.0158	0.0018
24	0.1113	0.0113	7.2881	8.9847	0.1015	65.4813	88.4973	9.8497	644.9733	0.1372	0.0153	0.0016
25	0.1102	0.0102	7.4580	9.0770	0.0923	67.6964	98.3471	10.8347	733.4706	0.1341	0.0148	0.0014
26	0.1092	0.0092	7.6186	9.1609	0.0839	69.7940	109.1818	11.9182	831.8177	0.1313	0.0143	0.0012
27	0.1083	0.0083	7.7704	9.2372	0.0763	71.7773	121.0999	13.1100	940.9994	0.1287	0.0139	0.0011
28	0.1075	0.0075	7.9137	9.3066	0.0693	73.6495	134.2099	14.4210	1062.0994	0.1264	0.0136	0.0009
29	0.1067	0.0067	8.0489	9.3696	0.0630	75.4146	148.6309	15.8631	1196.3093	0.1242	0.0133	0.0008
30	0.1061	0.0061	8.1762	9.4269	0.0573	77.0766	164.4940	17.4494	1344.9402	0.1223	0.0130	0.0007
31	0.1055	0.0055	8.2962	9.4790	0.0521	78.6395	181.9434	19.1943	1509.4342	0.1205	0.0127	0.0007
32	0.1050	0.0050	8.4091	9.5264	0.0474	80.1078	201.1378	21.1138	1691.3777	0.1189	0.0125	0.0006
33	0.1045	0.0045	8.5152	9.5694	0.0431	81.4856	222.2515	23.2252	1892.5154	0.1174	0.0123	0.0005
34	0.1041	0.0041	8.6149	9.6086	0.0391	82.7773	245.4767	25.5477	2114.7670	0.1161	0.0121	0.0005
35	0.1037	0.0037	8.7086	9.6442	0.0356	83.9872	271.0244	28.1024	2360.2437	0.1148	0.0119	0.0004
36	0.1033	0.0033	8.7965	9.6765	0.0323	85.1194	299.1268	30.9127	2631.2681	0.1137	0.0117	0.0004
37	0.1030	0.0030	8.8789	9.7059	0.0294	86.1781	330.0395	34.0039	2930.3949	0.1126	0.0116	0.0003
38	0.1027	0.0027	8.9562	9.7327	0.0267	87.1673	364.0434	37.4043	3260.4343	0.1117	0.0115	0.0003
39	0.1025	0.0025	9.0285	9.7570	0.0243	88.0908	401.4478	41.1448	3624.4778	0.1108	0.0114	0.0003
40	0.1023	0.0023	9.0962	9.7791	0.0221	88.9525	442.5926	45.2593	4025.9256	0.1099	0.0112	0.0002
∞	0.1000	0.0000	10.0000	10.0000	0.0000	100.0000	∞	∞	∞	0.1000	0.0100	0.0000

Table 501.11 Economics Factors for $i = 11\%$

n	A/P	A/F	A/G	P/A	P/F	P/G	F/A	F/P	F/G	G/A	G/P	G/F
1	1.1100	1.0000	0.0000	0.9009	0.9009	0.0000	1.0000	1.1100	0.0000			
2	0.5839	0.4739	0.4739	1.7125	0.8116	0.8116	2.1100	1.2321	1.0000	2.1100	1.2321	1.0000
3	0.4092	0.2992	0.9306	2.4437	0.7312	2.2740	3.3421	1.3676	3.1100	1.0746	0.4398	0.3215
4	0.3223	0.2123	1.3700	3.1024	0.6587	4.2502	4.7097	1.5181	6.4521	0.7300	0.2353	0.1550
5	0.2706	0.1606	1.7923	3.6959	0.5935	6.6240	6.2278	1.6851	11.1618	0.5580	0.1510	0.0896
6	0.2364	0.1264	2.1976	4.2305	0.5346	9.2972	7.9129	1.8704	17.3896	0.4550	0.1076	0.0575
7	0.2122	0.1022	2.5863	4.7122	0.4817	12.1872	9.7833	2.0762	25.3025	0.3867	0.0821	0.0395
8	0.1943	0.0843	2.9585	5.1461	0.4339	15.2246	11.8594	2.3045	35.0858	0.3380	0.0657	0.0285
9	0.1806	0.0706	3.3144	5.5370	0.3909	18.3520	14.1640	2.5580	46.9452	0.3017	0.0545	0.0213
10	0.1698	0.0598	3.6544	5.8892	0.3522	21.5217	16.7220	2.8394	61.1092	0.2736	0.0465	0.0164
11	0.1611	0.0511	3.9788	6.2065	0.3173	24.6945	19.5614	3.1518	77.8312	0.2513	0.0405	0.0128
12	0.1540	0.0440	4.2879	6.4924	0.2858	27.8388	22.7132	3.4985	97.3926	0.2332	0.0359	0.0103
13	0.1482	0.0382	4.5822	6.7499	0.2575	30.9290	26.2116	3.8833	120.1058	0.2182	0.0323	0.0083
14	0.1432	0.0332	4.8619	6.9819	0.2320	33.9449	30.0949	4.3104	146.3174	0.2057	0.0295	0.0068
15	0.1391	0.0291	5.1275	7.1909	0.2090	36.8709	34.4054	4.7846	176.4124	0.1950	0.0271	0.0057
16	0.1355	0.0255	5.3794	7.3792	0.1883	39.6953	39.1899	5.3109	210.8177	0.1859	0.0252	0.0047
17	0.1325	0.0225	5.6180	7.5488	0.1696	42.4095	44.5008	5.8951	250.0077	0.1780	0.0236	0.0040
18	0.1298	0.0198	5.8439	7.7016	0.1528	45.0074	50.3959	6.5436	294.5085	0.1711	0.0222	0.0034
19	0.1276	0.0176	6.0574	7.8393	0.1377	47.4856	56.9395	7.2633	344.9044	0.1651	0.0211	0.0029
20	0.1256	0.0156	6.2590	7.9633	0.1240	49.8423	64.2028	8.0623	401.8439	0.1598	0.0201	0.0025
21	0.1238	0.0138	6.4491	8.0751	0.1117	52.0771	72.2651	8.9492	466.0468	0.1551	0.0192	0.0021
22	0.1223	0.0123	6.6283	8.1757	0.1007	54.1912	81.2143	9.9336	538.3119	0.1509	0.0185	0.0019
23	0.1210	0.0110	6.7969	8.2664	0.0907	56.1864	91.1479	11.0263	619.5262	0.1471	0.0178	0.0016
24	0.1198	0.0098	6.9555	8.3481	0.0817	58.0656	102.1742	12.2392	710.6741	0.1438	0.0172	0.0014
25	0.1187	0.0087	7.1045	8.4217	0.0736	59.8322	114.4133	13.5855	812.8482	0.1408	0.0167	0.0012
26	0.1178	0.0078	7.2443	8.4881	0.0663	61.4900	127.9988	15.0799	927.2616	0.1380	0.0163	0.0011
27	0.1170	0.0070	7.3754	8.5478	0.0597	63.0433	143.0786	16.7386	1055.2603	0.1356	0.0159	0.0009
28	0.1163	0.0063	7.4982	8.6016	0.0538	64.4965	159.8173	18.5799	1198.3390	0.1334	0.0155	0.0008
29	0.1156	0.0056	7.6131	8.6501	0.0485	65.8542	178.3972	20.6237	1358.1562	0.1314	0.0152	0.0007
30	0.1150	0.0050	7.7206	8.6938	0.0437	67.1210	199.0209	22.8923	1536.5534	0.1295	0.0149	0.0007
31	0.1145	0.0045	7.8210	8.7331	0.0394	68.3016	221.9132	25.4104	1735.5743	0.1279	0.0146	0.0006
32	0.1140	0.0040	7.9147	8.7686	0.0355	69.4007	247.3236	28.2056	1957.4875	0.1263	0.0144	0.0005
33	0.1136	0.0036	8.0021	8.8005	0.0319	70.4228	275.5292	31.3082	2204.8111	0.1250	0.0142	0.0005
34	0.1133	0.0033	8.0836	8.8293	0.0288	71.3724	306.8374	34.7521	2480.3403	0.1237	0.0140	0.0004
35	0.1129	0.0029	8.1594	8.8552	0.0259	72.2538	341.5896	38.5749	2787.1778	0.1226	0.0138	0.0004
36	0.1126	0.0026	8.2300	8.8786	0.0234	73.0712	380.1644	42.8181	3128.7673	0.1215	0.0137	0.0003
37	0.1124	0.0024	8.2957	8.8996	0.0210	73.8286	422.9825	47.5281	3508.9317	0.1205	0.0135	0.0003
38	0.1121	0.0021	8.3567	8.9186	0.0190	74.5300	470.5106	52.7562	3931.9142	0.1197	0.0134	0.0003
39	0.1119	0.0019	8.4133	8.9357	0.0171	75.1789	523.2667	58.5593	4402.4248	0.1189	0.0133	0.0002
40	0.1117	0.0017	8.4659	8.9511	0.0154	75.7789	581.8261	65.0009	4925.6915	0.1181	0.0132	0.0002
∞	0.1100	0.0000	9.0909	9.0909	0.0000	82.6446	∞	∞	∞	0.1100	0.0121	0.0000

Table 501.12 Economics Factors for $i = 12\%$

n	A/P	A/F	A/G	P/A	P/F	P/G	F/A	F/P	F/G	G/A	G/P	G/F
1	1.1200	1.0000	0.0000	0.8929	0.8929	0.0000	1.0000	1.1200	0.0000			
2	0.5917	0.4717	0.4717	1.6901	0.7972	0.7972	2.1200	1.2544	1.0000	2.1200	1.2544	1.0000
3	0.4163	0.2963	0.9246	2.4018	0.7118	2.2208	3.3744	1.4049	3.1200	1.0815	0.4503	0.3205
4	0.3292	0.2092	1.3589	3.0373	0.6355	4.1273	4.7793	1.5735	6.4944	0.7359	0.2423	0.1540
5	0.2774	0.1574	1.7746	3.6048	0.5674	6.3970	6.3528	1.7623	11.2737	0.5635	0.1563	0.0887
6	0.2432	0.1232	2.1720	4.1114	0.5066	8.9302	8.1152	1.9738	17.6266	0.4604	0.1120	0.0567
7	0.2191	0.0991	2.5515	4.5638	0.4523	11.6443	10.0890	2.2107	25.7418	0.3919	0.0859	0.0388
8	0.2013	0.0813	2.9131	4.9676	0.4039	14.4714	12.2997	2.4760	35.8308	0.3433	0.0691	0.0279
9	0.1877	0.0677	3.2574	5.3282	0.3606	17.3563	14.7757	2.7731	48.1305	0.3070	0.0576	0.0208
10	0.1770	0.0570	3.5847	5.6502	0.3220	20.2541	17.5487	3.1058	62.9061	0.2790	0.0494	0.0159
11	0.1684	0.0484	3.8953	5.9377	0.2875	23.1288	20.6546	3.4785	80.4549	0.2567	0.0432	0.0124
12	0.1614	0.0414	4.1897	6.1944	0.2567	25.9523	24.1331	3.8960	101.1094	0.2387	0.0385	0.0099
13	0.1557	0.0357	4.4683	6.4235	0.2292	28.7024	28.0291	4.3635	125.2426	0.2238	0.0348	0.0080
14	0.1509	0.0309	4.7317	6.6282	0.2046	31.3624	32.3926	4.8871	153.2717	0.2113	0.0319	0.0065
15	0.1468	0.0268	4.9803	6.8109	0.1827	33.9202	37.2797	5.4736	185.6643	0.2008	0.0295	0.0054
16	0.1434	0.0234	5.2147	6.9740	0.1631	36.3670	42.7533	6.1304	222.9440	0.1918	0.0275	0.0045
17	0.1405	0.0205	5.4353	7.1196	0.1456	38.6973	48.8837	6.8660	265.6973	0.1840	0.0258	0.0038
18	0.1379	0.0179	5.6427	7.2497	0.1300	40.9080	55.7497	7.6900	314.5810	0.1772	0.0244	0.0032
19	0.1358	0.0158	5.8375	7.3658	0.1161	42.9979	63.4397	8.6128	370.3307	0.1713	0.0233	0.0027
20	0.1339	0.0139	6.0202	7.4694	0.1037	44.9676	72.0524	9.6463	433.7704	0.1661	0.0222	0.0023
21	0.1322	0.0122	6.1913	7.5620	0.0926	46.8188	81.6987	10.8038	505.8228	0.1615	0.0214	0.0020
22	0.1308	0.0108	6.3514	7.6446	0.0826	48.5543	92.5026	12.1003	587.5215	0.1574	0.0206	0.0017
23	0.1296	0.0096	6.5010	7.7184	0.0738	50.1776	104.6029	13.5523	680.0241	0.1538	0.0199	0.0015
24	0.1285	0.0085	6.6406	7.7843	0.0659	51.6929	118.1552	15.1786	784.6270	0.1506	0.0193	0.0013
25	0.1275	0.0075	6.7708	7.8431	0.0588	53.1046	133.3339	17.0001	902.7823	0.1477	0.0188	0.0011
26	0.1267	0.0067	6.8921	7.8957	0.0525	54.4177	150.3339	19.0401	1036.1161	0.1451	0.0184	0.0010
27	0.1259	0.0059	7.0049	7.9426	0.0469	55.6369	169.3740	21.3249	1186.4501	0.1428	0.0180	0.0008
28	0.1252	0.0052	7.1098	7.9844	0.0419	56.7674	190.6989	23.8839	1355.8241	0.1407	0.0176	0.0007
29	0.1247	0.0047	7.2071	8.0218	0.0374	57.8141	214.5828	26.7499	1546.5229	0.1388	0.0173	0.0006
30	0.1241	0.0041	7.2974	8.0552	0.0334	58.7821	241.3327	29.9599	1761.1057	0.1370	0.0170	0.0006
31	0.1237	0.0037	7.3811	8.0850	0.0298	59.6761	271.2926	33.5551	2002.4384	0.1355	0.0168	0.0005
32	0.1233	0.0033	7.4586	8.1116	0.0266	60.5010	304.8477	37.5817	2273.7310	0.1341	0.0165	0.0004
33	0.1229	0.0029	7.5302	8.1354	0.0238	61.2612	342.4294	42.0915	2578.5787	0.1328	0.0163	0.0004
34	0.1226	0.0026	7.5965	8.1566	0.0212	61.9612	384.5210	47.1425	2921.0082	0.1316	0.0161	0.0003
35	0.1223	0.0023	7.6577	8.1755	0.0189	62.6052	431.6635	52.7996	3305.5291	0.1306	0.0160	0.0003
36	0.1221	0.0021	7.7141	8.1924	0.0169	63.1970	484.4631	59.1356	3737.1926	0.1296	0.0158	0.0003
37	0.1218	0.0018	7.7661	8.2075	0.0151	63.7406	543.5987	66.2318	4221.6558	0.1288	0.0157	0.0002
38	0.1216	0.0016	7.8141	8.2210	0.0135	64.2394	609.8305	74.1797	4765.2544	0.1280	0.0156	0.0002
39	0.1215	0.0015	7.8582	8.2330	0.0120	64.6967	684.0102	83.0812	5375.0850	0.1273	0.0155	0.0002
40	0.1213	0.0013	7.8988	8.2438	0.0107	65.1159	767.0914	93.0510	6059.0952	0.1266	0.0154	0.0002
∞	0.1200	0.0000	8.3333	8.3333	0.0000	69.4444	∞	∞	∞	0.1200	0.0144	0.0000

The following sections present various types of engineering economics problems.

Present Worth

Example 501.1

What amount must be invested in a fund with a 10% effective annual interest rate to yield $2000.00 at the end of 3 years, assuming that interest is compounded half yearly?

Solution Let the half yearly interest rate be i.

$$\text{Then } (1+i)^2 = 1 + 0.1 = 1.1 \Rightarrow i = 0.04881$$

3 years = 6 half-year periods

$$P/F(i\%, n \text{ periods}) = (1+i)^{-n} = 1.04881^{-6} = 0.7513$$

$$P = (P/F) \times F = 0.7513 \times \$2000 = \$1502.62$$

Note that since the interest rate is 10% effective annual rate, we can also do the following:
From the tables: $(P/F, 3 \text{ years}, 10\%) = 0.7513$

Therefore, $P = \$2000 \times 0.7513 = \$1502.60$.

If the rate had been specified as a 10% nominal annual interest rate, the first approach would have been correct, but the second approach would have yielded a different and incorrect answer.

Example 501.2

Investment A costs $45,000 up-front investment and pays back $58,000 after 3 years. Investment B costs $25,000 up front and pays back $12,000 each year for 3 years. If the MARR is 6%, which investment is superior?

Solution Converting all cash flows to present worth—costs as negative cash flow and revenues as positive.

$$PW_A = -45,000 + 58,000 \times (P/F, 6\%, 3 \text{ years}) = -45,000 + 58,000 \times 0.8396 = 3,696.80$$

$$PW_B = -25,000 + 12,000 \times (P/A, 6\%, 3 \text{ years}) = -25,000 + 12,000 \times 2.6730 = 7,076.00$$

Therefore, investment B is superior.

Principal in a Sinking Fund

Example 501.3

To create a capital fund for future expenditures, a company invests $125,000 annually into a fund accumulating interest rate of 5% (compounded annually). What is the value of the fund immediately after the 5th payment?

Solution   The future worth of the first 5 payments is

$$125,000 \times (F/A, 5\%, 5 \text{ years}) = 125,000 \times 5.5256 = \$690,700$$

Nonannual Compounding

Example 501.4

A bank offers a 6% nominal interest rate per annum, compounded daily. What is the effective annual rate?

Solution   Nominal daily rate $= 6\% \div 365 = 0.0164\% = 1.6438 \times 10^{-4}$
The F/P factor is given by

$$F/P = (1 + i)^n = (1 + 0.00016438)^{365} = 1.0618$$

Therefore, effective annual interest rate $= 6.18\%$.

Capitalized Cost

Example 501.5

What is the capitalized cost of a project that has an initial cost of $50 million and requires annual maintenance cost of $2.5 million? Assume that the effective interest rate is 10%.

Solution   The capitalization cost of an annuity is the principal that generates the annuity to perpetuity. Therefore, the capitalization cost factor is given by (P/A) as $n \to \infty$ and is given by

$$(P/A, n \to \infty) = \frac{1}{i} \Rightarrow P_{cap} = \frac{A}{i}$$

$$\text{Capitalization cost} = \text{initial cost} + \text{annual cost} \times (P/A, 10\%, \infty)$$

$$= \$50M + \$2.5M \times (1/0.1) = \$75M$$

Therefore, the capitalization cost for the project is $75 million.

Equivalent Uniform Annual Cost

When two alternatives are being compared and they have unequal time lines, i.e., unequal useful life, they cannot be compared based on present worth or future worth. They must be compared in terms of annual cost. All costs must then be converted to annuities using the appropriate length of time. This cost is called the equivalent uniform annual cost (EUAC).

Example 501.6

A sidewalk construction project has two design alternatives—brick and concrete. Brick will last 12 years, have an initial cost of $600,000 and annual maintenance cost of $2000. Concrete will last 20 years, have an initial cost of $1,000,000 and annual maintenance cost of $1000. Which option is superior from an economy point of view? The interest rate is 6%.

Solution

$$EUAC_{brick} = 600,000(A/P, 6\%, 12 \text{ years}) + 2000 = 600,000 \times 0.1193 + 2000 = 73,580$$

$$EUAC_{concrete} = 1,000,000(A/P, 6\%, 20 \text{ years}) + 1000 = 1,000,000 \times 0.0872 + 1000 = 88,200$$

Therefore, the annual cost of constructing in brick is 17% cheaper than concrete.

Depreciation

Depreciation is an accounting device that is used to reduce the book value of capital assets. The reduction in book value represents the wear and tear on the equipment. The primary reason for using the device of depreciation is to lower the value of capital assets on the balance sheet in order to lower tax liabilities.

Of the various models for depreciation, the four most commonly used for equipment placed in service before 1981 are (1) straight line method, (2) declining balance method, (3) sum of years digits method, and (4) units of production method. For property placed in service after 1986, the MACRS (modified accelerated cost recovery system) is to be used.

Straight Line Method

Using the straight line method, the annual depreciation claimed on a property is given by

$$D = \frac{C - S}{n} \qquad (501.10)$$

where C = initial value of the property
S = salvage value at the end of design life
n = design life

Declining Balance Method

Under the declining balance method, the annual depreciation claimed on a property is calculated as

Annual depreciation = depreciation rate × book value at the beginning of the year

The most common depreciation rate used is double the straight line rate. For this reason, this technique is also referred to as the double-declining balance method.

For example, if the initial cost of the property is $1000 and the salvage value after a 5-year useful life is $100, then the depreciable value is $900. If the straight line approach were taken, the annual depreciation would be $180 (18%). Let us use a rate of 36% for the declining balance method. The following numbers are calculated. For comparison, the very last column in the table gives the end of yearbook value using straight line depreciation.

Year	Book value at beginning of year	Depreciation	Book value at end of year	Book value (straight line)
1	1000	360	640	820
2	640	230	410	640
3	410	148	262	460
4	262	94	168	280
5	168	60	108	100

Sum of Years Digits Method

Under this method, annual depreciation is determined by multiplying the depreciable cost by a schedule of fractions. If the useful life is n years, the sum of years' digits represents the sum $1 + 2 + 3 + \cdots + n = n(n + 1)/2$. The depreciation fractions are then

$$\frac{n}{\frac{n(n+1)}{2}}, \frac{n-1}{\frac{n(n+1)}{2}}, \frac{n-2}{\frac{n(n+1)}{2}}, \cdots, \frac{1}{\frac{n(n+1)}{2}}$$

For example, if the depreciation period is 5 years, the sum of years' digits is $1 + 2 + 3 + 4 + 5 = 15$. The fractions are therefore 5/15 for the 1st year, 4/15 for the 2nd year, 3/15 for the 3rd year, 2/15 for the 4th year, and 1/15 for the 5th year.

Units of Production Method

Under this method, annual depreciation is computed based on the number of units produced during the year. A depreciation rate per unit of production is first computed by dividing the depreciable cost by the total number of units expected to be produced over the useful life.

Modified Accelerated Cost Recovery System

Under the MACRS, the owner of the property is allowed to depreciate the property by published percentages that are given in Table 501.13.

Table 501.13   MACRS Depreciation Schedules

Recovery year	3-year property	5-year property	7-year property	10-year property	15-year property	20-year property
1	33.33	20.00	14.29	10.00	5.00	3.75
2	44.45	32.00	24.49	18.00	9.50	7.219
3	14.81	19.20	17.49	14.40	8.55	6.677
4	7.41	11.52	12.49	11.52	7.70	6.177
5		11.52	8.93	9.22	6.93	5.713
6		5.76	8.92	7.37	6.23	5.285
7			8.93	6.55	5.90	4.888
8			4.46	6.55	5.90	4.522
9				6.56	5.91	4.462
10				6.55	5.9	4.461
11				3.28	5.91	4.462
12					5.9	4.461
13					5.91	4.462
14					5.9	4.461
15					5.91	4.462
16					2.95	4.461
17						4.462
18						4.461
19						4.462
20						4.461

Example 501.7

An asset has a purchase price of $25,000. It has a 10-year useful life and a salvage value of $8000. Find the book value at the end of year 3 using straight line depreciation.

Solution

$$\text{Annual depreciation } D = (25{,}000 - 8000)/10 = \$1700 \text{ per year}$$

$$\text{Book value} = 25{,}000 - 3 \times 1700 = \$19{,}900$$

Example 501.8

What would be the book value at the end of 3 years for the above-named asset if the MACRS (modified accelerated cost recovery system) method of depreciation was used?

Solution   Depreciation percentages in years 1, 2, and 3 = 10%, 18%, and 14.4%, respectively

$$\text{Book value} = 25{,}000 \times [1 - 0.1 - 0.18 - 0.144] = 25{,}000 \times 0.576 = \$14{,}400$$

Tax Issues

Example 501.9

A corporation pays 48% in income tax on profits. It purchases an asset that will produce a revenue stream of $5000 for 10 years. The cost of the asset is $30,000. Annual expenses are $1500. Salvage value (at the end of year 10) of the equipment is $1200. Ignore depreciation. Use an MARR of 8%. What is the after-tax present worth of the asset?

Solution

Annual revenue = $5000

Annual expenses = $1500

Annual profit = $3500

Annual taxes = $0.48 \times $3500 = 1680

NET annual profit = $1820

$$\text{Present worth } = -30{,}000 + 1820(P/A, 8\%, 10 \text{ years}) + 1200(P/F, 8\%, 10 \text{ years})$$
$$= -30{,}000 + 1820 \times 6.7101 + 1200 \times 0.4632 = -17{,}231.78$$

Bonds

Example 501.10

What is the maximum an investor should pay for a 25-year bond with a $20,000 face value and an 8% coupon rate (with interest paid semiannually)? The bond will be held to maturity. The effective annual rate for comparison is 10%.

Solution If the interest is paid semiannually, the payment is 4% of $20,000 = 800. The effective MARR = 10%. This must be converted to an equivalent 6-month rate using

$$(1+i)^2 - 1 = 0.10 \Rightarrow i = 0.0488$$

Bond is held to maturity, which is 25 years, or 50 six-month periods. The present worth of 50 $800 payments at an interest rate 0.0488

$$\$800 \times \left(\frac{P}{A}, 4.88\%, 50\right) = \$800 \times \frac{1.0488^{50} - 1}{0.0488(1.0488^{50})} = \$800 \times 18.6 = 14{,}879.76$$

The present worth of a future sum of $20,000 is

$$\$20{,}000 \times \left(\frac{P}{F}, 4.88\%, 50\right) = \$20{,}000 \times \frac{1}{1.0488^{50}} = \$20{,}000 \times 0.0923 = 1846.70$$

The total present worth of all sources of income from the bond, if held to maturity, is $16,726.46.

Break-Even Analysis

Example 501.11

A company has a choice between two processes to manufacture a product. The data for both options (A and B) are given below. How many units must be produced annually to justify process A? Assume the MARR = 7%.

	Process A	Process B
Initial cost ($)	50,000	28,000
Fixed annual cost ($)	2,500	1,300
Annual production cost ($/unit)	20	30
Life (years)	10	8
Salvage value of equipment ($)	4,000	0

Solution   Let annual production be x units.

Since length of timeline is different for options A and B, use EUAC (equivalent uniform annual cost).

EUAC of all costs (option A) in thousands of dollars

$$50(A/P, 7\%, 10 \text{ years}) + 2.5 + 0.020x - 4(A/F, 7\%, 10 \text{ years}) = 50(0.1424)$$

$$+ 2.5 + 0.020x - 4(0.0724) = 9.3304 + 0.02x$$

EUAC of all costs (option B) in thousands of dollars

$$28(A/P, 7\%, 8 \text{ years}) + 1.3 + 0.030x = 28(0.1675) + 1.3 + 0.030x = 5.99 + 0.03x$$

To justify option A, we must have

$$9.3304 + 0.02x \le 5.99 + 0.03x \Rightarrow x \ge 334.04$$

At least 335 units must be produced annually in order to justify option A.

Benefit-Cost Analysis

A project is considered beneficial if benefits exceed costs. This may be expressed mathematically as

$$\frac{B}{C} > 1 \quad \text{or} \quad B - C > 0 \tag{501.11}$$

If the first approach is taken, all benefits and costs should be converted to equivalent quantities (i.e., all expressed as present worth, or all converted to annuities). For the second approach, it is typical to express the differential as the net present worth of all benefits and costs.

Example 501.12

For an existing bridge, the costs related to performing repairs are shown in the table below. All costs are in thousands of dollars. Calculate the benefit-cost ratio of performing repairs. Use MARR = 8%.

	Existing	Repair
Initial cost	—	120
Salvage value	50	180
Maintenance costs:		
Years 1–10	12	5
Years 11–20	18	7
Design life	20 years	

Solution   The added cost of repairing the bridge is $120,000. This is a present value P. The offset to the cost is the extra $180 − $50 = $130 in salvage. This can be considered a negative cost. This is a future value F.

By performing the repairs, the maintenance costs are reduced by $7000 throughout the 20 years plus an additional savings of $4000 in years 11 to 20. The first is an annuity A. The second is a shifted annuity A. We can visualize the shifted annuity (1) as the difference between two annuities as shown below: (2) − (3). Annuities 2 and 3 are easier to deal with because their origin is at $t = 0$, and therefore the tables can be used.

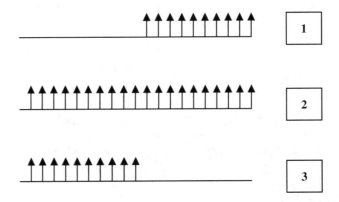

Converting all quantities to present worth (all sums in thousands of dollars)

Cost $= 120 - 130 \times (P/F, 8\%, 20 \text{ years}) = 120 - 130 \times 0.2145 = \$92.115$

Benefit $= 7 \times (P/A, 8\%, 20 \text{ years}) + 4 \times [(P/A, 8\%, 20 \text{ years}) - (P/A, 8\%, 10 \text{ years})] =$
$7 \times 9.8181 + 4 \times (9.8181 - 6.7101) = \$81.159$

Benefit cost ratio $= 0.88$

Life-Cycle Cost

Life-cycle cost of a capital asset is the lifetime cost to purchase, install, operate, maintain, and dispose of it. The capital expenditure, usually consisting of the purchase price and cost of installation, is often a fairly small fraction of the life-cycle cost. In the example below, annual energy and maintenance costs have been included in the life-cycle cost calculations.

Example 501.13

A pump is used to transport a fluid from a storage tank to a pressurized tank. A flow control valve (FCV) regulates the rate of flow at 400 gal/min. The plant has experienced a recurring problem involving the FCV, which tends to fail approximately once a year due to cavitation. The cost of repair for each instance of failure is $4000.

The following alternatives have been suggested:

Option 1 – Leave system as is, and incur the annual expense of the repair to the FCV. Annual energy cost for this option = $11,000. Routine annual maintenance = $800

Option 2 – Replace the FCV with one that can resist cavitation. The cost of a new valve is $15,000. For this option, annual energy cost = $9500. Routine annual maintenance = $1500

Life time (for each option) = 8 years

Interest rate = 8%

Inflation cost (affecting fuel prices) = 4%

Compare the life-cycle cost for the two alternatives.

Solution
OPTION 1

Actual energy costs (including 4% inflation) are: $11,000, $11,440, $11,897.60, $12,373.50, $12,868.44, $13,383.18, $13,918.51, $14,475.25

Adjusted to current dollars (present worth) = $11,000, $10,592.59, $10,200.27, $9,822.49, $9,458.69, $9108.37, $8771.02, $8446.17. TOTAL present worth of fuel costs = $77,399.60

Annual expense = cost of new FCV + routine annual maintenance = $4800

For $i = 8\%$, n = 8 years, P/A = 5.7466

Present worth of annual expenses (excluding fuel) = $4800 \times 5.7466 = $27,583.68

Life-cycle cost for option 1 = $77,399.60 + $27,583.68 = $104,983.28

OPTION 2

Using scaling (9500/11000) adjusted to current dollars (present worth) of fuel costs = $77,399.60 × 9.5/11 = $66,845.11

Annual expense = routine annual maintenance = $1500

Present worth of annual expenses (excluding fuel) = $1500 × 5.7466 = $8619.90

Capital expenditure = cost of new FCV = $15,000

Life-cycle cost for option 2 = $66,845.11 + $8,619.90 + $15,000.00 = $90,465.01

Therefore, option 2 (replacing the FCV) reduces the life-cycle cost by $14,518.27.

Probability and Statistics

Probabilistic Basis for Design

When designing facilities for probabilistic loads, such as those caused by wind, earthquake, and flood, it is common practice to define a design event, such as a *25-year flood* or *500-year earthquake*. The time is the *return period* of the design event. For example, a 25-year flood is defined to be one which has a return period of 25 years, which implies the magnitude has an *annual probability* of occurrence of $1 \div 25 = 0.04\%$ or 4%.

Let the return period or recurrence interval $= N$ years

Then, the annual probability of occurrence $= 1/N$

The annual probability of nonoccurrence $= 1 - \dfrac{1}{N}$

The probability of nonoccurrence over a sequence of M years $= \left(1 - \dfrac{1}{N}\right)^M$

Probability of the event occurring *at least once* over a period of

M years $= 1 - \left(1 - \dfrac{1}{N}\right)^M$

Based on these results, we may construct Table 502.1.

Example 502.1

A dam has a design life of 25 years. What is the probability of the dam being overtopped by a 20-year flood level during its design life?

Solution A 20-year flood level is defined as the high-water elevation that has a $1/20 = 5\%$ annual probability. To use Table 502.1, the input variables are

$M = 25$

$N = 20$

The corresponding probability from Table 502.1 is 72.3%.

Table 502.1 Probability of Occurrence During a Design Period

Return period N (years)	Duration M (years) over which probability of occurrence is calculated							
	1	**2**	**5**	**10**	**20**	**25**	**50**	**100**
2	50.0%	75.0%	96.9%	99.9%	100.0%	100.0%	100.0%	100.0%
5	20.0%	36.0%	67.2%	89.3%	98.8%	99.6%	100.0%	100.0%
10	10.0%	19.0%	41.0%	65.1%	87.8%	92.8%	99.5%	100.0%
20	5.0%	9.8%	22.6%	40.1%	64.2%	72.3%	92.3%	99.4%
25	4.0%	7.8%	18.5%	33.5%	55.8%	64.0%	87.0%	98.3%
50	2.0%	4.0%	9.6%	18.3%	33.2%	39.7%	63.6%	86.7%
100	1.0%	2.0%	4.9%	9.6%	18.2%	22.2%	39.5%	63.4%

If the given M and N values are not found in the table, this probability can be calculated from

$$P = 1 - \left(1 - \frac{1}{20}\right)^{25} = 0.7226$$

Design Recurrence Interval

Also, if a certain maximum probability of failure is specified (note that system reliability = 1 − probability of failure), then one may calculate the appropriate return period (N_o) over a design life of M years. This calculation is based on the formula

$$N_o = \frac{1}{1 - (1 - P_f)^{1/M}} \tag{502.1}$$

where N_o = required return period for the design event
M = design life (or duration being considered)
P_f = probability of failure

These results are summarized in Table 502.2. For example, if a maximum failure probability of 10% ($P_f = 0.1$) is acceptable over a design life of 50 years ($M = 50$), then the return period of the design event should be 475 years.

$$N_o = \frac{1}{1 - (1 - 0.1)^{1/50}} = 475.06$$

Table 502.2 Design Return Period for Given Failure Probability

Probability of failure	Duration M (years) over which probability of occurrence is calculated							
	1	2	5	10	20	25	50	100
0.0001	10000	19999.5	49998	99995.5	199990.5	249988	499975.5	999950.5
0.001	1000	1999.5	4998	9995.5	19990.5	24988	49975.5	99950.5
0.01	100	199.5	497.99	995.5	1990.5	2487.98	4975.46	9950.4
0.1	10	19.5	47.96	95.4	190.3	237.8	475.06	949.6
0.2	5	9.5	22.9	45.3	90.1	112.5	224.6	448.6
0.25	4	7.5	17.9	35.3	70.0	87.4	174.3	348.11

System Reliability

The reliability (R) of a system is defined as

$$R = 1 - P_f \tag{502.2}$$

where P_f = probability of failure.

Series System

For a system that can be modeled as a cascade of n processes in series, as shown in Fig. 502.1, the system reliability is calculated as

$$R_{system} = R_1 R_2 \ldots R_n \tag{502.3}$$

The failure probabilities of individual processes are designated P_{f1}, P_{f2}, and so on. For such a system, the entire system fails if any one of the component processes fails.

Example 502.2

A wastewater treatment plant can be modeled as a sequence of three processes, each accomplishing various levels of percent removal of suspended solids. What is the overall suspended solids removal efficiency of the plant?

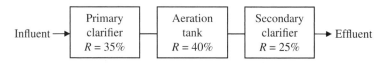

Solution It may be tempting to say that the total removal fraction is $35 + 40 + 25 = 100\%$, but it must be remembered that the removal fraction effected by the second process is on the

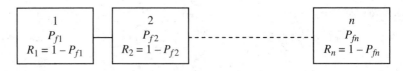

Figure 502.1 Reliability of a series system.

transmitted fraction from the first, and so on. The overall transmission (throughput) of suspended solids is therefore

$$T = (1 - 0.35) \times (1 - 0.40) \times (1 - 0.25) = 0.293$$

Therefore, the overall removal efficiency $= 1 - 0.293 = 0.706$ (70%).

Parallel System

For a system that can be modeled as a cascade of n processes in parallel, as shown in Fig. 502.2, the system reliability is calculated as

$$R = 1 - (1 - R_1)(1 - R_2) \cdots (1 - R_n) \tag{502.4}$$

Such a system has several redundant paths through it and therefore fails only if all component processes fail.

Figure 502.2
Reliability of a
parallel system.

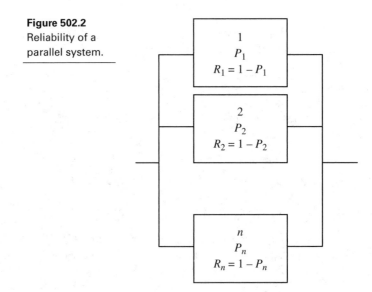

Normal Distribution

The outcome of many independent experiments (typically greater than 30) on a random variable usually follows the normal or Gaussian probability distribution. The Gaussian probability density function (bell-shaped curve) is given by

$$f(x) = \frac{1}{\sqrt{2\pi}\sigma}\exp\left[-\frac{1}{2}\left(\frac{x-\mu}{\sigma}\right)^2\right] = \frac{1}{\sqrt{2\pi}\sigma}\exp\left[-\frac{1}{2}Z^2\right] \qquad (502.5)$$

where Z is the standard normal variable, which results from the shifting and scaling of the random variable X, as given below

$$Z = \frac{X-\mu}{\sigma} \qquad (502.6)$$

In Eq. (502.6), μ is the mean and σ is the standard deviation of the normal variable X. The standard normal variable Z has mean 0 and standard deviation 1. For any random variable, which is expected to follow the normal distribution, the first step is to transform the relevant values of the random variable to a corresponding value of Z. Note that the mean value of the X variable will transform to a Z value of 0.0. Table 502.3 gives the area under the Gaussian probability density function from $-\infty$ to a particular value of Z. Because of the symmetry of the distribution about $Z = 0.0$, only the positive values of Z are listed.

Note that, due to the symmetry of the Gaussian curve, the area corresponding to a negative Z value (i.e., $X < \mu$) is given by (see Fig. 502.3)

$$P(z < -Z) = 1 - P(z < Z) \qquad (502.7)$$

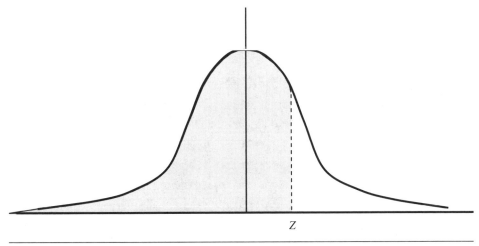

Figure 502.3 Cumulative probability for a normal distribution.

Table 502.3 Cumulative Probability for a Normal Distribution $P(z \leq Z)$

Z	0.00	0.01	0.02	0.03	0.04	0.05	0.06	0.07	0.08	0.09
0.0	0.5000	0.5040	0.5080	0.5120	0.5160	0.5199	0.5239	0.5279	0.5319	0.5359
0.1	0.5398	0.5438	0.5478	0.5517	0.5557	0.5596	0.5636	0.5675	0.5714	0.5753
0.2	0.5793	0.5832	0.5871	0.5910	0.5948	0.5987	0.6026	0.6064	0.6103	0.6141
0.3	0.6179	0.6217	0.6255	0.6293	0.6331	0.6368	0.6406	0.6443	0.6480	0.6517
0.4	0.6554	0.6591	0.6628	0.6664	0.6700	0.6736	0.6772	0.6808	0.6844	0.6879
0.5	0.6915	0.6950	0.6985	0.7019	0.7054	0.7088	0.7123	0.7157	0.7190	0.7224
0.6	0.7257	0.7291	0.7324	0.7357	0.7389	0.7422	0.7454	0.7486	0.7517	0.7549
0.7	0.7580	0.7611	0.7642	0.7673	0.7704	0.7734	0.7764	0.7794	0.7823	0.7852
0.8	0.7881	0.7910	0.7939	0.7967	0.7995	0.8023	0.8051	0.8078	0.8106	0.8133
0.9	0.8159	0.8186	0.8212	0.8238	0.8264	0.8289	0.8315	0.8340	0.8365	0.8389
1.0	0.8413	0.8438	0.8461	0.8485	0.8508	0.8531	0.8554	0.8577	0.8599	0.8621
1.1	0.8643	0.8665	0.8686	0.8708	0.8729	0.8749	0.8770	0.8790	0.8810	0.8830
1.2	0.8849	0.8869	0.8888	0.8907	0.8925	0.8944	0.8962	0.8980	0.8997	0.9015
1.3	0.9032	0.9049	0.9066	0.9082	0.9099	0.9115	0.9131	0.9147	0.9162	0.9177
1.4	0.9192	0.9207	0.9222	0.9236	0.9251	0.9265	0.9279	0.9292	0.9306	0.9319
1.5	0.9332	0.9345	0.9357	0.9370	0.9382	0.9394	0.9406	0.9418	0.9429	0.9441
1.6	0.9452	0.9463	0.9474	0.9484	0.9495	0.9505	0.9515	0.9525	0.9535	0.9545
1.7	0.9554	0.9564	0.9573	0.9582	0.9591	0.9599	0.9608	0.9616	0.9625	0.9633
1.8	0.9641	0.9649	0.9656	0.9664	0.9671	0.9678	0.9686	0.9693	0.9699	0.9706
1.9	0.9713	0.9719	0.9726	0.9732	0.9738	0.9744	0.9750	0.9756	0.9761	0.9767
2.0	0.9772	0.9778	0.9783	0.9788	0.9793	0.9798	0.9803	0.9808	0.9812	0.9817
2.1	0.9821	0.9826	0.9830	0.9834	0.9838	0.9842	0.9846	0.9850	0.9854	0.9857
2.2	0.9861	0.9864	0.9868	0.9871	0.9875	0.9878	0.9881	0.9884	0.9887	0.9890
2.3	0.9893	0.9896	0.9898	0.9901	0.9904	0.9906	0.9909	0.9911	0.9913	0.9916
2.4	0.9918	0.9920	0.9922	0.9925	0.9927	0.9929	0.9931	0.9932	0.9934	0.9936
2.5	0.9938	0.9940	0.9941	0.9943	0.9945	0.9946	0.9948	0.9949	0.9951	0.9952
2.6	0.9953	0.9955	0.9956	0.9957	0.9959	0.9960	0.9961	0.9962	0.9963	0.9964
2.7	0.9965	0.9966	0.9967	0.9968	0.9969	0.9970	0.9971	0.9972	0.9973	0.9974
2.8	0.9974	0.9975	0.9976	0.9977	0.9977	0.9978	0.9979	0.9979	0.9980	0.9981
2.9	0.9981	0.9982	0.9982	0.9983	0.9984	0.9984	0.9985	0.9985	0.9986	0.9986
3.0	0.9987	0.9987	0.9987	0.9988	0.9988	0.9989	0.9989	0.9989	0.9990	0.9990

Example 502.3

A long record ($N > 30$) of damage data to a small jurisdiction from coastal flooding due to hurricanes is assumed to follow the Gaussian (normal) distribution. The mean cost (thousands of dollars) is 680 and the standard deviation is 45.

(a) What is the probability of a storm causing more than $750,000 of damage?

(b) What is the probability that the damage is between $600,000 and $700,000?

Solution

(a) The specific values of X need to be converted to their corresponding Z values.

$$X_1 = 750 \Rightarrow Z_1 = \frac{750 - 680}{45} = +1.556$$

From the table, $P(Z \leq 1.556) = 0.9401$ (interpolating between 0.9394 and 0.9406). Therefore, $P(Z > 1.556) = 1.0 - 0.9401 = 0.0599$ (5.99%).

(b) Again, transforming the specific values of X to their corresponding Z values, the probability that X is between 600,000 and 700,000 is given by $P(-1.778 \leq Z \leq +0.444)$. However, since Table 502.3 does not contain negative values of Z, one needs to use the symmetry properties of the normal distribution.

$$P(-1.778 \leq Z \leq +0.444) = P(0.444) - P(-1.778) = P(0.444) - [1 - P(+1.778)]$$
$$= P(+0.444) + P(+1.778) - 1 = 0.6714 + 0.9623 - 1 = 0.6337$$

Therefore, $P(-1.778 \leq Z \leq +0.444) = 0.6337$ (63.37%)

Example 502.4

The cost function (thousands of dollars) for a construction project has been established empirically as

$$C = 50X_1 + 30X_2 - 20X_3$$

where X_1, X_2, and X_3 are relevant random variables (e.g., number of labor hours, number of material units consumed, number of embedded safety features, and the like). These three random variables have the following parameters:

X_1: normally distributed – mean = 250, standard deviation = 12

X_2: normally distributed – mean = 200, standard deviation = 10

X_3: normally distributed – mean = 60, standard deviation = 5

What is the probability that the overall project cost exceeds $18.5 million?

Solution Since the cost C is a linear combination of three normally distributed random variables, it will be normally distributed as well. The parameters of the distribution of C are

$$\mu_C = a_1\mu_{X_1} + a_2\mu_{X_2} + a_3\mu_{X_3} = 50 \times 250 + 30 \times 200 - 20 \times 60 = 17{,}300$$

Mean cost = 17.3 million dollars

$$\sigma_C^2 = a_1^2\sigma_{X_1}^2 + a_2^2\sigma_{X_2}^2 + a_3^2\sigma_{X_3}^2 = 50^2 \times 12^2 + 30^2 \times 10^2 + 20^2 \times 5^2 = 460{,}000$$

Standard deviation of cost = 678 thousand dollars
The probability that $C > 18{,}500$ is given by calculating the standard normal variable

$$Z = \frac{18{,}500 - 17{,}300}{678} = 1.77$$

The cumulative area under the standard normal curve to $Z = 1.77$ is 0.9616 (96.16%). Therefore, the probability of cost exceeding \$18.5 million is 3.84%.

Student's *t*-Distribution

For small samples, the normal distribution is not quite appropriate. In that case, the *t*-distribution should be used. Table 502.4 shows the *t*-value corresponding to α, the area in the upper tail of the *t*-distribution density function and the number of degrees of freedom $n = N - 1$ (see Fig. 502.4).

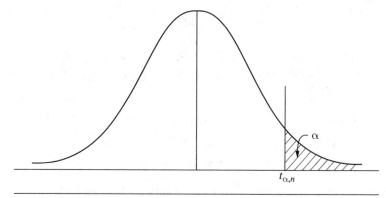

Figure 502.4 Area in the upper tail of the student's *t*-distribution.

Table 502.4 Values of $t_{n,\alpha}$ for the Student's t-Distribution

DOF	$\alpha = 0.25$	$\alpha = 0.20$	$\alpha = 0.15$	$\alpha = 0.10$	$\alpha = 0.05$	$\alpha = 0.025$	$\alpha = 0.01$	$\alpha = 0.005$
1	1.000	1.376	1.963	3.078	6.314	12.706	31.821	63.657
2	0.816	1.061	1.386	1.886	2.920	4.303	6.965	9.925
3	0.765	0.978	1.350	1.638	2.353	3.182	4.541	5.841
4	0.741	0.941	1.190	1.533	2.132	2.776	3.747	4.604
5	0.727	0.920	1.156	1.476	2.015	2.571	3.365	4.032
6	0.718	0.906	1.134	1.440	1.943	2.447	3.143	3.707
7	0.711	0.896	1.119	1.415	1.895	2.365	2.998	3.499
8	0.706	0.889	1.108	1.397	1.860	2.306	2.896	3.355
9	0.703	0.883	1.100	1.383	1.833	2.262	2.821	3.250
10	0.700	0.879	1.093	1.372	1.812	2.228	2.764	3.169
11	0.697	0.876	1.088	1.363	1.796	2.201	2.718	3.106
12	0.695	0.873	1.083	1.356	1.782	2.179	2.681	3.055
13	0.694	0.870	1.079	1.350	1.771	2.160	2.650	3.012
14	0.692	0.868	1.076	1.345	1.761	2.145	2.624	2.977
15	0.691	0.866	1.074	1.341	1.753	2.131	2.602	2.947
16	0.690	0.865	1.071	1.337	1.746	2.120	2.583	2.921
17	0.689	0.863	1.069	1.333	1.740	2.110	2.567	2.898
18	0.688	0.862	1.067	1.330	1.734	2.101	2.552	2.878
19	0.688	0.861	1.066	1.328	1.729	2.093	2.539	2.861
20	0.687	0.860	1.064	1.325	1.725	2.086	2.528	2.845
21	0.686	0.859	1.063	1.323	1.721	2.080	2.518	2.831
22	0.686	0.858	1.061	1.321	1.717	2.074	2.508	2.819
23	0.685	0.858	1.060	1.319	1.714	2.069	2.500	2.807
24	0.685	0.857	1.050	1.318	1.711	2.064	2.492	2.707
25	0.684	0.856	1.058	1.316	1.708	2.060	2.485	2.787
26	0.684	0.856	1.058	1.315	1.706	2.056	2.479	2.779
27	0.684	0.855	1.057	1.314	1.703	2.052	2.473	2.771
28	0.683	0.855	1.056	1.313	1.701	2.048	2.467	2.763
29	0.683	0.854	1.055	1.311	1.699	2.045	2.462	2.756
30	0.683	0.854	1.055	1.310	1.697	2.042	2.457	2.750
∞	0.674	0.842	1.036	1.282	1.645	1.960	2.326	2.576

Example 502.5

The following values were obtained for the failure load of eight test specimens. Determine the 95% confidence interval for the expected failure load.

Sample no.	Failure load (lb)	Mean	Deviation d	d^2
1	20,138		−1,250.75	1,564,375.6
2	24,123		2,734.25	7,476,123.1
3	19,045		−2,343.75	5,493,164.1
4	23,781	21,388.75	2,392.25	5,722,860.1
5	22,904		1,515.25	2,295,982.6
6	19,905		−1,483.75	2,201,514.1
7	20,004		−1,384.75	1,917,532.6
8	21,210		−178.75	31,951.6
				26,703,503.8

Solution

The sample size is $N = 8$

The (arithmetic) sample mean is $\bar{X} = \dfrac{\sum_{i=1}^{N} x_i}{N} = \dfrac{171,110}{8} = 21,388.75$

The sum of squares of deviations about the mean $= 26,703,503.8$

Sample variance is $s^2 = \dfrac{\sum_{i=1}^{N} (x_i - \bar{X})^2}{N-1} = \dfrac{26,703,503.8}{7} = 3,814,786.1$

Sample standard deviation is $s = \sqrt{3,814,786.1} = 1953.15$

95% confidence interval (two-sided) implies $\alpha = (1 - 0.95)/2 = 0.025$

From the t-distribution table (chosen because we are dealing with a small sample, $N < 30$), degrees of freedom $= N - 1 = 7$ and $\alpha = 0.025$, the t-value is 2.365

Therefore, the 95% confidence interval is $\bar{X} \pm \dfrac{ts}{\sqrt{n}} = 21,388.75 \pm \dfrac{2.365 \times 1953.15}{\sqrt{8}}$

The 95% confidence interval is therefore 19,755.62 to 23,021.88.

Binomial Distribution

The binomial distribution may be applied whenever a probabilistic event has only two possible mutually exclusive outcomes, each with a well-defined probability. For example, if the occurrence of a specific design flood has an annual probability P, then the annual probability of "nonoccurrence" is $(1 - P)$. The probability of x occurrences in N trials is given by

$$P(x,N) = C(N,x)P^x(1 - P)^{N-x} \tag{502.8}$$

where $C(N,x)$ is the combination function given by

$$C(N,x) = \frac{N!}{x!(N-x)!}$$ (502.9)

Example 502.6

What is the probability of three 20-year floods occurring during the next 10 years?

Solution   Since the 20-year flood has an annual probability of 0.05 (by definition), we can designate $P = 0.05$ and $(1 - P) = 0.95$. The probability of three floods over the next 10 years, is therefore

$$P(3,10) = C(10,3)(0.05)^3(0.95)^7 = 120 \times (0.05)^3(0.95)^7 = 0.01$$

Thus, there is a 1% probability of three 20-year floods occurring during the next 10 years.

Quality Control

There are various statistically sound methods which can be used by a quality control specialist. An acceptable quality limit (AQL) must be established before referring to these methods. For example, AQL $= 0.1\%$ means that the probability of defective samples is 0.1%.

AQL—Acceptable Quality Limit

In every batch of production—whether it be a machine part, or a garment or a batch of concrete poured for a construction project, there will be defects. The buyer has to set a limit on the proportion of defects that is acceptable. This is the AQL (acceptable quality limit). For example, the client may say "I want no more than 2% of all concrete cylinders prepared from a batch of concrete to test below the target compressive strength." Thus, the AQL for this sampling protocol is 2%. The client can then set a minimum sample size and the maximum number of failures in that sample that is acceptable. In other words, if the number of failures is less than this number, the entire batch is deemed acceptable.

The AQL tables are statistical tools used in the QA/QC process to answer the following questions:

- How many samples should be inspected?
- What is the threshold between "acceptable" and "refused"?

Three types of defects are usually recognized. The AQL values associated with each type of defect can vary.

Critical defects (AQL $= 0\%$). These are defects that are totally unacceptable, because these defects mean safety violations and serious adverse impacts.

Major defects (AQL $= 2.5\%$). Products having these defects do not necessarily pose safety violations but are serious enough that they would be unacceptable to the end user.

Table 502.5 Sample-Size Code Letters by Lot Size

Lot size	General inspection level		
	I	II	III
2 to 8	A	A	B
9 to 15	A	B	C
16 to 25	B	C	D
26 to 50	C	D	E
51 to 90	C	E	F
91 to 150	D	F	G
151 to 280	E	G	H
281 to 500	F	H	J
501 to 1200	G	J	K
1201 to 3200	H	K	L
3201 to 10,000	J	L	M
10,001 to 35,000	K	M	N
35,001 to 150,000	L	N	P
150,001 to 500,000	M	P	Q
500,001 and over	N	Q	R

Minor defects (AQL = 4%). Products having these defects would be violating some specifications, but the product is likely to be acceptable to the end user.

Lot Size and Inspection Level

The lot size is the total order quantity. There are three inspection levels—level I (reduced severity), level II (normal severity), or level III (increased severity). Based on the lot size and the inspection level, an ISO table yields a sample-size code letter (ranging from A to R). See Table 502.5.

A second table (selected according to level of severity) yields the required sample size and (based on the AQL) an acceptable number of defects. Table 502.6 is for level of severity II.

For a hypothetical inspection of a production with 4000 units, with an Inspection Level of II, Table 502.5 indicates a code letter level of "L."

Referring to row L of Table 502.6, a sample size of 200 should be used. If the AQL is 2.5%, no more than 10 units may fail for the entire batch to be acceptable.

Example 502.7

Consider the example of testing of a daily batch of concrete cylinders at a construction site. If the AQL = 3%, what is the probability of having zero defective (strength below target strength) samples in a batch of 5 cylinders?

Table 502.6 Sample Size for Various AQL Levels

Code letter	Sample size	AQL 1.0%	1.5%	2.5%	4.0%	6.5%
A	2	≤ 0	≤ 0	≤ 0	≤ 0	≤ 0
B	3	≤ 0	≤ 0	≤ 0	≤ 0	≤ 0
C	5	≤ 0	≤ 0	≤ 0	≤ 0	≤ 1
D	8	≤ 0	≤ 0	≤ 0	≤ 1	≤ 1
E	13	≤ 0	≤ 0	≤ 1	≤ 1	≤ 2
F	20	≤ 0	≤ 1	≤ 1	≤ 2	≤ 3
G	32	≤ 1	≤ 1	≤ 2	≤ 3	≤ 5
H	50	≤ 1	≤ 2	≤ 3	≤ 5	≤ 7
J	80	≤ 2	≤ 3	≤ 5	≤ 7	≤ 10
K	125	≤ 3	≤ 5	≤ 7	≤ 10	≤ 14
L	200	≤ 5	≤ 7	≤ 10	≤ 14	≤ 21
M	315	≤ 7	≤ 10	≤ 14	≤ 21	≤ 21
N	500	≤ 10	≤ 14	≤ 21	≤ 21	≤ 21
P	800	≤ 14	≤ 21	≤ 21	≤ 21	≤ 21
Q	1250	≤ 21	≤ 21	≤ 21	≤ 21	≤ 21
R	2000	≤ 21	≤ 21	≤ 21	≤ 21	≤ 21

Solution

Probability of one cylinder being defective $= 0.03$

Probability of one cylinder being acceptable $= 0.97$

Probability of zero defects in a batch of 5 cylinders $= 0.97^5 = 0.859$

Thus, given an AQL of 3%, the probability of having a defect-free batch of five samples is about 86%. Conversely, having a defect-free batch also only implies a certain limited confidence about the quality of the entire population.

Minimum Sample Size

Conversely, the following question may be asked: What is the minimum sample size to test defect free in order to have 90% confidence in the quality of the entire lot? This question is addressed in Example 502.8.

Given the inherent probability of an item being defective (AQL) $= p$, the probability of getting n defects in a sample of size N is given by

$$C(N,n)p^n(1-p)^{N-n}$$

where $C(N,n)$ is the combination function, given by

$$C(N,n) = \frac{N!}{n!(N-n)!}$$

Therefore, in order to have a certain level of confidence in a sample, the overall probability of n defects in N items is given by

$$C(N,n)p^n(1-p)^{N-n} = 1 - \text{ confidence level}$$

For zero defects ($n = 0$), this becomes

$$(1-p)^N = 1 - \text{ confidence level}$$

Therefore, the smallest defect-free sample size (N) that yields a confidence level C is

$$N = \frac{\log(1-C)}{\log(1-p)} \qquad (502.10)$$

Example 502.8

If an inspector at a construction site needs to have 90% confidence in the quality of the concrete used, what is the minimum number of randomly selected concrete cylinders that must all test satisfactory if the acceptable quality level is 8%?

Solution

The probability of a single failure (deficient cylinder) $= 0.08$

The probability of no failures in N cylinders $= (1 - 0.08)^N$

For 90% confidence in results, this must be equal to $1 - 0.90 = 0.1$

Therefore, $0.92^N = 0.1 \Rightarrow N = \dfrac{\log 0.1}{\log 0.92} = 27.6$

Therefore, given the underlying quality level of 8%, it requires a sample of at least 28 cylinders to test defect free in order for the engineer to have 90% confidence in the quality of the concrete.

Linear Regression

Given a data set containing x, y paired data, linear regression analysis uses the concept of minimum least square error to estimate the parameters of the "line of best fit." The equation of this line of best fit is designated

$$y = \hat{a} + \hat{b}x \qquad (502.11)$$

The two parameters slope and y-intercept of this line of best fit are given by

$$\hat{b} = \frac{S_{xy}}{S_{xx}} = \frac{\sum x_i y_i - 1/N \sum x_i \sum y_i}{\sum x_i^2 - 1/N \left(\sum x_i\right)^2} = \frac{1/N \sum x_i y_i - \left(\overline{X}\right)\left(\overline{Y}\right)}{1/N \sum x_i^2 - \left(\overline{X}\right)^2} \qquad (502.12)$$

$$\hat{a} = \overline{Y} - \hat{b}\overline{X} \qquad (502.13)$$

where the computed arithmetic average of the series x and y are

$$\overline{X} = \frac{\sum x_i}{N} \qquad \overline{Y} = \frac{\sum y_i}{N}$$

The standard error (S_e) of this estimate is given by

$$S_e^2 = \frac{S_{xx}S_{yy} - S_{xy}^2}{S_{xx}(n-2)} \qquad (502.14)$$

The correlation coefficient (Pearson) is given by

$$r = \frac{\sum x_i y_i - 1/N \sum x_i \sum y_i}{\sqrt{\sum x_i^2 - 1/N \left(\sum x_i\right)^2} \sqrt{\sum y_i^2 - 1/N \left(\sum y_i\right)^2}} \qquad (502.15)$$

Example 502.9

Given the following data on two random variables X and Y, determine the equation of the line of best fit and the correlation coefficient.

Solution

X	Y	XY	X²	Y²
5	10	50	25	100
12	32	384	144	1024
23	45	1035	529	2025
30	56	1680	900	3136
32	65	2080	1024	4225
$\Sigma X = 102$	$\Sigma Y = 208$	$\Sigma XY = 5229$	$\Sigma X^2 = 2622$	$\Sigma Y^2 = 10{,}510$

$$\hat{b} = \frac{1/N \sum x_i y_i - \overline{X}\,\overline{Y}}{1/N \sum x_i^2 - \left(\overline{X}\right)^2} = \frac{\sum x_i y_i - N\overline{X}\,\overline{Y}}{\sum x_i^2 - \left(\overline{X}\right)^2} = \frac{5229 - 5 \times 20.4 \times 41.6}{2622 - 5 \times 20.4^2} = 1.822$$

$$\hat{a} = \overline{Y} - \hat{b}\overline{X} = 41.6 - 1.822 \times 20.4 = 4.44$$

Therefore, the line of best fit is given by

$$y = 4.44 + 1.822x$$

The correlation coefficient (Pearson) is given by

$$r = \frac{5229 - 1/5 \times 102 \times 208}{\sqrt{2622 - 1/5 \times 102^2} \times \sqrt{10510 - 1/5 \times 208^2}} = \frac{985.8}{23.2637 \times 43.0952} = 0.983$$

The standard error (S_e) of this estimate is given by

$$S_e^2 = \frac{\left(\sum x_i^2 - 1/N\left(\sum x_i\right)^2\right) \times \left(\sum y_i^2 - 1/N\left(\sum x_i\right)^2\right) - \left(\sum x_i y_i - 1/N\sum x_i \sum y_i\right)^2}{\left(\sum x_i^2 - 1/N\left(\sum x_i\right)^2\right) \times (n-2)}$$

$$S_e^2 = \frac{(2622 - 102^2 \div 5) \times (10510 - 208^2 \div 5) - (5229 - 102 \times 208 \div 5)^2}{(2622 - 102^2 \div 5) \times (5-2)} = 20.52$$

Project Scheduling

Gantt Charts

A Gantt chart is a bar chart that is often used to track the scheduling of activities in a project. It can also be used to track budget, resources, and equipment usage. A Gantt chart, when updated regularly, can give an accurate snapshot of the status (percent completion) of activities, as well as incurred expenditures and equipment allocations. The major advantage of the Gantt chart is that time relationships between tasks is immediately apparent. It is also quite simple to tweak the information displayed on a Gantt chart to ask "what-if" questions.

A major disadvantage is that precedence relationships are difficult to display, and are therefore typically not shown on a Gantt chart. This disadvantage can be overcome by drawing a time-scaled network. An example of a time-scaled network is shown in Fig. 503.1.

Critical Path Analysis

The benefit of using critical path analysis (CPA) over Gantt charts is that CPA formally identifies tasks which *must* be completed on time for the whole project to be completed on time, and also identifies which tasks can be delayed for a while if resources need to be reallocated to catch up on missed tasks. A further benefit of CPA is that it helps you identify the minimum length of time needed to complete a project. The disadvantage of CPA is that the relation of tasks to time is not as immediately obvious as with Gantt charts. The following conventions and abbreviations are used in this chapter:

ES   early start

EF   early finish

LS   late start

LF   late finish

TF   total float

FF   free float

D   duration

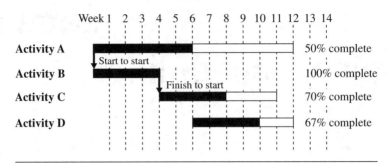

Figure 503.1 Time-scaled network.

The start date for the project is designated as zero. The first activities (i.e., those without predecessors) therefore have ES = 0. The finish date for an activity is calculated as

$$EF = ES + D \tag{503.1}$$

$$LF = LS + D \tag{503.2}$$

The early start date for any following activity is taken to be equal to the latest finish date of all predecessor activities.

Total Float, Free Float, and Interfering Float

Float or slack is the amount of time that a task in a project network can be delayed without causing a delay to subsequent tasks (free float) or project completion date (total float).

The total float of an activity is calculated as the difference between late start and early start dates, or as the difference between late finish and early finish dates

$$TF = LS - ES = LF - EF \tag{503.3}$$

An activity that has a total float equal to zero is said to be a "critical activity," which means that a delay in the finish time of this activity will cause the entire project to be delayed by the same amount of time. A critical activity typically has free float equal to zero, but an activity that has zero free float may not be on the critical path.

The free float of an activity is the maximum time that it can be delayed without delaying the start of its successors. It is calculated as the latest time the activity could start but still finish before the early start of the next activity minus the earliest start time for the activity. In other words, the free float of an activity can be expressed as

$$FF_i = \min(ES_{i+1}) - EF_i \tag{503.4}$$

where $\min(ES_{i+1})$ is the smallest of early start times of all successors of activity i
   EF_i is the early finish time of activity i

NOTE   For a terminal activity in the network, there are no successor activities. For such activities, the $\min(ES_{i+1})$ is taken as the *project duration*.

The interfering float is that portion of the total float that interferes with following activities. It is given by

$$\text{Interfering float} = \text{total float} - \text{free float} \qquad (503.5)$$

For an activity-on-arrow network, if there is only one activity link going into a node, the free float for that activity is zero. All the links going into the same node have the same value of interfering float. Also, all activities on an activity chain have the same value of total float.

Activity on Node and Activity on Arrow Networks

An activity network may be represented using either activity-on-node (AON) or activity-on-arrow (AOA) diagrams. In an AON diagram, the nodes represent the activities and the arrows indicate the predecessor-successor relationships between these activities.

In an AOA diagram, the arrows represent the activities and the nodes represent instants in time indicating the start and end of each activity. In the AOA notation, arrows are scaled to represent activity durations. An activity connecting nodes i and j is represented by duration t_{ij} and scaled accordingly. If these activities are summarized in a table, the precedence relationships may be *read* from the table as follows: an activity described as PQ will have preceding activities whose descriptor (label) ends in a P (such as AP, BP, CP, etc.) and will be followed by activities whose descriptor (label) starts with a Q (such as QR, QS, QT, etc.). A major disadvantage of the AOA notation is that spurious relationships between activities may arise. The use of dummy arrows is common with AOA networks to avoid such misinformation.

NOTE   The more precedence relationships are defined for a project, the longer it gets. In fact, the longest time a set of activities can take is simply the sum of all activity durations, as when they occur in sequence along a single path, with no parallelism between activities. On the other hand, if all activities are parallel (i.e., there are NO precedence relationships), then the project duration is a minimum and is equal to the duration of the longest activity.

Logic Dummies

When the activity-on-arrow diagram is used to represent a project, there may arise cases where showing a sequence with a single arrow is inadequate, since the arrow represents both the sequence and duration of an activity.

Arrow diagramming has problems representing a sequence of activities where sets of activities share some, but not all, of the prior activities. In this example (see table below), Activity C has A as a predecessor and Activity D has both A and B as predecessors. One way to know if a logic dummy is needed is to look at the activity list and find those activities that share some, *but not all* of the entire set of prior activities.

Activity	Predecessors
A	-
B	-
C	A
D	A, B

Figure 503.2 Use of dummy activity in arrow diagrams.

When we try to create an Arrow Diagram for this table, the standard notation shown in the figure on the left is incorrect, because activity B does not precede activity C. To solve this problem, the developers of the Arrow Diagram created the "logic dummy" (Fig.503.2). The logic dummy, an activity with zero duration, links together activities whose sequence would otherwise not be shown. To designate the dummy activity, a dashed arrow is used. As shown in the figure on the right, the logic dummy is used to show the relationship between activity A and D without the problem of linking Activity B with Activity C.

Example 503.1

Construct the flow diagram for the project described in the table below. Identify the critical path and parameters such as early start, late start, early finish, late finish for each task.

Task	Predecessors	Successors	Duration
A	—	B, C	2
B	A	D, E	3
C	A	G	1
D	B	F	3
E	B	F	4
F	D, E	J	1
G	C	H, I	2
H	G	J	1
I	G	K	4
J	F, H	L	3
K	I	L	2
L	J, K	—	2

The activities networks for the project are shown in Figs. 503.3 and 503.4.

Solution The *forward pass* through the network can be solved as follows:

For activity A, the early start date is 0 and the early finish date is $0 + \text{duration}_A = 2$.

Since activity B has the single predecessor (A), the early start date for B is the same as the early finish date for A (2). The early finish date for B is equal to $2 + \text{duration}_B = 5$.

Similarly, for activity C (single predecessor A), the early start date for C is the same as the early finish date for A (2). The early finish date for C is equal to $2 + \text{duration}_C = 3$.

For activity G (single predecessor C), the early start date for G is the same as the early finish date for C (3). The early finish date for G is equal to $3 + \text{duration}_G = 5$.

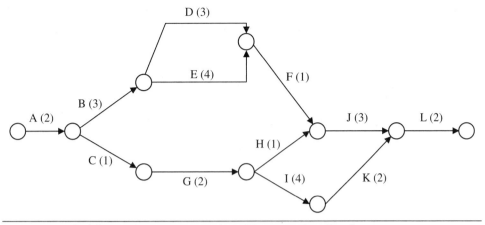

Figure 503.3 Activity-on-arrow representation of the example network.

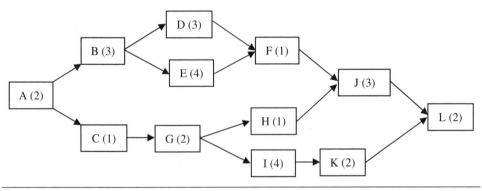

Figure 503.4 Activity-on-node representation of the example network.

Since activity H has the single predecessor (G), the early start date for H is the same as the early finish date for G (5). The early finish date for H is equal to $5 + \text{duration}_H = 6$.

Similarly, for activity I (single predecessor G), the early start date for I is the same as the early finish date for G (5). The early finish date for I is equal to $5 + \text{duration}_H = 9$.

Similarly, for activity K (single predecessor I), the early start date for K is the same as the early finish date for I (9). The early finish date for K is equal to $9 + \text{duration}_K = 11$.

Activity D has the single predecessor (B), the early start date for D is the same as the early finish date for B (5). The early finish date for D is equal to $5 + \text{duration}_D = 8$.

Similarly, for activity E (single predecessor B), the early start date for E is the same as the early finish date for B (5). The early finish date for C is equal to $5 + \text{duration}_E = 9$.

Activity F has two predecessors (D and E). Therefore, the early start date for F is the *larger* of the two early finish dates (9) and the early finish date for F is equal to $9 + \text{duration}_F = 10$.

Activity J has two predecessors (F and H). Therefore, the early start date for J is the larger of the two early finish dates (10) and the early finish date for J is equal to $10 + \text{duration}_J = 13$.

Activity L has two predecessors (J and K). Therefore, the early start date for L is the larger of the two early finish dates (13) and the early finish date for L is equal to $13 + \text{duration}_L = 15$.

The project duration is therefore 15 days.

The *backward pass* through the project progresses as follows:

Activity L is the terminal task of the project and its early finish date is the same as the project duration. If there were multiple ends to parallel paths, the activity with the latest finish date (same as the project duration) would be on the critical path and therefore have late finish date equal to the early finish date. *All terminal tasks would then have their late finish date set equal to the project duration.* Therefore, the late finish date for L is 15 and the late start date for L is equal to $15 - \text{duration}_L = 13$.

Both J and K have a single successor and therefore their late finish date is equal to the late start date of L. Thus, for activity J, late finish date is 13 and late start date is $13 - \text{duration}_J = 10$ and for activity K, late finish date is 13 and late start date is $13 - \text{duration}_K = 11$.

Activities F and H precede J. Thus, their late finish date is equal to the late start date of J. Thus, for activity F, late finish date is 10 and late start date is $10 - \text{duration}_F = 9$ and for activity H, late finish date is 10 and late start date is $10 - \text{duration}_H = 9$.

Activity I precedes K. Thus, the late finish date for I is equal to the late start date of K = 11 and late start date is $11 - \text{duration}_I = 7$.

Activity G has two successors—H and I. Thus, the late finish date for G is the *smaller* of their late start dates. Thus, the late finish date for G is equal to 7 and the late start date is $7 - \text{duration}_G = 5$.

Both D and E have a single successor and therefore their late finish date is equal to the late start date of F. Thus, for activity D, late finish date is 9 and late start date is $9 - \text{duration}_D = 6$ and for activity E, late finish date is 9 and late start date is $9 - \text{duration}_E = 5$.

Activity B has two successors—D and E. Thus, the late finish date for B is the smaller of their late start dates. Thus, the late finish date for B is equal to 5 and the late start date is $5 - \text{duration}_B = 2$.

Activity C precedes G. Thus, the late finish date for C is equal to the late start date of $G = 5$ and late start date is $5 - \text{duration}_C = 4$.

Activity A has two successors—B and C. Thus, the late finish date for A is the smaller of their late start dates. Thus, the late finish date for A is equal to 2 and the late start date is $2 - \text{duration}_A = 0$.

If all the details are filled in for each activity using the following convention, the following table may be constructed.

Task	D_i	ES_i	EF_i	LS_i	LF_i	TF_i	Successors	$min(ES_{i+1})$	FF_i
A	2	0	2	0	2	0	B, C	2	0
B	3	2	5	2	5	0	D, E	5	0
C	1	2	3	4	5	2	G	3	0
D	3	5	8	6	9	1	F	9	1
E	4	5	9	5	9	0	F	9	0
F	1	9	10	9	10	0	J	10	0
G	2	3	5	5	7	2	H, I	5	0
H	1	5	6	9	10	4	J	10	4
I	4	5	9	7	11	2	K	9	0
J	3	10	13	10	13	0	L	13	0
K	2	9	11	11	13	2	L	13	2
L	2	13	15	13	15	0	–	15	0

In the table above, free float for each activity has been calculated using the formula

$$FF_i = min(ES_{i+1}) - EF_i$$

The critical path for this project is therefore A-B-E-F-J-L, which has a total duration of 15 days. This is the longest path through the project and is therefore the minimum time in which the project can be completed. Note that every activity on the critical path has zero total float and zero free float. However, there are noncritical activities (C, G, I) that have zero free float. This simply means that delaying these activities will delay the start time for their successors, but will not necessarily delay the entire project. The critical path is highlighted in Fig. 503.5.

Note that if the available time for the entire project is the same as the duration of the longest path, then the float time or slack on the critical path will be zero. In the table above, note that

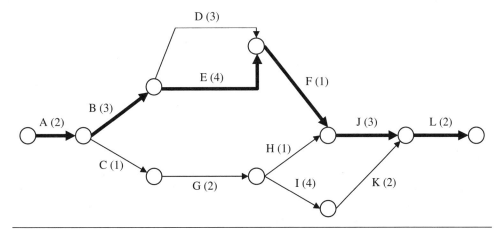

Figure 503.5 Critical path for the example network.

the late finish time for the last activity (L) is set to be the same as the early finish time for that activity. The backward sweep through the network starts from this value.

However, if the time available for the project were different (say 17 days), then that value would be the start value for the backward sweep and would change all the values in the *late finish* and *late start* columns. The float time along the critical path would then no longer be *zero*, but equal to the difference between the time available and the duration of the longest path.

Note that if we look at the original task diagram, we see that for this relatively simple project, there are only four task chains from project start to project finish. These are

      ABDFJL   duration 14 days

      ABEFJL   duration 15 days

      ACGHJL   duration 11 days

      ACGIKL   duration 13 days

Of these, ABEFJL is the longest and therefore is the critical path. All tasks on this path are critical and therefore must have total float and free float equal to zero. Depending on the question asked, such a quick approach may be adequate.

Lag Information

The previous discussion has assumed that there is no specified lag between activities. All predecessor-successor relationships have been assumed to have zero FS (finish to start) lag. In other words, as soon as a predecessor activity finishes, the successor activity can start. Activities that share the same start time are designated as SS and activities that have the same finish time are designated FF (Fig. 503.6).

In many projects, two tasks are related but with a specified lag. A simple example of FS lag may be the time that must elapse between the end of a concrete pour and the construction of the next stage (thus allowing the concrete to reach a specified minimum strength). The FS lag can also be negative, which is referred to as *finish-to-start lead*.

The lag between two activities can also be an SS (start to start) lag. In such a case, the start time of one activity lags the start time for another. If, on the other hand, the lag between the two activities is specified as an FF (finish to finish) lag, then there is a specified lag between the finish times of the two activities. Symbolically, these relationships are shown in Fig. 503.6.

Time-Cost Tradeoff

The primary role of the project manager is to balance time versus cost. During the initial project planning phase, the project manager must assign a reasonable completion time to each activity as well as allocate resources (human resources, equipment, and materials) to each activity. It must be recognized that as the project progresses, it is often necessary to make changes to these initial allocations (project monitoring and control). In making

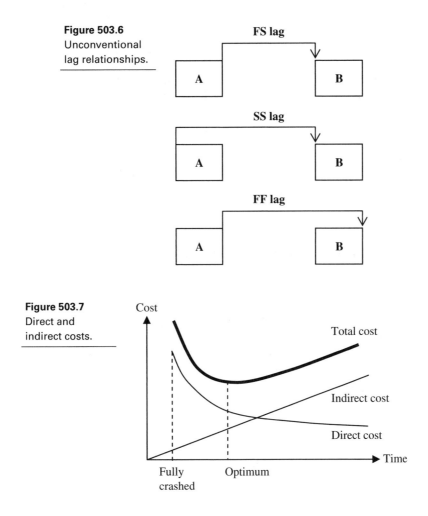

Figure 503.6
Unconventional lag relationships.

Figure 503.7
Direct and indirect costs.

these changes, the time versus cost tradeoff must be understood and incorporated. For example, if the project is behind schedule, the first question is—what are the benefit-cost implications (additional labor, fees and fines, additional equipment leasing, etc.) of corrective measures such as crashing certain activities? A typical cost versus time diagram is shown in Fig. 503.7. The indirect cost line is shown as a straight line because it is typically a linear function of time. The direct cost curve shows a maximum value at the "fully crashed" point (where the maximum resources have been allocated to reduce the duration to its minimum value).

Earned Value Management (EVM)

In project management, *budgeted cost of work performed* (BCWP) or *earned value* (EV) is the budgeted cost of work that has actually been performed. On the other hand, the *budgeted cost of work scheduled* (BCWS) is the approved budget that has been allocated to complete a scheduled task during a specific time period. Other names for BCWS are *Planned Value* (PV) and *Budget at Completion* (BAC). *Actual Cost of Work Performed* (ACWP) is the actual cost that has been spent, rather than the budgeted cost.

As an illustration of these terms, assume that a schedule contains a task that is budgeted to cost $10,000 to perform, and is expected to begin on January 1st and complete at the end of January 10th.

At the end of January 5th, the work is scheduled to be 50% complete (5 days of the scheduled 10 days). So, at the end of January 5th, the BCWS is $10,000 (the budgeted cost) times 50% (the scheduled completion percentage), or $5000.

Suppose that by the end of January 5th, the work is actually only 30% complete. In this case, the BCWP would be $10,000 (budgeted cost) times 30% (the actual completion percentage), or $3000.

Also, suppose that to reach the 30% complete level at the end of January 5, $2700 was actually spent. Then, the ACWP would be $2700.

The cost variance (CV) is defined as the difference between BCWP and ACWP. Thus, in this example, CV = BCWP − ACWP = 3000 − 2700 = $300 (positive value indicates under budget).

The schedule variance (SV) is defined as the difference between BCWP and BCWS. Thus, in this example, SV = BCWP − BCWS = 3000 − 5000 = − $2000 (negative value indicates behind schedule).

Cost performance index (CPI) is defined as the ratio of the BCWP to the ACWP. CPI greater than 1.0 means "under budget." CPI less than 1 means that the cost of completing the work is higher than planned.

In this example, CPI = BCWP/ACWP = 3000/2700 = 1.11

Schedule performance index (SPI) is defined as the ratio of the BCWP to the BCWS. SPI greater than 1.0 means "ahead of schedule."

In this example, SPI = BCWP/BCWS = 3000/5000 = 0.6 (indicating behind schedule).

The *estimate to complete* (ETC) is the estimate (additional funds) to complete the remaining part of the project. It is calculated as

$$ETC = \frac{BCWS - BCWP}{CPI} \tag{503.6}$$

In this example,

$$ETC = \frac{5000 - 3000}{1.11} = 1800$$

The estimate at complete (EAC) represents the total funds to complete the project. It is calculated as

$$EAC = ACWP + ETC \tag{503.7}$$

In this example, EAC = 2700 + 1800 = $4500.

Project Crashing

Identification of the critical path allows the project manager to make effective decisions that impact project completion. For example, the overall project duration may be shortened by allocating additional resources to certain tasks on the critical path, thereby shortening it. The length of the critical path may also be shortened by pruning activities on the critical path or by fast tracking (i.e., performing more activities in parallel). In fact, the first action the project manager must consider when trying to reduce project duration is to examine all precedence relationships along the critical path to see if there are any activities that can be taken off the critical path and performed in parallel. The following must be true for an activity to be a good candidate for crashing:

1. It must be on the critical path.
2. It must precede multiple activities, so that if shortened, there is widespread effect on several activities.
3. It must be of long duration—it is generally cheaper to reduce longer tasks by a set amount of time than short ones.
4. It must have a low cost per period gained.
5. It must be relatively early in the project, so that crashing it has widespread impact.
6. It must be low-skilled labor intensive. It is both easier and cheaper to crash such activities, simply by allocating extra resources.

Consider a scenario where the previously described (Example 503.1) project must be completed in 13 days. Currently, the normal length of the critical path A-B-E-F-J-L is 15 days. Additional resources must be allocated to individual tasks in order to shorten their duration. Consider the cost data in the following table outlining normal costs and crash costs for each activity in the project.

Task	Normal Duration	Cost	Crashed Duration	Cost	Cost differential
A	2	1000	1	1800	800
B	3	2500	2	3000	500
C	1	1200	1	1200	0
D	3	2200	2	2800	600
E	4	5400	3	6000	600
F	1	550	1	550	0
G	2	1300	1	2100	800
H	1	700	1	700	0
I	4	6500	3	9000	2500
J	3	2100	2	2600	500
K	2	1200	1	1900	700
L	2	1000	1	1600	600

To reduce the overall project duration by 2 days, the most inexpensive option is to allocate additional resources (crash) to activities B and J, since these have the lowest cost differential. With these additional resources, the new path durations become

ABDFJL   duration 12 days

ABEFJL   duration 13 days (critical)

ACGHJL   duration 10 days

ACGIKL   duration 13 days (critical)

Resource Leveling

When limited resources must be shared by various activities in the project, the network must be optimized using processes called *activity-based resource assignments* and *resource leveling*. Once resource leveling is conducted, it may become possible for a previously noncritical path to become the longest or "resource critical" path. CPM allows continuous monitoring of the project schedule, allows the project manager to track the critical activities, and alerts the project manager to the possibility that noncritical activities may be delayed beyond their total float, thus creating a new critical path and delaying project completion.

As an example, consider a time-scaled representation of some overlapping activities in a project, as shown in Fig. 503.1. The time-scaled network is a hybrid of a Gantt chart and a CPM network, sometimes known as a Plannet chart. Essentially, it is a Gantt chart that includes network logic. Activities are shown as bars between their ES and EF times, as in a Gantt chart, but precedence relationships are also shown using arrows.

In Fig. 503.8, activities A, B, and C are shown on a scaled time line. Each activity label is accompanied by the number of resource units consumed by that activity. As may be seen, some activities (Z) which precede A are taking place up to week 5, consuming 3 resource units. A starts in week 5 and over the period week 5 to week 12, A and Z occur simultaneously, consuming 7 resource units. Between weeks 12 and 17, A and B are concurrent (8 resource units). Between weeks 17 and 19, B and C are concurrent (9 units) and after week 19, only activity C is occurring, consuming 5 units of resources. The resource demand according to this schedule is plotted on the gray-shaded diagram in the figure.

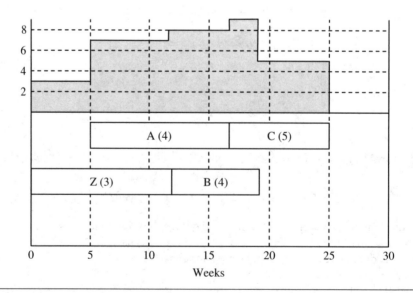

Figure 503.8   Resource demand in a project.

If the resources are constrained to 8 units, it is apparent from the figure that the period 17 to 19 is problematic. There are various alternatives that a project manager may consider:

1. Evaluate the dependence of B on its stated predecessors (Z) and examine the possibility of advancing the start date of B to week 10. Then B will occupy weeks 10 to 17 and the maximum resource demand will be 8 units.

2. Evaluate the cost associated with crashing activity B (normal duration 7 weeks, crashed duration 5 days) by 2 weeks, thereby eliminating the conflict with activity C.

3. Delay activity C by 2 weeks to eliminate the concurrence of B and C. If C is a critical activity, this has to be accompanied by crashing activity C by 2 weeks so that the overall project is not delayed.

Machine Production

When using construction equipment, concepts such as capacity, cycle time, efficiency, production rate are used to calculate the overall rate of production of that equipment. When calculating earth excavation operations, adjustments must be made for volume changes due to shrinkage and swell.

Example 503.2

A dump truck having a capacity of 15 yd^3 is used at a job site to haul excavated material. The distance from the excavation pit to the dump site is 5 mi and the average speed of the truck is 25 mph. The total time to load and unload a single load off the dump truck is 10 min. The overall efficiency factor for the dump truck is 0.85. The swell factor for the excavated soil is 1.25.

1. What are the number of trips the dump truck needs to make to move material from a pit with in-place volume = 120,000 ft^3?

2. How many 8-h working days are needed if using a single truck?

3. If a fleet of trucks is used to transport the material, how many trucks are needed to finish the haul operation in 2 days?

Solution

1. In-place volume = 120,000 ft^3 = 120,000 ÷ 27 = 4444 yd^3.

   Hauled volume = 1.25 × 4444 = 5555 yd^3.

   At 15 yd^3 hauled per trip, the number of trips needed at 100% efficiency = 370.4 trips.

   Operating at 85% efficiency, number of trips = 370.4 ÷ 0.85 = 435.7 (436 trips).

2. Each trip requires 10 min (load unload) + 12 min (travel) = 22 min = 0.367 h.

   In an 8-h day (operating at 85% efficiency), the number of daily trips = 8 ÷ 0.367 = 21.8 trips (21 complete trips).

   Therefore, a single truck needs 436 ÷ 21 = 20.8 days (21 days).

3. To finish hauling 5555 yd^3 in 2 days, daily haul = 2778 yd^3.

Each truck, operating at 85% efficiency, makes $0.85 \times 21.8 = 18$ trips per day = 270 yd^3.

Therefore, number of trucks needed = $2778 \div 270 = 10.3$ trucks (use 11 trucks).

Wright's Cumulative Average Model

The concept of the learning curve was introduced to the aircraft industry in 1936 by T. P. Wright. Wright described a basic theory for obtaining cost estimates based on repetitive production of airplane assemblies. Learning curves (also known as progress functions) have been applied to all types of work of various complexities.

In the theory of learning, it is recognized that repetition of the same operation results in less time or effort expended on that operation. For the Wright learning curve, the underlying hypothesis is that the direct labor man-hours necessary to complete a unit of production will decrease by a constant percentage each time the production quantity is doubled. If the rate of improvement is 20% between doubled quantities, then the **learning percent** would be 80% ($100 - 20 = 80$). While the learning curve emphasizes time, it can be easily extended to cost as well.

In Wright's model, the learning curve function is defined as follows:

$$Y = aX^b \tag{503.8}$$

where:

Y = the cumulative average time (or cost) per unit
X = the cumulative number of units produced
a = time (or cost) required to produce the first unit
b = slope of the function when plotted on log-log paper = log of the learning rate $\div$ log 2

Example 503.3

The pipefitter's installation of a 12-in diameter underground steam line takes 3.4 man-hours to complete the first weld joint. Based on the contractor's historic data, a learning curve rate of 80% is expected. The man-hours required to weld the 18th pipe joint is most nearly:

A. 0.75

B. 1.34

C. 3.44

D. 4.55

Solution

$X = 18$

$a = 3.4$

$b = \log 0.8 \div \log 2 = -0.322$

$Y = aX^b - 3.4 \times 18^{-0.322} = 1.34$

Answer is **B**.

Production Cycle Time

Every piece of construction equipment has an ideal capacity, an overall operational efficiency and a cycle time. The number of work cycles in a working day is obtained by dividing the length of the working day by the cycle length. If the capacity of the equipment is then multiplied by the number of daily work cycles, we obtain the daily production rate of the equipment.

Example 503.4

An excavator with a capacity of 3 yd^3 has a cycle time of 45 s. The overall efficiency of the operation is 0.84. What is the soil export (yd^3) from the job site in an 8-h day?

Solution

Number of daily cycles $= 8$ h $\div$ 45 s $= 640$

Daily production (ideal) $= 640 \times 3 = 1920$ yd^3

Daily production (real) $= 0.84 \times 1920 = 1613$ yd^3

Equipment Balancing

When several pieces of equipment need to work in tandem for accomplishing a task, the overall productivity will depend on the equipment with the longer cycle time, while the other equipment will have idle time. Consider the following example.

Example 503.5

A front-end loader has an output of 220 yd^3 per hour (bank measure). It is used to load a fleet of 7 trucks, each of which has a capacity of 15 yd^3 (loose). The truck cycle time, which includes load time, travel time, and unload time, is 18 min. The excavated soil has 20% swell and 10% shrinkage. The project requires 25,000 yd^3 of compacted soil. How many hours are needed to complete the job?

Solution

Overall requirement $= 25,000$ yd^3 (compacted) represents 90% of the bank measure.

Bank measure $= 25,000 \div 0.9 = 27,778$ yd^3.

Due to 20% swell, this soil will occupy $1.2 \times 27,778 = 33,333$ yd^3 in the loose condition.

Loader productivity $= 220$ yd^3/h (bank).

Each truck makes 3.33 trips per hour (cycle time $= 18$ min). Therefore, the fleet of 7 trucks makes 23.33 trips per hour, transporting $23.33 \times 15 = 350$ yd^3 loose soil, which is equivalent to $350 \div 1.2 = 291.67$ yd^3 bank soil per hour. This is higher than the loader's capacity and so the loader's capacity will determine overall productivity.

Therefore, the time required $= 27,778 \div 220 = 126.3$ h.

Under these conditions, the fleet of 7 trucks has some idle time.

Example 503.6

For the previous example, what is the size of the largest truck fleet for which trucks have no idle time?

Solution

As before, we compute loader productivity $= 220$ yd³/h (bank).

Each truck has productivity $= 3.33 \times 15 = 50$ yd³ loose soil, which is equivalent to $50 \div 1.2 = 41.67$ yd³ bank soil per hour.

For the productivity of the trucks to equal that of the loader, number of trucks $= 220 \div 41.67 = 5.28$.

Therefore, if 5 trucks are used, they will have no idle time.

PERT

PERT, which stands for Program Evaluation and Review Technique, was developed by Booz, Allen & Hamilton and Lockheed Corporation for the Polaris missile project for the U.S. Navy in 1958. As a project evaluation technique, PERT is similar to critical path analysis (CPA). However, whereas CPA uses a single deterministic value for each activity duration, PERT uses a probabilistic approach that allows three-time estimates for the duration of each activity since accurate estimates of task require good historical data.

Optimistic time a: Time an activity will take if everything goes perfectly.

Most likely time m: Most realistic time estimate to complete the activity.

Pessimistic time b: Time an activity takes if everything goes wrong.

From these we calculate the *expected time t* for the task.

The time estimates are often assumed to follow the beta probability distribution. These three-time estimates can be used to compute the expected time and the standard deviation for each activity.

$$\text{Expected time } t = \frac{a + 4m + b}{6} \tag{503.9}$$

$$\sqrt{\text{Variance}} = \left[\frac{(b - a)}{6} \right] \tag{503.10}$$

When estimates are too optimistic, the distribution curve is skewed to correct for this. In that case, the formula below might be used:

$$\text{Expected time } t = \frac{a + 3m + 2b}{6} \tag{503.11}$$

Thus, if a path within the project comprises the tasks A, E, H, and K, the expected time for the path may be given by

$$T = t_A + t_E + t_H + t_K \tag{503.12}$$

where t_A, t_E, t_H, and t_K are mean time estimate for activities A, E, H, and K, respectively, and the variance for the path is given by

$$\sigma^2 = \sigma_A^2 + \sigma_E^2 + \sigma_H^2 + \sigma_K^2 \tag{503.13}$$

The project variance is the sum of the variances of each of the tasks on the critical path. The square root of this is the project standard deviation.

From the normal distribution equation:

$$z = \frac{\text{due date} - \text{expected date of project completion}}{\text{project standard deviation}}$$

Looking this value up in the normal distribution tables gives the probability of the project being completed on our due date. At the end of this process we have

- An expected completion date for the project.
- A knowledge of tasks critical to the project, i.e., if they are delayed the delivery of the final product is delayed.
- A knowledge of tasks not that critical, that is, they can be delayed to some extent (the float of the task) without affecting the overall delivery of the final product. Resources from these tasks could be diverted to the critical tasks if they start to fall behind.
- An estimate of how likely it is that the project will be finished by the deadline.

Example 503.7
Consider the project described in the following table:

Activity	Predecessor	a	m	b	Mean μ	Variance ρ²
A	—	1	2	4	2.17	0.25
B	—	5	6	7	6.00	0.11
C	—	2	4	5	3.83	0.25
D	A	1	3	4	2.83	0.25
E	C	4	5	7	5.17	0.25
F	A	3	4	5	4.00	0.11
G	B, D, E	1	2	3	2.00	0.11

What is the probability that the project completion time will be between 10 and 13 days?

Solution The mean and variance of the activity durations are calculated and tabulated in the last two columns of the table.

$$\mu = \frac{a + 4m + b}{6}$$
$$\sigma = \frac{b - a}{6}$$

The activity on node network for the project is shown below.

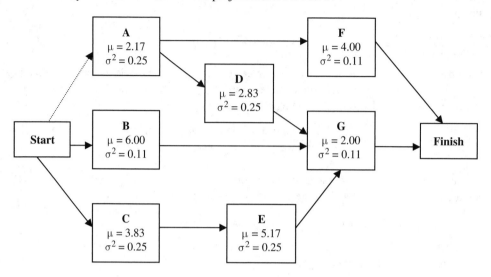

The network is simple enough that the critical path is seen as C-E-G (mean duration 11 days).

$$\mu = \mu_C + \mu_E + \mu_G = 3.83 + 5.17 + 2.00 = 11.0$$

The variance of this path is calculated as

$$\sigma^2 = \sigma_C^2 + \sigma_E^2 + \sigma_G^2 = 0.25 + 0.25 + 0.11 = 0.61$$
$$\sigma = 0.782$$

The probability that the project duration is between 10 and 13 days is given by

$$P(10 \leq T \leq 13) = P\left(z \leq \frac{13-11}{0.782}\right) - P\left(z \leq \frac{10-11}{0.782}\right)$$
$$= P(z \leq 2.558) - P(z \leq -1.279) = P(z \leq 2.558) - [1 - P(z \leq 1.279)]$$
$$= 0.9948 - (1 - 0.8997) = 0.8945$$

Design Loads during Construction

Load Combinations

Some of the provisions from ASCE 37 (2002), *Design Loads on Structures during Construction,* Reston, Virginia, are summarized in this chapter.

The following load types are considered during construction—dead, live, construction, environmental, and lateral earth pressure. The LRFD load combinations to be used during construction are given below:

$$1.4D + 1.4C_D + 1.2C_{FML} + 1.2C_{VML} \qquad (504.1a)$$

$$1.2D + 1.2C_D + 1.2C_{FML} + 1.4C_{VML} + 1.6C_P + 1.6C_H + 0.5L \qquad (504.1b)$$

$$1.2D + 1.2C_D + 1.2C_{FML} + 1.3W + 1.4C_{VML} + 0.5C_P + 0.5L \qquad (504.1c)$$

$$1.2D + 1.2C_D + 1.2C_{FML} + 1.0E + 1.4C_{VML} + 0.5C_P + 0.5L \qquad (504.1d)$$

$$0.9D + 0.9C_D + (1.3W \text{ or } 1.0E) \qquad (504.1e)$$

where C_D = dead loads
C_{FML} = fixed material loads
C_{VML} = variable material loads
C_P = personnel and equipment loads
C_H = horizontal construction loads

Personnel and Vehicle Loads (C_P)

When needed, an individual personnel load is to be taken as 250 lb. This load may be distributed over a 12×12 in area. A wheel of a manually powered vehicle is to be taken as 500 lb. A wheel of a powered vehicle is to be taken as 2000 lb.

Material Loads

Any material being lifted into place is treated as part of the equipment load. When these materials are discharged to their destination, they are treated as material load.

Horizontal Construction Load (C_H)

Horizontal construction load C_H is to be taken as the largest of the following:

1. For wheeled vehicles transporting materials, C_H is the larger of 0.2 times the fully loaded weight (for a single vehicle) or 0.1 times the fully loaded weight (for multiple vehicles).

2. 50 lb per person at the level of the platform.

3. 2% of the total dead load.

4. The calculated horizontal reaction.

Form Pressure

Freshly poured concrete in forms is considered as a material load. Once the concrete reaches the specified target strength, it should be included in the dead load.

Personnel and Equipment Load Reduction (ASCE 37)

For an influence area A_I greater than 400 ft^2, the live loads from personnel and equipment acting on temporary structures during construction may be taken as

$$C_p = L_o \left(0.25 + \frac{15}{\sqrt{A_I}} \right) \tag{504.2}$$

where L_o is the nominal live load. The reduced live load C_p must not be less than $0.5L_o$ for members supporting loads from one level and must not be less than $0.4L_o$ for members supporting loads from two or more levels. The exception to these limits is that if L_o is less than 25 psf, then L_o must not be less than $0.6L_o$.

The live load reduction accounts for the fact that, statistically, the probability of an area being fully loaded with the full specified live load becomes less and less as the area increases. Concrete floor live loads (other than vehicle loads) can be distributed over a 2.5 × 2.5 ft area. Vehicular live (wheel) loads are to be distributed over and area of 20 in^2.

Roof Live Load Reduction during Construction

The nominal value of the construction load on sloping roofs may be reduced by a reduction factor (R) which depends on the slope (pitch) of the roof, expressed as F (inches per foot). The reduction factor R is given by

$$R = 1.2 - 0.05F \qquad 0.6 \leq R \leq 1.0 \tag{504.3}$$

Table 504.1 Operational Class Designations

Load	Classification	Comments
20 psf	Very light duty	Sparsely populated with personnel with hand tools, very little construction materials
25 psf	Light duty	Sparsely populated with personnel operating equipment, staging of materials for light weight construction
50 psf	Medium duty	Concentrations of personnel, staging of materials for average construction
75 psf	Heavy duty	Material placement by motorized buggies, staging of materials for heavy construction

Operational Class of Working Surfaces

The classifications of working surfaces on temporary structures during construction are shown in Table 504.1.

Importance Factor

During construction, the importance factor (I) for all environmental loads shall be 1.0. The design wind speed during construction shall be the design wind speed from ASCE 7 multiplied by the appropriate factor from Table 504.2.

The ground snow load during construction shall be the ground snow loads from ASCE 7 multiplied by the appropriate factor from Table 504.3.

Lifting and Rigging

In this section, various examples of lifting devices are reviewed. Some of these employ arrays of vertical cables, which share the load in accordance with principles of equilibrium and compatibility

Table 504.2 Importance Factor for Wind Loads during Construction

Construction period	Factor
Less than 6 weeks	0.75
6 weeks–1 year	0.80
1–2 years	0.85
2–5 years	0.90

Table 504.3 Modification Factor for Ground Snow Loads during Construction

Construction period	Factor
Up to 5 years	0.80
More than 5 years	1.00

(see Chap. 101 on *Strength of Materials*). If only vertical cables are used, the system is somewhat lacking in stability. For example, visualize a load suspended by vertical cables only. Even though the system is in equilibrium (if the cables can provide a total vertical force equal to the weight of the suspended object), it is susceptible to uncontrolled lateral swinging under the effect of wind. This inherent instability can be eliminated by utilizing cables that are inclined to the vertical.

OSHA 29 CFR Section 1926.753 specifies criteria for hoisting and rigging. It specifies the following:

Cranes being used in steel erection activities shall be visually inspected prior to each shift by a competent person. The inspection shall include (but not be limited to) control mechanisms, drive mechanisms, safety devices, pressurized lines (air, hydraulic, etc.), hooks and latches, wire ropes, tires, and ground conditions.

If any deficiency is identified, an immediate determination shall be made by the competent person as to whether the deficiency constitutes a hazard. If the deficiency is determined to constitute a hazard, the hoisting equipment shall be removed from service until the deficiency has been corrected. The operator shall be responsible for those operations under the operator's direct control. Whenever there is any doubt as to safety, the operator shall have the authority to stop and refuse to handle loads until safety has been assured.

Safety latches on hooks shall not be deactivated or made inoperable except:

1. When a qualified rigger has determined that the hoisting and placing of purlins and single joists can be performed more safely by doing so; or

2. When equivalent protection is provided in a site-specific erection plan.

Range versus Load Capacity of Cranes

The operating range of a crane is limited by the following three factors:

1. Length of the boom – this is purely a geometric limitation. The boom length is measured from the boom pivot (or "foot") to the point at the tip of the boom where the lifting cables are deployed. Figure 504.1 shows a plot of vertical versus horizontal projections of the boom, for various boom lengths of a particular crane.

2. Crane load tables – these tables plot Radius versus Lifting capacity. The radius is measured outward from the boom foot to the boom tip along the horizontal projection of the boom. In other words, the radius belongs to the circle on the ground created by the boom tip if the crane were to be swung around in a 360 degree circle. This table is based upon overturning analysis of the crane (usually 75% of the tip load). Table 504.4 is an example of such a table, for the crane mentioned in item 1.

3. Site specific limitation – local codes and specifications may specify a boom standoff distance, which is a minimum prescribed clearance between the crane and the building edge.

An example of a load table for a crane with a 200 ft boom is shown below. The height (vertical) versus radius (horizontal) reach diagram for the same crane (for variable boom length) is shown in the figure following the table.

For an illustration of the use of Table 504.4 and Fig. 504.1, consider a situation where the crane is positioned in front of a 90-ft high building, such that the boom foot is 60 ft from the face of the building. The outline of the building is shown inserted into Fig. 504.1.

Table 504.4 Radius versus Load Capacity for a Crane with a 200 ft Boom

Radius (ft)	Capacity (lb)	Radius (ft)	Capacity (lb)	Radius (ft)	Capacity (lb)
32	146,300	80	39,200	130	17,900
36	122,900	85	35,800	135	16,700
40	105,500	90	32,800	140	15,500
45	89,200	95	30,200	145	14,500
50	76,900	100	27,900	150	13,600
55	67,200	105	25,800	155	12,700
60	59,400	110	23,900	160	11,800
65	53,000	115	22,200	165	11,100
70	47,600	120	20,600	170	10,300
75	43,100	125	19,200	175	9,600

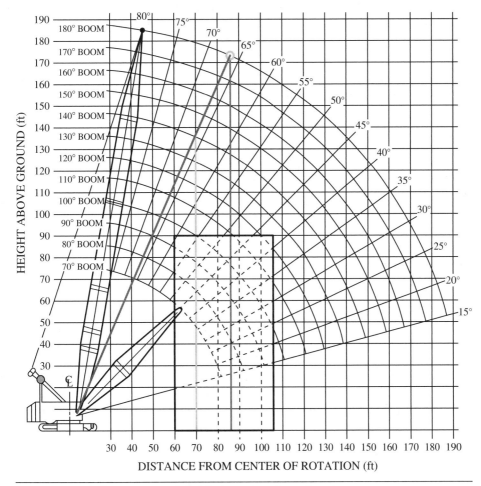

Figure 504.1 Radius versus reach height for various boom lengths.

Let's say the standoff distance (to allow for boom width and safe clearance) is specified as 10 ft. Based on this, locate the point 10 ft from building corner and draw a line from boom C.L. and through this point. This represents the geometric limit of boom position. This line corresponds to a boom inclination angle of approximately 66 degrees.

Next, if the maximum boom extension is given as 180 ft, we end the line drawn in the previous step at the circle for the 180-ft boom length. This is the physical limit of reach of the boom. If we drop a horizontal projection from this point (marked with a circle), this corresponds to a radius of about 86 ft. Since the face of the building is at 60 ft, this means that the maximum horizontal reach of the crane is about 26 ft behind the face of the building. From the load table (Table 504.4), the limit load for a radius of 86 ft is about 35,000 lb. In other words, for any load less than 35,000 lb, the boom can be extended fully and the load placed approximately 26 ft behind the front edge of the building.

For any load greater than 35,000 lb, the load table must be consulted to find the corresponding radius. For example, if the load (including rigging) is 46,000 lb, the table gives a radius of 70 ft. (Interpolation is not allowed for the load table. If a load falls between two entries, choose the more conservative). This corresponds to an offset (from the building face) = 70 − 60 = 10 ft. This also means that to place a load of 46,000 lb, the boom extension needs to be only 140 ft.

Working under Loads

Routes for suspended loads shall be preplanned to ensure that no employee is required to work directly below a suspended load except for:

1. Employees engaged in the initial connection of the steel; or
2. Employees necessary for the hooking or unhooking of the load.

When working under suspended loads, the following criteria shall be met:

1. Materials being hoisted shall be rigged to prevent unintentional displacement;
2. Hooks with self-closing safety latches or their equivalent shall be used to prevent components from slipping out of the hook; and
3. All loads shall be rigged by a qualified rigger.

Multiple Lifts

A multiple lift (a maximum of five members hoisted per lift) shall only be performed if a multiple lift rigging assembly is used. Only beams and similar structural members are permitted to be lifted in a multiple lift. Components of the multiple lift rigging assembly shall be specifically designed and assembled with a maximum capacity based on the manufacturer's specifications with a 5 to 1 safety factor for all components.

The multiple lift rigging assembly shall be rigged with members attached at their center of gravity and maintained reasonably level, rigged from top down, and rigged at least 7-ft (2.1-m) apart.

Example 504.1

The precast concrete element shown below weighs 4000 lb. During erection, it is lifted using two straight cables as shown below. The center of gravity of the load is located at a distance 14 ft from the left edge, as shown below. What is most nearly the tension in the cable on the left?

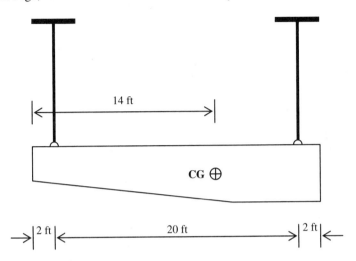

Solution   Taking moments about the cable on the right, we have

$$T_1 \times 20 = 4000 \times 8 \Rightarrow T_1 = 1600 \text{ lb}$$

Example 504.2

The precast concrete element described in the previous example is lifted using two inclined cables as shown. The center of gravity of the load is located at a distance 14 ft from the left edge, as shown. What is most nearly the tension in the cable on the left?

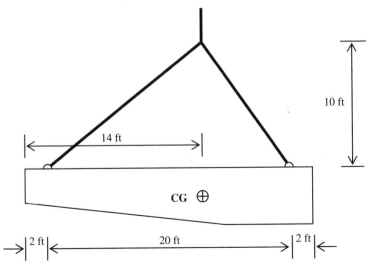

Solution

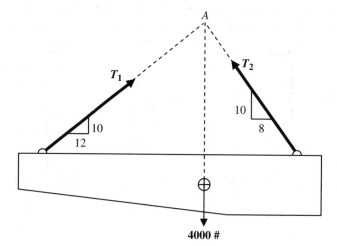

The equilibrium equations are $\Sigma F_x = 0$, $\Sigma F_y = 0$, and $\Sigma M = 0$. Of these equations, the moment equation is satisfied automatically if the springing point for the cables (point A in the figure above) is directly above the center of gravity of the load. This is a requirement for overall stability. Thus, the two surviving equations ($\Sigma F_x = 0$ and $\Sigma F_y = 0$) can be used to solve for two cable tensions. The use of more than two cables to suspend a load in a two-dimensional system creates a statically indeterminate system of forces. For such systems, compatibility equations must be used in addition to equilibrium equations to obtain a solution. The equilibrium equations are shown below:

$$\sum F_x = 0 \Rightarrow T_1 \times \frac{12}{\sqrt{244}} = T_2 \times \frac{8}{\sqrt{164}} \Rightarrow T_1 = 0.813 T_2$$

$$\sum F_y = 0 \Rightarrow T_1 \times \frac{10}{\sqrt{244}} + T_2 \times \frac{10}{\sqrt{164}} = 4000 \Rightarrow T_1 = 2499 \qquad T_2 = 3074 \text{ lb}$$

Example 504.3

A symmetric box (10-ft wide × 10-ft long × 5-ft high) is lifted by a crane utilizing four cables attached to each corner of the box as shown in the diagram below. The cables are connected to the vertical hoisting cable at a point which is 15 ft directly above the center of gravity of the load. If the weight of the box is 10,000 lb, what is most nearly the tension in the cables?

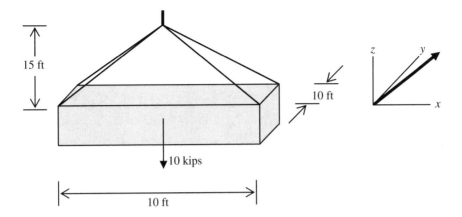

Solution   The solution to this problem is greatly simplified by the symmetry. If the vertical component of each of these (identical) cable forces can be calculated, one may simply write the vertical equilibrium equation.

$$\sum F_z = 4T_z - 10 = 0$$

For example, the cable tension on the front left corner (shown in the figure on the right) can be written as the vector

$$\mathbf{T} = T \frac{5\underline{i} + 5\underline{j} + 15\underline{k}}{\sqrt{5^2 + 5^2 + 15^2}} = T(0.302\underline{i} + 0.302\underline{j} + 0.905\underline{k})$$

Therefore, the vertical component is $0.905T$ and the equation of vertical equilibrium reduces to

$$\sum F_z = 4 \times 0.905T - 10 = 0 \Rightarrow T = 2.76 \text{ kips}$$

Scaffolding[1]

Definitions

A *scaffold* is any temporary elevated platform (supported or suspended) and its supporting structure (including points of anchorage) used for supporting employees and/or materials. Scaffolding can be fixed in place or be adjustable, being equipped with a hoist that can be operated by employee(s) on the scaffold.

A *supported scaffold* has one or more platforms supported by outrigger beams, brackets, poles, legs, uprights, posts, frames, or similar rigid supports. A *suspension scaffold* has one or more platforms suspended by ropes or by other nonrigid means from an overhead structure.

Supported scaffolds can be of various types, such as *lean-to scaffold* (kept erect by tilting it toward and resting it against a building or structure), *shore scaffold* (placed against a building or structure and held in place with props), *form scaffold* (a platform supported by brackets attached to formwork), and *ladder jack scaffold* (a platform resting on brackets attached to ladders).

An example of a suspension scaffold is a *catenary scaffold*, which is a platform supported by two essentially horizontal and parallel ropes attached to structural members of a building or other structure.

A *tube and coupler scaffold* is a supported or suspended scaffold consisting of a platform(s) supported by tubing, erected with coupling devices connecting uprights, braces, bearers, and runners.

Outrigger refers to a structural member of a supported scaffold used to increase the base width of a scaffold in order to provide support for and increase the stability of the scaffold. An *outrigger beam* (also called a *thrustout*) refers to a structural member of a suspension scaffold or outrigger scaffold which provides support for the scaffold by extending the scaffold point of attachment to a point out and away from the structure or building. An *outrigger scaffold* is a supported scaffold consisting of a platform resting on outrigger beams projecting beyond the wall or face of the building or structure, the inboard ends of which are secured inside the building or structure.

---

[1]Much of the contents of this chapter are based on the provisions of Code of Federal Regulations (CFR 29, part 1926).

A *bearer* (also known as a *putlog*) is a transverse horizontal scaffold member upon which the scaffold platform rests and which joins scaffold uprights, posts, poles, and similar members. A *runner* (also known as a *ledger*) is a longitudinal member which may support the bearers.

A *brace* is a rigid connection that holds one scaffold member in a fixed position with respect to another member, or to a building or structure. A *cleat* is a structural block used at the end of a platform to prevent the platform from slipping off its supports. A *coupler* is a device for locking together the tubes of a tube and coupler scaffold.

The *stall load* is the load at which the prime mover of a power-operated hoist stalls or the power to the prime mover is automatically disconnected.

A *single-pole scaffold* is a supported scaffold consisting of a platform resting on bearers, the outside ends of which are supported on runners secured to a single row of posts or uprights, and the inner ends of which are supported on or in a structure or building wall.

A *double-pole scaffold* is a supported scaffold consisting of a platform resting on cross beams (bearers) supported by ledgers and a double row of uprights independent of support (except ties, guys, braces) from any structure.

A *top plate bracket scaffold* is a scaffold supported by brackets that hook over or are attached to the top of a wall.

A *deceleration device* is any mechanism, such as a rope grab, rip-stitch lanyard, specially woven lanyard, tearing or deforming lanyard, or automatic self-retracting lifeline lanyard, which dissipates a substantial amount of energy during a fall arrest or limits the energy imposed on an employee during fall arrest.

Open sides and ends means the edges of a platform that are more than 14 in (36 cm) away horizontally from a sturdy, continuous, vertical surface (such as a building wall), or a sturdy, continuous horizontal surface (such as a floor), or a point of access.

Capacity of Scaffold Components

Each scaffold and scaffold component shall be capable of supporting, without failure, its own weight and at least four times the maximum intended load applied or transmitted to it.

Direct connections to roofs and floors, and counterweights used to balance adjustable suspension scaffolds, shall be capable of resisting at least four times the tipping moment imposed by the scaffold operating at the rated load of the hoist, or at least 1.5 times the tipping moment imposed by the scaffold operating at the stall load of the hoist, whichever is greater.

Each suspension rope, including connecting hardware, used on nonadjustable suspension scaffolds shall be capable of supporting, without failure, at least six times the maximum intended load applied or transmitted to that rope.

Each suspension rope, including connecting hardware, used on adjustable suspension scaffolds shall be capable of supporting, without failure, at least six times the maximum intended load applied or transmitted to that rope with the scaffold operating at either the rated load of the hoist, or two times the stall load of the hoist, whichever is greater.

The stall load of any scaffold hoist shall not exceed three times its rated load. Platforms shall not deflect more than 1/60 of the span when loaded.

Required Width of Scaffold Platforms

Each scaffold platform and walkway shall be at least 18-in (46-cm) wide. Each ladder jack scaffold, top plate bracket scaffold, roof bracket scaffold, and pump jack scaffold shall be at least 12-in (30-cm) wide. There is no minimum width requirement for boatswain's (bosun's) chairs. In narrow areas where these minimums cannot be achieved, platforms and walkways shall be as wide as feasible, and employees on those platforms and walkways shall be protected from fall hazards by the use of guardrails and/or personal fall arrest systems.

The front edge of all platforms shall not be more than 14 in (36 cm) from the face of the work, unless guardrail systems are erected along the front edge and/or personal fall arrest systems are used to protect employees from falling. The maximum distance from the face for outrigger scaffolds shall be 3 in (8 cm). The maximum distance from the face for plastering and lathing operations shall be 18 in (46 cm).

Allowable Span and Overhang Specifications for Platforms

Allowable spans for 2×10 in (nominal) or 2×9 in (rough) solid sawn wood planks are shown in Table 505.1.

The maximum permissible span for $1\ 1/4 \times 9$ in or wider wood plank of full thickness with a maximum intended load of 50 lb/ft^2 shall be 4 ft.

 NOTE Platform units used to make scaffold platforms intended for light-duty use shall be capable of supporting at least 25 lb/ft^2 applied uniformly over the entire unit-span area, or a 250-lb point load placed on the unit at the center of the span, whichever load produces the greater shear force.

Overhang Specifications

Unless cleated or otherwise restrained by hooks or equivalent means, each end of a platform shall extend over the centerline of its support at least 6 in (15 cm). Unless the platform is designed and installed so that the cantilevered portion of the platform is able to support employees and/or materials without tipping, or has guardrails which prevent access to the cantilevered end, platforms 10 ft or less in length shall not extend over its support more than 12 in (30 cm). For platforms greater than 10 ft in length, this overhang shall not exceed 18 in (46 cm).

Table 505.1 Allowable Span for Wood Planks

Maximum intended nominal load (psf)	Maximum permissible span using full thickness undressed lumber (ft)	Maximum permissible span using nominal thickness lumber (ft)
25	10	8
50	8	6
75	6	—

On scaffolds where platforms are overlapped to create a long platform, the overlap shall occur only over supports, and shall not be less than 12 in (30 cm) unless the platforms are nailed together or otherwise restrained to prevent movement.

At all points of a scaffold where the platform changes direction, such as turning a corner, any platform that rests on a bearer at an angle other than a right angle shall be laid first, and platforms which rest at right angles over the same bearer shall be laid second, on top of the first platform.

Criteria for Supported Scaffolds

Supported scaffolds with a height to base width (including outrigger supports, if used) ratio of more than four to one (4:1) shall be restrained from tipping by guying, tying, bracing, or equivalent means.

Supported scaffold poles, legs, posts, frames, and uprights shall bear on base plates and mud sills or other adequate firm foundation. Footings shall be level, sound, rigid, and capable of supporting the loaded scaffold without settling or displacement. Supported scaffold poles, legs, posts, frames, and uprights shall be plumb and braced to prevent swaying and displacement.

Criteria for Suspension Scaffolds

All suspension scaffold support devices, such as outrigger beams, cornice hooks, parapet clamps, and similar devices, shall rest on surfaces capable of supporting at least four times the load imposed on them by the scaffold operating at the rated load of the hoist or at least 1.5 times the load imposed on them by the scaffold at the stall capacity of the hoist, whichever is greater.

The inboard ends of suspension scaffold outrigger beams shall be stabilized by bolts or other direct connections to the floor or roof deck, or they shall have their inboard ends stabilized by counterweights, except mason's multipoint adjustable suspension scaffold outrigger beams shall not be stabilized by counterweights.

Outrigger beams which are not stabilized by bolts or other direct connections to the floor or roof deck shall be secured by tiebacks. Tiebacks shall be equivalent in strength to the suspension ropes.

Suspension scaffold outrigger beams shall be provided with stop bolts or shackles at both ends, securely fastened together with the flanges turned out when channel iron beams are used in place of I-beams, installed with all bearing supports perpendicular to the beam centerline, set and maintained with the web in a vertical position; and when an outrigger beam is used, the shackle or clevis with which the rope is attached to the outrigger beam shall be placed directly over the centerline of the stirrup.

When winding drum hoists are used on a suspension scaffold, they shall contain not less than four wraps of the suspension rope at the lowest point of scaffold travel. When other types of hoists are used, the suspension ropes shall be long enough to allow the scaffold to be lowered to the level below without the rope end passing through the hoist, or the rope end shall be configured or provided with means to prevent the end from passing through the hoist.

Access

When scaffold platforms are more than 2 ft (0.6 m) above or below a point of access, portable ladders, hook-on ladders, attachable ladders, stair towers, stairway-type ladders, ramps, walkways, integral prefabricated scaffold access, or direct access from another surface shall be used. Cross braces shall not be used as a means of access.

Hook-on and attachable ladders shall be positioned so that their bottom rung is not more than 24 in (61 cm) above the scaffold supporting level. When hook-on and attachable ladders are used on a supported scaffold more than 35-ft (10.7-m) high, they shall have rest platforms at no more than 35-ft (10.7-m) vertical intervals. Hook-on and attachable ladders shall be specifically designed for use with the type of scaffold used. Hook-on and attachable ladders shall have a minimum rung length of 11 1/2 in (29 cm) and shall have uniformly spaced rungs with a maximum spacing between rungs of 16 3/4 in.

Stairway-type ladders must have slip-resistant treads on all steps and landings and shall be positioned such that their bottom step is not more than 24 in (61 cm) above the scaffold supporting level. These ladders shall be provided with rest platforms at vertical intervals no exceeding 12 ft (3.7 m). Minimum step width on these ladders shall be 16 in (41 cm), except that mobile scaffold stairway-type ladders shall have a minimum step width of 11 1/2 in (30 cm). A landing platform at least 18-in wide by 18 in long shall be provided at each level. Each scaffold stairway shall be at least 18-in (45.7-cm) wide between stairrails. Stairways shall be installed between 40° and 60° from the horizontal.

Ramps and Walkways

Ramps and walkways 6 ft (1.8 m) or more above lower levels shall have a guardrail system installed. No ramp or walkway shall be inclined more than a slope of $1V{:}3H$. If the slope of a ramp or a walkway is steeper than $1V{:}8H$, the ramp or walkway shall have cleats not more than 14-in (35-cm) apart which are securely fastened to the planks to provide footing.

Integral prefabricated scaffold access frames shall be specifically designed and constructed for use as ladder rungs. The minimum rung length shall be 8 in (20 cm) and shall not be used as work platforms when rungs are less than 11 1/2 in in length, unless each affected employee uses fall protection, or a positioning device. These frames should be uniformly spaced within each frame section and be provided with rest platforms at 35-ft (10.7-m) maximum vertical intervals on all supported scaffolds more than 35-ft (10.7-m) high and have a maximum spacing between rungs of 16 3/4 in (43 cm). Nonuniform rung spacing caused by joining end frames together is allowed, provided the resulting spacing does not exceed 16 3/4 in (43 cm). Steps and rungs of ladder and stairway type access should line up vertically with each other between rest platforms.

Direct access to or from another surface shall be used only when the scaffold is not more than 14 in (36 cm) horizontally and not more than 24 in (61 cm) vertically from the other surface. Hook-on or attachable ladders shall be installed as soon as scaffold erection has progressed to a point that permits safe installation and use. Cross braces on tubular welded frame scaffolds shall not be used as a means of access or egress.

Clearance between Scaffolds and Power Lines

For insulated lines carrying voltage less than 300 V, the minimum lateral distance is 3 ft (0.9 m). For voltage between 300 V and 50 kV, the minimum lateral distance is to be 10 ft (3.1 m). For voltage exceeding 50 kV, the lateral distance is to be 10 ft plus 0.4 in for each kV over 50 kV.

For uninsulated lines, for voltage less than 50 kV, the minimum lateral distance is to be 10 ft (3.1 m). For voltage exceeding 50 kV, the lateral distance is to be 10 ft plus 0.4 in for each kV over 50 kV.

Scaffolds and materials may be closer to power lines than specified above where such clearance is necessary for performance of work, and only after the utility company, or electrical system operator, has been notified of the need to work closer and the utility company, or electrical system operator, has deenergized the lines, relocated the lines, or installed protective coverings to prevent accidental contact with the lines.

Fall Protection

Each employee on a scaffold more than 10 ft (3.1 m) above a lower level shall be protected from falling to that lower level. Each employee on a boatswain's chair, catenary scaffold, float scaffold, needle beam scaffold, or ladder jack scaffold shall be protected by a personal fall arrest system. Each employee on a single-point or two-point adjustable suspension scaffold shall be protected by both a personal fall arrest system and guardrail system. Each employee on a crawling board (chicken ladder) shall be protected by a personal fall arrest system, a guardrail system (with minimum 200-lb toprail capacity), or by a 3/4-in (1.9-cm) diameter grabline or equivalent handhold securely fastened beside each crawling board. Each employee on a self-contained adjustable scaffold shall be protected by a guardrail system (with minimum 200-lb toprail capacity) when the platform is supported by the frame structure, and by both a personal fall arrest system and a guardrail system (with minimum 200-lb toprail capacity) when the platform is supported by ropes. Each employee on a walkway located within a scaffold shall be protected by a guardrail system (with minimum 200-lb toprail capacity) installed within 9 1/2 in (24.1 cm) of and along at least one side of the walkway. Each employee performing overhand bricklaying operations from a supported scaffold shall be protected from falling from all open sides and ends of the scaffold (except at the side next to the wall being laid) by the use of a personal fall arrest system or guardrail system (with minimum 200-lb toprail capacity).

Personal fall arrest systems used on scaffolds shall be attached by lanyard to a vertical lifeline, horizontal lifeline, or scaffold structural member. When vertical lifelines are used, they shall be fastened to a fixed safe point of anchorage, shall be independent of the scaffold, and shall be protected from sharp edges and abrasion. Safe points of anchorage include structural members of buildings, but do not include standpipes, vents, other piping systems, electrical conduit, outrigger beams, or counterweights. When horizontal lifelines are used, they shall be secured to two or more structural members of the scaffold, or they may be looped around both suspension and independent suspension lines (on scaffolds so equipped) above the hoist and brake attached to the end of the scaffold. Horizontal lifelines shall not be attached only to the suspension ropes.

When lanyards are connected to horizontal lifelines or structural members on a single-point or two-point adjustable suspension scaffold, the scaffold shall be equipped with additional independent support lines and automatic locking devices capable of stopping the fall of the scaffold in the event one or both of the suspension ropes fail. The independent support lines shall be equal in number and strength to the suspension ropes.

Vertical lifelines, independent support lines, and suspension ropes shall not be attached to each other, nor shall they be attached to or use the same point of anchorage, nor shall they be attached to the same point on the scaffold or personal fall arrest system.

Guardrail systems shall be installed along all open sides and ends of platforms. Guardrail systems shall be installed before the scaffold is released for use by employees other than erection/dismantling crews.

The top edge height of toprails or equivalent member on supported scaffolds manufactured or placed in service after January 1, 2000, shall be installed between 38 in (0.97 m) and 45 in (1.2 m) above the platform surface. The top edge height on supported scaffolds manufactured and placed in service before January 1, 2000, and on all suspended scaffolds where both a guardrail and a personal fall arrest system are required shall be between 36 in (0.9 m) and 45 in (1.2 m).

Each toprail or equivalent member of a guardrail system shall be capable of withstanding, without failure, a force applied in any downward or horizontal direction at any point along its top edge of at least 100 lb (445 N) for guardrail systems installed on single-point adjustable suspension scaffolds or two-point adjustable suspension scaffolds, and at least 200 lb (890 N) for guardrail systems installed on all other scaffolds.

Midrails, screens, mesh, intermediate vertical members, solid panels, and equivalent structural members of a guardrail system shall be capable of withstanding, without failure, a force applied in any downward or horizontal direction at any point along the midrail or other member of at least 75 lb (333 N) for guardrail systems with a minimum 100-lb toprail capacity, and at least 150 lb (666 N) for guardrail systems with a minimum 200-lb toprail capacity.

Protection from Falling Objects

In addition to wearing hardhats each employee on a scaffold shall be provided with additional protection from falling hand tools, debris, and other small objects through the installation of toeboards, screens, or guardrail systems, or through the erection of debris nets, catch platforms, or canopy structures that contain or deflect the falling objects.

Where there is a danger of tools, materials, or equipment falling from a scaffold and striking employees below, the area below the scaffold to which objects can fall shall be barricaded, and employees shall not be permitted to enter the hazard area. Otherwise, a toeboard shall be erected along the edge of platforms more than 10 ft (3.1 m) above lower levels for a distance sufficient to protect employees below, except on float (ship) scaffolds where an edging of 3/4- × 1 1/2-in (2- × 4-cm) wood or equivalent may be used in lieu of toeboards. Where tools, materials, or equipment are piled to a height higher than the top edge of the toeboard, employees shall be protected by either (1) paneling or screening extending from the toeboard or platform to the top of the guardrail; (2) a guardrail system with openings small enough to prevent passage of potential falling objects; or (3) a canopy structure, debris net, or catch platform strong enough to withstand the impact forces of the potential falling objects.

Specifications for Guardrails

Guardrails shall satisfy the following criteria:

1. Toprails shall be equivalent in strength to 2- by 4-in lumber; or 1 1/4- × 1 1/4- × 1/8-in structural angle iron; or 1- × 0.070-in wall steel tubing; or 1.990- × 0.058-in wall aluminum tubing.

2. Midrails shall be equivalent in strength to 1- by 6-in lumber; or 1 1/4- × 1 1/4- × 1/8-in structural angle iron; or 1- × 0.070-in wall steel tubing; or 1.990- × 0.058-in wall aluminum tubing.

3. Toeboards shall be equivalent in strength to 1- by 4-in lumber; or 1 1/4- × 1 1/4- × 1/8-in structural angle iron; or 1- × 0.070-in wall steel tubing; or 1.990- × 0.058-in wall aluminum tubing.

4. Posts shall be equivalent in strength to 2- by 4-in lumber; or 1 1/4- × 1 1/4- × 1/8-in structural angle iron; or 1- × 0.070-in wall steel tubing; or 1.990- × 0.058-in wall aluminum tubing.

5. Distance between posts shall not exceed 8 ft.

6. Overhead protection shall consist of 2-in nominal planking laid tight, or 3/4-in plywood.

7. Screen installed between toeboards and midrails or toprails shall consist of No. 18 gauge U.S. Standard wire 1-in mesh.

Examples of Scaffolds

Suspended Scaffolds: Single-Point Adjustable

A single-point adjustable scaffold (see Fig. 505.1) consists of a platform suspended by one rope from an overhead support and equipped with means to permit the movement of the platform to desired work levels. The most common among these is the scaffold used by window washers to clean the outside of a skyscraper (also known as a boatswain's chair).

The supporting rope between the scaffold and the suspension device must be kept vertical unless the rigging has been designed by a qualified person, the scaffold is accessible to rescuers,

Figure 505.1
Single-point
suspended
scaffold.

the support rope is protected from rubbing during direction changes, and the scaffold is positioned so swinging cannot bring it into contact with other surfaces. The maximum intended load for these single-point adjustable suspension scaffolds is 250 lb.

Suspended Scaffolds: Two-Point (Swing Stage)

Two-point adjustable suspension scaffolds, also known as swing-stage scaffolds, are perhaps the most common type of suspended scaffold. Hung by ropes or cables connected to stirrups at each end of the platform, they are typically used by window washers on skyscrapers, but play a prominent role in high-rise construction as well (see Fig. 505.2).

Figure 505.2
Two-point
suspended
scaffolds.

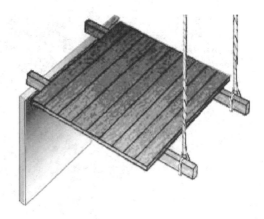

Figure 505.3
Suspended
scaffold on
needle beams.

Suspended Scaffolds: Needle Beams

This simple type of scaffold consists of a platform suspended from needle beams, usually attached on one end to a permanent structural member (see Fig. 505.3). Scaffold support beams must be installed on edge. Ropes or hangers must be used for supports. (Exception: One end of the scaffold may be supported by a permanent structural member). Ropes must be securely attached to needle beams. Support connections must be arranged to prevent the needle beam from rolling or becoming displaced. Platform units must be attached by bolts or equivalent means. Cleats and overhang are not considered adequate means of attachment.

For a maximum intended load of 25 lb/ft^2, beams must be 4- × 6-in in cross section, with a maximum beam span of 10 ft, and the platform span must be no more than 8 ft. Must be attached to the needle beam by a scaffold hitch or eye splice. The loose end of the rope must be tied by a bowline knot or a round turn and a half hitch.

Rope strength must at least be equal to 1-in-diameter, first-grade manila rope.

Suspended Scaffolds: Catenary

A catenary scaffold is a scaffold consisting of a platform supported by two essentially horizontal and parallel ropes attached to structural members of a building or other structure (see Fig. 505.4).

Catenary scaffolds may not have more than one platform between consecutive vertical pickups, and more than two platforms altogether. Platforms supported by wire rope must have hook-shaped stops on each of the platform to prevent them from slipping off the wire ropes. These hooks must be positioned so that they prevent the platform from falling if one of the horizontal wire ropes breaks.

Wire ropes must not be overtightened to the point that a scaffold load will overstress them. Wire ropes must be continuous and without splices between anchors. Each employee on a catenary scaffold must be protected by a personal fall-arrest system.

Catenary scaffolds have a maximum intended load of 500 lb. No more than two employees at a time are permitted on a catenary scaffold. Maximum capacity of come-along is 2000 lb. Vertical pickups must be spaced no more than 50-ft apart. Ropes must be equivalent in strength to at least 1/2-in-diameter improved plow steel wire rope.

Figure 505.4
Suspended
scaffold—
catenary.

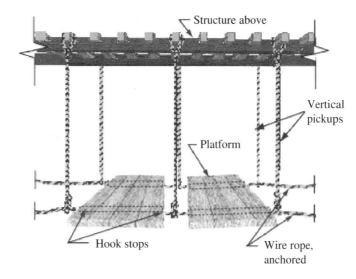

Suspended Scaffolds: Interior Hung

An interior hung suspension scaffold consists of a platform suspended from the ceiling or roof structure by fixed length supports (see Fig. 505.5).

Interior hung scaffolds must be suspended from roof structures (e.g., ceiling beams). Roof structures must be inspected for strength before scaffolds are erected. Suspension ropes/cables

Figure 505.5
Suspended
scaffold—interior
hung.

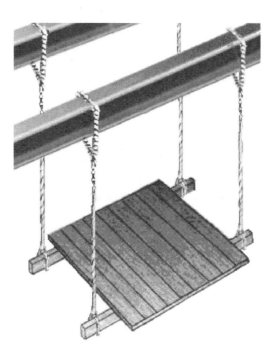

Chapter 505

1098

Figure 505.6
Suspended
scaffold—
multilevel.

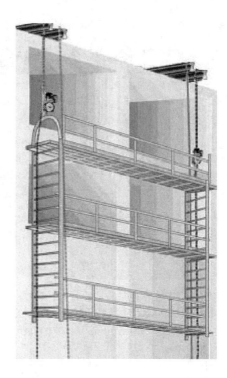

must be connected to overhead supports by shackles, clips, thimbles, or equivalent means. Bearers must have dimensions of 2×10 in, and be used on edge.

For an intended maximum load of 25 to 50 lb/ft^2, the maximum span is 10 ft. For an intended maximum load of 75 lb/ft^2, the maximum span is 7 ft.

Suspended Scaffolds: Multilevel

A multilevel scaffold is a two-point or multipoint adjustable suspension scaffold with a series of platforms at various levels resting on common stirrups (see Fig. 505.6).

Multilevel suspended scaffolds must be equipped with additional independent support lines that are equal in number to the number of points supported, equal in strength to the suspension ropes, and rigged to support the scaffold if the suspension ropes fail. Independent support lines and suspension ropes must not be anchored to the same points. Supports for platforms must be attached directly to support stirrups (not to other platforms).

Formwork for Concrete

Terminology

Formwork is the total system of support for freshly placed or partially cured concrete, including the form that is in contact with the concrete as well as all supporting members including shores, reshores, hardware, braces, and related hardware.

Shoring is the name given to supporting members that resist a compressive force imposed by a load. Reshoring is the operation in which shoring equipment (also called reshores or reshoring equipment) is placed, as the original forms and shores are removed, in order to support partially cured concrete and construction loads.

Lift slab process refers to a method of concrete construction in which floor and roof slabs are cast on or at ground level and lifted into position using jacks.

Vertical slip forms are jacked vertically during the placement of concrete. *Jacking* refers to the operation of lifting a slab or group of slabs vertically from one location to another during the construction of a building/structure where the lift-slab process is being used.

Limited access zone refers to an area alongside a masonry wall, which is under construction, and which is clearly demarcated to limit access by employees.

General

For cast-in-place (CIP) concrete construction, forms are used to mold the concrete to desired size and shape and to control its position and alignment. Formwork is a temporary structure that must support its own weight and the vertical and lateral loads imposed by the freshly placed concrete in addition to construction live loads (including materials, equipment, and personnel).

Well-designed formwork must be erected quickly, permit high loads for a few hours during the concrete placement and be disassembled after a short period for future use. Form building has three basic objectives:

Safety: The formwork must have a stable configuration that provides adequate safety for workers and the structure.

Quality: The formwork must have adequate strength to support all loads (materials, personnel, and equipment) and have accuracy of position and dimensions to ensure a quality structure.

Economy: Economy is a major concern since formwork costs constitute approximately half the total cost of concrete work in a typical project. In designing and building formwork, the contractor should aim for maximum economy without sacrificing quality or safety.

Materials

Lumber was once the predominant form material, with formwork being destroyed after a single use. Recent developments in the use of plywood, metal, plastics, and other materials and accessories have led to increased prefabrication of formwork. Today, modular panel forming is the norm. Increased reuse of prefabricated forms has led to significant economy.

Size, shape, and alignment of slabs, beams, and other concrete structural elements depend on accurate construction of the forms. The forms must be sufficiently rigid under the construction loads to maintain the designed shape of the concrete, stable and strong enough to maintain large members in alignment, and substantially constructed to withstand handling and reuse without losing their dimensional integrity. The formwork must remain in place until the concrete is strong enough to carry its own weight, or the finished structure may be damaged.

Causes of Formwork Failure

Formwork failures are the cause of many large accidents and failures that occur during concrete construction. The period during which formwork is most susceptible is during the placement of fresh concrete. Generally some unexpected event causes a localized failure and the resulting redistribution of loads causes other members to become overloaded, leading to progressive collapse of the entire formwork structure. Some of the main causes of formwork failure are as follows:

1. *Inadequate bracing*: The use of diagonal bracing, guys, and struts improves the overall stability of the shoring forms, particularly those of significant height. Inadequate bracing may lead to a local collapse to extend to a large portion of the structure.

2. *Improper stripping and shore removal*: Premature stripping of forms and premature removal of shores can cause catastrophic failures.

3. *Vibration*: When supporting shores or jacks are displaced by vibration caused by traffic, concrete placement, or movement of workers and equipment, formwork failures can occur. Diagonal bracing can help prevent failure due to vibration.

4. *Unstable soil under mudsills, shoring not plumb*: A mudsill is a plank, frame, or small footing on the ground used as a base for a shore or post in formwork. Excessive settlement of weak soil, thawing of frozen ground under the mudsill, resulting in additional settlement and loss of support, inadequate site drainage leading to washing out of supporting soil, can all be causes of form failure.

5. *Temperature and rate*: The temperature and rate of vertical placement of concrete are factors influencing the development of lateral pressures that act on the forms. If temperature drops during construction operations, rate of concreting often has to be slowed down to prevent a build up of lateral pressure overloading the forms.

6. *Poor quality detailing*: Insufficient nailing, failure to tighten the locking devices on metal shoring, inadequate provisions to prevent rotation of beam forms where slabs frame into them on the side, inadequate anchorage against uplift for sloping form faces, lack of bracing or tying of corners, bulkheads, or other places where unequal pressure is found, can all lead to form failure.

Shoring/Reshoring of Concrete Multistory Buildings

This section summarizes the guidelines of Committee ACI-347 (Guide for Shoring/Reshoring of Concrete Multistory Buildings—ACI 347.2R-05). The report uses the following definitions:

Preshores: These are added shores placed snugly under selected panels of a deck-forming system before any primary (original) shores are removed. Preshores and the panels they support remain in place until the remainder of the complete bay has been stripped and back-shored, a small area at a time.

Reshores: These are shores placed snugly under a stripped concrete slab or other structural member after the original forms and shores have been removed from a large area, requiring the new slab or structural member to deflect and support its own weight and existing construction loads applied before the installation of the reshores.

Shores: These are vertical or inclined support members designed to carry the weight of the formwork, concrete, and construction loads above.

ACI Committee 347 recommends that both vertical supports and horizontal framing components of formwork be designed for a minimum live load of 50 lb/ft² (2.4 kPa) of horizontal projection to provide for weight of workers, runways, screeds, and other equipment. When motorized carts are used, the minimum live load should be 75 lb/ft² (3.6 kPa). The minimum design value for combined dead and live loads should be 100 lb/ft² (4.8 kPa) or 125 lb/ft² (6.0 kPa) when motorized carts are used.

In a typical construction cycle for a multistory cast-in-place concrete building where both shores and reshores are used, there are four construction phases:

- Phase 1: Installation of the shores and formwork followed by the casting of the floor slab;
- Phase 2: Once the concrete slab has reached a target strength threshold, removal of the shores and formwork allowing the slab to deflect and carry its own weight;
- Phase 3: Removal of reshores at the lowest interconnected level; and
- Phase 4: Placement of reshores in the story from which the shores and forms were removed.

The reshores are placed snugly without initially carrying any load.

If only shores are used, then the third and fourth phases are eliminated. According to ACI Committee 347, the reshores should be installed snugly under the slab just stripped so that they are relatively load-free upon installation. This stripping procedure allows the slab to deflect under its own weight and the reshores are installed without preload.

Construction Load Distribution

The construction load distribution between the concrete slabs and the shoring/reshoring system is evaluated by using the simplified method. Although this example utilizes a wood shoring/reshoring system, it is assumed that the compressibility of the shoring/reshoring system does not significantly impact construction load redistribution.

Example 506.1

In the following example, one level of shores and two levels of reshores are used for construction of a multistory concrete building. The following symbols are assumed and used for this example:

D = weight of slab

Construction live load = $0.4D$

Shore and form weight = $0.05D$

Reshores weight negligible

The table below outlines the steps in the shoring/reshoring operation. The following symbols are used:

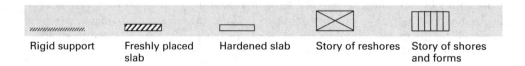

Rigid support   Freshly placed slab   Hardened slab   Story of reshores   Story of shores and forms

	Description of operation	Structure view	Load on slab			Load in shores/ reshores at the end of operation
			At start of operation	Change during operation	At the end of operation	
1	Place level 1 concrete. Floor is supported by forms and shores.		0	0	0	1.45 D
2	Level 1 slab hardens and carries own weight. Remove level 1 shores, allow slab to deflect, place reshores snug but not loaded. Construction load is gone.		0	+ 1.00 D	1.00 D	0
3	Form, shore, and place level 2 concrete. Slab 1 cannot deflect further, so all load goes through reshores to ground.		0 1.00 D	0 + 0	0 1.00 D	1.45 D 1.45 D
4	Slab 2 hardens. Construction live load is gone. Remove level 2 forms and shores, allow slab 2 to deflect. Reshore slab 2 snugly without loading reshores.		0 1.00 D	+ 1.00 D 0	1.00 D 1.00 D	0 0
5	Form, shore, and place level 3 concrete. All added load on slab 3 is carried to ground through shore and reshores.		0 1.00 D 1.00 D	0 0 0	0 1.00 D 1.00 D	1.45 D 1.45 D 1.45 D
6	Slab 3 hardens. Construction load is gone. Level 1 reshores are flown up to level 3.		0 1.00 D 1.00 D	+1.00 D 0 0	1.00 D 1.00 D 1.00 D	0 0
7	Form, shore, and place level 4 concrete. The total load is distributed equally to the three interconnected slabs. (1.45D÷3 = 0.48D)		0 1.00 D 1.00 D 1.00 D	0 +0.48 D +0.48 D +0.48 D	0 1.48 D 1.48 D 1.48 D	1.45 D 0.97 D 0.49 D 0
8	Slab 4 hardens. Construction live load (0.45D) is removed equally from interconnected slabs.		0 1.48 D 1.48 D 1.48 D	0 −0.15 D −0.15 D −0.15 D	0 1.33 D 1.33 D 1.33 D	1.05 D 0.72 D 0.39 D 0
9	Remove forms and shores supporting slab 4, allowing it to carry its own weight. The load in the shores is equally removed from the interconnected slabs. This step is now identical to step 6 and can repeat from step 7.		0 1.33 D 1.33 D 1.33 D	+1 D −0.35 D −0.35 D −0.35 D	1 D 0.98 D 0.98 D 0.98 D	0 0 0

Design Loads on Concrete Formwork

Vertical Loads

Gravity loads for formwork are dead loads and live loads. These are typically given as load per unit form contact area (lb/ft²). The dead load is defined as the weight of the reinforced concrete plus the weight of the formwork. The *live load* is defined as additional loads imposed during the process of construction such as material storage, personnel, and equipment. Some typical values are given below:

> *Dead loads:* Concrete and rebar (130–160 lb/ft³), formwork (at least 10 lb/ft²).

> *Live loads*: Minimum 50 psf (75 psf if motorized carts are used) includes the following loads—personnel, equipment, mounting of concrete, impact. Formwork impact load is a resulting load from dumping of concrete or the starting and stopping of construction equipment on the formwork. An impact load may be several times a design load.

The sum of dead load and live load shall not be taken less than 100 psf (125 psf if motorized carts are used).

Lateral Loads

Lateral loads shall be taken as given below:

> *For slab forms*: The greater of 100 lb/ft of slab edge or 2% of total dead load.

> *For wall forms*: The greater of 15 psf or local code requirements for wind load or 100 lb/ft of wall (for wall height >8 ft).

Lateral Pressure from Fresh Concrete

The lateral pressure imposed by concrete on a wall form is a function of the following primary factors: (1) density of concrete, (2) temperature of the concrete at the time of placing, (3) rate of concrete placement (feet of height per hour), and (4) height of concrete placement. For walls and columns poured rapidly, the lateral pressure from wet concrete is to be taken as the hydrostatic pressure:

$$p_{max} = w \times h \qquad\qquad (506.1)$$

where w = unit weight of the fresh concrete
h = full height of the form or the distance between horizontal construction joints

According to ACI 347[1], the lateral pressure C_c (from fresh concrete) on wall and column forms is given by the following empirical relationships, subject to the following conditions:

- Concrete unit weight is 150 lb/ft³.
- Concrete slump does not exceed 7 in.

---

[1] ACI 347 (2004), *A Guide to Formwork for Concrete*, Detroit, Michigan.

- Concrete is subjected to normal internal vibration to a depth of 4 ft or less.
- Concrete temperature is in the range of 40 to 90°F.

Walls

For rate of concrete placement $R < 7$ ft/h and placement height not exceeding 14 ft, the base value of lateral pressure is given by

$$p_{max} = C_w C_c \left(150 + 9000 \frac{R}{T} \right)$$ (506.2)

walls with placement rate less than 7 ft/h where placement height exceeds 14 ft and for all walls with placement rate 7 to 15 ft/h

$$p_{max} = C_w C_c \left(150 + \frac{43400}{T} + 2800 \frac{R}{T} \right)$$ (506.3)

where R = rate of concrete placement (ft/h)
T = temperature of the concrete in the forms (°F)

The pressure computed by this equation is subject to a minimum value of 600 psf and a maximum value of 2000 psf, and should never be taken to be greater than the hydrostatic pressure $p = 150h$. Values of lateral pressure on wall forms are summarized in Table 506.1.

Table 506.1 Concrete Pressure on Wall Forms

Rate of placement, ft/h	Maximum lateral pressure (psf) for temperature indicated					
	40°F	50°F	60°F	70°F	80°F	90°F
1	600	600	600	600	600	600
2	600	600	600	600	600	600
3	825	690	600	600	600	600
4	1050	870	750	664	600	600
5	1275	1050	900	793	713	650
6	1500	1230	1050	921	825	750
7	1725	1410	1200	1050	938	850
8	1795	1466	1247	1090	973	881
9	1865	1522	1293	1130	1008	912
10	1935	1578	1340	1170	1043	943

Slip-Form Construction

For slip-form concrete operation, the lateral pressure of fresh concrete in designing forms, braces, and wales is to be taken as

$$p_{max} = C_w C_c \left(C + 6000 \frac{R}{T} \right)$$ (506.4)

where $C = 100$ psf for concrete placed in 6- to 10-in lifts with slight or no vibration
$C = 150$ psf for concrete that requires additional vibration

If concrete is pumped from the base of the form, the lateral pressure is given by

$$p_{max} = 1.25 \times w \times h$$ (506.5)

The 25% increase is a default value to account for pump surge pressure, unless specifically indicated as a higher value by the pump manufacturer.

Columns

For columns (defined as a vertical element with no plan dimension greater than 6.5 ft), concrete with slump less than 7 in. and placed with normal internal vibration to a depth of 4 ft or less the base value of lateral pressure is given by

$$p_{max} = C_w C_c \left(150 + 9000 \frac{R}{T} \right)$$ (506.6)

where R = rate of concrete placement (ft/h)
T = temperature of the concrete in the forms (°F)

The pressure computed by this equation is subject to a minimum value of 600 psf and a maximum value of 3000 psf, and should never be taken to be greater than $150h$. Values of lateral pressure on column forms are summarized in Table 506.2.

Modifications to Form Pressure

The base values of lateral design pressure within forms, as given by the equations above and summarized in Tables 506.1 and 506.2, should be modified for the unit weight of the concrete and for concrete chemistry. The modification is effected via multiplication by the factors in Tables 506.3 and 506.4.

Example 506.2

Determine the lateral design pressure for the placement of a 12-ft-high wall in which light-weight concrete is placed at a rate of 4 ft/h. The temperature of the concrete is 60°F and the unit weight is 130 pcf. The concrete is cast with Type I cement without any retarding admixtures.

Table 506.2 Concrete Pressure on Column Forms

Rate of placement, ft/h	Maximum lateral pressure (psf) for temperature indicated					
	40°F	50°F	60°F	70°F	80°F	90°F
1	600	600	600	600	600	600
2	600	600	600	600	600	600
3	825	690	600	600	600	600
4	1050	870	750	664	600	600
5	1275	1050	900	793	713	650
6	1500	1230	1050	921	825	750
7	1725	1410	1200	1050	938	850
8	1950	1590	1350	1179	1050	950
9	2175	1770	1500	1307	1163	1050
10	2400	1950	1650	1436	1275	1150
11	2625	2130	1800	1564	1388	1250
12	2850	2310	1950	1693	1500	1350
13	3000	2490	2100	1821	1613	1450
14	3000	2670	2250	1950	1725	1550
15	3000	2850	2400	2079	1838	1650
16	3000	3000	2550	2207	1950	1750
17	3000	3000	2700	2336	2063	1850
18	3000	3000	2850	2464	2175	1950
19	3000	3000	3000	2593	2288	2050
20	3000	3000	3000	2721	2400	2150
21	3000	3000	3000	2850	2513	2250
22	3000	3000	3000	2979	2625	2350
23	3000	3000	3000	3000	2738	2450
24	3000	3000	3000	3000	2850	2550
25	3000	3000	3000	3000	2963	2650
26	3000	3000	3000	3000	3000	2750
27	3000	3000	3000	3000	3000	2850
28	3000	3000	3000	3000	3000	2950
29	3000	3000	3000	3000	3000	3000

Table 506.3 Modification to Form Pressure due to Concrete Weight (C_w)

Multiplication factor for unit weight of concrete not to be less than 0.8	
$\gamma < 140$ pcf	$0.5\left(1 + \dfrac{\gamma}{145}\right)$
140 pcf $< \gamma < 150$ pcf	1.0
$\gamma > 150$ pcf	$\left(\dfrac{\gamma}{145}\right)$

Table 506.4 Modification to Form Pressure due to Concrete Chemistry (C_c)

Multiplication factor for concrete chemistry	
Types I, II, and III cement without retarders	1.0
Types I, II, and III cement with a retarder	1.2
Other types or blends containing less than 70% slag or 40% fly ash without retarders	1.2
Other types or blends containing less than 70% slag or 40% fly ash with a retarders	1.4
Blends containing more than 70% slag or more than 40% fly ash	1.4

Solution The base value for lateral pressure for wall forms for concrete temperature $=$ 60°F and $R = 4$ ft/h is 750 psf. The modification factor for concrete unit weight is

$$0.5\left(1 + \frac{130}{145}\right) = 0.95$$

Therefore, the lateral pressure on the forms is $0.95 \times 750 = 712$ psf.

Example 506.3

Normal weight concrete ($\gamma = 150$ pcf) is poured into a wall form at the rate of 5 ft/h. The average temperature during the pour is 60°F. The form uses transverse ties that are spaced every 3 ft horizontally and vertically. Each tie rod is 3/4-in diameter and has ultimate stress $\sigma_u = 80$ ksi. Determine the factor of safety of the tie rods.

Solution For a wall form, the lateral pressure from fresh concrete at 60°F and $R = 5$ ft/h is $p = 900$ psf. Therefore, the total axial tension in each tie rod (having a tributary area $= 3$ ft $\times$ 3 ft $= 9$ ft²) is 8100 lb. The ultimate axial tension is

$$F_u = \sigma_u \times A = 80 \times \frac{\pi \times 0.75^2}{4} = 35.34 \text{ kips}$$

$$\text{FS} = \frac{35.34}{8.1} = 4.4$$

Table 506.5 Minimum Factors of Safety for Formwork Accessories (ACI 347)

Component	Minimum required FS
Form ties	2.0
Form anchor	2.0 (if formwork supports form weight and concrete pressures only)
	3.0 (if formwork supports form weight, concrete pressures, construction live load, and impact)
Form hangers	2.0
Anchoring inserts used as form ties for precast concrete panels used as forms	2.0

Table 506.5 lists minimum factors of safety for various types of formwork accessories, as prescribed by ACI 347.

Formwork Components

In Fig. 506.1, a typical arrangement of vertical (studs) and horizontal members (wales) is shown. The plyform spans between the studs spaced a distance L_1 between centers.

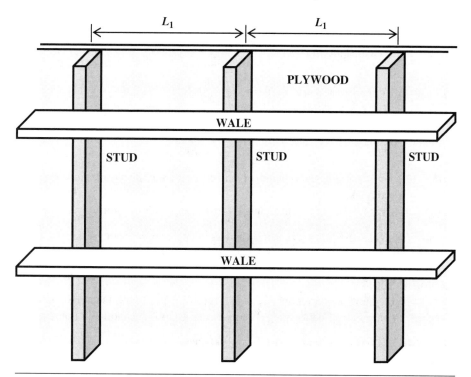

Figure 506.1 Parts of a typical wall form.

For a plyform supporting a uniform pressure behind it and supported by uniformly spaced supports, the required effective section modulus S (in³/ft) is given by

For one or two spans: $S = \dfrac{wL_1^2}{96F_b}$

For three or more spans: $S = \dfrac{wL_1^2}{120F_b}$

where w = uniform pressure (lb/ft²)
F_b = allowable bending stress (lb/in²)
S = required effective section modulus (in³/ft)
L_1 = span, center to center of supports (in)

When section choice is governed by shear stress, the required rolling shear constant Ib/Q (in²/ft) is given by

For one or two spans: $\dfrac{Ib}{Q} = \dfrac{wL_2}{19.2F_s}$

For three or more spans: $\dfrac{Ib}{Q} = \dfrac{wL_2}{20F_s}$

where w = uniform pressure (lb/ft²)
F_s = allowable rolling shear stress (lb/in²)
Ib/Q = required rolling shear constant (in²/ft)
L_2 = clear span (in)

Shear deflection is calculated as $\Delta_s = \dfrac{CwL_2^2 t^2}{1270EI}$

Bending deflection is calculated as $\Delta_b = \dfrac{wL_3^4}{1743EI}$

$C = 120$ for face grain across supports, and 60 for face grain parallel to supports
t = plywood thickness (in)
L_3 = clear span + ¼ in for 2-in framing (clear span + 5/8 in for 4-in framing)

Table 506.6 lists modulus of elasticity and allowable stresses for class I, class II, and structural plyform. See Table 114.6 for section properties of plyform.

Example 506.4

What is the recommended pressure for ¾-in Plyform Class I with face grain across supports spaced 16 in on center, if deflection is not to exceed $L/360$? Assume 2-in nominal framing.

Solution   For Class I ¾-in-thick plyform (see Table 114.6)
Moment of inertia, $I = 0.199$ in⁴/ft
Effective section modulus, $S = 0.455$ in³/ft

Table 506.6 Modulus of Elasticity and Allowable Stresses for Plyform

	Modulus of elasticity (lb/in²)		
	Plyform Class I	Plyform Class II	Structural I plyform
Use for bending deflection calculation	1,650,000	1,430,000	1,650,000
Use for shear deflection calculation	1,500,000	1,300,000	1,500,000
	Allowable Stress (lb/in²)		
Bending stress, F_b	1930	1330	1930
Rolling shear stress, F_s	72	72	102

Rolling shear constant, $Ib/Q = 7.187$ in²/ft
Allowable bending stress, $F_b = 1930$ psi
Allowable rolling shear stress, $F_s = 72$ psi

Based on bending stress, maximum pressure (assuming three spans),

$$w_b = \frac{120F_bS}{L_1^2} = \frac{120 \times 1930 \times 0.455}{16^2} = 412 \text{ psf}$$

Based on shear stress, maximum pressure (assuming three spans), using clear span, $L_2 = 16 - 1.5 = 14.5$ in

$$w_s = \frac{20F_s(Ib/Q)}{L_2} = \frac{20 \times 72 \times 7.187}{14.5} = 714 \text{ psf}$$

For load $w = 1$ psf, bending deflections is calculated as

$$\Delta_b = \frac{wL_3^4}{1743EI} = \frac{1.0 \times 14.75^4}{1743 \times 1,650,000 \times 0.199} = 8.3 \times 10^{-5}$$

For load $w = 1$ psf, shear deflections is calculated as

$$\Delta_s = \frac{120wL_2^2t^2}{1270EI} = \frac{120 \times 1.0 \times 14.5^2 \times 0.75^2}{1270 \times 1,500,000 \times 0.199} = 3.7 \times 10^{-5}$$

Total deflection for $w = 1$ psf is $\Delta = 1.2 \times 10^{-4}$ in.
Allowable deflection is $L/360 = 16/360 = 0.0444$ in
Therefore, allowable load based on maximum deflection is $0.0444 \div 1.2 \times 10^{-4} = 370$ psf
Governing load (based on deflection) = smallest of 412, 714, 370 = 370 psf

Concrete Maturity

The concept of concrete maturity is based on the principle that concrete's properties (such as in-place strength) are a function of time (age) and its temperature history. The complete temperature profile of a concrete during the process of curing gives an idea about the "maturity" of the concrete, which may be expressed in terms of a maturity index

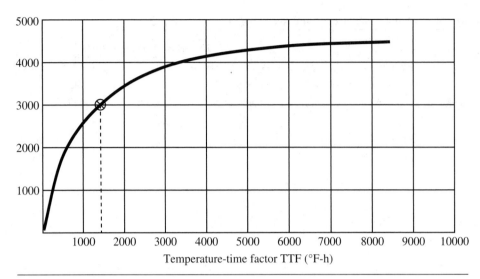

Figure 506.2 Representative concrete strength maturity curve (Nurse-Saul).

(equivalent age). Traditionally, testing of field-cured cylinders, cured under the same conditions as the structure, are used for process control, such as removal of forms, back-filling walls, opening of pavements, or bridges to traffic, and the like.

The main advantage of the concrete maturity concept is that it uses the actual tempera-ture profile (TTF or time-temperature function) of the concrete to predict strength and other parameters. Depending on the method used, this can be used to predict a maturity index of the concrete. The procedure for estimating concrete strength using the concept of concrete maturity is described in ASTM C1074[2]. The two methods recognized in the manual are the Nurse-Saul method and the Arrhenius method.

The Nurse-Saul function assumes that rate of strength development is directly proportional to the temperature. The maturity index is described in terms of a temperature-time factor (TTF) whose typical units are °C-h or °C-days, where the temperature is measured above an assumed datum temperature below which it is assumed no hydration occurs. ASTM 1074 suggests a default value of the datum temperature to be 0°C. A typical example is shown in Fig. 506.2.

Nurse-Saul Function

$$M = (T_a - T_o)\Delta t \tag{506.7}$$

where M = maturity (TTF) at age t
$\quad T_a$ = average concrete temperature during time interval Δt
$\quad T_o$ = datum temperature

---

[2]American Society for Testing and Materials, ASTM C1074: *Standard Practice for Estimating Concrete Strength by the Maturity Method,* 2004.

The *Arrhenius function* (which accounts for nonlinearity in the rate of cement hydration) assumes that rate of strength development is an exponential function of the temperature. The maturity index is described in terms of an "equivalent age" (which represents the equivalent duration of curing at the reference temperature that would result in the same value of maturity as the curing period for the given average temperature). The value of activation energy is needed for this method. ASTM 1074 suggests a default range of the activation energy of 40,000 to 45,000 J/mol (Type I cement).

$$t_e = \sum \left\{ e^{-\frac{E}{R}\left(\frac{1}{T_a} - \frac{1}{T_r}\right)} \Delta t \right\} \qquad (506.8)$$

where t_e = equivalent age at reference curing temperature
E = activation energy = 33,500 J/mol for $T \geq 20°C$
= 33,500 + 1472 (20–T) for $T < 20°C$
R = universal gas constant = 8.3144 J/(mol K)
T_a = average concrete temperature during time interval Δt, K
T_r = reference temperature, K
Δt = time interval

According to Carino (2004), the Arrhenius equation is a better representation of time-temperature function than the Nurse-Saul equation when a wide variation in concrete temperature is expected. Furthermore, the Nurse-Saul approach is limited in that it assumes that the rate of strength gain is a linear function. Nevertheless, the Nurse-Saul methodology is more widely used by many state highway agencies, largely because of its simplicity. Both maturity functions are outlined in ASTM C1074 (ASTM 2005a).

Example 506.5

Specifications for a construction job state that wall forms can be removed once the concrete reaches a compressive strength of 3000 psi. The maturity curve (Nurse-Saul) for the concrete is shown in Fig. 506.2. The cement hydration is assumed to cease below a temperature of 20°F. The temperature of the concrete at the time of the opening is 70°F. Estimate the length of time in days that the contractor has to wait before removing forms.

Solution From the curve, corresponding to a strength of 3000 psi, we obtain TTF = 1400°F-h.

Therefore, the time interval for sufficient maturity is given by

$$\Delta t = \frac{\text{TTF}}{T_f - T_o} = \frac{1400}{70 - 20} = 28 \text{ h} = 1.167 \text{ days}$$

The contractor has to wait at least 28 h after the concrete is placed before removing the forms.

Table 506.7 General Guidelines for Form Removal (ACI 347)

Element	Time to strip forms	
Walls	12 h	
Columns	12 h	
Sides of beams and girders	12 h	

Pan joist slabs
  Width $\leq$30 in
  Width >30 in
	3 days
	4 days	

Bottoms of joists, beams, and girders	Where design live loads are	
	< dead load	> dead load
Clear span <10 ft Clear span 10–20 ft Clear span >20 ft	7 days 14 days 21 days	4 days 7 days 14 days

One-way floor slabs
  Clear span <10 ft
  Clear span 10–20 ft
  Clear span >20 ft
	4 days
	7 days
	10 days
		3 days
		4 days
		7 days

Mass Concrete

ACI 116R[3] defines mass concrete as "any volume of concrete with dimensions large enough to require that measures be taken to cope with generation of heat from hydration of the cement and attendant volume change to minimize cracking." Mass concrete occurs in heavy civil engineering construction, such as in gravity dams, arch dams, gravity-retaining walls, lock walls, power-plant structures, and large building foundations. For such structures, heat of hydration is controlled by using cement or cementitious material possessing low or moderate heat generating characteristics, by postcooling (cooling the fresh concrete) or by placing sequence. Heat rise in mass concrete is most often controlled by replacement of cement with pozzolans, particularly fly ash.

Mass concrete falls into two distinct categories, namely, low-lift (heights of 5 to 10 ft) and high lift. Low-lift formwork usually consists of multiuse steel cantilever form units that incorporate their own work platforms. High-lift formwork is similar to the single-use wood forms used extensively for structural concrete.

Guidelines for Form Removal

In the absence of specific guidelines from the engineer in charge, forms may be stripped according to the timelines in Table 506.7.

Labor Costs for Forming

The various component costs in concrete work are material costs for concrete and reinforcement, cost of new and reused forms, and labor costs for form erection, concrete placement, and form stripping. In Example 506.5, these costs are computed for the construction of a concrete wall.

---

[3]Cement and Concrete Terminology, American Concrete Institute, Detroit, Michigan.

Example 506.6

A concrete wall that is 96-ft long × 12-ft high × 1-ft thick is to be built in four equal pours. The wall forms are 24-ft long × 12-ft high and will be reused. Assume 10% waste of concrete. The following unit costs are given:

Labor Rates

Carpenter	$36/h
Laborer	$26/h
Supervisor	$40/h

Materials

Formwork, initial erection	$3.10/ft²
Formwork, reuse	$0.76/ft²
Concrete	$100.00/yd³
Reinforcement	$120.00/yd³

The composition of crews and crew productivity for various operations are given below:

Crews

Erect and strip forms	4 carpenters + 2 laborers + 1 supervisor (working)
Place concrete	3 laborers + 1 carpenter + 1 supervisor (working)

Productivity

Erect forms	6.0 ft²/LH
Strip forms	15.0 ft²/LH
Place concrete	2.4 yd³/LH

What is the total cost of the wall?

Solution

$$\text{Total wall area} = 96 \times 12 = 1152 \text{ ft}^2$$

$$\text{Form area} = 2 \times 1152 = 2304 \text{ ft}^2$$

Of this area, since the wall is constructed in four equal pours, the first quarter (576 ft²) has a unit cost of $3.10/ft² and the remaining three quarter (1728 ft²) has a unit cost of $0.76/ft².

Volume of concrete placed = $96 \times 12 \times 1 = 1152 \text{ ft}^3 = 42.67 \text{ yd}^3$

Including 10% waste, concrete volume = 46.93 yd³

Hourly cost to erect and strip forms

$4 \times 36 + 2 \times 26 + 1 \times 40 = \$236/h = \$218.45 \div 7 = \$33.71/LH$

Hourly cost to place concrete

$$1 \times 36 + 3 \times 26 + 1 \times 40 = \$154/h = \$154 \div 5 = \$30.80/LH$$

Cost to erect forms = $\$33.71/LH \times 2304$ ft$^2 \div 6$ ft^2/LH = $\qquad$ \$12,945
Cost to strip forms = $\$33.71/LH \times 2304$ ft$^2 \div 15$ ft^2/LH = $\qquad$ \$ 5178
Cost to place concrete = $\$30.80/LH \times 42.67$ yd$^3 \div 2.4$ yd^3/LH = $\qquad$ \$ 548
Cost of materials: $\quad$ Formwork $\qquad \$3.10 \times 576 + 0.76 \times 1728 =$ \$ 3099
$\qquad\qquad\qquad\quad$ Reinforcement $\quad \$120.00 \times 42.67 =$ $\qquad$ \$ 5120
$\qquad\qquad\qquad\quad$ Concrete $\qquad\quad \$100.00 \times 46.93 =$ $\qquad$ \$ 4693
$\quad$ TOTAL $\qquad\qquad\qquad\qquad\qquad\qquad\qquad\qquad\qquad\qquad$ \$31,583

Falsework

In construction, falsework consists of temporary structures used to support structures and structural components in order to hold them in place until they are sufficiently rigid to support themselves. Falsework may include temporary support structures for formwork and scaffolding to give workers access to the structure being constructed.

There is some variation in the use of the terms "formwork" and "falsework." For example, ACI considers falsework to be included in formwork, but in some DOT literature, they are considered as distinct. In general, falsework is the combination of temporary structural elements that support formwork, while the primary purpose of formwork is to retain and support fresh concrete. Vertical forms are primarily used to resist pressure (from fresh concrete behaving like a fluid) and horizontal forms are used to resist weight.

Example 506.7

The figure below shows a cantilevered portion of a bridge deck. During construction, the formwork for the concrete deck is supported by the exterior girder (vertical support only) and a system of uniformly spaced braces as shown. The maximum allowable load for each brace is 5000 lb. Both ends of the brace may be assumed to be pinned. The cantilevered concrete slab has a weight of 1200 lb/ft and the parapet has a weight of 400 lb/ft. What is the maximum longitudinal spacing between braces?

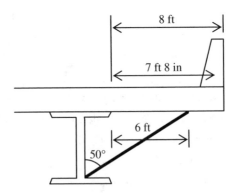

Solution A simple free-body diagram of the falsework is shown below:

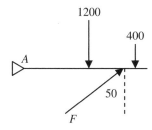

Taking moments about the assumed hinge at A, we have

$$\sum M_A = 1200 \times 4 + 400 \times 7.67 - F\cos 50 \times 6 = 0$$

Solving this, we get the compression force in the brace $F = 2040$ lb/ft. Therefore, the required maximum spacing on the braces is given by

$$S = \frac{5000 \text{ lb}}{2040 \text{ lb/ft}} = 2.45 \text{ ft}$$

Slipforming

Slipforming is a forming process in which a special form assembly slips or moves in a particular direction leaving the formed concrete in place. The vertical or horizontal movement of forms can be a continuous process or a sequence of discrete placements. The rate of movement of the forms is regulated so that the concrete is left unsupported only after it is stiff enough to retain its shape while supporting its own weight and the incidental loads during construction.

Slipforming can be used for vertical structures such as silos, storage bins, piers, chimneys, and other similar structures. Horizontal slipforming can be used for structures such as tunnel linings, water conduits, drainage channels, canal linings, pavements, curbs, and retaining walls.

Concrete to fill otherwise inaccessible areas can be placed pneumatically or by positive displacement pump and pipeline. When using slipforms for underground structures, rock is sometimes used as a form backing, permitting the use of rock anchors and tie rods instead of external bracing and shores.

Formwork design loads for underground structures are similar to those for above-ground structures, except for unusual vertical loads occurring near the crown of arch or tunnel forms and flotation or buoyancy effect beneath tunnel forms. In placing concrete in the crowns of tunnel forms, pressures up to 3000 psf have been induced in areas of overbreak and near vertical bulkheads from concrete placed pneumatically or by positive displacement pump. The assumed pressure should not be less than 1500 psf acting normally to the form plus the dead weight of the concrete.

Overbreak is the excess removal of rock or other excavated material above the forms beyond the required tunnel lining thickness.

Tunnel Placement Techniques

The two basic methods of placing a tunnel arch are commonly known as the *bulkhead method* and the *continuously advancing slope method*. The advancing slope method, a continuous method of placement, is usually preferred for tunnels driven through competent rock, ranging between 10 and 25 ft in diameter and at least 1 mile in length.

The bulkhead method is used exclusively where poor ground conditions exist, requiring the lining to be placed concurrently with tunnel driving operations. It is also used when some site-specific factor precludes the advancing slope method.

Materials

Usually, tunnel and shaft forms are made of steel or a composite of wood and steel. Experience is important in the design and fabrication of a satisfactory tunnel form due to the nature of the pressures developed by the concrete, placing techniques and the high degree of mobility required.

When reuse is not a factor, plywood and tongue-and-groove lumber are sometimes used for exposed surface finishes. High humidity in underground construction alleviates normal shrinkage and warping.

Excavations[1]

Site Layout and Control

For construction projects, it is very important to lay out a site survey. Such a survey must include control points that are used as reference points for both horizontal and vertical control of the project. Permanent bench marks must be set up for vertical control and well-marked points must be used for horizontal control.

The purpose of horizontal control is to accurately determine points for the various facilities of an engineering project. On a large facility, one should establish a grid network and use it for this control.

Vertical control methods determine the difference in elevation between points. If available, a level reference surface or datum must be established and referenced from a known bench mark. Differences in elevation, with corrections, are subtracted from or added to this known elevation, resulting in the elevation of the points.

The construction layout survey is the final preconstruction operation. It provides alignments, grades, and locations that guide construction operations. The survey includes determining exact placement of the centerline, laying out curves, setting all remaining stakes, grades, and shoulders, staking out necessary structures, laying out culvert sites and performing other work required to begin construction.

Construction Stakes

Construction stakes are used for centerline, slope, offset, shoulder, grade, reference, ditch, culvert, and intermediate slakes and for temporary bench marks. Typically, these stakes are approximately 1 in $\times$ 3 in in cross section and 2 ft in length. Finished grade stakes and temporary bench marks are 2 in $\times$ 2 in by 12 in.

The primary functions of construction stakes are to indicate facility alignment control elevations, guide equipment operators, and eliminate unnecessary work. They also determine the width of clearing required by indicating the limits of the cut and fill at right angles to the centerline of a road.

---

[1]For a detailed treatment of compaction, see Chap. 202. For earthwork topics, including the use of mass diagrams, see Chap. 212.

Survey crews mark and place construction stakes to conform to the planned line and grade of the proposed facility. A uniform system must be used so the information on the stakes can be properly interpreted by the construction crew.

Construction stakes indicate:

- The stationing or location of any part of the facility in relation to its starting point. If the stake is located at a critical point such as a point of curvature (PC), point of intersection (PI), or point of tangency (PT) of a curve, this should be noted on the stake.

- The height of cut or fill from the existing ground surface to the top of the subgrade for centerline stakes or to the shoulder grade for shoulder or slope stakes.

- The horizontal distance from the centerline to the stake location.

- The side-slope ratio used on slope stakes.

Ditch Stake

The example of a ditch stake shown in Fig. 507.1 is marked with the following information:

1 Offset distance to intercept point of back slope with natural ground

2 Cut to bottom of ditch from stake

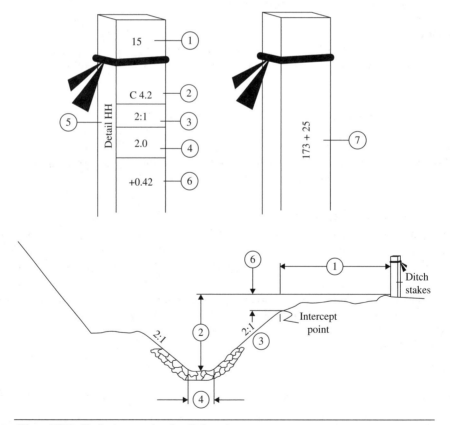

Figure 507.1 Typical example of a ditch stake.

3 Back slope of ditch

4 Base width

5 Ditch typical from plans

6 Vertical distance (±) between intercept point and offset ditch stake point + offset ditch stake point higher than intercept point – offset ditch stake point lower than intercept point

7 Station

Slope Stake

The example of slope stake shown in Fig. 507.2 is marked with the following information:

1 Offset distance (horizontal distance between catch point and slope stake)

2 Total fill or cut from base point to intercept point with natural ground

3 Total horizontal distance of slope

4 Rate of slope

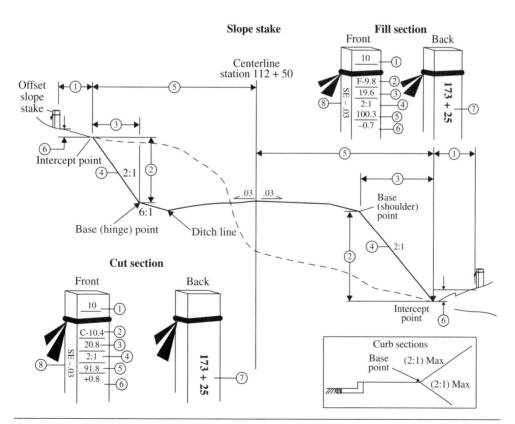

Figure 507.2 Typical examples of slope stakes—cut and fill sections.

5 Total distance from centerline to intercept point

6 Vertical distance ($\pm$) between intercept point and offset stake point + offset stake point higher than intercept point – offset stake point lower than intercept point

7 Station

8 Superelevation

Terminology

Benching: Excavating the sides of an excavation to form one or a series of horizontal levels or steps, usually with vertical or near-vertical surfaces between levels, as a method of protecting employees from cave-ins.

Cave-in: The separation of a mass of soil or rock material from the side of an excavation, or the loss of soil from under a trench shield or support system, in sufficient quantity so that it could entrap, bury, or otherwise injure and immobilize a person.

Cross braces: The horizontal members of a shoring system installed perpendicular to the sides of the excavation, the ends of which bear against either uprights or wales.

Kickout: The accidental release or failure of a cross brace.

Sheeting: The interconnected or closely spaced members of a shoring system that retain the earth in position and in turn are supported by other members of the shoring system.

Shoring: A structure that supports the sides of an excavation, preventing cave-ins.

Trench: A narrow excavation (in relation to its length) made below the surface of the ground. In general, the depth is greater than the width, but the width of a trench (measured at the bottom) is not greater than 15 ft.

A stairway, ladder, ramp or, other safe means of egress shall be located in trench excavations that are 4 ft or more in depth so as to require no more than 25 ft of lateral travel for employees.

Uprights: The vertical members of a trench shoring system placed in contact with the earth and usually positioned so that individual members do not contact each other. Uprights placed so that individual members are closely spaced, in contact with or interconnected to each other, are often called "sheeting."

Wales: Horizontal members of a shoring system placed parallel to the excavation face whose sides bear against the vertical members of the shoring system or earth.

Classification of Soil and Rock Deposits

For evaluation of excavations each soil and rock deposit should be classified as stable rock, type A, type B, or type C in accordance with the definitions set forth below:

Stable rock: This classification is used for natural solid mineral matter that can be excavated with vertical sides and remains intact while exposed.

Type A soil: This classification is used for cohesive soils with an unconfined compressive strength of 1.5 tons/ft^2 (144 kPa) or greater. Examples of cohesive soils are clay, silty clay, sandy clay, clay loam and, in some cases, silty clay loam, and sandy clay loam.

No soil may be classified as type A if (1) the soil is fissured, or (2) the soil is subject to vibration from heavy traffic, pile driving, or similar effects, or (3) the soil has been previously disturbed, or (4) the soil is part of a sloped, layered system where the layers dip into the excavation on a slope of 4H:1V or greater, or (5) the material is subject to other factors that would require it to be classified as a less stable material.

Type B soil: This classification is used for (1) cohesive soil with an unconfined compressive strength greater than 0.5 tons/ft^2 (48 kPa) but less than 1.5 tons/ft^2 (144 kPa), or (2) granular cohesionless soils including angular gravel (similar to crushed rock), silt, silt loam, sandy loam and, in some cases, silty clay loam, and sandy clay loam, (3) previously disturbed soils except those which would otherwise be classed as type C soil, (4) soil that meets the unconfined compressive strength or cementation requirements for type A, but is fissured or subject to vibration, or (5) dry rock that is not stable, or (6) material that is part of a sloped, layered system where the layers dip into the excavation on a slope less steep than 4H:1V, but only if the material would otherwise be classified as type B.

Type C soil: This classification is used for (1) cohesive soil with an unconfined compressive strength of 0.5 tons/ft^2 (48 kPa) or less, or (2) granular soils including gravel, sand, and loamy sand, or (3) submerged soil or soil from which water is freely seeping, or (4) submerged rock that is not stable, or (5) material in a sloped, layered system where the layers dip into the excavation or a slope of 4H:1V or steeper.

Maximum Allowable Slopes

Table 507.1 shows maximum allowable slopes for excavations less than 20-ft deep.

Exceptions

A short-term maximum allowable slope of 1H:2V (63°) is allowed in excavations in type A soil that are 12 ft or less in depth.

Short-term maximum allowable slopes for excavations greater than 12 ft in depth shall be 3/4H:1V (53°).

Table 507.1   Maximum Allowable Slopes for Excavations less than 20 ft

Soil or rock type	Maximum allowable slopes (H:V) for excavations less than 20-ft deep
Stable rock	Vertical (90°)
Type A	3/4:1 (53°)
Type B	1:1 (45°)
Type C	1 1/2:1 (34°)

Excavations in Type A Soil

All simple slope excavation 20 ft or less in depth shall have a maximum allowable slope of 3/4:1. Exception: Simple slope excavations which are open 24 h or less (short term) and which are 12 ft or less in depth shall have a maximum allowable slope of 1/2:1.

All benched excavations 20 ft or less in depth shall have a maximum allowable slope of 3/4 to 1. These slopes may have simple or multiple benches.

All excavations 8 ft or less in depth which have unsupported vertically sided lower portions shall have a maximum vertical side of 3 1/2 ft.

All excavations more than 8 ft but not more than 12 ft in depth with unsupported vertically sided lower portions shall have a maximum allowable slope of 1:1 and a maximum vertical side of 3 1/2 ft.

All excavations 20 ft or less in depth which have vertically sided lower portions that are supported or shielded shall have a maximum allowable slope of 3/4:1. The support or shield system must extend at least 18 in above the top of the vertical side.

Excavations in Type B Soil

All simple slope excavations 20 ft or less in depth shall have a maximum allowable slope of 1:1.

All benched excavations 20 ft or less in depth shall have a maximum allowable slope of 1:1. These slopes may have simple or multiple bench.

All excavations 20 ft or less in depth which have vertically sided lower portions shall be shielded or supported to a height at least 18 in above the top of the vertical side.

All such excavations shall have a maximum allowable slope of 1:1.

Excavations in Type C Soil

All simple slope excavations 20 ft or less in depth shall have a maximum allowable slope of 1 1/2:1.

All excavations 20 ft or less in depth which have vertically sided lower portions shall be shielded or supported to a height at least 18 in above the top of the vertical side.

All such excavations shall have a maximum allowable slope of 1 1/2:1.

Underpinning

Structures sometimes have foundations that require rehabilitation because of either increased loads from the superstructure or localized damage to parts of the foundation. To achieve greater bearing capacity, it may be necessary for foundation components to reach a deeper, stronger soil stratum. The most cost-effective alternative may be an underpinning of the structure. Underpinning is the installation of temporary or permanent supports to an existing foundation to provide either additional depth or an increase in bearing capacity. Other conditions which may necessitate underpinning are construction of a new project with a deeper foundation adjacent to an existing building, excessive settlement, or addition of basement below an existing structure.

Settlement may be caused by a lowering of the water table, caused by drawdown due to pumping. On the other hand, a rise of the water table can cause a lowering of bearing capacity of certain soils, causing additional settlement.

There are two basic concepts of underpinning: (1) mass concrete underpinning, which is suitable for relatively shallow depths of underpinning in dry stable ground conditions. Bases are excavated and constructed in a trench excavated directly under the footing, and (2) beam and base method of underpinning in which a reinforced concrete beam is constructed below, above, or in replacement of the existing footing. The beam then transfers the load of the building to mass concrete bases, which are constructed at designed strategic locations.

Underpinning of a Damaged Pile Foundation

When the tops of timber piles are damaged (decay), the underpinning sequence looks as shown in Fig. 507.3.

Step 1: Shore existing construction, excavate approach pit and expose existing piles.

Step 2: Remove top portion of piles and cut piles at new cutoff elevation.

Step 3: Install steel plates and wedge strut.

Step 4: Transfer load to piles using wedges.

Step 5: Pour encasement concrete and backfill approach pit.

Temporary Support with Maintenance Jacking

Light structures (e.g., wood-frame garages) that fall within the influence line of the adjacent excavation and which do not warrant the expense of an underpinning installation may be supported on timber or concrete mats (see Fig. 507.4).

If settlement occurs, the structure will be kept at the same level by means of mechanical or hydraulic jacks. At completion of the work in the adjacent lot, the jacks are replaced with short steel columns, and the void is filled with concrete.

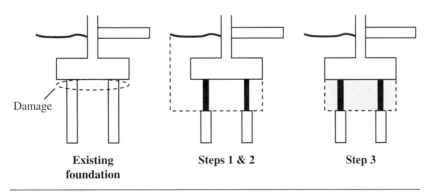

Damage   Existing foundation   Steps 1 & 2   Step 3

Figure 507.3 Underpinning of a foundation.

Figure 507.4
Temporary
support with
jacks.

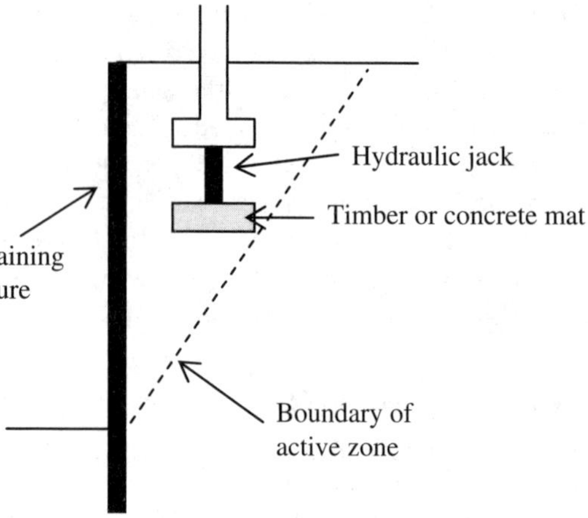

Hydraulic jack

Timber or concrete mat

Earth retaining
structure

Boundary of
active zone

Figure 507.5
Bracket pile
underpinning.

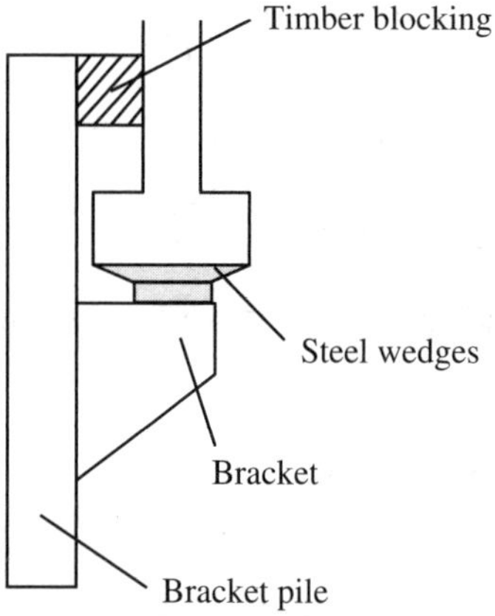

Timber blocking

Steel wedges

Bracket

Bracket pile

Bracket Pile Underpinning

When both the existing and future structures belong to the same owner, the use of bracket piles is very economical (most municipal building codes do not allow a building to be supported on the foundation that is located on someone else's property).

The steel bracket piles are driven or placed adjacent to the future structure in preaugered holes which are then backfilled with a lean sand-cement mix. The load is transferred from the structure into the pile through a steel bracket welded to the side of the pile (see Fig. 507.5).

This type of underpinning can be utilized for structures up to two stories high, depending on the weight of the building and the quality of the bearing material at subgrade or the new structure. The toe penetration of the piles is determined by the vertical load distribution of the bracket pile.

The spacing of the piles depends on the load distribution in the existing structure. The maximum spacing should not exceed 8 ft.

Erosion Control

General

Soil erosion can be caused by various natural forces such as water, wind, ice, and gravity. Soil erosion at unprotected construction sites can result in rills, gullies, sheet erosion, damaged slopes, eroded ditches, plugged drainage structures and culverts, and flooded work areas. Stream channels can be filled with sediment causing flooding of adjacent areas. Transported sediment can cause damage by covering agricultural lands and filling ditches and other drainage systems. Sediment deposited into reservoirs reduces their storage capacity and damages aquatic life. Sediment that reaches navigable waterways requires the intermittent dredging of these navigation channels.

Storm Water Discharge Permits

According to the U.S. Environmental Protection Agency (EPA), effective March 10, 2003, any land-disturbing activity that will "disturb" an area of 1 or more acres is required to have an NPDES (National Pollutant Discharge Elimination System) permit for its storm water discharge (see Chap. 306, "Water and Wastewater Treatment"). A land-disturbing activity includes actions that alter the surface of the land. Actions such as clearing, grading, excavation, building construction access roads may be designated as "land-disturbing."

The 1-ac limit is based on the "common plan of development." Thus, breaking down a large project into numerous projects of less than 1 ac does not eliminate the need for a storm water discharge permit. If the overall common plan of development will involve the disturbance of 1 or more acres, construction on any portion of the project needs to be covered by a storm water discharge permit.

The NPDES permit is required for storm water discharge from any disturbed areas. The term *storm water discharge* refers to any surface runoff from the disturbed areas of the construction site. Construction site storm water runoff results from rainfall runoff or snow melt. The need for a permit applies regardless of the location of the land-disturbing activity.

Pollution Prevention Plan

Every project must have a pollution prevention plan developed before the notice of intent is submitted to the permitting authority. The storm water pollution prevention (SWPP) plan contains a listing of all planned erosion and sediment control practices on site and addresses inspection and maintenance procedures. The SWPP plan must be signed by all contractors and subcontractors involved in land-disturbing activities and kept on the project site along with the record of inspection forms. The SWPP plan needs to include the following: (1) an identification of the selected erosion and sediment controls at the site, (2) the sequence of construction activities and an explanation of how the erosion control practices will be phased in with the construction activities, (3) an outline of each contractor's and subcontractor's role and responsibility in the project and toward erosion and sediment controls, (4) identification of the parties responsible for inspection, maintenance, and continuous evaluation of erosion and sediment control practices, and (5) recordkeeping of the inspection and maintenance of the sediment and erosion controls.

Storm Water Management

The storm water management plan must identify the selected measures that will be installed during construction to control pollutants in storm water discharges that will occur after completion of construction. The design of storm water management practices must be based on specific goals for sediment removal. Permanent controls may be needed to control erosion from the site. Increased runoff from the site may cause increased erosion downstream. A sediment basin of adequate size (satisfying a requirement specified by local agency) must be provided. If a sediment basin is not feasible, comparable erosion control measures on the downslope and sideslope of the project perimeter must be provided.

Velocity dissipation devices shall be placed at the discharge location and along the length on any outfall grass channel as necessary to provide a nonerosive flow from the structure to a receiving watercourse.

Energy Dissipaters

Energy dissipaters are permanent structural devices used to reduce high velocities at the outlet of pipes to prevent scour. On steep slopes and downstream of conduits carrying high flow rates, such energy dissipation mechanisms are warranted. Some of the most commonly used methods are forced hydraulic jumps, stilling basins, drop structures, riprap basins, and the like.

Natural Methods of Erosion Control

Some of the most effective vegetation and soil stabilization methods for erosion control are (1) compost blanket—a surface treatment of 1 to 4 in of compost or mulch, (2) grass channels—to act as diversion of runoff around an area, (3) rolled erosion control products (RECP), which are prefabricated blankets formed from natural and synthetic materials, (4) vegetative filter strip, which is a strip of grass planted at right angles to the runoff flow, and (5) seeding and fertilizing, which is probably the most effective natural method of erosion control.

Structural Methods of Erosion Control

Some of the most effective structural erosion control measures are (1) bench—a slightly reverse sloping step on a back slope, (2) temporary slope drain—a temporary pipe to carry runoff flow from the top of the slope to the bottom, (3) silt fence—a geotextile fabric barrier to intercept sediment, (4) sediment trap—a depression that slows the flow and allows silt to settle, and (5) sediment basin—usually consisting of a dam, a pipe outlet, and an emergency spillway.

Table 508.1 provides general guidance for the selection of the most appropriate temporary erosion control measures. Since every possible situation cannot be addressed by these general guidelines, the selection of temporary erosion control measures for some situations must be based upon good judgment and past experience under similar conditions.

Bale Ditch Check

The purpose of bale ditch checks is to intercept runoff. Checks should not be placed in ditches where high flows are expected or where the slope is steeper than 6%. Rock checks should be used instead. Bale ditch checks should be placed perpendicular to the flowline of the ditch. The sediment-laden runoff will pond behind the bales, slowing the runoff velocity and allowing most of the suspended sediment to drop out. Water is intended to flow over the bales, not around the side. The ditch check should extend far enough so that the ground level at the ends

Table 508.1 Best Management Practice for Erosion Control

Area to protect	Condition	Best management practice
Ditches	Grade ≤6%	Bale ditch check Silt fence ditch check TSD ditch check
	Grade >6%	Rock ditch check Erosion control blankets
	High flows expected	Rock ditch check Erosion control blankets
Slopes	Erosion control	Temporary seeding Erosion control blankets
	Sediment control	Bale slope barriers Silt fence slope barriers
Drop inlet protection	No decision needed	Bale drop-inlet barrier Silt fence drop-inlet barrier TSD drop-inlet barrier

Table 508.2 Maximum Spacing of Bale Ditch Checks

Ditch grade (%)	Check spacing ft (m)
1	200 (60)
2	100 (30)
3	65 (20)
4	50 (15)
5	40 (12)
6	10 (3)
>6	Do not use bales

of the check is higher than the top of the lowest center bale. This prevents water from flowing around the check. In ditches with high flows or steep slopes, erosion protection mechanisms such as erosion control blanket or riprap will be required on the downstream side of the bales.

Bale ditch checks may be constructed of wheat straw, oat straw, prairie hay, and the like. The stakes used to anchor the bales should be made of a hardwood material. Twine should be used to bind bales. The use of wire binding is discouraged because it does not biodegrade readily. Table 508.2 shows approximate spacing of bale ditch checks as a function of the ditch grade.

Triangular Silt Dikes

Triangular silt dikes (TSDs) work on the same principle as bale ditch checks; they intercept and pond sediment-laden runoff. Ponding the water reduces the velocity of any incoming flow and allows most of the suspended sediment to settle. Water exits the TSD by flowing over the top. The geotextile apron on the downstream side of the dike helps prevent scour caused by this flowing water. Because TSD installations require a minimal depth for anchoring, they are well suited to ditches with shallow soils underlain by rock.

Silt Fence Ditch Check

Silt fence ditch checks operate by intercepting, ponding, and filtering sediment-laden runoff. Checks should not be placed in ditches where high flows are expected or on slopes exceeding 6%. Rock checks should be used instead. Silt fence ditch checks should be placed perpendicular to the flowline of the ditch. The silt fence should extend far enough so that the ground level at the ends of the fence is higher than the top of the low point of the fence. This prevents water from flowing around the check. Silt fences must be placed in ditches where it is unlikely to be overtopped. Water should flow through a silt fence ditch check, not over it. Silt fence ditch checks often fail when overtopped.

Ponding the water reduces the velocity of the incoming flow and allows most of the suspended sediment to settle. As the ponded water percolates through the silt fence fabric, much of the remaining suspended sediment is filtered out. Silt fence ditch checks work well in ditches

Table 508.3   Maximum Spacing of Silt Fence Ditch Checks

Ditch grade (%)	Check spacing ft (m)
1	200 (60)
2	100 (30)
3	65 (20)
4	50 (15)
5	40 (12)
6	10 (3)
>6	Do not use silt fence

with low flows and moderate slopes. Table 508.3 shows approximate spacing of silt fence ditch checks as a function of the ditch grade.

Rock Ditch Checks

Rock ditch checks operate by intercepting and ponding sediment-laden runoff. Rock ditch checks should be perpendicular to the flowline of the ditch. Rock ditches must be designed so that water can flow over them, not around them. The ditch check should extend far enough so that the ground level at the ends of the check is higher than the low point on the crest of the check. Ponding the water dissipates the energy of any incoming flow and allows a large portion of the suspended sediment to settle. Water exits the ditch check by flowing over its crest. Rock ditch checks are ideal for ditches that eventually will have a riprap lining. Upon completion of the project, rock ditch checks can be spread out to form a riprap lined channel. Rock ditch checks should be constructed of stone that is between 4 in and 8 in (100 mm to 200 mm) in size. Table 508.4 shows approximate spacing of rock ditch checks as a function of the ditch grade.

Table 508.4   Maximum Spacing of Rock Ditch Checks

Ditch grade (%)	Check spacing ft (m)
5	60 (18)
6	50 (15)
7	43 (13)
8	36 (11)
9	33 (10)
10	30 (9)

Bale Slope Barriers

Bale slope barriers may be constructed of wheat straw, oat straw, prairie hay, and the like. Twine should be used to bind bales. The use of wire binding is discouraged because it does not biodegrade readily. Bale slope barriers operate by intercepting and ponding sediment-laden runoff. Ponding the water dissipates the energy of the incoming flow and allows much of the suspended sediment to settle. Water exits the bale slope barrier by flowing over the bales.

A slope barrier should be used at the toe of a slope when a ditch does not exist. The slope barrier should be placed on nearly level ground 5 ft to 10 ft (1.5 m to 3.0 m) away from the toe of a slope. The barrier is placed away from the toe of the slope to provide adequate storage for settling sediment. When possible, bale slope barriers should be placed along contours to avoid a concentration of flow.

Silt Fence Slope Barriers

Silt fence slope barriers operate by intercepting and ponding sediment-laden slope runoff. Ponding the water reduces the velocity of the incoming flow and allows most of the suspended sediment to settle. Water exits the silt fence slope barrier by percolating through the silt fence.

A slope barrier should be used at the toe of a slope when a ditch does not exist. The slope barrier should be placed on nearly level ground 5 ft to 10 ft (1.5 m to 3.0 m) away from the toe of a slope to provide adequate storage for settling sediment. When practicable, silt fence slope barriers should be placed along contours to avoid concentrated flows. Silt fence slope barriers also can be placed along right-of-way fence lines to keep sediment from crossing onto adjacent property. When placed in this manner, the slope barrier will not likely follow contours.

Drop Inlets

A drop inlet spillway is a mechanical system which lowers low-to-medium volumes of water over a sharp incline (20°) through a box or pipe structure. The incline height is typically 4 ft or greater. Such a system dissipates most of the energy contained in the water. Concrete catch basins, plastic drop pipes, and steel-sloped culverts are all examples of drop inlet spillways. Broadly, a drop inlet is of the drop-pipe or the sloped-pipe type. In a drop inlet, the kinetic energy is dissipated in the following three locations:

1. The bottom of the vertical pipe must break the energy of the falling water.

2. Energy is dissipated via friction in the horizontal pipe.

3. Energy is also dissipated below the horizontal pipe outlet. Under most circumstances, dissipaters such as rock rip rap, gabion mattresses, or equivalent should be installed.

Occupational Safety

OSHA Regulations for Construction Projects

Provisions in this chapter are based on Chap. 1926 of the Code of Federal Regulations CFR29, which is compliant with section 107 of the Contract Work Hours and Safety Standards Act, commonly known as the Construction Safety Act.

Rules of Construction

By contracting for full performance of a contract, the prime contractor assumes all obligations prescribed as employer obligations, whether or not he subcontracts any part of the work. The prime contractor assumes the entire responsibility under the contract and the subcontractor assumes responsibility with respect to his portion of the work. With respect to subcontracted work, the prime contractor and any subcontractor or subcontractors shall be deemed to have joint responsibility.

All employees required to enter into confined or enclosed spaces shall be instructed as to the nature of the hazards involved, the necessary precautions to be taken, and in the use of protective and emergency equipment required. The employer shall comply with any specific regulations that apply to work in dangerous or potentially dangerous areas.

Confined or enclosed space means any space having a limited means of egress, which is subject to the accumulation of toxic or flammable contaminants or has an oxygen deficient atmosphere. Confined or enclosed spaces include, but are not limited to, storage tanks, process vessels, bins, boilers, ventilation or exhaust ducts, sewers, underground utility vaults, tunnels, pipelines, and open top spaces more than 4 ft in depth such as pits, tubs, vaults, and vessels.

Hazardous substance means a substance which, due to being explosive, flammable, poisonous, corrosive, oxidizing, irritating, or otherwise harmful, is likely to cause death or injury.

Personal Protective Equipment (PPE)

The following are guidelines which an employer can use to begin the selection of the appropriate personal protective equipment (PPE). The site information may suggest the use of combinations of PPE selected from the different protection levels (A, B, C, or D) as being more suitable to the hazards of the work.

Personal protective equipment is divided into four categories based on the degree of protection afforded.

Level A

The greatest level of skin, respiratory, and eye protection is required. Level A protection should be used when:

1. The hazardous substance has been identified and requires the highest level of protection for skin, eyes, and the respiratory system based on either the measured (or potential for) high concentration of atmospheric vapors, gases, or particulates; or the site operations and work functions involve a high potential for splash, immersion, or exposure to unexpected vapors, gases, or particulates of materials that are harmful to skin or capable of being absorbed through the skin;

2. Substances with a high degree of hazard to the skin are known or suspected to be present, and skin contact is possible; or

3. Operations are being conducted in confined, poorly ventilated areas, and it has not been determined that conditions requiring Level A protection do not exist.

Level A Equipment

The following constitute Level A equipment:

1. Positive pressure, full face-piece self-contained breathing apparatus (SCBA), or positive pressure supplied air respirator with escape SCBA, approved by the National Institute for Occupational Safety and Health (NIOSH).

2. Totally encapsulating chemical-protective suit.

3. Gloves, outer, chemical-resistant.

4. Gloves, inner, chemical-resistant.

5. Boots, chemical-resistant, steel toe and shank.

6. Disposable protective suit, gloves and boots (depending on suit construction, may be worn over totally encapsulating suit).

Optional equipment include: Coveralls, long underwear, hard hat (under suit).

Level B

The highest level of respiratory protection is necessary but a lesser level of skin protection is needed. Level B protection should be used when:

1. The type and atmospheric concentration of substances have been identified and require a high level of respiratory protection, but less skin protection;

2. The atmosphere contains less than 19.5% oxygen; or

3. The presence of incompletely identified vapors or gases is indicated by a direct-reading organic vapor detection instrument, but vapors and gases are not suspected of containing high levels of chemicals harmful to skin or capable of being absorbed through the skin.

 NOTE This involves atmospheres with IDLH (immediately dangerous to life or health) concentrations of specific substances that present severe inhalation hazards and that do not represent a severe skin hazard; or that do not meet the criteria for use of air-purifying respirators.

Level B Equipment

The following constitute Level B equipment:

1. Positive pressure, full-facepiece self-contained breathing apparatus (SCBA), or positive pressure supplied air respirator with escape SCBA (NIOSH approved).
2. Hooded chemical-resistant clothing (overalls and long-sleeved jacket; coveralls; one or two-piece chemical-splash suit; disposable chemical-resistant overalls).
3. Gloves, outer, chemical-resistant.
4. Gloves, inner, chemical-resistant.
5. Boots, outer, chemical-resistant steel toe and shank.

Optional equipment include: Coveralls, chemical-resistant boot-covers (disposable), hard hat, face shield.

Level C

The concentration(s) and type(s) of airborne substance(s) is known and the criteria for using air-purifying respirators are met. Level C protection should be used when:

1. The atmospheric contaminants, liquid splashes, or other direct contact will not adversely affect or be absorbed through any exposed skin;
2. The types of air contaminants have been identified, concentrations measured, and an air-purifying respirator is available that can remove the contaminants; and
3. All criteria for the use of air-purifying respirators are met.

Level C Equipment

The following constitute Level C equipment:

1. Full-face or half-mask, air purifying respirators (NIOSH approved).
2. Hooded chemical-resistant clothing (overalls; two-piece chemical-splash suit; disposable chemical-resistant overalls).

3. Gloves, outer, chemical-resistant.

4. Gloves, inner, chemical-resistant.

Optional equipment include: Coveralls, chemical-resistant steel toe and shank boots, chemical-resistant boot-covers (disposable), hard hat, escape mask, face shield.

Level D

A work uniform affording minimal protection, used for nuisance contamination only. Level D protection should be used when:

1. The atmosphere contains no known hazard; and

2. Work functions preclude splashes, immersion, or the potential for unexpected inhalation of or contact with hazardous levels of any chemicals.

Level D Equipment

1. Coveralls.

2. Boots/shoes, chemical-resistant steel toe and shank.

3. Safety glasses or chemical splash goggles.

Optional equipment include: Gloves, chemical-resistant boots (disposable), hard hat, escape mask, face shield.

Exposure to Lead

During construction, employees may be exposed to lead from demolition, salvage, and removal of materials containing lead. For lead, action level (AL) means employee exposure, without regard to the use of respirators, to an airborne lead concentration of 30 $\mu g/m^3$, calculated as an 8-hour time-weighted average.

Permissible Exposure Limit (PEL)

The employer shall assure that no employee is exposed to lead at concentrations greater than 50 micrograms per cubic meter of air ($\mu g/m^3$) averaged over an 8-hour period. If an employee is exposed to lead for more than 8 hours in any work day, the employees' allowable exposure, as a time weighted average (TWA) for that day, shall be reduced according to the following formula:

Allowable employee exposure (in $\mu g/m^3$) = 400 divided by hours worked in the day.

Where lead is present, until the employer performs an employee exposure assessment and documents that the employee performing any of the listed tasks is not exposed above the permissible exposure limit (PEL), the employer shall treat the employee as if the employee were exposed above the PEL, and not in excess of 10 times the PEL, and shall implement appropriate employee protective measures. Where the employer establishes that the employee is exposed to levels of lead above the PEL but below 500 $\mu g/m^3$, the employer may provide the exposed employee with the appropriate respirator prescribed for such use at such lower exposures, in accordance with Table 509.1.

Monitoring

The employer shall use a method of monitoring and analysis which has an accuracy (to a confidence level of 95%) of not less than plus or minus 25% for airborne concentrations of lead equal to or greater than 30 $\mu g/m^3$.

If the initial determination or subsequent determination reveals employee exposure to be at or above the action level but at or below the PEL the employer shall perform monitoring in accordance with this paragraph at least every 6 months. The employer shall continue monitoring at the required frequency until at least two consecutive measurements, taken at least 7 days apart, are below the action level at which time the employer may discontinue monitoring for that employee.

Different Types of Respirators

Table 509.2 reproduces the APF guidelines from 29 CFR 1910.134 (Table 1) about the various types of respirators named in Table 509.1.

Table 509.1 Respiratory Protection for Lead Aeorsols

Airborne concentration of lead or condition of use	Required respirator
Not in excess of 500 $\mu g/m^3$	• Half-mask air purifying respirator with high-efficiency filters. • Half-mask supplied air respirator operated in demand (negative pressure) mode
Not in excess of 1250 $\mu g/m^3$	• Loose fitting hood or helmet powered air purifying respirator with high-efficiency filters • Hood or helmet supplied air respirator operated in continuous flow mode (e.g., type CE abrasive blasting respirators operated in continuous-flow mode)
Not in excess of 2500 $\mu g/m^3$	• Full-facepiece air purifying respirator with high-efficiency filters • Tight fitting powered air purifying respirator with high-efficiency filters • Full-facepiece supplied air respirator operated in demand mode • Half-mask or full-facepiece supplied air respirator operated in continuous flow mode • Full-facepiece self-contained breathing apparatus (SCBA) operated in demand mode
Not in excess of 50,000 $\mu g/m^3$	• Half-mask supplied air respirator operated in pressure demand or other positive-pressure mode
Not in excess of 100,000 $\mu g/m^3$	• Full-facepiece supplied air respirator operated in pressure demand or other positive pressure mode (e.g., type CE abrasive blasting respirators operated in a positive-pressure mode)
Greater than 100,000 $\mu g/m^3$ unknown concentration, or fire fighting	• Full facepiece SCBA operated in pressure demand or other positive pressure mode

NOTES:
- Respirators specified for higher concentrations can be used at lower concentrations of lead.
- Full facepiece is required if the lead aerosols cause eye or skin irritation at the use concentrations.
- A high-efficiency particulate filter (HEPA) means a filter that is 99.97% efficient against particles larger than 0.3 microns.

Table 509.2 Assigned Protection Factors

Type of respirator	Quarter mask	Half mask	Full facepiece	Helmet/ hood	Loose fitting facepiece
Air-Purifying Respirator	5	10	15	-	-
Powered Air-Purifying Respirator (PAPR)	-	50	1000	25/1000	25
Supplied-Air Respirator (SAR) or Airline Respirator					
• Demand Mode	-	10	50	-	-
• Continuous Flow Mode	-	50	1000	25/1000	25
• Pressure-demand or other positive-pressure mode		50	1000	-	-
Self-Contained Breathing Apparatus (SCBA)					
• Demand mode	-	10	50	50	-
• Pressure-demand or other positive-pressure mode	-	-	10,000	10,000	-

APF (Assigned Protection Factor)

An APF is a term used by OSHA to determine how well a respirator/filter combination will protect an individual from external contaminants. It is an estimate of the level of protection a respirator provides. APFs are used to select the appropriate class of respirators that will provide the necessary level of protection. There are certain levels used for different types of masks. The APF is based on the type of mask and size. An APF of 10 means that no more than one-tenth of the contaminants to which the worker is exposed will leak into the inside of the mask. An APF of 100 means only an one percent leakage.

MUC (Maximum Use Concentration)

The MUC is a term used by OSHA for the upper limit at which the class of respirators is expected to provide protection. The MUC can be calculated by multiplying the APF in Table 509.2, by the permissible exposure limit (PEL), short-term exposure limit (STEL), or ceiling limit of the contaminant.

If an exposure ever approaches the MUC, then the employer should select the next highest level of respirator. The respirator can be used up to this concentration as long as the MUC does not exceed the immediately dangerous to life or health (IDLH) level. When no OSHA PEL is available for a hazardous substance, the MUC must be determined using available information and professional judgment.

IDLH (Immediately Dangerous to Life or Health)

The term immediately dangerous to life or health (IDLH) is defined by the U.S. National Institute for Occupational Safety and Health (NIOSH) as exposure to airborne contaminants that is "likely to cause death or immediate or delayed permanent adverse health effects or prevent escape from such an environment." Some examples are smoke or other poisonous gases at sufficiently high concentrations.

Example: If the permissible exposure limit for a contaminant such as Toluene* is 200 ppm, then a half-faced mask with an APF of 10 should protect the employee up to 2000 ppm. 10 APF × 200 ppm = 2000 ppm (MUC).

Emergency Response

The employer shall establish an emergency action plan which should list the types of evacuation to be used in emergency situations. Before implementing the emergency action plan, the employer shall designate and train a sufficient number of persons to assist in the safe and orderly emergency evacuation of employees.

The employer shall review the plan with each employee covered by the plan at the following times: initially when the plan is developed, whenever the employee's responsibilities or designated actions under the plan change, and whenever the plan is changed.

The employer shall make provisions for prompt medical attention in case of serious injury. In addition, the employer shall provide proper equipment for prompt transportation of the injured person to a physician or hospital, or a communication system for contacting necessary ambulance service. In areas where 911 service is not available, the telephone numbers of the physicians, hospitals, or ambulances shall be conspicuously posted.

First aid supplies shall be easily accessible when required. The contents of the first-aid kit shall be placed in a weatherproof container with individual sealed packages for each type of item, and shall be checked by the employer before being sent out on each job and at least weekly on each job to ensure that the expended items are replaced. Where the eyes or body of any person may be exposed to injurious corrosive materials, suitable facilities for quick drenching or flushing of the eyes and body shall be provided within the work.

If it is reasonably anticipated employees will be exposed to blood or other potentially infectious materials while using first-aid supplies, employers should provide personal protective equipment (PPE). Appropriate PPE includes gloves, gowns, face shields, masks, and eye protection.

Every company must measure its safety performance, using experience modification rate (EMR), OSHA safety statistics, etc., and communicate these results to its employees. The EMR indicates whether the company is at, below, or above average (EMR <1.0 better than average, EMR >1.0 worse than average) for number of injuries for its particular industry. OSHA uses EMR <1.0 as a criteria for enrollment in its voluntary protection program (VPP). In addition, OSHA uses EMR as enrollment criteria in some of its partnership programs.

The incident rate is an OSHA term for the number of recordable injuries and illnesses for a facility in a year, normalized to 100 employees working 2000 h/year. Thus, the incident rate is calculated as

$$\text{Incident rate} = \frac{\text{number of recordable injuries (annual)}}{\text{total annual hours worked by all employees}} \times 200{,}000 \qquad (509.1)$$

Toilets

At the construction site, toilets shall be provided for employees according to Table 509.3.

Table 509.3 Number of Toilet Facilities at Construction Sites

Number of employees	Facilities
20 or less	1
More than 20	1 toilet seat and 1 urinal per 40 workers
200 or more	1 toilet seat and 1 urinal per 50 workers

Noise Exposure

Protection against the effects of noise exposure shall be provided when noise levels exceed those shown in Table 509.4.

Exposure to impulsive or impact noise should not exceed 140 dB peak sound pressure level.

When employees are subjected to sound levels exceeding those listed in Table 509.4, feasible administrative or engineering controls shall be utilized. If such controls fail to reduce sound levels within the levels of the table, personal protective equipment shall be provided and used to reduce sound levels within the levels of the table. If the variations in noise level involve maxima at intervals of 1 s or less, it is to be considered continuous.

When the daily noise exposure is composed of two or more periods of noise exposure of different levels, their combined effect should be considered, rather than the individual effect of each. Exposure to different levels for various periods of time shall be computed according to the formula for $F(e)$, the equivalent noise exposure factor

$$F(e) = \frac{T_1}{L_1} + \frac{T_2}{L_2} + \cdots + \frac{T_n}{L_n} \tag{509.2}$$

where T_i = period of noise exposure at any essentially constant level
L_i = duration of the permissible noise exposure at the constant level

If the value of $F(e)$ exceeds 1.0, the exposure exceeds permissible levels.

Table 509.4 Permissible Noise Exposure at Construction Sites

Permissible noise exposure	
Duration per day (hours)	Sound level (dB)
8	90
6	92
4	95
3	97
2	100
1 1/2	102
1	105
1/2	110
1/4 or less	115

Example 509.1

An employee is exposed to sound at these levels for these periods:

105 dB for 15 min

100 dB for 30 min

95 dB for 90 min

What is the noise exposure factor?

Solution   The equivalent noise exposure factor is calculated as

$$F(e) = \frac{0.25}{1.0} + \frac{0.5}{2.0} + \frac{1.5}{4.0} = 0.88$$

Since the value of $F(e)$ is less than 1.0, the exposure is within permissible limits.

Wherever it is not feasible to reduce the noise levels or duration of exposures to those specified in Table 509.4, ear protective devices shall be provided and used. Ear protective devices inserted in the ear shall be fitted or determined individually by competent persons. Plain cotton is not an acceptable protective device.

Light Exposure

Employees, when working in areas in which a potential exposure to direct or reflected laser light greater than 5mW exists, shall be provided with antilaser eye protection devices as specified in 29 CFR Part 1926 Subpart E. Employees shall not be exposed to microwave power densities greater than 10 mW/cm^2.

Employees shall not be exposed to light intensities above the following limits:

Direct staring: 1 μW/cm^2

Incidental observing: 1 mW/cm^2

Diffused reflected light: 2.5 W/cm^2

Table 509.5 shall be used as a guide for the selection of the proper shade numbers of filter lenses or plates used in welding. Shades more dense than those listed may be used to suit the individual's needs.

While work is in progress, construction areas, ramps, runways, corridors, offices, shops, and storage areas shall be lit to not less than the minimum illumination intensities listed in Table 509.6.

Signs, Signals, and Barricades

Signs and symbols required by this section shall be visible at all times when work is being performed, and shall be removed or covered promptly when the hazards no longer exist.

Table 509.5   Filter Lens Shade Numbers for Protection against Radiant Energy

Welding operation	Shade number
Shielded metal-arc welding 1/16-, 3/32-, 1/8-, 5/32-in diameter electrodes	10
Gas-shielded arc welding (nonferrous) 1/16-, 3/32-, 1/8-, 5/32-in diameter electrodes	11
Gas-shielded arc welding (ferrous) 1/16-, 3/32-, 1/8-, 5/32-in diameter electrodes	12
Shielded metal-arc welding 3/16-, 7/32-, 1/4-in diameter electrodes	12
5/16-, 3/8-in diameter electrodes	14
Atomic hydrogen welding	10–14
Carbon-arc welding	14
Soldering	2
Torch brazing	3 or 4
Light cutting, up to 1 in	3 or 4
Medium cutting, 1 in to 6 in	4 or 5
Heavy cutting, over 6 in	5 or 6
Gas welding (light), up to 1/8 in	4 or 5
Gas welding (medium), 1/8 inch to ½ in	5 or 6
Gas welding (heavy), over ½ in	6 or 8

Table 509.6   Minimum Illumination Intensities (Foot-Candles)

Foot-candles	Area of operation
5	General construction area lighting
3	General construction areas, concrete placement, excavation and waste areas, access ways, active storage areas, loading platforms, refueling, and field maintenance areas
5	Indoors: warehouses, corridors, hallways, and exitways.
5	Tunnels, shafts, and general underground work areas (Exception: minimum of 10-foot-candles is required at tunnel and shaft heading during drilling, mucking, and scaling. Bureau of Mines approved cap lights shall be acceptable for use in the tunnel heading.)
10	General construction plant and shops (e.g., batch plants, screening plants, mechanical and electrical equipment rooms, carpenter shops, rigging lofts and active store rooms, mess halls, and indoor toilets and workrooms)
30	First aid stations, infirmaries, and offices

Danger Signs

Danger signs shall be used only where an immediate hazard exists. Danger signs shall have red as the predominating color for the upper panel; black outline on the borders; and a white lower panel for additional sign wording.

Caution Signs

Caution signs shall be used only to warn against potential hazards or to caution against unsafe practices. Caution signs shall have yellow as the predominating color; black upper panel and borders; yellow lettering of "caution" on the black panel; and the lower yellow panel for additional sign wording (in black lettering).

Exit Signs

Exit signs, when required, shall be lettered in legible red letters, not less than 6-in high, on a white field and the principal stroke of the letters shall be at least 3/4 in. in width.

Safety Instruction Signs

Safety instruction signs, when used, shall be white with green upper panel with white letters to convey the principal message. Any additional wording on the sign shall be black letters on the white background.

Directional Signs

Directional signs, other than automotive traffic signs, shall be white with a black panel and a white directional symbol. Any additional wording on the sign shall be black letters on the white background.

Traffic Signs

Construction areas shall be posted with legible traffic signs at points of hazard.

All traffic control signs or devices used for protection of construction workers shall conform to Part VI of the Manual on Uniform Traffic Control Devices[1], Millennium Edition.

Accident Prevention Tags

Accident prevention tags shall be used as a temporary means of warning employees of an existing hazard, such as defective tools, equipment, etc. They shall not be used in place of, or as a substitute for, accident prevention signs.

Fall Protection

The critical fall height (requiring fall protection) at a construction site is 6 ft. Any potential fall hazard shall be protected by passive fall protection devices such as guardrails, or employees shall be provided with fall arrest harnesses or other safety devices.

---

[1]"Manual of Uniform Traffic Control Devices," Federal Highway Administration, Washington, DC, 2003.

Guardrails

Top edge height of toprails, or equivalent guardrail system members, shall be 42 in (1.1 m) plus or minus 3 in (8 cm) above the walking/working level. When conditions warrant, the height of the top edge may exceed the 45-in height, provided the guardrail system meets all other criteria of this paragraph.

 NOTE When employees are using stilts, the top edge height of the toprail, or equivalent member, shall be increased an amount equal to the height of the stilts. Guardrail systems shall be capable of withstanding, without failure, a force of at least 200 lb (890 N) applied within 2 in (5.1 cm) of the top edge, in any outward or downward direction, at any point along the top edge. When the 200 lb (890 N) test load is applied in a downward direction, the top edge of the guardrail shall not deflect to a height less than 39 in (1.0 m) above the walking/working level.

Midrails

Midrails, screens, mesh, intermediate vertical members, or equivalent intermediate structural members shall be installed between the top edge of the guardrail system and the walking/working surface when there is no wall or parapet wall at least 21 in (53 cm) high. When used, midrails shall be installed at a height midway between the top edge of the guardrail system and the walking/working level. Screens and mesh, when used, shall extend from the toprail to the walking/working level and along the entire opening between toprail supports. Intermediate members (such as balusters), when used between posts, shall be not more than 19-in (48-cm) apart. Other structural members (such as additional midrails and architectural panels) shall be installed such that there are no openings in the guardrail system that are more than 19-in (0.5-m) wide.

Safety Nets

Safety nets shall be installed as close as practicable under the walking/working surface on which employees are working, but in no case more than 30 ft (9.1 m) below such level.

When nets are used on bridges, the potential fall area from the walking/working surface to the net shall be unobstructed. Safety nets shall extend outward from the outermost projection of the work surface according to provisions in Table 509.7.

The mesh size of nets shall not exceed 6 inches by 6 inches. All new nets shall meet accepted performance standards of 17,500 foot-pounds minimum impact resistance as determined and certified by the manufacturers, and shall bear a label of proof test. Edge ropes shall provide a minimum breaking strength of 5,000 pounds.

Lanyards

Lanyards and vertical lifelines shall have a minimum breaking strength of 5000 lb (22.2 kN). Self-retracting lifelines and lanyards which automatically limit free fall distance to 2 ft (0.61 m)

Table 509.7 Safety Net Provisions in OSHA

Vertical distance from working level to horizontal plane of net	Minimum required horizontal distance of outer edge of net from the edge of the working surface
≤5 ft	8 ft
>5–10 ft	10 ft
>10 ft	13 ft

or less shall be capable of sustaining a minimum tensile load of 3000 lb (13.3 kN) applied to the device with the lifeline or lanyard in the fully extended position.

Self-retracting lifelines and lanyards which do not limit free fall distance to 2 ft (0.61 m) or less, ripstitch lanyards, and tearing and deforming lanyards shall be capable of sustaining a minimum tensile load of 5000 lb (22.2 kN) applied to the device with the lifeline or lanyard in the fully extended position. Personal fall arrest systems, when stopping a fall, shall satisfy the following criteria:

1. Limit maximum arresting force on an employee to 900 lb (4 kN) when used with a body belt.

2. Limit maximum arresting force on an employee to 1800 lb (8 kN) when used with a body harness.

3. Be rigged such that an employee can neither free fall more than 6 ft (1.8 m), nor contact any lower level.

4. Bring an employee to a complete stop and limit maximum deceleration distance an employee travels to 3.5 ft (1.07 m).

5. Have sufficient strength to withstand twice the potential impact energy of an employee free falling a distance of 6 ft (1.8 m), or the free fall distance permitted by the system, whichever is less.

Bracing of Masonry Walls

A masonry wall that is under construction lacks the lateral support usually provided by roofs, floors, pilasters, and walls. Until the mortar cures and gains adequate strength and floors and roof are installed, masonry walls must be braced laterally. The bracing plan is usually a function of the expected maximum wind speed and the self-weight of the wall.

OSHA (29 CFR 1926.706a) and bracing (29 CFR 1926.706b) requirements for Masonry Construction (enacted August 15, 1988) state the following:

a. A limited access zone shall be established whenever a masonry wall is being constructed. The limited access zone shall conform to the following.

   1. The limited access zone shall be established prior to the start of construction of the wall.

   2. The width of the limited access zone shall be equal to the height of the wall to be constructed plus 4 ft and shall run the entire length of the wall.

   3. The limited access zone shall be established on the side of the wall, which will be unscaffolded.

   4. The limited access zone shall be restricted to entry by employees actively engaged in constructing the wall. No other employees shall be permitted to enter the zone.

   5. The limited access zone shall remain in place until the wall is adequately supported to prevent overturning and to prevent collapse unless the height of the wall is over 8 ft, in which case the limited access zone shall remain in place until the requirements of paragraph (b) of this section have been met. OSHA considers bracing as adequate support, so once the wall is braced the limited access zone may be removed.

b. All masonry walls over 8 ft in height shall be adequately braced to prevent overturning and to prevent collapse unless the wall is adequately supported so that it will not overturn or collapse. The bracing shall remain in place until permanent supporting elements of the structure are in place.

For walls 8 ft or lower, the limited access zone must stay in place "until the wall is adequately supported to prevent overturning and to prevent collapse."

Walls higher than 8 ft must be braced until they are permanently supported by the structure, "unless the wall is adequately supported so that it will not overturn or collapse." OSHA considers self-supporting walls that have reached their design strength adequately supported. Thus, it seems logical that a wall may be built higher than 8 ft without bracing if the weight of the wall can resist the wind speeds expected on the wall

Some rules of thumb about wall bracing are given below:

- Usually, walls which are 10 times higher than their thickness should be braced. They should be braced on both sides, usually for 7 days.

- Horizontal spacing of vertical bracing (ft) must be less than 1.5 times the wall thickness (inches). Therefore, for an 8-in-thick masonry wall, the maximum horizontal spacing between vertical bracing members is 12 ft.

Figure 509.1 shows maximum unbraced wall height, calculated from wind speed (mph) and self-weight of the wall (psf). For example, a 12-in-thick masonry hollow block wall that weighs 65 psf and is subject to maximum wind speed of 60 mph must be braced at a height of 6 ft.

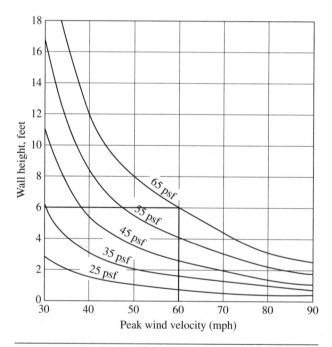

Figure 509.1 Maximum unbraced wall height as function of wind speed.

Steel Erection

A steel erection contractor shall not erect steel unless it has received written notification that the concrete in the footings, piers, and walls or the mortar in the masonry piers and walls has attained, on the basis of an appropriate ASTM standard test method of field-cured samples, either 75% of the intended minimum compressive design strength or sufficient strength to support the loads imposed during steel erection.

Site Layout

The controlling contractor shall ensure that the following is provided and maintained:

1. Adequate access roads into and through the site for the safe delivery and movement of derricks, cranes, trucks, other necessary equipment, and the material to be erected and means and methods for pedestrian and vehicular control.

2. A firm, properly graded, drained area, readily accessible to the work with adequate space for the safe storage of materials and the safe operation of the erector's equipment.

General Requirements for Erection Stability

All columns shall be anchored by a minimum of four anchor rods (anchor bolts). Each column anchor rod (anchor bolt) assembly, including the column-to-base plate weld and the column foundation, shall be designed to resist a minimum eccentric gravity load of 300 lb (136.2 kg) located 18 in (0.46 m) from the extreme outer face of the column in each direction at the top of the column shaft. Columns shall be set on level finished floors, pregrouted leveling plates, leveling nuts, or shim packs which are adequate to transfer the construction loads. The perimeter columns should extend a minimum of 48 in (1.2 m) above the finished floor to permit installation of perimeter safety cables prior to erection of the next tier, except where constructability does not allow such extension.

Anchor rods (anchor bolts) shall not be repaired, replaced, or field modified without the approval of the project structural engineer of record. Prior to the erection of a column, the controlling contractor shall provide written notification to the steel erector if there has been any repair, replacement, or modification of the anchor rods (anchor bolts) of that column.

Except where steel joists are used and columns are not framed in at least two directions with solid web structural steel members, a steel joist shall be field bolted at the column to provide lateral stability to the column during erection.

A vertical stabilizer plate shall be provided on each column for steel joists. The plate shall be a minimum of 6 by 6 in (152 by 152 mm) and shall extend at least 3 in (76 mm) below the bottom chord of the joist with a 13/16-in (21-mm) hole to provide an attachment point for guying or plumbing cables.

The bottom chords of steel joists at columns shall be stabilized to prevent rotation during erection.

Field-Bolted Joists

Except for steel joists that have been preassembled into panels, connections of individual steel joists to steel structures in bays of 40 ft (12.2 m) or more shall be fabricated to allow for field bolting during erection.

Steel joists and steel joist girders shall not be used as anchorage points for a fall arrest system unless written approval to do so is obtained from a qualified person.

Attachment of Steel Joists and Steel Joist Girders

Each end of K series steel joists shall be attached to the support structure with a minimum of two 1/8-in (3-mm) fillet welds 1 in (25 mm) long or with two 1/2-in (13-mm) bolts, or the equivalent.

Each end of LH and DLH series steel joists and steel joist girders shall be attached to the support structure with a minimum of two 1/4-in (6-mm) fillet welds 2-in (51-mm) long, or with two 3/4-in (19-mm) bolts, or the equivalent.

Panels that have been preassembled from steel joists with bridging shall be attached to the structure at each corner before the hoisting cables are released.

Quantity Estimating

A very important step in the process of preparing a bid is preparing an estimate of quantities (materials, labor, and equipment) involved in the proposed solution. Two significant components of this estimate are (1) estimating quantities accurately (quantity take-off) and (2) estimating unit prices as accurately as possible. The bill of quantities (BOQ) is an itemized list of materials, parts, and labor, and associated costs required to construct, maintain, or repair a structure.

There are various types of entities that require estimating. The unit costs for these entities are consistent with the entity type. For example, excavation quantities are estimated as a volume (yd^3 typical) and the unit cost must be specified in $/yd^3$.

The most common types of entities are listed in Table 510.1. Current material costs according to *Engineering News Review* (as of September 2014):

Cement	$114.81/ton
Steel	$49.60/ton
Lumber	$454.10/ton

For example, in the cost estimating of the construction of a free-standing concrete wall, the following categories must be considered:

1. Material cost of concrete (including waste)

    a. Calculate volume of wall in yd^3.

    b. Obtain the unit cost of concrete per yd^3 from a source such as RS Means.

2. Material cost of reinforcement.

    a. Calculate weight of reinforcement (ton) from structural drawings.

    b. Obtain unit cost of steel reinforcement per ton from a source such as RS Means.

3. Labor cost of placing concrete.

    a. Estimate productivity (crew hours per 100 yd^3).

    b. Estimate total crew hours for the total volume of concrete to be placed.

    c. Calculate labor cost using hourly labor cost data.

Table 510.1 Units of Measure of Various Activities

Unit measure	Typical units	Examples
Each	–	Doors, windows
Length	ft	Pipes
Area	ft² sheet	Forms, painting, plastering Plywood is sold in 4-ft × 8-ft sheets
Volume	yd³ board-foot	Excavation, backfilling, concrete Lumber (equivalent to a 12-in × 12-in × 1-in volume)
Weight	ton	Reinforcement steel
Effort	LH (labor hour)	Equipment, labor

4. Formwork material cost. If forms are reused, the cost of initial form erection ($ per ft²) will be different from the cost of reuse of forms (the latter is considerably smaller).

   a. Estimate total wall surface area to be formed. Estimate how much area will apply to initial erection and how much area will apply to form reuse.

   b. Apply unit material costs to calculate total form material costs.

5. Formwork labor cost. The operations of form erection and stripping usually have different values of productivity associated with them, so must be calculated separately.

Table 510.2 lists standard (ASTM) reinforcing bar sizes. Table 510.3 lists steel wire data for smooth and deformed wires used for welded wire fabric. Welded wire fabric is designated by two numbers and two letter-number combinations. An example is 6 × 10 – W8.0 × W4.0. The first number (6) gives the spacing in inches of the longitudinal wires and the second (10) gives

Table 510.2 Standard (ASTM) Reinforcing Bar Sizes

US size #	Metric size #	Weight lb/ft	Weight kg/m	Diameter in	Diameter mm	Area in²	Area mm²
3	10	0.376	0.560	0.375	9.52	0.11	71
4	13	0.668	0.994	0.500	12.70	0.20	129
5	16	1.043	1.552	0.625	15.88	0.31	200
6	19	1.502	2.235	0.750	19.05	0.44	284
7	22	2.044	3.042	0.875	22.22	0.60	387
8	25	2.670	3.973	1.000	25.40	0.79	510
9	29	3.400	5.059	1.128	18.65	1.00	645
10	32	4.303	6.403	1.270	32.36	1.27	819
11	36	5.313	7.906	1.410	35.81	1.56	1006
14	43	7.650	11.384	1.693	43.00	2.25	1452
18	57	13.600	20.238	2.257	57.33	4.00	2581

Table 510.3 Wire Sizes Used for Welded Wire Fabric

Wire size number		Diameter		Area		Weight	
Smooth	Deformed	in	mm	in²	mm²	lb/ft	kg/m
W31	D31	0.628	16.0	0.310	200	1.054	1.568
W28	D28	0.597	15.2	0.280	181	0.952	1.417
W26	D26	0.575	14.6	0.260	168	0.884	1.315
W24	D24	0.553	14.1	0.240	155	0.816	1.214
W22	D22	0.529	13.4	0.220	142	0.748	1.113
W20	D20	0.505	12.8	0.200	129	0.680	1.012
W18	D18	0.479	12.2	0.180	116	0.612	0.911
W16	D16	0.451	11.5	0.160	103	0.544	0.810
W14	D14	0.422	10.7	0.140	90	0.476	0.708
W12	D12	0.391	9.9	0.120	77	0.408	0.607
W11	D11	0.374	9.5	0.110	71	0.374	0.557
W10	D10	0.357	9.1	0.100	65	0.340	0.506
W9.5		0.348	8.8	0.095	61	0.323	0.481
W9	D9	0.338	8.6	0.090	58	0.306	0.455
W8.5		0.329	8.4	0.085	55	0.289	0.430
W8	D8	0.319	8.1	0.080	52	0.272	0.405
W7.5		0.309	7.8	0.075	48	0.255	0.379
W7	D7	0.299	7.6	0.070	45	0.238	0.354
W6.5		0.288	7.3	0.065	42	0.221	0.329
W6	D6	0.276	7.0	0.060	39	0.204	0.304
W5.5		0.265	6.7	0.055	36	0.187	0.278
W5	D5	0.252	6.4	0.050	32	0.170	0.253
W4.5		0.239	6.1	0.045	29	0.153	0.228
W4	D4	0.226	5.7	0.040	26	0.136	0.202
W3.5		0.211	5.4	0.035	23	0.119	0.177
W2.9		0.192	4.9	0.029	19	0.099	0.147
W2.5		0.178	4.5	0.025	16	0.085	0.126
W2		0.160	4.1	0.020	13	0.068	0.101
W1.4		0.134	3.4	0.014	9	0.048	0.071

the spacing in inches of the transverse wires. The first letter-number combination gives the type (W: smooth) and size (8.0) of the longitudinal wire while the second gives the same for the transverse wires (W4.0). Thus, the designation indicates a mesh in which the longitudinal wires are W8 ($A = 0.08$ in²) spaced 6 in apart and the transverse wires are W4 ($A = 0.04$ in²) spaced 10 in apart.

Welded wire fabric is sold in sheets. Normally, fabric sheets with a width of 8.5 ft or less can be transported on a truck. Widths for rail shipment are generally limited to 11.5 ft. In both cases, the length of sheets is limited to 40 ft.

Example 510.1

The (outer) plan dimensions of a wall footing for a building are 54 ft × 72 ft as shown below. The 18-in-thick footing is reinforced concrete and the wall is masonry (12-in blocks). The width of the footing is 3 ft and the bottom of the footing is located 30 in below grade. The wall is 1-ft thick and 14-ft high (measured from the top of the footing).

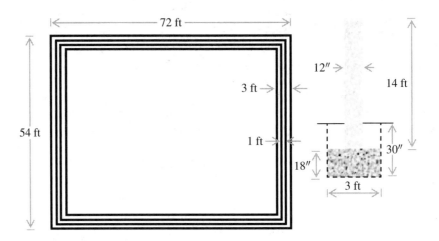

Estimate the following quantities: (1) total volume (yd³) of concrete used for the footings, (2) the total number of 12-in masonry blocks (12D × 8H × 16L) used to construct the wall (extending 14 ft above the footing), and (3) the total excavation volume (yd³).

Solution Dimensions of building (along footing centerline) = 69 ft × 51 ft
Perimeter (along footing centerline) = 2 × (69 + 51) = 240 ft

(1) Cross section of footing = 1.5 × 3 = 4.5 ft²
Volume of concrete in footing = 240 × 4.5 = 1080 ft³ = 40 yd³
(2) Total wall "surface" area (centerline) = perimeter × height = 240 × 14 = 3360 ft²
Each 12-in-wide masonry block is 8-in high and 16-in long.
Wall surface area covered by one block = 8 × 16 = 128 in² = 0.89 ft²
Number of masonry blocks required = 3360 ÷ 0.89 = 3780
(3) Cross section of trench = 2.5 ft × 3 ft = 7.5 ft²
Volume of excavation = 7.5 × 240 = 1800 ft³ = 66.7 yd³

Quantity Estimation for Excavations

For excavations in soil, we must account for the fact that once the soil is excavated and released from the confining pressures *in situ*, it expands to a larger volume. This expansion is quantified in terms of the bulking factor (or swell) and typically expressed either as a percentage or a decimal. A bank volume V expands to a loose volume $= V(1+ \text{bulking})$ upon being excavated. When this soil is placed and compacted into place, it usually occupies a lesser volume than the original bank volume. The shrinkage factor is also expressed as a decimal or percentage. The final compacted volume = $V(1 - \text{shrinkage})$.

Example 510.2

A contractor is excavating the trench shown below. The trench should be 5-ft deep × 3-ft wide. A 12-in-diameter water pipe is to be placed in the trench and then backfilled with the soil that was removed. All dimensions shown are on centerline.

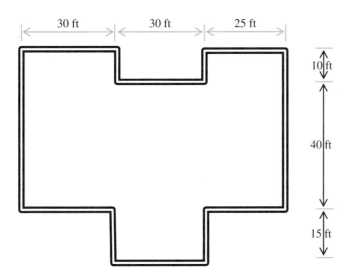

Does the contractor have enough soil to backfill the trench, or will some have to be borrowed? If borrow soil needs to be brought in, how much (LCY)?

Assume swell = 15% and shrinkage = 12%

Solution   Perimeter = 30 + 10 + 30 + 10 + 25 + 50 + 25 + 15 + 30 + 15 + 30 + 50 = 320 ft

Cross section of trench = 3 × 5 = 15 ft²
Volume of soil originally in the trench = 15 × 320 = 4800 ft³
After compaction, excavated soil will fill a volume = 0.88 × 4800 = 4224 ft³
Area of pipe = $\pi/4(1)^2 = 0.785$ ft²
Area to be filled with soil = 15 − 0.785 = 14.215 ft²
Required soil (compacted) volume = 14.215 × 320 = 4549 ft³
Deficit = 4549 − 4225 = 324 ft³ (compacted volume)

This soil is equivalent to 324 ÷ 0.88 ft³ = 368.2 ft³ (bank volume) = 368.2 × 1.15 = 423.4 ft³ (loose volume) = 15.68 LCY (loose cubic yards).

Estimating Mortar Quantity between Bricks or Masonry Blocks

A brick having dimensions $L \times W \times H$ is shown below. The front view of the wall is shown on the right, with the mortar joints between the brick and its adjacent units shaded in gray.

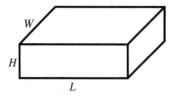 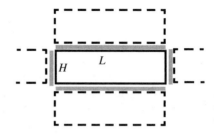

The top and bottom ($L \times W$) and left and right ($H \times W$) faces have mortar adjacent to them. If the mortar joint thickness is assumed to be t, then half of that thickness is assumed to be associated with each brick. The total surface area (of 4 faces) covered by a thickness $t/2$ is then given by

$$S = 2W\left(L + \frac{1}{2}t + H + \frac{1}{2}t\right) = 2W(L + H + t)$$

When multiplied by the mortar thickness of $t/2$, the mortar volume is calculated as

$$V = 2W(L + H + t)\frac{t}{2} = W(L + H + t)t$$

Example 510.3

If mortar joint thickness in problem 510.1 is ½ in, what is the volume of mortar required for the entire wall?

Solution   Thus, for the masonry blocks in Example 510.1, $L = 16$ in. $W = 12$ in, $H = 8$ in. If we assume mortar thickness $= 0.5$ in, the volume of mortar associated with each brick is

$$V = W(L + H + t)t = 12 \times (16 + 8 + 0.5) \times 0.5 = 147\,\text{in}^3$$

For Problem 510.1, total mortar volume required (for 3780 bricks) $= 555,660$ in^3 $= 321.6$ ft^3 $=$ 11.91 yd^3

Example 510.4

For the wall in problem 510.1, the wall is reinforced with vertical reinforcement in the form of a no. 5 bar every 16 in. The wall height is 14 ft and the perimeter measured along the center-line is 240 ft. What is the total weight of reinforcement steel needed for the wall?

Solution   Wall perimeter $= 240$ ft
Number of rebars $= 240 \times 12 \div 16 = 180$
Since each-bar is 14-ft long, total length of reinforcement $= 180 \times 14 = 2520$ ft
No. 5 bars have weight $= 1.043$ lb/ft (Table 510.2)
Total weight of reinforcement $= 2520 \times 1.043 = 2628$ lb

Example 510.5

For the 18-in-diameter column, the main reinforcement consists of 6 no. 10 bars confined laterally by no. 3 ties every 10 in. If the length of the column is 20 ft, what is the total weight of steel reinforcement used in the column?

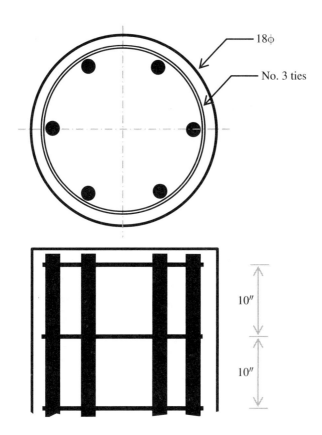

18ϕ

No. 3 ties

10″

10″

Solution   Using the ACI recommended clear cover of 1.5 in, the diameter of the ties $= 18 - 2 \times 1.5 - 0.375 = 14.625$ in

Circumference of each tie $= \pi \times 14.625 = 45.95$ in $= 3.83$ ft

Number of ties in a 20-ft-long column (using a spacing of 10 in) $= 240/10 + 1 = 25$

Total length of ties (no. 3) $= 25 \times 3.83 = 96$ ft

Total length of main reinforcement (no. 10) $= 20 \times 6 = 120$ ft

Total weight of reinforcement (using values from Table 510.2) $= 0.376 \times 96 + 4.303 \times 120 = 552.5$ lb.

Example 510.6

A reinforcing steel quantity take-off needs to be done for a 32-foot-long wall footing. The cross section shown below shows all steel groups as designed. The footing thickness is 30 in and the width is 8 ft. Compute the total weight of steel (lb) required.

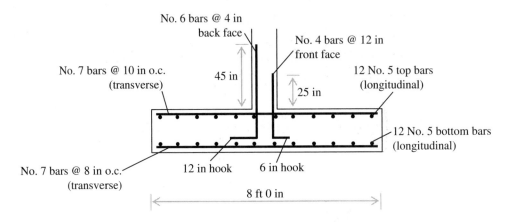

Solution The table below summarizes weight of reinforcement in each of the 6 categories.

Category	Size/spacing	Number	Length (ft)	Weight (lb/ft)	Total weight (lb)
Top transverse	No. 7 @ 10 in	384/10 ≈ 39	39 × 8 in = 312	2.044	637.7
Bottom transverse	No. 7 @ 8 in	384/8 ≈ 48	48 × 8 in = 384	2.044	784.9
Top longitudinal	12 No. 5	12	12 × 32 in = 384	1.043	400.5
Bottom longitudinal	12 No. 5	12	12 × 32 in = 384	1.043	400.5
Front face	No. 4 @ 12 in	384/12 ≈ 32	32 × 57 in = 152	0.668	101.5
Back face	No. 6 @ 4 in	384/4 ≈ 96	96 × 83 in = 664	1.502	997.3
					3322.4

Therefore, total weight of reinforcement required for the wall = 3322.4 lb

Note that length estimates for the various bars can be more precise than what is shown in the table. For example, the transverse bars that run the entire footing width will be a little shorter than 8 ft because of clear cover requirements in the ACI specifications. Similarly, the top transverse bars (@ 10-in spacing o.c.) will span a little less than the 32 ft (384 in) length of the wall. This will lead to 38 spaces (380 in ÷ 10 in) and therefore 39 bars. This is the same as that obtained by dividing 384 in. by the 10-in spacing and rounding it up to 39.

Example 510.7

A building has horizontal dimensions 100 ft by 80 ft. It has a hipped roof with a pitch of 6 in per foot, and it has a 3-foot overhang.

How many sheets of plywood will be needed to cover the roof?

Solution If the eaves are 3-ft horizontally out from each wall, the horizontal dimensions of the roof are 106 ft × 86 ft. The plan view looks like this:

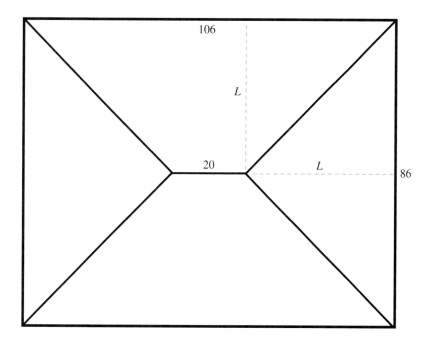

In the north-south direction, the horizontal projection from the eaves to the ridge = 43 ft. At a 1:2 pitch the height of the roof is 0.5 × 43 = 21.5 ft. The inclined length of the lines shown as dashed lines on the figure is given by the Pythagorean formula (using horizontal projection = 43 ft and vertical projection = 21.5 ft)

$$L = \sqrt{43^2 + 21.5^2} = 48.08 \text{ ft}$$

For both the triangular and the trapezoidal roof sections, L is the distance from the base to apex. The total area of the roof is equal to 2 each of these shapes.

$$\text{Area of the triangle: } A_1 = \frac{1}{2} \times 86 \times 48.08 = 2067.4 \text{ sq ft}$$

$$\text{Area of the trapezoid: } A_2 = \frac{1}{2} \times (106 + 20) \times 48.08 = 2029.0 \text{ sq ft}$$

The total roof area is twice the sum of these, since the roof consists of two of each type of section.

$$\text{Total area of roof: } A = 2(A_1 + A_2) = 10{,}193 \text{ sq ft}$$

Assuming 10% waste and each standard sheet of plywood being 4 ft × 8 ft in size, the number of sheets required is calculated as

$$N = \frac{10193 \times 1.1}{4 \times 8} = 350$$

Suggested shortcut: A quick estimate can be made by noting that because of the pitch of the roof, a plan dimension of 43 ft corresponds to an inclined dimension of 48.08 ft. We can use this ratio (48.08/43) to estimate the roof area from the corresponding plan area as follows:

$$A = 106 \times 86 \times \frac{48.08}{43} = 10{,}193 \text{ sq ft}$$

Quantity Estimation of a Stockpile Volume

For a stockpile of material with stable side slopes, the pile takes on the shape of a truncated pyramid, as shown in Fig. 510.1. The sideslope (H:V) is related to the internal stability of the material. For a cohesionless material, the "angle of repose" θ is related to slope m according to

$$m = \tan\theta \qquad\qquad (510.1)$$

If the side slope is given as 1:1, that should be taken as the slope normal to the edges, not at the corners. If the slope at the edges is maintained at 1:1, the slope at the corners will be less than that (same rise over a longer run) and that is OK from a stability point of view. So, for a height of H and a slope parameter $m = \tan\theta$ (rise over run), the horizontal dimension will "come in"

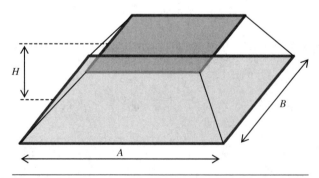

Figure 510.1 Geometric elements of a truncated pyramid.

H/m from either edge, resulting in the dimensions of the top reducing from the corresponding base dimensions A and B according to

$$A - \frac{2H}{m} \quad \text{and} \quad B - \frac{2H}{m} \tag{510.2}$$

So, for a pyramid with base dimensions 350 ft × 150 ft and height 15 ft and side slope = 1:1, the top dimensions will be 320 ft × 120 ft.

The inclined (hypotenuse) of trapezoids on each face will be

$$H\sqrt{1 + 1/m^2} \tag{510.3}$$

And the surface area of the inclined face rising from edge A will be

$$\left(A - \frac{H}{m}\right) H\sqrt{1 + 1/m^2} \tag{510.4}$$

Therefore, the total surface area of all four inclined sides is given by

$$2H\sqrt{1 + 1/m^2} \left(A + B - \frac{2H}{m}\right) \tag{510.5}$$

Area of the top (flat) surface is

$$\left(A - \frac{2H}{m}\right)\left(B - \frac{2H}{m}\right) \tag{510.6}$$

The volume of the truncated pyramid is

$$V = \frac{H}{6}\left[6AB - \frac{6H}{m}(A + B) + \frac{8H^3}{m^2}\right] \tag{510.7}$$

Example 510.8

Sand is stored at a construction site in the form of a pyramid as shown in Fig. 510.1. The angle of internal friction is 34°. If the maximum base dimension is 200 ft × 60 ft and the height of the pile is 12 ft, the total volume of sand (yd³) in the stockpile is most nearly:

Solution

The side-slope parameter

$$m = \tan 34 = 0.675$$

For $A = 200$ ft, $B = 60$ ft, $H = 12$ ft, and $m = 0.675$, the volume is given by Eq. (510.7) is 93,557.1 ft³ = 3465 yd³

End-of-Chapter Practice Problems

Chapter 101: Strength of Materials

101-001

The I-section shown below is to be used as a beam to withstand bending about the horizontal axis. Find the elastic section modulus of the section. All dimensions are in mm.

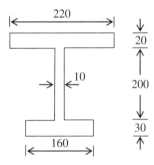

101-002

The I-section shown in Prob. 101-001 is to be used as a beam to withstand bending about the horizontal axis. Find the plastic section modulus of the section. All dimensions are in mm.

101-003

For the beam shown below, the maximum bending moment (kip-ft) is most nearly:

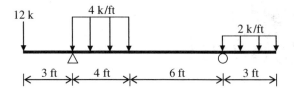

Chapter 102: Statically Determinate Structures

102-001

A cable system supports two vertical forces as shown. Node D is located 1 ft below node A. If the breaking strength of the cable is 40 tons, what is the factor of safety?

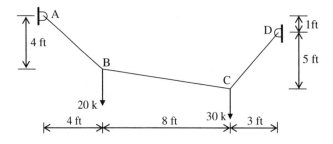

102-002

The following service loads apply to the timber-framed bearing wall building shown in transverse section:

Floor DL = 8 psf

Roof DL = 10 psf (on horizontal projection)

Floor LL = 40 psf

Roof LL = 20 psf (on horizontal projection)

Wall DL = 7 psf

Wind pressure on stud walls = 25 psf

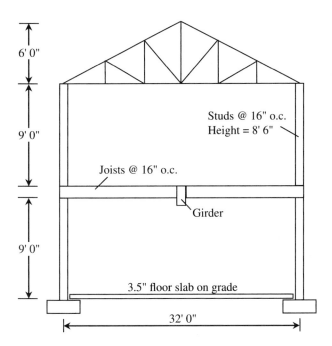

The service gravity load applied to each wall footing (lb/ft) is most nearly:

(A) 764

(B) 934

(C) 1000

(D) 1400

102-003

For the beam shown below, the dead load is a UDL = 2 kips/ft and the live load is a UDL = 5 kips/ft. Internal hinges exist at B and D. The maximum possible negative moment at support C (kip-ft) is most nearly:

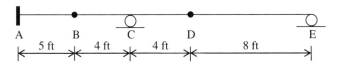

Chapter 103: Introduction to Indeterminate Structures

103-001

Calculate the maximum vertical deflection (inches) for the cantilever beam shown: $E = 29,000$ kips/in². Moments of inertia for segments AB and BC are $I_{AB} = 1200$ in⁴ and $I_{BC} = 700$ in⁴.

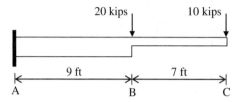

103-002

The propped cantilever shown below supports the uniformly distributed load as shown: $E = 29,000$ kips/in². Moment of inertia of beam section: $I = 1200$ in⁴. The vertical reaction (kips) at the roller support is most nearly:

(A) 17

(B) 35

(C) 16

(D) 32

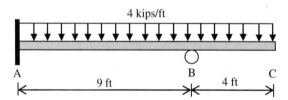

103-003

For the steel frame shown below, assume that the vertical members (columns) all have moment of inertia $I = 800$ in⁴ and all horizontal members (beams) have moment of inertia $I = 1200$ in⁴. All joints are to be considered rigid. The moment distribution factor for member EH at node E is most nearly:

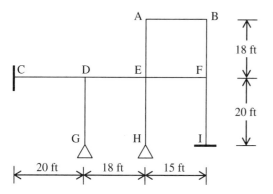

Chapter 104: Concrete Fundamentals

104-001

A 4-in-wide × 8-in-high rectangular unreinforced concrete beam is simply supported by supports 75-in apart and tested in third point loading as shown in the figure below. The load P at which the beam fails by flexural cracking near midspan is 560 lb. What is most nearly the 28-day compressive strength of the concrete?

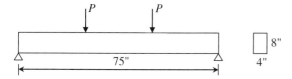

104-002

A 6-in-diameter × 12-in-high concrete cylinder is tested in a split-cylinder test as shown in the figure below. The load P at which the cylinder beam fails by tensile splitting is 35,600 lb. What is most nearly the 28-day compressive strength of the concrete?

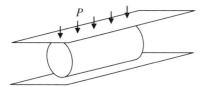

104-003

The table below shows characteristics of a concrete mix. If the total unit weight of the concrete is 142 lb/ft³, what is the air content (percent)?

Component	Weight (lb)	SG
Cement	450	3.15
Sand	870	2.65 (SSD)
Stone aggregate	1106	2.50 (SSD)
Water	205	1.0

Chapter 105: Reinforced Concrete Beams

105-001

A system of parallel T-shaped reinforced concrete beams support a floor with the following superimposed service loads: DL = 0.085 ksf, LL = 0.15 ksf. The beams are simply supported by girders 32-ft apart and the lateral spacing between beams is 8 ft. Twenty-eight-day compression strength of the concrete = 4000 psi and reinforcement is grade-60 deformed bars.

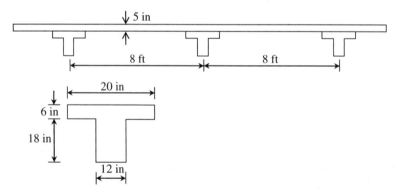

The required tension reinforcement (in²) in the T-beams is most nearly:

(A) 3.0

(B) 4.0

(C) 5.0

(D) 6.0

105-002

What is the maximum permitted spacing of no. 3 stirrups in the maximum shear zone near the supports for the T-shaped reinforced concrete beam, described in Prob. 105-001?

105-003

The cantilever beam shown is reinforced with tension steel only (three no. 9 bars) as shown. At a section where it is no longer required, it is proposed to cut off the central bar. What is the development length for this bar? The longitudinal bars are epoxy coated. Stirrup is no. 3 size. Use $f_c' = 4000$ psi, $f_y = 60,000$ psi.

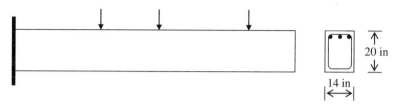

Chapter 106: Reinforced Concrete Slabs

106-001

A building floor system consists of a reinforced concrete slab ($f_c' = 4000$ psi, $f_y = 60,000$ psi) supported by a system of parallel beams spaced at 8 ft on center. The beams span 20 ft and are supported by a girder using simple connections. If explicit deflection calculations are not performed for the concrete slab, what is the minimum required thickness according to ACI 318?

106-002

A reinforced concrete floor consists of a 5-in-thick slab cast monolithically with beams as shown. Twenty-eight-day compressive strength = 4000 psi. Steel reinforcement is grade 60 deformed bars. Beams have simple span = 28 ft. The floor loads are superimposed dead load = 25 psf and live load = 85 psf. Beams are to be designed with tension steel only. The required reinforcement (in²) of the tensile reinforcement is most nearly:

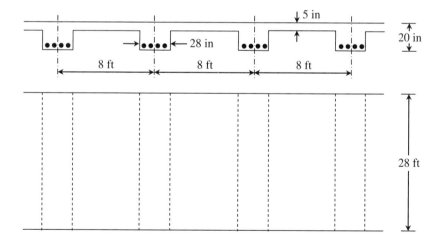

106-003

What is the reinforcement requirement for the longitudinal direction (perpendicular to plane of drawing) of the slab in Prob. 106-002?

Chapter 107: Reinforced Concrete Columns

107-001

What is the maximum pitch of a 3/8-in-diameter spiral to be used for a reinforced concrete (f'_c = 4500 psi, 60,000 psi) column with diameter = 18 in?

107-002

What is the appropriate arrangement of ties for a rectangular (15 in × 20 in) reinforced concrete (f'_c = 4500 psi, 60,000 psi) column? The longitudinal reinforcement is eight no. 9 bars, arranged on all four faces.

107-003

A reinforced concrete column (f'_c = 4000 psi, 60,000 psi) is loaded eccentrically with a factored load P_u as shown. The reinforcement is to be arranged on two opposing faces of the rectangular section only. What is the required reinforcement (in²)?

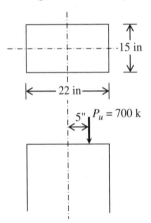

Chapter 108: Prestressed Concrete

108-001

For the rectangular prestressed concrete beam shown below, what is the most effective tendon profile? The beam is expected to be loaded with uniformly distributed gravity loads.

108-002

A prestressed concrete beam ($f'_c = 6000$ psi) is to be post-tensioned using seven-wire strand with total cross-sectional area = 3.67 in² and initial prestress = 75% of the ultimate stress $f_{pu} = 270$ ksi. The elastic section properties of the beam are shown below. Eccentricity of the effective prestressing force is 10.25 in. Calculate the bottom fiber stress for the following conditions:

1. Dead load + live load bending moment (causing tension on the bottom fiber) = 1200 kip-ft
2. Prestress loss = 32 ksi

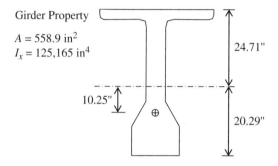

Girder Property
$A = 558.9$ in²
$I_x = 125,165$ in⁴

24.71"

10.25"

20.29"

108-003

For the beam in Prob. 108-002, what is the allowable tension according to PCI?

Chapter 109: Steel Tension Members

109-001

A36 steel ($F_y = 36$ ksi, $F_u = 58$ ksi) is used for a truss. What is the minimum gross area needed for the bottom chord. Assume effective net area = $0.7A_g$ (load is 30% DL + 70% LL).

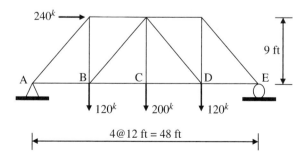

240^k

9 ft

A   B   C   D   E

120^k 200^k 120^k

4@12 ft = 48 ft

109-002

A C15 × 50 channel is used as a tension member. Steel grade is A36 ($F_y = 36$ ksi, $F_u = 58$ ksi). The member is connected using two lines of ¾-in-diameter bolts in the web as shown below. What is the design member capacity in tension?

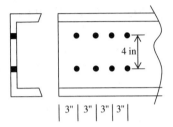

109-003

What is the recommended maximum slenderness ratio of a tension member according to AISC?

Chapter 110: Steel Compression Members

110-001

The legs of a transmission tower are designed as a built-up section under concentric compression load. The cross-section consists of four L6 × 4 × 7/8 angles arranged as shown below. The four angles are "stitched together" by a lattice of plates, represented by the dashed lines in the section. If the KL for the member is 32 ft, determine the governing slenderness ratio.

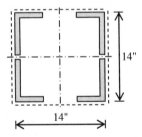

110-002

A steel ($F_y = 50$ ksi, $F_u = 65$ ksi) column is 50-ft long with pinned ends (rotation permitted, translation prevented). The following service loads produce axial force in the column (compression positive):

Dead load: 120 kips

Live load: 250 kips

Wind load: ± 340 kips

The weak axis of the column is laterally braced at midheight. What is the lightest W12 section that is adequate?

Chapter 111: Steel Beams

111-001

A simply supported steel beam ($L = 30$ ft) carries the following uniformly distributed loads: dead load = 2 kips/ft (not including self-weight) and live load = 4 kips/ft. The compression flange has lateral support only at the supports. What is the lightest W-shape that is adequate? Use $C_b = 1.0$ and $F_y = 50$ ksi.

111-002

What is the nominal shear capacity (V_n) of a M12.5 × 12.4 beam for $F_y = 36$ ksi and $F_u = 58$ ksi. The section is used as a beam without any stiffeners.

Chapter 112: Bolted and Welded Connections

112-001

What is the available shear strength (kips) of 1½-in-diameter A325 high-strength bolts in single shear? Bolts derive their strength from bearing. Assume that threads are included in the shear plane.

112-002

What is the available shear strength (kips) of 1½-in-diameter A325 high-strength bolts in single shear, if bolts are pretensioned adequately to prevent slip. Assume bolt holes are long slotted in the transverse direction. Assume slip is a strength limit state.

112-003

What is the maximum permitted factored load for the scenario shown below? Bolts are ¾-in-diameter A307 bolts in single shear. The eccentric vertical load is equally transmitted to two flange plates bolted to the flanges of the column as shown below.

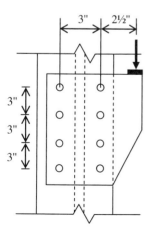

112-004

What is the maximum permitted factored load for the scenario shown below? Welds are E70XX electrodes. Weld size = ¼ in. The eccentric vertical load is equally transmitted to two flange plates welded to the flanges of the column as shown below.

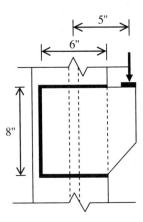

Chapter 113: Bridge Design (AASHTO LRFD)

113-001

A composite concrete-deck–steel-girder system for a simple span ($L = 80$ ft) bridge is shown. The beam is a doubly symmetric plate girder ($A_s = 143.0$ in²; $I_{xx} = 92,140$ in⁴; depth = 64 in). Slab thickness = 10 in.

Twenty-eight-day compressive strength of the concrete = 4000 psi. Yield stress for steel is 60 ksi. Load on the deck (including weight of slab plus asphalt overlay and equivalent traffic load) may be taken as 1.8 kips/ft². The maximum bending stress (kips/in²) in the steel girders is most nearly:

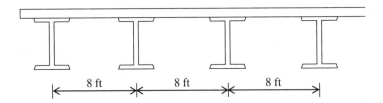

113-002

For the bridge shown above, what is the distribution factor for bending moment in an interior girder for a single lane loaded? Use AASHTO LRFD 4th edition.

113-003

For a steel stringer-concrete deck bridge, fully composite behavior is to be achieved by ¾-in-diameter shear studs (two per line) as shown below. If the number of stress cycles = 2×10^6, what is the required spacing for these shear studs in the high shear zone adjacent to the bearings?

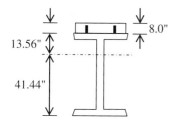

Moment of inertia of transformed section = 50,230 in⁴
Centroid distance from bottom flange = 41.44 in
Transformed slab = 8-in deep × 10.7-in wide

The following reactions are given:
Maximum reaction due to live load plus impact = 53.7 kips
Maximum reaction due to dead load = 38.6 kips

Chapter 114: Timber Design

114-001

What is the axial load capacity (ASD) of a solid sawn no. 1 Hem-Fir column—nominal size 6 in × 6 in. Pinned-pinned ends, length = 14 ft? Moisture condition = 22%. Surroundings are at room temperature.

114-002

A 2-in × 10-in floor joist spans 15 ft. Uniformly distributed loads are $D = 30$ lb/ft and $L = 60$ lb/ft. Timber is no. 2 Southern Pine. The joist is notched to a dressed depth of 7.25 in at the supports. Joists support a floor system, spaced 24 in on center. Top of joist has continuous lateral support by virtue of being nailed to wood panels. Joists are in building interior and subject to dry use. Check the shear capacity of the joist at the notch.

Chapter 115: Masonry Design

115-001

A 10-in-wide × 40-in-deep masonry beam ($f_m' = 1500$ psi) is reinforced with two no. 6 bars ($f_y = 40$ ksi) placed with effective cover = 3.5 in. Assume the beam is fully grouted lightweight concrete masonry with 8-in-high blocks. What is the moment capacity based on allowable compressive stress in masonry? What is the allowable moment based on the allowable stress in the steel?

115-002

A 16-ft-high 10-in × 18-in masonry column ($f_m' = 2000$ psi) has to carry an eccentric axial load of 20 kips. Eccentricity $e = 3$ in, measured parallel to the short direction. The column is reinforced with four no. 9 bars ($f_y = 40$ ksi). Check the adequacy of the column if (a) special inspection is not provided and (b) special inspection is provided.

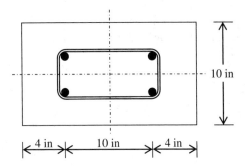

Chapter 201: Phase Relationships for Soils

201-001

A clayey sand (SC) was compacted into a 1/30th cubic foot cylindrical Proctor steel mold with the following results:

Weight of wet soil: 4.18 lb

Weight of dry soil: 3.67 lb

Specific gravity: 2.73

The percent saturation of this compacted sample is most nearly:

(A) 70%

(B) 50%

(C) 15%

(D) 85%

201-002

A sample of saturated soil has volume = 1050 cm³ and weighs 4.58 lb. If the specific gravity of solids = 2.60, what is the water content (%)?

201-003

A highway project requires 20,000 yd³ of fill compacted to 95% of maximum Proctor dry density. Laboratory tests on samples from the borrow pit show the following properties:

Unit weight = 124 pcf

Water content = 18%

Void ratio = 0.45

Maximum Proctor dry unit weight = 108 pcf

Optimum moisture content = 22%

What volume (yd³) of borrow soil is needed?

Chapter 202: Soil Sampling and Testing

202-001

An undrained triaxial test is conducted on a 2-in-diameter, 4-in-long cylindrical sample of silty clay. At shear failure, the following data is recorded:

Chamber pressure = 12.6 psi

Pore pressure = 4.5 psi

Added axial load = 93.2 lb

Ignoring internal friction, the cohesion (lb/ft²) of the soil is most nearly:

(A) 1100

(B) 1800

(C) 2200

(D) 4500

202-002

The figure below shows a rock recovery from a depth of 100 to 115 ft below the surface. The vertical lines on the sample represent the presence of natural fractures in the rock. What is the RQD of the sample?

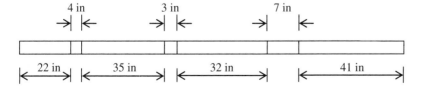

202-003

A clay soil is tested in the Atterberg test where a groove etched in the surface of the sample closes as a result of repeated blows delivered to the sample. In the table below, the related data from the five samples is presented.

Water content (%)	No. of blows to close groove
21	39
32	33
43	27
54	20
65	16

Which of the following is true?

  (A) The plasticity index of the soil is 32.

  (B) The liquid limit of the soil is 45.

  (C) The plastic limit of the soil is 23.

  (D) The liquidity index of the soil is 56.

202-004

The soil profile shown below consists of a sandy fill layer overlying a layer of normally consolidated clay with the soil properties shown. The soil below the elevation of -20 ft is stiff clay.

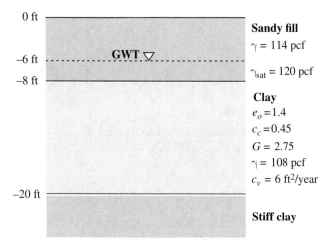

The GWT is lowered to the top of the clay layer and a mat foundation (bottom of mat at depth of 42 in below ground) of plan dimensions 190 ft × 250 ft is constructed. Total vertical (concentric) load on the mat = 20,000 tons.

What is the settlement caused by the consolidation of the clay layer (inches) immediately underlying the sandy fill?

202-005

A 20-ft-thick layer of clay is situated between layers of sand. The ultimate settlement is calculated to be 8 in. How much time (years) is needed for the first 6 in of settlement to occur? The following data are given:

Initial void ratio = 1.6

Void ratio after 4 years = 1.3

Consolidation index = 0.42

Recompression index = 0.10

Coefficient of consolidation = 5 ft²/year

Chapter 203: Soil Classification

203-001, 002, 003

The sieve analysis result summary for a soil sample is shown in the table below. The total mass of the soil sample is 250 g. In addition, hydrometer analysis of the fines indicates that 24 g is finer than 0.05 mm and 12 g is finer than 0.002 mm. Atterberg limits are LL = 32 and PL = 23.

Sieve size/opening (mm)	Weight retained (g)
50	12
25	15
12.5	21
No. 4 (4.75)	23
No. 10 (2.0)	16
No. 20 (0.85)	34
No. 40 (425)	27
No. 60 (0.25)	32
No. 100 (0.15)	12
No. 140 (0.106)	12
No. 200 (0.075)	15
Pan	31

This data is to be used for problems 203-001 to 203-003.

203-001

What is the textural description of the soil according to the USDA Soil Texture Classification System?

203-002

What is the classification of the soil according to the USCS Soil Classification System?

203-003

What is the classification of the soil according to the AASHTO Soil Classification system? Include the group index in your answer.

Chapter 204: Vertical Stress Increase at Depth

204-001

A circular footing carries a column load, $P = 105,000$ lb. The diameter of the footing is 5 ft and the footing depth is 4 ft. What is the vertical stress (psf) induced by the column load at a point 4 ft below the edge of the footing?

204-002

A rectangular footing is 6 ft $\times$ 4 ft in plan dimensions. The concentric column load on the footing is 80 kips. What is the stress increase (psf) at a point 5 ft below the center of the footing? Use the influence diagram on the next page.

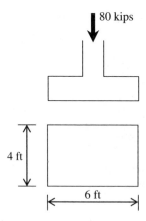

Vertical stress at a point directly below the comer
of a uniformly loaded rectangular area

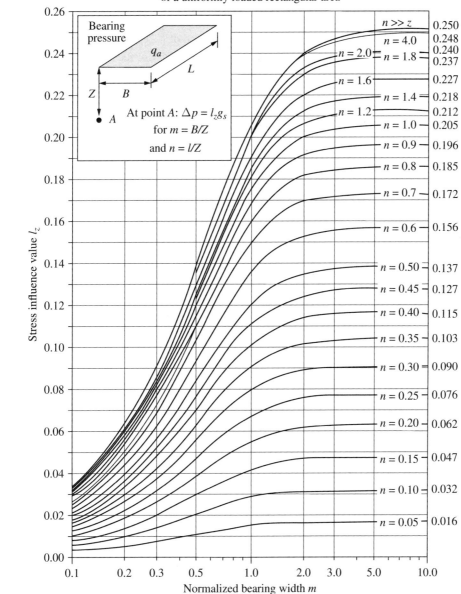

Chapter 205: Flow through Porous Media

205-001

Seepage occurs through the soil sample due to the hydraulic conditions shown below. The void ratio of the soil sample $= 0.45$. The hydraulic conductivity of the soil is 1.8×10^{-4} ft/s. At the instant shown, what is the seepage velocity (in/s)?

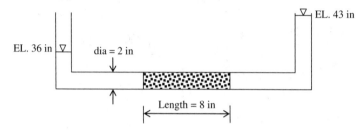

205-002

A confined aquifer has the following properties:

Thickness $= 34$ ft

Storativity $= 0.6$

Transmissivity $= 1.2$ ft^2/s

Elevation of piezometric surface $= 239.37$ ft above sea level.

A 9-in-diameter screened well is used to pump water from this aquifer at the rate of 750 gpm. What is the elevation of the piezometric surface at the well 2 h after pumping begins?

205-003

Using the flow net shown below, (1) determine the seepage underneath the 1000-ft-wide concrete dam, and (2) the velocity at point "a" in ft/h, where the height of the net's square is 19 ft. The soil has a $G_S = 2.67$, $D_{10} = 0.01$ mm. Estimate the flow by using Hazen's coefficient $C = 15$ to determine the hydraulic conductivity K.

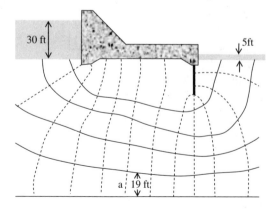

Chapter 206: Shallow Foundations

206-001

A wall carries a total load of 20 kips/ft. It is supported by a continuous footing with the bottom of footing founded in a sandy soil with negligible cohesion and an angle of internal friction = 32°. The dry unit weight of the soil is 94 pcf. The groundwater table is located at the same elevation as the bottom of the footing. The saturated soil below the water table has a water content of 26%. Depth of footing = 3.0 ft. The minimum factor of safety based on ultimate bearing capacity is 2.8. What is the minimum required width of footing?

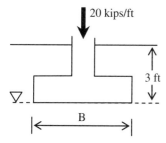

206-002

A square footing is to carry a concentric column load of 75 kips. The maximum allowable settlement is 0.8 in. Depth of footing = 34 in. The soil has the following properties:

Description: Coarse sand with traces of silt. Cohesion = 0. Angle of internal friction = 34°. Moisture content = 12%. Dry unit weight = 95 pcf. Corrected SPT N value = 25.

What is the minimum footing size based on settlement?

206-003

A rectangular footing (4.7 ft × 9.7 ft) carries a 12-in-square column as shown. The column carries a total concentric load of 130 kips. What is the approximate maximum soil pressure under the footing?

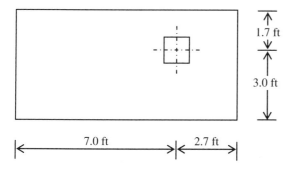

Chapter 207: Deep Foundations

207-001

The figure below shows a pile cap supported by a group of five piles supporting the following superstructure loads: vertical load: $V = 300$ kips, horizontal load: $H = 120$ kips.

Horizontal spacing between pile lines = 6 ft. Thickness of pile cap = 3 ft.

Effective location of the loads should be at the centroid of the pile group. Use one line of vertical piles and one line of batter piles. Determine the batter angle. Determine the axial force per pile.

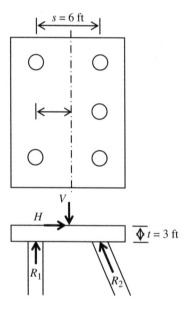

207-002

What is the ultimate capacity of a timber pile (diameter = 14 in) driven to a depth of 40 ft below grade through the soil layers shown in the figure below? The bottom of the pile cap is 6 ft below grade.

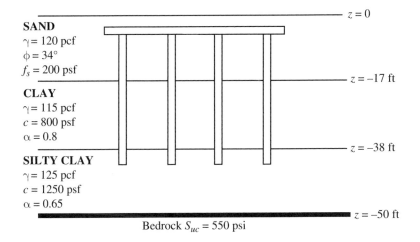

SAND
$\gamma = 120$ pcf
$\phi = 34°$
$f_s = 200$ psf

$z = 0$

$z = -17$ ft

CLAY
$\gamma = 115$ pcf
$c = 800$ psf
$\alpha = 0.8$

$z = -38$ ft

SILTY CLAY
$\gamma = 125$ pcf
$c = 1250$ psf
$\alpha = 0.65$

$z = -50$ ft

Bedrock $S_{uc} = 550$ psi

207-003

For the scenario shown in Prob. 207-002, the vertical load on the pile cap is 400 tons distributed over the pile cap dimensions of 30 ft × 15 ft. Prior to building the foundation, the water table is lowered from a depth of 10 to 17 ft. What is the settlement caused by the consolidation of the two clay layers between depths of 17 and 50 ft? Assume initial void ratio $e = 0.55$ and compression index $C_c = 0.45$ for both clay layers.

Chapter 208: Retaining Walls

208-001

A 12-in-thick cantilever wall is used to retain a backfill soil ($\gamma = 120$ lb/ft³, angle of internal friction $= 32°$) subject to surcharge load $= 400$ psf. Friction angle between base of wall footing and soil $= 24°$. The elevation of the top of the backfill is 150 ft and the bottom of the 3-ft-thick wall footing is at elevation 132 ft. What is the factor of safety against sliding?

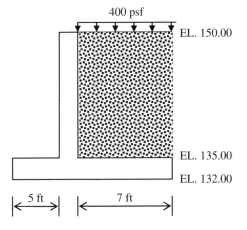

400 psf

EL. 150.00

EL. 135.00

EL. 132.00

5 ft

7 ft

208-002

For the wall shown above, what is the factor of safety against overturning?

208-003

The 12-in-thick retaining wall uses a 20-in key directly below the stem as shown. The wall is used to retain a backfill soil ($\gamma = 120$ lb/ft³, angle of internal friction $= 32°$) subject to surcharge load $= 400$ psf. Friction angle between base of wall footing and soil $= 24°$. Undrained cohesion of the soil below the footing $= 900$ psf and the adhesion factor for the concrete-soil interface $= 0.8$. The elevation of the top of the backfill is 150 ft and the bottom of the 3-ft-thick wall footing is at elevation 132 ft. What is the factor of safety against sliding?

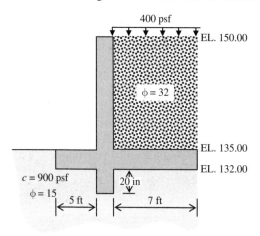

Chapter 209: Support of Excavation

209-001

A 12-ft-wide, 22-ft-deep trench is excavated in a sandy soil ($\gamma = 123$ pcf, $w = 12\%$, $e = 0.35$, $\phi = 34°$) and supported by a timber sheet piling which is in turn supported by a system of horizontal wales (vertical spacing $= 6$ ft) and struts (longitudinal spacing $= 20$ ft) as shown.

If the allowable bending stress in the sheet pile is 2200 psi, what is the required thickness (in) of the planks?

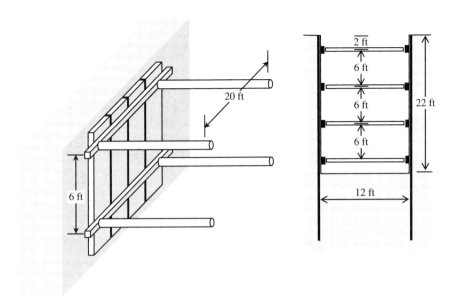

209-002

For the situation described in Prob. 209-001, what is the axial load in the second strut from the top?

209-003

A retaining wall is restrained by a steel anchor rod attached to a deadman behind the wall as shown. The axial force in the anchor rod has been calculated as 10,500 lb. At the location of the tieback, the minimum wall movement necessary to create the active condition is 0.25 in. The allowable stress in the anchor rod is 24,000 psi. What is the minimum required diameter and length (L) of the anchor rod?

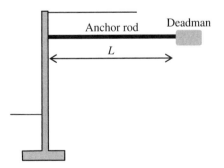

Chapter 210: Slope Stability

210-001

The stability of a slope is being investigated using the Bishop method. The region defined by the assumed failure circle is divided into vertical slices as shown. The radius of the assumed circle is 50 ft. The fourth slice (shown shaded) has a radial line which is inclined to the vertical at 25°. The (horizontal) width of each slice is 6 ft and the weight of the slice is 20,000 lb/ft. What is the tangential resisting force preventing sliding along the failure surface?

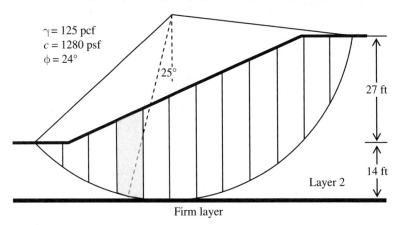

210-002

An embankment of c-ϕ soil is protected by riprap as shown below. Groundwater is at significant depth below the toe of the slope. What is the factor of safety against sliding along the face of the slope?

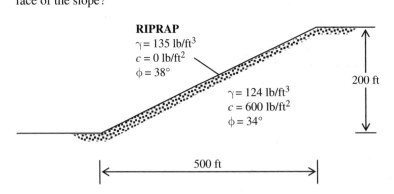

Chapter 211: Seismic Topics in Geotechnical Engineering

211-001

For the design earthquake at a particular site, the cyclic resistance ratio (CRR) is 0.25. The average cyclic shear stress induced at a depth of 23 ft below the surface is 425 psf. What is the factor of safety against liquefaction at this depth?

SILTY SAND $z = 0$
$\gamma = 125$ pcf
$w = 10\%$
$z = -10$ ft

GWT at $z = -13$ ft   COARSE SAND
$\gamma = 120$ pcf
$w = 20\%$

$z = -22$ ft

FINE SAND
$\gamma = 122$ pcf
$w = 15\%$

$z = -37$ ft

CLAY
$\gamma = 118$ pcf
$c_u = 1400$ psf
$w = 32\%$

211-002

According to the IBC, for calculating seismic loads on a foundation, the site class is determined using an average of the properties of what depth below the surface?

(A) 25 ft

(B) 50 ft

(C) 100 ft

(D) 200 ft

Chapter 212: Earthwork

212-001

The table below shows cut-and-fill sections at 100-ft stations over a project length of 600 ft. Assuming 8% shrinkage, calculate the cumulative earthwork volume between stations 0 + 0.00 and 6 + 0.00.

Station	Cut area (ft²)	Fill area (ft²)
0 + 0.00	456.12	45.32
1 + 0.00	634.54	120.54
2 + 0.00	342.23	563.20
3 + 0.00	123.45	547.23
4 + 0.00	84.23	231.70
5 + 0.00	0	102.71
6 + 0.00	123.90	58.45

212-002

The mass diagram shown below plots cumulative earthwork from A (station 0 + 0.00) to E (station 12 + 0.00). The vertical ordinates are cumulative earthwork volume (yd³) with each grid increment being 100 yd³. Assuming that the limit of free haul is 300 ft, what is the overhaul (yd³-sta) between C and E?

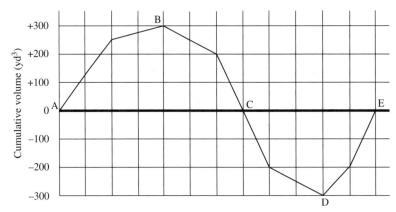

212-003

For the mass haul diagram shown above (Prob. 212-002), which of the following statements is *true*?

(A) Points A and C are called grade points

(B) If the free haul distance is 200 ft, then the limits of free haul are stations 6 + 00 to 8 + 00

(C) Stations 7 + 50.00 to 11 + 50.00 are called match points

(D) The net earthwork volume between stations 2 + 0.00 and 6 + 50.00 is 150 yd³ (cut)

Chapter 301: Basic Fluid Mechanics

301-001

A rectangular gate 10-ft wide × 6-ft high is hinged at the top edge A and held closed by a single bracket at B. The gate is used to control the depth of water in the reservoir to a depth of 10 ft as shown. What is the horizontal force (lb) that the bracket experiences as a result?

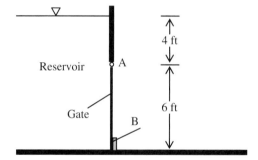

301-002

A 1:25 scale model of a spillway is built to simulate flow in the laboratory. The field conditions are

Spillway crest elevation = 234.12 ft above sea level

Spillway length = 120 ft

Head = 3.6 ft

Spillway weir coefficient = 3.3

Assuming that the model is perfectly geometrically similar to the prototype, what must be the model flow rate (ft³/s) in order to achieve the Froude number similarity?

301-003

Water flows at the rate of 3.2 cfs through a 6-in-diameter cast-iron pipe. The viscosity of the water can be taken as 1.0×10^{-5} ft²/s. What is expected to be the maximum flow velocity in the pipe?

301-004

Water flows at the rate of 3.2 cfs through a 6-in-diameter cast-iron pipe. The figure shows the plan view of the pipe layout near a 90° bend. Assume that elevation of the plane of the pipe network is 156.00 ft. The elevation of the HGL is 235.4 ft. An anchor bracket is used to support the pipe bend at A. What is the force exerted by the fluid on the bracket?

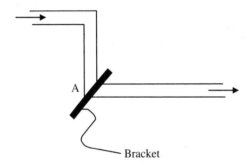

Bracket

Chapter 302: Closed Conduit Hydraulics

302-001

A segment of an 8-in-diameter cast-iron pipe conveys water (temperature 16°C) at a flow rate $Q = 3$ cfs. A valve located 300 ft from a bend is closed suddenly (closure time = 0.2 s). What is the excess pressure (psi) created by the valve closure?

302-002

A 3200-ft-long cast-iron pipe, 12-in-diameter, conveys flow ($Q = 2000$ gal/min) with the aid of a centrifugal pump which operates at an efficiency of 85%. The elevation of the HGL just upstream of the pump is 302.45 ft. The desired pressure at the downstream end of the pipe (elevation 167.5 ft) is 70 psig. What is the required rated power of the pump (horsepower)?

302-003

A centrifugal pump is used to pump water at a flow rate $Q = 2000$ gal/min. The water temperature is 15°C. The suction line diameter is 8 in and the discharge line diameter is 12 in. The hydraulic grade line elevation at the pump inlet is 231.9 ft and the pump elevation is 156.5 ft. What is the maximum permissible velocity of water inside the pump casing?

302-004

A reservoir with surface elevation at 435 ft serves a community with 160 households. The mean elevation at the community is 245 ft. Each household has one main water connection and the peak demand is 50 gpm per connection, what is the available gage pressure "at the tap" during peak conditions. Assume the following properties for the pipe from the reservoir to the distribution point in the community: length 3200 ft, diameter = 24 in, Hazen Williams $C = 110$. Assume minor losses in the distribution system to be 4 in of head loss per connection.

302-005

The figure below shows the system curve (H vs. Q) for a water delivery pipeline. Also shown on the figure is the manufacturer's curve (H vs. Q) for a centrifugal pump. An array of three of these pumps is used in series configuration to convey flow in this system. What is the operating discharge (gal/min)?

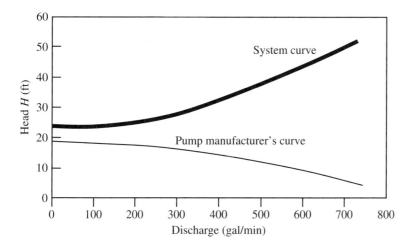

Chapter 303: Open Channel Hydraulics

303-001

A trapezoidal open channel (Manning's $n = 0.014$ constant with depth, bottom width = 10 ft, side slopes 1:1) conveys a flow rate $Q = 130$ cfs. The longitudinal slope of the channel bed is 0.5%. What is the normal depth of flow?

303-002

What is the critical depth of flow in a circular conduit (diameter = 48 in) that conveys a flow rate of 10,000 gal/min?

303-003

A circular conduit of diameter 36 in made of corrugated metal ($n = 0.024$) is used to convey a peak flow rate of 32 ft³/s. What is the critical slope of the conduit?

303-004

Following the design storm, a stormwater detention pond has surface elevation = 341.90 ft above sea level. A sharp crested rectangular weir (width = 20 ft) is used to control outflow from the pond. If it is desired to release water at the rate of 10,000 gal/min, what should be the weir crest elevation (ft)?

303-005

A rectangular open channel (width = 24 ft) conveys flow $Q = 700$ cfs at a depth of 1.6 ft. A hydraulic jump occurs in the channel at this location. What is the rate of energy loss (hp) in the hydraulic jump?

303-006

A rectangular open channel (width = 20 ft, longitudinally sloping down at 0.7%) conveys flow ($Q = 840$ cfs) at depth of 5.6 ft. The critical depth in this section of the channel is 3.8 ft. The normal depth for this section of the channel is 2.95 ft. What is the classification of the flow?

Choices: M1, M2, M3, C1, C3, S1, S2, S3, H2, H3, A2, A3

Chapter 304: Hydrology

304-001

A 180-ac watershed receives gross rainfall of 3.2 in during a 2-h storm. If the average infiltration into the prevalent soil is 0.5 in/h and evaporation losses are 5000 gal/ac-h, what is the net runoff volume in million gallons?

304-002

The 1-h unit hydrograph of excess precipitation is shown below. What is the peak runoff discharge (cfs) produced by a storm which produces 1.2 in of excess precipitation during the first hour, 0.9 in during the second hour, and 0.3 in during the third hour?

Time (h)	0.0	1.0	2.0	3.0	4.0	5.0	6.0
Runoff (cfs/in)	0	23	47	82	32	12	0

304-003

A 180-ac watershed has the following classifications:

Zone	Area (ac)	Description	Soil type
1	50	Townhomes	D
2	20	Paved areas: parking lots	C
3	70	Single-family homes on 1-ac lots	A
4	40	Grassy areas in good condition	A

Using the NRCS procedure, what is the estimated runoff depth (inches) from this watershed after a storm with gross rainfall of 3.2 in during the peak of the rainy season?

304-004

A 180-ac watershed has the following subareas (zones). The table indicates hydrologic properties of these zones (1-4).

Zone	Area (ac)	Average slope (%)	CN	Rational C	Time for overland flow (min)
1	50	0.7	90	0.85	45
2	20	1.2	67	0.70	32
3	70	0.8	54	0.67	50
4	40	1.0	89	0.82	38

The figure below shows intensity-duration-frequency curves based on historical hydrologic data for watersheds similar to this one.

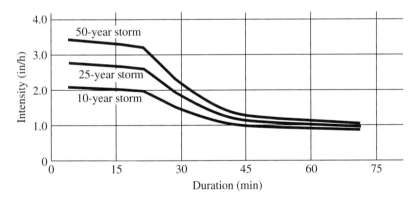

Using the rational method, what is the peak runoff (ft³/s) from this watershed due to a storm event with annual probability of occurrence = 4%?

304-005

Runoff from a watershed (area = 225 ac) drains into a stream. A monitoring station records the following discharges over the 6-h period following a 2-h storm. What would be the peak discharge (cfs) in the stream following a storm event with 2.3 in of runoff?

Time (h)	0.0	1.0	2.0	3.0	4.0	5.0	6.0
Runoff (cfs)	23	45	95	125	76	51	25

Chapter 305: Water Supply Quantity and Quality

305-001

A water sample shows the following data:

> Sample volume = 12 mL
>
> Initial dissolved oxygen concentration = 7.4 mg/L
>
> Incubation temperature = 25°C
>
> Dissolved oxygen after 5 days = 2.3 mg/L

If the deoxygenation rate constant (base 10 at temperature 20°C) = 0.25/day^{-1}, what is the ultimate BOD (mg/L)?

305-002

A water sample shows the following results from analysis:

> Temperature = 63°F
>
> pH = 7.9
>
> Ca^{++} = 34 mg/L
>
> Mg^{++} = 15 mg/L
>
> TDS = 410 mg/L
>
> Alkalinity = 110 mg/L as $CaCO_3$

What is the noncarbonate hardness (mg/L as $CaCO_3$) of this water sample?

305-003

A small stream has the following characteristics:

> Discharge Q = 200 MGD
>
> Temperature = 15°C
>
> Ultimate BOD negligible
>
> Dissolved oxygen = 6.3 mg/L

Velocity $= 5$ mph

Deoxygenation rate constant (base 10 at 15°C) $= 0.23$/day^{-1}

Reoxygenation rate constant (base 10 at 15°C) $= 0.32$/day^{-1}

A point source empties wastewater into the stream. The wastewater has the following characteristics:

Discharge $Q = 1000$ gpm

Temperature $= 35$°C

Ultimate BOD $= 1250$ lb/day

Dissolved oxygen $= 1.3$ mg/L

What is the oxygen deficit at a point 10 miles from the point at which the wastewater mixes with the river?

305-004

A community, with a population of 12,000, generates wastewater at an average rate of 100 gpcd. What is approximately the peak flow rate (ft^3/s) to be used for the design of sanitary sewers?

305-005

Drinking water is expected to account for approximately 75% of the total daily intake of a particular contaminant. The reference dose for this contaminant is 0.18 mg/kg-day. Assuming typical ([Environmental Protection Agency (EPA)] recommended) values, what is the MCLG (mg/L) for an adult?

Chapter 306: Water and Wastewater Treatment

306-001

A rotating biological contactor (RBC) unit consists of 12 polyurethane disks mounted on a shaft. Each disk is of diameter 10 ft. The shaft rotates at an angular velocity of 2 rpm. The disks are submerged in a tank where wastewater flows through with a velocity $= 3$ ft/s. The influent has flow rate $= 4$ MGD with ultimate BOD $= 230$ mg/L. The depth of immersion of the disks is 3 ft. The BOD removal of the RBC process is given by

$$\frac{S_o}{S_e} = \left(1 + \frac{kA}{Q}\right)^n$$

$k = 15{,}000$ gpd/ft^2

$n = 2.6$

$S_o =$ influent BOD$_5$ (mg/L)

$S_e =$ effluent BOD$_5$ (mg/L)

What is the ultimate BOD (mg/L) in the effluent?

306-002

Disinfection of a water supply is via chlorination. The chlorine demand of the water is 3.6 mg/L. The flow rate is 4 MGD. A 99.9% removal of Giardia requires 3-log inactivation which requires a CT value of 102 mg/L-min at the specified temperature and pH. If the chlorine dose is 5.0 mg/L, what is the minimum volume (gallons) of the chlorination chamber? Assume plug flow.

306-003

The following characteristics are given for an activated sludge process which treats primary effluent from a primary clarifier:

Influent flow rate = 3.8 MGD

MLVSS = 2600 mg/L

Volume of aeration tank = 100,000 ft^3

Solids concentration in RAS = 8500 mg/L

Waste-activated sludge volume = 50,000 gpd

What is the solids retention time (days)?

306-004

Type I settling occurs in a grit chamber whose dimensions are as follows: length = 40 ft, width = 10 ft, depth = 8 ft. The flow through velocity (in the long direction) is 4 ft/s. What is the percent settling rate for a particle whose setting velocity is 3 in/s?

306-005

Alum, $Al_2(SO_4)_3 \cdot 14H_2O$, is being used to induce coagulation in a water supply. The flow rate is $Q = 4$ MGD. The following data is known:

pH = 7.6

Temperature = 21°C

Total hardness = 210 mg/L as $CaCO_3$

Alkalinity = 194 mg/L as $CaCO_3$

If the alum is 80% pure, what dose (lb/day) would be required to treat the water supply?

306-006

A 10-MGD water treatment plant is to utilize eight identical rectangular clarifiers in parallel configuration. Clarifiers are 15-ft deep and have length:width ratio 4:1. The plant maintenance manual requires that the plant not be overloaded even when one unit is taken offline for cleaning and repairs. The maximum flow through velocity in the tank is 0.005 ft/s. What is the required length of each clarifier unit?

Chapter 401: Capacity Analysis

401-001

The highest hourly volume for a highway lane was observed to be 1340 vph. The peak hour factor, calculated from the 15-min peak flow, is 0.88. The average space mean speed is 56 mph. If the average vehicle length is 19 ft, what is the mean headway (seconds) during the peak period?

401-002

A car's deceleration characteristics are as follows: for speeds exceeding 40 mph, the deceleration rate is 6 mph/s and for lower speeds, the deceleration rate is 4 mph/s. If driver perception-reaction time is 2 s, what is the stopping sight distance for a car traveling at 70 mph.

401-003

A speed detector measures the following vehicle speeds (mph) over a stretch of highway. What is the space mean speed (mph)?

43, 23, 56, 32, 56, 34, 41, 67, 45, 72

Chapter 402: Highway Safety

402-001

A two-lane minor approach intersects with a major road at a T-junction as shown below. If left turns from the westbound approach are disallowed, how many conflict points are eliminated? Consider all types of vehicular conflict.

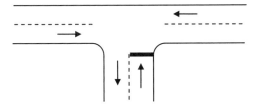

402-002

Several highway segments are being evaluated for allocation of limited funding. Statistics from the past 3 years have been summarized in the table below. From the point of view of safety, which segment should receive highest priority in the allocation of funding?

Segment	No. of accidents	Length (miles)	ADT (vpd)
1	46	6	20,000
2	67	4	12,000
3	43	5	8000
4	60	4	10,000

Chapter 403: Sight Distance

403-001

What is the passing sight distance on a parabolic vertical curve that connects an upgrade of 5% to a downgrade of 3%. The PVC is at station $23 + 45.60$ and the PVT is at station $38 + 12.54$.

403-002

According to the *AASHTO Green Book*, what is the vehicle design speed for which the passing sight distance of 715 ft is adequate on a two-lane highway (one lane in each direction)?

403-003

What is the maximum design speed for a parabolic vertical curve that connects an upgrade of 5% to a downgrade of 3%. The PVC is at station $23 + 45.60$ and the PVT is at station $38 + 12.54$.

Chapter 404: Highway Curves

404-001

A horizontal circular curve transition is required between two tangents with a deflection angle 67°30'45" (right). The tangents intersect at station $12 + 10.50$ and the PC is at station $7 + 14.06$. It has been decided to use two circular curves with radii = 800 and 500 ft. What is the length of longer curve?

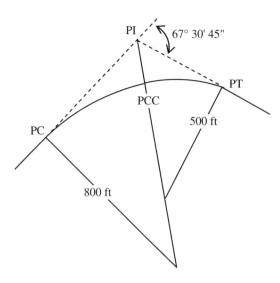

404-002

Two tangents intersect at station 23 + 34.50. The deflection angle between tangents is 56°45'. A spiral horizontal curve of length 140 ft is to be used as a transition from/to the tangents to a circular curve of radius 850 ft. What is the station of the tangent to spiral station?

404-003

A parabolic vertical curve must connect grades $G_1 = + 5\%$ and $G_2 = - 3\%$, intersecting at PVI station 23 + 34.87 (elevation 235.76 ft). If the design speed on the road is 65 mph, what is the elevation of the PVC, based on adequate stopping sight distance?

Chapter 405: Superelevation

405-001

A highway ramp is laid out as a circular curve with radius = 450 ft. If the design speed on the ramp is 40 mph and the side friction factor at this speed is 0.20, what is the minimum required superelevation?

405-002

A horizontal curve on a rural highway with a design speed of 75 mph is to be superelevated according to the guidelines in the *AASHTO Green Book*. Snow and ice is expected to be present during the worst winter months. What is the minimum recommended radius (ft)?

405-003

A circular curve (radius = 760 ft) requires a design superelevation of 6%. The cross section of the roadway consists of four travel lanes (lane width = 12 ft) crowned at the centerline with drainage cross slope = 2%. The transition to superelevation is achieved by rotating the cross section through the centerline. The superelevation runoff rate is 1:300. What is the length of the superelevation runoff?

Chapter 406: Freeways

406-001

A 1-mile segment of a rural freeway has the following data:

No. of lanes in each direction = 4

Hourly volume in design direction = 4670

Lane width = 11 ft

Right shoulder width = 6 ft

PHF = 0.88

Average grade = +5%

Trucks = 6%

RVs = 3%

Ramp density = 3 ramps per mile

What is the level of service?

406-002

Given the following peak flows (pcphpl) for a freeway off ramp, calculate the LOS in the diverge area.

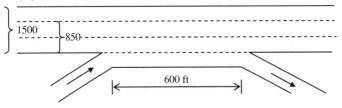

Chapter 407: Multilane Highways

407-001

A six-lane divided highway in rolling terrain has the following data:

Free flow speed = 52 mph

No. of lanes in each direction = 3

Hourly volume in design direction = 3420

Lane width = 11 ft

Right shoulder width = 5 ft

Average grade = +5%

Trucks = 6%

RVs = 3%

PHF = 0.90

Average spacing of driveways = 500 ft

What is the level of service?

407-002

What is the free flow speed for a four-lane undivided rural highway with the following specifications?

No. of lanes in each direction = 2

Hourly volume in design direction = 2310

Lane width = 12 ft

Shoulder width = 7 ft

Average grade = 5%

Trucks = 8%

RVs = 6%

PHF = 0.90

Chapter 408: Two-Lane Highways

408-001

A two-lane class I highway carries a two-way peak hourly volume of 2450 vph. The directional split is 65:35. The highway is on level grade. PHF = 0.90. Heavy vehicles are negligible. Access points are spaced approximately 500-ft apart, lanes width = 12 ft, shoulder width = 6 ft. What is the level of service?

408-002

The following data is given for a two-lane highway:

Lane width = 12 ft

Hourly volume = 850 vph in each direction

PHF = 0.88

40% no passing zones

Level terrain

6% trucks

What is the percent time spent following?

Chapter 409: Signalization Warrants

409-001

An unsignalized intersection is being evaluated for signalization according to MUTCD guidelines. The north-south highway has two lanes in each direction and the east-west street has one lane in each direction. The following data, collected during the peak hour, are given:

Northbound approach: Peak-hour volume, $V = 470$ mph, design speed $= 40$ mph

Southbound approach: Peak-hour volume, $V = 560$ mph, design speed $= 40$ mph

Eastbound and westbound approaches: design speed $= 35$ mph

What is the minimum hourly volume (vph) on the minor approach for which this criterion meets the "peak-hour" warrant for signalization?

409-002

An intersection is shown in the figure. The stop line on the minor street (one lane in each direction) is located a distance of 105 ft from the near-side rail. The hourly volume on the major street (sum of both approaches) is 200 vph. What is the minimum approach volume on the minor street, for which this intersection meets the signalization warrant in the MUTCD?

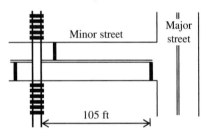

Chapter 410: Intersections

410-001, 002, 003

A right-angle signalized intersection between two four-lane highways (Main St. and South St.) in a non-CBD area is shown as follows. Assume a four-phase signal system, with signal cycle length $= 70$ s, lost time $= 3$ s per phase, and all-red time $= 1$ s per phase. Phases are (1) EBL and WBL, (2) EBTH/R and WBTH/R, (3) NBL and SBL, and (4) NBTH/R and SBTH/R. Lane group peak flow rates are shown on the figure.

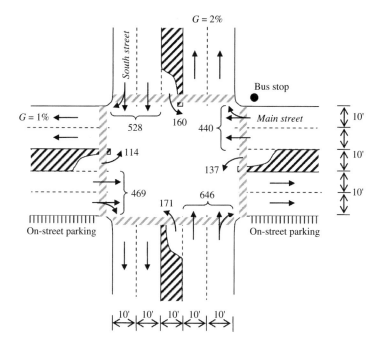

Design vehicle length = 40 ft

Deceleration rate = 10 ft/s^2

20 parking maneuvers per hour for EBTH/R approach

30 parking maneuvers per hour for WBTH/R approach

10 buses per hour for both EBTH/R and WBTH/R approaches

Cross-walk width = 8 ft

410-001

What is the saturation flow rate for the NBTH/R lane group?

410-002

What is the required yellow time for the N-S phase? Assume design vehicle length = 40 ft, perception-reaction time = 1.0 s, and deceleration rate = 10 ft/s^2.

410-003

If the pedestrian flow rate in the north-south crosswalk is 400 pedestrians per hour, then what is the pedestrian green time for the north-south signal phase?

Chapter 411: Design of Highway Pavements

411-001

An asphalt mix weighs 140 lb/ft^3. The mix contains 6.5% asphalt by weight. The specific gravity of the asphalt is 1.08. The following aggregates are used in the mix:

Aggregate	% by weight	Effective specific gravity
Coarse aggregate	64	2.70
Fine aggregate	24	2.60
Filler (dust)	12	2.65

The air content of the asphalt mix is most nearly.

411-002

A highway bridge has ADT = 23,000 in year 2001. Traffic is expected to grow at 4% per year. Truck traffic = 16%. Typical 2001 axle load data (for a representative set of 1000 trucks) is shown in the table below. What is the cumulative ESAL (W_{18}) during the 10-year period 2001–2010?

Axle category	Axle load (kips)	Frequency	Load equivalence factor	NxLEF
Single axle	4	340	0.002	0.68
	6	450	0.010	4.50
	8	150	0.032	48.0
	12	580	0.175	101.5
Double axle	6	230	0.002	0.46
	10	380	0.012	4.56
	16	120	0.081	9.72
		2250		169.42

Chapter 501: Engineering Economics

501-001

The purchase price of a piece of construction equipment is $50,000. The estimated useful life of the equipment is 10,000 h, after which the salvage value is $8000. Fuel and minor maintenance costs are $12.00 per hour of use. Tires are replaced after every 2000 h of use (at a cost of $3000). Major repairs are to be conducted every 4000 h at a cost of $9000. If the estimated use of this equipment is 2000 h per year, what is the minimum hourly charge for this equipment? Assume annual interest rate of 8%.

501-002

The declining balance method (rate = 30%) is used to depreciate the value of a piece of capital equipment. If the purchase price is $50,000, what is the book value at the end of year 3?

501-003

The following economic data is available for a construction project that involves doing repairs to an existing 2-mile stretch of highway:

Current annual user costs (environmental impact) = $24,000

Cost of repaving = $74,000

Annual user costs of newly paved road = $5000

Useful life of new pavement = 5 years

Annual interest rate = 8%

What is the benefit:cost ratio of the paving project?

Chapter 502: Probability and Statistics

502-001

A bridge being constructed is estimated to have a useful design life of 35 years. The magnitude of the design earthquake is to be such that its probability of exceedance over the life of the bridge (lifetime risk tolerance) is to be 10%. What should be the return period of the design earthquake?

502-002

A wastewater treatment plant can be represented by the system diagram composed of four processes as shown below. The operational efficiency of each process is shown. What is the overall efficiency of the plant?

502-003

Historical flood elevation data for a floodplain are summarized in the table below. The total length of the dataset is 100 years, with each year represented by the high water elevation for that year. The data is then presented in the form of grouped frequency table.

Elevation (ft)	Number of occurrences
356.00–356.50	1
356.50–357.00	3
357.00–357.50	4
357.50–358.00	9
358.00–358.50	13
358.50–359.00	21
359.00–360.00	17
360.00–360.50	12
360.50–361.00	11
361.00–361.50	6
361.50–362.00	2
362.00–362.50	1
	100

Based on this data, what is the annual probability that the flood elevation will exceed 360 ft 4 in?

Chapter 503: Project Scheduling

503-001

A project consists of eight activities (A through H). The activity-on-node diagram below shows predecessor-successor relationships between activities, which are all "finish-to-start" with the exception of B and E, which have an FF lag = 6 weeks. What is the minimum time to complete the project?

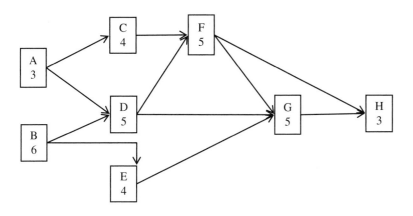

503-002

For the schedule shown in Fig. 503-001, the following data is given:

Activity	Normal duration (weeks)	Crashed duration (weeks)	Normal cost ($,000)	Crashed cost ($,000)
A	3	2	7	10
B	6	5	8	10
C	4	3	5	7
D	5	4	6	7
E	4	3	7	9
F	5	4	7	8
G	5	4	6	8
H	3	2	6	9

The critical path is BDFGH. If this path were to be shortened by 3 weeks, what is the minimum cost for which this can be accomplished?

503-003

For the activity network shown in Prob. 503-001, what is the free float of activity A?

Chapter 504: Design Loads during Construction

504-001

The following figure shows the range diagram for an articulated boom lift which is used to place a piece of HVAC equipment (weight 1000 lb) onto the roof of an 80-ft-tall building.

What is the farthest horizontal distance (feet) from the face of the building to the front axle of the loader?

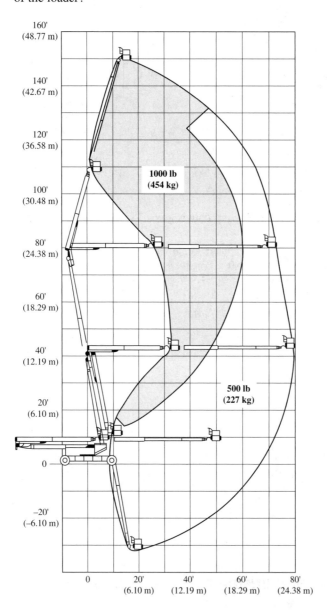

504-002

A truck, with fully loaded gross weight = 54,000 lb, has a 450-hp engine. When climbing a gravel road (slope = 5%, rolling resistance = 60 lb/ton) with the engine operating at full power and 85% efficiency, what is the velocity (mph)?

504-003

A scaffold platform is 10-ft long × 3-ft wide. The maximum permitted occupancy of the platform is four workpeople. Dead load of the platform and connected hardware is 400 lb. What is the horizontal construction load that must be used for the design of scaffold components?

Chapter 505: Scaffolding

505-001

A scaffold platform is supported by outrigger beams which are 12-ft apart. According to OSHA standards, what is the maximum overhang of the platform beyond the outriggers if the scaffold is not specifically designed to overcome tipping forces on the cantilevered portions?

505-002

What is the maximum vertical clearance between a scaffold platform and any point of access for which ladders or other type of access are not required according to OSHA standards?

505-003

A suspension scaffold is suspended by a symmetric arrangement of four suspension cables. The scaffold hoist has a stall load of 30,000 lb, and a rated load of 12,000 lb. What is the required breaking strength of each cable?

Chapter 506: Formwork for Concrete

506-001

Steel ties are used for 14-ft-high wall forms in which a concrete (unit weight = 140 pcf, type I cement with no retarders) is poured at a rate of 8 ft/h. Concrete temperature = 65°F. If the ties are spaced every 3 ft both vertically and horizontally and the allowable tensile stress in the ties is 24,000 psi, what is the required tie diameter?

506-002

Fresh concrete in a wall form is instrumented with probes that yield the following time-temperature data for the 45-h period following pour (at $t = 0$). If the threshold temperature for concrete curing is 32°F and the target TTF value for forms to be removed is 2000 °F-h, then the forms can be safely removed at $t = ?$

Time (h)	Temp (°F)
00:00–05:00	108
05:00–10:00	102
10:00–15:00	99
15:00–20:00	94
20:00–25:00	84
25:00–30:00	83
30:00–35:00	73
35:00–40:00	72
40:00–45:00	67

506-003

Using the temperature profile shown in Prob. 506-002, the equivalent age (h) after 15 h of curing is most nearly:

Chapter 507: Excavations

507-001

A trench is to be excavated to a depth of 18 ft. The site may be subjected to vibration from handheld compacting equipment. The soil has the following parameters: cohesive soil; unit weight = 124 lb/ft³; S_{uc} = 1600 lb/ft². What is the maximum slope angle (with respect to the horizontal) that is permitted in this excavation, according to OSHA standards?

507-002

According to OSHA standards, what is the maximum depth of an excavated trench for which ladders are not required to provide workers with egress?

Chapter 508: Erosion Control

508-001

Which of the following strategies has the highest probability of success in stabilizing an erodible slope which has a slope of 1:1?

(A) Mulching

(B) Revegetation

(C) Retaining structure

(D) Erosion control blanket

508-002

Which of the following is not an example of an energy dissipater?

(A) Hydraulic jump

(B) Ditch check

(C) Drop inlet

(D) Headwall

Chapter 509: Occupational Safety

509-001

What is the maximum permissible span for a 2 in × 10 in full-thickness undressed lumber platform with a working load of 25 psf?

(A) 6 ft

(B) 8 ft

(C) 10 ft

(D) 12 ft

509-002

Storing masonry blocks in stacks higher than 6 ft is permissible as long as

(A) The stack is braced every 6 ft

(B) The stack is braced every 8 ft

(C) The stack is tapered back one-half block per tier above the 6-ft level

(D) The stack is on level ground

509-003

At a construction site, what is the maximum permissible exposure to lead (μg/m^3 air), averaged over an 8-h period?

(A) 10

(B) 20

(C) 50

(D) 100

509-004

A 440-kV transmission line passes over a contractor's work area. If the line is not de-energized, what is the minimum distance (ft) between the line and any part of a crane?

(A) 11

(B) 14

(C) 17

(D) 23

509-005

According to OSHA, what is the minimum requirement of toilet facilities at a job site with 35 employees?

(A) 2 toilets

(B) 1 toilet seat + 1 urinal

(C) 1 toilet seat + 2 urinals

(D) 2 toilet seats + 2 urinals

509-006

According to OSHA regulations, employees shall be provided with anti-laser eye protection devices when working in areas in which a potential exposure to reflected laser light is greater than

(A) 5 mW

(B) 4 mW

(C) 3 mW

(D) 2 mW

509-007

According to OSHA regulations, the minimum illumination required for electrical equipment rooms is

(A) 3 foot-candles

(B) 5 foot-candles

(C) 10 foot-candles

(D) 30 foot-candles

509-008

According to OSHA, the maximum intensity of impulsive or impact noise (dB) is

(A) 92

(B) 110

(C) 140

(D) 188

509-009

Safety nets, where required, shall be provided when workplaces are more than __ ft above the ground or water surface.

(A) 100

(B) 75

(C) 50

(D) 25

509-010

According to OSHA, lifelines shall be secured above the point of operation to an anchorage or structural member capable of supporting a minimum dead weight (lb) of

(A) 4,200 lb

(B) 4,800 lb

(C) 5,400 lb

(D) 6,000 lb

509-011

When safety nets are required to be provided, how far beyond (ft) the edge of the work surface shall these nets extend?

(A) 4

(B) 6

(C) 8

(D) 10

509-012

When a safety net is required on a construction site, the net shall meet the minimum performance standard of

(A) 15,000 ft·lb impact

(B) 17,500 ft·lb impact

(C) 20,000 ft·lb impact

(D) 22,500 ft·lb impact

509-013

The mesh size of safety nets shall not exceed

(A) 6 in × 6 in

(B) 8 in × 8 in

(C) 10 in × 10 in

(D) 12 in × 12 in

509-014

What is the maximum volume (gal) of flammable and combustible liquids that can be stored outside an approved storage cabinet?

(A) 60

(B) 25

(C) 15

(D) 10

509-015

According to OSHA regulations, what is the highest stack (ft) allowed when bricks are being stored?

(A) 5

(B) 7

(C) 9

(D) 10

509-016

The maximum intended load for a metal tubular frame scaffold including its components is 1,500 lb. According to OSHA, the scaffold shall be designed to support a minimum load (tons) of

(A) 1.0

(B) 1.5

(C) 2.0

(D) 3.0

509-017

Each end of a scaffold platform which is not cleated or otherwise restrained shall extend over the centerline of its support a distance (in) at least

(A) 2

(B) 4

(C) 6

(D) 12

509-018

According to OSHA, what is the appropriate height of the required guardrails for a scaffold platform built after January 1, 2012?

(A) 36 in

(B) 42 in

(C) 48 in

(D) 54 in

509-019

When toeboards are used as a protection against falling objects, they shall have a minimum vertical height of

(A) 3.5 in

(B) 4.0 in

(C) 4.5 in

(D) 5.0 in

509-020

What is the OSHA requirement for the permissible span (ft) of full-thickness 2-in × 10-in undressed scaffolding lumber with a 50-psf working load?

(A) 4

(B) 5

(C) 6

(D) 8

509-021

A scaffold is located 12 ft above the ground. According to OSHA, what is the minimum required thickness for the screen between the toeboard and guardrail on the open side of the scaffold?

(A) 16 gauge

(B) 18 gauge

(C) 20 gauge

(D) 22 gauge

509-022

Ropes that are used to define control access areas shall have a minimum breaking strength (lb) of

(A) 50

(B) 100

(C) 200

(D) 300

509-023

According to OSHA, where oxygen deficiency (less than 19.5 percent oxygen) or a hazardous atmosphere exists or could be expected to exist, the atmosphere in excavations deeper than _____ must be tested before employees are allowed to enter the excavation.

(A) 3 ft

(B) 4 ft

(C) 5 ft

(D) 6 ft

509-024

In Type A soils, excavations 8 ft or less in depth which have unsupported, vertically sided lower portions shall have a maximum vertical side (ft) of

(A) 3.0

(B) 3.5

(C) 4.0

(D) 5.0

509-025

Whenever a masonry wall is being constructed, the width (ft) of the limited access zone that runs the length of the wall (on the unscaffolded side) is

(A) The height of the wall constructed

(B) The height of the wall constructed plus 2 ft

(C) The height of the wall constructed plus 4 ft

(D) The height of the wall constructed plus 6 ft

509-026

If a masonry wall is not adequately supported so that it will not overturn or collapse, what is the maximum height (ft) that can be left unbraced?

(A) 8

(B) 12

(C) 16

(D) 20

509-027

A non-self-supporting ladder has a working length of 16 ft. According to OSHA, the horizontal distance from the top support to the foot of the ladder is approximately

(A) 3 in

(B) 4 ft

(C) 5 ft

(D) 6 ft

509-028

According to OSHA, what is the maximum airborne concentration of asbestos (fibers per cm^3) when averaged over 30 min?

(A) 0.1

(B) 0.5

(C) 1.0

(D) 2.0

509-029

According to OSHA, cohesive soil with an unconfined compressive strength of less than 2.5 tons/ft^2 but greater than 1.5 tons/ft^2 is defined as

(A) Type A

(B) Type B

(C) Type C

(D) Type D

509-030

The short-term maximum allowable slope for excavations less than 12 ft in Type A soil shall be

(A) 1H:4V

(B) 1H:2V

(C) 3H:4V

(D) 1H:1V

509-031

Lumber that is handled manually shall not be stacked more than __ ft high.

(A) 14

(B) 16

(C) 18

(D) 20

509-032

A stairway, ladder, ramp, or other safe means of egress shall be located in trench excavations that are 4 ft or more in depth so as to require no more than __ ft of lateral travel for employees.

(A) 15

(B) 20

(C) 25

(D) 30

509-033

For lift-slab construction operations, what is the maximum number of manually controlled jacks allowed for the lifting of one slab?

(A) 8

(B) 10

(C) 12

(D) 14

509-034

When lifting concrete slabs by jacks, all points of the slab support shall be kept level within

(A) 0.5 in

(B) 1.0 in

(C) 1.5 in

(D) 2.0 in

509-035

At no time during steel erection shall there be more than _____ floors of unfinished bolting or welding above the foundation or uppermost permanently secured floor.

(A) 1

(B) 2

(C) 3

(D) 4

Solutions to End-of-Chapter Practice Problems

Chapter 101: Strength of Materials

101-001

The vertical coordinate of the centroid, measured from the bottom edge of the section, is

$$\bar{y} = \frac{\sum y_i A_i}{\sum A_i} = \frac{15 \times 4800 + 130 \times 2000 + 240 \times 4400}{4800 + 2000 + 4400} = 123.93 \, \text{mm}$$

Total depth = 250. Therefore, distance from NA to top fiber = 250 – 123.93 = 126.07 mm

The moment of inertia with respect to this centroidal axis is

$$I_{NA} = \sum \left(\frac{1}{12} bh^3 + Ad^2 \right) = \frac{1}{12} \times 160 \times 30^3 + 4800 \times (15 - 123.93)^2 + \frac{1}{12} \times 10 \times 200^3$$

$$+ 2000 \times (130 - 123.93)^2 + \frac{1}{12} \times 220 \times 20^3 + 4400 \times (240 - 123.93)^2$$

$$= 123.48 \times 10^6 \, \text{mm}^4$$

Elastic section modulus is

$$S_x = \frac{I_{NA}}{y_{\text{max}}} = \frac{123.48 \times 10^6}{126.07} = 9.795 \times 10^5 \, \text{mm}^3$$

101-002

The plastic neutral axis (PNA) can be found by dividing the area (11,200 mm²) into two equal halves (5600 mm²). The bottom flange area is 4800 mm². Area necessary from the web = 800 mm². Depth of web (to PNA) = 800/10 = 80 mm.

Therefore, vertical coordinate of the PNA, measured from the bottom edge of the section, is 110 mm.

The plastic section modulus Z_x is the first moment of all area components about the PNA:

$$Z_x = \sum Ad = 4800 \times (110 - 15) + 800 \times \frac{80}{2} + 1200 \times \frac{120}{2} + 4400 \times (240 - 110)$$
$$= 1.132 \times 10^6 \, \text{mm}^3$$

101-003

Taking moments (counterclockwise positive) about the roller support (B):

$$M_B = 0 \Rightarrow 12 \times 13 - 10 A_y + 16 \times 8 - 6 \times 1.5 = 0 \Rightarrow A_y = 27.5 \text{kips}$$

Since the total vertical load is 34 kips, the reaction at the roller support $B_y = 6.5$ kips.

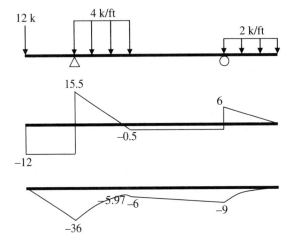

The entire bending moment diagram is negative (convex on top). Maximum bending moment magnitude = 36 kip-ft.

Chapter 102: Statically Determinate Structures

102-001

C is 6 ft below A. B is 4 ft below A. Therefore, C is 2 ft below B.

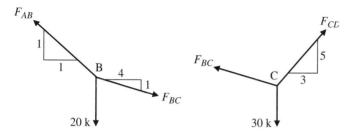

Solving the equilibrium equations at B:

$$-F_{AB}\frac{1}{\sqrt{2}} + F_{BC}\frac{4}{\sqrt{17}} = 0 \Rightarrow F_{AB} = 1.372F_{BC}$$

$$F_{AB}\frac{1}{\sqrt{2}} - F_{BC}\frac{1}{\sqrt{17}} - 20 = 0 \Rightarrow 0.7276F_{BC} = 20 \Rightarrow F_{BC} = 27.5\,\text{k}\,\&\,F_{AB} = 37.7\,\text{k}$$

Solving the equilibrium equations at C:

$$-F_{BC}\frac{4}{\sqrt{17}} + F_{CD}\frac{3}{\sqrt{34}} = 0 \Rightarrow F_{CD} = 1.886F_{BC} = 51.9\text{k}$$

Maximum cable tension = 51.9 k = 26 tons. Therefore, the factor of safety = 40/26 = 1.54.

102-002

The roof load is incident on 32 ft (horizontal projection). Each wall carries half of that (16-ft tributary width).

Total load from roof × 16 ft = 30 × 16 = 480 lb/ft

The floor load is incident on 32 ft (horizontal projection). Each 16-ft-wide half is carried equally by the girder and one wall. Therefore, the girder receives load from 16-ft-wide strip and each wall receives load from an 8-ft floor width.

Total load from floor × 8 ft = 48 × 8 = 384 lb/ft

Self weight of wall = 7 × 18 = 126 lb/ft

Total load to footing — 480 + 384 + 126 = 990 lb/ft

102-003

The influence diagram for the moment at C, drawn using the principle of Muller-Breslau, is shown below:

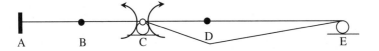

If the unit load is at D, then the bending moment at C is 1 × −4 ft = − 4 ft. Therefore, the ordinate at D is −4 ft. The bending moment at C is then calculated as the product of the distributed load and the area under the load. Placing the total load (UDL = 2 + 5 = 7 kips/ft) over CE, the maximum negative moment at C is:

$$M_{C,\text{max}} = 7\frac{\text{kips}}{\text{ft}} \times \frac{1}{2} \times 12\text{ft} \times -4\text{ft} = -168\,\text{kip-ft}$$

Chapter 103: Introduction to Indeterminate Structures

103-001

The maximum deflection for the cantilever beam will be at the free end C. The unit load method can be used, which gives the deflection as

$$\delta = \int \frac{M}{EI} m dx$$

where M = the bending moment diagram of the beam under given loads
m = the bending moment diagram of the beam under the unit load (vertical unit load at C)

For segment AB, flexural rigidity $EI = 3.48 \times 10^7$ kip-in^2 = 2.417×10^5 kip-ft^2

For segment BC, flexural rigidity $EI = 2.03 \times 10^7$ kip-in^2 = 1.41×10^5 kip-ft^2

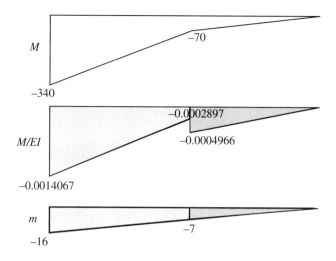

Utilizing the table for $\int Mmdx$, we get:

$$\delta = \int Mmdx = \frac{1}{6}\left[-0.0014067(2 \times -16 - 7) - 0.0002897(2 \times -7 - 16)\right] \times 9$$

$$+ \frac{1}{3}[-0.0004966 \times -7] \times 7 = 0.0953 + 0.0081 = 0.1034 \text{ft} = 1.24 \text{in}$$

103-002

The first-order indeterminate beam can be partitioned into a primary beam and a secondary beam using the vertical reaction at B as the redundant.

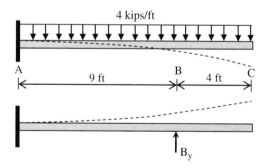

The vertical deflection at B for the primary beam is calculated from the standard result:

$$y_P(x) = \frac{wx^2}{24EI}(4Lx - x^2 - 6L^2) = \frac{4 \times 9^2}{24EI}(4 \times 13 \times 9 - 9^2 - 6 \times 13^2) = \frac{-8464.5}{EI}$$

For the secondary beam, the deflection at B is:

$$y_s(x) = \frac{PL^3}{3EI} = \frac{B_y \times 9^3}{3EI} = \frac{243B_y}{EI}$$

Compatibility at B is satisfied if vertical deflection at B is zero.

$$y(x) = \frac{243B_y}{EI} - \frac{8464.5}{EI} = 0 \Rightarrow B_y = 34.83$$

103-003

At node E, there are four members—EA (far end rigid, stiffness $= 4EI/L = 4 \times E \times 800/18 = 177.78E$), ED (far end rigid, stiffness $= 4EI/L = 4 \times E \times 1200/18 = 266.67E$), EH (far end pinned, stiffness $= 3EI/L = 3 \times E \times 800/20 = 120E$), and EF (far end rigid, stiffness $= 4EI/L = 4 \times E \times 1200/15 = 320E$).

The relative stiffness of member EH at node E is therefore

$$d_{EH} = \frac{K_{EH}}{\sum K} = \frac{120}{177.78 + 266.67 + 120 + 320} = 0.136$$

Chapter 104: Concrete Fundamentals

104-001

The maximum moment is given by $M = P\dfrac{L}{3} = \dfrac{560 \times 75}{3} = 14,000\text{lb-in}$

Section modulus of the beam section is $S = \dfrac{bh^2}{6} = \dfrac{4 \times 8^2}{6} = 42.67\text{in}^3$

Maximum bending stress (modulus of rupture) is $\sigma_{max} = \dfrac{M}{S} = \dfrac{14,000}{42.67} = 328.13\text{psi}$

ACI gives the following for modulus of rupture: $f_r = 7.5\sqrt{f_c'}$

Therefore, $f_c' = 1914$ psi.

28-day compressive strength is nearly 2000 psi.

104-002

Split cylinder test tensile strength: $f_{ct} = \dfrac{2P}{\pi DL} = \dfrac{2 \times 35,600}{\pi \times 6 \times 12} = 314.8\text{psi}$

$$f_{ct} = 6.7\sqrt{f_c}$$

Therefore, $f_c' = 2207$ psi.

104-003

The total weight of all components = 2606 lb. The total volume = 2606/142 = 18.3521 ft³.

Volumes are tabulated in the table below:

Component	Weight (lb)	SG	Volume (ft³)
Cement	450	3.15	2.2894
Sand	870	2.65 (SSD)	5.2612
Stone aggregate	1106	2.50 (SSD)	6.0897
Water	205	1.0	3.2853
Total	2606		16.9256

Therefore, air volume = 18.3521 − 16.9256 = 1.4265 ft³, which is 7.8% of the total volume.

Chapter 105: Reinforced Concrete Beams

105-001

Self weight of floor slab = $0.15 \times 5/12 = 0.0625$ ksf.

Cross-sectional area of T-beam = 336 in² = 2.33 ft². Self-weight of T-beam = 0.35 kip/ft.

Factored load on floor:

$$w_u = 1.2w_D + 1.6w_L = 1.2 \times (0.085 + 0.0625) + 1.6 \times 0.15 = 0.417 \text{ksf}$$

Factored load on beam: $w_u = 8 \times 0.417 + 0.35 = 3.686 \text{klf}$

Maximum moment: $M_u = \dfrac{w_u L^2}{8} = \dfrac{3.686 \times 32^2}{8} = 471.8 \text{kip-ft} = 5662 \text{kip-in}$

Assuming rectangular behavior, strength coefficient:

$$X = \frac{M_u}{\phi f_c' b d^2} = \frac{5662}{0.9 \times 4 \times 20 \times 21.5^2} = 0.170$$

Reinforcement parameter, $w = 0.192$. (See Table 105.5.)

Reinforcement ratio: $\rho = w \dfrac{f_c'}{f_y} = \dfrac{0.192 \times 4}{60} = 0.0128$

Required reinforcement: $A_s = \rho b d = 0.0128 \times 20 \times 21.5 = 5.5 \text{in}^2$

Depth of neutral axis: $c = \dfrac{A_s f_y}{0.85 f_c' b \beta} = \dfrac{5.5 \times 60}{0.85 \times 4 \times 20 \times 0.85} = 5.7 \text{in} < 6 \text{in}$

Rectangular behavior assumption is justified.

105-002

Factored load on beam: $w_u = 8 \times 0.417 + 0.35 = 3.686 \, \text{klf}$

Maximum shear: $V_u = \dfrac{w_u L}{2} = \dfrac{3.686 \times 32}{2} = 58.98 \, \text{kip}$

Nominal shear capacity of concrete section:

$$V_c = 2\sqrt{f_c'}\, b_w d = 2 \times \sqrt{4000} \times 20 \times 21.5 = 54391 \, \text{lb}$$

Since $V_u > \phi V_c'$ shear reinforcement is required.

Required capacity: $V_s = \dfrac{V_u}{\phi} - V_c = \dfrac{58.98}{0.75} - 54.4 = 24.25 \, \text{kips}$

Required spacing: $s = \dfrac{A_v f_y d}{V_s} = \dfrac{0.22 \times 60 \times 21.5}{24.25} = 11.7 \, \text{in}$

Use no. 3 stirrups at 11 in o.c.

105-003

No. 9 bars: Bar diameter $d_b = 1.128$ in. Clear spacing between bars $= (14 - 2 \times 1.5 - 2 \times 0.375 - 3 \times 1.128)/2 = 3.43$ in, which is less than $6d_b$.

Top bars $- \Psi_t = 1.3$, epoxy coated $- \Psi_e = 1.5$, bars larger than no. 6 $- \Psi_s = 1.0$, normal weight concrete $- \lambda = 1.0$

Area of transverse reinforcement within $s = 6$ in of the splitting plane, $A_{tr} = 2 \times 0.11 = 0.22 \, \text{in}^2$.

Number of bars being developed within splitting plane, $n = 1$.

$$K_{tr} = \dfrac{40 A_{tr}}{sn} = \dfrac{40 \times 0.22}{6 \times 1} = 1.467$$

Without adequate information about bar arrangement, c_b cannot be calculated. The confinement term is taken to be 2.5.

$$L_d = \left\{ \dfrac{3}{40 \times 1} \dfrac{60{,}000}{\sqrt{4000}} \dfrac{1.3 \times 1.5 \times 1.0}{2.5} \right\} \times 1.128 = 62.6 \, \text{in}$$

Chapter 106: Reinforced Concrete Slabs

106-001

The slab has support from the beams, every 8 ft. Both ends of the slab are continuous over the supporting beams. According to ACI 318, for solid one-way slabs with both ends continuous, minimum required slab thickness (based on deflection) = $L/28 = 8 \times 12/28 = 3.43$ in; use 3.5 in.

106-002

According to ACI 318, the effective width of flange = smallest of ($L/4$, c.c spacing, $b_w + 12t$) = min(7 ft, 8 ft, 88 in) = 7 ft = 84 in.

Cross section of beam and slab = $96 \times 5 + 28 \times 15 = 900$ in² = 6.25 ft²

Self weight of beam and slab = 6.25 ft² $\times 0.15$ kip/ft³ = 0.94 k/ft

Total dead load = $0.94 + 8 \times 0.025 = 1.14$ k/ft

Floor live load = 85 psf $\times$ 8 ft = 680 lb/ft = 0.68 k/ft

Total factored live load: $w_u = 1.2 \times 1.14 + 1.6 \times 0.68 = 2.456$ k/ft

Maximum moment: $M_u = \dfrac{w_u L^2}{8} = \dfrac{2.456 \times 28^2}{8} = 240.7\,\text{kip-ft} = 2888.3\,\text{kip-in}$

For a rectangular beam with $b = 84$, $M_u = 2888.3$ kip-in, $f_c' = 4$ ksi, and $f_y = 60$ ksi

Strength parameter: $X = \dfrac{M_u}{\phi f_c' b d^2} = \dfrac{2888.3}{0.9 \times 4 \times 84 \times 17.5^2} = 0.125$

From Table 105.5, corresponding value of the reinforcement parameter: $w = \rho \dfrac{f_y}{f_c'} = 0.135$

$$\rho = w\dfrac{f_c'}{f_y} = \dfrac{0.135 \times 4}{60} = 0.009$$

Required reinforcement: $A_s = \rho b d = 0.009 \times 84 \times 17.5 = 13.23\,\text{in}^2$

Use 9 no. 11 bars ($A_s = 14.0$ in², requires width = 27.8 in according to Table 105.4)

106-003

The slab panel is 28 ft $\times$ 8 ft and has support from the beams on parallel faces. Therefore, the slab behaves as a continuous one-way slab. For the long direction (28 ft), the code requires temperature and shrinkage steel. That area (per foot width) is

$$A_s = 0.0018bh = 0.0018 \times 12 \times 5 = 0.108\,\text{in}^2/\text{ft}$$

Spacing of these bars not to exceed 5 h, nor 18 in. Use no. 4 bars at 18 in spacing (0.13 in²/ft).

Chapter 107: Reinforced Concrete Columns

107-001

Assuming 1.5-in clear cover, the diameter of the core:

$$D_c = 18 - 2 \times (1.5 + 0.375) = 14.25 \, in$$

Gross area $A_g = 254.5 \, in^2$

Concrete core area $A_c = 159.5 \, in^2$

Area of spiral $A_{sp} = 0.11 \, in^2$

Spiral volume ratio: $\rho_s = 0.45 \left(\dfrac{A_g}{A_c} - 1 \right) \left(\dfrac{f'_c}{f_y} \right) = 0.45 \times \left(\dfrac{254.5}{159.5} - 1 \right) \left(\dfrac{4.5}{60} \right) = 0.02$

Maximum permitted spacing: $s = \dfrac{4 A_{sp}}{\rho_s D_c} = \dfrac{4 \times 0.11}{0.02 \times 14.25} = 1.54 \, in$

Use a spiral pitch of 1.5 in (this is within the limits of 1 in to 3 in). Therefore, OK.

107-002

For no. 10 and smaller bars, ACI recommends no. 3 bars to be used as lateral ties. For larger bars, no. 4 bars are required. If no. 3 bars are used, their spacing is not to exceed the smallest of (least lateral dimension of column = 15 in, 16 times the diameter of the longitudinal bars = $16 \times 1.128 = 18$ in and 48 times the diameter of the ties = $48 \times 0.375 = 18$ in).

Therefore, no. 3 ties should be spaced at 15 in oc.

107-003

The normalized eccentricity $e/h = 5/22 = 0.23$.

Assuming 1.5 in of clear cover, no. 3 ties and no. 9 bars, we get the center to center distance between the two lines of reinforcement = $22 - 2 \times (1.5 + 0.375 + 1.128/2) = 17.12$ in.

Therefore, the parameter $\gamma = 17.12/22 = 0.78$.

Using the appropriate chart—Fig. 107.5 ($f'_c = 4000$ psi, 60,000 psi, $\gamma = 0.78$), for $e/h = 0.23$ and the parameter

$$K_n = \dfrac{P_u}{\phi_c f'_c A_g} = \dfrac{700}{0.65 \times 4 \times 330} = 0.816$$

The required reinforcement ratio $\rho_g = 3.4\%$.

Therefore, reinforcement required = $0.034 \times 15 \times 22 = 11.22 \, in^2$ (12 no. 9 bars arranged with 6 on each of two opposing faces).

Chapter 108: Prestressed Concrete

108-001

The expected curvature (under the gravity loads) and the bending moment diagram are shown below. Accordingly, to minimize tension, the cable must be draped as shown in the third figure.

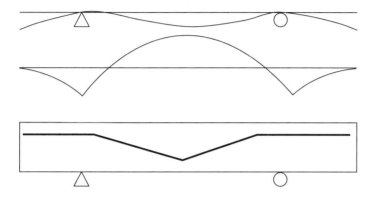

108-002

Prestress force (after losses): $F_p = (0.75 \times 270 - 32) \times 3.67 = 625.74 \, \text{kips}$

Moment due to prestress force: $M_p = F_p e = 625.74 \times 10.25 = 6413.8 \, \text{kip-in} = 534.5 \text{kip-ft}$. This moment causes compression on bottom fiber, so it is opposite to the moment due to dead and live loads.

Net moment $= 1200 - 534.5 = 665.5$ kip-ft

Bending stress on bottom fiber (compression positive):

$$\sigma_b = +\frac{625.74}{558.9} - \frac{665.5 \times 12 \times 20.29}{125,165} = -0.175 \, \text{ksi}$$

Bottom fiber stress $= 175$ psi (tension)

108-003

For stemmed deck members and beams, maximum tensile stress at service loads, after all losses, is

$$12\sqrt{f_c'} = 12 \times \sqrt{6000} = 929.5 \, \text{psi}$$

Chapter 109: Steel Tension Members

109-001

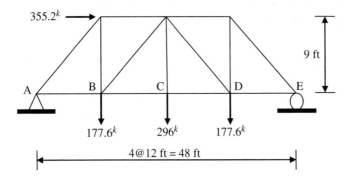

The (LRFD) factored loads are shown above. For example, the lateral load of 240 kips is 30% dead load + 70% live load = 72 k DL + 168 k LL. The factored load = $1.2 \times 72 + 1.6 \times 168 = 355.2$ k. The vertical reaction at A due to these loads is 259 kips. The maximum axial force will be in members BC and CD. The axial force in these members (using method of sections) is 809 kips.

YIELD: $0.9F_y A_g \geq 809 \rightarrow A_g \geq 25 \text{ in}^2$

FRACTURE: $0.75F_u A_e = (0.75F_u)(0.7A_g) \geq 809 \rightarrow A_g \geq 26.6 \text{ in}^2$

Minimum gross area required $= 26.6 \text{ in}^2$

109-002

$C15 \times 50 - A_g = 14.7 \text{ in}^2, t_w = 0.673 \text{ in}, x = 0.799 \text{ in}$

Net area: $A_{net} = 14.7 - 2 \times \dfrac{7}{8} \times 0.673 = 13.5 \text{ in}^2$

Shear lag factor: $U = 1 - \dfrac{\bar{x}}{L} = 1 - \dfrac{0.799}{9} = 0.91$

Design strength (yield): $\phi_t P_n = 0.9 \times 36 \times 14.7 = 476.3 \text{ kips}$

Design strength (fracture): $\phi_t P_n = 0.75 \times 58 \times 0.91 = 13.5 = 476.3 \text{ kips}$

109-003

Even though slenderness limit is not a design criterion for tension members, AISC recommends a maximum slenderness ratio of 300 to protect against incidental compression.

Chapter 110: Steel Compression Members

110-001

The centroid of the section is as shown, by symmetry.

The moment of inertia I_x is given by: $I_x = 4 \times [27.7 + 8 \times (7 - 2.12)^2] = 872.9 \, \text{in}^4$

The moment of inertia I_y is given by: $I_y = 4 \times [9.7 + 8 \times (7 - 1.12)^2] = 1145.2 \, \text{in}^4$

Total area: $A = 4 \times 8 = 32 \, \text{in}^2$

Smaller radius of gyration: $r = \sqrt{\dfrac{I}{A}} = \sqrt{\dfrac{872.9}{32}} = 5.22$

Governing slenderness ratio: $\dfrac{KL}{r} = \dfrac{32 \times 12}{5.22} = 73.6$

110-002

LRFD: factored load P_u = largest of 1.4D (168 k), 1.2D + 1.6L (544 k), 1.2D + 1.6W + 0.5L (813 k) = 813 kips

For KL_y = 25 ft and P_u = 813 k, lightest W12 is W12 × 120. r_x/r_y = 1.76. Convert the KL_x using 50′/1.76 = 28.4 ft. For KL_y = 28.4 ft and P_u = 813, lightest W12 is W12 × 152 ($\phi_c P_n$ = 875).

ASD: P_a = largest of D (120 k), D + L (370 k), D + 0.75L + 0.75 W (637.5 k) = 637.5 kips

For KL_y = 25 ft and P_a = 637 k, lightest W12 is W12 × 152. r_x/r_y = 1.77. Convert the KL_x using 50′/1.77 = 28.25 ft. For KL_y = 28.25 ft and P_a = 637, lightest W12 is W12 × 170 (P_n/Ω_c = 666).

Chapter 111: Steel Beams

111-001

LRFD factored load: $w_u = 1.2w_D + 1.6w_L = 1.2 \times 2 + 1.6 \times 4 = 8.8 \dfrac{\text{kips}}{\text{ft}}$

Design moment: $M_u = \dfrac{w_u L^2}{8} = \dfrac{8.8 \times 30^2}{8} = 990 \, \text{kip} \cdot \text{ft}$

For F_y = 50 ksi, L_b = 30 ft, and M_u = 990 kip-ft, the section chosen is W24 × 146. Increased dead load = 2.146 k/ft. New factored load, w_u = 8.975 k/ft, M_u = 1009.7 k-ft. $\phi_b M_n$ = 1068 k-ft > 990. Therefore, OK.

111-002

For M12.5 × 12.4, depth $d = 12.5$ in, $t_w = 0.155$ in, $h/t_w = 74.8$

$$2.24\sqrt{\frac{E}{F_y}} = 2.24 \times \sqrt{\frac{29,000}{36}} = 63.6$$

Since $h/t_w > 63.6$, Section G2.1(a) does not apply.

$$1.10\sqrt{\frac{k_v E}{F_y}} = 1.10 \times \sqrt{\frac{5 \times 29,000}{36}} = 69.8$$

$$1.37\sqrt{\frac{k_v E}{F_y}} = 1.37 \times \sqrt{\frac{5 \times 29,000}{36}} = 86.95$$

Since $69.8 < h/t_w < 86.95$

$$C_v = \frac{1.10\sqrt{k_v E/F_y}}{h/t_w} = \frac{69.8}{74.8} = 0.933$$

Nominal shear capacity: $V_n = 0.6F_y A_w C_v = 0.6 \times 36 \times 12.5 \times 0.155 \times 0.933 = 39.05\,\text{kips}$

Chapter 112: Bolted and Welded Connections

112-001

From Table 7-1 in the AISC Manual, A325 bolts (diameter 1.5 in) in single shear (S) with threads included in the shear plane (N), the available shear strength is $r_n/\Omega_v = r_n/2.0 = 42.4$ kips (ASD) or $\phi_v r_n = 0.75r_n = 63.6$ kips (LRFD).

112-002

From Table 7-4 in the AISC Manual, A325 bolts (diameter 1.5 in) in single shear (S) with long slotted holes (LSLT), the available shear strength is $r_n/\Omega_v = r_n/1.75 = 16.3$ kips (ASD) or $\phi_v r_n = 0.85r_n = 24.2$ kips (LRFD). The minimum bolt pretension for slip-critical behavior is 103 kips.

112-003

From Table 7-1 (AISC Manual of Steel Construction), $r_n/\Omega_v = r_n/2.0 = 5.3$ kips. Therefore, $r_n = 10.6$ kips. From Table 7-8, for eccentricity $e_x = 4$ in, vertical spacing $s = 3$ in and number of bolts per line $n = 4$, coefficient $C = 4.86$. Nominal capacity of the group $= R_n = Cr_n = 4.86 \times 10.6 = 51.5$ kips.

For ASD: $P_a \leq \dfrac{R_n}{\Omega} = \dfrac{51.5}{2.0} = 25.7$ kips

For LRFD: $P_u \leq \phi R_n = 0.75 \times 51.5 = 38.6$ kips

The total load is double the load capacity of each bolt group.

112-004

The centroid of the weld group is located at $\bar{x} = \dfrac{2 \times 6 \times 3 + 8 \times 0}{2 \times 6 + 8} = 1.8$ in . Therefore, eccentricity of the load with respect to the centroid $e_x = 8 - 1.8 = 6.2$ in. Parameter $a = e_x/l = 6.2/8 = 0.775$. Parameter $k = 6/8 = 0.75$. For these two parameters, for vertical load P,

parameters for use of AISC Table 8-8: $a = 0.775$ and $k = 0.75$. Therefore, $C = 2.82$.

For E70XX electrodes, $C_1 = 1.0$, D = no. of sixteenths of weld size = 4, $l = 8$ in

Nominal capacity of the weld group $= R_n = CC_1Dl = 2.82 \times 1.0 \times 4 \times 8 = 90.24$ kips.

For ASD: $P_a \leq \dfrac{R_n}{\Omega} = \dfrac{90.24}{2.0} = 45.1$ kips

For LRFD: $P_u \leq \phi R_n = 0.75 \times 90.24 = 67.7$ kips

The total load is double the load capacity of each weld group.

Chapter 113: Bridge Design (AASHTO LRFD)

113-001

According to AASHTO, effective width of slab = c/c distance between beams = 8 ft = 96 in.

Modular ratio: $n = \dfrac{E_s}{E_c} = \dfrac{29{,}000 \text{ ksi}}{1820\sqrt{4} \text{ ksi}} = 7.97 = 8$

Equivalent width of slab = 96/8 = 12 in.

Height of centroid (measured from bottom of steel): $\bar{y} = \dfrac{143 \times 32 + 12 \times 8 \times 69}{136 + 96} = 46.9''$

Moment of inertia of equivalent section:

$$I_{NA} = 92,140 + 143 \times (46.9 - 32)^2 + \frac{1}{12} \times 12 \times 8^3 + 96 \times (46.9 - 69)^2 = 171,287 \, in^4$$

Maximum bending moment in composite section:

$$M = \frac{wL^2}{8} = \frac{1 \cdot 8 \times 8 \times 80^2}{8} = 11,520 \, k \cdot ft = 138,240 \, k \cdot in$$

Bending stress in steel (tensile): $\sigma = \dfrac{My}{I} = \dfrac{138,240 \times 46 \cdot 9}{171,287} = 37.9 \, ksi$

113-002

Span $L = 80$ ft, lateral spacing $S = 8$ ft, girder moment of inertia $I_g = 92,140$ in^4, girder area $A = 143$ in^2, girder depth $d = 64$ in; slab thickness $t_s = 10$ in, modular ratio $n = 8$.

Stiffness parameter: $K_g = n\left(I_g + e_g^2 A\right) = 8 \times (92,140 + 37^2 \times 143) = 2.3 \times 10^6 \, in^4$

Distribution factor:

$$mg = 0.06 + \left(\frac{S}{14}\right)^{0.4} \left(\frac{S}{L}\right)^{0.3} \left(\frac{K_g}{12Lt_s^3}\right)^{0.1} = 0.06 + \left(\frac{8}{14}\right)^{0.4} \left(\frac{8}{80}\right)^{0.3} \left(\frac{2.3 \times 10^6}{12 \times 80 \times 10^3}\right)^{0.1} = 0.5$$

113-003

Maximum shear due to live load plus impact = 53.7 kips

Maximum shear force at the support for the fatigue limit state:

$$V = 0.75V_{max} = 0.75 \times 53.7 = 40.3 \, kips$$

Minimum shear force at the support (for a simply supported beam) = 0

Range of shear at the support: $V_r = 40.3 - 0 = 40.3$ kips

Allowable range of shear: $Z_r = \alpha d^2 = 7.53 \times 0.75^2 = 4.24$ kips

Statical moment of transformed slab about the centroid:

$$Q = 10.7 \times 8 \times 17.56 = 1503.14 \, in^3$$

Range of horizontal shear (at the interface between the slab and the top of the flange):

$$V_{sr} = \frac{V_r Q}{I} = \frac{40.3 \times 1503.14}{50,230} = 1.206 \frac{kip}{in}$$

With two studs per row, the required stud spacing is: $p = \dfrac{nZ_r}{V_{sr}} = \dfrac{2 \times 4.24}{1.206} = 7.03 \, in$

Chapter 114: Timber Design

114-001

Nominal size: 6 in $\times$ 6 in $-$ dressed size 5.5 in $\times$ 5.5 in $-$ area $= 30.25$ in^2

For no. 1 Hem-Fir, reference values are: $F_c = 850$ psi, $E = 1.3 \times 10^6$ psi, $E_{min} = 4.7 \times 10^5$ psi

Adjustment factors: For M.C. $= 22\%$, $C_t = 0.95$ (for F_c), $C_D = 1.15$

For pinned-pinned ends, $K = 1.0$

Slenderness ratio: $\dfrac{l_e}{d} = \dfrac{Kl}{d} = \dfrac{1.0 \times 14 \times 12}{5.5} = 30.6$

$$E'_{min} = E_{min} \times C_m \times C_t \times C_i = 4.7 \times 10^5 \times 1.0 \times 1.0 \times 1.0 = 4.7 \times 10^5 \, \text{psi}$$

$$F_{cE} = \dfrac{0.822 E_{min}}{(l_e/d)^2} = \dfrac{0.822 \times 4.7 \times 10^5}{30.6^2} = 412.6 \, \text{psi}$$

$$F_c^* = F_c \times C_F \times C_D \times C_i \times C_t \times C_M = 850 \times 1.0 \times 1.15 \times 1.0 \times 1.0 \times 0.95 = 928 \, \text{psi}$$

For sawn lumber, $c = 0.8$

$$C_P = \dfrac{1 + \dfrac{F_{cE}}{F_c^*}}{2c} - \sqrt{\left(\dfrac{1 + \dfrac{F_{cE}}{F_c^*}}{2c}\right)^2 - \dfrac{\dfrac{F_{cE}}{F_c^*}}{c}} = \dfrac{1 + 0.444}{2 \times 0.8} - \sqrt{\left(\dfrac{1 + 0.444}{2 \times 0.8}\right)^2 - \dfrac{0.444}{0.8}} = 0.393$$

$$F_c' = F_c \times C_F \times C_D \times C_M \times C_P = 850 \times 1.0 \times 1.15 \times 0.95 \times 0.39 = 362 \, \text{psi}$$

Allowable load: $P = F_c' \times A = 362 \times 30.25 = 10{,}950 \, \text{lb}$

114-002

Nominal size $-$ 2 in $\times$ 10 in $-$ dressed size 1.5 in $\times$ 9.25 in

For no. 2 Southern Pine, reference values are: $F_v = 175$ psi

Adjustment factors: Load duration factor $C_D = 1.0$, Wet service factor $C_M = 1.0$, Temperature factor $C_t = 1.0$, Incising factor $C_i = 1.0$

Distributed load: $w = 30 + 60 = 90$ lb/ft

Maximum shear: $V_u = \dfrac{w_u L}{2} = \dfrac{90 \times 15}{2} = 675 \, \text{lb}$

Allowable shear stress: $F_v' = F_v C_D C_M C_t C_i = 175 \times 1.0 \times 1.0 \times 1.0 \times 1.0 = 175\,\text{psi}$

Depth of beam at notched end: $d_n = 7.25\,\text{in}$

Maximum allowable shear force at notch:

$$V_r' = \left(\frac{2}{3}F_v' b d_n\right)\left(\frac{d_n}{d}\right)^2 = \left(\frac{2}{3}\times 175 \times 1.5 \times 7.25\right)\left(\frac{7.25}{9.25}\right)^2 = 779.4\,\text{lb} > 675\,\text{lb}$$

Therefore, beam has adequate capacity.

Chapter 115: Masonry Design

115-001

Effective depth $= 40 - 3.5 = 36.5$ in

Area of steel: $A_s = 0.88\,\text{in}^2$

For width $= 10$ in, actual width $= 9.63$ in

Reinforcement ratio: $\rho = \dfrac{A_s}{bd} = \dfrac{0.88}{9.63 \times 36.5} = 0.0025$

Modulus of elasticity: $E_m = 900 f_m' = 900 \times 1500 = 1.35 \times 10^6\,\text{psi}$

Modular ratio: $n = E_s/E_m = 29 \times 10^6/1.35 \times 10^6 = 21.5$

$n\rho = 21.5 \times 0.0025 = 0.0538$

$k = \sqrt{(n\rho)^2 + 2n\rho} - n\rho = 0.279$

$j = 1 - k/3 = 1 - 0.093 = 0.907$

Allowable compressive stress in masonry: $F_m = 0.45 f_m' = 675\,\text{psi}$

Allowable moment based on compressive stress in masonry:

$$M_m = \frac{1}{2}F_m kjbd^2 = \frac{1}{2}\times 675 \times 0.279 \times 0.907 \times 9.63 \times 36.5^2 = 1{,}095{,}714\,\text{lb}\cdot\text{in} = 91.3\,\text{kip}\cdot\text{ft}$$

Allowable tensile stress in steel: $F_s = 0.5 f_y = 20{,}000 \, \text{psi}$

Allowable moment based on tensile stress in steel:

$$M_s = A_s F_s jd = 0.88 \times 20{,}000 \times 0.907 \times 36.5 = 582{,}657 \, \text{lb} \cdot \text{in} = 48.6 \, \text{kip} \cdot \text{ft}$$

115-002

Cross section: $A = 18 \times 10 = 180 \, \text{in}^2$

Area of steel: $A_s = 4 \text{ no. } 9 \text{ bars} = 4.0 \, \text{in}^2$

Reinforcement ratio: $\rho = \dfrac{A_s}{A_g} = \dfrac{4}{180} = 0.022$

Radius of gyration: $r = 0.289 \times 10 = 2.89 \, \text{in}$

Slenderness ratio: $\dfrac{h}{r} = \dfrac{16 \times 12}{2.89} = 66.4 < 99$

Reduction factor: $R = 1 - \left(\dfrac{h}{140r} \right)^2 = 1 - \left(\dfrac{66.4}{140} \right)^2 = 0.775$

If special inspection is not provided, the allowable compressive stress in the masonry is reduced by half. Accordingly, the allowable axial load is given by

$$
\begin{aligned}
P_a &= \left(0.5 \times 0.25 f'_m + 0.65 \rho F_{sc} \right) R A_g \\
&= (0.5 \times 0.25 \times 2000 + 0.65 \times 0.0222 \times 16{,}000) \times 0.775 \times 180 \\
&= 67{,}082 \, \text{lb} = 67.1 \, \text{k}
\end{aligned}
$$

If special inspection is provided, the allowable axial load is given by

$$
\begin{aligned}
P_a &= (0.25 f'_m + 0.65 \rho F_{sc}) R A_g \\
&= (0.25 \times 2000 + 0.65 \times 0.0222 \times 16{,}000) \times 0.775 \times 180 \\
&= 101{,}958 \, \text{lb} = 102.0 \, \text{k}
\end{aligned}
$$

Chapter 201: Phase Relationships for Soils

201-001

Weight of soil solids = 3.67 lb

Weight of water = 4.18 − 3.67 = 0.51 lb

Water content w = 0.51/3.67 = 0.139

Total unit weight = 4.18/(1/30) = 125.4 pcf

Specific gravity: G_s = 2.73

Using the formula: $\gamma = \dfrac{(1+w)SG_s}{wG_s + S}\gamma_w \Rightarrow 125.4 = \dfrac{1.139 \times 2.73 \times S}{0.139 \times 2.73 + S} \times 62.4 \Rightarrow S = 0.693$

201-002

Volume = 1050 cc = 0.037 ft³

Unit weight = 4.58/0.037 = 123.5 lb/ft³

Degree of saturation S = 1.0

$$\gamma = \frac{(1+w)SG_s}{wG_s + S}\gamma_w \Rightarrow w = \frac{S(G_s\gamma_w - \gamma)}{G_s(\gamma - S\gamma_w)} = \frac{1 \times (2.6 \times 62.4 - 123.5)}{2.6 \times (123.5 - 1 \times 62.4)} = 0.244$$

Water content = 24%

201-003

In the fill, dry density = 0.95 × 108 = 102.6 pcf

20,000 cy of this soil contains weight of solids W_s = 102.6 × 20,000 × 27 = 5.54 × 10⁷ lb

In situ, the dry unit weight in the borrow pit: $\gamma_d = \dfrac{124}{1.18} = 105.1$ pcf

In order to get the same weight of solids, one needs the volume: $V = \dfrac{W_s}{\gamma_d} = \dfrac{5.54 \times 10^7}{105.1} =$

$5.27 \times 10^5\,\text{ft}^3 = 19{,}526\,\text{yd}^3$

Chapter 202: Soil Sampling and Testing

202-001

The total lateral pressure, $\sigma_3 = 12.6$ psi

Effective lateral pressure, $\sigma'_3 = 12.6 - 4.5 = 7.1$ psi

Added axial pressure $= 93.2/(\pi/4 \times 2^2) = 29.7$ psi

The total axial pressure $\sigma_1 = 12.6 + 29.7 = 42.3$ psi

Effective axial pressure, $\sigma'_1 = 42.3 - 4.5 = 37.8$ psi

Cohesion $= \frac{1}{2}(\sigma'_1 - \sigma'_3) = 15.35$ psi $= 2210.4$ psf

202-002

RQD is defined as the fraction of unfractured core pieces longer than 4 in.

Total length of the core run $= 115 - 100 = 15$ ft $= 180$ in.

Total length of pieces longer than 4 in $= 22 + 35 + 32 + 7 + 41 = 137$ in

RQD $= 137/180 = 0.76$

202-003

This test sounds like the liquid limit test, where a hemispherical sample is placed in a cup and the cup is dropped on a plate (delivering "blows" to the soil) repeatedly. The water content at which it takes 25 blows to close the groove is defined as the liquid limit. According to the data, this would occur very close to sample 3. The liquid limit is therefore approximately 46. A graphical solution is to plot log N versus water content (w) and to read LL as the water content corresponding to N = 25

202-004

Effective stress at the center of the clay layer (depth $= 14$ ft) is calculated as the sum of the z products for the soil layers (submerged where appropriate)

$$p_1 = 114 \times 6 + (120 - 62.4) \times 2 + (108 - 62.4) \times 6 = 1073 \text{psf}$$

The uniform pressure directly under the mat (depth $= 3.5$ ft) is: $\Delta p = \dfrac{20,000 \times 2000}{190 \times 250} = 842$ psf

The pressure increase at depth of 14 ft (10.5 ft below bottom of mat) can be considered to be 842 psf (no dissipation, since the load is exerted over relatively large plan dimensions). The pressure increase due to lowering of water table, application of the foundation load, and the stress relief due to the excavation for the mat is given by

$$p_2 = 114 \times 8 + (108 - 62.4) \times 6 + 842 - 3.5 \times 114 = 1629 \text{psf}$$

Primary settlement of the 12-ft-thick clay layer is calculated as

$$s = \frac{C_c H \log_{10}\left(\frac{p_2'}{p_1'}\right)}{1 + e_0} = \frac{0.45 \times 144}{1 + 1.4} \times \log_{10}\left(\frac{1629}{1073}\right) = 4.9\text{in}$$

202-005

The clay layer can be considered to be "doubly drained" since it can drain through the sand layers above and below. Therefore, the drainage thickness H_d = half the full layer thickness = 10 ft.

For degree of consolidation U = 6 in/8 in = 75%, time factor $T_v = 0.477$.

The time for 75% settlement to occur is given by

$$t = \frac{T_v H_d^2}{c_v} = \frac{0 \cdot 477 \times 10^2}{5} = 9.54\text{yrs}$$

Chapter 203: Soil Classification

203-001

The critical sizes for the USDA classification system are 2.0 mm (no. 10 sieve), 0.05 mm, and 0.002 mm.

According to the data, 87 g is coarser than the no. 10 sieve. This is the gravel fraction. Since 24 g is finer than 0.05 mm, 250 – 24 = 226 g is coarser than 0.05 mm and therefore, the total weight of the soil in the size range 2.0 mm > D > 0.05 mm is 226 – 87 = 139 g. The weight of the soil in the size range 0.05 mm > D > 0.002 mm is 24 – 12 = 12 g and the weight of the soil finer than 0.002 mm is 12 g.

Separating the gravel fraction, we get the following percentages for the sand, silt, and clay fractions:

Sand: 139/163 = 85.2%

Silt: 12/163 = 7.4%

Clay: 12/163 = 7.4%

According to the USDA Soil Texture Triangle, this soil can be classified as loamy sand.

203-002

For the USCS, first identify $F_{200} = 31/250 = 12.4\%$. Since this is less than 50%, the soil is predominantly coarse grained. For coarse-grained soils, one must identify how the coarse fraction separates into larger particles (G: gravel) and smaller particles (S: sand). The division between these is at the no. 4 sieve. In the data, 71 g (28.4%) is retained on the no. 4 sieve while the coarse fraction = $100 - F_{200} = 87.6\%$. Since the gravel fraction (28.4%) is less than half the coarse fraction (87.6%), the soil is predominantly sand and the first letter is "S".

To identify the second letter, one looks at F_{200}. Since F_{200} is greater than 12%, plasticity characteristics (Casagrande Plasticity Chart) must be used rather than particle size distribution. Since $LL = 32$ and $PL = 23$, the plasticity index $PI = 9$. This plots the point just above the A-line. The soil is therefore SC.

203-003

The AASHTO soil classification uses the no. 200 sieve to separate coarse-grained soils from fine-grained soils. $F_{200} = 31/250 = 12.4\%$. Since this is less than 35%, the soil is predominantly coarse grained.

According to the data, 87 g is coarser than the no. 10 sieve. Therefore, % retained on no. 10 sieve = $87/250 = 34.8\%$. Therefore, % passing no. 10 sieve = 65.2%. This does not meet the criteria for group A-1-a.

According to the data, 148 g is coarser than the no. 40 sieve. Therefore, % retained on no. 40 sieve = $148/250 = 59.2\%$. Therefore, % passing no. 40 sieve = 40.8%. This meets both gradation criteria for group A-1-b (% passing no. 40 sieve $\leq$ 50 and % passing no. 200 sieve $\leq$ 25). However, $PI = 9$ does not.

On the other hand, the soil meets ALL criteria of group A-2-4 (% passing no. 200 sieve $\leq$ 35, $LL \leq 40$, $PI \leq 10$)

Group Index: $GI = (F_{200} - 35)[0.2 + 0.005(LL - 40)] + 0.01(F_{200} - 15)(PI - 10)$
$$= (12.4 - 35)[0.2 + 0.005(32 - 40)] + 0.01(12.4 - 15)(9 - 10)$$
$$= -3.616 + 0.026 = -3.59$$

Negative answer for GI must be reported as zero. Therefore, the AASHTO classification is A-2-4 (0).

Chapter 204: Vertical Stress Increase at Depth

204-001

Vertical pressure directly under the footing: $q = P/A = 105/19.635 = 5.35$ ksf.

Using the Boussinesq stress contours for a circular footing, $x/r = 2.5/2.5 = 1.0$ and $z/r = 4/2.5 = 1.6$, the influence factor $= 0.27$. Therefore, the pressure increase $= 0.27 \times 5.35 = 1.444$ ksf $= 1444$ psf.

204-002

Vertical pressure directly under the footing $= 80,000$ lb/24 ft$^2 = 3333$ lb/ft^2.

Using the tool supplied with the problem, one has to subdivide the loaded area (4 ft $\times$ 6 ft) into four identical quarters (2 ft $\times$ 3 ft). For each of these areas, the normalized dimensions are $m = B/z = 2/5 = 0.8$ and $n = L/z = 3/5 = 0.6$. For $m = 0.8$ and $n = 0.6$, the influence coefficient $I = 0.12$.

Therefore, vertical stress increase directly under center $= 4 \times 0.12 \times 3333 = 1600$ lb/ft^2.

Chapter 205: Flow through Porous Media

205-001

At the instant shown, the hydraulic head difference $H = 43 - 36 = 7$ in

The hydraulic gradient: $i = \dfrac{H}{L} = \dfrac{7}{8} = 0.875$ in/in

Porosity: $n = \dfrac{e}{1+e} = \dfrac{0.45}{1.45} = 0.31$

Seepage velocity: $V = \dfrac{Ki}{n} = \dfrac{1.8 \times 10^{-4} \times 0.875}{0.31} = 5.1 \times 10^{-4} \dfrac{\text{ft}}{\text{sec}} = 6.1 \times 10^{-3} \dfrac{\text{in}}{\text{sec}}$

205-002

Pumping rate: $Q = 750$ gpm $= 1.674$ cfs

Pump radius $r = 4.5$ in $= 0.375$ ft

Dimensionless parameter: $u = \dfrac{r^2 S}{4Tt} = \dfrac{0.375^2 \times 0.6}{4 \times 1.2 \times 2} = 0.0088$

Value of Well function: $W(u) = 4.16$ (Table 205.1)

$s = \dfrac{Q}{4\pi T} W(u) = \dfrac{1.674 \times 4.16}{4\pi \times 1.2} = 0.46$ h

Elevation of piezometric surface at the well $= 239.37 - 0.46 = 238.91$ ft

205-003

According to Hazen's formula: $K = CD^2_{10} = 15 \times 0.01^2 = 0.0015$ mm/s $= 0.0177$ ft/h

$$q = K\frac{N_f}{N_e}H = 0.0177 \times \frac{5}{12} \times 25 = 0.1845\,\text{ft}^2/\text{h}$$

Total flow rate: $Q = qL = 0.1845 \times 1000 = 184.5$ ft³/h

Flow per channel $= 0.1845/5 = 0.0369$ ft²/h

Average discharge velocity in 19-ft-wide cell $= 0.0369/(19 \times 1) = 0.00194$ ft/h

Chapter 206: Shallow Foundations

206-001

For water table exactly at the bottom of the footing, the bearing capacity equation is modified to:

$$q_{ult} = cN_c + \gamma D_f N_q + \frac{1}{2}(\gamma_{sat} - \gamma_w)BN_\gamma$$

Saturated unit weight $= 94 \times 1.26 = 118.44$ pcf

For $\phi = 32°$, $N_q = 28.52$ and $N_\gamma = 26.87$ (Table 206.1)

The expression for bearing capacity becomes:

$$q_{ult} = 118.44 \times 3 \times 28.52 + \frac{1}{2}(118.44 - 62.4) \times 26.87B = 10134 + 753B$$

Allowable bearing stress: $q_{all} = \dfrac{q_{ult}}{FS} = \dfrac{10134 + 753B}{2.8} = 3620 + 269B$

Therefore, the load capacity equation can be written as: $(3620 + 269B)B > 20,000$

Solving this quadratic equation, $B \geq 4.21$ ft

206-002

Unit weight of soil: $\gamma = \gamma_d(1 + w) = 95 \times 1.12 = 106.4\text{pcf}$

According to Bowles-Meyerhof, the net allowable soil pressure (for footing width less than 4 ft) is given by:

$$q_{net} = \frac{N}{2.5}F_d S$$

Depth factor: $F_d = 1 + 0.33\dfrac{D_f}{B} \leq 1.33$

Try $B = 4$ ft, depth factor $= 1.23$, $q_{net} = 9.8$ ksf. Actual net soil pressure $= 75/16 = 4.7$ ksf. PASS

Try $B = 3$ ft, depth factor $= 1.31$, $q_{net} = 10.5$ ksf. Actual net soil pressure $= 75/9 = 8.3$ ksf. PASS

Try $B = 2.5$ ft, depth factor $= 1.37$, limited to 1.33, $q_{net} = 10.6$ ksf. Actual net soil pressure $= 75/6.25 = 12$ ksf. Therefore, FAIL

Try $B = 2.75$ ft, depth factor $= 1.34$, limited to 1.33, $q_{net} = 10.6$ ksf. Actual net soil pressure $= 75/7.56 = 9.9$ ksf. PASS. Therefore, a 2 ft 9 in square footing will work.

206-003

Eccentricity of the load—in the long direction, $e_x = 4.85 - 2.7 = 2.15$ ft; in the short direction, $e_y = 3.0 - 2.35 = 0.65$ ft.

The effective dimensions of the footing are:

$L_x = L - 2e_x = 9.7 - 2 \times 2.15 = 5.4$ ft

$L_y = B - 2e_y = 4.7 - 2 \times 0.65 = 3.4$ ft

Maximum soil pressure: $q_{max} \approx \dfrac{P}{L_x L_y} = \dfrac{130}{5.4 \times 3.4} = 7.08$ ksf

Chapter 207: Deep Foundations

207-001

Centroid distance: $h_v = \dfrac{3A \times 6 + 2A \times 0}{5A} = 3.6$ ft

Taking moments about the top of the battered pile: $M_B = 0 \Rightarrow R_1 s + Ht - V(s - h_v) = 0$

$$R_1 = \dfrac{V(s - h_v) - Ht}{s} = \dfrac{300 \times (6 - 3.6) - 120 \times 3}{6} = 60 \text{ kips}$$

$$R_{2v} = V - R_1 = \dfrac{Vh_v + Ht}{s} = \dfrac{300 \times 3.6 + 120 \times 3}{6} = 240 \text{ kips}$$

$$R_{2h} = H = 120 \text{ kips}$$

$$R_2 = \sqrt{R_{2h}^2 + R_{2v}^2} = \sqrt{120^2 + 240^2} = 268.3 \text{ kips}$$

Batter $= 240{:}120 = 2V{:}1H$

R_1 per pile $= 60/2 = 30$ kips, R_2 per pile $= 268.3/3 = 89.5$ kips

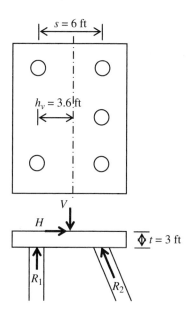

207-002

Point bearing capacity (layer 3): $Q_p = cN_cA_p = 1250 \times 9 \times \dfrac{\pi}{4} \times \left(\dfrac{14}{12}\right)^2 = 12{,}026\,\text{lb}$

Side friction capacity is calculated as the sum of contributions from the three layers of embedment (11 ft in sand, 21 ft in clay, and 2 ft in silty clay). Pile length $= 40 - 6 = 34$ ft.

Side friction capacity (ultimate):

$$Q_s = f_sL_1P + \alpha_2c_2L_2P + \alpha_3c_3L_3P$$
$$= 200 \times 11 \times \pi \times \frac{14}{12} + 0.8 \times 800 \times 21 \times \pi \times \frac{14}{12} + 0.65 \times 1250 \times 2 \times \pi \times \frac{14}{12}$$
$$= 63{,}280\,\text{lb}$$

Ultimate capacity: $Q_{ult} = Q_p + Q_s = 12.026 + 63.280 = 75.3\,\text{kips}$

207-003

Pile length $= 40 - 6 = 34$ ft.

Load is transmitted from a point which is 2/3 the length of the pile, i.e., at a depth of $6 + 22.67 = 28.67$ ft. The thickness of layer 2 below this depth is 9.33 ft, with center at 33.33 ft and the thickness of layer 3 is 12 ft, with center at 44 ft.

Effective vertical pressure (before and after) at a depth of 33.33 ft are:

$$p_1' = 120 \times 10 + (120 - 62.4) \times 7 + (115 - 62.4) \times (33.33 - 17) = 2462\,\text{psf}$$

$$p_2' = \frac{800{,}000}{(30 + 27.33) \times (15 + 27.33)} + 120 \times 17 + (115 - 62.4) \times (33.33 - 17) = 3229\,\text{psf}$$

Effective vertical pressure (before and after) at a depth of 44.0 ft are:

$$p_1' = 120 \times 10 + (120 - 62.4) \times 7 + (115 - 62.4) \times 21 + (125 - 62.4) \times (44 - 38) = 3083\,\text{psf}$$

$$p_2' = \frac{800{,}000}{(30 + 38) \times (15 + 38)} + 120 \times 17 + (115 - 62.4) \times 21 + (125 - 62.4) \times (44 - 38)$$
$$= 3742\,\text{psf}$$

Settlement of layer 2: $s = \dfrac{HC_c}{1 + e_o}\log_{10}\left(\dfrac{p_2'}{p_1'}\right) = \dfrac{9.33 \times 12 \times 0.45}{1.55} \times \log_{10}\left(\dfrac{3229}{2462}\right) = 3.83\,\text{in}$

Settlement of layer 3: $s = \dfrac{HC_c}{1 + e_o}\log_{10}\left(\dfrac{p_2'}{p_1'}\right) = \dfrac{12 \times 12 \times 0.45}{1.55} \times \log_{10}\left(\dfrac{3742}{3083}\right) = 3.52\,\text{in}$

Total settlement $= 3.893 + 3.52 = 7.35\,\text{in}$

Chapter 208: Retaining Walls

208-001

The active earth pressure coefficient is

$$K_a = \frac{1 - \sin 32}{1 + \sin 32} = 0.307$$

The resultant earth pressure has two components—due to overburden and due to surcharge:

$$R_a = \frac{1}{2}K_a\gamma H^2 + K_a q_s H = 0.5 \times 0.307 \times 120 \times 18^2 + 0.307 \times 400 \times 18 = 8178.5\,\frac{\text{lb}}{\text{ft}}$$

The total weight of concrete stem, concrete footing and backfill soil are:

$$W = 15 \times 1 \times 150 + 13 \times 3 \times 150 + 7 \times 15 \times 120 = 20{,}700\,\frac{\text{lb}}{\text{ft}}$$

Coefficient of friction (below footing) $= \tan 24° = 0.445$

Factor of safety against sliding: $FS = \dfrac{0.445 \times 20,700}{8178.5} = 1.13$

208-002

The overturning moment due to the two components of the earth pressure are:

$$M_{OT} = \frac{1}{6} K_a \gamma H^3 + \frac{1}{2} K_a q_s H^2 = \frac{1}{6} \times 0.307 \times 120 \times 18^3 + \frac{1}{2} \times 0.307 \times 400 \times 18^2$$

$$= 55,702 \frac{\text{lb} \cdot \text{ft}}{\text{ft}}$$

Stabilizing moment due to the weight components (concrete stem, concrete footing, and backfill soil):

$$M_s = \sum W_i x_i = 15 \times 1 \times 150 \times 5.5 + 13 \times 3 \times 150 \times 6.5 + 7 \times 15 \times 120 \times 9.5 = 170,100 \frac{\text{lb} \cdot \text{ft}}{\text{ft}}$$

Factor of safety against overturning: $FS = \dfrac{170,100}{55,702} = 3.05$

208-003

The active earth pressure coefficient in the backfill is

$$K_a = \frac{1 - \sin 32}{1 + \sin 32} = 0.307$$

The resultant earth pressure has two components—due to overburden and due to surcharge:

$$R_a = \frac{1}{2} K_a \gamma H^2 + K_a q_s H = 0.5 \times 0.307 \times 120 \times 18^2 + 0.307 \times 400 \times 18 = 8178.5 \frac{\text{lb}}{\text{ft}}$$

The total weight of concrete stem, concrete footing, and backfill soil are:

$$W = 15 \times 1 \times 150 + 13 \times 3 \times 150 + 7 \times 15 \times 120 = 20,700 \frac{\text{lb}}{\text{ft}}$$

Coefficient of friction (below footing) $= \tan 24° = 0.445$

The passive earth pressure coefficient of the soil on the passive side ($\phi = 15$) is

$$K_p = \frac{1 + \sin 15}{1 - \sin 15} = 1.7$$

The resultant earth pressure force is calculated as the resultant force on 4.67-ft depth of passive soil (top of footing to bottom of key):

$$R_p = \frac{1}{2} K_p \gamma H^2 = 0.5 \times 1.7 \times 120 \times 4.67^2 = 2224.5 \frac{\text{lb}}{\text{ft}}$$

Resistance due to cohesion along a length of 5 ft and adhesion along a length of 8 ft is calculated as:

$$F = 1200 \times 5 + 0.8 \times 1200 \times 8 = 13680 \frac{\text{lb}}{\text{ft}}$$

Factor of safety against sliding: $FS = \dfrac{0.445 \times 20,700 + 2224.5 + 13680}{8178.5} = 3.07$

Chapter 209: Support of Excavation

209-001

The active pressure behind the sheet pile wall: $p = 0.65 \gamma H \tan^2 (45 - \phi/2) = 497.3 \, \text{psf}$

Assuming continuous support from several evenly spaced supports:

$$M_{max} = \frac{wL^2}{10} = \frac{497.3 \times 6^2}{10} = 1790.2 \, \text{lb} \cdot \text{ft/ft} = 21,482 \, \text{lb} \cdot \text{in/ft}$$

Using the bending stress formula:

$$\frac{M}{S} \leq \sigma_{all} \Rightarrow S \geq \frac{M}{\sigma_{all}} \Rightarrow \frac{bt^2}{6} \geq \frac{M}{\sigma_{all}} \Rightarrow t \geq \sqrt{\frac{6M}{\sigma_{all} b}} = \sqrt{\frac{6 \times 21482}{2200 \times 12}} = 2.21 \text{in}$$

209-002

The active pressure behind the sheet pile wall: $p = 0.65 \gamma H \tan^2 (45 - \phi/2) = 497.3 \, \text{psf}$

Assuming a hinge at the second strut, and equating the moment of the earth pressure of the 8 ft above strut 2 to the moment of the S1 force, one gets:

$$497.3 \times 8 \times 20 \times 4 = F_1 \times 6 \Rightarrow F_1 = 53,045 \, \text{lb} = 53 \, \text{kips}$$

Repeating the process, assuming a hinge at the third strut, and equating the moment of the earth pressure of the 14 ft above strut 3 to the moment of the S1 and S2 forces, one gets:

$$497.3 \times 14 \times 20 \times 7 = F_1 \times 12 + F_2 \times 6 \Rightarrow F_2 = 56,361 \text{lb} = 56.4 \, \text{kips}$$

209-003

Allowable stress: $\dfrac{P}{A} \leq 24,000 \Rightarrow A \geq \dfrac{10,500}{24,000} \Rightarrow d \geq 0.746 \, \text{in}$

Use 0.75-in-diameter rod, $A = 0.44 \, \text{in}^2$

Minimum elongation: $\Delta = \dfrac{PL}{AE} \geq 0.25 \Rightarrow L \geq \dfrac{0.44 \times 29 \times 10^6 \times 0.25}{10,500} \Rightarrow L \geq 304 \, \text{in} = 25 \, \text{ft}$

Chapter 210: Slope Stability

210-001

Since the radial line makes an angle of 25° with the vertical, the tangent must make 25° with the horizontal. Assuming the short arc length (bottom edge of the shaded slice) to be approximately equal to the tangent, arc length = 6/cos 25 = 6.62 ft

$$F_t = W_T \cos\alpha \tan\phi + cL = 20{,}000 \times \cos25 \times \tan24 + 1280 \times 6.62$$
$$= 8070 + 8474 = 16{,}543\,\text{lb / ft}$$

210-002

Because of the significant length of the slope (540 ft), it can be treated as an "infinite" slope.

Slope angle: $\beta = \tan^{-1}\left(\dfrac{200}{500}\right) = 21.8$

For the embankment soil:

$$FS = \frac{c}{\gamma H \cos^2\beta\tan\beta} + \frac{\tan\varphi}{\tan\beta} = \frac{600}{124 \times 200 \times \cos^2 21.8 \times \tan21.8} + \frac{\tan34}{\tan21.8} = 1.74$$

For riprap: $FS = \dfrac{c}{\gamma H \cos^2\beta\tan\beta} + \dfrac{\tan\varphi}{\tan\beta} = 0 + \dfrac{\tan38}{\tan21.8} = 1.95$

Overall factor of safety (lower) is 1.74.

Chapter 211: Seismic Topics in Geotechnical Engineering

211-001

Effective vertical stress at a depth of 23 ft:

$$\sigma'_v = 125 \times 10 + 120 \times 12 + 122 \times 1 - 62.4 \times 10 = 2188\,\text{psf}$$

Since CRR = 0.25, the maximum allowable cyclic shear stress (resistance) = 0.25 × 2188 = 547 psf

FS = 547/425 = 1.287

211-002

Answer is C. IBC recommends using an average of soil shear wave velocity, standard penetration resistance, or undrained shear strength of the underlying 100 ft of soil.

Chapter 212: Earthwork

212-001

Station	Cut area (ft²)	Cut volume (ft³)	Fill area (ft²)	Fill volume/0.9 (ft³)
0 + 0.00	456.12		45.32	
1 + 0.00	634.54	54,533	120.54	9215
2 + 0.00	342.23	48,839	563.20	37,993
3 + 0.00	123.45	23,284	547.23	61,690
4 + 0.00	84.23	10,384	231.70	43,274
5 + 0.00	0	4212	102.71	18,578
6 + 0.00	123.90	6195	58.45	8953
		+ 147,446		− 179,703

Net earthwork volume = 147,446 − 179,703 = − 32,257 ft³ = 1,195 yd³ (fill)

212-002

The distance from station 8 + 0.00 to 11 + 0.00 is the 300-ft-long free haul vector between stations C and E. Therefore, the desired answer is calculated as the moment of the outlying segments about the edges of this free haul interval.

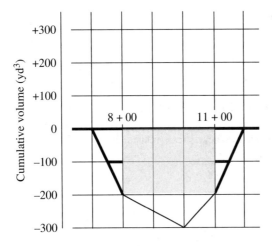

Overhaul between stations C and E = 200 × 0.5 + 200 × 0.5 = 200 yd³-sta

212-003

Points of transition between cut and fill (such as B ad D) are called grade points (A is false). At stations $6 + 00$ and $8 + 00$, the mass diagram ordinate does not match (B is false). The mass diagram ordinate at station $2 + 0.00$ is $+250$ yd^3 and at station $6 + 50.00$ is $+100$ yd^3. This means the net earthwork between these stations is $100 - 250 = -150$ yd^3 (which is a FILL of 150 yd^3) (D is false).

Stations $7 + 50.00$ to $11 + 50.00$ have identical mass diagram ordinate and therefore are called match points.

Chapter 301: Basic Fluid Mechanics

301-001

Pressure at centroid of gate (depth $= 7$ ft): $p_{CG} = 62.4 \times 7 = 436.8$ psf

Horizontal resultant force due to fluid pressure $= 436.8 \times 6 \times 10 = 26{,}208$ lb

Effective location (depth) of the resultant from A is found as the depth of the centroid of the trapezoidal pressure diagram: $\quad y = \dfrac{h}{3}\left(\dfrac{b_1 + 2b_2}{b_1 + b_2}\right) = \dfrac{6}{3}\left(\dfrac{249.6 + 2 \times 624}{249.6 + 624}\right) = 3.43$ ft

Taking moments about A: $26{,}208 \times 3.43 = F \times 6 \rightarrow F = 14{,}976$ lb $= 15$ kips

301-002

Spillway flow rate: $\quad Q_1 = 3.3b_1 H_1^{3/2} = 3.3 \times 120 \times 3.6^{3/2} = 2705 \dfrac{\text{ft}^3}{\text{sec}}$

$$Fr_1 = Fr_2 \Rightarrow \frac{V_1}{\sqrt{gH_1}} = \frac{V_2}{\sqrt{gH_2}} \Rightarrow \frac{V_1}{V_2} = \sqrt{\frac{H_1}{H_2}}$$

Since $Q = VbH$:

$$\frac{Q_1}{Q_2} \Rightarrow \frac{V_1 b_1 H_1}{V_2 b_2 H_2} = \left(\frac{H_1}{H_2}\right)^{2.5} = 20^{2.5} = 1789$$

Therefore, for the scale model, the flow rate must be $2705/1789 = 1.51$ ft^3/sec

301-003

Average velocity: $V = \dfrac{Q}{A} = \dfrac{3.2}{\dfrac{\pi}{4} \times 0.5^2} = 16.3$ fps

Reynolds Number: $\quad Re = \dfrac{VD}{\nu} = \dfrac{16.3 \times 0.5}{1.0 \times 10^{-5}} = 8.1 \times 10^5$

This is high enough for the flow to be considered "fully turbulent." For fully turbulent flow, the maximum flow velocity is about 18% higher than the average velocity.

$$V_{max} = 1.18 \times 16.3 = 19.2 \text{ fps}$$

301-004

Assuming negligible pressure loss at the bend, pressure head in the pipe = 235.4 − 156.0 = 79.4 ft. This corresponds to a pressure = 79.4 × 62.4 = 4954.6 psf.

Cross-sectional area of the pipe: $A = \dfrac{\pi}{4}\left(\dfrac{6}{12}\right)^2 = 0.196 \text{ ft}^2$

Average flow velocity: $V = \dfrac{Q}{A} = \dfrac{3.2}{0.196} = 16.3 \text{ fps}$

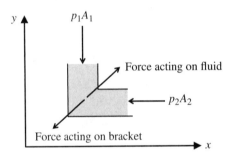

Writing the force-momentum relationship for x and y directions:

$$F_x - p_1 A_1 = \rho QV - 0 \Rightarrow F_x = p_1 A_1 + \rho QV = 4954.6 \times 0.196 + \dfrac{62.4}{32.2} \times 3.2 \times 16.3 = 1072.2 \text{ lb}$$

$$F_y - p_2 A_2 = 0 - \rho QV \Rightarrow F_y = p_2 A_2 - \rho QV = 4954.6 \times 0.196 - \dfrac{62.4}{32.2} \times 3.2 \times 16.3 = 870 \text{ lb}$$

Resultant force on bracket: $F = \sqrt{1070^2 + 870^2} = 1379 \text{ lb}$

Chapter 302: Closed Conduit Hydraulics

302-001

Flow velocity: $V = \dfrac{Q}{A} = \dfrac{3}{\dfrac{\pi}{4} \times \left(\dfrac{8}{12}\right)^2} = 8.6 \text{ fps}$

The excess pressure is given by

$$\Delta p = \frac{0.07VL}{t} = \frac{0.07 \times 8.6 \times 300}{0.2} = 903 \text{psi}$$

302-002

Head loss due to friction: $h_f = \dfrac{10.429 Q_{gpm}^{1.85} L_{ft}}{C^{1.85} D_{in}^{4.865}} = \dfrac{10.429 \times 2000^{1.85} \times 3200}{100^{1.85} \times 12^{4.865}} = 47.9 \text{ ft}$

Flow rate $Q = 2000 \text{ gpm} = 4.464 \text{ cfs}$

The gage pressure of 70 psi corresponds to an absolute pressure of $70 + 14.7 = 84.7 \text{ psi} = 195.5$ ft of water.

Writing the Bernoulli energy balance between a point just upstream of the pump and the outlet (we can ignore the velocity head at both locations, since the pipe diameter remains constant:

$$302.45 + h_p - 47.9 = 167.5 + 195.5 = 363.0 \text{ ft}$$

Therefore, pump head, $h_p = 108.4 \text{ ft}$

Pump power: $P = \dfrac{\gamma Q h_p}{\eta} = \dfrac{62.4 \times 4.464 \times 108.4}{0.85} = 35{,}524 \text{ lb} \cdot \dfrac{\text{ft}}{\text{sec}} = 65 \text{ hp}$

302-003

Pressure head at the inlet $= 231.9 - 156.5 = 75.4 \text{ ft.}$

Velocity in suction line: $V_1 = \dfrac{2000/448}{\dfrac{\pi}{4}\left(\dfrac{8}{12}\right)^2} = 12.79 \text{ fps}$

In order to avoid cavitation, the pressure inside the pump casing should not fall below the vapor pressure, which is 0.26 psi (at $T = 15°C$), which is equivalent to a pressure head of 0.6 ft

$$\frac{p_1}{\gamma} + \frac{V_1^2}{2g} = \frac{p_2}{\gamma} + \frac{V_2^2}{2g} \Rightarrow \frac{V_2^2}{2g} = \frac{V_1^2}{2g} + \frac{p_1 - p_2}{\gamma} = \frac{12.79^2}{2 \times 32.2} + 75.4 - 0.6 = 77.3 \Rightarrow V_2 = 70.6 \text{ fps}$$

302-004

At peak conditions, flow rate in the 24-in pipe $= 160 \times 50 = 8000 \text{ gpm} = 17.86 \text{ cfs}$

Velocity in pipe $= 5.684 \text{ fps}$

Head loss due to friction: $h_f = \dfrac{3.022 \times 3200 \times 5.684^{1.85}}{100^{1.85} \times 2^{1.165}} = 18 \text{ ft}$

Minor loss at all connections $= 4 \text{ in} \times 160 = 640 \text{ in} = 53.33 \text{ ft}$

Total head loss $= 18 + 53.33 \text{ ft} = 71.33 \text{ ft}$

Static head $= 435 - 245 = 190 \text{ ft}$

Available pressure head at community $= 118.67 \text{ ft} = 51.4 \text{ psi}$

302-005

By creating the H–Q curve for two pump in series, we see that the intersection with the system curve is around 360 pgm.

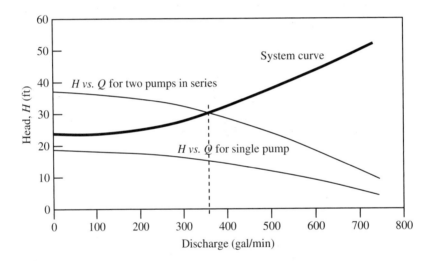

Chapter 303: Open Channel Hydraulics

303-001

As this is an unknown depth problem, calculate the parameter:

$$K = \frac{Qn}{kb^{8/3}S^{1/2}} = \frac{130 \times 0.014}{1.486 \times 10^{8/3} \times 0.005^{1/2}} = 0.0373$$

From Table 303.3, the depth ratio $d/b = 0.14$. Therefore, depth of flow, $d = 0.14 \times 10 = 1.4 \text{ ft}$.

303-002

$Q = 10,000$ gpm $= 22.3$ cfs

Flow parameter: $\dfrac{Q}{\sqrt{gD^5}} = \dfrac{22.3}{\sqrt{32.2 \times 4^5}} = 0.123$

From Fig. 303.8, corresponding depth ratio: $d_c/D = 0.34$. Therefore, critical depth $=$ $0.34 \times 48 = 16.3$ in

303-003

$Q = 32$ cfs

Flow parameter: $\dfrac{Q}{\sqrt{gD^5}} = \dfrac{32}{\sqrt{32.2 \times 3^5}} = 0.362$

From Fig. 303.8, corresponding depth ratio: $d_c/D = 0.64$. For this depth ratio, Table 303.3 yields $Q/Q_f = 0.7396$. Therefore, flow rate for pipe flowing full, $Q_f = 32/0.7396 = 43.3$ cfs.

Using Eq. (303.34), we solve for longitudinal slope, $S = 0.014$ (1.4%).

303-004

Desired flow rate $Q = 10,000$ gpm $= 22.3$ cfs

Assuming the weir contractions are suppressed, $Q = 3.33bH^{3/2} = 22.3 \Rightarrow H = 0.48$ ft

Now, if the width is adjusted for end contractions, $b_{eff} = 20 - 0.1 \times 2 \times 0.48 = 19.9$ ft

Recalculating H: $Q = 3.33 \times 19.9 \times H^{3/2} = 22.3 \Rightarrow H = 0.484$ ft

Weir crest elevation $= 341.90 - 0.48 = 341.42$ ft

303-005

Approach velocity: $V_1 = \dfrac{700}{24 \times 1.6} = 18.23$ fps

Downstream depth: $d_2 = -\dfrac{1}{2}d_1 + \sqrt{\dfrac{2V_1^2 d_1}{g} + \dfrac{d_1^2}{4}} = -\dfrac{1.6}{2} + \sqrt{\dfrac{2 \times 18.23^2 \times 1.6}{32.2} + \dfrac{1.6^2}{4}} = 5.0$ ft

Downstream velocity: $V_2 = \dfrac{700}{24 \times 5} = 5.83$ fps

Energy (head) loss: $\Delta E = \left(d_1 + \dfrac{V_1^2}{2g} \right) - \left(d_2 + \dfrac{V_2^2}{2g} \right) = 1.6 + \dfrac{18.23^2}{2 \times 32.2} - 5.0 - \dfrac{5.83^2}{2 \times 32.2} = 1.233$ ft

Rate of energy loss (power): $P = \gamma Q \Delta E = 62.4 \times 700 \times 1.233 = 53{,}857 \text{ lb} - \dfrac{\text{ft}}{\text{s}} = 98 \text{ hp}$

Figure 303.15 can be used to solve this problem, but the resolution of the h_L/d_1 axis is too coarse for the numbers in this problem.

303-006

Since the slope is not critical, horizontal, or adverse, that rules out C1, C3, H2, H3, A2, and A3

Normal depth $y_o = 2.95$ ft

Critical depth $y_c = 3.8$ ft

Actual depth $y = 5.6$ ft

Since $y_o < y_c < y$, the flow is classified as M3 (see Table 303.8)

Chapter 304: Hydrology

304-001

Total infiltration depth $= 0.5 \times 2 = 1.0$ in

Net (Gross − infiltration) rainfall volume $= 180 \times 43{,}560 \times 2.2/12 = 1{,}437{,}480 \text{ ft}^3 = 10.75$ million gallons

Evaporation loss $= 5000 \times 180 \times 2 = 1{,}800{,}000$ gallons $= 1.8$ million gallons

Net runoff volume $= 10.75 - 1.8 = 8.95$ million gallons

304-002

The table below shows the pattern of runoff produced by the three periods of runoff.

Time (h)	0.0	1.0	2.0	3.0	4.0	5.0	6.0	
Hour 1 (1.2 in)	0	27.6	56.4	98.4	38.4	1.44	0	
Hour 2 (0.9 in)		0	20.7	42.3	73.8	28.8	10.8	0
Hour 3 (0.3 in)			0	6.9	14.1	24.6	9.6	3.6	0
Total	0	27.6	77.1	147.6	126.3	54.84	20.4	3.6	0

Peak discharge $= 147.6$ cfs

304-003

Assuming that the soil is oversaturated condition (antecedent moisture condition III) when the storm occurs, the given curve numbers (AMC II) to be converted to CN_{III} values.

$$CN_{III} = \frac{23CN_{II}}{10 + 0.13CN_{II}}$$

The curve numbers (CN_{II}) and adjusted curve numbers (CN_{III}) for the four zones are shown in the table below:

Zone	Area (ac)	CN_{II}	CN_{III}
1	50	92	96
2	20	98	99
3	70	51	71
4	40	39	60

The weighted average curve number (CN_{III}) is then 78. For gross rainfall $= 3.2$ in and $CN = 78$, runoff depth [Table 304.8 or Fig. 304.11 or Eq. (304.26)] is 1.3 in.

304-004

Annual recurrence interval $= 1/0.04 = 25$ years.

Maximum time of concentration $= 50$ min. From the I-D-F curves, for $N = 25$ years and duration $= 50$ min, design intensity $= 1.2$ in/h.

Therefore, the total runoff from the four zones is calculated as:

$$Q = \sum C_i i A_i = i \sum C_i A_i = 1.2 \times (0.85 \times 50 + 0.7 \times 20 + 0.67 \times 70 + 0.82 \times 40)$$

$$= 163.4 \, ac - \frac{in}{hr} = 164.8 \, cfs$$

304-005

The table below shows the net discharge pattern of runoff (obtained by subtracting the base flow of 23 cfs)

Time (h)	0.0	1.0	2.0	3.0	4.0	5.0	6.0
Net Q (cfs)	0	22	72	102	53	28	2

Total runoff volume: $V = \sum Q_{net} \Delta t = \left(\sum Q_{net} \right) \Delta t = 279 \times 3600 = 1.0 \times 10^6 \, \text{ft}^3$

Average depth of runoff: $d = \dfrac{V}{A} = \dfrac{1.0 \times 10^6}{225 \times 43{,}560} = 0.1025 \, \text{ft} = 1.23 \, \text{in}$

Peak discharge due to runoff from a 2.3 in runoff $= 2.3/1.23 \times 102 = 190.7 \, \text{cfs}$

Adding in the base flow, the peak discharge that would be recorded would be $190.7 + 23 = 213.7 \, \text{cfs}$

Chapter 305: Water Supply Quantity and Quality

305-001

Adjust the temperature coefficient for temperature: $T_{25} = T_{20} \times 1.047^{25-20} = 0.315$

$$BOD_5 = \frac{DO_i - DO_f}{\left(\dfrac{V}{300} \right)} = \frac{7.4 - 2.3}{\left(\dfrac{12}{300} \right)} = 127.5 \frac{\text{mg}}{\text{L}}$$

$$BOD_{ult} = \frac{BOD_5}{1 - 10^{-kt}} = \frac{127.5}{1 - 10^{-0.315 \times 5}} = 131 \frac{\text{mg}}{\text{L}}$$

305-002

Using the conversion factors based on equivalent weight, the total hardness is calculated as:

$$H = 2.5 \times 34 + 4.1 \times 15 = 146.5 \frac{\text{mg}}{\text{L}}$$

Since the alkalinity is less than the total hardness, the difference is equal to the noncarbonated hardness. Therefore, noncarbonated hardness $= 146.5 - 110 = 36.5 \, \text{mg/L}$ as $CaCO_3$.

305-003

Stream discharge: $Q = 200 \, \text{MGD} = 200 \times 1.5472 = 309.4 \, \text{cfs}$

Wastewater discharge: $Q = 1000 \, \text{gpm} = 1.44 \, \text{MGD} = 1000/448.8 = 2.23 \, \text{cfs}$

Ultimate BOD concentration in wastewater $= 1250/(8.3454 \times 1.44) = 10.4 \, \text{mg/L}$

Assuming the stream velocity (5 mph) is not altered significantly, 10 miles is equivalent to time from mixing $= 2\,\text{h} = 0.083$ days

Dissolved oxygen at the instant of mixing: $\overline{DO} = \dfrac{309.4 \times 6.3 + 2.23 \times 1.3}{309.4 + 2.23} = 6.26\,\dfrac{\text{mg}}{\text{L}}$

Initial oxygen deficit $= \text{DO}_{\text{sat}} - \text{DO} = 10.07 - 6.26 = 3.81$ mg/L

Oxygen deficit at $t = 0.83$ h after mixing:

$$D_t = \frac{0.23 \times 10.4}{0.32 - 0.23}(10^{-0.23 \times 0.083} - 10^{-0.32 \times 0.083}) + 3.81 \times 10^{-0.32 \times 0.025} = 4.02\,\frac{\text{mg}}{\text{L}}$$

305-004

The peak factor is calculated according to: $\dfrac{Q_{\text{peak}}}{Q_{\text{ave}}} = \dfrac{18 + \sqrt{P}}{4 + \sqrt{P}} = \dfrac{18 + \sqrt{12}}{4 + \sqrt{12}} = 2.876$

Therefore, peak flow:

$$Q_{\text{peak}} = 2.876 \times 100 \times 12{,}000 = 3.45 \times 10^6\,\text{gpd} = 3.45\,MGD = 5.34\,\text{cfs}$$

305-005

Adult body weight $= 70$ kg

Water intake $= 2$ L/day

$$\text{MCLG} = \frac{R_f D \times \text{body weight}}{\text{Water intake}} \times \text{drinking water contribution} = \frac{0.18 \times 70}{2} \times 0.75 = 4.73\,\frac{\text{mg}}{\text{L}}$$

Chapter 306: Water and Wastewater Treatment

306-001

$Q = 4\,\text{MGD} = 4 \times 10^6$ gpd

For a circle of diameter $D = 10$ ft, submerged to a depth $d = 3$ ft, i.e., $d/D = 0.3$, Table 303.2 yields $A/D^2 = 0.1982$. Therefore, submerged area $= 0.1982 \times 10^2 = 19.82\,\text{ft}^2$

Total area (10 disks $\times$ 2 sides per disk): $A = 20 \times 19.82 = 396.4\,\text{ft}^2$

$$\frac{S_o}{S_e} = \left(1 + \frac{kA}{Q}\right)^n = \left(1 + \frac{15{,}000 \times 396.4}{4 \times 10^6}\right)^{2.6} = 10.68$$

Therefore, ultimate BOD in effluent: $S_e = \dfrac{S_o}{10.68} = \dfrac{230}{10.68} = 21.5\,\dfrac{\text{mg}}{\text{L}}$

306-002

Free chlorine, C = Chlorine dose – chlorine demand = $5.0 - 3.6 = 1.4$ mg/L

$t_{10} = 102/1.4 = 72.9$ min = 0.0506 day

For plug flow, baffling factor BF = 1.0. Therefore detention time = 0.0506.

Using the concept of detention time: $V = tQ = 0.0506 \times 4 \times 10^6 = 202,380$ gallons

306-003

Aeration tank volume, $V_a = 100,000$ ft^3 = 748,000 gal = 0.748 MG

Waste activated sludge flow rate, $Q_w = 0.05$ MGD

$$SRT = \frac{V_a X}{(Q_o - Q_w)X_e + Q_w X_r} \approx \frac{V_a X}{Q_w X_r} = \frac{0.748 \times 2600}{0.05 \times 8500} = 4.58 \text{ days}$$

306-004

Flow rate: $Q = VA = 4 \times 10 \times 8 = 320 \dfrac{\text{ft}^3}{\text{sec}}$

Surface loading velocity: $V^* = \dfrac{Q}{A_s} = \dfrac{320}{40 \times 10} = 0.8 \dfrac{\text{ft}}{\text{sec}} = 9.6 \dfrac{\text{in}}{\text{sec}}$

% removal of these particles = $3.0/9.6 \times 100\% = 31\%$

306-005

1 mg/L of pure alum (molecular weight = 594 mg/L, equivalent weight = 99 mg/L) treats alkalinity = 0.5 mg/L as $CaCO_3$.

Therefore, to treat alkalinity = 194 mg/L as $CaCO_3$, we need 388 mg/L of pure alum = 388/0.80 = 485 mg/L of stock alum.

Daily dose of alum = $485 \times 8.3454 \times 4 = 16,190$ lb/day

306-006

With one unit offline, each clarifier unit must be able to handle a load of 10/7 = 1.429 MGD = 2.21 ft^3/sec

If the width is W, then the length is $4W$. The flow through velocity:

$$V = \frac{Q}{15W} = \frac{2.21}{15W} \leq 0.005$$

This leads to a minimum width: $W \geq 29.5$ ft

Therefore, the tanks must be dimensioned 120-ft long $\times$ 30-ft wide $\times$ 15-ft deep.

Chapter 401: Capacity Analysis

401-001

Peak flow rate: $v_p = \dfrac{V}{PHF} = \dfrac{1340}{0.88} = 1523\text{vph}$

Headway (sec) $= \dfrac{3600\,\text{s/h}}{1523\,\text{veh/h}} = 2.36\,\text{sec/veh}$

401-002

Reaction distance (at speed of 70 mph for 2 sec) $= 1.47 \times 2 \times 70 = 205.8$ ft

From 70 to 40 mph, with a deceleration rate of 6 mph/sec, time required $= 30/6 = 5$ sec.

Distance traveled $= \text{ut} + \tfrac{1}{2}at^2 = 70 \times 5 - 0.5 \times 6 \times 5^2 = 275$ mph-sec $= 1.47 \times 275$ ft $= 404.25$ ft

From 40 mph to rest, with a deceleration rate of 4 mph/sec, time required $= 40/4 = 10$ sec.

Distance traveled $= \tfrac{1}{2}at^2 = 0.5 \times 4 \times 10^2 = 200$ mph-sec $= 1.47 \times 200$ ft $= 294$ ft

Stopping sight distance $= 205.8 + 404.25 + 294 = 904$ ft

401-003

The space mean speed is calculated as the harmonic mean of the given speeds:

$$\overline{V} = \frac{N}{\sum_{1}^{N} \dfrac{1}{V_i}} = \frac{10}{0.2385} = 41.9\,\text{mph}$$

Chapter 402: Highway Safety

402-001

In the two scenarios shown, the hollow circle is a diverging conflict, the solid circle is a merging conflict, and the square is a crossing conflict.

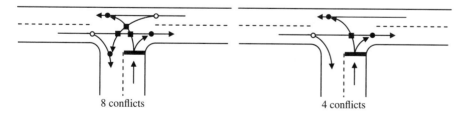

8 conflicts   4 conflicts

402-001

The accident rate for each highway segment is calculated according to:

$$\text{Accidents per hundred million vehicle miles} = \frac{\text{Accidents} \times 10^8}{\text{N} \times 365 \times \text{L} \times \text{ADT}}$$

The rates (for segments 1–4) are: 35, 127.5, 98.2, and 137.0, respectively. Therefore, the segment most in need of rehabilitation is segment 4.

Chapter 403: Sight Distance

403-001

The algebraic grade difference is: $A = 5 - (-3) = 8$

Length of curve: $L = 3812.54 - 2345.60 = 1466.94$ ft

According to AASHTO Green Book: $L = \dfrac{AS^2}{2800}$ for $S \leq L$

This yields: $S = 716.5$ ft. Since this fits the criterion $S \leq L$, it is correct.

403-002

According to Table 3-4 of the AASHTO Green Book, required passing sight distance for a design speed of 45 mph is 700 ft, and for 50 mph is 800 ft. Therefore, answer is 45 mph.

403-003

The algebraic grade difference is: $A = 5 - (-3) = 8$

Length of curve: $L = 3812.54 - 2345.60 = 1466.94$ ft

According to Eqn. (3-44) of the AASHTO Green Book, for a crest vertical curve, the stopping sight distance (S)

$$L = \frac{AS^2}{2158} \quad \text{for} \quad S \leq L$$

This yields: $S = 629.1$ ft. Since this fits the criterion $S \leq L$, it is correct. Using Table 403.1, the corresponding design speed (assuming level ground) is 64 mph. If choices of 60 and 65 mph are available, one must round down to 60 mph.

Chapter 404: Highway Curves

404-001

Length of back tangent $T_a = 1210.50 - 714.06 = 496.44$ ft

Deflection angle: $I = 67°30'45'' = 67.5125°$

$$T_a = \frac{R_2 - R_1 \cos I + (R_1 - R_2)\cos\Delta_2}{\sin I}$$

This equation can be solved for the deflection angles Δ_1 and Δ_2 for curves 1 and 2.

$$496.44 = \frac{500 - 800 \times \cos 67.5125 + (800 - 500)\cos\Delta_2}{\sin 67.5125} \Rightarrow \Delta_2 = 28.08, \Delta_1 = 39.43$$

$$L_1 = 100\frac{\Delta_1}{D_1} = 100\frac{\Delta_1 R_1}{5729.578} = 550.55\,\text{ft}$$

404-002

Degree of curve: $D_c = \frac{5729.578}{850} = 6.74°$

Spiral angle: $I_s = \frac{L_s D_c}{200} = \frac{140 \times 6.74}{200} = 4.7185$

$$x_c = L_s\left(1 - \frac{I_s^2}{32828}\right) = 140 \times \left(1 - \frac{4.7185^2}{32828}\right) = 139.91\,\text{ft}$$

$$k = x_c - R_c \sin I_s = 139.91 - 850 \times \sin 4.7185 = 69.99\,\text{ft}$$

$$y_c = \frac{L_s^2}{6R_c} = \frac{140^2}{6 \times 850} = 3.84\,\text{ft}$$

$$p = y_c - R_c(1 - \cos I_s) = 3.84 - 850 \times (1 - \cos 4.7185) = 0.96\,\text{ft}$$

$$T_s = k + (R_c + p)\tan\frac{I}{2} = 69.99 + (850 + 0.96)\tan\frac{56.75}{2} = 529.62\,\text{ft}$$

Therefore, T_S is at station $2334.50 - 529.62 = 1874.87$ (station $18 + 74.87$). Note that the major contributor to the difference between the original tangent length ($R_c \tan I/2 = 459.11$ ft without the spiral transition) and the new tangent length (from PI to TS) is the dimension k. So, a coarse estimate for T_s is $459.11 + 69.99 = 529.10$ ft, which is within 0.5 ft of the exact answer.

404-003

For design speed of 65 mph, level grade stopping sight distance = 644.1 ft.

AASHTO GDHS: For $S > L, L = 2S - \dfrac{2158}{A} = 2 \times 644.1 - \dfrac{2158}{8} = 1018.5\,\text{ft}$. This does not meet the criterion ($S > L$). So, we try the other equation:

For $S \leq L, L = \dfrac{AS^2}{2158} = \dfrac{8 \times 644.1^2}{2158} = 1538.0\,\text{ft}$.

Elevation of the PVC = elevation of the PVI $- G_1 L/2 = 235.76 - 0.05 \times 769 = 197.31$ ft

Chapter 405: Superelevation

405-001

Superelevation factor: $e + f \geq \dfrac{V^2}{15R} = \dfrac{40^2}{15 \times 450} = 0.237$

Required superelevation: $e = 0.237 - 0.20 = 0.037$. Use 4% superelevation.

405-002

AASHTO Green Book recommends maximum superelevation = 8% for regions with snow/ice. Adopting a maximum superelevation rate of 8%, Table 3-10 yields $R_{min} = 2210$ ft.

405-003

The superelevation runoff (2 lanes rotated) is calculated as:

$$L_r = \frac{b_w \times n \times w \times e_d}{SRR} = \frac{0.75 \times 2 \times 12 \times 0.06}{1/300} = 324\,\text{ft}$$

Chapter 406: Freeways

406-001

For the conditions given (1 mile segment on 5% upgrade), $E_T = 3.0$

Heavy vehicle factor:

$$f_{HV} = \frac{1}{1 + P_T(E_T - 1)} = \frac{1}{1 + 0.09 \times (3.0 - 1)} = 0.847$$

Peak flow rate: $v_p = \dfrac{V}{PHF \times N \times f_{HV} \times f_p} = \dfrac{4670}{0.88 \times 4 \times 0.806 \times 1.0} = 1646 \text{pcphpl}$

$FFS = 75.4 - f_{LW} - f_{LC} - 3.22 TRD^{0.84} = 75.4 - 1.9 - 0.0 - 3.22 \times 3^{0.84} = 65.4$

For $FFS = 65$ mph and $v_p = 1646$ pcphpl, LOS is C

406-002

The density (pc/mile/lane) in the diverge influence area is calculated from Eq. 14.23 of the Highway Capacity Manual:

$D_R = 4.252 + 0.0086 v_{12} - 0.009 L_D = 4.252 + 0.0086 \times 850 - 0.009 \times 600 = 6.16 \text{pcpmpl}$

According to Exhibit 14-3, LOS is A ($D_R < 10$ pcpmpl).

Chapter 407: Multilane Highways

407-001

Since the FFS is given, local adjustments, such as $f_A, f_M, f_{LW},$ and f_{LC} are not needed.

For the conditions given (rolling terrain), $E_T = 3.0$

Heavy vehicle factor:

$f_{HV} = \dfrac{1}{1 + P_T(E_T - 1)} = \dfrac{1}{1 + 0.06 \times (2.5 - 1)} = 0.847$

Peak flow rate: $v_p = \dfrac{V}{PHF \times N \times f_{HV} \times f_p} = \dfrac{3420}{0.90 \times 3 \times 0.847 \times 1.0} = 1495 \text{pcphpl}$

For $FFS = 52$ mph and $v_p = 1495$ pcphpl, LOS is D (range 1352 to 1766)

407-002

For a rural highway, default value for $BFFS = 60$ mph, access points per mile $= 8$.

Adjustment for access point density: $f_A = 2.0$ mph

For undivided design, $f_M = 1.6$ mph

For 12-ft lanes, $f_{LW} = 0.0$

For undivided design, $LC_L = 6$ ft and $LC_R = 7$ ft, total lateral clearance $= 12$ ft (max), $f_{LC} = 0.0$.

$FFS = 60 - 2.0 - 1.6 - 0 - 0 = 56.4$ mph

Chapter 408: Two-Lane Highways

408-001

Heavy vehicle factor: $f_{HV} = 1.0$

Hourly volume for design direction: $V = 0.65 \times 2450 = 1593$ vph

Flow rate: $v_p = \dfrac{V}{PHF \times f_G \times f_{HV}} = \dfrac{1593}{0.90 \times 1.0 \times 1.0} = 1769$ pcphpl

Since this is greater than capacity of 1700 pc/h, LOS is F.

408-002

For level terrain and significant two-way flow rate, equivalence factor $E_T = 1$ and $f_{HV} = 1.0$. Grade adjustment factor $f_G = 1.0$.

Peak flow rate: $v_p = \dfrac{V}{PHF \times f_G \times f_{HV}} = \dfrac{850}{0.88} = 966$

For 50:50 directional split, 40% no passing zones and flow rate $= 966, f_{np} = 35.8$ (by interpolation).

For $v_p = 966, a = -0.0048$ and $b = 0.83$

$$BPTSF_d = 100(1 - e^{-0.0048 \times 966^{0.83}}) = 76$$

$$PTSF_d = BPTSF_d + f_{np}\frac{V_d}{V_d + V_o} = 76 + 35.8 \times \frac{966}{966 + 966} = 94$$

Chapter 409: Signalization Warrants

409-001

Total volume (both approaches combined) on the major street (2 lanes each direction) $= 1030$ vph. From Fig. 4C-3 (MUTCD), corresponding point on the curve (for two-lane major intersecting one-lane minor), minor street high volume approach $= 280$ vph.

409-002

In Fig. 4C-9, the dimension $D = 105 - 6 = 99$ ft. Visually interpolating between the curves for $D = 90$ and $D = 110$, for major street volume $= 200$ vph, minor street approach volume $= 110$ vph.

Chapter 410: Intersections

410-001

Saturation flow rate for a lane group is given by: $s = s_0 N f_w f_{HV} f_g f_p f_{bb} f_a f_{RT} f_{LT}$

For lane width $= 10$ ft, $f_w = 0.933$

For 6% heavy vehicles, $f_{HV} = 0.943$

For 25% RIGHT TURNS, $f_{RT} = 0.9625$

NBTH/R: $s = 1900 \times 2 \times 0.933 \times 0.943 \times 0.99 \times 1.0 \times 1.0 \times 1.0 \times 0.9625 \times 1.0 = 3185\,\text{pcph}$

410-002

The length of the clearance interval is given by: $Y = t_R + \dfrac{V}{2(a \pm gG)} + \dfrac{W+L}{V}$

For N-S phase, width of intersection $= 50$ ft; grade $= 2\%$; speed $= 35$ mph, vehicle length $= 40$ ft

$$Y = 1.0 + \frac{1.47 \times 35}{2(10 - 32.2 \times .02)} + \frac{50 + 40}{1.47 \times 35} = 4.5\,\text{sec}$$

410-003

Cycle length $= 70$ sec

Number of cycles per hour $= 3600/70 = 51.43$

Pedestrians per cycle $= 400/51.43 = 7.78$

For crosswalk width < 10 ft, pedestrian green time is given by:

$$G_p = 3.2 + \frac{L}{S_p} + 0.27 N_{\text{ped}} = 3.2 + \frac{50}{4.0} + 0.27 \times 7.78 = 17.8\,\text{sec}$$

Chapter 411: Design of Highway Pavements

411-001

Assume total volume $= 1.0$ ft^3

Total weight $= 140$ lb. Of this, 6.5% is asphalt weight $= 9.1$ lb. The rest of the weight (130.9 lb) is aggregates. This can be further broken down into coarse aggregate (64%) $= 83.78$ lb, fine aggregate (24%) $= 31.42$ lb and filler (12%) $= 15.71$ lb.

Using the specific gravities, the volumes can be calculated as:

Asphalt volume = 0.135 ft³

Coarse aggregate volume = 0.497 ft³

Fine aggregate volume = 0.194 ft³

Filler volume = 0.095 ft³

Therefore, remaining volume = 0.079 ft³ is air content

Therefore, air content = 8% (approx.)

411-002

Annual ESAL (2001):

$$W_{18} = \frac{ADT \times 365 \times P_T \times \sum(N_i LEF_i)}{N} = \frac{23,000 \times 365 \times 0.16 \times 169.42}{1000} = 227,565$$

Sum of geometric series (first term = a, growth rate = r) is:

$$S = \frac{a(r^n - 1)}{r - 1} = \frac{a(1.04^{10} - 1)}{1.04 - 1} = 12.006$$

Thus, the cumulative ESAL = 12 × first year ESAL = 12 × 227,565 = 2.73 million

Chapter 501: Engineering Economics

501-001

At annual usage of 2000 h, useful life of 10,000 h = 5 years

Converting everything to present worth,

Initial cost: $P = 50,000$

Annual maintenance and fuel cost = $12 × 2000 = $24,000

Annual cost: $A = 24,000 \times \left(\frac{P}{A}, 5\ years, 8\%\right) = 24,000 \times 3.9920 = 95,808$

Tires are replaced every year: $T = 3000 \times \left(\dfrac{P}{A}, 5\,years, 8\%\right) = 3000 \times 3.9920 = 11,976$

Major maintenance after year 2 and year 4:

$$T = 9000 \times \left[\left(\dfrac{P}{F}, 2\,years, 8\%\right) + \left(\dfrac{P}{F}, 4\,years, 8\%\right)\right] = 9000 \times (0.8573 + 0.7350) = 14,330.70$$

Salvage value: $S = 8000 \times \left(\dfrac{P}{F}, 5\,years, 8\%\right) = 8000 \times 0.6806 = 5444.80$

Total present worth (thousands) $= 50 + 95.808 + 11.976 + 14.331 - 5.445 = 166.67$

Converted to an annuity, this is $166,670/3.992 = \$41,751$

Hourly charge to break even $= \$41,751/2000 = \$20.87$

501-002

The declining balance method calculates the depreciation based on the book value.

Purchase price $= \$50,000$

Year 1: Depreciation $= 30\%$ of $\$50,000 = \$15,000$. Book value at the end of year 1 $= \$35,000$

Year 2: Depreciation $= 30\%$ of $\$35,000 = \$10,500$. Book value at the end of year 2 $= \$24,500$

Year 3: Depreciation $= 30\%$ of $\$24,500 = \$7350$.

Book value at the end of year 3 $= \$17,150$

501-003

The capital cost of the project can be converted to an annuity:

$$A = 74,000 \times \left(\dfrac{A}{P}, 5\,years, 8\%\right) = 74,000 \times 0.2505 = 18,537$$

Annual benefit $= \$19,000$ in savings

Benefit:cost ratio $= 19,000/18,537 = 1.025$

Chapter 502: Probability and Statistics

502-001

If the return period is N, then the annual probability of occurrence $= 1/N$, and therefore, the probability of nonoccurrence is $1 - 1/N$.

Over a period of 35 years, the probability of nonoccurrence is

$$q = \left(1 - \frac{1}{N}\right)^{35}$$

Therefore, the probability of occurrence over the life of the structure is

$$p = 1 - q = 1 - \left(1 - \frac{1}{N}\right)^{35} = 0.1$$

Solving, we get: $\left(1 - \frac{1}{N}\right)^{35} = 0.9 \Rightarrow \left(1 - \frac{1}{N}\right) = 0.997 \Rightarrow \frac{1}{N} = 0.003 \Rightarrow N = 333$

The bridge should be designed for the 333-year earthquake.

502-002

Assuming the efficiencies represent some kind of removal efficiency, the "transmit" fractions of these processes are 33%, 30%, 75%, and 67%, respectively. That means the effective transmission fraction is $0.33 \times 0.30 \times 0.75 \times 0.67 = 0.0497$ (5%).

Therefore, the removal efficiency of the plant (cascade of processes $1 - 4$) = 95%

502-003

360 ft 4 in $= 360.33$ ft. This is within the interval $360.00 - 360.50$, which has a frequency of 12. Therefore, by interpolation, frequency below $360.33 = 2/3 \times 12 = 8$. Therefore, the cumulative frequency for $360.33 = 76$.

Annual probability of exceeding this elevation = 24%

Chapter 503: Project Scheduling

503-001

A: $ES_A = 0$, $EF_A = 3$

B: $ES_B = 0$, $EF_B = 6$

C: $ES_C = 3$, $EF_C = 7$

D: $ES_D = \max(EF_A, EF_B) = 6$, $EF_D = 11$

E: $EF_E = EF_B + FF\ lag = 6 + 6 = 12$

F: $EF_F = \max(EF_C, EF_D) = 11$, $EF_F = 16$

G: $\max(EF_D, EF_E, EF_F) = \max(11, 12, 16) = 16$, $EF_G = 21$

H: $\max(EF_F, EF_G) = 21$, $EF_H = 24$

Project duration = 24 weeks

503-002

The path can be reduced by 3 weeks by crashing the cheapest activities (smallest cost differentials). These are B, D, F, or G. The minimum additional cost to reduce 3 weeks = $1 + 1 + 2 = \$4000$. Cost of original schedule = \$52,000. Therefore, minimum crashed cost (21-week completion) is \$56,000.

503-003

The forward and backward passes are completed below. Activity A has two successors—C and D. The free float is defined as

$$FF_i = \min ES_{i+1} - EF_i$$

Therefore, $FF_A = \min(ES_C, ES_D) - EF_A = \min(3,6) - 3 = 0$

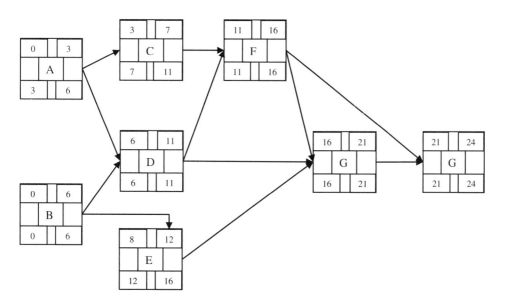

Chapter 504: Design Loads during Construction

504-001

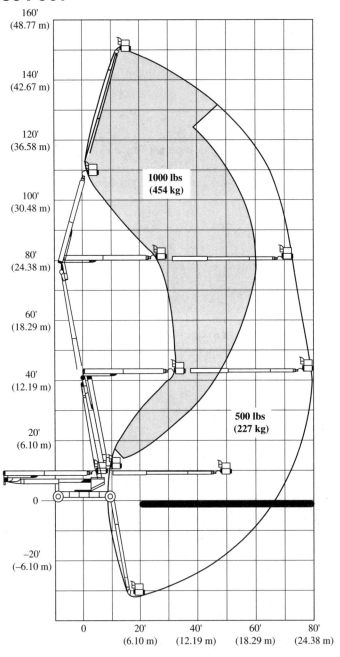

The 80-ft building is shown in semitransparent gray, positioned such that the roof is just accessible to the shaded 1000 lb zone of the reach diagram. At this position, the front axle is at 10 ft while the front of the building is at 55 ft. Therefore, the distance from the front axle to the face of the building is 45 ft.

504-002

Grade resistance = 5% = 0.05 = 100 lb/2000 lb = 100 lb/ton

Total resistance = rolling resistance + grade resistance = 160 lb/ton

Gross weight = 54,000 lb = 27 tons

Required rimpull = 27 × 160 = 4320 lb

$$V_{mph} = \frac{375 \times \eta \times HP}{Rimpull_{lb}} = \frac{375 \times 0.85 \times 450}{4320} = 33.2\,mph$$

504-003

According to ACI 347, horizontal construction load is equal to 50 lb per person = 4 × 50 = 200 lb.

Chapter 505: Scaffolding

505-001

For platforms greater than 10 ft in length, this overhang shall not exceed 18 in (46 cm).

505-002

When scaffold platforms are more than 2 ft (0.6 m) above or below a point of access, portable ladders, hook-on ladders, attachable ladders, stair towers, stairway-type ladders, ramps, walkways, integral prefabricated scaffold access, or direct access from another surface shall be used. Cross braces shall not be used as a means of access.

505-003

"Each suspension rope, including connecting hardware, used on adjustable suspension scaffolds shall be capable of supporting, without failure, at least six times the maximum intended load applied or transmitted to that rope with the scaffold operating at either the rated load of the hoist, or two times the stall load of the hoist, whichever is greater."

Scaffold operating at greater of the rated load of the hoist (12,000 lb) or two times the stall load (30,000 lb), i.e., for scaffold operating at 60,000 lb, each cable carries 15,000 lb. Therefore, each cable must have a breaking strength exceeding 90,000 lb.

Chapter 506: Formwork for Concrete

506-001

For $T = 65°F$ and pour rate $= 8$ ft/h, the base value of the lateral pressure $= 1168.5$ psf (by interpolation). Normal weight concrete – no adjustment. Type I cement without retarders – no adjustment. Maximum hydrostatic pressure $= \gamma h = 140 \times 14 = 1960$ psf.

Each tie has a tributary area $= 3 \times 3 = 9$ ft². Therefore, axial tension $= 9 \times 1168.5 = 10,517$ lb.

Required tie area $= 10.52$ k/24 ksi $= 0.44$ in². Required tie diameter $= 0.747$ in. Use ¾ in diameter ties.

506-002

The table below is constructed from the given data. Cumulative TTF $= 2000$ is reached approximately at midpoint of the 30:00 to 35:00 interval, i.e., at $t \approx 32.5$ h. Forms can be removed at $t = 33$ h.

Time (h)	Temp (°F)	T - To (°F)	TTF (°F-h)	Cumulative TTF
00:00–05:00	108	76	380	380
05:00–10:00	102	70	350	730
10:00–15:00	99	67	335	1065
15:00–20:00	94	62	310	1375
20:00–25:00	84	52	260	1635
25:00–30:00	83	51	255	1890
30:00–35:00	73	41	205	2095
35:00–40:00	72	40	200	2295
40:00–45:00	67	35	175	2470

506-003

The *Arrhenius function* for concrete maturity is given below:

$$t_{20} = \sum e^{-\frac{E}{R}\left(\frac{1}{T_i} - \frac{1}{T_o}\right)} \Delta t$$

where: t_{20} = time required, when curing at 20°C, to reach an equivalent maturity
E = activation energy = 41,500 J/mol for $T \geq 20°C$
R = universal gas constant = 8.3144 J/(mol-K)
T_i = average concrete temperature during time interval Δt, K
T_0 = reference temperature = 293 K

Time (h)	Temp (°F)	T (K)	$e^{-\frac{E}{R}\left(\frac{1}{T_i}-\frac{1}{T_o}\right)}\Delta t$
00:00–05:00	108	315.4	16.8
05:00–10:00	102	312.0	14.1
10:00–15:00	99	310.4	13.0
			43.9

Equivalent age after 15 h of curing = 43.9 h.

Chapter 507: Excavations

507-001

The unconfined compression strength $= 1600$ lb/ft$^2 = 0.8$ ton/ft^2. This classifies the soil as Type B and the steepest slope is 1:1 (45°).

507-002

Where employees are required to be in trenches 3 ft deep or more, ladders shall be provided, which extend from the floor of the trench excavation to at least 3 ft above the top of the excavation. They shall be located to provide means of exit without more than 25 ft of lateral travel.

Chapter 508: Erosion Control

508-001

Mulching typically works well for slopes upto 33%, while vegetation and erosion control mats are effective upto about 50%. For a steep slope (1:1, 45°), the most attractive way to prevent erosion on slopes over 50% in steepness is to install retaining walls to create nearly flat terraces.

508-002

The forced hydraulic jump is commonly used as an energy dissipater. A ditch check is typically used as an interceptor for sediment, but is not very effective as an energy dissipater. A drop structure is effective for energy dissipation. An impact-type dissipater such as a headwall is common.

Answer is B.

Chapter 509: Occupational Safety

509-001

1926 Subpart L Appendix A: Table gives maximum permissible span for full-thickness undressed lumber and nominal-thickness lumber.

Answer is C.

509-002

1926.250(b)(7): When masonry blocks are stacked higher than 6 ft, the stack shall be tapered back one-half block per tier above the 6-ft level.

Answer is C.

509-003

1926:62(c)(1): The employer shall assure that no employee is exposed to lead at concentrations greater than 50 micrograms per cubic meter (50 $\mu g/m^3$) of air averaged over an 8-h period.

Answer is C.

509-004

For transmission lines carrying over 50 kV, the OSHA rule for minimum clearance is 10 ft + 0.4 in for each kV over 50 kV. Therefore, for a 440-kV line, the minimum clearance is 10 ft + 0.4 in · 390 = 23 ft.

Answer is D.

509-005

1926.51(c)(1) Table D-1: For 20–200 employees, 1 urinal and 1 toilet seat per 40 workers.

Answer is B.

509-006

1926.54(c): Employees, when working in areas in which a potential exposure to direct or reflected laser light greater than 0.005 W (5 mW) exists, shall be provided with antilaser eye protection devices as specified in Subpart E of this part.

Answer is A.

509-007

1926.56 Table D-3: 10 foot-candles required for "General construction plant and shops (e.g., batch plants, screening plants, mechanical and electrical equipment rooms, carpenter shops, rigging lofts and active store rooms, mess halls and indoor toilets and workrooms."

Answer is C.

509-008

1926.52(e): Exposure to impulsive or impact noise should not exceed 140-dB peak sound pressure level.

Answer is C.

509-009

1926.105(a): Safety nets shall be provided when workplaces are more than 25 ft above the ground or water surface, or other surfaces where the use of ladders, scaffolds, catch platforms, temporary floors, safety lines, or safety belts is impractical.

Answer is D.

509-010

1926.104 (b): Lifelines shall be secured above the point of operation to an anchorage or structural member capable of supporting a minimum dead weight of 5400 lb.

Answer is C.

509-011

1926.105(c)(1): Nets shall extend 8 ft beyond the edge of the work surface where employees are exposed and shall be installed as close under the work surface as practical but in no case more than 25 ft below such work surface. Nets shall be hung with sufficient clearance to prevent user's contact with the surfaces or structures below. Such clearances shall be determined by impact load testing.

Answer is C.

509-012

1926.105(d): All new nets shall meet accepted performance standards of 17,500 ft·lb minimum impact resistance as determined and certified by the manufacturers, and shall bear a label of proof test.

Answer is B.

509.013

1926.105(d): The mesh size of nets shall not exceed 6 in by 6 in.

Answer is A.

509-014

1926.152(b)(1): No more than 25 gal of flammable liquids shall be stored in a room outside of an approved storage cabinet.

Answer is B.

509-015

1926.250(b)(6): Brick stacks shall not be more than 7 ft in height. When a loose brick stack reaches a height of 4 ft, it shall be tapered back 2 in in every foot of height above the 4-ft level.

Answer is B.

509-016

1926.451(a)(1): Except as provided in paragraphs (a)(2), (a)(3), (a)(4), (a)(5), and (g) of this section, each scaffold and scaffold component shall be capable of supporting, without failure, its own weight and at least four times the maximum intended load applied or transmitted to it.

$$1500 \times 4 = 6000 \text{ lb} = 3 \text{ tons}$$

Answer is D.

509-017

1926.451(b)(4): Each end of a platform, unless cleated or otherwise restrained by hooks or equivalent means, shall extend over the centerline of its support at least 6 in (15 cm).

Answer is C.

509-018

1926.451(g)(4)(ii): The top edge height of toprails or equivalent member on supported scaffolds manufactured or placed in service after January 1, 2000, shall be installed between 38 and 45 in above the platform surface.

Answer is B.

509-019

1926.451(h)(4)(ii): Where used, toeboards shall be at least 3½ in (9 cm) high from the top edge of the toeboard to the level of the walking/working surface.

Answer is A.

509-020

1926 Subpart L Appendix A: Table gives maximum permissible span for full-thickness undressed lumber and nominal-thickness lumber. For working load = 50 psf, maximum permissible span = 8 ft.

Answer is D.

509-021

1926 Subpart L Appendix A (f): Screen installed between toeboards and midrails or toprails shall consist of No. 18 gauge U.S. Standard wire 1-in mesh.

Answer is B.

509-022

1926.502(g)(3)(iii): Control lines shall consist of ropes, wires, tapes, or equivalent materials, and supporting stanchions as follows: Each line shall have a minimum breaking strength of 200 lb (0.88 kN).

Answer is C.

509-023

1926.651(g)(1)(i): Where oxygen deficiency (atmospheres containing less than 19.5% oxygen) or a hazardous atmosphere exists or could reasonably be expected to exist, such as in excavations in landfill areas or excavations in areas where hazardous substances are stored nearby, the atmospheres in the excavation shall be tested before employees enter excavations greater than 4 ft (1.22 m) in depth.

Answer is B.

509-024

1926 Subpart P Appendix B Fig. B-1: All excavations 8 ft or less in depth which have unsupported vertically sided lower portions shall have a maximum vertical side of 3½ ft.

Answer is B.

509-025

1926.706(a)(2): A limited access zone shall be established whenever a masonry wall is being constructed. The limited access zone shall be equal to the height of the wall to be constructed plus 4 ft, and shall run the entire length of the wall.

Answer is C.

509-026

1926.706(b): All masonry walls over 8 ft in height shall be adequately braced to prevent overturning and to prevent collapse unless the wall is adequately supported so that it will not overturn or collapse. The bracing shall remain in place until permanent supporting elements of the structure are in place.

Answer is A.

509.027

1926.1053(b)(5)(i): Non-self-supporting ladders shall be used at an angle such that the horizontal distance from the top support to the foot of the ladder is approximately one-quarter of the working length of the ladder (the distance along the ladder between the foot and the top support).

Answer is B.

509.028

1926.1101(c)(2): The employer shall ensure that no employee is exposed to an airborne concentration of asbestos in excess of 0.1 fiber per cm^3 of air as an 8-h time-weighted average (TWA), or 1.0 fiber per cm^3 of air as averaged over a sampling period of 30 min.

Answer is C.

509.029

1926 Subpart P Appendix A: "Type A" means cohesive soils with an unconfined, compressive strength of 1.5 tons per square foot (tsf) (144 kPa) or greater.

Answer is A.

509-030

1926 Subpart P Appendix B Fig. B-1: Exception: Simple slope excavations which are open 24 h or less (short term) and which are 12 ft or less in depth shall have a maximum allowable slope of ½:1.

Answer is B.

509-031

1926.250(b)(8)(iv): Lumber piles shall not exceed 20 ft in height provided that lumber to be handled manually shall not be stacked more than 16 ft high.

Answer is B.

509-032

1926.651(c)(2): Means of egress from trench excavations: A stairway, ladder, ramp, or other safe means of egress shall be located in trench excavations that are 4 ft (1.22 m) or more in depth so as to require no more than 25 ft (7.62 m) of lateral travel for employees.

Answer is C.

509-033

1926.705(j): The maximum number of manually controlled jacks/lifting units on one slab shall be limited to a number that will permit the operator to maintain the slab level within specified tolerances of paragraph (g) of this section, but in no case shall that number exceed 14.

Answer is D.

509-034

1926.705(g): Jacking operations shall be synchronized in such a manner to ensure even and uniform lifting of the slab. During lifting, all points at which the slab is supported shall be kept within ½ in of that needed to maintain the slab in a level position.

Answer is A.

509-035

1926.754(b)(2): At no time shall there be more than four floors or 48 ft (14.6 m), whichever is less, of unfinished bolting or welding above the foundation or uppermost permanently secured floor, except where the structural integrity is maintained as a result of the design.

Answer is D.

INDEX

Y

Z